Jens-Rainer Ohm

Multimedia Communication Technology

Springer-Verlag Berlin Heidelberg GmbH

Jens-Rainer Ohm

Multimedia Communication Technology

Representation, Transmission and Identification of Multimedia Signals

With 441 Figures

 Springer

Professor Jens-Rainer Ohm
RWTH Aachen University
Chair and Institute of Communications Engineering
Melatener Str. 23
52074 Aachen
Germany

Cataloging-in-Publication Data applied for

ISBN 978-3-642-62277-9   ISBN 978-3-642-18750-6 (eBook)
DOI 10.1007/978-3-642-18750-6

Typesetting: Digital data supplied by author
Cover-Design: Design & Production, Heidelberg
Printed on acid-free paper   62/3020 Rw 5 4 3 2 1 0

Preface

Information technology provides a plenty of new ways to process, store, distribute and access audiovisual information. Beyond traditional broadcast and telephone channels and analog storage media like film or tapes, the emerging Internet, mobile networks and digital storage are going to revolutionize the terms of distribution and access. This development is ruled by the convergence of audiovisual media technology, information technology and telecommunications technology. By capabilities of digital processing, established media like photography, movie, television and radio are changing their roles and are becoming subsumed by new integrated services which are *mobile*, *interactive*, *pervasive*, usable from anywhere, giving freedom to play with, and penetrating everyday life. Multimedia communication establishes new forms of communication between people, between people and machines, allows also communication between machines using audiovisual information or related feature parameters. Intelligent media interfaces are becoming increasingly important, and machine assistance in accessing media, in acquiring, organizing, distributing, manipulating and consuming audiovisual information becomes inevitable in the future.

This book intends to provide a deep insight into important enabling technologies of multimedia communication systems, which are methods of multimedia signal processing, analysis, identification and recognition, and schemes for multimedia signal representation, compression and expression by features or other properties. All these are lively and highly innovative areas at present, where this book reviews state-of-the-art technology and its scientific foundations, but shall primarily support systematic understanding of underlying methods, algorithms and their theoretical foundations. It is strongly believed that this is the best approach to contribute to future improvements in the field.

In part, the book is a substantially upgraded translation of my German language textbook on digital image and video coding, which was published by the mid '90s. Since then, the progress that was made in compression of audiovisual data has been breath-taking, and consequently newest developments are reflected, including the Advanced Video Coding standard and motion-compensated Wavelet coding. The second basis for this book are my lectures on topics of multimedia communications held regularly at RWTH Aachen University. These treat all aspects of image, video and audio compression, including networking interfaces, and also include multimedia signal identification and recognition. These latter aspects, topically related to the MPEG-7 multimedia content description standard, establish a profound basis for intelligent multimedia systems.

Most chapters are supplemented by homework problems, for which solutions are available from **http://www.ient.rwth-aachen.de**.

The book would not have been possible without contributions of numerous students and many other people who have worked with me on topics of image, video and audio processing, encoding and recognition over more than 15 years. These are (in alphabetical order) Sven Bauer, Michael Becker, Markus Beermann, Sven Brandau, Nicole Brandenburg, Michael Brünig, Ferry Bunjamin, Kai Clüver, Emmanuelle Corne, Holger Crysandt, Sila Ekmekci, Christoph Fehn, Ingo Feldmann, Oliver Fromm, Karsten Grüneberg, Karsten Grünheit, Jens Güther, Hafez Hadinejad, Konstantin Hanke, Guido Heising, Hans Dieter Höhne, Michael Höynck, Laetitia Hue, Ebroul Izquierdo, Peter Kauff, Jörg Kramer, Silko Kruse, Patrick Laurent, Thomas Ledworuski, Wolfram Liebsch, Oliver Lietz, Phuong Ma, Bela Makai, Claudia Mayer, Bernd Menser, Domingo Mery, Karsten Müller, Patrick Ndjiki-Nya, Bernhard Pasewaldt, Andreas Praatz, Lars Prokop, Oliver Rockinger, Katrin Rümmler, Thomas Rusert, Mihaela van der Schaar, Ansgar Schiffler, Oliver Schreer, Holger Schulz, Aljoscha Smolic, Frank Sperling, Peter Stammnitz, Jens Wellhausen, Mathias Wien and Detlef Zier. Please forgive me if I forgot anybody.

Very special thanks are also directed to my scientific mentors Peter Noll, Hans Dieter Lüke and Irmfried Hartmann, all people of IENT and to my family.

Aachen, August 15, 2003
Jens-Rainer Ohm

Table of Contents

1 Introduction

Multimedia communication systems are a flagship of the information technology revolution. The combination of multiple information types, particularly audiovisual information (speech/audio/sound/image/video/graphics) with abstracted (text), smelled or tactile information provides new degrees of freedom in exchange, distribution and acquisition of information. Communication includes exchange of information between different persons, between persons and machines, or between machines only. Sufficient perceptual quality must be provided, which is related to the compression and its interrelationship with transmission by networks. Advanced methodologies are based on content analysis and identification, which is of high importance for automatic user assistance and interactivity. In multimedia communication, concepts and methods from signal processing, systems and communications theory play a dominant role, where audiovisual signals are a primary challenge regarding transmission, storage and processing complexity. This chapter introduces basic concepts and terminology related to multimedia communication systems, and gives an overview about acquisition and common formats of audiovisual signal sources.

1.1 Concepts and Terminology

A basic paradigm of communication systems development is to convey information at lowest cost possible, while observing highest possible quality and maximum benefit for the user. This can directly be related to the value of services which can be implemented on such systems. In multimedia communication, a challenge is imposed by those types of information which require high bandwidth and high processing speed due to their large volume. From this point of view, *audio, image, video* and *graphics* signals are most critical; multimedia communication systems are reaching their limits in real time processing for these types of data, such that the need for processing of audiovisual signals is one important driving force in development of computer systems with ever increasing processing power.

The classical model of a communication system is illustrated in Fig. 1.1a. This goes back to the foundations of modern communications theory as introduced in [SHANNON 1949]. The path of transmission is *symmetric*, where the blocks on the left-hand side (transmitter) have their complementary parts on the right-hand side (receiver). Source coder and decoder, channel coder and decoder must understand each other. The goal of this system is to convey as much information as possible with a quality as high as possible over an existing channel. This is achieved by optimization of the source encoder, which compresses the source to lowest possible number of bits, the channel encoder, which protects the bit stream from possible losses, and the modulator which adapts to the physical channel, also taking into consideration loss and bandwidth characteristics.

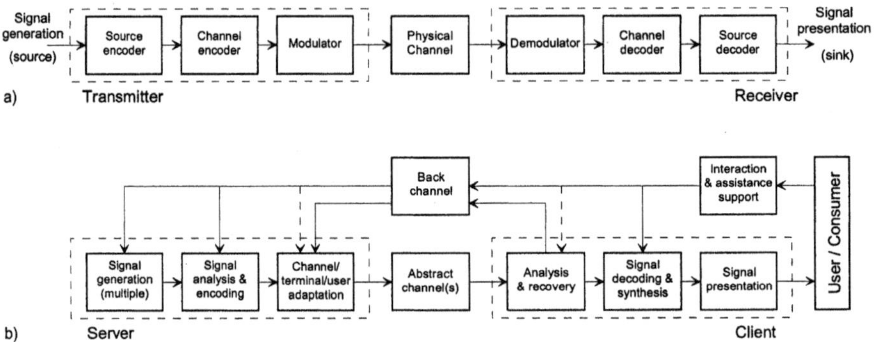

Fig. 1.1. **a** Classical model of a communication system   **b** Model of a multimedia communication system

The concept of the classical model in principle assumes transmission of linear media over a single path. Even though this is fully valid in many situations, it is by many aspects too simplistic for multimedia communication systems. The block diagram in Fig. 1.1b illustrates some important new aspects:

- The classical model assumes independent optimization of source and channel coding for best performance as the optimum solution. As will be shown at different places in this book, a source coding method which achieves optimum compression can be extremely sensitive against errors occurring in the channel, e.g. due to feedback of previous reconstruction errors into the decoding process. This requires joint optimization of the entire chain, such that in fact the best quality is retained for the user while the rate to be transmitted over the physical channel is made as low as possible.

- The classical model assumes a passive receiver ('sink'), which is very much related to broadcast services. In multimedia systems, the user can interact, and can take influence on any part of the chain, even back on the signal generation; this is reflected by providing a *back channel*, which can also be used by automatic mechanisms serving the user by best quality services. Instead of transmitter and receiver, denotation of devices at the front and back ends as *server* and *client* better reflects this new paradigm.

- The classical model assumes one monolithic channel for which the optimization of source coding, channel coding and modulation is made once. Multimedia communication mostly uses *heterogeneous networks*, which typically have largely varying characteristics; as a consequence, it is desirable to consider the channels more by an abstract level and perform proper adaptation to the instantaneous channel characteristics. Channels can be networks or storage devices. Recovery at the client side may include analysis which is far beyond traditional channel coding, e.g. by conveying loss characteristics to the server via the back channel.
- Multimedia services are becoming more 'intelligent', including elements of signal content analysis to assist the user. This includes support for content related interaction, support in finding the multimedia information which best serves the needs of the user. Hence, the information source is not just encoded at the front end, but more abstract analysis can be performed in addition; the encoding part itself may also include meta information about the content.
- Multimedia communication systems typically are *distributed systems*, which means that the actual processing steps involved are performed at different places. Elements of *adaptation* of the content to the needs of the network, to the client configuration, or to the user's needs can be found anywhere in the chain. Finally, temporary or permanent *storage* of content can also reside anywhere, as storage elements are a specific type of channel, intended for the purpose of later review instead of instantaneous transmission.

Multimedia communication systems are judged by the value they provide to the user, which is interpreted in terms of quality over cost. Different aspects of quality are hidden behind the block diagram of Fig. 1.1b, contributing to the overall *Quality of Service* (QoS). This term, even though is widely used, is subject to potential misunderstanding as different aspects are subsumed:

- The QoS relating to *network transmission* includes aspects like transmission bandwidth, delay and losses. It indirectly contributes to the perceived quality. This will be denoted as *Network QoS*.
- The QoS relating to *perceived signal quality* includes the entire transmission chain, including the compression performance of source encoding/decoding, and the inter-relationship with the channel characteristics. This is denoted as *Perceptual QoS*. An overview over methods for measurement is given in Appendix A.1.
- The QoS relating to the *overall service quality* is at the highest level. It includes aspects like the level of user satisfaction with the content itself, but also the satisfaction concerning additional services, e.g. how good an adaptation to the user's needs is made. Some methods that are used to express this category of QoS with regard to content identification are described in Appendix A.2. This may be denoted as the *Semantic QoS*.

The cost imposed to the user is related to transmission/storage cost and cost that is due to the complexity of systems. From the engineering point of view, a system of less complexity fulfilling the same functionality is better serving the user's needs.

This book concentrates on the topics of *representation, transmission and identification* of multimedia signals. The remaining part of this section will give a high-level overview about concepts and terminology related to each of these three aspects.

1.1.1   Signal Representation by Source Coding

By *multimedia signal compression*, systems for transmission and storage of multimedia signals shall generate the most compact representation, such that the highest possible perceptual quality is achieved. Immediately after capturing, the signal is converted into a digital representation having a finite number of samples and amplitude levels. This step already influences the final quality. If the range of rates that a prospective channel can convey, or the resolution required by an application are not known at the time of acquisition, it is advisable to capture the signal by highest possible quality, and scale it later.

In the *source coder*, the data rate needed for digital representation shall be reduced as much as possible. Properties of the signal which allow reduction of the rate can be expressed in terms of *redundancy* (which is e.g. the typically expected similarity of samples from the signal). The opinion about the quality of the overall system is ruled by the purpose of the *consuming* at the end of the chain. If the sink is a human observer, it is useful to adapt the source coding method to *perceptual properties of humans,* as it would be useless to convey a finer granularity of quality than the user can (or would desire to) perceive. Some artifacts occurring during the process of source coding and decoding may not be visible or audible, such that it is not necessary to suppress them. This is the *irrelevant information* contained in the original signal, which in fact must not be encoded either[1]. If however only a narrow-bandwidth channel is available, such that neither exploitation of redundancy nor of irrelevance can lead to the desired goal, it is unavoidable to accept distortions. The nature of these distortions can be different, e.g. reduction of resolution in time and space, or insertion of coding noise. Herein, it is still advisable to regard the perceptual properties or usage purpose of the consumer. In advanced methods of source coding, *content-related* properties can also be taken into consideration. This can e.g. be done by putting more emphasis on parts or pieces of the signal in which the user is expected to be most interested.

---

[1] In the case where consumption is made by a machine, the paradigm of irrelevance would also be derived from the purpose by which the machine is intending to use the signal. For example, if the goal is content analysis of the signal, it should not be unacceptably distorted by source coding such that the intended analysis would fail.

The encoded information is usually represented in form of binary digits (*bits*). The bit rate is measured either in *bit/sample*[1], or *bit per second* (*bit/s*), where the latter results from the *bit/sample* ratio, multiplied by the *samples/s* (the sampling rate). An important criterion to judge the performance of a source coding scheme is the *compression ratio*. This is the ratio between the bit rate necessary for representation of the uncompressed source and its compressed counterpart. If e.g. for digital TV the uncompressed source requires 165 *Mbit/s* [2], and the rate after compression is 4 *Mbit/s*, the compression ratio is 165:4=41.25. If compressed signal streams are stored as files on computer discs, the file size can be evaluated to judge the compression performance. When translating into bit rates, it must be observed that file sizes are often measured in *KByte*, *MByte* etc., where one *Byte* consists of 8 *bit*, 1 *KByte*=1,024 *Byte*, 1 *MByte*=1,024 *KByte* etc.

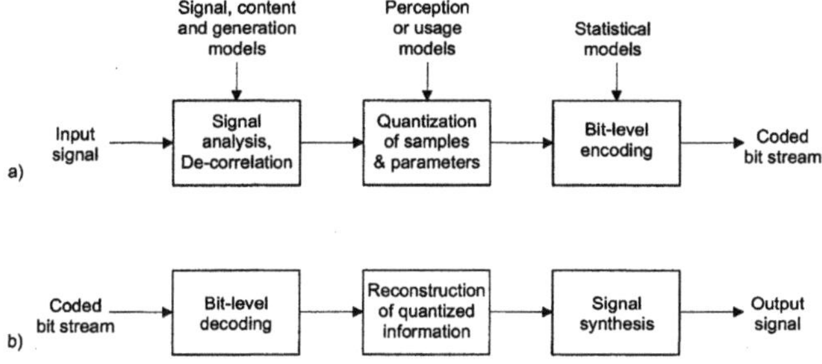

Fig. 1.2. Basic principle of a coding system for multimedia signals: **a** Encoder **b** Decoder

Fig. 1.2 shows the basic block diagram of a multimedia signal coding and decoding system. The first step of encoding is *signal analysis*. Important principles for this are *prediction of signals* and *frequency analysis* by transforms. In coding applications, the analysis step shall be reversible; by a complementary synthesis performed at the decoder, the signal shall be reconstructed achieving as much fidelity as possible. Hence, typical approaches of signal analysis used in coding are reversible transformations of the signal into *equivalent forms*, by which the encoded representation is as much as possible *free of redundancy*. If linear systems or transforms are used for this purpose, the removal of redundancy is also called *de-correlation*, as correlation expresses linear statistical dependencies between signal samples. To optimize such systems, availability of good and simple models reflecting the properties of the signal is crucial. Methods of signal analysis can also be related to the generation (e.g. properties of the acquisition process) and to the content of signals. Besides the samples of the signal or its equivalent representation, additional *side information* parameters can be generated by the analysis stage, such as adaptation parameters which are needed during decoding and syn-

---

[1] For image signals also *bit/pixel* (*bpp*), where 'pixel' is a widely-used acronym for *picture element*, an image sample.

[2] *Mbit/s*: 10^6 bit per second, also *kbit/s*: 10^3 bit per second etc.

thesis. The second step of encoding is *quantization*, which maps the signal, its equivalent representation or additional parameters into a discrete form. If the required compression ratio does not allow lossless reconstruction of the signal at the decoder output, perceptual properties or circumstances of usage should be considered during quantization to retain as much as possible the relevant information. The final step is *bit-level encoding*, which has the goal to represent the discrete set of quantized values by lowest possible rate. The optimization of encoding is mostly performed on basis of statistical criteria.

Important parameters to optimize a source coding algorithm are *rate, distortion, latency* and *complexity*. These parameters have mutual influence on each other. The relationship between rate and distortion is determined by the *rate distortion function* (see sec. 3.5 and 11.2) which gives a lower bound of the rate if a certain maximum distortion limit is required. Improved rate/distortion performance (which means improved compression ratio while keeping distortion constant) can usually be achieved by increasing the complexity of the encoding/decoding algorithm. Alternatively, increased latency also helps to increase compression performance; if for example an encoder is allowed to look ahead about consequences of present decisions on future encoding steps, this provides an advantage.

1.1.2   Optimization of Transmission

The interface between the source coder and the channel is also of high importance for the overall Perceptual QoS. While the source encoder removes redundancy from the signal, the channel encoder *adds redundancy* to the bit stream for the purpose of protection and recovery in case of losses. At the receiver side, the channel decoder removes the redundancy inserted by the channel encoder, while the source decoder supplements the redundancy which was removed by the source encoder. From this point of view, the operation of source encoding and channel decoding is similar and vice versa. Actually, the more complex part is usually on the side where redundancy is removed, which means finding the *relevant information* within an overcomplete representation. In fact, source and channel encoding play counteracting roles and should be optimized jointly for optimum performance. For example, it is not useful to add redundancy by channel encoding for parts of bit streams which are less relevant from a user perspective. The transmission over the channel also includes modulation, where the combination of channel coding principles and modulation at the physical level is one of the commonly used methods in communications technology to approach the limits of channel capacity.

Nevertheless, in the context of multimedia systems it often is advantageous to view the channel as a 'black box' for which a model exists. This in particular concerns error/loss characteristics, bandwidth, delay (latency) etc., which are the most important parameters of Network QoS. When parameters of Network QoS are guaranteed by the network, adaptation between source coding and the network transmission can be made in an almost optimum way. This is usually done by negotiation protocols. If no Network QoS is supported, specific mechanisms can be introduced for adaptation at the server and client sides. This includes application-specific error protection based on estimated network quality or usage of retransmission protocols. Introduction of latency is also a viable method to improve the transmission quality, e.g. by optimization of transmission schedules, temporary buffering of information at the receiver side before presentation is started, or scrambling/interleaving of streams when bursty losses are expected.

It becomes obvious that also for the channel part, improved quality can be expected when *higher latency* or *higher complexity* of the systems are introduced. Unfortunately, such concepts may not be suitable for all applications. For example, real-time conversational services (like video telephony) only allow minimum latency; the same is true for some types of interactive applications, where a quick reaction is expected based on user actions. For mobile devices, where battery capacity is critical, the overall complexity of systems must be kept low.

Physically, the abstract channel will often consist of a chain of several network paths having different QoS characteristics. Herein, the entire chain is only as strong as the weakest element. In heterogeneous networks, capabilities for easy adaptation of media streams to the moving target of network characteristics play an important role. In some cases, it may be necessary to *transcode* the streams into a different format that is better suitable for the specific network characteristics. From this point of view, source coding methods producing *scalable streams*, which are adaptable and can be truncated independent of the encoding process, are highly advantageous. Another approach is the usage of *multiple descriptions* of the content, which are conveyed to the receiver by different transmission paths.

Today's digital communication networks as used for multimedia signal transmission are based on the definition of distinct layers with clearly defined interfaces. On top of the physical transmission layer, a hierarchy of protocol stacks performs the adaptation up to the application layers. In such a configuration, optimization over the entire transmission chain could only be achieved by cross-layer signaling, which however imposes additional complexity to the transmission.

1.1.3   Content Identification

Advanced multimedia communication systems include components for *recognition of signal content*. This is an interdisciplinary area, which – besides the aspects treated in this book – is related to neurosciences and psychology. As technical systems for signal content recognition are intended to serve humans, e.g. by enabling communication between humans and automatic systems, they must be compatible with human cognition and take into regard possible human reactions. Such concepts are important in interactive multimedia systems. The principles and algorithms introduced in this book can be interpreted at a high level by the schematic diagram shown in Fig. 1.3. After acquisition and digitization, *preprocessing* is often applied, which shall improve the signal. In this context, *linear or nonlinear filters* are employed and methods to increase the resolution by *interpolation* are used when precision of quasi continuous signals is required. The next step is *feature extraction*. Examples of multimedia signal features are color, texture, shape, geometry, motion, depth for images and video; spectral characteristics, pitch, tempo, melody, phonemes etc. for audio and speech. If multiple features are used, a *feature transform* is useful, which is compacting the features in a different feature space or a sub-space which is more suitable for the subsequent *classification*. The last step is the classification itself, which consists of a suitable concatenation, weighting and comparison of the extracted features, which are usually matched against feature constellations which are known a priori. By this, mapping into semantic classes can be performed, which then allows to regard the signal at a higher level of abstraction.

Most of the processing steps described here are based on signal models or statistical models, and can involve estimation methods. On the other hand, when features are extracted and grouped into classes or categories, this knowledge also helps for better adaptation of the models and estimators. Hence, the four basic building blocks shown in the figure should not simply be regarded as forward connected. A recursive and iterative view is more appropriate as indicated by the dotted lines. For example, a classification hypothesis can be used to perform a feature extraction once again on the basis of an enhanced model, which is then expected to improve the result. This is actually very similar to models of human cognition which proceeds by a helix, where a better interpretation is achieved at a higher level, reflecting about the original input again on the basis of an initial hypothesis which is then either verified or rejected. The result of classification is abstracted information about the signal content. Even though it does not allow *reconstruction* of the signal any more, it may also be used to improve methods intended for reconstruction, such as source coding. In a more generalized view, the information gained by the different stages of content identification systems can be viewed as additional components describing multimedia signals, which are also denoted as *metadata information*.

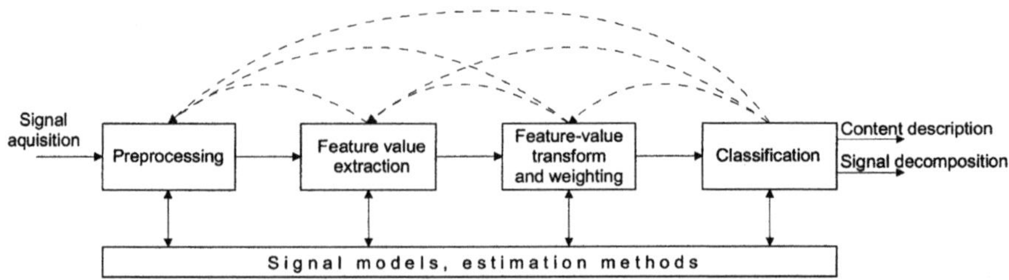

Fig. 1.3. Signal processing chain for multimedia content analysis and recognition

Metadata information can be used for many purposes in a multimedia communication system. Some examples are

–   Separation of relevant information on a semantic level (instead of the perceptual quality level), e.g. to put emphasis on more important objects like a per-

son's face within a video, or identification of an iterated melody in an audio signal;
–   Signal decomposition (e.g. segmentation) of signals, where the segments can uniquely be differentiated by specific features indicating a semantic context;
–   Automatic analysis of events, which can be used to trigger transmission in a real-time application;
–   Usage for adaptation of transmission and storage to user's preferences and needs, to achieve optimum usage of the available channel capacity;
–   Assist users to quickly find and identify specific content, e.g. by searching for similar images in a data base, similar scenes or specific events in a movie.

1.2   Signal Sources and Acquisition

Multimedia systems mainly process digital representations of signals, while the acquisition and generation of *natural signals* will in many cases not directly be performed by a digital device; electro-magnetic (microphone), optical (lens), chemical (film) media may be involved. In such cases, the properties of the digital signal are influenced by the signal conversion process during acquisition. The analog-to-digital conversion itself consists of a sampling step which maps a spatio-temporally continuous signal into discrete samples, and a quantization step which maps an amplitude-continuous signal into numerical values. Some acquisition devices like CCD or CMOS cameras inherently provide sampled signals. Sources of multimedia signals can also be synthetic, like graphics or sounds, usually generated by a computer or other synthesis device.
If natural signals are captured, part of the information originally available in the outside (three-dimensional) world is lost due to

–   limited bandwidth or resolution of the acquisition device;
–   "Non-pervasiveness" of the acquisition device, which resides at a singular position in the 3D exterior world, such that the properties of the signal are available only for this specific view or listening point; a possible solution is the usage of multiple cameras or microphones, where however acquisition of 3D spatial information will always be *incomplete*.

This relationship between exterior world and the incomplete projection into the signal shall now be studied for the example of a camera system. The global exterior 'world coordinate system' denote locations $\mathbf{W}=[W_1,W_2,W_3]^{\mathrm{T}}$, where an imaging property $I(W_1,W_2,W_3,\theta_1,\theta_2,\theta_3,t)$ shall be defined at a specific location $\mathbf{W}$ and time t. This imaging property I can be interpreted in a very general sense, e.g. it could be the spectral color and energy of light arriving at any location $\mathbf{W}$ by any angular orientation $\Theta=[\theta_1\ \theta_2\ \theta_3]^{\mathrm{T}}$. A camera is now assumed to be positioned at one position $\mathbf{W}$ with the optical axis ('ray of view') orientated by a specific direction Θ. For simplicity, a new coordinate system $\mathbf{X}=[X\ Y\ Z]^{\mathrm{T}}$ is defined with its origin exactly at the position of the camera and orientation such that the optical axis is the Z axis. A simple model is the *pin hole camera* shown in Fig. 1.4. A 2-dimensional image captured at time t is interpreted as a projection from the 3D exterior world into the image plane of the camera, regarding this singular view. The image plane can be a film, a CCD chip etc. In the pin hole model, only *one light ray* from the exterior world, emanating from some object in the exterior world, targets at one point of the image plane[1]. The relationship between a point P at exterior world coordinate $\mathbf{X}_P=[X_P\ Y_P\ Z_P]^{\mathrm{T}}$ and its projection into the point $\mathbf{r}_P=[r_P\ s_P]^{\mathrm{T}}$ of the image plane is characterized by the *central projection equation*, where F is the focal length of the camera,

$$r_P = F \cdot \frac{X_P}{Z_P} \quad ; \quad s = F \cdot \frac{Y_P}{Z_P} . \tag{1.1}$$

The origin of the world coordinate system is at the position of the 'pin hole', the focal point of the camera. The optical axis is orientated perpendicular to the image plane. The image plane itself is positioned at $Z=-F$, with center at $X=0$, $Y=0$, r-axis orientated at $Y=0$ by $-X$, s-axis at $X=0$ by $-Y$. The same projection equation holds where the rays traverse a mirrored image plane positioned at $Z=F$, r-axis orientated as X, s-axis as Y. The image plane itself has a horizontal width H and a vertical height V. The horizontal and vertical view angles spanning symmetrically around the optical axis are

$$\varphi_H = \pm\arctan\frac{H}{2F} \quad ; \quad \varphi_V = \pm\arctan\frac{V}{2F} . \tag{1.2}$$

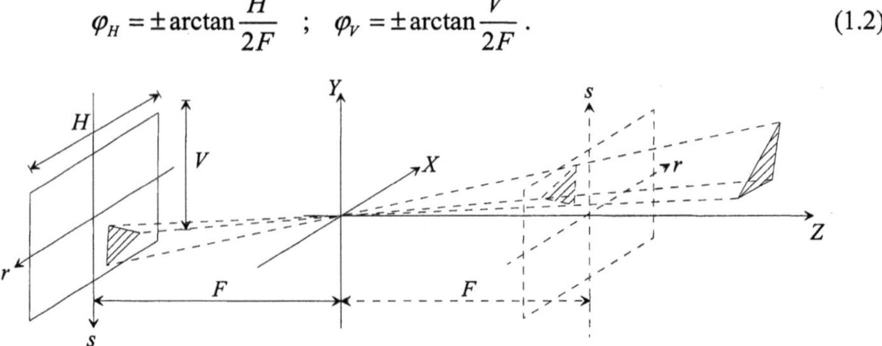

Fig. 1.4. Pin hole camera model

In electronic cameras and scanners, the image plane is usually scanned and sampled sequentially left-right / top-down row by row. In analog video cameras, the output is a voltage, where a unique location in the image plane can be associated

---

[1] In lens cameras this is different, as a number of rays are bundled by a lens all targeting for one point; if the lens is in focus, ideally all these rays will also emanate from the same point of an object in the exterior world.

with a precise time dependency of the signal. The change from one line to the next is indicated by the horizontal blanking interval (Fig. 1.5a), the change from one frame to the next by the vertical blanking interval. The instantaneous position on the image plane (Fig. 1.5b), either during acquisition or display, can be interpreted by the horizontal and vertical control voltages (Fig. 1.5c).

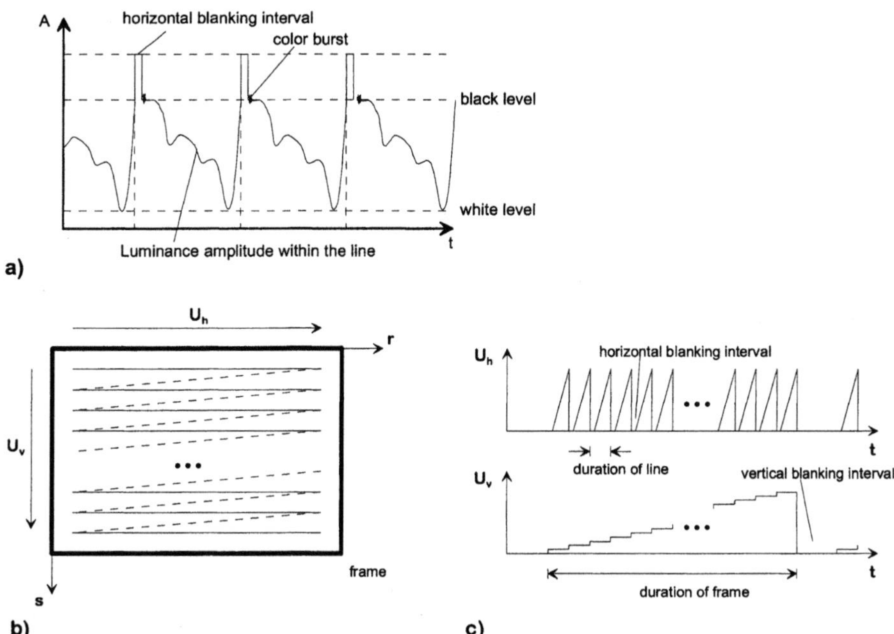

Fig. 1.5. a Analog TV signal (FBAS) **b** Row-wise scan sequence
c Horizontal and vertical control voltages, displayed as functions over time

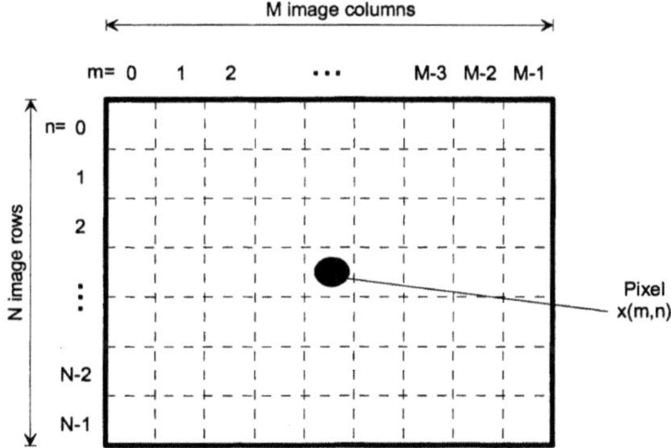

Fig. 1.6. Definition of the discrete image matrix

The time-sequential scanning of information in a video signal enables sampling by A-to-D converters applied to a one-dimensional signal, where only the time of sampling must be synchronized with the respective spatio-temporal location. Historically, camera image sampling evolved first over the temporal dimension in movie film cameras. By development of analog video technology, vertical row-by-row spatial sampling was introduced, while the signal within each row is still of continuous nature. Finally, in digital imaging the signal is sampled in the remaining horizontal dimension, and is converted (quantized) into numerical values instead of electrical signals. The image plane of width H and height V is mapped into M and N discrete sampling locations and represents a frame sample within a time-dependent sequence. Sampled and spatially bounded images can be expressed as matrices. Usually, it is assumed that the top left pixel of the image has coordinate $(0,0)$ and is the top left element of the matrix as well. This is illustrated in Fig. 1.6. A total of $M{\times}N$ pixels are mapped into a matrix of M columns and N rows,

$$\mathbf{X} = \begin{bmatrix} x(0,0) & x(1,0) & x(2,0) & \cdots & & \cdots & & x(M-1,0) \\ x(0,1) & x(1,1) & & & & & & x(M-1,1) \\ x(0,2) & & & \ddots & \ddots & & & \vdots \\ \vdots & & & & & & & x(M-1,N-2) \\ x(0,N-1) & x(1,N-1) & \cdots & & \cdots & x(M-2,N-1) & & x(M-1,N-1) \end{bmatrix}. \quad (1.3)$$

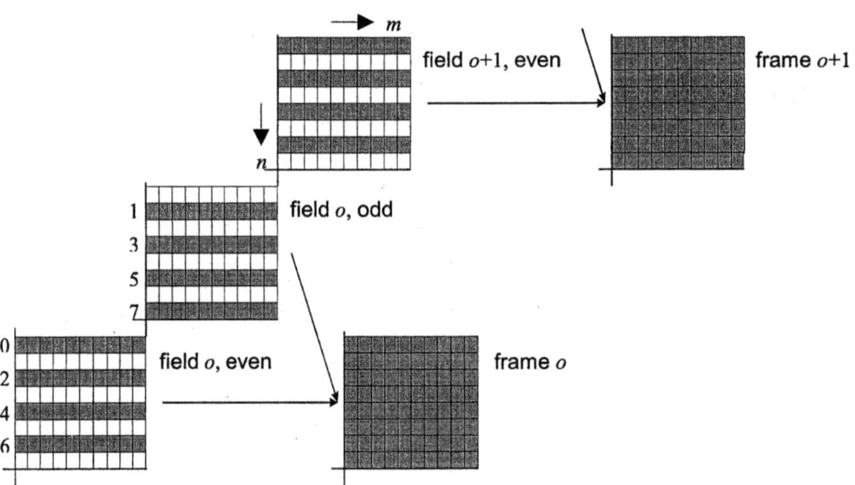

Fig. 1.7. Interlaced video sampling

In some cases (e.g. if values from different rows of the matrix shall be included in one single arithmetic operation) it is more convenient to re-arrange all samples from the image matrix into one single vector, which contains all pixels sequentially row-wise scanned, starting at the top left pixel of the image:

$$\mathbf{x} = [x(0,0) \quad x(1,0) \quad \cdots \quad x(M-1,0) \quad x(0,1) \quad x(1,1) \quad \cdots \quad x(0,N-1) \quad \cdots \quad x(M-1,N-1)]^{\mathsf{T}}. \quad (1.4)$$

In analog video technology, *interlaced* acquisition is widely used, where the even and odd lines are captured at different time instances, (see Fig. 1.7). Here, a *video frame* consists of *two fields*, each containing only half number of lines. When the entire frame is captured simultaneously, the acquisition is *progressive*, as it is the case in movie films.

1.3 Digital Representation of Multimedia Signals

The process of digitization of a signal consists of sampling (see sec. 2.2) and quantization (see chapter 11). The resultant 'raw' digital format is denoted as *Pulse Code Modulation* (PCM) representation. These formats are often regarded as the original references in digital multimedia signal processing applications; specific characteristics related to such formats for visual and audio signals are explained in this section.

1.3.1 Image and Video Signals

When image and video signals are represented digitally by B bit per component[1] sample, 2^B different levels can be represented by PCM. $B=8$ (256 levels) is typically regarded to be sufficient for representation of signals acquired by video cameras, scanners or acquisition chips in digital cameras. Video cameras usually perform a nonlinear mapping of illumination values into amplitude values, providing a finer scale of amplitudes for darker values. This is denoted as the *gamma characteristic* of the acquisition device and can be described by the transfer equation

$$V = c_1 \cdot \Phi^{\gamma} + c_2 , \qquad (1.5)$$

where Φ is the luminous flux normalized to a maximum level, V the amplitude (voltage) of the signal, c_1 the sensitivity of the camera and c_2 an offset value[2].

These nonlinear characteristics are often ignored when digital representations are processed, but are important either for the purpose of presentation (printing, display) or for exact analysis. If a higher dynamic range shall be processed, more

---

[1] Components are colors, luminance or chrominance, see below.

[2] For example, the ITU-R Rec. BT.709 defines the transfer characteristic in two ranges, where the lower range has a linear characteristic:

$c_1 = 1.099$; $c_2 = -0.099$; $\gamma = 0.45$ for $1 \geq \Phi \geq 0.018$

$c_1 = 4.500$; $c_2 = 0.000$; $\gamma = 1.00$ for $0 \leq \Phi < 0.018$.

than 8 bit per sample are required; film sampling is mostly done using 10 or 12 bit. When image signals are mapped onto a linear amplitude scale (e.g. in color graphics), up to 16 bit are used for representation of each component. This is still relatively low as compared to the available scale of visually distinguishable natural illumination levels, which covers a range of nine decades in light intensity (in comparison, the dynamic range of audible signals covers only a range of around 7 decades in sound pressure). Hence, human eyes and also cameras are equipped with regulation mechanisms, which are the iris, the diaphragm and the sensitivity of the receptor medium; in cameras also the exposure time can be adapted to the intensity of light.

To capture and represent color images, the most common representation consists of three primary components of active light, red (R), green (G) and blue (B). These components are separately acquired and sampled. This results in a count of samples which is higher by a factor of three as compared to monochrome images. True representation of color may even require more components in a multi-spectral representation (see also sec. 6.1.3 and 7.1).

Color images and video are often represented by a *luminance* component Y and two *chrominance* (color difference) components. For the transformation between R,G,B and luminance/chrominance representations, different definitions exist, depending on the particular application domain. For example, in standard TV resolution video, the following transform is mainly used[1]:

$$Y = 0.299 \cdot R + 0.587 \cdot G + 0.114 \cdot B \,;$$

$$C_b = \frac{0.5}{0.866} \cdot (B - Y) \quad ; \quad C_r = \frac{0.5}{0.701} \cdot (R - Y). \tag{1.6}$$

For high definition (HD) video formats, the transform

$$Y = 0.2126 \cdot R + 0.7152 \cdot G + 0.0722 \cdot B \,;$$

$$C_b = \frac{0.5}{0.9278} \cdot (B - Y) \quad ; \quad C_r = \frac{0.5}{0.7874} \cdot (R - Y) \tag{1.7}$$

is more commonly used. By the transformations (1.6) and (1.7), difference components are limited to maximum ranges $\pm 0.5A$. The possible color variations in the R,G,B color space are restricted such that perceptually and statistically more important colors are represented more accurately. Chrominance components are in addition usually sub-sampled, which is reasonable as the human visual sense is not capable to perceive differences in color by the same high spatial resolution as for the luminance component.

---

[1] It is assumed here that the component Y cover the same amplitude range $0...A$ as the original components R,G,B, and the chrominance components cover a range $-A/2...+A/2$. In some cases, it is useful to restrict the ranges to $A_{min}...A_{max}$. This is easily achieved by modification of the components, $Y'=Y(A_{max}-A_{min})/A+A_{min}$; $C_{b/r}'=C_{b/r}(A_{max}-A_{min})/A$.

In interlaced sampling, sub-sampling of chrominances is mostly performed only in horizontal direction to avoid color artifacts in case of motion, while for progressive sampling both horizontal and vertical directions of chrominance can be sub-sampled into lower resolution. Component sampling ratios are often expressed in a notation $C_1:C_2:C_3$ to express the relative numbers of samples[1]. For example,

- when the same number of samples is used for all three components like in R,G,B, the expression is '4:4:4';
- a Y,C_b,C_r sampling structure with horizontal-only sub-sampling of the two chrominances is expressed by the notation '4:2:2', while '4:1:1' indicates horizontal sub-sampling by a factor 4;
- if sub-sampling is performed in both directions, i.e. half number of samples in chrominances along both horizontal and vertical directions, the notation is '4:2:0'.

The respective source format standards also specify the sub-sampled component sample positions in relation to the luminance sample positions.

Table 1.1. Typical resolution formats of digital still images

	Resolution						
	320x240	480x320	640x480	1024x768	1536x1024	2048x1536	3072x2048
Number of pixels	76,800	153,600	307,200	786,432	1.572,864	3.145,728	6.291,456
Number of bits per component, 8 bit precision	614,400	1228,800	2.457,600	6.291,456	12.582,912	25.165,824	50.331,648
Number of bits total (3 comp.)	1.843,200	3.686,400	7.372,800	18.874,368	37.748,736	75.497,472	150.994,944

Table 1.1 shows pixel resolutions (*width* x *height*) commonly used for still image acquisition by digital cameras; numbers of pixels, number of bits per component image using $B=8$ bit precision, and total number of bits resulting for all three components are also listed. The required data volume for a raw data representation can be calculated by multiplication of the numbers from up to five contributing dimensions: width, height, number of components, bit depth per component, and number of frames (in case of image sequences).

---

[1] The rationale behind some of the following notations is not fully consistent. Actually, the expression by only three numbers is not complete enough to cover all possible sampling ratios in horizontal and vertical directions.

For video representation, besides the total number of bits e.g. required to store a movie, the *number of bits per second* is more interesting for transmission over channels. It is straightforward then to multiply by the number of frames per second instead of total number of frames. Table 1.2 describes parameters of commonly used video formats. In some cases, numbers of pixels, frame rates etc. allow variations.

Table 1.2. Digital video formats and their properties

	Format						
	QCIF	CIF/ SIF	HHR	ITU-R BT.601 (SDTV)	ITU-R BT.709 (HDTV)	SMPTE 296M (720p)	DC
Sampling frequency (Y) [*MHz*]	--	--	--	13.55	74.25 ... 148.5	74.25	var.
Sampling structure*)	P	P	I	I	P/I	P	P
Pixels/line (Y)	176	352	360	720	1920	1280	≤ 4000
Number of lines (Y)	144	288 (240)	576 (480)	576 (480)	1080	720	≤ 3000
Color sampling	4:2:0	4:2:0	4:2:2	4:2:2	4:2:2 4:4:4	4:2:2 4:4:4	4:4:4
Pixel aspect ratio	4:3	4:3	4:3	4:3	16:9	16:9	var.
Frame rate [*Hz*]	5-15	10-30	25 (30)	25 (30)	24-60	24-60	24-72
PCM bit depth [*b*]	8	8	8	8	8-10	8-10	10-12
	Data rates						
per frame [*kbyte*]	38.02	126.7 ... 152.1	414.7 (345.6)	829.4 (691.2)	4147 ... 7776	1843 ... 3456	Up to 54000
per second [*Mbit/s*]	1.52 ... 9.12	10.13 ... 36.5	82.95	165,9	796.2 ... 3732	353.9 ... 1659	Up to 3.888 x 10^6

*) P/I: Progressive/Interlaced

For standard TV resolution, the source of the digital TV signal is the analog TV signal of 625 lines in Europe (525 lines in Japan or US), sampled by an interlaced schema. These signals are sampled by a rate of 13.55 MHz for the luminance. After removal of vertical blanking intervals, 575 (480) active lines remain. The horizontal blanking intervals are also removed, which gives around 704 active pixels per line. The digital formats listed in Tab. 1.2 are storing only those active pixels with a very small overhead of few surplus pixels from the blanking intervals. Japanese and US (NTSC) formats are traditionally using 60 fields per second (30 frames per second)[1], while in Europe, 50 fields per second (25 frames per second) is used in analog TV (PAL, SECAM). The digital standards defining HD formats are more flexible in terms of frame and field rates, allowing ranges of 24, 25, 30, 50 or 60 frames/second, 50 or 60 fields/second; movie material, interlaced and progressive video are supported. For higher resolutions, the '720p' format (720 lines progressive) is widely used in professional digital video cameras. All 'true' HDTV formats have 1080 lines in the digital signal.

There are other commonly-used formats, some of which are generated by digitally down-converting the standard TV resolution, e.g. the *half horizontal resolution* (HHR), the *Common Intermediate Format* (CIF), *Standard Intermediate Format* (SIF) and the *Quarter CIF* (QCIF). Higher resolutions beyond HD are often denoted as *Digital Cinema* (DC) formats. Extrapolating from formats presently used in digital movie production, a set of standardized image resolutions can be expected in the future which will have a size of up to 4000x3000 pixel; it can be expected that also higher frame rates must be supported.

Fig. 1.8 gives a coarse impression of the sampled image areas supported in formats between QCIF and HDTV. An increased number of samples can either be used to increase the resolution (spatial detail), or to display scenes by a wider angle. For example, in a cinema movie close-up views of human faces are rarely shown. Movies displayed on a cinema screen allow the observer's eye to explore the scene, while on standard definition TV screens and even more for the smaller formats, this capability is very limited.

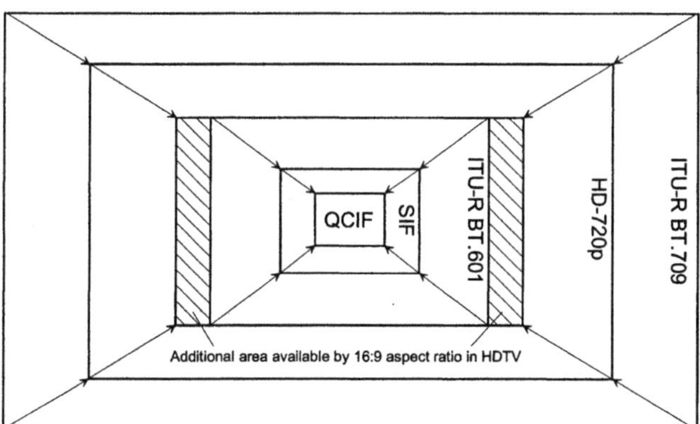

Fig. 1.8. Sizes and aspect ratios of digital video/image formats

For medical and scientific purpose, digital images with much higher resolution than in movie production are used, resolutions of up to 10,000x10,000 =

---

[1] 29.94 frames/second will exactly meet analog sampling rates.

100.000,000 pixels are quite common. Such formats are not realistic yet for real-time acquisition by digital video cameras, as the clock rates for sampling will be extremely high. In general, as it can be observed from the HDTV formats listed above, a certain tendency exists to finally replace the interlaced sampling structure by higher frame rate progressive sampling[1]. This is also important under the aspect of convergence between information technology and TV services, as progressive scanning is the usual operation mode of computer displays.

1.3.2   Speech and Audio Signals

For audio signals, parameters as sampling rate and precision (bit depth) take most influence on the resulting data rates of the digital representation. These parameters highly depend on the properties of the signals, and on the requirements for quality. In speech signal quantization, nonlinear mappings using logarithmic amplitude compression are used[2], which for the case of low amplitudes provides an equivalently low quantization noise as in 12 bit quantization, even though only 8 bit/sample are used. For music signals to be acquired by audio CD quality, linear 16 bit representation is most commonly necessary. For some specialized applications, even higher bit-depths and higher sampling rates than for CD are used. Table 1.3 shows typical parameters in the areas of speech and broadband-audio digital representation. The required bit rates (per second) are calculated by multiplication of the sampling rate by the bit depths and the number of audio channels. The number of bits required for storage is the rate per second multiplied by the duration of the audio track.

Table 1.3. Sampling and raw bit rates for audio signals

	Frequency range [Hz]	Sampling rate [KHz]	PCM resolution [bits/sample]	PCM rate [kb/s]
Telephony speech	300 – 3,400	8	8-16	64-128
Broadband speech	50 – 7,000	16	8-16	128-256
CD audio	10 – 20,000	44.1	2 x 16 (stereo)	1410
DVD multichannel	10 – 22,000	48	(5+1) x 16	$4.6 \cdot 10^3$
DVD audio	10 – 44,000	96	2 x 24	$4.6 \cdot 10^3$
Multichannel audio	10 – 44,000	96	(7+1) x 24	$18.4 \cdot 10^3$

---

[1] Interlaced sampling methods will nevertheless still play a dominant role in the future, as most material in TV archives is originally stored by these formats.

[2] Denoted as μ-law and A-law quantization characteristic

Due to the tremendous amount of rates necessary for representation of the original uncoded formats, the requirement for data compression by application of image, video and audio coding is permanently present, even though the available transmission bandwidth is further increasing by advances in communications technology. In general, the past experience has shown that multimedia traffic increases faster than new capacity is becoming available, and compressed transmission of data is inherently cheaper. If sufficient bandwidth is available, it is more efficiently used in terms of quality that serves the user, if the resolution of the signal is increased. Further, certain types of communication channels (in particular in mobile transmission) exist where the bandwidth is inherently expensive due to physical limitations.

1.4   Problems

Problem 1.1
A video movie of 150 minutes duration is digitally represented in format ITU-R 601 (720x576 pixels luminance, 2x360x576 pixels chrominances per frame, 25 frames/s, 8 bit/pixel PCM).
a)   Which hard disc capacity is required to store the entire movie in ITU-R 601 encoding?
b)   The movie shall be encoded using MPEG-2 compression, achieving a compression factor of 40:1 over PCM. Which file size is now required to store the stream on the hard disc?
c)   To transmit the movie over a satellite channel, error protection by channel coding is provided. This increases the rate by 10 %. Which transmission bandwidth (Mbit/s) of the channel is necessary?

Problem 1.2
Images from a Web camera shall be transmitted over a telephone modem, providing a constant data rate of 56 kbit/s. 16 kbit/s are needed for speech transmission. The frame rate shall be 10 frames/s.
a)   As the video coder does not output a constant-rate stream, it is necessary to buffer the information before transmitting over the channel. What maximum buffer size (measured in bits) is allowable, if the maximum delay caused by buffering shall be 200 ms?
b)   The output rate of the encoder can be changed by modification of the quantizer setting (the discrete-step 'Q factor'). As a rule of thumb, lowering the Q factor by one step may increase the rate by a factor 1.1. For one frame, it is measured that encoding has consumed 4.84 kbit. By how many steps must the Q factor be lowered or increased to avoid an over-run or under-run of the buffer?

Part A: Multimedia Signal Processing and Analysis

2 Signals and Sampling

One- or multidimensional sampling of sensor information is the origin of natural-source digital multimedia signals. The methodologies of sampling have an eminent impact on any subsequent processing steps, including encoding and content analysis of signals. In particular for video signals, full understanding of sampling phenomena is one key to further improve compression technology. To introduce the conditions behind multi-dimensional sampling, this chapter starts by a general introduction to multi-dimensional coordinate systems and related dependencies in the Fourier spectral domain. This also includes aspects of geometric modifications and their impact on the properties of the spectrum.

2.1 Signals and Fourier Spectra

Fourier spectral analysis is an important tool to explore the properties and perform processing of multimedia signals. It is assumed that the reader is familiar with processing of one-dimensional (1D), e.g. time-dependent (t) signals; therefore, concepts will be introduced in the subsequent sections primarily for two- and multi-dimensional signals. The *Fourier transform* of a 1D signal

$$X(j\omega) = \int_{-\infty}^{\infty} x(t)e^{-j\omega t}dt \qquad (2.1)$$

decomposes the signal into an infinite number of complex oscillations $exp(j2\pi t/\tau)$, which express sinusoids in their real and imaginary components. The *period* τ (in time-dependent signals) or *wavelength* λ (in spatially dependent signals) has a reciprocal relationship with *frequency* ω. In time-dependent signals the unit $Hz=s^{-1}$ is commonly used to express the frequency, while no explicit physical unit definition for spatially-dependent frequencies exists; a reciprocal value of a length unit, e.g. cm^{-1} is normally used, but other definitions which are normalized by the viewing distance exist as well (cf. sec. 6.1.2).

2.1.1   Spatial Signals and Two-dimensional Spectra

Rectangular coordinate systems. The amplitude of an image signal is dependent on the spatial position in two dimensions – horizontally and vertically. Spatially dependent frequency components shall be ω_1 (horizontally) and ω_2 (vertically). The two-dimensional Fourier transform of a spatially continuous signal is

$$X(j\omega_1, j\omega_2) = \int_{-\infty}^{\infty} \int_{-\infty}^{\infty} x(r,s) e^{-j\omega_1 r} e^{-j\omega_2 s}\, dr\, ds. \qquad (2.2)$$

(2.2) can be extended into a generic definition of κ-dimensional spectra associated with a κ-dimensional signal, when all frequency coordinates $\boldsymbol{\omega}=[\omega_1\ \omega_2\ ...\ \omega_\kappa]^T$ and signal coordinates in space and time $\mathbf{r}=[r\ s\ t\ ...]^T$ are expressed as vectors. Provided that the coordinate axes establish an orthogonal system,

$$X(j\boldsymbol{\omega}) = \int_{-\infty}^{\infty}..\int_{-\infty}^{\infty} x(\mathbf{r}) e^{-j\boldsymbol{\omega}^T \mathbf{r}}\, d\mathbf{r}. \qquad (2.3)$$

The multi-dimensional Fourier spectrum can alternatively be expressed by magnitude and phase,

$$X(\omega) = |X(j\omega)| = \sqrt{X(j\omega)\cdot X*(j\omega)}$$

$$\Phi(\omega) = \arctan\frac{\mathrm{Im}\{X(j\omega)\}}{\mathrm{Re}\{X(j\omega)\}} \pm k(\omega)\cdot\pi \ ; \ k(\omega) = \begin{cases} 1: \mathrm{Re}\{X(j\omega)\}<0 \\ 0: \mathrm{Re}\{X(j\omega)\}\ge 0. \end{cases} \qquad (2.4)$$

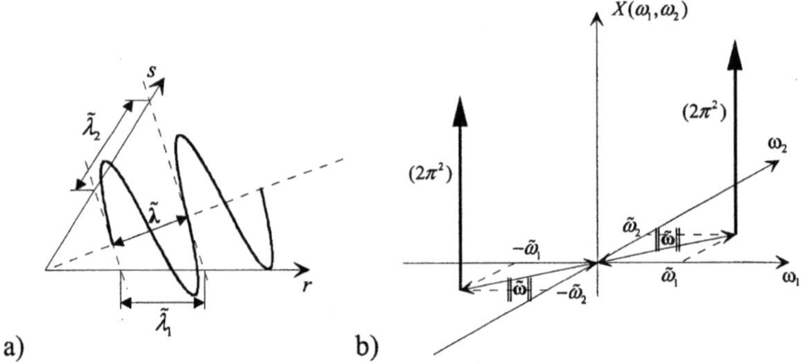

Fig. 2.1. a Directional orientation and wavelength of a sinusoid in the spatial plane **b** Directional orientation of the related Dirac impulses, representing the sinusoid spectrum in the 2D frequency domain

Example. A two-dimensional cosine signal is given as

$$x_c(r,s) = \cos\left(\tilde{\omega}_1 r + \tilde{\omega}_2 s\right). \qquad (2.5)$$

The solution for the Fourier integral of this signal is zero except for the frequencies $(\tilde{\omega}_1, \tilde{\omega}_2)$ and $(-\tilde{\omega}_1, -\tilde{\omega}_2)$ where singularities exist. This can be expressed by a pair of *Dirac impulses* (Fig. 2.1a):

$$X_c(j\omega_1, j\omega_2) = 2\pi^2 \left[\delta(\omega_1 + \tilde{\omega}_1, \omega_2 + \tilde{\omega}_2) + \delta(\omega_1 - \tilde{\omega}_1, \omega_2 - \tilde{\omega}_2) \right]. \tag{2.6}$$

In the 2-dimensional frequency plane, the distance of each of these impulses from the origin and the angular orientation (relative to the ω_1 axis) can be expressed as

$$\|\tilde{\omega}\| = \sqrt{\tilde{\omega}_1^{\,2} + \tilde{\omega}_2^{\,2}}\,; \quad \varphi = \arctan\frac{\tilde{\omega}_2}{\tilde{\omega}_1}\,; \quad \tilde{\omega}_1 \geq 0. \tag{2.7}$$

The two-dimensional sinusoid can be interpreted as a wave front with a propagation direction orientated by this angle. Sections of this wave front in the horizontal and vertical directions are sinusoids of frequencies $\tilde{\omega}_1$ or $\tilde{\omega}_2$, respectively. These correspond to the wavelengths (Fig. 2.1b)

$$\tilde{\lambda}_1 = \frac{2\pi}{\tilde{\omega}_1} \quad ; \quad \tilde{\lambda}_2 = \frac{2\pi}{\tilde{\omega}_2}. \tag{2.8}$$

Regard a horizontal section from the sinusoid of wave length $\tilde{\omega}_1$. The signal phase shall be shifted by $\phi(s)$, depending on the vertical position

$$x(r,s) = \sin\left(\tilde{\omega}_1 r + \phi(s)\right). \tag{2.9}$$

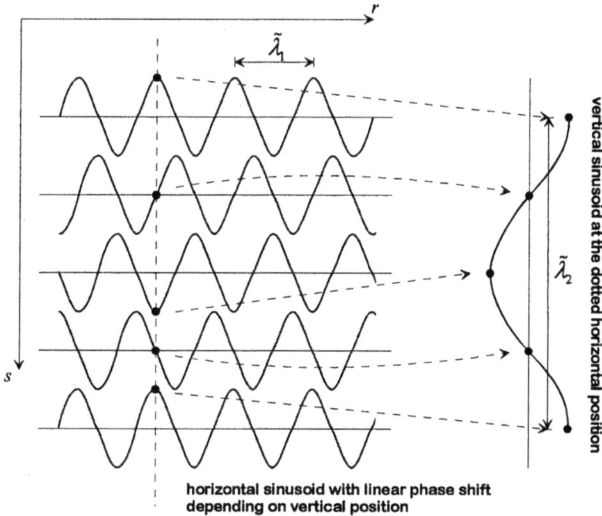

Fig. 2.2. Interpretation of a two-dimensional sinusoid: Linear phase shift depending on vertical position results in vertical wavelength and frequency

If the dependency on the vertical position is linear, $\phi(s) = \tilde{\omega}_2 s$. Then, for any $s = k\tilde{\lambda}_2$ (k an arbitrary integer number), $\phi(s) = 2\pi k$. This determines distances where the signal has equal amplitude, i.e. vertical periodicity. Thus, the 2-dimensional sinusoid can be interpreted as a 1-dimensional sinusoid having a linear phase shift depending on the second dimension. This is illustrated in Fig. 2.2, where any vertical section over the different phase-shifted versions will be a sinusoid of wave length $\tilde{\lambda}_2$. The wave length of the 2-dimensional sinusoid, measured by the direction of wave front propagation, can be determined from (2.7) and (2.8) as

$$\tilde{\lambda} = \frac{1}{\|\tilde{\omega}\|} = \frac{\tilde{\lambda}_1 \tilde{\lambda}_2}{\sqrt{\tilde{\lambda}_1^2 + \tilde{\lambda}_2^2}}. \tag{2.10}$$

Table 2.1 summarizes some of the most important properties of the κ-dimensional Fourier transform ($\kappa \geq 1$), which express the modification of the spectrum by certain operations applied to the signal.

Table 2.1. Properties of the κ-dimensional Fourier transform

	Signal domain	Fourier domain
Linearity	$a \cdot x_1(\mathbf{r}) + b \cdot x_2(\mathbf{r})$	$a \cdot X_1(j\omega) + b \cdot X_2(j\omega)$
Shift	$x(\mathbf{r} - \mathbf{r}')$	$X(j\omega) \cdot e^{-j\omega^T \mathbf{r}'}$
Convolution	$y(\mathbf{r}) = x(\mathbf{r}) * h(\mathbf{r})$ $= \int_{-\infty}^{\infty} .. \int_{-\infty}^{\infty} x(\mathbf{r} - \boldsymbol{\rho}) \cdot h(\boldsymbol{\rho}) \, d\boldsymbol{\rho}$	$Y(j\omega) = X(j\omega) \cdot H(j\omega)$
Multiplication	$y(\mathbf{r}) = x_1(\mathbf{r}) \cdot x_2(\mathbf{r})$	$Y(j\omega) = \frac{1}{(2\pi)^\kappa} X_1(j\omega) * X_2(j\omega)$
Separability	$x(\mathbf{r}) = x_1(r) \cdot x_2(s) \cdot ...$	$X(j\omega) = X_1(j\omega_1) \cdot X_2(j\omega_2) \cdot ...$
Inversion	$x(-\mathbf{r})$	$X*(j\omega)$
Complex modulation	$x(\mathbf{r}) \cdot e^{j\tilde{\omega}^T \mathbf{r}}$	$X[j(\omega - \tilde{\omega})]$
Scaling	$x(r \cdot U, s \cdot V, ...)$	$\frac{1}{\lvert U \cdot V \cdot ... \rvert} X(j\frac{\omega_1}{U}, j\frac{\omega_2}{V}, ...)$

By application of the *inverse Fourier transform*, the signal can be reconstructed from the Fourier spectrum:

$$x(\mathbf{r}) = \frac{1}{(2\pi)^\kappa} \int\limits_{-\infty}^{\infty} \dots \int\limits_{-\infty}^{\infty} X(j\boldsymbol{\omega}) \cdot e^{j\boldsymbol{\omega}^T \mathbf{r}} d\boldsymbol{\omega}. \tag{2.11}$$

Coordinate system mapping. Rectangular coordinate systems are only a special case for the description of two- and multidimensional signals. They allow expressing the relationships between the signal domain and the frequency domain by a multi-dimensional Fourier transform with separable integration over the different dimensions. The rectangular two-dimensional coordinate system is described by a reference system of unit vectors $\mathbf{e}_0 = [1\ 0]^T$ and $\mathbf{e}_1 = [0\ 1]^T$, which establish orientations of orthogonal axes. Any coordinate pair (r,s) can then be expressed as a vector $\mathbf{r} = r \cdot \mathbf{e}_0 + s \cdot \mathbf{e}_1$. The relationship with frequency vectors $\boldsymbol{\omega} = \omega_1 \cdot \mathbf{e}_0 + \omega_2 \cdot \mathbf{e}_1$ is given by (2.3), where the coordinate system describing the frequency plane must be based on orthogonal axes of same orientations (see Fig. 2.3a). Now, a coordinate mapping $\mathbf{r}' = r \cdot \mathbf{d}_0 + s \cdot \mathbf{d}_1$ shall be applied, which can be expressed in matrix notation $\mathbf{r}' = \mathbf{D} \cdot \mathbf{r}$ using the mapping matrix[1]

$$\mathbf{D} = \begin{bmatrix} d_{00} & d_{01} \\ d_{10} & d_{11} \end{bmatrix} = \begin{bmatrix} \mathbf{d}_0 & \mathbf{d}_1 \end{bmatrix}. \tag{2.12}$$

The vectors $\mathbf{d}_0$ und $\mathbf{d}_1$ are the *basis vectors* of this mapping. By application in the signal plane, a complementary mapping into frequency coordinates must occur, which shall be expressed similarly as $\boldsymbol{\omega}' = \mathbf{F} \cdot \boldsymbol{\omega}$ by using a mapping matrix (see Fig. 2.3b)

$$\mathbf{F} = \begin{bmatrix} f_{00} & f_{01} \\ f_{10} & f_{11} \end{bmatrix} = \begin{bmatrix} \mathbf{f}_0 & \mathbf{f}_1 \end{bmatrix}. \tag{2.13}$$

Unless the determinants of matrices $\mathbf{D}$ or $\mathbf{F}$ are zero, the mappings must be reversible, such that $\mathbf{r} = \mathbf{D}^{-1} \cdot \mathbf{r}'$ and $\boldsymbol{\omega} = \mathbf{F}^{-1} \cdot \boldsymbol{\omega}'$. The Fourier transform in the mapped coordinate system can then be expressed as

$$\tilde{X}(j\boldsymbol{\omega}') = \int\limits_{-\infty}^{\infty} \dots \int\limits_{-\infty}^{\infty} \tilde{x}(\mathbf{r}') \cdot e^{-j\boldsymbol{\omega}'^T \mathbf{r}'} d\mathbf{r}'. \tag{2.14}$$

In the Fourier integrals of (2.3) and (2.14), the coordinates describing positions in the signal and frequency domains can now be expressed by the respective mapping functions, such that

---

[1] It is assumed here that the origin of the coordinate transform is not changed by the mapping. A more general form is the mapping $\mathbf{r}' = \mathbf{D} \cdot \mathbf{r} + \mathbf{t}$, where $\mathbf{t}$ expresses a translation of the origin. This is the *affine transform* (7.108). Regarding the spectral relationships, the translation will only introduce a linear phase shift $e^{j\boldsymbol{\omega}^T \mathbf{t}}$ in the spectrum of $\boldsymbol{\omega}'$, which is not explicitly proven here; a similar effect occurs for the case of motion translation, see sec. 2.1.2.

$$X(j\omega) = \int_{-\infty}^{\infty} \cdots \int_{-\infty}^{\infty} x(\mathbf{D}^{-1}\mathbf{r}') \cdot e^{-j(\omega')^{\mathrm{T}}(\mathbf{F}^{-1})^{\mathrm{T}}\mathbf{D}^{-1}\mathbf{r}'} d(\mathbf{D}^{-1}\mathbf{r}') \tag{2.15}$$

and

$$\tilde{X}(j\omega') = \int_{-\infty}^{\infty} \cdots \int_{-\infty}^{\infty} \tilde{x}(\mathbf{D}\mathbf{r}) \cdot e^{-j\omega^{\mathrm{T}}\mathbf{F}^{\mathrm{T}}\mathbf{D}\mathbf{r}} d(\mathbf{D}\mathbf{r}). \tag{2.16}$$

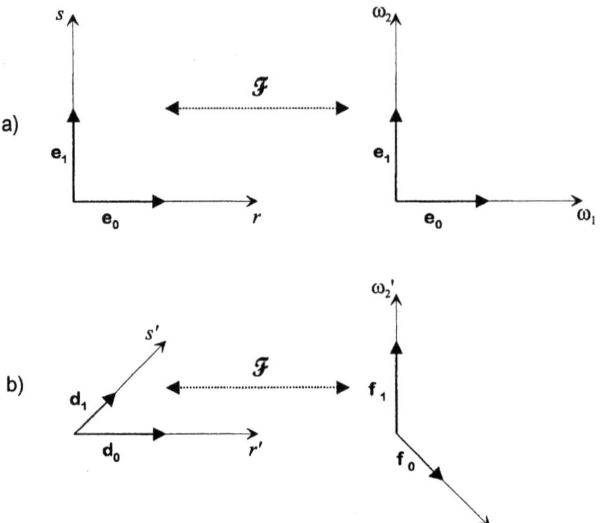

Fig. 2.3. Rectangular (**a**) and nonrectangular (**b**) coordinate systems in the spatial domain and the 2D frequency domain

The signal amplitudes shall not be changed by the coordinate mapping,

$$x(\mathbf{r}) = x(\mathbf{D}^{-1}\mathbf{r}') = \tilde{x}(\mathbf{r}') = \tilde{x}(\mathbf{D}\mathbf{r}). \tag{2.17}$$

Further, the spectral values must also uniquely be mapped, but may be subject to a constant scaling of amplitudes

$$X(j\omega) = X(j\mathbf{F}^{-1}\omega') = \frac{1}{c}\tilde{X}(j\omega') = \frac{1}{c}\tilde{X}(j\mathbf{F}\omega). \tag{2.18}$$

Then, comparing (2.3), (2.14)-(2.17), the following relationships must hold:

$$e^{-j\omega^{\mathrm{T}}\mathbf{r}} = e^{-j(\omega')^{\mathrm{T}}(\mathbf{F}^{-1})^{\mathrm{T}}\mathbf{D}^{-1}\mathbf{r}'} = e^{-j(\omega')^{\mathrm{T}}\mathbf{r}'} = e^{-j\omega^{\mathrm{T}}\mathbf{F}^{\mathrm{T}}\mathbf{D}\mathbf{r}}. \tag{2.19}$$

Further, using the inverse Fourier transform conditions to reconstruct the amplitudes at the origin of the respective coordinate system $x(\mathbf{0}) = \tilde{x}(\mathbf{0})$, the following relationship applies:

$$\int_{-\infty}^{\infty} \ldots \int_{-\infty}^{\infty} X(j\omega)d\omega = \int_{-\infty}^{\infty} \ldots \int_{-\infty}^{\infty} \tilde{X}(j\omega')d\omega' = \int_{-\infty}^{\infty} \ldots \int_{-\infty}^{\infty} \tilde{X}(j\omega')d(\mathbf{F}\omega). \qquad (2.20)$$

Hence,

$$X(j\omega) = \frac{1}{\det|\mathbf{D}|} \tilde{X}(j\omega'). \qquad (2.21)$$

From (2.19),

$$\mathbf{F}^{\mathrm{T}}\mathbf{D} = (\mathbf{F}^{-1})^{\mathrm{T}}\mathbf{D}^{-1} = \mathbf{I}, \qquad (2.22)$$

and as $\mathbf{A}^{-1}\mathbf{A}=\mathbf{I}$,

$$\mathbf{D}^{-1} = \mathbf{F}^{\mathrm{T}}; \quad \mathbf{F}^{-1} = \mathbf{D}^{\mathrm{T}} \quad \Rightarrow \quad \mathbf{F} = \left[\mathbf{D}^{-1}\right]^{\mathrm{T}}; \quad \mathbf{D} = \left[\mathbf{F}^{-1}\right]^{\mathrm{T}}. \qquad (2.23)$$

For the 2D case, $\mathbf{F}^{\mathrm{T}}\mathbf{D}=\mathbf{I}$ imposes four conditions on the basis vectors:

$$\mathbf{f}_0^{\mathrm{T}} \cdot \mathbf{d}_0 = 1; \quad \mathbf{f}_1^{\mathrm{T}} \cdot \mathbf{d}_0 = 0 ; \quad \mathbf{f}_0^{\mathrm{T}} \cdot \mathbf{d}_1 = 0; \quad \mathbf{f}_1^{\mathrm{T}} \cdot \mathbf{d}_1 = 1. \qquad (2.24)$$

It is evident that each basis vector in the spatial domain $\mathbf{d}_i$ has its orthogonal counterpart $\mathbf{f}_j$ in the frequency domain, such that $\mathbf{f}_j^{\mathrm{T}}\mathbf{d}_i=0$ when $i \neq j$; for $i=j$, the inner product of the vectors is unity[1]. The basis systems described by $\mathbf{D}$ and $\mathbf{F}$ are *biorthogonal* according to (B.21), which means that $\mathbf{F}$ is the dual matrix of $\mathbf{D}$. Fig. 2.4 shows the following examples :

- *Scaled rectangular* mapping:

$$\mathbf{D}_{scal} = \begin{bmatrix} U & 0 \\ 0 & V \end{bmatrix}; \; \mathbf{F} = \begin{bmatrix} \frac{1}{U} & 0 \\ 0 & \frac{1}{V} \end{bmatrix}; \; \mathbf{D}^{-1} = \mathbf{F}^{\mathrm{T}} = \begin{bmatrix} \frac{1}{U} & 0 \\ 0 & \frac{1}{V} \end{bmatrix}; \; \mathbf{F}^{-1} = \mathbf{D}^{\mathrm{T}} = \begin{bmatrix} U & 0 \\ 0 & V \end{bmatrix} \qquad (2.25)$$

- *Shear* mapping :

$$\mathbf{D}_{shear} = \begin{bmatrix} 1 & v \\ 0 & 1 \end{bmatrix}; \; \mathbf{F} = \begin{bmatrix} 1 & 0 \\ -v & 1 \end{bmatrix}; \; \mathbf{D}^{-1} = \mathbf{F}^{\mathrm{T}} = \begin{bmatrix} 1 & -v \\ 0 & 1 \end{bmatrix}; \; \mathbf{F}^{-1} = \mathbf{D}^{\mathrm{T}} = \begin{bmatrix} 1 & 0 \\ v & 1 \end{bmatrix} \qquad (2.26)$$

- *Quincunx* mapping :

$$\mathbf{D}_{quin} = \begin{bmatrix} 2 & 1 \\ 0 & 1 \end{bmatrix}; \; \mathbf{F} = \begin{bmatrix} \frac{1}{2} & 0 \\ -\frac{1}{2} & 1 \end{bmatrix}; \; \mathbf{D}^{-1} = \mathbf{F}^{\mathrm{T}} = \begin{bmatrix} \frac{1}{2} & -\frac{1}{2} \\ 0 & 1 \end{bmatrix}; \; \mathbf{F}^{-1} = \mathbf{D}^{\mathrm{T}} = \begin{bmatrix} 2 & 0 \\ 1 & 1 \end{bmatrix} \qquad (2.27)$$

- *Hexagonal* mapping :

$$\mathbf{D}_{hex} = \begin{bmatrix} \frac{2}{\sqrt{3}} & \frac{1}{\sqrt{3}} \\ 0 & 1 \end{bmatrix}; \; \mathbf{F} = \begin{bmatrix} \frac{\sqrt{3}}{2} & 0 \\ -\frac{1}{2} & 1 \end{bmatrix}; \; \mathbf{D}^{-1} = \mathbf{F}^{\mathrm{T}} = \begin{bmatrix} \frac{\sqrt{3}}{2} & -\frac{1}{2} \\ 0 & 1 \end{bmatrix}; \; \mathbf{F}^{-1} = \mathbf{D}^{\mathrm{T}} = \begin{bmatrix} \frac{2}{\sqrt{3}} & 0 \\ \frac{1}{\sqrt{3}} & 1 \end{bmatrix} \qquad (2.28)$$

- *Rotational* mapping, clockwise by an angle φ :

---

[1] These conditions apply in general for arbitrary number of dimensions.

$$\mathbf{D}_{rot} = \begin{bmatrix} \cos\varphi & \sin\varphi \\ -\sin\varphi & \cos\varphi \end{bmatrix} = \mathbf{F}\,; \quad \mathbf{D}^{-1} = \mathbf{F}^{\mathrm{T}} = \mathbf{F}^{-1} = \mathbf{D}^{\mathrm{T}} = \begin{bmatrix} \cos\varphi & -\sin\varphi \\ \sin\varphi & \cos\varphi \end{bmatrix}. \tag{2.29}$$

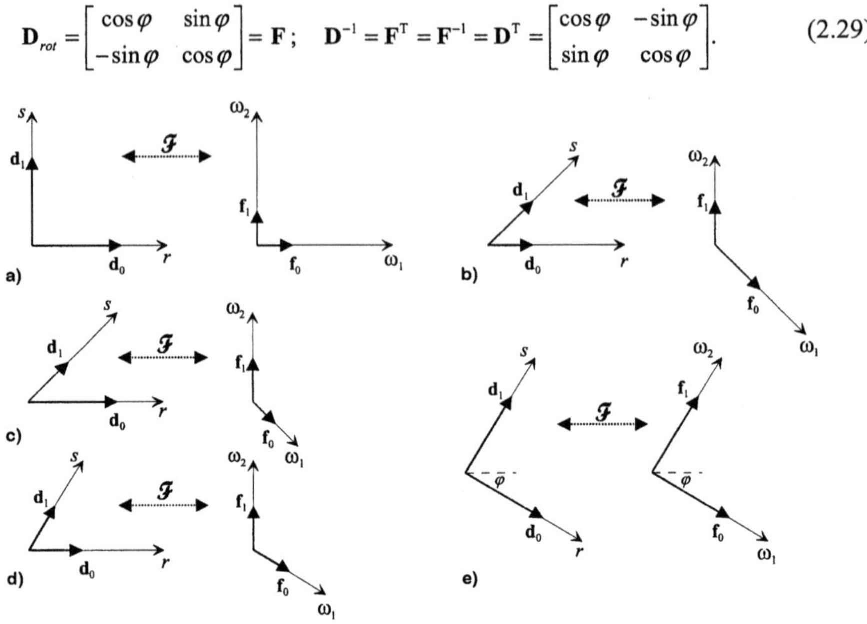

Fig. 2.4. Basis vectors of different geometric mapping systems:
a Scaled rectangular **b** Shear, v=1 **c** Quincunx **d** Hexagonal **e** Rotation

The scaled rectangular transformation describes the scaling property of the Fourier transform according to Table 2.1, where the amplitude factor results by (2.21) from the determinant of the transformation matrix. The shear mapping in principle allows re-defining the angular orientation of two-dimensional signals, while the horizontal spectral characteristics are retained (in case where only the vertical axis is sheared). When the signal is rotated, the spectrum rotates likewise. The coordinate transformations introduced here allow understanding what impact geometric modifications have on the spectra of two- and multi-dimensional signals. These relationships will also be used to analyze sampling of multidimensional signals.

2.1.2   Spatio-temporal Signals

In a video signal, two-dimensional images (frames) also vary over time. The time dependency t results in a 'temporal' frequency ω_3, where the Fourier spectrum is

$$X(j\omega_1, j\omega_2, j\omega_3) = \int\limits_{-\infty}^{\infty} \int\limits_{-\infty}^{\infty} \int\limits_{-\infty}^{\infty} x(r,s,t)e^{-j\omega_1 r} e^{-j\omega_2 s} e^{-j\omega_3 t}\, dr\,ds\,dt. \tag{2.30}$$

For the case of sinusoids, the spectral property resulting by temporal changes can be interpreted similarly to Fig. 2.3. In particular, if a constant motion (without local variations and without acceleration) occurs, the behavior of the signal can be expressed by a linear phase shift, depending on time t. Consider first the case of zero motion, $x(r,s,t)=x(r,s,0)$. Then, the three-dimensional Fourier spectrum (2.30) is

$$X(j\omega_1,j\omega_2,j\omega_3) = \int_{-\infty}^{\infty}\int_{-\infty}^{\infty} x(r,s,0)e^{-j\omega_1 r}e^{-j\omega_2 s}\,drds \cdot \int_{-\infty}^{\infty} e^{-j\omega_3 t}\,dt = X(j\omega_1,j\omega_2)\big|_{t=0} \cdot \delta(\omega_3).$$

$$(2.31)$$

The Dirac impulse $\delta(\omega_3)$ is the Fourier transform of a signal which is constant over time. Multiplication by Dirac impulses defines an ideal sampling. From (2.31), the 3D spectrum in case of unchanged signals just consists of *one sample* located at $\omega_3=0$:

$$X(j\omega_1,j\omega_2,j\omega_3) \sim \begin{cases} X(j\omega_1,j\omega_2)\big|_{t=0} & \text{when } \omega_3 = 0 \\ 0 & \text{when } \omega_3 \neq 0. \end{cases} \qquad (2.32)$$

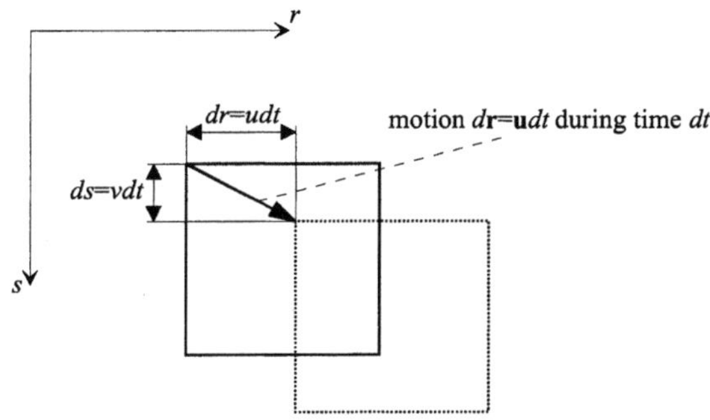

Fig. 2.5. Spatial shift caused by translational motion of velocities u and v

If motion is present in the signal, a spatial shift $dr=u\cdot dt$ in horizontal direction and $ds=v\cdot dt$ in vertical direction occurs within time interval dt, linearly dependent on the velocity vector $\mathbf{u}=[u,v]^{\mathrm{T}}$ (see Fig. 2.5). In case of constant-velocity translational motion, taking reference to the signal for $t=0$,

$$x(r,s,t) = x(r+ut,s+vt,0) \qquad (2.33)$$

$$\Rightarrow X(j\omega_1,j\omega_2,j\omega_3) = \int_{-\infty}^{\infty}\int_{-\infty}^{\infty}\int_{-\infty}^{\infty} x(r+ut,s+vt,0)\cdot e^{-j\omega_1 r}e^{-j\omega_2 s}e^{-j\omega_3 t}\,drdsdt. \qquad (2.34)$$

By substituting ($\eta=r+ut \Rightarrow d\eta=dr,\ r=\eta-ut$; $\xi=s+vt \Rightarrow d\xi=ds,\ s=\xi-vt$), the temporal dependency can be separated in the Fourier integration

$$X(j\omega_1,j\omega_2,j\omega_3) = \int_{-\infty}^{\infty}\int_{-\infty}^{\infty} x(\eta,\xi)\cdot e^{-j\omega_1 \eta}e^{-j\omega_2 \xi}\,d\eta d\xi \cdot \int_{-\infty}^{\infty} e^{-j(\omega_3-\omega_1 u-\omega_2 v)t}\,dt$$

$$= X(j\omega_1,j\omega_2)\big|_{t=0} \cdot \delta(\omega_3-\omega_1 u-\omega_2 v). \qquad (2.35)$$

Thus,

$$X(j\omega_1, j\omega_2, j\omega_3) \sim \begin{cases} X(j\omega_1, j\omega_2) & \text{when} \quad \omega_3 = \omega_1 u + \omega_2 v \\ 0 & \text{when} \quad \omega_3 \neq \omega_1 u + \omega_2 v. \end{cases} \tag{2.36}$$

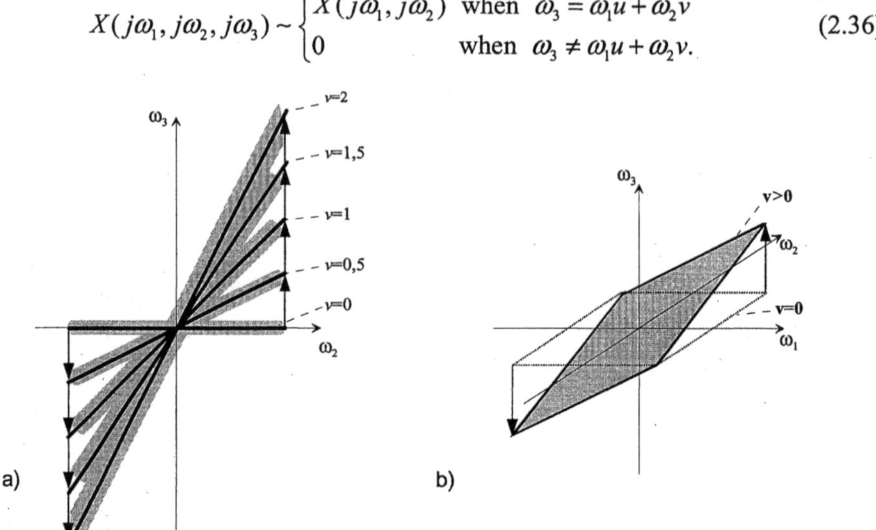

Fig. 2.6. **a** Shear of the non-zero spectral components by different translational motion velocities, shown in an (ω_2, ω_3) section of the 3D frequency domain **b** Position of the non-zero spectral components in cases of zero and non-zero 2D translational motion

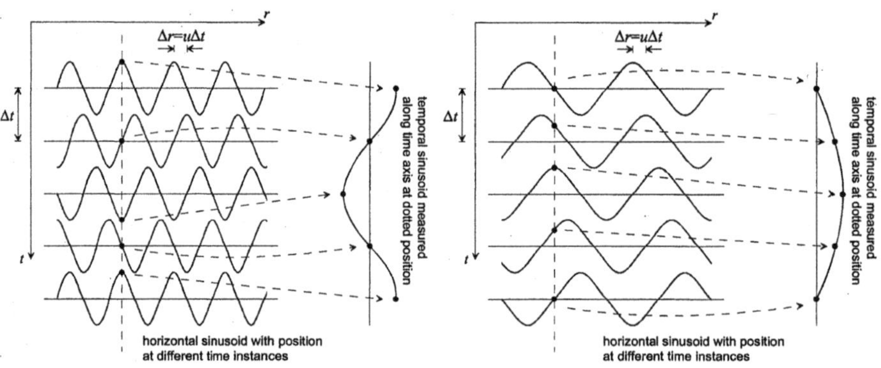

Fig. 2.7. Interpretation of the frequency in ω_3 for two sinusoids of different spatial frequencies, which are subject to the same translational motion

The spectrum $X(j\omega_1, j\omega_2)$ is now sampled on a plane $\omega_3 = \omega_1 u + \omega_2 v$ in the 3D frequency domain. Fig. 2.6a shows positions of non-zero spectrum planes for different normalized velocities v, where the (ω_2, ω_3) section is shown for $\omega_1 = 0$. Fig. 2.6b shows qualitatively the behavior in the full $(\omega_1, \omega_2, \omega_3)$ space for the zero-motion case and for motion by one constant velocity $\mathbf{u} > 0$, where the spectrum is assumed to be band-limited in ω_1 and ω_2.

From Fig. 2.6, it is obvious that the positions of non-zero spectral values in case of constant velocity establish a linear relationship between ω_3 and the frequencies relating to spatial coordinates. This effect can also be interpreted in the signal domain. Fig. 2.7 shows two sinusoids of different horizontal frequencies ω_1, both moving by the same velocity. The phase shift occurring due to the constant velocity motion linearly depends on the spatial frequency, which by a similar relationship as described in (2.9) effects a linear increase of the frequency ω_3 as well.

2.2 Sampling of Multimedia Signals

2.2.1 The Sampling Theorem

According to the Fourier transform relationships, any signal can be composed from an infinite number of sinusoids. The *sampling theorem* states that any sinusoidal component, having a period of length λ, can be reconstructed perfectly if the signal is sampled by an ideal sampling function with sample positions spaced regularly by $R < \lambda/2$. Defining the signal frequency $\omega = 2\pi/\lambda$ and the sampling frequency $\omega_R = 2\pi/R$, the signal must be band-limited prior to sampling into a frequency range $\omega_{max} < \omega_R/2$; otherwise, no perfect reconstruction from the samples will be possible. To illustrate this, Fig. 2.8 shows sampled sinusoids of frequencies $\omega < \omega_R/2$, $\omega = \omega_R/2$ and $\omega > \omega_R/2$. For the critical case $\omega = \omega_R/2$, it is not possible to reconstruct the *amplitude* of the sinusoid, unless the positions of the samples relative to the phase of the signal were known. For $\omega > \omega_R/2$, the amplitude is correct, but the *frequency* of the signal is interpreted wrongly. This effect is known as *aliasing*.

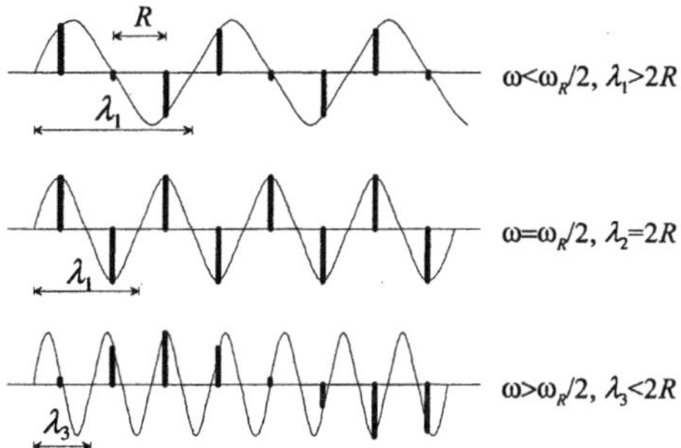

Fig. 2.8. Interpretation of the alias effect in the signal domain

The ideal sampling of a one-dimensional signal can mathematically be expressed as a multiplication by a train of Dirac impulses of period R,

$$\delta_R(r) = \sum_{m=-\infty}^{\infty} \delta(r - mR).$$

(2.37)

The Dirac impulse has the *sifting property*

$$x(\rho) = \int_{-\infty}^{\infty} x(r)\delta(r - \rho)dr,$$

(2.38)

from which

$$\int_{-\infty}^{\infty} e^{-j\omega r} \cdot \delta(r - mR)dr = e^{-j\omega mR}.$$

(2.39)

The Fourier transform of the Dirac impulse train is also a Dirac impulse train. Regarding the scaling property from Table 2.1, the solution is

$$\sum_{m=-\infty}^{\infty} e^{-jm\omega R} = \frac{2\pi}{R} \sum_{k=-\infty}^{\infty} \delta(\omega - k\omega_R) = \frac{2\pi}{R}\delta_{\omega_R}(\omega).$$

(2.40)

By the multiplication property of the Fourier transform (Table 2.1), and due to the fact that convolution by a single Dirac impulse effects a shift on the respective axis

$$X(j\omega) * \delta(\omega - k\omega_R) = X(j(\omega - k\omega_R)),$$

(2.41)

the spectrum of the sampled signal becomes an ω_R-periodic version of the original spectrum. Copies of the spectrum can be found at multiples of the sampling frequency (see Fig. 2.9):

$$X^{(R)}(j\omega) = \frac{1}{2\pi}\left[X(j\omega) * \frac{2\pi}{R}\delta_{\omega_R}(\omega) \right] = \frac{1}{R}\sum_{k=-\infty}^{\infty} X(j(\omega - k\omega_R)).$$

(2.42)

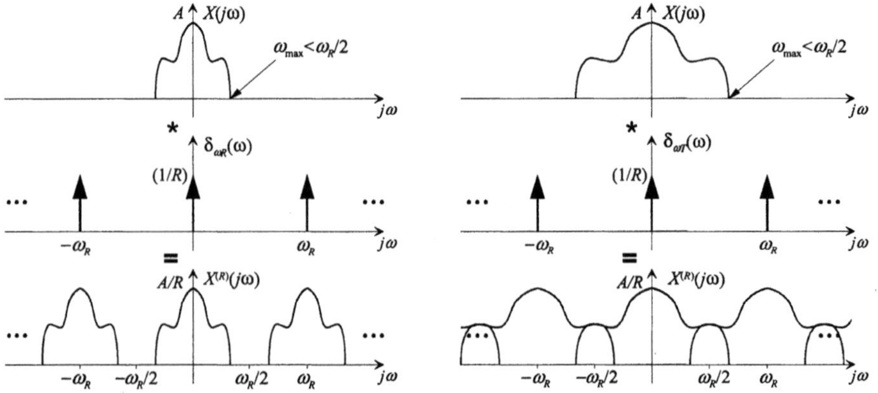

Fig. 2.9. Interpretation of alias in the frequency domain

If the spectrum of the signal contains components which are larger than half of the sampling frequency, the base band spectrum overlaps with its periodic copies (Fig. 2.9 right), which gives an interpretation of the alias effect in the frequency domain. It will no longer be possible to reconstruct the signal by a lowpass filter of cutoff frequency $\omega_R/2$. For a single sinusoid of frequency $\tilde{\omega}$, $\omega_R/2 \leq \tilde{\omega} \leq \omega_R$, the resulting alias frequency is $\omega_R - \tilde{\omega}$, which is a sinusoid of lower frequency as shown in Fig. 2.8 bottom.

When the sampled signal is processed in multimedia signal processing, it is mostly assumed that the sampling conditions have been properly observed. As then one period of the spectrum represents the entire information, the *normalized periodic frequency* Ω is introduced which is independent of R,

$$X(j\Omega) = \frac{1}{R} X(j\omega) * \delta_{\omega_R}(\omega) \quad \text{with} \quad \Omega = \frac{2\pi\omega}{\omega_R} = \omega R. \tag{2.43}$$

It is only necessary to regard the range $-\pi \leq \Omega < \pi$. Due to the sifting property (2.38), the Fourier transform (2.1) can be re-written as a sum, such that it is possible to compute the spectrum directly from the sequence of samples:

$$X(j\Omega) = \sum_{m=-\infty}^{\infty} x(m)e^{-jm\Omega}. \tag{2.44}$$

2.2.2   Separable Two-dimensional Sampling

Separable two- or multidimensional sampling is characterized by independent sampling positions being independent in the respective dimensions. Fig. 2.10 shows an example for the evolution of a two-dimensional, separable grid of Dirac impulses. Assume that sampling distances are S in vertical direction and R in horizontal direction. This can be characterized by *Dirac line impulses* $\delta_S(s)$ and $\delta_R(r)$, which are equally-spaced lines of Dirac property[1], the first being parallel with the r axis, the latter parallel with the s axis. Observe that each of these line impulses performs sampling in only one of the spatial dimensions; indeed $\delta_S(s)$ characterizes the line sampling structure of analog video systems. By multiplication of the two line impulses, which are perpendicular oriented, the rectangular two-dimensional Dirac impulse grid is defined as (Fig. 2.10d)

---

[1] The Dirac impulse $\delta(\mathbf{r})$ defined over a one- or multi-dimensional variable $\mathbf{r}$ is a function from distribution theory; formally it can be expressed as being zero for any point $\mathbf{r} \neq 0$, and the condition $\int_{-\infty}^{\infty} \cdots \int_{-\infty}^{\infty} \delta(\mathbf{r})d\mathbf{r} = 1$; see Fig. 2.10a.

$$\delta_{R,S}(r,s) = \delta_R(r) \cdot \delta_S(s) = \sum_{m=-\infty}^{\infty} \sum_{n=-\infty}^{\infty} \delta(r-mR, s-nS) \,. \tag{2.45}$$

In analogy with (2.40), using the separability property of the 2D Fourier transform, the rectangular impulse grid then has a two-dimensional spectrum

$$\sum_{m=-\infty}^{\infty} \sum_{n=-\infty}^{\infty} e^{-jm\omega_1 R} e^{-jn\omega_2 S} = \frac{2\pi}{R} \delta_{\omega_R}(\omega_1) \cdot \frac{2\pi}{S} \delta_{\omega_S}(\omega_2) = \frac{4\pi^2}{RS} \delta_{\omega_R,\omega_S}(\omega_1,\omega_2)$$

$$= \frac{4\pi^2}{RS} \sum_{k=-\infty}^{\infty} \sum_{l=-\infty}^{\infty} \delta(\omega_1 - k\omega_R, \omega_2 - l\omega_S). \tag{2.46}$$

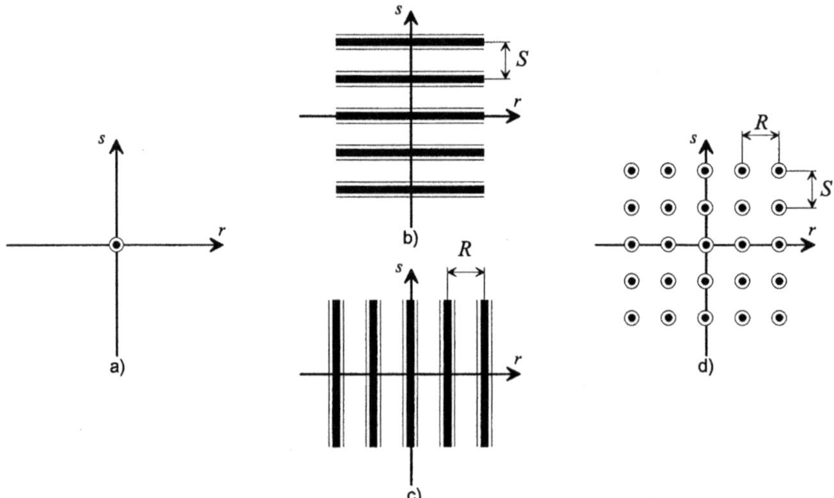

Fig. 2.10. 2D elementary sampling functions: **a** 2D Dirac impulse **b,c** Dirac line impulses **d** Dirac impulse grid

The operation of ideal *rectangular* sampling of a spatially-continuous 2D signal $x(r,s)$ is performed by multiplication with $\delta_{R,S}(r,s)$. The discrete signal $x(m,n)$ is generated by the amplitude samples $x(mR,nS)$. If the Fourier spectrum of the continuous image $x(r,s)$ is defined according to (2.1), the discrete spectrum results in analogy to (2.42) as

$$X^{(R,S)}(j\omega_1, j\omega_2) = \frac{1}{RS} X(j\omega_1, j\omega_2) * \delta_{\omega_R,\omega_S}(\omega_1,\omega_2)$$

$$= \frac{1}{RS} \sum_{k=-\infty}^{\infty} \sum_{l=-\infty}^{\infty} X[j(\omega_1 - k\omega_r), j(\omega_2 - l\omega_s)]. \tag{2.47}$$

Again, it is possible to define the periodic spectrum directly from the discrete signal, assuming normalized sampling distances $R=S=1$:

$$X(j\Omega_1, j\Omega_2) = X(j\omega_1, j\omega_2) * \delta_{\omega_R,\omega_S}(\omega_1,\omega_2) = \sum_{m=-\infty}^{\infty} \sum_{n-\infty}^{\infty} x(m,n) \cdot e^{-jm\Omega_1} \cdot e^{-jn\Omega_2} . \quad (2.48)$$

By the operation of 2D convolution, the pulse grid sampling results in periodic copies of the spectrum along *both directions*, where the respective sampling frequencies and normalized frequencies are defined as

$$\omega_R = \frac{2\pi}{R} \; ; \quad \omega_S = \frac{2\pi}{S} \; ; \quad \Omega_1 = 2\pi \cdot \frac{\omega_1}{\omega_R} \; ; \quad \Omega_2 = 2\pi \cdot \frac{\omega_2}{\omega_S} . \quad (2.49)$$

Examples of amplitude spectra $X(\omega_1,\omega_2)$ and $X^{(R,S)}(\omega_1,\omega_2)$ in case of rectangular sampling are shown in Fig. 2.11.

To avoid aliasing, $x(r,s)$ must be band limited before sampling. The 2D rectangular sampling theorem is then formulated as follows:

$$X(j\omega_1, j\omega_2) \overset{!}{=} 0 \quad \text{for} \quad |\omega_1| \geq \frac{\omega_R}{2} = \frac{\pi}{R} \quad \text{or} \quad |\omega_2| \geq \frac{\omega_S}{2} = \frac{\pi}{S} . \quad (2.50)$$

The conditions of alias-free sampling are in this case independently formulated for the horizontal and vertical spatial dimensions. In principle, this method of separable rectangular sampling can be extended to an arbitrary number of dimensions.

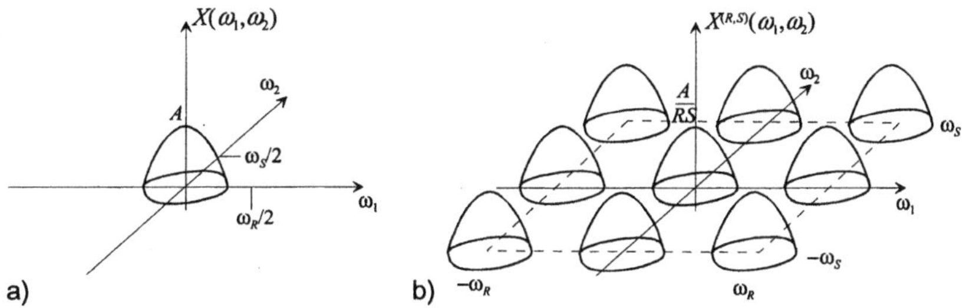

Fig. 2.11. Spectra of 2D image signals: **a** Continuous signal **b** sampled signal

2.2.3   Non-separable Two-dimensional Sampling

In contrast to regular one-dimensional sampling, sampling positions in one dimension can be shifted depending on the location(s) in other dimension(s) in the two- and multi-dimensional case. The separable rectangular sampling is a specific case where no such mutual dependency exists. Constraints may further be imposed by the acquisition devices; e.g. for signals generated by video cameras, sampling grids are restricted to a constant line spacing S. Different regular grids of 2D sampling are illustrated in Fig. 2.12. Regularity of grids means a certain periodicity of a basic structure and can be expressed by systems of *basis vectors* $\mathbf{d}_0 = [d_{00} \; d_{10}]^T$, $\mathbf{d}_1 = [d_{01} \; d_{11}]^T$. Linear combinations of these vectors, when multiplied by integer indexing vectors $[m,n]^T$, point to effective sampling positions $m \cdot \mathbf{d}_0 + n \cdot \mathbf{d}_1$. The basis vectors can be interpreted as the columns of a *sampling matrix* **D**, which maps the index values to the actual sampling positions of the spatially continuous signal:

$$\begin{bmatrix} r(m,n) \\ s(m,n) \end{bmatrix} = \begin{bmatrix} d_{00} & d_{01} \\ d_{10} & d_{11} \end{bmatrix} \cdot \begin{bmatrix} m \\ n \end{bmatrix} = \begin{bmatrix} \mathbf{d}_0 & \mathbf{d}_1 \end{bmatrix} \cdot \begin{bmatrix} m \\ n \end{bmatrix}. \tag{2.51}$$

For the rectangular case, the sampling distances R in horizontal and S in vertical direction are independent of each other. The corresponding sampling matrix is[1]

$$\mathbf{D}_{rect} = \begin{bmatrix} R & 0 \\ 0 & S \end{bmatrix} \Rightarrow \mathbf{F}_{rect} = \begin{bmatrix} 2\pi/R & 0 \\ 0 & 2\pi/S \end{bmatrix}. \tag{2.52}$$

For the case of *shear sampling*,

$$\mathbf{D}_{shear} = \begin{bmatrix} R & v \cdot R \\ 0 & S \end{bmatrix} \Rightarrow \mathbf{F}_{shear} = \begin{bmatrix} 2\pi/R & 0 \\ -2\pi v/S & 2\pi/S \end{bmatrix}, \tag{2.53}$$

the effective sampling grid can still be rectangular when v is an integer value (see Fig. 2.12b), however the neighborhood relationship of samples does no longer follow an orthogonal-axis scheme for any $v \neq 0$. Cases which can be deduced from shear sampling by non-integer shear factors are the hexagonal sampling scheme (Fig. 2.12c) and the quincunx sampling scheme (Fig. 2.12d). In these cases, the basis vectors are tuned such that each sample has same distances towards its six or four nearest neighbors, respectively. The only free parameter is the vertical sampling distance S but even this is pre-determined by the properties of the camera in case of analog video signals. The basis vectors as sketched in Fig. 2.12 influence the position shifts between samples in adjacent lines. For hexagonal and quincunx sampling, this is described by the matrices

$$\mathbf{D}_{hex} = S \cdot \begin{bmatrix} 2/\sqrt{3} & 1/\sqrt{3} \\ 0 & 1 \end{bmatrix} \Rightarrow \mathbf{F}_{hex} = \frac{2\pi}{S} \begin{bmatrix} \sqrt{3}/2 & 0 \\ -1/2 & 1 \end{bmatrix};$$

$$\mathbf{D}_{quin} = S \cdot \begin{bmatrix} 2 & 1 \\ 0 & 1 \end{bmatrix} \Rightarrow \mathbf{F}_{quin} = \frac{2\pi}{S} \begin{bmatrix} 1/2 & 0 \\ -1/2 & 1 \end{bmatrix}. \tag{2.54}$$

To determine the positions of periodic spectral copies in the case of non-rectangular sampling, a non-separable 2D Dirac impulse grid $\delta_D(\mathbf{r})$ is formally defined, corresponding to the positions addressed by the respective sampling ma-

---

[1] In analogy with (2.49), the relationships between matrices **D** and **F** must observe the factor 2π in the reciprocal relationship of sampling distances and sampling frequency, such that $\mathbf{F} = 2\pi[\mathbf{D}^{-1}]^{\mathrm{T}}$.

trix $\mathbf{D}$. By performing an inverse coordinate mapping, it is straightforward to transform $\delta_{\mathbf{D}}(\mathbf{r})$ into a separable sampling function, having unity sampling distances in both spatial directions:

$$\delta_{1,1}(\mathbf{r}') = \delta_{\mathbf{D}}(\mathbf{r}) \quad \text{where} \quad \mathbf{r}' = \mathbf{D}^{-1}\mathbf{r} \tag{2.55}$$

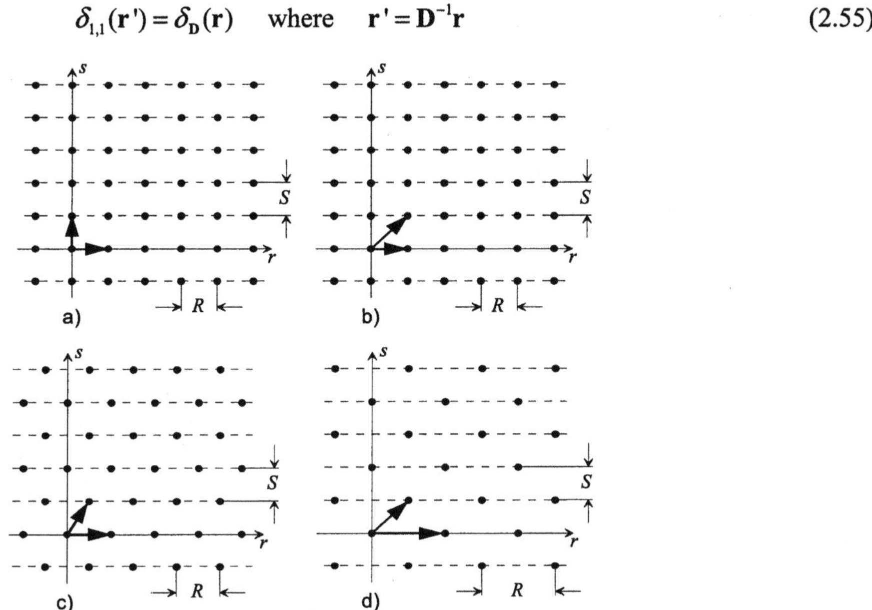

Fig. 2.12. 2D sampling grids: **a** Rectangular **b** Shear, $\nu{=}1$ **c** Hexagonal **d** Quincunx

The Fourier transform of $\delta_{\mathbf{D}}(\mathbf{r})$ must also be a non-separable Dirac impulse grid in the frequency domain, which can likewise be mapped into a separable impulse grid function as follows:

$$\mathscr{F}\{\delta_{\mathbf{D}}(\mathbf{r})\} = 4\pi^2 \delta_{\mathbf{F}}(\omega) = \frac{4\pi^2}{\det|\mathbf{D}|}\delta_{2\pi,2\pi}(\omega') \quad \text{where} \quad \omega = \mathbf{F}\cdot\omega'$$

$$= \frac{4\pi^2}{\det|\mathbf{D}|}\sum_{k=-\infty}^{\infty}\sum_{l=-\infty}^{\infty}\delta\left[\omega_1 - 2\pi k f_{00} - 2\pi l f_{01}, \omega_2 - 2\pi k f_{10} - 2\pi l f_{11}\right]. \tag{2.56}$$

The spectrum of the sampled signal then results as

$$X(j\omega) ** \delta_{\mathbf{F}}(\omega) = \frac{1}{\det|\mathbf{D}|}\sum_{k=-\infty}^{\infty}\sum_{l=-\infty}^{\infty}X\left[j(\omega_1 - 2\pi k f_{00} - 2\pi l f_{01}), j(\omega_2 - 2\pi k f_{10} - 2\pi l f_{11})\right]. \tag{2.57}$$

Assume that $\Omega{=}\mathbf{F}\omega$ defines the non-separable normalized frequency, and $x(\mathbf{n})$ describes a signal defined on a rectangular grid of samples, for which $\mathbf{Dn}$ are the original sampling positions. Then, a non-separable Fourier sum relating to the normalized sampling frequency can be expressed directly from the signal samples,

$$X(j\mathbf{\Omega}) = \sum_{m=-\infty}^{\infty} \sum_{n=-\infty}^{\infty} x(\mathbf{n}) \cdot e^{-j\mathbf{n}^{\mathrm{T}}\mathbf{D}^{\mathrm{T}}\mathbf{\Omega}} \ . \tag{2.58}$$

To generally describe the conditions of the sampling theorem for non-rectangular sampling cases, the mapping from ω to ω' according to (2.56) can be used:

$$\tilde{X}(j\omega_1',j\omega_2') = \tilde{X}(j\omega') \sim X(j\omega) \quad \text{where} \quad \boldsymbol{\omega}' = \begin{bmatrix} \omega_1' & \omega_2' \end{bmatrix}^{\mathrm{T}} = \mathbf{F}^{-1} \cdot \boldsymbol{\omega} \ . \tag{2.59}$$

To avoid alias and allow perfect reconstruction, the spectrum of the signal must be band-limited before sampling, following the condition

$$\tilde{X}(j\omega_1',j\omega_2') \overset{!}{=} 0 \quad \text{for} \quad |\omega_1'| \geq \pi \quad \text{or} \quad |\omega_2'| \geq \pi \ . \tag{2.60}$$

Here, (2.59) und (2.60) define the mapping of the non-rectangular system onto the system with a rectangular grid of periodic continuations of the spectrum. The sampling conditions can now be described separately in the single dimensions. For specific sampling geometries, it is also possible to define conditions without necessity to perform the mapping. For examples of sampling matrices in (2.53) and (2.54), the layouts of base bands and the periodic copies thereof are given in Fig. 2.13, and conditions are explicitly determined in the following paragraphs.

Quincunx sampling. In the first quadrant, the boundary of the base band is described by a line of slope $a=-1$ and intercept $b=\omega_S/2$. Generalization into all four quadrants gives the condition

$$X(j\omega_1,j\omega_2) \overset{!}{=} 0 \quad \text{for} \quad |\omega_1| + |\omega_2| \geq \frac{\pi}{S} \ . \tag{2.61}$$

In quincunx sampling, pure horizontal or vertical sinusoids can be reconstructed up to the same frequency as with quadrangular ($R=S$) sampling, though the number of samples is reduced by a factor of two. For sinusoids of diagonal orientation, the maximum allowable frequency is however lower.

Hexagonal sampling. The hexagonal shape of the base band requires piecewise definition. When $|\omega_1| \leq \omega_S \cdot \sqrt{3}/6$, the boundary of the base band is parallel with the ω_1 axis, while for higher frequencies $|\omega_1|$, lines of slope $a=\pm\sqrt{3}$ and intercept $b=\pm\omega_S$ define the boundary. This results in sampling conditions

$$X(j\omega_1,j\omega_2) \overset{!}{=} 0 \quad \text{for} \quad |\omega_2| \geq \frac{\pi}{S} \quad \text{or} \quad |\omega_1| + \frac{|\omega_2|}{\sqrt{3}} \geq \frac{2\pi}{\sqrt{3}S} \ . \tag{2.62}$$

For arbitrary orientation of sinusoids, hexagonal sampling schemes[1] guarantee

---

[1] The definition of the sampling matrix for hexagonal sampling in (2.54) is not unique, other definitions exist with different orientations of the hexagons. Refer to Problem 2.6 for an example.

omnidirectionally almost equal conditions. Regarding the same line spacing S, the effective number of samples compared to a square sampling approach (rectangular with $R{=}S$) can be reduced by a factor $\sqrt{3}/2 \approx 0{,}866$ [1]. On the other hand, no major disadvantage exists as in quincunx sampling, where stronger limitation of diagonal frequencies occurred.

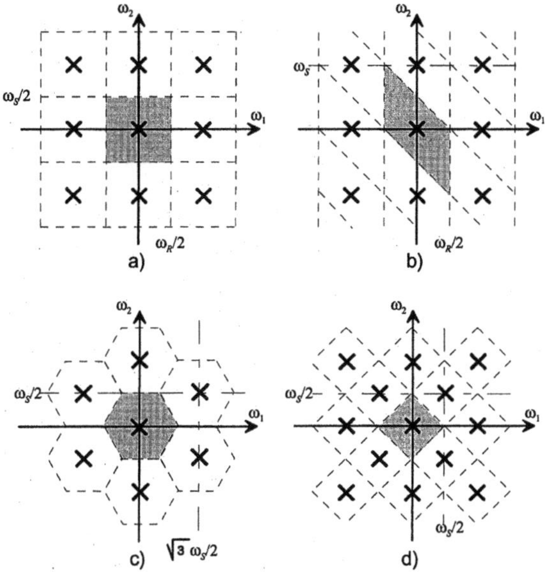

Fig. 2.13. Positions of base band and spectral copies for different 2D sampling grids.
a Rectangular **b** Shear, $v{=}1$ **c** Hexagonal **d** Quincunx

Shear sampling. The shear sampling is a specific case which physically has the same sampling positions as a rectangular grid, but artificially defines a neighborhood relationship between samples which may actually not be direct neighbors of each other. Regard the example (2.9) which was used to introduce 2D frequency orientations of sinusoids. When the phase shift $\phi(s)$ is *known*, it will be possible to define a sampling function and a base band shape in the related sampled spectrum, which allows alias-free reconstruction even when formally $\tilde{\omega}_2 > \pi/S$. The boundary of the base band relating to horizontal sinusoids is in parallel to the ω_2 axis, the same as in case of rectangular sampling. For the vertical frequency, two lines of slope $a{=}{-}v{\cdot}\omega_R/\omega_S$ and intercept $b{=}{\pm}\omega_S/2$ establish the sampling conditions, such that

---

[1] See Problem 2.2.

$$X(j\omega_1, j\omega_2) \overset{!}{=} 0 \quad \text{for} \quad |\omega_1| \geq \frac{\pi}{R} \quad \text{or} \quad \left| \omega_2 + \frac{v \cdot S}{R} \cdot \omega_1 \right| \geq \frac{\pi}{S}. \tag{2.63}$$

The reconstruction of a continuous signal by interpolation from the discrete (sampled) signal is typically performed by a lowpass filter having a frequency response which retains the base band. This means that the actual base-band layout and hence the conditions for alias-free reconstruction also depend on the pass-band shape of the lowpass filter to be used for reconstruction. Shear sampling allows reconstructing signals which clearly would violate the rigid sampling conditions of the equivalent rectangular grid; this is penalized by forbidding other frequency components, however. In total, the *bandwidth range* of signal frequencies which can be reconstructed alias-free is always identical, independent of the shear factor. For any sampling system, the bandwidth range (area of the base bands in Fig. 2.13) is identical to the determinant of the frequency sampling matrix $\mathbf{F}$. The definition of the base band allows certain degrees of freedom, indeed. As another example, quincunx sampling could still be realized combined with a reconstruction filter of horizontal pass-band cutoff $\pm\omega_S/4$, vertical cutoff $\pm\omega_S/2$ or vice versa. The question whether this makes sense can only be answered by an analysis of signal characteristics, and by the actual goal of sampling, e.g. the foreseen bandwidth of signals which shall be analyzed or reconstructed.

A digital image resulting from the sampling process has finite extension, when it originates from an acquisition device which has a finite image plane (camera, scanner). The spatially discrete image resulting after rectangular sampling can directly be arranged as an image matrix consisting of N rows and M columns (see Fig. 1.6). Typically, for the purpose of row-wise sequential processing, it is assumed that the top-left corner of the matrix establishes the origin of the discrete coordinate system ($m=0,n=0$), while at the bottom-right corner the pixel of coordinate ($m=M-1,n=N-1$) resides. Observe that the sampling matrices as introduced in this chapter assume orientations where the coordinate reference $(0,0)$ is at the bottom left position, by which the image plane is defined over the first quadrant in a common orthogonal coordinate system. In some processing and analysis tasks, in particular if projections from the 3D world into the image plane are considered, it is also advantageous to locate the origin at the center of the image plane. All these variants can be realized by simple transformations of coordinate systems.

2.2.4   Sampling of Video Signals

A discrete sequence of images can be interpreted as a three-dimensional (2D spatial+time) signal (see Fig. 2.14). A single image from a video or movie sequence is denoted as a *frame*; the sampling distance between subsequent frames shall be T. An extension of (2.51) to the third dimension leads to a mapping of sampling positions in the spatio-temporal continuum as

$$\begin{bmatrix} r(m,n,o) \\ s(m,n,o) \\ t(m,n,o) \end{bmatrix} = \mathbf{D} \cdot \begin{bmatrix} m \\ n \\ o \end{bmatrix}. \tag{2.64}$$

For the example of Fig. 2.14a, sampled pixels have identical spatial positions in any frame. Such a configuration is denoted as *progressive sampling*, which is shown in Fig. 2.14b over the vertical and temporal directions. The sampling matrix related to this fully-separable 3D sampling is given as

$$\mathbf{D}_{prog} = \begin{bmatrix} R & 0 & 0 \\ 0 & S & 0 \\ 0 & 0 & T \end{bmatrix}. \tag{2.65}$$

This type of sampling is in effect when movie frames are rectangular sampled, e.g. by film scanning. In analog video, *interlaced sampling* is typically used. For each time instance, only half number of lines is processed, and an alternation over time is made between even and odd line acquisition. The images containing only even or odd lines are the even and odd *fields*, respectively (see. Fig. 2.14c). The sampling matrix can in this case be defined as[1]

$$\mathbf{D}_{inter} = \begin{bmatrix} R & 0 & 0 \\ 0 & 2S & S \\ 0 & 0 & T/2 \end{bmatrix} \tag{2.66}$$

In principle, this scheme is very similar to a quincunx sampling grid applied to the vertical/temporal plane of the 3D coordinate system. The bottom-right 2x2 sub-matrix in (2.66) is indeed similar to (2.54) except for the fact that S and T which actually express different physical units. The effect is that higher vertical frequencies are allowable when no high temporal frequencies (e.g. caused by motion) are present.

In progressive sampling – as an extension of 2D rectangular sampling – the conditions of the sampling theorem can be formulated independently in the particular directions. In this case the sampling matrix is diagonal, such that no interrelationships occur between the three dimensions:

$$X(j\omega_1, j\omega_2, j\omega_3) \overset{!}{=} 0 \quad \text{when} \quad |\omega_1| \geq \frac{\pi}{R} \quad \text{or} \quad |\omega_2| \geq \frac{\pi}{S} \quad \text{or} \quad |\omega_3| \geq \frac{\pi}{T}. \tag{2.67}$$

In interlaced sampling, only the condition for the horizontal frequency can be separated from the other two, as the horizontal sampling positions are fixed and

---

[1] In Fig. 2.14c and in the sampling matrix (2.66) a configuration is shown where the top field (lines 0,2,4,..) is the field which is sampled first within the frame. In NTSC TV and digital 60 Hz interlaced video derived thereof, the bottom field is sampled first.

can be expressed by one single plane impulse $\delta_R(r)$, being independent of s and t. In analogy with (2.61),

$$X(j\omega_1, j\omega_2, j\omega_3) \overset{!}{=} 0 \quad \text{when} \quad |\omega_1| \geq \frac{\pi}{R} \quad \text{or} \quad \frac{|\omega_2|}{T} + \frac{|\omega_3|}{2S} \geq \frac{\pi}{S \cdot T}. \quad (2.68)$$

In video acquisition, spatial sampling is often assumed to be alias free, as the elements of the acquisition system (lenses etc.) naturally have a lowpass effect. As was shown in (2.36), the temporal frequency ω_3 depends on spatial frequency and the strength of motion. Assume that the signal could contain sinusoids of almost the maximum allowed spatial frequencies ($\omega_1 \approx \omega_R/2$, $\omega_2 \approx \omega_S/2$). Substituting the condition for ω_3 from (2.67) into (2.36), the following limiting condition must be imposed on the velocity to achieve alias-free sampling:

$$|u| \cdot \frac{T}{R} + |v| \cdot \frac{T}{S} \overset{!}{<} 1 \quad \text{resp.} \quad |k| + |l| \overset{!}{<} 1 \quad \text{with} \quad k = u \cdot \frac{T}{R}, \ l = v \cdot \frac{T}{S}. \quad (2.69)$$

Herein, k and l are discrete shift parameters, which express by how many pixels the signal is shifted horizontally/vertically from one frame to the next, if the velocity (u,v) is observed in the continuous signal. The strict limitation of only one pixel shift either in row or in column directions appears surprising at first sight, as it is common experience that humans are capable to watch moving pictures of much higher motion without any problem. It must be argued however, that the limitation in (2.69) is strictly valid if only *one* sinusoid of near-highest allowable spatial frequency is sampled. Natural video signals are typically mixtures of many different sinusoids, where those of low frequency will allow perceiving the motion reliably and alias-free. The fact that linear motion effects a linear phase shift of the signal allows the observer's eyes to track the motion. It is nevertheless much more difficult to reliably track smaller moving areas with high amount of spatial structure, where significant lower frequency components are obviously missing.

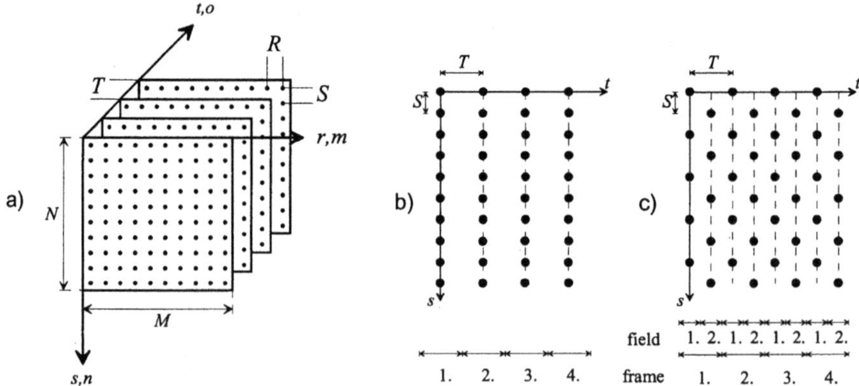

Fig. 2.14. **a** Progressively sampled image sequence **b/c** Video sampling in vertical/temporal directions: Progressive (**b**) and interlaced (**c**) schemes

To illustrate the effects of alias occurring in the case of progressive sampling, Fig. 2.15 shows the vertical/temporal planar section (ω_2, ω_3) of the 3D frequency space. A spatial sinusoid of nearly half vertical sampling frequency is assumed. Center frequencies of the periodic spectral copies are marked by '**x**'. Fig. 2.15a is the spectrum of the signal without motion. Fig. 2.15b shows the skewing of the position in direction of ω_3, when the signal is moved by half a sampling position ($v{=}{-}0.5{\cdot}S/T$) upwards, Fig. 2.15c illustrates the case of motion by 1.5 sampling positions ($v{=}{-}1.5{\cdot}S/T$). In the latter case, alias components appear in the base band, such that the viewer might interpret this as a motion by half a sample downwards ($v{=}0.5{\cdot}S/T$). The spatial frequency of the signal remains unchanged in any case, i.e. aliasing in ω_3 only effects wrong interpretation of motion in the case of progressive sampling. In movies, this is well-known as the "stage coach effect", where the wheels of the vehicle seem to move slowly, stand still or sometimes turn backwards.

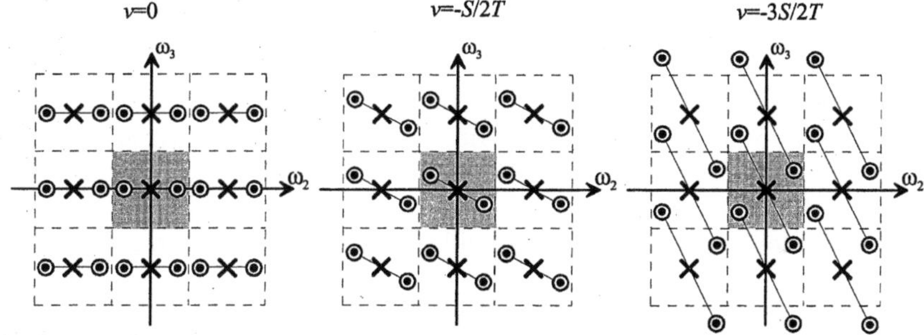

Fig. 2.15. Effect of alias, vertical motion of a progressively-sampled sinusoid

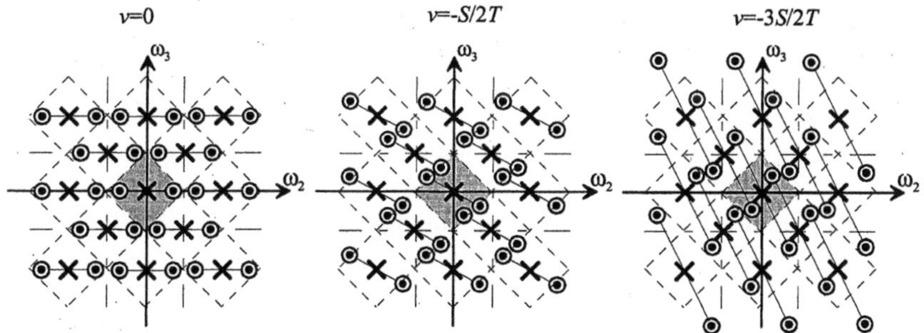

Fig. 2.16. Effect of alias, vertical motion of an interlaced-sampled sinusoid

Fig. 2.16 shows the effect for the case of interlace sampling of the same signal. First, it is obvious that aliasing occurs even when only quite low motion is present. Second, not only a wrong interpretation of the motion occurs; if occasionally the alias spectra originate from the diagonally-adjacent spectral copies, a vertical spatial frequency $\tilde{\omega}_2$ is mapped into the base band as $\omega_S/2 - \tilde{\omega}_2$. In particular when highly-detailed periodic stripes are present in the scene and moving, this results in appearance of strange sinusoidal components, most probably also having different orientations than the original.

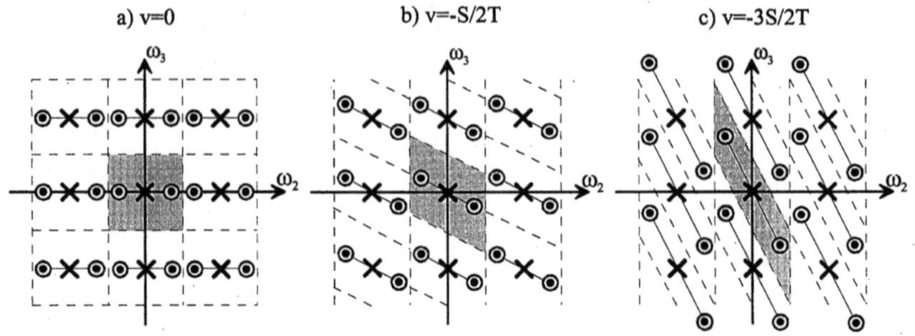

Fig. 2.17. Avoidance of alias by adaptation of the human visual system; tracking by the eyes effects correct reconstruction in sheared sampling.

As motion leads to well-defined non-zero spectral positions, perfect reconstruction and correct perception is in principle possible from sampled spectra when the motion is known to the observer. This relates to the case of shear sampling. If an observer tracks a detail within a video, this is only possible by knowing the velocity. Correspondences are set between subsequent frame coordinate positions considering the motion shift, such that implicitly a reconstruction from a base-band which is sheared in ω_3 is performed. Fig. 2.17 illustrates that a single sinusoid moving by higher velocity than strictly allowed could be interpreted correct; however, from a *single* sinusoid it is typically not possible to guess the motion which is actually performed, as the signal is periodic and no unique correspondences can be detected between the subsequent frames. Only for the case of structured signals with a broader spectrum of frequencies, an observer can recognize the true motion and track it accordingly. This means that in movie and video sampling, typically frequencies in ω_3 will appear which are much beyond the limit expressed by (2.67) for the case of progressive sampling. Motion-compensated processing, which is similar to the tracking by the eye, allows to analyze these signals along the temporal axis without running into the danger of merely processing alias components.

2.3   Problems

Problem 2.1

The vectors $\mathbf{a}=[2\ 0]^T$ and $\mathbf{b}=[2\ 1]^T$ establish a basis system $\mathbf{A}=[\ \mathbf{a}\ \mathbf{b}\ ]$.

a)   Determine the vectors of the dual basis $\widetilde{\mathbf{A}}$, such that the property of biorthogonality is fulfilled.

b)   Sketch the positions of all four vectors in a coordinate system.

c)   Compute the determinants of $\mathbf{A}$ and $\widetilde{\mathbf{A}}$.

Problem 2.2

a)   Prove the condition for hexagonal sampling (2.62).

b)   What is the ratio of the area of the base band in Fig. 2.13c (shaded), as compared to the rectangular sampling case with $\omega_R=\omega_S$?

c)   What is the ratio of the horizontal sampling distance R in Fig. 2.12c, as compared to the rectangular sampling case with $R=S$?

d)   Compute the determinant of the sampling matrix $\mathbf{D}_{hex}$, normalized by $S=1$. Compare the value against the results from parts b) and c).

e)   The CCD chip of a digital camera has an active area of 10 mm x 7,5 mm. The number of pixels is 4.000,000. Compute the approximate vertical sampling distances S for the cases of rectangular sampling ($R=S$) and hexagonal sampling.

Problem 2.3

A two-dimensional cosine of horizontal frequency $\omega_1=\omega_S/3$ is sampled by a quincunx grid (see Figs. 2.12d and 2.13d).

a)   Determine the upper limit for vertical frequency ω_2 (absolute value) guaranteeing alias free sampling.

b)   Which horizontal frequency becomes visible after ideal lowpass filter reconstruction from the sampled signal, if the vertical frequency is $\omega_2=\omega_S/3$?

Problem 2.4

By motion of a video camera using progressive sampling, a translational motion shift of $k=20$ pixels horizontally and $l=10$ pixels vertically results. The frame rate is 50 frames/s, the CCD chip of the camera has a size of 10 mm x 7.5 mm, and the number of pixels is 360 x 288.

a)   Determine the horizontal and vertical velocities u and v, by which the points move on the image plane.

b)   The camera shoots a cosine oscillation, where in the sampled image one period is 40 pixels. Which is the maximum allowable frequency in horizontal direction, such that the sampling theorem is not violated, i.e. the motion of the camera can be interpreted correctly ?

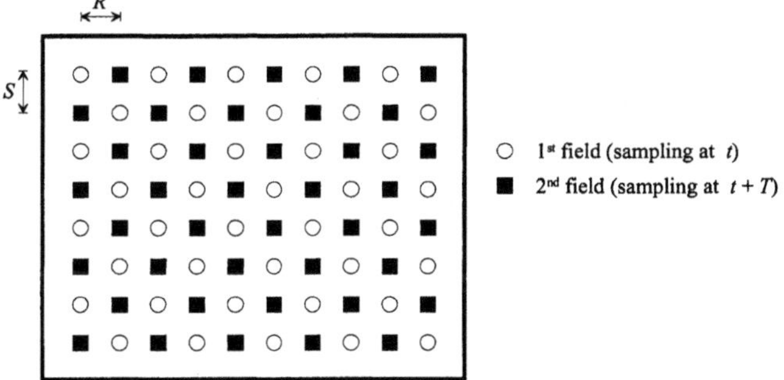

Fig. 2.18. Interlace sampling based on two quincunx sampled fields

Problem 2.5

In a variant of interlaced video sampling, the fields shall not be established by rectangular grids, but by quincunx grids (see Fig. 2.18).

a)   Determine a sampling matrix **D**.
b)   Compute the related frequency sampling matrix **F**.
c)   Which is the shape of the base band in the 3D frequency domain ?

Problem 2.6
A 2D signal shall be sampled by the grid shown in Fig. 2.19, where the black and white dots establish two rectangular sub-grids of equal sampling distances. The black grid is shifted horizontally and vertically by R' and S' relative to the white grid.

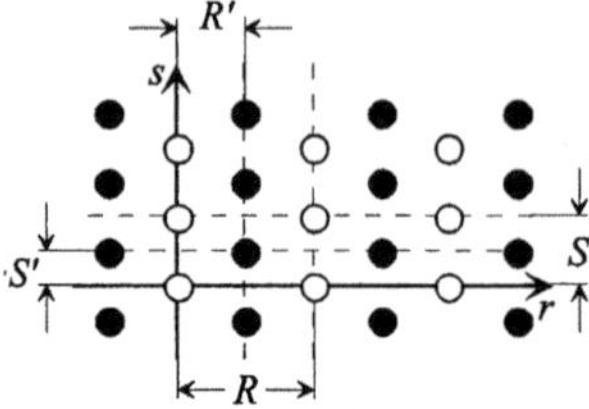

Fig. 2.19. Sampling grid composed by two rectangular sub-grids

a)   For a given vertical sampling distance S, determine values of R, R' and S', such that
      i) a quincunx grid *ii)* a hexagonal grid results.
b)   Describe the sampling of signal $x(r,s)$ by a sampling function $\delta_{\bullet\circ}(\mathbf{r})$ which is a super-
      position of two (black&white bullet) rectangular impulse grids $\delta_{R,S}$.
c)   Determine the Fourier spectrum of $\delta_{\bullet\circ}(\mathbf{r})$. For which cases will the number of periodic
      spectral copies be lower than for one of the rectangular-grid spectra?
d)   Values $0 \le R' \le R$, $0 \le S' \le S$ shall be allowed. Under which conditions will it not be possible
      to express the sampling function $\delta_{\bullet\circ}(\mathbf{r})$ by a closed-form expression, using a sampling
      matrix **D** ?
e)   For $R=S$, $R'=S'=S/2$ sketch the spectrum of a signal, which was sampled observing the
      sampling criteria (Sketched frequency layout shall at least include the base band and the
      eight nearest alias frequency bands).
f)   A signal $x_c(r,s) = A \cos[\, 2\pi/S\,(r+s)\,] + B \cos[\, 2\pi/S\,(r-s)\,]$ is sampled by the grid given
      in e). Which signal will be observed after reconstruction using an ideal lowpass filter
      with same pass band as the base-band shape?

3   Statistical Analysis of Multimedia Signals

Statistical analysis methods are inevitable in optimization of multimedia signal encoding schemes and for identification-related analysis. Of particular importance are signal properties that can be analyzed from probability distributions, which includes first-order probabilities as well as joint and conditional probabilities. Joint analysis of different signals or of signal samples from different locations in the same signal is also performed by correlation, covariance and higher-order moment computation. These functions can be set in direct relationships with probability distributions and with spectral analysis methods. The chapter further introduces statistical models and methods for statistical tests. Finally, the basic relationships of statistic properties within information-theoretic concepts are discussed.

3.1   Properties Related to Sample Statistics

In this chapter, mainly statistical analysis of *sampled* multimedia signals $x(\mathbf{n})$ is treated, whereas similar properties can be derived for signals which are continuous over time or space. For a κ-dimensional signal, the index is expressed by a vector $\mathbf{n}$ of κ dimensions, where sample-related analysis methods are in principle independent of the number of dimensions a signal actually has. Statistical properties of samples from signals can be characterized by the *Probability Density Function* (PDF) $p(x)$, interpreting continuous signal amplitudes x as *random variables*. Probabilities are normalized into values between 0 and 1. As signal samples have any value $-\infty < x < \infty$, the following condition must hold for the PDF:

$$\int_{-\infty}^{\infty} p(x)dx = 1. \tag{3.1}$$

The probability of a value $x \leq x_\mathrm{a}$ is defined by the *Cumulative Distribution Function*

$$P(x \leq x_a) = \int_{-\infty}^{x_a} p(x)dx \ . \tag{3.2}$$

The probability of a signal amplitude to fall into an interval $x_a < x \leq x_b$ is

$$P(x_a < x \leq x_b) = \int_{x_a}^{x_b} p(x)dx = P(x \leq x_b) - P(x \leq x_a) \ . \tag{3.3}$$

The *expected value* $E\{f[x]\}$ is the mean over an asymptotically infinite set of signal samples to which the function $f[x]$ is applied; it can directly be related to the PDF:

$$E\{f[x(\mathbf{n})]\} = \lim_{N \to \infty} \frac{1}{N} \sum_{\mathbf{n}} f[x(\mathbf{n})] = \int_{-\infty}^{\infty} f(x) \cdot p(x)dx \ . \tag{3.4}$$

From these definitions, the following fundamental parameters are defined for *sample statistics*:

-   Mean value $\mu_x = \int_{-\infty}^{\infty} x \cdot p(x)dx = E\{x(\mathbf{n})\}$ (3.5)

-   Quadratic mean $L_x = \int_{-\infty}^{\infty} x^2 \cdot p(x)dx = E\{x^2(\mathbf{n})\}$ (3.6)

-   Variance $\sigma_x^2 = \int_{-\infty}^{\infty} (x - \mu_x)^2 \cdot p(x)dx = E\{(x(\mathbf{n}) - \mu_x)^2\} = L_x - \mu_x^2$ (3.7)

-   P^{th} order moment[1] $m_x^{(P)} = \int_{-\infty}^{\infty} x^P \cdot p(x)dx = E\{x^P(\mathbf{n})\}$ (3.8)

-   Central moment $m'_x{}^{(P)} = \int_{-\infty}^{\infty} (x - \mu_x)^P \cdot p(x)dx = E\{(x(\mathbf{n}) - \mu_x)^P\}$ (3.9)

So far, only continuous-amplitude signals were regarded. For numeric digital processing, these signals are quantized, which means they are mapped into discrete-amplitude signals. Here, we assume the most simple case of uniform quantization (see section 11.1), where amplitude values are mapped into equally distant reconstruction values y_j. These are values at the mid positions of non-overlapping inter-

---

[1] Strictly spoken, (3.8) defines only the *central value* of Pth order moments. A general definition related to joint statistics will be given in sec. 3.2. From the entire series of Pth order moments, $P=1,...,\infty$, $p(x)$ can be reconstructed, following the *moment theorem* [PAPOULIS 1984]. From a finite set of Pth order moments, a truly continuous approximation of $p(x)$ can be made. Another way to approximate $p(x)$ is by measurement of the frequency of occurrence (e.g. histogram) from a large set of quantized signal samples, which approximates a discrete probability function as described on the subsequent pages. If the quantization is sufficiently fine, $p(x)$ can be approximated by interpolation.

vals of width Δ on the amplitude scale. The mapping function is the *quantizer characteristic*, which is a nonlinear staircase function (see Fig. 3.1). The *probability mass function* of a value y_j can be determined by integrating the PDF within the respective quantization interval,

$$p(y_j) = \int\limits_{y_j - \Delta/2}^{y_j + \Delta/2} p(x)dx \;.$$
(3.10)

Further, if the discrete amplitudes shall be represented by a finite number of bits, the number of different values y_j must be limited. Assume that the PDF of a discrete-amplitude signal shall have only J discrete non-zero-values. This is expressed by a sum of weighted Dirac impulses

$$p(x) = \sum_j p(y_j)\delta(x - y_j)$$
(3.11)

where further from (3.1),

$$\sum_j p(y_j) = 1 \;.$$
(3.12)

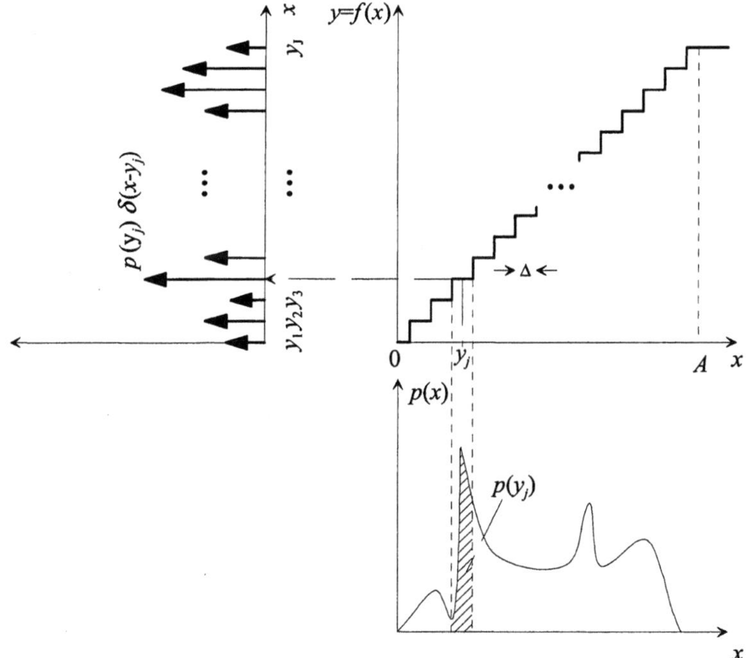

Fig. 3.1. Quantizer characteristic function and mapping of the PDF $p(x)$ of the continuous-amplitude signal to the probability mass function $p(y_j)$ of the quantized (discrete-amplitude) signal.

For a finite set of signal samples, the probability distribution of a discrete signal can be approximated from a *histogram H(j)*, which is the count of occurrences for discrete signal values falling into the single amplitude levels. If the number of samples is N, this gives:

$$p(y_j) = P(j) = \lim_{N \to \infty} \frac{H(j)}{N} \quad \text{and} \quad P(j) \approx \hat{P}(j) = \frac{H(j)}{N}. \tag{3.13}$$

In a strict statistical sense, only the left part of (3.13) is true; the right part could be a good approximation provided that a sufficient number of samples N is analyzed which are representative for the characteristic behavior of the signal[1]. As an analogy with the cumulative distribution function, the cumulative histogram $C(j)$ can be defined as a function which counts the number of values having amplitudes $y_i \leq y_j$ from a set of discrete-amplitude samples:

$$C(j) = \sum_{i=1}^{j} H(i) \Rightarrow P(y_i \leq y_j) = \lim_{N \to \infty} \frac{C(j)}{N}. \tag{3.14}$$

The histogram $H(j)$ or the approximation of $p(y_j)$ from (3.13) can also be used to compute an approximation of the statistic behavior from a finite signal or a finite set of signals in analogy with (3.5)-(3.9)[2]:

$$\mu_x = E\{x(\mathbf{n})\} = \sum_{j=1}^{J} x_j \cdot p(x_j)$$

$$\approx \hat{\mu}_x = \sum_{j=1}^{J} x_j \cdot \hat{P}(j) = \frac{1}{N} \sum_{j=1}^{J} x_j \cdot H(j) = \frac{1}{N} \sum_{\mathbf{n}} x(\mathbf{n}) \tag{3.15}$$

---

[1] Whether such an approximation will be reasonable or not, depends on the properties of *stationarity* and *ergodicity*. For stationary signals, the limit transitions in (3.4) and (3.13) will give identical results, irrespective of positions in time and space where samples are analyzed. In case of ergodicity, one single signal is representative for the entire class, such that measurements over a sufficiently large range can determine the statistical parameters. *Wide sense stationarity* and *ergodicity* can be claimed, if these properties only apply to mean and quadratic mean/variance, the moments up to order 2. For signals with statistically dependent samples, these conditions must also include joint sample statistics (see sec. 3.2).

[2] Probability mass functions $p(y_j)$ were used so far, expressing that distortion is introduced by the quantization of the continuous-value amplitudes x into discrete values y_j. In the sequel, the probabilities over discrete sets are mostly denoted as $p(x_j)$ or $P(j)$. This expresses that a signal of discrete amplitudes x_j is now regarded as the 'original' signal; the mapping $j \leftrightarrow x_j$ is unique, and for many tasks of statistical analysis, it is possible to abstract from the actual amplitude values behind, such that the probabilities $P(j)$ are sufficiently expressive.

$$L_x = E\{x^2(\mathbf{n})\} = \sum_{j=1}^{J} x_j^2 \cdot p(x_j)$$

$$\approx \hat{L}_x = \sum_{j=1}^{J} x_j^2 \cdot \hat{P}(j) = \frac{1}{N}\sum_{j=1}^{J} x_j^2 \cdot H(j) = \frac{1}{N}\sum_{\mathbf{n}} x^2(\mathbf{n}) \tag{3.16}$$

$$\sigma_x^2 = E\{[x(\mathbf{n}) - \hat{\mu}_x]^2\} = \sum_{j=1}^{J} (x_j - \hat{\mu}_x)^2 \cdot p(x_j)$$

$$\approx \hat{\sigma}_x^2 = \sum_{j=1}^{J} (x_j - \hat{\mu}_x)^2 \cdot \hat{P}(j) = \frac{1}{N}\sum_{j=1}^{J} (x_j - \hat{\mu}_x)^2 \cdot H(j) \tag{3.17}$$

$$= \frac{1}{N}\sum_{\mathbf{n}} [x(\mathbf{n}) - \hat{\mu}_x]^2 = \hat{L}_x - \hat{\mu}_x^2$$

$$m_x^{(P)} = E\{x^P(\mathbf{n})\} = \sum_{j=1}^{J} x_j^{P} \cdot p(x_j)$$

$$\approx \hat{m}_x^{(P)} = \sum_{j=1}^{J} x_j^{P} \cdot \hat{P}(j) = \frac{1}{N}\sum_{j=1}^{J} x_j^{P} \cdot H(j) = \frac{1}{N}\sum_{\mathbf{n}} x^P(\mathbf{n}) \tag{3.18}$$

$$m'^{(P)}_x = E\{(x(\mathbf{n}) - \mu_x)^P\} = \sum_{j=1}^{J} (x_j - \mu_x)^P \cdot p(x_j)$$

$$\approx \hat{m}_x^{(P)} = \sum_{j=1}^{J} (x_j - \mu_x)^P \cdot \hat{P}(j) \tag{3.19}$$

$$= \frac{1}{N}\sum_{j=1}^{J} (x_j - \mu_x)^P \cdot H(j) = \frac{1}{N}\sum_{\mathbf{n}} (x(\mathbf{n}) - \mu_x)^P$$

A model PDF can characterize the statistical behavior of a signal. For multimedia signals, the generalized Gaussian distribution is often used:

$$p(x) = a \cdot e^{-|b(x-\mu_x)|^{\gamma}} \quad ; \quad a = \frac{b\gamma}{2\Gamma\left(\frac{1}{\gamma}\right)} \quad ; \quad b = \frac{1}{\sigma_x}\sqrt{\frac{\Gamma\left(\frac{3}{\gamma}\right)}{\Gamma\left(\frac{1}{\gamma}\right)}} \cdot \tag{3.20}$$

For the gamma function $\Gamma(\cdot)$, tabulated solutions exist. For $\gamma=2$, (3.20) gives the Gaussian normal PDF

$$p(x) = \frac{1}{\sqrt{2\pi\sigma_x^2}} \cdot e^{-\frac{(x-\mu_x)^2}{2\sigma_x^2}}, \tag{3.21}$$

for which many optimization problems can be solved analytically by linear algebra. For $\gamma=1$, the Laplacian PDF results; both cases are also shown in see Fig. 3.2:

$$p(x) = \frac{1}{\sqrt{2}\sigma_x} \cdot e^{-\frac{\sqrt{2}|x-\mu_x|}{\sigma_x}} \ . \tag{3.22}$$

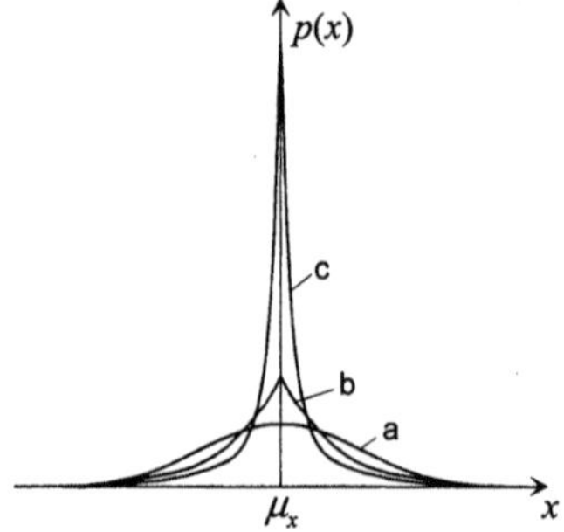

Fig. 3.2. Generalized Gaussian PDF for different parameters γ: $\gamma=2$, Gaussian (**a**); $\gamma=1$, Laplacian (**b**); $\gamma=0,5$ (**c**)

3.2   Joint Statistical Properties

If two continuous-amplitude signals $x(\mathbf{n})$ and $y(\mathbf{n})$ shall be analyzed with regard to dependent statistical behavior, the *joint PDF $p(x,y)$* or *joint probability mass function $p(x_i,y_j)$* are used, the latter one for discrete amplitude signals. These functions express probabilities of sample amplitudes (random variables) from a signal $x(\mathbf{n})$ and another signal $y(\mathbf{n})$ to occur jointly. The definition of 'joint' may e.g. relate to a co-occurrence at the same location in time, space, or also co-occurrence by a certain temporal or spatial distance. If e.g. $y(\mathbf{n})$ is a distorted version of $x(\mathbf{n})$, the joint probability characterizes the deviations and also the degree of similarity which still exists between both signals. If $y(\mathbf{n})=x(\mathbf{n+k})$ is a shifted version of $x(\mathbf{n})$, the joint probability can be used to express *self-similarities* between the samples within one signal.

The joint PDF $p(x,y)$ is a 2-dimensional continuous function, while the joint probability mass function $p(x_i,y_j)$ relates to discrete-amplitude signals. The basic rules which are given in this section are applicable likewise to both cases, but mostly expressions relating to $p(x_i,y_j)$ or the synonymous *joint discrete probability $P(i,j)$* will be used. The joint functions are symmetric,

$$p(x_i, y_j) = p(y_j, x_i) \ . \tag{3.23}$$

For identical signals, $p(x_i,y_j)=0$ when $x_i \neq y_j$, while for $x_i=y_j$, $p(x_i,y_j)=p(x_i)=p(y_j)$. If signals are statistically independent,

$$p(x_i, y_j) = p(x_i) \cdot p(y_j) \,. \tag{3.24}$$

If x_i und y_j allow J amplitude levels each, $p(x_i, y_j)$ consists of up to J^2 non-zero entries. Again, it is possible to approximate the joint probability by an estimate computed from a *joint histogram* under the same conditions formulated above for (3.13):

$$p(x_i, y_j) = P(i, j) = \lim_{N \to \infty} \frac{H(i, j)}{N} \quad ; \quad P(i, j) \approx \hat{P}(i, j) = \frac{H(i, j)}{N} \tag{3.25}$$

Fig. 3.3 shows a schema to compute a joint histogram from sequences of discrete-value samples x_i and y_j; in this example, each sample has one out of four discrete values 1...4. Hence, the joint histogram has $4^2=16$ values. For each combination of x_i- and y_j-values occurring jointly, the respective histogram bin is incremented.

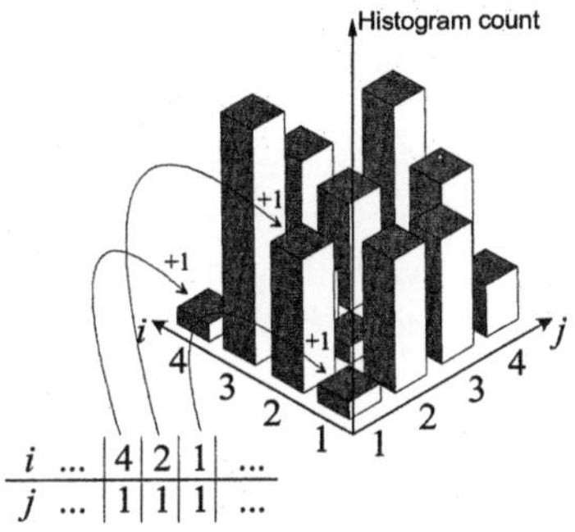

Fig. 3.3. Computation of a joint histogram.

Fig. 3.4 shows two more examples. In Fig. 3.4a, both signals are identical (all joint histogram values where $x_i \neq y_j$ are zero), while in Fig.3.4b, they are almost identical (joint histogram values where $x_i \neq y_j$ are very small).

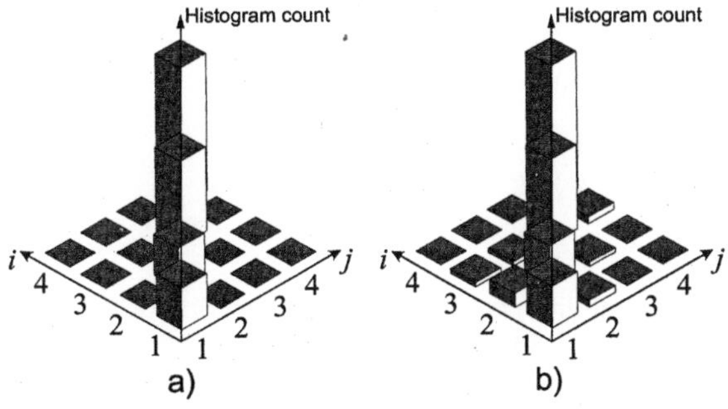

Fig. 3.4. Joint histogram examples:
a Case of equality   **b** Case of high similarity of signals x and y.

The *conditional probabilities* $p(x_i|y_j)$ and $p(y_j|x_i)$ allow to express the certainty by which the occurrence of a value x_i can be expected, if a related value y_j is already known: $p(x_i|y_j)$ is the *'probability of x_i given y_j'*, $p(y_j|x_i)$ the opposite case. The relationships between the joint probability and the conditional probabilities are:

$$p(x_i|y_j) = \frac{p(x_i, y_j)}{p(y_j)} \quad ; \quad p(y_j|x_i) = \frac{p(y_j, x_i)}{p(x_i)} . \tag{3.26}$$

The conditional probabilities are determined by normalizing the joint probability with the probability of the value which is setting the condition. If both signals are equal, $p(x_i|y_j){=}p(y_j|x_i){=}1$ for $x_i{=}y_j$, and all other values are 0. This expresses an absolute certainty, no deviation between the values can be expected. For statistically-independent signals, from (3.24) and (3.26) $p(x_i|y_j){=}p(x_i)$ and $p(y_j|x_i){=}p(y_j)$. A conditional probability mass function $p(x_i|y)$ expresses that a discrete amplitude signal x_i is conditioned by a continuous amplitude signal y. If x is continuous and y_j a discrete-value signal, the relationship is defined by a conditional PDF $p(x|y_j)$.

These concepts can likewise be extended to joint statistics of more than two signals or more than two samples from one signal. If e.g. K values from a continuous-amplitude signal are combined into a vector $\mathbf{x} = [x_1, x_2, \dots, x_K]^T$, the joint probability density becomes also K-dimensional and is denoted as *vector PDF*

$$p_K(\mathbf{x}) = p(x_1, x_2, \dots, x_K) , \tag{3.27}$$

where specifically for the case of statistical independency of the vector elements

$$p_K(\mathbf{x}) = p(x_1) \cdot p(x_2) \cdot \dots \cdot p(x_K) . \tag{3.28}$$

The conditional PDF of a sample x, provided that a vector $\mathbf{x}$ is given, is defined as

$$p(x|\mathbf{x}) = \frac{p(x; x_1, x_2, \dots, x_K)}{p_K(\mathbf{x})} , \tag{3.29}$$

which is a one-dimensional PDF over variable x[1]. For joint analysis, also the definition of the expected value must be extended to functions over several variables which are taken from distant positions in the signal, such that[2]

$$E\{f[x(\mathbf{n}), x(\mathbf{n}+\mathbf{k}), \dots]\} = \lim_{N \to \infty} \frac{1}{N} \sum_{\mathbf{n}} f[x(\mathbf{n}), x(\mathbf{n}+\mathbf{k}), \dots] . \tag{3.30}$$

---

[1] Similarly, it is possible to define a probability $p(x_j|\mathbf{x})$. The joint probability between the discrete x_j and the continuous vector $\mathbf{x}$ is a $K{+}1$ dimensional function $p(x_j; x_1, \dots x_K) = p_K(\mathbf{x}) \cdot p(x_j|\mathbf{x}) \cdot \delta(x - x_j)$, which behaves like a vector PDF $p_K(\mathbf{x})$ over the K dimensions of the vector, and like a sampled function over the dimension of x_j.

[2] For the example of 2D images, the distance vector $\mathbf{k}$ parametrizes the joint relationship between samples $x(m,n)$ and their respective neighbors $x(m+k,n+l)$ which are at a distance of $\mathbf{k}{=}[k,l]^T$ sampling positions.

For sample-related statistics, relationships were established between the probability function and expected values (see (3.5)-(3.9), or (3.15)-(3.19) for the discrete-probability case). The joint function $p(x_{j_1},x_{j_2};\mathbf{k})$ shall express the probability of a constellation where one sample has a value x_{j_1}, while another sample at a distance $\mathbf{k}$, has a value x_{j_2}. Linear statistical dependencies between the two samples are expressed by the *autocorrelation function* (ACF):

$$r_{xx}(\mathbf{k}) = \sum_{j1=1}^{J}\sum_{j2=1}^{J} x_{j1}\cdot x_{j2}\cdot p_{\mathbf{k}}(x_{j1},x_{j2}) = E\{x(\mathbf{n})\cdot x(\mathbf{n}+\mathbf{k})\}$$

$$= \lim_{N\to\infty}\frac{1}{N}\sum_{\mathbf{n}} x(\mathbf{n})\cdot x(\mathbf{n}+\mathbf{k}) \approx \hat{r}_{xx}(\mathbf{k}) = \frac{1}{N}\sum_{\mathbf{n}} x(\mathbf{n})\cdot x(\mathbf{n}+\mathbf{k}). \tag{3.31}$$

(3.31) is the quadratic mean value (3.16) when $\mathbf{k}=0$, which is the maximum value of $r_{xx}(\mathbf{k})$ over all $\mathbf{k}$. The *autocovariance function* results by separating the mean value (3.15):

$$r'_{xx}(\mathbf{k}) = E\{[x(\mathbf{n})-\mu_x]\cdot[x(\mathbf{n}+\mathbf{k})-\mu_x]\} = r_{xx}(\mathbf{k})-\mu_x^2$$

$$\approx \hat{r}'_{xx}(\mathbf{k}) = \frac{1}{N}\sum_{\mathbf{n}}[x(\mathbf{n})-\mu_x]\cdot[x(\mathbf{n}+\mathbf{k})-\mu_x]. \tag{3.32}$$

For $\mathbf{k}=0$, (3.32) gives the variance (3.17), which again is the maximum value of the function[1]. When ACF and autocovariance are normalized by their respective maxima, the *autocorrelation* and *autocovariance coefficients* have value ranges between -1 and $+1$:

$$\rho_{xx}(\mathbf{k}) = \frac{r_{xx}(\mathbf{k})}{r_{xx}(0)} = \frac{r_{xx}(\mathbf{k})}{L_x} \quad ; \quad \rho'_{xx}(\mathbf{k}) = \frac{r'_{xx}(\mathbf{k})}{r'_{xx}(0)} = \frac{r'_{xx}(\mathbf{k})}{\sigma_x^2}. \tag{3.33}$$

While the autocorrelation and autocovariance functions analyze the *self-similarity within a signal*, a similarity *between distinct signals* $x(\mathbf{n})$ und $y(\mathbf{n}+\mathbf{k})$ is analyzed by the *cross correlation function* (CCF) and the *cross covariance* function:

$$r_{xy}(\mathbf{k}) = E\{x(\mathbf{n})\cdot y(\mathbf{n}+\mathbf{k})\} = \lim_{N\to\infty}\frac{1}{N}\sum_{\mathbf{n}} x(\mathbf{n})\cdot y(\mathbf{n}+\mathbf{k}); \quad \rho_{xy}(\mathbf{k}) = \frac{r_{xy}(\mathbf{k})}{\sqrt{L_x L_y}}, \tag{3.34}$$

$$r'_{xy}(\mathbf{k}) = E\{[x(\mathbf{n})-\mu_x]\cdot[y(\mathbf{n}+\mathbf{k})-\mu_y]\} = r_{xy}(\mathbf{k})-\mu_x\mu_y; \quad \rho'_{xy}(\mathbf{k}) = \frac{r'_{xy}(\mathbf{k})}{\sigma_x\sigma_y}. \tag{3.35}$$

The correlation functions and covariance functions analyze *linear statistical dependencies* between two signals or two sampling positions within one signal. If two signals are *uncorrelated*, $r_{xy}(\mathbf{k})=\mu_x\mu_y$ and $r'_{xy}(\mathbf{k})=0$ over all $\mathbf{k}$. Unless periodic components are present in a signal, the following conditions hold for the ACF and covariance if $|\mathbf{k}|$ grows large[2]:

---

[1] For a proof, see Problem 3.1a.

[2] It is sufficient when one of the values in the vector $\mathbf{k}$ converges towards ∞.

$$\lim_{|\mathbf{k}|\to\infty} r_{xx}(\mathbf{k}) = \mu_x^2 \quad ; \quad \lim_{|\mathbf{k}|\to\infty} r'_{xx}(\mathbf{k}) = 0 \,. \tag{3.36}$$

It should be observed that uncorrelated signals or signal samples are not necessarily statistically independent. In particular, *nonlinear dependencies* can not be analyzed by the correlation functions.

Another important statistical property is *stationarity* (see footnote on p. 52). A signal is called *wide sense stationary*, when the mean value and the autocorrelation function are shift invariant (independent from the position of measurement):

$$\mu_x = E\{x(\mathbf{n})\} \stackrel{!}{=} E\{x(\mathbf{n}+\mathbf{l})\}$$

$$r_{xx}(\mathbf{k}) = E\{x(\mathbf{n})\cdot x(\mathbf{n}+\mathbf{k})\} \stackrel{!}{=} E\{x(\mathbf{n}+\mathbf{l})\cdot x(\mathbf{n}+\mathbf{l}+\mathbf{k})\} \qquad \text{over all } \mathbf{l}\,. \tag{3.37}$$

A test whether a signal has the property of wide-sense stationarity can be performed by the *structure function*

$$D_x(\mathbf{k}) = \frac{E\{[x(\mathbf{n}+\mathbf{k})-x(\mathbf{n})]^2\}}{E\{x^2(\mathbf{n})\}-\mu_x^2} = \frac{E\{x^2(\mathbf{n})\}+E\{x^2(\mathbf{n}+\mathbf{k})\}-2\cdot E\{x(\mathbf{n})\cdot x(\mathbf{n}+\mathbf{k})\}}{E\{x^2(\mathbf{n})\}-\mu_x^2}\,. \tag{3.38}$$

For $|\mathbf{k}|\to\infty$, the last term in the numerator converges towards $2\mu_x^2$, such that $D_x(\mathbf{k})$ will approach a value of 2 when the signal is wide-sense stationary. Such a signal may analytically be described by a statistical model with ideal stationarity assumption.

The Gaussian PDF can also be extended to express the joint statistics of two signal values, including their correlation properties. For a certain shift $\mathbf{k}$ between the signal value positions,

$$p(x,y;\mathbf{k}) = \frac{1}{\sqrt{(2\pi)^2\cdot|\mathbf{R}'_{xy}(\mathbf{k})|}}\cdot e^{-\frac{1}{2}\boldsymbol{\xi}^{\mathrm{T}}\mathbf{R}'_{xy}(\mathbf{k})^{-1}\boldsymbol{\xi}}$$

$$\boldsymbol{\xi} = \begin{bmatrix} x-\mu_x \\ y-\mu_y \end{bmatrix};\ \mathbf{R}'_{xy} = E\{\boldsymbol{\xi}\cdot\boldsymbol{\xi}^{\mathrm{T}}\} = \begin{bmatrix} \sigma_x^2 & r'_{xy}(\mathbf{k}) \\ r'_{yx}(\mathbf{k}) & \sigma_y^2 \end{bmatrix}. \tag{3.39}$$

Fig. 3.5a shows the shape of the 2D Gaussian PDF for the case of statistically independent values of identical variances $\sigma_x^2=\sigma_y^2$. In this case, the exponent in the 2D Gaussian function (3.39) is a circle equation, such that values of equal probability will be circles around the position $\boldsymbol{\mu}=[\mu_x\ \mu_y]^{\mathrm{T}}$; the width of the circles is determined by the variance[1]. Fig. 3.5b/c illustrates the influence of variances and co-variances. For uncorrelated signals of variances $\sigma_x^2\neq\sigma_y^2$, the circles are deformed into ellipsoids with principal axes in parallel to the coordinate axes x and y.

---

[1] The standard deviation σ_x describes the turning point of the Gaussian hull.

It can further be shown[1] that covariances $r'_{xy}(k)=r'_{yx}(-k)\neq 0$ effect an ellipsoid deformation, where the orientations of the principal axes will change into the orientation of the eigenvectors of $\mathbf{R'}_{xy}$, while the lengths of the principal axes are the square roots of the related eigenvalues. For example, when $\sigma_x^2=\sigma_y^2$, the principal axes will have 45^0 orientations relative to the coordinate axes x and y; for high correlation, they degenerate into very narrow functions (Fig. 3.5c).

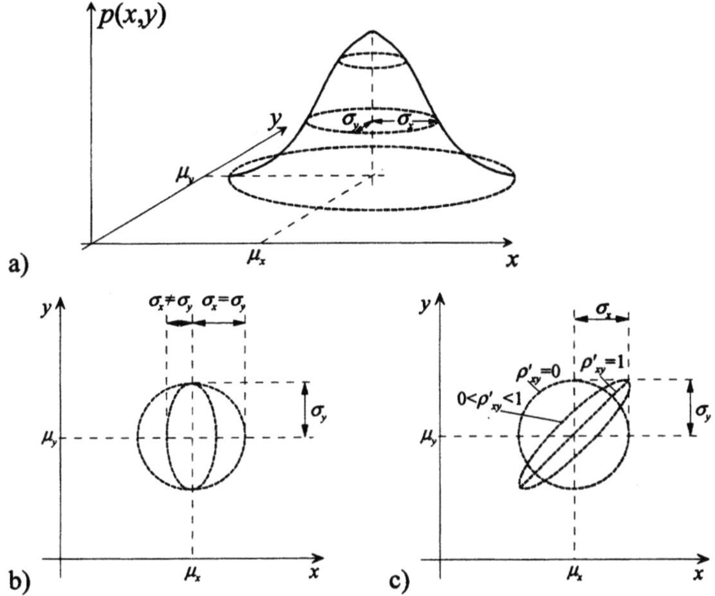

Fig. 3.5. Joint 2D Gaussian PDF $p(x,y)$ **a** for case of statistically independent signals
b modification of the shape by different variances
c modification of the shape by different co-variances

The joint Gaussian PDF can be extended into the *Vector Gaussian PDF* relating to vectors $\mathbf{x}$ constructed from K adjacent samples of a stationary signal,

$$p_K(\mathbf{x}) = \frac{1}{\sqrt{(2\pi)^K \cdot |\mathbf{R'}_{xx}|}} \cdot e^{-\frac{1}{2}[\mathbf{x}-\boldsymbol{\mu}]^T \mathbf{R'}_{xx}^{-1}[\mathbf{x}-\boldsymbol{\mu}]} . \tag{3.40}$$

The mean vector $\boldsymbol{\mu}$ is $\mu_x \cdot \mathbf{1}$. The *autocovariance matrix* $\mathbf{R'}_{xx}$ describes the covariances between the respective samples within the vector and possesses the following Toeplitz matrix structure[2]:

---

[1] For a proof, refer to (9.33)-(9.38).

[2] Due to stationarity, the variances and covariances at all sample positions must be equal, e.g. $E\{x(0)x(1)\}=E\{x(1)x(2)\}=\dots$, which leads to the Toeplitz structure. The Vector Gaussian PDF is not limited to this case, see e.g. (9.1)-(9.2).

$$
\mathbf{R'}_{xx} =
\begin{bmatrix}
r'_{xx}(0) & r'_{xx}(1) & & & \cdots & r'_{xx}(K-1) \\
r'_{xx}(1) & r'_{xx}(0) & r'_{xx}(1) & & & \cdots \\
& r'_{xx}(1) & r'_{xx}(0) & \ddots & & \\
\cdots & & \ddots & \ddots & & r'_{xx}(1) \\
r_{xx}'(K-1) & \cdots & & & r'_{xx}(1) & r'_{xx}(0)
\end{bmatrix}
$$

$$
= \sigma_x^{\,2} \cdot
\begin{bmatrix}
1 & \rho'_{xx}(1) & & & \cdots & \rho'_{xx}(K-1) \\
\rho'_{xx}(1) & 1 & \rho'_{xx}(1) & & & \cdots \\
& \rho'_{xx}(1) & 1 & \ddots & & \\
\cdots & & \ddots & \ddots & & \rho'_{xx}(1) \\
\rho'_{xx}(K-1) & \cdots & & & \rho'_{xx}(1) & 1
\end{bmatrix}.
\qquad (3.41)
$$

An interpretation of a vector value in the *signal space* $\mathcal{R}^K$ is given in Fig. 3.6 for the case of $K=3$. The signal space has orthogonal coordinate axes, such that the amplitude values of the vector $\mathbf{x}=[\ x(0)\ x(1)\ x(2)\ ]^T$ are linearly combined using three unity vectors $\mathbf{e}_1=[1\ 0\ 0]^T$, $\mathbf{e}_2=[0\ 1\ 0]^T$ and $\mathbf{e}_3=[0\ 0\ 1]^T$. The mean is expressed as expected vector value $\mu=E\{\mathbf{x}\}$. The autocovariance matrix is the expected outer product of the zero-mean vector values, $\mathbf{R'}_{xx}=E\{[\mathbf{x}\text{-}\mu][\mathbf{x}\text{-}\mu]^T\}$.

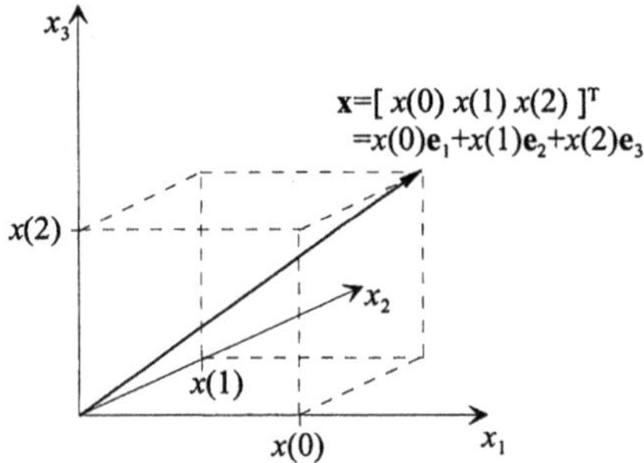

Fig. 3.6. Interpretation of $\mathbf{x}$ in the vector space $\mathcal{R}^K$, example of $K=3$.

Statistical dependencies between the elements of the vector are *linear*, if concentrations of the elements of a vector data set are found within a cloud around a *line* in the signal space. If the vector is composed of values (random variables) from the same signal, this line will be the axis spanned by an angle of 45^0 between the axes of the vector space (see Fig. 3.7a). An example of nonlinear statistical dependencies within vectors $\mathbf{x}$, where the cloud deviates from the line, is shown in Fig. 3.7b.

In advance to statistical analysis, it can be useful to perform an operation which combines signal samples, e.g. to reduce the influence of noise superimposed to the signal. One common way to achieve this is *convolution*. This is a linear operation which performs a weighted summation of samples within a window of size Π, positioned around the sample to be analyzed. If performed at all coordinate positions **n**, the result is an output signal $y(\mathbf{n})$, which can then be subject to statistical analysis[1]:

$$y(\mathbf{n}) = x(\mathbf{n}) * h(\mathbf{n}) = \sum_{\mathbf{k}\forall\{\mathbf{\Pi}\}} x(\mathbf{n}-\mathbf{k}) \cdot h(\mathbf{k}). \tag{3.42}$$

The convolution can be used to compute a local mean-value ('moving average') over a finite set of samples, or to perform frequency-selective analysis (see section 4.1). It is interesting to notify the duality with the operation of correlation (3.31): The computation of the convolution product differs from the correlation product by one inversion of sign[2].

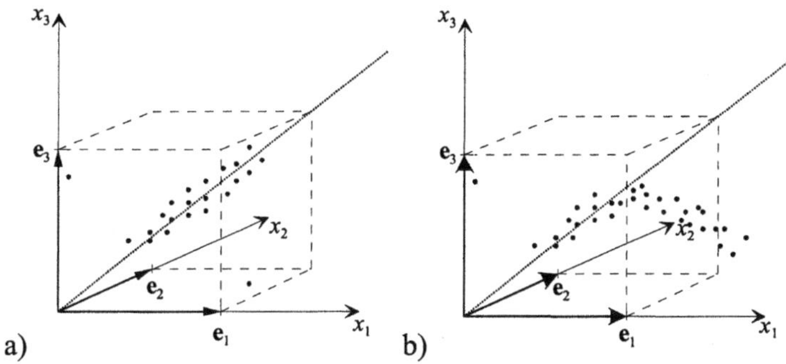

Fig. 3.7. Statistical dependencies within a vector data set plotted over locations in $\mathcal{R}^K$: **a** linear dependency (correlation) indicated by a cloud of values in the vicinity of a line **b** nonlinear dependency

The general definition of the joint expected value (3.30) is based on a function $f[\cdot]$ which combines samples from a one- or multidimensional signal $x(\mathbf{n})$. So far, we have only considered functions combining two values at most (ACF, covariance). An extension to higher-order functions gives the generalized definition of the *moment* of order P (based on joint statistics):

$$m_{x}^{(P)}(\mathbf{k}_1,...,\mathbf{k}_{P-1}) = E\{x(\mathbf{n}) \cdot x(\mathbf{n}+\mathbf{k}_1) \cdot ... \cdot x(\mathbf{n}+\mathbf{k}_{P-1})\}. \tag{3.43}$$

The related *central moment* is defined as

$$m_{x}^{\prime(P)}(\mathbf{k}_1,...,\mathbf{k}_{P-1}) = E\{(x(\mathbf{n})-\mu_x) \cdot (x(\mathbf{n}+\mathbf{k}_1)-\mu_x) \cdot ... \cdot (x(\mathbf{n}+\mathbf{k}_{P-1})-\mu_x)\}. \tag{3.44}$$

The definition of the Pth order moment is parametrized over P-1 shift vectors $\mathbf{k}_p$.

---

[1] To retain the mean value of $x(\mathbf{n})$, the summation over values $h(\mathbf{n})$ should be unity. In principle, $h(\mathbf{n})$ must not be of finite extension; e.g. Gaussian functions can be applied for such purposes.

[2] Correlation functions can indeed be expressed by a convolution operation, $r_{xx}(\mathbf{k}) = x(-\mathbf{k}) * x(\mathbf{k})$ and $r_{xy}(\mathbf{k}) = x(-\mathbf{k}) * y(\mathbf{k}) = y(\mathbf{k}) * x(-\mathbf{k}) = r_{yx}(-\mathbf{k})$.

The first and second order moments are the mean value and the autocorrelation function, respectively. Moments of orders $P>2$ are useful to describe or analyze nonlinear behavior of signals and systems. Likewise, the availability of all higher-order joint moments will allow reconstructing the related joint PDF (cf. footnote on p. 50). The Gaussian normal distribution (3.21) and its vector extension (3.40) also play an important role in this context, as they are *fully described by first and second order moments*. For Gaussian models, all moments related to orders higher than 2 can recursively be determined from the moments of first two orders. As an example, the 3rd order moment in the case of a Gaussian process is (where $m_G^{(1)}$ and $m_G^{(2)}$ express mean and ACF of the process) [NIKIAS, MENDEL 1993]

$$m_G^{(3)}(\mathbf{k}_1,\mathbf{k}_2) = m_G^{(1)}\left[m_G^{(2)}(\mathbf{k}_1)+m_G^{(2)}(\mathbf{k}_2)+m_G^{(2)}(\mathbf{k}_1-\mathbf{k}_2)\right]-2\cdot\left[m_G^{(1)}\right]^3 . \tag{3.45}$$

This enables a simple test whether a signal has Gaussian property, which is performed by the *cumulants*. To determine the cumulant of order P from the Pth order moment, the first and second order moments as measured from a signal are used to estimate the higher-order moments $\hat{m}_G^{(P)}$ of a Gaussian signal. This estimate is subtracted from the actually measured higher order moment[1]. It is then obvious that all cumulants of order >2 must be zero if the signal is indeed Gaussian:

$$c_x^{(P)}(\mathbf{k}_1,...,\mathbf{k}_{P-1}) = m_x^{(P)}(\mathbf{k}_1,...,\mathbf{k}_{P-1}) - \hat{m}_G^{(P)}(\mathbf{k}_1,...,\mathbf{k}_{P-1}) . \tag{3.46}$$

The computation of cumulants requires the same number of shifted samples as for the moments. In many cases however, it is fully sufficient to analyze the *central values* which are the zero-shift cases ($\mathbf{k}_p=\mathbf{0}$) both for moments and cumulants. The central values of the first and second-order cumulants are identical with the zero-shift moments, mean and variance, respectively. The central value of the third-order cumulant is the *skewness*

$$\begin{aligned}
c_x^{(3)}(\mathbf{0},\mathbf{0}) &= E\left\{ x^3(\mathbf{n}) \right\}-3m_x^{(1)}m_x^{(2)}(\mathbf{0})+2\left[m_x^{(1)}\right]^3 \\
&= E\left\{ (x(\mathbf{n})-\mu_x)^3 \right\}= m_x^{\prime(3)}(\mathbf{0},\mathbf{0}),
\end{aligned} \tag{3.47}$$

and the central value of the 4th order cumulant is the *kurtosis*

---

[1] The ad hoc definition of cumulants as given here is only valid for low orders as $P=3$ and $P=4$, however it is less usual to apply orders $P>4$ in multimedia signal analysis anyway. The mathematically correct definition of moments and cumulants is formulated by using the *characteristic function* which is the Fourier transform of the PDF [PAPOULIS 1984]. The zero position of the Pth derivative of the characteristic function is the central value of the Pth order moment as defined in (3.8). If the natural logarithm of the characteristic function is derived P times, the central value of the Pth order cumulant results likewise. For the generalized versions of moments (3.43) and cumulants (3.46), characteristic functions would have to be determined from P^{th} order joint PDFs, which turns out to be a problem of extreme complexity.

$$c_x^{(4)}(0,0,0) = E\{x(\mathbf{n})^4\} - 3\left[m_x^{(2)}(0)\right]^2 - 4m_x^{(1)}m_x^{(3)}(0,0)$$

$$+6\left[m_x^{(1)}\right]^2 m_x^{(2)}(0) - 6\left[m_x^{(1)}\right]^4 = E\{(x(\mathbf{n})-\mu_x)^4\} - 3\sigma_x^4 . \tag{3.48}$$

Both skewness and kurtosis are uniquely zero for the case of a Gaussian PDF.

3.3 Spectral Properties

As it is expected that the reader is familiar with continuous and discrete-time Fourier analysis, only extensions into multiple dimensions are developed here for the example of 2D (image) signals. A generalization for higher number of dimensions is straightforward. For signals having infinite support, the 2D Fourier transform was defined in (2.2) for the continuous and in (2.48) for the discrete (sampled) case. These transform relationships apply likewise to continuous-amplitude and discrete-amplitude signals. If an infinite, continuous signal is periodic, spectral properties can be analyzed over periods of length λ_i in the respective dimensions. The related *Fourier series analysis* for the two-dimensional case is defined by[1]

$$c_{u,v} = \frac{\tilde{\omega}_1 \cdot \tilde{\omega}_2}{4\pi^2} \int\limits_{-\pi/\tilde{\omega}_1}^{\pi/\tilde{\omega}_1} \int\limits_{-\pi/\tilde{\omega}_2}^{\pi/\tilde{\omega}_2} x(r,s) \cdot e^{-ju\tilde{\omega}_1 r} \cdot e^{-ju\tilde{\omega}_2 s} dr ds \quad ; \quad \tilde{\omega}_i = \frac{2\pi}{\lambda_i} . \tag{3.49}$$

The spectrum resulting from Fourier series analysis consists of discrete spectral lines, but extends to infinity. Non-zero spectral lines $c_{u,v}$ can be found equidistantly spaced at integer multiples $u \cdot \tilde{\omega}_1$ in horizontal and $v \cdot \tilde{\omega}_2$ in vertical direction.

The application to sampled signals of limited spatio-temporal extension allows to practically perform a Fourier analysis on a computer, signal or media processor. This is the *Discrete Fourier Transform* (DFT), which leads to a discrete (line) spectrum as in the case of Fourier series analysis (3.49), but is also periodic by the sampling frequency as in the case of Fourier transform for sampled signals (2.48)[2]:

$$\hat{X}(u,v) = \sum_{m=0}^{M-1}\sum_{n=0}^{N-1} x(m,n) \cdot W_M^{-mu} \cdot W_N^{-nv} \quad \text{with} \quad W_M = e^{j\frac{2\pi}{M}} \text{ and } W_N = e^{j\frac{2\pi}{N}} . \tag{3.50}$$

---

[1] In principle, it is irrelevant whether the analysis is performed over one period of a periodic signal, or over a finite signal of the same length; periodicity by λ_i will always implicitly be assumed due to the property of the basis functions

[2] The complex numbers W_i are introduced for a more convenient notation, and can be interpreted as angular step sizes on the unit circle in the complex plane, which can be mapped into distances between the spectral samples.

The discrete frequencies u in horizontal and v in vertical direction have the following relationship with the normalized frequencies describing the periodic spectrum of the sampled signal[1]:

$$\Omega_1 = u \cdot \frac{2\pi}{M} \quad ; \quad \Omega_2 = v \cdot \frac{2\pi}{N} . \tag{3.51}$$

The spectral resolution increases by increased horizontal/vertical lengths of the analysis window, M and N. Table 3.1 shows the basic differences in properties for the four methods of Fourier analysis. The implicit coupling of properties discrete↔periodic and continuous↔aperiodic in spectral and signal domains is obvious.

Table 3.1. Methods of Fourier analysis and associated properties of the signal and the spectrum

	Spectral domain	Signal domain
Fourier transform integral (2.2)	continuous aperiodic	aperiodic continuous
Fourier series analysis (3.49)	discrete aperiodic	periodic continuous
Fourier transform sum (2.48)	continuous periodic	aperiodic discrete
Discrete Fourier transform (3.50)	discrete periodic	periodic discrete

The limitation of the DFT analysis over a window of finite extension leads to the effect that both the spectrum and the signal must be regarded as being periodic (see Problem 3.4). In a two-dimensional signal, periodic copies exist both in horizontal and vertical directions. If an amplitude difference between opposite boundaries of the analysis window is high, high-frequency components will be analyzed which are not present in the non-periodic signal. Two-dimensional DFT spectra are often visualized as image matrices of size MN[2].

---

[1] It should be observed that any of the sampled complex exponential functions in case of the DFT has a period which is a fraction of the analysis window size over both dimensions (periods of $M/2$, $M/3$, $M/4,\ldots$ etc.). Whenever a sinusoid is analyzed which was not sampled such that the sampling distance is an *exact integer fraction* of the signal period, the DFT spectrum will never become an exact line spectrum; it will rather exhibit a certain spread of spectral energy over spectral lines in the vicinity of the actual signal frequency.

[2] It is possible to display the real and imaginary parts of the transform coefficients separately, or to display magnitude and phase. Usually, logarithmic magnitudes

The number of values in the real and imaginary DFT spectral matrices is doubled when compared to the number of samples in the original image matrix. Nevertheless, the number of *linearly independent* spectral values is still $M \cdot N$, when the signal is real-valued. Then, the condition $X[j(-\Omega)] = X*[j\Omega]$ holds, which likewise applies to the coefficients resulting from the DFT (see Problem 3.4). Due to $e^{jk\pi}=\pm 1$, spectral positions will have *real-valued* DFT coefficients, where (u,v) indices exactly map into values $\Omega_i \in [0,\pi]$ according to (3.51).

The DFT is an *orthogonal* transform (cf. sec. 4.3.1). Regarding the complex exponential function, the following condition applies:

$$\sum_{n=0}^{N-1} e^{-j\frac{nk}{N}2\pi} \cdot e^{j\frac{ni}{N}2\pi} = \sum_{n=0}^{N-1} W_N^{-nk} \cdot W_N^{ni} = \begin{cases} N & \text{for } k = i \\ 0 & \text{for } k \neq i. \end{cases} \tag{3.52}$$

The extension for the 2-dimensional case is

$$\sum_{m=0}^{M-1}\sum_{n=0}^{N-1} W_M^{-mk} W_N^{-nl} W_M^{mi} W_N^{nj} = \begin{cases} MN & \text{for } k = i \quad \text{and} \quad l = j \\ 0 & \text{for } k \neq i \quad \text{or} \quad l \neq j, \end{cases} \tag{3.53}$$

where the definition of the W_i is the same as in (3.50). By substituting (3.50) into (3.54) and using the orthogonality condition (3.52), it is possible to show that the 2D DFT is invertible (*inverse DFT*, IDFT):

$$x(m,n) = \frac{1}{MN}\sum_{u=0}^{M-1}\sum_{v=0}^{N-1} \hat{X}(u,v) \cdot W_M^{um} \cdot W_N^{vn} \quad ; \quad x(m,n) \mathrel{\overset{DFT}{\underset{IDFT}{\rightleftarrows}}} \hat{X}(u,v). \tag{3.54}$$

Most properties of the continuous Fourier transform (e.g. linearity, separability, see Table 2.1) apply likewise to the 2D DFT. The shifting condition

$$x(m-k,n-l) \mathrel{\overset{DFT}{\underset{IDFT}{\rightleftarrows}}} W_M^{-uk} W_N^{-vl} \hat{X}(u,v) \tag{3.55}$$

and the modulation

$$W_M^{m\tilde{u}} W_N^{n\tilde{v}} x(m,n) \mathrel{\overset{DFT}{\underset{IDFT}{\rightleftarrows}}} \hat{X}(u-\tilde{u},v-\tilde{v}) \tag{3.56}$$

can now be expressed by discrete coordinates. The conjugate-complex symmetry for the case of real-valued signals is modified due to the periodic continuation of the spectrum,

---

$x(m,n) = \log(|\hat{X}(m-M/2,n-N/2)|+1)$ should be shown, which emphasizes spectral components of low energy. The frequency $(0,0)$ then appears at the center position of the image, which is denoted as the *optical Fourier transform*.

$$x(-m,-n) = x(M-m, N-n) \underset{IDFT}{\overset{DFT}{\rightleftarrows}} \hat{X}^*(u,v) = \hat{X}(M-u, N-v). \qquad (3.57)$$

In the case of rectangular sampling, the DFT is separable, i.e. the 2D transform can be computed sequentially along the two dimensions (likewise for higher number of dimensions). By writing

$$\hat{X}(u,v) = \sum_{m=0}^{M-1} \left[\sum_{n=0}^{N-1} x(m,n) \cdot e^{-j2\pi\frac{nv}{N}} \right] e^{-j2\pi\frac{mu}{M}}, \qquad (3.58)$$

and defining

$$G(m,v) = \sum_{n=0}^{N-1} x(m,n) \cdot e^{-j2\pi\frac{nv}{N}} \qquad (3.59)$$

we get

$$\hat{X}(u,v) = \sum_{m=0}^{M-1} G(m,v) \cdot e^{-j2\pi\frac{mu}{M}}. \qquad (3.60)$$

Here, a one-dimensional DFT of length N is first computed along all columns of the image matrix, and subsequently a row-wise DFT of length M is applied on the intermediate result $G(m,v)$. This can also be processed by inverse sequence, as the transform is linear and separable.

Conjugate-complex multiplication of two DFT spectra having same sizes MxN results in

$$\hat{S}_{xy}(u,v) = \hat{X}^*(u,v) \cdot \hat{Y}(u,v) = \sum_{m=0}^{M-1}\sum_{n=0}^{N-1} x(m,n) \cdot W_M^{um} W_N^{vn} \cdot \sum_{k=0}^{M-1}\sum_{l=0}^{N-1} y(k,l) \cdot W_M^{-uk} W_N^{-vl}$$

$$= \sum_{m=0}^{M-1}\sum_{n=0}^{N-1}\sum_{k=0}^{M-1}\sum_{l=0}^{N-1} x(m,n) \cdot y(k,l) \cdot W_M^{-u(k-m)} \cdot W_N^{-v(l-n)}. \qquad (3.61)$$

By substituting $k'=k-m$, $l'=l-n$, and assuming periodic extension of the signal

$$\hat{S}_{xy}(u,v) = \sum_{m=0}^{M-1}\sum_{n=0}^{N-1}\sum_{k'=0}^{M-1}\sum_{l'=0}^{N-1} x(m,n) \cdot y(m+k',n+l') \cdot W_M^{-uk'} \cdot W_N^{-vl'}, \qquad (3.62)$$

this can be re-written as

$$\hat{S}_{xy}(u,v) = M \cdot N \sum_{k'=0}^{M-1}\sum_{l'=0}^{N-1} \underbrace{\left[\frac{1}{M \cdot N} \sum_{m=0}^{M-1}\sum_{n=0}^{N-1} x(m,n) \cdot y(m+k',n+l') \right]}_{=\hat{r}_{xy}(k',l')} \cdot W_M^{-uk'} \cdot W_N^{-vl'}. \qquad (3.63)$$

(3.63) is equivalent to the Fourier transform of the cross correlation estimate (3.34), however assuming a periodic extension of the signals. This means that the *circular cross correlation* (for all shifts **k**) can be computed by application of the DFT on the two images, complex-conjugate multiplication of spectra, and IDFT of the result. If another than periodic extension of the signal is desirable, it is necessary to extend the signal up to a size of $(2M-1)(2N-1)$.

As a special case of (3.63), the Fourier transform of the circular autocorrelation function estimate is the multiplication of the spectrum and its conjugate:

$$\hat{S}_{xx}(u,v) = \hat{X}^*(u,v) \cdot \hat{X}(u,v)$$

$$= M \cdot N \sum_{k=0}^{M-1}\sum_{l=0}^{N-1} \underbrace{\left[\frac{1}{M \cdot N} \sum_{m=0}^{M-1}\sum_{n=0}^{N-1} x(m+k,n+l) \cdot x(m,n) \right]}_{= \hat{r}_{xx}(k,l)} \cdot W_M^{-um} \cdot W_N^{-vn} \quad (3.64)$$

$$\hat{r}_{xx}(k,l) \quad \overset{DFT}{\underset{IDFT}{\overset{\rightarrow}{\leftarrow}}} \quad \frac{1}{M \cdot N}\hat{S}_{xx}(u,v)$$

$$\hat{r}'_{xx}(k,l) \quad \overset{DFT}{\underset{IDFT}{\overset{\rightarrow}{\leftarrow}}} \quad \frac{1}{M \cdot N}\hat{S}'_{xx}(u,v) \; ; \; \hat{S}'_{xx}(u,v) = \begin{cases} \hat{S}_{xx}(u,v) & \text{für } (u,v) \neq (0,0) \\ 0 & \text{für } (u,v) = (0,0). \end{cases} \quad (3.65)$$

Using (3.63), this can be generalized into a Fourier transform pair between the cross correlation function and the cross power spectrum,

$$\hat{r}_{xy}(k,l) \quad \overset{DFT}{\underset{IDFT}{\overset{\rightarrow}{\leftarrow}}} \quad \frac{1}{M \cdot N}\hat{S}_{xy}(u,v) = \frac{1}{M \cdot N}\hat{X}^*(u,v) \cdot \hat{Y}(u,v) . \quad (3.66)$$

If spectra are computed from a finite set of signals, an estimate of spectral properties for a certain signal class can be made. From the expected value over an asymptotically infinite set of spectra, the *DFT power spectrum* and *cross power spectrum* are statistically defined as expected values from spectrum instances i

$$S_{xx}(u,v) = \lim_{I \to \infty} \frac{1}{I}\sum_{i=1}^{I} \hat{S}_{xx}^{(i)}(u,v) \quad ; \quad S_{xy}(u,v) = \lim_{I \to \infty} \frac{1}{I}\sum_{i=1}^{I} \hat{S}_{xy}^{(i)}(u,v) . \quad (3.67)$$

DFT-based power spectra are always related to finite signal extensions, and can only be interpreted as Fourier transforms of finite or circular correlation functions. If signals are assumed to be infinitely extended, the correlation functions become infinite as well, and power spectra are continuous. The *Wiener-Khintchine theorem* establishes the Fourier transform relation between the autocorrelation function and the power spectrum, which is formally computed by the Fourier transform sum (2.48):

$$r_{xx}(\mathbf{k}) \quad \overset{\mathcal{F}}{\leftrightarrow} \quad S_{xx}(\Omega) = \sum_{k_1=-\infty}^{\infty} \cdots \sum_{k_\kappa=-\infty}^{\infty} r_{xx}(\mathbf{k}) \cdot e^{-j\mathbf{k}^T\Omega}$$

$$r_{xy}(\mathbf{k}) = r_{yx}(-\mathbf{k}) \quad \overset{\mathcal{F}}{\leftrightarrow} \quad S_{xy}(j\Omega) = S^*_{yx}(j\Omega). \quad (3.68)$$

The autocorrelation function of real-valued one- and multidimensional signals has an even symmetry $r_{xx}(\mathbf{k})=r_{xx}(-\mathbf{k})$ (see footnote on p. 61). This implies that the power spectrum $S_{xx}(\Omega)$ is real and positive-valued. In contrary, cross power spectra $S_{xy}(j\Omega)=S^*_{yx}(j\Omega)$ are complex-valued.

The following relationships hold between the estimates for central values of first and second order moments (3.15)-(3.17) and the DFT power spectrum:

$$\hat{\mu}_x^2 = \frac{1}{M \cdot N}\hat{S}_{xx}(0,0)\,;\; \hat{L}_x = \frac{1}{M \cdot N}\sum_{u=0}^{M-1}\sum_{v=0}^{N-1}\hat{S}_{xx}(u,v)\,;\; \hat{\sigma}_x^2 = \frac{1}{M \cdot N}\left[\sum_{\substack{u=0 \\ (u,v)\neq(0,0)}}^{M-1}\sum_{v=0}^{N-1}\hat{S}_{xx}(u,v)\right]$$

(3.69)

The relationship between the quadratic mean value and the sum over the power spectrum coefficients is expressed by *Parseval's theorem*, which can also consistently be defined for the case of continuous power spectra,

$$L_x = r_{xx}(0) = \frac{1}{(2\pi)^\kappa}\int_{-\pi}^{\pi}\cdots\int_{-\pi}^{\pi}S_{xx}(\Omega)d\Omega\,.$$

(3.70)

By defining a power spectrum for the zero-mean counterpart of a signal, it is further possible to establish the relationships for mean value and variance,

$$r'_{xx}(\mathbf{k}) \overset{\mathscr{F}}{\leftrightarrow} S'_{xx}(\Omega) \Rightarrow \sigma_x^2 = r'_{xx}(0) = \frac{1}{(2\pi)^\kappa}\int_{-\pi}^{\pi}\cdots\int_{-\pi}^{\pi}S'_{xx}(\Omega)d\Omega$$

(3.71)

$$S_{xx}(\Omega) = S'_{xx}(\Omega) + \mu_x^2 \cdot (2\pi)^\kappa \delta(\Omega).$$

Finally, also the Fourier transform of higher-order moments can be defined, which are *higher order spectra*, also an important tool for analysis of nonlinear signal behavior. The spectrum of order P results from the Pth order moment as follows[1]:

$$S_{xx}^{(P)}(\Omega_1,\ldots,\Omega_{P-1}) = \mathscr{F}\left\{m_x^{(P)}(\mathbf{k}_1,\ldots,\mathbf{k}_{P-1})\right\}$$

$$= \sum_{k_{1,1}=-\infty}^{\infty}\cdots\sum_{k_{\kappa,P}=-\infty}^{\infty}m_x^{(P)}(\mathbf{k}_1,\ldots,\mathbf{k}_{P-1})\cdot e^{-j\mathbf{k}_1^{\mathsf{T}}\Omega_1}\cdot\ldots\cdot e^{-j\mathbf{k}_{P-1}^{\mathsf{T}}\Omega_{P-1}}.$$

(3.72)

3.4 Statistical Modeling and Tests

To optimize multimedia signal compression and identification algorithms, analysis of statistical properties based on moments or probability distributions is often applied. In fact, both aspects are not completely independent of each other, as it is

---

[1] For a signal of κ dimensions, the dimensionality of the Fourier transform computation will be $(P\text{-}1)\cdot\kappa$

possible to reconstruct the PDF if the entire (infinite) series of moments is known and all moments are finite (*moment theorem*).

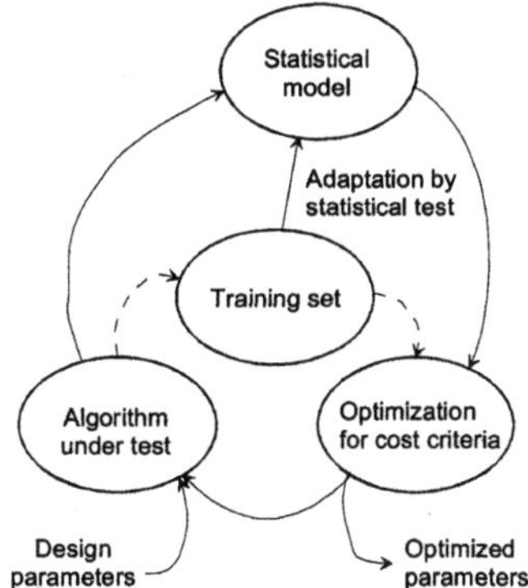

Fig. 3.8. Optimization procedure of an algorithm for multimedia signal processing

It can also be useful to define a model for the PDF or probability distribution directly, based on measurements of occurrences from signal amplitudes. For this purpose, typically a *training set* of samples is used which should be statistically representative of the data that shall later be processed by the algorithm. A typical process of algorithmic optimization is shown in Fig. 3.8. In many cases, such optimization procedures will iteratively loop until the desired result is achieved.

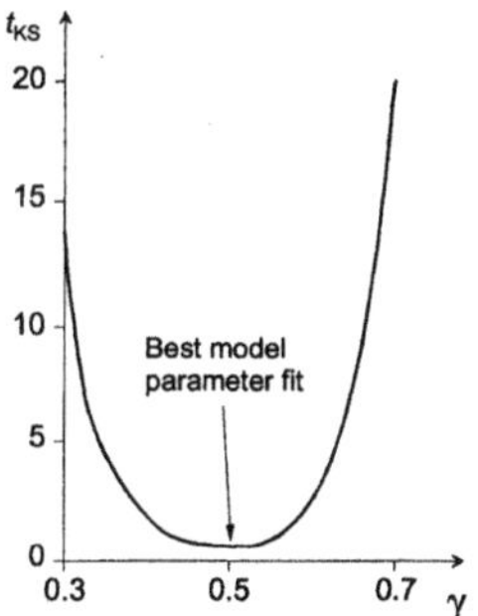

Fig. 3.9. Kolmogorov-Smirnov test, parametrized over γ of a generalized Gaussian model source

The algorithmic optimization could be run directly on the training set, which is actually done in some cases, e.g. for optimization of quantizers. It is however advantageous if a statistical model can be used which is described by a small number of parameters. In such an optimization, a typical goal is minimization of a cost parameter (e.g. distortion), which can be achieved by computing the derivative of an analytic function. In particular, Gaussian probability models will frequently be used for such purposes in the subsequent chapters. The Gaussian PDF is fully described by mean and autocovariance, which can be adapted by measurements from the training set.

For non-Gaussian signals, higher-order moments may be needed to express the statistical behavior, such that simple modeling by mean and autocovariance may not be sufficient. The purpose of statistical tests is a direct proof on how good the distribution characteristics of a signal can be modeled by an analytic distribution model. Assume that an estimate of a probability distribution $\hat{P}(j)$ has been measured by amplitude occurrences of a signal, and its similarity with a given model distribution $P_M(j)$ shall be tested. Both distributions shall relate to the same set of amplitude levels $\{x_j; j=1,...,J\}$ of a quantized signal; if the actual model is continuous, sufficiently accurate discrete probability values can be determined by (3.10). The *Chi-Square test* computes the quadratic deviation over all levels, normalized by the model probability of the respective level:

$$t_{\chi^2} = J \sum_{j=1}^{J} \frac{\left[\hat{P}(j) - P_M(j)\right]^2}{P_M(j)} .$$
(3.73)

The best fit is given where (3.73) is minimum. The *Kolmogorov-Smirnov test* is not based on probability functions, but on the maximum deviation in the cumulative probability (3.14) according to the formula

$$t_{KS} = \sqrt{J} \arg\max_{j=1,...,J} \left|\hat{P}(x \leq x_j) - P_M(x \leq x_j)\right| .$$
(3.74)

Again the best model fit is given for a minimum value t_{KS}. An example how such tests can be used to adapt a parametric PDF model is given in Fig. 3.9. The model distribution is a generalized Gaussian PDF (3.20), where mean and variance are directly adapted by the measurements from a set of signals (transform coefficients), but the parameter γ must be adapted by fitting the measured distribution with the model. In this example, the test indicates a best fit for $\gamma=0.5$.

Markov chain models. In some cases, it is not necessary to model a large range of amplitude random variables, but rather the *state change behavior* of signals. This applies to modeling of binary signals $b(n) \in \{0,1\}$ (e.g. bit streams, two-tone images), or to modeling of features on a more abstract level, e.g. segment transitions in space or time, where a segment relates to a semantic unit (spoken word, video scene, region in an image). A simple model to define states of binary signals with memory is the *2-state Markov chain* shown in Fig. 3.10a[1]. As the signal $b(n)$ has only two levels, S_1='0' and S_2='1' establish the two states of the model, which is fully defined by two transition probabilities $P(0|1)$ ('0' follows a '1'), and $P(1|0)$

---

[1] The problem will be discussed here mainly for binary sequences $b(n)$, but it can formally be extended to any sequences of events $\mathcal{E}(n) \in \{S_j; j=1,...,J\}$. Extensions to multi-level signals $x(n)$ are also made by *Markov Random Fields* (cf. sec. 7.2.1).

('1' follows a '0'). The remaining probabilities $P(0|0)$ and $P(1|1)$, which signify sequences of two equal values, result as

$$P(0|0) = 1 - P(1|0) \;;\; P(1|1) = 1 - P(0|1) . \qquad (3.75)$$

The model shall fulfill two conditions:
-   The probability to be in a state is only dependent on the transition probabilities leading into this state, coupled with the respective probability of the state from which the transition is made;
-   The model shall be stationary, which means that the probability of states shall be invariant against the time or location of observation.

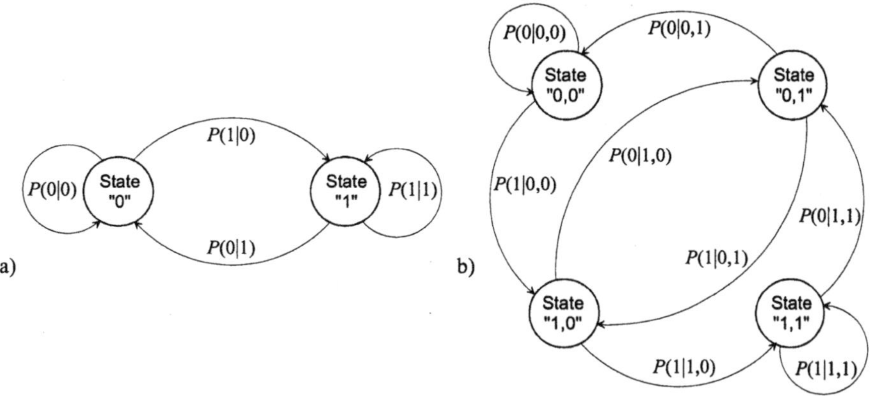

Fig. 3.10. Binary sequences modeled by Markov chain of **a** two states **b** four states

This can be formulated as follows for the two-state model, based on a state transition matrix **P**:

$$\begin{bmatrix} P(0) \\ P(1) \end{bmatrix} = \underbrace{\begin{bmatrix} P(0|0) & P(0|1) \\ P(1|0) & P(1|1) \end{bmatrix}}_{\mathbf{P}} \cdot \begin{bmatrix} P(0) \\ P(1) \end{bmatrix} . \qquad (3.76)$$

The global probabilities of states '1' and '0' then must result as

$$P(1) = \frac{P(1|0)}{P(0|1) + P(1|0)} = 1 - P(0) \;;\; P(0) = \frac{P(0|1)}{P(0|1) + P(1|0)} = 1 - P(1) \qquad (3.77)$$

The probability of '0'- or '1'-sequences of length l can be determined by the fact that the model rests in state '0' for l-1 cycles, and then changes to state '1',

$$P_0(l) = P(1|0) \cdot [1 - P(1|0)]^{l-1} \;;\; P_1(l) = P(0|1) \cdot [1 - P(0|1)]^{l-1} . \qquad (3.78)$$

These probabilities decay exponentially by increasing length l.

The subsequent binary samples will be statistically independent for the case where $P(0|1)=P(0)$ and $P(1|0)=P(1)$. Markov chains with more than two states can be defined accordingly, where again only the transition probabilities between all of these states will define the model. This general formulation of the Markov chain transitions for a model of J states can be written as an extension of (3.76)

$$\begin{bmatrix} P(S_1) \\ P(S_2) \\ \vdots \\ P(S_J) \end{bmatrix} = \begin{bmatrix} P(S_1|S_1) & P(S_1|S_2) & \cdots & P(S_1|S_J) \\ P(S_2|S_1) & P(S_2|S_2) & & \vdots \\ \vdots & & \ddots & \\ P(S_J|S_1) & \cdots & & P(S_J|S_J) \end{bmatrix} \cdot \begin{bmatrix} P(S_1) \\ P(S_2) \\ \vdots \\ P(S_J) \end{bmatrix}. \qquad (3.79)$$

In Markov chains, the probability of transition into a state only depends on *one previous state*, no dependencies with pre-previous states shall exist, such that for the binary 2-state model

$$P\big(b(n)=\beta\,|\,b(n-1),b(n-2),b(n-3),...\big) \equiv P\big(b(n)=\beta\,|\,b(n-1)\big) \quad ; \quad \beta \in \{0,1\} \quad (3.80)$$

Now, if a binary sequence $b(n)$ shall be defined where the state of each sample shall indeed depend on *two predecessors*, the transition probabilities must be expressed as $P[b(n)=\beta\,|\,b(n-1),b(n-2)]$. This can not be realized by the 2-state model, as it is in contradiction with (3.80). Hence, it is necessary to define a Markov chain of four states, relating to the four configurations of $[b(n-1),b(n-2)]$. As however the group $[b(n),b(n-1)]$ will be mapped for the follow-up state into $[b(n-1),b(n-2)]$, certain transitions must be prohibited, which can be interpreted such that the respective transition probability will be zero. The related state diagram is shown in Fig. 3.10b. This model straightforwardly extends to the case where a sample $b(\mathbf{n})$ is conditioned by a K-dimensional vector $\mathbf{b}$ of previous values, which can be established from a one- or multi-dimensional neighborhood context $\mathcal{N}(\mathbf{n})$ of K members, not including the current position $\mathbf{n}$. The model will then be based on 2^K different states, and is fully described by 2^{K+1} state transitions

$$P\big(b(\mathbf{n})=\beta\,|\,\mathbf{b}\big) \quad ; \quad \mathbf{b} = \big\{[b(\mathbf{i})]\,|\,\mathbf{i} \in \mathcal{N}(\mathbf{n}); \mathbf{i} \neq \mathbf{n}\big\} \quad ; \quad \beta \in \{0,1\}. \qquad (3.81)$$

However, only 2^K state transitions are freely selectable, as clearly

$$P\big(b(\mathbf{n})=0\,|\,\mathbf{b}\big) = 1 - P\big(b(\mathbf{n})=0\,|\,\mathbf{b}\big). \qquad (3.82)$$

The follow-up state can also be constrained by zero-probability transitions, as in the example above.

Even though in the cases discussed so far the number of states is finite, the complete sequences $b(n)$ or $b(\mathbf{n})$ can be regarded as infinite. If a Markov chain model allows to transit with a non-zero probability from any state to any other state within a finite number of steps, it is said to be *irreducible*. This would not be the case for chains where one or several states S_i exist with all outgoing transition probabilities $P(S_j|S_i)=0$, but any $P(S_i|S_j)>0$. In such a case, S_i will be a terminating state which is once reached and can never again be left. Such models can be useful in cases where finite sequences of states shall be modeled (cf. sec. 9.4.6).

3.5   Statistical Foundations of Information Theory

Statistical considerations on certainty and uncertainty establish the most important foundations of *information theory*. In general, information is intended to reduce the *uncertainty* about an event, a letter, the state of a signal etc. Assume a discrete set $\mathcal{S}$ of possible events, which are characterized by J different states j. Each state shall have a given probability $P(j)$. The goal is to define a measure for the information $i(j)$ which is related to the knowledge that the event is in state j. Consequently, the mean of information over all states will be $H(\mathcal{S})=\Sigma\, P(j)i(j)$. Availability of *complete information* means that any uncertainty is removed about the state of the event. The function $H(\mathcal{S})$ must however still be consistent, if the amount of certainty is varied, e.g. if it stays uncertain whether the state is $j=1$ or $j=2$, while it is already certain that the state will not be $j=3...J$. This implies that the function must be defined such that it is separable into the information contained in states 1,2 and the remaining states. Assuming that $i(j)$ is somehow related to the probabilities of the states, the following condition must be observed:

$$
\begin{aligned}
H(\mathcal{S}) &= H\left\{P(1),P(2),P(3),...,P(J)\right\} \\
&\overset{!}{=} H\left\{P(1)+P(2),P(3),...,P(J)\right\}+\left(P(1)+P(2)\right)\cdot H\left\{P(1),P(2)\right\}
\end{aligned}
\tag{3.83}
$$

If (3.83) is valid, an arbitrary separation of the information into certainty and uncertainty about any of the states of the event is possible. It can be shown that the only function fulfilling (3.83) is the *self information* of a discrete event of state j, defined as[1]

$$
i(j) = \log_2 \frac{1}{P(j)} = -\log_2 P(j) .
\tag{3.84}
$$

The mean value of the self information over all possible events is denoted as the *entropy*

---

[1] In (3.84), any base of the logarithm can be selected, but the base relates to the unit of information, which is 'bit' in the case of base 2. A probability $P(j)=0$ will lead to an infinite self information; in the subsequent definitions of entropy this is not a problem, as $\lim\limits_{x\to 0}\left(x\cdot\log\frac{1}{x}\right)=0$.

$$H(\mathcal{S}) = -\sum_{j=1}^{J} P(j) \cdot \log_2 P(j) \, . \tag{3.85}$$

If two distinct events defined over sets $\mathcal{S}_1$ and $\mathcal{S}_2$ occur, their *joint information* and *joint entropy* can be defined from the joint probability

$$H(\mathcal{S}_1, \mathcal{S}_2) = \sum_{j_1=1}^{J_1} \sum_{j_2=1}^{J_2} P(j_1, j_2) \cdot i(j_1, j_2) \, ; \quad i(j_1, j_2) = -\log_2 P(j_1, j_2) \, . \tag{3.86}$$

The concept of conditional probability allows to introduce the *conditional information* of an event state j_2, provided that the state j_1 of the other event is already known: This allows defining a criterion for the predictability of state j2 from j1, and considering statistical dependencies in the information theoretic concepts:

$$i(j_2 \,|\, j_1) = -\log_2 P(j_2 \,|\, j_1) = -\log_2 \frac{P(j_1, j_2)}{P(j_1)} \, ; \quad P(j_2 \,|\, j_1) > 0 \, . \tag{3.87}$$

For statistically independent events, due to (3.24) and (3.26) $P(j_2|j_1)=P(j_2)$, which makes the conditional information identical to the self information $i(j_2)$. The difference between the self information and the conditional information is the *mutual information*. It signifies the amount of information in event j_2, which was already provided by j_1. Likewise, this can be interpreted as the amount of information which could possibly be saved (e.g. needs not to be transmitted) when the knowledge about statistical dependency is exploited:

$$i(j_2; j_1) = i(j_2) - i(j_2 \,|\, j_1) \, . \tag{3.88}$$

Combining (3.84) and (3.87) into (3.88), further considering (3.26) gives

$$i(j_2; j_1) = \log_2 \frac{P(j_2 \,|\, j_1)}{P(j_2)} = \log_2 \frac{P(j_1, j_2)}{P(j_1) \cdot P(j_2)} = \log_2 \frac{P(j_1 \,|\, j_2)}{P(j_1)} = i(j_1; j_2) \, . \tag{3.89}$$

This shows the symmetry property of mutual information. If two events are statistically independent, the mutual information becomes zero. This is a very strict condition for statistical independency, which even allows to test for presence or absence of nonlinear dependencies, being a much more rigid criterion than the cross correlation.

The mean value of conditional information over all j_1 and j_2 is the *conditional entropy*, considering all statistical dependencies between j_1 and j_2 that can be exploited:

$$H(\mathcal{S}_2|\mathcal{S}_1) = -\sum_{j_1=1}^{J}\sum_{j_2=1}^{J} P(j_1)\cdot P(j_2|j_1)\cdot\log_2 P(j_2|j_1)$$

$$= -\sum_{j_1=1}^{J}\sum_{j_2=1}^{J} P(j_1,j_2)\cdot\log_2 P(j_2|j_1).$$

(3.90)

The *mean of mutual information* can be expressed from (3.88) and (3.89) as follows[1]:

$$I(\mathcal{S}_2;\mathcal{S}_1) = I(\mathcal{S}_1;\mathcal{S}_2) = \sum_{j_1=1}^{J}\sum_{j_2=1}^{J} P(j_1,j_2)\cdot\log_2 \frac{P(j_1,j_2)}{P(j_1)\cdot P(j_2)}$$

$$= H(\mathcal{S}_2) - H(\mathcal{S}_2|\mathcal{S}_1) = H(\mathcal{S}_1) - H(\mathcal{S}_1|\mathcal{S}_2).$$

(3.91)

The general relationships between entropy, conditional entropy and mutual information are shown in Fig. 3.11a by a diagram of information flow. In principle, the whole schema is invertible, i.e. the event states j_1 and j_2 can change their roles, while the mutual information will not change. In addition, Fig. 3.11b shows an interpretation from the background of set theory, where the circles indicate the information over event sets $\mathcal{S}_1$ and $\mathcal{S}_2$. If an intersection appears, a part of the information is shared, such that at least some statistical dependency between the events is in effect.

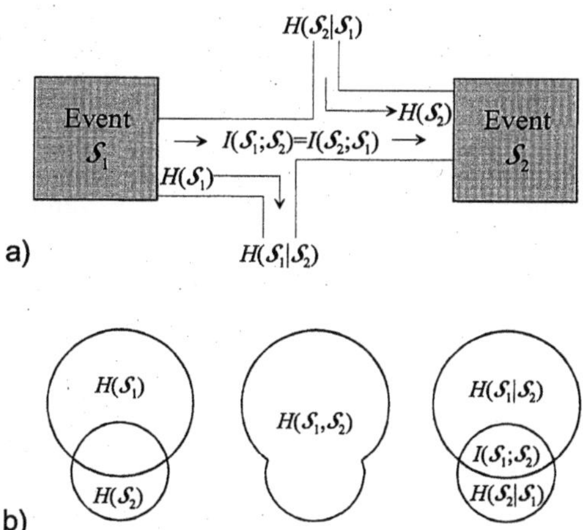

Fig. 3.11. Graphical interpretation of information-theoretic statistical parameters in terms of information flow (*a*) and in terms of set theory (*b*)

In a more abstract way, entropy, conditional entropy and mutual information can be used to express the problem of encoding information by a *discrete alphabet*. Typical examples of discrete alphabets are sets of alphanumeric letters, or sets of

---

[1] (3.91) also is often denoted as *mutual information*

reconstruction values in case of signal quantization. Let a source alphabet $\mathcal{A}$ be defined, which contains all distinct letters that a discrete source could ever produce. Further, a reconstruction alphabet $\mathcal{B}$ is given. Both alphabets need not necessarily be identical (however only if $\mathcal{A}$ is identical with $\mathcal{B}$ or a subset thereof, it is possible to perform *lossless coding*). The mapping of values from $\mathcal{A}$ into values from $\mathcal{B}$ is defined by a code $\mathcal{C}$. Then,

$$I(\mathcal{A};\mathcal{B})\big|_{\mathcal{C}} = H(\mathcal{A}) - H(\mathcal{A}\,|\,\mathcal{B})\big|_{\mathcal{C}} \tag{3.92}$$

As the mutual information cannot become negative,

$$0 \le H(\mathcal{A}\,|\,\mathcal{B})\big|_{\mathcal{C}} \le H(\mathcal{A}), \tag{3.93}$$

where for $H(\mathcal{A}\,|\,\mathcal{B})\big|_{\mathcal{C}} = 0$ it is possible to perform lossless decoding, while for $H(\mathcal{A}\,|\,\mathcal{B})\big|_{\mathcal{C}} = H(\mathcal{A})$ *nothing is known* after decoding, what the state of the source has been. For any values of $H(\mathcal{A}\,|\,\mathcal{B})\big|_{\mathcal{C}}$ between these extrema, lossy decoding will be performed, such that *distortion* occurs. Let $\mathcal{C}_D$ define the set of all codes, which are capable to perform the mapping from $\mathcal{A}$ onto $\mathcal{B}$ by introducing a distortion D[1]. The best possible code among all $\mathcal{C}_D$ is obviously one which needs lowest bit rate for its representation, which is the code providing least mutual information when the mapping from $\mathcal{A}$ into $\mathcal{B}$ is performed. The lowest bound for the rate by a given distortion D will then be

$$R(D) = \min_{\mathcal{C} \in \mathcal{C}_D} I(\mathcal{A};\mathcal{B})\big|_{\mathcal{C}} . \tag{3.94}$$

$R(D)$ is the *Rate Distortion Function* (RDF), which defines an interrelationship between rate R and distortion D. In this abstract form, the definition is valid for arbitrary source alphabets and arbitrary definitions of distortion. From (3.92)-(3.94) and the related reasoning, the following conclusions can be drawn:

- Lossless coding of a source, which generates letters from a discrete source alphabet $\mathcal{A}$, can be achieved by investing a minimum rate $R=H(\mathcal{A})$.
- The minimum achievable rate is zero. Actually, in this case, *nothing* can be known at the receiver end about the state of the source. In this case, a maximum distortion D occurs which shall never be superseded at any other rate.
- If the source is continuous by amplitude, the number of letters in the source alphabet $\mathcal{A}$ will grow towards infinity. Hence, it is not possible to achieve lossless encoding by a finite rate. However if the reconstruction alphabet $\mathcal{B}$ is sufficiently large, the distortion may become negligibly small.

Qualitative graphs of rate distortion functions for both cases of continuous-amplitude and discrete-amplitude sources are shown in Fig. 3.12. Typically, the

---

[1] At this point, D shall be introduced in a quite abstract way, more concrete definitions will be used in sect. 11.2

rate distortion function is convex and continuously decreasing until the maximum distortion (for rate zero) is reached.

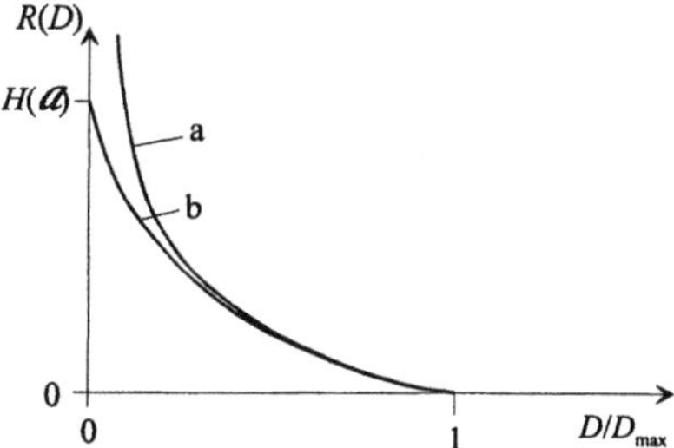

Fig. 3.12. Examples of $R(D)$ for sampled continuous (**a**) and discrete (**b**) sources

Example: Entropy of a Markov process. A Markov chain of J states is defined by the transition probability definitions in (3.79). Due to the property that the probabilities of next-state transitions are independent of history, the entropy of each state can first be defined independently by the respective probabilities of the next-state transitions

$$H(S_j) = -\sum_{i=1}^{J} P(S_i \mid S_j) \log_2 P(S_i \mid S_j) \quad ; \quad j = 1,...,J . \tag{3.95}$$

The overall entropy of the Markov process can then be computed as the probability-weighted average over all states,

$$H(\mathcal{S}) = \sum_{j=1}^{J} P(S_j) H(S_j) . \tag{3.96}$$

3.6 Problems

Problem 3.1

Show that the following conditions hold for stationary signals:

a) $E\left\{ [x(n) \pm x(n+k)]^2 \right\} \geq 0 \quad \Rightarrow \quad \sigma_x^2 \geq |r'_{xx}(k)|$

b) $r_{xx}(k) = r_{xx}(-k)$

c) $r'_{xy}(k) = E\left\{ [x(n) - \mu_x] \cdot [y(n+k) - \mu_y] \right\} = r_{xy}(k) - \mu_x \mu_y$

d) $\dfrac{d}{dy}\left[E\left\{ (x(n) - y)^2 \right\} \right] = 0 \quad \Rightarrow \quad y = E\{x(n)\}$

e) $\mathbf{R}_{xx} = \begin{bmatrix} r_{xx}(0) & r_{xx}(1) & r_{xx}(2) \\ r_{xx}(1) & r_{xx}(0) & r_{xx}(1) \\ r_{xx}(2) & r_{xx}(1) & r_{xx}(0) \end{bmatrix} = E\left\{ \mathbf{x} \cdot \mathbf{x}^T \right\}$ with $\mathbf{x} = \begin{bmatrix} x(n) \\ x(n+1) \\ x(n+2) \end{bmatrix}$

Problem 3.2

For the generalized Gaussian PDF (3.20),
a) show that $\gamma=2$ results in the Gaussian normal PDF (3.21).
b) show that $\gamma=1$ results in the Laplacian PDF (3.22).
c) with $\Gamma(c)=\Gamma(c+1)/c$, which PDF can asymptotically be expected for $\gamma\rightarrow\infty$?

   [use values $\Gamma(3)=2$; $\Gamma(1)=1$; $\Gamma(1.5)=\sqrt{\pi}/2$; $\Gamma(0.5)=\sqrt{\pi}$].

Problem 3.3

A one-dimensional, stationary zero-mean signal with Gaussian PDF has an autocovariance function $r'_{xx}(k)=\sigma_x^2\cdot\rho^{|k|}$.
a) Construct the autocovariance matrix $\mathbf{R}'_{xx}$ of size 3x3 .
b) Show that for $\rho=0$: $p_3(\mathbf{x})=p_G(x_1)\cdot p_G(x_2)\cdot p_G(x_3)$. Here, $p_3(\mathbf{x})$ is a vector Gaussian PDF (3.40) for vector random variables $\mathbf{x}=[\,x_1\ x_2\ x_3\,]^T$, and $p_G(x_i)$ shall be Gaussian normal distributions (3.21).

Problem 3.4

a) The Fourier transform of multidimensional sampled signals is defined as

$$X(j\mathbf{\Omega})=\sum_n x(\mathbf{n})\cdot e^{-j\mathbf{n}^T\mathbf{\Omega}}\quad;\quad \mathbf{n}=\begin{bmatrix}m&n&...\end{bmatrix}^T\quad;\quad \mathbf{\Omega}=\begin{bmatrix}\Omega_1&\Omega_2&...\end{bmatrix}^T.$$

   For real signals, spectral values are also real at those frequencies where over all dimensions i the condition $\Omega_i=k_i\cdot\pi$ (k_i integer) holds. Determine the numbers and positions of real spectral values in 2D Discrete Fourier transform, were the block sizes of the transform in horizontal and vertical direction are *i)* M and N both even; *ii)* M even and N odd; *iii)* M odd and N even; *iv)* M and N both odd.
b) For the DFT, show the relationship $\hat{X}^*(u,v)=\hat{X}(M-u,N-v)$. Sketch the positions of conjugate complex spectral value pairs for a 2D transform over even lengths M and N.

Problem 3.5

a) Two events j_1 and j_2 shall be statistically independent, i.e. $P(j_1,j_2)=P(j_1)\cdot P(j_2)$. Prove the following relationships: $H(j_1|j_2)=H(j_1)$; $H(j_2|j_1)=H(j_2)$; $I(j_1;j_2)=0$.
b) The events j_1 and j_2 shall be identical. Show that $H(j_1|j_2)=0$; $H(j_2|j_1)=0$; $I(j_1;j_2)=H(j_1)=H(j_2)$.

Problem 3.6

The joint PDF of two correlated Gaussian signals x and y shall be defined by (3.39), assuming **k=0**.
a) Determine the joint PDF for the cases of uncorrelated signals ($r'_{xy}=0$) and identical signals ($x=y$).
b) Determine the conditional PDF $p(y|x)$ for the general case first, then specifically for the case of uncorrelated signals.

4 Linear Systems and Transforms

In multimedia signal analysis for compression and identification, linear systems and transforms play an important role. This chapter starts by an introduction into two- and multi-dimensional linear systems, which are characterized by their impulse responses and by transfer functions in the Fourier or z domain. Linear prediction and autoregressive modeling are developed from linear filters. Signal analysis based on sets of orthogonal basis functions is another important concept for linear multimedia signal analysis. The classical approach are block transforms, which is then extended into block-overlapping and subband transforms. In fact, transforms using filter banks show the synergies and similarities between the signal analysis concepts by linear filters and linear transforms. The Wavelet transform applies the view of multi-resolution signal processing which partially allows releasing the orthogonality constraint imposed on the sets of analysis basis functions.

4.1 Two- and Multi-dimensional Linear Systems

4.1.1 Properties of Two-dimensional Filters

Linear systems realize the convolution operation (3.42). We concentrate here on specific aspects of two- and multidimensional linear systems for image and video processing. The convolution performs a *linear superposition* of a sample with its neighbors. In principle, the neighborhood could be infinite, but finite neighborhoods are regarded first, which must then be defined over multiple dimensions. In image processing, symmetric neighborhoods and linear functions which are symmetric around the origin are often used. This implements zero phase-shift systems, retaining an unchanged position of the signal after filtering. Typical examples of symmetric 2D windows are defined by *homogeneous neighborhood systems*, where image samples at positions (i,j) establish the neighborhood of a sample at position (m,n) according to a maximum distance norm of order P[1]:

$$\mathcal{N}_c^{(P)}(m,n) = \left\{ (i,j) \ : \ 0 < |i-m|^P + |j-n|^P \le c \right\}. \tag{4.1}$$

The parameter c influences the size of the neighborhood system, while P influences the shape. Fig. 4.1 shows examples for distance norms $P=1$, the 'diamond shaped' neighborhood, and $P=2$, the circular neighborhood. The trivial case $c=0$ means that no neighborhood is defined, while $P=0$ extends the neighborhood to infinity for any value $c>0$. Various values of c are shown, the position (m,n) is marked by '•'.

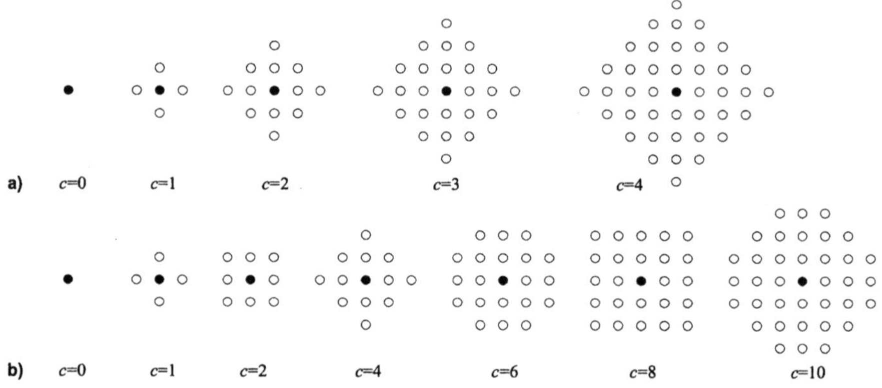

Fig. 4.1. Homogeneous neighborhood systems $\mathcal{N}_c^{(P)}(m,n)$ with $P=1$ (**a**) and $P=2$ (**b**) for various values of c.

The computation of the output value is performed by a combination function which could in principle be linear or nonlinear[2]

$$y(m,n) \ = f\left[x(i,j) \in \mathcal{N}_c^{(P)}(m,n) \right]. \tag{4.2}$$

If the function is linear, the combination is a weighted sum (superposition) of samples from the neighborhood. In the case of *linear systems*, the weighting function is denoted as *impulse response*. An eminent advantage is the dual interpretation of the effect of linear systems by a *frequency transfer function*, which is the Fourier transform of the impulse response.

The discrete *2D convolution* of a signal $x(m,n)$ by the impulse response $h(m,n)$ of a linear system is defined as[1]

---

[1] Homogeneous neighborhood systems are *symmetric*, which means that the current sample at position (m,n) is also a member of the neighborhood systems of any of its neighbors (i,j).

[2] For nonlinear functions, see chapter 5.

$$y(m,n) = \sum_{k=-\infty}^{\infty} \sum_{l=-\infty}^{\infty} x(m-k,n-l) \cdot h(k,l) = x(m,n) * h(m,n) \,. \tag{4.3}$$

(4.3) supports impulse responses having either finite or infinite extension. In the case of a homogeneous neighborhood system, $h(m,n)$ is finite, which means $h(k,l)=0$ for all $(k,l) \notin \mathcal{N}_c^{(P)}(0,0)$. Then, the 2D impulse response can be arranged within a filter matrix $\mathbf{H}$. Filters of this type are denoted as *Finite Impulse Response* (FIR) filters.

The discrete impulse response is the response of a discrete system when a *2D unit impulse* $\delta(m,n)^2$ is fed to the input. The system with the response[3]

$$h(m,n) = L\{\delta(m,n)\} \tag{4.4}$$

is called *linear* if for arbitrary complex constants a and b the following conditions hold:

$$y_1(m,n) = L\big[x_1(m,n)\big] \quad \text{and} \quad y_2(m,n) = L\big[x_2(m,n)\big]$$
$$\Rightarrow a \cdot y_1(m,n) + b \cdot y_2(m,n) = L\big[a \cdot x_1(m,n) + b \cdot x_2(m,n)\big]. \tag{4.5}$$

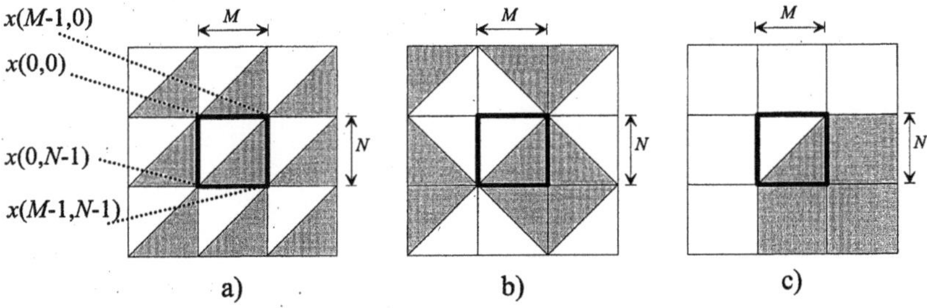

Fig. 4.2. Boundary extensions of finite image signals.
a periodic **b** symmetric (antiperiodic) **c** constant value

A system is *shift invariant* if

$$y(m,n) = L\big[x(m,n)\big] \quad \Rightarrow \quad y(m-k,n-l) = L\big[x(m-k,n-l)\big]. \tag{4.6}$$

A system of both properties is an *LSI system* (*L*inear *S*hift *I*nvariant). The operation of convolution underlies similar mathematical rules as a multiplication: Commutative, associative and distributive properties apply. Accordingly, if several LSI

---

[1] This is the 2D formulation of the more general form (3.42) which applies to an arbitrary number of dimensions.

[2] Also denoted as *Kronecker delta function*. This is a signal of value 1 for $(m,n)=(0,0)$, value 0 at all other positions; it plays a role as the discrete counterpart of the Dirac impulse.

[3] $L\{\cdot\}$ is the system transfer function, which describes and input-output relationship of the system.

systems operate sequentially, any sequence of processing can be chosen, while the result remains identical. Further, impulse responses of several systems running sequential or in parallel can be combined.

Image signals are typically of finite spatial extension, and the output of filtering shall in most cases have the same size as the input e.g. for display compatibility. Indices *m-k* and *n-l* in (4.3) can however have values less than zero or larger than *M*-1 or *N*-1, when an output near the image boundary shall be processed. Hence, it is necessary to define a signal extension beyond the boundaries of the input signal to consistently compute the convolution. Zero-setting of values will not always be useful, in particular if the signal has non-zero mean and can be expected to have non-zero amplitudes at the boundaries. The following methods of boundary extension are frequently employed (see Figs. 4.2a-c) :

– *Periodic extension*:

$$x(m,n)=x(m+M,n) \text{ for } m<0\,; \quad x(m,n)=x(m-M,n) \text{ for } m\ge M$$
$$x(m,n)=x(m,n+N) \text{ for } n<0\,; \quad x(m,n)=x(m,n-N) \text{ for } n\ge N$$
$$\tag{4.7}$$

– *Symmetric extension*:

$$x(m,n)=x(-m-1,n) \text{ for } m<0\,; \quad x(m,n)=x(2M-m-1,n) \text{ for } m\ge M$$
$$x(m,n)=x(m,-n-1) \text{ for } n<0\,; \quad x(m,n)=x(m,2N-n-1) \text{ for } n\ge N$$
$$\tag{4.8}$$

– *Constant-value extension*:

$$x(m,n)=x(0,n) \text{ for } m<0\,; \quad x(m,n)=x(M-1,n) \text{ for } m\ge M$$
$$x(m,n)=x(m,0) \text{ for } n<0\,; \quad x(m,n)=x(m,N-1) \text{ for } n\ge N$$
$$\tag{4.9}$$

A 2D system is *separable* if

$$h(k,l)=h_h(k)\cdot h_v(l)\,, \tag{4.10}$$

where $h_h(k)$ and $h_v(l)$ are impulse responses of 1D systems operating in horizontal and vertical directions, respectively. Substituting (4.10) into (4.3) results in

$$y(m,n)=\sum_{k=-\infty}^{\infty}\sum_{l=-\infty}^{\infty}x(m-k,n-l)h(k,l)$$
$$=\sum_{k=-\infty}^{\infty}h_h(k)\sum_{l=-\infty}^{\infty}x(m-k,n-l)h_v(l). \tag{4.11}$$

By defining

$$y_v(m,n)=\sum_{l=-\infty}^{\infty}x(m,n-l)h_v(l)\,, \tag{4.12}$$

(4.11) can be re-written as

$$y(m,n)=\sum_{k=-\infty}^{\infty}y_v(m-k,n)h_h(k)\,. \tag{4.13}$$

Fig. 4.3a shows the principle, where first a vertical 1D convolution by impulse response $h_v(n)$ is performed along each column, resulting in $y_v(m,n)$. In the next step, $y(m,n)$ is computed by row-wise convolution of $y_v(m,n)*h_h(m)$. As both 1D systems are linear, the result is identical, when the sequence of row-wise and column-wise processing is interchanged (Fig. 4.3b).

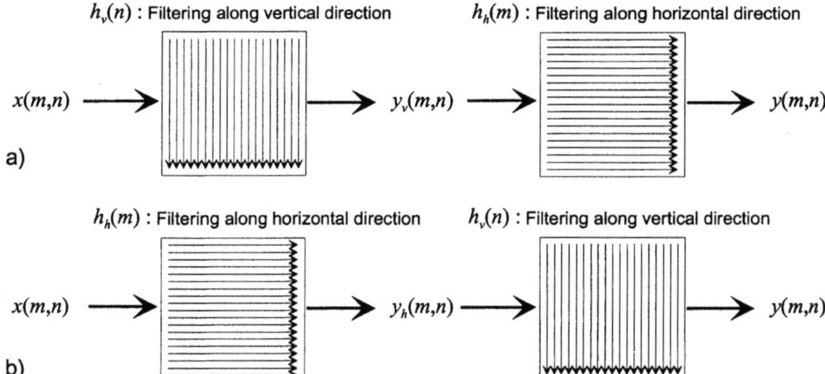

Fig. 4.3. Principle of a separable linear system with **a** first vertical, then horizontal filter step **b** first horizontal, then vertical filter step

Example: Lowpass filter. Let impulse responses of horizontal/vertical 1D filters be

$$h_h(k) := \{h_h(-1) = 0.2 \; ; \; h_h(0) = 0.6 \; ; \; h_h(1) = 0.2 \; ; \; h_h(k) = 0 \;\; \text{for} \;\; |k| > 1\}$$
$$h_v(l) := \{h_v(-1) = 0.3 \; ; \; h_v(0) = 0.4 \; ; \; h_v(1) = 0.3 \; ; \; h_v(l) = 0 \;\; \text{for} \;\; |l| > 1\}. \tag{4.14}$$

From (4.10), this results in a 2D filter with the following matrix:

$$\mathbf{H} = \begin{bmatrix} h(1,1) & h(0,1) & h(-1,1) \\ h(1,0) & h(0,0) & h(-1,0) \\ h(1,-1) & h(0,-1) & h(-1,-1) \end{bmatrix} = \begin{bmatrix} .06 & .18 & .06 \\ .08 & .24 & .08 \\ .06 & .18 & .06 \end{bmatrix} = \mathbf{h}_v \cdot \mathbf{h}_h^{\mathrm{T}}$$

$$\text{where} \quad \begin{aligned} \mathbf{h}_v &= \begin{bmatrix} 0.2 & 0.6 & 0.2 \end{bmatrix}^{\mathrm{T}} \\ \mathbf{h}_h &= \begin{bmatrix} 0.3 & 0.4 & 0.3 \end{bmatrix}^{\mathrm{T}}. \end{aligned} \tag{4.15}$$

The center value $h(0,0)$ is applied as a weight for the current pixel position. Following (4.3), the value $x(m-k,n-l)$ is multiplied by $h(k,l)$. In (4.15) and all subsequent examples, the filter matrix is constructed as it will be overlaid to the image, i.e. the top-left value is the coefficient multiplied by the top-left neighbor of the current pixel etc. If the 2D convolution is applied directly, a number of 9 multiplications/additions per pixel must be computed. If the convolution is performed separably as in (4.11)-(4.13), only 6 multiplications/additions are necessary. It is however not possible to separate any filter; further, separable filters do not allow all degrees of freedom that a non-separable 2D filter could have, in particular for the diagonal orientations. The following matrices give examples for non-separable 2D filters (both describe lowpass filters):

$$\mathbf{H} = \begin{bmatrix} 0 & .2 & 0 \\ .2 & .2 & .2 \\ 0 & .2 & 0 \end{bmatrix} \quad ; \quad \mathbf{H} = \begin{bmatrix} .25 & 0 & 0 \\ 0 & .5 & 0 \\ 0 & 0 & .25 \end{bmatrix}. \tag{4.16}$$

A filter matrix is separable, if all rows and all columns are linearly dependent. If the columns of the filter matrix are written as vectors such that

$$\mathbf{H} = \begin{bmatrix} \mathbf{h}_0 & \mathbf{h}_1 & \mathbf{h}_2 & \cdots & \mathbf{h}_K \end{bmatrix}, \tag{4.17}$$

the following condition must hold for separability:

$$a_0\mathbf{h}_0 = a_1\mathbf{h}_1 = a_2\mathbf{h}_2 = \ldots = a_K\mathbf{h}_K . \tag{4.18}$$

The coefficients of the horizontal filter will then be reciprocally dependent on the factors a_k. Then, a separable 2D filter matrix can uniquely be factorized into the underlying 1D impulse response vectors $\mathbf{h}_h$ and $\mathbf{h}_v$[1].

Fig. 4.4. Discrete 2D convolutions by filter matrices from (4.19).
a Original image b Convolution by $\mathbf{H}_T$ c Convolution by $\mathbf{H}_H$

Results of a 2D convolution employing an averaging lowpass filter $\mathbf{H}_T$ and a highpass filter $\mathbf{H}_H$,

$$\mathbf{H}_T = \frac{1}{9}\begin{bmatrix} 1 & 1 & 1 \\ 1 & 1 & 1 \\ 1 & 1 & 1 \end{bmatrix} \quad ; \quad \mathbf{H}_H = \begin{bmatrix} -1 & -1 & -1 \\ -1 & 8 & -1 \\ -1 & -1 & -1 \end{bmatrix} \tag{4.19}$$

are illustrated in Fig. 4.4. For a symmetric odd-length FIR filter of matrix size $(K+1)\mathrm{x}(L+1)$, the output is computed as[2]

---

[1] The gain factors of the 1D filters may not be unique, e.g. the 1D filters $\mathbf{h}_v=[0.1\ 0.3\ 0.1]^T$ and $\mathbf{h}_h=[0.6\ 0.8\ 0.6]^T$ would result in the same 2D filter (4.15).
[2] Filter coefficients $a(k,l)$ are identical with the impulse response values $h(k,l)$ in the FIR case.

$$y(m,n) = \sum_{k=-K/2}^{K/2} \sum_{l=-L/2}^{L/2} a(k,l) \cdot x(m-k,n-l) \,. \tag{4.20}$$

Infinite Impulse Response (IIR) filters are not realized by direct implementation of the convolution equation (4.3), but introduce feedback from previous output values $y(m,n)$. For IIR systems, a strict sequence of processing must be observed due to the recursive relationship in the feedback loop. For a 2D geometry, all positions which must have been readily processed to provide the feedback for the current position establish the *Region of Support* (RoS). Fig. 4.5 shows three different causal IIR filter geometries with their respective RoS geometries: The *wedge plane filter*, the *quarter plane filter* and the *asymmetric half plane filter*. For the cases of quarter-plane and wedge-plane filter masks, either row-wise or column-wise recursion scans would be possible; these filters also allow diagonal or zigzag scans. For the asymmetric half plane filter, row-wise processing (starting at the top left position) is the only possible sequence of recursion[1].

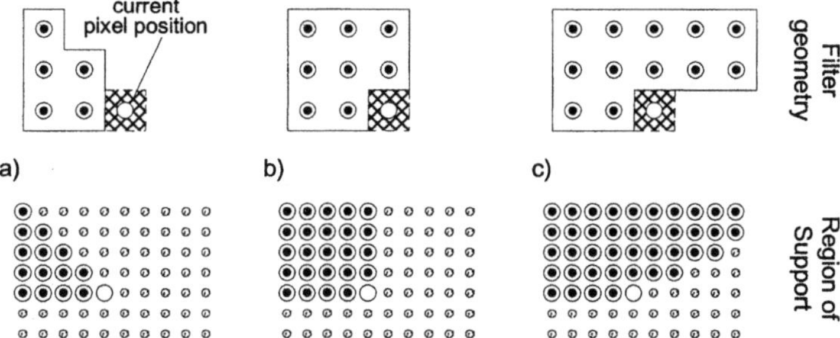

Fig. 4.5. Causal 2D filter masks and geometries of their support regions:
a Wedge plane **b** Quarter plane **c** Asymmetric half plane

A recursive quarter-plane filter, where the filter mask defines the feedback from $(P+1)\cdot(Q+1)-1$ previous output pixels, generates the output signal

$$y(m,n) = x(m,n) + \sum_{\substack{p=0 \\ (p,q)\neq(0,0)}}^{P} \sum_{q=0}^{Q} b(p,q) \cdot y(m-p,n-q) \,. \tag{4.21}$$

4.1.2   Frequency Transfer Functions of Multi-dimensional Filters

Fourier transfer functions. The frequency transfer function of a two-dimensional linear system can be computed as the Fourier transform of the impulse response. This transfer function is continuous and periodic,

---

[1] See also footnote on p. 100; for finite-extension signals, more general filter geometries, including truly noncausal IIR systems, can be realized.

$$H(j\Omega_1, j\Omega_2) = \sum_{k=-\infty}^{\infty} \sum_{l=-\infty}^{\infty} h(k,l) \cdot e^{-jk\Omega_1} e^{-jl\Omega_2} \ . \tag{4.22}$$

If the system has FIR property, the summation limits can be bounded, such that the complex transfer function can directly be computed. For a system with a symmetric filter matrix as in (4.20),

$$A(j\Omega_1, j\Omega_2) = \sum_{k=-K/2}^{K/2} \sum_{l=-L/2}^{L/2} a(k,l) \cdot e^{-jk\Omega_1} e^{-jl\Omega_2} \ . \tag{4.23}$$

For the case of an IIR system, (4.21) gives

$$x(m,n) = y(m,n) - \sum_{\substack{p=0 \\ q=0 \\ (p,q)\neq(0,0)}}^{P \quad Q} b(p,q) \cdot y(m-p, n-q) \ . \tag{4.24}$$

Comparing (4.20)/(4.23) and (4.24), the transfer function is determined as follows:

$$X(j\Omega_1, j\Omega_2) = Y(j\Omega_1, j\Omega_2) \cdot \left[1 - \sum_{\substack{p=0 \\ q=0 \\ (p,q)\neq(0,0)}}^{P \quad Q} b(p,q) \cdot e^{-jp\Omega_1} e^{-jq\Omega_2} \right]$$

$$\Rightarrow H(j\Omega_1, j\Omega_2) = \frac{Y(j\Omega_1, j\Omega_2)}{X(j\Omega_1, j\Omega_2)} = \frac{1}{1 - \displaystyle\sum_{\substack{p=0 \\ q=0 \\ (p,q)\neq(0,0)}}^{P \quad Q} b(p,q) \cdot e^{-jp\Omega_1} e^{-jq\Omega_2}} \ . \tag{4.25}$$

A system with FIR *and* IIR parts has a transfer function which combines (4.23) and (4.25),

$$H(j\Omega_1, j\Omega_2) = \frac{A(j\Omega_1, j\Omega_2)}{1 - B(j\Omega_1, j\Omega_2)} = \frac{\displaystyle\sum_{k=-K/2}^{K/2} \sum_{l=-L/2}^{L/2} a(k,l) \cdot e^{-jk\Omega_1} e^{-jl\Omega_2}}{1 - \displaystyle\sum_{\substack{p=0 \\ q=0 \\ (p,q)\neq(0,0)}}^{P \quad Q} b(p,q) \cdot e^{-jp\Omega_1} e^{-jq\Omega_2}} \ . \tag{4.26}$$

The complex transfer function can either be represented by its real and imaginary parts, or by magnitude $F(\Omega_1,\Omega_2)$ and phase $\Phi(\Omega_1,\Omega_2)$:

$$F(\Omega_1,\Omega_2) = |H(j\Omega_1, j\Omega_2)| = \sqrt{\mathrm{Re}\{H(j\Omega_1, j\Omega_2)\}^2 + \mathrm{Im}\{H(j\Omega_1, j\Omega_2)\}^2}$$

$$\Phi(\Omega_1,\Omega_2) = \arctan \frac{\mathrm{Im}\{H(j\Omega_1, j\Omega_2)\}}{\mathrm{Re}\{H(j\Omega_1, j\Omega_2)\}} \pm k(\Omega_1,\Omega_2) \cdot \pi \tag{4.27}$$

$$\text{with } k(\Omega_1,\Omega_2) = \begin{cases} 0 & \text{for } \text{Re}\{H(j\Omega_1,j\Omega_2)\} \geq 0 \\ 1 & \text{for } \text{Re}\{H(j\Omega_1,j\Omega_2)\} < 0. \end{cases}$$

Linear transform based frequency analysis is related to *eigenfunctions*, which is a class of functions being unchanged when fed into a linear system, except for a possible scaling of the amplitude and a phase shift. The amplitude/phase modification applied to one eigenfunction is equivalent with the frequency transfer function for *one* frequency value. In the case of the Fourier transform, the eigenfunctions are complex exponentials of unity magnitude, $e^{jn^{\mathrm{T}}\Omega}$. In principle, the Fourier transform performs an analysis of a signal (in this case an impulse response), expressing it by a weighted superposition of sinusoids. If the impulse response is not absolutely summable (see condition on BIBO stability, p. 89), the existence of a Fourier transfer function is not guaranteed, because the sum (4.22) may not converge[1]. The *z transform* analyses a broader class of signals and impulse responses, and converge in many cases where the Fourier transform does not exist [OPPENHEIM, WILLSKY, YOUNG 1997].

Multi-dimensional z transform. The z transform has eigenfunctions of the type

$$x(\mathbf{n}) = z_1^{-k} z_2^{-l}... \quad \text{with } z_i = \rho_i \cdot e^{j\Omega_i} \text{ and } \rho_i > 0. \tag{4.28}$$

In (4.28), i is the dimension index of a multi-dimensional transform, e.g. $i=1,2,3$ for horizontal, vertical and temporal axis analysis of an image or video signal. The multi-dimensional z transform of a signal $x(\mathbf{n})$ is defined by

$$X(\mathbf{z}) = X(z_1,z_2,...) = \sum_{m=-\infty}^{\infty} ... \sum_{n=-\infty}^{\infty} x(m,n,...) \cdot z_1^{-m} z_2^{-n}.... \tag{4.29}$$

The z transform for $\rho_i=1$ is identical to the discrete Fourier sum (2.48). For $\rho_i \neq 1$, the eigenfunctions of the z transform are more generic, as they also allow describing exponentially decaying or growing behavior over time.

The Fourier transform of a signal shows a singularity $\pi \cdot \delta\left(\Omega_i - \tilde{\Omega}_i\right)$ contributed in dimension i when the signal bears a component $e^{jn^{\mathrm{T}}\tilde{\Omega}_i}$, which means that the signal oscillates by a frequency $\tilde{\Omega}_i$. For the case of the z transform, a singularity (*pole*, which is equivalent to the Dirac impulse in the Fourier spectrum) appears at position $z_i = \tilde{\rho}_i e^{-j\tilde{\Omega}_i}$, when the signal bears a component $\tilde{\rho}_i^n e^{jn\tilde{\Omega}_i}$. The location of this singularity allows checking whether the signal amplitude is decaying or growing by increased n, while the oscillation has a frequency $\tilde{\Omega}_i$. By the locations of poles in the complex z plane, it can be analyzed whether a recursive filter is stable. If the

---

[1] An exceptional case where a system is not BIBO stable but a Fourier transform exists is given for impulse responses which can be expressed as a superposition of infinitely extended sinusoidal oscillations; then, Dirac impulses appear in the Fourier spectrum at the respective frequency positions.

summation needs to be performed only over the positive values of n, the *one-sided* z transform is sufficient, where the pole locations should only be found at positions $\tilde{p}_i < 1$[1]:

$$X_1(\mathbf{z}) = X_1(z_1, z_2, \ldots) = \sum_{m=0}^{\infty} \ldots \sum_{n=0}^{\infty} x(m, n, \ldots) \cdot z_1^{-m} z_2^{-n} \ldots . \tag{4.30}$$

Properties of the z transform. Properties of the multidimensional z transform are very similar to those of the Fourier transform[2]:

$$\text{Linearity:} \qquad a \cdot x_1(\mathbf{n}) + b \cdot x_2(\mathbf{n}) \quad \overset{z}{\longleftrightarrow} \quad a \cdot X_1(\mathbf{z}) + b \cdot X_2(\mathbf{z}) \tag{4.31}$$

$$\text{Shift:} \qquad x(\mathbf{n} - \mathbf{k}) \overset{z}{\longleftrightarrow} z_1^{-k} \cdot z_2^{-l} \cdot \ldots \cdot X(\mathbf{z}) \tag{4.32}$$

$$\text{Convolution:} \qquad y(\mathbf{n}) = x(\mathbf{n}) * h(\mathbf{n}) \quad \overset{z}{\longleftrightarrow} \quad Y(\mathbf{z}) = X(\mathbf{z}) \cdot H(\mathbf{z}) \tag{4.33}$$

$$\text{Separability:}$$

$$x(\mathbf{n}) = x_1(m) \cdot x_2(n) \cdot \ldots \quad \overset{z}{\longleftrightarrow} \quad X(\mathbf{z}) = X_1(z_1) \cdot X_2(z_2) \cdot \ldots \tag{4.34}$$

$$\text{Inversion:} \qquad x(-\mathbf{n}) \quad \overset{z}{\longleftrightarrow} \quad X(z_1^{-1}, z_2^{-1}, \ldots) \tag{4.35}$$

$$\text{Scaling}[3]:$$

$$x(m', n') = x_{(U,V)}(m' \cdot U, n' \cdot V) \quad \overset{z}{\longleftrightarrow} \quad X(z_1, z_2) = X_{(U,V)}(z_1^{1/U}, z_2^{1/V}) \tag{4.36}$$

$$\text{Expansion:} \qquad x_{(U,V)}(m,n) = \begin{cases} x(m',n'), & m = m'U \text{ and } n = n'V \\ 0, & \text{else} \end{cases} \tag{4.37}$$

$$\overset{z}{\longleftrightarrow} \quad X_{(U,V)}(z_1, z_2) = X(z_1^{U}, z_2^{V})$$

$$\text{Modulation:} \qquad x(\mathbf{n}) \cdot e^{jm\tilde{\Omega}_1} \cdot e^{jn\tilde{\Omega}_2} \cdot \ldots \quad \overset{z}{\longleftrightarrow} \quad X(z_1 \cdot e^{-j\tilde{\Omega}_1}, z_2 \cdot e^{-j\tilde{\Omega}_2}, \ldots) . \tag{4.38}$$

Special cases of (4.38) are the modulation by cosines :

$$x(m,n) \cdot \cos m\tilde{\Omega}_1 \cdot \cos n\tilde{\Omega}_2 \quad \overset{z}{\longleftrightarrow} \quad \frac{1}{4} \sum_{p,q \in \{-1,1\}} X(z_1 \cdot e^{-jp\tilde{\Omega}_1}, z_2 \cdot e^{-jq\tilde{\Omega}_2}) , \tag{4.39}$$

and the modulation by a 2D grid of unit impulses[4] of periods U und V

---

[1] The one-sided two-dimensional z transform covers the first quadrant of an infinitely extended 2D plane, and is hence applicable e.g. to quarter-plane or wedge-plane filters.

[2] Only those properties, which are used in the forthcoming sections and chapters are listed here. For more complete tables of properties, see e.g. [OPPENHEIM, WILLSKY, YOUNG 1997].

[3] Scaling is a sub-sampling operation. The z transform mapping as expressed in (4.36) is strictly valid when all samples in $x_{(U,V)}(m,n)$ except for $m=m'U$ and $n=n'V$ are zero. Scaling is typically used in combination with (4.40).

[4] The unit pulse grid $\delta_{U,V}(m,n)$ is a discrete signal which has a value of 1 at positions $(m,n)=(m'U, n'V)$, where m' and n' are integer numbers; all other positions are zero.

$$x_{(U,V)}(m,n) = x(m,n) \cdot \delta_{U,V}(m,n) \quad \overset{z}{\longleftrightarrow} \quad \frac{1}{U \cdot V} \sum_{u=0}^{U-1} \sum_{v=0}^{V-1} X(z_1 \cdot e^{-j\frac{2\pi u}{U}}, z_2 \cdot e^{-j\frac{2\pi v}{V}}) \quad (4.40)$$

Stability analysis in multi-dimensional IIR systems. The z transform expresses whether a signal (or an impulse response) tends to oscillate by increasing or decreasing amplitude. This is often used for stability analysis. For a 1D z transform, the usual approach is to determine the *Region of Convergence* (RoC) within the z plane, which is the range where the transform can be computed giving finite values. On the unit circle of the complex z plane, the z transform is identical to the Fourier transform (2.48). If the unit circle is included in the RoC, the system will be BIBO stable (*bounded input, bounded output*), as the existence of the Fourier transform without singularities (Dirac impulses) is a clear indicator that the area under the impulse response is finite and the system is BIBO stable. The z transfer function of a discrete linear 2D system having a symmetric mask of the FIR part and a quarter-plane mask of the IIR part is

$$H(z_1,z_2) = \frac{A(z_1,z_2)}{1 - B(z_1,z_2)} = \frac{\displaystyle\sum_{k=-K/2}^{K/2} \sum_{l=-L/2}^{L/2} a(k,l) \cdot z_1^{-k} z_2^{-l}}{1 - \displaystyle\sum_{\substack{p=0 \\ (p,q)\neq(0,0)}}^{P} \sum_{q=0}^{Q} b(p,q) \cdot z_1^{-p} z_2^{-q}} \; . \tag{4.41}$$

In 1D systems, the roots of the numerator polynomial are the *zeros*, the roots of the denominator polynomial are the *poles* of the z transfer function. For *causal impulse responses* ($h(m,n)=0$ for $m<0$, $n<0$), the RoC is an area outside the radius described by the outermost pole, while for *anti-causal impulse responses* ($h(m,n)=0$ for $m\geq0$, $n\geq0$) the RoC is an area inside the radius of the innermost pole. Hence, the analysis of pole locations, which must be outside the unit circle for anti-causal components and inside the unit circle for causal components, is sufficient for stability analysis. For separable multi-dimensional systems, stability analysis can be applied separately to the 1D functions; if all of them are stable, the entire system will be stable likewise.

The quarter-plane filter from Fig. 4.5b is *strictly causal* and can be analyzed by the one-sided z transform (4.30). For this case, the roots of the denominator polynomial *must* be at positions $|z|>1$ to fulfill stability criteria. For more general two- and multidimensional filters, factorization into causal and anti-causal parts may be necessary. The asymmetric half-plane filter from Fig. 4.5c has indeed *anticausal* components and can not be analyzed by the one-sided z transform. Such impulse responses must be factorized into components which are either *strictly causal* or *strictly anti-causal* in the single dimensions. Depending on the respective configuration, existence of poles for $|z|>1$ or $|z|<1$ must then be tested.

In the two- and multidimensional case, the RoC becomes a 'hypervolume of convergence', e.g. for two signal dimensions the z domain is a 4-dimensional hyperspace of two real and two imaginary axes. Roots of the denominator polynomial are then no longer singular locations, but are described by two-dimensional surfaces within this 4D space. A simple stability test is based on the analysis of *root maps* separately in the dimensions z_1 and z_2. The root map in the z_2 plane shows the movements of singularities (poles) in z_2 as they occur when the other variable z_1 is varied along the unit circle, $e^{j\Omega_1}$. Three cases can be in effect (see Fig. 4.6): The root maps are *entirely within* the unit circle in z_2, are *intersecting* the unit circle, or are *entirely outside*. The first case indicates instability for anti-causal systems, the latter case for causal systems, and the intersection case for both types of systems. From this, the following stability theorems are derived for causal systems; in the case of anticausal systems, the respective conditions must be interchanged to test for existence of singularities inside the unit circle.

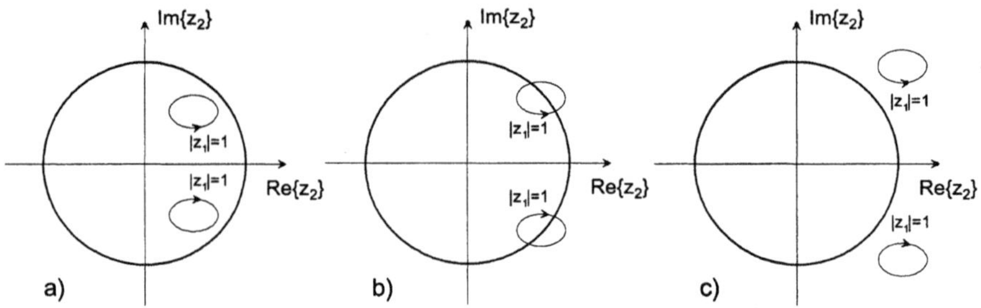

Fig. 4.6. Positions of root maps in relation to the unit circle of the z_2 plane
a within **b** intersecting **c** outside

Stability theorem [STRINTZIS 1977]. $H(z_1,z_2) = 1/[1-B(z_1,z_2)]$ shall be a recursive 2D filter in the first quadrant. This filter is stable under following conditions:

a. $1-B(z_1,z_2) \neq 0$ for $|z_1| = 1$ and $|z_2| = 1$
     (on the common unit hypersphere of both coordinates).
b. $1-B(a,z_2) \neq 0$ for $|z_2| \geq 1$ and any a with $|a|=1$.
c. $1-B(z_1,b) \neq 0$ for $|z_1| \geq 1$ and any b with $|b|=1$.

Condition a. can be computed by the Fourier transform (e.g. approximated by a 2D DFT of sufficiently dense spectral lines), while b. and c. are usual conditions as applied in stability analysis based on 1D z transform; for simplicity, set $a=b=1$.

   Proof : For case of instability, root maps and the related 2D root surfaces either intersect the unit circle / hypersphere (checked densely by condition a.), or they are completely outside (which is proven for any value by conditions b. and c.). The stability theorem extends as follows to the κ-dimensional case, where $\kappa+1$ conditions have to be checked, where a. can again be computed via a κ-dimensional DFT:

a. $1-B(z_1,z_2,...,z_\kappa) \neq 0$ for $|z_1| = 1, |z_2| = 1,...,|z_\kappa| = 1$
b. $1-B(1,1,...,z_\kappa,...,1,1) \neq 0$ for $|z_\kappa| \geq 1, k=1,2,..., \kappa.$

4.1.3   Image filtering by Matrix Operations

The convolution of one row from an image matrix by a 1D FIR filter can be expressed in matrix notation. This is shown here for the case of a symmetric 3-tap filter:

$$
\underbrace{\begin{bmatrix}
y(0,n) \\
y(1,n) \\
y(2,n \\
\vdots \\
\vdots \\
y(M-2,n) \\
y(M-1,n)
\end{bmatrix}}_{\mathbf{y(n)}}
=
\underbrace{\begin{bmatrix}
h(0) & h(-1) & 0 & \cdots & \cdots & 0 & h(1) \\
h(1) & h(0) & h(-1) & 0 & \cdots & \cdots & 0 \\
0 & h(1) & h(0) & h(-1) & 0 & \cdots & 0 \\
\vdots & \ddots & \ddots & \ddots & \ddots & \ddots & \vdots \\
\vdots & & 0 & h(1) & h(0) & h(-1) & 0 \\
0 & & & 0 & h(1) & h(0) & h(-1) \\
h(-1) & 0 & 0 & \cdots & 0 & h(1) & h(0)
\end{bmatrix}}_{\mathbf{H}_{\text{period}}}
\cdot
\underbrace{\begin{bmatrix}
x(0,n) \\
x(1,n) \\
x(2,n \\
\vdots \\
\vdots \\
x(M-2,n) \\
x(M-1,n)
\end{bmatrix}}_{\mathbf{x(n)}}.
\tag{4.42}
$$

The filter matrix $\mathbf{H}$ in (4.41) shows the case of *periodic* boundary extension (4.7). Alternatively, the cases of constant-value extension (4.9) and symmetric extension (4.8) can be implemented in the matrix:

$$
\mathbf{H}_{\text{const}} =
\begin{bmatrix}
h(0)+h(1) & h(-1) & 0 & \cdots & \cdots & 0 & 0 \\
h(1) & h(0) & h(-1) & 0 & \cdots & \cdots & 0 \\
0 & h(1) & h(0) & h(-1) & 0 & \cdots & 0 \\
\vdots & \ddots & \ddots & \ddots & \ddots & \ddots & \vdots \\
\vdots & & 0 & h(1) & h(0) & h(-1) & 0 \\
0 & & & & 0 & h(1) & h(0) & h(-1) \\
0 & 0 & 0 & \cdots & 0 & h(1) & h(0)+h(-1)
\end{bmatrix};
\tag{4.43}
$$

$$
\mathbf{H}_{\text{symm}} =
\begin{bmatrix}
h(0) & h(1)+h(-1) & 0 & \cdots & \cdots & 0 & 0 \\
h(1) & h(0) & h(-1) & 0 & \cdots & \cdots & 0 \\
0 & h(1) & h(0) & h(-1) & 0 & \cdots & 0 \\
\vdots & \ddots & \ddots & \ddots & \ddots & \ddots & \vdots \\
\vdots & & 0 & h(1) & h(0) & h(-1) & 0 \\
0 & & & & 0 & h(1) & h(0) & h(-1) \\
0 & 0 & 0 & \cdots & 0 & h(1)+h(-1) & h(0)
\end{bmatrix}.
\tag{4.44}
$$

If the horizontal filter shall process all rows of the image, this can be written by the following matrix multiplication (see (1.3) for definition of the image matrix):

$$Y = \left[\mathbf{H} \cdot \mathbf{X}^{\mathrm{T}} \right]^{\mathrm{T}} = \mathbf{X} \cdot \mathbf{H}^{\mathrm{T}} \quad ; \quad \mathbf{X} = \begin{bmatrix} \mathbf{x}(0)^{\mathrm{T}} \\ \mathbf{x}(1)^{\mathrm{T}} \\ \vdots \\ \mathbf{x}(N-1)^{\mathrm{T}} \end{bmatrix}. \tag{4.45}$$

Observe that in row-wise filtering, $\mathbf{H}$ is of size $M{\times}M$, while matrices $\mathbf{X}$ and $\mathbf{Y}$ are of size $M{\times}N$. If a separable filter operation on the image shall be expressed, a second $N{\times}N$ matrix $\mathbf{H}_v$ must be defined for the vertical filter, while $\mathbf{H}_h$ is identical with the previous horizontal filter. In the following equations, the first step is the vertical filter, where the intermediate step of transposing the matrix $\mathbf{X}$ is not necessary in the expression on the right-hand side:

$$\mathbf{Y} = \left[\mathbf{H}_h \cdot \left[\mathbf{H}_v \cdot \mathbf{X} \right]^{\mathrm{T}} \right]^{\mathrm{T}} = \left[\mathbf{H}_v \cdot \mathbf{X} \right] \cdot \mathbf{H}_h^{\mathrm{T}} . \tag{4.46}$$

If the filter is not separable, it is no longer possible to express the operation by two subsequent steps of matrix multiplication. The matrix multiplication by $\mathbf{H}$ must then perform linear operations on adjacent horizontal and vertical samples simultaneously. This can be expressed by input and output images arranged in a row-wise sequence as one-dimensional vectors referring to the scheme in (1.4). The operation then is a one-step matrix-vector multiplication:

$$\begin{bmatrix} y(0,0) \\ y(1,0) \\ \vdots \\ y(M-1,0) \\ y(0,1) \\ \vdots \\ y(0,N-1) \\ \vdots \\ y(M-2,N-1) \\ y(M-1,N-1) \end{bmatrix} = \mathbf{H} \cdot \begin{bmatrix} x(0,0) \\ x(1,0) \\ \vdots \\ x(M-1,0) \\ x(0,1) \\ \vdots \\ x(0,N-1) \\ \vdots \\ x(M-2,N-1) \\ x(M-1,N-1) \end{bmatrix}. \tag{4.47}$$

The matrix $\mathbf{H}$, apparently still has identical coefficients populating the diagonals, but now several disparate diagonal non-zero stripes appear:

$$\mathbf{H} = \begin{bmatrix} \ddots & \ddots & \ddots & \ddots & \ddots & \ddots & \ddots & \ddots & \ddots & \ddots & \ddots & \ddots & \ddots & \ddots & \ddots \\ \cdots 0 & h(1,1) & \mathbf{h(0,1)} & h(-1,1) & 0\cdots 0 & h(1,0) & \mathbf{h(0,0)} & h(-1,0) & 0\cdots 0 & h(1,-1) & \mathbf{h(0,-1)} & h(-1,-1) & 0\cdots & \cdots & \cdots \\ \cdots & \cdots 0 & h(1,1) & \mathbf{h(0,1)} & h(-1,1) & 0\cdots 0 & h(1,0) & \mathbf{h(0,0)} & h(-1,0) & 0\cdots 0 & h(1,-1) & \mathbf{h(0,-1)} & h(-1,-1) & 0\cdots & \cdots \\ \cdots & \cdots & \cdots 0 & h(1,1) & \mathbf{h(0,1)} & h(-1,1) & 0\cdots 0 & h(1,0) & \mathbf{h(0,0)} & h(-1,0) & 0\cdots 0 & h(1,-1) & \mathbf{h(0,-1)} & h(-1,-1) & 0\cdots \\ \ddots & \ddots & \ddots & \ddots & \ddots & \ddots & \ddots & \ddots & \ddots & \ddots & \ddots & \ddots & \ddots & \ddots & \ddots \end{bmatrix} \tag{4.48}$$

In (4.48), the example of a 2D filter of size $K{\times}L{=}3{\times}3$ pixels is shown. The matrix $\mathbf{H}$ is only sparsely populated by non-zero values. Boundary extensions can be defined similar to the methods in (4.42)-(4.44). The full matrix $\mathbf{H}$ is of size $M\cdot N\times M\cdot N$. In the case of separable filters, it is also possible to construct the matrix (4.48) by computing the outer product of the two horizontal and vertical matrices:

$$\mathbf{H} = \mathbf{H}_v \times \mathbf{H}_h . \tag{4.49}$$

As filter matrices both for the separable and non-separable cases are quadratic, they are invertible if no singularities occur[1]. Then, it is possible to reconstruct the signal $\mathbf{X}$ from $\mathbf{Y}$:

$$\mathbf{Y} = \mathbf{H}\cdot\mathbf{X} \quad \Rightarrow \quad \mathbf{X} = \mathbf{H}^{-1}\cdot\mathbf{Y} . \tag{4.50}$$

For the separable filter operation, the inverse matrix notation is

$$\mathbf{X} = \left[\mathbf{H}_h^{-1} \cdot \left[\mathbf{H}_v^{-1} \cdot \mathbf{Y} \right]^{\mathrm{T}} \right]^{\mathrm{T}} = \left[\mathbf{H}_v^{-1} \cdot \mathbf{Y} \right] \cdot \left[\mathbf{H}_h^{-1} \right]^{\mathrm{T}} . \tag{4.51}$$

4.1.4   Realization of Two-dimensional Filters

Convolution requires storage of pixels. Assume row-wise processing starting from the top-left pixel, the common image scan. Depending on the geometry of the filter, memory for single pixels (to access samples from the same row), for complete rows (to access samples from previous rows) or for complete images (to access samples from previous frames) may be necessary. A symmetric filter matrix of size[2] $(K+1)\mathrm{x}(L+1)$ shall be centered around the current position, such that it extends by $K/2$ left and right, and $L/2$ top and down (see Fig. 4.7a). The output signal $y(m,n)$ can then only be computed after $L/2$ future rows plus another $K/2$ pixels have become available. The same amount of pixels from past rows must be retained in memory, and in total it is necessary to provide a memory of $M\cdot L+K$ pixels (Fig. 4.7b).

It is possible to implement this 2D filter by using a one-dimensional shift-register memory structure. Following the notation of the z transform, delay elements are expressed by z^{-1}. The structure of a one-dimensional symmetric FIR filter is shown in Fig. 4.8a, the related structure of a 2D filter in Fig. 4.8b[3]. The input signal to the 2D filter is the sequence of row-wise scanned image pixels. Observe that in each filter step only $(K+1)(L+1)$ pixels are invoked, whereas the memory consists of $M\cdot L+K$ delay elements. For each of the L rows, $M-K$ pixels must be held in memory for subsequent filtering steps, which are not used in the

---

[1] Typically, the non-zero filter coefficients are found along diagonals of the matrix. Hence, it is unlikely that the determinant could become zero.

[2] K und L are even in these examples.

[3] Specific boundary extensions are neglected here. In particular for the structures shown in Figs. 4.8 and 4.9, the rows will 'wrap around' in memory, which is a similar effect as periodic boundary extension (4.7), however by an offset of one line vertically.

current operation. Due to the usage of 'future' pixels a delay occurs, whereby the output signal $y(m,n)$ can only be computed when the signal $x(m+K/2,n+L/2)$ appears at the input of the system.

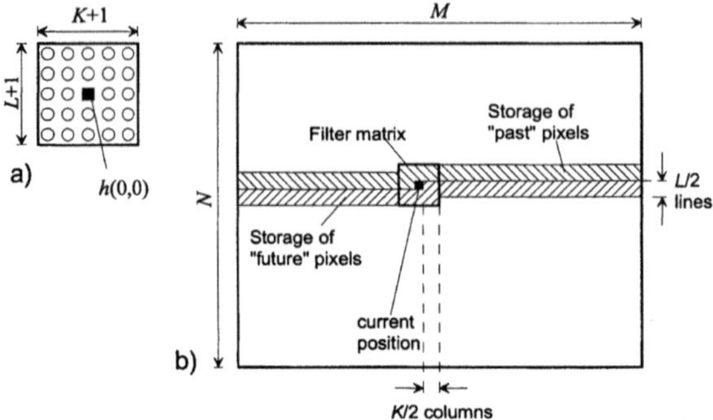

Fig. 4.7. a Filter matrix of a symmetric FIR filter **b** Positions which have to be stored to perform the 2D convolution by this filter matrix

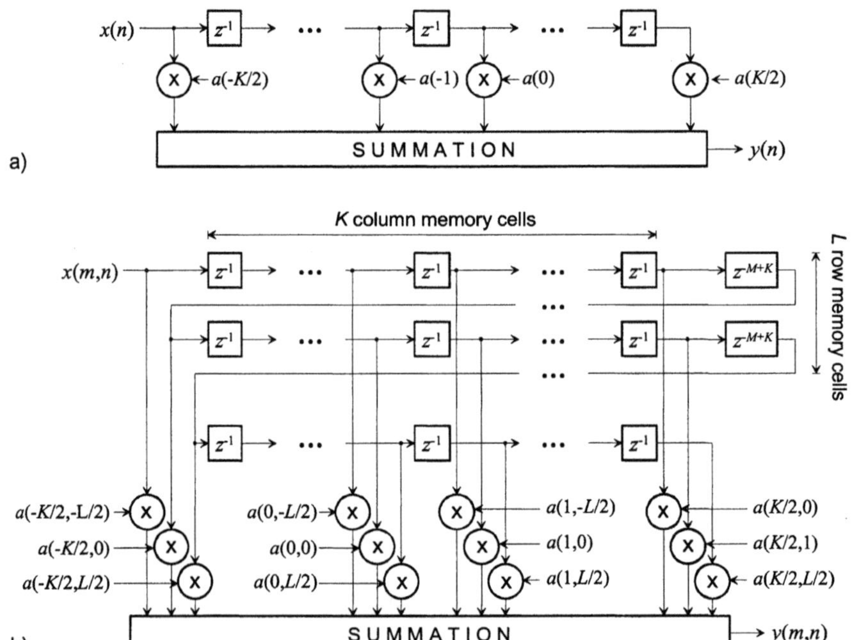

Fig. 4.8. Structures of a one-dimensional (**a**) and two-dimensional (**b**) symmetric FIR filter. The result is available with a delay of $K/2$ horizontally, $L/2$ vertically.

Fig. 4.9a shows the structure of a 1D IIR system, Fig. 4.9b is the 2D counterpart. In the 2D case, it is also possible to arrange the memory structure of the feedback loop as a one-dimensional shift register. Systems containing both IIR and FIR components can be combined from Figs. 4.8 and 4.9.

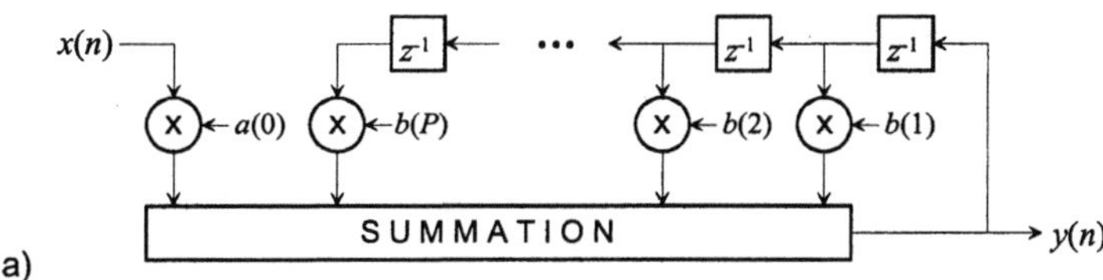

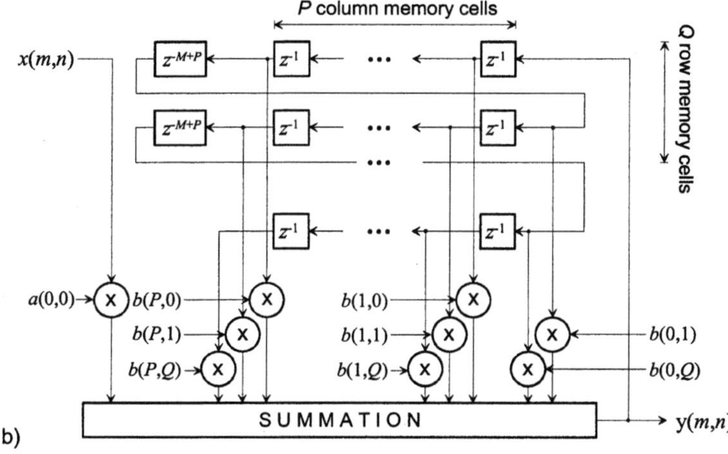

Fig. 4.9. Structures of a one-dimensional (**a**) and two-dimensional (**b**) IIR filter

Discrete 2D convolution in the DFT domain. Multiplication of DFT spectra from an image $x(m,n)$ of size MxN and from a symmetrically-centered finite impulse response $h(k,l)$ of odd non-zero size $(K+1)x(L+1)$ gives

$$Y(u,v) = X(u,v) \cdot H(u,v) = \sum_{m=0}^{M-1}\sum_{n=0}^{N-1} x(m,n) W_M^{-um} W_N^{-vn} \cdot \sum_{k=-K/2}^{K/2}\sum_{l=-L/2}^{L/2} h(k,l) W_M^{-uk} W_N^{-vl}$$

$$= \sum_{m=0}^{M-1}\sum_{n=0}^{N-1}\sum_{k=-K/2}^{K/2}\sum_{l=-L/2}^{L/2} x(m,n) \cdot h(k,l) \cdot W_M^{-u(m+k)} \cdot W_N^{-v(n+l)} . \tag{4.52}$$

By substitutions $m'=m+k$, $n'=n+l$, the following expression results:

$$Y(u,v) = \sum_{k=-K/2}^{K/2}\sum_{l=-L/2}^{L/2}\sum_{m'=k}^{M-1+k}\sum_{n'=l}^{N-1+l} x(m'-k,n'-l) \cdot h(k,l) \cdot W_M^{-um'} \cdot W_N^{-vn'} , \tag{4.53}$$

which due to the periodicity can be written as

$$Y(u,v) = \sum_{m'=0}^{M-1} \sum_{n'=0}^{N-1} \left[\sum_{k=-K/2}^{K/2} \sum_{l=-L/2}^{L/2} x(m'-k,n'-l) \cdot h(k,l) \right] \cdot W_M^{-um'} \cdot W_N^{-vn'} . \qquad (4.54)$$

The multiplication of the 2D DFT spectra is identical to the DFT of the discrete 2D convolution sum (4.3), provided that periodic ('cyclic') extension of the signal and impulse response is performed beyond the boundaries. If the convolution sum shall be computed by the DFT/IDFT using another extension than the periodic method, it is necessary to enlarge the analysis size of the DFT up to a size $(M+K)$x$(N+L)$. For this case, Fig. 4.11 shows the arrangements of the image matrix, the filter impulse response[1] and the convolution result in the respective extended windows.

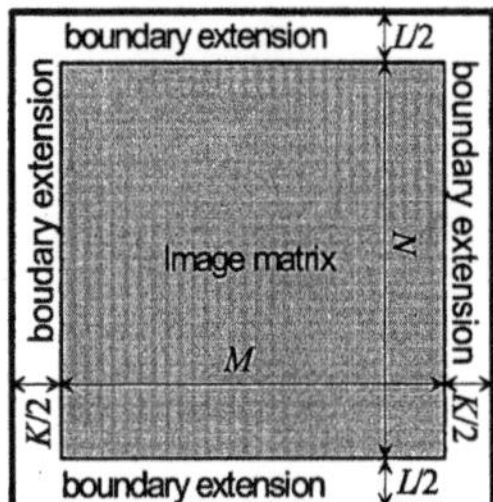

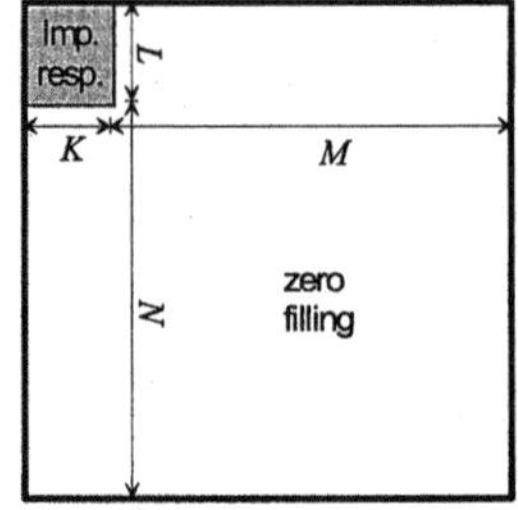

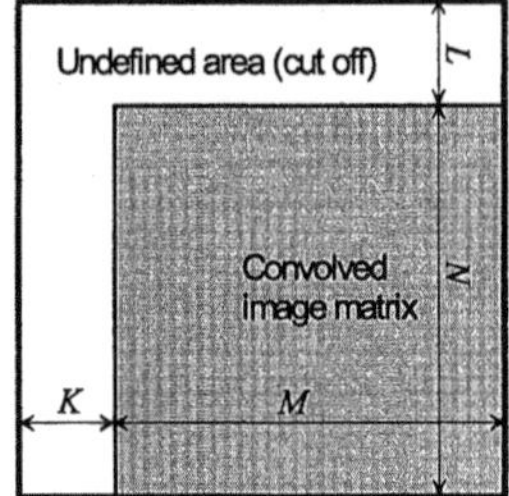

Fig. 4.10. Positions of signals and impulse response for 2D convolution via DFT

4.2 Linear Prediction

4.2.1 One- and Two-dimensional Autoregressive Models

Algorithms of multimedia signal processing often require a model about the statistical properties of signals for analytic optimization (cf. sec. 3.4). If statistical assumptions are made which go beyond sample statistics, modeling of statistical dependencies between samples is required. The ACF or autocovariance function are sufficient to optimize *linear systems* for a given purpose, as they can fully characterize *linear statistical dependencies* between samples.

The *autoregressive (AR-) model* (Fig. 4.11) is generated by a recursive filter with z transfer function $A(\mathbf{z})=1/(1-H(\mathbf{z}))$, which is fed by a white Gaussian noise process $z(\mathbf{n})$. The process $x_{AR}(\mathbf{n})$ from the filter output possesses spectral properties which are determined by the amplitude transfer function of the filter. The PDF of this stationary process is also Gaussian. As the property of stationarity does not

---

[1] For the case of symmetric filters also the *filter matrix*, which is the spatially inverted impulse response as shown in (4.14)/(4.15), can be transformed directly. Alternatively, it is also possible to transform the filter matrix and perform complex conjugate multiplication of spectra.

necessarily apply to natural signals, there is a certain deficiency by mapping signals to this model. Usually, a high degree of variation is observed in the local properties of image, speech and audio signals, for which a local adaptation of models is necessary. In general, the model helps to simplify problems of optimization due to its analytic properties. As the AR model is Gaussian, it is fully described by moments up to the order two. AR model signals perfectly follow the vector Gaussian PDF (3.40).

The autoregressive model of first order [AR(1)] is frequently used to model the global statistics of image signals. For the 1D case, a Gaussian zero-mean white-noise signal $z(n)$ (*innovation signal*) of variance σ_z^2 is fed into a recursive filter of transfer function

$$A(z) = \frac{1}{1-\rho z^{-1}} . \tag{4.55}$$

The computation of the signal can be expressed as

$$x_{AR}(n) = \rho \cdot x(n-1) + z(n) . \tag{4.56}$$

The AR(1) process has an autocorrelation function[1]

$$r_{xx}(k) = \sigma_x^2 \rho^{|k|} \quad ; \quad \sigma_x^2 = \frac{\sigma_z^2}{1-\rho^2} \tag{4.57}$$

and a power spectrum

$$S_{xx}(\Omega) = \left|X(j\Omega)\right|^2 = \sigma_x^2 \sum_{k=-\infty}^{\infty} \rho^{|k|} e^{-jk\Omega} = \frac{\sigma_x^2(1-\rho^2)}{1-2\rho\cos\Omega+\rho^2} . \tag{4.58}$$

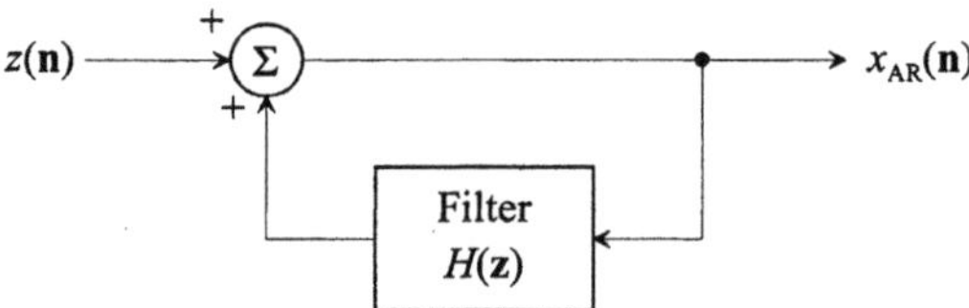

Fig. 4.11. System for generation of an autoregressive model process

The autoregressive process has zero mean property. Obviously, the model is fully characterized by the filter parameter ρ, which is identical to the autocovariance coefficient[2] $\rho(1)$, and the variance σ_x^2. Typical values of $\rho(1)$ for natural images are between .9 and .98. Fig. 4.12 shows row-wise power spectrum estimates (3.67)

---

[1] For a proof on (4.57) and (4.58), see Problem 4.2.

[2] For simplicity, we write $\rho(k)$ instead of the usual notation $\rho'(k)$ for the autocovariance; due to the zero mean property of the AR process, autocorrelation and autocovariance functions are identical.

computed from two natural images and related spectra of AR(1) models, adapted by setting ρ equal to the estimated autocovariance values (3.32)/(3.33).

This 1D model can be extended into two and multiple dimensions in different ways, where the methods differ by the properties of the 2D autocovariance function. We first develop some models which are only parametrized by the autocovariance coefficients $\rho_h \equiv \rho_h(1)$ and $\rho_v \equiv \rho_v(1)$ of horizontal and vertical directions. The properties can be illustrated by 2D plots in the (k,l) plane, showing lines where autocovariance values are constant (Fig. 4.13).

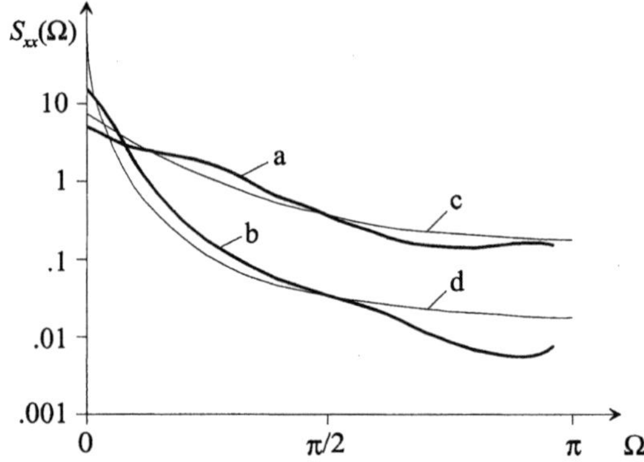

Fig. 4.12. Row-wise spectra of two 2D image signals (a,b) and model spectra of AR(1) processes adapted by autocovariance parameters $\rho=0.70$ (c) and $\rho=0.96$ (d).

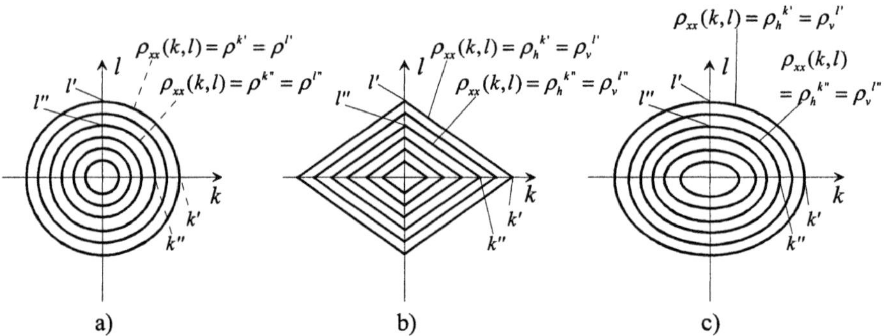

Fig. 4.13. Lines of constant autocovariance in 2D AR(1) models.
a isotropic b separable c elliptic

The *isotropic* model has an autocovariance function

$$r_{xx}(k,l) = \sigma_x^2 \rho^{\sqrt{k^2+l^2}} , \tag{4.59}$$

expressing circular-symmetric values independent of the direction, $\rho_h = \rho_v$ is inherently assumed. Constant values appear on circles of radius $r = \sqrt{k^2 + l^2}$ (see Fig. 4.13a). The two-dimensional power spectrum of the isotropic model is then also circular-symmetric,

$$S_{xx}(\Omega_1, \Omega_2) = \frac{\sigma_x^2 (1 - \rho^2)}{1 - 2 \cdot \rho \cdot \cos\sqrt{\Omega_1^2 + \Omega_2^2} + \rho^2} . \tag{4.60}$$

For other models, autocovariance values are defined truly separate for the horizontal and vertical directions. In natural images, it can be observed that autocovariance statistics differ significantly for different orientations. It is often observed that the covariance along the vertical axis is lower than along the horizontal axis. The *separable model* with autocovariance function

$$r_{xx}(k,l) = \sigma_x^2 \cdot \rho_h^{|k|} \cdot \rho_v^{|l|} \quad ; \quad \sigma_x^2 = \frac{\sigma_z^2}{(1 - \rho_h^2) \cdot (1 - \rho_v^2)} , \tag{4.61}$$

shows straight lines of constant autocovariance[1]. These lines intersect with axes k and l at positions k' and l' where $\rho_h^{|k'|} = \rho_v^{|l'|}$ (see Fig. 4.13b). The generation of the discrete 2D signal can be implemented by a separable recursive filter, which is numerically expressed by the equation

$$x_{AR}(m,n) = \rho_h x_{AR}(m-1,n) + \rho_v x_{AR}(m,n-1) - \rho_h \rho_v x_{AR}(m-1,n-1) + z(m,n) . \tag{4.62}$$

The related power spectrum is

$$S_{xx}(\Omega_1, \Omega_2) = \sigma_x^2 \cdot \frac{1 - \rho_h^2}{1 - 2 \cdot \rho_h \cdot \cos\Omega_1 + \rho_h^2} \cdot \frac{1 - \rho_v^2}{1 - 2 \cdot \rho_v \cdot \cos\Omega_2 + \rho_v^2} . \tag{4.63}$$

The *elliptic model* has an autocovariance function

$$r_{xx}(k,l) = \sigma_x^2 \cdot e^{-\sqrt{(\beta_h k)^2 + (\beta_v l)^2}} \quad ; \quad \beta_h = -\ln\rho_h \quad ; \quad \beta_v = -\ln\rho_v . \tag{4.64}$$

It shows constant autocovariance lines of elliptic shape due to the elliptic equation in the exponent (Fig. 4.13c). This model can also be interpreted as an extension of the isotropic model, which is given for the special case of $\beta_h = \beta_v$. Intersections of constant autocovariance graphs with the coordinate axes are identical to the case of the separable model. The power spectrum of the elliptic model is

$$S_{xx}(\Omega_1, \Omega_2) = \frac{2\pi\sigma_x^2}{\beta_h \beta_v} \cdot \frac{1}{\left[1 + \left(\dfrac{\Omega_1}{\beta_h}\right)^2 + \left(\dfrac{\Omega_2}{\beta_v}\right)^2 \right]} . \tag{4.65}$$

---

[1] If the two exponential expressions in (4.61) are modified for a common basis, a line equation over absolute values appears in the exponent, see (4.66).

All models introduced so far can be interpreted as special cases of a generalized 2D AR(1) model with autocovariance

$$r_{xx}(k,l) = \sigma_x^2 \cdot e^{-\left[(\beta_h \cdot |k|)^\gamma + (\beta_v \cdot |l|)^\gamma\right]^{\frac{1}{\gamma}}} . \tag{4.66}$$

For the isotropic and for the elliptic model, $\gamma=2$; for the separable model $\gamma=1$. When $\gamma<1$, the lines of constant autocovariance are convex, for $\gamma>1$ concavity is observed. The intersections with the axes remain identical as in the case of separable and elliptic models, irrespective of γ. The relation of factors β with the autocovariance coefficients is given in (4.64).

Autoregressive models of higher order are frequently applied in speech analysis and for texture analysis of images. As natural signals are however assumed to be instationary, the autocovariance function is usually estimated over segments (finite windows) of samples. Then, at least local invariance of statistical behavior is assumed. The coefficients $a(\mathbf{p})$ shall define the generation of the AR model process[1]

$$x_{AR}(\mathbf{n}) = \sum_{\substack{\mathbf{p}\in\Pi \\ \mathbf{p}\neq 0}} a(\mathbf{p}) \cdot x_{AR}(\mathbf{n}-\mathbf{p}) + z(\mathbf{n}) . \tag{4.67}$$

Within the range Π (defining the neighborhood of output samples which are fed back into the recursive generator filter), it is assumed that the process and the signal to be modeled shall possess identical autocovariance values $r_{xx}(\mathbf{k})$. As the autocovariance function establishes a Fourier transform pair with the power spectrum, the model has a spectrum

$$S_{xx}(\mathbf{\Omega}) = \frac{\sigma_z^2}{\left|1 - \sum_{\substack{\mathbf{p}\in\Pi \\ \mathbf{p}\neq 0}} a(\mathbf{p}) \cdot e^{-j\mathbf{p}^T\mathbf{\Omega}}\right|^2} . \tag{4.68}$$

First, a set of coefficients $a(\mathbf{p})=a_1(p)\cdot a_2(q)\cdot...$ of a separable AR model shall be determined. For this purpose, it is necessary to perform the autocovariance analysis separately along the particular directions of the multidimensional coordinate

---

[1] Observe that the definitions (4.68) and (4.69) do not implicitly postulate the *causality* of the AR synthesis filters. In fact, all following deductions can likewise be made for non-causal filter sets without any limitation. Non-causal recursive filtering is indeed practically applicable for signals of finite extension, e.g. image signals. In such a case, the non-causal mask must be factorized into causal and anticausal components. While the causal parts can be processed recursively left-right and top-down, the anticausal parts are processed by right-left and bottom-up recursions. As still all filters are linear systems, the output is finally generated by superposition of all partial components. Only the current position $\mathbf{n}$ must not be used, which means that $a(\mathbf{0})=0$. For more detail on non-causal AR modeling of images, see e.g. [JAIN 1989].

system. For each 1D component, estimation of the autocovariance is performed by analysis over a window of N signal values[1],

$$\hat{r}_{xx}(k) = \frac{1}{N} \sum_{n=0}^{N-1} x(n) \cdot x(n+k) \; ; \; -P \leq k \leq P \qquad (4.69)$$

To optimize the model, it is postulated that the white-noise signal $z(\mathbf{n})$ to be fed into the AR synthesis filter shall have lowest possible variance:

$$\sigma_z^2 = E\{z^2(n)\} = E\left\{\left[x(n) - \sum_{p=1}^{P} a(p) \cdot x(n-p)\right]^2\right\} \overset{!}{=} \min$$

$$= E\{x^2(n)\} - 2 \cdot E\left\{\left[x(n) \cdot \sum_{p=1}^{P} a(p) \cdot x(n-p)\right]\right\} + E\left\{\left[\sum_{p=1}^{P} a(p) \cdot x(n-p)\right]^2\right\}. \qquad (4.70)$$

The minimization is achieved by partial derivations, performed separately for each filter coefficient:

$$\frac{\partial z^2(n)}{\partial a(k)} \overset{!}{=} 0 \Rightarrow E\{x(n) \cdot x(n-k)\} = \sum_{p=1}^{P} a(p) \cdot E\{x(n-p) \cdot x(n-k)\} \; ; \; 1 \leq k \leq P. \qquad (4.71)$$

This gives a Pth order equation system, known as the *Wiener-Hopf equation*, where the optimum filter coefficients must fulfill the following conditions:

$$r_{xx}(k) = \sum_{p=1}^{P} a(p) \cdot r_{xx}(k-p) \quad ; \quad 1 \leq k \leq P. \qquad (4.72)$$

Due to the symmetry of the autocovariance, $r_{xx}(k\text{-}p) = r_{xx}(p\text{-}k)$, the problem can be simplified in a more regular matrix structure, and the Wiener-Hopf equation can be written as follows, where $\mathbf{R}_{xx}$ is the autocovariance matrix (3.41):

$$\underbrace{\begin{bmatrix} r_{xx}(1) \\ r_{xx}(2) \\ \vdots \\ \vdots \\ r_{xx}(P) \end{bmatrix}}_{\mathbf{r}_{xx}} = \underbrace{\begin{bmatrix} r_{xx}(0) & r_{xx}(1) & \cdots & \cdots & r_{xx}(P-1) \\ r_{xx}(1) & r_{xx}(0) & r_{xx}(1) & \cdots & r_{xx}(P-2) \\ \vdots & r_{xx}(1) & r_{xx}(0) & \ddots & \vdots \\ \vdots & \vdots & \ddots & \ddots & r_{xx}(1) \\ r_{xx}(P-1) & r_{xx}(P-2) & \cdots & r_{xx}(1) & r_{xx}(0) \end{bmatrix}}_{\mathbf{R}_{xx}} \cdot \underbrace{\begin{bmatrix} a(1) \\ a(2) \\ \vdots \\ \vdots \\ a(P) \end{bmatrix}}_{\mathbf{a}}. \qquad (4.73)$$

The solution is obtained when the vector $\mathbf{r}_{xx}$ and the matrix $\mathbf{R}_{xx}$ are filled by the autocovariance estimates (4.69), and the matrix is inverted:

---

[1] Zero-mean signals are assumed here; for signals which do not have this property, the mean should be subtracted before analysis. If the mean by itself varies locally, a *moving average* filter can be used to model this behavior (see sec. 7.2.1).

$$\mathbf{a} = \hat{\mathbf{R}}_{xx}^{-1} \cdot \hat{\mathbf{r}}_{xx} \ .$$

(4.74)

$\mathbf{R}_{xx}$ has a *Toeplitz structure*, which means that it is diagonally symmetric, and each single diagonal is only filled by one identical value. Even though the matrix is not sparse, it is highly regular, and fast algorithms exist to resolve the problem of matrix inversion [RABINER, SCHAFER 1978]. This is mainly due to the fact that the lower and upper triangular matrices, as well as many sub-matrices are identical, such that many sub-matrix inversions need to be computed only once (cf. (B.14)). The variance of the innovation signal for the case of a 1D AR(P) model is[1]

$$\sigma_z^2 = E\{z^2(n)\} = E\left\{\left(x_{AR}(n) - \sum_{p=1}^{P} a(p)x_{AR}(n-p)\right)^2\right\} = \underbrace{E\{(x_{AR}(n))^2\}}_{=\sigma_x^2}$$

$$-2\sum_{p=1}^{P} a(p)\underbrace{E\{x_{AR}(n)x_{AR}(n-p)\}}_{=r_{xx}(-p)} + \sum_{p=1}^{P}\sum_{q=1}^{P} a(p)a(q)\underbrace{E\{x_{AR}(n-p)x_{AR}(n-q)\}}_{=r_{xx}(p-q)}$$

(4.75)

$$= \sigma_x^2 - 2\sum_{p=1}^{P} a(p)r_{xx}(p) + \sum_{p=1}^{P} a(p)\underbrace{\sum_{q=1}^{P} a(q)r_{xx}(p-q)}_{=r_{xx}(p) \text{ acc. to W-H eq.}} = \sigma_x^2 - \sum_{p=1}^{P} a(p)r_{xx}(p)$$

This leads to an alternative formulation of the Wiener-Hopf equation, where the computation of the innovation signal variance is included by the first row of the matrix:

$$\sigma_z^2\delta(k) = r_{xx}(k) - \sum_{p=0}^{P} a(p)\cdot r_{xx}(k-p) \quad ; \quad 0 \le k \le P$$

$$\underbrace{\begin{bmatrix} \sigma_z^2 \\ 0 \\ \vdots \\ \vdots \\ 0 \end{bmatrix}}_{\mathbf{r}_{xx}} = \underbrace{\begin{bmatrix} r_{xx}(0) & r_{xx}(1) & r_{xx}(2) & \cdots & r_{xx}(P) \\ r_{xx}(1) & r_{xx}(0) & r_{xx}(1) & \cdots & r_{xx}(P-1) \\ r_{xx}(2) & r_{xx}(1) & \ddots & & \vdots \\ \vdots & \vdots & & \ddots & \vdots \\ r_{xx}(P) & r_{xx}(P-1) & \cdots & \cdots & r_{xx}(0) \end{bmatrix}}_{\mathbf{R}_{xx}} \cdot \underbrace{\begin{bmatrix} 1 \\ -a(1) \\ \vdots \\ \vdots \\ -a(P) \end{bmatrix}}_{\mathbf{a}}$$

(4.76)

For a separable model of κ dimensions, filter coefficients are optimized independently by solving 1D Wiener-Hopf equations over the different coordinate axes. The variance σ_z^2 of the innovation signal to be fed into the κ-dimensional model can be computed as follows by concatenation of the synthesis systems for dimensions i, where each has a certain ratio of input and output variances according to (4.75):

---

[1] This is a generalization of (4.57) and also covers the AR(1) case.

$$\sigma_{z,i}^{2} = r_{xx,i}(0) - \sum_{p=1}^{P_i} a_i(p) \cdot r_{xx,i}(p) \;\Rightarrow\; \sigma_z^{2} = \frac{\prod_{i=1}^{K} \sigma_{z,i}^{2}}{\prod_{i=1}^{K-1} r_{xx,i}(0)}. \tag{4.77}$$

Separable models do not allow optimum adaptation to the properties of multidimensional signals. For example in the two-dimensional case, the diagonal orientations of autocovariances are not considered. When non-separable autocovariance functions are used, non-separable IIR filters must also be defined as synthesis filters. Using the 2D autocovariance function (3.31), the solution results by the 2D Wiener-Hopf equation, which is a 2D extension of (4.76). For the case of quarter-plane filter optimization this gives [MARAGOS, SCHAFER, MERSEREAU 1984]

$$\sigma_z^{2}\delta(k,l) = r_{xx}(k,l) - \sum_{\substack{p=0 \\ (p,q)\neq(0,0)}}^{P} \sum_{q=0}^{Q} a(p,q) \cdot r_{xx}(k-p,l-q), \tag{4.78}$$

which can again be written as $\mathbf{r}_{xx} = \mathbf{R}_{xx} \cdot \mathbf{a}$. $\mathbf{R}_{xx}$ is a *Toeplitz block matrix*

$$\mathbf{R}_{xx} = \begin{bmatrix} \phi_0 & \phi_1 & \cdots & \cdots & \phi_P \\ \phi_1 & \phi_0 & \cdots & \cdots & \phi_{P-1} \\ \vdots & \vdots & \ddots & & \vdots \\ \vdots & \vdots & & \ddots & \vdots \\ \phi_P & \phi_{P-1} & \cdots & \cdots & \phi_0 \end{bmatrix} \tag{4.79}$$

with associated sub-matrices

$$\phi_p = \begin{bmatrix} r_{xx}(0,p) & r_{xx}(-1,p) & \cdots & \cdots & r_{xx}(-Q,p) \\ r_{xx}(1,p) & r_{xx}(0,p) & \cdots & \cdots & r_{xx}(-Q+1,p) \\ \vdots & \vdots & \ddots & & \vdots \\ \vdots & \vdots & & \ddots & \vdots \\ r_{xx}(Q,p) & r_{xx}(Q-1,p) & \cdots & \cdots & r_{xx}(0,p) \end{bmatrix}. \tag{4.80}$$

The vector of coefficients is arranged by row-wise order

$$\mathbf{a} = \left[1,-a(1,0),..,-a(Q,0),-a(0,1),...,-a(Q,P)\right]^{\mathrm{T}}, \tag{4.81}$$

and the autocorrelation vector is

$$\mathbf{r}_{xx} = \left[\sigma_z^{2},0,0,...,0\right]^{\mathrm{T}}. \tag{4.82}$$

The lengths of the vectors and the row/column lengths of the quadratic matrix are $(P{+}1)\cdot(Q{+}1)$. The task is to determine $(P{+}1)\cdot(Q{+}1){-}1$ unknown coefficients in $\mathbf{a}$. This is achieved as in (4.74), inverting the autocovariance matrix $\mathbf{R}_{xx}$. The matrix of the 2D formulation does however no longer have a Toeplitz structure, and not even the sub-matrices (4.80) are always diagonally symmetric, since $r_{xx}(k,p) \neq r_{xx}(-k,p)$. As a consequence, the solution imposes higher computational complexity, and also the number of covariance values to be used in the optimization is larger than the number of filter coefficients to be determined. This may indeed result in computation of unstable synthesis filters.

Example. For the case $P=2$, $Q=2$, the separable method based on (4.73) and (4.74) gives the optimum predictor coefficients

$$r_{xx}(1) = r_{xx}(0) \cdot a(1) \quad \Rightarrow \quad a_{h,opt} = \rho_h \quad ; \quad a_{v,opt} = \rho_v \ . \tag{4.83}$$

In the non-separable case, the matrix $\mathbf{R}_{xx}$ from (4.79) and (4.80) is

$$\mathbf{R}_{xx} = \begin{bmatrix} r_{xx}(0,0) & r_{xx}(-1,0) & r_{xx}(0,1) & r_{xx}(-1,1) \\ r_{xx}(1,0) & r_{xx}(0,0) & r_{xx}(1,1) & r_{xx}(0,1) \\ r_{xx}(0,1) & r_{xx}(-1,1) & r_{xx}(0,0) & r_{xx}(-1,0) \\ r_{xx}(1,1) & r_{xx}(0,1) & r_{xx}(1,0) & r_{xx}(0,0) \end{bmatrix} . \tag{4.84}$$

For the computation of the 3 filter coefficients, the first row and first column relating to the computation of the innovation signal variance can be discarded, which reduces to the following matrix equation system of normalized autocovariances; all inherent symmetries of the 2D autocovariance have been utilized here:

$$\begin{bmatrix} \rho_{xx}(1,0) \\ \rho_{xx}(0,1) \\ \rho_{xx}(1,1) \end{bmatrix} = \begin{bmatrix} 1 & \rho_{xx}(1,1) & \rho_{xx}(0,1) \\ \rho_{xx}(-1,1) & 1 & \rho_{xx}(1,0) \\ \rho_{xx}(0,1) & \rho_{xx}(1,0) & 1 \end{bmatrix} \cdot \begin{bmatrix} a(1,0) \\ a(0,1) \\ a(1,1) \end{bmatrix} . \tag{4.85}$$

It is evident that even for this simple case, the matrix $\mathbf{R}_{xx}$ does not have a Toeplitz structure any more, and four linearly independent autocovariance values must be used to determine three filter coefficients. As a consequence, a unique bi-directional mapping between the 2D autocovariance function and non-separable AR model parameters is not possible in general, which is a major difference as compared to the 1D case. This phenomenon applies for any number of dimensions $\kappa > 1$.

4.2.2   Linear Prediction

Autoregressive modeling of signals has a direct relationship with *linear prediction*, in which a linear *predictor filter* computes an estimate $\hat{x}(\mathbf{n})$ for the signal value $x(\mathbf{n})$. The difference is the *prediction error*

$$e(\mathbf{n}) = x(\mathbf{n}) - \hat{x}(\mathbf{n}) \ . \tag{4.86}$$

The signal can be reconstructed by using the prediction error and the estimate,

$$y(\mathbf{n}) \overset{!}{=} x(\mathbf{n}) = e(\mathbf{n}) + \hat{x}(\mathbf{n}) . \tag{4.87}$$

If estimates $\hat{x}(\mathbf{n})$ are exclusively computed by past values of the signal, the prediction error $e(\mathbf{n})$ is a unique equivalent of $x(\mathbf{n})$[1]. The prediction is typically performed by an FIR filter of transfer function $H(\mathbf{z})$ (Fig. 4.14a); the *prediction error filter* (Fig. 4.14b), performing the operation described in (4.86), has a transfer function

$$A(\mathbf{z}) = 1 - H(\mathbf{z}) . \tag{4.88}$$

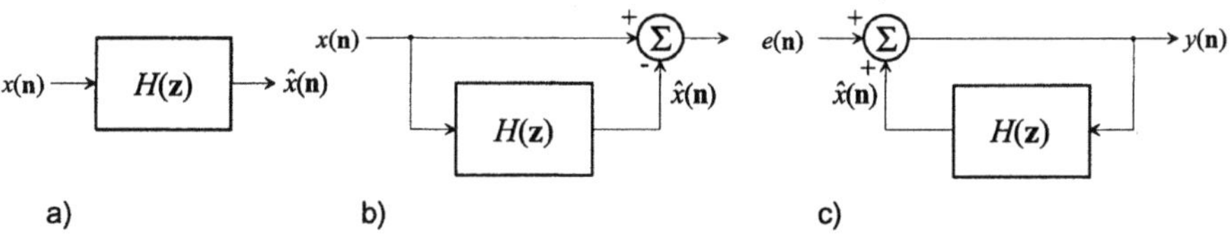

Fig. 4.14. System elements in linear prediction: **a** Predictor filter $H(\mathbf{z})$ **b** Prediction error filter (analysis filter) $A(\mathbf{z})$ **c** inverse prediction error filter (synthesis filter) $B(\mathbf{z})$

The *inverse prediction error filter* (synthesis filter, Fig. 4.14c) performs the operation (4.87). It is a recursive filter of transfer function (cf. (4.24)/(4.25))

$$B(\mathbf{z}) = \frac{1}{A(\mathbf{z})} = \frac{1}{1 - H(\mathbf{z})} . \tag{4.89}$$

The filter (4.89) is identical to the synthesis filter of an AR model, from which it can be concluded that the prediction error signal is actually Gaussian white noise if an AR process is optimally predicted (i.e. predicted using the same coefficients as in the complementary synthesis filter by which the process was generated). In the context of linear prediction, the ratio of signal variance and prediction error variance is denoted as the *prediction gain*

$$G = \frac{\sigma_x^2}{\sigma_e^2} , \tag{4.90}$$

which can be determined from (4.75) for the case of an AR model.

Two-dimensional prediction. The prediction equation in the case of a 2D quarter-plane predictor filter of order $(P+1) \cdot (Q+1)-1$ is

---

[1] See also footnote on p. 100; for finitely extended signals like images, noncausal prediction can also be performed by using 'future' samples, while the mapping $e(\mathbf{n}) \leftrightarrow x(\mathbf{n})$ is still unique and reversible. An example of non-causal prediction is used in the so-called 'fast KLT', see p. 124.

$$\hat{x}(m,n) = \sum_{\substack{p=0 \\ q=0 \\ (p,q)\neq(0,0)}}^{\substack{P \\ Q}} a(p,q)\cdot x(m-p,n-q) . \tag{4.91}$$

The z transfer function of this filter is

$$H(z_1,z_2) = a(1,0)\cdot z_1^{-1} + ... + a(P,0)\cdot z_1^{-P} + a(0,1)\cdot z_2^{-1} + ... + a(P,1)\cdot z_1^{-P}\cdot z_2^{-1}$$
$$+ ... + a(0,Q)\cdot z_2^{-Q} + ... + a(P,Q)\cdot z_1^{-P}\cdot z_2^{-Q}. \tag{4.92}$$

For the case of 2D signals, 2D prediction can be expected to better minimize the variance of the prediction error signal, as compared to 1D (horizontal or vertical) prediction. Assume that 2D prediction is applied to a separable 2D AR(1) model, where the same prediction filter $H(z_1,z_2)$ is used as in the feedback loop of the model generator. Hence, the prediction error filter $A(z_1,z_2)=1-H(z_1,z_2)$ will exactly reproduce the Gaussian white noise fed into the generator of the AR process. For the 2D separable AR(1) model, the optimum predictor filter results from the two (horizontal and vertical) 1D filters as follows:

$$H(z_1) = \rho_h\cdot z_1^{-1} \Rightarrow A(z_1) = 1 - \rho_h\cdot z_1^{-1}$$
$$H(z_2) = \rho_v\cdot z_2^{-1} \Rightarrow A(z_2) = 1 - \rho_v\cdot z_2^{-1}$$
$$A(z_1,z_2) = A(z_1)\cdot A(z_2) = 1 - \rho_h\cdot z_1^{-1} - \rho_v\cdot z_2^{-1} + \rho_h\cdot\rho_v\cdot z_1^{-1}\cdot z_2^{-1} . \tag{4.93}$$
$$\Rightarrow H(z_1,z_2) = 1 - A(z_1,z_2) = \rho_h\cdot z_1^{-1} + \rho_v\cdot z_2^{-1} - \rho_h\cdot\rho_v\cdot z_1^{-1}\cdot z_2^{-1}$$

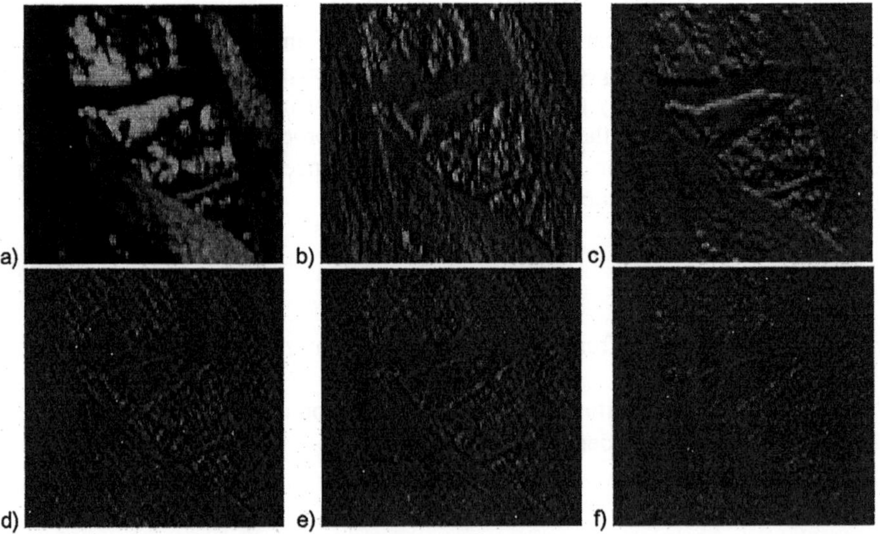

Fig. 4.15. Original image (**a**) and prediction error images: **b-c** 1D prediction row-wise (b) and column-wise (c) **d-f** 2D prediction with fixed coefficients (*d*), separable adaptation (*e*) and non-separable adaptation (*f*)

Fig. 4.15 shows an original image (*a*), prediction error images obtained by horizontal (*b*) and vertical (*c*) 1D prediction, and by separable 2D prediction (*d*). In these examples, fixed predictor coefficients were used, determined from the global horizontal and vertical one-step covariances $\rho'_{xx}(1,0)$ and $\rho'_{xx}(0,1)$ of the given image. While the horizontal prediction is not capable to predict vertical edges, the vertical filter fails at horizontal edges, but the 2D prediction performs reasonably well in both cases. Specifically in areas showing regular textured structures (e.g. grass, water, hairs, etc.) the usage of higher-order 2D predictors can be advantageous, if adapted properly to the signal. Typically, the spatial correlation is lower within structured or edge regions than it is in flat areas; it may also be direction-dependent.

Forward-adaptive prediction. Fig. 4.15e/f shows prediction error images for cases of separable and non-separable adaptive prediction. In both cases, the adaptation block size is 16x16 pixels, quarter-plane prediction filters of size 3x3 are optimized by solving the Wiener-Hopf equation system. This method is *forward-adaptive*, as the parameters of the filters are estimated from the statistics of the areas where they shall actually be used, which is only possible if a delay is introduced in the process; in usage for coding, the predictor parameters must be sent to the receiver as side information. The prediction error is obviously lower than in the cases of fixed settings of predictor filters from Fig. 4.15b-d; the non-separable method effects a better prediction in particular along diagonal structures.

Backward-adaptive prediction. Methods of predictor filter adaptation which use only *past* samples for the optimization of the prediction, are *backward-adaptive*. The *Least Mean Square algorithm* (LMS) is often applied for adaptation of separable or non-separable predictors[1]. In the separable case, $a_{h,m}(p)$ shall be the pth horizontal, and $a_{v,n}(q)$ the qth vertical predictor coefficient at the image position (m,n). This results in the prediction equation

$$\hat{x}(m,n) = \sum_{\substack{p=0 \\ (p,q)\neq(0,0)}}^{P-1} \sum_{q=0}^{Q-1} a_{h,m}(p) \cdot a_{v,n}(q) \cdot x(m-p,n-q) \, . \tag{4.94}$$

The LMS update of the coefficients [ALEXANDER, RAJALA 1985]is performed by

$$a_{h,m+1}(p) = a_{h,m}(p) + \varepsilon \cdot e(m,n) \cdot x(m-p,n)$$
$$a_{v,n+1}(q) = a_{v,n}(q) + \varepsilon \cdot e(m,n) \cdot x(m,n-q) . \tag{4.95}$$

If an image is processed line-sequentially, the LMS adaptation can also be formulated in two dimensions, which allows to perform backward-adaptive prediction of two-dimensional predictor filters [CHUNG, KANEFSKY 1992]:

---

[1] This is a gradient-based approach very similar to the methods described in sec. 8.3.

$$a_{m+1,n}(p,q) = a_{m,n}(p,q) + \varepsilon \cdot e(m,n) \cdot x(m-p,n-q). \tag{4.96}$$

Useful values for the step size factor ε are around 10^{-5}. The prediction and predictor adaptation can however fail entirely at signal discontinuities in edge regions in images, which can hardly be foreseen from previous samples. Nevertheless, the prediction gain is globally in the same order of magnitude for cases of forward and backward adaptation.

Motion compensated prediction. When temporal prediction from previous frame(s) of a video signal shall be performed, it has to be observed that the position of similar areas, which would be suitable as prediction estimates, highly depends on motion. In *motion compensated prediction*, predictor adaptation is performed by *motion estimation* (see sec. 7.6). If frame o shall be predicted, and the best-matching position in the reference frame (e.g. the previous frame of index $o-1$) is found by a shift of k pixels horizontally and l pixels vertically, the prediction equation is

$$e(m,n,o) = x(m,n,o) - \hat{x}(m,n,o) \quad \text{with} \quad \hat{x}(m,n,o) = x(m+k,n+l,o-1). \tag{4.97}$$

The motion compensated predictor filter can be characterized by a 3D z transfer function

$$H(z_1,z_2,z_3) = z_1^{k} \cdot z_2^{l} \cdot z_3^{-1}, \tag{4.98}$$

which obviously describes a linear system; motion-compensated prediction indeed is a specific type of linear prediction. This simple type of filter uses a copy of pixels from one previous frame and shifts them by an integer number of sample positions. If the brightness of the signal changes, it will be more appropriate to multiply by an additional factor; in a more generalized approach, values from different previous frames can be superimposed for prediction: Each could then be weighted individually by a weighting factor a_t, the 'temporal predictor coefficient'. If prediction shall further support sub-pixel motion shifts, a spatial interpolation filter must be included in the predictor filter. The estimate is then computed by using R previous frames[1]

$$\hat{x}(m,n,o) = \sum_{r=1}^{R} a_t(r) \sum_{p=-(P+1)/2}^{(P-1)/2} \sum_{q=-(Q+1)/2}^{(Q-1)/2} a_r(p,q) \cdot x[m+k(r)-p,n+l(r)-q,o-r]. \tag{4.99}$$

The interpolation filter $A_r(z_1,z_2)$ is typically of lowpass characteristic. The z transfer function of the predictor filter is then described by

---

[1] Here, 2D interpolation filters of even length (odd orders P and Q) are used for the horizontal and vertical sub-pixel interpolations. For a more exact definition, refer to (5.47). As sub-pixel shifts can be different for each previous frame used in the prediction, the interpolation filters are defined individually. Also, the integer motion shifts are defined individually for each of the previous frames. More realistically, motion shifts will also vary locally, such that the predictor filter is also a shift-variant system.

$$H(z_1, z_2, z_3) = \sum_{r=1}^{R} a_t(r) \cdot A_r(z_1, z_2) \cdot z_1^{k(r)} \cdot z_2^{l(r)} \cdot z_3^{-r} . \tag{4.100}$$

Fig. 4.16 shows results of a frame predicted without and with inclusion of motion compensation, the latter case also with additional interpolation filtering for sub-pixel accuracy.

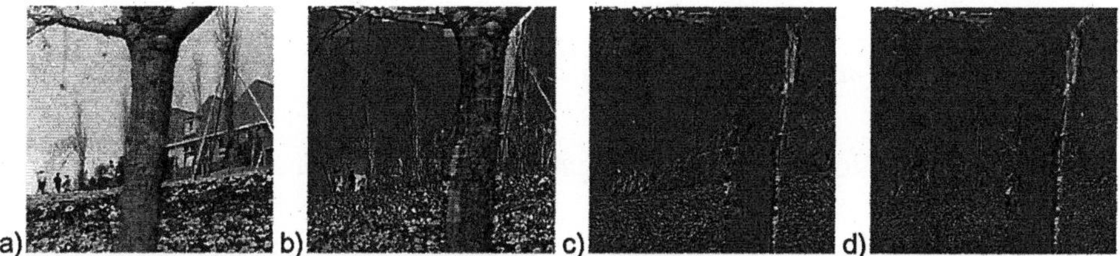

Fig. 4.16. Original frame from a video sequence (**a**) and prediction error frames:
b without motion compensation **c** with motion compensation, full-pixel shift accuracy
d with motion compensation, half-pixel shift accuracy

4.3 Linear Block Transforms

4.3.1 Orthogonal Basis Functions

In the DFT (sec. 3.4), spectral analysis is performed by multiplication of the signal by a series of complex exponential functions. The property of orthogonality (3.52) is important in this context. Other sets of basis functions can be defined where the members are orthogonal as well. In general, two finite discrete complex functions $t_0(m)$ and $t_1(m)$ of length M are called *orthogonal*, if their linear combination gives zero,

$$\sum_{m=0}^{M-1} t_0(m) t_1(m)^* = 0 \quad \text{or} \quad \mathbf{t}_0^{\mathrm{T}} \mathbf{t}_1^* = 0 \quad \text{with} \quad \mathbf{t}_u = \left[t_u(0) \;\; \cdots \;\; t_u(M-1) \right]^{\mathrm{T}} . \tag{4.101}$$

If the functions $t_u(m)$ are impulse responses of linear filters, and the operations $y_0(m)=x(m)*t_0(m)$, $y_1(m)=x(m)*t_1(m)^*$ are performed, it can be shown that the cross correlation between the two unshifted outputs becomes zero, i.e. $r_{y_0y_1}(0)=0$. This is the *decorrelating effect* of orthogonal basis functions.

For an *orthogonal set* of basis functions, each member of the set is orthogonal with any other. It is also possible to define such sets from real-valued functions. Instead of performing convolution of a signal by the members of the set, the method of *block transform* extracts block segments (vectors) $\mathbf{x}$ of length M from the signal, and applies the operations $c_u=\mathbf{x}^{\mathrm{T}}\mathbf{t}_u$. In principle, as for the example of the DFT, this is a mapping from the signal domain into a transform domain, where the values c_u

are the transform coefficients. If it is possible to reconstruct the signal from the transform coefficients, these establish an equivalent representation of the signal. In the sequel, this problem will first be studied for the case of one-dimensional transforms, from which related two- and multidimensional transforms can mostly be derived by a separable operation similar to the case of separable filters. A segment from the signal, consisting of M subsequent samples, shall be mapped into a set of transform coefficients

$$c_u = \sum_{m=0}^{M-1} x(m) \cdot t_u(m) \quad ; \quad 0 \le u < U . \tag{4.102}$$

It shall be possible to reconstruct the signal by an inverse transform synthesis, such that

$$x(m) = \frac{1}{A} \sum_{u=0}^{U-1} c_u \cdot s_u(m) ; \quad 0 \le m < M . \tag{4.103}$$

Substituting (4.103) into (4.102) gives

$$\sum_{m=0}^{M-1} t_u(m) \cdot \sum_{p=0}^{U-1} \frac{c_p}{c_u} \cdot s_p(m) = A \quad ; \quad 0 \le u < U . \tag{4.104}$$

This condition can only be valid for all u, if the factor c_p/c_u is zero for $u \ne p$, such that

$$\sum_{m=0}^{M-1} t_u(m) \cdot s_p(m) = \begin{cases} A & \text{for} \quad u = p \\ 0 & \text{for} \quad u \ne p \end{cases} \quad ; \quad 0 \le (p,u) < U \tag{4.105}$$

For a real valued A, the analysis and synthesis functions must be conjugates of each other, but in addition the product must be equal for any of the analysis/synthesis function pairs, such that in fact the vector norms of the analysis basis functions must be constant over the entire set:

$$\mathbf{s}_u = \mathbf{t}_u^* \quad \text{and} \quad \mathbf{t}_u^{\mathrm{T}} \mathbf{t}_u^* = A . \tag{4.106}$$

By combining (4.105) and (4.106), the condition of orthogonality results as follows for the set of basis functions:

$$\mathbf{t}_u^{\mathrm{T}} \mathbf{t}_p^* = \sum_{m=0}^{M-1} t_u(m) \cdot t_p(m)^* = \begin{cases} A & \text{for} \quad u = p \\ 0 & \text{for} \quad u \ne p \end{cases} \quad ; \quad 0 \le (p,u) < U \tag{4.107}$$

For the example of the DFT (3.50) and IDFT (3.54), $t_u(m) = e^{-j2\pi mu/M}$, $s_u(m) = e^{j2\pi mu/M}$ and $A=M$, which apparently fulfills these conditions. A discrete transform of M signal values into U coefficients can be formulated by matrix notation

$$
\begin{bmatrix} c_0 \\ c_1 \\ \vdots \\ \vdots \\ c_{U-1} \end{bmatrix} = \begin{bmatrix} t_0(0) & t_0(1) & \cdots & \cdots & t_0(M-1) \\ t_1(0) & t_1(1) & \cdots & \cdots & t_1(M-1) \\ \vdots & & \ddots & & \vdots \\ \vdots & & & \ddots & \vdots \\ t_{U-1}(0) & t_{U-1}(1) & \cdots & \cdots & t_{U-1}(M-1) \end{bmatrix} \cdot \begin{bmatrix} x(0) \\ x(1) \\ \vdots \\ \vdots \\ x(M-1) \end{bmatrix} ,
\qquad (4.108)
$$

$$
\underbrace{}_{\mathbf{c}} \qquad\qquad \underbrace{}_{\mathbf{T}} \qquad\qquad \underbrace{}_{\mathbf{x}}
$$

where the signal vector $\mathbf{x}$ is M-dimensional, the transform matrix $\mathbf{T}$ is of size $M{\times}U$, and the resulting vector $\mathbf{c}$ holds the U transform coefficients c_u. The rows of $\mathbf{T}=[\mathbf{t}_0\ \mathbf{t}_1\ \ldots\ \mathbf{t}_{U-1}]^T$ are the *basis functions* of the transform. Following (4.108), the values of $x(m)$ can uniquely be reconstructed from the coefficients c_u if $\mathbf{T}$ is invertible, which will be the case if the matrix is quadratic, i.e. $U=M$[1], and the determinant is $\neq 0$. Then,

$$
\mathbf{x} = \mathbf{T}^{-1}\mathbf{c} .
\qquad (4.109)
$$

As a special case of orthogonality, a transform is called *orthonormal*, if $A=1$ in (4.107). Analysis and synthesis vectors in (4.106) are identical for real-valued orthonormal bases. For an arbitrary complex, orthogonal transform matrix the following condition holds:

$$
\mathbf{T}^{-1} = \frac{1}{A}\left[\mathbf{T}^*\right]^T .
\qquad (4.110)
$$

The synthesis functions $\mathbf{s}_p$ from (4.105) are the columns of $[\mathbf{T}^*]^T$; in combination with (4.110), this gives the relationship $\mathbf{T}{\cdot}\mathbf{T}^{-1}=\mathbf{I}$. In orthonormal linear transforms, the quadratic norm (energy) of a signal vectors can directly be computed without any normalization from the coefficient vectors, otherwise

$$
\mathbf{x}^T\mathbf{x} = \mathbf{x}^T \underbrace{\frac{1}{A}\left[\mathbf{T}^*\right]^T \mathbf{T}}_{\mathbf{I}} \mathbf{x} = \frac{1}{A}\left[\mathbf{c}^*\right]^T \mathbf{c} \ \text{ or } \ \|\mathbf{x}\|^2 = \frac{1}{A}\|\mathbf{c}\|^2 .
\qquad (4.111)
$$

A *separable two-dimensional transform* can be expressed by matrix notation

$$
\begin{bmatrix} c_{0,0} & c_{1,0} & \cdots & c_{U-1,1} \\ c_{0,1} & c_{1,1} & \cdots & c_{U-1,1} \\ \vdots & \vdots & \ddots & \vdots \\ c_{0,V-1} & c_{1,V-1} & \cdots & c_{U-1,V-1} \end{bmatrix} = \begin{bmatrix} t_0(0) & t_0(1) & \cdots & t_0(N-1) \\ t_1(0) & t_1(1) & \cdots & t_1(N-1) \\ \vdots & \vdots & \ddots & \vdots \\ t_{V-1}(0) & t_{V-1}(1) & \cdots & t_{V-1}(N-1) \end{bmatrix} .
$$

---

[1] The proof given earlier by substituting (4.103) into (4.102) only showed the necessary condition of orthogonality of the transform; in fact, perfect reconstruction also requires a *complete transform*, it is not possible if $U{<}M$.

$$\cdot \begin{bmatrix} x(0,0) & x(1,0) & \cdots & x(M-1,0) \\ x(0,1) & x(1,1) & \cdots & x(M-1,1) \\ \vdots & \vdots & \ddots & \vdots \\ x(0,N-1) & x(1,N-1) & \cdots & x(M-1,N-1) \end{bmatrix} \cdot \begin{bmatrix} t_0(0) & t_1(0) & \cdots & t_{U-1}(0) \\ t_0(1) & t_1(1) & \cdots & t_{U-1}(1) \\ \vdots & \vdots & \ddots & \vdots \\ t_0(M-1) & t_1(M-1) & \cdots & t_{U-1}(M-1) \end{bmatrix} \tag{4.112}$$

which is very similar to the formulation of separable filters in (4.51). In a first step, all columns (length N) of an image matrix $\mathbf{X}$ are transformed separately by the multiplication $\mathbf{C}_v = \mathbf{T}_v \cdot \mathbf{X}$. The matrix $\mathbf{C}_v$ then holds the result of the vertical transform step. The subsequent horizontal transform of $\mathbf{C}_v$ is rather performed by multiplying the transposed transform matrix $\mathbf{T}_h^{\mathrm{T}}$ instead of transposing the matrix $\mathbf{C}_v$ [1]. The matrix equations for the separable 2D transform and the related inverse transform are as follows:

$$\mathbf{C} = \underbrace{\left[\mathbf{T}_v \cdot \mathbf{X}\right]}_{\mathbf{C}_v} \cdot \mathbf{T}_h^{\mathrm{T}} \quad \Rightarrow \quad \mathbf{X} = \left[\mathbf{T}_v^{-1} \cdot \mathbf{C}\right] \cdot \left[\mathbf{T}_h^{-1}\right]^{\mathrm{T}} \tag{4.113}$$

The basis functions relating to $U \cdot V$ coefficients of the separable 2D transform are

$$t_{u,v}(m,n) = t_u(m) \cdot t_v(n) \quad ; \quad 0 \le u < U \quad ; \quad 0 \le v < V . \tag{4.114}$$

2D basis functions are also denoted as *basis images*. As in the case of a non-separable 2D filter, any 2D transform can be executed in one single step of matrix multiplication, which is necessary if the transform is non-separable. To express this, the image and the resulting transform coefficient block are arranged by row-wise sequence as 1D vectors (cf. (4.47) und (4.48)):

$$\begin{bmatrix} c_{0,0} \\ c_{1,0} \\ \vdots \\ c_{U-1,0} \\ c_{0,1} \\ \vdots \\ c_{U-1,V-1} \end{bmatrix} = \begin{bmatrix} t_{0,0}(0,0) & t_{0,0}(1,0) & \cdots & t_{0,0}(M-1,0) & t_{0,0}(0,1) & \cdots & t_{0,0}(M-1,N-1) \\ t_{1,0}(0,0) & t_{1,0}(1,0) & \cdots & t_{1,0}(M-1,0) & t_{1,0}(0,1) & \cdots & t_{1,0}(M-1,N-1) \\ \vdots & \vdots & \ddots & \vdots & \vdots & \ddots & \vdots \\ t_{U-1,0}(0,0) & t_{U-1,0}(1,0) & \cdots & t_{U-1,0}(M-1,0) & t_{U-1,0}(0,1) & \cdots & t_{U-1,0}(M-1,N-1) \\ t_{0,1}(0,0) & t_{0,1}(1,0) & \cdots & t_{0,1}(M-1,0) & t_{0,1}(0,1) & \cdots & t_{0,1}(M-1,N-1) \\ \vdots & \vdots & \ddots & \vdots & \vdots & \ddots & \vdots \\ t_{U-1,V-1}(0,0) & t_{U-1,V-1}(1,0) & \cdots & t_{U-1,V-1}(M-1,0) & t_{U-1,V-1}(0,1) & \cdots & t_{U-1,V-1}(M-1,N-1) \end{bmatrix} \cdot \begin{bmatrix} x(0,0) \\ x(1,0) \\ \vdots \\ x(M-1,0) \\ x(0,1) \\ \vdots \\ x(M-1,N-1) \end{bmatrix}$$

$$\underbrace{\phantom{c_{0,0}}}_{\mathbf{c}_{2D}} \qquad \underbrace{\phantom{t_{0,0}(0,0)}}_{\mathbf{T}_{2D}} \qquad \underbrace{}_{\mathbf{x}_{2D}}$$

$$\tag{4.115}$$

For the case of a separable transform, the 2D transform matrix from (4.115) can be expressed as the outer product of the two 1D matrices,

$$\mathbf{T}_{2D} = \mathbf{T}_v \times \mathbf{T}_h . \tag{4.116}$$

---

[1] Alternatively, the second step could be $\mathbf{C} = [\mathbf{T}_h \cdot \mathbf{C}_v^{\mathrm{T}}]^{\mathrm{T}}$, however the other formulation is more elegant as the output is in correct (not transposed) order right away. It is of course also possible to perform the horizontal transform first, achieving the same final result.

4.3.2   Basis Functions of Orthogonal Transforms

In this section, basis functions of some important transforms are introduced mostly by their one-dimensional versions. They extend to the case of two-dimensional separable transforms according to (4.113)-(4.115).

Rectangular basis functions. The analysis block lengths M of the following rectangular basis function transforms are powers of 2 ($M=2^b$, $b \in \mathbb{N}$). The simplest version is the set of *Rademacher functions*. These are periodic rectangular functions defined by

$$t_u^{\mathrm{Rad}}(m) = \frac{1}{\sqrt{M}} \cdot (-1)^{\mathrm{int}\left(\frac{m \cdot 2^u}{M}\right)} \tag{4.117}$$

The factor $1/\sqrt{M}$ guarantees a unity norm of the basis functions, which would establish an orthonormal system. In total, the Rademacher set consists of $U=\log_2 M+1$ orthonormal basis functions. The 4 functions for the case $M=8$ are shown in Fig. 4.18a.

For $U>2$, the transform matrix is non-quadratic ($U<M$). The following example shows the Rademacher transform matrix for $M=8$ ($U=4$):

$$T^{\mathrm{Rad}}(8) = \frac{1}{2\sqrt{2}} \begin{bmatrix} 1 & 1 & 1 & 1 & 1 & 1 & 1 & 1 \\ 1 & 1 & 1 & 1 & -1 & -1 & -1 & -1 \\ 1 & 1 & -1 & -1 & 1 & 1 & -1 & -1 \\ 1 & -1 & 1 & -1 & 1 & -1 & 1 & -1 \end{bmatrix}. \tag{4.118}$$

As a consequence, no unique inverse transform exists. Such types of transform are denoted as *incomplete*. The goal is now to find additional basis functions, which would extend the Rademacher set into a complete orthogonal set consisting of M orthonormal rectangular functions. One such extension leads to the set of *Haar functions*. All those Rademacher functions, which contain more than one transition between positive and negative values, are split into two or more non-overlapping partial functions. The Haar transform still has $U^*=\log_2 M+1$ 'basis types', which are subsequently described by an index $u^*=0,1,...,\log_2 M$. From each basis type, M^* basis functions are developed, each of which has only one flip of the sign. M^* is 1 for $u^*=0,1$ and 2^{u^*-1} for the other basis types. Basis functions determined from the same basis type are indexed by $r=0,1,..., M^*-1$. The orthonormal transform basis set is then defined by[1]

---

[1] The following definition of the Haar basis is made to make it interpretable as a block transform. The definition and interpretation as a Wavelet transform (see sec. 4.4.4) is much more clear and obvious.

$$t_u^{\text{Haar}}(m) = \begin{cases} \text{haar}(u^*, m - r \cdot \frac{M}{M^*}, r) & \text{for } r \cdot \frac{M}{M^*} \leq m < (r+1) \cdot \frac{M}{M^*} \\ 0, & \text{else} \end{cases}$$
(4.119)

with

$$u = \begin{cases} u^* & \text{for} \quad u^* = 0,1 \\ M^* + r & \text{for} \quad u^* > 1 \end{cases}$$
(4.120)

and

$$\text{haar}(u^*, m, r) = \sqrt{\frac{M^*}{M}} \cdot (-1)^{\text{int}\left(\frac{m \cdot 2^{u^*}}{M}\right)} \quad \text{for} \quad r \cdot \frac{M}{M^*} \leq m < (r+1) \cdot \frac{M}{M^*}$$
(4.121)

The transform matrix of a Haar transform with block length $M=U=8$ is

$$\mathbf{T}^{\text{Haar}}(8) = \frac{1}{2\sqrt{2}} \begin{bmatrix} 1 & 1 & 1 & 1 & 1 & 1 & 1 & 1 \\ 1 & 1 & 1 & 1 & -1 & -1 & -1 & -1 \\ \sqrt{2} & \sqrt{2} & -\sqrt{2} & -\sqrt{2} & 0 & 0 & 0 & 0 \\ 0 & 0 & 0 & 0 & \sqrt{2} & \sqrt{2} & -\sqrt{2} & -\sqrt{2} \\ 2 & -2 & 0 & 0 & 0 & 0 & 0 & 0 \\ 0 & 0 & 2 & -2 & 0 & 0 & 0 & 0 \\ 0 & 0 & 0 & 0 & 2 & -2 & 0 & 0 \\ 0 & 0 & 0 & 0 & 0 & 0 & 2 & -2 \end{bmatrix}.$$
(4.122)

The scaling factors for the different basis types vary. This is necessary to realize an orthonormal transform by unity vector norms, since the lengths of the different basis functions are not identical. The basis functions for the case $M=8$ are shown in Fig. 4.18b.

The *Walsh basis* also consists of $U=M$ functions, and can be interpreted as another extension of the Rademacher set. The constant length M is retained for all basis functions, however it is postulated that the basis function u should have exactly u flips between positive and negative values, which could then straightforwardly be interpreted as a frequency. This is satisfied by the Rademacher functions of $u=0$ and $u=1$, which are also the first two basis functions of the Walsh system. The development of remaining Walsh functions is then performed recursively, starting from $t_1(m)$. The number of recursions necessary to generate all basis functions is $\log_2 M - 1$. During one recursion step, all basis functions developed in the previous step are scaled (which is done by eliminating each second sample), and then combined into new basis functions once periodically and once antiperiodically. The number of new basis functions is doubled by each iteration step, and with v flips in the scaled function, two new functions can be generated, one of which has $2v-1$ sign flips, the other $2v$ flips. The process of recursion can be described as follows:

Let

$$t_u^{\text{Walsh}}(m) = t_u^{\text{Rad}}(m) \quad \text{for} \quad 0 \le u < 2, \quad 0 \le m < M$$

$u^*=1$, $M^*=2$, $U^*=\log_2 M$, $P(0)=-1$

While $u^*<U^*$

{

For $0 \le r < \log_2 M^*$:

$$t_{\text{scal}}(m,r) = t_{(M^*+2r)/2}^{\text{Walsh}}(2m,r)$$

$$t_{M^*+2r+i}^{\text{Walsh}}(m) = \begin{cases} t_{\text{scal}}(m,r) & \text{for} \quad 0 \le m < M/2 \\ P(r)^{i+1} \cdot t_{\text{scal}}(m-M/2,r) & \text{for} \quad m \ge M/2 \end{cases} \quad \text{for} \quad i=0,1$$

For the next step, set $P(2r+i)=-P(r)^{i+1}$, $M^*=2M^*$, $u^*=u^*+1$

} $\hspace{10cm}$ (4.123)

The factor $P(r)$ effects that from a periodic basis function (having even number of flips), an antiperiodic function is generated at first in the next step and vice versa. This guarantees an ever increasing number of flips, equal to the index of the function. The recursion process is illustrated for the case $M=8$ in Fig. 4.17, the whole set of basis functions for $M=8$ is shown in Fig. 4.18c. The transform matrix of the Walsh transform is then

$$\mathbf{T}^{\text{Walsh}}(8) = \frac{1}{2\sqrt{2}} \begin{bmatrix} 1 & 1 & 1 & 1 & 1 & 1 & 1 & 1 \\ 1 & 1 & 1 & 1 & -1 & -1 & -1 & -1 \\ 1 & 1 & -1 & -1 & -1 & -1 & 1 & 1 \\ 1 & 1 & -1 & -1 & 1 & 1 & -1 & -1 \\ 1 & -1 & -1 & 1 & 1 & -1 & -1 & 1 \\ 1 & -1 & -1 & 1 & -1 & 1 & 1 & -1 \\ 1 & -1 & 1 & -1 & -1 & 1 & 1 & -1 \\ 1 & -1 & 1 & -1 & 1 & -1 & 1 & -1 \end{bmatrix} \quad (4.124)$$

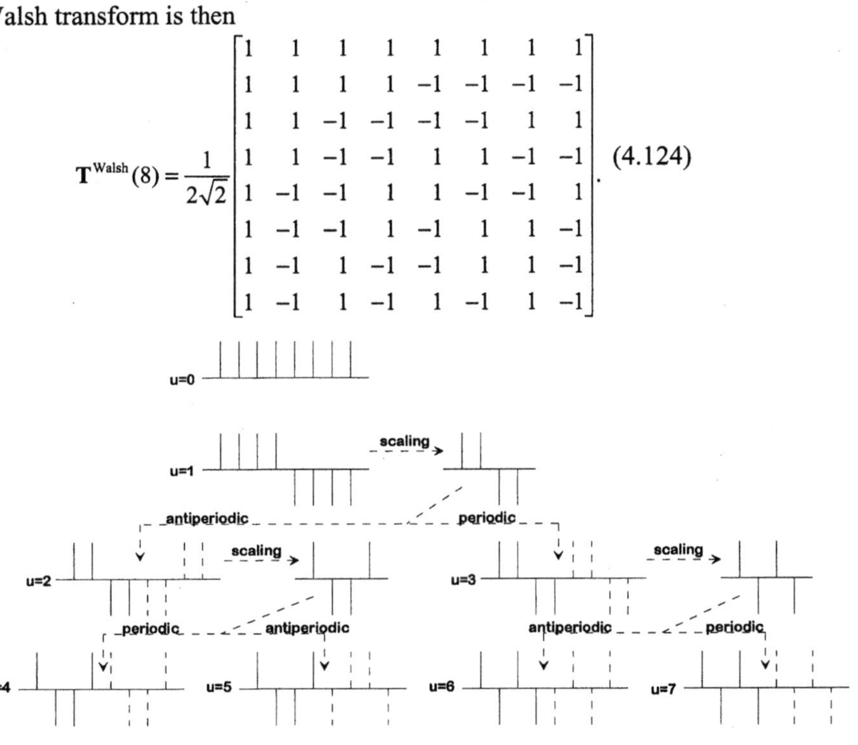

Fig. 4.17. Development of the Walsh transform basis

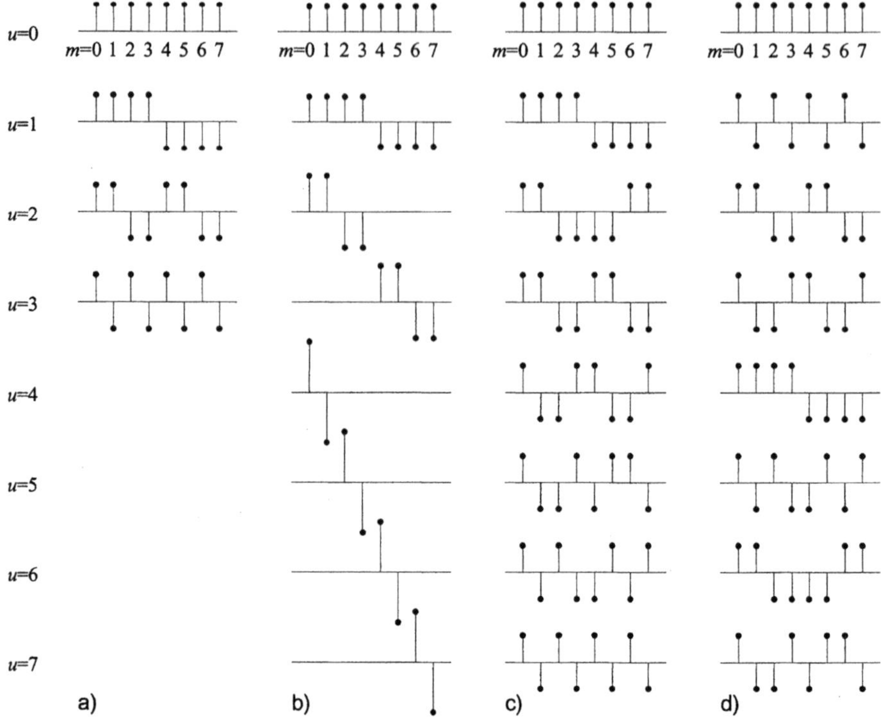

Fig. 4.18. Rectangular basis function systems
a Rademacher **b** Haar **c** Walsh **d** Hadamard

The *Hadamard transform* has in principle the same basis vectors as the Walsh system, however the ordering within the matrix (or the indexing of basis functions) is different. Hadamard coefficients can not be interpreted to express 'increasing frequency' by their order. On the other hand, the recursive construction of the Hadamard transform matrix is much simpler to describe. For a given length $M=2M^*$, which shall be a power of two, the definition of the transform matrix is made from the matrix of the next lower power of two, M^*. Starting from a trivial 1x1 unity matrix of $M^*=1$, the recursion is as follows

$$\mathbf{T}^{\text{Had}}(1) = \begin{bmatrix} 1 \end{bmatrix}$$

$$\mathbf{T}^{\text{Had}}(2M^*) = \frac{1}{\sqrt{2}} \begin{bmatrix} \mathbf{T}^{\text{Had}}(M^*) & \mathbf{T}^{\text{Had}}(M^*) \\ \mathbf{T}^{\text{Had}}(M^*) & -\mathbf{T}^{\text{Had}}(M^*) \end{bmatrix} \quad \text{for } M^* = 1, 2, 4, \ldots, M/2. \tag{4.125}$$

The Hadamard transform matrix for the case $M=8$ is (see also Fig. 4.18d) :

$$\mathbf{T}^{\text{Had}}(8) = \frac{1}{2\sqrt{2}}
\begin{bmatrix}
1 & 1 & 1 & 1 & 1 & 1 & 1 & 1 \\
1 & -1 & 1 & -1 & 1 & -1 & 1 & -1 \\
1 & 1 & -1 & -1 & 1 & 1 & -1 & -1 \\
1 & -1 & -1 & 1 & 1 & -1 & -1 & 1 \\
1 & 1 & 1 & 1 & -1 & -1 & -1 & -1 \\
1 & -1 & 1 & -1 & -1 & 1 & -1 & 1 \\
1 & 1 & -1 & -1 & -1 & -1 & 1 & 1 \\
1 & -1 & -1 & 1 & -1 & 1 & 1 & -1
\end{bmatrix} \tag{4.126}$$

Sinusoidal basis functions. The DFT is an orthogonal transform of sinusoidal basis, where the complex exponentials can be expressed by sinusoids of specific frequency and phase. The transform matrix of the DFT is

$$\mathbf{T}^{\text{DFT}} =
\begin{bmatrix}
1 & 1 & 1 & 1 & 1 & \cdots & 1 \\
1 & e^{-j\frac{2\pi}{M}} & e^{-j\frac{4\pi}{M}} & e^{-j\frac{6\pi}{M}} & \cdots & & e^{-j\frac{2\pi(M-1)}{M}} \\
1 & e^{-j\frac{4\pi}{M}} & e^{-j\frac{8\pi}{M}} & & & \ddots & e^{-j\frac{4\pi(M-1)}{M}} \\
1 & e^{-j\frac{6\pi}{M}} & & & & & \\
1 & \vdots & & \ddots & & \ddots & \vdots \\
\vdots & & & & & & \\
1 & e^{-j\frac{2\pi(U-1)}{M}} & e^{-j\frac{4\pi(U-1)}{M}} & & \cdots & & e^{-j\frac{2\pi(U-1)(M-1)}{M}}
\end{bmatrix}. \tag{4.127}$$

It must be noted that this version of the DFT is not orthonormal. From (4.107), $A=M$ for a 1D transform and $A=MN$ for a separable 2D transform[1]. Further, the DFT possesses a property which is highly disadvantageous in analysis of limited segments of a signal: As the signal is implicitly regarded as periodic, occasional amplitude differences between the left and right boundaries of the analysis segment are interpreted as high frequencies (see Fig. 4.19a). This can be avoided, if a symmetric (mirror) extension of the signal is made. Then, the Fourier transform must be computed over a length of $2M$ (see Fig. 4.19b). If the first sample of the analysis block has coordinate position $m=0$, the point of symmetry is at $m=-\frac{1}{2}$ to define a perfectly symmetric signal. This can be realized by a modification of the DFT basis function, where the exponential function can provide a phase shift by half a sample. Then,

$$c_u = \sum_{m=-M}^{M-1} x(m) \cdot e^{-j2\pi\frac{u}{2M}\left(m+\frac{1}{2}\right)} \quad \text{with} \quad x(m) = x(-m-1) \quad \text{for} \quad m < 0, \tag{4.128}$$

---

[1] For an orthonormal version of the 1D DFT, normalization by a factor $1/\sqrt{M}$ must be applied both in the analysis and synthesis (IDFT). Likewise, for a 2D transform, the normalization must use a factor $1/\sqrt{MN}$.

which can be re-written by

$$c_u = \sum_{m=0}^{M-1} x(m) \cdot \left[e^{-j2\pi\frac{u}{2M}\left(-m-1+\frac{1}{2}\right)} + e^{-j2\pi\frac{u}{2M}\left(m+\frac{1}{2}\right)} \right] = 2 \cdot \sum_{m=0}^{M-1} x(m) \cdot \cos\left[u \cdot \left(m+\frac{1}{2}\right) \cdot \frac{\pi}{M} \right].$$

$$(4.129)$$

The result of this transform are $U=M$ linearly-independent real-valued coefficients (note that the real part of the Fourier transform corresponds to the symmetric part of a signal). However, the transform (4.129) does not fulfill (4.107), as the norm of the basis vector $\mathbf{t}_0$ is higher than for the other. By a slight modification, orthonormality is achieved; the result is the *Discrete Cosine Transform* (DCT) with basis vectors

$$t_u^{\cos}(m) = \sqrt{\frac{2}{M}} \cdot C_0 \cdot \cos\left[u \cdot \left(m+\frac{1}{2}\right) \cdot \frac{\pi}{M} \right]$$

$$(4.130)$$

$$\text{with} \quad C_0 = \frac{1}{\sqrt{2}} \quad \text{for} \quad u=0 \quad ; \quad C_0 = 1 \quad \text{for} \quad u \neq 0.$$

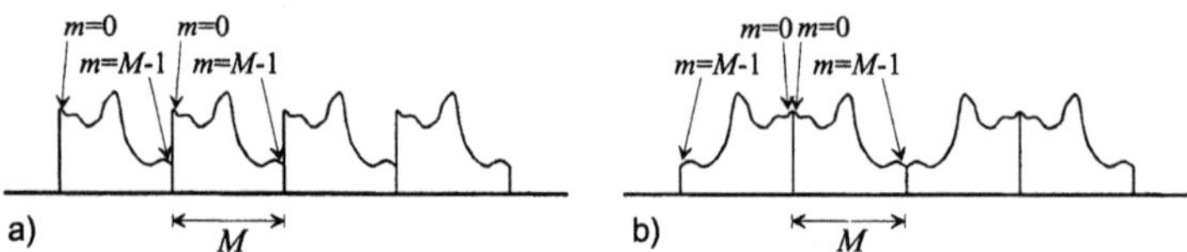

Fig. 4.19. Continuation of the signal at boundaries of a finite analysis segment
a periodic **b** symmetric

The two-dimensional DCT is widely used in image and video compression (e.g. in standards like MPEG, JPEG, H.261/2/3). The respective formulations of the 2D DCT and the inverse 2D DCT over a rectangular signal block of size $M \cdot N$ are

$$c_{u,v} = \frac{2}{\sqrt{M \cdot N}} C_0^{(u)} \cdot C_0^{(v)} \sum_{m=0}^{M-1}\sum_{n=0}^{N-1} x(m,n) \cdot \cos\left[u \cdot \left(m+\tfrac{1}{2}\right) \cdot \tfrac{\pi}{M} \right] \cos\left[v \cdot \left(n+\tfrac{1}{2}\right) \cdot \tfrac{\pi}{N} \right] \quad (4.131)$$

$$x(m,n) = \frac{2}{\sqrt{M \cdot N}} \sum_{m=0}^{M-1}\sum_{n=0}^{N-1} C_0^{(u)} \cdot C_0^{(v)} \cdot c_{u,v} \cdot \cos\left[u \cdot \left(m+\tfrac{1}{2}\right) \cdot \tfrac{\pi}{M} \right] \cos\left[v \cdot \left(n+\tfrac{1}{2}\right) \cdot \tfrac{\pi}{N} \right]. \quad (4.132)$$

It is also possible to extend the signal *antisymmetrically*, even though this would only be useful for signals which tend towards zero at the block boundaries. Then, a DFT of length $2 \cdot (M+1)$ can be applied to the extended signal $[0,-x(M-1),...,-x(0),0,x(0),...,x(M-1)]$. The imaginary part of the result is the *Discrete Sine Transform* (DST) which has basis functions[1]

---

[1] Even though formally the imaginary part of the DFT, DST coefficients can be regarded as real-valued if they are use standalone. The DST is used as one step in the so-called 'fast KLT', see p. 124.

$$t_u^{\sin}(m) = \sqrt{\frac{2}{M+1}} \cdot \sin\left[(m+1)\cdot(u+1)\cdot\frac{\pi}{M+1}\right] \quad ; \quad u = 0,\ldots,M-1, \qquad (4.133)$$

such that again M signal samples are transformed into $U{=}M$ linearly-independent coefficients. Other real-valued orthogonal sinusoidal transforms are based on mixtures of sines and cosines in their basis functions. An example is the *Discrete Hartley transform* (DHT), which for $M{=}4$ would have the same basis functions as the Walsh and Hadamard transforms, but for larger transform block lengths attains more of sinusoidal behavior,

$$t_u^{\text{Hart}}(m) = \text{cas}\left[2\pi\frac{um}{M}\right] \quad \text{with} \ \ \text{cas}(x) = \cos(x) + \sin(x). \qquad (4.134)$$

A *non-separable* 2D variant of this transform, which allows a more efficient compression of image patterns having different diagonal orientations, is defined by basis images

$$t_{u,v}^{\text{Hart}}(m,n) = \text{cas}\left[2\pi\left(\frac{um}{M}+\frac{vn}{N}\right)\right]. \qquad (4.135)$$

1D basis functions of sinusoidal transforms are shown in Fig. 4.20. Fig. 4.21 shows *basis images* of different separable 2D transforms, as defined by (4.114).

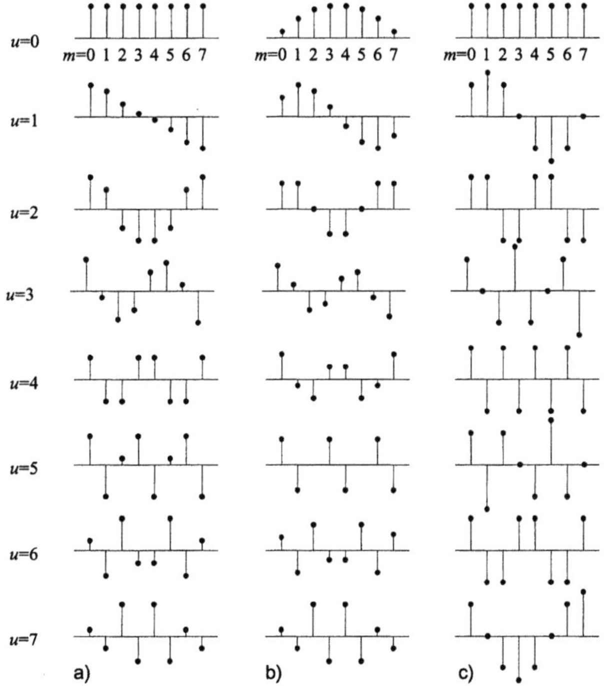

Fig. 4.20. Basis vectors of sinusoidal transforms a DCT b DST c DHT

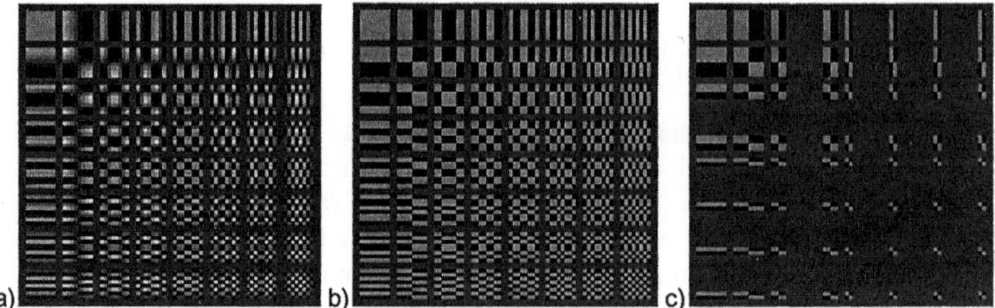

Fig. 4.21. 2D basis images of transforms. **a** DCT **b** Walsh **c** Haar

Integer transforms. Rectangular-basis transforms are attractive due to their low complexity, as they can be computed by multiplication-free integer operations. This is fully achieved if quantization is performed using step sizes corresponding to the respective normalization factors; likewise, this allows perfect reconstruction of signals from a transform coefficient representation of finite bit precision. On the other hand, the basis vectors of these transforms give only a poor representation of typical signal behavior such as smooth increase of amplitudes. Sinusoidal transforms are better capable to approximate such types of signals by minimum error, but can only be implemented by float precision due to the usage of trigonometric functions; they are further affected by rounding errors when the arithmetic precision is limited. As an alternative, integer transform bases were developed, allowing to capture smooth signal behavior, but can be implemented by low-precision integer arithmetic. These basis functions establish orthonormal sets, when appropriate normalization is performed, which can also be done by proper definition of step sizes in quantization. One example for this class is the following transform of block length $M=4$, which is used in the Advanced Video Coding standard (cf. sec. 17.4.4), having a transform matrix [MALVAR ET AL. 2002],

$$\mathbf{T}^{\text{int}}(4) = \begin{bmatrix} 1 & 1 & 1 & 1 \\ 2 & 1 & -1 & -2 \\ 1 & -1 & -1 & 1 \\ 1 & -2 & 2 & -1 \end{bmatrix}. \tag{4.136}$$

Different row normalization factors $1/2, 1/\sqrt{10}, 1/2, 1/\sqrt{10}$ must be applied for orthonormality of the respective basis vectors of (4.136). In fact, (4.136) was designed as a simplification of a truly orthogonal transform in which one unique normalization can be applied to achieve orthonormality of all basis vectors:

$$\mathbf{T}^{\text{int}}(4) = \frac{1}{\sqrt{676}} \begin{bmatrix} 13 & 13 & 13 & 13 \\ 17 & 7 & -7 & -17 \\ 13 & -13 & -13 & 13 \\ 7 & -17 & 17 & -7 \end{bmatrix}. \tag{4.137}$$

(4.137) can be interpreted as a reasonably good integer-precision approximation of the DCT basis, where it is an important property in applications of coding that orthogonality is still guaranteed. A similar orthogonal integer transform of block length $M=8$ [WIEN 2003] is defined by the following transform matrix; observe that the even indexed basis functions are exactly mirror-symmetric extensions of (4.137):

$$\mathbf{T}^{int}(8) = \frac{1}{\sqrt{1352}} \begin{bmatrix} 13 & 13 & 13 & 13 & 13 & 13 & 13 & 13 \\ 19 & 15 & 9 & 3 & -3 & -9 & -15 & -19 \\ 17 & 7 & -7 & -17 & -17 & -7 & 7 & 17 \\ 9 & 3 & -19 & -15 & 15 & 19 & -3 & -9 \\ 13 & -13 & -13 & 13 & 13 & -13 & -13 & 13 \\ 15 & -19 & -3 & 9 & -9 & 3 & 19 & -15 \\ 7 & -17 & 17 & -7 & -7 & 17 & -17 & 7 \\ 3 & -9 & 15 & -19 & 19 & -15 & 9 & -3 \end{bmatrix} . \tag{4.138}$$

An optimum transform – KLT. One important goal in using linear transforms for multimedia signal processing is representation of a signal or its properties as accurate as possible, using lowest possible number of transform coefficients. Typically, the energy of the reconstruction error is the criterion for optimality. Assuming that only T out of U coefficients shall be used to represent the signal, the reconstruction error of a one-dimensional transform is

$$e(m) = x(m) - \frac{1}{A} \sum_{u=0}^{T-1} c_u \cdot t_u{}^*(m) . \tag{4.139}$$

This results in an energy of the error over the complete block

$$\|\mathbf{e}\|^2 = \sum_{m=0}^{M-1} e^2(m)$$

$$= \sum_{m=0}^{M-1} \left[x^2(m) - 2 \cdot x(m) \cdot \frac{1}{A} \sum_{u=0}^{T-1} c_u \cdot t_u{}^*(m) + \left| \frac{1}{A} \sum_{u=0}^{T-1} c_u \cdot t_u{}^*(m) \right|^2 \right] . \tag{4.140}$$

Substituting in (4.140)

$$x(m) = \frac{1}{A} \sum_{u'=0}^{U-1} c_{u'} \cdot t_{u'}{}^*(m) = \left[\frac{1}{A} \sum_{u'=0}^{U-1} c_{u'}{}^* \cdot t_{u'}(m) \right]^* = \frac{1}{A} \sum_{u'=0}^{U-1} c_{u'}{}^* \cdot t_{u'}(m) \tag{4.141}$$

gives

$$\|\mathbf{e}\|^2 = \sum_{m=0}^{M-1} \left[x^2(m) - \frac{2}{A^2} \sum_{u'=0}^{U-1} \sum_{u=0}^{T-1} c_u \cdot t_u{}^*(m) \cdot c_{u'}{}^* \cdot t_{u'}(m) + \left| \frac{1}{A} \sum_{u=0}^{T-1} c_u{}^* \cdot t_u(m) \right|^2 \right] , \tag{4.142}$$

by which (using the orthogonality condition (4.107)),

$$\|e\|^2 = \sum_{m=0}^{M-1} x^2(m) - \frac{1}{A} \sum_{u=0}^{T-1} |c_u|^2 . \tag{4.143}$$

(4.143) shows that for $T=U$ where $\|e\|^2=0$, the energy of a finite signal can be determined from the coefficients of a linear orthogonal transform. This corresponds to the condition (4.111), however it is more general now as it shows that by omitting transform coefficients, the error energy in the reconstruction is exactly the energy of these coefficients scaled by the normalization factor A.

Consequently, the optimum transform can be derived under the condition that the energy of the error shall be *minimized* if reconstruction from a finite number of coefficients is performed. This is equivalent with a condition to *maximize* the energy of the first T coefficients. From the expected energy values of the first T coefficients, the following condition can be formulated (for simplification here, orthonormality $A=1$ is assumed):

$$\sum_{u=0}^{T-1} E\left\{|c_u|^2\right\} = \sum_{u=0}^{T-1} E\left\{c_u \cdot c_u *\right\} = \sum_{u=0}^{T-1} E\left\{\sum_{m=0}^{M-1} x(m) \cdot t_u(m) \cdot \sum_{k=0}^{M-1} x(k) \cdot t_u *(k)\right\}$$

$$= \sum_{u=0}^{T-1} E\left\{\sum_{k=0}^{M-1}\left(\sum_{m=0}^{M-1} x(m) \cdot x(k) \cdot t_u(m)\right) \cdot t_u *(k)\right\} \tag{4.144}$$

$$= \sum_{u=0}^{T-1}\left[\sum_{k=0}^{M-1}\left(\sum_{m=0}^{M-1} r_{xx}(|m-k|) \cdot t_u(m)\right) \cdot t_u *(k)\right].$$

The maximum of this expression is approached if the innermost parenthesis fulfills the condition

$$\sum_{m=0}^{M-1} r_{xx}(|m-k|) \cdot t_u(m) = \lambda_u \cdot t_u(k), \tag{4.145}$$

which means the optimum basis functions will be the set of eigenvectors of the discrete autocorrelation sequence. Substituting (4.145) into (4.144) gives

$$\sum_{u=0}^{T-1} E\left\{|c_u|^2\right\} = \sum_{u=0}^{T-1} \lambda_u \left[\sum_{k=0}^{M-1} t_u(k) \cdot t_u *(k)\right], \tag{4.146}$$

which shows that the related eigenvalues λ_u are the expected values of the coefficient energies. This optimum, best-decorrelating transform is the *Karhunen-Loève transform* (KLT). The basis functions of this transform must specifically be adapted to the autocorrelation statistics of a given signal, and are globally optimum for Gaussian (autoregressive processes) of stationarity property. Formulating the condition (4.145) for the entire set of basis functions, these can be computed as eigenvectors Φ_u of the autocovariance matrix (3.41)[1], which must then be con-

---

[1] Assuming zero-mean property here and in the subsequent equations; if however the signal is not zero-mean, the expected block-energy component $M \cdot \mu_x^2$ related to the mean is typi-

structed as an MxM matrix containing the covariance function samples $r_{xx}(0)$ ... $r_{xx}(M-1)$:

$$\mathbf{R}_{xx}\cdot\mathbf{\Phi}_u = \lambda_u\cdot\mathbf{\Phi}_u$$

$$\text{with}\quad \mathbf{\Phi}_u = \left[\phi_u(0)\quad \phi_u(1)\quad \cdots\quad \phi_u(M-1)\right]^{\mathrm{T}}\quad;\quad 0\le u\le U-1. \tag{4.147}$$

Alternatively, the conjugates[1] of the eigenvectors $\phi_u{}^{*}$ are defined to establish the rows of transform matrix $\mathbf{T}^{\mathrm{KLT}}$; (4.147) can then be written in matrix notation

$$
\begin{bmatrix}
\phi_0^{*}(0) & \phi_0^{*}(1) & \cdots & \phi_0^{*}(M-1)\\
\phi_1^{*}(0) & \ddots & \ddots & \\
\vdots & \ddots & & \vdots\\
\phi_{U-1}^{*}(0) & & \cdots & \phi_{U-1}^{*}(M-1)
\end{bmatrix}
\cdot
\begin{bmatrix}
r_{xx}(0) & r_{xx}(1) & \cdots & r_{xx}(M-1)\\
r_{xx}(1) & r_{xx}(0) & r_{xx}(1) & \\
\vdots & r_{xx}(1) & \ddots & \ddots & \vdots\\
 & & & \ddots\\
r_{xx}(M-1) & & \cdots & r_{xx}(0)
\end{bmatrix}
$$

$$
=
\begin{bmatrix}
\lambda_0 & 0 & 0 & \cdots & 0\\
0 & \lambda_1 & 0 & & \\
0 & 0 & \lambda_2 & \ddots & \vdots\\
\vdots & & \ddots & \ddots & 0\\
0 & & \cdots & 0 & \lambda_{U-1}
\end{bmatrix}
\cdot
\begin{bmatrix}
\phi_0^{*}(0) & \phi_0^{*}(1) & \cdots & \phi_0^{*}(M-1)\\
\phi_1^{*}(0) & \ddots & \ddots & \\
\vdots & \ddots & & \vdots\\
\phi_{U-1}^{*}(0) & & \cdots & \phi_{U-1}^{*}(M-1)
\end{bmatrix}
$$

$$\Leftrightarrow \mathbf{T}^{\mathrm{KLT}}\cdot\mathbf{R}_{xx} = \mathbf{\Lambda}\cdot\mathbf{T}^{\mathrm{KLT}}. \tag{4.148}$$

Multiplying both sides of (4.148) by the inverse transform matrix retains the diagonal matrix $\mathbf{\Lambda}$ on the right side, which is occupied by the eigenvalues λ_u:

$$\mathbf{T}^{\mathrm{KLT}}\cdot\mathbf{R}_{xx}\cdot\left[\mathbf{T}^{\mathrm{KLT}*}\right]^{\mathrm{T}} = \mathbf{\Lambda} = \mathrm{Diag}\{\lambda_u\}\quad\Rightarrow\sum_{u=0}^{U-1}E\left\{\left|c_u\right|^2\right\} = \mathrm{tr}\{\mathbf{\Lambda}\} = M\cdot\sigma_x^2. \tag{4.149}$$

$\mathbf{\Lambda}$ in (4.149) can also be regarded as the 'optimum transform' of the autocovariance matrix, and can be interpreted as a special case of the Wiener-Khintchine theorem (3.68). By transforming the autocovariance matrix, a statistical representation of the signal by discrete spectral power samples λ_u is generated. While the correlation inherent in the signal is indicated by the fact that $\mathbf{R}_{xx}$ is not a diagonal matrix, the diagonal type of $\mathbf{\Lambda}$ indicates that no correlation is present anymore between the spectral samples.

Example: KLT for an AR(1) process. For the AR(1) process with autocovariance (4.57), the normalized eigenvalues are

---

cally concentrating in coefficient c_0, such that the basic proof for the optimality of the KLT is not affected.

[1] Definition by conjugates of eigenvectors makes the KLT consistent with the DFT notation; actually, the DFT is the optimum transform for perfect cyclic signals, such as signals composed from sinusoids, each with a period being an exact fraction of the analysis block length.

$$\lambda_u = \frac{1-\rho^2}{1-2\rho\cos\Omega_u+\rho^2}\,. \tag{4.150}$$

These are discrete samples of the power spectrum (4.58) at frequencies Ω_u. The eigenvectors are [JAIN 1989]:

$$\phi_u(p) = \left(\frac{2}{U+\lambda_u}\right)^{1/2}\cdot\sin\left(\Omega_u\left(p+1-\frac{U+1}{2}\right)+\frac{(u+1)\cdot\pi}{2}\right) \quad ; \quad 0\le p\le U-1. \tag{4.151}$$

In the AR(1) case, eigenvectors are real-valued, however no diagonal symmetry appears in the transform matrix. The frequencies Ω_u are determined as the positive solutions of the transcendental equation system[1]

$$\tan(U\Omega) = -\frac{(1-\rho^2)\cdot\sin\Omega}{\cos\Omega-2\rho+\rho^2\cos\Omega}\,. \tag{4.152}$$

Even for this simple case, the analysis frequencies are not located at equidistant positions of the frequency axis. This results in a *non-harmonic allocation* of the discrete frequencies Ω_u. Due to this fact, it is not possible to factor the transform matrix into a chain of sparse product matrices, which is a usual condition to design a fast algorithm for a transform (see sec. 4.3.4). The KLT is also important in analysis and classification of correlated feature values, see chapter 9.

'Fast KLT' for AR(1) processes. Assume a finite segment (vector) is extracted from an AR(1) process $x_b(m)$ which has zero values beyond the boundaries. If a *noncausal two-sided prediction*

$$e(m) = x_b(m)-\alpha\cdot x_b(m-1)-\alpha\cdot x_b(m+1) \quad ; \quad \alpha\le 0{,}5 \tag{4.153}$$

is applied to this signal, the following autocovariance matrix results for the prediction error signal:

$$\mathbf{R}_{ee} = \begin{bmatrix} 1 & -\alpha & 0 & \cdots & 0 \\ -\alpha & 1 & \ddots & \ddots & \vdots \\ 0 & \ddots & \ddots & \ddots & 0 \\ \vdots & \ddots & \ddots & \ddots & -\alpha \\ 0 & \cdots & 0 & -\alpha & 1 \end{bmatrix}. \tag{4.154}$$

The eigenvectors of the matrix (4.154) are the basis vectors of the Discrete Sine transform (4.133). Since the DST can be computed as the imaginary part of the DFT by a Fast Fourier Transform (FFT), an algorithm has proposed on this basis which is denoted as *fast KLT* [JAIN 1976]. Observe however, that this does not relate to a direct computation of a KLT from the signal $x(m)$, but restricts to the equiva-

---

[1] For $U=2$, the solutions are $\Omega_0=0$ and $\Omega_1=\pi$. In this case, the KLT is identical to most other block transforms, like Rademacher, Haar, Walsh, DCT; see also Problem 4.1.

lent residual signal $e(m)$, and the coefficients are not an approximation of (4.150). The method of prediction, in particular the choice of the coefficient α, will influence the performance, but nevertheless the method leads to an efficiently decorrelating transform with good data compaction properties.

The fast KLT algorithm. In a first step, the signal must be transformed into finite segments of zero-value boundary conditions. This is achieved by defining *boundary control samples* $x(m'U)$, $m'=\text{int}(m/U)$ which establish a signal sampled at each Uth position of $x(m)$. The signal segments between two adjacent control samples (length $U+1$) are now transformed into the equivalent signal $x_b(m)$. This is generated from an estimate made by linear interpolation between the two control samples at the tails of the segment, where the prediction errors at these tails are naturally zero:

$$x_b(m) = x(m) - \hat{x}(m) \quad ; \quad \hat{x}(m) = \frac{1-(m-m'U)}{U}x(m'U) + \frac{m-m'U}{U}x(m'U+1). \quad (4.155)$$

Using $x_b(m)$, the residual $e(m)$ is computed according to (4.153), which has then $U-1$ non-zero values in the current segment. The last step is a DST of length $U-1$. In addition to the coefficients, the boundary control samples are an integral part of the representation, which must be handled independently. Reconstruction of the signal is achieved by inverting the procedure: Inverse DST, reconstruction of the signal $x_b(m)$ from $e(m)$[1], reconstruction of $x(m)$ from $x_b(m)$, where only for the latter step the boundary control samples are required.

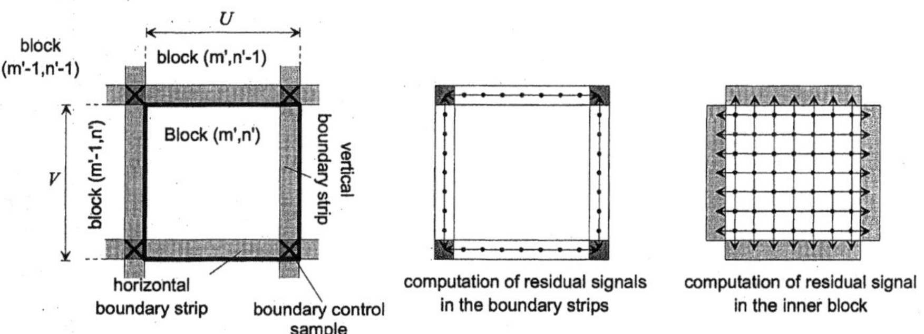

Fig. 4.22. Principle of two-dimensional 'fast KLT'

For a 2D transform of block size $U{\times}V$, 4 control samples at the four corners are required to process a block (see Fig. 4.22 left). First, the 1D algorithm as described above is performed over horizontal boundary strips of length U-1 and

---

[1] The process of non-causal prediction of a finite signal in (4.153) can be written as a vector-matrix multiplication (4.42), where $\mathbf{e}=\mathbf{H}\cdot\mathbf{x}_b$. The vectors are of length U-1, and the $(U\text{-}1){\times}(U\text{-}1)$ matrix $\mathbf{H}$ has the same structure as $\mathbf{R}_{ee}$ in (4.154). The inverse operation is $\mathbf{x}_b=\mathbf{H}^{-1}\cdot\mathbf{e}$.

vertical strips of length V-1, as shown in Fig. 4.22 center. Finally, the inner block of size $(U$-1$)$x$(V$-1$)$ is processed by separable horizontal/vertical application of the operations (4.155), (4.153) using the samples from boundary strips for prediction (see Fig. 4.22 right). Finally, a 2D DST is applied to the inner block. Each one of the horizontal/vertical boundary strips and three of the corner positions should already be available from previously processed blocks. Logically, each block relates to an image area of size UxV and is represented by one boundary control sample, two boundary strips of lengths U-1 and V-1, and the inner block. The entire number of coefficients and corner values is then still equal to the number of pixels in an image. Similar methods has been proposed under the names of *Pinned Sine Transform* (PST) [MEIRI, YUDILEVICH 1981] and *Recursive Block Coding* (RBC) [FARRELLE, JAIN 1986].

4.3.3   Efficiency of Transforms

An important criterion related to the efficiency of a linear orthogonal transform is concentration of as much signal energy as possible into the lowest possible number of transform coefficients. The related *energy packing efficiency* η_e of a transform [CLARKE 1985] is the normalized ratio of expected energy values, contained within the first T out of U transform coefficients:

$$\eta_e(T) = \frac{\sum_{t=0}^{T-1} E\{c_t^2\}}{\sum_{u=0}^{U-1} E\{c_u^2\}} .$$

(4.156)

Obviously, the KLT maximizes the energy packing efficiency, as it is optimized regarding this criterion, see (4.144). The *decorrelation efficiency* η_c is determined from a UxU=MxM size autocovariance matrix $\mathbf{R}_{xx}$ and its transform

$$\mathbf{R}_{cc} = E\{\mathbf{c}\cdot[\mathbf{c}*]^{\mathrm{T}}\} = E\{(\mathbf{T}\cdot\mathbf{x})\cdot[(\mathbf{T}\cdot\mathbf{x})*]^{\mathrm{T}}\}$$
$$= \mathbf{T}\cdot E\{\mathbf{x}\cdot\mathbf{x}^{\mathrm{T}}\}\cdot[\mathbf{T}*]^{\mathrm{T}} = [\mathbf{T}\cdot\mathbf{R}_{xx}]\cdot[\mathbf{T}*]^{\mathrm{T}} .$$

(4.157)

It is obvious that for the case of the KLT, $\mathbf{R}_{cc}$ is identical to the diagonal eigenvalue matrix Λ in (4.149). The decorrelation efficiency is then defined as

$$\eta_c = 1 - \frac{\sum_{\substack{k=0 \\ (k\neq l)}}^{U-1}\sum_{l=0}^{U-1} |r_{cc}(k,l)|}{\sum_{\substack{k=0 \\ (k\neq l)}}^{U-1}\sum_{l=0}^{U-1} |r_{xx}(k,l)|} .$$

(4.158)

For the case of the optimized KLT, η_c achieves the maximum value of 1, as all elements contributing to the numerator sum, where the trace of the matrix is excluded, must be zero. This means that the KLT achieves optimum decorrelation if optimized for a signal of given statistics. However only for the case of Gaussian signals (like AR processes), this also means that coefficients are fully statistically independent, as otherwise nonlinear dependencies may exist additionally. For other transforms than the KLT, also linear statistical dependencies (correlations) between coefficients of the discrete transform representation may be present.

Fig. 4.23 shows the decorrelation efficiency of different transforms for the example of an AR(1) model which is parametrized over the correlation ρ. The KLT is shown once optimized for $\rho=.91$, once for $\rho=.36$. The DCT (4.130) exhibits very good decorrelation and energy packing efficiency for highly-correlated signals ($\rho>.9$), close to the performance of the KLT. It can be observed in this range of relatively high correlation that for case of a slight deviation of the AR process statistics from the optimization target of the KLT, the DCT even performs better.

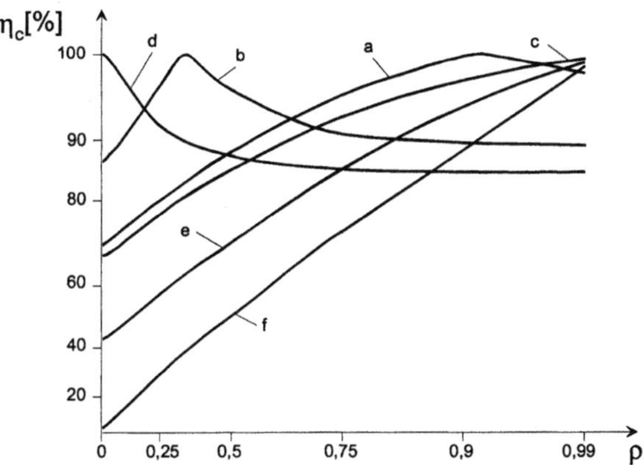

Fig. 4.23. Efficiency η_c as function of ρ for an AR(1) model, transform block size $U=8$ (*a*) KLT with $\rho=.91$, (*b*) KLT with $\rho=.36$, (*c*) DCT, (*d*) DST, (*e*) WHT, (*f*) HT
[adopted from CLARKE 1985]

Example. $\mathbf{R}_{xx}$ and the transformed matrix $\mathbf{R}_{cc}$ for an AR(1) model with $\rho=0,91$; a DCT of block length $M=8$ is applied:

$$\mathbf{R}_{xx} = \sigma_x^2 \cdot \begin{bmatrix} 1.000 & 0.910 & 0.828 & 0.754 & 0.686 & 0.624 & 0.568 & 0.517 \\ 0.910 & 1.000 & \ddots & \ddots & \ddots & \ddots & \ddots & 0.568 \\ 0.828 & \ddots & 1.000 & \ddots & \ddots & \ddots & \ddots & 0.624 \\ 0.754 & \ddots & \ddots & 1.000 & \ddots & \ddots & \ddots & 0.686 \\ 0.686 & \ddots & \ddots & \ddots & 1.000 & \ddots & \ddots & 0.754 \\ 0.624 & \ddots & \ddots & \ddots & \ddots & 1.000 & \ddots & 0.828 \\ 0.568 & \ddots & \ddots & \ddots & \ddots & \ddots & 1.000 & 0.910 \\ 0.517 & 0.568 & 0.624 & 0.686 & 0.754 & 0.828 & 0.910 & 1.000 \end{bmatrix} \qquad (4.159)$$

$$\mathbf{R}_{cc} = \sigma_x^{2} \cdot \begin{bmatrix} 6.344 & 0.000 & \mathbf{-0.291} & 0.000 & \mathbf{-0.066} & 0.000 & \mathbf{-0.021} & 0.000 \\ 0.000 & 0.930 & 0.000 & \mathbf{-0.027} & 0.000 & \mathbf{-0.008} & 0.000 & \mathbf{-0.002} \\ \mathbf{-0.291} & 0.000 & 0.312 & 0.000 & \mathbf{-0.001} & 0.000 & \mathbf{0.000} & 0.000 \\ 0.000 & \mathbf{-0.027} & 0.000 & 0.149 & 0.000 & \mathbf{-0.001} & 0.000 & \mathbf{0.000} \\ \mathbf{-0.066} & 0.000 & \mathbf{-0.001} & 0.000 & 0.094 & 0.000 & \mathbf{0.000} & 0.000 \\ 0.000 & \mathbf{-0.008} & 0.000 & \mathbf{-0.001} & 0.000 & 0.068 & 0.000 & \mathbf{0.000} \\ \mathbf{-0.021} & 0.000 & \mathbf{0.000} & 0.000 & \mathbf{0.000} & 0.000 & 0.055 & 0.000 \\ 0.000 & \mathbf{-0.002} & 0.000 & \mathbf{0.000} & 0.000 & \mathbf{0.000} & 0.000 & 0.049 \end{bmatrix} \quad (4.160)$$

Correlation is observed from the matrix $\mathbf{R}_{cc}$ between pairs of even coefficients, and pairs of odd coefficients, which are typed boldface in (4.160). More interpretation of this behavior will be given in sec. 4.3.5 and 13.4.1.

An important application of linear transforms is encoding. In image and video coding, the area of the signal is usually first partitioned into relatively small blocks, often quadratic blocks of size 8x8 pixels or smaller. In audio coding, the signal is partitioned into segments, with lengths of typically up to 1024 samples. After application of a one- or multidimensional transform, the coefficients are quantized and encoded. To reconstruct the signal, the inverse transform is applied on quantized coefficients (see Fig. 4.24). The resulting error energy between the original signal $\mathbf{x}$ and the reconstructed signal $\mathbf{y}$ is linearly dependent by the energy of the quantization error $\mathbf{q}$, which is the difference between the coefficient vector $\mathbf{c}$ and its quantized counterpart $\mathbf{c'}$[1]:

$$\sigma_e^2 = E\left\{(\mathbf{x}-\mathbf{y})^{\mathrm{T}}(\mathbf{x}-\mathbf{y})\right\} = \frac{1}{A}E\left\{(\mathbf{x}-\mathbf{y})^{\mathrm{T}}\mathbf{T}^{\mathrm{T}}\mathbf{T}(\mathbf{x}-\mathbf{y})\right\}$$
$$= \frac{1}{A}E\left\{(\mathbf{c}-\mathbf{c'})^{\mathrm{T}}(\mathbf{c}-\mathbf{c'})\right\} = \frac{1}{A}E\left\{\mathbf{q}^{\mathrm{T}}\mathbf{q}\right\} = \frac{1}{A}\sigma_q^2 \qquad (4.161)$$

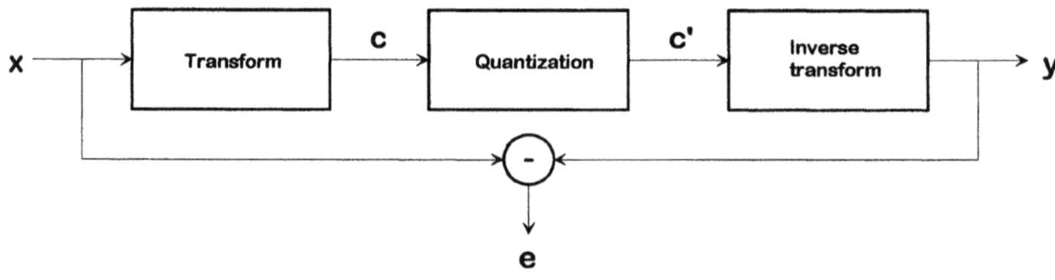

Fig. 4.24. Principle of data compression by transform coding

---

[1] Remark that the following proof is based on the condition $\mathbf{T}^{\mathrm{T}}\mathbf{T}\sim\mathbf{I}$, such that the linear dependency of the quantization error in the transformed domain and the reconstruction error is strictly valid only for the case of orthogonal transforms.

4.3.4   Fast Transform Algorithms

In implementations of linear transforms, the direct computation of vector/matrix multiplication (4.108) would be overly complex. Algorithms for fast computation exist, most of which obtain exactly the same result as the matrix form when sufficient arithmetic precision is used. The essential idea behind these algorithms is the factorization of the transform matrix into the product of several sparsely populated matrices. Usually this is possible if the basis functions of the transform are harmonically related, or if one is a scaled version of another. As an example, the Hadamard transform of block length $M=4$ is for the case of direct computation:

$$\begin{bmatrix} c_0 \\ c_1 \\ c_2 \\ c_3 \end{bmatrix} = \begin{bmatrix} 1 & 1 & 1 & 1 \\ 1 & -1 & 1 & -1 \\ 1 & 1 & -1 & -1 \\ 1 & -1 & -1 & 1 \end{bmatrix} \cdot \begin{bmatrix} x(0) \\ x(1) \\ x(2) \\ x(3) \end{bmatrix}. \tag{4.162}$$

This transform matrix can be factorized as follows:

$$\begin{bmatrix} 1 & 1 & 1 & 1 \\ 1 & -1 & 1 & -1 \\ 1 & 1 & -1 & -1 \\ 1 & -1 & -1 & 1 \end{bmatrix} = \begin{bmatrix} 1 & 0 & 1 & 0 \\ 0 & 1 & 0 & 1 \\ 1 & 0 & -1 & 0 \\ 0 & 1 & 0 & -1 \end{bmatrix} \cdot \begin{bmatrix} 1 & 1 & 0 & 0 \\ 1 & -1 & 0 & 0 \\ 0 & 0 & 1 & 1 \\ 0 & 0 & 1 & -1 \end{bmatrix}. \tag{4.163}$$

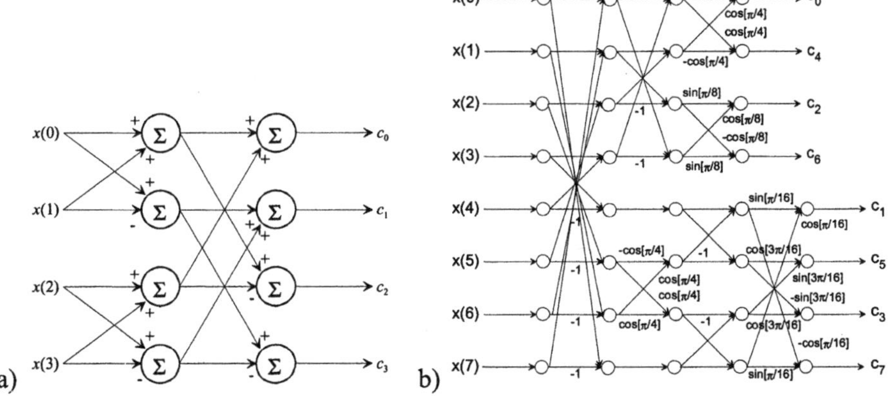

Fig. 4.25. Signal flow diagrams **a** Fast Hadamard transform, $M=4$ **b** Fast DCT, $M=8$

In the first step, the signal vector is multiplied by the right matrix, which requires 4 additions or subtractions. Multiplying this intermediate result by the left matrix requires another 4 additions or subtractions. The direct computation (4.162) would however require 12 additions or subtractions. Regarding a more general case (e.g. including the *Fast Fourier Transform*, FFT), perfect factorization of an $M\times M$ transform matrix is possible into a concatenation of $\log_2 M$ matrices, where any row or column contains only two factors unequal zero. For transforms over non-rectangular bases, also multiplications are required. By proper factorization, only

$M \cdot \log_2 M$ non-zero operations remain, while for the direct computation almost M^2 operations would have to be performed.

This principle can also be interpreted by *signal flow diagrams*. Here, intermediate results are used systematically for as many inputs to the next stage as possible. If multiplications using the same factors are applied to several samples, these are performed only *after* summation of the samples. Fig. 4.25 shows the diagram for the example of the Hadamard transform (4.163), and a signal flow diagram of a fast algorithm for computation of the DCT with M=8 [CHEN, SMITH, FRALICK 1977].

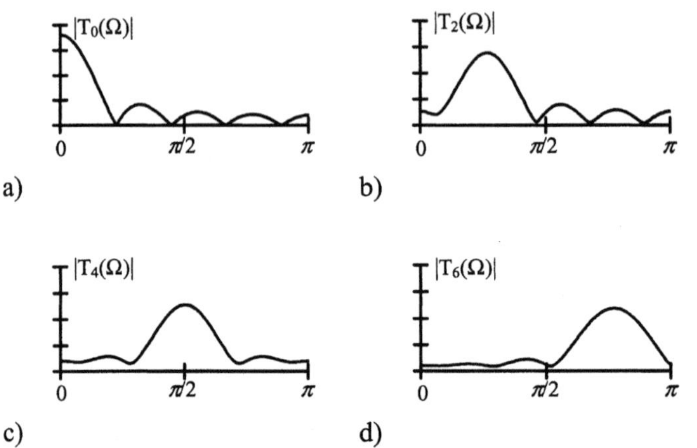

Fig. 4.26. Transfer functions of DCT basis vectors t_0 (*a*), t_2 (*b*), t_4 (*c*), t_6 (*d*)

4.3.5   Transforms with Block Overlap

Fig. 4.26 shows the amplitude transfer functions computed from several basis vectors of the DCT, block length M=8. The transform could in principle be interpreted as a parallel convolution of the signal, using impulse responses which are the reversed sequences of the values in the basis vectors. The computation of transform coefficients is however not performed at each sampling position, but only at M-distant positions, once for each block. Nevertheless, the operation of the transform can be interpreted as multiplication of the signal spectrum by the transfer functions of the basis vectors. The transform coefficients are then identified to carry frequency information related to *all frequencies* which are passing their respective transfer functions. In the case of the DCT, it can be observed that not only one unique pass-bands exists which is related to the typical band of frequencies that a coefficient is assumed to represent; additionally, significant side lobes are present. This may lead to the conclusion that the frequency separation performed by the transform is not optimum.

The center frequencies of the pass bands related to DCT basis functions appear at equidistant locations on the frequency axis, and the spectra of the even basis functions roughly have minima where the odd functions have maxima and vice versa. On the other hand, for pairs of even-indexed functions, one may have the maximum pass band amplitude where another has a maximum of a side lobe; the same applies for pairs of odd-indexed functions. This is one reason to observe correlations in (4.160) between coefficient pairs of both even and odd parity.

By using basis functions which have smaller amplitudes of the side lobes in their transfer spectra, it will be possible to design transforms which achieve better de-correlation and energy compaction, but still have frequency bands at equidistant positions. This can be achieved by application of *longer basis functions*, in particular if weighting windows are used which let the basis functions diminish more smoothly towards their tails. It shall still be possible to realize an orthogonal basis system by such an approach; further, the number of transform coefficients shall be the same as with a comparable block transform. This can be achieved if the basis functions are extended into neighbored blocks, but the distance of block positions is equal to the number of transform coefficients as before. This principle of *block-overlapping transforms*, has mainly been realized by a combination of cosine basis functions and appropriate window functions. Cosine-modulated functions are e.g. used in the *Lapped Orthogonal Transform* (LOT) [MALVAR, STAELIN 1989] and in the TDAC transform (*Time Domain Aliasing Cancellation*) [PRINCEN, BRADLEY 1986]; more generally, this family of transforms is denoted as *Cosine Modulated Filter-banks* or *Modified DCT*; these are e.g. used for audio signal waveform coding (cf. sec. 14.2).

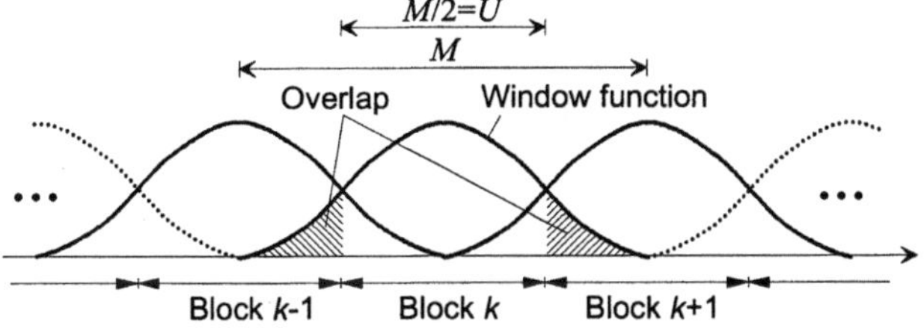

Fig. 4.27. Positions of analysis blocks with their overlapping window functions in a block-overlapping transform, $M=2U$

Example: TDAC transform. Decomposition is performed into U frequency bands, the basis function is of length $M=2U$ and defined by

$$t_u(m) = \sqrt{\frac{4}{M}} \cdot w(m) \cdot \cos\left[\frac{2\pi}{M} \cdot (u+0.5) \cdot (m+0.5-M/4)\right]. \qquad (4.164)$$

The squared window function $w^2(m)$ shall result in a unity value, when superimposed from neighboring blocks (see Fig. 4.27 and (4.165)), and the window shall be symmetric. As the window is applied to an orthogonal set of basis functions once during analysis and once during synthesis, (4.105)-(4.107) are still fulfilled and a perfect reconstruction of the signal is guaranteed:

$$w(m) = w(M-1+m)$$
$$w^2(m+M/2) + w^2(m) = 1 \tag{4.165}$$

An example is the sine window

$$w(m) = \sin\left(\frac{\pi}{M} \cdot (m+0,5)\right). \tag{4.166}$$

Fig. 4.28 shows the amplitude transfer functions of different TDAC basis functions in with $U=8$, $M=16$. The side lobes of the spectra are largely reduced as compared to the DCT case in Fig 4.26.

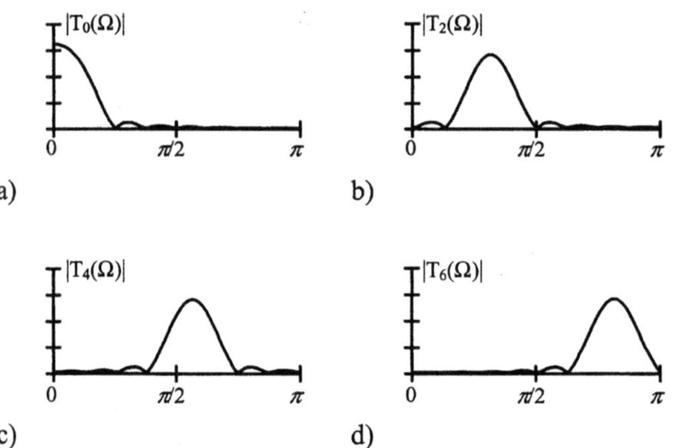

Fig. 4.28. Transfer functions of TDAC basis vectors t_0 (*a*), t_2 (*b*), t_4 (*c*), t_6 (*d*)

As $U<M$, the transform of a *single* transform block over length M is formally incomplete; this is however only the case if the coefficients belonging to the single blocks are regarded as independent, which can in fact no longer be the case. The reconstruction from coefficients of an isolated block would be a signal of length M with a window-dependent roll-off of amplitudes towards both tails; only by combining with the missing information which is carried by the coefficients of neighboring blocks (see Fig. 4.27), full reconstruction is possible. Regarding the entire signal (all blocks), the total number of coefficients is again identical to the number of original samples, such that the transform is complete. Block-overlapping transforms over finite signals can also be described by a matrix operation, which now covers the entire signal being formally split into a number of overlapping blocks; the index m' denotes the assignment of the resulting transform coefficients to particular block locations:

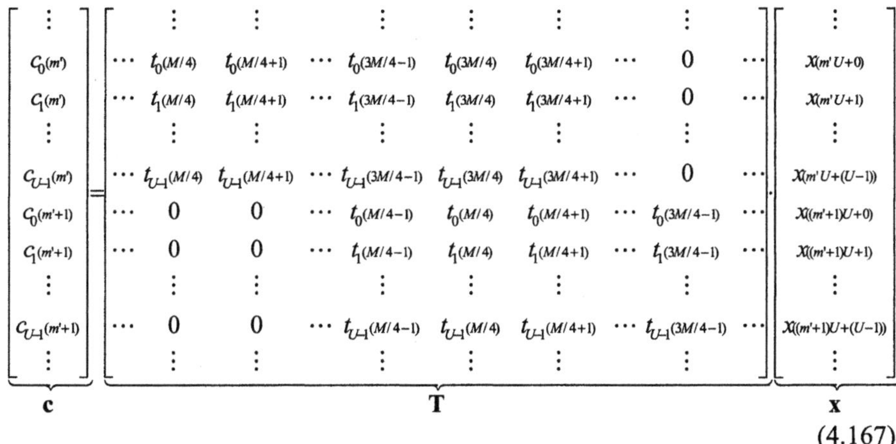

$$(4.167)$$

As the number of transform coefficients in **c** is equal to the number of samples in **x**, the matrix **T** is quadratic and invertible, and the signal can be perfectly reconstructed from the coefficients. Fast algorithms to perform the block-overlapping transform re-use results of intermediate operations from previous blocks. As a core, a fast DCT algorithm can be used, where the window function is embedded accordingly. A DCT block transform can even be interpreted as a special case of the block-overlapping transform (4.164)-(4.166), using a (non-overlapping) rectangular function of length $M=U$ as a window function $w(m)$.

4.4 Filterbank Transforms

Block-overlapping transforms establish a link between the block transforms and a broader class of transforms which are based on filter operations for frequency analysis. The principle of *filterbank transforms*, in which the linear block transforms can also be interpreted as a special case, is shown in Fig. 4.29. The basis functions of the linear transform would now indeed be applied in convolution operations during analysis and synthesis. As the convolution is by nature an operation which slides over any position of a signal, properties of the basis functions can be formulated much more flexible, as the strict processing of separate block segments is inherently given up.

Frequency analysis is typically performed using U parallel filters, which are one lowpass filter, U-2 bandpass filters and one highpass filter. However, application of U parallel filters to the signal will lead to a representation of filter outputs (transform coefficients) that is *overcomplete* by a factor of U. Hence, the output signals are down-sampled (decimated). If the factor of down-sampling is equal to the number of subband signals (which will be assumed here), the decimation is *critical*. This is the limit case which can still lead to a complete representation of the signal and guarantee perfect reconstruction. The entire number of coefficients is then again equal to the number of samples in the signal. During synthesis, the signal is reconstructed by *interpolation* of subband signals and subsequent superposition of all frequency components.

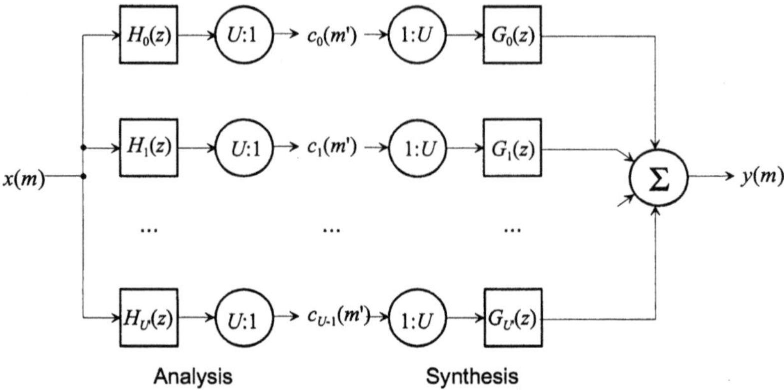

Fig. 4.29. Subband analysis and synthesis system, U frequency bands ($U'=U-1$)

Fig. 4.30 illustrates the elementary operations of decimation and interpolation for the case of a sub-sampling factor $U=2$. Decimation is the operation of retaining only each Uth sample from the original signal (Fig. 4.30a $\rightarrow$ 4.30b). This can be interpreted to consist of two steps, where samples to be discarded are first replaced by zeros (Fig. 4.30c) before they are actually removed. During interpolation, the step from Fig 4.30b $\rightarrow$ 4.30c expands the signal by filling zero samples, by which no information is supplemented. By subsequent *interpolation filtering*, the zeros are finally eliminated (Fig. 4.30d). Typically, signals in Figs. 4.30a and 4.30d could only be identical, if the signal $x(m)$ was band-limited such that no alias occurs by the decimation. This means that frequency components for $\Omega \geq \pi/U$ should be zero before decimation is performed.

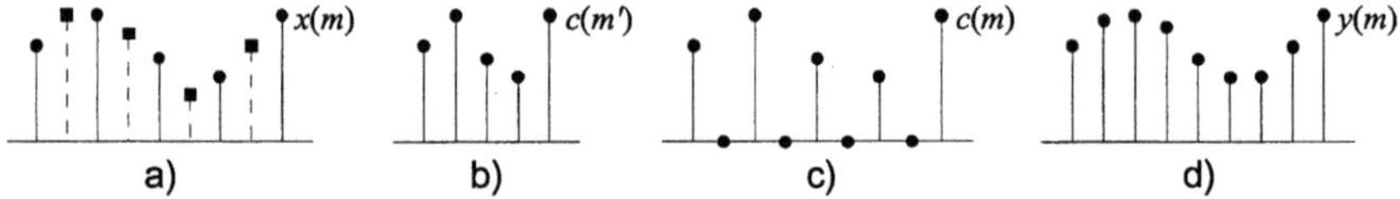

Fig. 4.30. Decimation and interpolation: **a** Original signal **b** decimated Signal **c** first interpolation step (filling by zeros) **d** second interpolation step (filtering)

In comparing the DCT and block overlapping transforms, the aspect of spectral separation properties of transforms was discussed. If a set of ideal lowpass, bandpass and highpass filters would be realizable, U non-overlapping frequency bands would exactly cover a bandwidth of $\Delta\Omega=\pi/U$ each. This is impossible if filters of finite impulse response length shall be used[1]. With non-ideal filters and critical sampling, an overlap of frequency bands always occurs as shown in Fig. 4.31. In Fig. 4.31a, the amplitude transfer function of a lowpass filter is shown[2], of which the bandpass and highpass filters are modulated versions. The resulting layout of the spectrum is shown in Fig. 4.31b.

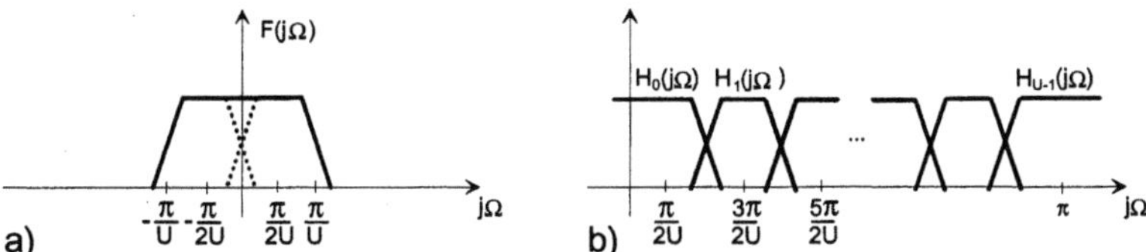

Fig. 4.31. a Lowpass filter **b** overlapping modulated bandpass filters

4.4.1   Decimation and Interpolation

The signal $c(m')$ is generated by sub-sampling a full-rate signal $x(m)$ by a factor U, which is in principle performed by eliminating samples. The first step can be interpreted as a multiplication (modulation) of the signal by a delta impulse train $\delta_U(m)$ of period U[3]:

$$c(m) = x(m)\delta_U(m) = \begin{cases} x(m) & \text{if } m' = m/U \in \mathbb{Z} \\ 0 & \text{else.} \end{cases} \tag{4.168}$$

In the spectral domain, additional periodic copies of the signal spectrum appear, such that the effect is the same as if the continuous signal had been sampled by ω_R/U right away; ω_R is the sampling frequency by which the signal $x(m)$ was originally sampled. The normalized spectral frequency Ω relates to the sampling frequency ω_R. U copies of the base-band spectrum will appear due to sub-sampling at integer multiples of $2\pi/U$:

$$C(j\Omega) = \frac{1}{U}\sum_{u=0}^{U-1} X\left[j\left(\Omega - \frac{2\pi u}{U} \right) \right] \tag{4.169}$$

---

[1] The impulse responses of ideal filters would be modulated versions of infinitely extended sinc functions $h_u(m) = \text{si}\left(\dfrac{\pi}{U}m \right)\cdot\cos\left(\dfrac{\pi(2u+1)}{2U}m \right)$ with $\text{si}(x) = \dfrac{\sin x}{x}$.

[2] This lowpass filter can actually be interpreted as a superposition of two symmetric bandpass filters at center frequencies $\pi/2U$, which seamlessly combine. The same approach is taken to define the modulated sinc functions in the previous footnote.

[3] Unlike the Dirac impulse train introduced in ideal sampling, the delta impulse train is a sampled function itself. Its pulses have a value of 1.

In the second step of decimation, zero samples are discarded from $c(m)$, which effects a scaling of the m-axis, resulting in

$$c(m') = x(m'U) \tag{4.170}$$

As m' now relates to a different sampling frequency ω_R/U, a different normalized frequency $\Omega^* = U \cdot \Omega$ is defined, where the scaling of the m-axis leads to an expansion of the frequency axis:

$$C(j\Omega^*) = \frac{1}{U} \sum_{u=0}^{U-1} X\left(j\frac{\Omega^* - 2\pi u}{U} \right). \tag{4.171}$$

This can also be expressed in the domain of the z transform, where the effects of sampling (4.40) and scaling (4.36) can altogether be expressed in one step without re-definition of the z variable:

$$C(z) = \frac{1}{U} \sum_{u=0}^{U-1} X\left(e^{-j\frac{2\pi u}{U}} \cdot z^{\frac{1}{U}} \right). \tag{4.172}$$

Indices u in (4.169)-(4.172) characterize the different modulated spectral components in the normalized frequency range between $\Omega=0$ and $\Omega=2\pi$. If these spectra overlap, alias occurs as shown in Fig. 4.32. If this is the case, the decimation is *not reversible*.

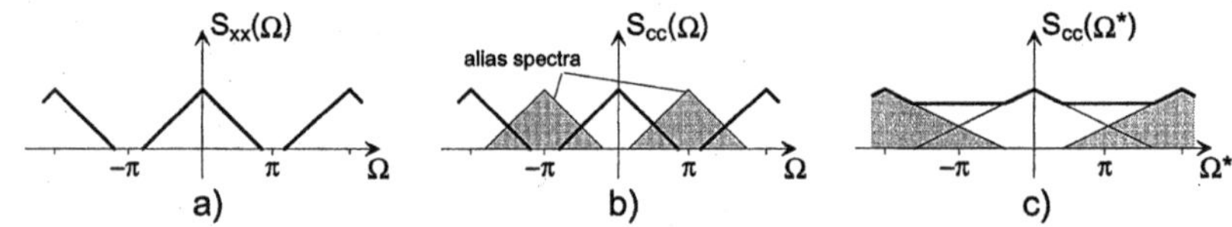

Fig. 4.32. Alias introduced by decimation 2:1
a Original spectrum **b** Spectrum after sub-sampling **c** Spectrum after scaling

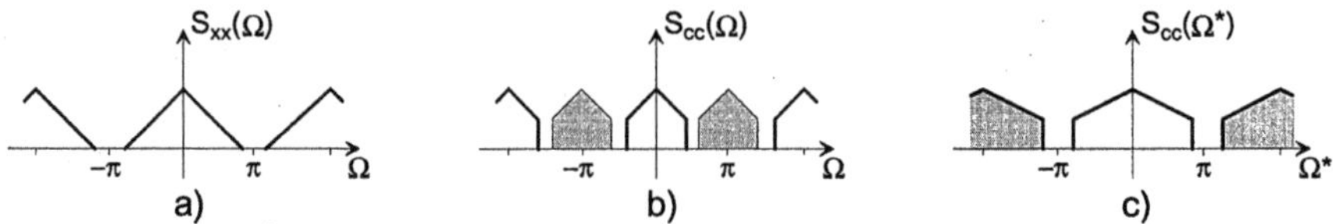

Fig. 4.33. Alias free sub-sampling 2:1 after lowpass filtering
a Original spectrum **b** Spectrum after filtering **c** Spectrum after sub-sampling

Fig. 4.33 shows the principle of an alias-free decimation with $U=2$. If a lowpass filter is applied before the decimation, it is at least possible to reconstruct the band-limited version from the sub-sampled signal according to (4.170) as follows:

$$y(m) = \begin{cases} c(m') & \text{if } m' = \dfrac{m}{U} \in \mathbb{Z} \\[2ex] 0 & \text{else.} \end{cases} \tag{4.173}$$

From (4.37), the expansion in the z and Ω spectra gives

$$C\!\left(z^{U}\right) \quad \Rightarrow \quad C(j\Omega) = C\!\left(jU\Omega^{*}\right). \tag{4.174}$$

Now, the spectrum is compressed, while the m' axis is expanded into the scale of the m axis. No spectral overlaps can occur by this step. Fig. 4.34 shows the spectrum after filling zeros into the signal as in Fig. 4.30c. Finally, application of a lowpass filter eliminates the unwanted spectral alias component around $\Omega=\pi$. This last step effects the elimination of the zero values by application of an *interpolation filter* (Fig. 4.30c $\rightarrow$ 4.30d). If interpolation filtering shall reverse the effect of the previous decimation, an amplification by a factor U must be performed to compensate for the respective amplitude scaling factor in (4.169) and (4.171).

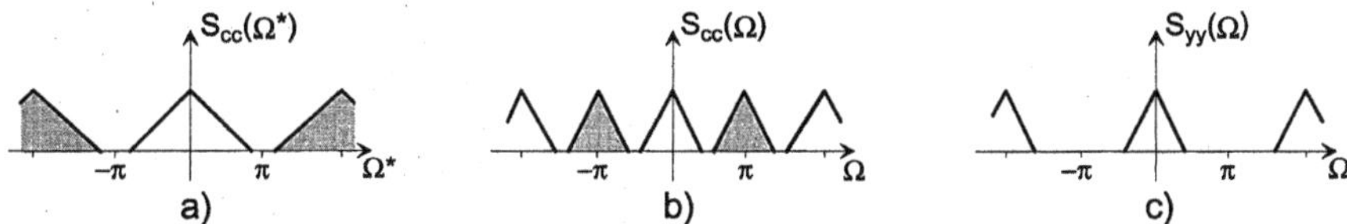

Fig. 4.34. Interpolation 1:2
a Original spectrum **b** Spectrum after expansion **c** Spectrum after filtering

Fig. 4.35 illustrates the application of the decimation principle to the *high frequency* component of a signal. Fig. 4.35a shows the spectrum of the original, Fig. 4.35b the result of highpass filtering with subsequent zero-setting of each second sample. By sub-sampling ($U=2$, Fig. 4.35c) the spectrum is expanded again. Observe that after sub-sampling the signal appears over an *inverted frequency axis*. Those components of the signal which in original resolution were found at $\Omega=\pi$ are now mapped to $\Omega=0$, while the components which were originally near $\Omega=\pi/2$ are found near $\Omega=\pi$ after sub-sampling.

If the sub-sampling after highpass filtering does not incur spectral overlaps (alias), it is likewise possible to reconstruct the original highpass part of the signal by interpolation, such that the spectral positions are correct. To achieve this, it is necessary to use a *highpass filter* as interpolator. This principle can be extended to any frequency components where bandpass filters are employed in a system using $U>2$ bands. Assume the frequency range up to half of the original sampling frequency is uniformly partitioned into U frequency bands. Then, after filtering and critical sub-sampling, each *even-indexed* band will be mapped by correct frequency order to the sub-sampled axis, while each *odd-indexed* band is frequency inverted[1].

So far, only the case of alias free sub-sampling was regarded, which is unrealistic, as no ideal filter can be applied. In the following sub-sections, it will be shown that perfect reconstruction can indeed be achieved even if the sub-sampling of the particular bands can not be performed alias-free. It is however necessary to design the filter banks for the analysis and synthesis stages such that elimination of alias components occurs when the interpolated signals are combined.

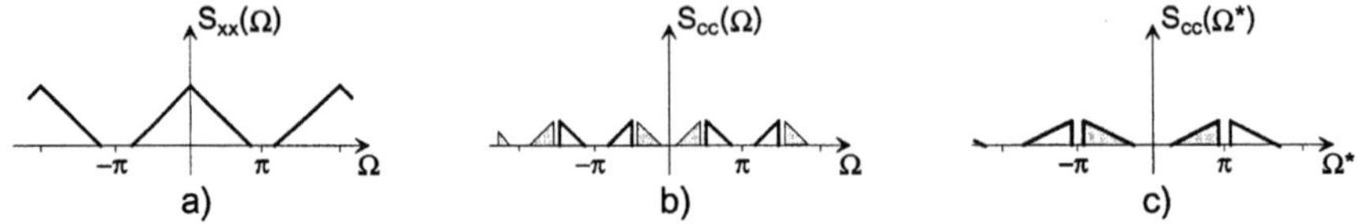

Fig. 4.35. Decimation of a high-frequency signal component
a Signal spectrum **b** Spectrum after highpass filtering and zero-setting of each second value   **c** Spectrum after sub-sampling

4.4.2   Properties of Subband Filters

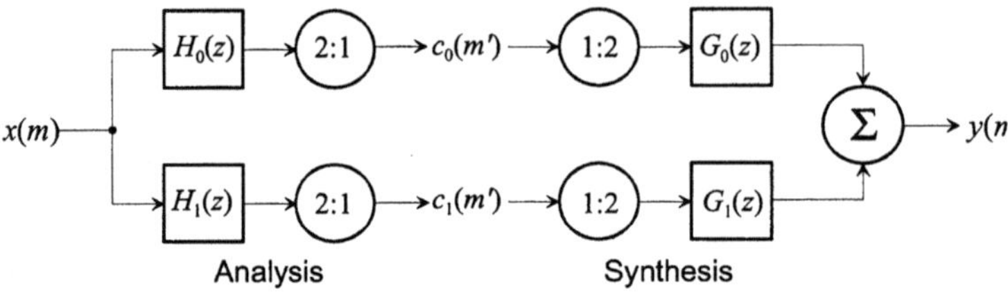

Fig. 4.36. Subband analysis system with U=2 frequency bands

The frequency transfer functions of the analysis filters of a critically-sampled filterbank overlap in case of finite impulse responses. Signal components will appear beyond the normalized bandwidth π/U, which means that alias occurs by subsampling. We generally define $H_u(z)$ as the transfer functions of the analysis filters, $G_u(z)$ are those of synthesis (interpolation) filters. For the case of U=2, which refers to the subband system in Fig. 4.36, only one lowpass band and one highpass band are generated. Hence, from (4.171), only one alias spectrum will appear at $\Omega=\pi$ or $z=-1$. Alias occurs in both branches (lowpass and highpass) during subsampling, which means that the components after filtering $H_u(z)X(z)$, are copied.

---

[1] This gives ground for yet another interpretation about the correlations between even-indexed and odd-indexed coefficients that was observed in block transforms (which can fully be interpreted as subband systems). The bands overlap in frequency, which is one of the causes for correlation after sub-sampling. On the other hand, as even and odd bands appear by reverse frequency order, the correlation is cancelled out.

Regarding the complete system chain, the following spectrum results after synthesis:

$$Y(z) = \frac{1}{2}\big[H_0(z)\cdot G_0(z) + H_1(z)\cdot G_1(z)\big]\cdot X(z)$$
$$+ \frac{1}{2}\big[H_0(-z)\cdot G_0(z) + H_1(-z)\cdot G_1(z)\big]\cdot X(-z). \tag{4.175}$$

Quadrature Mirror Filters (QMF). In (4.175), the upper term expresses the usable signal, while the lower term is the alias signal, which should be eliminated. This can be achieved if the expression within the lower parenthesis in (4.175) has a value of zero. This condition is fulfilled by different classes of filters. In QMF, the lowpass analysis filter H_0 and the highpass filter H_1 establish a system of *orthogonal quadrature components*: The highpass filter is determined from the lowpass filter by modulating the impulse response by a cosine of half sampling frequency ($\Omega=\pi$ or $z=-1$). This discrete modulation sequence alternates between $+1$ and -1 by each sample. The relationships of the different filters in the signal domain and the spectral domains of Ω- and z-transfer functions are listed in Table 4.1. Some mapping relationships between the impulse responses, the Ω- and z- transfer functions are also given in the lower part of the table.

Table 4.1. Definition of Quadrature Mirror Filters (QMF): Relationships of orthogonal lowpass and highpass analysis and synthesis filters, expressed by impulse responses, z- and Ω-spectra

	$f(m)$	$F(\Omega)$	$F(z)$
H_0	$h_0(m)=f(m)$	$H_0(\Omega)=F(\Omega)$	$H_0(z)=F(z)$
H_1	$h_1(m)=(-1)^{1-m}\cdot f(1-m)$	$H_1(\Omega)=e^{-j\Omega}\cdot F(\pi-\Omega)$	$H_1(z)=z^{-1}\cdot F(-z^{-1})$
G_0	$g_0(m)=f(-m)$	$G_0(\Omega)=F(-\Omega)$	$G_0(z)=F(z^{-1})$
G_1	$g_1(m)=(-1)^{m-1}\cdot f(m-1)$	$G_1(\Omega)=e^{j\Omega}\cdot F(\Omega-\pi)$	$G_1(z)=z\cdot F(-z)$
Inversion	$h(-m)$	$H(-\Omega)=H^*(\Omega)$	$H(z^{-1})$
Modulation	$(-1)^m h(m)$	$H(\Omega-\pi)=H^*(\pi-\Omega)$	$H(-z)=H(z\cdot e^{j\pi})$

Substituting Ω-frequencies into (4.175) gives

$$Y(\Omega) = \frac{1}{2}\big[H_0(\Omega)\cdot G_0(\Omega) + H_1(\Omega)\cdot G_1(\Omega)\big]\cdot X(\Omega)$$
$$+ \frac{1}{2}\big[H_0(\Omega-\pi)\cdot G_0(\Omega) + H_1(\Omega-\pi)\cdot G_1(\Omega)\big]\cdot X(\Omega-\pi). \tag{4.176}$$

A simpler expression results, if the definition of all four filters is related to the common model filter $F(\Omega)$ as parametrized in Table 4.1, which is identical to the analysis lowpass,

$$Y(\Omega) = \frac{1}{2}\left[F(\Omega)\cdot F(-\Omega)+F(\pi-\Omega)\cdot F(\Omega-\pi)\right]\cdot X(\Omega)$$
$$+\frac{1}{2}\left[F(\Omega-\pi)\cdot F(-\Omega)+e^{j\pi}F(-\Omega)\cdot F(\Omega-\pi)\right]\cdot X(\Omega-\pi). \tag{4.177}$$

With $e^{j\pi}=-1$, the alias term is eliminated, while for the output signal the condition

$$F(\Omega)\cdot F^{*}(\Omega)+F(\pi-\Omega)\cdot F^{*}(\pi-\Omega)=\left|F(\Omega)\right|^{2}+\left|F(\pi-\Omega)\right|^{2}=2 \tag{4.178}$$

will give perfect reconstruction. (4.178) is generalized to the case of arbitrary number U of subbands by

$$\sum_{u=1}^{U}\left|H_{u}(\Omega)\right|^{2}\overset{!}{=}U. \tag{4.179}$$

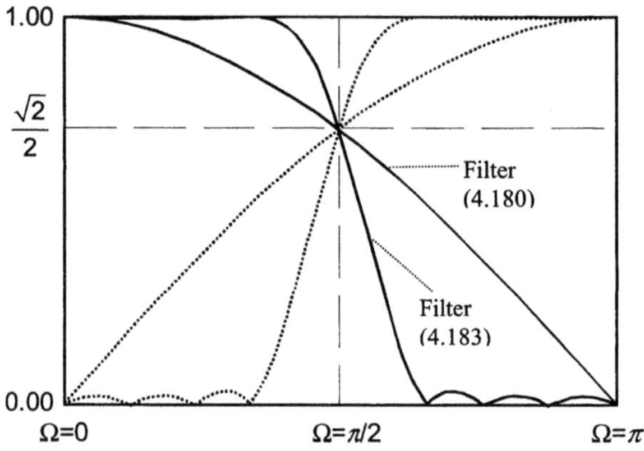

Fig. 4.37. Amplitude transfer functions of filters from (4.180) and (4.183)
[— lowpass $\cdots$ highpass]

Observe that the alias term is eliminated perfectly. Unfortunately, the condition (4.179) for perfect reconstruction of the signal can for the case of orthogonal filters (symmetry of lowpass and highpass transfer functions) only be fulfilled for two cases:

- Impulse response lengths are identical to the number of subbands U (which is the special case of block transforms);
- Ideal filters, which would have an infinite impulse response (modulated sinc functions, see footnote on p. 135).

Example: Haar filter basis. The Haar filter pair represents the basis functions of almost any orthonormal transforms for the case $U=M=2$, which includes DCT,

Walsh, Hadamard, Rademacher, Haar and the KLT as optimized for an AR(1) process. The z transfer functions according to Table 4.1 are[1]

$$H_0(z) = F(z) = \frac{\sqrt{2}}{2} + \frac{\sqrt{2}}{2} \cdot z^{-1}$$

$$H_1(z) = z \cdot F(-z^{-1}) = -\frac{\sqrt{2}}{2} + \frac{\sqrt{2}}{2} \cdot z^{-1}. \tag{4.180}$$

From the Fourier spectrum, we get

$$\left|H_0(j\Omega)\right|^2 + \left|H_1(j\Omega)\right|^2 = \left(\frac{\sqrt{2}}{2}\right)^2 \cdot \left(2\cos\frac{\Omega}{2}\right)^2 + \left(\frac{\sqrt{2}}{2}\right)^2 \cdot \left(2\sin\frac{\Omega}{2}\right)^2 = 2, \tag{4.181}$$

which fulfills condition (4.179). The disadvantage however is the flat decay of the amplitude transfer function, due to the low order of the filters. The Haar filters have a very poor frequency separation property, which will cause strong alias components in the sub-sampled signals. Other finite-length filters constructed by the conditions of Table 4.1 will not approach (4.179) perfectly. The design of such filters is made as a compromise between frequency separation properties for alias suppression in the subbands, and a reconstruction error which should be kept as low as possible, such that

$$U - \sum_{u=1}^{U} \left|H_u(\Omega)\right|^2 \overset{!}{=} \min. \tag{4.182}$$

Example. A symmetric 16 tap QMF pair [JOHNSTON 1980] following the relationships in Table 4.1 and optimized by the constraint (4.182) has lowpass and highpass impulse responses

$$\begin{aligned}
H_0(z) = {} & 0.007 \cdot z^7 - 0.02 \cdot z^6 + 0.002 \cdot z^5 + 0.046 \cdot z^4 \\
& -0.026 \cdot z^3 - 0.099 \cdot z^2 + 0.118 \cdot z + 0.472 \\
& +0.472 \cdot z^{-1} + 0.118 \cdot z^{-2} - 0.099 \cdot z^{-3} - 0.026 \cdot z^{-4} \\
& +0.046 \cdot z^{-5} + 0.002 \cdot z^{-6} - 0.02 \cdot z^{-7} + 0.007 \cdot z^{-8}
\end{aligned} \tag{4.183}$$

$$\begin{aligned}
H_1(z) = {} & 0.007 \cdot z^7 + 0.02 \cdot z^6 + 0.002 \cdot z^5 - 0.046 \cdot z^4 \\
& -0.026 \cdot z^3 + 0.099 \cdot z^2 + 0.118 \cdot z - 0.472 \\
& +0.472 \cdot z^{-1} - 0.118 \cdot z^{-2} - 0.099 \cdot z^{-3} + 0.026 \cdot z^{-4} \\
& +0.046 \cdot z^{-5} - 0.002 \cdot z^{-6} - 0.02 \cdot z^{-7} - 0.007 \cdot z^{-8}.
\end{aligned}$$

The filter pair establishes an orthogonal basis system according to (4.107); to achieve *orthonormality*, all coefficients or the filter output have to be multiplied

---

[1] Another definition $H_1(z) = \frac{\sqrt{2}}{2} - \frac{\sqrt{2}}{2} \cdot z^{-1}$ may alternatively be used.

by a factor $\sqrt{2}/2$. Fig. 4.37 shows the amplitude transfer functions of the filters (4.180) and (4.183). The amplitude transfer functions of lowpass and highpass filters are mirrored versions of each other, with symmetry about the frequency $\Omega = \pm\pi/2$.

To define more general conditions for alias-free *and* lossless reconstruction, the constraint that transfer functions H_0 and H_1 should be mirror-symmetric shall now be released. From (4.175), the elimination of the alias component is also achieved if the following conditions are observed:

$$G_0(z) = z^k \cdot H_1(-z)$$
$$G_1(z) = -z^k \cdot H_0(-z)$$

(4.184)

Substituting (4.184) into (4.175) gives

$$Y(z) = \frac{1}{2}\left[H_0(z) \cdot H_1(-z) - H_1(z) \cdot H_0(-z)\right] \cdot X(z) \cdot z^k ,$$

(4.185)

which results in a condition for perfect reconstruction

$$P(z) - P(-z) = 2 \cdot z^{-k} \quad \text{with} \quad P(z) = H_0(z) \cdot H_1(-z).$$

(4.186)

The term z^{-k} expresses an arbitrary phase shift which may occur anywhere in the analysis/synthesis chain. It is common practice to perform the process of filtering such that the current sample relates to the center of the impulse response (see sec. 4.1). As filters are mainly of FIR type[1], this can simply be resolved by a temporal delay or spatial shift. Two types of filters, which are determined from (4.184)-(4.186), are introduced in the following sub-sections.

Perfect Reconstruction Filters (PRF). For this type of filter, the basis functions of lowpass and highpass can be orthogonal, but the highpass impulse response must no longer be a modulated version of the lowpass response, and the transfer functions no longer obey the mirror symmetry as in the QMF case. Typically, the resulting frequency bands have unequal widths. Besides the property of guaranteed perfect reconstruction, the filters have linear phase property, and can be implemented by integer arithmetic through appropriate selection of the filter coefficients. (4.186) can be expressed by matrix notation

$$\det|\mathbf{P}(z)| = 2z^{-k} \quad \text{with} \quad \mathbf{P}(z) = \begin{bmatrix} H_0(z) & H_0(-z) \\ H_1(z) & H_1(-z) \end{bmatrix}.$$

(4.187)

The factorization of $P(z)$ into $H_0(z)$ and $H_1(-z)$ is now reduced into a problem to factorize the matrix, which shall have a determinant expressing a shift by k samples and multiplication by a factor of 2. The factorization is simplified, if the z

---

[1] For IIR subband filters, see e.g. [SMITH 1991].

polynomials are decomposed into polyphase components, where sub-responses of subscripts A and B are the even and odd samples of the impulse response, respectively:

$$H_u(z) = 2\left[H_{u,A}(z^2) + z^{-1}H_{u,B}(z^2)\right].$$

$$(4.188)$$

Writing the polyphase components of the filter pair into the polyphase matrix[1],

$$\mathbf{H}(z) = \begin{bmatrix} H_{0,A}(z) & H_{0,B}(z) \\ H_{1,A}(z) & H_{1,B}(z) \end{bmatrix},$$

$$(4.189)$$

(4.187) will be fulfilled if (4.189) has a determinant which is $2z^{-k+1}$. A possible construction of polyphase matrices fulfilling the constant-shift condition is made by

$$\mathbf{H}(z) = \begin{bmatrix} 1 & 1 \\ 1 & -1 \end{bmatrix} \cdot \prod_{p=1}^{P-1} \begin{bmatrix} 1 & 0 \\ 0 & z^{-1} \end{bmatrix} \cdot \begin{bmatrix} 1 & \alpha_p \\ \alpha_p & 1 \end{bmatrix}$$

$$(4.190)$$

[VETTERLI 1991]. $\mathbf{H}(z)$ is complemented by its inverse, which represents the polyphase components of the synthesis filters,

$$\begin{aligned}
\mathbf{G}(z) &= \begin{bmatrix} G_{0,A}(z) & G_{1,A}(z) \\ G_{0,B}(z) & G_{1,B}(z) \end{bmatrix} \\
&= \frac{1}{2}\begin{bmatrix} 1 & 1 \\ 1 & -1 \end{bmatrix} \cdot \prod_{p=1}^{P-1}\left[\begin{bmatrix} z^{-1} & 0 \\ 0 & 1 \end{bmatrix} \cdot \begin{bmatrix} 1 & -\alpha_p \\ -\alpha_p & 1 \end{bmatrix} \cdot \frac{1}{1-\alpha_p^2}\right].
\end{aligned}$$

$$(4.191)$$

The length of the filters $H_0(z)$ and $H_1(z)$ will then be $2P$ samples.

Examples. For $P=1$, the result from (4.190) and (4.191) is the Haar filter pair (4.192), using the alternative form of $H_1(z)$ as defined in the footnote on p. 141. For P=2, the following filters result, which are by the short length of their impulse responses also often denoted as *short kernel filters*:

$$\begin{aligned}
H_0(z) &= \frac{1}{\sqrt{2\left(\alpha_1^2-1\right)}}(1+\alpha z^{-1}+\alpha z^{-2}+z^{-3}) \\
H_1(z) &= \frac{1}{\sqrt{2\left(\alpha_1^2-1\right)}}(-1-\alpha z^{-1}+\alpha z^{-2}+z^{-3}) \\
G_0(z) &= H_1(-z) = \frac{1}{\sqrt{2\left(\alpha_1^2-1\right)}}(-1+\alpha z^{-1}+\alpha z^{-2}-z^{-3}) \\
G_1(z) &= -H_0(-z) = \frac{1}{\sqrt{2\left(\alpha_1^2-1\right)}}(1-\alpha z^{-1}+\alpha z^{-2}-z^{-3})
\end{aligned}$$

$$(4.193)$$

---

[1] For an introduction to polyphase systems, see sec. 4.4.3.

As example, the normalization factor is 1/4 for $\alpha_1=3$, which enables an implementation by integer arithmetic precision. Substituting into (4.186), the result is $P(z)-P(-z)=2 \cdot z^{-3}$.

Biorthogonal filters. In the PRF construction, lowpass and highpass filter kernels are of same length, and orthogonality applies due to the symmetry in (4.190). Even these relationships between the bases H_0 and H_1 can be released[1]. Condition (4.185) only postulates that the analysis highpass H_1 shall be a '$-z$'-modulated version of the synthesis lowpass G_0, and the synthesis highpass G_1 shall be a '$-z$'-modulated version of the analysis lowpass H_0. Hence, a cross-orthogonal relationship exists between the analysis highpass / synthesis lowpass and analysis lowpass / synthesis highpass pairs. If impulse responses (basis functions) are interpreted as vectors, this is equivalent to the property of *bi-orthogonality* (B.21) of a basis system. As an example, for the following sets of filters, analysis lowpass and highpass impulse responses have unequal lengths. Linear-phase properties are retained here, but are not required in general[2]:

$$H_0^{(5/3)}(z)=\frac{1}{8}\left(-z^2+2\cdot z^1+6+2\cdot z^{-1}-z^{-2}\right); \quad H_1^{(5/3)}(z)=\frac{1}{2}\left(-1+2\cdot z^{-1}-z^{-2}\right)$$

$$G_0^{(5/3)}(z)=\frac{1}{2}\left(z^{-1}+2+z^{-1}\right); \quad G_1^{(5/3)}(z)=\frac{1}{8}\left(-z^1+2+6\cdot z^{-1}+2\cdot z^{-2}-z^{-3}\right) \tag{4.194}$$

$$H_0^{(9/7)}(z)=0.027z^{-4}-0.016z^{-3}-0.078z^{-2}+0.267z^{-1}$$
$$+0.603+0.267z^1-0.078z^{-2}-0.016z^{-3}+0.027z^4$$
$$H_1^{(9/7)}(z)=0.091z^{-3}-0.057z^{-2}-0.591z^{-1}$$
$$+1.115-0.591z^1-0.057z^2+0.091z^3. \tag{4.195}$$

Biorthogonal filters are often used in the *Discrete Wavelet transform* (see sec. 4.4.4). Certain constraints must be regarded in this context, in particular an iterative application of the lowpass filter on scaled (sub-sampled) signals must still have the effect of a lowpass filter. Such criteria are important if biorthogonal filters are designed.

---

[1] Orthogonality is however an important property regarding encoding of frequency coefficients, see (4.161).

[2] By the lengths of their lowpass/highpass analysis filter kernels, these two filters are denoted as 5/3 and 9/7 filters, respectively. Both filters are sometimes modified, multiplying H_0 by $\sqrt{2}$ and dividing H_1 by $\sqrt{2}$, which almost approaches orthonormality at least for the case of the 9/7 filter.

4.4.3 Implementation of Filterbank Structures

If a direct implementation of filterbanks according to the structures previously introduced is performed, the complexity of realization for subband analysis and synthesis is considerably higher than for fast transform algorithms. Methods exist however to reduce the computational complexity considerably. A certain similarity between the subsequently described methods and fast transform algorithms as described in sec. 4.3.4 will become obvious. Actually, fast transform algorithms can be interpreted as cascaded subband systems ($\Leftrightarrow$ factorization of the transform matrix) in polyphase realization ($\Leftrightarrow$ computation of the transform at block-shifted positions), using all equalities of multiplication factors ($\Leftrightarrow$ perform multiplications after summations in the signal-flow diagrams). In particular, fast block transforms of rectangular basis systems can directly be interpreted as subband systems based on Haar filter kernels (4.180).

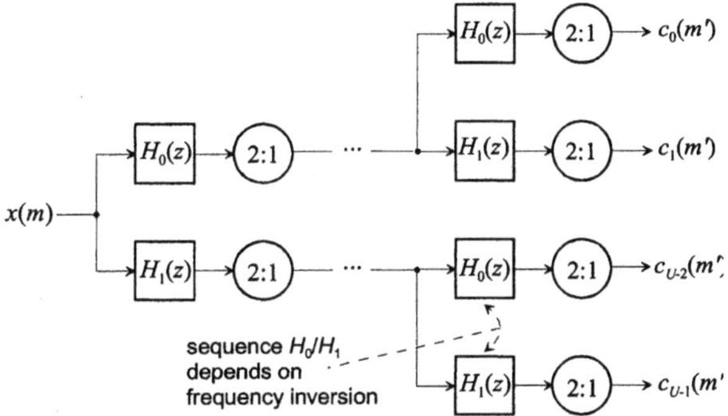

Fig. 4.38. Realization of subband analysis filter in a cascade from 2-band systems

Cascaded two-band systems. If T two-band systems from Fig. 4.36 are configured in a cascade, the system is operated in a chain of subsequent levels, which folds up like a tree. Each output signal of a preceding level of the chain is again decomposed into two more narrow subbands, and a complete decomposition into $U=2^T$ subbands can be realized as shown in Fig. 4.38. The impulse response of a bandpass filter along a specific path of the chain results by the iterative convolution of the respective lowpass and highpass filters involved. The computational effort is considerably lower than in a system with parallel processing of bandpass outputs, as intermediate results are used multiple times, and processing of higher levels uses increasingly sub-sampled signals. Due to the frequency inversion occurring in highpass band sub-sampling (see Fig. 4.35), any frequency band that results from an odd number of highpass filter / decimation steps will be frequency inverted. For the subsequent level, it is then necessary to exchange the sequence of filters H_0 and H_1 for an arrangement of subbands by increasing frequency order[1].

Exploitation of filter symmetries. If symmetric (linear phase) filters are used, duplicate multiplications can be avoided, where samples are multiplied by same factors several times. If the highpass basis function is a modulated version of the lowpass or uses the same multiplication factors (as in the cases of QMF and PRF types), the reduction of multiplications can be up to a factor of 4 (see also Problem 4.10).

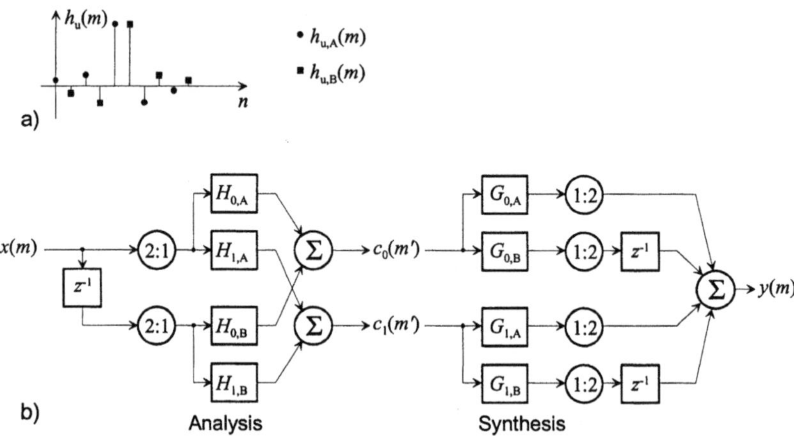

Fig. 4.39. Realization of subband analysis and synthesis by polyphase systems, $U{=}2$
a Separation of impulse response into partial terms $h_A(n)$ und $h_B(n)$
b Structure of the overall polyphase system

Polyphase systems. Only each Uth sample will be retained by subband analysis after the filtering and sub-sampling steps. Hence, the convolution must not be performed at positions which are discarded anyway. This will lead to a reduction of the computational complexity by a factor of U. The structure of a *polyphase system* is shown in Fig. 4.39 for the example $U{=}2$. Sub-sampling is performed in *advance to filtering*, whereby the signal is decomposed into U *polyphase components*, which are a set of distinct sequences sub-sampled at all possible phase positions. Likewise, it is necessary to decompose the filter impulse responses into polyphase components, such that instead of a length-P filter, U partial filters of lengths int(P/U) or int$(P/U{+}1)$ are obtained. If the subband filter impulse response h_i is decomposed into U polyphase components $h_{i,A}(m)$, $h_{i,B}(m)$, ... , U partial filters of transfer functions $H_{i,A}(z)$, $H_{i,B}(z)$, ... are given (see Fig. 4.39a for the case $U{=}2$). Similarly, it is not necessary to apply multiplications on zero-padded values

---

[1] This shows an analogy with the distinction between Walsh and Hadamard transforms (see Fig. 4.17).

during synthesis. This can be realized by performing the interpolation filtering step within the polyphase components, and compose the different phase positions into the reconstructed signal only by the last step. Again, this means that the expansion of the signal by zero-filling is performed *after filtering*, but no zeros will remain, as the polyphase components from all partial filters will complement each other by filling the gaps. A reduction of implementation complexity by a factor of U is achieved in synthesis as well (Fig. 4.39b).

For a system with $U=2$, the polyphase components of a signal $x(n)$ are sequences of even samples $x(2m')$ and odd samples $x(2m'+1)$. In the z transform domain, the following relationships apply:

$$x(2m') \overset{z}{\leftrightarrow} X_A(z); \quad x(2m'+1) \overset{z}{\leftrightarrow} X_B(z);$$

$$x(m) \overset{z}{\leftrightarrow} X(z) = 2\left[X_A(z^2) + z^{-1} X_B(z^2) \right],$$

(4.196)

where the subscripts A and B relate to the even and odd polyphase components, respectively. Formally, the components of the z transform related to even and odd polyphase components can be written in vector notation

$$\mathbf{X}(z) = \begin{bmatrix} X_A(z) \\ z^{-1} X_B(z) \end{bmatrix}.$$

(4.197)

The same procedure can be applied to the z polynomials of the filter impulse responses. As the convolution in the signal domain corresponds to a multiplication in the z domain, the filtering of the even/odd signal spectra by the respective filter transfer functions can be expressed as

$$\underbrace{\begin{bmatrix} C_0(z) \\ C_1(z) \end{bmatrix}}_{\mathbf{C}(z)} = \underbrace{\begin{bmatrix} H_{0,A}(z) & H_{0,B}(z) \\ H_{1,A}(z) & H_{1,B}(z) \end{bmatrix}}_{\mathbf{H}(z)} \underbrace{\begin{bmatrix} X_A(z) \\ z^{-1} X_B(z) \end{bmatrix}}_{\mathbf{X}(z)}.$$

(4.198)

For the synthesis part, a similar view applies. Writing the reconstructed signal by

$$\mathbf{Y}(z) = \begin{bmatrix} Y_A(z) \\ z \cdot Y_B(z) \end{bmatrix} \;;\; Y(z) = 2\left[Y_A(z^2) + z \cdot Y_B(z^2) \right],$$

(4.199)

the synthesis filter step can be expressed as

$$\underbrace{\begin{bmatrix} Y_A(z) \\ z \cdot Y_B(z) \end{bmatrix}}_{\mathbf{Y}(z)} = \underbrace{\begin{bmatrix} G_{0,A}(z) & G_{1,A}(z) \\ G_{0,B}(z) & G_{1,B}(z) \end{bmatrix}}_{\mathbf{G}(z)} \underbrace{\begin{bmatrix} C_0(z) \\ C_1(z) \end{bmatrix}}_{\mathbf{C}(z)}.$$

(4.200)

Combining (4.198) and (4.200), the condition for perfect reconstruction is

$$\begin{bmatrix} G_{0,A}(z) & G_{1,A}(z) \\ z^{-1}G_{0,B}(z) & z^{-1}G_{1,B}(z) \end{bmatrix}\begin{bmatrix} H_{0,A}(z) & z \cdot H_{0,B}(z) \\ H_{1,A}(z) & z \cdot H_{1,B}(z) \end{bmatrix} = \mathbf{I} , \tag{4.201}$$

from which the following relationships result:

$$
\begin{aligned}
G_{0,A}(z)H_{0,A}(z) + G_{1,A}(z)H_{1,A}(z) &= 1 \\
z \cdot G_{0,A}(z)H_{0,B}(z) + z \cdot G_{1,A}(z)H_{1,B}(z) &= 0 \\
z^{-1}G_{0,B}(z)H_{0,A}(z) + z^{-1}G_{1,B}(z)H_{1,A}(z) &= 0 \\
G_{0,B}(z)H_{0,B}(z) + G_{1,B}(z)H_{1,B}(z) &= 1.
\end{aligned}
\tag{4.202}
$$

These are fulfilled by the constellation of filters

$$
\begin{aligned}
H_{0,A}(z) &= G_{1,B}(z) \quad ; \quad H_{0,B}(z) = -G_{1,A}(z); \\
H_{1,A}(z) &= -G_{0,B}(z) \quad ; \quad H_{1,B}(z) = G_{0,A}(z),
\end{aligned}
\tag{4.203}
$$

which by substitution into (4.202) gives the additional condition

$$H_{0,A}(z)H_{1,B}(z) - H_{0,B}(z)H_{1,A}(z) = \det|\mathbf{H}(z)| = 1 . \tag{4.204}$$

Using (4.197) to express the polyphase filters in the (not down-sampled) z domain, (4.202) is equivalent to (4.184) with shift $k=-1$, while (4.204) is equivalent to (4.186). A special case of the polyphase transform is observed for $\mathbf{H}(z)=\mathbf{G}(z)=\mathbf{I}$, which is the so-called *lazy transform* where the 'subband' signals c_0 and c_1 would just be the polyphase components generated without any lowpass or highpass filtering.

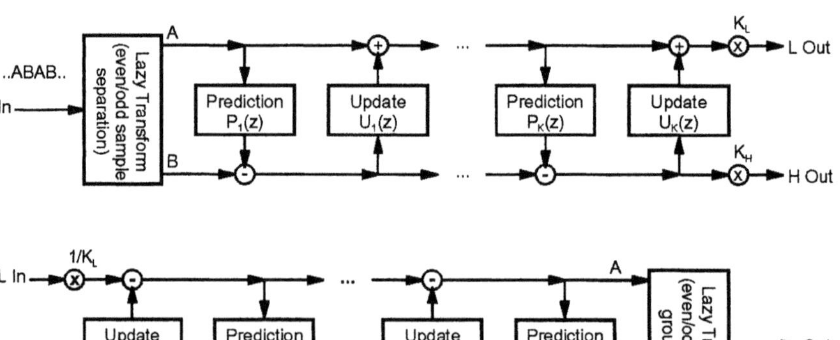

Fig. 4.40. Lifting structure of a biorthogonal filter.

Lifting implementation. Any pair of biorthogonal filters can be implemented in a *lifting structure* as shown in Fig. 4.40 [SWELDENS 1995]. The first step of the lifting filter is a decomposition of the signal into its even- and odd-indexed polyphase components, by the lazy transform. Then, the two basic operations are *prediction* *steps $P(z)$* and *update steps $U(z)$*. The prediction and update filters are primitive kernels typically of 2 or 3 taps each; the number of steps necessary and the values of coefficients in each step are determined by a factorization of biorthogonal filter pairs. Finally, normalization by factors K_{Low} and K_{High} is applied to achieve orthonormal decomposition.

The determination of the lifting and update filters can best be started from the polyphase representation. Assume that decomposition of a signal has been performed by a polyphase filter matrix $\mathbf{H}^0(z)$ (which could be the identity matrix $\mathbf{I}$ for the lazy transform in the beginning). If a *prediction step* is performed using the filter transfer function $P(z)$, the result is identical to a filter expressed by the polyphase matrix

$$\mathbf{H}^{\text{pr}}(z) = \underbrace{\begin{bmatrix} 1 & 0 \\ -P(z) & 1 \end{bmatrix}}_{\mathbf{P}(z)} \cdot \mathbf{H}^0(z)$$

$$= \begin{bmatrix} H_{0,A}(z) & H_{0,B}(z) \\ H_{1,A}(z) - P(z)H_{0,A}(z) & H_{1,B}(z) - P(z)H_{0,B}(z) \end{bmatrix}. \tag{4.205}$$

The complementary synthesis filter guaranteeing perfect reconstruction such that $\mathbf{G}^{\text{pr}}(z)\mathbf{H}^{\text{pr}}(z){=}\mathbf{I}$ when $\mathbf{G}^0(z)\mathbf{H}^0(z){=}\mathbf{I}$ is expressed as

$$\mathbf{G}^{\text{pr}}(z) = \mathbf{G}^0(z) \cdot \begin{bmatrix} 1 & 0 \\ P(z) & 1 \end{bmatrix} = \begin{bmatrix} G_{0,A}(z) + P(z)G_{1,A}(z) & G_{1,A}(z) \\ G_{0,B}(z) + P(z)G_{1,B}(z) & G_{1,B}(z) \end{bmatrix}. \tag{4.206}$$

Similarly, a single *update step* can be formulated as

$$\mathbf{H}^{\text{up}}(z) = \underbrace{\begin{bmatrix} 1 & U(z) \\ 0 & 1 \end{bmatrix}}_{\mathbf{U}(z)} \cdot \mathbf{H}^0(z)$$

$$= \begin{bmatrix} H_{0,A}(z) + U(z)H_{1,A}(z) & H_{0,B}(z) + U(z)H_{1,B}(z) \\ H_{1,A}(z) & H_{1,B}(z) \end{bmatrix}, \tag{4.207}$$

where the complementary synthesis filter is

$$\mathbf{G}^{\text{up}}(z) = \mathbf{G}^0(z) \cdot \begin{bmatrix} 1 & -U(z) \\ 0 & 1 \end{bmatrix} = \begin{bmatrix} G_{0,A}(z) & G_{1,A}(z) - U(z)G_{0,A}(z) \\ G_{0,B}(z) & G_{1,B}(z) - U(z)G_{0,B}(z) \end{bmatrix}. \tag{4.208}$$

Using (4.205)-(4.208) iteratively starting by a lazy transform, the equivalent polyphase matrix after a number of subsequent prediction and update steps is the product of all matrices, e.g. for a number of L subsequent prediction and update steps

$$\mathbf{H}(z) = \begin{bmatrix} K_{\mathrm{L}} & 0 \\ 0 & K_{\mathrm{H}} \end{bmatrix} \prod_{l=1}^{L} \begin{bmatrix} 1 & U_l(z) \\ 0 & 1 \end{bmatrix} \begin{bmatrix} 1 & 0 \\ -P_l(z) & 1 \end{bmatrix}. \tag{4.209}$$

Vice versa, it is possible to *factor* any given polyphase matrix into a number of primitive prediction/update matrices. Separation of single prediction and update steps from a given (complete) polyphase matrix $\mathbf{H}(z)$ will result in the following expression according to (4.205) and (4.207):

$$\mathbf{H}(z) = \begin{bmatrix} 1 & 0 \\ -P(z) & 1 \end{bmatrix} \cdot \mathbf{H}^{-\mathrm{pr}}(z) \quad ; \quad \mathbf{H}(z) = \begin{bmatrix} 1 & U(z) \\ 0 & 1 \end{bmatrix} \cdot \mathbf{H}^{-\mathrm{up}}(z) \tag{4.210}$$

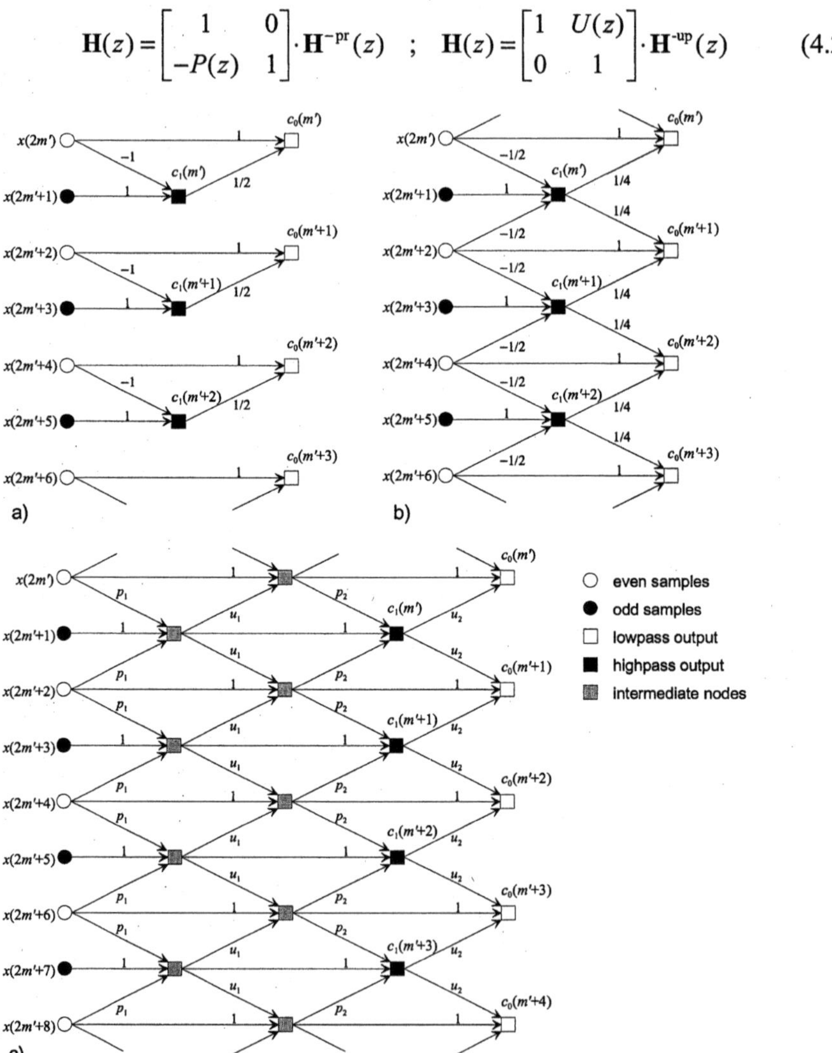

Fig. 4.41. Lifting signal flows implementing **a** Haar filter **b** biorthogonal 5/3 filter **c** biorthogonal 9/7 filter with parameters p_1=-1.586134342059924; u_1=-.052980118572961; p_2=.882911075530934; u_2=.443506852043971; K_{H}=1/K_{L}=1.230174104914001

The factorization is possible, as the determinant of any of the single prediction and update matrices is one, and hence inversion is possible. By polynomial division, the result can be computed step by step, and the factorization typically terminates when only a diagonal matrix of normalization factors K_L and K_H is retained.

Examples. The biorthogonal 5/3 filter from (4.194) can be expressed by the following polyphase matrix, which is further factored into normalization factors, prediction and update matrices

$$\mathbf{H}(z) = \begin{bmatrix} -\frac{1}{8}z^{-1} + \frac{3}{4} - \frac{1}{8}z & \frac{1}{4} + \frac{1}{4}z \\ -\frac{1}{2}z^{-1} - \frac{1}{2} & 1 \end{bmatrix} = \underbrace{\begin{bmatrix} 1 & 0 \\ 0 & 1 \end{bmatrix}}_{\mathbf{K}} \cdot \underbrace{\begin{bmatrix} 1 & \frac{1}{4} + \frac{1}{4}z \\ 0 & 1 \end{bmatrix}}_{\mathbf{U}(z)} \cdot \underbrace{\begin{bmatrix} 1 & 0 \\ -\frac{1}{2}z^{-1} - \frac{1}{2} & 1 \end{bmatrix}}_{\mathbf{P}(z)}. \quad (4.211)$$

Here, $K_H = K_L = 1$, $P(z) = \frac{1}{2}(z^{-1} + 1)$ and $U(z) = \frac{1}{4}(1 + z)$. Another example is for the Haar filter[1], where $K_L = \sqrt{2}$, $K_H = \sqrt{2}/2$, $P(z) = -1$ and $U(z) = \frac{1}{2}$:

$$\mathbf{H}(z) = \frac{\sqrt{2}}{2} \begin{bmatrix} 1 & 1 \\ -1 & 1 \end{bmatrix} = \underbrace{\begin{bmatrix} \sqrt{2} & 0 \\ 0 & \sqrt{2}/2 \end{bmatrix}}_{\mathbf{K}} \underbrace{\begin{bmatrix} 1 & \frac{1}{2} \\ 0 & 1 \end{bmatrix}}_{\mathbf{U}(z)} \underbrace{\begin{bmatrix} 1 & 0 \\ -1 & 1 \end{bmatrix}}_{\mathbf{P}(z)}. \quad (4.212)$$

The lifting structure can also be interpreted by a signal flow diagram, which is shown in Fig. 4.41 for the examples of a Haar filter (4.212) (without considering the normalization factors), the biorthogonal 5/3 filter (4.211) and the biorthogonal 9/7 filter (4.195).

The lifting structure also allows definition of *nonlinear subband filters*. A simple example is usage of rank-order filters like median or weighted median filters (cf. sec. 5.1.1) in prediction and update steps.

4.4.4   Wavelet Transform

Frequency signals established by orthogonal decomposition have a number of advantages for an efficient signal analysis and synthesis; in particular, they achieve a good decorrelation effect, and offer to directly relate signal or error energies in the signal and frequency domains. This can be accomplished by applying orthogonal basis functions as filter impulse responses. The *Wavelet Transform (WT)* is another type of orthogonal transform, applicable to continuous signals $x(t)$, which results likewise in a continuous time-dependent spectrum

---

[1] In the case of the Haar filter, the usage of the lifting approach seems not to give an advantage in terms of complexity for signal decomposition. This method becomes however a key approach in motion-compensated temporal-axis wavelet filtering, cf. sec. 13.4.2.

$$WT_x^{(\psi)}(t,f) = \int_{-\infty}^{\infty} x(\tau) \cdot \psi_f(t,\tau)d\tau \,. \tag{4.213}$$

The WT is based on bandpass filter kernels using basis functions

$$\psi_f(t,\tau) = \frac{1}{\sqrt{\alpha}} \cdot \psi\left(\frac{\tau-t}{\alpha}\right) \quad \text{with} \quad \alpha = \frac{f_0}{f} \tag{4.214}$$

The function $\psi(\cdot)$ is the *mother wavelet*, which is a bandpass filter of center frequency f_0. As it is also a time-dependent function, the WT allows analyzing which components are contained in a signal at frequencies around f and around temporal position t.

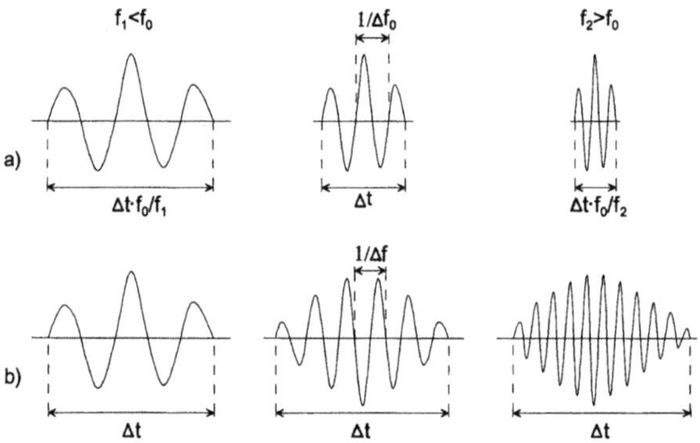

Fig. 4.42. Construction of analysis functions in WT (**a**) and STFT (**b**)

Another method of spectrum analysis which establishes a time and frequency dependency is the *Short Time Fourier Transform* (STFT)[1]

$$F(j\omega,t) = \int_{-\infty}^{\infty} x(\tau) \cdot w(\tau-t) \cdot e^{-j\omega\tau}d\tau \,. \tag{4.215}$$

The STFT analyzes the frequency behavior of a signal using a finite analysis window $w(\tau)$ centered at position t. In contrast to the WT, the time and frequency resolution of the STFT is constant over all locations and frequencies, as the length of the analysis window is fixed. In WT, the mother wavelet is scaled by a factor α in length, reciprocally depending on the analyzed frequency. As a result, the analysis window is scaled down for higher frequencies, while it is expanded for lower frequencies as shown in Fig. 4.42. In principle, the product of analyzed frequency and length of the analysis window is constant. The continuous WT in (4.213) is however not directly useful for digital signal analysis. It is highly overcomplete, being defined for infinite instances both of time and frequency positions. In the

---

[1] The notion of STFT is well established, but shall not limit the scope of this transform; the usage is not restricted to time-dependent signals.

Discrete Wavelet Transform (DWT), the analysis shall only be performed for discrete (sampled) signal positions, and only for a discrete set of frequencies. The common method is definition of a set of basis functions by a *dyadic scheme*, where the center frequencies f_i and the distances of sampling positions t_i used for the respective frequency bands are defined by powers of two relationships:

$$\alpha(i) = 2^i \quad ; \quad f_i = \frac{f_0}{\alpha(i)} \quad ; \quad i \geq 0$$

$$t_i(m) = m \cdot 2^i \cdot 2T$$

(4.216)

The distances $\Delta f_i = f_0 \cdot [\alpha(i) - \alpha(i\text{-}1)]$ between discrete analysis frequencies are no longer constant, and the width of the frequency bands is scaled down by a factor of 2 with each incremental step i. Simultaneously, the distance between analysis positions $\Delta t_i = [t_i(m) - t_i(m\text{-}1)]$ increases by a factor of two. This means that for higher frequency bands (lower i), the resolution by which the frequency expansion of the signal is sampled becomes more precise, while broader resolution bandwidth of the frequency representation is effected. This is illustrated in Fig. 4.43 for both cases of WT and STFT. Using the definitions in (4.216), the dyadic basis functions can be expressed as follows:

$$\psi_{i,m}(\tau) = 2^{-i/2} \cdot \psi\left(2^{-i} \cdot \tau - 2mT\right) \; ; \; WT^{(\psi)}(i,m) = \int_{-\infty}^{\infty} x(\tau) \cdot \psi_{i,m}(\tau)d\tau \qquad (4.217)$$

Remark that these basis functions of the DWT are still time-continuous signals. They can be interpreted as a kind of interpolation functions which are used to reconstruct a signal of given frequency-bandwidth resolution from the set of discrete wavelet coefficients. As in (4.216), T is the smallest sampling distance or highest resolution accuracy in the signal domain that can be represented by the DWT.

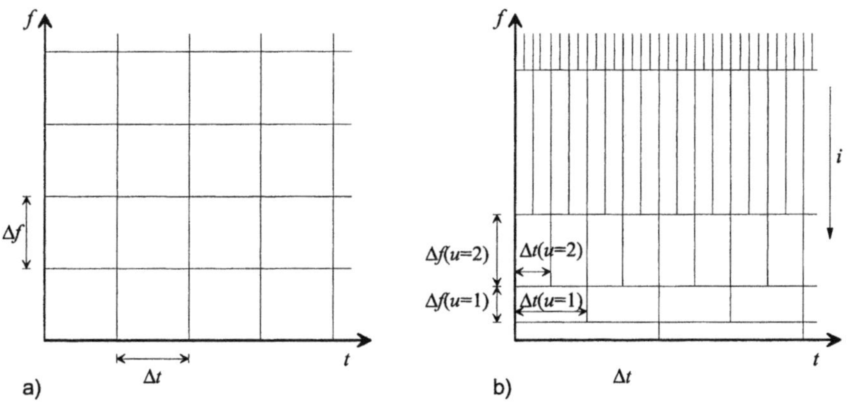

Fig. 4.43. Resolution accuracy in signal and frequency domains **a** for STFT **b** for DWT

Consequently, the DWT allows to reconstruct the signal by different resolution levels (scales). In a more abstract sense, the frequency domain representation up to half-sampling frequency can be constructed from a set of *scale spaces* and a set of *wavelet spaces*, each of which is related to one of the dyadic resolution levels (see Fig. 4.44). When the scale space $\mathcal{V}_{i-1}$ represents a certain resolution of the sampled signal, the next-higher scale space $\mathcal{V}_i$ represents the same signal by half number of samples and half bandwidth. The scale space $\mathcal{V}_{-1}$ shall represent a signal by the maximum resolution relating to a sampling distance T, which then fills the frequency bandwidth $0 \leq \Omega \leq \pi$. The wavelet space $\mathcal{W}_i$ is an orthogonal complement which couples two adjacent scale spaces:

$$\mathcal{V}_{i-1} = \mathcal{V}_i \oplus \mathcal{W}_i \quad \text{and} \quad \mathcal{V}_i \perp \mathcal{W}_i \tag{4.218}$$

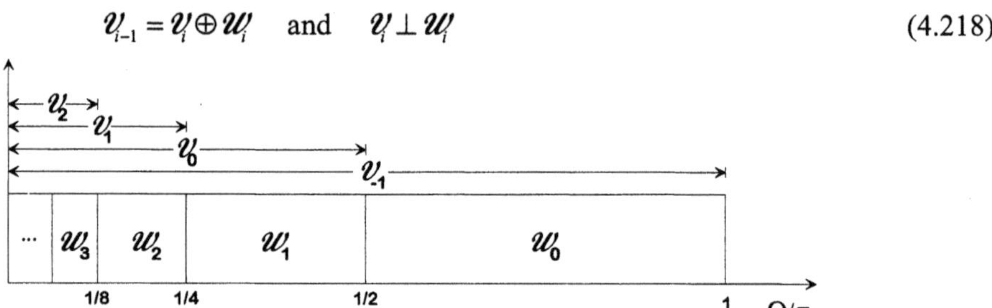

Fig. 4.44. Definition of scale and wavelet spaces

If the conditions in (4.218) hold true, it is obvious that also all subsequent wavelet spaces must be orthogonal. All details which are lost when reducing the resolution from $\mathcal{V}_{i-1}$ to $\mathcal{V}_i$ are mapped into $\mathcal{W}_i$. By iterative application it follows that an arbitrary scale space can be expressed as a direct sum of all higher-indexed wavelet spaces, where the summation is either infinite or terminated by a single low-resolution scale space:

$$\mathcal{V}_i = \mathcal{W}_{i+1} \oplus \mathcal{W}_{i+2} \oplus \mathcal{W}_{i+3} \oplus \mathcal{W}_{i+4} \ldots \tag{4.219}$$

The termination by a scale space is in particular necessary if the analyzed signal is finite, or if the delay occurring by the analysis shall be finite. The analysis of the signal, i.e. the decomposition into components which relate to the respective scale and wavelet spaces, is performed by *scaling functions* $\varphi(\tau)$ and *wavelet functions* $\psi(\tau)$. The scaling function is in principle a lowpass filter which is used to generate a lower-resolution representation, e.g. to construct $\mathcal{V}_i$ out of $\mathcal{V}_{i-1}$)

$\mathcal{V}_{-1}$ is the scale space representing the resolution of the sampled original signal. As $\mathcal{V}_0 \subset \mathcal{V}_{-1}$, any function in $\mathcal{V}_0$ can be expressed as a linear combination of basis functions $\varphi_{-1,k}(\tau)$ related to the scale space $\mathcal{V}_{-1}$. In particular, any scaling function related to $\mathcal{V}_0$ can be described by the *refinement equation* expressing a superposition of scaling functions related to $\mathcal{V}_{-1}$:

$$\varphi_{0,l}(\tau) = \sum_k h_0(k-2l) \cdot \varphi_{-1,k}(2\tau).$$ (4.220)

As for the wavelet space $\mathcal{W}_0 \subset \mathcal{V}_1$ is also valid, an associated wavelet function can be generated similarly by the *wavelet equation*

$$\psi_{0,l}(\tau) = \sum_k h_1(k-2l) \cdot \varphi_{-1,k}(2\tau).$$ (4.221)

(4.220) and (4.221) are the key equations of the DWT. They are used to determine the discrete lowpass and highpass analysis filter coefficients $h_0(k)$ and $h_1(k)$ of a filter bank system, which are then applicable in the discrete computation of the transform. Remark that continuous scaling and wavelet functions were regarded so far which are orthogonal by definition. If these can be constructed iteratively using discrete filter coefficients $h_0(k)$ and $h_1(k)$, the signal decomposition performed by these coefficients will be orthogonal as well, even if the sequences of discrete coefficients (impulse responses) $h_0(k)$ and $h_1(k)$ may not be orthogonal. This leads to constraints which e.g. can be used to design biorthogonal filter pairs. They have to be constructed under the constraint that by iterative application convergence towards an orthogonal set of continuous wavelet and scaling functions is achieved. The iterative development of scaling and wavelet functions can be expressed as

$$\varphi_{i,l}(\tau) = \sum_k h_0(k-2l) \cdot \varphi_{i-1,k}(2\tau); \quad \psi_{i,l}(\tau) = \sum_k h_1(k-2l) \cdot \varphi_{i-1,k}(2\tau).$$ (4.222)

The orthogonality of the decomposition is guaranteed if the following condition holds true:

$$\varphi(\tau) = \lim_{i \to \infty} \varphi_i(\tau) \quad ; \quad \psi(\tau) = \lim_{i \to \infty} \psi_i(\tau) \quad \text{and} \quad \int_{-\infty}^{\infty} \varphi(\tau)\psi(\tau)d\tau = 0.$$ (4.223)

Likewise, the operations (4.222) can be reversed, the next-lower scaling function (representing a signal of next-higher resolution) can be reconstructed from a current level of scaling and wavelet functions as

$$\varphi_{i,k}(\tau) = \sum_l g_0(k-2l) \cdot \varphi_{i+1,l}(\tau) + g_1(k-2l) \cdot \psi_{i+1,l}(\tau).$$ (4.224)

The operations of (4.222) and (4.224) are supporting shift variance to keep the definition as general as possible. Practically, it is more useful to define shift-invariant scaling and wavelet functions, such that

$$\varphi_i(\tau) = \sum_k h_0(k) \cdot \varphi_{i-1,k}(2\tau); \quad \psi_i(\tau) = \sum_k h_1(k) \cdot \varphi_{i-1,k}(2\tau)$$ (4.225)

$$\varphi_i(\tau) = \sum_l g_0(-2l) \cdot \varphi_{i+1,l}(\tau) + g_1(-2l) \cdot \psi_{i+1,l}(\tau).$$ (4.226)

The iterative development of scaling and wavelet functions shall now be illustrated for the shift-invariant case using the simplest orthogonal wavelet basis, which is the Haar basis. The refinement and wavelet equations to perform the mapping from $\mathcal{V}_1$ into $\mathcal{V}_0$ and $\mathcal{W}_0$ are

$$\varphi_{0,0}(\tau) = \underbrace{\frac{\sqrt{2}}{2}}_{h_0(0)}\varphi_{-1,0}(\tau) + \underbrace{\frac{\sqrt{2}}{2}}_{h_0(1)}\varphi_{-1,1}(\tau); \quad \psi_{0,0}(\tau) = \underbrace{\frac{\sqrt{2}}{2}}_{h_1(0)}\varphi_{-1,0}(\tau) - \underbrace{\frac{\sqrt{2}}{2}}_{h_1(1)}\varphi_{-1,1}(\tau). \quad (4.227)$$

The scaling function in $\mathcal{V}_1$ is a hold element of length T. Fig. 4.45 shows the weighted superposition of two copies of the scaling function, resulting in the scaling and wavelet functions of the next level.

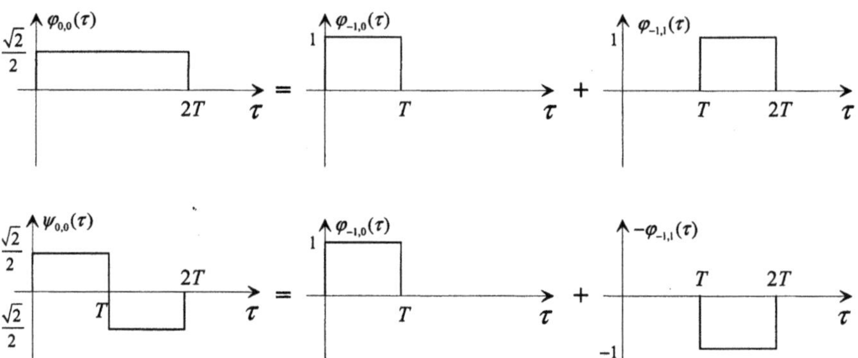

Fig. 4.45. Development of next-level scaling and wavelet functions for the Haar basis

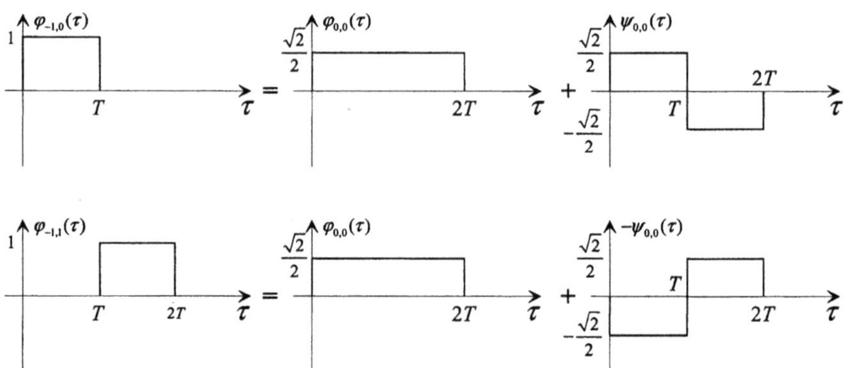

Fig. 4.46. Reconstruction of next-lower level scaling functions for the Haar basis

Now, reconstruction of the different copies of the scaling function in $\mathcal{V}_1$ shall be made from the scaling and wavelet functions in $\mathcal{V}_0$ and $\mathcal{W}_0$. The related equations are

$$\varphi_{-1,0}(\tau) = \underbrace{\frac{\sqrt{2}}{2}}_{g_0(0)}\varphi_{0,0}(\tau) + \underbrace{\frac{\sqrt{2}}{2}}_{g_1(0)}\psi_{0,0}(\tau) \; ; \quad \varphi_{-1,1}(\tau) = \underbrace{\frac{\sqrt{2}}{2}}_{g_0(1)}\varphi_{0,0}(\tau) - \underbrace{\frac{\sqrt{2}}{2}}_{g_1(1)}\psi_{0,0}(\tau) . \qquad (4.228)$$

This process of reconstruction is shown in Fig. 4.46.

The Haar basis converges towards a continuous scaling function which is a hold element. The ideal scaling function would be the *sinc* function (ideal lowpass impulse response, which is not realizable as it is infinite); the corresponding wavelet basis functions would be cosine-modulated versions thereof. An alternative approach is usage of *Gabor functions* as wavelet bases, which are products of Gaussian functions and complex sinusoids. Even though having infinite impulse responses as well, they have a faster decay than the *sinc* and can better be approximated by finite functions. A general form can be written as

$$g(\tau;\omega_C,\sigma) = \frac{1}{\sqrt{2\pi\sigma^2}} \cdot e^{-\frac{\tau^2}{2\sigma^2}} \cdot e^{j\omega_C\tau} \quad \Rightarrow \quad G(\omega;\omega_C,\sigma) = e^{-\frac{\sigma^2(\omega-\omega_C)^2}{2}} , \qquad (4.229)$$

where ω_C relates to the center frequency, and σ is a parameter to determine the bandwidth. Gabor functions are impulse responses of bandpass filters, and establish complete, non-orthogonal basis sets; it can be shown that any function $x(\tau)$ can be reconstructed exactly from an infinite series of discrete coefficients, which are available at regularly-spaced sampling positions in signal and frequency domains [GABOR 1946][1]. For cases of finite signals and finite frequency ranges, the number of coefficients will then become finite as well. When Gabor functions shall be used for an octave-band representation as in the DWT, the neighbored center frequencies must fulfill the condition of constant octave-band spacing $\Delta=\log_2[\omega_C(i+1)/\omega_C(i)]$ over all i (cf. (4.216)). The center frequencies will thus be positioned at

$$\omega_C(i) = \omega_0 \cdot 2^{-i\Delta} , \qquad (4.230)$$

where ω_0 is the center frequency of the highest wavelet band. Further, the bandwidth must be proportional with the center frequency on the octave-band scale, such that

$$\sigma(i) = \frac{1}{\kappa\omega_C(i)} \quad \text{with} \quad \kappa = \frac{2B-1}{\sqrt{2\ln 2}\cdot(2B+1)} . \qquad (4.231)$$

κ is a constant which can be derived by the following condition, where the ratio of the filter transfer amplitudes at the center frequency ω_C and the frequency $\omega=0$ is constant irrespective of the frequency band i [HALEY, MANJUNATH 1999],

---

[1] This is a specific version of discrete STFT based on Gaussian shaped window functions, which is denoted as Gabor transform.

$$\frac{|G(0\,;\omega_c,\sigma)|}{|G(\omega_c\,;\omega_c,\sigma)|} = 2^{-\gamma} \quad \text{with} \quad \gamma = \frac{2^{B+1}}{2^{B-1}}. \tag{4.232}$$

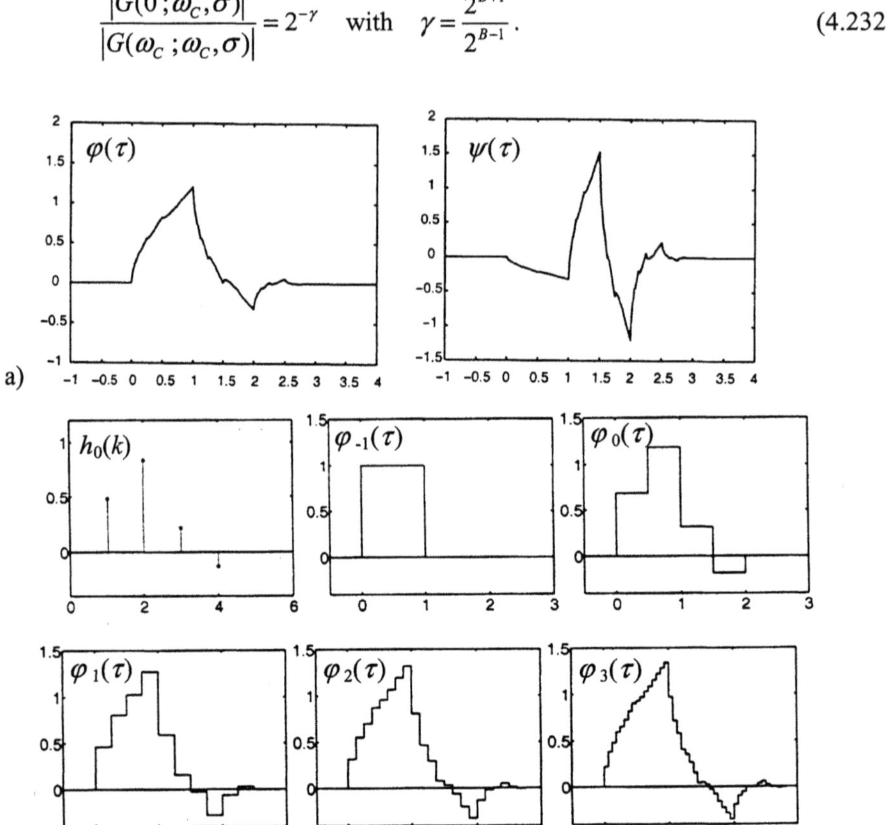

Fig. 4.47. a Daubechies 'D4' scaling and wavelet functions **b** first iteration to develop the continuous scaling function from 4 discrete coefficients

Types of orthogonal continuous wavelet and scaling functions with *finite support* are also known. Some of these functions look rather strange. As an example, the continuous scaling and wavelet filter responses of the *Daubechies function set* 'D4' [DAUBECHIES 1988] are shown in Fig. 4.47a. The first steps of iterative development of the scaling function by four discrete coefficients $h_0(k)$ is shown in Fig. 4.47b.

The discrete coefficients according to (4.220) and (4.221) can now directly be used to decompose a discrete signal $x(m)$ by a DWT. Assume that an approximation of the signal is available by a scaled representation $s_i(m)$. The scaling coefficients of the next-coarser dyadic approximation (representing a signal of half resolution or half number of samples) are then computed as

$$s_{i+1}(m) = \sum_k h_0(k) \cdot s_i(2m-k) \,, \tag{4.233}$$

and the complementary wavelet coefficients are

$$c_{i+1}(m) = \sum_k h_1(k) \cdot s_i(2m-k) \,. \tag{4.234}$$

This decomposition can be computed iteratively, starting by $s_{-1}(m){\equiv}x(m)$. Actually, each level of this decomposition is identical to the decomposition of a signal into low- and high-frequency subbands as introduced in sec. 4.4.2. However, in contrast to the cascaded system from Fig. 4.38, only the low frequency output (the next lower scale) is subject to further decomposition. It is straightforward to compute the reconstruction of the signal by inverting the sequence of recursion (*inverse DWT*, IDWT):

$$s_i(m) = \sum_k g_0(m-2k) \cdot s_{i+1}(k) + \sum_k g_1(m-2k) \cdot c_{i+1}(k) \,. \tag{4.235}$$

As shown earlier, perfect reconstruction is possible if the synthesis coefficients $g_0(k)$ and $g_1(k)$ are related to $h_0(k)$ and $h_1(k)$ by biorthogonality (4.170). By the orthogonality properties of the scale space and the complementary wavelet spaces, it can be concluded that the sequences of scaling and wavelet coefficients will approach mutual orthogonality, even though the filter basis may only be biorthogonal.

Fig. 4.48 shows the block diagram of a DWT analysis/synthesis filter bank, and a schematic layout of the resulting frequency decomposition, which can be described as an *octave-band structure*. If the Haar basis (4.180) is used, the resulting decomposition is exactly the same for the Haar transform (4.119)-(4.122). If longer filter impulse responses are realized, which can in principle be done by selecting a different function $\psi(t)$, a better frequency separation and also better alias suppression in the scaled signal versions can be achieved.

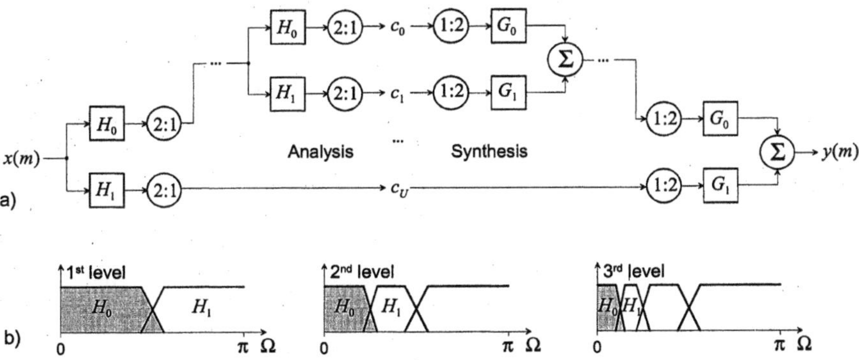

Fig. 4.48. **a** Octave-band filter bank system for DWT and IDWT **b** Octave-band frequency layout (3 levels of analysis)

For many classes of signals, in particular for natural image signals, the higher accuracy of frequency resolution for the lower-frequency bands provides a good fit with signal models. According to the AR(1) model, significantly more low-frequency than high-frequency components can be expected, which are relevant when larger areas of the signal are smoothly changing. For the high-frequency components, accurate frequency analysis is less important than an *accurate localization of detail*, in particular if a signal potentially exhibits discontinuities, as it is the case in edge areas.

4.4.5   Two- and Multi-dimensional Filter Banks

The simplest realization of a two- or multi-dimensional filter bank is the *separable* method, where the analysis and synthesis filters are a product of horizontal and vertical filters. For the 2D case, the basis functions for the frequency band of index u in horizontal and v in vertical direction can then be described as

$$h_{u,v}(m,n) = h_u(m) \cdot h_v(n)$$

$$g_{u,v}(m,n) = g_u(m) \cdot g_v(n).$$

(4.236)

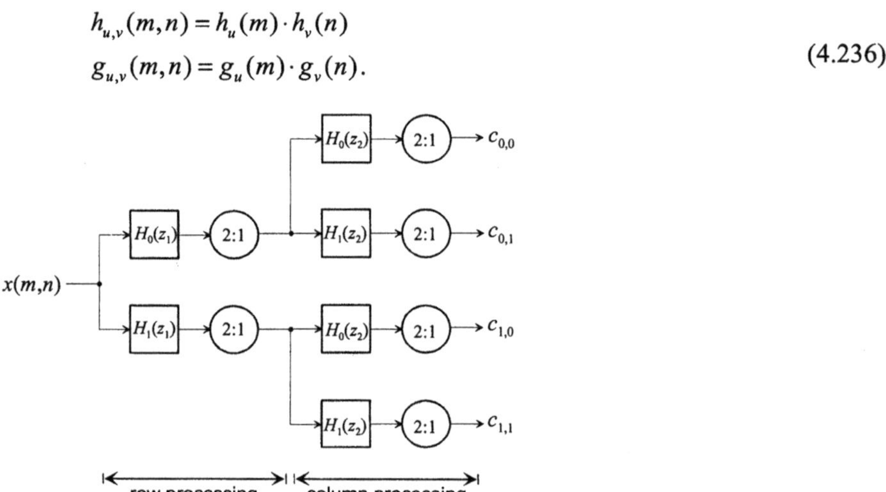

Fig 4.49. Separable 2D system for decomposition into four frequency bands

For a total of $U \cdot V$ bands, the total sub-sampling factor is $U^* = U \cdot V$. 2D systems with 2-band decomposition structures shall be applied to each of the directions ($U=2$, $V=2$, $U^*=4$), which can best be realized sequentially, such that filtering and sub-sampling is first performed over one dimension. Only a reduced number of samples then needs to be fed into the second directional decomposition stage. A block diagram is shown in Fig. 4.49. Fig. 4.50a depicts the related layout of subbands in the 2D frequency domain. This basic 4-band decomposition structure of Fig. 4.49 can be applied iteratively. For a case where all 4 subbands are equally decomposed in the next level, Fig. 4.50b shows a layout with sub-division into 16 bands. Fig. 4.50c is an example where an octave-band decomposition is applied fully separable over both dimensions, which matches with the case of the Haar transform basis in Fig. 4.21c. Fig 4.50d shows the layout which is commonly denoted as *2D DWT*, where only the lowpass output c_{00} of the 4-band system is subject to further 4-band decomposition etc. In Fig. 4.50c/d, '*S*' denotes the scaling band of lowest resolution.

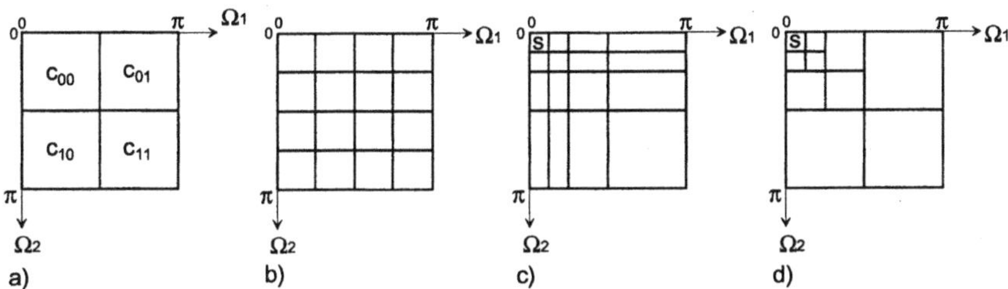

Fig. 4.50. Layout of 2D frequency bands. **a** 4 band elementary decomposition **b** 16 bands of equal bandwidth   **c** Separable octave-band, 16 bands   **d** 2D DWT, 10 bands

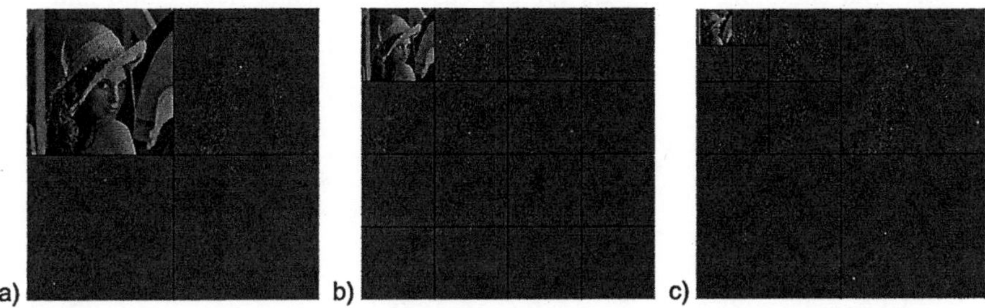

Fig. 4.51. Decomposition of an image into subbands (amplified by factor 4, except c_{00}) **a** relating to Fig. 4.49a **b** relating to Fig. 4.49b **c** relating to Fig. 4.49d

Fig. 4.51 shows examples of subband and wavelet decompositions applied to an image signal.

It is also possible to realize non-separable 2D filter banks. Fig. 4.52 shows an example of a 2D decimation by a factor of 2, where a subband system decomposes a rectangular sampled signal into two components of quincunx sampling. To describe such systems, principles introduced in the context of multi-dimensional sampling can be used. In general, if $x(m,n)$ is the original signal and $c(m',n')$ the sub-sampled signal, the relationship between the indices can be expressed by the sampling matrix $\mathbf{D}$ such that

$$\begin{bmatrix} m \\ n \end{bmatrix} = \begin{bmatrix} d_{00} & d_{01} \\ d_{10} & d_{11} \end{bmatrix} \cdot \begin{bmatrix} m' \\ n' \end{bmatrix} \quad ; \quad \mathbf{n} = \mathbf{D} \cdot \mathbf{n'} . \tag{4.237}$$

The factor of sub-sampling, and hence the number of spectral copies (original plus alias spectra) is

$$U^* = \det[\mathbf{D}] = d_{00} \cdot d_{11} - d_{10} \cdot d_{01} . \tag{4.238}$$

The related frequency sampling matrix $\mathbf{F}=[\mathbf{D}^{-1}]^{\mathrm{T}}$ points to the positions of periodic spectral copies, by which alias may occur. In analogy with (4.172), the z transform of the decimated signal is

$$C(z_1,z_2) = \frac{1}{U^*} \sum_{u=0}^{(d_{00}-1)} \sum_{v=0}^{(d_{11}-1)} X\left(W^{-(f_{00}u+f_{01}v)} \cdot z_1^{f_{00}} z_2^{f_{01}}, W^{-(f_{10}u+f_{11}v)} \cdot z_1^{f_{10}} \cdot z_2^{f_{11}} \right) \tag{4.239}$$

$$\text{with} \quad W = e^{j2\pi}.$$

The reverse operation is an interpolation by factor U^*, which results by modification of (4.174) using the parameters in $\mathbf{D}$

$$Y(z_1,z_2) = C(z_1^{d_{00}} \cdot z_2^{d_{10}}, z_1^{d_{01}} \cdot z_2^{d_{11}}) . \tag{4.240}$$

Example. The sampling matrix $\mathbf{D}_q$ for the case of quincunx decimation and the related frequency sampling matrix $\mathbf{F}_q$ in analogy with (2.54) and Fig. 4.52 are expressed as

$$\mathbf{D}_q = \begin{bmatrix} 2 & 1 \\ 0 & 1 \end{bmatrix} \quad ; \quad \mathbf{F}_q = [\mathbf{D}_q^{-1}]^{\mathrm{T}} = \begin{bmatrix} \tfrac{1}{2} & 0 \\ -\tfrac{1}{2} & 1 \end{bmatrix}. \tag{4.241}$$

The z transform of the decimated signal is

$$C(z_1,z_2) = \frac{1}{2} \sum_{u=0}^{1} X\left(W^{-\frac{1}{2}u} \cdot z_1^{\frac{1}{2}}, W^{\frac{1}{2}u} \cdot z_1^{-\frac{1}{2}} \cdot z_2 \right). \tag{4.242}$$

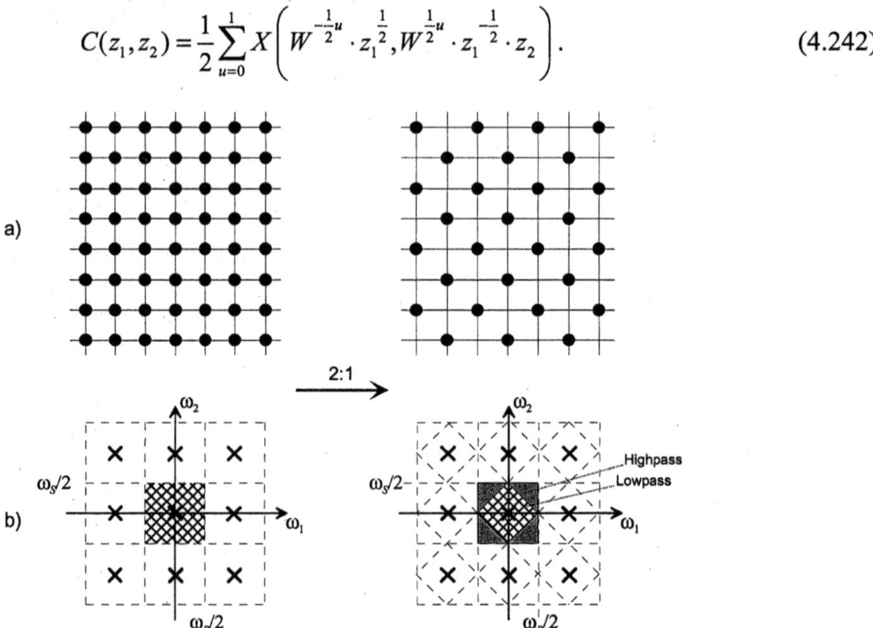

Fig. 4.52. Non-separable 2D system with 2:1 quincunx decimation
a Sub-sampling schema in the spatial domain **b** Layout of frequency bands

To realize a non-separable decimation, it is typically necessary to use non-separable filters. The quincunx decimation can be performed using the biorthogonal pair of 2D filter matrices [KOVACEVIC, VETTERLI 1992]

$$\mathbf{H}_0 = \frac{1}{32}\begin{bmatrix} 0 & 0 & -1 & 0 & 0 \\ 0 & -2 & 4 & -2 & 0 \\ -1 & 4 & 28 & 4 & -1 \\ 0 & -2 & 4 & -2 & 0 \\ 0 & 0 & -1 & 0 & 0 \end{bmatrix} \quad ; \quad \mathbf{H}_1 = \frac{1}{4}\begin{bmatrix} 0 & -1 & 0 \\ -1 & 4 & -1 \\ 0 & -1 & 0 \end{bmatrix} \tag{4.243}$$

As in (4.194), the kernels of lowpass and highpass filters are of different size. Applying the relationships $G_0(\mathbf{z})=H_1(-\mathbf{z})$ and $G_1(\mathbf{z})=-H_0(-\mathbf{z})$ from (4.185), the synthesis filters are determined by multiplication (modulation) with alternating signs. For symmetric 2D filters, this is realized such that impulse response values with an odd sum of indices are multiplied by -1, i.e.

$$g_0(m,n) = (-1)^{m+n} h_1(m,n)$$
$$g_1(m,n) = (-1)^{m+n+1} h_0(m,n)$$

$$\tag{4.244}$$

The resulting synthesis filter matrices are

$$\mathbf{G}_0 = \frac{1}{4}\begin{bmatrix} 0 & 1 & 0 \\ 1 & 4 & 1 \\ 0 & 1 & 0 \end{bmatrix} \quad ; \quad \mathbf{G}_1 = \frac{1}{32}\begin{bmatrix} 0 & 0 & 1 & 0 & 0 \\ 0 & 2 & 4 & 2 & 0 \\ 1 & 4 & -28 & 4 & 1 \\ 0 & 2 & 4 & 2 & 0 \\ 0 & 0 & 1 & 0 & 0 \end{bmatrix}. \tag{4.245}$$

Polar forms of 2D Wavelet functions. For rotation-invariant analysis of 2D signals, separable filters are not well suitable, as the cut-off frequencies of 'diagonal' frequency bands are different compared to the horizontal/vertical bands. Further, the common 2D DWT as introduced above does not allow to differentiate between the 45^0 and 135^0 diagonal orientations. To overcome this problem, analysis methods based on continuous Gabor functions (4.229)-(4.232) have been proposed, which allow to define 2D wavelet bases by polar coordinates. The center frequencies of the bands are located in an octave-band schema at different radial (scale) orientations, where in addition a number of regularly-spaced angular (directional) orientations is introduced at each scale (for an example see Fig. 7.13). From (4.229), the frequency ω maps the radial orientation (distance from the origin of the 2D frequency plane), and the other center frequency θ_C relates to the angle. The parameters σ_ρ and σ_θ define to the bandwidths in radial and angular orientations, respectively [HALEY, MANJUNATH 1999]:

$$G_P(\omega,\theta\,;\omega_C,\theta_C,\sigma_\rho,\sigma_\theta) = G(\omega;\omega_C,\sigma_\rho)e^{-\frac{\sigma_\theta^2(\theta-\theta_C)^2}{2}}.\tag{4.246}$$

By mapping the polar coordinates into Cartesian coordinates of the frequency plane, the 2D continuous impulse responses of the direction/scale oriented Gabor filters can be expressed as (omitting the parametrization over $(\omega_C,\theta_C,\sigma_\rho,\sigma_\theta)$ for simplicity):

$$g_P(r,s) = \int\limits_{-\infty}^{\infty}\int\limits_{-\infty}^{\infty} G_P(\sqrt{\omega_1^2+\omega_2^2},\arctan\frac{\omega_2}{\omega_1}) \cdot e^{j(\omega_1 r+\omega_2 s)}\, d\omega_1 d\omega_2.\tag{4.247}$$

4.4.6 Pyramid Decomposition

The DWT is a *multi-resolution scheme* for signal representation. This means that by using more wavelet bands, the resolution of the signal is increased. Another type of multi-resolution methods are the so-called *pyramid representation schemes*. A number of image planes is arranged by levels of a pyramid, where the area of the images (number of pixels) decreases towards the tip of the pyramid, and full resolution relates to the base (see Fig. 4.53). The following methods are applied, which are typically also based on dyadic levels of scale spaces[1]:

- *Gaussian Pyramid*: All resolution levels can be used independently, i.e. the next coarser level must not be available if a finer level shall be used. The generation of the Gaussian pyramid representation is performed by elementary building blocks consisting of lowpass filtering followed by decimation. This is performed in a sequential cascade through all levels of the pyramid, starting from the base and terminating at the top (see Fig. 4.54). By cascading T elementary building blocks, a total of $T+1$ resolution levels (including the original resolution) is generated.
- *Laplacian Pyramid*: Each resolution level (except for the smallest scale image) is represented by a difference signal. The principle as applied for generation of the difference signals is shown in Fig. 4.55a. For reconstruction, the difference is added to a prediction generated by spatial interpolation of the next-coarser signal (Fig. 4.55b). The generation of the Laplacian pyramid representation starts also from the base (full resolution), while the reconstruction starts from the tip. If T elementary building blocks are arranged in a cascade structure, a total of $T+1$ resolution levels is represented by T difference signals and one strongly-scaled image. The reconstructed signals are equivalent to the output of the Gaussian pyramid.

---

[1] This means that for the next-finer resolution level (next-lower layer of the pyramid) the number of pixels is doubled in each spatial direction of 2D signals, which gives in total a quadruple number of pixels.

The first method is denoted as *Gaussian* pyramid, because filters with a Gaussian shape or a discrete approximation thereof are widely used as impulse response of the lowpass filter for the purpose of resolution reduction prior to decimation. The convolution of two Gaussian functions results in a Gaussian of extended length. Hence, the effect of the cascaded system is similar to the application of just one Gaussian filter with a very long impulse response (strong lowpass) to the original signal. The implementation complexity in the cascaded pyramid system is however much lower due to the intermediate sub-sampling operations.

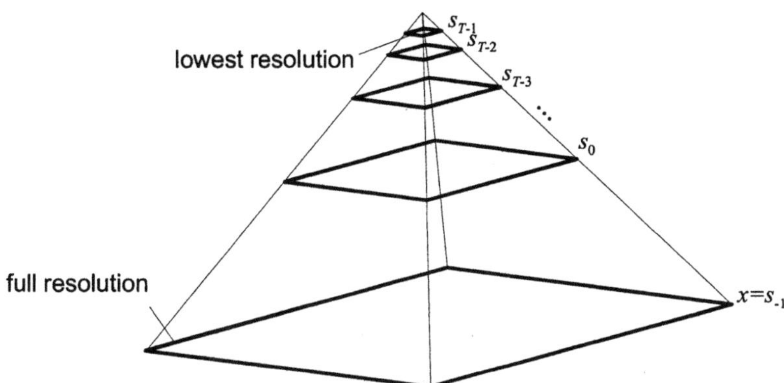

Fig. 4.53. Resolution levels of an image signal in pyramid representation

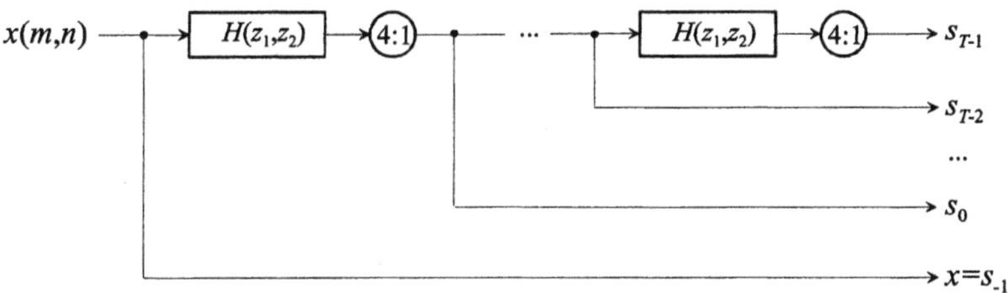

Fig. 4.54. Generation of the Gaussian pyramid representation

An example for a very simple approximation of a non-separable 2D Gaussian by a short filter kernel is given by the matrix[1]

$$\mathbf{H}_G = \frac{1}{8}\begin{bmatrix} 0 & 1 & 0 \\ 1 & 4 & 1 \\ 0 & 1 & 0 \end{bmatrix}. \tag{4.248}$$

---

[1] A typical primitive 1D approximation of a Gaussian filter function is the binomial filter with the coefficient vector $\mathbf{h}=[\ ¼\ ½\ ¼\ ]^T$. (4.248) represents a superposition combination of one horizontal and one vertical binomial filter.

The difference between the original signal and the signal filtered by (4.248) could in one step be generated by the filter

$$\mathbf{H}_L = \begin{bmatrix} 0 & 0 & 0 \\ 0 & 1 & 0 \\ 0 & 0 & 0 \end{bmatrix} - \frac{1}{8}\begin{bmatrix} 0 & 1 & 0 \\ 1 & 4 & 1 \\ 0 & 1 & 0 \end{bmatrix} = \frac{1}{8}\begin{bmatrix} 0 & -1 & 0 \\ -1 & 4 & -1 \\ 0 & -1 & 0 \end{bmatrix}. \tag{4.249}$$

This filter kernel provides a raw approximation of the second derivative of the signal, and is denoted as *Laplacian filter operator*, which is also employed for the purpose of edge detection (see 7.3.2). Hence, the differential pyramid is denoted as *Laplacian* pyramid. In principle, this pyramid represents the second derivatives of the signal within different scale spaces.

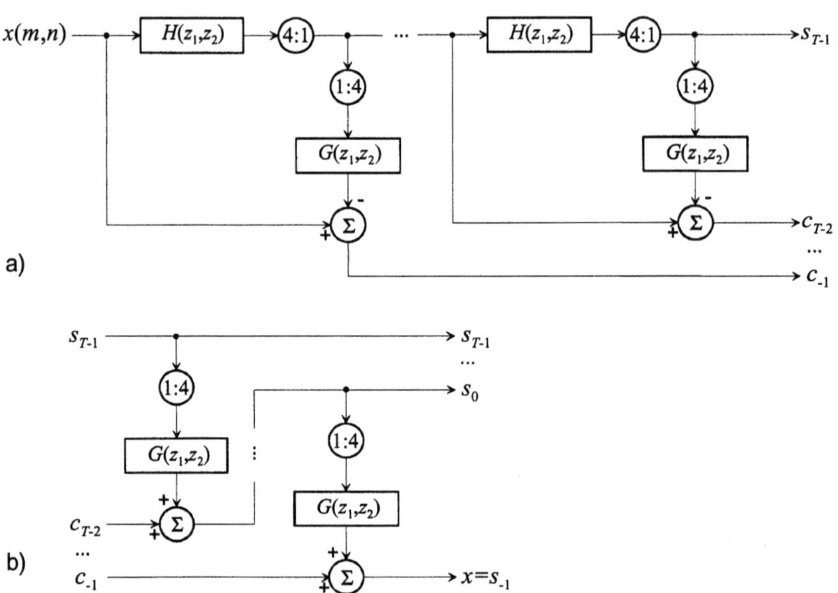

Fig. 4.55. Laplacian pyramid representation: **a** Analysis **b** Synthesis

The resolution levels of the Gaussian pyramid can roughly be seen to be identical to the scale spaces of the wavelet transform, however the differential channels of the Laplacian pyramid are not orthogonal complements as the wavelet channels indeed are. An over-completeness is inherent to the pyramid schemes: If T pyramid levels are used, the number of samples to be represented grows by a factor of

$$\sum_{t=1}^{T-1}\left(\frac{1}{4}\right)^t \tag{4.250}$$

as compared to the number of samples in the original image signal. In subband or DWT transforms with critical sampling, the number of signal samples is always identical to the entire number of scaling and wavelet coefficients. Comparison of (4.249) and (4.243) even shows that for a biorthogonal wavelet transform an identical choice of highpass (difference) basis functions is possible as for the case of the Laplacian pyramid, without the penalty of over-completeness.

4.5   Problems

Problem 4.1

The eigenvalues of the matrix $\mathbf{R} = \begin{bmatrix} 1 & \rho \\ \rho & 1 \end{bmatrix}$ are $\lambda_0=1+\rho$ and $\lambda_1=1-\rho$.

a)   Following (B.16), determine Φ_0 und Φ_1 of $\mathbf{R}$ such that they establish an orthonormal base $\Psi=[\ \Phi_0\ \Phi_1\ ]$ according to (B.20).
b)   Sketch the eigenvectors within a coordinate system of axes x_1, x_2.
c)   Determine the inverse Ψ^{-1}.
d)   Compute the determinant of $\mathbf{R}$, and compare the result against the product of the eigenvalues.

Problem 4.2
For the AR(1) model from (4.55) and (4.56), prove the validity of the autocorrelation and variance properties (4.57) and of the spectral properties (4.58).

Problem 4.3
For statistical modeling of a 1D signal, an AR(1) model of variance σ_x^2 and correlation coefficient $\rho=0.95$ is used. By using the autocorrelation function of the model, a linear predictor shall be optimized. Determine the coefficients $a(1)$ and $a(2)$ for a predictor filter of order $P=2$ by solving the Wiener-Hopf equation (4.73).

Problem 4.4
For linear 2D prediction of a separable AR(1) process with $\rho_h=\rho_v=0.95$, a non-separable predictor filter is used which implements the prediction equation $\hat{x}(m,n) = 0.5\cdot x(m-1,n)+0.5x(m,n-1)$. Determine the variance of the prediction error signal, and compare against the variance of the innovation signal $z(n)$ from (4.61).

Problem 4.5
A video sequence consists of image frames which are unchanged except for global translational motion. The picture information is an output from a 2D separable AR(1) model generator. Parameters are $\rho_h=\rho_v=0.95$. From one frame to the next, translational shift by $k=7$ horizontally and $l=3$ vertically occurs. Different methods of linear prediction shall be compared by the criterion of prediction error variance:
a)   Spatial prediction, separable predictor according to (4.93);
b)   Temporal prediction $\hat{x}(m,n,o) = x(m,n,o-1)$;
c)   Motion-compensated temporal prediction $\hat{x}(m,n,o) = x(m-k,n-l,o-1)$.

Problem 4.6
a)   Construct the transform matrices of 1D Haar and Walsh transforms for $M=4$.
b)   Transform the following image matrix by the related separable 2D transforms using (4.113):

$$X = \begin{bmatrix} 18 & 4 & 2 & 4 \\ 18 & 4 & 2 & 4 \\ 2 & 4 & 2 & 4 \\ 2 & 4 & 2 & 4 \end{bmatrix}$$

c)   Interpret the results. Discuss in particular which transform better compacts the given image.

Problem 4.7
a)   Determine the transform matrix of a 1D DCT for $M=3$, and show that the transform is orthonormal.
b)   Set up the autocorrelation matrix (3.41) of size 3x3 for an AR(1) model. The matrix shall then be transformed by the 1D DCT of a), using (4.157).
c)   For a model of variance σ_x^2, two different cases $\rho=0.9$ and $\rho=0.5$ shall be considered to fill the matrix $\mathbf{R}_{cc}$. Give an interpretation of the differences you observe in the matrix entries.
d)   For both cases from c), compute the trace (B.17) of the autocorrelation matrices and their transformed counterparts. Give an interpretation of the result.

Problem 4.8
a) Determine the Fourier transfer functions $\mathcal{F}\{t_u\}$ for the basis vectors of a transform

$$t_0 = \left[\frac{\sqrt{2}}{2} \quad \frac{\sqrt{2}}{2} \right]^T \qquad t_1 = \left[\frac{\sqrt{2}}{2} \quad -\frac{\sqrt{2}}{2} \right]^T$$

b) Prove the orthogonality of this basis system.
c) Show that the amplitude transfer functions of both basis vectors have a mirror-symmetric relationship around the frequency $\Omega=\pi/2$.
d) Show that $|\mathcal{F}\{t_0\}|^2 + |\mathcal{F}\{t_1\}|^2 = \text{const}$.

Problem 4.9
a)   Determine the basis vectors of the block-overlapping transform according to (4.164)-(4.166) with settings $U=2$, $M=4$. [Hint : To simplify expressions of the trigonometric functions, use constants $\cos(3\pi/8)=\sin(\pi/8)=A$ and $\cos(\pi/8)=\sin(3\pi/8)=B$; reflect for which other angles values $\pm A$ or $\pm B$ would result.]
b)   Show the orthogonality of the basis system.
c)   Determine the Fourier transfer functions $\mathcal{F}\{t_u\}$. Do the basis functions have a linear-phase property ?
d)   Would a realization of this transform by a fast algorithm be possible ?

Problem 4.10
a)   Show the validity of the orthogonality property for linear-phase QMF systems for the following cases of filters:

    i) $H_0(z) = A \cdot z^2 + B \cdot z + C + C \cdot z^{-1} + B \cdot z^{-2} + A \cdot z^{-3}$

    ii) $H_0(z) = A \cdot z^2 + B \cdot z + C + B \cdot z^{-1} + A \cdot z^{-2}$

b)   Determine the z transform representations of polyphase filters $H_{0,A}$, $H_{0,B}$, $H_{1,A}$ and $H_{1,B}$ according to Fig. 4.39 for both filter configurations from a). Which number of multiplications per sample would be necessary at minimum ?

Problem 4.11

This problem relates to the generation of sub-sampled signals in the Gaussian pyramid representation (Fig. 4.54). By self-convolution $h(n)*h(n)$ of a Gaussian-shaped impulse response kernel of width σ, a kernel of width 2σ is generated. This will have half cut-off frequency as compared to the previous. Filtering a signal by the kernel of width 2σ will give the same result as first filtering by the kernel of width σ, sub-sample the result and again filter by the same kernel (possible alias effects neglected). The solution of this problem will in particular show that these properties hold likewise for the 2D case. In particular, a shorter-width kernel inherits its geometric shape, as expressed by a type of neighborhood system, to the broader-width kernels which are developed by the iterated convolution.

a)   Compute the 2D self-convolution of the Gaussian filter kernel approximation (4.248).

b)   Determine the orders c of a homogeneous neighborhood system based on the absolute norm $\mathcal{N}c^{(1)}(m,n)$ according to (4.1): *i*) for the filter (4.248) *ii*) for the result from a). Which order will result by yet another self-convolution of the result from a) ?

c)   The 2D 'variance' of a Gaussian filter kernel shall be defined as

$$\sigma^2 = \sum_i \sum_j \, (i^2 + j^2) \cdot h(i,j).$$
$$\scriptstyle (i,j)\in\mathcal{N}_c(m,n)$$

Determine σ^2 for the filter (4.248) and the result from a)

Problem 4.12

A one-dimensional, zero-mean signal shall be modeled by an AR(1) process. To describe the parameters of the process, the value of spectral power density $S_{xx}(\Omega=\pi) = \sigma_x^2/9$ is given.

a)   Determine the correlation parameter ρ, and the variance σ_z^2 of the spectrally-white innovation signal in dependency of σ_x^2.

The signal shall be decomposed into two polyphase components, where the sequences of even- and odd-indexed samples shall be processed independently by predictor filters of first order, $H(z)=az^{-1}$, as shown in Fig. 4.56.

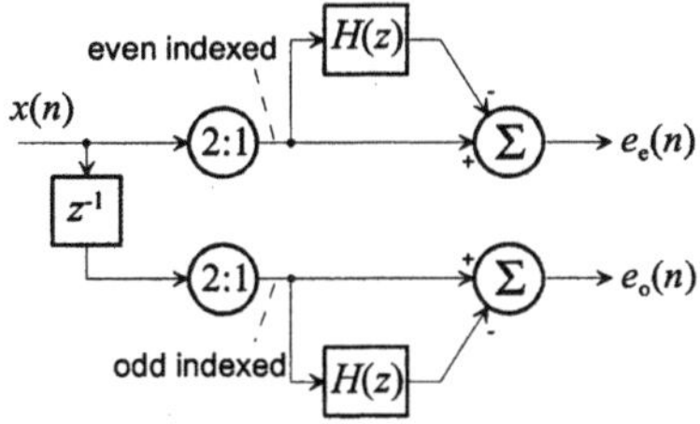

Fig. 4.56. Prediction within polyphase components

b)   Determine the optimum predictor coefficient a.

c)   Determine the variance of the prediction error signal $e_e(n)$ of the even-indexed samples when the optimum a is used. From this, compute the coding gain $G=\sigma_x^2/\sigma_{e_e}^2$. By which factor will this gain be smaller, as compared to the optimum case of prediction for the AR(1) model (without polyphase decomposition) ?

d)   Compute the covariance between the signals $e_e(n)$ und $e_o(n)$ when the optimum a is used.

Problem 4.13

An AR(1) process $x_{AR}(n)$ is characterized by the correlation parameter $\rho=0,75$ and the variance of the Gaussian innovation signal, $\sigma_z^2=7$.

a)   Determine the variance of the AR process.

For linear prediction of $x_{AR}(n)$, a falsely adapted prediction error filter of transfer function $A(z) = 1 - z^{-1}$ is used (see Fig. 4.57).

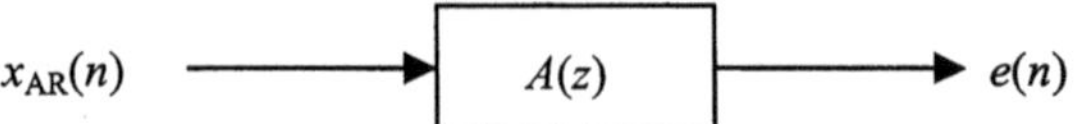

Fig. 4.57. Prediction of an AR(1) process

b)   Compute the variance of the prediction error signal $e(n)$ and the coding gain.
c)   Which would be the coding gain in case of optimum prediction? By which factor is the coding gain worse in the case of the falsely-adapted prediction from b)?
d)   Determine the power density spectrum $S_{ee}(\Omega)$. Is $e(n)$ a white noise signal?
e)   Determine a system (z transfer function and block diagram), which generates the optimum prediction error signal of lowest possible variance from $e(n)$.

Problem 4.14

A linear block transform of block size $M=U=2$ is defined by the following transform matrix:

$$\mathbf{T}(2) = \begin{bmatrix} 1/2 & 1/2 \\ 1 & -1 \end{bmatrix}$$

a)   Determine the inverse transform matrix $\mathbf{T}^{-1}$.
b)   Are the basis vectors of the transform orthogonal? Describe a transform matrix of an orthonormal transform $\mathbf{S}=[\mathbf{s}_0^T \, \mathbf{s}_1^T]^T$, such that $\mathbf{s}_0{\sim}\mathbf{t}_0$ and $\mathbf{s}_1{\sim}\mathbf{t}_1$.
c)   An AR(1) process is characterized by σ_x^2 and ρ. It shall be transformed block-wise by $\mathbf{T}$. Determine the variances of the transform coefficients $E\{c_0^2\}$ and $E\{c_1^2\}$.
d)   A quantization error vector $\mathbf{q}=[\, q_0 \; q_1 \,]^T$ is superimposed to the transform coefficients. Compute the resulting error $\mathbf{y}$-$\mathbf{x}$ in the reconstructed signal $\mathbf{y}$. How must the quantization step sizes of uniform quantizers be chosen, such that the contributions of q_0 and q_1 to the energy of the reconstruction error will become equal?
e)   The transform $\mathbf{T}$ is used as basis for a wavelet transform using 2 decomposition steps, i.e. depth 2 of the wavelet tree. This can then be interpreted as a block transform with $M=4$. Sketch a signal flow diagram related to the analysis of a single block. Also, express the related transform matrix $\mathbf{T}(4)$. Is this transform orthogonal or orthonormal?

5 Pre- and Postprocessing

Preprocessing methods are applied in multimedia signal analysis, coding and identification. Nonlinear filtering methods are often applied in preprocessing of image signals, as they can preserve characteristic signal properties like edges much better than linear filtering. In coding of multimedia signal, post-processing methods play an additional role for improving the perceptual quality of signals to be presented. Linear filter kernels are frequently used in signal enhancement in combination with adaptation mechanisms, which take into regard specific local properties of the signal or employ mechanisms which are specifically tuned to the expected distortions. Amplitude transforms establish another group of signal modifications, which are mainly based on manipulations of sample statistics. Finally, different interpolation methods are introduced which must be used if signal samples are needed at positions not available from the given sampling grid.

5.1 Nonlinear Filters

In filtering operations, the value of the pixel at position (m,n) is set in relationship with values from a neighborhood as e.g. described in (4.1) and (4.2). The geometry of a neighborhood system $\mathcal{N}_c(m,n)$ determines the shape of a *filter mask*, by which a combination of a pixel with its neighbors is performed. The output is computed by application of the combining function to any position of the input image. It shall be assumed here that the output has the same number of samples as the input. If the combining function $f[\cdot]$ is nonlinear, the output is the result of nonlinear filtering[1], as shown in Fig. 5.1. Unlike impulse responses of linear sys-

---

[1] Nonlinear filters are of particular importance in image processing applications, and are explained here mainly for the example of 2D images. One- or multi-dimensional nonlinear filtering can be developed by definition of the neighborhood systems over only one or more than two coordinate axes.

tems, nonlinear function behavior can not be mapped into a frequency transfer function.

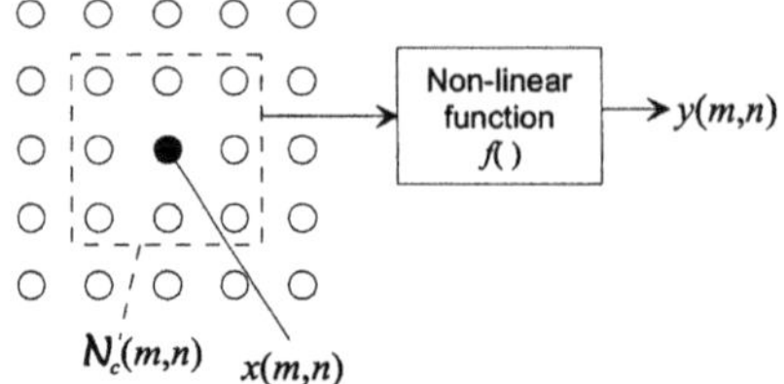

Fig. 5.1. Typical method of linear 2D filtering using a finite neighborhood system

Image signals show some properties which can hardly be modeled or analyzed by linear systems, in particular at amplitude discontinuities (edges). Consequently, methods of nonlinear filtering are widely used in image processing applications. Nonlinear filters can be grouped into different categories as shown in Fig. 5.2.

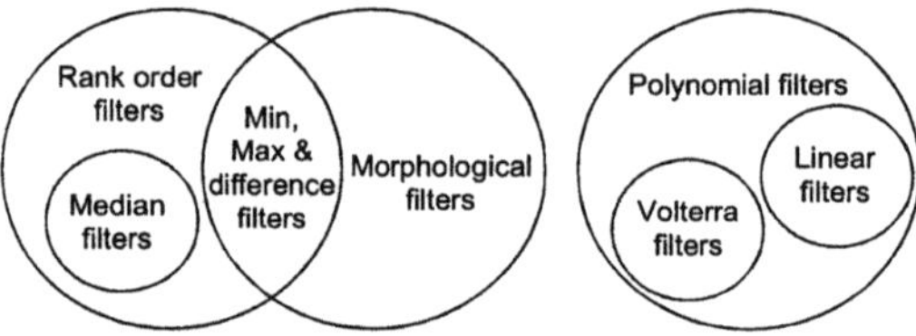

Fig. 5.2. Categories of nonlinear filters

While polynomial filters perform the combination of pixels in the neighborhood by a weighted sum of elements – which can then also include quadratic and higher-order terms –, rank-order filters and morphological filters implement combination functions which are based on value comparisons and logical operations within the neighborhood.

5.1.1   Median Filters and Rank Order Filters

The median is a value from a set, for which at least half the number of values is less or equal, and at least half the number of values is larger or equal. In median filtering of image signals, the values positioned under the geometry of the finite filter mask are forming the set, which must consist of an odd number of values (e.g. 3x3 or 5x5 pixels), as otherwise it would not be possible to determine a unique median value[1]. The median value usually is filled at the output position relating to the center of the mask. Median filtering effects an elimination of iso-

---

[1] Methods which allow usage of even number of input values are weighted median filtering (see below) and averaging of the two values at the center of the ordered set.

lated outlier values. It is also applied for purposes of nonlinear prediction and interpolation, where it simply replaces the linear lowpass filters typically used.

Example: Median filter over neighborhood $\mathcal{N}_2^{(2)}(m,n)$. Regard the image matrix **X** given in (5.1), assuming a constant-value extension (4.9) when the filter mask accesses a pixel from outside **X**. If the 3x3 filter mask is centered at the second pixel in the second row, the set of values **m**=[10,10,20,20,10,20,10,10,10] is the filter input. Re-ordering by amplitudes gives **m'**=[10,10,10,10,**10**,10,20,20,20], resulting in the value MED[**m**]=10. The pixel remains unchanged, as the majority of values from the neighborhood has the same amplitude. For the third pixel in the third row, the set is **m**=[10,20,20,10,10,20,10,20,20]. Re-ordering of the set gives **m'**=[10,10,10,10,**20**,20,20,20,20], such that MED[**m**]=20. This output value is not identical to the input of the current position, but it is an original value from the neighborhood. Application of the same operation to any position gives the output matrix **Y**, from which it is obvious that the median filter eliminates single, isolated amplitude values and straightens edge boundaries between areas of constant amplitude:

$$\mathbf{X} = \begin{bmatrix} 10 & 10 & 20 & 20 \\ 20 & \underline{10} & 20 & 20 \\ 10 & 10 & \underline{10} & 20 \\ 10 & 10 & 20 & 20 \end{bmatrix} \quad ; \quad \mathbf{Y} = \mathrm{MED}[\mathbf{X}] = \begin{bmatrix} 10 & 10 & 20 & 20 \\ 10 & 10 & 20 & 20 \\ 10 & 10 & 20 & 20 \\ 10 & 10 & 20 & 20 \end{bmatrix} \tag{5.1}$$

The *root signal* of a median filter with a certain mask geometry is the smallest neighborhood constellation of pixels with identical amplitudes that will remain unchanged in case of an iterated application of the filter. For any position of the root signal, the majority of elements under the median filter mask must then also be a member of the root signal. The shape of the root signal relates to the resolution preservation capability of a median filter; any detail structures that are 'smaller' than the root signal would possibly be affected by the filter. Examples of median filter geometries and their root signals are shown in Fig. 5.3.

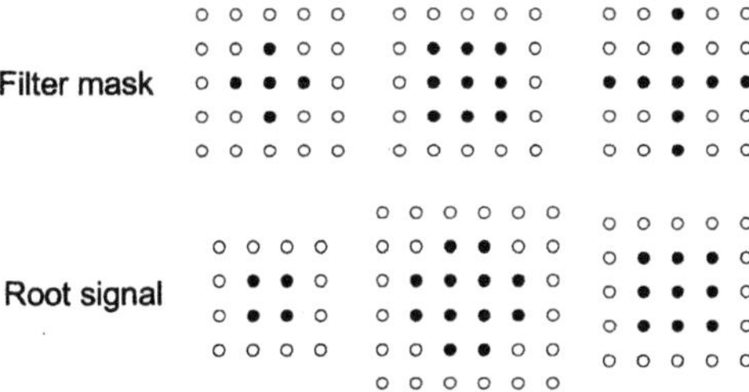

Fig. 5.3. Median filter geometries and related root signals

Median filters have an effect of equalization, they tend to limit the number of distinct amplitudes within a local environment. Nevertheless, amplitude discontinuities which signify characteristic properties of the signal are fully preserved. Fig. 5.4a shows the effects of a median filter and of a linear mean-value filter applied to an idealized edge (amplitude discontinuity). Edge ringing[1] is also eliminated by median filters (Fig. 5.4b).

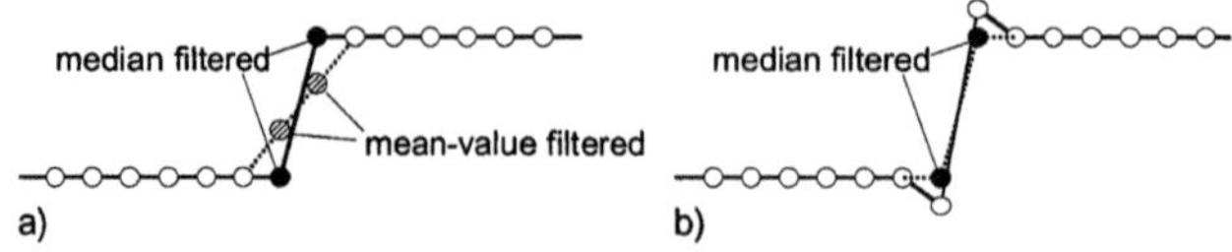

Fig. 5.4. Effects of filters with mask width 3 **a** Median filter and mean-value filter at a flat amplitude discontinuity **b** Median filter at an amplitude discontinuity with ringing

Variants of median filters are:

– Weighted median filters: For each position under the filter mask, an integer-numbered weighting factor M is defined. In the sets **m** and **m'**, the value of the respective pixel is copied M times. Usually, the pixel positioned at the center of the mask is given the highest weight. The root signal will then cover a smaller area and also retain smaller structures[2]. It is still important that the sum of weighting factors over the entire mask is an uneven number. By application of weighted median filters, thin lines in images can be preserved, while single isolated pixels of different amplitude are discarded. This is illustrated in the example (5.2), where output values of '10' would be produced at all positions by an unweighted median filter of mask size 3x3. If the center pixel is weighted by M=5, the single isolated value '20' in the second column of **X** is erased, but the column of values '20' is preserved, as the weight effects a majority for this value:

$$
\mathbf{X} = \begin{bmatrix} 10 & 10 & 10 & 20 & 10 \\ 10 & 10 & 10 & 20 & 10 \\ 10 & 20 & 10 & 20 & 10 \\ 10 & 10 & 10 & 20 & 10 \\ 10 & 10 & 10 & 20 & 10 \end{bmatrix} ; \ \mathbf{Y} = \mathrm{MED}_{\mathbf{G}}[\mathbf{X}] = \begin{bmatrix} 10 & 10 & 10 & 20 & 10 \\ 10 & 10 & 10 & 20 & 10 \\ 10 & 10 & 10 & 20 & 10 \\ 10 & 10 & 10 & 20 & 10 \\ 10 & 10 & 10 & 20 & 10 \end{bmatrix} ; \ \mathbf{G} = \begin{bmatrix} 1 & 1 & 1 \\ 1 & 5 & 1 \\ 1 & 1 & 1 \end{bmatrix} \qquad (5.2)
$$

– Hybrid linear/median filters: Output signals of different linear filters establish the set for median computation.

---

[1] Edge ringing is a typical effect which occurs by application of linear highpass or high-emphasis filters at amplitude discontinuities. Edge ringing is also a problem if a signal with an amplitude discontinuity is reconstructed from an incomplete set of transform basis functions; this is known as the *Gibbs phenomenon.*

[2] For definition of root signals for weighted median filters, see Problem 5.1.

These generalized types of median filters are also categorized as *rank order filters*. Other types of rank order filters are

– Minimum-value filters, which produce as output the *minimum amplitude* from the neighborhood set,

$$y_{\min}(m,n) = \min_{x(i,j)\in\mathcal{N}_c(m,n)}\left[x(i,j)\right];$$

(5.3)

– Maximum-value filters, which produce as output the *maximum amplitude* from the neighborhood set,

$$y_{\max}(m,n) = \max_{x(i,j)\in\mathcal{N}_c(m,n)}\left[x(i,j)\right];$$

(5.4)

– Difference filters, which produce as output the *maximum difference between any two values* from the neighborhood set, which in fact is the always-positive difference,

$$y_{\text{diff}}(m,n) = y_{\max}(m,n) - y_{\min}(m,n);$$

(5.5)

Example. The image matrix **X** is transformed into the following output images if minimum, maximum and difference filters of mask size 3x3 are applied:

$$\mathbf{X} = \begin{bmatrix} 10 & 10 & 10 & 20 & 20 \\ 10 & 10 & 10 & 20 & 20 \\ 10 & 10 & 10 & 20 & 20 \\ 10 & 10 & 10 & 20 & 20 \\ 10 & 10 & 10 & 20 & 20 \end{bmatrix} \Rightarrow \mathbf{Y}_{\min} = \begin{bmatrix} 10 & 10 & 10 & 10 & 20 \\ 10 & 10 & 10 & 10 & 20 \\ 10 & 10 & 10 & 10 & 20 \\ 10 & 10 & 10 & 10 & 20 \\ 10 & 10 & 10 & 10 & 20 \end{bmatrix}; \ \mathbf{Y}_{\max} = \begin{bmatrix} 10 & 10 & 20 & 20 & 20 \\ 10 & 10 & 20 & 20 & 20 \\ 10 & 10 & 20 & 20 & 20 \\ 10 & 10 & 20 & 20 & 20 \\ 10 & 10 & 20 & 20 & 20 \end{bmatrix}; \ \mathbf{Y}_{\text{diff}} = \begin{bmatrix} 0 & 0 & 10 & 10 & 0 \\ 0 & 0 & 10 & 10 & 0 \\ 0 & 0 & 10 & 10 & 0 \\ 0 & 0 & 10 & 10 & 0 \\ 0 & 0 & 10 & 10 & 0 \end{bmatrix}$$

(5.6)

The effect of a minimum-value filter *erodes* isolated peaks or amplitude plateaus from a signal, while a maximum-value filter discards minima or fills up troughs of the amplitude shape. The difference filter allows to analyze a kind of *nonlinear gradient* within an image signal. These latter types of rank order filters also have an interpretation by types of *morphological filters* (see the subsequent section). According to their effect, the maximum-value filter is then denoted as *erosion filter*, while the minimum-value filter is the *dilation filter*.

5.1.2   Morphological Filters

The term morphology is deduced from an ancient Greek word for 'shape' or 'figure'. Morphological filtering is often applied to manipulate geometric shapes expressed as binary signals by '1' value constellations. By generalization, they can be applied to signals and functions of multiple-level amplitude values, gray-level or color image signals. They can also be used for nonlinear contrast enhancement, elimination of local details or detection of characteristic points such as edges and corners in image signals.

Application to Binary Signals. The two basic operations in morphological filtering are *erosion* and *dilation*. Fig. 5.5a shows the example of a binary shape of an object, Fig. 5.5b is a *structure element* S of size 3x3 pixels, which plays the role of a filter mask. The structure element is shifted over the object A, where the current shift position is aligned by the center of S. Only those positions are retained as part of the output object shape, for which *only* pixels of A are found under S (Fig. 5.5c). The counterpart operation is *dilation*, where all those pixels are included in the output object shape, where *at least one* pixel of A is found under S at the respective position (Fig. 5.5d).

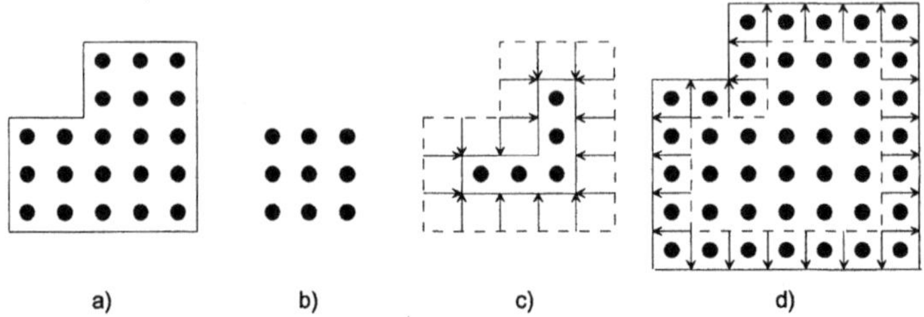

a)   b)   c)   d)

Abb. 5.5. Morphological basic operations: **a** Binary object shape A **b** Structure element S **c** Shape after erosion **d** Shape after dilation

Let by definition pixels belonging to an object shape A and pixels which characterize the shape of the structure element S be described by a logical "1". Mathematically, dilation is the *Minkovsky addition*

$$A \oplus S = \left\{ (m,n) \middle| [S+(m,n)] \cap A \neq 0 \right\}. \tag{5.7}$$

The expression '$S+(m,n)$' characterizes a translation (shift) of the structure element such that its center is at position (m,n). Similarly, erosion is the *Minkovsky subtraction*

$$A \odot S = \left\{ (m,n) \middle| [S+(m,n)] \subseteq A \neq 0 \right\}. \tag{5.8}$$

In the example shown by Fig. 5.5, the operations of erosion and dilation are *reversible*, such that from the shape of Fig. 5.5c the original object (Fig. 5.5a) is reconstructed by dilation; from the shape in Fig. 5.5d it is reconstructed by erosion. Reversibility is not generally guaranteed, in particular single holes in the object shape, noses or troughs in the object boundaries would usually be lost. From the basic operations erosion and dilation, additional morphological features can be defined. The *inner contour* of an object is given by the *exclusive-or* combination of the original and eroded signals,

$$A - (A \odot S), \tag{5.9}$$

while the *outer contour* results by *exclusive-or* combination of original and dilated signals,

$$(A \oplus S) - A. \tag{5.10}$$

By appropriate choice of the structure element's shape or additional criteria (e.g. minimum or maximum number of pixels that must belong to the object when the set under the structure element is analyzed), further features like corner pixels of an object shape can be extracted. The operation of *opening*,

$$A \circ S = (A \odot S) \oplus S \tag{5.11}$$

is defined as erosion followed by dilation, which straightens convex shapes and eliminates thin noses. The counterpart is *closing*

$$A \bullet S = (A \oplus S) \odot S, \tag{5.12}$$

which is defined as dilation followed by erosion, effecting a straightening of concave shapes and elimination of holes, channels and troughs. Effects of opening and closing are illustrated in Fig. 5.6, where again a structuring element of size 3x3 pixels was used.

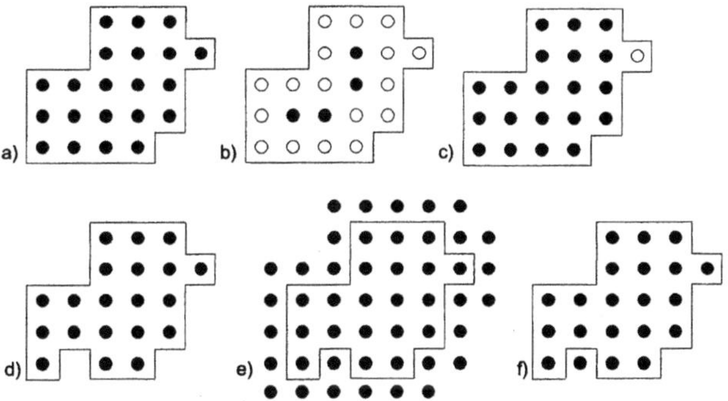

Fig. 5.6. a Original shape **b** erosion **c** opening (dilation of result in b)
d Original shape **e** dilation **f** closing (erosion of result in e)

In principle, mainly the size of the structure element influences the strength of the effect that morphological filters have. Alternatively, filters defined by small structure elements can also be applied iteratively to achieve a stronger effect. In some cases, it is also desirable to adapt the effect by the size of the object in general. If an object resists to (is not eliminated by) opening even though a large structure element or iterated opening is used, it can be retained in the original shape through an operation denoted as *opening by reconstruction*. Similarly, holes within an object that resist against closing can be retained by performing *closing by reconstruction*.

Application to non-binary Signals. A non-binary signal has a larger number of amplitude levels. For 2D signals, this can be interpreted as an amplitude surface, where the amplitudes represent a height map of the surface. The volume under this surface is stacked by a number of 'amplitude layers', each of which possesses a binary 2D shape. At one pixel position which has an integer amplitude value j, the stack consists of j layers. A one-dimensional section of this stack, which may e.g. be the amplitudes of an image in one row, is shown in Fig. 5.7. The width of a binary shape in layer j is larger if more pixels in the neighborhood have an amplitude which is larger or equal to j.

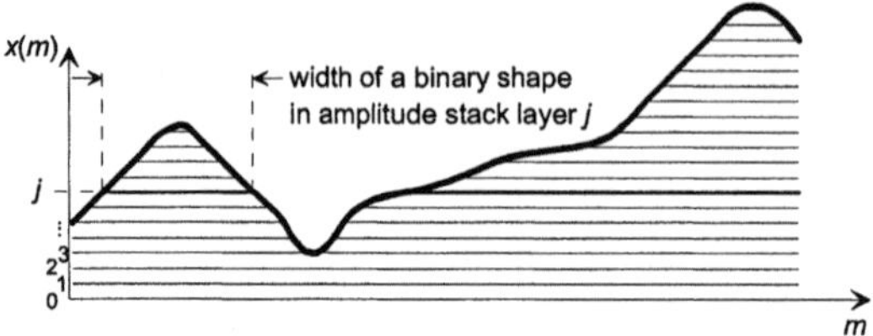

Fig. 5.7. Signal of multiple amplitude levels, composed from 'amplitude layers'

The operations of erosion and dilation can now be interpreted such that they are executed separately within each layer. The eroded or dilated shapes retain their original stacking order, however as by guarantee the upper layers are smaller or equal in width compared to the lower amplitude layers, no 'caves' ever appear, such that the amplitude volume is eroded like a mountain or is filled up (in dilation) as if sand is dispersed over the surface. The amplitude resulting by the entire stack of eroded or dilated layers is the eroded or dilated amplitude function (Fig. 5.8a).

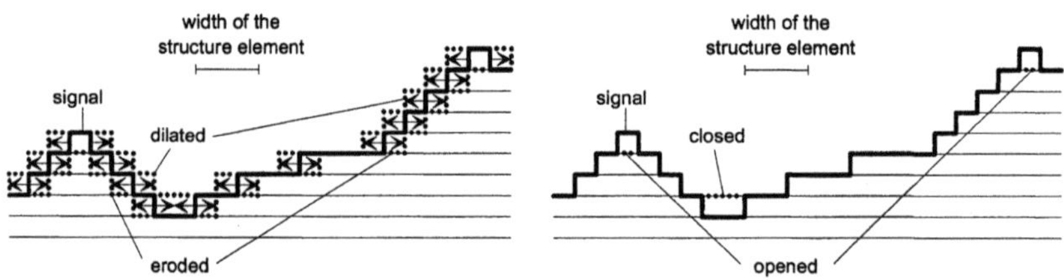

Fig. 5.8. Application of morphological operations to a non-binary signal
a Erosion and dilation **b** opening and closing

The results of dilation and erosion are exactly equivalent to the respective effects of maximum-value filters and minimum-value filters defined in (5.3) and (5.4). Hence, it is in fact not necessary to apply level-wise processing, as the much simpler operation of rank-order filtering fulfills the same effect.

By tendency, the dilated image will have an increased mean value of amplitude, while for the eroded image the mean value is lowered. Opening and closing are defined as before by subsequent processing of erosion and dilation or vice versa. Both of these functions effect a nonlinear equalization of signal amplitudes (Fig. 5.8b), where the opening eliminates peaks, and closing fills up troughs of the amplitude volume. By tendency, this modifies the signal such that plateaus of equal amplitudes are generated.

Analogous with the inner and outer contour definitions in (5.9) and (5.10), *morphological gradients* can be defined for gray-value signals. A typical goal of gradient analysis is finding of contour points. In the binary case, contours were defined by exclusive-or combinations of a shape and its dilated (for outer contour) or eroded versions (for inner contour). Accordingly, gradients of multiple-level amplitude signals are defined by differencing the values of the eroded or dilated signals with the original signal. These are described as erosion or dilation gradients[1]. The *morphological gradient* is the difference between the dilated and eroded signals; this will be the result of the min-max difference filter (5.5).

5.1.3   Polynomial Filters

Polynomial filters are conceptionally a superset of linear filters (characterized by impulse responses), but include also nonlinear combinations of samples, e.g. by concatenating multiplications of several samples from the input signal. Practically, only polynomial filters up to an order of two, the *Volterra filters*, appear to be relevant. The transfer equation of a one-or multidimensional Volterra filter is defined as

$$y(\mathbf{n}) = \overline{h}_1\left[x(\mathbf{n})\right] + \overline{h}_2\left[x(\mathbf{n})\right], \tag{5.13}$$

where the linear (FIR) term is

$$\overline{h}_1\left[x(\mathbf{n})\right] = \sum_{\mathbf{p}} a(\mathbf{p}) \cdot x(\mathbf{n}-\mathbf{p}), \tag{5.14}$$

and the nonlinear (quadratic) term is

$$\overline{h}_2\left[x(\mathbf{n})\right] = \sum_{\mathbf{p}}\sum_{\mathbf{q}} b(\mathbf{p},\mathbf{q}) \cdot x(\mathbf{n}-\mathbf{p}) \cdot x(\mathbf{n}-\mathbf{q}). \tag{5.15}$$

Recursive structures can be defined similarly, where however stability of the higher-order components is not as straightforward to test as for the case of linear filters. Volterra filters can e.g. be applied for nonlinear prediction[2] and noise reduction. The computation of the output signal in polynomial filters of higher order is similar to the computation of higher-order moments (sec. 3.3). This is an analogy to the linear case, where a high similarity between the convolution (linear filter) and the computation of correlation functions (moments of second order) can be observed. To optimize Volterra filters, moments or central moments of third order (3.43) must be used for the non-linear terms in addition to the autocorrela-

---

[1] For strictly positive values, the erosion gradient is defined by $x(m,n)$-$y_{min}(m,n)$, the dilation gradient by $y_{max}(m,n)$-$x(m,n)$.

[2] While the prediction estimate is computed by a non-recursive filter, the synthesis filter will be recursive, as usual in linear prediction.

tion or autocovariance function for the linear terms. More viable solutions could be realized by performing backward-adaptation of Volterra predictors [LIN, UNBEHAUEN 1992], which give in analogy with (4.94)-(4.96)

$$a_{n++}(\mathbf{p}) = a_n(\mathbf{p}) + \varepsilon_1 \cdot e(\mathbf{n}) \cdot y(\mathbf{n}-\mathbf{p})$$
$$b_{n++}(\mathbf{p},\mathbf{q}) = b_n(\mathbf{p},\mathbf{q}) + \varepsilon_2 \cdot e(\mathbf{n}) \cdot y(\mathbf{n}-\mathbf{p}) \cdot y(\mathbf{n}-\mathbf{q}).$$

$$(5.16)$$

5.2   Signal Enhancement

Signal enhancement is often used to improve the presentation of multimedia signals (postprocessing) for the comfort of the human observer, or to facilitate the analysis of signals in the context of coding or recognition (preprocessing). In particular for images and video, a significant effect to eliminate deficiencies which occur during acquisition, encoding and transmission can be achieved by relatively simple methods of linear or nonlinear filtering, often trying to utilize the interrelationships between the different available dimensions of the signal. Subsequently, some methods used in image and video enhancement are described.

Blurring filters. The goal of blurring is to cancel out high frequencies from the signal. Even though not optimum in a statistical sense (cf. chapter 8), this can subjectively reduce the impression of noise, in particular when the signal has low-frequent characteristics, where the noise is not hidden by signal components at higher frequencies. If the goal is to keep the content of the image unaffected, while noise or other artifacts shall be reduced, *direction-adaptive* lowpass filters can be applied, e.g. the following set of binomial filters for horizontal, vertical and the two diagonal directions,

$$\mathbf{H}_h = \frac{1}{4}\begin{bmatrix} 0 & 0 & 0 \\ 1 & 2 & 1 \\ 0 & 0 & 0 \end{bmatrix}; \; \mathbf{H}_{d^+} = \frac{1}{4}\begin{bmatrix} 1 & 0 & 0 \\ 0 & 2 & 0 \\ 0 & 0 & 1 \end{bmatrix}; \; \mathbf{H}_v = \frac{1}{4}\begin{bmatrix} 0 & 1 & 0 \\ 0 & 2 & 0 \\ 0 & 1 & 0 \end{bmatrix}; \; \mathbf{H}_{d^-} = \frac{1}{4}\begin{bmatrix} 0 & 0 & 1 \\ 0 & 2 & 0 \\ 1 & 0 & 0 \end{bmatrix}. \quad (5.17)$$

If these filters are adapted to local edge directions (which can be achieved by edge analysis, sec. 7.3), the relevant information of the image is sharply retained, while noise or other interferences are blurred out. Median filters or morphological filters such as open/close methods can also be used as blurring filters with good edge preservation properties.

De-blurring (contrast enhancement) filters. Linear filters can also be applied to *enhance the contrast* by enlarging the gradient of the amplitude. An example for such a filter is given by the filter matrix

$$H = \begin{bmatrix} -\alpha/8 & -\alpha/8 & -\alpha/8 \\ -\alpha/8 & \alpha+1 & -\alpha/8 \\ -\alpha/8 & -\alpha/8 & -\alpha/8 \end{bmatrix}, \tag{5.18}$$

where $\alpha \geq 0$ regulates the strength of the contrast enhancement, useful values being in the range between 1 and 4.

Noise insertion. Even though a typical goal of signal enhancement is *reduction of noise*, in some case *addition of noise* can have an advantageous effect on improved subjective impression, which is also known as *dithering* or *comfort noise insertion*. Regard a case where a signal is quantized into a distinguishable number of amplitude levels. If the signal itself varies smoothly, quantization can effect the appearance of staircase functions. This is likewise true for images where quantization into a small number of gray levels produces unnatural plateaus of identical amplitudes, or for audio/speech, where in case of low amplitudes occasional switching between quantization levels becomes perceptible. Addition of noise from an independent noise source can then mask the unnatural quantization noise components and improves the subjective impression.

Deblocking and deringing filters. In cases of lossy coding of multimedia signals, certain types of errors in the decoded signal are *deterministic*. For example, if block-based transforms are used for compression, *blocking artifacts* are encountered. Such undesirable components in the reconstructed signal can be reduced by *post processing*, taking into account the known block grid structure. This can be achieved by application of 1D lowpass filters across the block edges (cf. Fig. 5.9a); at the corner positions, 2D filtering is applied.

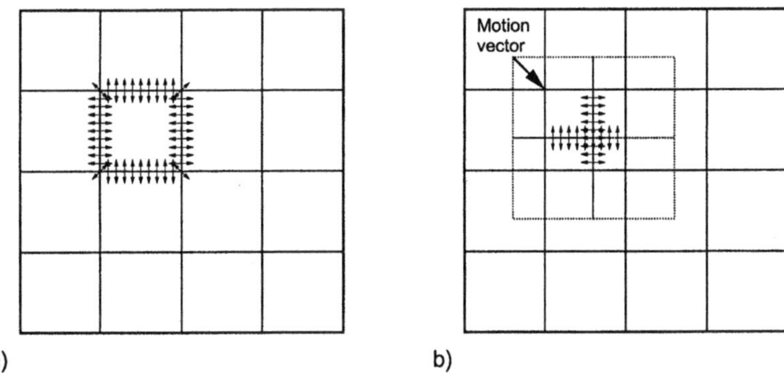

a)   b)

Fig. 5.9. Deblocking filter **a** for one block of a static block grid **b** in the context of motion compensation

Unconditional application of lowpass deblocking filters across any block boundaries will however also smooth out details of the images; even worse, as filtering is applied exclusively at boundaries, a 'smearing grid' artifact could become visible replacing the blocking artifact. This can be resolved by filter adaptation. Typical methods applied in this context are the analysis of the coded information (number and positions of coefficients retained, identification of blocks where intraframe coding is applied, estimation of expected distortion based on quantizer step size, edge direction analysis, analysis about presence of unnatural edge plateaus at block boundaries etc.)[1]

The problem becomes more severe, when motion-compensated processing is involved, as shown in Fig. 5.9b. Depending on the motion vector, blocking artifacts which were present in a previous decoded frame can be fed back into the current frame by motion-compensated prediction. Here, they would appear at an inner location of the block, depending on the value of the motion vector. It is then necessary to analyze the motion vectors and apply deblocking filtering accordingly at the respective positions where artifacts are expected[2].

A second effect that often affects the quality of decoded signals are *ringing artifacts* at signal discontinuities (such as edges in images). These are caused by the discarding or strong quantization of frequency coefficients. To reduce ringing, it is necessary to predict the expected ringing characteristics. Partially, this can be determined from the map of decoded coefficients, but directional edge analysis is very effective as well.

5.3   Amplitude-value transformations

Amplitude-value transformations define mappings of input amplitudes to output amplitudes. This can be interpreted as manipulation of probability distributions, e.g. a modification of the histogram for discrete-amplitude or of the PDF for continuous-amplitude signals. Amplitude mapping can either be performed globally or locally within small segments of a signal.

Contrast enhancement is a typical goal of amplitude value transformations applied to images. In audio signals, it is often desirable to *compress* amplitudes into a pre-defined range, such that dynamic fluctuations are limited. For example, it is convenient for the user of a radio, if the loudness stays relatively constant in audio broadcasting applications, such that it is not necessary to change the volume settings depending on local signal behavior. Another application of amplitude mapping is *companding* of signals, which is often done in transmission systems where noise is expected to interfere with the signal; when the low amplitudes are in-

---

[1] It is beyond the scope here to give a detailed insight into such algorithms which are often optimized on a heuristic basis. Reasoning behind different methods can e.g. be found in [PARK, LEE 1999].

[2] This effect does not occur, if deblocking filtering is applied *within the prediction loop*, as it is e.g. defined in the Advanced Video Coding standard (cf. sec. 17.4.4). In this case, prediction is made from an already de-blocked signal.

creased prior to transmission, the noise will take less effect, as an *expansion* (the reverse principle of compression), applied after receiving, reduces the noise amplitude and reconstructs the signal into its original amplitude range. This principle is e.g. applied in PCM encoding of speech signals where it provides suppression of quantization noise in cases of low amplitude levels.

For images, amplitude mapping can be applied either to the luminance or to the color component amplitudes. In color mapping, it is not straightforward to define objective contrast enhancement functions, as the subjective color impression could be falsified undesirably. An extreme example is the usage of *color lookup tables*, which are used to systematically highlight specific image content without being coherent with the original natural color any more. Color lookup tables are also used when the number of different colors that can be displayed is limited; in such a case, the lookup table can be determined by a vector quantizer codebook design (cf. sec. 11.5.3). *Quantization* and *re-quantization* can indeed be regarded as specific optimized cases of non-linear amplitude mapping.

5.3.1   Amplitude Mapping Functions

The amplitude value of a sample $x(\mathbf{n})$ shall be mapped into the output value $y(\mathbf{n})$. This relationship can be described by a mapping characteristic function $f(\cdot)$, which could be linear or nonlinear:

$$y(\mathbf{n}) = f\left[x(\mathbf{n})\right]. \tag{5.19}$$

If the mapping characteristic is steady, unique and monotonous, the mapping is reversible, i.e.

$$x(\mathbf{n}) = f^{-1}\left[y(\mathbf{n})\right]. \tag{5.20}$$

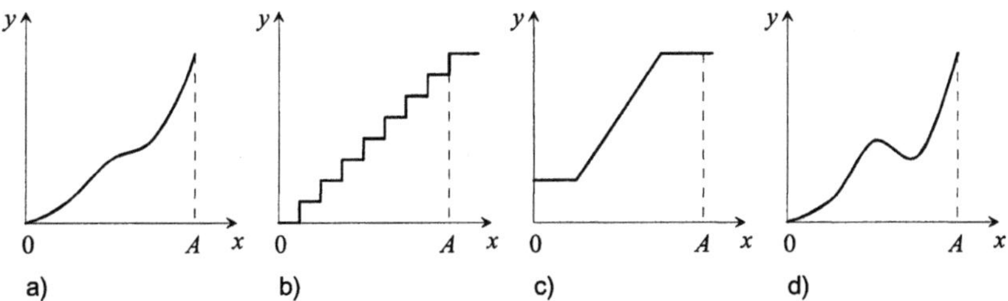

Fig. 5.10. Examples of mapping characteristics (explanations see text below)

Examples of mapping functions are shown in Fig. 5.10. The function in Fig. 5.10a would be reversible. Irreversible functions are shown in Fig. 5.10b (quantizer characteristic, which is unsteady), Fig. 5.10c (clipping characteristic, which is not unique in certain ranges) and Fig. 5.10d (non-monotonous function). Some important invertible mapping characteristics are:

–   *Linear characteristic* (Fig. 5.11a) :

$$y(\mathbf{n}) = \alpha \cdot x(\mathbf{n}) + y_a \quad ; \quad x(\mathbf{n}) = \frac{1}{\alpha} \cdot y(\mathbf{n}) - \frac{y_a}{\alpha}, \tag{5.21}$$

which includes the case ($\alpha=-1$) of negative amplitude mapping (Fig. 5.11b).

–   *Piecewise-linear characteristic* (Fig. 5.11c):

$$y(\mathbf{n}) = \begin{cases} \alpha \cdot x(\mathbf{n}) & \text{for} \quad x(\mathbf{n}) \le x_a \\ \beta \cdot \left[x(\mathbf{n}) - x_a \right] + y_a & \text{for} \quad x_a \le x(\mathbf{n}) \le x_b \\ \gamma \cdot \left[x(\mathbf{n}) - x_b \right] + y_b & \text{for} \quad x_b \le x(\mathbf{n}) \end{cases} \tag{5.22}$$

with $y_a = \alpha \cdot x_a$, $y_b = \beta \cdot [x_b - x_a] + y_a$; this can likewise be extended to more than three pieces, and is invertible if all slopes ($\alpha, \beta, \gamma, ..$) are non-zero and of equal sign.

–   *Root and quadratic characteristics*, which are examples of reversible compression/expansion function pairs as illustrated in Figs. 5.11d/e

$$y(\mathbf{n}) = \sqrt{\alpha |x(\mathbf{n})|} \cdot \operatorname{sgn}(x(\mathbf{n})) \quad ; \quad x(\mathbf{n}) = \frac{y^2(\mathbf{n})}{\alpha} \cdot \operatorname{sgn}(y(\mathbf{n})), \ \alpha > 0, \tag{5.23}$$

–   Another compression/expansion pair are the *logarithmic and exponential characteristics*

$$\begin{aligned} y(\mathbf{n}) &= \log_\alpha \left(1 + |x(\mathbf{n})| \right) \cdot \operatorname{sgn}(x(\mathbf{n})) \\ x(\mathbf{n}) &= \left(\alpha^{|x(\mathbf{n})|} - 1 \right) \cdot \operatorname{sgn}(y(\mathbf{n})) \end{aligned} \quad , \ \alpha > 1. \tag{5.24}$$

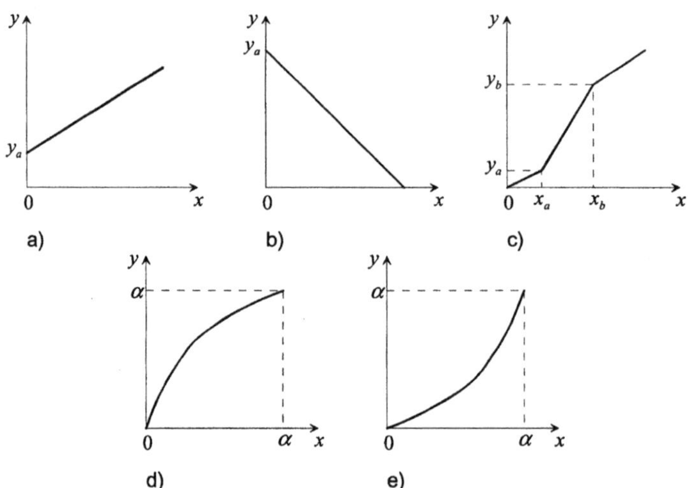

Fig. 5.11. Examples of reversible mapping characteristics (5.21)-(5.24)

5.3.2   Probability Distribution Modification and Equalization

Mapping functions can be determined systematically, provided that criteria for optimization are given. As an example, the goal of a mapping might be

- to obtain a desired probability distribution at the output, e.g. to maximize the contrast of a signal or achieve a uniform distribution of probabilities in a discrete representation;
- to minimize the resulting error in the mapping from a continuous-value to a discrete-value signal (quantization)[1].

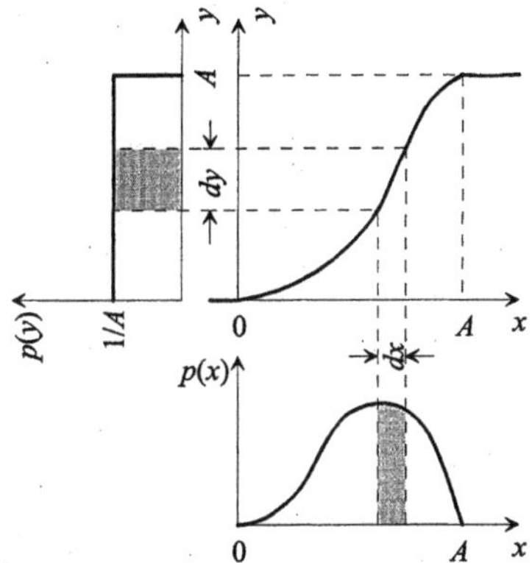

Fig. 5.12. Relationships of PDFs for input and output signals in amplitude mapping

For steady, monotonous functions, the areas under the PDFs within a differential amplitude range of the input and the corresponding range of the output must be identical (see Fig. 5.12):

$$p(x)\cdot dx = p(y)\cdot dy \quad\Rightarrow\quad \frac{df(x)}{dx}=\frac{p(x)}{p(y)} \quad\text{or}\quad \frac{df^{-1}(y)}{dy}=\frac{p(y)}{p(x)}. \tag{5.25}$$

Further, the probability of samples within an amplitude interval $[x_a, x_b]$ of the input must be identical to the probability within the corresponding output interval $[f(x_a), f(x_b)]$:

$$\int_{x_a}^{x_b} p(x)dx = \int_{f(x_a)}^{f(x_b)} p(y)dy. \tag{5.26}$$

The cumulative distribution functions of input and output signals must hence be related by

---

[1] See optimization of non-uniform quantizer characteristics, sec. 11.1

$$P(x \leq x_a) = \int_{-\infty}^{x_a} p(x)dx = \int_{-\infty}^{f(x_a)} p(y)dy = P[f(x) \leq f(x_a)]. \tag{5.27}$$

Of particular interest is the case where the mapping characteristic shall result in a uniform distribution of the output in the amplitude range $0 \leq x \leq A_{max}$, which is a solution to maximize the contrast of an image signal. Here, $p(y)=1/A_{max}$ such that $P(y \leq y_a)=y_a/A_{max}$. Assuming that the input is restricted to amplitudes $0 \leq x \leq A_{max}$, and using (5.27) results in the mapping characteristic

$$f(x) \cdot \frac{1}{A_{max}} = \int_0^x p(\tilde{x})d\tilde{x} \quad \Rightarrow \quad f(x) = A_{max} \cdot \int_0^x p(\tilde{x})d\tilde{x}. \tag{5.28}$$

In principle, any PDF can be targeted for the output signal, but the solution is more complicated for the case of a non-uniform target, as the linear dependency of $f(x)$ on the left side of (5.28) would be replaced by an integral condition.

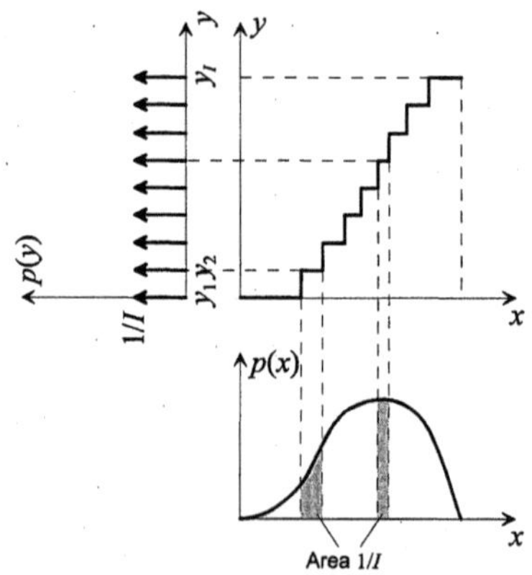

Fig. 5.13. Non-uniform quantizer characteristic to obtain a quantized signal with uniform distribution for reconstruction of maximum contrast

These methods can be adopted for the case of direct mapping into discrete probability distributions $P(x_j)$ or histograms $H(j)$. Assume that a quantization of an image shall result in a uniform discrete probability distribution. With a number of $M \cdot N$ pixels and I discrete amplitude values, $(M \cdot N)/I$ samples must fall into each bin of the histogram. The lower and upper boundaries of quantization intervals can then be determined recursively (starting with $x_{lo,1}=0$) such that

$$P(x \leq x_{up,i}) = \frac{i}{I} = \int_0^{x_{up,i}} p(x)dx = \frac{1}{MN}\sum_{\tilde{i}=1}^{i} H(\tilde{i}) = C(i) \tag{5.29}$$

$$\text{and } x_{lo,i+1} = x_{up,i} \quad ; \quad i \leftarrow i+1.$$

The mapping function will be

$$y = f(x) = y_i \quad \text{for} \quad x_{\text{lo},i} \leq x < x_{\text{up},i} \,, \tag{5.30}$$

which typically gives a non-uniform quantizer characteristic. Remark that – unlike the quantization approaches introduced in sec. 11.1 – this quantizer is not optimized for *minimum distortion*, but for *maximization of contrast* or *maximization of entropy* in the digital signal. The signal of maximum amplitude contrast is generated when the quantized signal is reconstructed by an inverse quantizer of uniform step size (see Fig. 5.13):

$$y_i = (i-1) \cdot \frac{A_{\max}}{I-1} \quad ; \quad i = 1, 2, \dots, I \,. \tag{5.31}$$

5.4   Interpolation

In signal analysis and signal presentation, it is often necessary to estimate values of signals *between known sampling positions*. The ultimate goal of this estimation is the reconstruction of a continuous signal from the discrete samples, which is achieved by interpolation. Control positions[1] can in principle be defined on arbitrary sampling grids, including irregular grids with non-equidistant positions (see Fig. 5.14). In the latter case, interpolation is more complicated, and high-quality lowpass filters[2] are not usable for this purpose. Further, it will no longer be possible to interpret interpolation by signal expansion and filtering (cf. Fig. 4.34), as the mapping of an irregular grid into a regular one can not directly be interpreted by spectral operations.

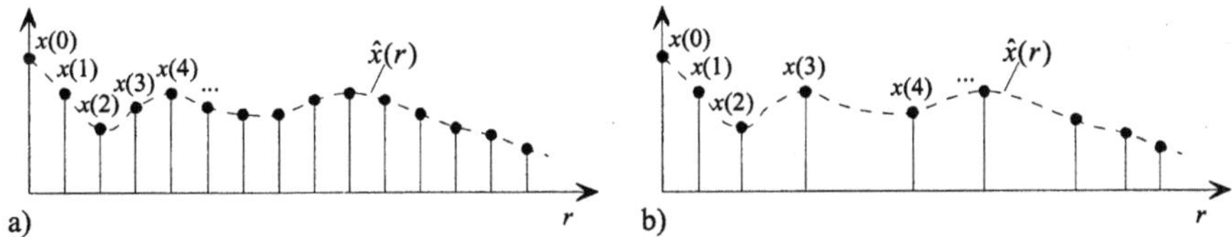

Fig. 5.14. Interpolation of a signal in cases of equidistant (*a*) and non-equidistant (*b*) positions of support.

---

[1] Control positions are the positions in time or space for which actually samples are available. In the simplest case, interpolation is then performed directly from the samples; in some interpolation methods, control coefficients are computed first.

[2] For a broader background on the interpretation of interpolation using linear systems, also refer to sec. 4.4.1.

5.4.1   Zero- and First-order Interpolators

Let values of the signal x be known at uniquely indexed support positions $r(m)$[1]. The most simple interpolator is the *hold element* (zero-order interpolator), giving the estimated value (Fig. 5.16a)

$$\hat{x}(r) = x\big[r(m)\big] \text{ for } r(m) \le r < r(m+1).$$ (5.32)

Similarly, a separable 2D hold element can be defined as

$$\hat{x}(r,s) = x\big[r(m),s(n)\big] \text{ for } r(m) \le r < r(m+1) \text{ and } s(n) \le s < s(n+1).$$ (5.33)

The simplest 'true' interpolation (using more than one sample) is performed by the first-order linear interpolator (Fig. 5.16b)

$$\hat{x}(r) = x\big[r(m)\big] \cdot \frac{r(m+1)-r}{r(m+1)-r(m)} + x\big[r(m+1)\big] \cdot \frac{r-r(m)}{r(m+1)-r(m)}.$$ (5.34)

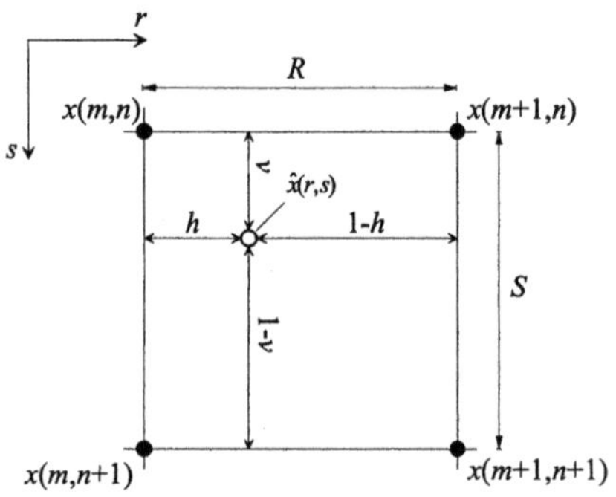

Fig. 5.15. Bilinear interpolation

In case of rectangular sampling, the 2D separable version of (5.34) is *bilinear interpolation*. The principle is illustrated in Fig. 5.15. The value to be estimated at position (r,s) results by weighted averaging of four neighboring positions, which are used for the interpolation within the range $m{\cdot}R \le r < (m+1){\cdot}R$, $n{\cdot}S \le s < (m+1){\cdot}S$. The horizontal and vertical fractional distances h and v are normalized by the sampling distances:

$$\hat{x}(r,s) = x(m,n) \cdot (1-h) \cdot (1-v) + x(m+1,n) \cdot h \cdot (1-v)$$
$$+ x(m,n+1) \cdot (1-h) \cdot v + x(m+1,n+1) \cdot h \cdot v$$

---

[1] In case of equidistant sampling, the condition $r(m)=m{\cdot}R$ holds, where R is the sampling distance.

$$\text{with} \quad h = \frac{r - mR}{R} \; ; \; v = \frac{s - nS}{S}. \tag{5.35}$$

Bilinear interpolation can be performed separably as follows :

a) *Horizontal interpolation step:*

$$\hat{x}(r,n) = (1-h) \cdot x(m,n) + h \cdot x(m+1,n)$$
$$\hat{x}(r,n+1) = (1-h) \cdot x(m,n+1) + h \cdot x(m+1,n+1) \tag{5.36}$$

b) *Vertical interpolation step:*

$$\hat{x}(r,s) = (1-v) \cdot \hat{x}(r,n) + v \cdot \hat{x}(r,n+1) \tag{5.37}$$

The result of interpolation using systems of order zero (hold element) and one (linear interpolator) is shown in Fig. 5.16. It is obvious that none of these systems is capable to approximate the original signal accurately, as no discontinuities or corners would be expected in the signal amplitudes. A better approximation can only be achieved by higher-order interpolators, using information from more than two control positions. The elements to be combined for the computation of the interpolation result need not necessarily be original samples from the signal. In a generalized approach, these can be *control coefficients* $c(k)$. If the interpolation is defined piecewise as in (5.32) and (5.33), the continuous basis functions used for the interpolation are shift variant and valid only locally. If $P+1$ control coefficients are involved in the interpolation at any position r, the interpolation system is of order P.

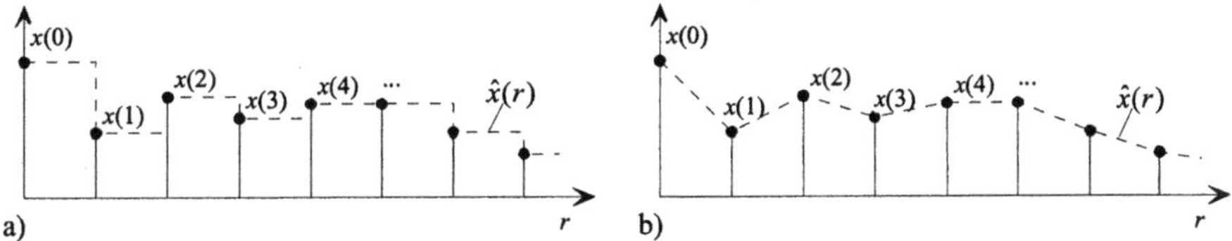

Fig. 5.16. Interpolation using systems **a** of order zero (hold element) **b** of order one (linear interpolator)

A generic notation for interpolation of order P based on control coefficients can be expressed for the one-dimensional case as

$$\hat{x}(r) = \sum_{k=0}^{P} c_k \cdot \varphi_k^{(P)}(r) . \tag{5.38}$$

$\varphi_k^{(P)}(r)$ is a member from a system of continuous basis function, which is used to determine the contribution of coefficient c_k to the interpolation at position r. For the case of a 2D system, (5.38) can be extended as follows, where the right side of (5.39) defines the case of a separable 2D basis function:

$$\hat{x}(r,s) = \sum_{k=0}^{P}\sum_{l=0}^{Q} c_{k,l} \cdot \varphi_{k,l}{}^{(P,Q)}(r,s) \; ; \; \varphi_{k,l}{}^{(P,Q)}(r,s) = \varphi_k{}^{(P)}(r) \cdot \varphi_l{}^{(Q)}(s) \, . \qquad (5.39)$$

In the sequel, interpolations using linear filters, frequency coefficient expansions and spline interpolation are described as realizations of this generic interpolation definition.

5.4.2   Interpolation using linear Filters

For interpolation by linear filters, the control coefficients c_k are identical to signal samples,

$$c_k = \begin{cases} x(m-P/2+k), & P \quad \text{even} \\ x(m-(P-1)/2+k), & P \quad \text{odd} \end{cases} \quad \text{for} \quad r(m) \le r < r(m+1) \, . \qquad (5.40)$$

The functions $\varphi_k(r)$ represent continuous interpolation filter impulse responses. In case of shift-invariant interpolation functions, this can be expressed by a *parent interpolation function* $\varphi^{(P)}(r)$ which is symmetric around position $r=0$[1]

$$\varphi_k{}^{(P)}(r) = \begin{cases} \varphi^{(P)}[r-r(m-P/2+k)-R/2], & P \quad \text{even} \\ \varphi^{(P)}[r-r(m-(P-1)/2+k)], & P \quad \text{odd} \end{cases} \quad \text{for} \quad r(m) \le r < r(m+1) \qquad (5.41)$$

The superposition of the different shifted functions, weighted by the respective c_k, results in the interpolated signal at position r. The continuous function systems for cases $P=0$ (hold) and $P=1$ (first-order linear) are shown in Fig. 5.17a/b,

$$\varphi^{(0)}(r) = \begin{cases} 1, & 0 \le |r| < \dfrac{R}{2} \\ 0, & \text{else} \end{cases} \; ; \; \varphi^{(1)}(r) = \begin{cases} 1-\left|\dfrac{r}{R}\right|, & 0 \le |r| < R \\ 0, & \text{else} \end{cases} \qquad (5.42)$$

These are impulse responses of finite extension, where the number of samples to be used for interpolation at any position is $P+1$, except for the positions $r(m)$, where only the original sample of this position is used. The *ideal interpolator* is the *sinc function* which has an infinite impulse response (Fig. 5.17c)[2]

$$\varphi^{(\infty)}(r) = \text{si}(\omega_c r) \quad \text{with} \quad \text{si}(x) = \frac{\sin x}{x}; \quad \omega_c = \frac{\pi}{R} \, . \qquad (5.43)$$

---

[1] The additional shifts depending on P in (5.40) and (5.41) are necessary by the definition that the value $\varphi^{(P)}(0)$ shall give the interpolation weight of the control value at position $r(m)$.

[2] For the case of the ideal interpolator, it would be necessary to modify the summation limits in (5.38) and (5.44) to run from $-\infty$ to $+\infty$. Then, the additional shifts by $P/2$, $R/2$ in (5.40) and (5.41) would not apply.

The frequency transfer function has ideal lowpass characteristics with cut-off frequency ω_c. This will however be an interpolator of order $P=\infty$, which cannot be realized. A set of Pth order interpolation basis functions must fulfill the following condition at any position r:

$$\sum_{k=0}^{P} \varphi_k(r) = 1 .$$

(5.44)

Different methods exist to design best linear filter interpolation functions of finite extension. One common approach is *windowing* of the sinc function, where specific windowing functions are designed to minimize ripples in the transfer function of the interpolation filter and give a sharp cutoff transition. An example is the *Hamming window* of length W,

$$w(r) = 0.54 + 0.46 \cos\left(\frac{2\pi r}{W}\right) \text{ for } |r| \le \frac{W}{2} .$$

(5.45)

Depending on the choice of W in relationship with the sampling distance R, the order of the interpolator can be adjusted. Another class of optimized linear interpolation functions can be derived from Wiener filters (see sec. 8.2.2), which are also adaptable to properties of the signal, and can be optimized in particular when additional noise is present[1].

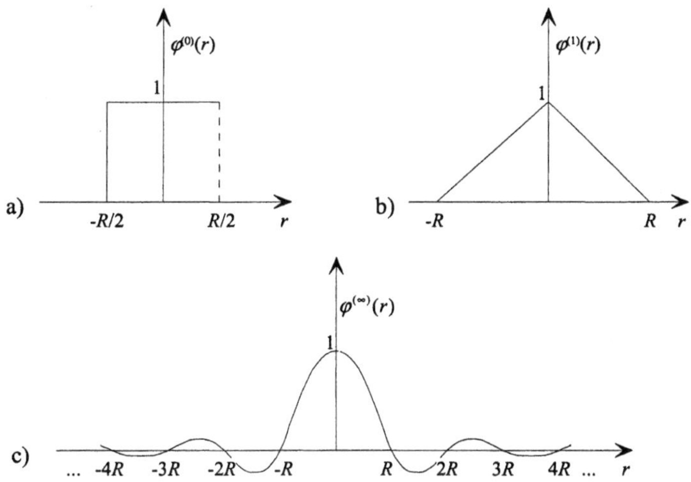

Fig. 5.17. Linear function systems $\varphi^{(P)}(r)$ for interpolation: **a** Hold element, $P=0$ **b** Linear interpolator, $P=1$ **c** Ideal interpolator (sinc function), $P=\infty$

---

[1] Most of the signal estimation methods described in chapter 8 can indeed be used for signal interpolation.

For some applications (e.g. in motion estimation), interpolation must be performed only on a discrete grid of sub-pixel positions. In such cases, it is sufficient to define a discrete set of interpolation filters, one for each fractional position; the interpolation operation can then be interpreted as a discrete convolution (see sec. 4.1 and 4.4.1)[1]. Assume an example where only interpolation of one additional sample at any position $r=(m+r')\cdot R$, $0\leq r'<1$ between two control positions is required. A linear FIR filter with an even number $P+1$ of coefficients can best be used for this purpose (see Fig. 5.18 for the example of $r'=0.5$). For the 1D case, the intermediate values are directly computed using a digital filter of odd order P:

$$\hat{x}\big((m+r')\cdot R\big) = \sum_{p=-(P+1)/2}^{(P-1)/2} x(m-p)\cdot h_{r'}(p).\tag{5.46}$$

(5.46) is directly determined from the general form (5.38), setting c_k according to (5.40) and $h_{r'}(p)=\varphi^{(P)}[(m+r')R]$ in (5.41)-(5.43).

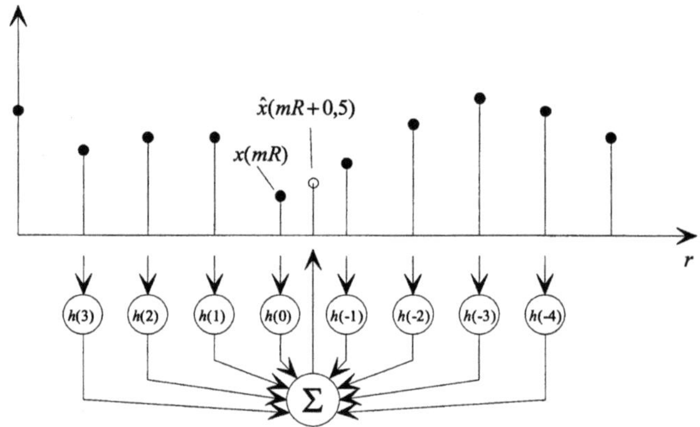

Fig. 5.18. Interpolation of an intermediate value at a mid position between two samples, using a discrete linear filter

A separable 2D FIR interpolation of order P for the horizontal and order Q for the vertical interpolation filter is

$$\hat{x}\big((m+r')\cdot R,(n+s')\cdot S\big) = \sum_{p=-(P+1)/2}^{(P-1)/2}\left(\sum_{q=-(Q+1)/2}^{(Q-1)/2} x(m-p,n-q)\cdot h_{s'}(q)\right)\cdot h_{r'}(p).\tag{5.47}$$

In some cases, *only* interpolated samples will be used. This is for example the case in block-based motion compensation, when a motion vector of sub-pixel accuracy is defined for the entire block. Then, the interpolation filter will imprint its fre-

---

[1] Nevertheless, the discrete coefficients in this case are typically determined from samples of continuous functions such as the windowed sinc.

quency cut-off properties to the interpolated signal. This effect is shown in Fig. 5.19 for different filters which shift the signal by half of the sampling distance. Observe that only the ideal lowpass is capable to retain the information of the signal perfectly up to the resolution which is in fact available from the known samples. The linear interpolator (which is a two-sample averaging filter in this case) has a 3 dB attenuation at $\Omega=\pi/2$. Better interpolation quality is generally possible when higher-order filters are used, but the design must be made carefully as a compromise between ripple in the pass-band and highest possible cut-off frequency.

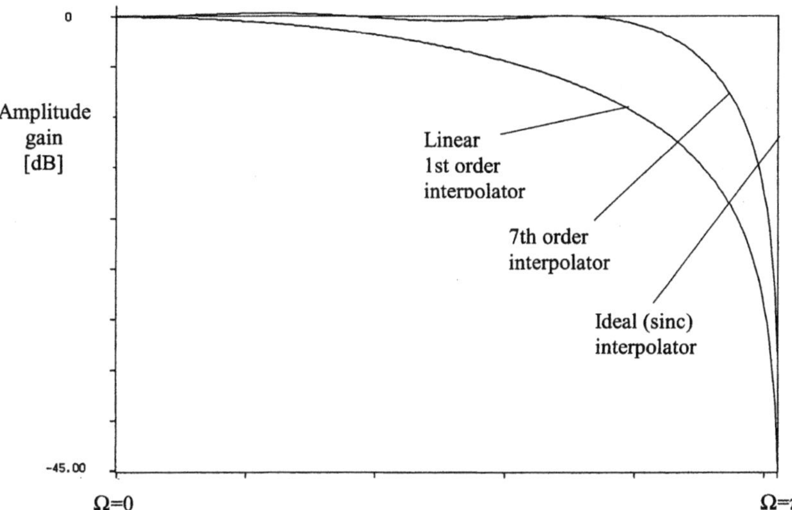

Fig. 5.19. Amplitude transfer functions of 1D linear interpolation filters with $P=1$, $P=7$ (windowed sinc) and ideal interpolator ($P=\infty$), interpolation of position $r'=0.5$.

5.4.3   Interpolation based on Frequency Extension

Transform coefficients can also be used as control coefficients for interpolation. The continuous functions $\varphi_k(r)$ are then the basis functions of a transform synthesis (e.g. the scaling functions of a DWT or complex exponential functions of a DFT). For this purpose, the discrete transform representation of a sampled signal is extended by values representing higher frequencies, which relates to an increased signal resolution. This could be zero-values in the simplest case, or estimates of higher-frequency components when applicable, e.g. for signals with harmonic properties. Synthesis is then performed using the corresponding set of synthesis basis functions of the extended transform representation. It must be observed in this context that relationships between signals, transform coefficients and basis functions are often shift variant; further, effects which occur due to the processing of finite signal segments must be compensated. For example, in the case of block transforms, the interpolation result which is gained by a synthesis of double block size or double-length basis functions (providing a signal of double number of samples per dimension) is typically of high quality at the center of the block and will tend to become erroneous towards the block boundaries. Block-overlapping transforms (due to the smoothing effect of windowing functions) and in particular the wavelet transform (due to its inherent scalability properties) are better suitable for signal interpolation.

5.4.4   Spline and Lagrangian Interpolation

B-splines are another class of continuous interpolation functions of finite length, where at any position r only a finite number of control values is used for interpolation. The following definition (5.48) allows an application for interpolation from any regular or irregular sampling grid; the value of the basis function of order P can be computed using the formula[1]

$$\varphi_k^{(P)}(r) = (-1)^{P+1} \cdot (P+1) \cdot \sum_{i=0}^{P+1} \frac{[r-r(i)]^P \cdot \varepsilon[r-r(i)]}{\prod_{\substack{j=0 \\ i \neq j}}^{P+1} [r(i)-r(j)]} \; ; \; P \geq 0 \tag{5.48}$$

This function is non-zero in the range $r(m) \leq r < r(m+P+1)$, which would be an equivalent definition as for a *causal filter*, where the response shall be zero before an input is available.

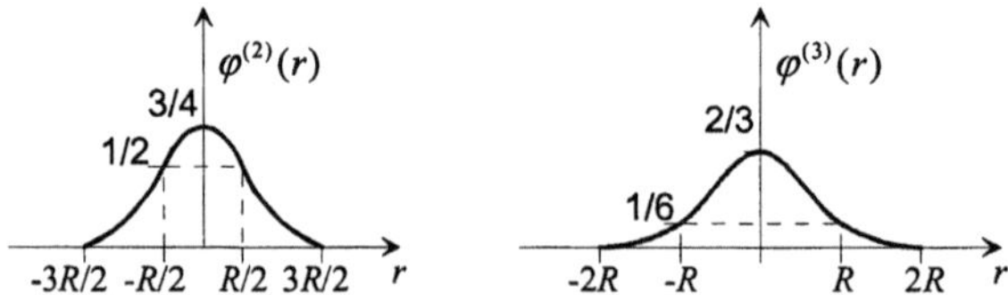

Fig. 5.20. Centered *B*-spline functions of orders $P=2$ and $P=3$ for the case of a regular grid of control positions

Even though spline interpolation is generically applicable to arbitrary sampling grids, we will now mainly study the case of regular grids with control positions $r(m)=mR$. Further, similar to the case of linear filters, parent functions $\varphi^{(P)}(r)$ shall be defined which are symmetric around position $r=0$. Mapping into the interpolation functions relating to positions of coefficients c_k is then again performed by (5.41). In the regular-grid case, B-spline functions can be interpreted as an iterated (P-fold) convolution of the hold element $\varphi^{(0)}(r)$ of (5.42). Obviously, the B-spline $\varphi^{(1)}(r)$ is then also identical to the basis function of the linear interpolator. Fig.

---

[1] $\varepsilon(r)$ is the *unit step function* [$\varepsilon(r)=0$ for $r<0$, $\varepsilon(r)=1$ for $r>0$]

5.20 shows the *B*-spline functions of orders $P=2$ and $P=3$. For $P\rightarrow\infty$, the iterated convolution converges towards a Gaussian function. The commonly-used B-spline functions of orders $P=2$ and $P=3$ are *quadratic* and *cubic splines*, which can directly be defined as follows:

$$\varphi^{(2)}(r) = \begin{cases} \dfrac{3}{4}-\left(\dfrac{r}{R}\right)^2 & \text{for } |r| \leq \dfrac{R}{2} \\[2ex] \dfrac{\left(1.5-|r/R|\right)^2}{2} & \text{for } \dfrac{R}{2} < |r| \leq \dfrac{3}{2}R \\[2ex] 0 \ \text{ for } |r| > \dfrac{3}{2}R \end{cases} \tag{5.49}$$

$$\varphi^{(3)}(r) = \begin{cases} \dfrac{4+3|r/R|^3-6|r/R|^2}{6} & \text{for } |r| \leq R \\[2ex] \dfrac{\left(2-|r/R|\right)^3}{6} & \text{for } R < |r| \leq 2R \\[2ex] 0 \ \text{ for } |r| > 2R \end{cases} \tag{5.50}$$

Fig. 5.21 illustrates the computation of a cubic spline interpolation ($P=3$) within an interval $r(m)\leq r<r(m+1)$. A total of 4 control coefficients is used to weight the respective interpolation functions, which have their maxima at $r(m-1)$, $r(m)$, $r(m+1)$ und $r(m+2)$. The result of interpolation is determined from the general formulation (5.38), using the basis function (5.50) mapped into position r by (5.41). The argument of (5.41) is expressed here as $r'=r-r[m-(P-1)/2+k]$, such that for $P=3$ the control coefficient c_1 corresponds to the position $r(m)$. Normalization of sampling distances $R=1$ is now assumed. This gives the interpolated result within the range $r(m)\leq r<r(m+1)$, $r'=r-r(m)$

$$\begin{aligned} \hat{x}(r) &= c_0 \cdot \frac{\left(2-(r'+1)\right)^3}{6} + c_1 \cdot \frac{4+3r'^3-6r'^2}{6} \\[2ex] &\quad +c_2 \cdot \frac{4-3(r'-1)^3-6(r'-1)^2}{6} + c_3 \cdot \frac{\left(2+(r'-2)\right)^3}{6} \\[2ex] &= c_0 \cdot \frac{-r'^3+3r'^2-3r'+1}{6} + c_1 \cdot \frac{3r'^3-6r'^2+4}{6} + c_2 \cdot \frac{-3r'^3+3r'^2+3r'+1}{6} + c_3 \cdot \frac{r'^3}{6} \end{aligned} \tag{5.51}$$

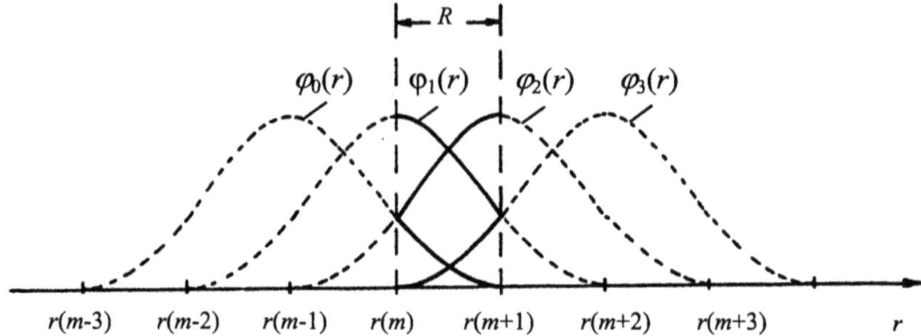

Fig. 5.21. Cubic spline interpolation and basis functions which are combined for interpolation within an interval $r(m) \leq r < r(m+1)$

(5.51) can be re-written by the following matrix expression:

$$\hat{x}(r) = \frac{1}{6} \cdot \begin{bmatrix} r'^3 & r'^2 & r' & 1 \end{bmatrix} \cdot \begin{bmatrix} -1 & 3 & -3 & 1 \\ 3 & -6 & 3 & 0 \\ -3 & 0 & 3 & 0 \\ 1 & 4 & 1 & 0 \end{bmatrix} \cdot \begin{bmatrix} c_0 \\ c_1 \\ c_2 \\ c_3 \end{bmatrix} \tag{5.52}$$

The remaining problem is optimization of the control coefficients c_k. At the control positions, $r'=0$, such that

$$\hat{x}[r(m)] = \frac{1}{6}[c_0 + 4c_1 + c_2]. \tag{5.53}$$

Even though formally computation of spline interpolation has been regarded exactly like linear interpolation so far, (5.53) shows a fundamental difference. In linear interpolation of order $P=3$, only the coefficient $c_1=x[r(m)]$ would be used to interpolate the control position itself, linear interpolation functions are *interference free*, no other samples will contribute to the interpolation of the originally available positions. On the other hand, the condition

$$\hat{x}[r(m)] \overset{!}{=} x(m) \tag{5.54}$$

can now be used to determine the unknown spline control coefficients. This is an incomplete problem if only one condition (5.53) is used. Any coefficient however takes influence on different intervals, such that the c_k defined for interval $r(m) \leq r < r(m+1)$ will be identical with c_{k-1} of interval $r(m+1) \leq r < r(m+1+1)$. Regarding all those control coefficients jointly which are used within a finite signal segment of length M, the operation of interpolation at the control positions can be written by a vector-matrix system, where $c(1)$ maps to the coefficient c_1 of the first interval $r(0) \leq r < r(1)$. The following definition assumes a circular (periodic) extension[1] of the sequence of control coefficients:

$$\underbrace{\begin{bmatrix} \hat{x}[r(0)] \\ \hat{x}[r(1)] \\ \hat{x}[r(2)] \\ \vdots \\ \\ \hat{x}[r(M-1)] \end{bmatrix}}_{\hat{\mathbf{x}}} = \frac{1}{6} \underbrace{\begin{bmatrix} 4 & 1 & 0 & \cdots & 0 & 1 \\ 1 & 4 & 1 & 0 & & 0 \\ 0 & 1 & 4 & 1 & \ddots & \vdots \\ \vdots & \ddots & \ddots & \ddots & \ddots & \\ 0 & & 0 & 1 & 4 & 1 \\ 1 & 0 & \cdots & 0 & 1 & 4 \end{bmatrix}}_{\mathbf{H}} \underbrace{\begin{bmatrix} c(1) \\ c(2) \\ c(3) \\ \vdots \\ \\ c(M) \end{bmatrix}}_{\mathbf{c}} \tag{5.55}$$

Using condition (5.54), the control coefficients can be computed by multiplication of the inverted matrix $\mathbf{H}$ from (5.55) by the vector of original signal values:

$$\begin{bmatrix} c(1) \\ c(2) \\ c(3) \\ \vdots \\ \\ c(M) \end{bmatrix} = 6 \cdot \begin{bmatrix} 4 & 1 & 0 & \cdots & 0 & 1 \\ 1 & 4 & 1 & 0 & & 0 \\ 0 & 1 & 4 & 1 & \ddots & \vdots \\ \vdots & \ddots & \ddots & \ddots & \ddots & \\ 0 & & 0 & 1 & 4 & 1 \\ 1 & 0 & \cdots & 0 & 1 & 4 \end{bmatrix}^{-1} \begin{bmatrix} x(0) \\ x(1) \\ x(2) \\ \vdots \\ \\ x(M-1) \end{bmatrix}. \tag{5.56}$$

The computation of the control coefficients does not necessarily require inversion of the matrix; in fact, even though the matrix is sparsely populated, the inversion for cases of large segment sizes M is not trivial. Observe however that $\mathbf{H}$ is a filter matrix similar to (4.42) relating to a non-recursive filter $H(z)=(z+4+z^{-1})/6$. Apparently, the inverse filter is a non-causal recursive filter of transfer function

$$H^{(-1)}(z) = \frac{6}{z+4+z^{-1}} = \frac{6a}{(1+az^{-1})\cdot(1+az)} \; ; a = 2 - \sqrt{3}. \tag{5.57}$$

This filter has a 'causal' pole at $z_1=a$ within and an 'anticausal' pole at $z_1=1/a$ outside the unit circle, such that it is stable. Factorization of (5.57) gives:

$$\begin{aligned} H^{(-1)}(z) &= \frac{6a}{1-a^2} \cdot \left(\frac{1+az}{(1+az^{-1})\cdot(1+az)} + \frac{1+az^{-1}}{(1+az^{-1})\cdot(1+az)} - \frac{1+a^2+az+az^{-1}}{(1+az^{-1})\cdot(1+az)} \right) \\ &= \sqrt{3} \cdot \left(\frac{1}{1+az^{-1}} + \frac{1}{1+az} - 1 \right). \end{aligned} \tag{5.58}$$

The transfer function is factorized into additive components in (5.58). This means that the segment of length M (e.g. a row of an image) can be processed by super-

---

[1] For other types of extension, refer to (4.43)-(4.44).

position of the results from two independent FIR filters. Herein, the recursion of the 'causal' filter H^+ and 'anticausal' filter H^-

$$H^+(z) = \frac{1}{1 + az^{-1}} \; ; \quad H^-(z) = \frac{1}{1 + az} \tag{5.59}$$

proceeds from left to right and right to left, respectively. Each of the partial filter outputs can be computed recursively, provided that initial boundary conditions are set. The corresponding signal processing equations are

$$y^+(m) = x(m) - a \cdot y^+(m-1) \quad ; \quad 0 \leq m \leq M - 1$$

$$y^-(m) = x(m) - a \cdot y^-(m+1) \quad ; \quad M - 1 \geq m \geq 0 \tag{5.60}$$

$$y(m) = \sqrt{3} \cdot \left(y^+(m) + y^-(m) - x(m) \right).$$

Apparently $y(m)$ maps to $c(m+1)$, such that the control coefficients can be computed recursively by this procedure; it is then only necessary to define boundary conditions for $c^+(0)$ and $c^-(M+1)$, which can e.g. both be set to zero.

Lagrangian interpolation. B-spline interpolation can also be interpreted as an approach of polynomial fitting, where the control coefficients are determined such that for the available positions an exact match of the estimate is achieved, while the estimated function in between should be as smooth as possible. Another common approach of polynomial fitting is *Lagrangian interpolation*, which can again be interpreted by the generic interpolation formula (5.38). The definition of the basis functions as given below in (5.61) implicitly guarantees $\hat{x}[r(m)] = x(m)$.

Lagrangian interpolation, though computationally attractive, has a certain deficiency regarding smoothness of the interpolated signal, where discontinuities of the slope can appear at the control positions. This effect is similar to the first-order interpolator (5.42), which also is the Lagrangian interpolator for $P=1$. The control coefficients are signal samples at the same positions as defined in (5.40), where the basis functions for interpolation of order P are defined as

$$\varphi_k^{(P)}(r) = \prod_{\substack{j=0 \\ j \neq k}}^{P} \frac{r - r(m+j)}{r(m+k) - r(m+j)} \quad \text{for} \quad r(m) \leq r \leq r(m+1). \tag{5.61}$$

5.4.5   Interpolation on Irregular 2D Grids

When *projection* of images is performed, typically a rectangular 2D grid is mapped into an irregular warping grid. If the warped image shall be represented again by a rectangular sampled image matrix, it is necessary to interpolate intermediate values from samples which are positioned irregularly in a 2D plane. Extensions of bilinear interpolation are mostly used in this context, which are in principle based on back-projection of the irregular grid into a rectangular grid. Interpolation is performed from the nearest four samples which span a circumscribing quadrangular polygon; the point to be interpolated should be inside this shape (see Fig. 5.20). Let the difference between the positition to be interpolated, (r,s), and one of the surrounding positions of index p, $p=1,...,P$, $(P=4)$ be defined as the Euclidean distance,

$$\mathbf{d}_p = \begin{bmatrix} r-r(p) & s-s(p) \end{bmatrix}^{\mathrm{T}} \quad \Rightarrow \left\| \mathbf{d}_p \right\| = \sqrt{\mathbf{d}_p{}^{\mathrm{T}} \mathbf{d}_p} \ . \tag{5.62}$$

The interpolated value is then computed as

$$\hat{x}(r,s) = \frac{\displaystyle\sum_{p=1}^{P} \frac{x\big[r(p),s(p)\big]}{\left\|\mathbf{d}_p\right\|}}{\displaystyle\sum_{p=1}^{P} \frac{1}{\left\|\mathbf{d}_p\right\|}} \quad ; \quad \left\|\mathbf{d}_p\right\| \neq 0 \ . \tag{5.63}$$

This interpolation can only be performed if the position to be interpolated is not one of the four (corner) control position coordinates, and if it is included within the area of the polygon spanned by the control positions. The latter condition can be tested, if the difference vector $\mathbf{d}_p$ relating to a control position is right-sided of the vector $\mathbf{v}_p$, which clock-wise connects the current control position with the next:

$$\mathbf{v}_p = \begin{bmatrix} r\big(\mathrm{mod}(p,P)+1\big)-r(p) & s\big(\mathrm{mod}(p,P)+1\big)-s(p) \end{bmatrix}^{\mathrm{T}}$$
$$\Rightarrow \det\begin{vmatrix} \mathbf{v}_p & \mathbf{d}_p \end{vmatrix} \overset{!}{\geq} 0, \quad p=1,...,P. \tag{5.64}$$

(5.64) can also be used to determine suitable control positions. In principle, (5.62)-(5.64) are not restricted to the usage of $P=4$ control positions, it is however difficult to determine what the optimum number P would be, as this highly depends on the geometric constellation. Typically, the quality of the interpolation result is higher, if control positions are available as near as possible to the position that is to be interpolated.

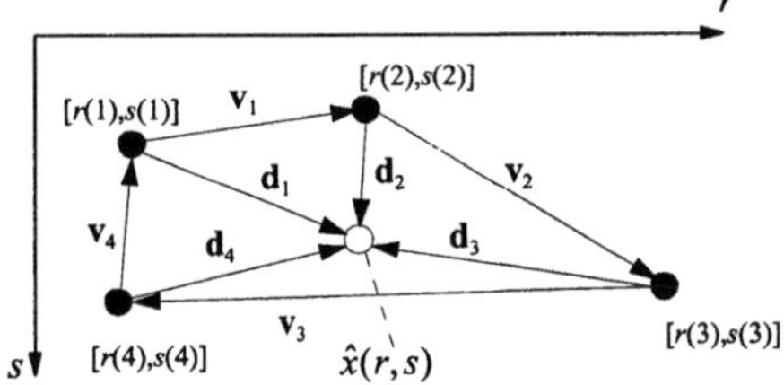

Fig. 5.22. Interpolation of a position $\hat{x}(r,s)$ from the vertices of a quadrangular polygon

Another approach which is directly related to bilinear mapping from four corner control positions was proposed in [WOLBERG 1990] and is based on 2D amplitude surface fitting within the area defined by (5.64),

$$\hat{x}(r,s) = \alpha_0 + \alpha_1 r + \alpha_2 s + \alpha_3 rs \;. \tag{5.65}$$

To determine the coefficients α_i, the following equation system relating to the boundary (corner) conditions must be solved:

$$\underbrace{\begin{bmatrix} x[r(1),s(1)] \\ x[r(2),s(2)] \\ x[r(3),s(3)] \\ x[r(4),s(4)] \end{bmatrix}}_{\mathbf{x}} = \underbrace{\begin{bmatrix} 1 & r(1) & s(1) & r(1)s(1) \\ 1 & r(2) & s(2) & r(2)s(2) \\ 1 & r(3) & s(3) & r(3)s(3) \\ 1 & r(4) & s(4) & r(4)s(4) \end{bmatrix}}_{\mathbf{R}} \cdot \underbrace{\begin{bmatrix} \alpha_0 \\ \alpha_1 \\ \alpha_2 \\ \alpha_3 \end{bmatrix}}_{\mathbf{a}} \quad \Rightarrow \mathbf{a} = \mathbf{R}^{-1}\mathbf{x}\;. \tag{5.66}$$

This method is also extensible to use lower or higher number of control positions and use higher order surface-fitting polynomials. If the number of control positions is higher than the number of coefficients α_i, the solution can be achieved by pseudo inversion (8.20) of the matrix, which gives good results unless the available samples at control positions are subject to untypical variations.

Finally, median filters (see sec. 5.5.1 and Problem 5.7) can also be used for interpolation, where simply the median value is computed from a set consisting of an uneven number of control position values in the neighborhood. If a weighted median filter is used, distance dependent weighting functions can also be included.

5.5   Problems

Problem 5.1
The following image signal shall be given.

$$\begin{bmatrix} 5 & 5 & 15 & 15 \\ 10 & 30 & 25 & 20 \\ 5 & 20 & 25 & 20 \\ 5 & 5 & 10 & 20 \end{bmatrix}$$

a)   At the highlighted positions, perform median filtering using the following unweighted and weighted configurations:
   i) 4-neighborhood $\mathcal{N}_1^{(1)}(m,n)$
   ii) 8-neighborhood $\mathcal{N}_2^{(2)}(m,n)$
   iii) Neighborhood $\mathcal{N}_1^{(1)}(m,n)$, center pixel weighted three-fold
   iv) Neighborhood $\mathcal{N}_2^{(2)}(m,n)$, center pixel weighted three-fold
b)   Sketch the root signals for the median filters of iii) and iv).

Problem 5.2
Sketch the root signals of the median filter geometries shown in Fig. 5.23. Black dots indicate positions of values belonging to the filter masks.

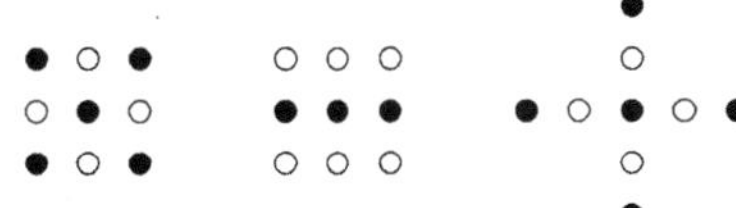

Fig. 5.23. Median filter masks for which root signals shall be found

Problem 5.3

Within the area marked in the following image matrix, perform nonlinear filter operations using 3x3 filter masks. Whenever necessary, use constant-value extensions of the images:
a) Median   b) Maximum value (dilation)   c) Minimum value (erosion)
d) Maximum-difference   e) opening   f) closing

$$\mathbf{X} = \begin{bmatrix} 10 & 10 & 10 & 10 & 20 & 20 & 20 \\ 10 & 10 & 10 & 20 & 20 & 20 & 20 \\ 10 & 10 & 10 & 10 & 20 & 20 & 20 \\ 10 & 10 & 10 & 10 & 10 & 20 & 20 \\ 10 & 10 & 10 & 10 & 20 & 20 & 20 \end{bmatrix}$$

Problem 5.4

Apply morphological filters to the binary images (black=1, white=0) shown in Figure 5.24. Use a neighborhood system $\mathcal{N}_2^{(2)}(m,n)$ from (4.1) for the shape of the structure element.
a)   Filtering by an erosion filter. Erode the image until only one pixel is retained. Which will be the coordinate position of this pixel? How many iterations of erosion must be performed?
b)   Filtering by a closing filter. Which effect do you observe?

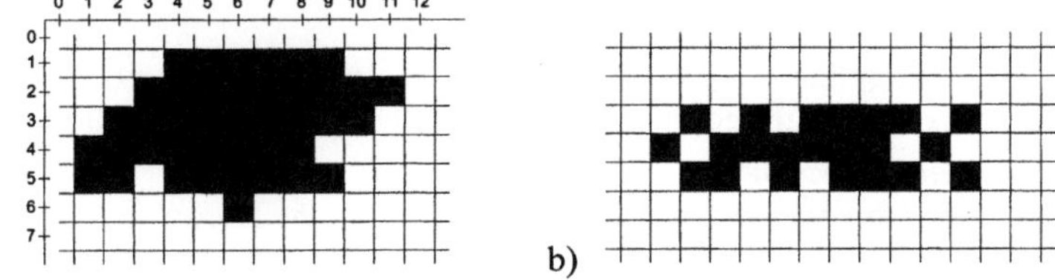

Fig. 5.24. Binary image shapes to be eroded (a) and closed (b)

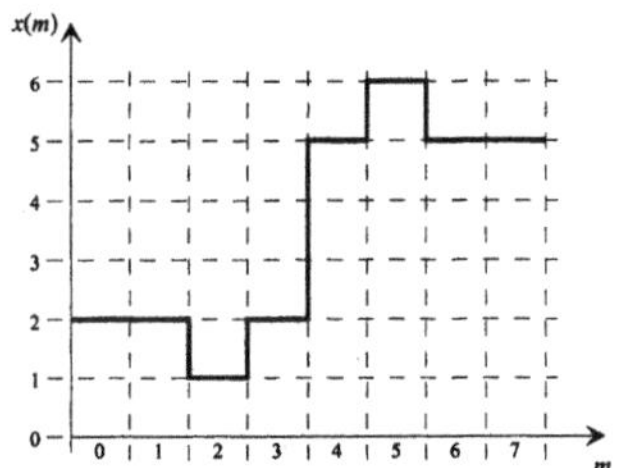

Fig. 5.25. 1D amplitude shape

Problem 5.5

For the 1D signal $x(m)$ shown in Fig. 5.25, sketch the results by the following nonlinear filter operations. Use a structure element of length 3 and assume constant-value extension at the boundary:

i) Median filter   ii) Maximum-difference filter   iii) Erosion filter
iv) Dilation filter   v) Opening filter   vi) Closing filter

Problem 5.6
Determine the inverse of the mapping characteristic (5.22).

Problem 5.7
The amplitude values of an image signal (2,3,4,5) shall be known at the positions shown in Fig. 5.26.

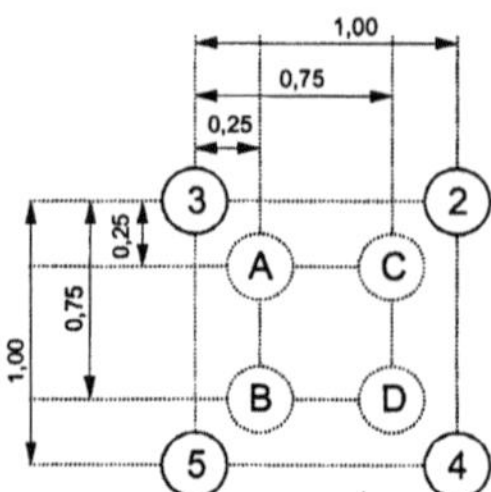

Fig. 5.26. 1D amplitude shape

a)   Determine the intermediate values **A,B,C,D** by median filter interpolation. Herein, the median values shall be computed from each three sampling positions which are nearest to the interpolated position.
b)   At positions **A** and **B**, determine the deviations of median-interpolated values from the values of bilinear interpolation (5.35).

Problem 5.9
Basis functions of quadratic spline interpolation are piecewise defined as in (5.49).
a)   Construct the matrix form of the equation to determine the value $\hat{x}(r)$ from the coefficients c_0, c_1, c_2 in dependency of r'.
b)   Determine conditions to compute the coefficients $c(m)$ from the known samples of a signal.

Part B: Content-related Multimedia Signal Analysis

6 Perceptual Properties of Vision and Hearing

Optimization of multimedia signal coding algorithms for best subjective quality, and adaptation of signal identification methods for behavior similar to human senses requires knowledge about perceptual properties of vision and hearing. Both in coding and recognition it is important to distinguish the information contained in the signals into categories of relevant and irrelevant components. Physiological properties of the human perception system, e.g. characteristic properties of eyes and ears have a direct impact here. In particular masking effects, which are well known to be in effect both in vision and hearing, will be explained in this chapter. Further, it should be noticed that certain elements of human vision and hearing have a high similarity to the technical systems presented in this book, and could as well be regarded as methods of signal and information analysis and processing.

6.1 Properties of Vision

6.1.1 Physiology of the Eye

Light rays intruding into the eye (Fig. 6.1) reflect a natural scene from the outside world. These are bundled towards the retina by a lens which is protected by the cornea. Upon the retina, light-sensitive receptors perform a sampling of the image that is projected from the outside. Two types of receptors exist, which are the *rod cells* (around 120 million units), and the *cone cells* (around 6 million units). Both types of receptors contain light-sensitive pigments, where the rods are specialized for acquisition of the brightness, while the cones are capable to perceive brightness *and* color; three different types of cones are tuned to perceive red, green and blue primaries of the color spectrum. The distribution of the receptors is not uniform over the retina. In the area with highest density of receptors, the *fovea centralis*, mostly rods are present. At the periphery of the retina, mostly cones can be found.

The information from the receptors has to be forwarded towards the brain. For this purpose, the light information is transformed into electrical information. This is done by an electro-chemical reaction influenced by the intensity of the light, which modulates the frequency of sharp electrical pulses. Approximately 800,000 nerve fibers perform the transport of visual information from the eyes to the brain, which means that a mechanism of data compression is executed within the retina, as the number of receptors is higher by a factor of ≈ 150. This is in fact done by conveying only differences appearing in the information that is acquired by adjacent receptor cells. Four different types of neurons in the retina perform an analysis of local similarity between the information propagated by the receptor cells: *horizontal cells*, *bipolar cells*, *amacrine cells* and *ganglion cells*[1]. The first layer of neurons only propagates an excitation signal to the next layer when the difference between the neighbored receptor inputs is *above a threshold*. Further layers proceed similarly by evaluating inputs from neurons at the previous layer[2].

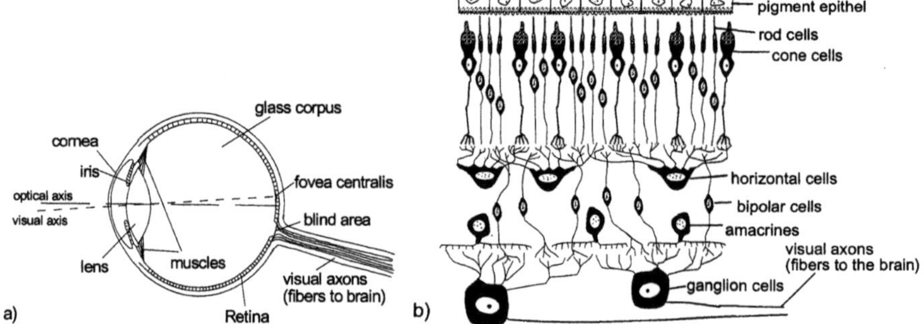

Fig. 6.1. Physiology of the human eye **a** Overall structure **b** Structure of the retina

The visual nerves propagate the information towards the visual center of the brain, the *cortex praestriata*. It is assumed that certain precincts of the visual cortex are responsible to process and analyze the information originating from related areas of the retina. Depending on the information which is input, specialized structures in each precinct are responsible to detect edges, lines, slits etc. of different orientations and resolutions. This theory of vision supposes that the early visual processing stages of the brain perform similar operations as technical computer vision or image processing systems, e.g. oriented edge analysis and detail (frequency) analysis to discover the width of structures. To justify this theory, experiments applied on apes and cats have indicated that specialized areas of the brain (denoted as *hypercolumns*) are primarily active if horizontal or vertical stripes are shown to the subject [HUBEL, WIESEL 1962].

---

[1] The schematic drawing in Fig. 6.1b is physiologically not fully correct, as the horizontal and bipolar cells are rather found within the outer layers of the retina.

[2] This structure of the retina has been a model for a specific structure of *artificial neural networks*, the *multi layer perceptron* (see sec. 9.4.5).

Further, specialized areas of the visual cortex are responsible to process information and detect basic structures separately from the inputs of the left and right eye. Presumably, further processing of the visual information is performed in the brain areas below the cortex. This includes detection and recognition of more complex shapes and objects, stereoscopic vision etc. Finally, as *differences of excitations* are processed by priority, objects moving over time draw higher attention than objects which are static within a scene. This is related to the necessity to protect the animal or human from danger or any threat occurring in the outside world.

In principle, due to the extremely high volume of visual data, only massive parallel processing performed both in the retina and in the brain can enable the challenging task of analyzing visual information. Until today, machine vision appears to be quite immature as compared to the capabilities of the human or animal vision and recognition systems.

6.1.2   Sensitivity Functions

The amplitude sensitivity of the *Human Visual System* (HVS) follows *Weber's law*, which states that a perceptible difference in brightness ΔL has a constant ratio compared to absolute brightness:

$$\frac{\Delta L}{L} = const. \approx 0,02 \qquad\qquad (6.1)$$

This law has frequently been utilized in technical systems for processing and display of visual information. An example is the nonlinear amplitude transfer characteristic of video cameras, cathode ray tubes and film material, which implicitly reduces the visibility of noise in dark areas. In addition to the global ratio $\Delta L/L$, noise fluctuations and other local differences are masked in areas of high detail, while they are more clearly perceived in flat areas. This property is however somewhat congruent to sensitivity functions dependent on spatial frequency, which will be introduced below.

Subjective perception of contrast is dependent on angular orientation and on spatial frequency (or detail) of a structure. Regarding angular orientation, experiments have shown that humans are less capable to recognize amplitude differences of diagonal stripe structures, as compared to horizontal and vertical structures of same stripe period. Due to this effect, more preference should be given to horizontal and vertical structure orientations, both in image coding and recognition.

The visual sensitivity as depending on spatial frequency and directionality is typically measured by showing periodic signals (e.g. sinusoids or stripes) to test subjects. Just noticeable amplitude differences with a given stripe period (width) and orientation are then mapped into the value of a frequency-dependent sensitivity function. The spatial frequency measurement unit used in the observation of a regular stripe or sinusoid pattern should be normalized, such that it becomes independent of the viewing distance. An appropriate unit to achieve this is *cycles per degree* of the viewing angle (*cy/deg*), as illustrated in Fig. 6.2a. Assume a stripe structure is viewed from a distance d, such that the width of one stripe, the period λ, maps into an angle φ. For half viewing distance $d/2$, the width of the same stripes as mapped to the retina would be doubled, or the same width would be perceived if a stripe structure has only half period $\lambda/2$. The spatial frequency f_{cy} is defined as the reciprocal value of φ,

$$\varphi = 2\arctan\frac{\lambda}{2d} \quad \Rightarrow f_{cy} = \frac{1}{\varphi}\left[\frac{cy}{deg}\right]. \tag{6.2}$$

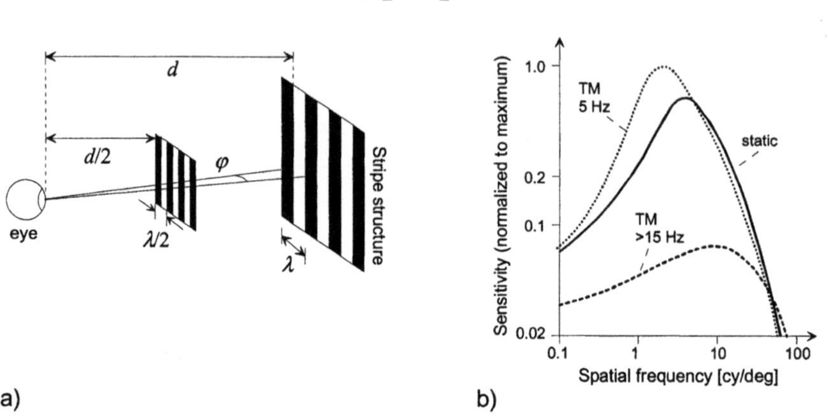

Fig. 6.2. **a** Definition of spatial frequency by *cy/deg*   **b** Contrast sensitivity depending on spatial frequency for static case and with two different temporal modulations (TM)

In Fig. 6.2b, typical sensitivity curves parametrized over the spatial frequency are shown. Slightly different results on this are reported in the literature, whereas it is justified to assume that the qualitative properties as shown in Fig. 6.2b are applicable. In particular, the unit of measurement is often vague, such that it should be normalized for a maximum sensitivity level. For static stripe patterns (no change over time), the maximum sensitivity is reported to be around 3-5 *cy/deg*, and drastic decay of the sensitivity function is found towards lower and higher frequencies; the lower sensitivity towards higher frequencies can directly be interpreted by the resolution limitation of the HVS, where the number of receptor cells on the retina is finite. The decay towards lower frequencies may also be explained by the physiology of the retina, where excitation difference processing seems to work more effectively in the small scale investigated by the early neuronal processing stages.

Example. 5 *cy/deg* according to (6.2) approximately relates to a 2 cm wavelength of a sinusoid when the viewing distance is 2.90 m. The optimum viewing distance towards a TV screen is around a factor of 4-6 over the screen diagonal size, such that here the signal should be displayed on a screen not smaller than 49 cm diagonal, which gives a horizontal size of 39 cm when the aspect ratio is 4:3. Now assume a digital signal of standard TV resolution, which has 704 active pixels horizontally. In this case, the horizontal sampling distance of pixels on the screen would be $\approx$0.554 mm. Assume now that a DCT block transform of size 8x8 pixels is applied to the signal. The length of the block sides would be 4.43 mm. As the first basis function of the DCT consists of half a cosine period over the block size (cf. Fig. 4.20), the coefficient $c_{1,0}$ represents a sinusoid of roughly 1 cm wavelength, which is already beyond the frequency of maximum sensitivity. For half of this viewing distance, the coefficient will roughly represent the frequency of maximum sensitivity. This means that for any higher-frequency coefficients, the frequency sensitivity graph is yet in the range of strong roll-off characteristic; consequently, higher coding errors can be allowed for higher-frequency coefficients, as these are masked by the HVS. This property is widely exploited in still image coding.

Spatial sensitivity is not independent of temporal sensitivity. For example, the effect of *line flicker* is well-known in interlaced video: The slight decay of amplitudes that happens from one field period to the next is recognized as spatial variation between adjacent lines, which becomes visible in large uniform areas. This effect is however quite specific for interlaced display devices, and is indeed rather an aliasing problem. If lines of one field are black and lines of the other field are white, the impression is even more annoying to the eye. To measure spatio-temporal transfer functions of the visual system, a spatial pattern of certain frequency (e.g. a sinusoid or stripe) must be modulated over time. Fig. 6.2b illustrates this effect, showing in addition a spatial frequency sensitivity function for a value near the maximum of temporal sensitivity, which is reported to be between 5 and 10 Hz; the qualitative behavior is the same as for the static case, but the maximum is shifted towards slightly lower spatial frequency. The temporal sensitivity function decays drastically for temporal frequencies above 10 Hz[1], however in this case, even though globally at a lower level, the maximum of spatial sensitivity is again shifted towards a higher spatial frequency [ROBSON 1966].

Effects of spatio-temporal sensitivity functions have hardly been utilized in video data compression so far, which is partially due to their strong dependency on viewing conditions. It was pointed out in [GIROD 1993] that the sensitivity for temporal frequency changes dramatically, if the observer is allowed to *track the motion*. The eye then performs a *motion compensation*, and is hence able to detect much higher spatio-temporal frequency components; in principle, the temporal spectral component is 'de-skewed', the effects discussed in sec. 2.1.2 are reversed. The structure of a fast moving object in a movie or video can indeed only be perceived when it is tracked; staring at only one position on the screen will lead to very strange perceptual effects. In particular, it can be concluded that for parts of the scene which can well be tracked, such as smooth-moving objects, coding errors will have a higher visibility than in case of 'chaotic' motion, e.g. objects with fast changes in illumination and shape, deformations etc.

---

[1] Without this decay, it would not be possible to perceive fluent motion in movies.

6.1.3   Color Vision

The retina holds three different types of cones which are specifically tuned to the acquisition of the red, green and blue primaries. The entire color spectrum however consists of visible light components of many different tones (frequencies). The visible spectrum covers wavelengths between approximately 400 nm (boundary of ultraviolet range) and 700 nm (boundary of infrared range). Mixtures of spectral components appear as unique (but non-primary) color tones. Perception of such mixtures by the visual system is possible, as the receptors are not performing narrowband analysis of the primaries; the wavelengths of the primaries can be interpreted as center frequencies of bandpass systems with spectral overlap. Fig. 6.3 shows the receptor potentials, which relate to the bandpass functions of red, green and blue receptor sensitivities, normalized by their respective maxima. In general, due to the large overlaps, information about the three primaries enables the observer to perceive pure color tones very clear, and also provides sufficient means to distinguish color mixtures as well. Some special cases exist where color tones can not uniquely be identified. An example are spectra which consist of multiple narrow peaks with sharp cut-off. This is one reason why perception of certain colors, e.g. golden and brown tones, are subject to rather subjective impression.

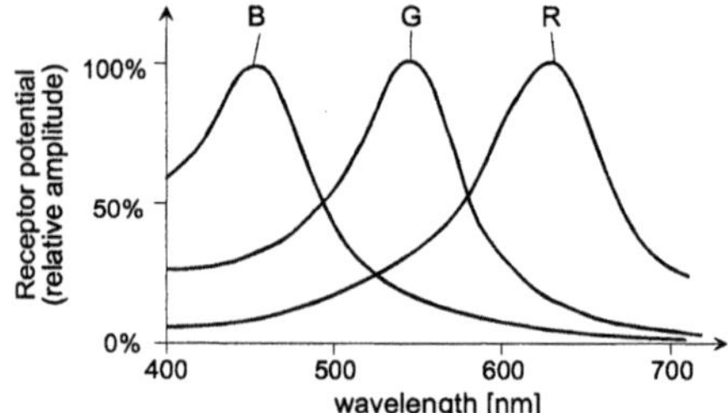

Fig. 6.3. Normalized sensitivity transfer functions of the three types of cone receptors (schematically)

It can be assumed that a preliminary interpretation of color is performed by an early stage of vision, in the neurons of the retina. As the number of cones is lower than the number of rods, the spatial resolution of color perception is lower than it is the case for brightness reception. As a consequence, encoding of chrominance components in images can usually be done by lower resolution than for the luminance (cf. sec. 1.3.1). In the multiplication factors used to transform (R,G,B) into luminance and chrominance components, additional properties of the HVS are taken into account. It can be observed that one dominant component may suppress the influence of the other two. This is specifically true for the green component, which falls into the spectral range where the sensitivity of the HVS is highest, while for the blue component it is clearly lowest. In particular, highly saturated green tones of different hue can much better be differentiated than it is the case for blue tones.

Color perception is one of the most subjectively biased components of the HVS, and can easily be jeopardized. Even if relatively false colors are presented to an observer, the visual sense adapts after a short period of time. The human observer is also able to adapt on illuminant conditions, such that objects for which the color is known are perceived subjectively as *having* this color, which is probably mainly due to associative processes in the brain. Further, the color sensitivity curves will largely vary among different individuals, not even considering exceptionalities like color blindness. In the interest of a "physically objective" representation of colors, it could be necessary to use more than 3 color channels for representation of the visible color spectrum, and also use a physical reference relating to illumination conditions (see sec. 7.1.1).

6.2   Properties of Hearing

6.2.1   Physiology of the Ear

A global schematic diagram of the human ear is shown in Fig. 6.4. It can roughly be described by the parts of *outer ear*, *middle ear* and *inner ear*. The task of the outer ear is to receive acoustic waves from the outside world and forward them by the *auditory canal* towards the *eardrum*. Length and shape influences the resonance frequency of the canal, taking influence on the frequency sensitivity peak of the auditory sense which is between 1 KHz and 4 KHz. The main task of the middle ear, starting at the eardrum, is a transformation of the air sound pressure into the hydrodynamic pressure which is processed by the inner ear. This is achieved by the three *ossicles* (*hammer*, *anvil*, *stirrup*), which direct the pressure from the eardrum towards the *oval window*, the entrance of the inner ear. The inner ear is entirely filled by fluid, and all operations are effected by hydro-mechanic oscillations. The central part of the inner ear is the *cochlea* (which looks like a snail's house of about 2½ windings). It consists of three parallel channels, also denoted as *scales*. These are separated by membranes, where the *scala media* (the middle one) is filled by another fluid than the *scala vestibuli* and the *scala tympani*.

The *basilar membrane*, which plays a central role in audible perception, is connected to the scala tympani. It is further connected to *Corti's organ*, which contains the receptor cells. By the construction of the scales, which have different widths from the input towards the inner parts of the cochlea, the basilar membrane performs a *frequency-space-transformation*. This means that different tone frequencies will effect maximum resonance at different locations of the basilar membrane, which can be interpreted like a filter bank of bandpass systems [ZWICKER 1982]. The oscillations of the basilar membrane are received by Corti's organ through *hair cells*, where the attached synapses perform a transformation into nervous pulses by a chemical reaction. The transmission towards the brain is performed by around 30,000 nerve fibers. It is interesting to note that a significant amount of audible information processing is already done before the information is directed towards the brain. In particular the relatively raw frequency analysis which is performed by the basilar membrane has a significant effect on psychoacoustic sensitivity functions discussed in the following sections.

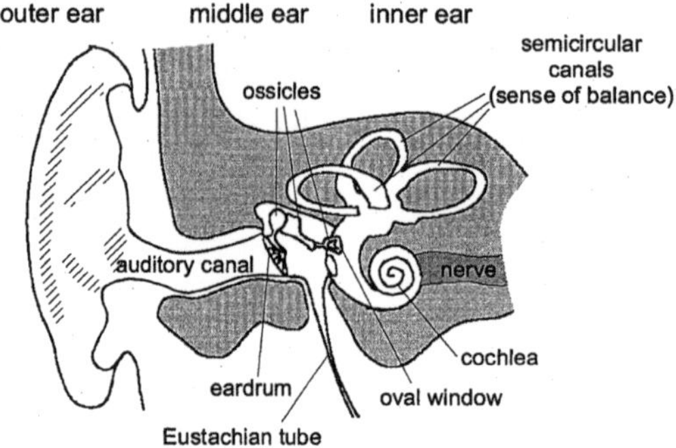

Fig. 6.4. Schematic diagram of the human ear

6.2.2   Sensitivity Functions

In hearing, spectral properties (related to analysis of the 1D temporally-dependent signal) and amplitude (sound level or loudness) are the most important parameters for subjective impression. While in spectral color mixtures, only single color tones are perceived, the sense of hearing is able to finely separate single audible tones or sounds according to spectral and temporal properties. This is in particular true for *harmonic sounds*, where spectra are composed by a pitch frequency and its harmonics, a case that applies to many types of musical instruments and also to voiced segments of speech. An important criterion for the physical intensity of audio signals is the *sound pressure level*

$$L = 20 \cdot \log \frac{p}{p_0} \; dB \, , \tag{6.3}$$

where the normalization factor $p_0 = 20 \mu Pa$ relates to a sound intensity of 10^{-12} W/m^2, which references to the human *threshold of hearing* at a frequency of 1 KHz. In general, the threshold of hearing is frequency dependent; in particular, very low and very high frequencies can only be perceived when the sound pressure level is considerably higher. Fig. 6.5 shows the frequency-dependent threshold of hearing, the upper *threshold of pain*, and typical ranges of frequency and sound levels of music and speech. The spectral range of sounds audible by humans approximately covers the range between 16 Hz and 16 KHz, which are more than 3 decades on a frequency scale.

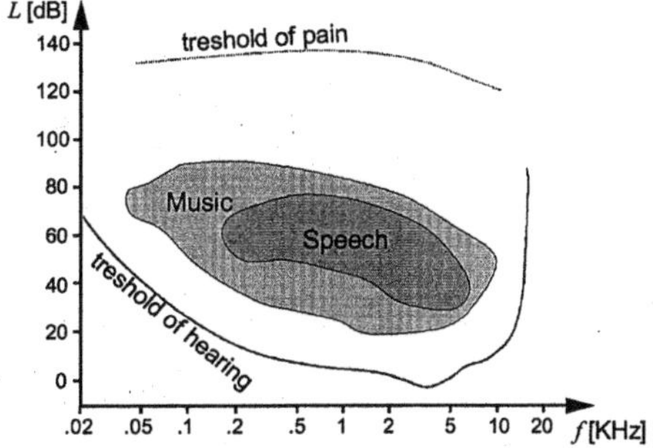

Fig. 6.5. Parameters of human hearing depending on sound pressure level and frequency (adopted from ZWICKER)

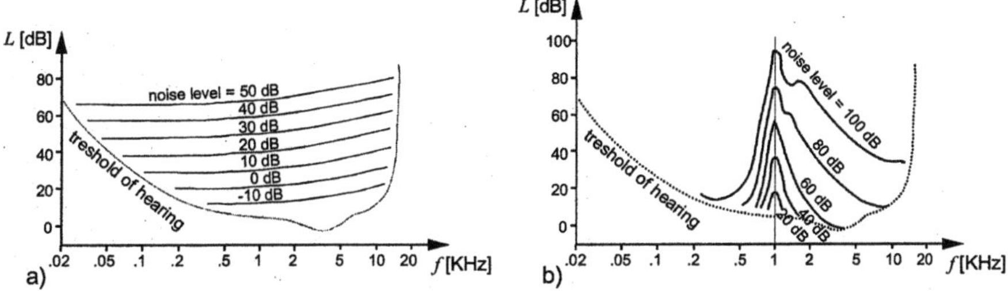

Fig. 6.6. Masked threshold of hearing with interfering signals of different sound pressure levels: **a** White noise **b** Narrow-bandwidth noise of mid frequency 1 KHz (adopted from ZWICKER)

Important effects that are exploited in audio signal compression are based on the *masking properties* of human hearing. In principle, distinct sounds are differentiated by their spectral properties. This is however only true if not another sound source is present which acts as interference or *masking signal*. Further, the strength of the masking effect depends on the sound pressure level of the masking signal. Fig. 6.6a shows the frequency-dependent *masked threshold of hearing*, where the masking signal is broadband white noise of sound pressure levels between −10 dB and +50 dB. It is evident that the threshold for audibility of other tones is increased substantially above the original threshold of hearing. White noise as masking signal is not very realistic, however. To determine effects of mutual masking for natural signals, it is more useful to investigate the masking threshold(s) when the masking signal is narrow-bandwidth noise. This is shown in Fig. 6.6b for the example of an interfering narrow-bandwidth signal of mid frequency 1 KHz and different sound pressure levels. It is obvious that the masking takes higher effect on frequencies which are above the frequency of the interfering signal.

The masking effect is also noticeable when the interfering signal is not of narrow-bandwidth nature, but a harmonic signal. Fig. 6.7 shows that masking is in effect over the entire range of the harmonic spectrum, starting from the pitch frequency.

The effect of higher disturbance towards higher frequencies can be observed likewise.

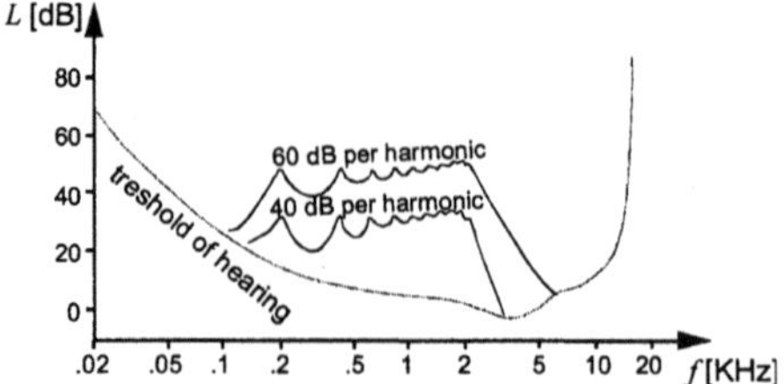

Fig. 6.7. Frequency-dependent masking in case of harmonic signal as interference, pitch frequency 200 Hz (adopted from ZWICKER)

It has been shown in psycho-acoustic experiments that a group of pure sines within a narrow frequency band is perceived like a narrow-bandwidth noise signal in this frequency band. It was further found that two pure tones must be distant by a minimum frequency offset to be perceived as separate. This leads to the conclusion that the frequency resolution of the audible sense is finite. For criteria which better reflect the psycho-acoustic properties, the definition of *frequency groups* was introduced [ZWICKER 1966]. It is assumed in this context that the human hearing system does not analyze sounds of discrete frequency, but is rather operating like a filter bank of bandpass systems. Each filter is tuned to a specific mid frequency f_0, having bandwidth Δf_G. The bandwidths shall be defined such that discrete frequencies (sinusoids) within the same frequency group are not perceived as disparate sounds. It turns out that neither the mid frequencies f_0 establish a linear scale on the frequency axis, nor are the bandwidths Δf_G equal over all groups. Fig. 6.8a shows that the width of the frequency groups is approximately constant up to f=500 Hz ($\Delta f_G \approx 100$ Hz), and is approximately proportional (linearly increasing) with the frequency within the range above this frequency. The overall conclusion was made that the range of the audible spectrum should be separated into 24 distinguishable frequency groups, assuming that the audible sense judges the spectrum of a broad-band sound (e.g. a complex music signal) according to the spectral energy or power appearing within these frequency groups. Based on this assumption, the *Bark scale* was defined, where the artificial unit Bark[1] expresses the *toneness* of a sound. The basis is the frequency-group partitioning as illustrated in Fig. 6.8a, describing the nonlinear mapping from the linear frequency axis f onto the linear (in the perceptually adjusted domain) Bark scale. An approximation of this transformation is expressed by the formula

$$\frac{f}{Bark} = 13 \arctan\left(0.76 \frac{f}{KHZ}\right) + 3.5 \arctan\left(\frac{1}{7.5} \frac{f}{KHZ}\right). \tag{6.4}$$

---

[1] named after the acoustic scientist Barkhausen.

Fig. 6.8b shows the mapping of the audible range of frequencies onto the discrete values of the Bark scale, ranging from 0-24, which define the lower and upper bounds of the 24 frequency groups. Comparing two sounds with regard to their Bark-scale related spectral energy distribution leads to a similarity measure which much better approximates the properties of the human hearing system.

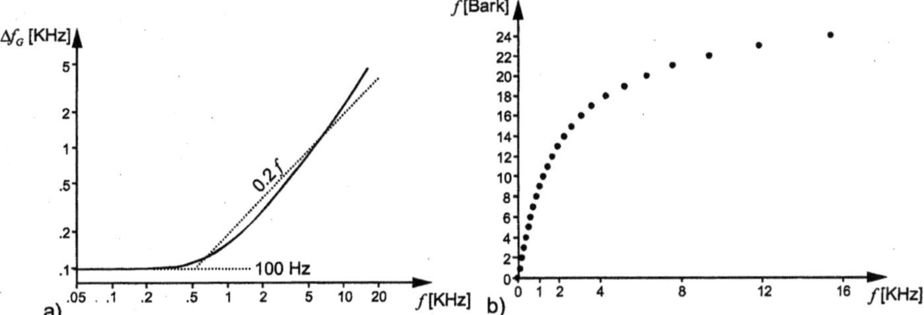

Fig. 6.8. **a** Relationship between frequency and frequency group width within the audible spectral range **b** Definition of discrete frequencies on the *Bark scale* (adopted from ZWICKER)

The *static* properties of frequency-dependent audibility effects can primarily be characterized in the frequency domain as done so far. In addition, a *dynamic* behavior of masking can be shown, which is in particular effective if masking sounds are switched on or off. This can only be analyzed in the temporal domain. Fig. 6.9 shows the temporal *pre and post masking effects* (which are dynamic) and the *simultaneous masking effect* (which is static). Here, an interfering sound (noise signal) was switched on for a duration of 200 ms. The effect of post masking is quite plausible, as the hearing system has adapted itself to the masking sound, such that it takes a short period until another tone which was masked before is detected. The pre masking effect is more surprising at first sight. Apparently, this is related to the 'processing time' within the chain from the ear to the brain. A loud sound event suddenly arriving seems to interrupt the processing of previous events by its higher priority.

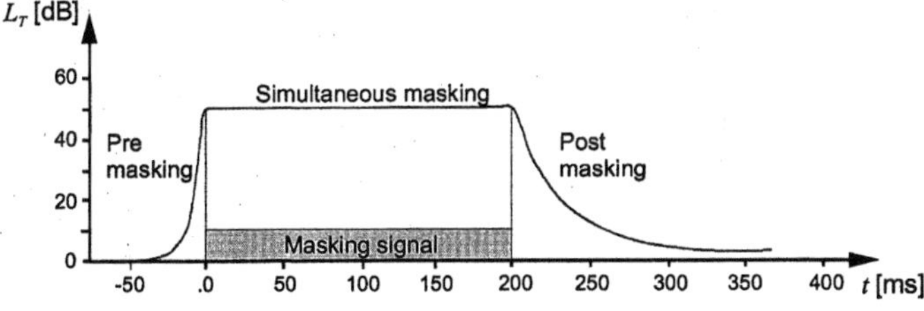

Fig. 6.9. Pre and post masking effects when an interfering signal (white noise) is switched on and off (adopted from ZWICKER)

7 Features of Multimedia Signals

Feature extraction is the first step towards an abstraction, to gain content-related information from the signal or make two signals comparable by criteria beyond sample-wise comparison. Usually, the output of feature extraction are sets of parameters which allow classifying the signal based on feature properties. Feature description parameters should in many application cases be invariant against typical distortions and variations that can occur to the signal, like geometric modifications, noise, illumination changes etc. Many feature description methods are hence based on statistical models, or related to statistical parametrization of the feature properties. Important features of image and video signals are color, texture, shape, geometry, motion, depth and spatial (2D/3D) structure; for audio signals, envelopes (hulls) of the temporal or spectral representations, timbre, modulation of tones, loudness, tempo, harmonic structure, melody and structure of music pieces are important features. Most of the feature extraction and representation methods introduced in this chapter have direct or indirect relationships to the physical generation of signals or to perceptual relevance; some features can directly be related to semantic meaning.

7.1 Color

For human visual perception, color is an important distinction criterion to recognize exterior-world objects; similarly, content-based analysis of image and video signals often relies on color criteria. The spectrum of visible light (wavelengths ~400 nm…700 nm) is roughly sampled by the camera acquisition, when typically only the primary color components R (red), G (green) and B (blue) are output. For certain applications, e.g. satellite and night image analysis, multi-spectral sensors and infrared cameras are also used, which actually provide more information than distinguishable by the human eye. Image sensors possess a bandpass characteristic related to color spectrum sensitivity. In case of RGB sensors, and even more for multi-spectral sensors, bandpass transfer functions are overlapping. Hence, it can be expected that different color channel signals are highly correlated and redundant. If the number of color channels is K, a *color space* of K dimensions is spanned; the coordinate axes of this space relate to the amplitude levels of the respective components. Conceptually, this is not different from vector spaces signifying combined signal samples (cf. sec. 3.2) or features (cf. sec. 9.1). The property of a color pixel at a given spatio-temporal position can then be represented by a color vector $\mathbf{x(n)}$ of dimension K, with the position defined over κ dimensions by the index vector $\mathbf{n}$.

7.1.1   Color Space Transformations

Due to following reasons, the color representations provided by image sensors are often not directly suitable for color classification:

–   Redundancies between the different color components exist, which in particular for the case of multispectral images leads to an unnecessarily high dimensionality of the color space;
–   The color space of capture may allow color combinations which barely are present in natural images or may be irrelevant to distinguish colors;
–   The (linear) Euclidean distance between two colors in the acquisition color space is not linearly related to the subjective perception about the dissimilarity between these colors.

Hence, it is advantageous for color feature extraction to use color spaces where the components are statistically independent, which support the natural occurrence of colors, providing as compact representation as possible, and which are adapted to the human color perception properties. Indeed, a large number of different color space representations have been proposed, which by the relationship between the acquisition color space and the representation color space can roughly be categorized into linear transforms and nonlinear transforms. Linear color transforms can be described by a matrix operation. The number of output components must not necessarily be identical to the number of input components actually acquired; a very generic form of a linear color transform can be expressed by the following mapping of K input components x into L output components x':

$$\begin{bmatrix} x'_1(m,n) \\ x'_2(m,n) \\ \vdots \\ x'_L(m,n) \end{bmatrix} = \begin{bmatrix} f_{11} & f_{12} & \cdots & f_{1K} \\ f_{21} & f_{22} & \ddots & \vdots \\ \vdots & & \ddots & \\ f_{L1} & \cdots & & f_{LK} \end{bmatrix} \cdot \begin{bmatrix} x_1(m,n) \\ x_2(m,n) \\ \vdots \\ x_K(m,n) \end{bmatrix} \qquad (7.1)$$

With $L<K$, the dimensionality of the color space to be processed is reduced; the goal would be to transform the information into a small number of statistically independent components, retaining most information about the input color properties. For multispectral images (e.g. satellite images), this can be achieved by appli-

cation of the *Karhunen-Loève Transform* (4.147) or by independent component analysis (cf. sec. 9.1.2).

The source representation by (R,G,B) color triplets is often decomposed into one luminance component (Y) and two color difference or chrominance components. As an example, the (Y,C_b,C_r) transform given in (1.6) can be expressed as a matrix transform

$$\begin{bmatrix} Y \\ C_b \\ C_r \end{bmatrix} = \begin{bmatrix} .299 & .587 & .114 \\ -.169 & -.331 & .5 \\ .5 & -.419 & -.081 \end{bmatrix} \cdot \begin{bmatrix} R \\ G \\ B \end{bmatrix}. \tag{7.2}$$

This type of transforms also enables a compatibility between color and monochrome (Y) representations by supplementing the chrominances as difference information, which can even be represented by a reduced sampling rate (cf. sec. 1.3.1).

Definition of color references. In the color representations introduced so far, any relationship of primary components with physical color references was disregarded. Such a relationship could only be set, if the lighting conditions during the capturing of an image were known, because objects will appear by different colors if illuminated differently. Firstly, it is necessary to define the contributions of the primary color components for the case when a narrow-bandwidth light (one pure color tone) shall be output. The spectral mid frequencies and spectral transfer functions of the primaries were specified in the standard CIE 15.2, related to the physical characteristics of cathode-ray display devices (luminescent phosphor). It was found however that it is not possible to synthesize *arbitrary* single color tones of the visible spectrum by mixing these primary components; a certain range of wavelengths between blue and green tones exists where indeed a negative excitation of the red component would be necessary, which is in fact not achievable physically.

As a consequence, it is also not possible to generate a white reference light (containing all color tones of the visible spectral range such that they appear perceptually by same brightness) from (R,G,B) components. To circumvent this problem, a system of primary components was to be defined, which would allow to generate a white reference without negative excitation of any component. This resulted in the definition of the CIE (X,Y,Z) primary component system, where Y still expresses the luminance component as in (7.2):

$$\begin{bmatrix} X \\ Y \\ Z \end{bmatrix} = \begin{bmatrix} .607 & .174 & .200 \\ .299 & .587 & .114 \\ .000 & .066 & 1.116 \end{bmatrix} \cdot \begin{bmatrix} R \\ G \\ B \end{bmatrix} \tag{7.3}$$

A reference to a physical color can now be set by defining the weighting factors for (X,Y,Z) primaries which will produce a unique white light reference. The perception of white color is however highly subjective, but for a technical system it is sufficient to rely on a normative definition. As an example, the white reference D65 (related to a *color temperature* of 6500 K) is defined by the values $X_0=0.3127$, $Y_0=0.3290$, $Z_0=0.3583$. Irrespective of the actual illumination color, it can be characterized by an (X,Y,Z) triplet in relationship to the standardized white reference light source. This can be used to analyze the actual physical colors within an image as if it had been captured under the standardized illumination source.

The transform (7.2) is not orthogonal (cf. Problem 7.1). Hence, (4.161) is not applicable and the Euclidean distance between two color triplets in the (R,G,B) color space, (R_1,G_1,B_1) and (R_2,G_2,B_2),

$$d = \sqrt{(R_1 - R_2)^2 + (G_1 - G_2)^2 + (B_1 - B_2)^2} \, , \tag{7.4}$$

cannot directly be mapped into the Euclidean distance between the corresponding triplets in (Y,C_b,C_r). If further the Euclidean distance is only computed over the chrominance subspace (C_b,C_r), the result would be quite different from the distance in (R,G,B) space. Abstraction from possible differences in brightness will in particular be better related to the human perception about differences between colors, which are expressed primarily by the chrominance components. In general however, linear transforms are not fully suitable to resolve the nonlinear relationship between (Euclidean) measured and perceived differences. An example for a nonlinear transform is the (H,S,V) transform[1]; in the following definition, $H=0$ expresses the red primary color. $A_{\max}$ is the maximum amplitude range of the components in the (R,G,B) space. The outputs are normalized for values $0 \leq (V,S) \leq 1$, $0 \leq H \leq 2\pi$:

$$R' = R / A_{\max}; \quad G' = G / A_{\max}; \quad B' = B / A_{\max}$$

$$MAX = \max(R',G',B') \, ; \quad MIN = \min(R',G',B')$$

$$V = MAX; \; S = 1 - \frac{MIN}{MAX} \; \text{if } V > 0 \tag{7.5}$$

$$H' = \arccos \frac{\frac{1}{2}\left[(R'-G')+(R'-B')\right]}{\left[(R'-G')^2 + (R'-B')(G'-B')\right]^{\frac{1}{2}}} \; \text{if } S > 0$$

$$H = \begin{cases} H' & \text{if } B' \leq G' \\ 2\pi - H' & \text{if } B' > G' \end{cases}$$

The (H,S,V) transform is nonlinear, but invertible (see Problem 7.2). As the (R,G,B) input values are bounded in amplitude, the transformation into polar coordinates results in an outer boundary of the (H,S,V) color space formed by a cylinder (Fig. 7.1a), where the rotation axis is V, and sections perpendicular with this

---

[1] *H*ue, *S*aturation, *V*alue. Other similar transforms exist, e.g. *H,S,I* (*I* for intensity) and *H,L,S*. For an overview on these definitions, see [PLATANIOTIS, VENETSANOPOULOS 2000].

axis are circles (Fig. 7.1b). The angular position is represented by H and the radial position by S. $S=0$ on the central axis corresponds to a gray value, while $S=1$ is a fully saturated color, which means at least one of the three components is entirely missing. On the hue circle, three pure *secondary colors* Yellow, Cyan and Magenta can be found, which are equally-balanced mixtures of *only two* adjacent primaries. Primary and secondary colors are differentiated by angular orientation steps of $\pi/3$ each.

Fast algorithms exist, which allow approximating H without computation of squares, square roots and trigonometric functions. One example is the following definition, which has been used for the specification of the (H,S,V) color transformation in the MPEG-7 standard:

$$H = \begin{cases} \dfrac{G-B}{MAX-MIN} \cdot \dfrac{\pi}{3} & \text{if } R=MAX \ \wedge \ B=MIN \\[3ex] 2\pi - \dfrac{B-G}{MAX-MIN} \cdot \dfrac{\pi}{3} & \text{if } R=MAX \ \wedge \ G=MIN \\[3ex] \left[2+\dfrac{B-R}{MAX-MIN}\right] \cdot \dfrac{\pi}{3} & \text{if } G=MAX \\[3ex] \left[4+\dfrac{R-G}{MAX-MIN}\right] \cdot \dfrac{\pi}{3} & \text{if } B=MAX \end{cases} \qquad \text{if } S>0 \qquad (7.6)$$

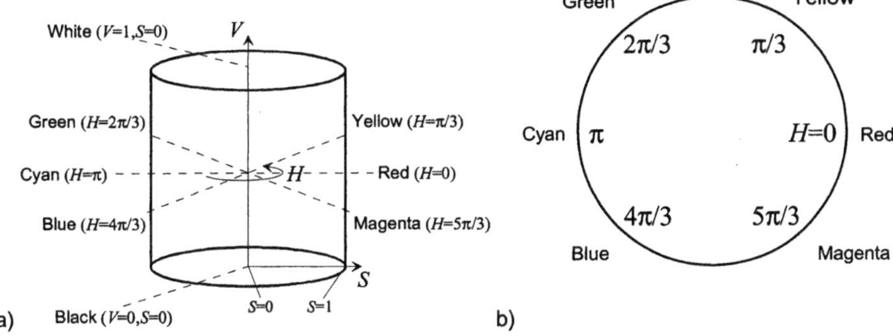

Fig. 7.1. a Representation of the (H,S,V) color cylinder, positions of primary and secondary colors.

The normalized Euclidean distance between color values (H_1, S_1, V_1) and (H_2, S_2, V_2) is computed as[1]

$$d = \sqrt{\frac{(V_1 - V_2)^2 + (S_1 \cdot \cos H_1 - S_2 \cdot \cos H_2)^2 + (S_1 \cdot \sin H_1 - S_2 \cdot \sin H_2)^2}{5}}. \qquad (7.7)$$

Best fit with subjective rating of color differences. The Euclidean distance in the (H,S,V) color space is better adjusted to human perception than it was the case for the (R,G,B) space. H and S are important characteristics to judge the subjective impression about a color. Nevertheless, the (H,S,V) color space is still not perfect yet for brightness-independent color comparison. To get an even better adaptation to subjective perception of color under changing illumination conditions, it is appropriate to go back to the (X,Y,Z) components. A good fit with the human color difference perception is provided by the definition of the CIE (L^*,a^*,b^*) color space[2],

$$L^* = \begin{cases} 25 \cdot \left[100 \dfrac{Y}{Y_0} \right]^{\frac{1}{3}} - 16 & \text{if} \quad \dfrac{Y}{Y_0} > 8.856 \cdot 10^{-3} \\[4mm] 903.3 \dfrac{Y}{Y_0} & \text{if} \quad \dfrac{Y}{Y_0} \le 8.856 \cdot 10^{-3} \end{cases}$$

$$a^* = 500 \cdot \left[\left[\frac{X}{X_0} \right]^{\frac{1}{3}} - \left[\frac{Y}{Y_0} \right]^{\frac{1}{3}} \right]; \quad b^* = 200 \cdot \left[\left[\frac{X}{X_0} \right]^{\frac{1}{3}} - \left[\frac{Z}{Z_0} \right]^{\frac{1}{3}} \right]. \qquad (7.8)$$

This transform is also reversible by[3],

$$X = X_0 \left[\frac{a^*}{500} + \left[\frac{1}{100} \right]^{\frac{1}{3}} \left[\frac{L^*+16}{25} \right] \right]^3; \quad Y = Y_0 \left[\left[\frac{1}{100} \right]^{\frac{1}{3}} \left[\frac{L^*+16}{25} \right] \right]^3$$

$$Z = Z_0 \left[\frac{a^*}{500} + \left[\frac{1}{100} \right]^{\frac{1}{3}} \left[\frac{L^*+16}{25} \right] - \frac{b^*}{200} \right]^3. \qquad (7.9)$$

The Euclidean distance between two colors (L_1^*, a_1^*, b_1^*) and (L_2^*, a_2^*, b_2^*) gives a color difference criterion which reasonably matches subjective perception,

---

[1] By the factor 5, normalization is made for $0 \le d \le 1$.

[2] Another similar definition is made for the CIE (L^*,u^*,v^*) color space. For exact definitions of diverse color spaces, see [PLATANIOTIS, VENETSANOPOULOS 2000].

[3] Shown here only for the case of the higher luminance range.

$$\Delta E_{ab}{}^* = \sqrt{\left|L^*_1 - L^*_2\right|^2 + \left|a^*_1 - a^*_2\right|^2 + \left|b^*_1 - b^*_2\right|^2} \; . \tag{7.10}$$

To achieve an illumination-independent color comparison, it is also possible to only regard the Euclidean distance in a^* and b^*. An almost perfect adaptation to human color perception is achieved by the following derived definition [CIE 1995]:

$$\Delta C_{ab}{}^* = \left| \sqrt{\left(a^*_1{}^2 + b^*_1{}^2\right)} - \sqrt{\left(a^*_2{}^2 + b^*_2{}^2\right)} \right| , \tag{7.11}$$

$$\Delta H_{ab}{}^* = \sqrt{\left(\Delta E_{ab}{}^*\right)^2 - \left|L^*_1 - L^*_2\right|^2 - \left(\Delta C_{ab}{}^*\right)^2} \; , \tag{7.12}$$

$$\Delta E_{94}{}^* = \sqrt{\left|L^*_1 - L^*_2\right|^2 + \left(\frac{\Delta C_{ab}{}^*}{S_C}\right)^2 + \left(\frac{\Delta H_{ab}{}^*}{S_H}\right)^2} \; , \tag{7.13}$$

with $\quad S_C = 1 + .045 \cdot C_{ab}{}^* \; ; \; S_H = 1 + .015 \cdot C_{ab}{}^* \; ;$

$$\text{and} \quad C_{ab}{}^* = \sqrt{\left(a^*_1{}^2 + b^*_1{}^2\right) \cdot \left(a^*_2{}^2 + b^*_2{}^2\right)} . \tag{7.14}$$

7.1.2 Representation of Color Features

The color transform is pixel-wise applied to the image signal. To represent the color of image and video signals, the following feature criteria are commonly used:

-   Statistical color features: Probability distributions or histograms based on pixel color occurrences; identification of dominant colors by searching for concentrations (clusters) in the probability distribution;
-   Properties of color localization in time and space, or structures of color within a neighborhood.

Color Histograms. A histogram is a discrete approximation of a probability distribution, based on the analysis of occurrences of quantized signal values (see sec. 3.1). In the case of color analysis, the dimensionality of the signal space is related to the number of color components. As histograms are computed over sets of discrete samples, a quantization[1] is necessary which should be performed such that a sufficient number of distinguishable color classes is available. The result of quantization is expressed by indices, where all color values which fall into the same partition of the color space are annotated by the same index value. Fig. 7.2 shows an example of partitions in the (H,S,V) color space. First, the cylinder is partitioned into 4 slices, where the identification of the slice is performed by uniform

---

[1] For more details on methodologies of quantization, refer to sec. 11.1 and 11.5.

scalar quantization of V (Fig. 7.2a). Each of these slices is then further partitioned in H and S (Fig. 7.2b)[1]. First, a scalar quantization in S must be applied to identify in which ring of the slice the partition will be located. Finally, H is quantized into a number of steps which depends on S, the radius of the ring. This scheme results in partitions which have approximately equal sizes. For low saturation values, no color tones (H) can be differentiated anyway, while for high saturation, the centers of outermost ring partitions are exactly at the six primary and secondary color tones as shown in Fig. 7.1b. In histogram computation, these quantization partitions are denoted as *bins*. In the example shown, the total number of bins is 4x10=40.

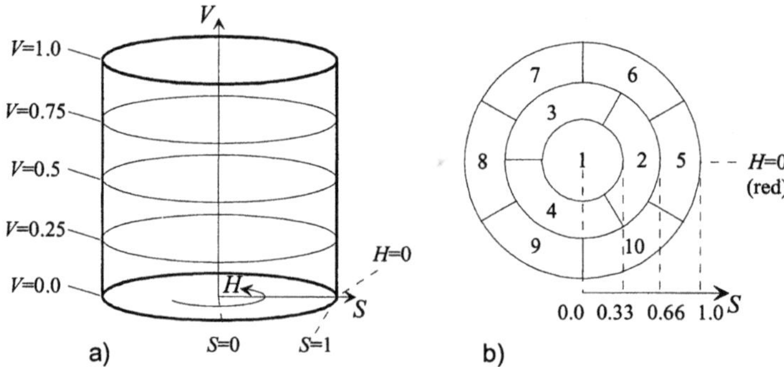

Fig. 7.2. Example of a color quantization for the (H,S,V) color space: **a** Slice-wise quantization of the color cylinder in V **b** Quantization of a slice in S and H

Color histograms can likewise be extracted from full images, from image regions, from a sequence of video frames or from a series of still images, provided that the set of samples that is analyzed expresses the color properties of a visual item in a reasonable way. In any case, it is necessary to normalize by the count of samples over which the histogram was computed (cf. (3.13)). The resulting set of values, expressing the frequency of occurrence for each color, can be interpreted as a feature vector of a dimensionality that is equal to the number of bins. A number of 30-40 bins should typically be provided at minimum to achieve sufficient distinction between color histograms of different images; useful maximum numbers depend on the precision required for a specific application, but a saturation of expressiveness can often be observed beyond 300-500 bins [OHM ET AL. 2002]. It can further be expected that occurrence values of adjacent histogram bins are statistically dependent. This can be explained by the fact that characteristic colors appearing within images are not unique, but show statistical variations (e.g. due to illumination variations, shadows etc.). Occurrences of such colors will occasionally spread over several neighbored histogram bins, as the boundaries were defined without any knowledge about color distributions in a specific image. For a more

---

[1] Even though the quantization of H and S is not independent, it is not a typical vector quantization (see sec. 11.5), the partitions are not Voronoi cells.

dense representation of color histograms, linear transforms can be applied, which are also advantageous for simplified feature comparison (sec. 9.1 and 9.3). Fig. 7.3 shows the application of a Haar transform (cf. (4.122) and (4.180)) on the occurrence values of a color histogram, a method that is implemented in the *Scalable Color Descriptor* of the MPEG-7 standard. The transform systematically computes sums and differences of adjacent histogram bins, such that a scalable representation of the histogram is achieved[1].

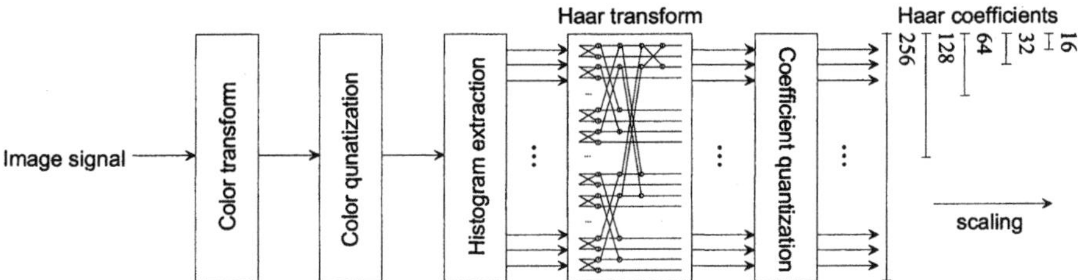

Fig. 7.3. Color histogram extraction with scalable representation of the color feature in the Haar transform domain

Dominant color clusters. The histogram characterizes the distribution of *all colors* occurring in an image. Alternatively, it can be interesting to describe only *some colors*, which can also directly be related to semantic properties of the content (e.g. blue of sky, red of fire, skin color of face). The extraction of dominant colors is a classification problem, where concentrations of values (clusters) in the color feature space must be identified[2]. The characterization of the properties of dominant colors can then be done by statistical analysis for the set of color samples x which are allocated to the cluster S. The simplest approach is made by characterization of mean values and variances in the single color components, combined into mean and variance vectors[3]

$$\mathbf{\mu}_S = \int \cdots \int \mathbf{x} \cdot p(\mathbf{x} \mid S) d\mathbf{x} \; ; \; \mathbf{\sigma}_S^2 = \int \cdots \int (\mathbf{x} - \mathbf{\mu}_S)^2 \cdot p(\mathbf{x} \mid S) d\mathbf{x}. \tag{7.15}$$

As an extension, covariances between the color components within each class are characterized by the covariance matrix

$$\mathbf{C}_S = \int \cdots \int (\mathbf{x} - \mathbf{\mu}_S) \cdot (\mathbf{x} - \mathbf{\mu}_S)^{\mathrm{T}} p(\mathbf{x} \mid S) d\mathbf{x}. \tag{7.16}$$

---

[1] Note that summing each two adjacent histogram bins in the computation of the 'lowpass' coefficient is equivalent to the computation of a histogram holding half number of bins. Scaling of the multi-dimensional color feature space is done by dyadic reduction of the number of 'samples'≡color partitions over the different dimensions H, S and V.

[2] Methods to identify and define clusters are described in sec. 9.4.3.

[3] Color amplitudes and their statistical parameters are described here by continuous values. They will normally be determined by numerical computation, where however the underlying quantization uses much finer levels than in histogram computation.

The statistical descriptions introduced so far are practically invariant against geometric modifications of images. In analyzing the color properties of images, it is often relevant to determine *where* a color is found in an image, and if it is *locally concentrated* (dominant visible in a certain area) or randomly distributed and mixed with other colors over the entire area of an image.

Color localization. The color localization properties can be described by analyzing dominant colors locally and associate them with their position in an image. Within smaller areas of an image, it is also more likely that features are homogeneous, such that only one or very few dominant colors will be found. A simple approach is the analysis of dominant colors within a pre-defined block grid[1]; more sophisticated approaches can perform a region segmentation (see sec. 10.1), which however requires combination with features describing the region shape. With a pre-defined block partition, it is likely that adjacent blocks exhibit similar dominant colors. This will in particular occur if larger areas of homogeneous color are present in an image, such as regions showing sky, water, meadows etc. To exploit such dependencies in the localized feature information for a more compact representation, 2D transforms can be applied on small images assembled from the block-related feature values. In particular for the case of dominant color, any types of transforms are suitable for this purpose which are also suitable for 2D images in general, e.g. the DCT (4.131). The set of transform coefficients is then directly used as feature vector, where the de-correlating properties of the transform allow using only a sub-set of coefficients for feature representation (see sec. 9.1.1). This results in a very compact feature vector representation of an image color layout. A diagram of a system is shown in Fig. 7.4, where dominant colors are extracted from blocks of a grid layout. The matrix of dominant block colors is processed by a 2D block transform. The feature is then represented by quantized coefficients in the transformed space[2].

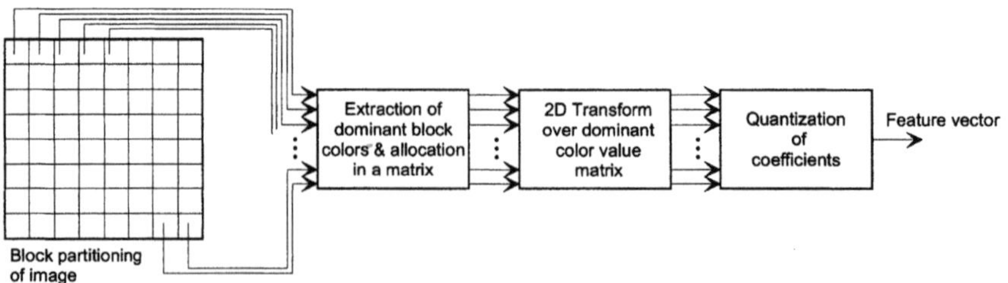

Fig. 7.4. Extraction of the dominant color feature over a block grid with feature representation in a transformed space

---

[1] Note that the same approach described here for the example of dominant colors can be used for spatial localization of many other features like textures, edges, motion, depth etc. as well. In the MPEG-7 standard, this is defined as a *Grid Layout*.

[2] This principle is used by the *Color Layout Descriptor* of the MPEG-7 standard.

Color structure. Subjective impression and semantic meaning will be very different for cases where pixels of the same color are locally concentrated as *blobs*, or where they are largely scattered and mixed with other colors over large areas of an image, appearing as a kind of 'color noise'. In the latter case, a human observer will not be able to identify a unique color. The histogram as computed by first-order pixel occurrences could however be identical for both cases. By application of morphological pre-processing, the difference can be analyzed as follows. Assume that color quantization into J partitions (bins) has been applied to an image, such that for each pixel a mapping is made into the respective index value, $\mathbf{x}(m,n) \rightarrow i(m,n)$. Now, J binary images can be defined, where

$$b_j(m,n) = \begin{cases} 1 & \text{if} \quad i(m,n) = j \\ 0 & \text{else} \end{cases} \quad \text{for } j = 1,\ldots,J. \tag{7.17}$$

Examples of images quantized into three colors are shown in Fig. 7.5. After extraction of the images $b_j(m,n)$, a morphological operation is performed. Typically, an opening operation is most useful as it will discard isolated samples of color indices, but operations of erosion or dilation can also be applied which will compress or expand any structures. In Fig. 7.5, application of a structure element with a 4-nearest-neighbor shape $\mathcal{N}_1^{(1)}$ is used. The histogram is now computed such that for each bin j, the number of values $b_j(m,n)=1$ is counted. This gives value counts of $[H(1),H(2),H(3)]=[54,53,57]$ for the scattered case, and $[31,28,34]$ for the blob-like case. In comparison, a histogram count applied directly to $i(m,n)$ will result by $[20,19,27]$ in both cases. In the practical realization, all operations are run in parallel, such that it is simply analyzed which color indices are found at a given position of the structure element in $i(m,n)$. For example, if the morphological operation is a dilation, all those histogram counts are incremented in parallel for which a matching index is found under the current position of the structure element. This latter method is used in the *Color Structure Descriptor* defined by the MPEG-7 standard.

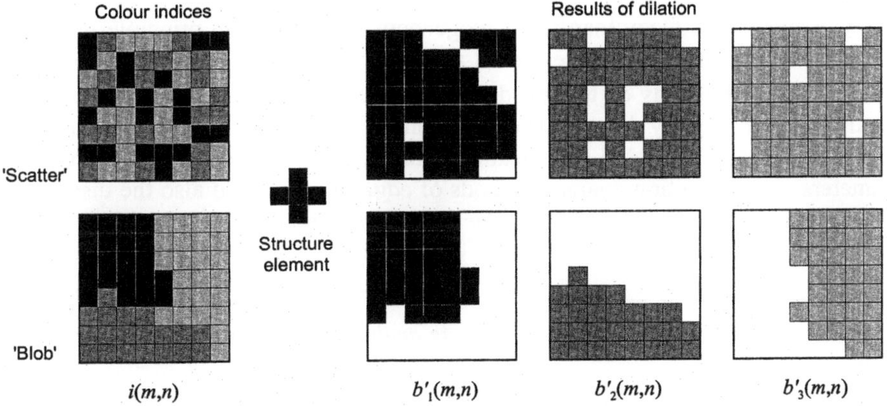

Fig. 7.5. Color structure analysis by morphological dilation of color index values

7.2   Texture

Textures are fine localized structures which are appearing due to the surface properties of exterior-world objects and light reflections emanating from these surfaces. This results in a local variation mostly of luminance, in some cases also of color information[1]. Locally *homogeneous textures*, like images of hair, water surfaces etc. are probably those components of images which can best be characterized by stationary statistical processes like AR models. Some typical examples of textures are shown in Fig. 7.6. With regard to rough structure, two extreme cases of homogeneous textures can be distinguished:

- Regular textures, which exhibit strong periodic or quasi-periodic behavior (Fig. 7.6 top left) : *Exact* periodicity is a very rare case mostly found in synthetic images. In natural images, regular structures are often quasi-periodic, which means that periodic patterns can clearly be recognized, but they have slight variations of periods. This is the case for many object structures occurring in nature, e.g. waves, cristalline structures, honeycombs, or structures built by humans. In any case, peaks can be found in the power spectrum and in the ACF, signifying the frequency and length of the periods.
- Irregular or quasi-irregular textures (Fig. 7.6 bottom right) : These typically result if a texture shows rudimentary elements which may come by different sizes and shapes, but nevertheless can uniquely be identified as members of a homogeneous class, like a pile erected from stones of different sizes. Those smaller objects can better be characterized by their stochastic variational behavior.

In fact, these categorizations are not fully transparent, and boundaries between regular and irregular texture categories are quite smooth, as indicated by the remaining examples from Fig. 7.6.

Heterogeneous textures are characterized by rather instationary behavior. This view is even more fuzzy than for homogeneous textures, as in principle the luminance structure of entire images could always be denoted as a heterogeneous texture[2]. Certain methods of texture description can however clearly be applied to analyze images e.g. by directional orientations of their global structures, amount of detail in average by global and local structures etc. For these types of texture parameters, the distinction against methods of *edge analysis*, and also the distinction against luminance-related parts of *color analysis* are not clearly definable.

---

[1] See also footnote on p. 228. Mostly, texture *analysis* is performed on to the luminance component only.

[2] The notion of 'texture' is also used in a more abstracted sense for the entire pixel-based image information including color. An example for this is the usage of the term 'texture encoding' in the Visual part of the MPEG-4 standard.

For analysis and description of texture features, mostly methods of *statistical parametrization* and *spectral analysis* are employed, which are described in more detail in the subsequent sections.

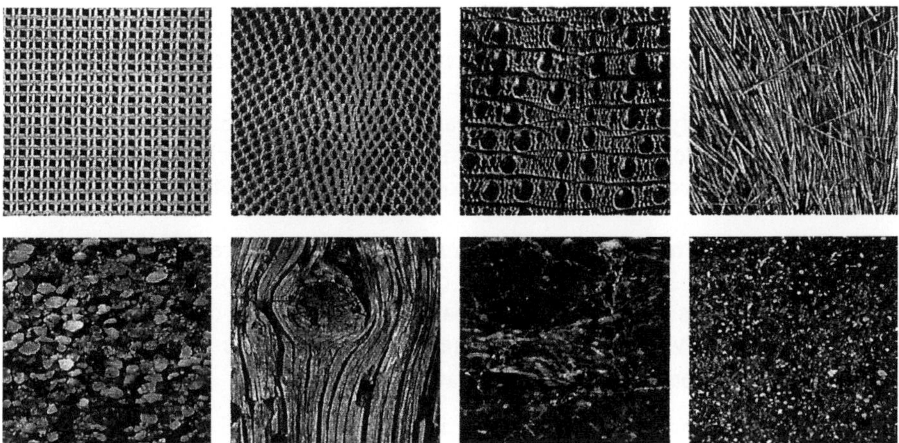

Fig. 7.6. Examples of textures with regularity decreasing from top-left to bottom-right (from the BRODATZ texture database)

7.2.1   Statistical Texture Analysis

The most simple statistical features of textures relate to local measurements of mean values and variances (first and second order moments). This can be performed by either using an analysis window applied over a neighborhood at each pixel position, or by partitioning the image into a pre-defined grid of non-overlapping blocks. The moments are then computed locally for each block. Observe that the global mean and variance values are not very expressive for the texture feature unless it is extremely homogeneous; an analysis of the local variations in mean and variance over the entire analysis areas much better captures the properties in the general case.

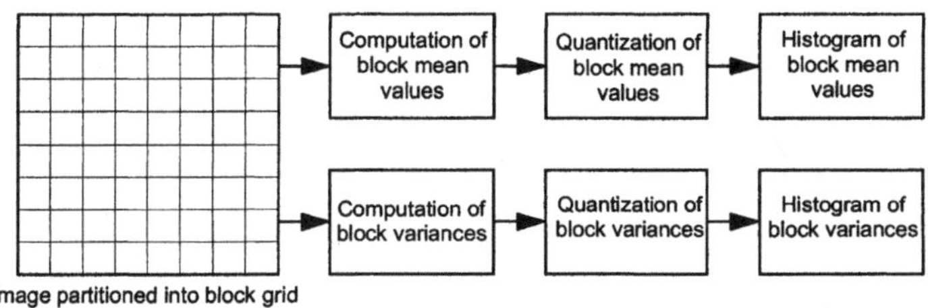

Fig. 7.7. Characterization of first- and second-order moment statistics by histograms

An example for a histogram-based analysis of moment statistics is shown in the block diagram of Fig. 7.11, which results in a very compact representation of the texture feature. Mean and variance values are quantized into a small number of bins each [OHM ET AL. 2000]. The two histogram distributions establish the feature values and can be used for comparison of different textures; the texture will usually be inhomogeneous when the histogram distribution does not show significant peaks.

Pixel-based analysis features like mean and variance are however not applicable to characterize textures by their directional properties, which require to analyze joint probabilities (co-occurrences) of pixels for different directional orientations.

Co-occurrence matrices. The co-occurrence matrix is a representation of a *joint histogram* (see Fig. 3.3), computed over combinations of amplitude values, where the joint analysis is made for sample pairs over a distance of k pixel positions horizontally and l vertically. If J different amplitude levels are analyzed, joint histograms and co-occurrence matrices have a size of $J {\times} J$[1].

Example. The following image matrix $\mathbf{X}$ is quantized into $J{=}3$ different amplitude levels 0-2.

$$\mathbf{X} = \begin{bmatrix} \underline{0} & \underline{0} & \underline{0} & 1 & 2 \\ \underline{1} & \underline{1} & \underline{0} & \underline{1} & 1 \\ \underline{2} & \underline{2} & \underline{1} & \underline{0} & 0 \\ \underline{1} & \underline{1} & \underline{0} & \underline{2} & 0 \\ 0 & 0 & 1 & 0 & 1 \end{bmatrix} \tag{7.18}$$

The co-occurrence matrix is initialized by zero values. Then, for each position (m,n) the amplitude x_i is analyzed jointly with amplitude x_j at position $(m+k,n+l)$, and the matrix element $c_{i,j}$ is incremented. In this example, only the 16 underlined values are analyzed to avoid access of corresponding pixels which are outside the image[2]. The matrices $\mathbf{C}^{(k,l)}$ resulting for the cases of horizontal, vertical and diagonal neighborhood relationships are:

$$\mathbf{C}^{(1,0)} = \begin{bmatrix} 3 & 3 & 1 \\ 2 & 3 & 1 \\ 1 & 1 & 1 \end{bmatrix} ; \ \mathbf{C}^{(0,1)} = \begin{bmatrix} 1 & 4 & 1 \\ 4 & 1 & 2 \\ 1 & 2 & 0 \end{bmatrix} ; \ \mathbf{C}^{(1,1)} = \begin{bmatrix} 4 & 2 & 1 \\ 2 & 3 & 2 \\ 0 & 2 & 0 \end{bmatrix} . \tag{7.19}$$

To establish criteria which are independent of the image size, the results are normalized by the pixel count:

---

[1] The number of amplitude levels must not necessarily be the same as for the original representation of the signal. To keep co-occurrence analysis simple, it is advisable to perform a quantization of the signal amplitude levels into a coarser scale prior to the extraction of the joint histogram.

[2] In Problem 7.4, the same example is used with a periodic boundary extension.

$$\hat{\mathbf{P}}^{(1,0)} = \frac{1}{16}\begin{bmatrix} 3 & 3 & 1 \\ 2 & 3 & 1 \\ 1 & 1 & 1 \end{bmatrix}; \ \hat{\mathbf{P}}^{(0,1)} = \frac{1}{16}\begin{bmatrix} 1 & 4 & 1 \\ 4 & 1 & 2 \\ 1 & 2 & 0 \end{bmatrix}; \ \hat{\mathbf{P}}^{(1,1)} = \frac{1}{16}\begin{bmatrix} 4 & 2 & 1 \\ 2 & 3 & 2 \\ 0 & 2 & 0 \end{bmatrix}. \quad (7.20)$$

The co-occurrence matrices by themselves can be rather large (e.g. 16x16=256 in case of 2^4 gray levels), and must be computed for several pixel shift distances to achieve sufficient expressiveness about the texture feature. Hence, it is impractical to use them directly for the representation. Reasonable feature criteria to characterize the result of co-occurrence analysis (7.20) are listed subsequently, which can then directly be used to compare features of different textures by applying common distance measures:

- Maximum occurrence and its index $\qquad \max_{i,j}(\hat{P}^{(k,l)}_{i,j}) \ ; \ \arg\max_{i,j}(\hat{P}^{(k,l)}_{i,j}) \qquad (7.21)$

- Order p difference moment $\qquad\qquad \sum_i \sum_j |x_i - x_j|^p \ \hat{P}^{(k,l)}_{i,j} \qquad (7.22)$

- Order p inverse difference moment $\qquad \displaystyle\sum_i \sum_{\substack{j \\ i \ne j}} \frac{\hat{P}^{(k,l)}_{i,j}}{|x_i - x_j|^q} \qquad (7.23)$

- Entropy[1] $\qquad\qquad\qquad\qquad -\sum_i \sum_j \hat{P}^{(k,l)}_{i,j} \cdot \log \hat{P}^{(k,l)}_{i,j} \qquad (7.24)$

- Uniformity $\qquad\qquad\qquad\qquad \sum_i \sum_j \left(\hat{P}^{(k,l)}_{i,j} \right)^2 \qquad (7.25)$

Co-occurrence matrices can also relate to model distributions. For example, let a binary signal be modeled by a Markov chain of transition probabilities $P(0|1)$ and $P(1|0)$ between horizontally neighbored samples of distance $(k,l)=(1,0)$ (see sec. 3.4). Then, the normalized co-occurrence matrix can be written as

$$\mathbf{P}^{(1,0)} = \begin{bmatrix} P^{(1,0)}(0,0) & P^{(1,0)}(0,1) \\ P^{(1,0)}(1,0) & P^{(1,0)}(1,1) \end{bmatrix} = \begin{bmatrix} P^{(1,0)}(0|0) & P^{(1,0)}(0|1) \\ P^{(1,0)}(1|0) & P^{(1,0)}(1|1) \end{bmatrix} \cdot \begin{bmatrix} 1/P(0) \\ 1/P(1) \end{bmatrix}. \quad (7.26)$$

Autocorrelation analysis. The co-occurrence matrix $\hat{\mathbf{P}}^{(k,l)}$ as computed for a certain distance (k,l) between two pixels is an approximation (estimate) of joint probability values. This can be used directly to compute an estimate of the autocorrelation function (3.31); observe that equality in (7.27) will only hold if the same boundary value extension is used in the computation of the autocorrelation and of the co-occurrence values:

---

[1] The entropy as defined here can use any base of the logarithm, as it is not directly related to units of information

$$\hat{r}_{xx}(k,l) = \sum_{i=1}^{J}\sum_{j=1}^{J} x_i \cdot x_j \cdot \hat{P}^{(k,l)}_{i,j} \approx \frac{1}{M \cdot N}\sum_{m=0}^{M-1}\sum_{n=0}^{N-1} x(m,n)\cdot x(m+k,n+l) \ . \quad (7.27)$$

By a sequence of autocorrelation values (i.e. over a range $-P \le k \le P$, $-Q \le l \le Q$), parameters of an autoregressive (AR) image model can be determined using the Wiener-Hopf equation (4.74) and its 2D extensions. This model is then described by the coefficients $\mathbf{a}$ of a recursive filter, which can be computed from the values of the autocorrelation or autocovariance function by solving a linear equation system

$$\mathbf{r}_{xx} = \mathbf{R}_{xx}\cdot\mathbf{a} \Rightarrow \mathbf{a} = \mathbf{R}_{xx}^{-1}\cdot\mathbf{r}_{xx} \ . \qquad (7.28)$$

An AR model is suitable for *homogeneous* texture feature description, which could be regarded as quasi-stationary. Either the AR model parameters or the autocorrelation function are equivalent approximations of the texture's power spectrum. A texture is characterized by the vector of filter coefficients $\mathbf{a}_x$, in addition the autocorrelation matrix $\mathbf{R}_{xx}$ is known. It shall be compared against a reference texture represented by filter coefficients $\mathbf{a}_r$. The best approximation of the spectrum is given, if the difference[1]

$$d = \left[\mathbf{a}_x - \mathbf{a}_r\right]^{\mathrm{T}}\cdot\mathbf{R}_{xx}\cdot\left[\mathbf{a}_x - \mathbf{a}_r\right] \qquad (7.29)$$

becomes minimum. In addition, the filter $\mathbf{a}_x$ can be used for linear prediction of the texture signal. Then, statistical analysis of the prediction error signal $e(\mathbf{n})$, which must not necessarily be zero-mean Gaussian white noise, can further be used to describe the texture as shown in Fig. 7.8. This includes

- Probability distribution of $e(\mathbf{n})$
- Variance and mean values of $e(\mathbf{n})$
- Power spectrum of $e(\mathbf{n})$

The variance is pre-determined by the AR analysis. If the prediction error signal is not a zero-mean signal, this is a clear indicator that the signal itself does not have zero mean. Addition of a mean (bias) value can in fact be applied either before or behind the AR synthesis filter, if a non-zero mean process shall be generated. More interesting is the property of the prediction error spectrum in case that it is not white noise. In particular, a lowpass characteristic of the prediction error spectrum indicates the tendency that subsequent prediction error values are correlated which leads to slow variations of average values measured over local groups of samples. Such a characteristic can well be modeled by a non-recursive filter which

---

[1] The autocorrelation matrix and the coefficient vector are here by the form of (4.76) and (4.79)-(4.82). As $\mathbf{R}_{xx}\mathbf{a}=[\sigma_z^2\ 0\ \dots\ 0]^{\mathrm{T}}$, $\mathbf{a}^{\mathrm{T}}\mathbf{R}_{xx}\mathbf{a}=\sigma_z^2$. Apparently, (7.29) expresses the difference between the variances of two prediction error signals, where the analyzed texture is once predicted by the correct filter $\mathbf{a}_x$ and once by the reference filter $\mathbf{a}_r$. It can be shown that minimization of this criterion is equivalent to the best approximation between two power spectra represented by AR model parameters.

must be adapted to the spectral properties of the prediction error signal; this filter is denoted as *Moving Average* (MA) filter. It can be characterized to be complementary with the recursive AR synthesis filter, which always produces a zero-mean signal globally and locally when the input is Gaussian white noise $z(\mathbf{n})$. A process which is generated by feeding $z(\mathbf{n})$ first into a 'coloring' MA filter, and subsequently into the AR synthesis filter, is called an ARMA process. Optionally, a global mean value addition can be performed as well.

ARMA modeling can also be used for ARMA synthesis of textures (Fig. 7.8b) : A white noise signal, optionally adapted to the PDF that was measured in case when the prediction error signal is not Gaussian, is first 'colored' by the MA filter, and finally fed into the AR synthesis filter. If statistical parameters at each stage of this process correspond with those gained by the analysis, very naturally-looking textures can be synthesized by such a method [CHELAPPA, KASHYAP 1985].

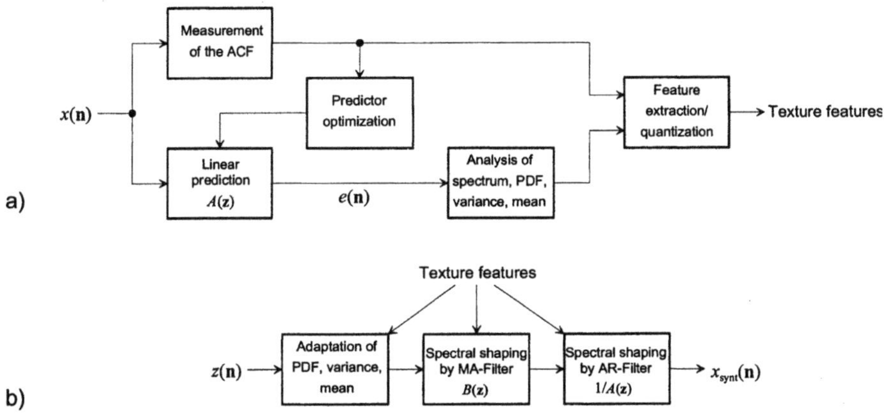

Fig. 7.8. a Analysis of texture features by AR modeling and statistical analysis of the prediction error signal   **b** ARMA synthesis of textures

Markov Random Fields. In the computation of co-occurrence matrices and in the computation of the autocorrelation function, the statistical relationships between pixels and their respective neighbors are determined *separately* for each particular shift distance. This is sufficient if the statistical dependencies between the amplitude values are linear. Markov Random Field (MRF) models[1] are more generally applicable, as they are based on *conditional probabilities* covering *complete sets of neighborhood constellations*, which bears the capability to model nonlinear dependencies as well. To implement this, it is necessary to determine the probabilities for a pixel $x(m,n)$ to possess a certain amplitude value, if the pixels from a neighborhood system show an amplitude pattern $\mathbf{x}(m,n)$. In the case of a homogeneous neighborhood system $\mathcal{N}_c^{(2)}(m,n)$ of (4.1), this can be expressed as

---

[1] For a reference on the topic, see [CHELAPPA, JAIN 1996].

$$P\left[x(m,n) = x_i \mid \mathbf{x}(m,n) = \mathbf{x}_j \right]$$
$$\text{with} \quad \mathbf{x}(m,n) = \left\{ x(k,l) \ : \ 0 < (k-m)^2 + (l-n)^2 \leq c \right\}. \tag{7.30}$$

Due to the high number of possible configurations $\mathbf{x}(m,n)$, the definition of the probability values can be a problem of extreme high complexity. Even for the most simple homogeneous four-neighbor system ($c=1$) and $J=16$ amplitude levels, $\mathbf{x}(m,n)$ can be instantiated by 16^4 different values. Practically, if analysis of the transition probabilities is made from a given texture image, most of the state transitions will never occur, such that a conditional probability of zero can be assigned. Another approach could be made by fit the available probability estimates by a vector Gaussian distribution (3.40), which however will limit the statistical dependencies between pixels to linear cases. Then, the expressiveness of MRF models will not go beyond the degrees of freedom offered by autocorrelation analysis. In general, in MRF modeling it must be assumed that the neighborhood relationships are symmetrical, and the following condition must hold:

$$x(m,n) \in \mathcal{N}_c(k,l) \Leftrightarrow x(k,l) \in \mathcal{N}_c(m,n) . \tag{7.31}$$

The MRF model has another interpretation as an extension of state models like the Markov chain concepts (see sec. 3.4). With J different amplitude values per pixel, J different transitions are possible towards the next state, but the entire number of states will be J^K when K is the number of pixels in the neighborhood vector. However, in the most general case described by (7.30), the next-state transition probabilities can no longer be characterized as memoryless or independent of previous-state transitions. For case of memoryless state models (like the Markov chain), the probabilities of a number of neighbored pixels to have identical amplitude values will follow a negative exponential distribution (cf. (3.78)); this can be modeled by a Gibbs distribution. For more discussion of these aspects, refer to the usage of MRF models in the context of image segmentation in sec. 10.1.4.

It is also possible to perform *texture synthesis* based on the MRF approach [CROSS, JAIN 1983]. First, conditional probability parameters must be estimated, which will typically leave many of the states unresolved in a concrete texture analysis; these do not fit the actual texture, such that they will never be entered by the synthesis model. A random number generator can then be used to determine the respective next state; sub-dividing the interval $[0,1]$ into sub-intervals of widths which are equal to the different transition probabilities, the next state is determined by the number the generator produces. Depending on the local properties of the texture, the size of the neighborhood system will be of high importance for natural synthesis. In the case of quasi-periodic regular textures, the neighborhood must at least span one entire period [PAGET/LONGSTAFF 1998].

7.2.2   Spectral Features of Texture

Texture analysis based on Fourier spectra. The power spectrum is the Fourier transform of the autocorrelation function. Hence, it can be expected that the texture features can likewise be expressed by features in the spatial domain or in the spectral domain. Spectral representations relating to the DFT can be computed by the Fast Fourier Transform (FFT).

Power spectrum values contain no information about phase, they are invariant against statistical variations caused by small-scale location changes of the signal. A straightforward approach to measure the similarity of a texture image $x(m,n)$ against a reference texture $x_{ref}(m,n)$ can use the distance of DFT power spectra estimates (3.64)[1]

$$d = \sum_u \sum_v \left| \hat{S}_{xx}(u,v) - \hat{S}_{xx,ref}(u,v) \right| = \sum_u \sum_v \left| \left| \hat{X}(u,v) \right|^2 - \left| \hat{X}_{ref}(u,v) \right|^2 \right| . \quad (7.32)$$

If the image is rotated or scaled against the reference texture, it will no longer be possible to perform this comparison:

- In case of rotation of an image, the power spectrum rotates likewise, see (2.29);
- If the image is captured from a different distance than the reference texture, the spectrum will be down-scaled over all dimensions (in case of spatial expansion) or expanded (in case of spatial down-scaling); the entire spectral energy (normalized by the area in units of pixels) will however be identical in both cases, if measured along a ray of certain orientation in the spectral domain.

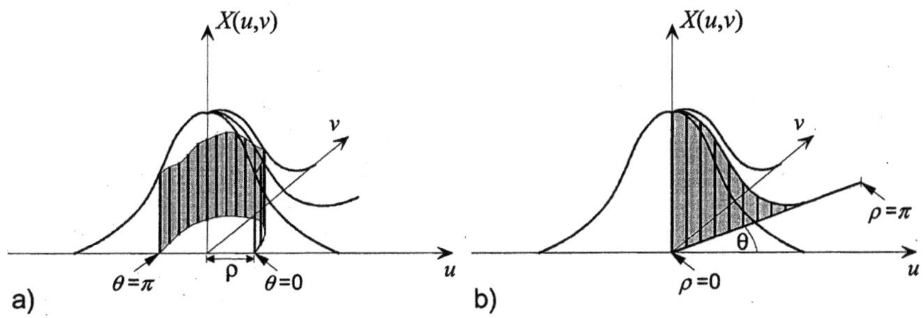

Fig. 7.9. Definition of ring spectra (a) and ray spectra (b)

---

[1] Alternatively, a similarity criterion based on maximization of the normalized magnitude of cross power spectra (3.62) would be applicable. This could be interpreted as a 'spectral correlation coefficient' which also abstracts from phase shifts between the textures,

$$c = \sum_u \sum_v \left| \hat{X}^*(u,v)\hat{X}_{ref}(u,v) \right| \Big/ \sqrt{\sum_u \sum_v \hat{S}_{xx}(u,v)\hat{S}_{xx,ref}(u,v)} .$$

Both properties can be utilized, when first the spectral values (Ω_1, Ω_2) or their discrete counterparts (u,v) according to (3.51) are transformed into discrete polar coordinates (ρ_r, θ_q), and then spectral power values along R rings (constant ρ) or Q rays (constant θ) are computed (see Fig. 7.9) The results are $S_{xx}(r)$ as a rotation-invariant *ring spectrum* and $S_{xx}(q)$ as a scaling-invariant *ray spectrum*:

$$\hat{S}_{xx}(r) = \sum_{q=0}^{Q-1} \hat{S}_{xx}(\rho_r, \theta_q) \quad ; \quad \rho_r = \frac{r\pi}{R} \tag{7.33}$$

$$\hat{S}_{xx}(q) = \sum_{r=0}^{R-1} \hat{S}_{xx}(\rho_r, \theta_q) \quad ; \quad \theta_q = \frac{q\pi}{Q}. \tag{7.34}$$

For the angular orientation q, it is sufficient to regard the angular range between 0 and π due to the symmetry properties of 2D spectra computed from real valued signals (3.57). For the radial orientation r, the entire spectral information is also contained in the range between 0 and π (half of sampling frequency). As discrete DFT spectra are used, an appropriate mapping onto the discrete radial and angular orientations is necessary[1].

Using the methods described, it is possible to reduce the complexity of comparison and classification, as the number of parameters to be compared is less than the number of coefficients in the original power spectrum. Two textures can be assumed to be similar except for a rotation angle, if the expression

$$d = \sum_{r=0}^{R-1} \left| \hat{S}_{xx}(r) - \hat{S}_{xx,ref}(r) \right| \tag{7.35}$$

becomes minimum. They are similar except for a scale factor, if the expression

$$d = \sum_{q=0}^{Q-1} \left| \hat{S}_{xx}(q) - \hat{S}_{xx,ref}(q) \right| \tag{7.36}$$

is minimized. If textures have to be compared which may be both rotated and scaled compared to the reference, it is still necessary to perform comparison by determining the minimum over multiple angular orientations[2],

$$d = \min_{\Delta q} \sum_{q=0}^{Q-1} \left| \hat{S}_{xx}(q) - \hat{S}_{xx,ref}(q + \Delta q) \right| \; ; \; 0 \le \Delta q < Q \tag{7.37}$$

or the minimum over multiple scaling factors s[3]

---

[1] Even though approximations of discrete 2D polar Fourier transforms exist, none of these can be accurate on rectangular grids. It is either necessary to interpolate a discrete polar representation over the image plane, which loses resolution for radial distances far from the origin, or to interpolate coefficients in the frequency domain.

[2] Under the assumption of radial symmetry $\hat{S}_{xx,ref}(\tilde{q}) = \hat{S}_{xx,ref}(\tilde{q} - Q)$ for cases $\tilde{q} \ge Q$.

[3] For scaling factors $s<1$, it is necessary to interpolate values in the reference spectrum.

$$d = \min_s \sum_{r=0}^{\min(R-1,\,s\cdot r)} \left| \hat{S}_{xx}(r) - \hat{S}_{xx,ref}(s\cdot r) \right| \; ; \; s_{\min} \le s \le s_{\max} \,. \tag{7.38}$$

Texture analysis based on filter banks. Frequency analysis for texture feature extraction must not necessarily be performed by the Fourier transform. One problem of Fourier transform is the (rectangular) block size to be determined a priori, while in natural images it may not be clear whether the entire block is covered by a homogeneous texture; e.g. this will no be the case for arbitrary-shaped texture objects. Frequency analysis based on filter banks, in particular by the Wavelet transform, have successfully been applied to texture analysis, providing the advantage not to be restricted to block-based processing. The variable-resolution properties of the Wavelet transform are beneficial due to the following reasons (cf. sec. 4.4.4): High-frequency components are analyzed almost by an accuracy of pixel positions, which is favorable at the boundaries of objects. On the other hand, accurate frequency resolution is not necessary for high frequencies, except eventually for narrow-band analysis of harmonic spectra. Compared to block-based transforms, an octave-band analysis significantly reduces the number of parameters which are required to express the texture feature.

Depending on the depth of the wavelet tree and the size of the texture area, the result of filtering will be a certain number of coefficients expressing each frequency band. As in Fourier analysis, a representation shall be defined which is invariant against local variations and shift of the texture. This can be achieved if absolute values of the coefficients are statistically analyzed. If a number of $N_{u,v}$ coefficients $c_{u,v}$ have been computed in one frequency band, the *absolute moments* of first and second order are defined as[1]

$$\mu_{u,v} = \frac{1}{N_{u,v}} \sum_{m'} \sum_{n'} \left| c_{u,v}(m',n') \right| \; ; \; \sigma^2_{u,v} = \left[\frac{1}{N_{u,v}} \sum_{m'} \sum_{n'} \left| c_{u,v}(m',n') \right|^2 \right] - \mu^2_{u,v} \tag{7.39}$$

The discrete WT (DWT) is orthogonal or biorthogonal, and hence is suitable to achieve characterization of textures by a low number of feature parameters. If the typical critically-sampled 2D wavelet transform is used (see Fig. 4.51c), restrictions however apply regarding directional analysis, which may constrain the accuracy of texture description. For more details, refer to sec. 4.4.4:

– The typical frequency layout schema of 2D DWT enables a unique differentiation between the horizontal and vertical orientations within a texture, however not likewise for diagonal orientations[2];

---

[1] Alternatively, frequency bands can be established by combining second order moments (energies) from adjacent DFT power spectrum coefficients. In principle, octave-band frequency partitions can be defined from such a method as well.

[2] E.g. 45° and 135° orientations can not be distinguished, unless the signs of co-located horizontal and vertical wavelet coefficients are evaluated.

–   Orthogonality and biorthogonality are compromising the effect of alias suppression, such that frequency separation may not be optimum, depending on the type of filters chosen.

Nevertheless, it is attractive to use the 2D DWT for texture analysis, in particular if a wavelet representation is generated anyway, e.g. for the purpose of encoding. An example of a 2D wavelet transform frequency layout with one scaling band S and wavelet bands 1-9 is shown in Fig. 7.10. Analysis is based on a 2D wavelet tree of depth 3. The dotted circles indicate how the frequency bands[1] can be combined for rotation-invariant (scale diversifying) spectral channel groups. The straight lines indicate combination of bands into scale-invariant (angular orientation diversifying) spectral channel groups.

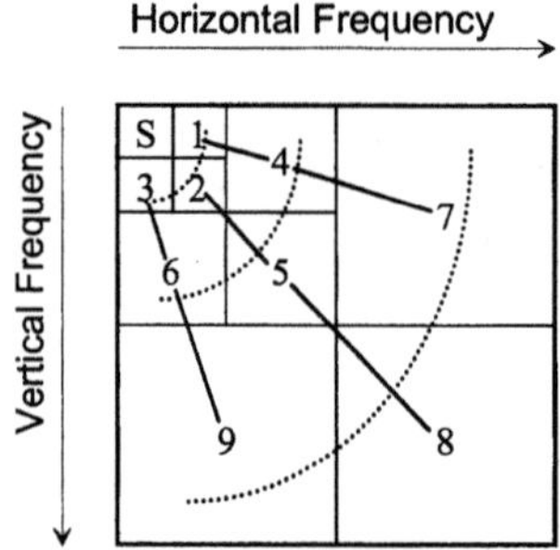

Fig. 7.10. Frequency layout for a 2D wavelet tree of depth 3, and combination of wavelet channel groups for rotation-invariant (···) and scale-invariant (——) texture criteria

In contrast to the Fourier transform, which generates exactly *one* coefficient per frequency over the entire analysis range, wavelet coefficients retain the *localization* property, where the accuracy of spatial localization increases dyadically towards the higher frequency bands. This has the following advantages:

–   By the wavelet transform, localized texture criteria can be separately analyzed, which can e.g. be used for the purpose of a segmentation, and allows to perform region-related texture analysis *after* computation of wavelet coefficients[2];
–   Statistical analysis can be applied to the different frequency bands, each of which consists of a number of coefficient samples. The statistical parameters are then used as feature criteria.

---

[1] Wavelet frequency bands are often called *channels* in the context of texture analysis, where not the coefficient samples, but rather the statistical description of information flowing by a certain frequency is seen as relevant.

[2] Region-oriented analysis is possible by the 2D Fourier transform as well, but the region shape must exactly be known *before* the transform, such that the region mask can be padded accordingly.

Example of wavelet-based texture feature analysis. A method of wavelet-based texture feature extraction which has been proposed in [OHM ET AL. 2000] is shown in Fig. 7.11. It is suitable for texture characterization both for homogeneous and inhomogeneous texture types. The 2D wavelet transform is computed by the common critically sampled method, which could directly be subject to the statistical analysis described below. Another attractive method, providing rotation- and scale-invariant texture analysis and reduced number of feature parameters, is realized by the combination of channels according to Fig. 7.10.

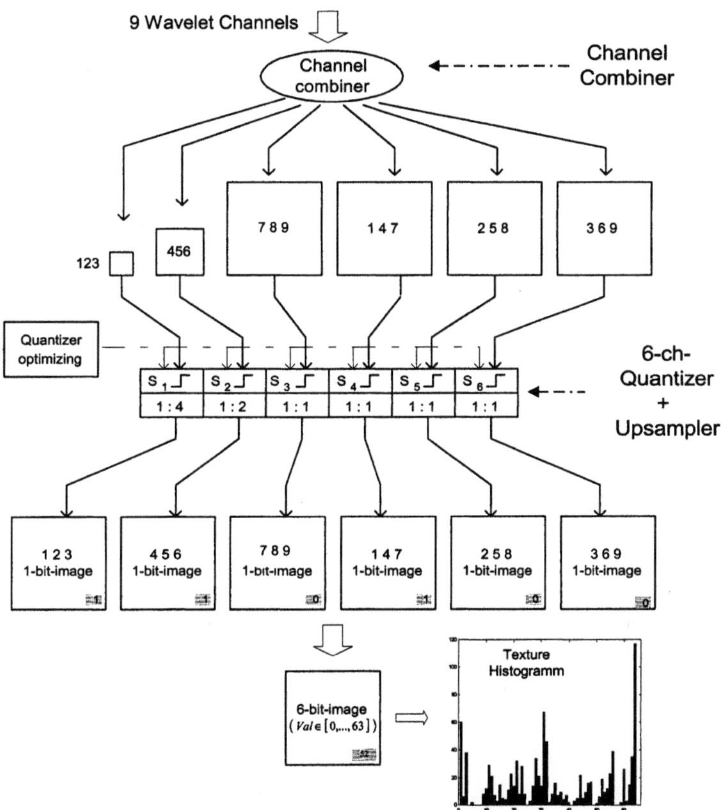

Fig. 7.11. Extraction of a frequency pattern histogram from the wavelet coefficients

At each sample position (m',n') relating to the first level of the transform, coefficients shall be combined into one ring 7-8-9 and three ray channel groups 1–4–7, 2–5–8 and 3–6–9. This requires using the lower-frequency coefficients multiple times at 4 or 16 neighbored positions. The remaining ring channel groups 1–2–3 and 4–5–6 can be computed at the respective resolution of 2nd and 3rd transform levels. By the schema shown, coefficients $c_u(m',n')$ from a total of 9 wavelet channels are mapped into 6 combined channels, three of which represent scale-diversifying ring spectra, while three relate to angular-orientation diversifying ray spectra. Linear sums of channel energies are used by the channel combiners.

The texture description can directly be made by the absolute-moment analysis (7.39). This however assumes quasi-stationarity, and will only be an appropriate method for the case of homogeneous textures. Another description method which is capable to better characterize inhomogeneous textures, is based on a histogram description of channel properties. For this, the coefficient energy values from the wavelet channels or combination channels are subject to a binary (threshold) quantization, indicating whether relevant components are present in the respective channel at position (m',n')[1]. This gives a very compact feature representation by a *frequency pattern vector* $\mathbf{b}(m',n')=[\ b_1\ b_2\ \dots\ b_6\ ]^\mathrm{T}$, containing the six bits of the binary decision related to the combination channels. Only $J=2^6=64$ different frequency patterns are available. Optimum threshold values of the binary decision are determined such that approximately half of the samples fall below the threshold, which maximizes the entropy of the binary representation. A histogram analysis is then performed related to the frequency patterns $\mathbf{b}(m',n')$, where the length-J vector of histogram values is the descriptor of the texture feature.

The information contained in the scaling channel, which is a sub-sampled representation of the original image, can further be used to extract additional features like mean and variance distributions as described earlier. Such features can then be seen as an orthogonal complement with regard to the wavelet-based features. Fig. 7.12 shows an example of similarity search in a retrieval application, where the texture feature representation based on wavelet frequency patterns as described above was used. The 10 images at the right-hand side are the most similar items from a large database, which were found when the feature comparison was made relative to the left image. The rotation and scale invariance of the method is obvious.

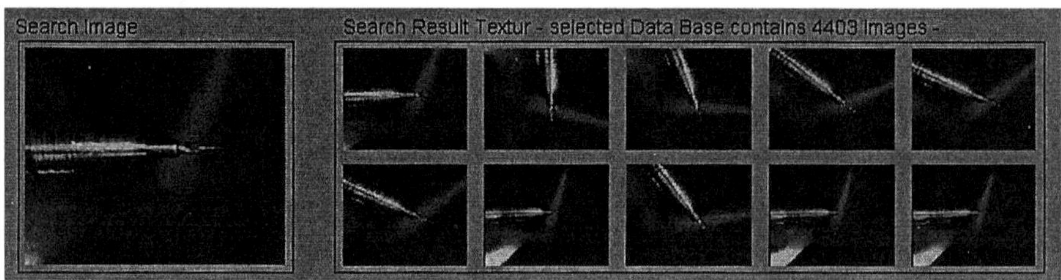

Fig. 7.12. Example of a set of images classified as similar, based on the wavelet texture feature (source: HHI)

---

[1] The sub-sampled positions (m',n') depend on the depth of the wavelet tree. For high localization accuracy of the texture feature, it may be useful to use no sub-sampling at all, which means usage of an overcomplete wavelet representation in the computation of the texture feature. This would for example be the case when the feature values shall be applied for pixel-accurate texture-based region segmentation.

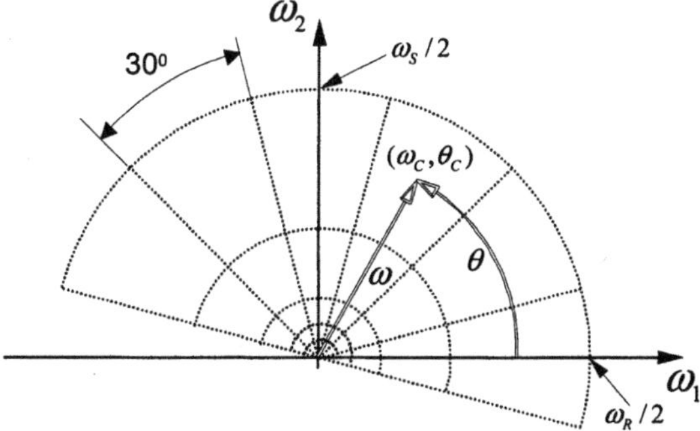

Fig. 7.13. Frequency layout for texture analysis based on Gabor Wavelet function

If the requirement for a frequency representation without over-completeness is released, it is also possible to use Wavelet filters having finer granularity of direction and/or scale analysis. Due to their properties in orientation and scale adaptability, polar Gabor filters as introduced in (4.246) and (4.247) can favorably be used for this purpose. Fig. 7.13 shows an example of a frequency layout which is used for homogeneous texture description in the MPEG-7 standard based on polar Gabor filters. The signal is analyzed within five octave-band resolution scales and 6 angular orientations distant by 30 degree steps. The other half of the frequency plane is redundant due to the conjugate symmetry of spectra. The meaning of the radial frequency ω and angular frequency θ is shown for the example of one center frequency (ω_C, θ_C). With five radial scales and 6 angles, the total resulting number of frequency channels is 5x6=30. Filters used for analysis are non-orthogonal, such that information from neighbored frequency channels is likely to be redundant if the 30 frequency channels are represented by absolute-moment features from (7.39). Up to 60 parameters are necessary to describe a texture; comparison of two textures is made by analysis of the difference between the moments.

High-frequency components contained in texture characterize high local variations of image amplitudes. These can either be related to detailed structures, or to discontinuities which indicate the boundaries of objects. If the analysis aims at the identification of amplitude changes of the latter type, it could rather be denoted as *edge analysis*. On the other hand, taking a global view onto an image by a coarse scale, edges may appear very similar to local texture structures, which is somehow related to the high similarity of microscopic and macroscopic structures of images, which can be interpreted as self-similarity or fractal behavior of images over different resolution scales. From this points of view, the methods of signal processing applied in texture and edge analysis can not be regarded fully independent. Typical edge analysis methods may be used to characterize textures, and texture analysis will typically also capture edge characteristics. Texture analysis methods, if applied to sub-sampled images, can be used to analyze the global structure of images, in particular capturing the variations of luminance at a global level.

7.3 Edge Analysis

A high amount of semantic information about image content is conveyed by shapes of objects. The analysis of object edges, which are the physical basis to perceive contours and shapes, plays an important role in the human visual system. In a simplistic view, an edge is a discontinuity of amplitude. In natural images, it will barely happen that an edge sharply separates two distinct plateaus of amplitude. This type of *step edges* (Fig. 7.14a) can primarily be found in synthetically-generated graphics images, and even there it is undesirable as it does not provide a photo-realistic impression. In natural images, due to shadows and reflections, the type of *ramp edges* (Fig. 7.14b) is a better model, which is characterized by an *edge width a* and an *edge slope b*. Natural edges are even more *smooth* as shown in the model of Fig. 7.14c; the slope is not constant over the edge, such that a point of *maximum edge slope* can be identified.

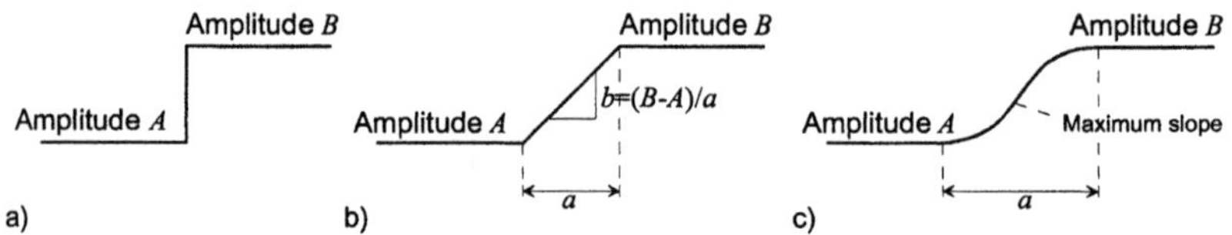

Fig. 7.14. Edge type models **a** Step edge **b** Ramp edge **c** Smooth edge

7.3.1   Edge Detection by Gradient Operators

The gradient of local amplitudes is an important criterion to detect the presence of an edge within an image. It is necessary to apply a *vertical* gradient analysis to detect *horizontal* edges and vice versa; the gradient is always relevant *perpendicular with the edge orientation*. A discrete approximation of the gradient is the difference between neighboring pixels,

$$\frac{\partial x(r,s)}{\partial r} \approx \frac{1}{R} \cdot \left[x(m,n) - x(m-1,n) \right] \quad \text{for} \quad r = mR, s = nS$$

$$\frac{\partial x(r,s)}{\partial s} \approx \frac{1}{S} \cdot \left[x(m,n) - x(m,n-1) \right].$$

$$(7.40)$$

If the normalization regarding the horizontal and vertical sampling distances R and S is omitted assuming unity sampling intervals, the gradient detection is performed by horizontal and vertical filtering operations using the filter matrices[1]

---

[1] The subscripts 'h' and 'v' are not related to gradient directions, but here indicate the orientation of the edge to be analyzed by the respective filter.

$$\mathbf{G}_v = \begin{bmatrix} -1 & 1 \end{bmatrix} \quad ; \quad \mathbf{G}_h = \begin{bmatrix} -1 \\ 1 \end{bmatrix}. \tag{7.41}$$

These simple gradient filters are quite sensitive against noise. This can be avoided if the gradient filter is combined with a local mean or lowpass filtering operation, where the mean values are computed *in parallel* with the edge orientation, or perpendicular to the orientation of the respective gradient filter:

$$\mathbf{M}_v = \begin{bmatrix} 1 \\ 1 \end{bmatrix} \quad ; \quad \mathbf{M}_h = \begin{bmatrix} 1 & 1 \end{bmatrix}. \tag{7.42}$$

Simple edge detection operators for horizontal or vertical edges can then be interpreted as a separable combination of 1D gradient and mean-value filters:

$$\mathbf{K}_v = \mathbf{G}_v \times \mathbf{M}_v = \begin{bmatrix} -1 & 1 \\ -1 & 1 \end{bmatrix} \quad ; \quad \mathbf{K}_h = \mathbf{G}_h \times \mathbf{M}_h = \begin{bmatrix} -1 & -1 \\ 1 & 1 \end{bmatrix}. \tag{7.43}$$

In a similar way, gradients can be computed for diagonal directions. With gradient and mean-value filters defined as

$$\mathbf{G}_{d^+} = \begin{bmatrix} 0 & 1 \\ -1 & 0 \end{bmatrix} \quad ; \quad \mathbf{G}_{d^-} = \begin{bmatrix} -1 & 0 \\ 0 & 1 \end{bmatrix} \quad ; \quad \mathbf{M}_{d^+} = \begin{bmatrix} 1 & 0 \\ 0 & 1 \end{bmatrix} \quad ; \quad \mathbf{M}_{d^-} = \begin{bmatrix} 0 & 1 \\ 1 & 0 \end{bmatrix}, \tag{7.44}$$

the convolution of filter kernel pairs results in the following edge detection filter matrices:

$$\mathbf{K}_{d^+} = \mathbf{G}_{d^+} * \mathbf{M}_{d^+} = \begin{bmatrix} 0 & 1 & 0 \\ -1 & 0 & 1 \\ 0 & -1 & 0 \end{bmatrix} \quad ; \quad \mathbf{K}_{d^-} = \mathbf{G}_{d^-} * \mathbf{M}_{d^-} = \begin{bmatrix} 0 & -1 & 0 \\ -1 & 0 & 1 \\ 0 & 1 & 0 \end{bmatrix}. \tag{7.45}$$

These latter filters have an output which is maximum for edges exactly crossing the position of the center pixel – unlike the cases of (7.44), where the edge would be positioned *between* two pixels. The previous horizontal/vertical filters can be modified to analyze centered edges as well, where simultaneously more stability results due to the usage of longer mean-value filters:

$$\mathbf{G}_v = \begin{bmatrix} -1 & 0 & 1 \end{bmatrix} ; \mathbf{G}_h = \begin{bmatrix} -1 \\ 0 \\ 1 \end{bmatrix} ; \quad \mathbf{M}_v = \begin{bmatrix} 1 \\ 1 \\ 1 \end{bmatrix} ; \quad \mathbf{M}_h = \begin{bmatrix} 1 & 1 & 1 \end{bmatrix}. \tag{7.46}$$

If also the mean value filters used for diagonal edge detection are modified into

$$\mathbf{M}_{d^+} = \begin{bmatrix} 1 & 1 \\ 1 & 1 \end{bmatrix} \quad ; \quad \mathbf{M}_{d^-} = \begin{bmatrix} 1 & 1 \\ 1 & 1 \end{bmatrix}, \tag{7.47}$$

an ensemble of edge-detection filter masks results by four directional orientations:

$$\mathbf{K}_v = \begin{bmatrix} -1 & 0 & 1 \\ -1 & 0 & 1 \\ -1 & 0 & 1 \end{bmatrix} \; ; \; \mathbf{K}_h = \begin{bmatrix} -1 & -1 & -1 \\ 0 & 0 & 0 \\ 1 & 1 & 1 \end{bmatrix} \; ; \; \mathbf{K}_{d^+} = \begin{bmatrix} 0 & 1 & 1 \\ -1 & 0 & 1 \\ -1 & -1 & 0 \end{bmatrix} \; ; \; \mathbf{K}_{d^-} = \begin{bmatrix} -1 & -1 & 0 \\ -1 & 0 & 1 \\ 0 & 1 & 1 \end{bmatrix}.$$

$$(7.48)$$

If another set of filter matrices[1]

$$\mathbf{M}_v = \frac{1}{4} \cdot \begin{bmatrix} 1 \\ 2 \\ 1 \end{bmatrix} \; ; \; \mathbf{M}_h = \frac{1}{4} \cdot [1 \; 2 \; 1] \; ; \; \mathbf{M}_{d^+} = \frac{1}{4} \cdot \begin{bmatrix} 1 & 2 \\ 2 & 1 \end{bmatrix} \; ; \; \mathbf{M}_{d^-} = \frac{1}{4} \begin{bmatrix} 2 & 1 \\ 1 & 2 \end{bmatrix} \quad (7.49)$$

is used for lowpass filtering, the result is the *Sobel operator*:

$$\mathbf{K}_v = \frac{1}{4} \begin{bmatrix} -1 & 0 & 1 \\ -2 & 0 & 2 \\ -1 & 0 & 1 \end{bmatrix} \; ; \; \mathbf{K}_h = \frac{1}{4} \begin{bmatrix} -1 & -2 & -1 \\ 0 & 0 & 0 \\ 1 & 2 & 1 \end{bmatrix} \; ; \; \mathbf{K}_{d^+} = \frac{1}{4} \begin{bmatrix} 0 & 1 & 2 \\ -1 & 0 & 1 \\ -2 & -1 & 0 \end{bmatrix} \; ; \; \mathbf{K}_{d^-} = \frac{1}{4} \begin{bmatrix} -2 & -1 & 0 \\ -1 & 0 & 1 \\ 0 & 1 & 2 \end{bmatrix}.$$

$$(7.50)$$

Orthogonality of directional operators. Observe that in the edge operators introduced here, either the horizontal and vertical filters, or the two diagonal filters are establishing orthogonal basis systems according to (4.107); on the other hand, neither the horizontal nor the vertical filter are orthogonal with any of the diagonal filters. As a consequence, the vertical operator outputs a zero value in cases of pure horizontal edges, while neither of the two diagonal operators will do so; the output of the diagonal operators will nevertheless be lower here than the output of the horizontal operator. The system can be supplemented by further operator(s) which are orthogonal with any of the other operators. This can be useful to identify whether an area of the image shows high detail but with *no unique edge orientation*. An operator of this kind which is orthogonal with all directional filters of (7.48) and (7.50) is defined by the filter matrix

$$\mathbf{K}_\perp = \begin{bmatrix} -1 & 1 & -1 \\ 1 & 0 & 1 \\ -1 & 1 & -1 \end{bmatrix}.$$

$$(7.51)$$

7.3.2   Edge Characterization by second Derivative

The mathematical operation of second derivative $\partial^2 x/\partial r + \partial^2 x/\partial s$ is described by the *Laplacian operator* $\nabla^2 x$. The second derivative can be computed by derivation of the first derivative. As this is a linear operation, it is also possible to define a sec-

---

[1] For the horizontal and vertical orientations, these are *binomial filters*.

ond-derivative filter directly, which results by self-convolution of gradient filters (7.41),

$$\nabla^2{}_v = \frac{1}{2}\mathbf{G}_v * \mathbf{G}_v = \frac{1}{2}\begin{bmatrix} -1 & 2 & -1 \end{bmatrix} \quad ; \quad \nabla^2{}_h = \frac{1}{2}\mathbf{G}_h * \mathbf{G}_h = \frac{1}{2}\begin{bmatrix} -1 \\ 2 \\ -1 \end{bmatrix}. \quad (7.52)$$

Instead of separate directional analysis, the mean of horizontal and vertical second derivatives can be computed. This result is identical to the convolution by the subsequent filter matrix, which results by center-aligned addition of the two filter matrices in (7.52). This is the non-separable 2D Laplacian operator using a neighborhood system $\mathcal{N}_1^{(1)}$:

$$\nabla^2{}_{\mathcal{N}_1^{(1)}} = \frac{\nabla^2{}_h + \nabla^2{}_v}{2} = \frac{1}{4}\begin{bmatrix} 0 & -1 & 0 \\ -1 & 4 & -1 \\ 0 & -1 & 0 \end{bmatrix}. \quad (7.53)$$

If the horizontal and vertical second derivatives (7.52) are used as components of a separable filter, this is equivalent to a convolution by a 2D Laplacian operator matrix over an 8-value neighborhood,

$$\nabla^2{}_{\mathcal{N}_2^{(2)},sep} = \nabla^2{}_h \times \nabla^2{}_v = \frac{1}{4}\begin{bmatrix} 1 & -2 & 1 \\ -2 & 4 & -2 \\ 1 & -2 & 1 \end{bmatrix}. \quad (7.54)$$

Further defining diagonally-oriented second-derivative filters

$$\nabla^2{}_{d^-} = \frac{1}{2}\begin{bmatrix} 0 & 0 & -1 \\ 0 & 2 & 0 \\ -1 & 0 & 0 \end{bmatrix} \quad ; \quad \nabla^2{}_{d^+} = \frac{1}{2}\begin{bmatrix} -1 & 0 & 0 \\ 0 & 2 & 0 \\ 0 & 0 & -1 \end{bmatrix}, \quad (7.55)$$

another non-separable version of a 2D Laplacian operator results by center-aligned averaging of all four directional filters from (7.52) and (7.55),

$$\nabla^2{}_{\mathcal{N}_2^{(2)}} = \frac{\nabla^2{}_h + \nabla^2{}_v + \nabla^2{}_{d^-} + \nabla^2{}_{d^+}}{4} = \frac{1}{8}\begin{bmatrix} -1 & -1 & -1 \\ -1 & 8 & -1 \\ -1 & -1 & -1 \end{bmatrix}. \quad (7.56)$$

The second derivative characterizes the *change of the gradients*. Hence, zero crossings of the second derivative can be found where the gradient itself is maximum or minimum, which can be interpreted as the true edge position in cases of positive or negative edge slope, respectively. If the edge shall characterize an amplitude transition of sufficient strength, one clear maximum *and* one clear minimum must be found in the second derivative towards both sides of the zero crossing. The distance between maximum and minimum can then be interpreted as the *width* of the edge. Relevant zero crossings of the second derivative (being framed by a clear maximum and minimum) can be found by using maximum difference filters (5.6)[1]. Fig. 7.15 shows amplitudes of a signal, its first and second derivatives, where the models are a ramp edge and a smooth edge with a continuous change of the gradient. The simplistic model of the ramp edge does not show a unique zero crossing, but extreme maxima and minima indicate the edge width.

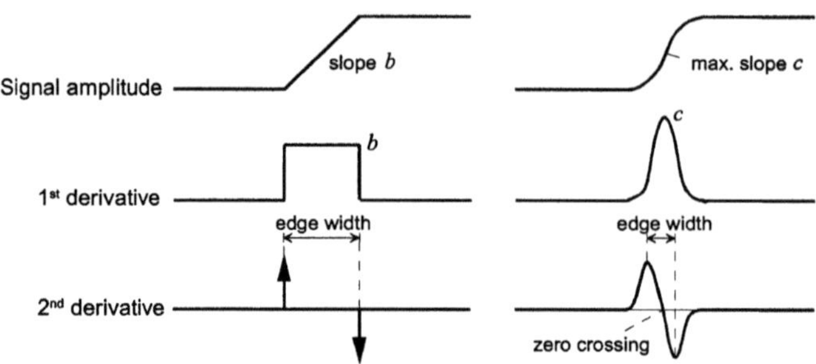

Fig. 7.15. Interpretation of first and second derivatives of the signal amplitude at image edges. Left: ramp edge model, right: smooth edge model with continuous change of slope

Laplacian operators are also sensitive against granular variations of amplitude and noise. This can be avoided by application of lowpass filtering before the second derivative is computed. Filters with a Gaussian shaped impulse response are frequently applied for this purpose. The discrete filter coefficients are derived from a continuous circular-symmetric 2D Gaussian function,

$$h(r,s) = \frac{1}{\sqrt{2\pi\tau^2}} \cdot e^{-\frac{r^2+s^2}{2\tau^2}} \ . \tag{7.57}$$

The parameter τ influences the width of the Gaussian shape and thereby the strength of lowpass filtering. The second derivative of this function is

$$\nabla_h^2(r,s) = \frac{1}{\pi^2\tau^4} \cdot \left[2 - \frac{r^2}{\tau^2} - \frac{s^2}{\tau^2} \right] \cdot e^{-\frac{r^2+s^2}{2\tau^2}} \ . \tag{7.58}$$

This radial-symmetric function is commonly denoted as *Mexican hat filter*, of which a 1D section is shown in Fig. 7.16. For a discrete approximation, this continuous function is sampled, such that the center of the discrete impulse response is at the maximum of the continuous function. The parameter τ can be set to 1, and the strength of lowpass filtering is simply varied by variation of the sampling distance, which is equivalent to scaling. This *Laplacian of Gaussian* (LoG) operator provides a very robust method for second derivative based edge detection, and is mostly applied in combination with multi-resolution edge analysis described in the next section.

---

[1] See also Problem 7.7.

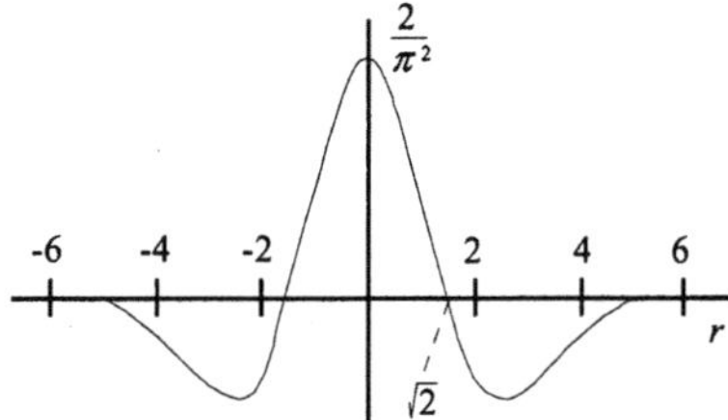

Fig. 7.16. Mexican hat filter impulse response

7.3.3   Edge Finding and Consistency Analysis

The gradient filters introduced in the previous section (for computation of first and second derivatives) are *linear* filters, and their application is only a first step of edge or shape feature extraction. If a set of I directional operators is used, a total of I output signals $y_i(m,n)$ is computed. Further processing highly depends on the goals of edge detection. Analysis steps can be as follows:

– To analyze the *edge direction* θ, find the filter output of highest amplitude at a given position, i.e.:
$$\theta(m,n) = \arg\max_{i=1\dots I} |y_i(m,n)| \tag{7.59}$$

– Generation of a *gray-value edge image*. The largest absolute value of all given directional operator outputs at a given position is selected, such that
$$y(m,n) = \max_{i=1..I} |y_i(m,n)| \tag{7.60}$$

Fig. 7.17 shows an example generated from combination of the directional outputs from (7.50).

– Generation of *binary edge maps*. For the decision *whether* an edge is present at an image position, a threshold criterion can be applied, by which a hypothesis about the presence of an edge is made. The resulting binary edge map can be computed separately as $b_i(m,n)$ for each edge orientation, or be combined from all directional operators as in (7.60):
$$b_i(m,n) = \begin{cases} 0 & \text{if } y_i(m,n) < \Theta \\ 1 & \text{if } y_i(m,n) \geq \Theta \end{cases} \;;\; b(m,n) = \begin{cases} 0 & \text{if } \max(y_i(m,n)) < \Theta \\ 1 & \text{if } \max(y_i(m,n)) \geq \Theta. \end{cases} \tag{7.61}$$

The simple threshold and directional criteria are quite sensitive against noise, and the optimum threshold values typically vary over a set of images. Further, it is difficult to distinguish semantically between true edges and high details of texture. To increase the reliability of edge detection, further criteria must be introduced. An example are neighborhood criteria, checking whether only isolated high gradient values are present, or whether the presence of consistently connected high values allows to conclude the presence of an edge or contour chain. Useful methods to analyze such consistency are *hierarchical edge analysis* and *edge tracking*.

a)   b)

Fig. 7.17. a Original image **b** Output image after combination of directional Sobel operator outputs (7.50) (high absolute values of gradients are shown dark)

Hierarchical edge analysis. This method investigates whether a detected edge is relevant globally (separating larger objects) or just related to small-scale local variations of amplitude. Edge analysis is performed at different resolution levels using lowpass filtering of different strength, eventually followed by sub-sampling. The reliability of edge detection is then judged by criteria which match the edge presence over different resolution levels, checking whether edges are consistently found over all resolutions. This can most efficiently be realized in a pyramid representation, where the resolution is varied by a factor of four from one level to the next (cf. sec. 4.4.6). Analysis typically starts at the tip of the pyramid, the lowest resolution level. Edges which are detected at this level are regarded to be globally relevant, as local fine structures will hardly be found at this level due to the strong lowpass filtering. The processing then proceeds to finer resolutions, which increases the accuracy of localization; on the other hand, additional edge positions which are only detected at finer resolution levels can be regarded as increasingly irrelevant, most probably only expressing local fine structures. If a Laplacian pyramid is used (see Fig. 4.55), the differential outputs can readily be interpreted as the outputs of the LoG-Operator at different resolution levels.

Edge tracking. The binary edge image as defined in (7.61) will most probably contain lines which are wider than one pixel. It is possible that no consistently connected edge contours are found, such that it will be necessary to link isolated contour segments. This requires to eliminate some pixels having a value '1' in $b(m,n)$, or for the case of incomplete edge lines, to change values from '0' to '1'. This can be achieved by *edge tracking* procedures as illustrated in Fig. 7.18. Fig. 7.18a shows an example output image $y(m,n)$ resulting by (7.60). Fig. 7.18b shows those pixels which are retained in an edge image $b(m,n)$ if a threshold value $\Theta_a=7$ is chosen in (7.61). These pixels are selected as initial points for hypothetical edges to be tracked. Each initial point is then connected with its neighbor of highest gradient value from Fig. 7.18a, this one again with the highest of the remaining 7 neighbors etc. The result is the contour image in Fig. 7.18c. In this example, three different edge contour chains are found, numbered as 1-3. An edge needs only to be continued if the highest-valued neighbor is at least higher than a threshold $\Theta_b < \Theta_a$. This algorithm can be realized by different variants, e.g. edge contours of straight lines can preferably be tracked; *edge paths* can be compared, where decisions are made by accumulated gradient values over chains of connected points. Fig. 7.18d shows a result from a decision based on edge chains of length 2. The value '9' in the lower-right quadrant could be connected either with '4-4' or with '3-7'. While in the scheme underlying Fig. 7.18c the decision for the first path is made directly based on the higher value of '4', the schema of Fig. 7.18d also investigates the '7' following the '3' in the other path. This further results in a merging of two contour chains.

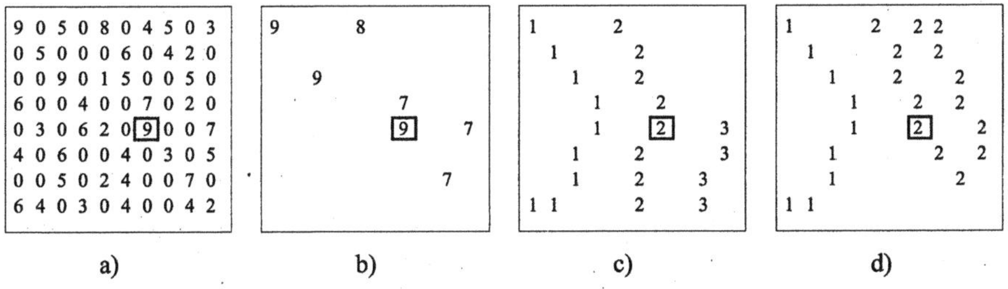

a)   b)   c)   d)

Fig. 7.18. Edge tracking (for explanation related to **a-d** refer to the text above)

7.3.4   Edge Model Fitting

For a straight step edge, a simple 2D edge model can be defined as shown in Fig. 7.19. Model parameters are the amplitude plateau values A and B on both sides of the edge, the edge position (r_{ed}, s_{ed}) and the orientation angle θ[1]:

$$\hat{y}(r,s) = \begin{cases} A & \text{if } (r-r_{ed})\cdot\sin\theta \geq (s-s_{ed})\cdot\cos\theta \\ B & \text{else} \end{cases} \tag{7.62}$$

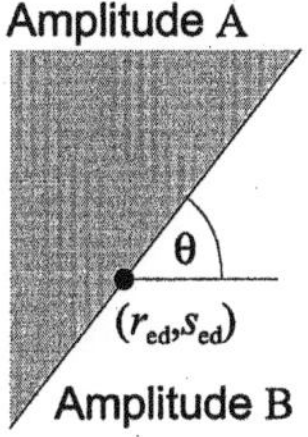

Fig. 7.19. Amplitudes and directional orientation in the step edge model (7.62).

By employing more complex models, e.g. including the width and/or the gradient of an edge, also parameters of ramp edges or other types of edges can be modeled. The goal is to adapt the edge model signal $\hat{y}(r,s)$, such that an error criterion (e.g. the mean squared error) is minimized when comparing the model to an image signal $x(m,n)$. A reasonable method in this context is decomposition of the image

---

[1] By definition here, $\theta=0$ shall relate to a horizontal edge with region A positioned above B.

signal and the edge model into orthogonal components. Basis functions are used for this purpose which are adapted to a given edge model. An example of two orthogonal basis images relating to a step-edge model is shown in Fig. 7.20. The area over which these functions extend should not be too small; even though this would simplify processing, the result would lack of reliability in detecting sufficiently large plateaus of clearly distinct amplitudes. Due to the properties of the basis functions, local amplitude variations within these plateaus are implicitly disregarded in the analysis. If the basis functions are defined over a quadratic image area of size PxP pixels[1], the following conditions hold :

$$\sum_{m=0}^{P-1}\sum_{n=0}^{P-1} t_0(m,n)\cdot t_1(m,n) = 0 \; ; \quad \sum_{m=0}^{P-1}\sum_{n=0}^{P-1} t_k(m,n)^2 = P\cdot(P-1) \; ; \; k = 0,1 . \qquad (7.63)$$

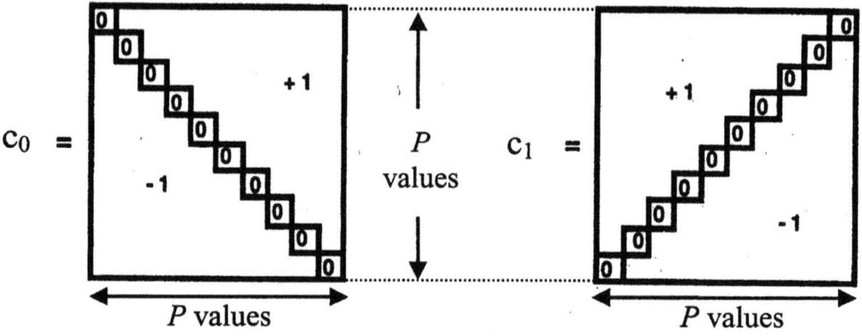

Fig. 7.20. Orthogonal basis images for edge model fitting (source: HARTMANN)

A (zero mean) approximation of the partial image within the given area is characterized as

$$x(m,n) = c_0 \cdot t_0(m,n) + c_1 \cdot t_1(m,n) , \qquad (7.64)$$

where the coefficients c_k due to the orthogonality condition result as

$$c_k = \frac{1}{P\cdot(P-1)}\sum_{m=0}^{P-1}\sum_{n=0}^{P-1} x(m,n)\cdot t_k(m,n) . \qquad (7.65)$$

The slope of the model edge, as related to the orientation of the basis functions, will be

$$\alpha = \arctan\frac{c_1}{c_0} , \qquad (7.66)$$

from which finally the angle of the edge as presented in Fig. 7.19 is computed as

$$\tan\theta = \frac{\tan\alpha - 1}{\tan\alpha + 1} = \frac{c_1 - c_0}{c_1 + c_0} \Rightarrow \theta = \arctan\frac{c_1 - c_0}{c_1 + c_0} . \qquad (7.67)$$

---

[1] Definitions over circular areas are also useful

At least one of the two values should be above a threshold value to distinguish between significant edges and local variations in amplitude. It is also possible to use basis images and related coefficients of common linear block transforms, the analysis will then however not be adapted to specific edge characteristics. If e.g. a DCT or a Walsh transform is applied, an edge orientation within the block can roughly be determined from the ratio of the coefficients which represent the first horizontal and vertical basis images[1],

$$\theta = \arctan \frac{c_{01}}{c_{10}} \, . \tag{7.68}$$

7.3.5   Description and Analysis of Edge Properties

Edge Histograms. Occurrences of edges or edge directions within an image can be described by an *edge histogram*. Depending on the accuracy of the underlying edge detection method, a number of distinct edge orientations has to be analyzed[2]. In addition, the number of occurrences for 'direction-less' edges as e.g. analyzed by the filter matrix (7.51) can be counted. An example where the resulting edge histogram vector consists of five bins is shown in Fig. 7.21. The edge histogram as shown here analyzes the frequency of occurrence for the different orientations, without taking into account the positions of edges. Combination with a localization method like a block grid structure (see sec. 7.1.2) is useful if the local distribution of edge directions within an image shall be characterized[3].

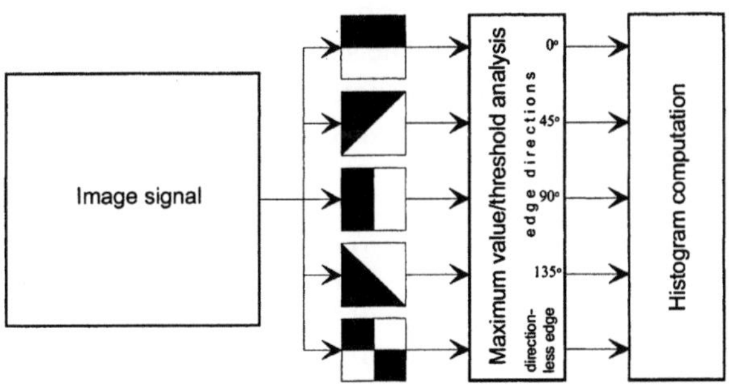

Fig. 7.21. System for extraction of edge histograms

---

[1] To analyze edge positions exactly, it would however not be sufficient to compute the transform over a set of non-overlapping blocks; in principle, the basis functions in edge model fitting are applied at any position of the image.

[2] Practically, this should at least be the four basic orientations, i.e. horizontally, vertically and two diagonal directions. In fact, more distinct orientations can only be differentiated of filter matrices or basis functions of sizes > 3x3 pixels are used.

[3] This is used by the *Edge Histogram Descriptor* of the MPEG-7 standard.

Hough Transform. The Hough transform is a parametric description, allowing to determine sets of edge pixels which establish straight lines. The following equation expressing line parameters by a polar form should hold for any points that are members of a straight line[1]:

$$\rho = m_P \cdot \cos\alpha + n_P \cdot \sin\alpha . \tag{7.69}$$

(7.69) allows to establish a relationship between the coordinates of edge points (e.g. points with $b(m_P,n_P)=1$ of (7.61)), which are potentially situated on the same line of distance ρ and angular orientation α (see Fig. 7.21a). The transformation into the *Hough space* (ρ,α) results by one graph for each edge point, expressing which straight lines characterized by (ρ,α) pairs might have this point as a member. If several graphs intersect within one single point in the Hough space, all related edge points will be on the same straight line, which is then characterized by the coordinates (ρ_S,α_S) of the intersection point (Fig. 7.21b). Practically, due to measurement inaccuracies in edge detection and due to sampling inaccuracies, the Hough graphs will typically not *exactly* intersect in unique points. It is then necessary to test for concentration of intersection points over certain neighborhood ranges within the Hough space (Fig. 7.21c). To capture potential positions of straight lines in an image, it is necessary to determine intersection points of Hough graphs pair-wise and apply a cluster analysis to find concentrations (see sec. 9.4.3). Practically, this can also be implemented by a fine uniform quantization of the Hough space, counting for each quantization cell the numbers of traversing graphs. In general, the Hough transform is not limited to the analysis of straight lines; if edge contours are described by higher-order polynomials, the dimensionality of the Hough space will increase, but in principle the same method is applicable to classify memberships on a model contour. The *Generalized Hough Transform* (sec. 9.1.3) extends the concept even further to analysis of arbitrary parametric descriptions of point features.

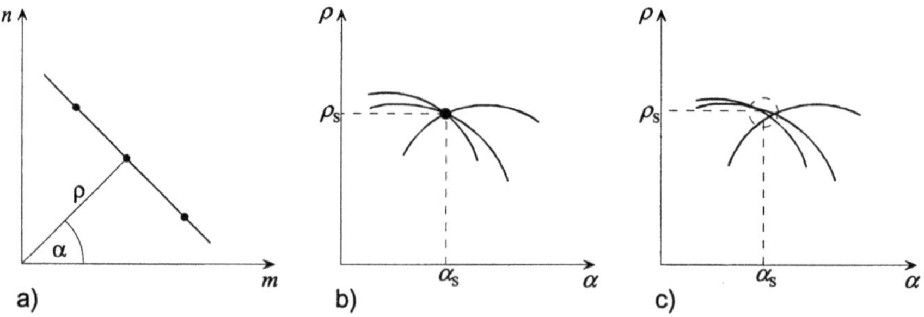

Fig. 7.22. **a** Parametrization of a straight line in polar coordinates **b** Representation of edge points by graphs in the Hough space in the ideal case and **c** in the realistic (non-ideal) case

---

[1] Refer to Problem 7.8 for a parametrization of the Hough transform over Cartesian coordinates.

7.4 Contour and Shape Analysis

7.4.1 Contour fitting

By *contour fitting* and *boundary fitting*, closed edge contours are generated implicitly. The underlying principle is approximation of the observed edge pixel positions by function systems. Among the methods which can be applied in general, the *Fourier*, *Wavelet*, *Spline* and *Polygon approximation* will be treated in detail.

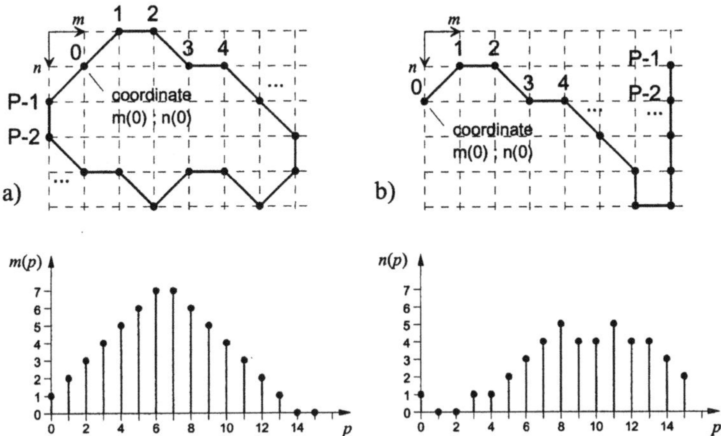

Fig. 7.23. Indexing of discrete contour graphs: **a** Closed contour **b** Open contour **c** Contour signals $m(p)$ and $n(p)$ relating to the closed contour

A discrete contour of P contour pixels indexed by $0 \leq p < P$ shall be given. This can uniquely be described by the m and n coordinates of the contour pixels on a discrete grid. For each contour pixel, values $m(p)$ and $n(p)$ describe the position (Fig. 7.23). The contour can be closed (Fig. 7.23a) or open (Fig. 7.23b). Contour pixels of indices p and $p+1$ can in principle have arbitrary distances. The coordinates $m(p)$ and $n(p)$ are interpreted as two *one-dimensional signals* in Fig. 7.23c. To realize a translation-invariant contour representation (describe the contour independent of its absolute position on the image grid), it is sufficient to subtract the contour mean value[1]:

$$m'(p) = m(p) - \bar{m} \ \text{ with } \ \bar{m} = \frac{1}{P}\sum_{p=0}^{P-1} m(p) \ ; \ n'(p) = n(p) - \bar{n} \ \text{ with } \ \bar{n} = \frac{1}{P}\sum_{p=0}^{P-1} n(p) \quad (7.70)$$

---

[1] Observe that the mean values $\bar{m}$ and $\bar{n}$ are not typically integer values. If the contour pixels shall still be at grid positions after mean subtraction, mean values must be rounded.

An alternative representation, split into two one-dimensional signals as well, can be gained by transforming the contour positions into polar coordinates,

$$\rho(p) = \sqrt{m^2(p) + n^2(p)}$$

$$\varphi(p) = \arctan\frac{n(p)}{m(p)} + k(p)\cdot\pi\;;\quad k(p) = \begin{cases} 0 \text{ if } m(p) \geq 0 \\ 1 \text{ if } m(p) < 0. \end{cases} \tag{7.71}$$

The subsequent descriptions of approximation techniques can be applied either to Cartesian or polar contour coordinates. The representation by polar coordinates is advantageous, as by neglecting $\varphi(p)$ a rotation-invariant, and by neglecting $\rho(p)$ a scaling-invariant contour description results[1]. Using zero-mean coordinates according to (7.70), the distance between each contour pixel and the center of the contour is determined as follows:

$$\rho'(p) \approx \sqrt{m'(p)^2 + n'(p)^2}\;. \tag{7.72}$$

An appropriate distance criterion to measure the similarity between two contours A and B is the *area* between the contours in the 2D image plane. If the angular differences $\varphi(p)-\varphi(p-1)$ are roughly constant over all p, the following simple criterion is approximately proportional with the difference area[2]:

$$d = \sum_p \left|\rho'_A(p) - \rho'_B(p)\right|. \tag{7.73}$$

All methods introduced in the forthcoming subsections can also be used for lossy or lossless encoding of continuous or discrete contours. It is then necessary to quantize and encode the resulting contour parameters (see also sec. 12.6.2).

Fourier approximation. To describe a contour by Fourier coefficients, the two 'coordinate signals' are interpreted as the real and imaginary parts of a complex number

$$t(p) = n(p) + j\cdot m(p)\quad;\; 0 \leq p < P\;. \tag{7.74}$$

The complex contour signal $t(p)$ is a periodic function if the contour is closed. Then, the DFT which implicitly interprets the analyzed segment of samples as being periodic (cf. sec. 3.3), is an optimum transform for $t(p)$[3]. The contour coefficients $T(q)$ typically are also complex numbers and periodic in any case; as the

---

[1] Under the assumption that contours are rotated or scaled with reference to the center coordinates.

[2] This would exactly be the case for the contour of a circle.

[3] The DFT is the optimum decorrelating transform for periodic signals. The contour which can be described most compactly is a circle, where the signals $m(p)$ and $n(p)$ are a sine and a cosine of period P, phase shifts depending on the definition of the contour starting point. One complex DFT coefficient is then sufficient for perfect description.

signal itself is complex, no complex-conjugate symmetry exists between coefficients $T(q)$ and $T(P-q)$. The number linearly independent values is $2P$ either in real and imaginary parts of the DFT representation or in the contour coordinate representation. The contour can be reconstructed by computing the inverse transform:

$$T(q) = \sum_{p=0}^{P-1} t(p) \cdot e^{-j\frac{2\pi pq}{P}} \quad \text{and} \quad t(p) = \frac{1}{P}\sum_{q=0}^{P-1} T(q) \cdot e^{j\frac{2\pi pq}{P}} . \tag{7.75}$$

By the properties of the DFT, the following conditions hold:

- A translation of the contour (adding constant offsets in m or n) will only change the coefficient $T(0)$, which expresses the mean value of the contour coordinates; $T(0)$ will be zero, if removed in advance as in (7.70).
- A scaling (change of size without changing the center point or shape) effects a linear scaling of all coefficients $T(q)$ with $q>0$.
- A rotation of the contour around the center point effects a linear-phase shift (proportional with the frequency) on all coefficients $T(q)$ with $q>0$, while the amplitude is not affected.

These properties are advantageous if the similarity of contours shall be compared independent of size, position and orientation. If the shape of the contour is relatively flat, the contour spectrum exhibits a compaction of 'energies' at low frequencies. As the transform is linear, the squared deviation between two contours can be computed directly in the transformed domain. The contour can also be approximated by a reduced number of transform coefficients[1] $P'<P$, such that the comparison of two contours can be simplified. A smoothing of contours can be performed by discarding high-frequency coefficients. If contours are not closed, or if the contour coordinates are extracted from positions which are not directly neighbored, the Fourier representation will be less efficient, as discontinuities appear which can not be approximated very well by sinusoids.

Spline approximation. Spline interpolation (sec. 5.4.3) approximates a continuous signal by discrete control coefficients. It is also frequently used to approximate contours. I control coefficients $\mathbf{c}_i=[r(i)\ s(i)]^T$ shall express contour sampling points that are known by their continuous coordinate positions. The approximation results in a continuous contour of values $\mathbf{r}(t)=[r(t)\ s(t)]^T$ by[2]

$$\mathbf{r}(t) = \sum_{i=0}^{I-1} \mathbf{c}_i \phi_i^{(Q)}(t). \tag{7.76}$$

---

[1] It is highly probable that truncating off high-frequency coefficients will lead to a relatively low approximation error, as the signals $m(p)$ and $n(p)$ are smooth, when extracted from direct neighbors.

[2] t is a one-dimensional continuous coordinate here, which is the running variable along the contour, where $t=0$ is the starting point of the contour.

The function $\phi_i^{(Q)}(t)$ is the *B spline* of order Q according to (5.48). If contour pixels are extracted from discrete images, the input will be a discrete contour of P observed points t_p. The control coefficients must be determined such that the following condition holds at the known positions of the contour t_p (cf. (5.55) and (5.56)):

$$\mathbf{r}(t_p) = \sum_{i=0}^{I-1} \mathbf{c}_i \phi_i^{(Q)}(t_p) \quad ; \quad 0 \le p < P. \tag{7.77}$$

(7.77) represents an equation system which can also be written in matrix notation as

$$\mathbf{R} = \mathbf{\Phi}^{(Q)} \cdot \mathbf{C} \tag{7.78}$$

with

$$\mathbf{R} = \begin{bmatrix} r(t_0) & s(t_0) \\ \vdots & \vdots \\ r(t_P) & s(t_P) \end{bmatrix} \quad ; \quad \mathbf{C} = \begin{bmatrix} r(0) & s(0) \\ \vdots & \vdots \\ r(I-1) & s(I-1) \end{bmatrix}. \tag{7.79}$$

$\mathbf{\Phi}^{(Q)}$ is a matrix of size $I \times P$. If $P > I$ (lower number of control points than original number of contour coordinates), (7.78) is an *overdetermined equation system*. The computation of the control point values can then be performed by pseudo inversion of the matrix $\mathbf{\Phi}^{(Q)}$ (for definition of the pseudo inverse, refer to sec. 8.3):

$$\mathbf{C} = \left[\mathbf{\Phi}^{(Q)} \right]^{\text{P}} \cdot \mathbf{R}, \tag{7.80}$$

which for the special case of $I = P$ is identical to the inverse of the matrix $[\mathbf{\Phi}^{(Q)}]^{-1}$.

Polygon Approximation. The principle of polygon approximation is a *linear interpolation* of the contour, which means that it is identical to spline interpolation of order $Q = 1$. The determination of control points is performed iteratively, where by each iteration step one more control point is set. The approximation starts with one line, which interconnects the terminating points A and B of the contour for case of open contours, or the pair of points having largest distance for the case of closed contours. By each further step, one position on the original contour must be found which has the largest geometric distance from the contour approximated by the polygon. At this position, a new control point is set. An example is shown in Fig. 7.24, where a/b illustrate the first two approximation steps, in which the control points C and D are selected. The approximation is terminated, if a fidelity criterion is fulfilled, e.g. the approximated contour shall not deviate by more than a certain maximum allowable distance $d_{\max}$ from the original contour (Fig. 7.24c). If a contour shape actually consists of straight lines and has corners, polygon approximation can achieve a better approximation than higher-order spline functions. The polygon is uniquely described by a sequence of *vertices*, which are the coordinate positions of the control points.

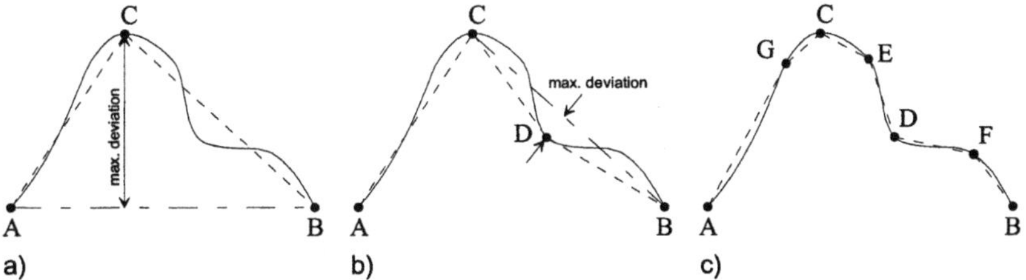

Fig. 7.24. Polygon approximation of a contour

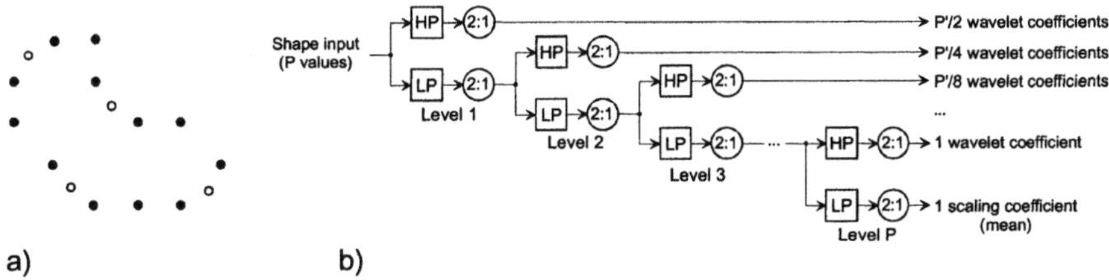

Fig. 7.25. a Interpolation of a contour into a number of samples P' that equals a power of 2 (white dots inserted) **b** Dyadic wavelet decomposition of the contour of length P'

Wavelet Approximation. For approximation of contours, wavelet basis functions are also applicable [CHUANG, KUO 1996]. These can represent the contour signal by a series of resolution scales represented by discrete coefficients. The coefficients are simply obtained by wavelet analysis of contour coordinate sequences.

In the case of a dyadic wavelet bases, the largest possible depth of a wavelet tree can be achieved if the number of contour points is a power of 2. Let P be the number of contour pixels, and P' the next-higher power-of-two value, such that $R=P'-P$ contour pixels are missing. This can be resolved, if additional points are inserted at each position p where the modulo value $\mathrm{mod}(p, P/R)=0$ (see Fig. 7.25a) [MÜLLER, OHM 1999]. If the new points are generated by interpolation, the related wavelet coefficients of the highest frequency band will be near zero. It is then possible to decompose the shape by $\log_2 P'$ levels of 2-band splits into $P'-1$ wavelet coefficients and one scaling coefficient (see Fig. 7.25b). The latter one represents the mean value of the contour.

Horizontal and vertical coordinates $m(p)$ and $n(p)$ can be transformed independently into scaling and wavelet coefficients $c_u^{(m)}(p')$ and $c_u^{(n)}(p')$, where the coordinates p' are the sub-sampled index variables. Due to the properties of the wavelet transform, the following relationships apply:

- A translation of the contour from one center $(\bar{m}_1, \bar{n}_1)$ to another $(\bar{m}_2, \bar{n}_2)$ only modifies the *scaling coefficients* $c_0^{(m)}(p')$ and $c_0^{(n)}(p')$.

- A scaling by a factor Θ, performed with reference to the contour center coordinate, effects a constant scaling of all *wavelet coefficients*:

$$\begin{bmatrix} \tilde{c}_u^{(m)} \\ \tilde{c}_u^{(n)} \end{bmatrix} = \Theta \cdot \begin{bmatrix} c_u^{(m)} \\ c_u^{(n)} \end{bmatrix} \quad \text{for all } u > 0. \tag{7.81}$$

– A rotation by an angle θ around the origin of the coordinate system effects a shift of coefficient magnitudes between the two directional components of *scaling and wavelet coefficients*, which can be expressed by a rotation matrix:

$$\begin{bmatrix} \tilde{c}_u^{(m)} \\ \tilde{c}_u^{(n)} \end{bmatrix} = \begin{bmatrix} \cos\theta & \sin\theta \\ -\sin\theta & \cos\theta \end{bmatrix} \begin{bmatrix} c_u^{(m)} \\ c_u^{(n)} \end{bmatrix}. \tag{7.82}$$

The scaling and localization properties of the wavelet transform allow identifying specific coefficients which have a more global or only local influence on the contour characteristics. For example, a change in a higher-frequency wavelet coefficient mainly effects local changes of the contour (Fig. 7.26). If higher-frequency bands are discarded, the remaining scaling and wavelet coefficients represent subsampled contours; this simplifies the contour feature comparison at different contour resolutions.

Fig. 7.26. Effect of localization property of the wavelet transform in contour modifications (source : MÜLLER)

The contour wavelet transform can also be applied to the polar coordinate representation (7.72), which is advantageous as rotation of a contour affects the phase coefficients, while scaling modifies the magnitude coefficients. For example, in a similarity-search algorithm, the ρ-component is most important and can be used as standalone criterion. Since wavelets are a naturally scalable representation, comparison of contours having different sizes is straightforward; when contours are scaled up to the next-higher power of two in the wavelet representation, it will always be possible to find one scale where two contours are of equal size. Thus, the dyadic wavelet representation is implicitly *scale invariant*. To obtain *rotation invariance*, it is necessary to modify (7.73) such that search is performed for

$$d = \min_{p'=0\ldots P'-1} \sum_{p=0}^{P'-1} \left| \rho_A(p) - \rho_B(p+p') \right|. \tag{7.83}$$

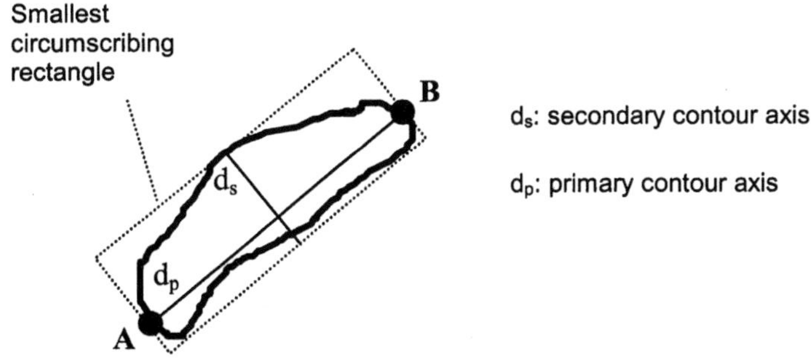

Fig. 7.27. Principal axes of the contour (source: MÜLLER)

Fig. 7.28. Examples of contours classified as similar by using a wavelet-based contour description

For a given resolution, P' comparisons are necessary. For contours with a clear difference between the lengths of the principal axes, the best energy compaction is achieved if the wavelet representation is constructed such that the first wavelet coefficient approximately represents the distance between points **A** and **B** on the primary axis (see Fig. 7.27). By defining the coordinate $p'=0$ at point **A** or **B**, and retaining the samples with even values p' in the iterated sub-sampling over the wavelet tree, the two points retained at the last level of the wavelet decomposition tree will approximately span the primary axis[1].

Fig. 7.28 shows examples of contours which were classified as being similar based on the wavelet contour feature in a database retrieval application.

7.4.2 Contour Description by Orientation and Curvature

If a starting point and the length of a closed contour are defined, it is possible to describe the contour by the tangential angle φ or its differential change $\Delta\varphi$ along the contour line. Fig. 7.29 shows this principle of *generalized chain coding*[2] for the case of a continuous contour. The tangential angle at position t of a *continuous contour* can be determined as follows:

---

[1] The two principal axes (primary and secondary) of a shape can be determined by eigen-vector analysis, cf. sec. 7.4.5

[2] Refer to the principles of *chain coding* and *differential chain coding* in the case of discrete contours, sec. 12.6.2; in these cases, discrete sets of angles are used.

$$\cos\left[\varphi(t)\right] = \frac{dr(t)}{dt} \; ; \; \sin\left[\varphi(t)\right] = \frac{ds(t)}{dt} \Rightarrow \varphi(t) = \arctan\frac{ds(t)}{dr(t)}. \tag{7.84}$$

If a *discrete contour* is given by P positions p, the following approximation of the contour coordinate derivatives can be used[1]:

$$\frac{dr(t)}{dt}\bigg|_{t=t(p)} \approx \Delta m(p) = m(p+1) - m(p)$$
$$\frac{ds(t)}{dt}\bigg|_{t=t(p)} \approx \Delta n(p) = n(p+1) - n(p) \tag{7.85}$$

An approximation of the contour length is then given by

$$L \approx \sum_{p=0}^{P-1} \sqrt{\Delta m^2(p) + \Delta n^2(p)} \tag{7.86}$$

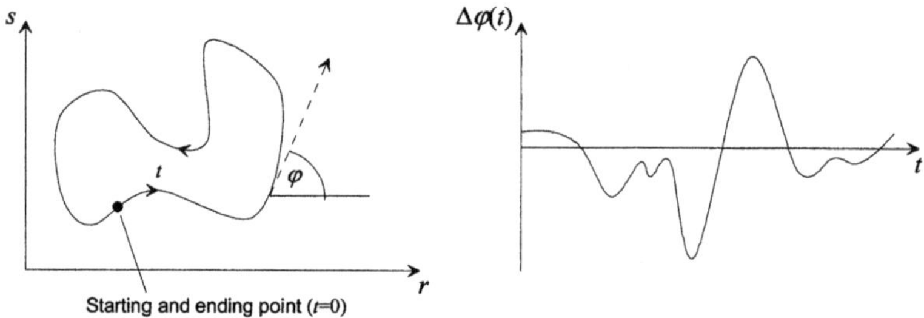

Abb. 7.29. Generalized chain coding of a continuous contour

Another contour feature is the *curvature*, which can be determined by computation of the *second derivatives* of the contour signals, the absolute value giving one single parameter for any given contour position:

$$\left|\kappa(t)\right| = \sqrt{\left(\frac{d^2r(t)}{dt^2}\right)^2 + \left(\frac{d^2s(t)}{dt^2}\right)^2}. \tag{7.87}$$

For discrete contours, the curvature is computed by the following approximation of the second derivatives at contour position p, which is analogous to the computation of second derivatives of a signal by the Laplacian operator (7.52),

$$\frac{d^2r(t)}{dt^2}\bigg|_{t=t(p)} \approx -m(p+1) + 2 \cdot m(p) - m(p-1)$$

---

[1] neglecting normalization by sampling distances R and S, which is strictly correct for case of directly neighbored contour points and 1:1 sample aspect ratio.

$$\left.\frac{d^2 s(t)}{dt^2}\right|_{t=t(p)} \approx -n(p+1)+2\cdot n(p)-n(p-1).\tag{7.88}$$

Fig. 7.30 qualitatively shows the graphs of curvature parameter $|\kappa(t)|$ for the examples of regular geometric shapes.

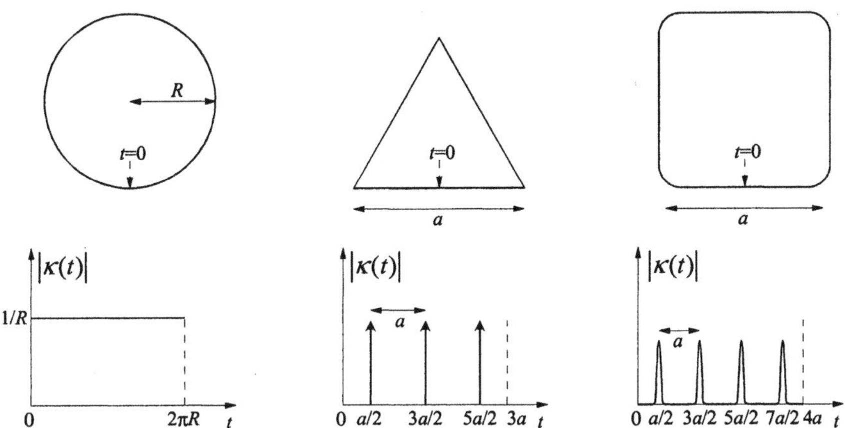

Fig. 7.30. Curvature parameters $|\kappa(t)|$ for the examples of a circle, a triangle and a rounded rectangle

Extreme turning points of the contour have a high significance to characterize a shape, in particular when their relative positions are regarded. Many natural objects which can be described by a contour are subject to geometric modifications of shape, e.g. a horse will have four legs, one head etc., but they may be positioned differently depending on the situation of acquisition. Nevertheless, the *presence* of a specific number of contour curvature extrema, their strengths and positions on the contour are highly expressive for the semantic interpretation. Extreme points of the contour curvature can be determined by the zero crossings of the following function, which could be interpreted as a direction-independent third derivative of a contour signal, such that the zero crossings of this function express the *turning points* of the contour:

$$\lambda(t)=\frac{\dfrac{dr(t)}{dt}\cdot\dfrac{d^2 s(t)}{dt^2}-\dfrac{ds(t)}{dt}\cdot\dfrac{d^2 r(t)}{dt^2}}{\sqrt{\left(\left(\dfrac{dr(t)}{dt}\right)^2+\left(\dfrac{ds(t)}{dt}\right)^2\right)^3}}.\tag{7.89}$$

To compute $\lambda(t)$ on discrete contours, the discrete approximations of second derivatives (7.88), and of first derivatives (7.85) are used. If two contours representing the shapes of objects are sampled by the same number P of discrete contour points (spaced equidistantly on the contour), information carried by $\lambda(t)$ can be used for comparison of contours, irrespective of the original (continuous) contour sizes. It will however be necessary to determine the most relevant zero crossings, which best express the global properties of the contour. The idea behind is very similar to hierarchical methods for edge detection (sec. 7.3.3)[1].

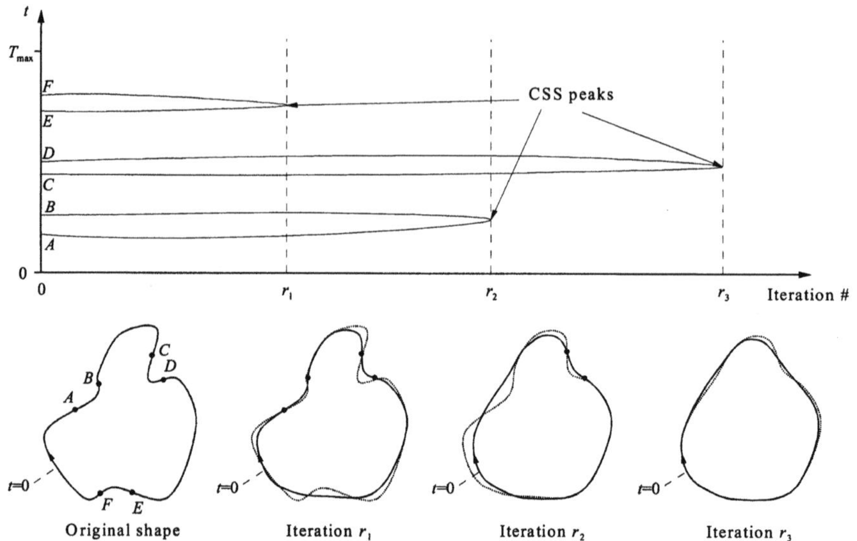

Fig. 7.31. CSS plot of zero-crossing positions in curvature parameter $\lambda(t)$ of an iteratively lowpass-filtered contour

Curvature scale space analysis. *Smoothing of contours* (elimination of local irregularities and deviations) can be performed by lowpass filtering of the contour signals $m(p)$ and $n(p)$[2]. Relevant zero crossings of $\lambda(t)$ (extrema of curvature or turning points of the contour direction) will however be preserved even after strong smoothing of the contour, as they signify important global characteristics of the shape. This process of smoothing can be applied iteratively, similar to a Gaussian pyramid approach (sec. 4.4.6); as a simple kernel to approximate a Gaussian-shaped impulse response, the binomial filter $\mathbf{h}=[1/4\ 1/2\ 1/4]^T$ can be used. After a certain number of iterated smoothing filter steps, any closed contour will become *convex* and *converge towards an ellipse* orientated by the principal axes of the contour. If the contour is convex, no contour-direction turning points will exist any more, such that no zero-crossings are found in $\lambda(t)$. Due to the analogy with scale-space analysis of signals (cf. sec. 4.4.4 and 4.4.6), the method is denoted as *Curvature Scale Space* (CSS) analysis [MOKHTARIAN, BOBER 2003]. To use this for an expression of contour properties, the *positions* and the *strengths* of the relevant zero

---

[1] In fact, the zero-crossings of $\lambda(t)$ characterize the discontinuities in the contour signals $r(t)$ and $s(t)$ or their discrete counterparts $m(p)$ and $n(p)$, as determined by gradient analysis.
[2] See Problem 7.9.

crossings must be described. After each step of lowpass filtering in the iterative smoothing of contours, the function (7.89) is computed and zero-crossings are detected. Obviously, the number of zero-crossings will become lower by tendency with stronger smoothing of contours. Hence, the *strength* of a curvature turning point can be characterized by the number of smoothing iterations which must be applied until it disappears by convexity. Likewise, the curvature peak *position* can be defined as the position where a zero crossing was last observed before it disappeared. This behavior can be visualized in a *CSS plot*, which shows the positions of zero crossings over the iteration number of the smoothing process (see Fig. 7.31). This method is used in the *Contour Shape Descriptor* of the MPEG-7 standard, where the *CSS peak* parameters express the strengths and positions of the most relevant zero crossings, computed as explained above. Fig. 7.31 shows the example of a contour which is iteratively lowpass-filtered, and the related CSS plot. Any concave segment in the contour is framed by two curvature turning points, which move towards each other during the iterated smoothing and will converge towards the same position when convexity is approached.

7.4.3   Geometric Features and Binary Shape Features

Methods for contour description are suitable to characterize the *outline shape* of an object. Different analysis methods are used to describe the *area-related* shape properties. A binary shape of an object shall be given by

$$b(m,n) = \begin{cases} 1: & \text{part of the object} \\ 0: & \text{not part of the object.} \end{cases} \tag{7.90}$$

This can be used to determine a number of *basic shape features*. The area of a shape (measured by the number of binary shape pixels) is

$$A = \sum_m \sum_n b(m,n). \tag{7.91}$$

The geometric *centroid* of a shape – this is the center of gravity, not being identical with the mean values of contour points (7.70) – is positioned at coordinate (r_S, s_S) where

$$r_S = \bar{m} = \frac{1}{A} \sum_m \sum_n b(m,n) \cdot m$$

$$s_S = \bar{n} = \frac{1}{A} \sum_m \sum_n b(m,n) \cdot n. \tag{7.92}$$

Observe that the centroid of the binary shape defined on a discrete grid is typically not an integer value, such that it may be positioned between points of the sampling grid. The *bounding rectangle* defines the size of the smallest rectangular image which could contain the entire shape by *original orientation*. It is defined by the side lengths (see Fig. 7.32a)

$$a_1 = \max\left[\left. r\right|_{b(r,s)=1}\right] - \min\left[\left. r\right|_{b(r,s)=1}\right]; \quad a_2 = \max\left[\left. s\right|_{b(r,s)=1}\right] - \min\left[\left. s\right|_{b(r,s)=1}\right] \qquad (7.93)$$

for a shape on continuous coordinates and

$$a_1 = \max\left[\left. m\right|_{b(m,n)=1}\right] - \min\left[\left. m\right|_{b(m,n)=1}\right] + 1; \quad a_2 = \max\left[\left. n\right|_{b(m,n)=1}\right] - \min\left[\left. n\right|_{b(m,n)=1}\right] + 1 \qquad (7.94)$$

for a shape on discrete coordinates[1]. Alternatively, a *smallest circumscribing rectangle* could be defined (see Fig. 7.27), where the orientations of sides a_1 and a_2 are determined by the principal axes[2]. Other circumscribing figures, e.g. *ellipses*, are useful as well.

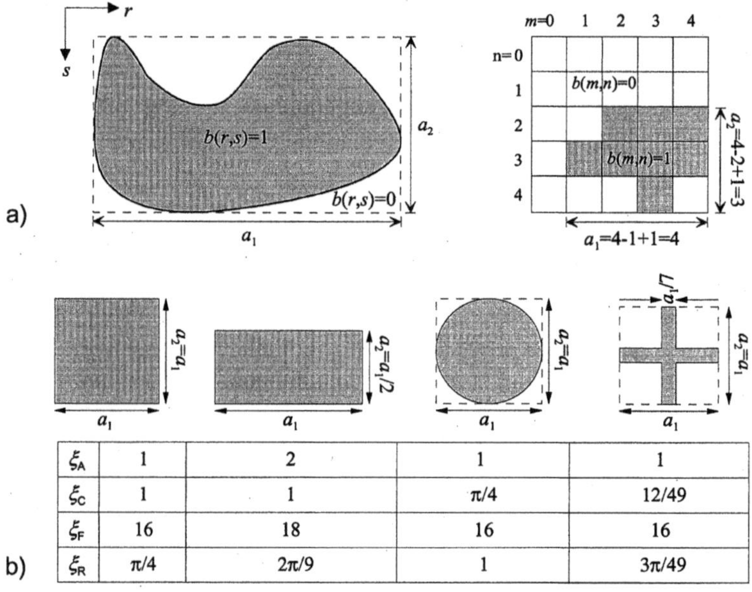

ξ_A	1	2	1	1
ξ_C	1	1	$\pi/4$	12/49
ξ_F	16	18	16	16
ξ_R	$\pi/4$	$2\pi/9$	1	$3\pi/49$

Fig. 7.32. a Definition of bounding rectangle in continuous (left) and discrete (right) binary shapes. **b** Examples of binary shapes and their basic shape parameters

The following basic feature parameters can then be determined for shape description:

---

[1] As the maximum and minimum coordinates in $b(m,n)$ describe samples of the *inner contour*, it is necessary to increment the difference values by 1 in (7.94). This is based on a geometric interpretation of a 2D image matrix to be a package of rectangular cells, where the sampling coordinates are the centers, and contour positions are at the boundaries of the rectangles.

[2] Principal axes can be determined by eigenvector analysis, cf. (7.125)-(7.133).

$$\text{Aspect ratio} \qquad \xi_A = \frac{a_1}{a_2} \qquad\qquad\qquad (7.95)$$

$$\text{Compactness} \qquad \xi_C = \frac{A}{a_1 \cdot a_2} \qquad\qquad\qquad (7.96)$$

$$\text{Form factor} \qquad \xi_F = \frac{4(a_1 + a_2)^2}{a_1 \cdot a_2} \qquad\qquad\qquad (7.97)$$

$$\text{Roundness}^{[1]} \qquad \xi_R = \frac{4\pi A}{L^2} \qquad\qquad\qquad (7.98)$$

The contour length L can be computed according to (7.86). Examples of these parameters are shown for different shapes in Fig. 7.32b. Except for the roundness, *no rotation invariance* applies. If rotation invariance is required, adequate parameters can be computed by using the *smallest circumscribing rectangle* which must be determined a priori[2].

The *projection profile* consists of a horizontal component Π_h and a vertical component Π_v, which analyze the numbers of shape pixels contained in each row or column of the circumscribing rectangle image (see Fig. 7.33). If a discrete shape has a circumscribing rectangle of side lengths a_1 and a_2, a total of a_1+a_2 values characterize the projection profiles, which are defined by

$$\Pi_h(n) = \sum_{m=0}^{a_1-1} b(m,n) \; ; \; 0 \le n < N$$

$$\Pi_v(m) = \sum_{n=0}^{a_2-1} b(m,n) \; ; \; 0 \le m < M \qquad\qquad (7.99)$$

Projection profiles only allow performing analysis of width and size of an object related to a given orientation and size, they are not scaling- and rotation-invariant.

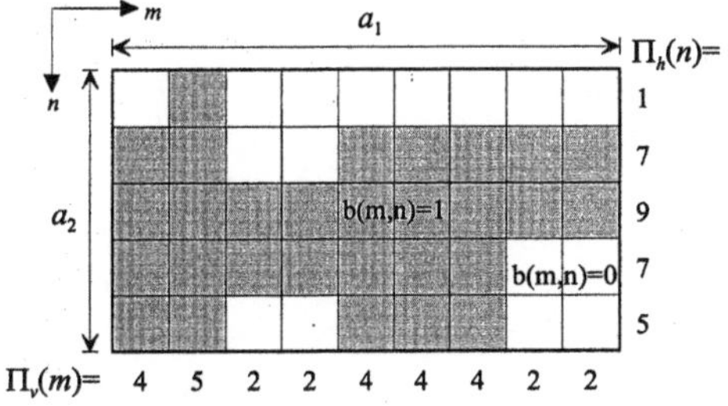

Fig. 7.33. Example of projection profiles

---

[1] A similar parameter denoted as *circularity* is defined as (perimeter)^2/area $= 4\pi/\xi_R$.

[2] A similar parameter as the aspect ratio relating to the circumscribing rectangle is the eccentricity (7.135)

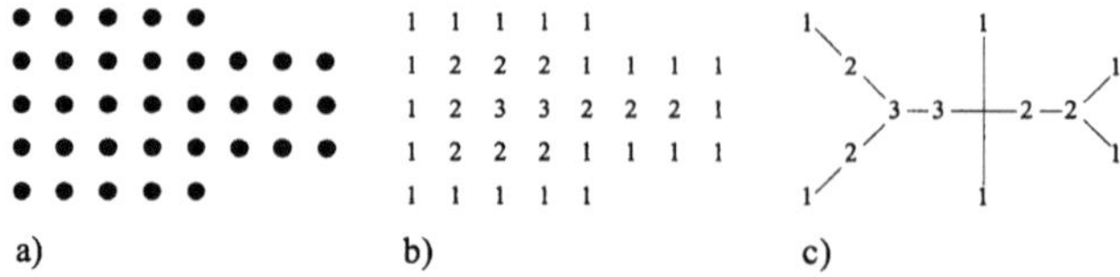

Fig. 7.34. Skeleton description   **a** Shape of an object **b** Values of distance transform **c** morphological skeleton

Distance transformation and skeleton description. By iterative application of morphological erosion (see sec. 5.1.2), the shape of an object can be *thinned* until only lines remain, which are denoted as the *skeleton* of the shape. This will possibly discard the information about corner positions which are relevant for description of the object's shape properties. If in addition to the skeleton the distance of its points from the object boundary is known, reconstruction of the shape is possible. The combination of skeleton and distances provides a compact representation of shapes, and the parameters can likewise be used as shape features. The distance information can be generated by a *distance transformation*. For each pixel of the binary shape (Fig. 7.34a), the shortest horizontal or vertical distance from the object boundary is determined (Abb. 7.34b). This information can be gained by iterative erosion using a nearest-neighbor structure element, where the distance is the iteration number of the erosion step by which the respective pixel will be discarded. The morphological skeleton is then the set of points, for which no member of the neighborhood system $\mathcal{N}_1^{(1)}(m,n)$ according to (4.1) will have a higher distance value (Fig. 7.34c). Within the skeleton, points of highest distance values have highest influence on description of the shape; comparing the constellations of these points gives an indication about similarity of two shapes. From the example in Fig. 7.34c, it is further obvious that the skeleton does not necessarily consist of connected pixels.

Symmetry axis transform (SAT). In SAT, also denoted as *medial axis transform* (MAT), the shape is described by a superposition of elements A_i with certain symmetry properties. The elements may be overlapping, which means that the entire shape A is a unification $A_1 \cup A_2 \cup \ldots \cup A_P$ for the case of P single binary elements. Fig. 7.35 shows an example, where the shape is described by $P=5$ squares of different positions and sizes. The SAT representation consists of the number P, and for each square the parameters of center position (cross-marked) and side length. Alternatively, rectangles, circles, ellipses etc. can be used as describing elements. The center position and size signify translation and scaling of the elements. Additionally, rotation, shear etc. could be parametrized, if useful by the nature of shapes to be described in a given application. Lossy SAT representations can straightforwardly be determined, where the largest elements have highest priority, and the shape distortion is minimized if only smallest elements or elements with lowest contribution to the shape (smallest non-overlapping areas) are

with lowest contribution to the shape (smallest non-overlapping areas) are discarded.

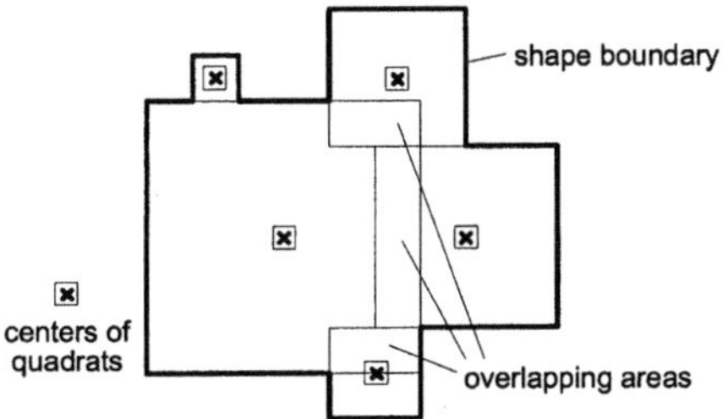

Fig. 7.35. Symmetry axis transform

7.4.4   Projection and geometric mapping

One ultimate goal in characterization and comparison of shapes is identification of exterior-world objects from image and video signals. The available projection into the image plane depends on the relative positions of the camera and the 3D world object which is captured. On the other hand, the features to be extracted for the description of shapes should be independent of the situation of capturing. Hence, it is useful to take into regard models for this projection process, to define criteria which are mostly invariant against position changes in the capturing process. For this analysis, the pin-hole camera model (Fig. 1.4) is used. Any point $\mathbf{P}=[X_P\ Y_P\ Z_P]^{\mathrm{T}}$ from the 3D world is mapped by the central projection equation (1.1) into image plane coordinates $\mathbf{r}_P=[r_P\ s_P]^{\mathrm{T}}$:

$$r_P = \frac{F}{Z_P} X_P\ ; \qquad s_P = \frac{F}{Z_P} Y_P\ . \tag{7.100}$$

If the 3D world object is at far distance from the camera, the *orthographic projection* (parallel projection) can be used as an approximation. In this case, only the horizontal/vertical 3D coordinates X and Y are projected into image coordinates, while focal length and depth (distance) are both compensated by a factor C which is constant over all points of an object:

$$r_P = \frac{X_P}{C}\ ; \qquad s_P = \frac{Y_P}{C}\ . \tag{7.101}$$

In the sequel, the projection relationships are described for the case of changing 3D position of a rigid object relative to the camera position. In principle, it is irrelevant if this change of relative position incurs due to camera movement or due to movement of the object. The change of a position of a rigid object can entirely be described by three translational parameters (change of the object centroid in X,Y,Z). The change of its orientation is expressed by three rotational parameters, the rotation angles φ_x, φ_y, φ_z around the axes of a coordinate system with origin at the mass centroid of the object. For a mass point belonging to the object, the resulting shift from position $\mathbf{P}=[X_P\ Y_P\ Z_P]^\mathrm{T}$ into position $\mathbf{P}'=[X_P\ Y_P\ Z_P]^\mathrm{T}$ can be described as

$$\mathbf{P}' = \mathbf{R}\cdot\mathbf{P}+\mathbf{t}\,. \tag{7.102}$$

Herein, the translation vector $\mathbf{t}=[\tau_x\ \tau_y\ \tau_z]^\mathrm{T}$ describes the shift of the centroid, and the rotation matrix is defined by

$$\mathbf{R} = \mathbf{f}\left(\varphi_x,\varphi_y,\varphi_z\right) = \begin{bmatrix} \rho_{11} & \rho_{12} & \rho_{13} \\ \rho_{21} & \rho_{22} & \rho_{23} \\ \rho_{31} & \rho_{32} & \rho_{33} \end{bmatrix}. \tag{7.103}$$

The components of the rotation matrix ρ_{ij} are in principle functions of the rotation angles φ_x, φ_y and φ_z, which define the rotation around the object's centroid. The relevant coordinate system of (7.103) has however its origin at the focal point of the camera. An exact description of the relationships between the φ-values and the matrix elements is beyond the scope of detail that can be given here; it shall be noted however, that due to the change of the centroid relative to the camera coordinate system, the ρ_{ij} are not only dependent on the angles, but are influenced by the τ-values as well. A similar problem will be treated below in the context of a 2D rotation. From the matrix system, the coordinates after the change of the relative position can be specified as

$$\begin{aligned} X'_P &= \rho_{11}X_P + \rho_{12}Y_P + \rho_{13}Z_P + \tau_x \\ Y'_P &= \rho_{21}X_P + \rho_{22}Y_P + \rho_{23}Z_P + \tau_y \\ Z'_P &= \rho_{31}X_P + \rho_{32}Y_P + \rho_{33}Z_P + \tau_z\,. \end{aligned} \tag{7.104}$$

Substituting (7.104) into (7.100), the projection onto the image plane gives

$$\begin{aligned} \frac{Z'_P}{F}r'_P &= \rho_{11}\frac{Z_P}{F}r_P + \rho_{12}\frac{Z_P}{F}s_P + \rho_{13}Z_P + \tau_x \\ \frac{Z'_P}{F}s'_P &= \rho_{21}\frac{Z_P}{F}r_P + \rho_{22}\frac{Z_P}{F}s_P + \rho_{23}Z_P + \tau_y\,. \end{aligned} \tag{7.105}$$

This can be resolved for direct specification of r'_P and s'_P by

$$r'_P = \frac{\rho_{11}r_P + \rho_{12}s_P + \rho_{13}F + F\dfrac{\tau_x}{Z_P}}{\rho_{31}\dfrac{1}{F}r_P + \rho_{32}\dfrac{1}{F}s_P + \rho_{33} + \dfrac{\tau_z}{Z_P}}\ ;\ s'_P = \frac{\rho_{21}r_P + \rho_{22}s_P + \rho_{23}F + F\dfrac{\tau_y}{Z_P}}{\rho_{31}\dfrac{1}{F}r_P + \rho_{32}\dfrac{1}{F}s_P + \rho_{33} + \dfrac{\tau_z}{Z_P}}\,. \tag{7.106}$$

The equations (7.106) define the image plane coordinates r'_P and s'_P after the movement as a function of the image plane coordinates r_P and s_P before the movement. Further dependencies from the parameters describing the movement, from the focal length F and the depth Z are in effect. For the case of orthographic projection (7.101), a much simpler relationship expresses the dependency of image plane coordinates before and after the movement:

$$r'_P = \rho_{11}r_P + \rho_{12}s_P + \rho_{13}Z_P + \tau_x$$
$$s'_P = \rho_{21}r_P + \rho_{22}s_P + \rho_{23}Z_P + \tau_y.$$

(7.107)

In this case, no dependency exists on the focal length and on the translation in Z direction τ_z. The entire mapping is described by linear relationships. According to (7.101), no depth-dependency is given, such that the condition must be $\rho_{13}=\rho_{23}=0$. These simplifications can however be seen as a tradeoff against the less accurate modeling of the projection, which is only valid for objects which are at far distance from the camera. A description of the relationships in (7.107) is then possible by a total number of 6 parameters, which is the *affine transform*, expresses a linear relationship of coordinates (r,s) and (r',s') by

$$
\begin{aligned}
r' &= a_1 \cdot r + a_2 \cdot s + t_1 \\
s' &= a_3 \cdot r + a_4 \cdot s + t_2
\end{aligned}
\quad \text{or} \quad
\underbrace{\begin{bmatrix} r' \\ s' \end{bmatrix}}_{\mathbf{r'}} = \underbrace{\begin{bmatrix} a_1 & a_2 \\ a_3 & a_4 \end{bmatrix}}_{\mathbf{A}} \cdot \underbrace{\begin{bmatrix} r \\ s \end{bmatrix}}_{\mathbf{r}} + \underbrace{\begin{bmatrix} t_1 \\ t_2 \end{bmatrix}}_{\mathbf{t}}.
$$

(7.108)

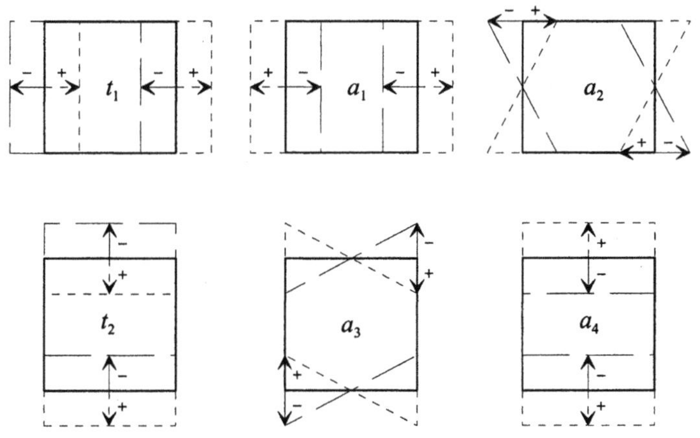

Fig. 7.36. Geometric modifications of squares as effected by single parameters of the affine transform (center of the coordinate system at the centers of the squares).

The affine transform is the most universal *linear* transform applicable to 2D coordinate relationships[1]. The single parameters of the affine transform have the following effects regarding modifications of a 2D geometry (see Fig. 7.36):

- a_1 / a_4 : Scaling in horizontal / vertical direction
- a_2 / a_3 : Shear in horizontal / vertical direction
- t_1 / t_2 : Translation in horizontal / vertical direction

---

[1] (7.102) plays a similar role as a *linear* 3D coordinate transformation.

Fig. 7.37 shows examples of geometric modifications which result by combinations of several affine transform parameters. These are in particular typical for the 2D projection from rigid 3D objects with planar surfaces. An overview is given in Table 7.1. These geometric modifications are directly related to effects of the 3D→2D projection. Regarding the example of a planar surface which is positioned in the 3D world in parallel to the image plane orientation, the following phenomena are observed:

- A 3D translation in parallel to the image plane is perceived as 2D translation;
- A 3D translation perpendicular to the image plane (distance change) is observed as 2D scaling;
- A 3D rotation around the Z coordinate axis (optical axis) is interpreted as 2D rotation;
- 3D rotations exclusively around one of the X or Y coordinate axes appear as a change of the aspect ratio;
- Mixed forms of rotations around several axes effect a mapping onto a non-orthogonal coordinate system, which can be described as shear.

Table 7.1. Relationship between geometric modifications and typical combinations of several parameters of the affine mapping

	a_1	a_2	a_3	a_4	t_1	t_2
Translation by k,l	1	0	0	1	k	l
Rotation by $\pm\theta$	$\cos\theta$	$\pm\sin\theta$	$\mp\sin\theta$	$\cos\theta$	0	0
Scaling by Θ	Θ	0	0	Θ	0	0
Change of aspect ratio by $\mu{:}\nu$	μ	0	0	ν	0	0
Shear of vertical coordinate axis by $\pm\theta$	1	$\pm\sin\theta$	0	1	0	0
Shear of horizontal coordinate axis by $\pm\theta$	1	0	$\mp\sin\theta$	1	0	0

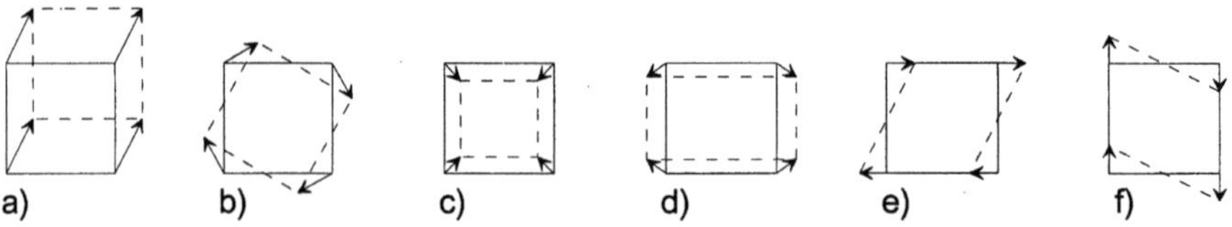

Fig. 7.37. Affine transform mappings of a quadratic area: **a** Translation **b** Rotation **c** Scaling **d** Change of aspect ratio **e** Shear of vertical coordinate axis **f** Shear of horizontal coordinate axis

Analogies are obvious when compared to the coordinate and sampling system transformations introduced in sec. 2.1.1 and 2.2.3. The spatial/frequency mapping relationships deduced there can directly be used to understand how the 2D spectrum changes in case of an affine transform: All linear geometric modifications can be interpreted as regular transformations into a coordinate system with different origin, scaling and orientation of the axes. The determinant of the matrix $\mathbf{A}$ in (7.108) represents the change of area, which is the effective scaling factor if all dimensions are regarded, and is e.g. related to a change of distance between the camera and the captured object. The translation corresponds to a change of the coordinate system reference. In total, the spectrum is modified with a mapping performed by the dual matrix $[\mathbf{A}^{-1}]^{\mathrm{T}}$ and a linear phase shift $e^{j(\omega_1 t_1 + \omega_2 t_2)}$ related to the translation.

The affine transform has been shown to be the perfect model for geometric mapping in 2D if orthographic projection (7.101) is assumed. For larger rigid 3D objects which are not too far from the camera, all changes in the projection could still be described by 6 parameters of 3D rotation and translation. This is no longer true for the changes of image coordinates after the central projection (1.1), as can be concluded from (7.106). For the special case of planar object surfaces, linear relationships exist between X_P, Y_P and Z_P[1], which can be used to eliminate the dependency on Z_P in (7.106). The result is the following 8-parameter model, which is also known as the *perspective mapping model*:

$$r' = \frac{a_1 \cdot r + a_2 \cdot s + t_1}{b_1 \cdot r + b_2 \cdot s + 1} \quad ; \quad s' = \frac{a_3 \cdot r + a_4 \cdot s + t_2}{b_1 \cdot r + b_2 \cdot s + 1}. \tag{7.109}$$

The geometric modifications which are enabled by (7.109) for a quadratic shape are shown in Fig. 7.38a in addition to the modifications by the affine model. The relationship with perspective mapping is obvious. Due to the division by variables r and s, the implementation of the perspective mapping is no longer linear, and has higher overall complexity than affine mapping. This also makes the estimation of parameters more difficult. A simplified (but not accurate) approximation of the perspective geometric mapping, using 8 parameters as well, is *bilinear mapping*[2]. Here, parameters a_1-a_4 are identical with those of the affine model, however b_1 and b_2 are *not* identical with those of the perspective model:

$$\begin{aligned} r' &= b_1 \cdot rs + a_1 \cdot r + a_2 \cdot s + t_1 \\ s' &= b_2 \cdot rs + a_3 \cdot r + a_4 \cdot s + t_2. \end{aligned} \tag{7.110}$$

---

[1] The following perfect mapping is also valid for the case when an arbitrary static 3D scene is captured and the *camera* is rotated without changing the position of the focal point. This will no longer be true in case of camera systems where optical (lens) distortions occur, as the central projection equations are then violated.

[2] Bilinear mapping is equivalent to bilinear interpolation of the mapping destinations (r',s') from the respective values of corner positions, cf. (5.65) and (5.66).

The differences between perspective and bilinear mapping are not visible by the exterior shape e.g. of a square modified into an arbitrary quadrangle. Rather, the inner structure of the quadratic area is mapped differently (see Fig. 7.38b-d). Both models effect nonlinear relationships between the original and mapped coordinates. If a 3D object has *non-planar* (curved) surfaces, nonlinear geometric distortions of higher order can occur in the mapping when objects rotate in the 3D world. For example, a line drawn onto a parabolic surface appears as a parabolic curve after the projection into the image plane, where the actual curvature depends on the orientation of the image plane relative to the surface. Such higher-order geometric distortions can be captured by higher-order polynomial models. Beyond the bilinear model[1], the *parabolic model* includes all polynomials up to a degree of two and has 12 parameters:

$$r' = c_1 \cdot r^2 + c_2 \cdot s^2 + b_1 \cdot rs + a_1 \cdot r + a_2 \cdot s + t_1$$
$$s' = c_3 \cdot r^2 + c_4 \cdot s^2 + b_2 \cdot rs + a_3 \cdot r + a_4 \cdot s + t_2. \tag{7.111}$$

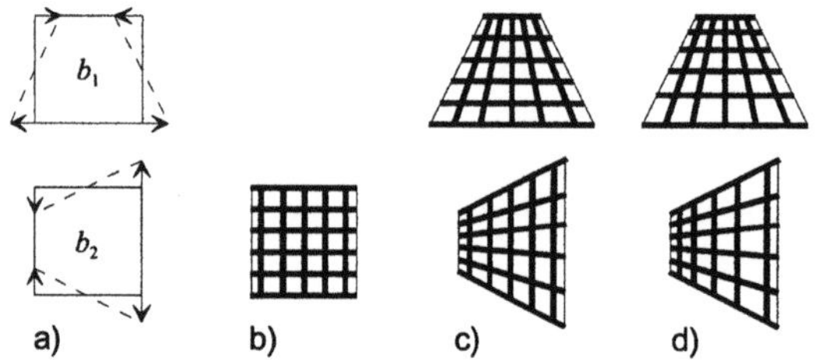

Fig. 7.38. a Modification of a quadratic area by parameters b_1 and b_2 of perspective and bilinear coordinate mapping, and **b-d** modification of a pattern (**b**) in bilinear mapping (**c**) and perspective mapping (**d**)

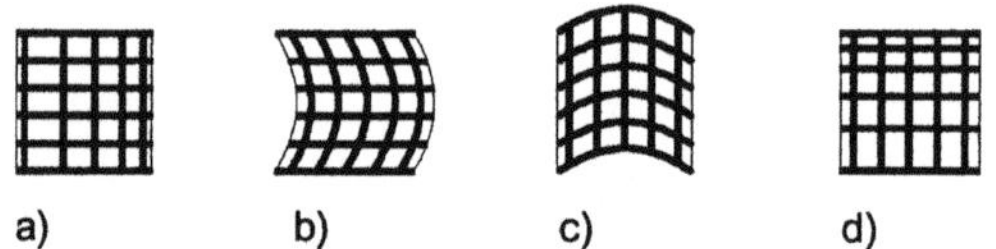

Fig. 7.39. Geometric modifications of the pattern from Fig. 7.38b by parabolic mapping **a** Parameter c_1 **b** Parameter c_2 **c** Parameter c_3 **d** Parameter c_4

The parameters b_1 and b_2 of the parabolic model have the same effect as for the bilinear model (Fig. 7.38c). The effects of the other nonlinear parameters c_1-c_4 are shown in Fig. 7.39. This model can also provide the effect of a perspective mapping model by a combination of parameters b_1/c_3 and b_2/c_1.

The effects of rotation and scaling were related to center coordinates so far (particularly in Tab. 7.1). This is correct, if an object turns around the center of the image, or if the camera performs rotation or zoom and the optical axis is exactly at

---

[1] Remark that the perspective model does not belong to this class of polynomial projection models

the center of the image plane. Scaling can more generally be described by the scaling factor Θ and the coordinate of the center of scaling (r_S, s_S)[1]. The position of a pixel (r,s) is then shifted towards (see Fig. 7.40a)

$$r' = \Theta \cdot (r - r_S) + r_S \quad ; \quad s' = \Theta \cdot (s - s_S) + s_S \; . \tag{7.112}$$

If a rotation by an angle θ is performed around a not-centered axis (r_R, s_R) in the image plane, the coordinate position (r,s) is shifted to the destination (Fig. 7.40b)

$$r' = \cos\theta(r - r_R) - \sin\theta(s - s_R) + r_R$$
$$s' = \sin\theta(r - r_R) + \cos\theta(s - s_R) + s_R \tag{7.113}$$

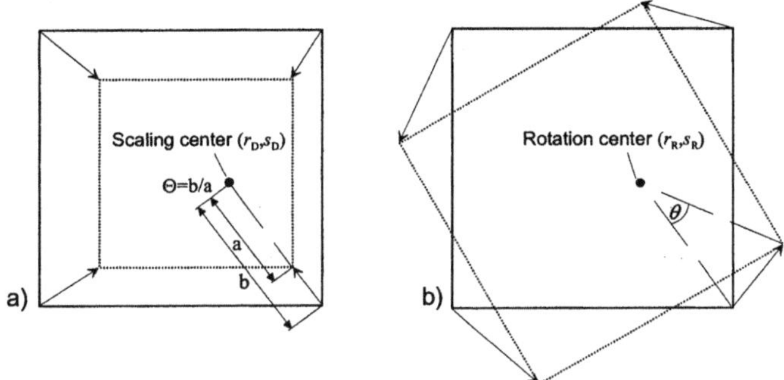

Fig. 7.40. Illustration of geometric mappings effected by non-centered operations of scaling (*a*) and rotation (*b*)

Beyond translation, rotation and scaling are the most important geometric modifications. This can be integrated into a simplified mapping model, excluding all other modifications of the affine transform. Substituting (7.112) and (7.113) into the affine equations (7.108) gives

$$r' = \Theta \cdot \cos\theta \cdot r - \Theta \cdot \sin\theta \cdot s - \Theta \cdot (\cos\theta - 1) \cdot r_R + \Theta \cdot \sin\theta \cdot s_R - (\Theta - 1) \cdot r_S + r_T$$
$$s' = \Theta \cdot \sin\theta \cdot r + \Theta \cdot \cos\theta \cdot s - \Theta \cdot \sin\theta \cdot r_R - \Theta \cdot (\cos\theta - 1) \cdot s_R - (\Theta - 1) \cdot s_S + s_T \; . \tag{7.114}$$

The values r_T and s_T represent pure translation here, as e.g. caused by a pan of the camera. This leads to a model of only 4 parameters

$$\begin{aligned} r' &= a_1 \cdot r - a_2 \cdot s + t_1 \\ s' &= a_2 \cdot r + a_1 \cdot s + t_2 \end{aligned} \quad \text{or} \quad \begin{bmatrix} r' \\ s' \end{bmatrix} = \begin{bmatrix} a_1 & -a_2 \\ a_2 & a_1 \end{bmatrix} \cdot \begin{bmatrix} r \\ s \end{bmatrix} + \begin{bmatrix} t_1 \\ t_2 \end{bmatrix} , \tag{7.115}$$

where[2]

---

[1] This is also denoted as the focus of expansion, the point at which straight extensions of all vectors $[\, r'\text{-}r \;\; s'\text{-}s \,]^T$ describing the mapping would intersect.

[2] Observe that parameters t_1 and t_2 now contain information about the centers of rotation and scaling, in addition to the original translation component.

$$a_1 = \Theta \cdot \cos\theta \quad t_1 = -\Theta \cdot (\cos\theta - 1) \cdot r_R + \Theta \cdot \sin\theta \cdot s_R - (\Theta - 1) \cdot r_S + r_T$$
$$a_2 = \Theta \cdot \sin\theta \;;\quad t_2 = -\Theta \cdot \sin\theta \cdot r_R - \Theta \cdot (\cos\theta - 1) \cdot s_R - (\Theta - 1) \cdot s_S + s_T. \tag{7.116}$$

Geometric transformations so far were related to continuous coordinates (r,s). In most cases of mapping, it will not be possible to map a sample directly with another sampling position. Even for the case where the original position is an integer value from a regular grid ($r=mR$, $s=nS$), the mapping (r',s') will typically not be at a regular sampling grid position, the mapping defines an *irregular sampling grid*. If reference between two regular grids is required (see e.g. motion estimation in sec. 7.6.4 and image warping in sec. 15.2), a signal interpolation must be performed. The mapping into discrete coordinates (m,n) further requires a normalization observing the actual sampling distances R and S or the sampling aspect ratio. For the cases of the affine and the parabolic mapping, where the parameters a_i, b_i, c_i, t_i shall be identical with those of the respective mappings in continuous coordinates, examples of discrete mapping can be described as follows:

a) Affine discrete mapping:

$$m' = a_1 \cdot m + a_2 \cdot \frac{S}{R} \cdot n + \frac{t_1}{R}$$
$$n' = a_3 \cdot \frac{R}{S} \cdot m + a_4 \cdot n + \frac{t_2}{S}. \tag{7.117}$$

b) Parabolic discrete mapping:

$$m' = c_1 \cdot R \cdot m^2 + c_2 \cdot \frac{S^2}{R} \cdot n^2 + b_1 \cdot S \cdot mn + a_1 \cdot m + a_2 \cdot \frac{S}{R} \cdot n + \frac{t_1}{R}$$
$$n' = c_3 \cdot \frac{R^2}{S} \cdot m^2 + c_4 \cdot S \cdot n^2 + b_2 \cdot R \cdot mn + a_3 \cdot \frac{R}{S} \cdot m + a_4 \cdot n + \frac{t_2}{S}. \tag{7.118}$$

7.4.5   Moment analysis

Moments were introduced to characterize statistical behavior of signal amplitudes (see sec. 3.1 and 3.2), which can be interpreted as an analysis of mass concentrations under probability density functions in a one- or multidimensional signal-amplitude space. If the same concept of moment analysis is formally applied to an image signal over spatial coordinates, the moments express criteria about mass distributions within the image amplitude. In case of a binary image, the mass density is assumed to be constant, such that the moments are useful criteria to characterize a binary shape. Nevertheless, moments over geometric figures can analyze the general case of multiple-amplitude signals such as gray-level images. The geometry-related *moment* of order $k+l$ for a two-dimensional discrete image signal $x(m,n)$ is defined as

$$\mu^{(k,l)} = \sum_{m=0}^{M-1}\sum_{n=0}^{N-1} m^k n^l \cdot x(m,n). \tag{7.119}$$

The center of mass is located at

$$\bar{r} = \frac{\mu^{(1,0)}}{\mu^{(0,0)}} \ ; \ \bar{s} = \frac{\mu^{(0,1)}}{\mu^{(0,0)}} . \tag{7.120}$$

By subtracting the center of mass coordinate, the (translation-invariant) *central moment* of order $k+l$ is defined by

$$\rho^{(k,l)} = \sum_{m=0}^{M-1}\sum_{n=0}^{N-1} (m-\bar{r})^k (n-\bar{s})^l \cdot x(m,n). \tag{7.121}$$

While the center of mass coordinate expresses the position of the object, the central moments can be used as parameters to express shape properties irrespective of the position where the shape is found in an image. The following relationships hold for the moments and central moments of low orders[1]:

$$\rho^{(0,0)} = \mu^{(0,0)} = \sum_{m=0}^{M-1}\sum_{n=0}^{N-1} x(m,n) \tag{7.122}$$

$$\begin{aligned}
\rho^{(1,0)} &= \sum_{m=0}^{M-1}\sum_{n=0}^{N-1} (m-\bar{r}) \cdot x(m,n) \\
&= \sum_{m=0}^{M-1}\sum_{n=0}^{N-1} m \cdot x(m,n) - \bar{r} \cdot \sum_{m=0}^{M-1}\sum_{n=0}^{N-1} x(m,n) \\
&= \mu^{(1,0)} - \bar{r}\mu^{(0,0)} = 0
\end{aligned} \tag{7.123}$$

$$\begin{aligned}
\rho^{(1,1)} &= \sum_{m=0}^{M-1}\sum_{n=0}^{N-1} (m-\bar{r})(n-\bar{s}) \cdot x(m,n) \\
&= \mu^{(1,1)} - \bar{r} \cdot \mu^{(0,1)} - \bar{s} \cdot \mu^{(1,0)} + \bar{r} \cdot \bar{s} \cdot \mu^{(0,0)} \\
&= \mu^{(1,1)} - \bar{s} \cdot \mu^{(1,0)} = \mu^{(1,1)} - \frac{\mu^{(0,1)} \cdot \mu^{(1,0)}}{\mu^{(0,0)}} .
\end{aligned} \tag{7.124}$$

Similarly, $\rho^{(0,1)}=0$ and $\rho^{(1,1)} = \mu^{(1,1)} - \bar{r} \cdot \mu^{(0,1)}$. In addition to the translation invariance, it would now be useful to establish the central moments as rotation-invariant criteria. Assume that rotation is performed around the center of mass. To simplify the following expressions, the reference of the coordinate system is defined by 'zero mean' at the center-of-mass coordinate ($\bar{r} = \bar{s} = 0$). If the object is rotated by an angle α, coordinates (m,n) are shifted to positions

---

[1] Also refer to Problems 7.11 and 7.12, where in addition relationships of moments and central moments with the projection profiles (7.99) are established.

$$\begin{bmatrix} \tilde{m} \\ \tilde{n} \end{bmatrix} = \begin{bmatrix} \cos\alpha & -\sin\alpha \\ \sin\alpha & \cos\alpha \end{bmatrix} \cdot \begin{bmatrix} m \\ n \end{bmatrix} \equiv \tilde{\mathbf{m}} = \mathbf{R}\cdot\mathbf{m}. \tag{7.125}$$

The rotation matrix $\mathbf{R}$ in (7.125) is an orthonormal basis system according to (B.20), with two orthogonal basis vectors $[\cos\alpha\ \sin\alpha]^{\mathrm{T}}$ and $[-\sin\alpha\ \cos\alpha]^{\mathrm{T}}$, and a unity determinant value. The whole system is then orthonormal, such that according to (B.22), the following relationships apply:

$$\mathbf{R}\cdot\mathbf{R}^{-1} = \mathbf{I}\ ;\ \mathbf{R}^{-1} = \mathbf{R}^{\mathrm{T}}\ ;\ \|\mathbf{R}\cdot\mathbf{m}\| = \|\mathbf{m}\|. \tag{7.126}$$

All central moments of order 2 are now formally combined into a matrix[1]

$$\mathbf{\Gamma} = \begin{bmatrix} \rho^{(2,0)} & \rho^{(1,1)} \\ \rho^{(1,1)} & \rho^{(0,2)} \end{bmatrix} = \sum_{m=0}^{M-1}\sum_{n=0}^{N-1} \begin{bmatrix} m^2 & mn \\ mn & n^2 \end{bmatrix}\cdot x(m,n). \tag{7.127}$$

The rotated object of same shape results by another matrix

$$\tilde{\mathbf{\Gamma}} = \begin{bmatrix} \tilde{\rho}^{(2,0)} & \tilde{\rho}^{(1,1)} \\ \tilde{\rho}^{(1,1)} & \tilde{\rho}^{(0,2)} \end{bmatrix} = \sum_{m=0}^{M-1}\sum_{n=0}^{N-1} \begin{bmatrix} \tilde{m}^2 & \tilde{m}\tilde{n} \\ \tilde{m}\tilde{n} & \tilde{n}^2 \end{bmatrix}\cdot x(\tilde{m},\tilde{n}). \tag{7.128}$$

Additionally,

$$\begin{bmatrix} \tilde{m}^2 & \tilde{m}\tilde{n} \\ \tilde{m}\tilde{n} & \tilde{n}^2 \end{bmatrix} = \begin{bmatrix} \tilde{m} \\ \tilde{n} \end{bmatrix}\cdot[\tilde{m}\ \ \tilde{n}] = \tilde{\mathbf{m}}\cdot\tilde{\mathbf{m}}^{\mathrm{T}} = \mathbf{R}\cdot\mathbf{m}\cdot\mathbf{m}^{\mathrm{T}}\cdot\mathbf{R}^{\mathrm{T}} = \mathbf{R}\cdot\begin{bmatrix} m^2 & mn \\ mn & n^2 \end{bmatrix}\cdot\mathbf{R}^{\mathrm{T}}. \tag{7.129}$$

Using (7.127) and (7.128),

$$\begin{aligned}
\tilde{\mathbf{\Gamma}} &= \sum_{m=0}^{M-1}\sum_{n=0}^{N-1}\mathbf{R}\cdot\begin{bmatrix} m^2 & mn \\ mn & n^2 \end{bmatrix}\cdot\mathbf{R}^{\mathrm{T}}\cdot x(\tilde{m},\tilde{n}) \\
&= \mathbf{R}\cdot\left[\sum_{m=0}^{M-1}\sum_{n=0}^{N-1}\begin{bmatrix} m^2 & mn \\ mn & n^2 \end{bmatrix}\cdot x(m,n)\right]\cdot\mathbf{R}^{\mathrm{T}} = \mathbf{R}\cdot\mathbf{\Gamma}\cdot\mathbf{R}^{\mathrm{T}}.
\end{aligned} \tag{7.130}$$

The rotation matrix $\mathbf{R}$ is equivalent to an orthonormal transform matrix. Hence, if applied to the matrix of central moments $\mathbf{\Gamma}$, the rotated matrix $\tilde{\mathbf{\Gamma}}$ must have identical *eigenvalues* and *determinant*. The two eigenvalues λ_1 and λ_2 can hence be used as rotation-invariant criteria. An even better manageable criterion results, if the *sum of eigenvalues* is used, which must be identical to the trace of any of the two matrices:

$$m_1 = \lambda_1 + \lambda_2 = \mathrm{tr}(\mathbf{\Gamma}) = \mathrm{tr}(\tilde{\mathbf{\Gamma}}) \Rightarrow \rho^{(2,0)} + \rho^{(0,2)} = \tilde{\rho}^{(2,0)} + \tilde{\rho}^{(0,2)}. \tag{7.131}$$

The eigenvalues are computed from the characteristic polynomial

---

[1] This matrix is in fact a geometric counterpart of the covariance matrix in (3.39).

$$\lambda^2 - \lambda \cdot \mathrm{tr}(\Gamma) + \mathrm{Det}(\Gamma) = 0, \tag{7.132}$$

which gives the solution

$$
\begin{aligned}
\lambda_{1,2} &= \frac{1}{2}\left[\mathrm{tr}(\Gamma) \pm \sqrt{\left(\mathrm{tr}(\Gamma)\right)^2 - 4 \cdot \mathrm{Det}(\Gamma)} \right] \\
&= \frac{1}{2}\left[\rho^{(0,2)} + \rho^{(2,0)} \pm \sqrt{\left(\rho^{(0,2)}\right)^2 + 2\rho^{(0,2)}\rho^{(2,0)} + \left(\rho^{(2,0)}\right)^2 - 4\rho^{(0,2)}\rho^{(2,0)} + 4 \cdot \left(\rho^{(1,1)}\right)^2} \right] \\
&= \frac{1}{2}\left[\rho^{(0,2)} + \rho^{(2,0)} \pm \sqrt{\left(\rho^{(0,2)} - \rho^{(2,0)}\right)^2 + 4 \cdot \left(\rho^{(1,1)}\right)^2} \right].
\end{aligned}
$$

$$\tag{7.133}$$

Hence, the *difference of eigenvalues* can be used as a second rotation-invariant criterion

$$m_2 = \lambda_1 - \lambda_2 = \sqrt{\left(\rho^{(0,2)} - \rho^{(2,0)}\right)^2 + 4 \cdot \left(\rho^{(1,1)}\right)^2}. \tag{7.134}$$

For this method of rotation-invariant analysis, the *eigenvectors* are not needed at all. They are however useful, as they implicitly determine the *principal axes* of the shape orientation. This can be interpreted in a way that moments up to the order of two fully characterize a Gaussian function, which forms an elliptical mass distribution (cf. sec. 3.2). Hence, the eigenvectors span an ellipse which matches the mass distribution of the shape as good as possible. The first eigenvector will be orientated over the main axis, which signifies the longitude extension of the shape, the second will be perpendicular with the first and indicates the direction of lateral extension. The ratio of eigenvalues could further be interpreted as a kind of shape aspect ratio, giving an indication whether the shape has narrow width[1]. This is denoted as the *eccentricity*

$$\xi_E = \sqrt{\frac{\lambda_1}{\lambda_2}}. \tag{7.135}$$

More rotation-invariant criteria can be defined for the central moments of higher orders. As an example, up to an order $k+l=3$, five additional criteria are useful (for an explanation on m_3-m_7, see [HU 1962]) :

$$m_3 = \left(\rho^{(3,0)} - 3\rho^{(1,2)}\right)^2 + \left(3\rho^{(2,1)} - \rho^{(0,3)}\right)^2 \tag{7.136}$$

$$m_4 = \left(\rho^{(3,0)} + \rho^{(1,2)}\right)^2 + \left(\rho^{(2,1)} + \rho^{(0,3)}\right)^2 \tag{7.137}$$

---

[1] Even for a horizontal or vertical orientation of a shape figure, this will not be identical to the aspect ratio of the circumscribing rectangle (7.95), as the latter one may be extended by any thin appendices of the shape, which hardly contribute to the mass distribution.

$$m_5 = \left(\rho^{(3,0)} - 3\rho^{(1,2)}\right) \cdot \left(\rho^{(3,0)} + \rho^{(1,2)}\right) \cdot \left[\left(\rho^{(3,0)} + \rho^{(1,2)}\right)^2 - 3\left(\rho^{(2,1)} + \rho^{(0,3)}\right)^2\right]$$
$$+ \left(3\rho^{(2,1)} - \rho^{(0,3)}\right) \cdot \left(\rho^{(2,1)} + \rho^{(0,3)}\right) \cdot \left[3\left(\rho^{(3,0)} + \rho^{(1,2)}\right)^2 - \left(\rho^{(2,1)} + \rho^{(0,3)}\right)^2\right] \tag{7.138}$$

$$m_6 = \left(\rho^{(2,0)} - \rho^{(0,2)}\right) \cdot \left[\left(\rho^{(3,0)} + \rho^{(1,2)}\right)^2 - \left(\rho^{(2,1)} + \rho^{(0,3)}\right)^2\right]$$
$$+ 4\rho^{(1,1)}\left(\rho^{(3,0)} + \rho^{(1,2)}\right) \cdot \left(\rho^{(2,1)} + \rho^{(0,3)}\right) \tag{7.139}$$

$$m_7 = \left(3\rho^{(2,1)} - \rho^{(0,3)}\right) \cdot \left(\rho^{(3,0)} + \rho^{(1,2)}\right) \cdot \left[\left(\rho^{(3,0)} + \rho^{(1,2)}\right)^2 - 3\left(\rho^{(2,1)} + \rho^{(0,3)}\right)^2\right]$$
$$+ \left(3\rho^{(2,1)} - \rho^{(0,3)}\right) \cdot \left(\rho^{(2,1)} + \rho^{(0,3)}\right) \cdot \left[3\left(\rho^{(3,0)} + \rho^{(1,2)}\right)^2 - \left(\rho^{(2,1)} + \rho^{(0,3)}\right)^2\right]. \tag{7.140}$$

Feature criteria based on central moments can have very different value ranges, such that they have no uniform influence on shape characterization. The aspect of feature-value normalization, which can solve this problem, is treated by more depth in sec. 9.2.1. If a normalization of the central moments is performed, the criteria defined in this section can be turned to be *scale invariant* as well. A simple normalization method which is specific for central moments of second order uses the trace of the matrix Γ,

$$\rho^{(2,0)*} = \frac{\rho^{(2,0)}}{\rho^2} ; \; \rho^{(0,2)*} = \frac{\rho^{(0,2)}}{\rho^2} \; \text{with} \; \rho^2 = \rho^{(2,0)} + \rho^{(0,2)} . \tag{7.141}$$

As the sum of eigenvalues is then normalized by $\lambda_1 + \lambda_2 = 1$, the remaining comparison criterion will be the ratio of eigenvalue differences and sums,

$$\xi_M = \frac{m_2}{m_1} = \frac{\lambda_1 - \lambda_2}{\lambda_1 + \lambda_2} . \tag{7.142}$$

The expressed meaning is similar to (7.135), but values are limited to a range $-1 \leq \xi_M \leq 1$.

7.4.6   Shape Analysis by Basis Functions

Moment analysis describes the spectrum in terms of mass distributions related to polynomials, where the distributions may become increasingly irregular when higher orders are used. An alternative way is representation by coefficient expansions related to basis functions, which can eventually better be adapted to specific shape properties. This method is used in the region shape descriptor of the MPEG-7 standard. To achieve invariance against scaling and translation, the binary shape is centered within a square window of normalized size, which means that the shape is scaled to the window size depending on the length of the primary axis. The specific transform basis is the *Angular Radial Transform* (ART), which is a 2D complex transform defined by the following continuous expression in polar coordinates on a unit disk, where also the binary shape image $b(m,n)$ must be transformed into the polar representation for computation:

$$c_{k,l} = \int_0^{2\pi} \int_0^1 t_{k,l}(\rho,\theta) \cdot b(\rho,\theta) \rho \, d\rho \, d\theta. \qquad (7.143)$$

In (7.143), $c_{k,l}$ is the discrete coefficient of order k and l, and $t_{k,l}(\rho,\theta)$ is the related basis function, which is separable over the angular and radial directions,

$$t_{k,l}(\rho,\theta) = a_k(\theta) r_l(\rho), \qquad (7.144)$$

with

$$a_k(\theta) = \frac{1}{2\pi} e^{-jk\theta}, \quad r_l(\rho) = \begin{cases} 1 & l = 0 \\ 2\cos(\pi l \rho) & l \neq 0. \end{cases} \qquad (7.145)$$

These are in principle basis functions related to a complex (in angular direction) and real-valued (in radial direction) Fourier/Cosine transform over polar coordinates. In the MPEG-7 Descriptor, twelve angular frequencies and three radial frequencies are analyzed (Fig. 7.41). The coefficient $c_{0,0}$ relates to the area of the shape that covers the unit disk. It may hence be omitted when other mechanisms are provided to describe the area or size of the shape. The total number of coefficients describing the shape will be 35 in this case. The first-order absolute moments (absolute values of coefficients) are used as feature values.

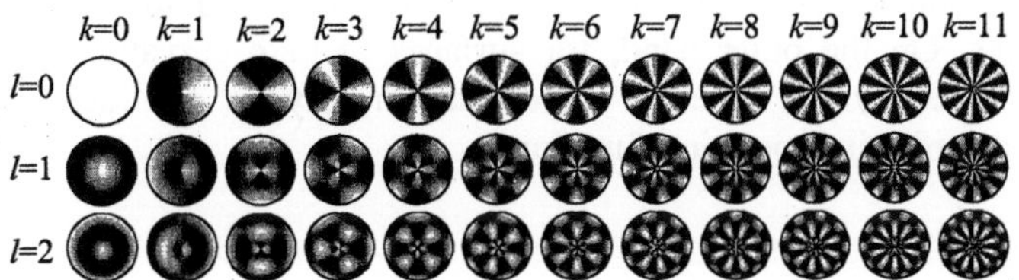

Fig. 7.41. Basis functions of ART as defined for the MPEG-7 Region Shape Descriptor

7.4.7   Three-dimensional Shapes

In principle, most methods for description of 3D volumetric shapes can be interpreted as extensions of the 2D methods. The 3D extension of contour shape representations are *surface shapes*, which relate to the *hull* of volumetric objects; the 3D extension of binary (area related) shape representations are *volumetric shapes*. A third approach for description of 3D shapes is made by *projection into 2D*. The extraction of three-dimensional shape parameters from still images or video frames can be highly inaccurate, as the projection into the image plane is generally lossy. More accurate approaches can be based on stereoscopic or multi-camera acquisition (cf. sec. 7.7); further, if illumination conditions are known, the reflections and shading occurring over convex or concave surfaces can be exploited as additional information to estimate the 3D shape [ZHANG ET AL. 1999].

3D surface shapes. One of the most widely used surface shape representations, in particular in computer graphics applications, are *wireframe* models (see Fig. 7.42). Wireframes represent the surface by a finite number of vertices with (X,Y,Z) coordinates, which have to be connected by the closest configuration of interconnections with their respective neighbors. In case of purely convex surfaces, this will be a Delaunay grid (cf. sec. 9.4.3). For objects with noses or holes, deviations from the global Delaunay topology occur; in such cases, the connectivity of vertices must be explicitly described. The interconnection of vertices results in planar triangular surface elements, for which the orientation in space is described by the three corner vertices. Within each triangular patch, the surface points can be interpreted as a linear interpolation result from the related vertices. This can indeed be interpreted as a 3D counterpart of polygon approximation for contours (sec. 7.4.1); both methods are based on linear interpolation, but planes are interpolated in 3D instead of lines in 2D. From a vertex-based representation, other methods for dense surface interpolation are possible, e.g. by *higher-order polynomial functions, spline functions, harmonic functions* [LI, LUNDMARK, FORCHHEIMER 1994] or *wavelets* [GROSS ET AL. 1996]; these will result in continuously curved surfaces instead of planar patches. Nevertheless, planar surface approximations are more widely used in computer graphics due to the fact that light reflection and projection parameters are constant over all points of the respective surface patch, which highly reduces the computational effort in rendering, as compared to curved-surface approximations. On the other hand, for surfaces of high curvature, the number of vertices necessary for accurate representation may be considerably higher than with higher-order interpolation functions.

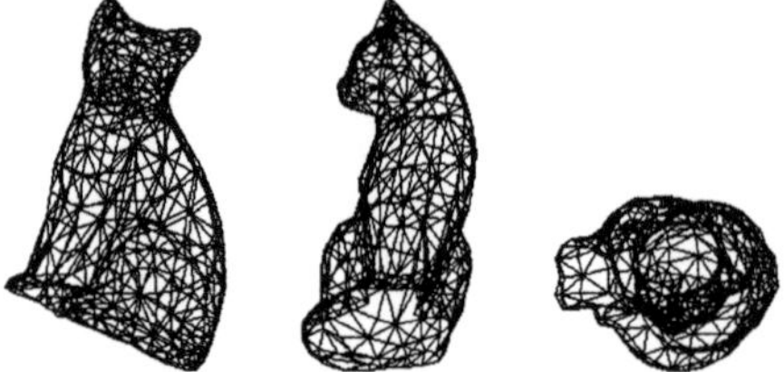

Fig. 7.42. Example of a wireframe model seen from three different view angles

Regarding the *feature description* of 3D surface shapes, very simple methods can be realized based on histograms of the surface angular orientations, denoted as *3D shape spectrum* analysis. This is a histogram of the curvatures of a 3D surface. The strength of the curvatures is analyzed by comparing the angles between orientations of neighbored triangular patches in the wireframe representation of a shape. This method is included for 3D shape description in the MPEG-7 standard [BOBER 2001].

3D volumetric shapes. An elementary description of volumetric information can be made by *voxels* (*volume elements*), which are small cube elements defining the shape. As an extension of the binary shape image (7.90), a binary volume $b(m,n,o)$ can be described, which is set to the value 1 if a voxel is present at the given location. The voxel representation is however not suitable for a characterization of the volume shape by a small number of parameters.

Generalizations of more compact 2D area-shape descriptions into 3D can be made as well. In analogy to the SAT method, 3D volumes can be represented by a superposition of cubes of different sizes, so-called *superquadrics*. This is more efficient, if the cube elements can be linearly deformed into *deformable superquadrics* [BARR 1981]. For volumes showing curved surfaces, other basic elements like cylinders are better suitable [MARR, NISHIHARA 1978]. Also morphological methods, including skeleton generation, can be extended for 3D volumes, if 3D structure elements are used.

The extension of moments and central moments (sec. 7.4.5) into 3D is straightforward, and can be used again to describe the set of voxels establishing the volume and their mass distribution in the 3D space. Of particular interest is the covariance matrix Γ, which is the 3D extension of (7.127). Let the P voxels of the volume be described by vectors $\mathbf{v}(p)=[m_p\ n_p\ o_p]^T$, $p=1,2,\ldots,P$, which contain the positions of the respective voxels. Then, the mass centroid of the volume and the covariance matrix can be written as

$$\overline{\mathbf{v}} = \frac{1}{P}\sum_{p=1}^{P}\mathbf{v}(p) \quad ; \quad \Gamma = \frac{1}{P}\sum_{p=1}^{P}[\mathbf{v}(p)-\overline{\mathbf{v}}][\mathbf{v}(p)-\overline{\mathbf{v}}]^T . \tag{7.146}$$

By performing eigenvector analysis of the covariance matrix, it is possible to determine the three principal axes of the volume, where the square roots of the eigenvalues are related to the deviations of mass concentrations from the centroid along the direction of the respective principal axis. The eigenvectors $\mathbf{r}_i$ are orthogonal and real-valued due to the symmetries in Γ:

$$\Lambda = \mathbf{R}^{-1}\Gamma\mathbf{R} \quad \text{with} \quad \Lambda = \begin{bmatrix} \lambda_1 & 0 & 0 \\ 0 & \lambda_2 & 0 \\ 0 & 0 & \lambda_3 \end{bmatrix} \quad \text{and} \quad \mathbf{R} = [\mathbf{r}_1\ \ \mathbf{r}_2\ \ \mathbf{r}_3] . \tag{7.147}$$

Description of 3D shapes based on projections. Volumetric shape extraction and description can also be achieved when a 3D object is acquired by multiple cameras or mapped into multiple 2D projections. For each of the projections, exact view directions and other parameters related to the projection itself, e.g. the focal length of a camera, have to be known[1]. An example is shown in Fig. 7.43 for an orthographic view projection. Apparently in this case, the 3D shape can perfectly be reconstructed from the mapped 2D shapes, by analyzing the set of voxels for

---

[1] These are extrinsic and intrinsic camera parameters, see further explanations in sec. 7.7.1.

which the back-projections of shapes from both camera views intersect. The example of Fig. 7.43 allows this by a minimum set of only two camera views, as it reflects a special case where the image planes are in parallel with exclusively planar surfaces of the 3D object, and the viewing rays of the cameras are orthogonal. In more general cases, including more complex objects with holes and concavities, perspective projection etc., two camera views will by far not be sufficient nor may a perfect reconstruction of the 3D shape from a finite number of camera views be guaranteed at all.

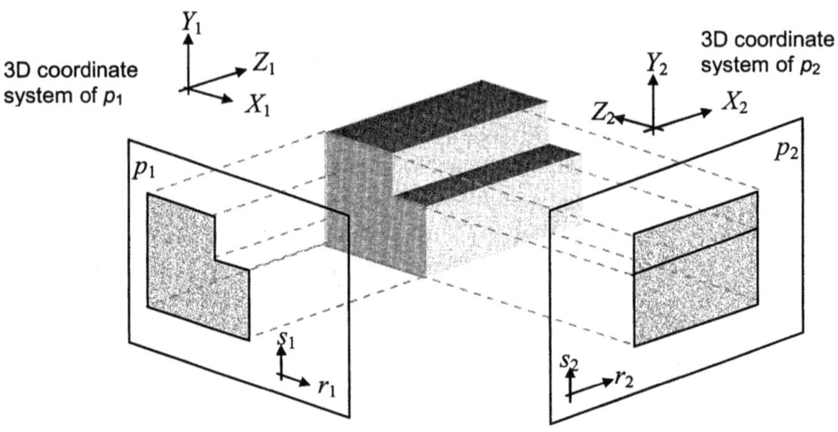

Fig. 7.43. Orthographic projection of a 3D shape into two different camera views (source: MÜLLER)

The *Radon Transform* is a very general tool to reconstruct a signal of higher dimension from a projection. Herein, the projection must not necessarily be binary, but can be dependent on the depth or density of the volume that is projected. The principle is illustrated in Fig. 7.44 for the case of a projection from a 2D slice of a volume, which is equivalent to a 2D shape. The equation of a line with a distance t_1 and an angular orientation θ relative to the origin of the coordinate system is

$$r\cos\theta + s\sin\theta = t_1 \, . \tag{7.148}$$

This line represents one projection ray passing through the volume. It is assumed that the ray 'accumulates' the voxels which are found along this way. This can mathematically be expressed by an integration along the line, for which the sifting property of the Dirac impulse is formally used,

$$\mathscr{R}_x(\theta, t_1) = \int_{-\infty}^{\infty} \int_{-\infty}^{\infty} x(r,s) \cdot \delta(r\cos\theta + s\sin\theta - t_1)\,dr\,ds \quad ; \quad 0 \le \theta < \pi \, . \tag{7.149}$$

If applied by using a variable t instead of only one fixed distance t_1, this defines the continuous, two-dimensional Radon transform $\mathscr{R}_x(\theta,t)$ of $x(r,s)$. It allows to analyze rays of any orientation and position on the slice plane. The first variant of the Radon transform is based on parallel (orthographic) projection (Fig. 7.45a), which relates to the definition in (7.149). Another definition relates to point-source (perspective) projection $\mathcal{R}_x(\beta,t)$, where the angle β has to be defined for each ray separately (Fig. 7.45b). The Radon transform has a number of properties which are very similar to the Fourier transform, in particular linearity, symmetry and scaling.

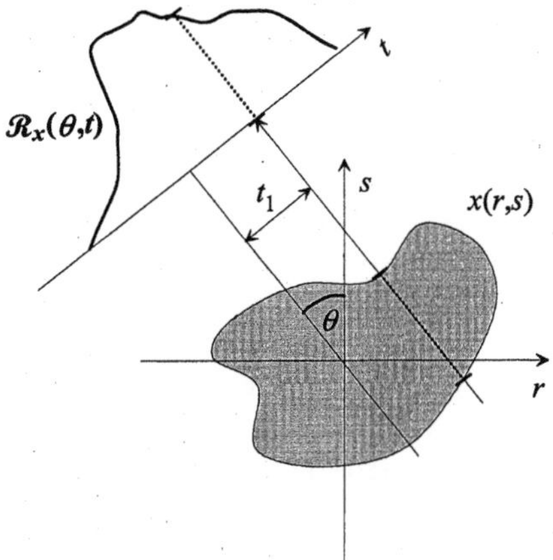

Fig. 7.44. Principle of the Radon Transform

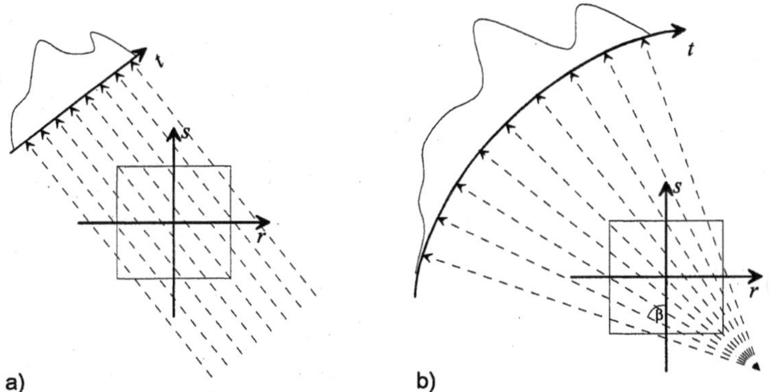

Fig. 7.45. Methods of projection in Radon Transform **a** Parallel **b** Point source

Further, it can be shown that 1D Fourier transforms performed on the Radon transform output along the direction of t (for any angles θ) are directly related with the 2D Fourier transform of the original slice image ('projection slice theorem'):

$$X(\omega_1,\omega_2) = \mathcal{F}_{2D}\{x(r,s)\} \quad ; \quad X(\theta,\rho) = \mathcal{F}_{1D}\{\mathcal{R}_x(\theta,t)\}$$
$$X(\omega_1,\omega_2) = X(\theta,\rho) \quad \text{for} \quad \omega_1 = \rho\cos\theta, \omega_2 = \rho\sin\theta \tag{7.150}$$

It turns out that $X(\theta,\rho)$ is the 2D Fourier transform in polar coordinates (see also footnote on p. 236). Consequently, it is possible to reconstruct the image from

$\mathcal{R}_x(\theta,t)$, which can be shown by the inverse Fourier transform relationship formulated over polar coordinates:

$$x(r,s) = \int\limits_{0}^{2\pi} \int\limits_{0}^{\infty} X(\theta,\rho)e^{j2\pi\rho(r\cos\theta + s\sin\theta)}\rho\,d\rho\,d\theta. \tag{7.151}$$

A perfect reconstruction is however only possible for the spatially continuous case; for sampled signals, the number of projection angles is finite. Even worse, it is hardly possible to define rays of arbitrary angular orientation consistently on a discrete sampling grid. This makes it necessary to interpolate positions, with the consequence that additional losses are introduced. Further, the resolution of 'ray sampling' will not be constant over the entire area of the plane. The integrals in (7.149) and (7.151) change into sums; the reconstruction improves when more projections are available, which means 'finer sampling' in t and θ. Even though explained here for the case of plane projections, the extension of the Radon transform into the third dimension is straightforward by defining projections depending on elevation angles φ. Due to its tight relationship with the polar-coordinate Fourier transform, the Radon transform is also closely related to other polar representations in the frequency domain, such as the Gabor wavelet filters (4.246)-(4.247).

7.5   Correspondence analysis

Correspondence analysis is based on *comparison of signal samples*, where in addition a modification *by a linear or nonlinear operation* (filtering, transform etc.) or a *geometric transform* may be made before the comparison is executed. The actual signal configuration to be compared is controlled by parameters from a parameter set, where the *parameter value giving optimum correspondence* is mapped to the feature value. Applications of correspondence analysis are

–   Comparison of signal samples or transformed representations of a signal to a 'catalogue' of patterns, e.g. for identification of objects, faces etc.;
–   Motion analysis by comparison of sample patterns from two or more video frames;
–   Disparity analysis in stereo or multiview image processing by comparison of sample patterns from different camera views;
–   Search for similar signal segments within one signal, e.g. for analysis of periodicities or structure analysis;
–   Phoneme analysis in speech recognition.

Correspondence analysis relies on a *cost function*, where optimization of the cost gives a hint to the best corresponding match. Typical examples of cost functions are *difference criteria* (to be minimized), but *correlation criteria* (maximum statistical dependency) or *information related criteria* (maximum mutual information, minimum bit rate) can be used as well.

To identify corresponding (similar) configurations of samples in one or several signals, these are interpreted as patterns from a neighborhood context. A *pattern* typically consists of the amplitudes from a group of neighbored signal values. It can also be a transformation thereof, e.g. amplitude gradients related to the sampling positions. The pattern comparison is usually performed sample-wise; if useful, a coordinate mapping or geometric modification can be applied prior to the comparison. As an example, let a pattern be extracted from a one- or multidimensional signal $x(\mathbf{n})$. The pattern to be compared shall have the shape Λ, where only samples of coordinates belonging to a set $\mathbf{n} \in \Lambda$ shall be used for comparison. This can e.g. be characterized by a binary mask as in (7.90), where $|\Lambda|$ is then the number of samples in the pattern:

$$\Lambda = \left\{ \; \mathbf{n} : b(\mathbf{n}) = 1 \; \right\} \; ; \; |\Lambda| = \sum_{\mathbf{n}} b(\mathbf{n}) \tag{7.152}$$

This pattern shall be set in correspondence with references from another signal or several other signals, wherein also additional mappings of coordinates may be allowed. The whole set of possible comparisons shall be finite, e.g. assuming that a set S of reference signals $y_s(\mathbf{n}) \in S$ is given, and for each of the signals different coordinate mappings (which could be a simple shift, or also more complex geometric mappings in case of images) $\gamma(\mathbf{n}) \in \mathcal{G}$, defined as members of a set $\mathcal{G}$, are allowed. It is now necessary to perform a comparison of the pattern against the different members of S, taking into account possible mappings of $\mathcal{G}$. A common criterion to perform such comparison is the normalized energy of pattern differences

$$\sigma_e^{2}(s,\gamma) = \frac{1}{|\Lambda|} \sum_{\mathbf{n} \in \Lambda} \left[x(\mathbf{n}) - y_s \left(\gamma(\mathbf{n}) \right) \right]^2$$
$$= \frac{1}{|\Lambda|} \left[\sum_{\mathbf{n} \in \Lambda} x^2(\mathbf{n}) + \sum_{\mathbf{n} \in \Lambda} y_s^{2} \left(\gamma(\mathbf{n}) \right) - 2 \sum_{\mathbf{n} \in \Lambda} x(\mathbf{n}) \cdot y_s \left(\gamma(\mathbf{n}) \right) \right]. \tag{7.153}$$

It is reasonable to assume the best correspondence where (7.153) becomes *minimum*. Then,

$$[s,\gamma]_{\text{opt}} = \arg \min_{s \in S, \gamma \in \mathcal{G}} \frac{1}{|\Lambda|} \sum_{\mathbf{n} \in \Lambda} \left[x(\mathbf{n}) - y_s \left(\gamma(\mathbf{n}) \right) \right]^2 . \tag{7.154}$$

The difference energy will also be minimized, when the last term in (7.153) approaches a *maximum* (provided that the first two terms are approximately constant for two given signals):

$$[s,\gamma]_{\text{opt}} = \arg \max_{s \in S, \gamma \in \mathcal{G}} \frac{1}{|\Lambda|} \sum_{\mathbf{n} \in \Lambda} x(\mathbf{n}) \cdot y_s \left(\gamma(\mathbf{n}) \right) . \tag{7.155}$$

(7.155) is the *cross correlation* between the signal pattern and a pattern from the reference set under a given geometric mapping. The result gained by (7.155) is however not necessarily identical to the result of (7.154). This is caused by the fact that $[y_s(\gamma(\mathbf{n}))]^2$ will typically not be constant neither over the different mappings $\gamma(\mathbf{n})$, nor over the different members from the set S. A better result can be obtained if the *normalized correlation* is used for comparison, where the normalization compensates variations:

$$[s,\gamma]_{opt} = \arg\max_{s \in S, \gamma \in \mathcal{G}} \frac{\sum_{\mathbf{n} \in \Lambda} x(\mathbf{n}) \cdot y_s(\gamma(\mathbf{n}))}{\sqrt{\sum_{\mathbf{n} \in \Lambda} x^2(\mathbf{n}) \cdot \sum_{\mathbf{n} \in \Lambda} y_s^2(\gamma(\mathbf{n}))}}. \tag{7.156}$$

The Cauchy-Schwarz inequality formulates an interdependency between the three terms in (7.153). From this, it also follows that (7.156) is absolutely less or equal than 1:

$$\left| \sum_{\mathbf{n} \in \Lambda} x(\mathbf{n}) \cdot y_s(\gamma(\mathbf{n})) \right| \le \sqrt{\sum_{\mathbf{n} \in \Lambda} x^2(\mathbf{n}) \cdot \sum_{\mathbf{n} \in \Lambda} y_s^2(\gamma(\mathbf{n}))}. \tag{7.157}$$

Equality in (7.157) holds exactly if $x(\mathbf{n})=c \cdot y_s(\gamma(\mathbf{n}))$, where c can be an arbitrary real-valued constant. It follows that correlation criteria such as (7.156) are more universally applicable to the problem of pattern matching than the difference energy criterion (7.154): If a signal pattern $x(\mathbf{n})$ is a linear amplitude-scaled version of a reference pattern $y_s(\gamma(\mathbf{n}))$[1], the normalized correlation will find the best match, while the difference criterion can give a confusing result whenever the energies of the image pattern and of the reference pattern deviate largely. In case of signals which do not have zero-mean, a better comparison criterion is the *normalized cross-covariance*:

$$[s,\gamma]_{opt} = \arg\max_{s \in S, \gamma \in \mathcal{G}} \frac{\sum_{\mathbf{n} \in \Lambda} [x(\mathbf{n}) - \hat{\mu}_x] \cdot [y_s(\gamma(\mathbf{n})) - \hat{\mu}_{y_s}]}{\sqrt{\sum_{\mathbf{n} \in \Lambda} [x(\mathbf{n}) - \hat{\mu}_x]^2 \cdot \sum_{\mathbf{n} \in \Lambda} [y_s(\gamma(\mathbf{n})) - \hat{\mu}_{y_s}]^2}} \tag{7.158}$$

$$= \arg\max_{s \in S, \gamma \in \mathcal{G}} \frac{\frac{1}{|\Lambda|} \sum_{\mathbf{n} \in \Lambda} x(\mathbf{n}) \cdot y_s(\gamma(\mathbf{n})) - \hat{\mu}_x \cdot \hat{\mu}_{y_s}}{\hat{\sigma}_x \cdot \hat{\sigma}_{y_s}}.$$

In (7.158), the μ- and σ-estimates represent mean and standard deviation of the signals $x(\mathbf{n})$ and $y_s(\gamma(\mathbf{n}))$, each measured over the area $\mathbf{n} \in \Lambda$.

---

[1] In images or video, the corresponding areas in $x(m,n)$ do not expose equal brightness with the references $y_s(\gamma(\mathbf{n}))$ if lighting conditions change. For an audio signal, a similar case would be for change of loudness.

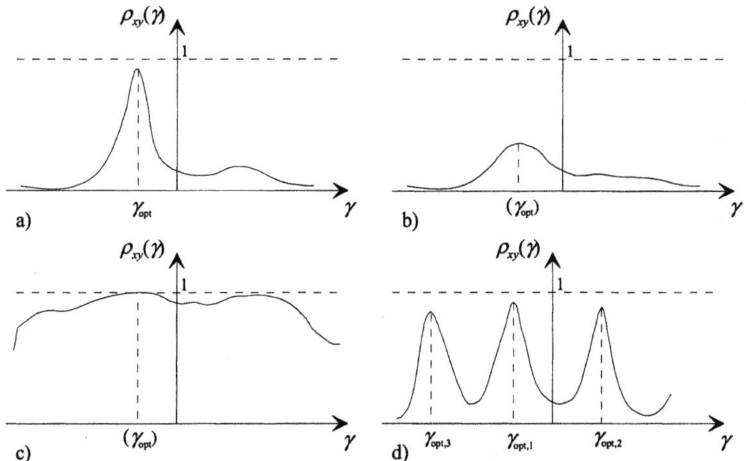

Fig. 7.46. Typical cost functions for normalized covariance criteria in correspondence analysis: **a** Unique maximum   **b** Lack of maximum   **c** Maximum, but diversified
d Several unique maxima

The cost function itself gives an indication for the reliability of the matching result. If a very unique minimum or maximum of the cost function is found, the decision can be regarded as more reliable. Typical graphs of cross-covariance based cost functions[1], some of which could cause wrong decisions, are shown in Fig. 7.46[2]. The ideal case is as in Fig. 7.46a, where *one* unique maximum exists; a decision for the parameter γ under this maximum can be identified as highly reliable. On contrary, if only a *weak* maximum exists (Fig. 7.46b), it can be concluded that no appropriately similar reference was available at all. If the cost function only decays *smoothly* from the maximum (Fig. 7.46c), the pattern is probably of insufficient structure, such that any match could be valid. If *several clearly distinct* maxima exist (Fig. 7.46d), several highly similar references exist, but there is no good indication for a decision which one is actually the best; this is a typical case, when periodic structures or other iterated copies appear in the signal pattern and/or in the reference pattern. The phenomenon of non-unique matches is reflected as the *correspondence problem*. If the cost function is *convex* and has *only one unique* maximum, iterative or hierarchical search for optimum parameters can be made

---

[1] For case of distance or difference criteria, the following statements would apply for 'minima' instead of 'maxima'.

[2] Parametrization is made here over dense variation of the geometric mapping γ, which is formally shown as a 1D function; in fact, the optimization over different parameters as they appear in geometric mapping functions will typically lead to multidimensional optimization problems, where the cost function would become a multi-dimensional surface instead of a 1D graph. Nevertheless, the typical cases discussed here apply likewise for one-and multidimensional parameter dependencies.

which will usually successfully find the global optimum; such methods play a central role in fast motion estimation algorithms as explained in the following section.

7.6   Motion Analysis

7.6.1   Mapping of motion into the image plane

In the pinhole camera model (Fig. 1.4), 3D world coordinates $\mathbf{P}=[X_\mathrm{P}\ Y_\mathrm{P}\ Z_\mathrm{P}]^\mathrm{T}$ which are representing the distance of a point P from the focal point, are mapped by the central projection equation (1.1) into the image plane coordinates $\mathbf{r}_P=[r_P\ s_P]^\mathrm{T}$.
The *speed* $\mathbf{V}(P)=[V_\mathrm{X}\ V_\mathrm{Y}\ V_\mathrm{Z}]$ of a mass point P in the 3D space is determined by the change of position $\mathbf{P}$ over time. By deriving (7.100), the movement observed in the image plane will then be of speed $\mathbf{u}(P)=[u\ v]^\mathrm{T}=d\mathbf{r}_P/dt$:

$$u = \frac{1}{Z_P}\left(FV_X - r_P V_Z\right) \quad ; \quad v = \frac{1}{Z_P}\left(FV_Y - s_P V_Z\right). \tag{7.159}$$

Substituting (7.100) into (7.159), it becomes obvious that motions in the 3D world can be invisible or be interpreted as not unique when observed in the camera image plane. For example, no motion will be recognized if $V_X/V_Z=X_P/Z_P$ and $V_Y/V_Z=Y_P/Z_P$. In this case, motion occurs along one of the rays of the projection towards the focal point. If temporal sampling of period T is applied (capturing a sequence of image frames), the following relationship applies between the *velocity vector* $\mathbf{u}$ and the (spatially continuous, but time-discrete) *displacement vector* $\mathbf{v}=[\Delta r\ \Delta s]^\mathrm{T}=\Delta\mathbf{r}$:

$$\mathbf{u} = \frac{\mathbf{v}}{T}. \tag{7.160}$$

Typically, velocity functions have a steady behavior over time due to mass inertia properties of exterior-world objects. The speed is either constant ($d\mathbf{u}/dt=0$) or continuously accelerated ($d\mathbf{u}/dt\neq0$). The acceleration (second derivation of spatial coordinates over time) can only be analyzed if at least three temporal positions of the point, to be connected by the *motion trajectory*, are known. In the context of identification and unique interpretation of displacements in the image plane, the following problems occur:

-   *Aperture problem*: If the window area for motion analysis only captures a part of an object which is homogeneous in amplitude, its motion cannot be determined uniquely, or eventually only along one direction (Fig. 7.47). In general, it is at least necessary that a signal contains non-zero frequency components orientated by the direction where the estimation of motion shall be performed.

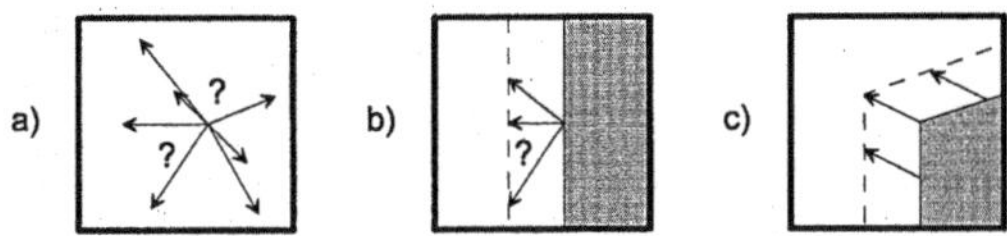

Fig. 7.47. Aperture problem **a** Impossibility to detect motion in a homogeneous area **b** Non-unique motion detection at a unidirectional edge **c** Unique motion detection at a corner

– *Correspondence problem*: In case of several equal objects or regular, periodic patterns, it may be impossible to identify the true correspondence (Fig. 7.48a/c). A unique correspondence may also be impossible to find in case of object deformations (Fig. 7.48b).

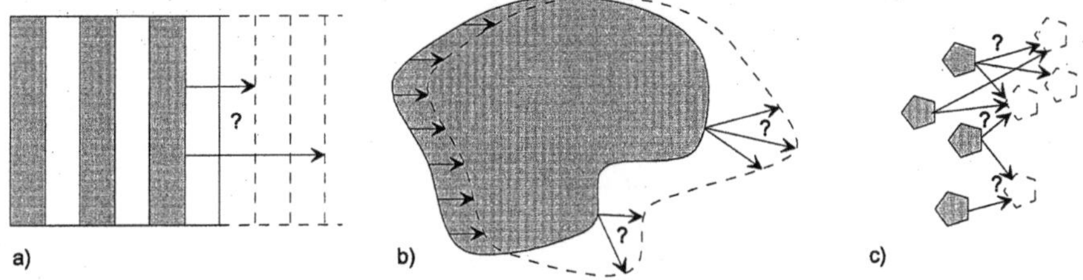

Fig. 7.48. Correspondence problem in cases of **a** periodic structure **b** object deformation **c** multiple equal objects

– *Occlusion problem*: Due to the projection, parts of scenes (background or objects) can be occluded in one frame, but will become visible if the relative position between camera and object changes (Fig. 7.49). If areas are covered or uncovered, they cannot uniquely be matched in motion estimation.

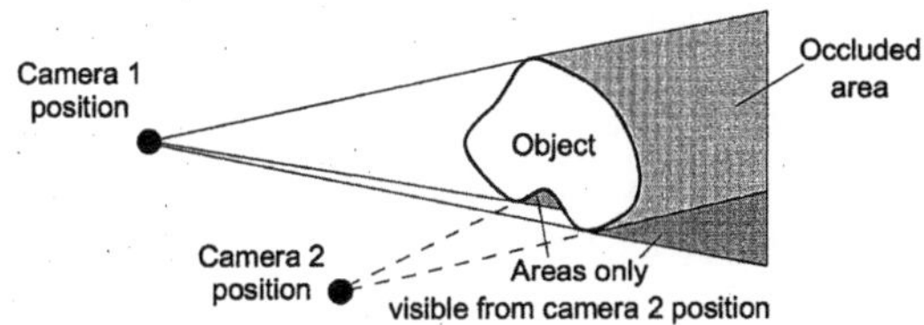

Fig. 7.49. Occlusions in case of different camera positions

Considerations as made above, concerning relationships between speed and displacement, are related to movements of mass points in the 3D space and the corresponding projected positions in the image plane. A point at position $\mathbf{r}$ shall be subject to a translational displacement $\mathbf{v}(\mathbf{r})$ between the sampling times of two frames. In most cases, the displacement $\mathbf{v}$ will not be constant over all positions $\mathbf{r}$. However, the motion of coherent objects in the 3D space typically effects only differential changes of local displacement shifts in the image. To investigate this effect, it is necessary to establish a relationship between neighbored displacement vector positions, which is characterized by the *displacement vector field* and its changes $\mathbf{v}(\mathbf{r}+d\mathbf{r})$. Fig. 7.50 shows examples which could be described by a parametric geometric transform, e.g. the affine transform (7.108).

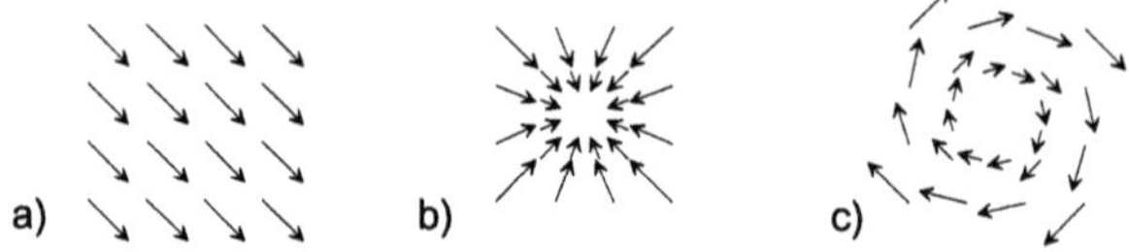

Fig. 7.50. Displacement vector fields in cases of different object motions
a Translation parallel to the image plane **b** Scaling (e.g. zoom)
c Rotation around an axis which is perpendicular with image plane

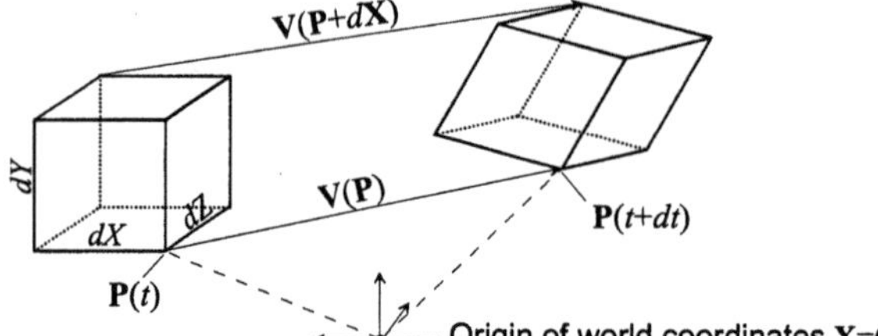

Fig. 7.51. Movement of a deformable infinitesimal volume element

Generic descriptions of volumetric movements are well-known from continuum mechanics. Herein, a rigid body is decomposed into an infinite number of an infinitesimal volume element $d\mathbf{X}=[dX,dY,dZ]^{\mathrm{T}}$, and the movement within time dt is described by the differential changes of these volume elements (Fig. 7.51). According to the fundamental theorems of kinematics, the movement of infinitesimal volume elements can uniquely be described as a sum of translation, rotation and deformation. Translation and rotation have *3 degrees of freedom* each[1], while deformation has another *6 degrees of freedom* (each 3 for the changes in lengths of the sides of the cube, 3 for the shear of the angles). The deformation of an infinitesimal volume element can fully be interpreted as a linear transformation in a 3D coordinate system.

By the projection into the image plane, only the *surfaces* of objects become visible (except for the case of transparent objects). A very similar analysis can then be performed using infinitesimal 2D area elements $d\mathbf{r}=[dr,ds]^{\mathrm{T}}$. The description of the displacement vector field can be described as a sum of translation (2 degrees of freedom), rotation (one degree of freedom) and deformation (3 degrees of freedom), which can be mapped into the 6 degrees of freedom supported by the affine transform (7.108). The change in the displacement vector field is then described as

---

[1] Those 6 parameters are sufficient to describe the motion of rigid bodies, cf. sec. 7.4.4

$$\mathbf{v}(\mathbf{r}+d\mathbf{r}) = \mathbf{v}(\mathbf{r}) + \mathbf{A}d\mathbf{r}, \tag{7.161}$$

where $\mathbf{A}$ would be identical to the matrix from (7.108) when $\mathbf{r}$ is at the center of the coordinate system. It is however clear from the discussion in sec. 7.4.4 that this model will not uniquely apply to larger, non-rigid objects of arbitrary shape; further, the *occlusion problem* is not considered at all. If occlusion occurs due to object motion, where parts of the background will become covered or uncovered, the assumption of continuity and differential changes of displacement vector fields is no longer valid. In particular, differentiation will not be possible at the positions of discontinuities. Fig. 3.41 shows possible configurations of displacement vector fields at the boundaries between two regions which are moving in opposite directions (e.g. foreground/background). In Fig. 7.52a/b, part of the background object is uncovered (the displacement vectors diverge at this position), while in Fig. 7.52c/d the case of covered background appears (the displacement vectors converge). It depends on the movement direction of the boundary, which of the two regions is the foreground. The boundary movement is always coupled to the movement of the foreground object. The width of the stripe which is covered or uncovered corresponds to the displacement component which is perpendicular to the boundary. In the example shown, the newly-uncovered or covered areas can be allocated either to region A (Fig. 7.52a/c) or to region B (Fig. 7.52b/d).

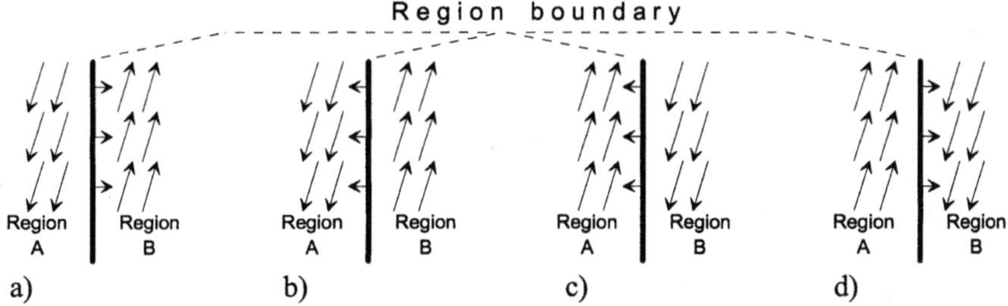

Fig. 7.52. Occlusion phenomena at region boundaries (for explanation, see text)

The displacement vector field $\mathbf{v}(r,s)$, which is still continuous in spatial coordinates, can be set in relationship to the *motion vector field* $\mathbf{k}(m,n)$, which is discrete in all three dimensions and expresses the motion shift by units of pixel coordinates. In digital video processing, usually only the motion vector field is directly processed, however it is necessary to understand its relationships with the displacement vector field and the velocity vector field if the result of analysis shall be used for semantic understanding and interpretation of the exterior-world motion. By substituting the following equation into (7.160), also the direct relationship between the spatially- and temporally-discrete motion vector $\mathbf{k}$ and the spatially- and temporally-continuous velocity vector $\mathbf{u}$ is given:

$$\mathbf{k} = \mathbf{D}\cdot\mathbf{v} \;\Rightarrow\; \mathbf{k} = T\cdot\mathbf{D}\cdot\mathbf{u} \;\Rightarrow\; \mathbf{u} = \frac{1}{T}\cdot\mathbf{D}^{-1}\cdot\mathbf{k}. \tag{7.162}$$

The motion vector $\mathbf{k}(m,n)=[k,l]^T$ represents the translational motion at position (m,n), the motion shift of one single pixel. $\mathbf{D}$ is the sampling matrix (2.51).

Motion feature extraction shall allow to describe the behavior of an object or also camera motion by a small set of motion parameters. According to a motion model selected (e.g. affine transform), parameters must be estimated to fit the motion vector field over smaller or larger areas, or even for the entire image plane if camera motion shall be estimated. In the following sections, the most widely used classes of motion estimators, based on the *optical flow* principle and on *matching* methods, are introduced. This is first done for the cases of translational motion parameter estimation; in sec. 7.6.4, the principles are then generalized to estimation of arbitrary non-translational motion parameters.

7.6.2   Motion Estimation by the Optical Flow Principle

Fig. 7.53 shows signal amplitudes at time points t and $t+dt$ for a continuous signal $x(r)$. The second amplitude is assumed to be a spatially-shifted but otherwise perfect copy of the first, such that a difference $dx=x(r,t)-x(r,t+dt)$ results at coordinate position r. If only the linear part of this change is considered, the difference dx can be linearly approximated from the differential slopes $\partial x/\partial r$ or $\partial x/\partial t$:

$$dx \cong \frac{\partial x}{\partial r} \cdot dr \cong -\frac{\partial x}{\partial t} \cdot dt . \qquad (7.163)$$

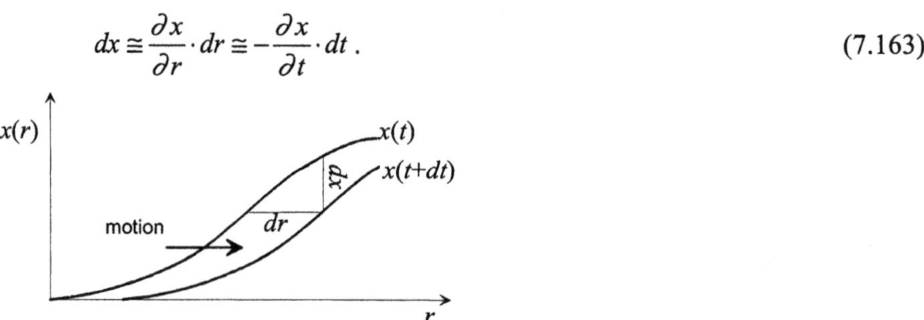

Fig. 7.53. Displacement shift dr of a signal occurring during time span dt, and relationship with the observed amplitude difference dx

If both spatial coordinates are regarded, the shift of the signal value $x(r,s,t)$ by dr and ds during time span dt can be expressed as

$$x(r,s,t) = x(r+dr,s+ds,t+dt). \qquad (7.164)$$

A Taylor series expansion applied to the right side of (7.164) gives

$$x(r,s,t) = x(r,s,t)+\frac{\partial x(r,s,t)}{\partial r} \cdot dr +\frac{\partial x(r,s,t)}{\partial s} \cdot ds +\frac{\partial x(r,s,t)}{\partial t} \cdot dt +\varepsilon, \qquad (7.165)$$

where ε are nonlinear approximation terms of orders two and higher. Further, the following interrelationships exist between the speed $\mathbf{u}=[u,v]^T$ and the shift $d\mathbf{r}=[dr,ds]^T$:

$$u = \frac{dr}{dt} \Rightarrow dr = u \cdot dt \quad ; \quad v = \frac{ds}{dt} \Rightarrow ds = v \cdot dt \, . \tag{7.166}$$

Neglecting the higher-order terms, the *continuity equation* known from hydrodynamics results from (7.165) and (7.166), which can then be used to determine the *flow speeds u* and *v* :

$$\frac{\partial x(r,s,t)}{\partial r} \cdot u(r,s,t) + \frac{\partial x(r,s,t)}{\partial s} \cdot v(r,s,t) + \frac{\partial x(r,s,t)}{\partial t} = 0 \, . \tag{7.167}$$

In analogy with the flow of fluids, the terminology *optical flow* is used to characterize the projected motion in the image plane. For this equation, no unique solution exists, as only one condition is given to determine two parameters u and v. Assuming that the velocity vector field – similar to a flow field – will be approximately continuous over a finite measurement window Λ, the following minimization problem can be formulated[1]:

$$\mathbf{u}_{\mathrm{opt}} = [u,v]_{\mathrm{opt}} = \arg\min_{[u,v]} \iint_{\Lambda} \left\| \frac{\partial x(r,s,t)}{\partial t} + u(r,s,t)\frac{\partial x(r,s,t)}{\partial r} + v(r,s,t)\frac{\partial x(r,s,t)}{\partial s} \right\|^2$$
$$+ c_1 \cdot \left\| \frac{\partial \mathbf{u}(r,s,t)}{\partial r} + \frac{\partial \mathbf{u}(r,s,t)}{\partial s} \right\|^2 + c_2 \cdot \left\| \frac{\partial \mathbf{u}(r,s,t)}{\partial t} \right\|^2 drds \, . \tag{7.168}$$

Constants c_1 and c_2 play the role of additional weighting factors. These will penalize strong spatial (c_1) or temporal (c_2) changes, fluctuations or discontinuities of the velocity field as unreliable results, evaluating differences with measurements at neighbored positions. (7.168) covers different solution approaches, which will be discussed in the sequel. First, a discrete formulation of the flow conditions will be introduced, where the constraints for continuity of the velocity field are not realized yet, which means that parameters c_1 and c_2 are set to zero.

For the case of sampled signals $x(m,n,o)$, the analysis relates to discrete positions $m=r/R$, $n=s/S$, $o=t/T$, where the following approximations of gradients can be made:

$$\frac{\partial x}{\partial r} \approx \frac{1}{R}\left[x(m,n,o) - x(m-1,n,o) \right] = \frac{1}{R} x_r(m,n)$$
$$\frac{\partial x}{\partial s} \approx \frac{1}{S}\left[x(m,n,o) - x(m,n-1,o) \right] = \frac{1}{S} x_s(m,n) \tag{7.169}$$
$$\frac{\partial x}{\partial t} \approx \frac{1}{T}\left[x(m,n,o) - x(m,n,o-1) \right] = \frac{1}{T} x_t(m,n) \, .$$

Using motion vectors $\mathbf{k}=[k,l]^{\mathrm{T}}$ as in (7.162), a *discrete* formulation of the continu-

---

[1] The first term of the following cost criterion is for minimization of the energy from flow condition violations over the measurement window. These are expressed as quadratic deviations of the left-hand side in (7.167) from the ideal value zero, when constant $\mathbf{u}$ is assumed over the window.

ity equation for position $x(m,n,o)$ results; by assuming that motion estimation is only performed in one frame at one time, the frame index variable o can be omitted for simplicity. The motion shift components k and l are still of *continuous value*, but only defined for discrete (sampled) positions in space and time:

$$k(m,n) \cdot x_r(m,n) + l(m,n) \cdot x_s(m,n) + x_t(m,n) = 0 \,. \tag{7.170}$$

This equation can be solved if differential values x_r, x_s and x_t are available for at least two pixel positions. In practice, due to effects of noise, errors can occur in the estimation. Hence, it is advisable to use more than two positions to achieve stability. A measurement area comprises P positions of coordinates $(m_1,n_1),...,(m_P,n_P)$, which are typically accessed from a neighborhood window centered around the position of current estimation. P realizations of (7.170) result in an *over-determined* equation system of order P, which is used to estimate the two unknown parameters k and l. In a matrix notation, this can be written as

$$\begin{bmatrix} x_r(1) & x_s(1) \\ x_r(2) & x_s(2) \\ \vdots & \vdots \\ \vdots & \vdots \\ x_r(P) & x_s(P) \end{bmatrix} \cdot \begin{bmatrix} k \\ l \end{bmatrix} = - \begin{bmatrix} x_t(1) \\ x_t(2) \\ \vdots \\ \vdots \\ x_t(P) \end{bmatrix} \qquad \Leftrightarrow \qquad \mathbf{G} \cdot \mathbf{k} = \mathbf{g}. \tag{7.171}$$

By (7.168), the following criterion applies,

$$\|e\|^2 = \|\mathbf{g} - \mathbf{G} \cdot \mathbf{k}\|^2 \overset{!}{=} \min \,, \tag{7.172}$$

for which a solution is achieved by computation of the *pseudo inverse* of the matrix $\mathbf{G}$[1],

$$\mathbf{k} = \mathbf{G}^p \cdot \mathbf{g} \quad ; \quad \mathbf{G}^p = \left(\mathbf{G}^T \mathbf{G} \right)^{-1} \mathbf{G}^T \,. \tag{7.173}$$

The error energy $\|e\|^2$ is used here as a criterion to judge the accuracy of the motion estimation. Specifically in case of large motion shifts and large fluctuations of local signal amplitudes, the result of (7.173) will only be a poor approximation of the true motion. This is partially due to neglecting the nonlinear terms from (7.165), which is only practical in a small spatial neighborhood, and also due to general sensitivity of the pseudo inverse solution against noise influences. An improvement can be made by using *recursive* or *iterative* motion estimation methods, where the minimum of the error energy (respectively the optimum motion shift) is gradually approached by *linear regression*. The *frame difference* represented by the vector $\mathbf{g}$ is then replaced by the *displaced frame difference* (DFD)

---

[1] More background on pseudo inversion of matrices will be given in the context of least-squares estimation in sec. 8.3.

vector $\mathbf{g'}$, which is taking into account the motion vectors computed in a previous step of estimation:

$$\mathbf{g'}(\hat{\mathbf{k}}) = \begin{bmatrix} x_t'(1) \\ x_t'(2) \\ \vdots \\ \vdots \\ x_t'(P) \end{bmatrix} \quad \text{with} \quad x_t'(m_p,n_p) = x(m_p,n_p,o) - x(m_p+\hat{k},n_p+\hat{l},o-1). \quad (7.174)$$

In *pixel-recursive* methods, $\hat{\mathbf{k}}$ is determined by values of previously-estimated motion vectors from neighbored positions in the same frame, in *frame-recursive* methods one or several vectors from the previous frame are used as initialization of the new estimation step. The spatio-temporal continuity of the motion vector field is an explicit condition for success of such methods. By using $\mathbf{g'}$ in the computation of (7.173), an update vector $\mathbf{k_u}$ is determined, which results in the finally estimated vector $\mathbf{k}$,

$$\mathbf{k_u} = \mathbf{G^p} \cdot \mathbf{g'}(\hat{\mathbf{k}}) \quad \Rightarrow \quad \mathbf{k} = \hat{\mathbf{k}} + \mathbf{k_u} . \quad (7.175)$$

In recursive methods, discontinuities of the motion vector field and undefined or erroneous vectors at the positions of occlusions can be highly problematic. This can partially be resolved by adaptation, e.g. selective switching of positions used to determine $\hat{\mathbf{k}}$. Further, criteria can be used which penalize divergences in the motion vector field as in (7.168).

In *iterative* flow estimation, the recursion is applied iteratively at the present position. The prediction vector $\hat{\mathbf{k}}$ is the result $\mathbf{k}^{(r)}=[k^{(r)},l^{(r)}]^{\mathrm{T}}$ of the rth iteration step. This is modified by an update vector $\mathbf{k_u}^{(r)}$,

$$\mathbf{k}^{(r+1)} = \mathbf{k}^{(r)} + \mathbf{k_u}^{(r)} . \quad (7.176)$$

The optimum update is typically found by a gradient-descent method, which is based on the derivative of the DFD energy $\mathbf{g'^T g'}$,

$$\mathbf{k_u}^{(r)} = -\frac{1}{2}\varepsilon \cdot \frac{\partial}{\partial \mathbf{k}^{(r)}}\left[\mathbf{g'}(\mathbf{k}^{(r)})^{\mathrm{T}}\mathbf{g'}(\mathbf{k}^{(r)})\right]. \quad (7.177)$$

Fig. 3.31 illustrates the ideal convergence towards the minimum of the cost criterion; the number of necessary iterations is dependent on the choice of the convergence factor ε, which can also variably be adapted by the instantaneous gradient.

The values of motion vectors estimated by the optical flow method are *continuous*. A spatial interpolation of intermediate values is necessary to compute $\mathbf{g'}$. Assuming that bilinear interpolation (5.35) is sufficient for this purpose, a computation of intermediate signal values can be performed by relatively low effort. Defining

$\underline{k}^{(r)}$ and $\underline{l}^{(r)}$ as integer-truncated values of the vector components, the derivation of (7.177) results in the discrete linear approximation of the update vector

$$\mathbf{k_u}^{(r)} = \varepsilon \cdot \sqrt{\left[\mathbf{g}'\right]^{\mathrm{T}} \cdot \left[\mathbf{g}'\right]} \cdot \begin{bmatrix} \dfrac{1}{P}\sum_{p=1}^{P} \dfrac{\hat{x}(m_p + k^{(r)}, n_p + l^{(r)}, o-1) - \hat{x}(m_p + \underline{k}^{(r)}, n_p + l^{(r)}, o-1)}{k^{(r)} - \underline{k}^{(r)}} \\ \dfrac{1}{P}\sum_{p=1}^{P} \dfrac{\hat{x}(m_p + k^{(r)}, n_p + l^{(r)}, o-1) - \hat{x}(m_p + k^{(r)}, n_p + \underline{l}^{(r)}, o-1)}{l^{(r)} - \underline{l}^{(r)}} \end{bmatrix}. \tag{7.178}$$

The iterative solution can now directly be computed without explicitly solving the pseudo inverse of matrix $\mathbf{G}$ in (7.173); it may be advantageous to use the pseudo inverse as the first step of estimation, but in principle it is also possible to use the steepest-descent method from the beginning, starting by a zero vector or initial estimate $\hat{\mathbf{k}}^{(0)}$ for the first iteration.

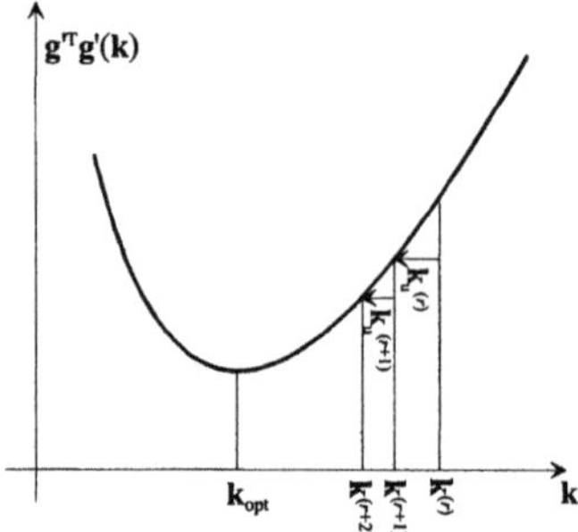

Fig. 7.54. Convergence towards the minimum of the cost function in iterative estimation

If a spatial continuity of the motion vector field shall be enforced by setting $c_1 > 0$ in (7.168), the iterations at the different pixel positions can no longer be regarded as independent. In such a case, it is best to perform any iteration step r simultaneously (or sequentially) over all positions before proceeding to the subsequent step $r+1$. Likewise, combinations of recursive and iterative optical flow estimation are possible. A more systematic approach to impose constraints on spatially-continuous recursive flow estimation is described in the next paragraph.

Recursive estimation with smoothness constraint. As an initial condition for the solution of the optical flow equation, the *smoothness of the motion vector field* can be postulated. Then, according to (7.168), the gradient between neighbored estimated motion vectors must be minimized simultaneously with the DFD energy by a joint cost criterion. The minimization of motion vector gradients is performed by postulating the second derivative to be zero. The second derivative is approximated by the Laplacian operator $\nabla^2 \mathbf{k} = \alpha \cdot (\mathbf{k} - \bar{\mathbf{k}})$, where $\bar{\mathbf{k}}$ is the mean or weighted mean over a discrete set of motion vectors from a neighborhood of the current pixel position, and α is a weighting factor. The following solution proposed by [HORN, SCHUNCK 1981] can be realized for iterative computation of motion parameters at position (m,n):

$$k^{(r+1)} = \overline{k}^{(r)} - x_r(m,n) \cdot \beta(m,n) \quad ; \quad l^{(r+1)} = \overline{l}^{(r)} - x_s(m,n) \cdot \beta(m,n) \qquad (7.179)$$

with

$$\beta(m,n) = \frac{x_r(m,n) \cdot \overline{k}^{(r)} + x_s(m,n) \cdot \overline{l}^{(r)} + x_t{'}(m,n)}{c_1 + x_r(m,n)^2 + x_s(m,n)^2}. \qquad (7.180)$$

Observe that only pixel-gradient approximations at the present pixel position and local mean values of motion parameters from the previous iteration influence the new result, which means that the DFD criterion is no longer computed over a larger measurement window. The parameter c_1 should be set approximately equal to the error occurring in gradient approximation $[x_r(m,n)^2 + x_s(m,n)^2]$, as effected by noise influences. A relevant influence of c_1 is therefore mainly effective in areas where the local amplitude is almost constant.

7.6.3   Motion Estimation by Matching

Optical flow based methods use the *difference* between two images as a criterion for motion estimation. The inherent assumption is made that the brightness of an object will not change between the two frames when the motion is estimated. Indeed, illumination changes can occur (a lamp is switched on, the sun hides behind clouds etc.), reflections from the surface of moving objects may change, or the object as a whole can move into a shadowed area. As discussed in sec. 7.5, the cross covariance (7.155) could be a more useful criterion for motion estimation in such cases. Nevertheless, frame difference criteria are also frequently applied in matching for motion estimation, as they are attractive in terms of complexity, and have turned out to give very stable results as well. In application of correspondence matching for motion estimation, the following aspects must be observed:

- *Matching areas* Λ must be defined, where for each area only one set of motion parameters shall be determined. If the matching areas are non-overlapping rectangular blocks of size $M'\mathrm{x}N'$ with start coordinate (upper left pixel) at (pM',qN'),

$$\Lambda(p,q) = \left\{ (m,n) : (pM' \leq m < (p+1)M') \wedge (qN' \leq n < (q+1)N') \right\}, \qquad (7.181)$$

   the method is denoted as *block matching*. In principle, matching areas of arbitrary shape can also be defined.
- Values of cost functions in matching can only be determined for a finite number of discrete motion shift constellations. Hence, in contrast to the optical flow method, resulting motion vectors are always given by *discrete values*. A very generic definition for this is made when motion parameters are defined as geometric mappings $\gamma(m,n)$ from a finite set $\mathcal{G}$ as in (7.154)ff. A very simple case occurs when the entire matching area shall be subject to *constant translational shift* by a motion vector $[k\ l]^T$. Then, the parameter space is often limited by defining a symmetric *search range*, which specifies maximum allow-

able shifts $\pm k_{max}$ and $\pm l_{max}$. The set of values in case of integer (full-pixel) accuracy of motion vectors can then formally be written as

$$\Pi = \left\{(k,l) : |k| \leq k_{max} \wedge |l| \leq l_{max} \wedge (k,l) \in \mathbb{Z}\right\}. \tag{7.182}$$

In addition, the number of elements in the set of allowable shift parameters can be modified by the *search step size s*, which is $s=1$ in (7.182), but could as well be larger or smaller than 1. The set of discrete motion shifts is then defined by

$$\Pi = \left\{(k,l) : (|k| \leq k_{max}) \wedge (|l| \leq l_{max}) \wedge (\mathrm{mod}(k,s) = 0) \wedge (\mathrm{mod}(l,s) = 0)\right\}. \tag{7.183}$$

Fig. 7.55 illustrates how this simple method of translational block matching works. In Fig. 7.55a, the terms 'matching area', 'search area' and 'search step size' are explained. Fig. 7.55b shows a principal inconsistency of the block matching concept assuming constant translational motion. The matching areas (blocks) are assumed to be non-overlapping in the current frame $x(m,n,o)$. As it can be expected that different values of vectors are estimated for different blocks, certain related best-matching areas will probably overlap in the *reference frame*[1]. Physically, this can be explained only in exceptional situations: The block-shift motion model is generally inadequate, whenever gradual changes of the motion vector field occur. Further, the block matching definition allows discontinuities only at pre-defined block boundary positions, which will hardly coincide with the real discontinuities of the motion vector field.

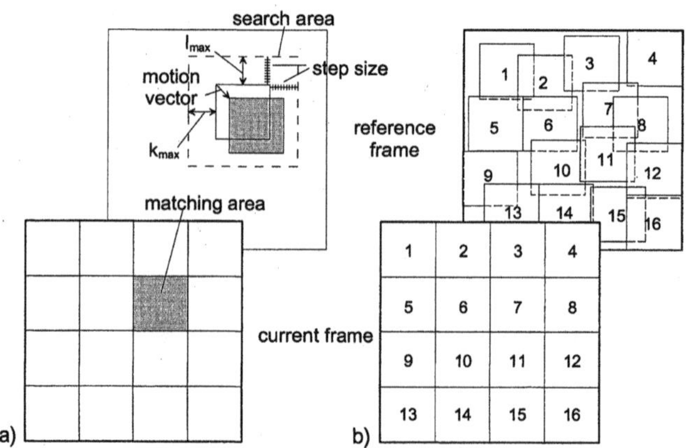

Fig. 7.55. Block matching motion estimation. **a** Definition of matching area, search range and step size in the current frame **b** Possible overlaps of best-matching blocks in the reference frame

---

[1] In most of the subsequent equations, we assume that the reference frame, which is the frame referred to by the motion vectors, is the previous frame $x(m,n,o\text{-}1)$. In principle, reference can be made to any past or future frame.

If large search ranges are used, cost functions based on cross correlation can efficiently be computed in the frequency domain, see (3.63), which generates the correlation results simultaneously for the entire search range. A special variant of frequency-domain processing is *phase correlation*, where only the phase spectrum is used in the correlation computation. This puts more emphasis on detail structures, edges etc. in the matching comparison. Difference criteria are nevertheless frequently used in block matching as well. To determine the optimum translational shift motion vector $[k,l]_{\text{opt}}$, a difference criterion of norm P can be used[1]

$$[k,l]_{opt} = \arg\min_{[k,l]\in\Pi}\left|\frac{1}{|\Lambda|}\sum\sum_{(m,n)\in\Lambda}\left|x(m,n,o)-\hat{x}(m+k,n+l,o-1)\right|^P\right|^{\frac{1}{P}}. \quad (7.184)$$

The *true motion* between two sampled frames of a sequence will usually not be by integer-pixel shifts. In the optical flow methods, this is no problem as continuous-value vectors are implicitly computed. To achieve a sub-pixel accurate comparison by block matching, it is necessary to generate additional discrete sub-pixel positions in the reference frame by interpolation (cf. sec 5.4), which is expressed in subsequent equations by marking the reference sample as $\hat{x}$. When the search step size s is decreased over both dimensions, the total number of positions to be compared in matching is increased by $1/s^2$. It is however not useful to test all possible sub-pixel positions over the entire search range, as it can be expected that the cost criterion changes smoothly for very similar motion vector parameters. Mostly, strategies are used which start by larger search step sizes and refine the estimated motion vector into sub-pixel accuracy only by the last step.

The complexity of block-matching motion estimation is mainly influenced by the size of the discrete parameter set, i.e. the different motion shift candidates which have to be compared by the cost criterion. If all candidates are checked, the number of operations is linearly dependent on the size of the search range, linear-reciprocally dependent on the search step size, and also linearly dependent on the complexity of the elementary cost function computation[2]. If matching areas are *non-overlapping*, the number of computations is in principle not changed if their size is modified. Small and irregular-shaped matching areas may however be disadvantageous regarding implementation of fast memory access.

Exhaustive block matching incorporates *full search* over all possible values of the parameter set. This guarantees that motion vectors are found which are optimum regarding the given cost function criterion, but must not necessarily coincide with the true motion. However, exhaustive-search matching is extremely complex due to the linear dependency on several parameters like search range, step size etc. Assume a rectangular search range, where all integer-shift positions up to $\pm k_{\text{max}}$ and $\pm l_{\text{max}}$ shall be compared. Then it is necessary to compute the cost criterion at

---

[1] $P{=}1$ for *minimum absolute difference* (MAD), $P{=}2$ for *minimum squared difference* (MSD); a more thorough introduction of the L_P norm in general is given in sec. 9.2.2.

[2] The L_1 norm or *Minimum Absolute Difference* (MAD) criterion is frequently used due to its low computational complexity.

$(2k_{\max}+1)\cdot(2l_{\max}+1)$ candidate positions, each implying at least one multiplication or absolute-value computation, additions/subtractions and comparison operations. If a maximum shift of only $k_{\max}=l_{\max}=15$ pixel is allowed, a total number of $31\cdot31=961$ candidate positions must be evaluated. In terms of arithmetic operations, this has to be multiplied by the number of pixels per second, which is around 10^7 for video of standard TV resolution. A dramatic reduction of computational complexity can be achieved by implementation of fast motion-vector search algorithms. Most of these algorithms do no longer guarantee that the global optimum of full search over the same range of the parameter space is reached. However, as the reduction of complexity may allow to increase the range of the parameter space, fast algorithms can often achieve better results at the same or lower computational complexity compared to exhaustive approaches. In one or the other way, fast motion estimation algorithms inherently exploit

- the smoothness of cost functions in case of slight variation of the parameter values, which allows to gradually optimize the result by iterative steps;
- the smoothness of motion vector fields, which allows to predict good initial estimates from previously estimated results;
- The scaling property of signals and motion vector fields, such that for lowpass filtered and spatially down-sampled frames both the number of sample-related operations, and the size of search ranges can be reduced[1].

Fast globally-optimum search. Under certain criteria, it is possible to reach the same global optimum as for the case of full search, but save a significant amount of computational effort nevertheless. As typically the cost function is computed sequentially over pixel positions within the matching area, it can be checked during the accumulation of cost contributions, whether the cost of the best match found so far was already superseded; in such a case, the computation can be interrupted and the respective motion vector is rejected. More advanced methods interpret the matching and reference block areas as vectors in the signal space $\mathcal{R}^K$. Then, certain candidates can immediately be excluded by analysis of the triangular inequality [BRÜNIG, NIEHSEN 2001], which is based on similar methods as described in (11.55)-(11.58) for the case of fast optimum search in vector quantization.

Multi-step search. Two principles of fast motion estimation algorithms are shown in Fig. 7.56a/b. Both are based on selecting a small subset of search positions which are tested out of the entire set of parameters, where an iterative multi-step search tracks the favorable direction for optimization of the cost criterion. In Fig.

---

[1] The scaling property also imposes an interesting relationship between the frame size and the complexity of motion estimation. If the frame size is doubled horizontally and vertically, the number of pixels is increased by a factor of 4. However, the size of the search range must also be doubled horizontally and vertically, as now any displacement vector $\mathbf{v}$ maps into a motion vector $\mathbf{k}$ of double length in (7.162). Considering exhaustive search, this leads to a complexity increase by a factor of 16 in case of doubled frame resolution.

7.56a/b, all positions tested in the particular steps are drawn as black dots, the steps are referenced by numbers, and the optimum as found in the respective step is marked by a circle. In both of these examples, the motion vector is finally found as k=-5, l=2. These two algorithms are typical representatives for a variety of similar approaches (see e.g. [GHANBARI 1990], [NAM ET AL. 1995],[PO, MA 1996]).

In the method of Fig. 7.56a, denoted as *three-step search* [MUSMANN ET AL. 1985], only a small search range of 9 positions is evaluated in each iteration step. Simultaneously, the search step size is decreased by each iteration (s=3, 2, 1 pixel for the three iterations in the example shown). The center of the search range in iteration step r is selected from the best-matching position of the previous iteration r-1, such that cost criteria need to be computed only for 8 new positions in iterations 2 and 3. In the example shown, a total of 9+8+8=25 candidate positions are compared in the three iterations; the possible parameter range is k_{max}=l_{max}=±6 pixel. A full search over the same search range would require comparison of 13·13=169 positions. The effective reduction of computational complexity is more drastic when the search range is larger and more iteration steps are implemented.

In the *logarithmic search* method shown in Fig. 7.56b, 5 different positions are compared in the first iteration step. The search starts from zero-shift position, the other positions which are examined are distant by two pixels horizontally or vertically. After finding the best match, it is compared in the next step against three more positions, which are the remaining horizontal and vertical neighbors at two-pixel distances that were not yet compared by the previous step. This process is continued until the best-match position remains unchanged, which indicates that a minimum over the cost function has been approached. Then, in a final step, all 8 positions around this optimum are once more checked as candidates. In the example shown, the cost function has to be computed for a total of 5+2·3+8=19 candidate positions.

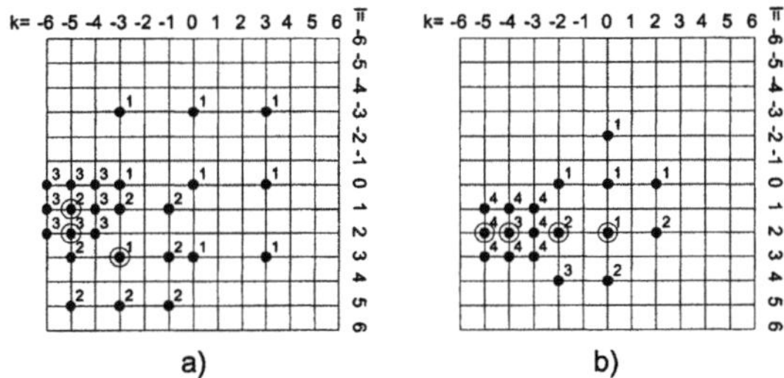

a)   b)

Fig. 7.56. Multi-step block-matching estimation methods
a 'Three-step search' **b** 'Logarithmic search'

These algorithms will actually reach the *globally-optimum* result (as in full search), if the cost function is truly convex, steadily improving over the parameter space up to the optimum position. If however local optima of the cost function exist, it is possible that the final result gets stuck in such a position. This can for example be the case when periodic structures appear.

Recursive block matching and usage of motion vector predictors. It is possible to utilize estimation results available from spatial or temporal neighbors of the current matching block. Due to the continuity of the motion vector field, it can be expected that correct vectors or at least good predictions for correct vectors are yet available from the neighborhood. In principle, this is the extension of the pixel-recursive methods described in sec. 7.6.2, now being applied over entire blocks. The block recursion can proceed along the temporal axis and/or along both spatial dimensions. Fig. 7.57 shows possible candidates of previously processed vectors, which may be used to predict an initial vector value for the current block. The temporal predictor should be selected such that the vector points from its position in the reference frame into the current block.

Different approaches are possible to determine the current estimate:

- The mean or median values of the previously-estimated vectors selected as predictor candidates are computed, and the new value is optimized within a search range around this initial hypothesis;
- The range between minimum and maximum of the predictor candidates is used as a search range, optionally extended by an additional margin;
- Initially, from several candidate vectors available, the best-matching predictor is selected by a block-matching criterion; the result is improved by testing an additional search range around the predictor position.

To perform block matching within a search range around the initially predicted vector value usually gives reliable results, but can still cost considerable effort, depending on the size of the search range to be tested around the predicted position. Much faster methods for updating the predicted vector value have been proposed[1], based on

- selection of *random values* (which are chosen whenever better than the initial vector according to the respective cost criterion) [DE HAAN ET AL. 1993];
- pixel-recursive gradient optimization, where also the final result is only selected if it is better than the initial estimate [BRAUN ET AL. 1997].

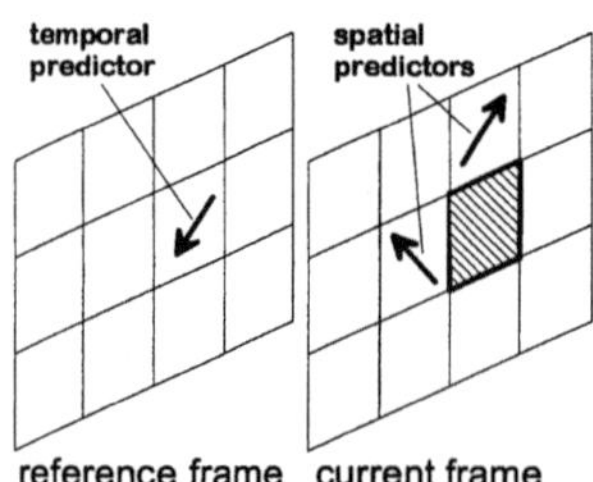

Fig. 7.57. Temporal and spatially adjacent previously computed vectors as possible candidates in recursive block matching

---

[1] These methods are optimized for block matching motion estimation of relatively small block sizes, and are intended to preferably estimate smooth motion vector fields.

Multi-resolution motion estimation. Like in multi-step methods, multi-resolution estimation increases the accuracy of the estimation result stepwise, which is accompanied by systematic modification of search parameters[1]. In both approaches, the size of the search range and the search step width are typically decreased during the iterative process. In multi-resolution estimation, also the *size of the matching area* is decreased, and the *resolution of the signal* is increased. This can ideally be combined with methods for hierarchical representation of signals, e.g. the Gaussian pyramid decomposition (sec. 4.4.6); due to the decreased matching area sizes, also a *pyramid of motion vector fields* of increasing resolution and precision is produced. If the estimation starts at the top of the pyramid, the resolution (number of samples in the signal) is increased by a factor of four with each iteration step. If e.g. at each resolution level same parameter ranges, step sizes and matching block sizes are used in the estimation process, the desired effect is inherently achieved: In the down-sampled image, the same block size covers a larger relative area than in the full resolution image; a search range of $\pm(k_{max}, l_{max})$ applied in an image down-scaled by a factor of U in each dimension is equivalent to a search range of $\pm(U \cdot k_{max}, U \cdot l_{max})$ in the original full-resolution image. A similar relationship concerns the search step size.

Tab. 7.2. Parameters of a multi-resolution motion estimation over four levels

Iteration step #	1	2	3	4
Frame size of pyramid level	88x72	176x144	352x288	704x576 (full)
Number of blocks / frame (horizontally x vertically)	11 x 9	22 x 18	44 x 36	88 x 72
Equivalent matching block size of full resolution	64 x 64	32 x 32	16 x 16	8 x 8
Equivalent search range size of full resolution	±64 pixel	±32 pixel	±16 pixel	±8 pixel
Equivalent search step width of full resolution	4 pixel	2 pixel	1 pixel	1/2 pixel
Maximum accumulated search range related to full resolution	64 pixel	64+32 =96 pixel	96+16 =112 pixel	112+8 =120 pixel

Tab. 7.2 shows example parameters for a pyramid representation of a frame, which is down-sampled over three levels. If at each level of the pyramid, block matching motion estimation is applied using a matching area size of 8x8 pixels, a search range of ±8 pixels and a search step size of 1/2 pixel, it becomes evident that the

---

[1] Both multi-step and multi-resolution methods are often denoted as *hierarchical motion estimation*.

respective equivalent parameters (related to full image resolution) are much larger. If the final result of motion estimation is generated by accumulation of the vectors from the single levels, the maximum range of vectors is significantly larger than the range used in each of the steps. From this point of view, multi-resolution estimation can be seen as a generalization of the fast estimation methods described earlier, but has an additional effect in reduced computational complexity, as the down-scaling of the image size for the first iteration steps also has a drastic effect on the reduction of arithmetic operations (cf. footnote on p. 299).

By simultaneously decreasing the equivalent reference and search area sizes, the results of motion estimation are regularized such that neighbored blocks at the highest resolution level can no longer be largely different[1]. Therefore, hierarchical estimation implicitly generates spatially-smooth motion vector fields, which is in harmony with the motion models described in sec. 7.6.1. The relationships between matching areas and estimated motion vectors at two different levels are illustrated in Fig. 7.58.

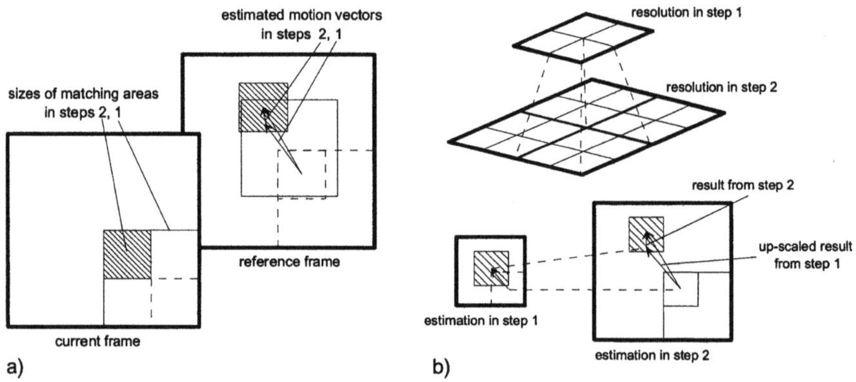

Fig. 7.58. Hierarchical motion estimation over two steps **a** Interpretation at full resolution **b** Principle of reduced resolution in the first step

Constrained estimation. In analogy to the *smoothness constraint* in optical-flow based methods, it is possible in block matching to establish interrelationships between adjacent blocks during estimation. An example for an optimization criterion in analogy with (7.184) is

$$[k,l]_{\text{opt}}^{(p,q)} = \arg\min_{[k,l]\in\Pi}\left[\frac{1}{|\Lambda|}\sum\sum_{(m,n)\in\Lambda}\left|x(m,n,o)-\hat{x}(m+k,n+l,o-1)\right|^P\right.$$

---

[1] It is also possible to allow larger search ranges, if adjacent blocks at the previous iteration step (coarser pyramid level) have largely different vectors. Such methods can be well combined with the constrained estimation described subsequently.

$$+\frac{c_1}{|\mathcal{N}|}\left(\sum_{(i,j)\in\mathcal{N}(p,q)}\left|k-\hat{k}^{(i,j)}\right|^P+\left|l-\hat{l}^{(i,j)}\right|^P\right) \tag{7.185}$$

$$+c_2\left(\left|k-\hat{k}_{-1}^{(p+k/M',q+l/N')}\right|^P+\left|l-\hat{l}_{-1}^{(p+k/M',q+l/N')}\right|^P\right)\Bigg]^{\frac{1}{P}},$$

where (p,q) is the index of the actual reference position and $\mathcal{N}(p,q)$ describes a neighborhood system related to a set of $|\mathcal{N}|$ neighbored blocks. Again, the parameters c_1 and c_2 weight for spatial and temporal smoothness of the motion vector field, respectively. Temporal reference is made to vectors $[k_{-1}\,l_{-1}]^{\mathrm{T}}$ estimated at the (previous) reference frame[1]. If the neighborhood system is *causal*, i.e. if it includes only blocks for which finally estimated motion parameters are available, the complexity is only marginally higher than with the conventional matching method (7.184). For non-causal neighborhood systems, which in case of row-wise sequential block processing could also include right-handed or lower neighbor blocks, optimization can only be done iteratively due to the mutual influences. In motion estimation for video coding, other criteria are applicable in constrained estimation, e.g. the rate $R(\mathbf{k})$ to be spent for encoding of the estimated motion vector,

$$[k,l]_{\mathrm{opt}}^{(p,q)}=\arg\min_{[k,l]\in\Pi}\left(\frac{1}{|\Lambda|}\sum_{(m,n)\in\Lambda}\left|x(m,n,o)-\hat{x}(m+k,n+l,o-1)\right|^P+\lambda\cdot R(\mathbf{k})\right). \tag{7.186}$$

This is a typical approach of Lagrangian optimization, where the optimum factor λ has to be determined from the operational goal of the optimization, which could e.g. be the minimization of distortion in motion compensated prediction coding with a given overall rate target.

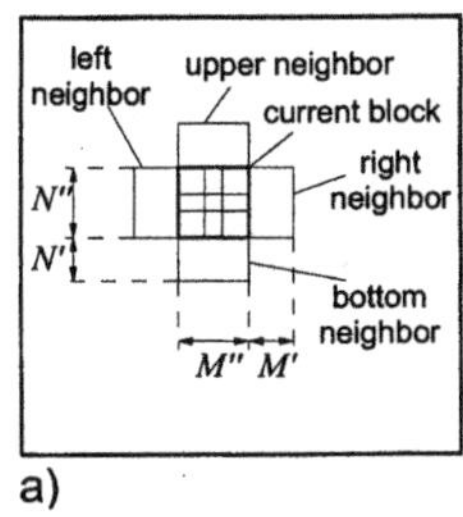

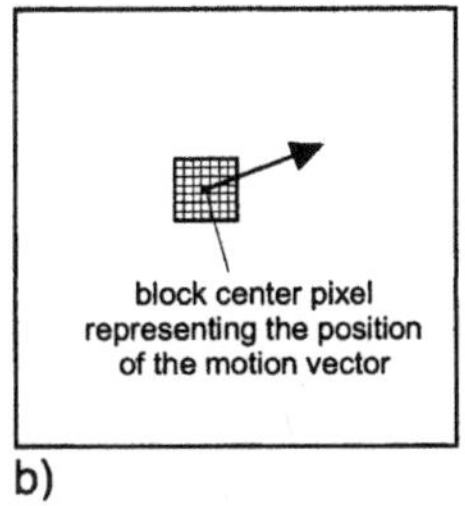

Fig. 7.59. Block matching with block overlap **a** Block positions having distances (M',N') horizontally and vertically **b** Interpretation of the motion vector to be valid at the center of the current block

---

[1] All vector differences are formally expressed referring to vectors $\hat{\mathbf{k}}$, which may be interpolated from different previous-frame estimation results, or may relate to preliminary estimates in the current frame.

Estimation with block overlap. Block matching was regarded so far to be based on non-overlapping matching areas or block partitions. The block reference areas can also overlap (Fig. 7.59a). Here, the block centers are positioned at distances M' and N' horizontally and vertically, the matching block areas are of sizes $M''\text{x}N''$, with $M''\geq M'$, $N''\geq N'$. The overlap towards each side is $(M''-M')/2$ horizontally and $(N''-N')/2$ vertically. To implement fast algorithms, computations of the cost function can be split into sub-blocks which can be re-used for estimation of the vectors in several blocks. This also leads to the conclusion that the estimation results for adjacent blocks will no longer be completely independent, such that a smoother motion vector field would be a side effect. In principle, for $M'=N'=1$ it becomes possible to apply block matching estimation *at any pixel position*, such that a dense motion vector field is generated. The matching area can then best be interpreted as a block with the current pixel at its center position (Fig. 7.59b). Conceptionally, this is not much different from solutions for the optical flow method, where also an area from the pixel neighborhood is often used to stabilize the estimation result. In estimation with block overlap, it can be useful to apply a weighting function $w(m,n)$ in the cost computation, which for the case of difference criteria can be implemented as

$$[k,l]_{opt} = \arg\min_{[k,l]\in\Pi} \left| \frac{1}{|\Lambda|} \sum \sum_{(m,n)\in\Lambda} \left| w(m,n)\left[x(m,n,o) - \hat{x}(m+k,n+l,o-1) \right] \right|^p \right|^{\frac{1}{p}} . \quad (7.187)$$

The function $w(m,n)$ will typically decrease towards the block boundaries, such that the motion estimation is more accurate for the center part of the block. Application of such a weighting function also establishes the optimum motion estimation for *overlapping block motion compensation* (cf. sec. 13.2.6, (13.25)/(13.26)), where the same weighting function must be used for the estimation step as in the compensation itself.

In methods to generate smooth motion vector fields, it should be observed that the motion vector field is by nature *unsteady* and *discontinuous* at object boundaries. Parameters in estimation should be chosen appropriately, e.g. by allowing sufficient deviation between the motion vectors of adjacent blocks in hierarchical search methods. In any case, it is most important to find a good compromise between homogeneity (continuity) of the vector field and the necessarily-occurring discontinuities. More exact solutions on this can only be achieved if motion estimation is directly coupled with region segmentation (see sec. 10.2.3).

Fig. 7.60 shows results of motion estimation, where different search strategies have been used in block matching. It can in particular be observed that the methods for generation of smooth motion vector fields are much better capable to capture the true motion. The full-search method objectively minimizes the criterion of DFD energy, which may not be the best choice from the viewpoint of true-motion estimation. In the case of the example shown, this is obviously caused by the correspondence problem, where different similar structures appear in the background area. This also indicates that smoothness constraints can partially resolve the correspondence problem by introducing dependencies over larger areas[1].

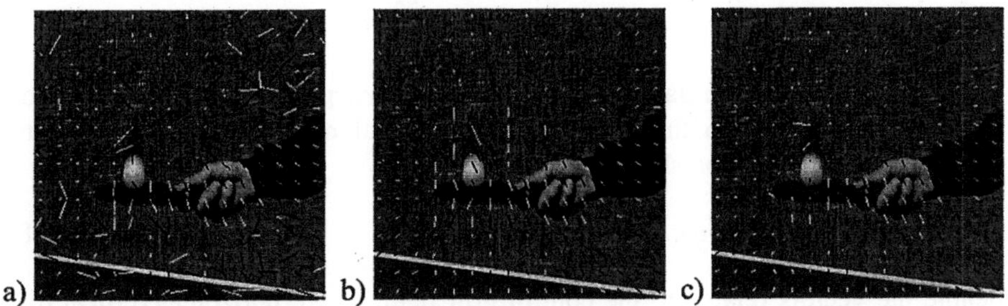

Fig. 7.60. Block matching motion estimation employing different search strategies
a Full search **b** Multi resolution search **c** Constrained search

7.6.4   Estimation of Parameters for Warping Grids

Both in overlapping and non-overlapping block matching motion estimation, one motion vector is computed for a relative area size M'xN'; this could also be interpreted as a *sub-sampled* approximation of a dense motion vector field, where sub-sampling factors are M' horizontally and N' vertically. The interpretation made for translational block matching is an interpolation of the dense field by a hold element, such that a constant motion vector is applied for the entire block. Alternatively, using the same factors of sub-sampling and same number of motion parameters, other interpolation functions could be used[2]. This can be interpreted in a way that the motion vector field is described by a grid of *control values*. Methods of this type have been introduced as *control grid interpolation* (CGI) [SULLIVAN, BAKER 1991] or *quadrangle-based motion compensation* (QBMC), where bilinear interpolation (5.35) is used component-wise to generate the dense motion vector field from the control values. In a more general view, any parametric *warping functions* can be used for the interpolation, which may be seen as more reasonable considering camera projection models (cf. sec. 7.4.4)[3]. If such an approach is e.g. used for the purpose of motion compensation, it is important to employ the same interpolation function during estimation yet, because otherwise the estimated vectors will not be optimized for the desired result. The case of a bilinear interpolation func-

---

[1] Another very simple approach to resolve the correspondence problem in unstructured areas can be made by a 'zero vector threshold': If the variance of the matching area is below a pre-defined threshold, the vector is forced into a zero value, or can be marked as unreliable.

[2] In fact, the dense motion vector field can be interpreted like a signal having two amplitude components, such that any conventional signal processing approach can be applied to increase resolution, eliminate outliers etc.

[3] Bilinear interpolation (5.35) of the motion vector field is fully equivalent to bilinear parametric mapping of coordinate positions (7.110).

tion will be discussed here, as it is complexity-wise the most simple approach; generalization to other functions is straightforward, as a one-to-one mapping between parametric warping models and a set of support position shifts with the same number of free parameters is always possible (cf. (7.192)).

Optimization of motion parameters for bilinear warping. Bilinear interpolation is applied separately to the horizontal and vertical motion vector components $k(m,n)$ and $l(m,n)$. Motion shifts

$$\breve{\mathbf{k}}(p,q) = \left[\breve{k}(p,q), \breve{l}(p,q)\right]^{\mathrm{T}} \tag{7.188}$$

denote the vectors related to the control grid field, where horizontal/vertical distances between the positions in a rectangular grid shall be M' and N'. The control-value coordinates are $(m_a, n_a) = (m'M', n'N')$[1]. The motion shift at any positions in the area spanned by four corner points is then computed according to (5.35),

$$\begin{aligned}
\hat{\mathbf{k}}(m,n) = {}&\breve{\mathbf{k}}(p,q) \cdot \left(1 - (m/M' - p)\right) \cdot \left(1 - (n/N' - q)\right) \\
&+ \breve{\mathbf{k}}(p+1,q) \cdot (m/M' - p) \cdot \left(1 - (n/N' - q)\right) \\
&+ \breve{\mathbf{k}}(p,q+1) \cdot \left(1 - (m/M' - p)\right) \cdot (n/N' - q) \\
&+ \breve{\mathbf{k}}(p+1,q+1) \cdot (m/M' - p) \cdot (n/N' - q)
\end{aligned} \tag{7.189}$$

with $m_a = p \cdot M' \le m \le (p+1) \cdot M'$ and $n_a = q \cdot N' \le n \le (q+1) \cdot N'$. If the control grid vector at position (p,q) shall be optimized by a matching criterion, the result shall be

$$\begin{aligned}
\breve{\mathbf{k}}_{\mathrm{opt}}(p,q) = \underset{[k,l] \in \Pi}{\arg\min} \sum_{m=m_a-M'+1}^{m_a+M'-1} \sum_{n=n_a-N'-1}^{n_a+N'-1} \Big| x(m,n,o) \\
- \hat{x}\left[m + \hat{k}(m,n), n + \hat{l}(m,n), o-1\right]\Big|^p,
\end{aligned} \tag{7.190}$$

where the summation limits exactly specify the range of influence for $\breve{\mathbf{k}}(p,q)$. In the current frame, this is a rectangular area of size $(2M'-1)(2N'-1)$, as shown in Fig. 7.61. Within this area, the values of the eight nearest-neighbor control positions will influence the result as well, which means that the optimum motion parameters can no longer be estimated *independently* of each other. As however in one step of the estimation process, only one vector is locally optimized according to (7.190), the global optimum for the motion vector field can only be obtained by an iterative procedure, where the results for the neighbored control positions from the previous iteration are used for the optimization of the current control point in the next iteration.

The speed of convergence in an iterative estimation for global optimization is a central problem in grid-based motion estimation. Starting from a grid field where the values $\breve{\mathbf{k}}(p,q)$ are initially set to zero would lead to a very slow process. Start-

---

[1] Optionally, as shown in Fig. 7.61, positions can be shifted by an offset of $(M'/2, N'/2)$.

ing from values that were obtained by conventional block matching gives reasonable initial values. Likewise, application of a *smoothness constraint* as in (7.185) is advantageous. By guarantee, no iteration step will ever deteriorate the cost criterion when the parameter set of the previous best result is one of the candidates for the next iteration. There is however a risk of getting stuck in local minima, depending on the surface of the cost function. This is in particular the case as the cost function value depends on a high number of parameters due to the interdependencies of adjacent grid positions, such that it is not straightforward to detect the presence of local minima on the function surface. In practical realization of grid-based motion estimation, convergence to *some* minimum of the cost function is often observed by two or three iterations, when a block-matching vector field computed by a smoothness constraint is used for initialization.

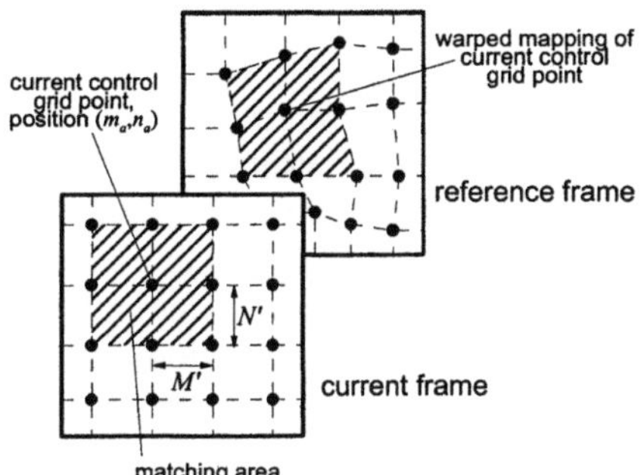

Fig. 7.61. Motion field interpolation based on bilinear warping from control positions

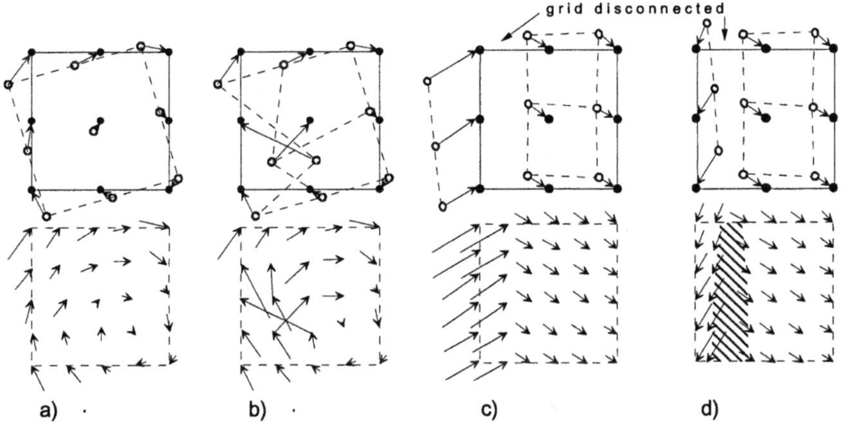

Fig. 7.62. Examples of warping grids. *Top*: Support grid in current frame (●) and reference frame (○); *Bottom*: Interpolated motion vector field
a Case of rotational motion **b** Unreasonable configuration with overlapping vectors
c/d Motion discontinuities in cases of a covered area (*c*) and an uncovered area (*d*)

Fig. 7.62a illustrates how rotational motion can be captured by a warping motion model. This is a clear advantage as compared to conventional block-independent matching, which only supports translational motion, even though the number of parameters to describe the motion vector field is equal in both cases. Warping grid interpolation has however severe limitations as well. First, *overlapping* motion vectors are unreasonable, as they would lead to mirrored coordinate relationships within partial areas (Fig. 7.62b). Such configurations can nevertheless occur, when an object is moving fast in front of a background; however in this case, a large occluded area is typically present. As motion vector fields must show discontinuities at object boundaries, the application of smooth interpolation across the boundaries is not reasonable under physical model considerations; warping interpolation is in fact only advantageous for motion vector field areas of smooth variation. A possible solution to solve this problem is shown in Fig. 7.62c/d. When large differences between adjacent grid vectors are observed, the grid is artificially disconnected, and extrapolation using a hold function is performed towards the object boundary, for which a position has to be defined between the two extrapolated areas. The determination of this boundary can also be performed as part of the estimation process [OHM 1994B], [OHM 1996] [HEISING 2002]. In the simplest case, the boundary is arbitrarily defined at the center between the diverging or converging grid positions. It is then possible to perform flexible switching between interpolative motion description and block-based (hold-element interpolation) motion description, where during the estimation process, the method which better minimizes the cost function is selected. As a by-product, overlapping motion vectors as in Fig. 7.62b, which are in fact reasonable in the case of fast moving objects, are fully supported. In [HEISING 2002], also a hybrid switching between warping interpolation and an overlapping block matching method has been investigated, which is reported to mostly resolve the problem of discontinuous motion vector fields.

Feature matching. When availability of motion vectors on a fixed discrete grid (e.g. block or control point grid) is not a prerequisite in an application, motion estimation based on *feature matching* can provide another factor of stability. This is for example the case when a global motion vector field shall be estimated from a finite set of local vectors. The first step is selection of unique *feature points* or *feature areas* which should have significant properties to make them unique matches. This can e.g. be edge points, corner points or points with highest available variation of neighborhood pixels, such that sufficient structure guarantees avoidance of the aperture problem. It is then possible either to employ matching over block areas centered around the feature points, or directly match feature data such as edges, which can be represented by gradients or as binary lines.

7.6.5   Estimation of non-translational Motion Parameters

When the purpose of motion estimation is extraction of motion-related semantic features or motion-compensated frame interpolation, the true motion must be found. Further, the number of parameters should be as low as possible. An example case are global motions, as effected by zoom or pan of a camera, which can be described extremely compact by only one set of parameters for the entire frame. Similar principles can be applied if the goal is to characterize non-translational motion properties of smaller regions. Non-translational parameters can be estimated by analysis of occurrences and relationships between local translational parameters. The positions where local motion parameters are available shall be denoted as *measurement points* which can best be determined by feature matching. By defining the motion shifts of measurement points as $k(m,n)=m'-m$ and $l(m,n)=n'-n$, using the discrete version of the affine mapping (7.117) as an example here, the parametrization is made by

$$k(m,n) = (a_1 - 1) \cdot m + a_2 \cdot \frac{S}{R} \cdot n + \frac{t_1}{R} \quad ; \quad l(m,n) = a_3 \cdot \frac{R}{S} \cdot m + (a_4 - 1) \cdot n + \frac{t_2}{S} \quad (7.191)$$

From the k- and l-values of one singular measurement point, it is not possible to compute non-translational parameters (cf. sec. 7.6.1). To determine the parameters of the affine model, information from at least 3 measurement points of coordinates (m_1,n_1), (m_2,n_2), (m_3,n_3) and their translational motion shifts (k_1,l_1), (k_2,l_2), (k_3,l_3) must be available[1]. This will result in a uniquely solvable equation system of six unknowns; in fact, this equation system can be separated into two systems of three unknowns each:

$$\begin{bmatrix} (a_1-1) \cdot R \\ a_2 \cdot S \\ t_1 \end{bmatrix} = R \cdot \begin{bmatrix} m_1 & n_1 & 1 \\ m_2 & n_2 & 1 \\ m_3 & n_3 & 1 \end{bmatrix}^{-1} \cdot \begin{bmatrix} k_1 \\ k_2 \\ k_3 \end{bmatrix} ; \quad \begin{bmatrix} a_3 \cdot R \\ (a_4-1) \cdot S \\ t_2 \end{bmatrix} = S \cdot \begin{bmatrix} m_1 & n_1 & 1 \\ m_2 & n_2 & 1 \\ m_3 & n_3 & 1 \end{bmatrix}^{-1} \cdot \begin{bmatrix} l_1 \\ l_2 \\ l_3 \end{bmatrix}. \quad (7.192)$$

If data from more than 3 measurement points are used, an overdetermined equation system results analogously to (7.171), for which a solution can e.g. be achieved by pseudo-inversion[2]:

$$\begin{bmatrix} (a_1-1) \cdot R \\ a_2 \cdot S \\ t_1 \end{bmatrix} = R \cdot \begin{bmatrix} m_1 & n_1 & 1 \\ m_2 & n_2 & 1 \\ \vdots & \vdots & \vdots \\ m_P & n_P & 1 \end{bmatrix}^{P} \cdot \begin{bmatrix} k_1 \\ k_2 \\ \vdots \\ k_P \end{bmatrix} ; \quad \begin{bmatrix} a_3 \cdot R \\ (a_4-1) \cdot S \\ t_2 \end{bmatrix} = S \cdot \begin{bmatrix} m_1 & n_1 & 1 \\ m_2 & n_2 & 1 \\ \vdots & \vdots & \vdots \\ m_P & n_P & 1 \end{bmatrix}^{P} \cdot \begin{bmatrix} l_1 \\ l_2 \\ \vdots \\ l_P \end{bmatrix}. \quad (7.193)$$

It is useful to select measurement points such that estimated values k_p,l_p shall become highly reliable. This can be achieved by additional criteria, e.g. selection

---

[1] As a generalization, for a parametric model with N free parameters, at least $N/2$ measurement points with unique motion shifts in both directions must be given.

[2] Pseudo inversion (cf. sec. 8.3) gives a reasonable result to fit measurement data by a function based on a least squares criterion, provided that the data set used is not affected by noise. For the problem of non-translational motion estimation, this means that the error in motion estimation shall be small for the measurement points. Methods to detect outliers in such a process are presented in sec. 8.7.

from positions of high spatial detail, or by checking the reliability of the motion estimation for the measurement points by analysis of the cost function.

Optical flow estimation. Parameters of a non-translational motion model can directly be estimated by an optical-flow method. Substitution of (7.191) into (7.170) gives

$$\left[(a_1-1)\cdot m+a_2\cdot\frac{S}{R}\cdot n+\frac{t_1}{R}\right]\cdot x_r(m,n)+\left[a_3\cdot\frac{R}{S}\cdot m+(a_4-1)\cdot n+\frac{t_2}{S}\right]\cdot x_s(m,n)=-x_t(m,n). \tag{7.194}$$

If P measurement points are used ($P{\geq}6$), the matrix formulation is

$$\begin{bmatrix} m\cdot x_r(1) & n\cdot x_r(1) & m\cdot x_s(1) & n\cdot x_s(1) & x_r(1) & x_s(1) \\ m\cdot x_r(2) & n\cdot x_r(2) & m\cdot x_s(2) & n\cdot x_s(2) & x_r(2) & x_s(2) \\ \vdots & \vdots & \vdots & \vdots & \vdots & \vdots \\ m\cdot x_r(P) & n\cdot x_r(P) & m\cdot x_s(P) & n\cdot x_s(P) & x_r(P) & x_s(P) \end{bmatrix}\cdot\begin{bmatrix} a_1-1 \\ a_2\cdot\frac{S}{R} \\ a_3\cdot\frac{R}{S} \\ a_4-1 \\ t_1\cdot\frac{1}{R} \\ t_2\cdot\frac{1}{S} \end{bmatrix}=-\begin{bmatrix} x_t(1) \\ x_t(2) \\ \vdots \\ x_t(P) \end{bmatrix}. \tag{7.195}$$

For $P{>}6$, this can again be solved by computation of the pseudo-inverse (7.172). Recursive and iterative methods for minimization of the estimation error can also be applied, as described in sec. 7.6.2.

All methods introduced here for estimation of non-translational parameters by the example of the affine model can be extended to other polynomial parametric models without any restrictions. For the perspective model however, the parameters influencing the denominator term do not allow direct expression by a matrix-form equation system[1].

Matching-based direct estimation. The matching principle can also directly be applied to estimate motion modeled by a parametric geometric transform. As matching always optimizes over a discrete parameter space, a finite number of 'quantized' parameter configurations must be defined in advance. Best definition of this discrete parameter set is not a trivial problem, as parameters of polynomial models will usually have a nonlinear effect on the geometric transform. If the parametric model is defined over a Q-dimensional parameter space (e.g. $Q{=}6$ for the case of an affine transform), each of which could be expressed by $J_1, J_2, ..., J_Q$ discrete values, a total of J_1 x J_2 x ... x J_Q constellations of the model is possible. The application of exhaustive search methods is almost impossible for parameter spaces of $Q{>}2$. If gradient methods such as multi-step search are employed, this will lead to similar solutions as in the case of the optical-flow approach. A very general notation for the matching problem based on arbitrary geometric transformations $\gamma(m,n)$ was already introduced in (7.153)-(7.158).

---

[1] Perspective mapping can be expressed by a matrix-form equation system, if mapping into *homogeneous coordinates* is performed (cf. (7.207)), which linearizes the problem in a higher-dimensional feature space.

7.6.6   Estimation of Motion Vector Fields at Object Boundaries

To resolve the occlusion problem, discontinuities of the motion vector field must be properly detected and resolved. However, no valid motion parameters can be estimated for these areas, when no correspondences exist. If it is tried to estimate motion nevertheless, the result will be erroneous and can even affect the parameters of neighbored not-occluded areas, e.g. if a matching area occasionally contains both occluded and non-occluded positions. If reliable information about positions of object boundaries was available, the motion estimation might become perfect; unfortunately, such information is usually not known by the time of motion estimation[1]. A hypothesis about the general *presence* of a discontinuity, and also a categorization of the type of occlusion (covered or uncovered area) can however be gained from unreliable motion information.

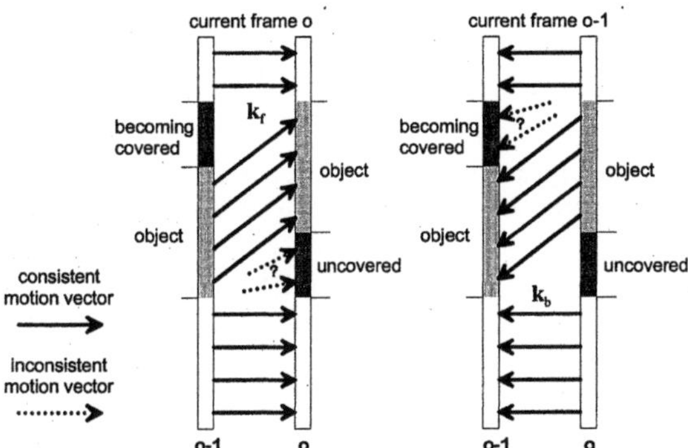

Fig. 7.63. Detection of occluded areas by forward/backward motion estimation

Forward/backward estimation. In forward-directed motion estimation, the reference frame precedes the current frame. In most motion estimation algorithms, motion vectors are estimated for each pixel position of the current frame, irrespective if they are valid or not. In this case, it can however be expected that motion vectors for areas *newly uncovered* are undefined, as no correspondence can then be established with the reference frame. Areas of the reference frame which are *becoming covered* in the current frame can eventually be identified, as these will not be referenced by a motion vector. In a backward estimation (where the subsequent frame is the reference frame), these conditions are reversed (Fig. 7.63). For a *detection of the presence* of occluded areas it is useful to compute the sum of related forward and backward motion vectors from a frame pair. The vector $k_f(m,n)$ is the forward vector estimated with frame o as 'current' frame, and $k_b(m,n)$ is the backward vector, where frame o-1 is the 'current' frame. Typically, vectors should

---

[1] This is a hen-egg problem: Reliable motion information may even be needed to give sufficient evidence about positions of object boundaries. Possible solutions are discussed in sec. 10.2.3.

be consistent, which means that they should uniquely connect two positions; as forward and backward vectors are only different by sign, the sum should be zero. Hence, a criterion to detect occluded positions can be gained by comparing the norm of the vector sum against a threshold Θ. A criterion for pixel position (m,n) in frame o-1 becoming covered will be

$$\left\| \mathbf{k}_b(m,n) + \mathbf{k}_f(m + k_b(m,n), n + l_b(m,n)) \right\| > \Theta, \tag{7.196}$$

while position (m,n) in frame o is potentially becoming uncovered if

$$\left\| \mathbf{k}_f(m,n) + \mathbf{k}_b(m + k_f(m,n), n + l_f(m,n)) \right\| > \Theta. \tag{7.197}$$

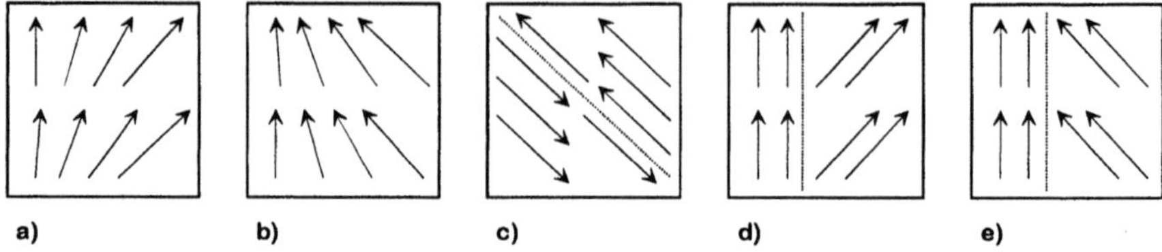

Fig. 7.64. Constellations in motion field divergence analysis: **a** Expansive, div>0 **b** Contractive, div<0 **c** Antipodal, div=0 **d** Uncovered area, div >0 **e** Covered area, div<0

Divergence analysis. The divergence of the discrete motion vector field can be defined at the pixel position of index (m,n) as[1]

$$\mathrm{div}\left(\mathbf{k}(m,n)\right) = l(m, n+1) - l(m,n) + k(m+1, n) - k(m,n). \tag{7.198}$$

The divergence is in general a criterion for geometric contraction or expansion between the two frames that are referenced. For example, the divergence will be

- >0 (Fig. 7.64a), when the reference frame scales up compared to the current frame, which will be denoted as an *expansive* motion vector field;
- <0 (Fig. 7.64b), when the reference frame is down-scaled, which means that the motion vector field is *contractive*.

At positions of discontinuities in the motion vector field, a divergence $\neq 0$ can be observed in case of occlusions. The difference as compared to largely contractive or expansive behavior is the abrupt change of the divergence, which means that the derivative of the divergence is $\neq 0$ at discontinuities. If however a discontinuity is such that two objects move *antipodal* in opposite directions without effecting occlusions, the divergence is zero (Fig. 7.64c). Otherwise, it can be concluded that for div>0 a new area is becoming uncovered (Fig. 7.64d), and for div<0, an area is becoming covered.

---

[1] For the case of block motion vectors, this analysis would relate to the pixel position at the bottom right corner of each block.

The divergence criterion is only valid if the estimation is correct, which is critical in the vicinity of motion discontinuities anyway. Divergence analysis should hence not be applied to vectors detected as unreliable, such that in fact only combination of different criteria, such as forward/backward estimation, cost-function analysis etc. is useful. To eliminate small noise influences, a threshold value Θ is defined to determine whether the motion vector field shows 'normal' behavior, whether it is 'contractive' (an area being covered) or 'expansive' (an area being uncovered)[1]:

$$a) \left| \mathrm{div} \left(\mathbf{k}(m,n) \right) \right| \leq \Theta \Rightarrow \text{'normal'}$$

$$b) \, \mathrm{div} \left(\mathbf{k}(m,n) \right) > \Theta \Rightarrow \text{'expansive' or 'uncovered area'} \tag{7.199}$$

$$c) \, \mathrm{div} \left(\mathbf{k}(m,n) \right) < -\Theta \Rightarrow \text{'contractive' or 'covered area'.}$$

The absolute value of the divergence can also be used to estimate the *width* of an occluded area.

Contractive or expansive behavior of motion vector fields will be detected under different scenarios, e.g. zoom, an object approaching or disappearing from the camera etc., but these will typically exhibit a smaller divergence than for cases of severe occlusions. At boundaries of frames, occlusions are also present in case of global camera motion.

7.6.7 Analysis of 3D Motion

Projection relationships, e.g. the central projection equation (1.1) and the related speed equations (7.159) can be integrated into the formulation of the motion estimation problem, where parametrization is made similar as in (7.191)-(7.195). This can be used to perform motion and speed analysis directly relating to the 3D world, provided that the distance Z is known. Due to the reciprocal dependency on Z, the problem formulation in the respective equation system will neither be linear nor can a polynomial form be described. In this case, it is possible to perform the analysis by a system of homogeneous coordinates (cf. (7.207)). In general, solutions are very similar to depth estimation, which will be described in sec. 7.7.

For 3D shapes described by grids of vertices, motion estimation can be performed similar as in warping grid estimation (cf. sec. 7.6.4). For the case of a wireframe description, it is either possible to project the vertices and their connectivity into a grid on the image plane, or to estimate 3D motion of the vertices and perform linear plane interpolation for the positions on the planar triangular patches (cf. Fig. 7.65). In case of extraordinary shape deformations, it may also happen that vertices are removed or added, or that a new interconnection topology of vertices has to be defined. It is however dependent on the application whether such changes are desirable; in particular if it is expected that the original 3D shape will

---

[1] Definitions in (7.199) are for the case of backward estimation; in forward estimation, the roles of 'covered' and 'uncovered' have to be interchanged.

appear once again, or if temporary geometric distortions are acceptable, it will not be useful to change the topology of vertex interconnections. In the case of moved or animated 3D objects, which can contain components of rigid or non-rigid body motion, the representation of motion will be more compact if the global (rigid body) motion which applies to all vertices is described independently; this requires 3 parameters for translation and 3 parameters for rotation. Then, the deviation between the vertex point movements and the prediction determined by global motion parameters characterizes the local deformations of the object.

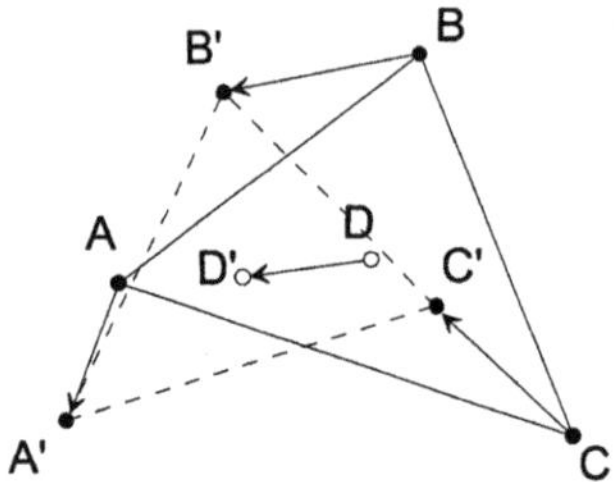

Fig. 7.65. Movement of vertices and surface points in a wireframe model

7.7   Disparity and Depth Analysis

Disparity analysis is also based on estimation of shift vectors by correspondence analysis, where the reference of estimation is the image from another camera simultaneously capturing the same exterior world scene. This is often used for the purpose of depth analysis by stereoscopic camera systems. The relationship of two images acquired by cameras from different positions I and II at the same time can be interpreted as equivalent to the relationship of two frames from a video signal, when the scene is static and the camera translates from position I to position II in the sampling difference time between two subsequent frames. From this observation, it can be expected that no fundamental difference exists between motion analysis and disparity analysis. It must however be observed that in video camera movement translations of the camera position are less frequently performed than rotations e.g. on a tripod. Occlusion effects are more severe and occluded areas can be much wider in cases of camera translation and in stereoscopic acquisition, which is in particular true when an object is at relatively short distance from the cameras. As a consequence, robust methods are needed for disparity estimation, which gives a favor to methods like multi-resolution block matching and feature matching. In stereoscopic analysis, mostly two equal cameras having a pre-defined distance and relative orientation are used to gain depth information about a scene or an object.

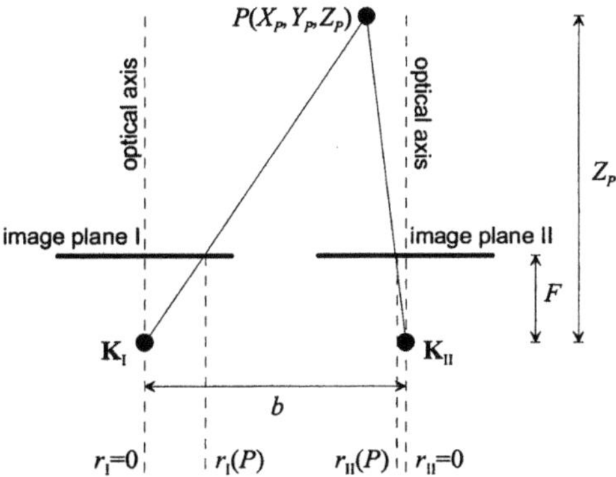

Fig. 7.66. Projection of a point **P** from the exterior world into the image planes of a stereoscopic camera system

The principle of stereoscopic acquisition is shown in Fig. 7.66. Regarding the central projection equation (7.100), the focal point of camera I is at $\mathbf{K_I}=[0,0,0]$, the focal point of camera II at $\mathbf{K_{II}}=[b_1,b_2,0]$. The distances b_1 and b_2 are the *baseline distances* between the cameras in horizontal and vertical directions, respectively. The optical axes are assumed to be parallel here, both cameras 'staring' into the same direction. The centers of the image planes are at $[0,0,F]$ and $[b_1,b_2,F]$, and the image planes are in parallel with the (X,Y)-plane of the global coordinate system. A point $\mathbf{P}=[X_P,Y_P,Z_P]$ from the exterior world is projected into the image plane of camera I at position $\mathbf{r_I}=[r_I(P)\ s_I(P)]^T$, into the plane of camera II at position $\mathbf{r_{II}}=[r_{II}(P)\ s_{II}(P)]^T$ (see Fig. 7.66) [1]. Substituting these values in (7.100) gives

$$X_P = r_I(P)\cdot\frac{Z_P}{F_I} = b_1 + r_{II}(P)\cdot\frac{Z_P}{F_{II}} \quad ; \quad Y_P = s_I(P)\cdot\frac{Z_P}{F_I} = b_2 + s_{II}(P)\cdot\frac{Z_P}{F_{II}}. \qquad (7.200)$$

Assuming that both cameras have identical focal lengths $F=F_I=F_{II}$, the depth distance of the point can be computed as

$$Z_P = F\cdot\frac{b_1}{r_I(P)-r_{II}(P)} = F\cdot\frac{b_2}{s_I(P)-s_{II}(P)}. \qquad (7.201)$$

The relative shift between the coordinates, at which the same point is observed in the image planes of cameras I and II, is the *stereoscopic parallax*. The discrete equivalent thereof, i.e. the shift by a certain number of pixels in the sampled images, is the *disparity* d_h (horizontally) or d_v (vertically); disparities relate to parallax values normalized by the sampling distances R and S:

---

[1] The image plane coordinates refer to an origin which is the center of the image plane of the respective camera.

$$d_h(P) = \frac{r_I(P) - r_{II}(P)}{R} = \frac{b_1 F}{R \cdot Z_P} \quad ; \quad d_v(P) = \frac{s_I(P) - s_{II}(P)}{S} = \frac{b_2 F}{S \cdot Z_P} . \quad (7.202)$$

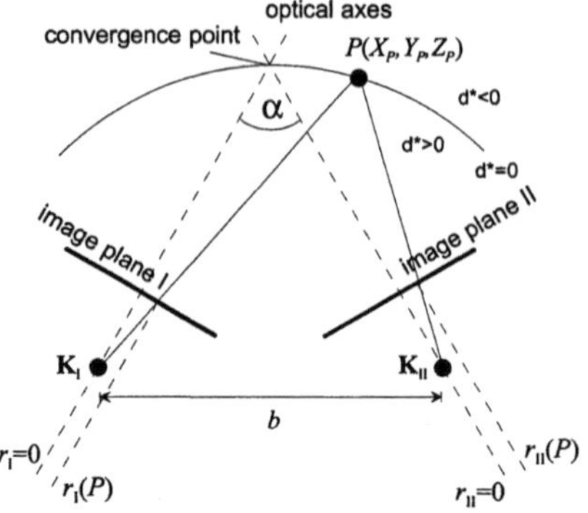

Fig. 7.67. Stereoscopic camera system with convergent camera axes, and position of world coordinate points at which the disparity will be zero

Parallaxes and disparities are reciprocally-dependent on the depth distance of **P**:

$$Z_P = F \cdot \frac{b_1}{r_I(P) - r_{II}(P)} = F \cdot \frac{b_1}{R \cdot d_h(P)} = F \cdot \frac{b_2}{s_I(P) - s_{II}(P)} = F \cdot \frac{b_2}{S \cdot d_v(P)} \quad (7.203)$$

In principle, the depth can be determined either from horizontal or from vertical disparities, but the result will be more reliable if the direction of higher baseline distance is used. Substituting (7.201) into (7.200), the coordinates X_P and Y_P of **P** can be computed as[1]

$$X_P = \frac{b_1 r_I(P)}{r_I(P) - r_{II}(P)} \quad ; \quad Y_P = \frac{b_2 s_I(P)}{s_I(P) - s_{II}(P)} . \quad (7.204)$$

By the given equations, it is possible to determine the position of a point in the 3D world, if the properties of the stereoscopic camera system are exactly known, if the baseline distances are greater than zero in horizontal and vertical dimensions[2] and if the parallax or disparity can be estimated. If only the depth Z_P shall be determined, it is sufficient to estimate either the horizontal or the vertical disparity. In general, (7.200)-(7.204) reflect only a specific case of stereoscopy, where both cameras have *parallel optical axes*, which is a *co-planar* stereoscopic system. The optical axes of the two cameras in this case only converge at infinite distance. In normal human vision and stereoscopic camera acquisition, it is typically desirable that an object at a given distance shall be visible at or near the center of the image

---

[1] Provided that $Z_P < \infty$, where the disparity converges towards a zero value.

[2] If e.g. $b_2 = 0$, the vertical distance Y_P can not be estimated.

planes in both cameras[1]. Then, the optical axes must be convergent. Assume the cameras are at the same vertical position (b_2=0), and the convergence is achieved only by a rotation of the image planes around axes which are parallel with the Y axis. The optical axes are tilted towards each other by a convergence angle α. Only the horizontal disparity $d_h{\equiv}d$ shall be estimated, and the horizontal baseline distance shall be b. Then, the disparity will not be zero for infinite distance, but at the *point of convergence*; it will also be zero for any points **P** which are positioned on an elliptic curve as shown in Fig. 7.67. Rough and simple depth estimation is possible if a correction factor is introduced in (7.203). In a coplanar geometry, the point of convergence at distance Z^0 of the cameras, would cause a disparity

$$R \cdot d_0 = F \cdot \frac{b}{Z^0} \approx F \cdot \alpha, \qquad (7.205)$$

with α measured in radians. Approximately, the case of convergent cameras can be interpreted such that only a constant offset of horizontal disparities applies as compared to the coplanar case, if the depth to be analyzed is not too much different of Z^0, and the point **P** is not too far from the optical axes. Then, for small convergence angles, the following relationship can be given for the disparities $d^*(P)$ measured in the convergent camera system, and the absolute depth position of the respective point **P**[2]:

$$Z_P \approx F \cdot \frac{b}{R \cdot \left(d*(P) + d_0 \right)} \approx \frac{b}{(R/F) \cdot d*(P) + \alpha}. \qquad (7.206)$$

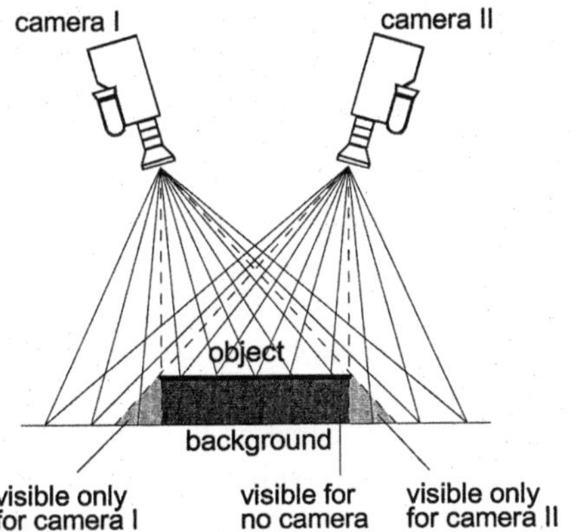

Fig. 7.68. Correspondence problem caused by occlusions in stereoscopic acquisition

---

[1] In human vision, an object which is focused by the eyes will also be projected to the *fovea centralis* of the retina where it appears by highest resolution (cf. sec. 6.1.1).

[2] It is very likely that the human visual system performs similar simplified methods of 'depth estimation'; all the constraints given here hold for the case where an object is focused by the eyes.

For an exact estimation of the 3D world coordinates and for cases where the constraints as given do not apply, a full analysis of the projection mapping into both cameras is necessary, which will be explained in detail in sections 7.7.1 and 7.7.2.

As in motion estimation, the underlying problem in disparity estimation is to determine exact point-wise correspondences between the projected points appearing in the image planes of cameras I and II. This can be problematic at boundaries of foreground objects, where parts of the background will be occluded in the image of one of the cameras: For these positions, it is impossible to determine disparity and distance (see Fig. 7.68). This phenomenon also applies to human vision, however the human subject will often utilize semantic knowledge to make a guess about the distance of these areas, e.g. by extrapolating from the depth of surrounding areas which are visible to both eyes[1].

The position at which a corresponding point can be found in the other camera image is not arbitrary, if the exact configuration of cameras is known as shown in Fig. 7.69. If $\mathbf{P}$ appears in image plane I at $\mathbf{p_I}=[r_I(\mathbf{P}),s_I(\mathbf{P}),F_I]^T$, it must be positioned in the 3D world on the line $(\mathbf{K_I p_I})$, which connects the focal point $\mathbf{K_I}$ with $\mathbf{p_I}$. This line is one of the 'rays of vision' of camera I, it is however not known at which distance (Z_P) $\mathbf{P}$ is positioned on the ray. The projection of $\mathbf{P}$ into image plane II must originate from some point on $(\mathbf{K_I p_I})$, and will be directed towards the focal point $\mathbf{K_{II}}$ of camera II. The *epipolar plane* is defined as a plane spanned by $(\mathbf{K_I p_I})$ and the base line $(\mathbf{K_I K_{II}})$ between the projection centers. All possible projection positions of $\mathbf{P}$ in image plane II must be on the *epipolar line*, which is the intersection between the epipolar plane and image plane II. By reversing this principle, an epipolar line in image plane I is defined. Corresponding points need only to be searched along the epipolar lines, which highly simplifies disparity estimation for the case where the stereoscopic camera configuration is known. The epipolar plane and also its intersection with the respective image plane (the epipolar line) depend on the position of $\mathbf{P}$. In correspondence estimation, for any point in the left plane, a unique epipolar line can be determined for the right plane and vice versa. In case of co-planar cameras with $b_2=0$, the epipolar lines are horizontal, i.e. for a given point in row n of the left image, the epipolar line is the row n of the right image and vice versa. Then, correspondences between the two images can exclusively be found in identical rows; consequently, disparity estimation for co-planar configurations is extremely simple. A very accurate adjustment of cameras is however required. By application of *camera calibration* [ZHANG, XU 1996], it is possible to determine the exact camera configuration which allows to compensate deviations from the perfect co-planar adjustment[2]. The following section gives a more gener-

---

[1] In general, for depth perception in human vision, the *stereoscopic parallax* is of higher importance only in the relatively near range ($<10m$). Other effects, in particular the *motion parallax* (where the viewpoint is slightly changed) play a similar important role.

[2] Camera calibration requires availability of 3D information assigned to a number of characteristic points from the 3D scene, which can uniquely be identified in the image plane;

alized view of projection relationships in stereoscopic and other multiple-view camera systems.

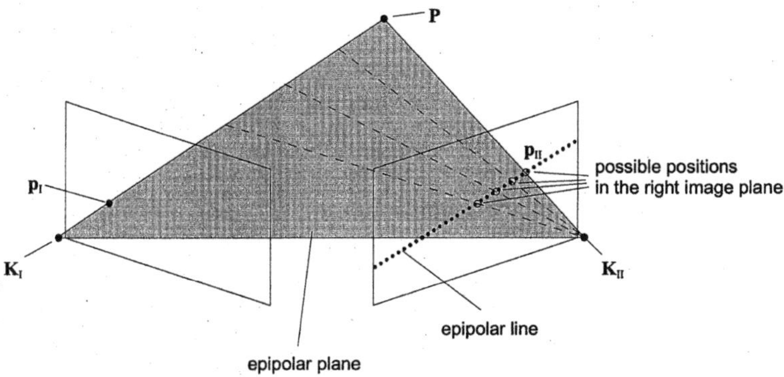

Fig. 7.69. Epipolar geometry in a convergent stereoscopic camera system

7.7.1   Central Projection in Stereoscopic and Multiple-camera Systems

In a system of two cameras I und II, the world coordinate system shall have its origin at the focal point $\mathbf{K_I}$, and the optical axis of camera I shall be the Z axis. The mapping of a point $\mathbf{P}$ from the 3D world into the point $\mathbf{p_I}$ of the first camera's image plane can according to (7.100) be described as

$$\mathbf{P} = \begin{bmatrix} X_P \\ Y_P \\ Z_P \end{bmatrix} \Rightarrow \mathbf{p}_I = \begin{bmatrix} r_1(P) \\ s_1(P) \end{bmatrix} = \begin{bmatrix} F_I \cdot X_P / Z_P \\ F_I \cdot Y_P / Z_P \end{bmatrix}. \tag{7.207}$$

This nonlinear relationship can be linearized if a homogeneous coordinate mapping is used. Homogeneous coordinates of a vector in a K-dimensional space are extended by one additional dimension, which expresses a 'scaling factor' for the original components:

$$\mathbf{p} = \begin{bmatrix} p_1 \\ p_2 \\ \vdots \\ p_K \end{bmatrix} \Leftrightarrow \widehat{\mathbf{p}} = \begin{bmatrix} \widehat{p}_1 \\ \widehat{p}_2 \\ \vdots \\ \widehat{p}_K \\ \widehat{p}_{K+1} \end{bmatrix} = \begin{bmatrix} \lambda \cdot \mathbf{p} \\ \lambda \end{bmatrix} = \lambda \cdot \begin{bmatrix} \mathbf{p} \\ 1 \end{bmatrix} \Rightarrow \mathbf{p} = \frac{1}{\widehat{p}_{K+1}} \cdot \begin{bmatrix} \widehat{p}_1 \\ \widehat{p}_2 \\ \vdots \\ \widehat{p}_n \end{bmatrix}. \tag{7.208}$$

Using homogeneous coordinates on both sides, the mapping of a point into plane I (7.207) can be re-written as follows:

---

the problem simplifies, if groups of reference points are found on planar surfaces which allow unique mapping into all three coordinate axis orientations of the 3D space.

$$\begin{bmatrix} \lambda_1 \cdot r_{\mathrm{I}}(P) \\ \lambda_1 \cdot s_{\mathrm{I}}(P) \\ \lambda_1 \end{bmatrix} = \underbrace{\begin{bmatrix} F_{\mathrm{I}} & 0 & 0 & 0 \\ 0 & F_{\mathrm{I}} & 0 & 0 \\ 0 & 0 & 1 & 0 \end{bmatrix}}_{\mathbf{L}} \cdot \begin{bmatrix} \lambda_2 \cdot X_P \\ \lambda_2 \cdot Y_P \\ \lambda_2 \cdot Z_P \\ \lambda_2 \end{bmatrix}$$

$$\Leftrightarrow \lambda_1 \cdot \hat{\mathbf{p}}_{\mathrm{I}} = \mathbf{L} \cdot \lambda_2 \cdot \hat{\mathbf{P}} \Leftrightarrow \lambda \cdot \hat{\mathbf{p}}_{\mathrm{I}} = \mathbf{L} \cdot \hat{\mathbf{P}} \Leftrightarrow \lambda \cdot \begin{bmatrix} r_{\mathrm{I}}(P) \\ s_{\mathrm{I}}(P) \\ 1 \end{bmatrix} = \mathbf{L} \cdot \begin{bmatrix} X_P \\ Y_P \\ Z_P \\ 1 \end{bmatrix}, \quad \lambda = \frac{\lambda_1}{\lambda_2}.$$

$$(7.209)$$

It is assumed that the origin of the image-plane coordinate system is at the center of the image plane, the system shall be equally scaled and axes (r,s) have the same orientations as the (X,Y) plane of the world coordinate system. To allow usage of discrete coordinates on the image plane, it is necessary to shift the origin into the top-left corner as for common addressing of image matrices (Fig. 1.6). Further, horizontal and vertical sampling distances R and S are included in the formulation. This is achieved by applying a coordinate transform $\mathbf{A}$, which describes a rectangular sampling[1] and a translational shift; in the new coordinate system, the center of the image plane is at position (m_0,n_0).

$$\mathbf{A} = \begin{bmatrix} \mathbf{D}_{rect}^{-1} & \mathbf{t} \\ \mathbf{0} & 1 \end{bmatrix} = \begin{bmatrix} 1/R & 0 & m_0 \\ 0 & 1/S & n_0 \\ 0 & 0 & 1 \end{bmatrix}.$$

$$(7.210)$$

For camera I, the following discrete image matrix positions reflect the projection of $\mathbf{P}$:

$$\lambda \cdot \hat{\mathbf{p}}_{\mathrm{I}} = \mathbf{A} \cdot \mathbf{L} \cdot \hat{\mathbf{P}} = \begin{bmatrix} 1/R^{(\mathrm{I})} & 0 & m_0^{(\mathrm{I})} \\ 0 & 1/S^{(\mathrm{I})} & n_0^{(\mathrm{I})} \\ 0 & 0 & 1 \end{bmatrix} \cdot \begin{bmatrix} F_{\mathrm{I}} & 0 & 0 & 0 \\ 0 & F_{\mathrm{I}} & 0 & 0 \\ 0 & 0 & 1 & 0 \end{bmatrix} \cdot \hat{\mathbf{P}}$$

$$= \underbrace{\begin{bmatrix} F_{\mathrm{I}}/R^{(\mathrm{I})} & 0 & m_0^{(\mathrm{I})} \\ 0 & F_{\mathrm{I}}/S^{(\mathrm{I})} & n_0^{(\mathrm{I})} \\ 0 & 0 & 1 \end{bmatrix}}_{\mathbf{H}^{(\mathrm{I})}} \cdot \underbrace{\begin{bmatrix} 1 & 0 & 0 & 0 \\ 0 & 1 & 0 & 0 \\ 0 & 0 & 1 & 0 \end{bmatrix}}_{\mathbf{L_N}} \cdot \hat{\mathbf{P}}.$$

$$(7.211)$$

The matrix $\mathbf{H}$ holds the *intrinsic camera parameters*, $\mathbf{L_N}$ is the normalized projection matrix. The scaling factor λ can in principle be chosen arbitrarily, and is not explicitly used in the subsequent deductions.

---

[1] Other sampling matrix definitions are possible.

The mapping process (7.211) shall now be described separately for the two cameras. Herein, it must be observed that the origin of the world coordinate system was defined by relationship with the orientation and focal point $\mathbf{K}_I$ of the first camera; regarding camera II, a 3D translation shift vector $\mathbf{t}$ and a rotation matrix $\mathbf{R}$ express the position and orientation. (7.102) can be mapped into homogeneous coordinates by

$$\mathbf{P}' = \mathbf{R} \cdot \mathbf{P} + \mathbf{t} \ \Rightarrow\ \hat{\mathbf{P}}' = \underbrace{\begin{bmatrix} \mathbf{R} & \mathbf{t} \\ \mathbf{0} & 1 \end{bmatrix}}_{\mathbf{D}} \cdot \hat{\mathbf{P}} . \tag{7.212}$$

The matrix $\mathbf{D}$ contains the *extrinsic parameters* of camera system II[1]; (7.212) expresses at which position $\mathbf{P}'$ of the exterior world a point must be located, if it shall be visible in the image plane of camera II at position $\mathbf{p}_I$. By reversion of this principle (camera is interpreted to be translated and rotated, while the point is fixed), it is straightforward to determine the actual projected position $\mathbf{p}_{II}$ which exactly corresponds to $\mathbf{p}_I$:

$$\hat{\mathbf{p}}_I = \mathbf{H}^{(I)} \cdot \mathbf{L}_N \cdot \hat{\mathbf{P}} \ \Rightarrow\ \hat{\mathbf{p}}_{II} = \mathbf{H}^{(II)} \cdot \mathbf{L}_N \cdot \mathbf{D} \cdot \hat{\mathbf{P}} = \mathbf{H}^{(II)} \cdot [\mathbf{R}\ \ \mathbf{t}] \cdot \hat{\mathbf{P}} . \tag{7.213}$$

7.7.2   Epipolar Geometry for arbitrary Camera Configurations

As shown in Fig. 7.69, the point $\mathbf{P}$ and the focal points $\mathbf{K}_I$, $\mathbf{K}_{II}$ span the epipolar plane Π_P, on which the projected points $\mathbf{p}_I$, $\mathbf{p}_{II}$ and the rays of projection are found. Π_P intersects the camera planes I and II by the epipolar lines $\mathbf{l}_P^{(I)}$ and $\mathbf{l}_P^{(II)}$. Epipolar planes Π_{P_i} which are spanned by points $\mathbf{P}_i$ and the focal points $\mathbf{K}_I$, $\mathbf{K}_{II}$, establish a 'set of planes'[2]. All members of this set intersect at the baseline connection $(\mathbf{K}_I\mathbf{K}_{II})$ between the two focal points of the cameras (see Fig. 7.70).

The intersecting lines of epipolar planes and image planes also establish a set of epipolar lines. All epipolar planes intersect in $(\mathbf{K}_I\mathbf{K}_{II})$, and further all epipolar lines must be on the respective image plane. Hence, it can be concluded that all epipolar lines related to an image plane will intersect in just one point, which is the intersection of $(\mathbf{K}_I\mathbf{K}_{II})$ with the image plane or its infinite extension; these unique points $\mathbf{e}_I$ for camera I and $\mathbf{e}_{II}$ for camera II are the *epipoles*[3]. The epipoles can be interpreted as the projection of the focal point $\mathbf{K}_I$ into image plane II (giving $\mathbf{e}_{II}$)

---

[1] The stereoscopic camera configuration is now fully described by the 12 extrinsic parameters and each 5 intrinsic parameters (7.211) for cameras I and II. If the world coordinate reference is *not* the coordinate system of camera I, also the extrinsic parameters of camera I must be given. For any additional camera, another 12 extrinsic and 5 intrinsic parameters must be provided.

[2] The number of epipolar planes is in fact much lower than the number of possible points $\mathbf{P}_i$, as many points will 'share' each epipolar plane.

[3] In case of a co-planar stereoscopic camera system, the epipoles are at an infinite distance, and all epipolar lines are in parallel.

and vice versa. The image plane coordinates of the two epipoles are determined in the sequel; when the epipoles are known, the computation of the epipolar line parameters is simple.

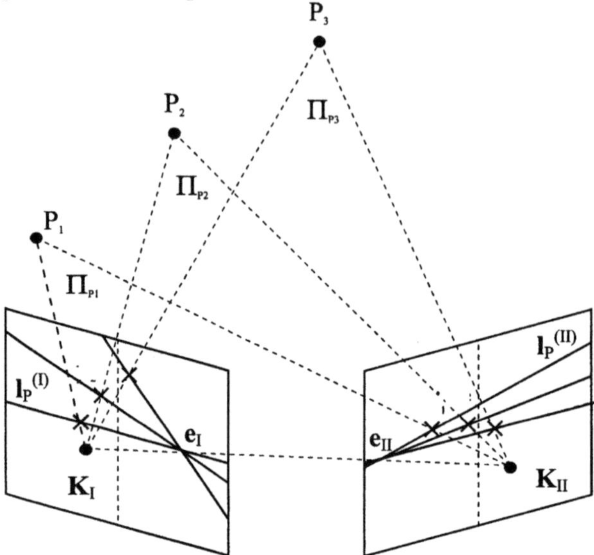

Fig. 7.70. Set of epipolar planes, epipolar lines and epipoles (source: PASEWALDT)

The focal point $\mathbf{K}_I$ of the first camera has homogeneous coordinates $[0\ 0\ 0\ 1]^T$ in its own coordinate system which is also the world coordinate system. Substituted in (7.213), the homogeneous image plane coordinates of the epipole $\mathbf{e}_{II}$ as projection of $\mathbf{K}_I$ into image plane II result by

$$\hat{\mathbf{e}}_{II} = \mathbf{H}^{(II)} \cdot [\mathbf{R}\quad \mathbf{t}] \cdot \begin{bmatrix} 0 \\ 0 \\ 0 \\ 1 \end{bmatrix} = \mathbf{H}^{(II)} \cdot \mathbf{t} = \begin{bmatrix} \mathbf{h}_1^T \\ \mathbf{h}_2^T \\ \mathbf{h}_3^T \end{bmatrix} \cdot \mathbf{t} \Rightarrow \mathbf{e}_{II} = \frac{1}{\mathbf{h}_3^T \cdot \mathbf{t}} \cdot \begin{bmatrix} \mathbf{h}_1^T \cdot \mathbf{t} \\ \mathbf{h}_2^T \cdot \mathbf{t} \end{bmatrix}. \tag{7.214}$$

The coordinates $\mathbf{K}_{II}$ in the (non-homogeneous) coordinate system of the second camera are $[0\ 0\ 0]^T$. The coordinate transformation (7.212) maps $\mathbf{K}_{II}$ into the system of camera I as

$$\mathbf{0} = \mathbf{R} \cdot \mathbf{K}_{II} + \mathbf{t} \Rightarrow \mathbf{K}_{II} = -\mathbf{R}^{-1} \cdot \mathbf{t}. \tag{7.215}$$

$\mathbf{R}$ describes a rotation along three orthogonal coordinate axes without scaling, which means that $\mathbf{R}$ represents an orthonormal and real-valued transform matrix, for which $\mathbf{R}^{-1} = \mathbf{R}^T$. Using (7.213), the epipole on image plane I is

$$\hat{\mathbf{e}}_I = \mathbf{H}^{(I)} \cdot \begin{bmatrix} \mathbf{I} & \mathbf{0} \end{bmatrix} \cdot \begin{bmatrix} -\mathbf{R}^T \cdot \mathbf{t} \\ 1 \end{bmatrix} = -\mathbf{H}^{(I)} \cdot \mathbf{R}^T \cdot \mathbf{t} = -\begin{bmatrix} \mathbf{h}_1^T \\ \mathbf{h}_2^T \\ \mathbf{h}_3^T \end{bmatrix} \cdot \mathbf{R}^T \cdot \mathbf{t}$$

$$\Rightarrow \mathbf{e}_I = \frac{1}{\mathbf{h}_3^T \cdot \mathbf{R}^T \cdot \mathbf{t}} \cdot \begin{bmatrix} \mathbf{h}_1^T \cdot \mathbf{R}^T \cdot \mathbf{t} \\ \mathbf{h}_2^T \cdot \mathbf{R}^T \cdot \mathbf{t} \end{bmatrix}.$$

(7.216)

A point $\mathbf{p}_I$ in image plane I corresponds with $\mathbf{p}_{II}$ in plane II which must be on the epipolar line $\mathbf{l}_P^{(II)}$. This can be described by the *epipolar equation* as follows. As the 3D world point $\mathbf{P}$ must be on the projection ray $(\mathbf{K}_I\mathbf{p}_I)$ (see Fig. 7.69), its coordinates can be determined from the coordinates of $\mathbf{p}_I$ except for a scaling factor c. The non-uniqueness follows from the arbitrary selection of the value λ when mapping into the homogeneous coordinate system is made, such that the reverse mapping is not unique. From (7.207) and (7.208),

$$\mathbf{P} \cong \begin{bmatrix} \mathbf{H}^{(I)} \end{bmatrix}^{-1} \cdot \hat{\mathbf{p}}_I \Rightarrow \hat{\mathbf{P}} = \begin{bmatrix} X_P \\ Y_P \\ Z_P \\ 1 \end{bmatrix} \cong \begin{bmatrix} \begin{bmatrix} \mathbf{H}^{(I)} \end{bmatrix}^{-1} \cdot \hat{\mathbf{p}}_I \\ c \end{bmatrix} = \begin{bmatrix} \begin{bmatrix} \mathbf{H}^{(I)} \end{bmatrix}^{-1} \cdot \begin{bmatrix} m_I(P) \\ n_I(P) \\ 1 \end{bmatrix} \\ c \end{bmatrix}$$

(7.217)

$$= \begin{bmatrix} r_I(P) \\ s_I(P) \\ 1 \\ c \end{bmatrix} = \frac{1}{Z_P} \cdot \begin{bmatrix} X_P \\ Y_P \\ Z_P \\ Z_P \cdot c \end{bmatrix} \Rightarrow c = 1/Z_P.$$

If (7.217) is substituted into the equations related to camera II from (7.213), the following relationship (epipolar equation) results:

$$\hat{\mathbf{p}}_{II} \cong \mathbf{H}^{(II)} \cdot \begin{bmatrix} \mathbf{R} & \mathbf{t} \end{bmatrix} \cdot \begin{bmatrix} \begin{bmatrix} \mathbf{H}^{(I)} \end{bmatrix}^{-1} \cdot \hat{\mathbf{p}}_I \\ c \end{bmatrix} = \mathbf{H}^{(II)} \cdot \mathbf{R} \cdot \begin{bmatrix} \mathbf{H}^{(I)} \end{bmatrix}^{-1} \cdot \hat{\mathbf{p}}_I + c \cdot \mathbf{H}^{(II)} \cdot \mathbf{t}.$$

(7.218)

It can be shown that the same factor c also applies for the reverse mapping $\mathbf{p}_{II} \rightarrow \mathbf{p}_I$. Therefore, if the correspondence between the positions of mappings $\mathbf{p}_I$ and $\mathbf{p}_{II}$ of $\mathbf{P}$ into the two image planes can be found (e.g. by disparity estimation), the factor c can be determined, and subsequently, by substituting into (7.217), the full position of $\mathbf{P}$ in the world coordinate system can be computed[1]. To achieve this, all intrinsic and extrinsic parameters of the camera system must be known and used in $\mathbf{H}^{(I)}$, $\mathbf{H}^{(II)}$, $\mathbf{R}$ and $\mathbf{t}$. (7.218) establishes a very general relationship between the image plane mappings of arbitrarily positioned cameras. It is straightforward to extend this to multi-camera configurations, where e.g. correspondences between points in

---

[1] Of course, this is not possible if no correspondence can be established between $\mathbf{p}_I$ and $\mathbf{p}_{II}$.

more than two camera planes are sought. The relationship is further applicable in
view point synthesis, e.g. to determine the mapping of a point into the image plane
of a virtual camera with desired position and orientation (cf. sec. 15.3 and 15.5). In
typical stereoscopic camera configurations with almost identical intrinsic parameters for two cameras of same type, $\mathbf{K}_I$ and $\mathbf{K}_{II}$ differ only by coordinate X, and
vector $\mathbf{t}$ equals zero except for the horizontal component, such that significant
simplifications are possible.

7.8   Mosaics

A mosaic is a large panoramic view which is made up of patches of image content
as appearing over a certain period of time in a video sequence, typically performing global motion. It can also be constructed from several still images shot under
different view angles with partially overlapping areas. A prerequisite condition
will be that no local motion occurs in the frames, or that at least locally moving
objects are excluded when the mosaic is generated. Mosaics are usually defined
using parametric motion models that describe the geometric transform (warping)
of the single frames into a common reference coordinate system. For the case of
mosaics generated from video sequences, it is also useful to retain the temporal
reference information of each of the contributing frames along with the sets of
motion parameters, which allows to use any spacing of frames in generating the
mosaic. This will in particular be useful if single frames shall be reconstructed
from the mosaic, by performing back projection into the image plane coordinates
for the original temporal position.

If the mosaic is defined on a plane, the key step in constructing is parametric
global motion estimation, which can be performed as described in sec. 7.6.5. Once
the global motion is known, the frames can be warped and blended into one common mosaic frame, which can be interpreted and further processed as a larger-size
image. For more general cases, mosaics can also be defined on cylindrical, spherical and adaptive manifold coordinate systems [PELEG ET AL. 2000]. However, these
transformations are mathematically more complex and require exact knowledge
about the intrinsic and extrinsic camera parameters related to each view [MCMILLAN,
BISHOP 1995]. It is important to note in this context that high-quality mosaics can
only be generated if no significant occlusion occurs between the different frames
of the video sequence or the different images from a series of stills. This can
strictly only be guaranteed if the relevant scene content is at far distance (e.g.
panoramic view over a city or landscape), or if the camera is strictly rotated
around the focal point. In both cases, when the pin hole camera model is used and
the central projection equation (1.1) applies, perspective mapping (7.109) is the
perfect warping model to generate the mosaic on a plane [SMOLIC, SIKORA, OHM 1999].
If however the camera acquisition bears geometric distortions such as lens distor-

tions, it can be advantageous to use higher-order parametric models like (7.111) to obtain a better global mapping in case of camera movements.

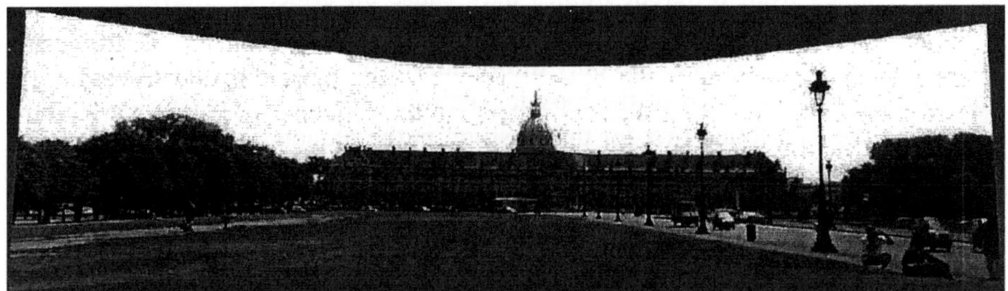

Fig. 7.71 Example of a mosaic image generated from a video sequence which was shot by a panning camera (source: SMOLIC)

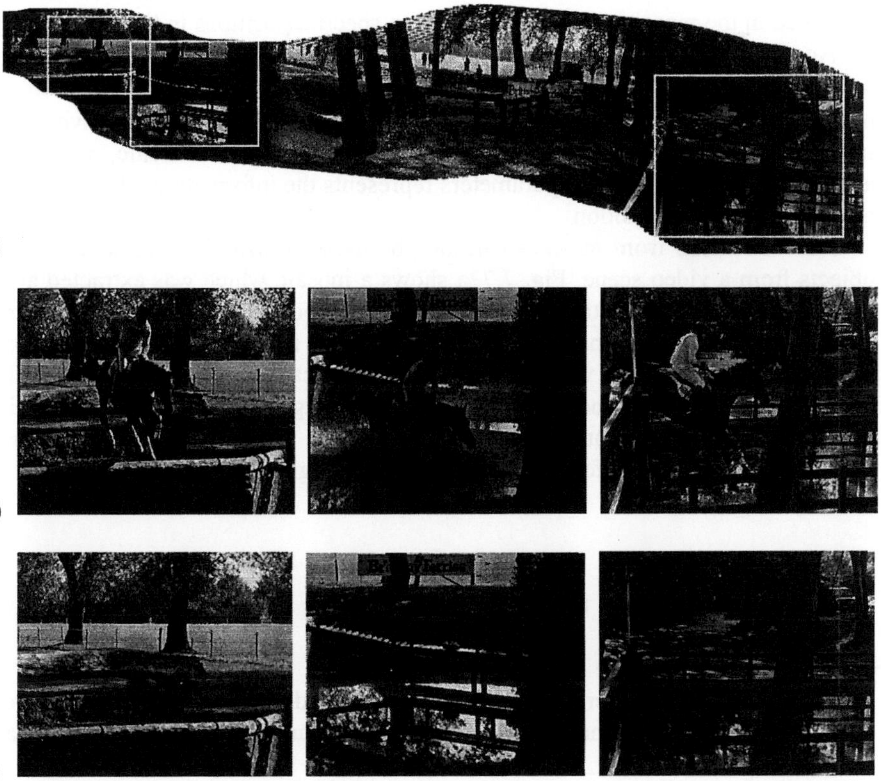

a)

b)

c)

Fig. 7.72. a Mosaic extracted from a video sequence **b** Three original frames from the sequence **c** Background-only frames reconstructed from the mosaic (source : SMOLIC)

Various schemes can be used for integrating and blending of the aligned frames in the construction of the mosaic. A regular or weighted temporal average of the intensity values from different frames may cause an effect of blurring; alternatively, a temporal (optionally weighted) median filtering of the intensity values can be applied. A third way is to use only information from nor more than one frame at one specific pixel position of the mosaic, such that the mosaic is constructed e.g. by pasting the data found first for this position in the sequence of pictures. Finally, if the motion warping is done with sub-pixel accuracy, it is also possible to generate mosaics which have *higher resolution* than the original frames of the sequence [SMOLIC, WIEGAND 2001]. An example of a mosaic generated from a sequence of video with slowly-panning camera is shown in Fig. 7.71.

Usage of mosaics. As mosaics contain the condensed image information either from a video sequence or from a series of still images, they can be used right away to extract any feature data that relate to an *entire scene*; in case where the motion information and temporal reference information are evaluated, localized features contained in the mosaic can also be related to specific positions in time.

Another usage of mosaics is for the purpose of *frame reconstruction*. This is done by back projection, inverting the motion equations and warping from the mosaic into the reference coordinate systems of the output images[1]. Conventional image compression techniques can be applied to represent the mosaic, which then along with the global motion parameters represents the information which is *static*, except for the global motion.

Reconstruction from mosaics can also be used to artificially remove moving objects from a video scene. Fig. 7.72a shows a mosaic which was extracted automatically from a scene with a camera panning and zooming, where the white boxes indicate the positions of three original frames shown in Fig. 7.72b. The horse rider in the foreground is not visible in the mosaic, because only areas are included which match the global motion parameter model. Fig. 7.72c shows the respective background parts, as reconstructed from the mosaic using the inverse global motion mapping parameters for back projection into original positions.

7.9   Face Detection and Description

Human faces are an important content of images and video sequences with direct relationship to semantic meaning. It is firstly important to analyze *where* faces are

---

[1] If this is done for the purpose of video encoding, the mosaic which establishes a large image memory containing pieces from several frames is also denoted as *sprite*. Actually, this denotation originated from computer games, where it was used for a large (graphical) background image from which only a projection of a smaller area is shown at each moment. The term 'sprite' was later taken up in the MPEG-4 standard.

in an image and eventually perform further processing aiming at *recognition of persons* or *analysis of facial actions*.

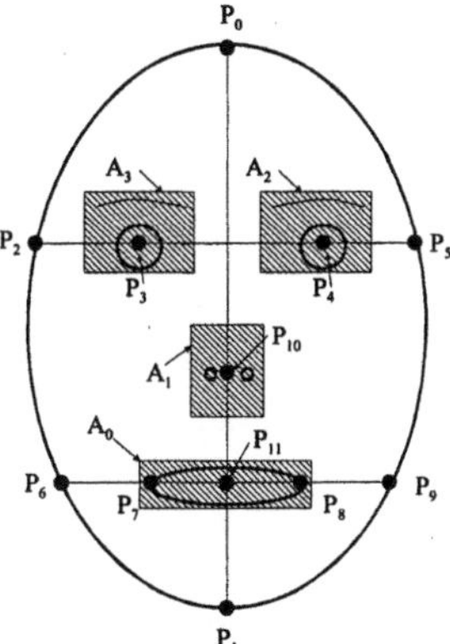

Fig. 7.73. Positions of feature points and feature areas in a simple 2D face structure model for frontal-view analysis.

For face detection, color is an important feature; colors of human skin are quite characteristic and can be modeled by vector Gaussian distributions in the color space, such that statistical criteria can be applied for confidence testing and classification (cf. chap. 9). It is however important to distinguish between face areas and other areas of same color, such as hands etc. This can be done by incorporating either *face structure models* or *face texture models*. Face texture models are also frequently used for face recognition as described below. A simple 2D face structure model is shown in Fig. 7.73. Typically, morphological operations are first applied to the patch areas which fulfill the color model, to fill holes, remove irregularities from the outer shape etc. Pre-selection of candidate face regions is made based on primitive form parameters such as (7.95)-(7.98). Edge detection is then applied to the texture in the candidate areas, where it is analyzed if e.g. characteristic edge shapes are found by the expected orientations as specified by feature points within typical feature areas like eyes, nose, mouth etc.

Face texture models. Face texture models are often used for face identification. As a first step, it is necessary to align a face by geometric mapping, such that eyes, mouth etc. are at the same positions as assumed by the texture model; such alignment can again be made on basis of a structure model as shown in Fig. 7.73. It is further necessary to compensate effects of different illumination conditions. For the case of frontal-view analysis, the face texture is typically defined within an elliptical shape. One of the most common approaches for face texture description is *eigenface analysis* [TURK, PENTLAND 1991]. The face eigenvectors are constructed from autocovariance analysis of a model face and are basis images of the optimum transform of the face texture model (cf. sec. 4.3.2). An example of first 10 eigenfaces extracted from a mean face texture model is shown in Fig. 7.74.

Fig. 7.74. 'Mean face' texture model computed from a set of geometrically adjusted face images and its first 10 eigenfaces (source: MENSER)

Eigenface texture models can either be generated for individual persons or as an average over the face texture of multiple persons; the first approach is useful for face identification, where it can be tested if a face of an acquired image fits with the pre-defined eigenfaces of a specific person. Alternative basis functions which are used for face texture analysis are Gabor wavelets [BUHMANN ET AL. 1992] (sec. 7.2.2) or bases from linear discriminant analysis (cf. sec. 9.2.5 and 17.6.3).

Facial action analysis. Facial action analysis is important for recognition of facial gestures and expressions, but can also be used for synthesis of graphical face models. Again, pre-defined models must be aligned to the captured images, e.g. by matching the face acquired from a camera to the correct position of a head model, such that the model can accordingly be adapted. Alternatively, 3D head models based on wireframe constructions (sec. 7.4.6) are used, which must then be matched by projection with an actual face image. Two examples of such models are shown in Fig. 7.75, which put explicit emphasis on important areas such as eyes or mouth. It can be assumed that a person's head is moved within the 3D world as an entire unit (globally). If the head is interpreted as *approximately rigid*, 6 parameters are sufficient to characterize the global head motion, e.g. if a person is nodding the head. The visible surface of the head is subdivided into smaller segments or patches, which can continuously be aligned with the actually observed image by properties of color, texture etc. The movements of some patches will show typical deviations from the global motion e.g. by mimic expressions of the face, by opening of the mouth or the eyes. As the patches of the model are pre-defined for specific areas of the face, a semantic relationship can be assigned, and an analysis of mimic expressions becomes possible.

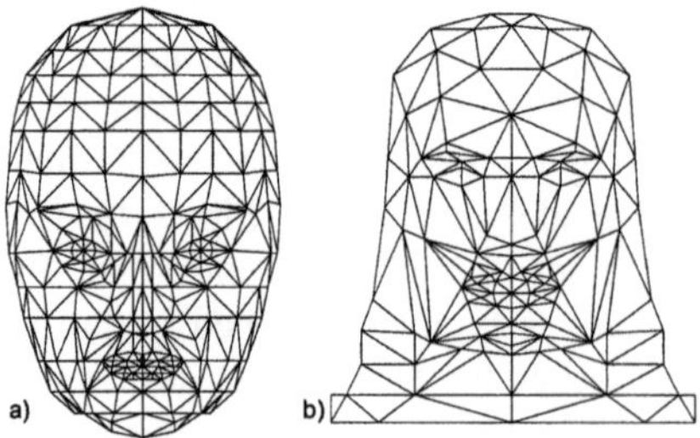

Fig. 7.75. Prototypic head models **a** as proposed by AIZAWA et. al.
b "Candide" after [RYDFALK 1987]

Most important mimic expressions can be related to movements of eyes, eyebrows and eyelids, of mouth and lips, and of the nose[1]. Fig. 7.76a shows face control points, which can be used to characterize the most important mimic expressions. In addition, Fig. 7.76b shows examples how typical lip configurations can be expressed by a very low number of control points. It is not even necessary to regard each of these control points independently, as usually symmetry is present, and groups of control points will be moved simultaneously in typical mimic expressions. Such groupings of control point movements are denoted as *facial action units* (FAU). Accurate description of mimic expressions can be achieved by definition of a finite set of FAUs[2], which can however hardly be analyzed by automatic systems today. These FAUs can further be structured and combined into *visemes*[3] which are rudimentary elements of *facial expressions*. The most relevant facial expressions are related to the moods of surprise, disgust, anger, fear, happiness and sadness.

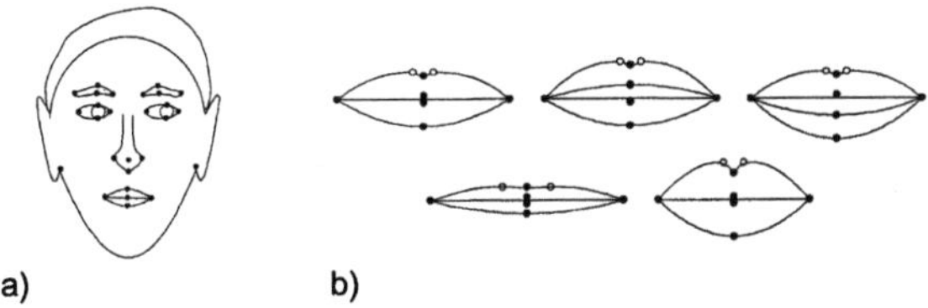

a)   b)

Fig. 7.76. a Control points for mimic characterization **b** Typical lip movements [after AIZAWA et. al.]

7.10 Audio Signal Features

In definition of reasonable features for characterization of audible signals, it is first useful to categorize the signals into classes such as speech, acoustic or sound events, music, single instruments etc. Certain basic features are however applicable to any type of these signals. Important aspects for audio signal categorization are also related to room acoustics, where a detailed treatise would however go beyond the scope of this book. In the following sections, an overview is given about the most important features for description of audible signals and related feature-extraction methods. For audio signals, it is in some cases straightforward to define features which are quite analogous to subjective perception, since the properties of

---

[1] Observe this is the same as for graphical characters drawn in comic books or movies.

[2] For facial description in the MPEG-4 standard, 66 FAUs are defined.

[3] The term *viseme* is borrowed from the denotation of *phonemes* which are often used in speech description and recognition.

the human hearing system are well-known. In other cases, such as speech recognition, a direct relationship with semantic meaning can be established; automatic extraction methods exist for such applications.

7.10.1 Basic Features

Time/frequency descriptions (spectral properties in dependency of time) are very important for audio signal analysis. By *spectrograms*, basic features like pitch frequency, harmonic properties, changes or fluctuations of tone frequency, and spectral amplitude distributions over time can be analyzed[1]. For this purpose, in principle any frequency transform method can be applied. To achieve a continuous change of the spectrum over time, it is advantageous to employ a transform with overlapping windows (see Fig. 7.77). This is equivalent to a short-time Fourier transform (cf. sec. 4.4.4), where now the DFT is used over a finite window of samples. To avoid errors in the spectrum estimation due to the finite analysis interval, windowing functions like e.g. the Hamming window (5.45) are applied prior to DFT computation.

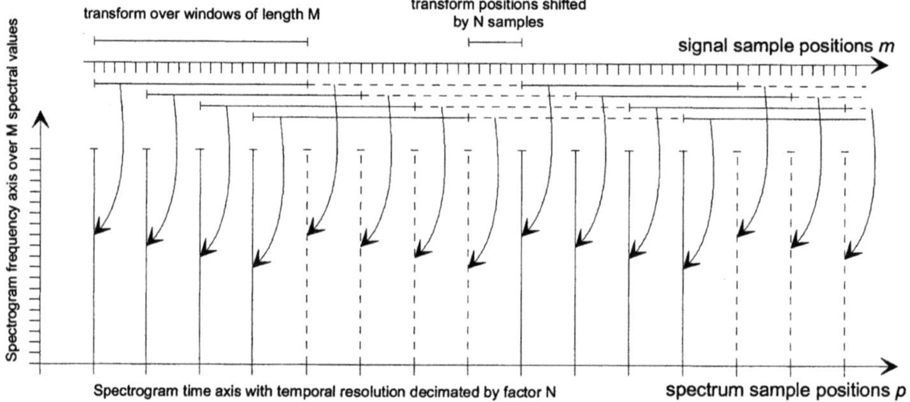

Fig. 7.77. Generation of a spectrogram representation for audio signals

Audio signals are often harmonic by nature, such that the spectrum can be interpreted by a pitch frequency and harmonics, which altogether establish a line spectrum. The analysis of pitch is an important task in audio signal analysis, applicable both to speech signals and music signals. If the pitch shall be detected from the spectrogram, the accuracy is however dependent of the DFT analysis length[2].

---

[1] In fact, spectrograms are often visualized by 2D images, where the horizontal position relates to the time axis, the vertical position to the frequency axis.

[2] Ideal spectral lines are often spread over a range of neighbored frequencies when spectral analysis is performed over finite segments. This occurs in particular when the analysis

Other methods of pitch analysis are based on autocorrelation analysis (where side maxima of the autocorrelation function are found at the pitch period and its multiples) or computation of the *cepstrum*[1]

$$c(m) = \sum_{u=0}^{U-1} \log\left[1 + \left|\hat{X}(u)\right|\right] \cdot e^{j2\pi\frac{mu}{U}} \tag{7.219}$$

In the signal $c(m)$, a pulse train appears for periodic signals which by the period between the pulses relates to the pitch frequency. In particular for speech signals, also the linear-prediction parameters can be utilized for raw pitch analysis (see sec. 4.2.2 and 14.1). The z-transform of an LPC synthesis filter exhibits poles, which in case of harmonic signals will be positioned near the unit circle in the complex z plane, at angular orientations which correspond to the frequencies of the pitch and its harmonics.

7.10.2   Speech Signal Analysis

Speech signals can be decomposed into voiced and unvoiced segments (see sec. 14.1). To classify the signal into its voiced/unvoiced parts, cepstrum analysis can be applied as well: In unvoiced segments, the cepstrum possesses just one peak at $n=0$, while for voiced segments a pulse train appears. One important application of speech analysis is *speech recognition*. It is however not possible to identify spoken letters directly by the sound of speech, as the context is important to determine how a letter is pronounced. An indirect mapping between speech sound and letters is enabled by the meta-unit of *phonemes*, which are combinations of letters in a certain pronounciation. Definition of phonemes highly depends on the language, typically the number of phonemes will be significantly higher (10-15x) than the number of letters.

Properties both of voiced and unvoiced phonemes can also be determined from the spectrum. In voiced speech, spectral distribution into a low number of spectral lines is characteristic. The dominant harmonics are called *formants*, where usually the largest three formants are sufficient to distinguish the different voiced speech sounds. In unvoiced speech, criteria like spectral maximum and roll-off characteristics (decay towards higher frequencies) also play an important role. In addition temporal properties like the amplitude envelope are important, which e.g. allows distinguishing between plosive and whizzing sounds.

To recognize speech reliably, a simple concatenation of recognized phonemes is not sufficient, either. This is in particular true, as the features of phonemes are subject to variations, while some phonemes have a high similarity (e.g. sounds

---

block length of the DFT is not a multiple of the pitch period. Only if the latter condition is perfectly fulfilled, a true line spectrum can be detected by DFT analysis.

[1] This artificial word denotes the inverse Fourier transform of a logarithmically-transformed amplitude spectrum. The following definition assumes usage of a DFT spectrum.

containing spoken letters 'b' or 'd', 't' or 'p'). Hence, the decision may be ambiguous; it is rather necessary to analyze the occurrence of *sequences of phonemes*, giving the context of words. This can be achieved by application of probability models. The most widely used methods for speech recognition are based on *Hidden Markov Models* (HMM, cf. sec. 9.4.6), which are described by state transition probabilities, with states characterizing the subsequent phonemes. On this basis, the sequence of highest probability is detected by comparing the likelihood of the analyzed phoneme sequence against different hypotheses of state sequences.

An important aspect is also the detection of speechless intervals. This is necessary to segment the signal into sections for further analysis, e.g. separate sentences into single words. In the most simple approach, this can be achieved by a threshold detection (suppression of small amplitudes), which will however fail if background noise is present. The separation of phonemes is more difficult, but often also a transition phase of smaller amplitude will be present; furthermore, the change of characteristic phoneme properties is a criterion by itself. The separation can then be achieved by first performing analysis over a fixed block-length raster, where a hypothesis about a change of the phoneme is made when the characteristics of two subsequent analysis blocks are significantly different. The exact boundary is then found by analysis on a more granular level. In general, the reliability of speech recognition will be clearly worse in an environment distorted by background noise.

Due to the generation by the human vocal tract, tones which are sung have similar characteristics as speech. The methods for speech analysis can hence partially be applied to analyze music pieces of song type. In general, tones in singing are much more pure and less subject to temporal variations than speech tones.

7.10.3 Musical Signals, Instruments and Sounds

Due to the physiological properties of the human ear (excitation of the basilar membrane by oscillations, cf. sec. 6.2.1) harmonic sounds are better distinguishable and are perceived as more comfortable. Traditionally, many musical instruments produce tones of harmonic characteristics, and also melodies and music pieces use sequences or an orchestration which preferably gives a harmonic impression. Similar to voiced and unvoiced speech, music sounds and the related instruments can be roughly characterized into *harmonic* and *percussive*. Harmonic sounds are typically also *sustained* and *coherent*, which means they keep similar properties over a longer time interval. Percussive sounds do not have a sustain phase, as percussive instruments are typically excited by pulse-like events (key of a piano, beat on a drum) instead of the permanent excitation of harmonic instruments (air blown through a flute). A sustain phase will however exist for instruments which are excited by pulses but tend to resonance, as e.g. bells. Such *inharmonic sounds* are also coherent. Further, *non-coherent sounds* exist. The spectral models for harmonic instruments are similar as for voiced speech, however for the identification of some instruments (e.g. violins), three formants are not sufficient.

Instrumental sounds, synthetic sounds and mixtures of any of these can be characterized either by the properties of the spectral envelope (peaks, roll-off characteristics) or by the envelope of temporal amplitudes (e.g. transitions where a tone starts, increases/decreases by amplitude or stops). Some of the features subsequently described can directly be mapped into semantic characterizations known from music science, and e.g. establish a solid basis for automatic annotation of music pieces or music genres. Most of the basic methods introduced in this section are also implemented as audio feature descriptors in the MPEG-7 standard (cf. sec. 17.6.4)

Tone frequency and loudness. The pitch of a tone and its harmonics are usually generated much cleaner by music instruments than it is the case for speech. For some instruments, the possible number of different tones is finite, i.e. except for the case of mis-tuning, the instrument will only be capable to generate tones from a discrete set (Examples: flute, piano; counter-examples: string instruments). For the overall impression of a sound, the pitch is rather important as well; the pitch distance between two tones is however not a good metric to be mapped into human capability to distinguish different sounds. As an example, a tone which is higher by one octave is perceived as being more similar than a tone which may only be slightly higher, but *not a harmonic* of the first tone's pitch. Related to the similarity and capability to distinguish between two tones by their height, the 'pitch helix' shown in Fig. 7.78 is an interesting concept [SHEPARD 1999], where tones which are different by one octave in pitch are stacked directly above each other. For reliable pitch analysis, a minimum number of periods must be available. Humans also typically need to hear at least 10-20 periods of a waveform until being able to recognize the frequency of a short tone burst.

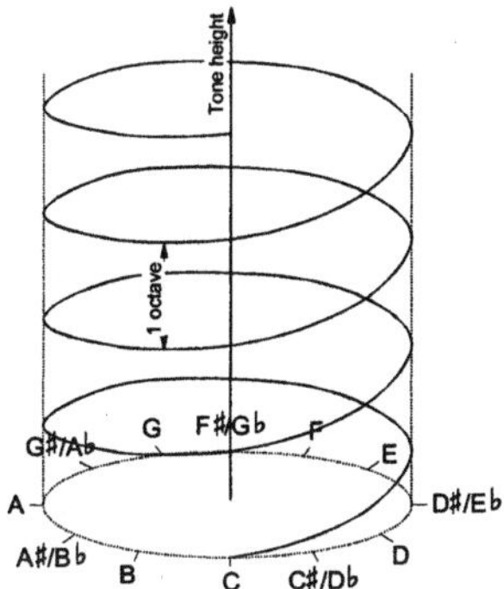

Fig. 7.78. Pitch helix [after SHEPARD]

A rough analysis of the instantaneous loudness can be gained by a windowed power analysis, taking the squares of signal amplitudes over a local neighborhood window of length $2M+1$,

$$P_x(m) = \frac{1}{2M+1} \sum_{k=m-M}^{m+M} x^2(m+k) \, . \qquad (7.220)$$

The signal $P_x(m)$ can be sub-sampled by a factor that corresponds to the window length; on the other hand, the window should be short enough to capture instantaneous power fluctuations whenever this is required by the target application of the analysis.

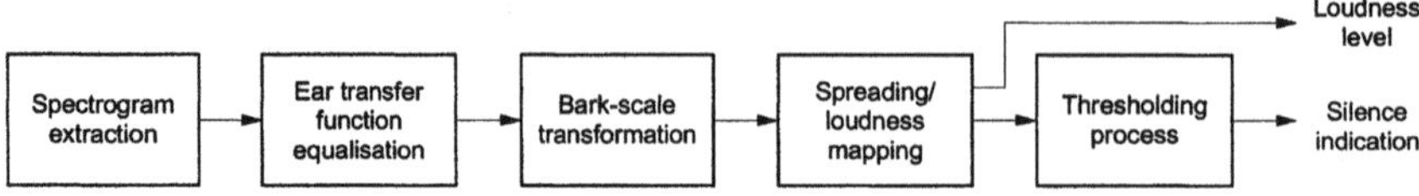

Fig. 7.79. Method to analyze loudness and silence

Loudness relates to the actually-perceived sound level of a music piece or sound event. Subjectively-adapted loudness measurement units are *phon* and *sone* (where the latter one is based on the psycho-acoustically adapted Bark scale, see sec. 6.2.2) [ZWICKER 1982]. Subjectively perceived 'doubling of loudness' does not correspond to a doubling of sound pressure level. Another important aspect of loudness measurements is related to detection of imperceptible or irrelevant sound events, which could be characterized as *perceptual silence*. A schema to analyze loudness and silence based on psycho-acoustic criteria is shown in Fig. 7.79. As loudness is frequency-dependent, the analysis is performed by application of a short-time Fourier transform over a given analysis window. The result is weighted by the psycho-acoustic transfer function, the mapping of spectral coefficients into the Bark scale and finally the loudness computation from the energy contributions of the frequency groups (in units of *sone*). Based on the loudness values, a threshold operation is applied to detect silence, which however also may need to take into account the temporal masking effects (cf. sec. 6.2.2)

Due to temporal masking, silence periods will only be relevant if they are longer than a certain minimum duration, which further depends on the previous loudness level. To avoid that the silence detection will frequently turn on and off in cases where low loudness levels fluctuate by a certain variation, a principle based on two threshold levels as shown in Fig. 7.80 can be applied[1]. The start of a silence period is detected when a *lower threshold value* is reached. Only if an *upper threshold level* is superseded, the silence period is terminated again. Switching to a silence period is performed only if the lower threshold criterion holds for a certain minimum duration t_{min}. A sliding-window mean filtering of loudness amplitudes can be performed in addition to stabilize the procedure. Different variants of such a scheme are possible, e.g. using more than 2 threshold levels, adaptive thresholds or context-dependent definitions of minimum durations.

---

[1] Similar dual-threshold methods are used in edge tracking (sec. 7.3.3) and video segmentation (sec. 10.2.1).

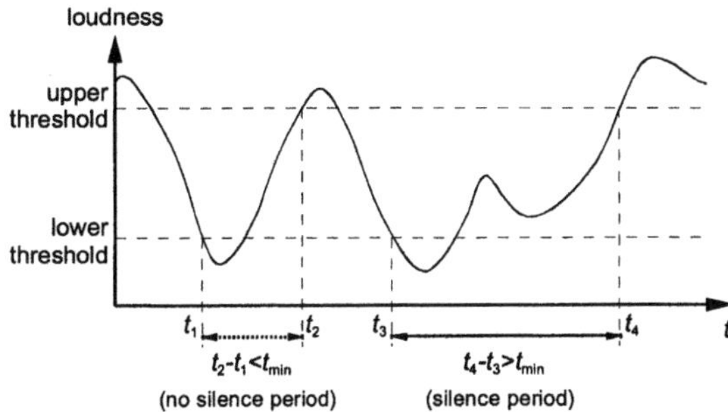

Fig. 7.80. Threshold analysis using two different thresholds to detect perceptual silence [after MPEG]

Basic spectral features. Direct spectral analysis based on (windowed) DFT coefficients or DFT power coefficients is not useful. Perceptually weighted frequency band partitions have rather non-uniform distribution. A typical approach in particular for music analysis is *octave band scaling*, where a logarithm of base 2 is used to perform a nonlinear mapping of the frequency axis. It is necessary to observe lower and upper band edges of the frequency range that shall be analyzed, typically bounding the range of the audible spectrum. Assume that $k=1$ is the lowest band within the relevant range, and K DFT coefficients $X(k)$ are available between lower and upper band edges. The *centroid* of spectral power distribution as related to an octave-scale frequency axis with an anchor frequency of 1 KHz is then computed as

$$C_x = \frac{\sum_{k=1}^{K} \log_2\left(f(k)/1000\right)\cdot\left|\hat{X}(k)\right|^2}{\sum_{k=1}^{K}\left|\hat{X}(k)\right|^2} \,, \qquad (7.221)$$

where $f(k)$ is the frequency [Hz] associated to coefficient k. The actual frequency associated with the centroid will be 2^C kHz. The *spectral spread* can be expressed on the same logarithmic scale as a 'standard deviation' of the power distribution around the centroid,

$$S_x = \sqrt{\frac{\sum_{k=1}^{K} \log_2\left(f(k)/1000-C_x\right)^2\cdot\left|\hat{X}(k)\right|^2}{\sum_{k=1}^{K}\left|\hat{X}(k)\right|^2}} \,. \qquad (7.222)$$

C_x and S_x are scalar parameters which only roughly express the spectral power distribution. On the other hand, using the linear frequency scale of the DFT does not provide a good match with the perceptual properties. More reasonable analysis is made by usage of a constant number of frequency bands per octave, or by using the nonlinear Bark scale (cf. sec. 6.2.2). In these cases, the numbers of DFT coefficients that fall into the different bands are not constant. Assume that within a band b on the nonlinear frequency axis, the lowest DFT frequency index will be $kl(b)$ and the highest index will be $kh(b)$. Then, the percentage of spectral power that falls into band b can be computed as[1]

$$P_x(b) = \frac{\displaystyle\sum_{k=kl(b)}^{kh(b)} \left|\hat{X}(k)\right|^2}{\displaystyle\sum_{k=1}^{K} \left|\hat{X}(k)\right|^2} \cdot 100 \quad [\%] \ . \tag{7.223}$$

Another feature related to single frequency bands b is the *spectral flatness*

$$\mathrm{SF}_x(b) = \frac{\sqrt[kh(b)-kl(b)+1]{\displaystyle\prod_{k=kl(b)}^{kh(b)} \left|\hat{X}(k)\right|^2}}{\frac{1}{kh(b)-kl(b)+1} \displaystyle\sum_{k=kl(b)}^{kh(b)} \left|\hat{X}(k)\right|^2} \ , \tag{7.224}$$

which is the ratio of geometric and arithmetic mean, expressing the homogeneity of the DFT power coefficients within the given frequency band; for a flat (white) spectrum, $\mathrm{SF}_x(b)=1$; otherwise values will be between 0 and 1[2].

Harmonicity. The measurement of harmonicity of an audio signal can – in addition to the methods introduced in sec. 7.10.1 – also be performed by *autocorrelation analysis* which will show a peak when the signal is shifted by a pitch period. The minimum reasonable pitch period is $P_{min}=2$. Analysis is performed over a segment of M adjacent samples starting at m_0:

$$f(P,m_0) = \frac{\displaystyle\sum_{m=m_0}^{m_0+M-1} x(m) \cdot x(m-P)}{\sqrt{\displaystyle\sum_{m=m_0}^{m_0+M-1} x^2(m) \cdot \sum_{m=m_o}^{m_0+M-1} x^2(m-P)}} \ ; \ P = P_{min}, \dots P_{max} \ . \tag{7.225}$$

After the maximum of $f(P)$ over the entire search range of P has been determined, a useful harmonicity criterion is given by the *harmonic ratio*

$$\mathrm{HR}(m_0) = \max_P(f(P,m_0)) \tag{7.226}$$

For a clean periodic signal, (7.226) gives HR=1, while for a totally aperiodic signal (e.g. white noise), HR=0. Now, for the case where harmonicity is detected by a

---

[1] If a DFT coefficient is positioned near the edge between two frequency bands, its power can also be spread into both bands using weights which should sum to unity.

[2] For the case of no power within the band (all coefficients zero), a limit transition also gives a value of 1.

sufficiently large HR value, a *comb-filtered signal* can be computed for the segment starting at m_0, using the period P as detected by (7.226)

$$y_{\text{comb}}(m) = x(m) - HR(m_0)x(m-P) \; ; \; m = m_0, ..., m_0 + M - 1 . \tag{7.227}$$

Then, the DFT can be computed over the segment both for $x(m)$ and for $y_{\text{comb}}(m)$. For each discrete frequency position $f(k)$, starting with $k=1$ at the lower band edge, the ratio of powers of the comb-filtered and original signal spectra is computed as

$$R_x(k) = \frac{\sum\limits_{l=k}^{K} \left| \hat{Y}(l) \right|^2}{\sum\limits_{l=k}^{K} \left| \hat{X}(l) \right|^2} \; ; \quad k = 1, ..., K . \tag{7.228}$$

The comb filter has the property to eliminate the pitch and its harmonic components. Thus, a low value of $R_x(k)$ will appear whenever a harmonic has been filtered out successfully. The frequency at *upper limit of harmonicity* can be defined as the highest frequency for which $R_x(k)$ is lower than a pre-defined threshold Θ.

Temporal envelope and timbre description. An envelope computed over the absolute amplitude of the temporal signal allows detecting the transient behavior of audio signals. To stabilize the result, averaging is performed over a number of adjacent samples. In (7.229), a window over $2M+1$ samples is used, which is centered on the position of the current value:

$$h_x(m) = \frac{1}{2M+1} \sum\limits_{k=n-M}^{m+M} \left| x(m+k) \right| . \tag{7.229}$$

To detect transient behavior reliably, M should be relatively low, typically corresponding to a duration of $\sim 10\ ms$. The temporal envelope of a tone is often separated into the following segments (see Fig. 7.81):

- the period of rising signal amplitude up to the maximum level (*attack*);
- the period of (fast) decreasing amplitude from the maximum level to a sustained level (*decay*)
- the period of (almost) constant signal amplitude (*sustain*);
- the period of finally decreasing signal amplitude (*release*).

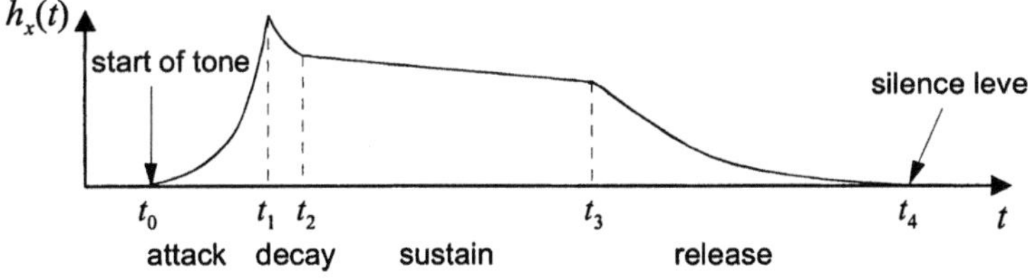

Fig. 7.81. Qualitative shape of the temporal envelope

The values of *attack time* $(t_1\text{-}t_0)$, *decay time* $(t_2\text{-}t_1)$, *sustain time* $(t_3\text{-}t_2)$ and *release time* $(t_4\text{-}t_3)$ are important parameters to characterize the temporal properties of tones generated by an instrument. These are also commonly used as control parameters in sound synthesis. Percussive sounds are typically characterized by a lack of the sustain phase. The temporal envelope as extracted from signals will usually not have the idealized characteristics indicated in Fig. 7.81.

Timbre is a semantic notion which could be interpreted vaguely as the *sharpness* of a sound; it is the criterion that makes two sounds of same tone height be perceived as different. Regarding the behavior over time, this is mainly related to the *attack phase* and the compactness of the envelope. Hence, good parameters describing timbre by its temporal characteristics are the *logarithmic attack time* (LAT)

$$\text{lat} = \log_{10}(t_1 - t_0) \tag{7.230}$$

and the *centroid of the temporal envelope*, with offset by the starting time t_0

$$\text{TC} = T \cdot \sum_{m=t_0/T}^{t_4/T} n \cdot h_x(m) - t_0 \Bigg/ \sum_{m=t_0/T}^{t_4/T} h_x(m) \ [s] . \tag{7.231}$$

Spectral timbre description. To express timbre for arbitrary monophonic (instrumental) or polyphonic (orchestral) sounds, and to allow timbre-based comparison of the characteristics of different instruments, it is necessary to use additional features from the spectral domain. For harmonic signals, timbre can closely be related with the *roll-off characteristics* of the spectrum, which express the relationship of the spectral amplitude levels at the pitch frequency and its harmonics. Assume that a number of L-1 harmonics exists at frequencies $f(l)\text{=}l{\cdot}f_p$, with pitch frequency $f_p\text{=}f(1)$. Let further $\hat{X}(l,p)$ express the DFT spectrum estimate of the harmonic l within analysis block p (one block of the short-term analysis over the spectrogram time axis, cf. Fig. 7.77)[1]. Then, a roll-off based timbre criterion can be computed from the spectral power values of B blocks at those harmonic frequencies as

$$\text{roll-off} = \frac{20}{L} \cdot \sum_{l=1}^{L} \frac{\log_{10}\left(\sum_{p=1}^{B} |\hat{X}(1,p)| \Bigg/ \sum_{p=1}^{B} |\hat{X}(l,p)| \right)}{\log_2 l} \ [dB\,/\,\text{octave}] . \tag{7.232}$$

$$|\hat{X}(l,p)| > 0.$$

---

[1] In case where the DFT analysis block length is a multiple of the pitch period, the harmonic frequencies will be at exact frequency positions of DFT coefficients. Otherwise, it may be necessary to interpolate the estimate from several neighbored spectrum coefficients.

Signals having a roll-off factor of 0 dB/octave are perceived as 'extremely sharp', those with a factor of 12 dB/octave are judged as 'less sharp' [MATHEWS 1999].

The *Harmonic Spectral Centroid* (HSC) is defined over a group of B analysis blocks of a sustained harmonic sound as follows:

$$\text{HSC}(p) = \frac{\sum_{l=1}^{L} f(u) \cdot |\hat{X}(l,p)|}{\sum_{l=1}^{L} |\hat{X}(l,p)|} \ [Hz] \quad \Rightarrow \quad \text{HSC} = \frac{1}{B} \sum_{p=1}^{B} \text{HSC}(p). \qquad (7.233)$$

The *Harmonic Spectral Deviation* (HSD) is the difference between the pure harmonic spectral lines and their envelope, allowing determining the 'pureness' of a tone[1]:

$$\text{HSD}(p) = \frac{\sum_{l=1}^{L} |\hat{X}(l,p)| - \bar{X}(l,p)}{\sum_{l=1}^{L} |\hat{X}(l,p)|} \quad \Rightarrow \quad \text{HSD} = \frac{1}{B} \sum_{p=1}^{B} \text{HSD}(p). \qquad (7.234)$$

The envelope values are typically determined by averaging neighbored harmonic amplitudes such as[2]

$$\bar{X}(l,p) = \frac{1}{2M+1} \sum_{k=l-M}^{l+M} |\hat{X}(l,p)|. \qquad (7.235)$$

The *Harmonic Spectral Spread* (HSS) characterizes the standard deviation of the spectral amplitude concentration around the centroid:

$$\text{HSS}(p) = \frac{1}{\text{HSC}(p)} \sqrt{\frac{\sum_{l=1}^{L} |\hat{X}(l,p)| \cdot (f(l) - \text{HSC}(p))^2}{\sum_{l=1}^{L} |\hat{X}(l,p)|}} \qquad (7.236)$$

$$\Rightarrow \quad \text{HSS} = \frac{1}{B} \sum_{p=1}^{B} \text{HSS}(p).$$

The *Harmonic Spectral Variation* (HSV) expresses the normalized cross correlation between spectral amplitudes from adjacent analysis windows:

---

[1] The HSD is more expressive, if logarithmic amplitudes are used for the harmonic spectrum and the envelope spectrum. In this case, either the expression $\log(1+X)$ should be used, or the normalization by the denominator should be skipped to avoid negative values. The latter approach is taken in the definition of the HSD in the MPEG-7 audio standard.

[2] At the boundaries of the spectral range, where (7.235) would access spectral lines with $l<1$ or $l>L$, a modification is necessary by averaging less values.

$$\text{HSV}(p) = 1 - \frac{\sum_{u=1}^{L}\left|\hat{X}(l,p)\right|\cdot\left|\hat{X}(l,p-1)\right|}{\sqrt{\sum_{l=1}^{L}\left|\hat{X}(l,p)\right|^2}\cdot\sqrt{\sum_{l=1}^{L}\left|\hat{X}(l,p-1)\right|^2}}$$

$$\Rightarrow \quad \text{HSV} = \frac{1}{B-1}\sum_{p=2}^{B}\text{HSV}(p). \tag{7.237}$$

For non-harmonic signals, criteria (7.232)-(7.237) are not applicable. Here, the *Spectral Centroid* (SC) can be computed from the DFT power spectrum coefficients (at all available frequency positions), from which first the spectral envelope is computed similarly to (7.235); alternatively, the values $\left|\overline{X}(k)\right|$ can be computed as averages from several analysis blocks p of the spectrogram:

$$\text{SC} = \frac{\sum_{k=1}^{K} f(k)\cdot\left|\overline{X}(k)\right|^2}{\sum_{k=1}^{K}\left|\overline{X}(k)\right|^2} \quad [Hz]. \tag{7.238}$$

For timbre classification of non-harmonic sounds, (7.238) can be combined with (7.230) and (7.231). The logarithmic attack time (7.230) is also applicable to harmonic sounds, along with the more specialized feature values introduced above. The temporal centroid (7.231) is not expressive to analyze the timbre of harmonic signals, as it may highly vary depending on the length of the sustain phase (length of the note that is played).

Modulations. Modulations are regular periodic fluctuations of a signal property. The most important types of modulations applicable to musical sounds are *vibrato*, which is a modulation of the pitch, and *tremolo*, which is a modulation of the amplitude envelope (hull). Vibrato is equivalent to *frequency modulation*, if the temporal variation of the tone frequency follows a sinusoidal function. In the example of Fig. 7.82a, the vibrato over a harmonic tone (with pitch frequency f_p and 2 harmonics) is shown; Fig. 7.82b illustrates tremolo applied to a sinusoid tone. If the amplitude is modulated by a cosine function of frequency $\tilde{f}$, the modulated signal and its spectrum will be

$$\tilde{x}(t) = x(t)\cdot\left[1 + A\cos\left(2\pi\tilde{f}t\right)\right]$$

$$\tilde{X}(f) = X(f) + \frac{A}{2}\left[X\left(f-\tilde{f}\right) + X\left(f+\tilde{f}\right)\right]. \tag{7.239}$$

This is equivalent to a *double-sideband amplitude modulation*, where typically $\tilde{f}$ is much smaller than the bandwidth of the signal, such that the two additional copies of the spectrum in (7.239) overlap.

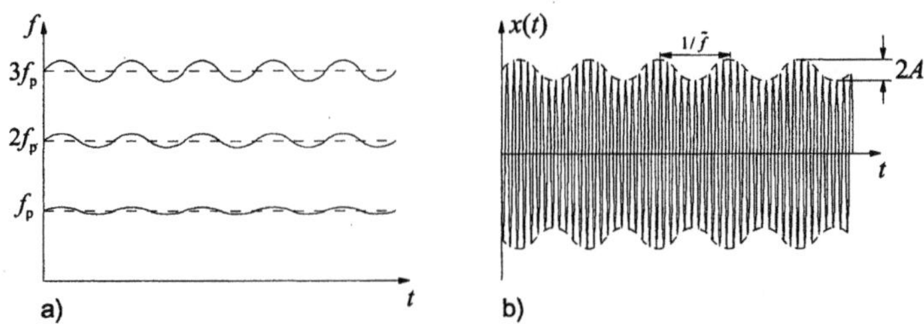

Fig. 7.82. a Spectrogram of a constant harmonic tone in case of a narrow-band vibrato
b Tremolo of a sinusoid tone

Tempo. The tempo of a melody sequence played can be analyzed by the rate of distinguishable tones, which are differentiated by their starting times t_0. If e.g. N tones of starting times $t_0(n)$ are found, the mean value of differences is

$$\Delta_{\text{Tone}} = \frac{1}{N-1} \sum_{n=2}^{N} t_0(n) - t_0(n-1), \tag{7.240}$$

and an analysis of tone length variations can be performed according to the following criterion:

$$\Delta_{\text{Var}} = \frac{1}{N-2} \sum_{n=2}^{N-1} \frac{t_0(n+1) - t_0(n)}{t_0(n) - t_0(n-1)}. \tag{7.241}$$

Music structure. The basic analysis elements described so far can be used to capture the structure of music pieces like songs etc. For this, it is necessary to perform a segmentation, to analyze periods with relatively constant properties of basic features in temporal context and variation. A time-dependent feature vector then has a structure

$$\mathbf{m}_{\text{segment}} = \begin{bmatrix} t_{\text{start}} & t_{\text{duration}} & \text{sound\_feature\_1} & \dots & \text{sound\_feature\_}K \end{bmatrix}^{\text{T}} \tag{7.242}$$

By analyzing a sequence of such vectors in their temporal variation, it is possible to characterize the structure of entire music pieces, including detection of verses, refrain etc. in songs. The problem of segmentation based on features is further discussed in sec. 10.3.

Melody. Melodies can be extracted from audio signals by analysis of tone-related segments. This must typically be done by a combination of temporal (power, envelope) analysis to best determine start and duration of tones as well as the different transition phases, and spectral analysis to determine the tone height. Melodies played with tone bindings by harmonic instruments can best be analyzed from spectral criteria. Spectral variation criteria can be used to detect sustained tones or transition phases, where however the temporal granularity of the analysis will highly depend on the transform block length that is used to compute the spectrum. In principle, the spectrum is sufficiently expressive to separate both monophonic and polyphonic sounds, where however for complex orchestra sounds it will be necessary to use a priori knowledge about characteristic properties of specific instruments, such as harmonic power distribution and typical attack characteristics.

7.10.4 Room Properties

In the description of audio signal properties, their features and underlying physical phenomena, the effects of sound propagation in the 3D exterior world were not considered so far. This is approximately correct for *reflection-free sound fields*, e.g. in an outside world without any obstacles, where effects of reflections and reverberations can be neglected. A distance-dependent amplitude transfer function will be in effect in any case, where higher frequencies are typically subject to stronger attenuation if the sound source is farther away. Reflections can also be neglected in reflection-free rooms which are specifically designed to test acoustical phenomena; recording rooms in sound studios for music production are designed similarly, because room effects can then be supplemented independent of the recording in a controllable way. Under normal situations, in particular when audio signals are recorded within rooms, the sound field is *diffuse*, and reflections occur. This plays an important role in the resultant characteristics of the signal, including their perception as being 'natural', but also makes signal analysis more difficult. Further, good modeling of room properties is important if a natural-sounding synthesis of audio signals shall be performed.

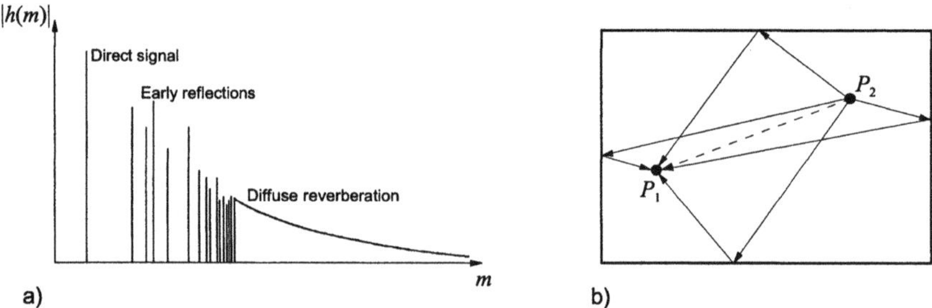

Fig. 7.83. a Qualitative properties of a room transfer function   **b** Physical interpretation of direct signal (...) and early reflections (—)

The reflection properties of a room can be characterized by the *room transfer function*, which is in fact the impulse response of the room. Typical components which are contributing are shown in Fig. 7.83a. The room transfer function corresponds strictly to one point P_1 of a room, where the sound is received, while the sound source is located at another point P_2 (see Fig. 7.83b). First, the *direct signal* arrives. Then, *early reflections* follow, where the delay depends on the room size, in particular the distance of the sound source from the walls or other obstacles. Early reflections can be observed as relatively discrete pulses in the transfer function. For the first reflection, exactly one *sound path* is possible from P_2, that is reflected by the nearest wall towards P_1. If W walls or other obstacles are present, the Nth reflection may be arriving over W^N different sound paths, all of which may not be much different by length. Hence, after a short time, the room transfer function will approach characteristics of *diffuse reverberation* instead of the pulse-like first echoes; single reflections can no longer be distinguished. The attenuation of the reflections depends on the absorbing capabilities of walls or other obstacles. The *reverberation time* is usually defined as the time span, by which after an impulse-like sound event[1] the sound pressure level decreases by 60 dB. For natural-like sound synthesis it is usually sufficient to simulate the reverberation in general as an exponentially decaying function, and insert in addition some early echoes. Deviations of single reflections by phase characteristics are hardly perceived, while the impression about the room size is highly influenced by the reverberation time and decay at large scale. This also implies that it is usually sufficient to record the room transfer function related to one single point P_1, if the reflection properties of the walls are not too much different; the amplitude transfer function of the room will then almost be invariant against changes in the position.

Room transfer functions can also be measured by feeding a signal from a noise generator[2] into a loudspeaker, recording the resulting sound field by a microphone. By measuring the power spectrum of the input noise and the cross-power spectrum between the input and the recorded signal, the overall transfer function of the system (including speaker and microphone) results as:

$$S_{xy}(j\Omega) = X^*(j\Omega) \cdot Y(j\Omega) = X^*(j\Omega) \cdot X(j\Omega) \cdot H(j\Omega)$$

$$= S_{xx}(\Omega) \cdot H(j\Omega) \Rightarrow H(j\Omega) = \frac{S_{xy}(j\Omega)}{S_{xx}(\Omega)}. \tag{7.243}$$

If the frequency transfer functions of loudspeaker and microphone are known, and if these have sufficient bandwidth such that no spectral zeros appear in the range of frequencies to be investigated, the frequency transfer function of the room can be computed by

$$H_{\text{Room}}(j\Omega) = \frac{H(j\Omega)}{H_{\text{Speaker}}(j\Omega) \cdot H_{\text{Microphone}}(j\Omega)}. \tag{7.244}$$

By discrete approximation and inverse DFT of sufficient length, it is then possible to determine the room transfer function (impulse response). It is also possible to regard the speaker system as part of the entire transfer function, which is in particular made if the goal is the optimization of a sound system within a room. The transfer function given in (7.244) is again related to exactly *one* position of the speaker as sound source, and *one* position of the microphone as receiver. For sound synthesis, analysis of reverberation time and other applications this will be fully sufficient. If the transfer function would *exactly* be known for any position, the influence of room acoustics to the analysis of signal properties could be com-

---

[1] E.g. a shot, which could be seen as the acoustic version of a Dirac impulse.

[2] E.g. flat-spectrum noise, band limited within the audible range of frequencies.

pletely eliminated. As it is almost impossible to capture the transfer function for any combination of speaker/microphone positions, it is more practical to use acoustical room models for this purpose. Two types of models are shown in Fig. 7.84:

–   The *ray tracing model* (Fig. 7.84a) assumes a point sound source with omnidirectional emission of sound waves. If the reflection properties of walls are known, 'sound rays' can be constructed. Considering the sound propagation speed, the transfer function can be determined for any location in the room.

–   In the *mirror image model* (Fig. 7.84b), virtual 'mirrored copies' of the room allocated in all three spatial dimensions are assumed, in which the sound rays propagate linearly. Each time when a wall is virtually passed, the respective sound ray is attenuated by a factor which is determined by the reflection properties of the wall. The impulse response of the room at any given position can then be determined by superposition of the sound field amplitudes of the room and all of its mirrored copies, where the time of arrival at a certain place is proportional with the factor of sound speed and the length of the ray. Fig. 7.84b shows the direct sound wave which arrives at the destination D, as well as the first four reflections which exactly traverse one wall.

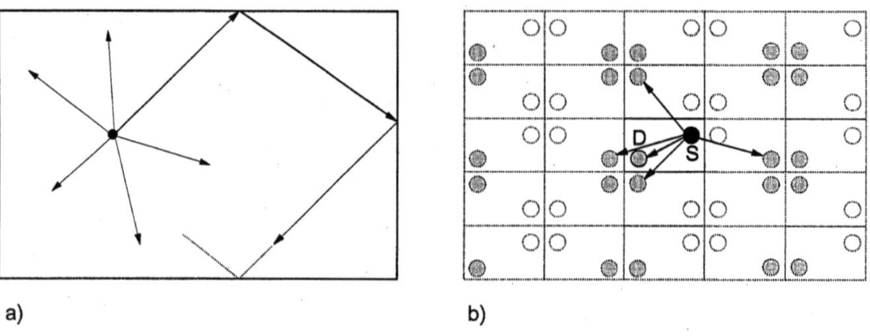

Fig. 7.84. Models to characterize room transfer functions **a** Ray tracing **b** Mirror image model with S=source, D=destination

7.11 Problems

Problem 7.1
Compute the angles between the basis vectors (rows of transform matrices) for the following color transforms. Are the basis vectors perpendicular? Is any of the transforms orthogonal? [The R,G,B space shall be interpreted as an orthogonal reference]
a) Y,C_b,C_r transform (7.2)
b) X,Y,Z transform (7.3)
c) I,K,L transform according to the following definition:

$$\begin{bmatrix} I \\ K \\ L \end{bmatrix} = \begin{bmatrix} \frac{1}{3} & \frac{1}{3} & \frac{1}{3} \\ -\frac{1}{\sqrt{6}} & -\frac{1}{\sqrt{6}} & \frac{2}{\sqrt{6}} \\ \frac{1}{\sqrt{6}} & -\frac{1}{\sqrt{6}} & 0 \end{bmatrix} \cdot \begin{bmatrix} R \\ G \\ B \end{bmatrix},$$

Note: The I,K,L transform is the linear first step of the I,H,S (intensity, hue, saturation) transform. It is followed by a transform of K and L into a polar coordinate system, where H expresses the hue angle and S the saturation by the distance from the origin. It is also obvious that this transform is fully invertible:

$$H = \arctan\left[\frac{L}{K}\right]; \ S = \sqrt{K^2 + L^2} \ \Rightarrow \ K = S\cos H \ ; \quad L = S\sin H .$$

Problem 7.2

a) Determine the inverse mapping of the H,S,V color transform (7.6).

b) Determine the color values in the R,G,B color space for the secondary colors yellow, cyan and magenta (see Fig. 7.1) with $S=1$, $V=0.5$ and $S=0.5$, $V=0.5$

c) Compute the distances between the black value ($V=S=R=G=B=0$) and the colors of b) in the H,S,V and R,G,B color spaces. Interpret the result.

Problem 7.3

A color transform from the primary components R,G,B shall be defined by

$$\begin{bmatrix} I \\ D_1 \\ D_2 \end{bmatrix} = \begin{bmatrix} 1/3 & 1/3 & 1/3 \\ 2/3 & -1/3 & -1/3 \\ -1/3 & -1/3 & 2/3 \end{bmatrix} \cdot \begin{bmatrix} R \\ G \\ B \end{bmatrix}.$$

a) Express the components D_1 and D_2 each as a function of I and *one* of the primary components.

b) Check if the transform is orthogonal.

c) Determine the inverse color transform.

d) Two pairs of R,G,B color triplets are defined as follows:

$$\mathbf{f}_{A,1}=[0\ 0\ 0]^T , \ \mathbf{f}_{B,1}=[1\ 1\ 1]^T \ ; \qquad \mathbf{f}_{A,2}=[1\ 0\ 0]^T , \ \mathbf{f}_{B,2}=[0\ 0\ 1]^T$$

Compute the Euclidean distances $d_2(\mathbf{f}_{A,i},\mathbf{f}_{B,i})$ in the R,G,B color space and in the I,D_1,D_2 color space. In which of the two color spaces will differences by intensity (I) take stronger influence?

e) Assume that the three primary components have a Gaussian PDF, are mutually uncorrelated and have mean values $\mu_R = 5$; $\mu_G = 8$; $\mu_B = 2$ and variances $\sigma_R^2 = 4$; $\sigma_G^2 = 3$; $\sigma_B^2 = 2$. Determine the mean value and the variance of the component I.

Problem 7.4

a) Determine for the image matrix $\mathbf{X}$ in (7.18) the co-occurrence matrix $\mathbf{C}^{(1,1)}$ assuming that the signal shall be periodically extended beyond the boundaries [cf. (4.7)]; all 25 pixels shall then be used to determine the co-occurrences.

b) Determine the co-occurrence-Matrix $\mathbf{C}^{(1,1)}$ for the following image matrix, also assuming periodic extension of the signal:

$$\mathbf{X} = \begin{bmatrix} 1 & 0 & 0 & 1 & 0 \\ 2 & 0 & 0 & 0 & 1 \\ 1 & 1 & 1 & 0 & 1 \\ 0 & 2 & 2 & 1 & 0 \\ 0 & 1 & 1 & 0 & 2 \end{bmatrix}$$

c)   For a signal of three amplitude levels 0,1,2, which will be the values of the criteria (7.22) with q=2, (7.24) and (7.25), when *i*) all entries in $\mathbf{P}^{(1,1)}$ are equal   *ii*) an image has a constant gray level ? Compute the same criteria from the co-occurrence matrices of a) and b), compare and interpret the results.

Problem 7.5

The texture feature of a binary image $b(m,n)$ shall be characterized. For this, co-occurrence matrices referencing horizontally and vertically adjacent neighbor pixels are available by the normalized form:

$$\mathbf{P}^{(1,0)} = \begin{bmatrix} 0.45 & 0.05 \\ 0.05 & 0.45 \end{bmatrix}; \quad \mathbf{P}^{(0,1)} = \begin{bmatrix} 0.4 & 0.1 \\ 0.1 & 0.4 \end{bmatrix} \quad \text{with} \quad \mathbf{P}^{(k,l)} = \begin{bmatrix} P^{(k,l)}(0,0) & P^{(k,l)}(0,1) \\ P^{(k,l)}(1,0) & P^{(k,l)}(1,1) \end{bmatrix}$$

Assume that the occurrence values are identical to the values of probability distribution for a signal model.

a)   Determine probabilities $P(0)$ and $P(1)$.
b)   Determine mean and variance of the binary texture.
c)   Compute the autocorrelation values $r_{bb}(0,1)$ and $r_{bb}(1,0)$, the autocovariance values $r'_{bb}(0,1)$ and $r'_{bb}(1,0)$, and the autocovariance coefficients $\rho'_{bb}(0,1)$ and $\rho'_{bb}(1,0)$.
d)   A similar texture shall be synthesized by the method according to the block diagram given in Fig. 7.85. Herein, $z(m,n)$ shall be white, zero-mean Gaussian noise of variance σ_z^2. It is assumed that correlation properties of the continuous-value process $x(m,n)$ are inherited by the binary process $b(m,n)$. **S** shall be a threshold decision circuit of following characteristics:

$$b(m,n) = \begin{cases} 0 & \text{if } x(m,n) \leq C \\ 1 & \text{if } x(m,n) > C \end{cases}$$

Determine parameters A, B and C.

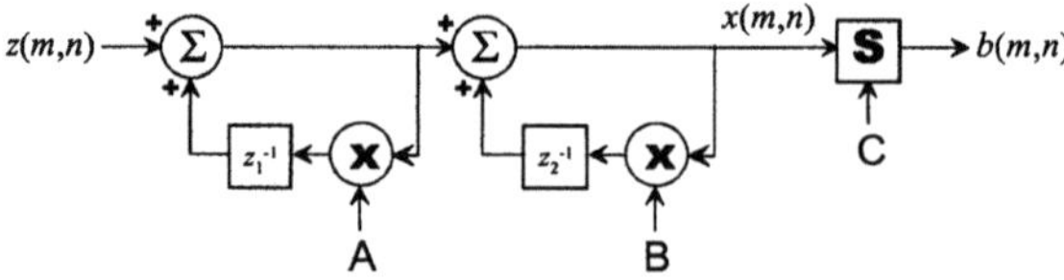

Fig. 7.85. Circuit for generation of a binary texture

Problem 7.6

A texture shall be described by the parameters of an isotropic first-order autoregressive 2D model. The power spectrum is given by (4.60). How is this spectrum changed in case of
a)   Rotation of the texture pattern by 45° or 90° ?

b)   Uniform scaling of the texture pattern by factors of 0.5 (compressed) or 2.0 (expanded) ?

Problem 7.7
a)   The image matrix **X** as shown below shall be filtered using the different directional masks of the operator (7.48). The analysis shall be limited to the bounded area. Determine the results.

$$\mathbf{X} = \begin{bmatrix} 5 & 5 & 5 & 5 & 10 & 10 \\ 5 & 5 & 5 & 5 & 10 & 10 \\ 5 & 5 & 5 & 10 & 10 & 10 \\ 5 & 5 & 10 & 10 & 10 & 10 \\ 5 & 5 & 10 & 10 & 10 & 10 \\ 10 & 10 & 10 & 10 & 10 & 10 \end{bmatrix}$$

b)   Using the results from a), determine an edge image by absolute-value computation and maximum search among the different directional filter results. Determine a threshold value by (7.61), such that an edge image with an edge width of one pixel is obtained.
c)   Now, the Laplace operator (7.53) shall be applied for edge detection in the bounded area. Determine the zero-crossings of the second derivative by application of a maximum difference filter (5.5) with a homogeneous neighborhood $\mathcal{N}_1^{(1)}$. Assume that the result of Laplace filtering is set to zero outside of the marked area.

Problem 7.8
An edge detector has identified the presence of following six (r,s) coordinate pairs of edge pixels within an image:

$$r(p) = [2.0 \quad 1.5 \quad 1.0 \quad -0.5 \quad 0 \quad 0.5\,]$$
$$s(p) = [1.0 \quad 1.5 \quad 2.0 \quad 0.5 \quad 1.0 \quad 1.5\,]$$

By application of the Hough transform, the parameters of two straight lines (angle α, distance of coordinate origin ρ) shall be determined. The Hough transform for continuous polar coordinates is $\rho = r \cdot \cos(\alpha) + s \cdot \sin(\alpha)$.
a)   Draw an image plane and sketch the positions of edge pixels.
b)   Compute and sketch the resulting curve sets in the Hough space, and determine the line parameters.
c)   Determine a parametrization of the Hough transform according to the slope/intercept form of a line equation. Sketch the result in the Hough space, based on this parametrization.
d)   How would it be possible to determine the end points of *line segments* ?

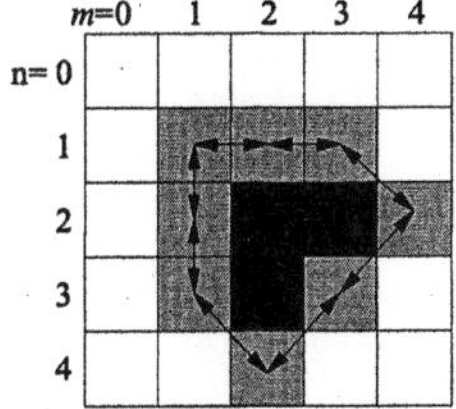

Fig. 7.86. Binary shape and its inner contour on a discrete grid

Problem 7.9
The *inner contour* of a binary shape is given in Fig. 7.86.
a)   The contour start position $p=0$ shall be the top-left pixel of the contour. Perform a mean-value filter operation of the contour coordinates $m(p)$ and $n(p)$ using each three neighbored values (contour pixel and its neighbors on both sides); observe the cyclic continuation of the contour, where the pixel at position $p=7$ is neighbored to $p=0$. Round the result to nearest integer and sketch the filtered contour.
b)   Compute the contour lengths, areas of the shape, compactness parameters ξ_K and form factors ξ_F for cases of original and filtered contours.

Problem 7.10
In Fig. 7.87, three continuous contours A, B and C are given. First, determine the lengths of the contours, assuming that the horizontal and vertical grid distances are unity. Now, the contours shall be sampled. The sampled contour shall be constructed such that lines interconnect the centers of adjacent cells; the interconnections shall be done such that the area between the continuous and the resulting discrete contour is minimized. Now, the lengths of the discrete contours shall be determined in the following two configurations of neighbor interconnections:
a)   Neighborhood system $\mathcal{N}_1^{(1)}$ of (4.1);
b)   Neighborhood system $\mathcal{N}_2^{(2)}$ of (4.1.
Compare the lengths of the continuous and discrete contours.

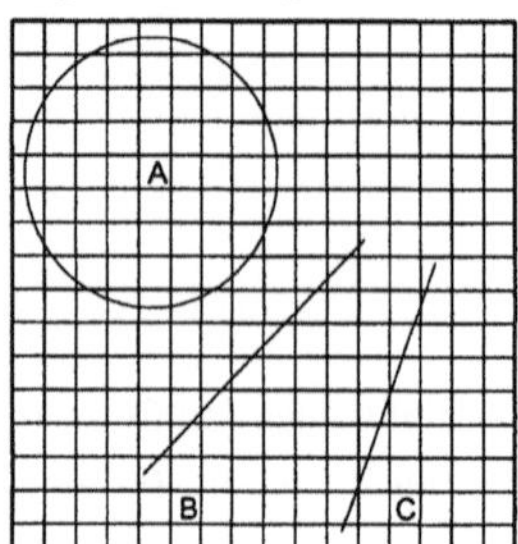

Fig. 7.87. Three continuous contours A, B, C drawn upon a discrete sampling grid

Problem 7.11
Determine relationships by which the central moments $\rho^{(2,0)}$ and $\rho^{(0,2)}$ can be computed from moments $\mu^{(k,l)}$ with $k+l\leq2$.

Problem 7.12
The binary shape shown in Fig. 7.88 is given (black : $b(m,n)=1$).
a)   Compute the distance transform. The shortest distances from the boundary shall be determined only horizontally and vertically. Sketch the skeleton of the shape.
b)   Compute the projection profiles.
c)   Determine the center of gravity from the values of the projection profiles.
d)   Compute the central moments $\rho^{(2,0)}$, $\rho^{(0,2)}$, und $\rho^{(1,1)}$. Construct the matrix Γ according to (7.127) and determine the rotation invariant criteria (7.131) and (7.134).
e)   Construct the matrix $\tilde{\Gamma}$ for the same form, rotated by 90° counterclockwise. Compare the rotation-invariant criteria to the result from d).

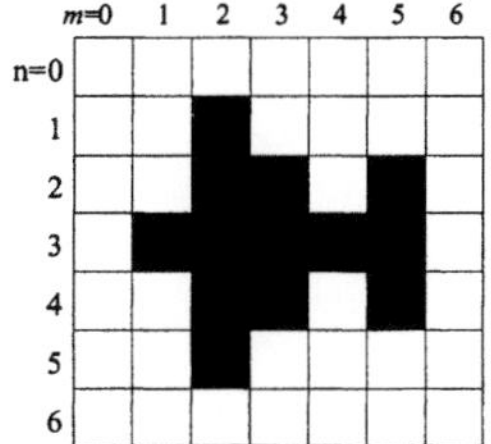

Fig. 7.88. Definition of a binary shape

Problem 7.13
The following image matrices express the values of two subsequent frames from a video sequence. The motion shift shall be determined for the bounded block of the reference frame $\mathbf{X}(o)$, using a block matching method with a search range of ± 1 pixel both horizontally and vertically.

$$\mathbf{X}(o) = \begin{bmatrix} 5 & 5 & 5 & 7 & 9 & 9 \\ 5 & 5 & 5 & 7 & 9 & 9 \\ 5 & 5 & 7 & 9 & 9 & 9 \\ 5 & 7 & 9 & 9 & 9 & 9 \\ 5 & 5 & 5 & 5 & 5 & 5 \\ 5 & 5 & 5 & 5 & 5 & 5 \end{bmatrix} \quad ; \quad \mathbf{X}(o-1) = \begin{bmatrix} 5 & 5 & 5 & 9 & 9 & 9 \\ 5 & 5 & 5 & 9 & 9 & 9 \\ 5 & 5 & 5 & 9 & 9 & 9 \\ 5 & 5 & 9 & 9 & 9 & 9 \\ 5 & 9 & 9 & 9 & 9 & 9 \\ 5 & 5 & 5 & 5 & 5 & 5 \end{bmatrix}$$

a) How many shift positions must be compared for the case of one-pixel accuracy?
b) Compute the values of the cost functions (matching criteria) over the search range for the cases of cross correlation between the frames (7.154) and mean squared error (MSE) of the frame difference (7.153).
c) Compute the value of both cost functions for the case of half-pixel accuracy at the shift position $k=-0.5$, $l=1$. Use bilinear interpolation to compute the intermediate values.

Problem 7.14
Construct the optical flow equation similar to (7.194) for the case of the parametric model with four parameters (7.115). What effect would pure rotation-and-zoom motion have on the parameters of the affine model, if the horizontal/vertical sampling distances are indeed $R \neq S$, but the motion estimator assumes equality $R=S$?

8   Signal and Parameter Estimation

Methods to estimate signal or parameter values are frequently needed in multimedia signal coding and analysis. In this chapter, a more general introduction into principles of linear and nonlinear estimation is given. Some examples and procedures of estimation were already explained in the context of linear prediction and motion estimation. This chapter is intended to provide a deeper understanding of the underlying methods and the related optimization procedures. This is partially done for examples from signal restoration, which can straightforwardly be extended to prediction and interpolation of incomplete signals; most of the methods introduced are not restricted to the estimation of signals, but likewise applicable to the estimation of parameters. This is in particular true for cases where parameters are used for optimum modeling, mapping or similarity comparison between signals. The most general applicability is given for nonlinear estimation methods, which do not rely on linear filter models. Finally, methods are discussed which are capable to reject unreliable data in estimation procedures.

8.1   Observation and Degradation Models

Methods for signal and parameter estimation rely on an observation (input), and an assumption which must be made, how this observation is related to an actual signal that shall be estimated either itself or be described by parameters. Usually, a statistical model of the signal (process) and a model for the degradation must be provided; optimization of parameters characterizing these models can also be part of the estimation process. On this basis, *objective criteria* (e.g. minimization of expected differences between original signal and estimated signal) are used to achieve optimum results. A degradation model which supports a variety of possible degradations that can occur in multimedia signals is shown in Fig. 8.1a. It consists of

-   a linear shift-invariant filter of impulse response $h(\cdot)$, which can be used to model degradations as smoothing or blurring which may e.g. be caused by the limited resolution capability of the acquisition device (camera lens focus, microphone), or inaccurate sampling (shutter time too long, motion blur).

- a nonlinear distortion $g(\cdot)$, which typically expresses the nonlinear behavior of the acquisition device, often expressed by an exponential function (where α, β are type-specific constants):

$$y(\mathbf{n}) = g[x(\mathbf{n})] = \alpha \cdot x(\mathbf{n})^{\beta} . \tag{8.1}$$

- a geometric distortion $\gamma(\cdot)$, which models the mapping into the camera plane, can include nonlinear optical distortions of the lens, but may even more generally be interpreted such that a camera has another position or orientation than would be desirable. Geometric distortions can be of linear nature (e.g. affine transform) or nonlinear. Some geometric distortions are globally valid for the signal, but effects of local motion in video can also fall under this category.
- a noise component $z_1(m,n)$, which is coupled to the signal by a nonlinearity $f(\cdot)$. This is e.g. useful to model granular film noise, noise occurring in electronic cameras, or model coding noise where signal-dependent (e.g. companding) quantization functions are used.
- an additive noise component $z_2(m,n)$ which is not correlated with the signal.

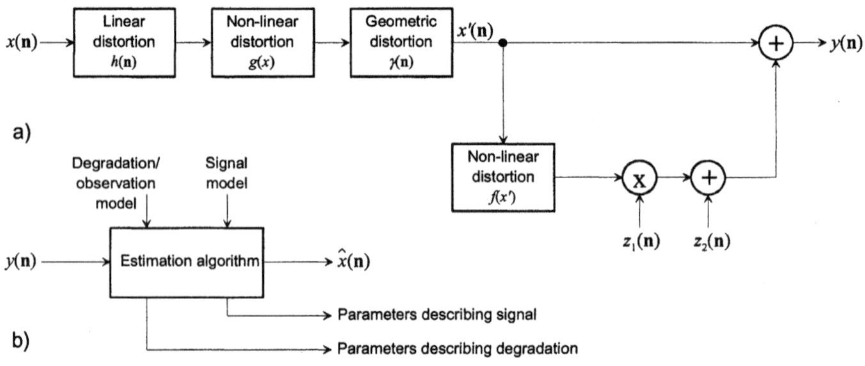

Fig. 8.1. a Degradation model including linear, nonlinear and geometric distortions, coupled and uncorrelated noise components **b** Estimation process inputs and outputs

A high-level view on the inputs and outputs of an estimation process is shown in Fig. 8.1b. The signal observation $y(\mathbf{n})$ is the input of the estimation algorithm, which is optimized based on signal models and degradation/observation models. The output can be an estimate of the signal, but parameters describing the signal or the degradation can also be provided.

8.2   Estimation based on linear filters

In this class of estimators, the goal is to use an optimized linear filter $h^{I}(\mathbf{n})$ for the estimation of the signal, such that the output is

$$\hat{x}(\mathbf{n}) = y(\mathbf{n}) * h^{I}(\mathbf{n}) \quad \Leftrightarrow \quad \hat{X}(j\Omega) = Y(j\Omega) \cdot H^{I}(j\Omega). \tag{8.2}$$

The assumption on degradation involves only the linear distortion component $h(\cdot)$ and the additive noise component $z_2(\mathbf{n})$ (subsequently denoted as $z(\mathbf{n})$), such that the process of degradation can be described as

$$y(\mathbf{n}) = x(\mathbf{n}) * h(\mathbf{n}) + z(\mathbf{n}) \quad \Leftrightarrow \quad Y(j\Omega) = X(j\Omega) \cdot H^{I}(j\Omega) + Z(j\Omega). \tag{8.3}$$

8.2.1   Inverse Filtering

A linear distortion of impulse response $h(\mathbf{n})$ has a transfer function $H(j\Omega)$. The inverse filter then has a reciprocal transfer function

$$H^{I}(j\Omega) = \frac{1}{H(j\Omega)}, \tag{8.4}$$

and will enable perfect elimination of the distortion, when no zero value is present in $H(j\Omega)$ at any frequency. Otherwise $H^{I}(j\Omega)$ would be unstable with an infinite gain (singularity) at certain frequency positions. A possible solution to this problem is the *pseudo-inverse filter*

$$H^{I}(j\Omega) = \begin{cases} \dfrac{1}{H(j\Omega)} & \text{for } H(j\Omega) \neq 0 \\ 0 & \text{for } H(j\Omega) = 0. \end{cases} \tag{8.5}$$

Pseudo-inverse filters can be realized very efficiently for finite signals like images by frequency-domain (FFT) processing. Inverse and pseudo-inverse filters are however quite sensitive in the case of additive noise distortions. The noise in the observation is not subject to the linear distortion $h(\mathbf{n})$, such that in cases where the noise component has higher energy than the linearly-distorted signal component, it may become unacceptably amplified by the inverse filter.

In principle, linear prediction can also be interpreted as an application of inverse filtering in signal estimation from the following point of view: It is assumed that a correlated AR process has been generated from a white noise innovation process by using a synthesis filter with transfer function $1/(1-H(\mathbf{z}))$. The inverse filter is the prediction error filter $1-H(\mathbf{z})$, which indeed perfectly estimates the innovation signal $z(\mathbf{n})$, providing the core of the information within the correlated process. This however is only perfect when the filter transfer function is known and the correlated process is indeed a stationary AR process. Otherwise, estimation of the optimum decorrelating filter must be performed as discussed in sec. 4.2.1, which is highly related to Wiener filtering, the topic of the subsequent section.

8.2.2   Wiener Filtering

The Wiener[1] filter is the optimum linear filter for the problem of signal estimation and reconstruction in case of an observation according to (8.3). The concept is closely related to the problem of linear prediction (cf. sec. 4.2). The goal is to determine an estimate $\hat{x}(\mathbf{n})$ for the (unknown) signal $x(\mathbf{n})$, suppressing the noise and eliminating the linear distortion as good as possible, such that the error variance

$$\sigma_e^2 = E\left\{ \left(x(\mathbf{n}) - \hat{x}(\mathbf{n}) \right)^2 \right\} \tag{8.6}$$

is minimized. When realized by FIR filters, the following relationships result:

$$y(\mathbf{n}) = \sum_{\mathbf{p}} h(\mathbf{p}) \cdot x(\mathbf{n}-\mathbf{p}) + z(\mathbf{n}) \tag{8.7}$$

$$\hat{x}(\mathbf{n}) = \sum_{\mathbf{p}} h^{\mathrm{I}}(\mathbf{p}) \cdot y(\mathbf{n}-\mathbf{p}) \tag{8.8}$$

Substituting (8.8) into (8.6) and performing a derivation over the Wiener filter coefficients gives:

$$\sigma_e^2 = E\left\{ x(\mathbf{n})^2 \right\} - 2E\left\{ x(\mathbf{n}) \cdot \sum_{\mathbf{p}} h^{\mathrm{I}}(\mathbf{p}) \cdot y(\mathbf{n}-\mathbf{p}) \right\} + E\left\{ \left[\sum_{\mathbf{p}} h^{\mathrm{I}}(\mathbf{p}) \cdot y(\mathbf{n}-\mathbf{p}) \right]^2 \right\} \tag{8.9}$$

$$\Rightarrow \frac{\partial \sigma_e^2}{\partial h^{\mathrm{I}}(\mathbf{k})} = -2E\left\{ x(\mathbf{n}) \cdot y(\mathbf{n}-\mathbf{k}) \right\} + 2E\left\{ \left[\sum_{\mathbf{p}} h^{\mathrm{I}}(\mathbf{p}) \cdot y(\mathbf{n}-\mathbf{p}) \right] \cdot y(\mathbf{n}-\mathbf{k}) \right\}. \tag{8.10}$$

The optimum set of coefficients is determined when the derivative (8.10) gives zero, which results in another form of the Wiener-Hopf equation

$$r_{xy}(\mathbf{k}) = \sum_{\mathbf{p}} h^{\mathrm{I}}(\mathbf{p}) \cdot r_{yy}(\mathbf{k}-\mathbf{p}) . \tag{8.11}$$

The order of the filter is equal to the order of the resulting linear equation system. Similar to (4.72)-(4.74), this can be expressed in matrix notation as

$$\mathbf{r}_{xy} = \mathbf{R}_{yy} \cdot \mathbf{h}^{\mathrm{I}} \Rightarrow \mathbf{h}^{\mathrm{I}} = \mathbf{R}_{yy}^{-1} \cdot \mathbf{r}_{xy} . \tag{8.12}$$

It turns out that only statistical model relationships expressed by the cross-correlation between original and observed signal have to be provided, while the statistics of the observed signal is known anyway. In (8.12), the impulse response

---

[1] Norbert WIENER was one of the fathers of system theory.

of the linear-distortion filter $h(\mathbf{p})$ is also hidden in the cross- and autocorrelation parameters. As the Fourier transforms of auto- and cross-correlation functions are power spectra and cross-power spectra (cf. sec. 3.3), the Wiener-Hopf equation can more elegantly be expressed in the frequency domain, where

$$S_{xy}(j\Omega) = S_{yy}(\Omega) \cdot H^{1}(j\Omega) \Rightarrow H^{1}(j\Omega) = \frac{S_{xy}(j\Omega)}{S_{yy}(\Omega)} . \tag{8.13}$$

Assuming that signal and noise are uncorrelated and using relationships[1]

$$S_{yy}(\Omega) = |H(j\Omega)|^{2} \cdot S_{xx}(\Omega) + S_{zz}(\Omega) \quad ; \quad S_{xy}(j\Omega) = H^{*}(j\Omega) \cdot S_{xx}(\Omega), \tag{8.14}$$

the following transfer function results for the optimum filter:

$$H^{1}(j\Omega) = \frac{H^{*}(j\Omega) \cdot S_{xx}(\Omega)}{|H(j\Omega)|^{2} \cdot S_{xx}(\Omega) + S_{zz}(\Omega)} = \frac{H^{*}(j\Omega)}{|H(j\Omega)|^{2} + \dfrac{S_{zz}(\Omega)}{S_{xx}(\Omega)}} . \tag{8.15}$$

The optimum Wiener filter is uniquely described by the transfer function of the distorting linear filter and the power spectra of the original signal and the additive noise. For S_{xx}, a model (for image and speech signals: e.g. autoregressive models) can be used which describes the expected statistical behavior of the original signal. In a restoration problem, Wiener filtering can also be applied iteratively, where the resulting estimate from one iteration is used to determine an improved approximation of the original power spectrum for the next iteration. Specifically for video sequences, the restored previous image can be used to compute a good estimate of S_{xx}, because the power spectrum is invariant against phase shifts of the signal occurring due to motion.

A comparison of (8.4) and (8.15) shows that the Wiener filter in the zero-noise case inherently reduces into the inverse filter; otherwise, compared to inverse filtering, the Wiener filter will influence the attenuation of a specific frequency component depending on the ratio of noise power and signal power at this frequency. Hence, even though the reconstruction quality highly depends on the power of the noise, the Wiener filter achieves at least the optimum solution in terms of minimization of the squared error between signal and estimate.

Wiener filtering can also be used for optimum linear interpolation in the presence of noise and by adaptation to the signal statistics. Even though this is not straightforward to see from the formulation of the problem in the spatial domain which assumes a fixed sampling rate, it is obvious that (8.15) can be combined with sampling-rate conversion, when the frequency transfer function is scaled by the desired factor of up-sampling, and the transfer function is combined with an ideal lowpass interpolation filter. Even though this will result in filters of infinite impulse response, approximations by causal filters can be used (cf. sec. 5.4.2).

---

[1] The equivalent expressions related to autocovariance matrices are
$\mathbf{R}_{yy}=\mathbf{h}[\mathbf{R}_{xx}\mathbf{h}]^{T}+\mathbf{R}_{zz}$ and $\mathbf{r}_{xy}=\mathbf{R}_{xx}\mathbf{h}$.

8.3   Least Squares Estimation

In Wiener filtering, the criterion is the minimization of error energy between the estimated (restored) signal and the original signal, provided that the linear distortion filter $h(\mathbf{n})$ and the statistical behavior of signal and noise are known. In *least squares estimation*, the criterion is the minimization of the energy between the observed signal $y(\mathbf{n})$ and the filtered estimate $\hat{x}(\mathbf{n}) * h(\mathbf{n})$, where $h(\mathbf{n})$ is again assumed to be known. This starting point (in contrast to Wiener filtering) does no longer require a statistical model for the original signal[1]:

$$\left\| y(\mathbf{n}) - \hat{x}(\mathbf{n}) * h(\mathbf{n}) \right\|^2 \overset{!}{=} \min . \tag{8.16}$$

At first sight, the problem formulation seems to be quite similar to the case of Wiener filtering when no noise is present, $E\{z^2(\mathbf{n})\}=0$. The resulting filter will in fact be identical to the case of inverse filtering, if the number of samples in the reconstructed signal $\hat{x}(\mathbf{n})$ *is exactly identical* to the number of samples in the observed signal $y(\mathbf{n})$. Least squares estimation inherently includes the problems of *interpolation* and *decimation*, where less or more samples shall be reconstructed than are available by the observation. This can be expressed in a matrix filter notation (cf. sec. 4.1.3). The observed signal is

$$\mathbf{y} = \mathbf{H} \cdot \mathbf{x} , \tag{8.17}$$

where the filter matrix $\mathbf{H}$ is a KxL matrix, with K the number of values in $\mathbf{x}$, L the number in $\mathbf{y}$. This means L samples are available, while K samples shall be reconstructed. The solution of this least squares problem targets for minimization of

$$\left\| \mathbf{e} \right\|^2 = \left\| \mathbf{y} - \mathbf{H} \cdot \hat{\mathbf{x}} \right\|^2 = \left[\mathbf{y} - \mathbf{H} \cdot \hat{\mathbf{x}} \right]^{\mathrm{T}} \left[\mathbf{y} - \mathbf{H} \cdot \hat{\mathbf{x}} \right] \overset{!}{=} \min . \tag{8.18}$$

A straightforward solution is provided by the *pseudo inverse matrix* $\mathbf{H}^{\mathrm{P}}$:

---

[1] Remark that this is a quite universal approach. As a variant, observe the case where the signals $x(\mathbf{n})$ and $y(\mathbf{n})$ are known and the function $h(\mathbf{n})$ is unknown. Then, the formulation of the estimation problem will be

$$\left\| y(\mathbf{n}) - x(\mathbf{n}) * \hat{h}(\mathbf{n}) \right\|^2 \overset{!}{=} \min ,$$

and subsequent optimization is done regarding estimation of the filter function. As convolution is commutative, all following steps can be performed by interchanging the role of x and h. Observe that this is exactly the approach taken in optical-flow motion estimation (sec. 7.6.2), where two video frames x and y are known, and the task is to estimate the motion shift, which gives in principle the parameters of a linear (phase-shift) filter. Even more, the concept is not limited to linear distortions, but can be extended to estimate parameters of nonlinear and geometric distortions, as long as the underlying processing applied to the signal can be expressed by a matrix-vector form, where the vector shall contain the parameters to be estimated. For an example, refer to (7.195).

$$\hat{\mathbf{x}} = \mathbf{H}^{\mathrm{P}} \cdot \mathbf{y} , \qquad (8.19)$$

where different cases have to be distinguished:

$$K < L \;:\; \mathbf{H}^{\mathrm{P}} = (\mathbf{H}^{\mathrm{T}} \cdot \mathbf{H})^{-1} \mathbf{H}^{\mathrm{T}} \quad ; \quad \mathbf{H}^{\mathrm{P}} \cdot \mathbf{H} = \mathbf{I} \;; \quad \mathbf{H} \cdot \mathbf{H}^{\mathrm{P}} \neq \mathbf{I}$$

$$K = L \;:\; \mathbf{H}^{\mathrm{P}} = \mathbf{H}^{-1} \quad ; \quad \mathbf{H}^{\mathrm{P}} \cdot \mathbf{H} = \mathbf{H} \cdot \mathbf{H}^{\mathrm{P}} = \mathbf{I} \qquad (8.20)$$

$$K > L \;:\; \mathbf{H}^{\mathrm{P}} = \mathbf{H}^{\mathrm{T}} (\mathbf{H} \cdot \mathbf{H}^{\mathrm{T}})^{-1} \quad ; \quad \mathbf{H} \cdot \mathbf{H}^{\mathrm{P}} = \mathbf{I} \quad ; \quad \mathbf{H}^{\mathrm{P}} \cdot \mathbf{H} \neq \mathbf{I} .$$

The pseudo inverse has a size LxK, and for the case of $K=L$ is identical to the conventional inverse of a matrix (B.14). For $K<L$, the inverse solution will be unique as well, which means that in both cases $\hat{\mathbf{x}} = \mathbf{x}$, whenever the inverse or pseudo-inverse can be computed. For $K>L$, the equation system is *underdetermined*, which means that less conditions (free parameters) than unknowns exist, and hence typically $\hat{\mathbf{x}} \neq \mathbf{x}$. Even in this case, the solution of the pseudo-inverse minimizes the error energy in (8.18), when no noise is present [JAIN 1989]. The pseudo inverse will however no longer be the optimum solution if noise is added to the degraded signal[1],

$$\mathbf{y} = \mathbf{H} \cdot \mathbf{x} + \mathbf{z} \;\Rightarrow\; E\left\{[\mathbf{y} - \mathbf{H} \cdot \mathbf{x}][\mathbf{y} - \mathbf{H} \cdot \mathbf{x}]^{\mathrm{T}}\right\} = \underbrace{E\left\{\mathbf{z} \cdot \mathbf{z}^{\mathrm{T}}\right\}}_{\mathbf{R}_{zz}} . \qquad (8.21)$$

In this case, it is necessary to observe the properties of the noise as well. Multiplying both sides of (8.21) by $\mathbf{R}_{zz}^{-1}$, it can be concluded that the following error energy shall be minimized now:

$$\|\mathbf{e}\|^2 = [\mathbf{y} - \mathbf{H} \cdot \hat{\mathbf{x}}]^{\mathrm{T}} \mathbf{R}_{zz}^{-1} [\mathbf{y} - \mathbf{H} \cdot \hat{\mathbf{x}}] \overset{!}{=} \min , \qquad (8.22)$$

where $\mathbf{R}_{zz}^{-1}$ is the inverse of the noise autocovariance matrix[2]. Assuming that the noise is not correlated with the signal, the derivative of the error function over the estimate will be:

$$\frac{\partial \|\mathbf{e}\|^2}{\partial \hat{\mathbf{x}}} = -2\mathbf{H}^{\mathrm{T}} \mathbf{R}_{zz}^{-1} [\mathbf{y} - \mathbf{H}\hat{\mathbf{x}}] . \qquad (8.23)$$

A common solution to this problem is to determine an initial estimate $\hat{\mathbf{x}}_0$ by pseudo-inversion, and then to optimize $\hat{\mathbf{x}}$ using the gradient of the error energy function (8.23). In an iterative process of *linear regression*, an estimate $\hat{\mathbf{x}}_r$ is computed by the rth iteration. Then, $\hat{\mathbf{x}}_{r+1}$ of the subsequent iteration is optimized

---

[1] (8.21) is the matrix formulation of (8.3).

[2] A zero-mean noise process is assumed here. Weighting the squared vector norm by the inverse of a covariance matrix is equivalent to a spectral weighting adapted to the spectral property that this covariance matrix expresses. This will more explicitly be shown in sec. 9.2 in the context of the Mahalanobis distance criterion; presently, we just conclude that the error energy is weighted by the spectral properties of the additive noise.

by the direction of the negative gradient vector $\mathbf{g}_r$. In the *steepest descent* method, the gradient is multiplied by a convergence factor ε_r, such that[1]

$$\hat{\mathbf{x}}_{r+1} = \hat{\mathbf{x}}_r + \varepsilon_r \mathbf{g}_r \quad \text{with} \quad \mathbf{g}_r = -\mathbf{H}^{\mathrm{T}}(\mathbf{y} - \mathbf{H} \cdot \hat{\mathbf{x}}_r). \tag{8.24}$$

From this,

$$\hat{\mathbf{x}}_{r+1} = \hat{\mathbf{x}}_r - \varepsilon_r \mathbf{H}^{\mathrm{T}} \left(\mathbf{y} - \mathbf{H} \cdot \left[\hat{\mathbf{x}}_{r-1} + \varepsilon_{r-1} \mathbf{g}_{r-1} \right] \right)$$
$$\Rightarrow \mathbf{g}_r = \mathbf{H}^{\mathrm{T}} \left(\mathbf{y} - \mathbf{H} \cdot \hat{\mathbf{x}}_{r-1} \right) + \varepsilon_{r-1} \cdot \mathbf{H}^{\mathrm{T}} \mathbf{H} \cdot \mathbf{g}_{r-1} = \mathbf{g}_{r-1} + \varepsilon_{r-1} \cdot \mathbf{H}^{\mathrm{T}} \mathbf{H} \cdot \mathbf{g}_{r-1}. \tag{8.25}$$

It turns out that $\mathbf{g}_r$ by itself can be computed recursively, which simplifies the procedure. The optimum convergence factor for iteration r in the case of the steepest-descent method is[2] [JAIN 1989]

$$\varepsilon_r = -\frac{\mathbf{g}_r^{\mathrm{T}} \mathbf{g}_r}{\sigma_z^2 \cdot \mathbf{g}_r^{\mathrm{T}} \mathbf{H}^{\mathrm{T}} \mathbf{H} \mathbf{g}_r}. \tag{8.26}$$

Fig. 8.2a illustrates the problem of iterative optimization, where $\|\mathbf{e}\|^2$ is the cost function which shall be minimized. In general, the gradient-descent optimization guarantees that the optimum is found, if the cost function is convex. If several local minima exist (Fig. 8.2b), the process may get stuck without reaching the global optimum.

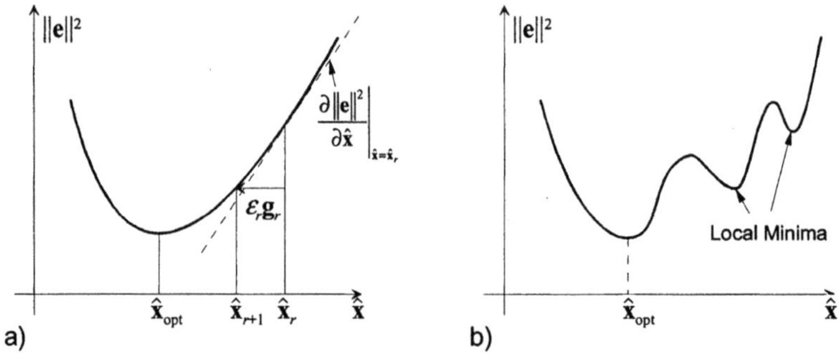

Fig. 8.2. a Convergence towards the minimum of the cost function in linear regression **b** multiple local minima

---

[1] under simplified assumption that the additive noise shall be spectrally white, i.e. $\mathbf{R}_{zz} = \sigma_z^2 \mathbf{I}$.
[2] The general form of this approach is commonly denoted as *Gauss-Newton method*.

8.4   Singular Value Decomposition

A generalization of pseudo-inversion is provided by *Singular Value Decomposition* (SVD) of the matrix $\mathbf{H}$. As $\mathbf{H}$ is non-quadratic (K columns and L rows), neither an inverse nor a determinant can be computed. It is however possible to define quadratic sub-matrices of size $P\mathrm{x}P$, $1\leq P\leq\min(K,L)$. The rank R of a non-quadratic matrix $\mathbf{H}$ is the size P of the largest quadratic sub-matrix having a non-zero determinant. According to the definition of determinant computation (B.9), all possible larger sub-matrices where $P>R$ must have a zero determinant. The method of pseudo-inversion as previously introduced allows the optimum solution of the estimation problem in the least-squares sense, if for the matrix $\mathbf{H}$, $R=\min(K,L)$.

An $L\mathrm{x}L$-Matrix $\mathbf{U}$ and a $K\mathrm{x}K$-Matrix $\mathbf{V}$ are defined, which shall have the following relationship with $\mathbf{H}$:

$$\mathbf{U}^{\mathrm{T}}\mathbf{H}\mathbf{V} = \mathbf{\Lambda}^{(1/2)} = \left[\begin{array}{ccccc} \sqrt{\lambda(1)} & 0 & \cdots & 0 & \vdots \\ 0 & \ddots & \ddots & \vdots & \mathbf{0} \\ \vdots & \ddots & \ddots & 0 & \vdots \\ 0 & \cdots & 0 & \sqrt{\lambda(R)} & \\ \cdots & \mathbf{0} & \cdots & & 0 \end{array}\right] \begin{array}{l} \left.\vphantom{\begin{array}{c}a\\a\\a\\a\end{array}}\right\}R \\ \left.\vphantom{a}\right\}L-R \end{array} \tag{8.27}$$

$$\underbrace{\hspace{3cm}}_{R} \quad \underbrace{\hspace{1cm}}_{K-R}$$

The elements in $\mathbf{\Lambda}^{(1/2)}$ are the R *singular values* $\lambda^{1/2}(r)$ of $\mathbf{H}$. They are square roots of the R non-zero eigenvalues of the $L\mathrm{x}L$-Matrix $\mathbf{H}\mathbf{H}^{\mathrm{T}}$ and the $K\mathrm{x}K$-Matrix $\mathbf{H}^{\mathrm{T}}\mathbf{H}$ (for $L>R$ or $K>R$, the remaining eigenvalues of $\mathbf{H}\mathbf{H}^{\mathrm{T}}$ or $\mathbf{H}^{\mathrm{T}}\mathbf{H}$ are zero, respectively). The columns of $\mathbf{U}$ are the eigenvectors $\mathbf{u}_r$ of $\mathbf{H}\mathbf{H}^{\mathrm{T}}$, the columns of $\mathbf{V}$ the eigenvectors $\mathbf{v}_r$ of $\mathbf{H}^{\mathrm{T}}\mathbf{H}$. The following conditions apply:

$$\mathbf{U}^{\mathrm{T}}\left[\mathbf{H}\mathbf{H}^{\mathrm{T}}\right]\mathbf{U} = \mathbf{\Lambda}^{(K)} = \left[\begin{array}{ccccc} \lambda(1) & 0 & \cdots & 0 & \vdots \\ 0 & \ddots & \ddots & \vdots & \mathbf{0} \\ \vdots & \ddots & \ddots & 0 & \vdots \\ 0 & \cdots & 0 & \lambda(R) & \\ \cdots & \mathbf{0} & \cdots & & 0 \end{array}\right] \begin{array}{l} \left.\vphantom{\begin{array}{c}a\\a\\a\\a\end{array}}\right\}R \\ \left.\vphantom{a}\right\}L-R \end{array} \tag{8.28}$$

$$\underbrace{\hspace{3cm}}_{R} \quad \underbrace{\hspace{1cm}}_{L-R}$$

$$\mathbf{V}^{\mathrm{T}}\left[\mathbf{H}^{\mathrm{T}}\mathbf{H}\right]\mathbf{V} = \mathbf{\Lambda}^{(L)} = \left[\begin{array}{ccccc} \lambda(1) & 0 & \cdots & 0 & \vdots \\ 0 & \ddots & \ddots & \vdots & \mathbf{0} \\ \vdots & \ddots & \ddots & 0 & \vdots \\ 0 & \cdots & 0 & \lambda(R) & \\ \cdots & \mathbf{0} & \cdots & & 0 \end{array}\right] \begin{array}{l} \left.\vphantom{\begin{array}{c}a\\a\\a\\a\end{array}}\right\}R \\ \left.\vphantom{a}\right\}K-R \end{array} \tag{8.29}$$

$$\underbrace{\hspace{3cm}}_{R} \quad \underbrace{\hspace{1cm}}_{K-R}$$

By reversing the principle in (8.27), it is possible to express $\mathbf{H}$ as follows; due to the diagonally populated matrix $\Lambda^{(1/2)}$, only R non-zero components are retained:

$$\mathbf{H} = \mathbf{U}\Lambda^{(1/2)}\mathbf{V}^{\mathrm{T}} = \sum_{r=1}^{R} \lambda^{1/2}(r) \cdot \mathbf{u}_r \mathbf{v}_r^{\mathrm{T}} \ . \tag{8.30}$$

Obviously, $\mathbf{H}$ can be expressed as a linear combination of outer products of eigenvectors $\mathbf{u}_r\mathbf{v}_r^{\mathrm{T}}$ from $\mathbf{U}$ and $\mathbf{V}$, weighted by the respective singular values $\lambda^{1/2}(r)$. As all eigenvectors are orthogonal, this set of matrices will be orthogonal as well. Their inverse will then be the transpose, while the inverse of $\Lambda^{(1/2)}$ is a matrix $\Lambda^{(-1/2)}$, which has a similar structure as $\Lambda^{(1/2)}$, but replaces singular values by their reciprocals $\lambda^{-1/2}(r)$. The *generalized inverse* $\mathbf{H}^{\mathrm{g}}$ of (8.30) is then the formal concatenation of the three inverted matrices, and is for the case $R=\min(K,L)$ identical to the pseudo-inverse $\mathbf{H}^{\mathrm{P}}$:

$$\mathbf{H}^{\mathrm{g}} = \mathbf{V}\Lambda^{(-1/2)}\mathbf{U}^{\mathrm{T}} = \sum_{r=1}^{R} \lambda^{-1/2}(r) \cdot \mathbf{v}_r \mathbf{u}_r^{\mathrm{T}} \ . \tag{8.31}$$

Similar to (8.19), the signal estimation can be performed by the pseudo inverse, again the solution minimizes – if no additive noise is involved – the error in estimation in the sense of a least-squares problem. Like the pseudo-inverse, $\mathbf{H}^{\mathrm{g}}$ is an LxK-Matrix such that

$$\hat{\mathbf{x}} = \mathbf{H}^{\mathrm{g}} \cdot \mathbf{y} \ . \tag{8.32}$$

By SVD, it is now possible to perform an estimation without using *all singular values*. SVD performs a decomposition of $\mathbf{H}$ into an optimally compact transformed expansion, similar to the Karhunen-Loève transform (4.148)[1]. Consequently, the always-positive singular values are ordered by sequence of decreasing amplitude or relevance. Hence, using only the first R' singular values $(R'<R)$ will give a sufficiently good approximation for the estimation as well. This can be advantageous for suppression of noise influences.

Applications of the SVD method are not limited to the restoration problem. If SVD is applied to matrices which bear statistically dependent values (e.g. feature vectors captured at different time instances), SVD provides an optimum compact representation by the singular values (cf. sec. 9.1.2).

---

[1] Actually, SVD could also be interpreted as a generalization of the KLT for the case of orthogonalization of non-quadratic matrices.

8.5 ML and MAP Estimation

The methods of signal estimation introduced so far were restricted to compensate linear degradations. Consequently, methods as linear filters, linear transforms and other linear optimization (e.g. regression) were applicable to solve the estimation problem. If however nonlinear relationships exist between the signal $\mathbf{x}$ to be estimated and the observed signal $\mathbf{y}$, the results achievable by linear methods are probably sub-optimum. Most commonly, methods for nonlinear estimation are based on statistical criteria of *conditional probabilities $P(\cdot|\cdot)$* for signals of discrete values, or *conditional probability density functions $p(\cdot|\cdot)$* for continuous-value signals. Here, the case of continuous values is regarded first, where the vector Gaussian distribution (3.40) shall be fitted as a model for the probability density of signals and noise [1]. Identical principles are applicable for the case of discrete-value signals, which are used by numerical approximations of the continuous-value case anyway. The conditional PDF $p(\mathbf{y}|\mathbf{x})$ expresses by which probability an observation $\mathbf{y}$ can be expected from a given signal value $\mathbf{x}$; $p(\mathbf{y}|\mathbf{x})$ is the 'a priori' conditional PDF, which can fully be determined by signal and degradation models. Assume that a nonlinear distortion $g(\cdot)$ is effective in addition to the linear distortion $h(\cdot)$, and a noise component $\mathbf{z}$ shall be added such that

$$\mathbf{y} = g(\mathbf{H} \cdot \mathbf{x}) + \mathbf{z}. \tag{8.33}$$

A vector Gaussian distribution may characterize the zero-mean noise process,

$$p(\mathbf{z}) = \left[\frac{1}{(2\pi)^K |\mathbf{R}_{zz}|} \right]^{1/2} e^{-1/2 \cdot \mathbf{z}^{\mathrm{T}} \mathbf{R}_{zz}^{-1} \mathbf{z}} \quad ; \quad \mathbf{R}_{zz} = E\{\mathbf{z} \cdot \mathbf{z}^{\mathrm{T}}\}. \tag{8.34}$$

The a priori PDF $p(\mathbf{y}|\mathbf{x})$ then describes the remaining uncertainty in the statistical description of the observed signal $\mathbf{y}$, provided that the distortion of $\mathbf{x}$ occurs by functions $h(\cdot)$ and $g(\cdot)$. This is characterized by the difference $\mathbf{y}-g(\mathbf{H}\cdot\mathbf{x})=\mathbf{z}$, where the additive noise causes a deviation as described by its covariance parameters[2]:

$$p(\mathbf{y}|\mathbf{x}) = \left[\frac{1}{(2\pi)^K |\mathbf{R}_{zz}|} \right]^{1/2} e^{-1/2 \cdot [\mathbf{y}-g(\mathbf{H}\cdot\mathbf{x})]^{\mathrm{T}} \mathbf{R}_{zz}^{-1} [\mathbf{y}-g(\mathbf{H}\cdot\mathbf{x})]}. \tag{8.35}$$

---

[1] Remark that processes of Gaussian distribution can fully be modeled by linear methods. This imposes for the moment a certain restriction to model nonlinear behavior; however, usage of Gaussian models is highly attractive as it allows analytic optimization and simplifies understanding of the principles behind the optimization.

[2] Note that the exponent of this conditional probability is identical to (8.22), when no nonlinear distortion is in effect.

The *Maximum Likelihood* (ML) estimation selects an estimate $\hat{\mathbf{x}}$ in the case when $\mathbf{y}$ is observed such that

$$\hat{\mathbf{x}} = \arg\max_{\mathbf{x}} p(\mathbf{y} \mid \mathbf{x}), \tag{8.36}$$

which means that just one signal constellation vector $\mathbf{x}$ is selected as estimate, for which the a priori conditional PDF is largest. A simple solution can be obtained by a logarithmic transformation of (8.35). The logarithm is a steady function, such that it is irrelevant for optimization whether the maximum is sought in $p(\mathbf{y}|\mathbf{x})$ or by the logarithmic mapping:

$$\ln p(\mathbf{y}|\mathbf{x}) = -\tfrac{1}{2}\big[\mathbf{y} - g(\mathbf{H}\cdot\mathbf{x})\big]^{\mathrm{T}} \mathbf{R}_{zz}^{-1} \big[\mathbf{y} - g(\mathbf{H}\cdot\mathbf{x})\big] - \tfrac{1}{2}\ln\big[(2\pi)^{K} |\mathbf{R}_{zz}|\big]. \tag{8.37}$$

The rightmost term is a constant not depending on $\mathbf{x}$, such that the optimum estimate for $\hat{\mathbf{x}}$ will be achieved by inverting the sign and *minimization* of the following cost function:

$$\Delta_{\mathrm{ML}}(\hat{\mathbf{x}}) = \tfrac{1}{2}\big[\mathbf{y} - g(\mathbf{H}\cdot\hat{\mathbf{x}})\big]^{\mathrm{T}} \mathbf{R}_{zz}^{-1} \big[\mathbf{y} - g(\mathbf{H}\cdot\hat{\mathbf{x}})\big]. \tag{8.38}$$

Derivation of (8.38) gives

$$\frac{\partial \Delta_{\mathrm{ML}}(\hat{\mathbf{x}})}{\partial \hat{\mathbf{x}}} = -\mathbf{H}^{\mathrm{T}} \cdot \mathbf{G}' \cdot \mathbf{R}_{zz}^{-1} \big[\mathbf{y} - g(\mathbf{H}\cdot\hat{\mathbf{x}})\big]. \tag{8.39}$$

The 'derivative' of the nonlinear function is expressed by a diagonal matrix $\mathbf{G}'$, which can be gained by linear approximation of partial derivatives around an actual value v, which is the input to the nonlinear distortion block from Fig. 8.1,

$$\mathbf{G}' = \begin{bmatrix} \left.\frac{\partial g(v)}{\partial v}\right|_{v=\hat{u}(1)} & 0 & \cdots & 0 \\ 0 & \left.\frac{\partial g(v)}{\partial v}\right|_{v=\hat{u}(2)} & \ddots & \vdots \\ \vdots & \ddots & \ddots & 0 \\ 0 & \cdots & 0 & \left.\frac{\partial g(v)}{\partial v}\right|_{v=\hat{u}(K)} \end{bmatrix} \;;\quad \hat{\mathbf{u}} = \mathbf{H}\hat{\mathbf{x}} = \begin{bmatrix} \hat{u}(1) & \hat{u}(2) & \cdots & \cdots & \hat{u}(K) \end{bmatrix}^{\mathrm{T}}. \tag{8.40}$$

Observe that for the case where only linear distortion is involved, $g(\mathbf{H}\hat{\mathbf{x}}) = \mathbf{H}\hat{\mathbf{x}}$, $\mathbf{G}'=\mathbf{I}$, such that the result (8.39) would be identical with the derivative of the least-squares cost function (8.23). In general, the ML estimation is performed by similar iterative solutions as used in least-squares estimation to find the minimum of (8.38). In such procedures, the linear approximation of $\mathbf{G}'$ can be computed by using the result of the previous iteration step.

The *Maximum a Posteriori* (MAP) method for signal estimation is based on the conditional PDF $p(\mathbf{x}|\mathbf{y})$, which is in analogy with the previous explanation the 'a posteriori' conditional PDF. It expresses how large the probability is that a certain value $\mathbf{x}$ has been the original signal, when a value $\mathbf{y}$ is observed. As an optimum value for $\hat{\mathbf{x}}$, it is then reasonable to choose an estimate

$$\hat{\mathbf{x}} = \arg\max_{\mathbf{x}} p(\mathbf{x}\,|\,\mathbf{y})\,. \tag{8.41}$$

According to (3.26), the following relationship exist between the two conditional PDFs and the first-order PDFs $p(\mathbf{x})$ and $p(\mathbf{y})$:

$$p(\mathbf{x}|\mathbf{y})\cdot p(\mathbf{y}) = p(\mathbf{y}|\mathbf{x})\cdot p(\mathbf{x})\,. \tag{8.42}$$

Re-formulation gives the *Bayes theorem*[1]

$$p(\mathbf{x}|\mathbf{y}) = \frac{p(\mathbf{y}|\mathbf{x})\cdot p(\mathbf{x})}{p(\mathbf{y})}\,. \tag{8.43}$$

This optimization can be performed independently of $p(\mathbf{y})$[2]. Substituting (3.40) and (8.35) into (8.43), and again performing a logarithmic transformation gives

$$\ln p(\mathbf{x}|\mathbf{y}) = -\tfrac{1}{2}\big[\mathbf{y} - g(\mathbf{H}\cdot\mathbf{x})\big]^{\mathrm{T}} \mathbf{R}_{zz}^{-1}\big[\mathbf{y} - g(\mathbf{H}\cdot\mathbf{x})\big] - \tfrac{1}{2}\ln\big[(2\pi)^{K}\,|\mathbf{R}_{zz}|\big]$$
$$-\tfrac{1}{2}\big[\mathbf{x} - \mathbf{x}_{m}\big]^{\mathrm{T}} \mathbf{R}'_{xx}{}^{-1}\big[\mathbf{x} - \mathbf{x}_{m}\big] - \tfrac{1}{2}\ln\big[(2\pi)^{K}\,|\mathbf{R}'_{xx}|\big] - \ln p(\mathbf{y})\,. \tag{8.44}$$

All logarithmic terms in (8.44) do not depend on the actual value of $\mathbf{x}$ and are hence constants, such that they can be neglected in the maximization of the posterior PDF. By reverting the signs in the remaining terms, the following function must be *minimized*:

$$\Delta_{\mathrm{MAP}}(\hat{\mathbf{x}}) = \tfrac{1}{2}\big[\hat{\mathbf{x}} - \mathbf{x}_{m}\big]^{\mathrm{T}} \mathbf{R}'_{xx}{}^{-1}\big[\hat{\mathbf{x}} - \mathbf{x}_{m}\big] + \tfrac{1}{2}\big[\mathbf{y} - g(\mathbf{H}\cdot\hat{\mathbf{x}})\big]^{\mathrm{T}} \mathbf{R}_{zz}^{-1}\big[\mathbf{y} - g(\mathbf{H}\cdot\hat{\mathbf{x}})\big]\,. \tag{8.45}$$

The derivative of (8.45) is

$$\frac{\partial\Delta_{\mathrm{MAP}}(\hat{\mathbf{x}})}{\partial\hat{\mathbf{x}}} = \mathbf{R}'_{xx}{}^{-1}\big[\hat{\mathbf{x}} - \mathbf{x}_{m}\big] - \mathbf{H}^{\mathrm{T}}\cdot\mathbf{G}'\cdot\mathbf{R}_{zz}^{-1}\big[\mathbf{y} - g(\mathbf{H}\cdot\hat{\mathbf{x}})\big]\,. \tag{8.46}$$

Comparing the cost functions (8.38) and (8.45), it becomes obvious that estimates $\hat{\mathbf{x}}$ which largely deviate from the mean $\mathbf{x}_{m}$ will have less chances to be considered as optimum values, unless high variance/covariance values of the source would support the presence of such outliers (cf. Problems 8.4 and 8.5). Similar to the iterative approach in least-squares estimation, the minimization of the cost function in ML or MAP estimation can be approached iteratively as follows:

---

[1] MAP estimation is often also denoted as *Bayesian estimation*.

[2] The irrelevance of *statistical expectation* about the observation was already shown for the case of Wiener filters. In fact, *standalone statistical knowledge* about the observation is useless in general, as it does not help to gain any knowledge about the source signal. Only statistical knowledge that helps to understand *how the source is mapped* into the observed signal is useful, such that conclusions about the signal can be drawn from the concrete observation. This is also in harmony with the information-theoretic as discussed in sec. 3.5: Only if the mutual information between source and receiver alphabets is larger than zero, useful knowledge about the source status can be available at the receiver side.

$$\hat{\mathbf{x}}_{r+1} = \hat{\mathbf{x}}_r + \varepsilon_r \cdot \frac{\partial \Delta(\hat{\mathbf{x}}_r)}{\partial \hat{\mathbf{x}}_r}. \tag{8.47}$$

The MAP concept is a widely used nonlinear optimization method in signal restoration, parameter estimation, signal classification and segmentation.

8.6   Kalman Estimation

Estimation methods considered so far were based on the optimization of the estimated result for an observed set of signal values (vector $\mathbf{y}$). In particular for signals which change over time or space, values or parameters must be estimated at various spatio-temporal instances (e.g. motion parameters to be estimated at different positions for each video frame). In such cases it can be advantageous if recursive estimation methods are applied, in which previous results are fed back into subsequent estimation steps. This follows by an assumption that the signal or the parameters to be estimated are only changing slowly[1]. Similar constraints were already introduced in the context of motion estimation (cf. sec. 7.6.2 and 7.6.3). An additional advantage of recursive estimation is the suppression of noise influences which may otherwise affect the reliability of the estimation result. To achieve this, a good model must be available about characteristics of such possible perturbations in the estimation process. The most widely used method to optimize recursive estimation in the presence of noise fluctuations is *Kalman filtering*, which is an estimation approach based on a state model. Again, the method is introduced based on an example from signal restoration, where a model for the signal to be estimated is needed. A vector state model is used here, where a vector consists of signal values $\mathbf{x}_r$ in state r,

$$\mathbf{x}_r = \mathbf{A}_r \mathbf{x}_{r-1} + \mathbf{B}_r \mathbf{e}_r, \tag{8.48}$$

which is described from the previous-state vector $\mathbf{x}_{r-1}$, and a zero-mean error vector $\mathbf{e}_r$, not to be correlated with the signal. The observation is however not the signal vector $\mathbf{x}_r$, but a disturbed version thereof, which is modeled here by a linear filter matrix $\mathbf{H}$ and an additive zero-mean noise component $\mathbf{z}_r$,

$$\mathbf{y}_r = \mathbf{H}_r \mathbf{x}_r + \mathbf{z}_r. \tag{8.49}$$

Further, the autocovariance matrices $\mathbf{R}_{ee,r} = E\{\mathbf{e}_r \mathbf{e}_r^{\mathrm{T}}\}$ and $\mathbf{R}_{zz,r} = E\{\mathbf{z}_r \mathbf{z}_r^{\mathrm{T}}\}$ shall be given, describing the statistical properties in state r, and the estimation result $\hat{\mathbf{x}}_{r-1}$ of the previous state r-1 is available. By this, a preliminary estimate for the signal in state r can be defined by the *state prediction equation*:

---

[1] As an example, the motion of a video sequence, due to mass inertia of the object that actually moves, follows a more or less continuous and steady trajectory.

$$\hat{\mathbf{x}}_r ' = \mathbf{A}_r \hat{\mathbf{x}}_{r-1} \; , \tag{8.50}$$

from which an estimation error

$$\varepsilon_r = \mathbf{x}_r - \hat{\mathbf{x}}_r ' \; , \tag{8.51}$$

characterized by the autocovariance matrix $\mathbf{R}_{\varepsilon\varepsilon,r} = E\{\varepsilon_r \varepsilon_r^{\mathrm{T}}\}$ will occur[1]. The estimate $\hat{\mathbf{x}}_r '$ is now used to compute an estimate for the *observed* vector:

$$\hat{\mathbf{y}}_r = \mathbf{H}_r \hat{\mathbf{x}}_r ' \; . \tag{8.52}$$

The difference

$$\mathbf{v}_r = \mathbf{y}_r - \hat{\mathbf{y}}_r = \mathbf{H}_r \mathbf{x}_r + \mathbf{z}_r - \mathbf{H}_r \hat{\mathbf{x}}_r ' = \mathbf{H}_r \varepsilon_r + \mathbf{z}_r \tag{8.53}$$

is the *Kalman innovation*, which reflects the uncertainties both about the new signal state and about the distortions. Assuming that ε and $\mathbf{z}$ are uncorrelated, $\mathbf{v}_r$ has an autocovariance matrix

$$\mathbf{R}_{vv,r} = E\{\mathbf{v}_r \mathbf{v}_r^{\mathrm{T}}\} = \mathbf{H}_r \mathbf{R}_{\varepsilon\varepsilon,r} \mathbf{H}_r^{\mathrm{T}} + \mathbf{R}_{zz,r} \; . \tag{8.54}$$

During the state recursions, the error (8.51) shall be minimized. Herein, the matrix $\mathbf{R}_{\varepsilon\varepsilon,r}$ accumulates all information about the covariance from all previous observations $\mathbf{y}_k$, where $k<r$. For a K-dimensional state vector $\mathbf{x}$, this matrix has the form

$$\mathbf{R}_{\varepsilon\varepsilon,r} = \begin{bmatrix} E\{\varepsilon_r^2(1)|\mathbf{y}_k,...,\mathbf{y}_1\} & E\{\varepsilon_r(1)\varepsilon_r(2)|\mathbf{y}_k,...,\mathbf{y}_1\} & \cdots & E\{\varepsilon_r(1)\varepsilon_r(K)|\mathbf{y}_k,...,\mathbf{y}_1\} \\ E\{\varepsilon_r(2)\varepsilon_r(1)|\mathbf{y}_k,...,\mathbf{y}_1\} & E\{\varepsilon_r^2(2)|\mathbf{y}_k,...,\mathbf{y}_1\} & & \\ \vdots & & \ddots & \vdots \\ & & & \ddots \\ E\{\varepsilon_r(K)\varepsilon_r(1)|\mathbf{y}_k,...,\mathbf{y}_1\} & & \cdots & E\{\varepsilon_r^2(K)|\mathbf{y}_k,...,\mathbf{y}_1\} \end{bmatrix} . \tag{8.55}$$

If the goal of optimization is minimization of the energy of ε, the trace of this matrix must be minimized. This is achieved by performing a prediction of the matrix, which describes the state change according to (8.48). This gives the *covariance prediction equation*:

$$\hat{\mathbf{R}}_{\varepsilon\varepsilon,r} = \mathbf{A}_r \mathbf{R}_{\varepsilon\varepsilon,r-1} \mathbf{A}_r + \mathbf{B}_r \mathbf{R}_{ee,r} \mathbf{B}_r \; , \tag{8.56}$$

by which the *Kalman gain* matrix is determined as

$$\mathbf{K}_r = \hat{\mathbf{R}}_{\varepsilon\varepsilon,r} \mathbf{H}_r^{\mathrm{T}} \mathbf{R}_{vv,r}^{-1} \; . \tag{8.57}$$

The Kalman gain is finally used to perform the *state update* and achieve the estimation result

---

[1] If for an initial state $\hat{\mathbf{x}}_0 = 0 \Rightarrow \hat{\mathbf{x}}_1 ' = 0$ is assumed, the autocovariance of ε is identical to the signal autocovariance, $\mathbf{R}_{\varepsilon\varepsilon,1} = E\{\mathbf{x}_0 \mathbf{x}_0^{\mathrm{T}}\}$.

$$\hat{\mathbf{x}}_r = \hat{\mathbf{x}}_r{'} + \mathbf{K}_r \mathbf{v}_r \, .$$

(8.58)

The last step is the update of the covariance matrix for the next state, also denoted as *Riccati equation*

$$\mathbf{R}_{\varepsilon\varepsilon,r} = \hat{\mathbf{R}}_{\varepsilon\varepsilon,r} - \mathbf{K}_r \mathbf{H}_r \hat{\mathbf{R}}_{\varepsilon\varepsilon,r} \, .$$

(8.59)

If $\mathbf{H}_r$ is a *KxL* matrix (where K is the length of $\mathbf{x}/\mathbf{e}/\boldsymbol{\varepsilon}$, L the length of $\mathbf{y}/\mathbf{z}/\mathbf{v}$), $\mathbf{K}_r$ will be an *LxK* matrix, and the Kalman estimator is also applicable to the case where the number of samples in the observed signal is not equal to the number of samples to be estimated.

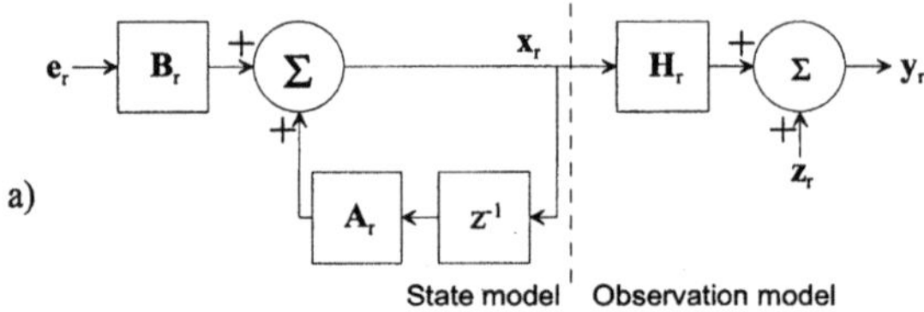

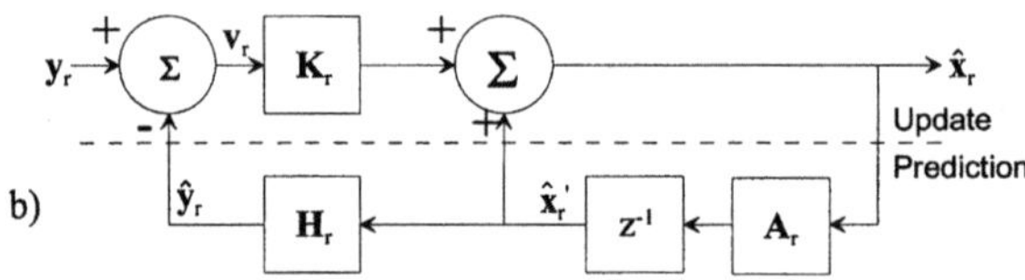

Fig. 8.3. Principle of a Kalman estimator

Fig. 8.3a shows the elements of the state model and the observation model. Fig. 8.3b illustrates the structure of a Kalman estimator as described above. If the update of all elements is performed within each step, Kalman estimation is computationally rather complex. Simplifications are achieved if an invariant signal-state model or invariant observation models are assumed. Kalman estimators are also applicable as *predictors* for signals with high additive noise components.

Kalman filtering for image restoration. The separable AR(1) model (4.62) can be written in row-wise vector form, when all elements of the signal relating to the nth row are brought to the left side of a difference equation

$$x(m,n) - \rho_h x(m-1,n) = \rho_v x(m,n-1) - \rho_h \rho_v x(m-1,n-1) + e(m,n),$$

(8.60)

where $e(m,n)$ is the non-predictable component. Each row characterizes one state ($r \equiv n$) of Kalman filter processing. The AR(1) model can then be expressed in the following vector state notation for row n:

$$\mathbf{S}_1 \mathbf{x}_n = \mathbf{S}_2 \mathbf{x}_{n-1} + \mathbf{e}_n \Rightarrow \mathbf{x}_n = \mathbf{S}_1^{-1} \mathbf{S}_2 \mathbf{x}_{n-1} + \mathbf{S}_1^{-1} \mathbf{e}_n \, .$$

(8.61)

Herein, $\mathbf{x}$ and $\mathbf{e}$ are vectors of M elements each, and the two filter matrices of size

MxM are organized such that they only process values from the same row (S_1) or from the previous row (S_2):

$$\mathbf{S}_1 = \begin{bmatrix} 1 & 0 & \cdots & \cdots & 0 \\ -\rho_h & \ddots & \ddots & & \vdots \\ 0 & \ddots & \ddots & \ddots & \vdots \\ \vdots & \ddots & \ddots & \ddots & 0 \\ 0 & \cdots & 0 & -\rho_h & 1 \end{bmatrix} ; \quad \mathbf{S}_2 = \begin{bmatrix} \rho_v & 0 & \cdots & \cdots & 0 \\ -\rho_h\rho_v & \ddots & \ddots & & \vdots \\ 0 & \ddots & \ddots & \ddots & \vdots \\ \vdots & \ddots & \ddots & \ddots & 0 \\ 0 & \cdots & 0 & -\rho_h\rho_v & \rho_v \end{bmatrix} \quad (8.62)$$

The observation is an image disturbed by noise $z(m,n)$. No linear distortion filter shall be present here ($\mathbf{H}=\mathbf{I}$), which gives the observation:

$$\mathbf{y}_n = \mathbf{x}_n + \mathbf{z}_n . \tag{8.63}$$

Then, the preliminary estimate is identical to the observation estimate, $\hat{\mathbf{y}}_n = \hat{\mathbf{x}}_n{}'$. Only results of the restoration up to the previous row (state) are required to compute this estimate

$$\hat{\mathbf{y}}_n = \mathbf{S}_1^{-1}\mathbf{S}_2\hat{\mathbf{x}}_{n-1} . \tag{8.64}$$

The output estimate $\hat{\mathbf{x}}_n$ is computed by the update step, using the actual observation **y**:

$$\hat{\mathbf{x}}_n = \hat{\mathbf{y}}_n + \mathbf{S}_1^{-1}\hat{\mathbf{e}}_n \quad ; \quad \hat{\mathbf{e}}_n = \mathbf{K'}_n\mathbf{v}_n \quad ; \quad \mathbf{v}_n = \mathbf{y}_n - \hat{\mathbf{y}}_n . \tag{8.65}$$

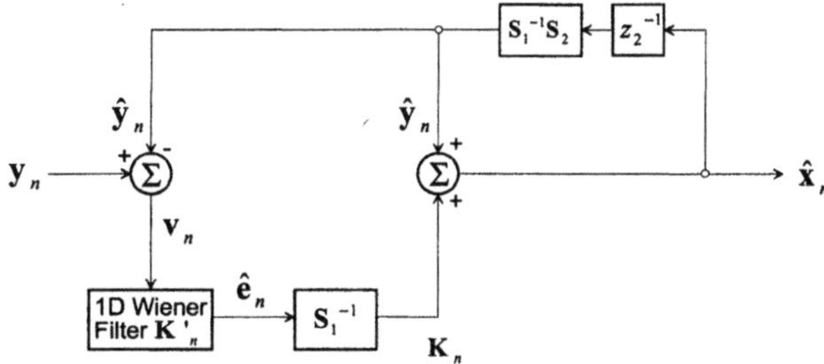

Fig. 8.4. Structure of a Kalman estimator for line-sequential restoration of images

A block diagram is shown in Fig. 8.4. Due to the signal model chosen, $\mathbf{B}_r=\mathbf{S}_1^{-1}$ and $\mathbf{A}_r=\mathbf{S}_1^{-1}\mathbf{S}_2$ are constant. As further the noise component $\mathbf{z}_n$ can be regarded as stationary, the update step can fully be interpreted as a computation of an estimate $\hat{\mathbf{e}}_n$ from the Kalman innovation $\mathbf{v}_n$, by which the noise component $\mathbf{z}_n$ should be suppressed as good as possible. The Kalman gain operation $\mathbf{K}_n=\mathbf{K'}_n\mathbf{A}_1^{-1}$ can then be realized as concatenation of a 1D Wiener filter optimized to suppress the noise and a matrix operation $\mathbf{A}_1^{-1}$ expressing the recursive AR synthesis in the nth row.

Kalman tracking of motion trajectories. Kalman filters are often used for temporal stabilization in feature extraction problems. For estimation of a motion trajectory, the state vector for a point P in a frame consists of the coordinate $\mathbf{r}_P$, the first derivative (speed) and the second derivative (acceleration); the state model is characterized by an assumption about motion continuity, for which an extrapolation of trajectory parameters from those of the previous frame is sufficient. The velocity state vectors $\mathbf{u}^{(o)} \equiv \mathbf{x}_r$ in a state ($r \equiv o$) can be formulated separately for horizontal and vertical coordinates r and s,

$$\hat{\mathbf{u}}_r^{(o)} = \begin{bmatrix} r_P^{(o)} \\ \frac{d}{dt} r_P^{(o)} \\ \frac{d^2}{dt^2} r_P^{(o)} \end{bmatrix} ; \; \hat{\mathbf{u}}_s^{(o)} = \begin{bmatrix} s_P^{(o)} \\ \frac{d}{dt} s_P^{(o)} \\ \frac{d^2}{dt^2} s_P^{(o)} \end{bmatrix}. \tag{8.66}$$

The prediction for the respective states of the vector (8.50) results from the previous state, consisting of the motion parameters estimated for the previous frame. Considering the temporal distance $\Delta t = T$ between two states,

$$\hat{\mathbf{u}}_{(r,s)}^{(o)} = \mathbf{A}\hat{\mathbf{u}}_{(r,s)}^{(o-1)} \;\text{ with }\; \mathbf{A} = \begin{bmatrix} 1 & \Delta t & \frac{1}{2}\Delta t^2 \\ 0 & 1 & \Delta t \\ 0 & 0 & 1 \end{bmatrix}. \tag{8.67}$$

The observation vector $\mathbf{y}_o$ results from a motion estimation applied separately, the vector $\mathbf{z}_o$ is the error occurring by this estimation, and $\mathbf{H}=\mathbf{I}$. Then, according to (8.49), (8.52) and (8.53) $\mathbf{v}_r = \mathbf{y}_r - \hat{\mathbf{x}}'_r$. The remaining processing in Kalman estimation follows the scheme described earlier, in particular the covariance matrix $\mathbf{R}_{\varepsilon\varepsilon,r}$ is continuously updated based on the estimated point positions. More parameters of a motion model (e.g. affine transform) can also be included in the state vector. The motion trajectory which finally results from the Kalman recursion is temporally stabilized, such that instantaneous fluctuations which otherwise occur in separate estimation of the motion vectors $\mathbf{y}_r$ are eliminated.

8.7   Outlier rejection in estimation

Measurement data used in estimation can be unreliable. As an example, motion vectors can be erroneous at positions of occlusions; also, if motion parameters shall be determined on the basis of parametric models (cf. sec. 7.6.5), disturbances may occur due to different local motion of small objects or deformation of object parts. Similar problems occur in disparity estimation, in estimation of shapes from edge data etc. Two different cases can be distinguished in this context:

- Additional information is available, which gives a classification about the reliability of measurement data at specific positions; for the example of motion estimation, this can be determined from the value of the matching cost function or other reliability checking mechanisms such as forward/backward estimation (cf. sec. 7.6.6). In such a case, unreliable data can a priori be given a lower weight in the global optimization of the estimate.
- No hint on reliability is available, except for the poor fit of certain measurement data with the hypothesis of the estimation. This is the more general case, applicable to many estimation problems. The remaining part of this section will rather concentrate on this aspect of the problem.

Levenberg-Marquardt algorithm. The steepest-descent criterion (7.20) is a specific formulation of the Gauss-Newton method, which determines the best update of an estimate $\hat{\mathbf{x}}_{r+1} = \hat{\mathbf{x}}_r + \mathbf{\Delta}_r$ as

$$\mathbf{\Delta}_r = -\frac{\mathbf{J}^T\left(\hat{\mathbf{x}}_r\right)\mathbf{e}\left(\hat{\mathbf{x}}_r\right)}{\mathbf{J}^T\left(\hat{\mathbf{x}}_r\right)\mathbf{J}\left(\hat{\mathbf{x}}_r\right)}, \tag{8.68}$$

where $\mathbf{e}$ is the residual error vector (as e.g. given by (8.18)) based on the estimate of iteration r, and $\mathbf{J}$ is the *Jacobian matrix* consisting of the partial derivatives of the residual error vectors $\mathbf{e}=[e_0\,e_2\,\dots\,e_{M-1}]^T$. With estimates relating to length-M vectors $\mathbf{x}=[x_0\,x_2\,\dots\,x_{M-1}]^T$ this can be written as

$$\mathbf{J}(\hat{\mathbf{x}}) = \begin{bmatrix} \dfrac{\partial e_0(\hat{\mathbf{x}})}{\partial \hat{x}_0} & \dfrac{\partial e_0(\hat{\mathbf{x}})}{\partial \hat{x}_1} & \dots & \dfrac{\partial e_0(\hat{\mathbf{x}})}{\partial \hat{x}_{M-1}} \\[2mm] \dfrac{\partial e_1(\hat{\mathbf{x}})}{\partial \hat{x}_0} & \ddots & & \vdots \\[2mm] \vdots & & \ddots & \vdots \\[2mm] \dfrac{\partial e_{M-1}(\hat{\mathbf{x}})}{\partial \hat{x}_0} & \dots & \dots & \dfrac{\partial e_{M-1}(\hat{\mathbf{x}})}{\partial \hat{x}_{M-1}} \end{bmatrix}. \tag{8.69}$$

The Gauss-Newton minimization is based on optimum curve-fitting of the quadratic error function, and is optimum in linear sense. The Levenberg-Marquardt method can be seen as a nonlinear extension of this approach, where the update vector is modified into

$$\mathbf{\Delta}_r = -\frac{\mathbf{J}^T\left(\hat{\mathbf{x}}_r\right)\mathbf{e}\left(\hat{\mathbf{x}}_r\right)}{\mathbf{J}^T\left(\hat{\mathbf{x}}_r\right)\mathbf{J}\left(\hat{\mathbf{x}}_r\right)+\lambda_r\mathbf{I}}. \tag{8.70}$$

The factor λ_r is adapted such that a *trust region* (or by negative expression: a *rejection zone*) is established which gives different weight to the different elements of the vectors, depending on the steepness of the individual gradients. The radius of the trust region is typically adjusted in the iterations according to an agreement between predicted and actual reduction in the error function. This way, fast convergence is guaranteed.

M-estimators. M-estimators are developed from maximum likelihood theory to minimize the influence of outliers on optimization problems [HUBER 1981]. Herein, each observation value is given an individual weighting factor which depends on an error criterion. The influence on the estimation result is minimized when the error criterion is increased.

In maximum likelihood estimation, an estimate is determined which is the most probable cause for the observed data. M-estimators work similar in principle, but eliminate outliers a posteriori; no prior knowledge about the properties of outliers is necessary. As a criterion, the residual error vectors $\mathbf{e}=[e_0\ e_2\ \dots\ e_{M-1}]^T$ can be evaluated at each position; in the case of parameter estimation, e.g. the deviation of the measurement data from the estimated model can be used as an error criterion. The error criterion should be positive, such that it is appropriate to use the absolute value $\varepsilon_m = |e_m|$. The mean of errors is

$$\mu_\varepsilon = \frac{1}{M} \sum_{m=0}^{M-1} \varepsilon_m \ . \tag{8.71}$$

In the subsequent iteration step of the gradient approach, the update vector is computed such that outliers with high values ε_m are given less importance by a weighting function. A robust function typically used in M-estimators is Tukey's bi-weight

$$w(m) = \begin{cases} \left(1 - \left(\dfrac{\lambda \varepsilon_m}{\mu_\varepsilon} \right)^2 \right)^2 & \varepsilon_m < \dfrac{\mu_\varepsilon}{\lambda} \\[2ex] 0 & \varepsilon_m > \dfrac{\mu_\varepsilon}{\lambda} \ . \end{cases} \tag{8.72}$$

The correct selection of the factor λ is crucial for the performance of the algorithm. For $\lambda=0$, no weighting is applied, such that the normal gradient-descent approach without weighting is performed. By increasing λ, outliers are given less and less weight, where in principle for $\lambda \to \infty$ all measurement data might be classified as outliers (except for perfectly-fitting samples); the weighting function converges into a unit impulse. Again, λ must be adjusted such that the fastest descent of the cost function is achieved, but criteria such as maximum percentage of outliers can be set additionally.

8.8 Problems

Problem 8.1
An image signal is distorted by motion blur during acquisition. The distortion can be described by a hold element, which performs averaging of three horizontally-adjacent pixels in the image matrix. The following operations for reconstruction shall be performed in horizontal direction:
a)   Compute the transfer function of the linear distortion.
b)   Determine the transfer function of an inverse filter and of a pseudo-inverse filter (for the case where inverse filtering leads to an unstable result). Would the potential instability still be critical, if the inverse filtering is performed in the DFT domain with horizontal size of the image *i)* M=30 pixels *ii)* M=32 pixels?
c)   The signal has a one-dimensional spectrum $X(j\Omega_1)=A\cdot e^{j\varphi_x(\Omega)}$. A noise of spectrum $Z(j\Omega_1)=(A/2)\cdot e^{j\varphi_z(\Omega)}$ is added to the blurred signal. How should the pseudo inverse filter be modified, when after reconstruction the spectral noise energy shall not be higher than the signal energy at any frequency?
d)   For the case of c), determine the transfer function of a Wiener filter and its gain at the cut-off frequency found in c).

Problem 8.2
An image signal is distorted during acquisition by an out-of-focus lens. This shall be modeled by a linear filter with a transfer function characterized as $H(z)=(z^1+2+z^{-1})/4$ in any radial direction.
a)   Compute the transfer function $H(j\Omega)$ of the linear distortion.
b)   Compute the transfer function $H^1(j\Omega)$ of an inverse filter. Will inverse filtering allow a perfect reconstruction in the frequency range $0\leq|\Omega|<\pi$?
c)   The sampled signal has a power spectrum $S_{xx}(\Omega)=\sin^{-2}\Omega$ and is distorted by the noise of the CCD chip with power spectrum $S_{nn}(\Omega)$=1/4. Determine the transfer function of the optimum Wiener filter $H^{1W}(j\Omega)$.
d) $H_{2D}(j\Omega_1, j\Omega_2)$ shall be a rotation-symmetric 2D transfer function, such that $H_{2D}(j\Omega_1, j\Omega_2)= H(j\Omega)$, where $|\Omega|$ is the distance of any frequency pair (Ω_1,Ω_2) from the origin of the 2D frequency plane. Express $H_{2D}(j\Omega_1, j\Omega_2)$ as a function of Ω_1 and Ω_2.
e)   Define the inverse and (in case of instability) the pseudo-inverse filter $H^1_{2D}(j\Omega_1, j\Omega_2)$ in the frequency range $0\leq|\Omega_1|<\pi,\ 0\leq|\Omega_2|<\pi$.

Problem 8.3
A signal vector $\mathbf{x}$=[3 9 15]^T is mapped by the following matrices into an observation $\mathbf{y}$:

$$\text{i) } \mathbf{H} = \begin{bmatrix} \frac{1}{2} & \frac{1}{2} & 0 \\ 0 & \frac{1}{2} & \frac{1}{2} \end{bmatrix} \quad \text{ii) } \mathbf{H} = \begin{bmatrix} 1 & 0 & 0 \\ \frac{1}{3} & \frac{2}{3} & 0 \\ 0 & \frac{2}{3} & \frac{1}{3} \\ 0 & 0 & 1 \end{bmatrix}$$

a)   Compute the values of $\mathbf{y}$ for both cases.
b)   Compute the pseudo-inverses of matrices i) and ii).
c)   Determine the reconstruction values $\hat{\mathbf{x}}$.

d)   For case i), determine the singular values of $\mathbf{H}\mathbf{H}^T$. Then, perform the reconstruction by the SVD method using only the first singular value.

Problem 8.4

A binary image signal x (amplitude values 0 and 1) with $P(0)=0.2$ and $P(1)=0.8$ is disturbed by white Gaussian noise of variance $\sigma_z^2=0.1$. Nothing is known about possible correlation in the signal. Values $y=0.3$; 0.5 ; 0.7 are observed. Determine the optimum reconstruction of the three values using
a)   the maximum-likelihood criterion;
b)   the maximum-a-posteriori criterion.

Problem 8.5

A signal $x(n)$ can be described as an AR(1) process with $\rho=0{,}75$ and $\sigma_x^2=16/7$. The signal is distorted by additive white Gaussian noise $z(n)$ of variance $\sigma_z^2=1$. Two adjacent samples are observed, which are combined into a vector $\mathbf{y} = [1 \quad 3]^T$.
a)   Determine the autocovariance matrices $\mathbf{R'}_{zz}$, $\mathbf{R'}_{xx}$ and their inverses.

Now, two hypotheses for estimates, $\hat{\mathbf{x}}_1 = [0 \quad 4]^T$ and $\hat{\mathbf{x}}_2 = [3 \quad 3]^T$ are given.

b)   Determine the best of the two estimates by a maximum-likelihood criterion.
c)   Determine the best of the two estimates by a maximum-a-posteriori criterion.
d)   By which factor should the variance of the signal be larger, such that ML and MAP estimation would lead to identical results?

9 Feature Transforms and Classification

One single feature is usually not sufficient to detect, characterize or classify the content of a multimedia signal. Decisions must often be derived from multiple features, which can be quite different in terms of physical meaning, but also statistically may be significantly distinct in terms of value ranges, mean values, variances etc. Further, linear or nonlinear statistical dependencies may exist between different features, which makes definition of subsequent classification criteria more difficult. In this chapter, methods of feature transforms are regarded first, which allow to reduce the set of feature values available to a possibly small number of independent, significant values. These must then be adapted (be normalized) in value ranges such that the combination of features in the subsequent classification step becomes as simple as possible. The classification itself requires similarity criteria by which a certain feature constellation can be matched against characteristic properties of a class, or by which similarity of two different signals is judged. Different similarity criteria are discussed, including the comparison of statistical distributions and comparison of entire sets. Finally, classification methods are introduced, where optimization typically relies on class definitions obtained by training sets. Linear classification, cluster-based and nearest neighbor methods, maximum a posteriori classification, artificial neural networks and hidden Markov models are representing state of the art classification approaches.

9.1 Feature Transforms

The goal of feature transforms is to analyze mutual interdependencies between the values of different features as extracted from a multimedia signal. This can then be used to generate a feature description which is as compact as possible. Interdependencies will as well exist between values expressing the same feature extracted at different locations or different time instances of one signal. In this section, *linear dependencies* are considered first, which can be captured by analysis of the covariance between different feature values. These can either be resolved by *ei-*

genvector analysis or by *independent component analysis* (ICA), which is a variant of singular value decomposition (SVD). Finally, as an example to resolve nonlinear dependencies, the usage of parametric models in combination with a transformation into generalized Hough parameter spaces is discussed.

9.1.1   Eigenvector Analysis of Feature Value Sets

Assume that feature vectors $\mathbf{m}=[m_1, m_2, \dots , m_K]^\mathrm{T}$, are compiled from K different feature values. Linear statistical dependencies between the elements of these vectors can be described by the *covariance matrix*

$$\mathbf{C}_{mm} = \begin{bmatrix} c_{mm}(1;1) & c_{mm}(1;2) & c_{mm}(1;3) & \cdots & c_{mm}(1;K) \\ c_{mm}(2;1) & c_{mm}(2;2) & c_{mm}(2;3) & & \vdots \\ c_{mm}(3;1) & c_{mm}(3;2) & \ddots & \ddots & \\ \vdots & & \ddots & \ddots & \vdots \\ c_{mm}(K;1) & \cdots & & \cdots & c_{mm}(K;K) \end{bmatrix} \tag{9.1}$$

with

$$\mathbf{C}_{mm} = E\left\{ [\mathbf{m}-\bar{\mathbf{m}}]\cdot[\mathbf{m}-\bar{\mathbf{m}}]^\mathrm{T} \right\} \text{ resp. } c_{mm}(k;l) = E\left\{ (m_k - \bar{m}_k)\cdot(m_k - \bar{m}_l) \right\}, \tag{9.2}$$

$$\text{where } \bar{\mathbf{m}} = E\{\mathbf{m}\} \text{ resp. } \bar{m}_k = E\{m_k\}.$$

As compared to the autocorrelation or autocovariance matrices of type (3.41), which are related to signals, this matrix directly expresses the covariances between any possible pair of feature values contained in the vector[1]. Like the autocovariance matrix, $\mathbf{C}_{mm}$ and the underlying set of feature vectors can be orthogonalized by an eigenvector transform (cf. (B.16) and (4.148)):

$$\mathbf{\Phi}^\mathrm{T}\cdot\mathbf{C}_{mm}\cdot\mathbf{\Phi} = \mathbf{\Lambda} = \begin{bmatrix} \lambda_1 & 0 & \cdots & & 0 \\ 0 & \lambda_2 & 0 & & \vdots \\ \vdots & 0 & \lambda_3 & \ddots & \\ & & \ddots & \ddots & 0 \\ 0 & \cdots & & 0 & \lambda_K \end{bmatrix} \tag{9.3}$$

In the transform matrix $\mathbf{\Phi}$, columns are the eigenvectors of $\mathbf{C}_{mm}$, which can be used as basis vectors of the transform. This results in K uncorrelated transform coefficients with variances expressed by the eigenvalues $\lambda_1 \dots \lambda_K$. For a given feature vector $\mathbf{m}$, the transform $\mathbf{f} = \mathbf{\Phi}^\mathrm{T}\cdot\mathbf{m}$ is applied for generation of the transform coefficients, which are also denoted as *principal components*. If the elements of the feature vector are highly correlated, only a negligible error occurs when the

---

[1] For an autocorrelation matrix of a stationary signal, $r_{xx}(k;l)=r_{xx}(|k-l|)$, which can be interpreted as a special case of (9.1).

first T components are retained (see (4.139)-(4.146)). A good indicator for the amount of covariance between the different feature values can be obtained by entropy analysis of the eigenvalues, which expresses the 'flatness' of the transformed representation[1]. First, the eigenvalues are normalized such that their sum will give the value of 1:

$$\lambda_k^* = \frac{\lambda_k}{\mathrm{tr}(\Lambda)} = \frac{\lambda_k}{\sum_k \lambda_k} . \tag{9.4}$$

Using the definition of entropy (3.85), substituting $\lambda_k^* \equiv P(k)$ gives

$$H^* = -\sum_k \lambda_k^* \cdot \log \lambda_k^* = -\frac{\sum_k \left[\lambda_k \cdot \log \lambda_k - \lambda_k \cdot \log \sum_j \lambda_j \right]}{\sum_j \lambda_j} = -\frac{\sum_k \lambda_k \cdot \log \lambda_k}{\sum_j \lambda_j} + \log \sum_j \lambda_j . \tag{9.5}$$

In case of a uniform distribution, $\lambda_k=1/K$. This would indicate that the feature values in $\mathbf{m}$ are uncorrelated, and the result will be $H^*=\log K$.

9.1.2   Independent Component Analysis

Statistical dependencies can be present not only *within* feature vectors, but also *between* feature vectors of same type, e.g. expressing the features of a signal for different locations in time or space. Assume that a number of L feature vectors $\mathbf{m}$ of same type are available, which can be arranged in a $K\mathrm{x}L$ matrix

$$\mathbf{M} = \begin{bmatrix} \mathbf{m}_1 & \mathbf{m}_2 & \cdots & \mathbf{m}_L \end{bmatrix}^{\mathrm{T}} . \tag{9.6}$$

As this matrix is non-quadratic except for the case $K=L$, it is not possible to compute eigenvectors; as an alternative solution for non-quadratic matrices, the SVD expansion was introduced in sec. 8.4. If (8.27) is applied to the matrix $\mathbf{M}$, the result is

$$\mathbf{U}^{\mathrm{T}}\mathbf{M}\mathbf{V} = \mathbf{F} , \tag{9.7}$$

where $\mathbf{F}$ is a diagonal matrix containing an *independent component representation* of $\mathbf{M}$. In principle, eigenvectors $\mathbf{u}_r$ of the $L\mathrm{x}L$-Matrix $\mathbf{M}\mathbf{M}^{\mathrm{T}}$ are used to remove the *temporal correlation* between the vectors, while the eigenvectors $\mathbf{v}_r$ of $\mathbf{M}^{\mathrm{T}}\mathbf{M}$ are used likewise to remove the *correlation between the elements* of the vectors. Even though it is not possible to *reconstruct* $\mathbf{M}$ when only the values in $\mathbf{F}$ are known, these establish a very compact representation summarizing the feature values expressed by the vectors $\mathbf{m}$. Alternatively, if $\mathbf{m}$ already contains uncorrelated ele-

---

[1] The notion 'flatness' is used here similar to the usage in the spectral flatness measure (11.19), however applied to the values of a discrete transform. The computation can be performed analogously with (3.85).

ments (e.g. values of a spectral representation or another decorrelating transform), it is only necessary to compute the eigenvectors $\mathbf{u}_r$ and apply the transformation $\mathbf{U}^T\mathbf{M}$. By retaining only components which are related to the first $L'{<}L$ eigenvectors, a very compact representation is generated, which represents the feature values from L vectors of length K by only $L'K$ independent component values.

9.1.3   Generalized Hough Transform

The Hough transform was originally introduced for parametric description of contour features (see sec. 7.3.5). The concept is much more universally applicable as a transformation of measurements related to any parametric feature description into a related (generalized Hough) feature space [BALLARD, BROWN 1985]. In the Hough space, local concentrations of parametrized functions (gained from measurement data) are searched to determine if a parametric model fits consistently. The generalized Hough space has to be defined over a number of dimensions which is equal to the number of free parameters in the parametric model.

Example: Generalized Hough analysis of parametric motion. The motion shift as described by the 4 parameter model (7.115) expresses translation, rotation by θ and scale change Θ between two frames of a video sequence by[1]

$$\begin{bmatrix} m' \\ n' \end{bmatrix} = \begin{bmatrix} a_1 & -a_2 \cdot \frac{S}{R} \\ a_2 \cdot \frac{R}{S} & a_1 \end{bmatrix} \cdot \begin{bmatrix} m \\ n \end{bmatrix} + \begin{bmatrix} t_1 \cdot \frac{1}{R} \\ t_2 \cdot \frac{1}{S} \end{bmatrix}; \ a_1 = \Theta \cdot \cos\theta; \ a_2 = \Theta \cdot \sin\theta. \quad (9.8)$$

If motion shifts $k=m'-m$ and $l=n'-n$ have been estimated for a number of pixel positions, the following relationships apply:

$$\begin{aligned} k &= (\Theta \cdot \cos\theta - 1) \cdot m - \Theta \cdot \tfrac{S}{R} \cdot \sin\theta \cdot n + t_1 \cdot \tfrac{1}{R} \\ l &= \Theta \cdot \tfrac{R}{S} \cdot \sin\theta \cdot m + \Theta \cdot (\cos\theta - 1) \cdot n + t_2 \cdot \tfrac{1}{S}, \end{aligned} \quad (9.9)$$

which can e.g. be re-formulated for dependency of the two translational parameters on the rotation and scale-change parameters,

$$\begin{aligned} t_1 &= (m+k) \cdot R + \Theta(n \cdot S \cdot \sin\theta - m \cdot R \cdot \cos\theta) \\ t_2 &= (n+l) \cdot S - \Theta(m \cdot R \cdot \sin\theta + n \cdot S \cdot \cos\theta). \end{aligned} \quad (9.10)$$

As k,l,m,n,R and S are constants for a given measurement point, (9.10) describes (hyper-) surface equations $t_1{=}f(\Theta,\theta)$ and $t_2{=}f(\Theta,\theta)$. A unique parametric motion behavior for a set of measurement points is detected if concentrations of plane intersections occur around given points $(\Theta_S,\theta_S;t_{1S},t_{2S})$ in the generalized Hough space. In principle, the positions of intersections in t_1 and t_2 are independent in this example, but must relate to the same parameters (Θ_S,θ_S) for validity of the motion

---

[1] Discrete coordinates are used here similar to (7.191).

model. Consequently, it can be concluded by presence of such intersection clusters that measurement points falling into one cluster belong to an object which follows a consistent motion model [KRUSE 1996].

9.2   Feature Value Normalization and Weighting

Normalization and weighting of multiple feature values of different meaning and sources is an important condition for successful application of classification methods introduced below in this chapter. Typically, features to be used for classification will have different value ranges. Second, it is necessary to define distance functions which allow determining the similarity of two distinct feature vectors in the feature space. The problems of feature weighting and normalization can in principle not be regarded independent of the subsequent classification. In particular, optimal methods will be gained by analysis of a *training set* of typical values, which are often selected as relevant for a given classification problem. In some cases it will further be necessary to build on *a priori knowledge* about the class separation to be expected, e.g. number of classes, statistical distribution of feature values within each class etc. At the first sight, this seems to be a 'hen-and-egg' problem. It should be considered however, that these a priori assumptions can be of preliminary nature, which can be updated dynamically based on classification decisions already made. This means that adaptive and learning systems can continuously improve their performance by iterative applications of such basic methods (cf. sec. 1.1.3). The subsequent discussions require the following definitions:

- $\mathbf{m}^* = [m_1{}^*, m_2{}^*, ..., m_K{}^*]^T$ is a feature vector consisting of K not-normalized feature values $m_k{}^*$.
- $\mathbf{m} = [m_1, m_2, ..., m_K]^T$ is a normalized version of the feature vector $\mathbf{m}^*$.
- $\mathbf{m}_q$ is *one* normalized feature vector from a training set $\mathcal{M}$, which has a total of Q members indexed by $q=1,2,...,Q$; $\mathbf{m}_q$ contains K feature values $m_{k,q}$.
- $\mathbf{m}_q{}^{(l)}$ is *one* normalized feature vector from the same training set, to which a class label $S(q)=S_l$ was assigned a priori, which designates one out of L classes. The entire training set $\mathcal{M}$ can be seen as a unification of class-related sub-sets $\mathcal{M}_l$, where the number of elements in each sub-set shall be Q_l, and $Q=\Sigma Q_l$.
- $\mathbf{z}^{(l)} = [z_1{}^{(l)}, z_2{}^{(l)}, ..., z_K{}^{(l)}]^T$ is a *centroid vector* which has the same meaning as $\mathbf{m}$; $\mathbf{z}^{(l)}$ contains K normalized mean feature values $z_k{}^{(l)}$ related to class S_l. Even though it is possible to define these centroids a priori by a statistical model, they can also be computed from the values of a representative training set as

$$\mathbf{z}^{(l)} = \frac{1}{Q_l} \sum_{q=1}^{Q} \mathbf{m}_q \cdot w(q) \quad \text{where} \quad w(q) = \begin{cases} 1 & \text{if} \quad S(q) = S_l \\ 0 & \text{else.} \end{cases} \tag{9.11}$$

9.2.1   Normalization of Feature Values

To adapt the *range* of feature values, the following methods are typically used:

–   Extreme value normalization. The normalization factor c_k is derived from the difference between any occurring minimum and maximum value over the entire training set; optionally, the minimum value can be subtracted as an offset, such that the normalized value will be strictly positive:

$$c_k = \left[\max_{q=1..Q}(m_{k,q}{}^*) - \min_{q=1..Q}(m_{k,q}{}^*) \right]; \quad m_k = \frac{m_k{}^* - \min_{q=1..Q}(m_{k,q}{}^*)}{c_k} \tag{9.12}$$

The method is disadvantageous in cases where single outliers occur as minima and maxima. It can then be improved by truncating the outliers in favor of more reasonable values by clipping, e.g. for minimum and maximum values which are representative for a sufficiently large percentage of the training set members.

–   Variance normalization. Feature values are normalized regarding the variance over the training set; by optional subtraction of the mean value, zero-mean feature vectors are generated:

$$\sigma_k{}^2 = \frac{1}{Q} \sum_{q=1}^{Q} (m_k{}^* - \overline{m_k}{}^*)^2; \quad m_k = \frac{m_k{}^* - \overline{m_k}{}^*}{\sigma_k}; \quad \overline{m_k}{}^* = \frac{1}{Q} \sum_{q=1}^{Q} m_k{}^* \tag{9.13}$$

Relevance normalization. This approach can strictly only be applied if a priori knowledge about the relevance of the given feature for a certain classification problem is available. This could be achieved by using additional *reliability information* gained during feature extraction, or by using methods of *relevance feedback*, where relevance information is assigned by users in an operating classification system. It is then possible to adjust the relevance of a given feature value in relation to the other feature values in a vector, e.g. by defining *weighting factors* w_k,

$$m_k = w_k \cdot m_k{}^* \tag{9.14}$$

In principle, relevance weights related to a given classification problem can also be derived from class likelihood functions, which express whether a feature is a good indicator for class separation in this case. Measures which indicate whether a classification problem is clearly solvable based on a given set of feature values are discussed in the subsequent sections, in particular sec. 9.2.5. Further, an interesting framework that can be used for relevance normalization and weighting has been developed by the *Dempster-Shafer evidence theory* [DEMPSTER 1968][DEMPSTER 1976][SHAFER 1976].   This is generalized from Bayes estimation theory, based on concepts of knowledge certainty. For different information components, parameters expressing *belief* and *plausibility* are assigned, which are used to develop the best *rule of combination* for different information components, which could actually be mapped into the weighting factors of (9.14).

9.2.2   Simple Distance Metrics

The distance measures introduced in the sequel are defined to express similarity or dissimilarity between a feature vector $\mathbf{m}$ and a class centroid $\mathbf{z}^{(l)}$ by one numerical value, as typically needed for classification. It is also possible to compute the distance between two distinct feature vectors, e.g. to express the similarity between two given signals. One of the most widely used distance metrics is the *Euclidean distance*, which is based on the quadratic norm of the difference between two vectors:

$$d_2(\mathbf{m},\mathbf{z}^{(l)}) = \left\|\mathbf{m}-\mathbf{z}^{(l)}\right\|_2 = \sqrt{\sum_{k=1}^{K}(m_k - z_k^{(l)})^2} = \sqrt{(\mathbf{m}-\mathbf{z}^{(l)})^{\mathrm{T}}\cdot(\mathbf{m}-\mathbf{z}^{(l)})}\,. \qquad (9.15)$$

This is a special case of the L_P *norm*[1]

$$d_P(\mathbf{m},\mathbf{z}^{(l)}) = \left\|\mathbf{m}-\mathbf{z}^{(l)}\right\|_P = \left(\sum_{k=1}^{K}\left|m_k - z_k^{(l)}\right|^P\right)^{\frac{1}{P}}\,. \qquad (9.16)$$

Two interesting cases of L_P norm are given for $P=1$, the sum of absolute differences over all feature values, and $P\rightarrow\infty$, which is dominated by only one largest absolute feature value difference[2]:

$$d_\infty(\mathbf{m},\mathbf{z}^{(l)}) = \max_{k=1..K}\left|m_k - z_k^{(l)}\right|\,. \qquad (9.17)$$

A geometric interpretation of distances L_1, L_2 and L_∞ in a 2-dimensional feature space is shown in Fig. 9.1.

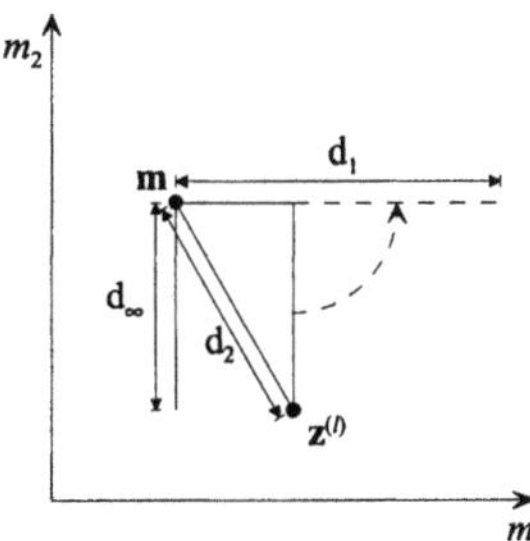

Fig. 9.1. Interpretation of L_P norms in a feature space with $K=2$ for $P=1,2,\infty$

---

[1] Throughout this book, if the subscript indicating the norm P is missing, assume the Euclidean norm $P=2$.

[2] In general, larger distances between feature values in the vector take higher influence on the result when P is larger.

In general, it is possible to weight the distances for the single values of the feature vector **m** individually, such that the relevance weighting introduced in the previous section becomes part of the distance measure:

$$d_{W,P}(\mathbf{m},\mathbf{z}^{(l)}) = \left(\sum_{k=1}^{K} w_k^{\,P}\left|m_k - z_k^{\,(l)}\right|^P\right)^{\frac{1}{P}}.$$

(9.18)

Specifically for the Euclidean norm, a weighting matrix can be introduced which also allows considering mutual interdependencies of feature values in the computation of the distance:

$$\begin{aligned}
d_{W,2}(\mathbf{m},\mathbf{z}^{(l)}) &= \sqrt{\sum_{j=1}^{K}\sum_{k=1}^{K} w_{j,k}\cdot(m_j - z_j^{(l)})\cdot(m_k - z_k^{(l)})} \\
&= \sqrt{(\mathbf{m}-\mathbf{z}^{(l)})^{\mathrm{T}}\cdot\mathbf{W}\cdot(\mathbf{m}-\mathbf{z}^{(l)})} \; ; \; \mathbf{W} = \left[w_{j,k}\right]
\end{aligned}$$

(9.19)

If no correlation exists between the different feature values in **m**, **W** is usually defined as a diagonal matrix, where only the values $w_{j,k}|_{j=k}$ are populated by the respective values w_k from (9.18).

9.2.3   Distance Metrics related to Statistical Distributions

Statistical moments such as mean, variance etc. (e.g. of a texture) can in principle be treated like any other feature value, and can directly be used for feature-based comparison. In some cases (e.g. color histograms, edge histograms), approximations of probability or occurrence-count functions are also used to characterize the properties of signals. Typical distance functions for probability distributions can be based on statistical tests such as the χ^2 test (3.73) or the Kolmogorov-Smirnov test (3.74). Another distance function, which is derived from information-theoretic concepts, is the *Kullback-Leibler distance*, expressing the distance between two probability density functions by

$$\Delta(p_A, p_B) = \int_{-\infty}^{\infty} p_A(x)\ln\frac{p_A(x)}{p_B(x)}\,dx\,.$$

(9.20)

For discrete probability distributions, this can also be interpreted as the *relative entropy*

$$\Delta(P_A, P_B) = \sum_{j=1}^{J} P_A(j)\log_2\frac{P_A(j)}{P_B(j)} = \left[-\sum_{j=1}^{J} P_A(j)\log_2 P_B(j)\right] - H(A)\,.$$

(9.21)

Observe that the Kullback-Leibler distance is not symmetric, $\Delta(p_A,p_B)\neq\Delta(p_B,p_A)$. If symmetry is required, the following function can be used:

$$\overline{\Delta}(p_A, p_B) = \frac{1}{2}\left[\Delta(p_A, p_B) + \Delta(p_B, p_A)\right]. \tag{9.22}$$

Distance functions as (3.73), (3.74) and (9.21) are suitable to judge the similarity of mass distribution functions which are statistically reliable and can in the best case be expressed by parameters of a model distribution. If approximations of statistical distributions are made by extraction of histograms (occurrence counts) from finite signals or signal segments, the functions may be much more irregular. While model distributions are typically smoothly changing for neighbored amplitude values, this is not necessarily the case for histograms. Alternatively, the occurrence values of histograms can directly be interpreted as a vector feature. The comparison of two signals is then e.g. performed by computation of the L_P norm difference between the normalized histogram values $H'(j)=H(j)/N$. In fact, the difference in the underlying feature (e.g. color) itself changes *linearly* with the difference in the probability distribution; also statistical tests are approximately computing the area between two distributions, which is rather related to the analysis of *absolute differences* of probability values. In practice, it usually turns out that the L_1 norm is most appropriate to compare features which are described by histograms. For example, comparing the histogram vectors $\mathbf{h}_A$ and $\mathbf{h}_B$ describing a feature of signals A and B, the L_1 difference function is

$$d_1(\mathbf{h}_A, \mathbf{h}_B) = \left\|\mathbf{h}_A - \mathbf{h}_B\right\|_1 = \sum_{j=1}^{J}\left|H'_A(j) - H'_B(j)\right| . \tag{9.23}$$

A histogram expresses the frequency of occurrence of quantized values $\mathbf{y}_j$, whereas (9.23) does not at all consider the distance of underlying *feature values* that are described by the different histogram lines: Assume similarity of two quantized values $\mathbf{y}_i \approx \mathbf{y}_j$. Then, two signals are indeed quite similar when $H_A(i)+H_A(j)\approx H_B(i)+H_B(j)$, but indeed much higher differences $H_A(i)\neq H_B(i)$ and $H_A(j)\neq H_B(j)$ may be found. To take this effect into regard, the distance computation can be performed considering sums over different histogram lines, when these represent similar feature-value amplitudes. Assuming that Π_j describes a set of other feature-value amplitudes $\mathbf{y}_i$ which are most similar with $\mathbf{y}_j$, a modified histogram distance considering these relationships is defined by

$$d(\mathbf{h}_A, \mathbf{h}_B) = \sum_{j=1}^{J}\left|\sum_{i\in\Pi_j} H'_A(i) - \sum_{i\in\Pi_j} H'_B(i)\right|. \tag{9.24}$$

For example, in case of a uniform scalar quantization of the respective feature, a useful set Π_j should include the values falling into neighbored quantizer cells. A more exact mapping of feature value similarities can be achieved by using a weighting function. Subsequently, a method is introduced which is based on the weighted Euclidean distance between quantized feature amplitude values in the feature space. This weighting function shall have a value of 1 in case of greatest similarity, and 0 in case of greatest dissimilarity. This can be realized by the following function, where $\Delta_{\max}$ expresses the maximum Euclidean distance of any possible pair $(\mathbf{y}_i, \mathbf{y}_j)$ of quantized values:

$$w_{i,j} = 1 - \frac{\left\| \mathbf{y}_i - \mathbf{y}_j \right\|_2}{\Delta_{\max}} \quad ; \quad \Delta_{\max} = \max_{i,j} \left\| \mathbf{y}_i - \mathbf{y}_j \right\|_2 . \tag{9.25}$$

As an example, regard a histogram in (H,S,V) color space with maximum-value normalization of Euclidean distance (7.7). Between two quantized color points indexed as i and j, the following weighting can then be applied:

$$w_{i,j} = 1 - \sqrt{\frac{(V(i)-V(j))^2 + (S(i)\cdot\cos H(i) - S(j)\cdot\cos H(j))^2 + (S(i)\cdot\sin H(i) - S(j)\cdot\sin H(j))^2}{5}} . \tag{9.26}$$

The computation of distances between two histograms having J bins each is then performed by using the weighted Euclidean distance (9.19),

$$\begin{aligned} d_{W,2}(\mathbf{h}_A, \mathbf{h}_B) &= \sqrt{(\mathbf{h}_A - \mathbf{h}_B)^{\mathrm{T}} \cdot \mathbf{W} \cdot (\mathbf{h}_A - \mathbf{h}_B)} \\ &= \sqrt{\sum_{i=1}^{J}\sum_{j=1}^{J} w_{i,j} \cdot (H'_A(i) - H'_B(i)) \cdot (H'_A(j) - H'_B(j))} \\ &= \sqrt{\sum_{i=1}^{J}\sum_{j=1}^{J} w_{i,j} \cdot [H'_A(i)\cdot H'_A(j) + H'_B(i)\cdot H'_B(j) - 2H'_A(i)\cdot H'_B(j)]}. \end{aligned} \tag{9.27}$$

By generalization of the last line in (9.27), it is also possible to compare histograms which represent the same feature by different underlying quantization of the feature, which means that numbers of bins J_A and J_B are different :

$$d(\mathbf{h}_A, \mathbf{h}_B) = \sqrt{\sum_{i=1}^{J_A}\sum_{j=1}^{J_A} w_{i,j} \cdot H'_A(i)\cdot H'_A(j) + \sum_{k=1}^{J_B}\sum_{l=1}^{J_B} w_{k,l} \cdot H'_B(k)\cdot H'_B(l) - 2\sum_{i=1}^{J_A}\sum_{k=1}^{J_B} w_{i,k} \cdot H'_A(i)\cdot H'_B(k)}. \tag{9.28}$$

Statistical representations of features can also be represented by parameters of a continuous PDF. This can for example be mean values, variances and co-variances of a scalar or vector Gaussian distribution. Two vector Gaussian distributions shall be indexed as i and j, defined by the mean-value vectors $\boldsymbol{\mu}^{(i)}$ and $\boldsymbol{\mu}^{(j)}$ and the co-variance matrices $\mathbf{R'}_{xx}^{(i)}$ and $\mathbf{R'}_{xx}^{(j)}$. A simple evaluation of similarity is defined by the following likelihood weighting, which approaches one for equality of the mean values and has a value of zero for the case of maximum dissimilarity:

$$w_{i,j} = e^{-\frac{1}{2}[\boldsymbol{\mu}_\Delta]^{\mathrm{T}} \bar{\mathbf{R}}'_{xx}{}^{-1}[\boldsymbol{\mu}_\Delta]} \quad \text{with} \quad \bar{\mathbf{R}}'_{xx} = \frac{1}{2}\left(\mathbf{R'}_{xx}^{(i)} + \mathbf{R'}_{xx}^{(j)}\right) \quad \text{and} \quad \boldsymbol{\mu}_\Delta = \boldsymbol{\mu}^{(i)} - \boldsymbol{\mu}^{(j)} . \tag{9.29}$$

In many cases, one single Gaussian hull is not sufficient to characterize the probability distribution of a feature. For example when an image shows several significant concentrations of dominant colors, modeling can better be performed by a *mixture of Gaussians*, which consists of J different Gaussian distributions, each characterized by a weighting factor $G^{(j)}$, mean value vector $\mu^{(j)}$ and Covariance matrix $\mathbf{R'}_{xx}^{(j)}$,

$$p_{\mathrm{mix}}(\mathbf{x}) = \sum_{j=1}^{J} G^{(j)} \cdot \frac{1}{\sqrt{(2\pi)^K \cdot \left|\mathbf{R'}_{xx}^{(j)}\right|}} \cdot e^{-\frac{1}{2}\left[\mathbf{x}-\mu^{(j)}\right]^{\mathrm{T}}\left[\mathbf{R'}_{xx}^{(j)}\right]^{-1}\left[\mathbf{x}-\mu^{(j)}\right]} \quad \text{with} \quad \sum_{j=1}^{J} G^{(j)} = 1. \quad (9.30)$$

Assume that two signals A and B are described by Gaussian mixtures consisting of J_A and J_B contributing Gaussians, which are defined by parameters $G_A^{(j)};G_B^{(j)}$, $\mu_A^{(j)};\mu_B^{(j)}$, $\mathbf{R'}_{xx,A}^{(j)};\mathbf{R'}_{xx,B}^{(j)}$. The following distance measure can be interpreted similar as (9.28)[1]; herein, the weighting function (9.29) is used:

$$d(p_{\mathrm{mix,A}}, p_{\mathrm{mix,B}}) = \sqrt{\sum_{i=1}^{J_A}\sum_{j=1}^{J_A} w_{i,j} \cdot G_A^{(i)} \cdot G_A^{(j)} + \sum_{k=1}^{J_B}\sum_{l=1}^{J_B} w_{k,l} \cdot G_B^{(k)} \cdot G_B^{(l)} - 2\sum_{i=1}^{J_A}\sum_{k=1}^{J_B} w_{i,k} \cdot G_A^{(i)} \cdot G_B^{(k)}}.$$

$$(9.31)$$

9.2.4   Distance Metrics based on Class Features

The Gaussian distribution is a widely used model to characterize statistical properties of signals and features, and is in particular suitable for analytic optimization. Assume that the PDF of those feature vectors, which have been assigned a priori to the set of class S_l, can be expressed as a vector Gaussian distribution

$$p(\mathbf{m}\,|\,S_l) = \frac{1}{\sqrt{(2\pi)^K \cdot \mathrm{Det}(\mathbf{C}_{mm}^{(l)})}} \cdot e^{-\frac{1}{2}(\mathbf{m}-\mathbf{z}^{(l)})^{\mathrm{T}}\left[\mathbf{C}_{mm}^{(l)}\right]^{-1}(\mathbf{m}-\mathbf{z}^{(l)})}. \quad (9.32)$$

Here, $\mathbf{C}_{mm}^{(l)}$ expresses the covariance matrix similar to (9.1), but only describing members of class S_l. When the number Q_l of class members is sufficiently large, the covariance matrix and the class centroid $\mathbf{z}^{(l)}$ can reliably be approximated from the training set,

$$\mathbf{C}_{mm}^{(l)} = \frac{1}{Q_l}\sum_{q=1}^{Q_l}(\mathbf{m}_q^{(l)} - \mathbf{z}^{(l)})\cdot(\mathbf{m}_q^{(l)} - \mathbf{z}^{(l)})^{\mathrm{T}} \;;\; \mathbf{z}^{(l)} = \frac{1}{Q_l}\sum_{q=1}^{Q_l}\mathbf{m}_q^{(l)}. \quad (9.33)$$

In analogy with (9.3), eigenvectors and eigenvalues of the covariance matrix can then be determined specifically for this class. By using the transpose of the eigenvector matrix $[\Phi^{(l)}]^{\mathrm{T}}$, the distance between a given feature vector $\mathbf{m}$ and the class centroid can be transformed into a decorrelated component space,

$$\mathbf{f}^{(l)} = \left[\Phi^{(l)}\right]^{\mathrm{T}}(\mathbf{m} - \mathbf{z}^{(l)}). \quad (9.34)$$

---

[1] The weights G_j play indeed a similar role as histogram values, but do not represent discrete amplitudes, but a discrete set of Gaussian hulls ($\mu^{(j)}$ describing the sampling position, and $\mathbf{R'}_{xx}^{(j)}$ the shape). As the shape influences the definition of the weights $w_{i,j}$, the comparison (9.31) is in principle comparing continuous mixture PDFs.

The eigenvalues $\lambda_k^{(l)}$ according to (1.14) express the expected within-class variances of transformed component vectors $\mathbf{f}^{(l)}$. If the Euclidean distance criterion (9.15) is applied to transformed components and normalization by variances (eigenvalues) is applied, the distance between the vector $\mathbf{m}$ and the class centroid can be expressed in the component space as follows:

$$d_l^2\left(\mathbf{m}, \mathbf{z}^{(l)}\right) = \sum_{k=1}^{K} \frac{\left(f_k^{(l)}\right)^2}{\lambda_k^{(l)}} = \left[\mathbf{f}^{(l)}\right]^{\mathrm{T}} \cdot \left[\Lambda^{(l)}\right]^{-1} \cdot \mathbf{f}^{(l)}. \tag{9.35}$$

Formulating the matrix equation on the right-hand side of (9.35) explicitly by relationship with the distance in the feature space gives

$$d_l^2\left(\mathbf{m}, \mathbf{z}^{(l)}\right) = \underbrace{\left[\mathbf{m} - \mathbf{z}^{(l)}\right]^{\mathrm{T}} \cdot \Phi^{(l)}}_{\left[\mathbf{f}^{(l)}\right]^{\mathrm{T}}} \cdot \underbrace{\left[\Phi^{(l)}\right]^{\mathrm{T}} \cdot \mathbf{C}_{mm}^{-1} \cdot \Phi^{(l)}}_{\left[\Lambda^{(l)}\right]^{-1}} \cdot \underbrace{\left[\Phi^{(l)}\right]^{\mathrm{T}} \cdot \left[\mathbf{m} - \mathbf{z}^{(l)}\right]}_{\mathbf{f}^{(l)}}. \tag{9.36}$$

Then, from the conditions of an orthonormal transform $[\Phi^{(l)}]^{\mathrm{T}} = \Phi^{-1}$ and $\Phi\Phi^{-1} = \mathbf{I}$

$$d_l^2\left(\mathbf{m}, \mathbf{z}^{(l)}\right) = \left[\mathbf{m} - \mathbf{z}^{(l)}\right]^{\mathrm{T}} \cdot \mathbf{C}^{(l)}_{mm}{}^{-1} \cdot \left[\mathbf{m} - \mathbf{z}^{(l)}\right]. \tag{9.37}$$

The distance measure (9.37) is the *Mahalanobis distance*. Compared to (9.19), it becomes obvious that the inverse of the covariance matrix plays a similar role here as a weighting matrix. This weighting is fully equivalent to a *decorrelation* and *variance normalization* in the related component space, while it is not even necessary to compute the transform itself. The given feature vector $\mathbf{m}$ can then be allocated to the class S_l where d_l^2 is minimum.

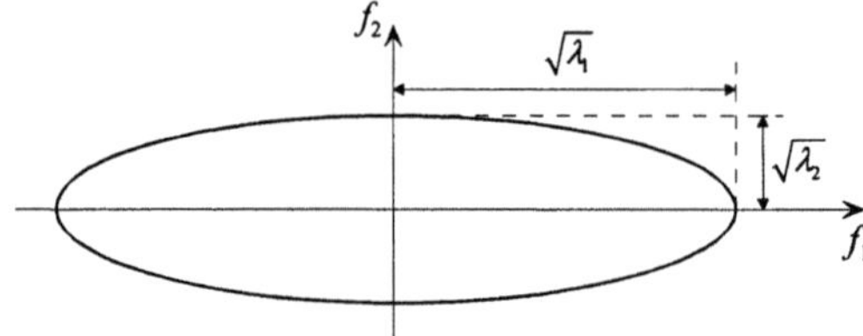

Fig. 9.2. Interpretation of the Mahalanobis distance, $K=2$

Example : Interpretation of the Mahalanobis distance for $K=2$. Obviously,

$$d_l^2\left(\mathbf{m}, \mathbf{z}^{(l)}\right) = \frac{\left[f_1^{(l)}\right]^2}{\lambda_1^{(l)}} + \frac{\left[f_2^{(l)}\right]^2}{\lambda_2^{(l)}} \tag{9.38}$$

is the equation of an ellipse, and the distance function will have a constant value on elliptic contours within the transformed component space. The transformed value of the centroid establishes the center, and the square roots of eigenvalues are the extensions of the principal axes (see Fig. 9.2). For a decision on assignment of a feature vector into a class, a higher deviation from the centroid is apparently allowed over those components which show high variation within the class. This is the case here for the direction of the primary axis. For $K>2$, the same interpretation applies to a K-dimensional ellipsoid hull indicating positions of constant Mahalanobis distance in the K-dimensional component space.

9.2.5   Reliability measures

While distance metrics establish criteria for classification, the question is important *how reliable* a decision can be when several classes have to be distinguished. In this context, the following observations can be regarded as relevant:

- the actual value of distance between a feature vector and the centroid of the chosen class S_l, in particular if compared to the distances of other class centroids;
- the a priori probability parameters of the selected class S_l, in particular the probability of the class in general and the expected variations of feature values around the centroid.

These criteria implicitly include the distances between the centroids of different classes and possible overlaps of class-related probability density functions. If for example centroids of two classes are very close, the decision about a feature vector in between will be unreliable, even more if a high variance is observed within the classes towards the centroid of the respective other class. To determine these relationships analytically, first the probabilities of classes S_l are estimated from a training set as

$$\hat{P}(l) = \frac{Q_l}{Q}, \tag{9.39}$$

where the global centroid of all feature vectors over the training set is

$$\mathbf{z} = \sum_{l=1}^{L} \hat{P}(l) \cdot \mathbf{z}^{(l)}. \tag{9.40}$$

Now, by defining a covariance matrix related to the deviations of the class centroids from the global centroid

$$\mathbf{C}_{zz} = \sum_{l=1}^{L} \hat{P}(l) \cdot \left[\mathbf{z}^{(l)} - \mathbf{z} \right] \cdot \left[\mathbf{z}^{(l)} - \mathbf{z} \right]^{\mathrm{T}}, \tag{9.41}$$

covariance parameters *'between classes'* can be expressed. In addition, the following *'within class'* covariance matrix is computed as a mean value over the class covariance matrices (9.33),

$$\mathbf{C}_{mm} = \sum_{l=1}^{L} \hat{P}(l) \cdot \mathbf{C}_{mm}^{(l)} . \tag{9.42}$$

To achieve a reliable separation of classes, it is advantageous if values of (9.41) get as large as possible, while values in (9.42) should be as small as possible[1]. Possible criteria are as follows, each expressing by one single (scalar) value the reliability of a certain set of features for a given classification problem:

$$c_1 = \mathrm{tr}\left\{\mathbf{C}_{mm}^{-1} \cdot \mathbf{C}_{zz}\right\} \quad \text{or} \quad c_1' = \frac{\mathrm{tr}\left\{\mathbf{C}_{zz}\right\}}{\mathrm{tr}\left\{\mathbf{C}_{mm}\right\}} \tag{9.43}$$

$$c_2 = \mathrm{tr}\left\{\left(\mathbf{C}_{mm} + \mathbf{C}_{zz}\right)^{-1} \cdot \mathbf{C}_{zz}\right\} \quad \text{or} \quad c_2' = \frac{\mathrm{tr}\left\{\mathbf{C}_{zz}\right\}}{\mathrm{tr}\left\{\mathbf{C}_{mm} + \mathbf{C}_{zz}\right\}} \tag{9.44}$$

$$c_3 = \frac{\det\left|\mathbf{C}_{zz}\right|}{\det\left|\mathbf{C}_{mm}\right|} \quad \text{or} \quad c_3' = \log\frac{\det\left|\mathbf{C}_{zz}\right|}{\det\left|\mathbf{C}_{mm}\right|} \tag{9.45}$$

$$c_4 = \frac{\det\left|\mathbf{C}_{zz}\right|}{\det\left|\mathbf{C}_{mm} + \mathbf{C}_{zz}\right|} \quad \text{or} \quad c_4' = \log\frac{\det\left|\mathbf{C}_{zz}\right|}{\det\left|\mathbf{C}_{mm} + \mathbf{C}_{zz}\right|} \tag{9.46}$$

In all criteria (9.43)-(9.46), larger values express larger reliability of class separation decisions. These criteria can e.g. be used to select optimum features for a given classification problem, providing a priori statistics which enables a reliable classification. If the centroid vector $\mathbf{z}$ in (9.41) is replaced by an instantiation of a feature vector $\mathbf{m}$, the same criteria can be used to judge how reliable a classification decision for this vector will be.

Another approach is to maximize the separability of classes in a classification problem by application of a linear transform $\mathbf{\Psi}$ to the feature vectors $\mathbf{m}$, which shall fulfill the condition that the ratio of the between-class and within-class covariances is maximized. Usage of a criterion similar to (9.45) gives

$$c_5 = \frac{\det\left|\mathbf{\Psi}^{\mathrm{T}}\mathbf{C}_{zz}\mathbf{\Psi}\right|}{\det\left|\mathbf{\Psi}^{\mathrm{T}}\mathbf{C}_{mm}\mathbf{\Psi}\right|} . \tag{9.47}$$

The transform $\mathbf{f} = \mathbf{\Phi}^{\mathrm{T}} \cdot \mathbf{m}$ determined from (9.3) describes a *Principal Component Analysis* (PCA) based on statistics of the entire set of vectors, which would not be very efficient in terms of decorrelation in the case where the set is a mixture of various classes. (9.34) uses a priori knowledge to apply PCA adapted to class properties for better discrimination. Maximization of (9.47) or similar criteria is achieved for a transform satisfying the condition

---

[1] A large 'between class' covariance expresses that class centroids are widely spread over the feature space, while a small 'within class' covariance expresses a narrow concentration of class-member feature vectors around the respective centroid.

$$\mathbf{\Psi}^{T} \cdot \left[\mathbf{C}_{mm}^{-1} \mathbf{C}_{zz} \right] \cdot \mathbf{\Psi} = \mathbf{\Lambda} \, , \tag{9.48}$$

where again the matrix $\mathbf{\Lambda}$ will be a diagonal matrix containing the expected values of the transformed components by order of relevance. The transform $\mathbf{f} = \mathbf{\Psi}^{T} \cdot \mathbf{m}$ applied to a feature vector is denoted as *Linear Discriminant Analysis* (LDA) [MUIRHEAD, CHEN 1994]. It maximizes the capability for class discrimination for any given feature statistics.

9.3   Feature-based Comparison

Single features or combinations of features can be used to compare two multimedia signals regarding their similarity. For a signal with features represented by a normalized feature vector $\mathbf{m}$, the most similar signal shall be found from a set of N other signals. Each signal is represented by a feature vector of same type, $\mathbf{m}_1...\mathbf{m}_N$. Based on a given distance criterion $d(\cdot)$, the index i of the most similar signal can be determined as

$$i = \underset{\mathbf{m}_n \in \{\mathbf{m}_1...\mathbf{m}_N\}}{\arg\min} \; d\left(\mathbf{m}, \mathbf{m}_n\right) . \tag{9.49}$$

In some applications (e.g. search and retrieval of similar signals from a database), it is further interesting to get as the result of similarity comparison a ranked set of M most similar signals, where usually $M \ll N$. Then, it is necessary to determine not only a best match, but in addition to compute indices $i_1...i_M$ ranked by decreasing similarity, such that[1]

$$d(\mathbf{m}, \mathbf{m}_{i_1}) \leq d(\mathbf{m}, \mathbf{m}_{i_2}) \leq ... \leq d(\mathbf{m}, \mathbf{m}_{i_M}) . \tag{9.50}$$

In value-wise comparison, a linear dependency exists between the complexity of comparison and the number K of values in the feature vector on one hand, the number N of feature vectors to be compared in total on the other hand; the computational effort is proportional with KN. If a difference measure is used which is based on a weighting matrix (e.g. the Mahalanobis distance or the weighted histogram difference), the complexity even increases by the squared number of values in the feature vector, but still increases linearly with N. Nevertheless, as usually $N \gg K$, the reduction of comparisons in N is a key to reduce complexity e.g. for search within huge data bases. Typically, most feature vectors of items in the database will be very dissimilar when compared to the target, such that after testing only some significant feature values, they can be rejected as candidates. This can be utilized to reduce the complexity of comparison: *Hierarchical search* strategies intend to use a small sub-set of feature values to achieve a preliminary selection of

---

[1] i_1 in (9.50) is identical with i in (9.49).

some good candidates for a more accurate comparison, and reject most of the other items. This can also be applied over several levels. A reduction of the feature space dimensionality in the pre-selection is important, while the retained features should be as expressive as possible.

Assume that $\mathbf{m}_n^{(t)}$ shall be a sub-vector of $\mathbf{m}_n$ with $K^{(t)} < K$ feature values. With a total of T steps in a hierarchical search, the following conditions apply:

$$\mathbf{m}_n^{(1)} \subset \mathbf{m}_n^{(2)} \subset \ldots \mathbf{m}_n^{(T-1)} \subset \mathbf{m}_n \; ; \; K^{(1)} < K^{(2)} < \ldots < K^{(T-1)} < K \, . \tag{9.51}$$

Only in the first step, the entire set of N feature sub-vectors of length $K^{(1)}$ is compared. Then, the $M^{(1)}$ best matches are selected, compared by more accurate criteria of $K^{(2)}$ feature values to select $M^{(2)}$ best matches etc.; by the final step, $M^{(T)}$ most similar items (compared by all $K^{(T)}$ feature values available) are retained in a ranked list as provided by (9.50). The number of items retained by each step will be ever decreasing, such that $M^{(1)} > M^{(2)} > \ldots > M^{(T)}$. The number of value-comparison operations (e.g. difference computations) is $NK^{(1)} + M^{(1)}K^{(2)} + \ldots + M^{(T-1)}K$. The principle is shown by the block diagram in Fig. 9.3.

The definition of sub-spaces for pre-selection is much simpler, if feature values are transformed into decorrelated component representations, e.g. by eigenvector analysis (9.3). The values in the component vectors $\mathbf{f} = \mathbf{\Phi}^T \cdot \mathbf{m}$ will inherently be ordered by a sequence of decreasing relevance, since the related eigenvalues (expected values) are ordered by decreasing size. The optimum sub-space partitioning can then simply be done by truncation of the transformed feature representation into a lower number of components, which are still as expressive as possible concerning the feature discrimination from the given reduced number of values. The vector $\mathbf{f}$ is partitioned into T sub-vectors according to the following schema:

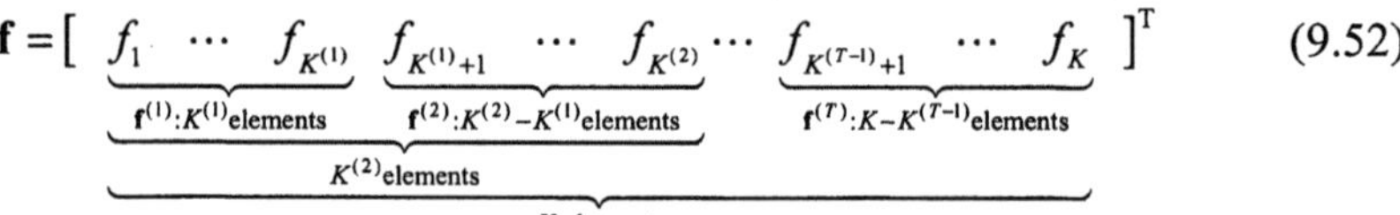

$$\mathbf{f} = \begin{bmatrix} \underbrace{f_1 \; \cdots \; f_{K^{(1)}}}_{\mathbf{f}^{(1)}:K^{(1)}\text{elements}} \; \underbrace{f_{K^{(1)}+1} \; \cdots \; f_{K^{(2)}}}_{\mathbf{f}^{(2)}:K^{(2)}-K^{(1)}\text{elements}} \cdots \underbrace{f_{K^{(T-1)}+1} \; \cdots \; f_K}_{\mathbf{f}^{(T)}:K-K^{(T-1)}\text{elements}} \end{bmatrix}^T \tag{9.52}$$

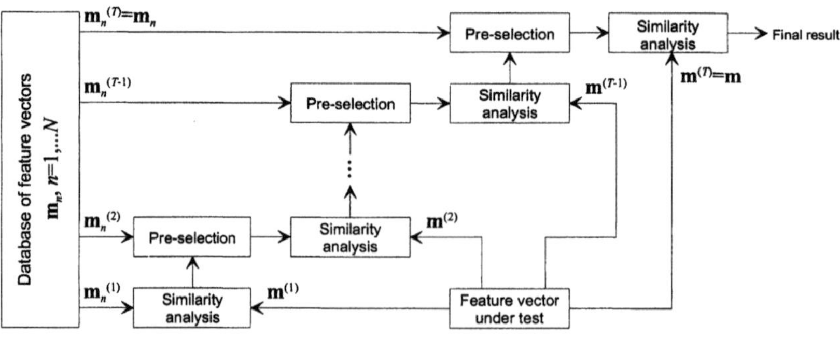

Fig. 9.3. Hierarchical comparison of feature vectors

The sub-vectors $\mathbf{f}^{(t)}$ establish orthogonal complements, and the full vector $\mathbf{f}$ is the concatenation $[\mathbf{f}^{(1)}\ \mathbf{f}^{(2)}\ ...\ \mathbf{f}^{(T)}]^{\mathrm{T}}$, which contains the same evidence about the feature as the original vector $\mathbf{m}$. The transformed component representation must only be computed once for the vectors contained in the database, and can then be used in any comparison operations. If the set of N reference feature vectors in the database is fixed, they can further be organized into a hierarchy of sub-space clusters for *tree-based decisions* which further improves and speeds up the pre-selection process.

Comparison of similarity between sets of data. When properties of a multimedia signal are expressed by multiple feature vector samples, a *similarity between sets* must be determined. A set could e.g. be a series of feature vectors extracted from frames of a video shot, from a collection of still images or from segments of a music piece. A reasonable criterion analyzing the similarity of two data sets $\mathcal{A}$ and $\mathcal{B}$ is the *Hausdorff distance*. Assume that a basic distance criterion $d(\mathbf{m}_a,\mathbf{m}_b)$ is defined, signifying the similarity between two feature vectors relating to any two items $a\in\mathcal{A}$ and $b\in\mathcal{B}$. This can e.g. be the Euclidean distance, which has to be computed for any pair-wise combination of members from $\mathcal{A}$ and $\mathcal{B}$. The Hausdorff distance d_h is then defined as follows:

$$\tilde{d}_h(\mathcal{A},\mathcal{B}) = \max_{a\in\mathcal{A}}\left\{\min_{b\in\mathcal{B}} d\left(\mathbf{m}_a,\mathbf{m}_b\right)\right\}; \quad \tilde{d}_h(\mathcal{B},\mathcal{A}) = \max_{b\in\mathcal{B}}\left\{\min_{a\in\mathcal{A}} d\left(\mathbf{m}_a,\mathbf{m}_b\right)\right\}$$
$$\Rightarrow\quad d_h(\mathcal{A},\mathcal{B}) = \max\left(\tilde{d}_h(\mathcal{A},\mathcal{B}),\tilde{d}_h(\mathcal{B},\mathcal{A})\right). \tag{9.53}$$

9.4 Feature-based Classification

The features of multimedia signals or a transformed equivalent thereof shall be classified, which in the ideal case shall result in semantic categories. This requires *a priori knowledge* which maps feature characteristics into content-related characteristics. Thereby, knowledge about typical feature constellations or rules will enable a system to conclude on specific content, and allow describing multimedia signals by a more abstract level. On the other hand, some decision criteria may not be unique, or feature data may be flagged as unreliable by mechanisms as described before. In such a case, a level of uncertainty must be associated with the classification decision, and it may be tried to obtain more evidence by additional features. Remark that cognition as performed by a human observer works quite similar: Conclusions are drawn which are coupled to reliability considerations. For example, even though an observer recognizes that a face is visible in an image, it may be difficult to actually recognize a person if due insufficient lighting conditions the face appears under low contrast. A subsequent step of verification can then be made checking an initial hypothesis by more specific features, e.g. investigation of very particular facial features for a specific person who is assumed to be visible in the scene.

Reliability of classification. For most classification methods covered in this chapter, optimization of the classification algorithms requires availability of a typical *training set* of features extracted from signal data. The a priori knowledge about class allocations can e.g. be gained by manual annotation of the training set, or by preparing the training set such that the optimum classification is inherently known. For example, in the problem of person recognition mentioned above, it would be necessary to provide a number of pictures for each person that shall be identified. Features are then extracted from this set which are used to train the classifier. The suitability of the training set in the foreseen purpose of classification must however be validated. To evaluate the performance of classifiers, criteria exist which are typically based on the percentages of right and wrong decisions (see Appendix A.2). A validation of the classification performance can be achieved when the training set is partitioned into several sub-sets. The algorithm is then trained only by some of the sub-sets, and verified using other sub-sets as *ground truth*, which were not used for the training. By using each subset once for training and once as ground truth, a mean value of performance over the entire set can be computed. From the reliability considerations discussed in sec. 9.2.5, the number of correct decisions will be high, if the classes are clearly separated in the feature space. The training set must however be compiled in a way that it reflects the situation of the envisaged classification purpose. This even means that preferably *most critical* data shall be included in the training set. For the example of face recognition mentioned above, representative feature data for one specific person should be selected from a variety of images showing the same person, which have a variety as expected by the application (e.g. photographs showing a person from different view angles, under different lighting conditions etc.). The general goal of classification will be correct detection of as many members of a given class as possible, while not allowing members of other classes to be falsely classified into this class. These two paradigms are reflected by the concepts of *precision* and *recall* rates (see Annex A.2). While often an optimum solution of this problem for one specific class can be formulated, multiple-class decisions will in many cases be compromising.

The class separation problem. The optimum separation of classes as expressed by feature constellations in the available training set is the basic problem to be solved. If feature vectors of dimension K are available, the classification is performed within a K-dimensional feature space, which must be partitioned into sub-spaces according to the number of classes L. This class separation problem is illustrated for the example of a feature space with $K=2$ features (dimensions) and $L=2$ classes in Fig. 9.4. It is necessary to define *boundaries* between the partitions of distinct classes in the feature space, such that the allocation of any feature vector to a class becomes straightforward. These boundaries should be determined such that the number of mis-classifications is minimized. In principle, this can be interpreted as a *quantization* of the feature space, which also has a high similarity to methods of vector quantization (cf. sec. 11.5)[1].

First, the *linear classifier* will be introduced, which is a good model to study the effects of class separation and distinction based on linear statistical properties of classes (e.g. mean values, variances, covariances). The linear classifier does however not provide a unique solution to the class separation problem in cases where more than two classes exist.

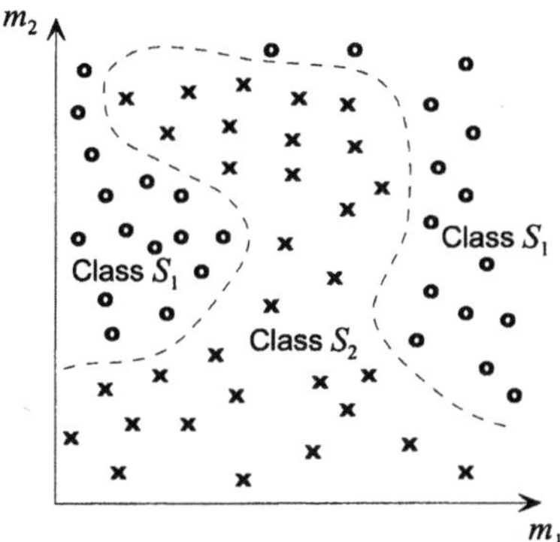

Fig. 9.4. Optimum sub-space partitioning in a classification problem with two classes

9.4.1   Linear Classification of two Classes

In linear classification, the decision boundary between two classes is described by a straight line in a two-dimensional feature space, by a plane or hyperplane in three or more-dimensional feature spaces. An illustration for the 2-dimensional feature space is given in Fig. 9.5. The decision line has an equation

$$w_0 + w_1 m_{1,0} + w_2 m_{2,0} = 0 . \tag{9.54}$$

Any feature vectors $\mathbf{m}_0 = [m_{1,0}, m_{2,0}]^T$ are on the decision line, where $\mathbf{w} = [w_1, w_2]^T$ is the normal vector of any length >0, perpendicular to the decision line. The distance between the decision line and the origin of the feature space is

$$d_0 = \frac{|w_0|}{\|\mathbf{w}\|} . \tag{9.55}$$

Normalization by the length of $\mathbf{w}$ gives the distance between any feature vector $\mathbf{m} = [m_1, m_2]^T$ and the decision line

---

[1] The criterion for optimum partitioning is however not the minimization of squared error as in VQ, but rather the correct association of feature-value constellations to semantic classes.

$$d(\mathbf{m}) = \left| \frac{w_0 + w_1 m_1 + w_2 m_2}{\|\mathbf{w}\|} \right| = \frac{\left| \mathbf{w}^{\mathrm{T}} \cdot \Delta \right|}{\|\mathbf{w}\|} \quad \text{with} \quad \Delta = \mathbf{m} - \mathbf{m}_0 . \tag{9.56}$$

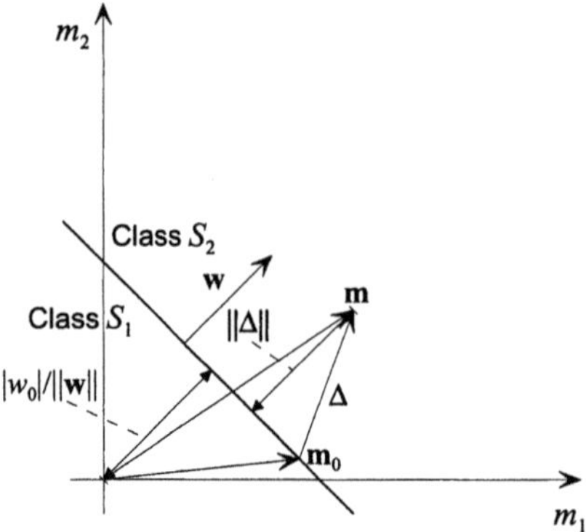

Fig. 9.5. On definition of a linear classifier for $K{=}2$ dimensions of the feature space

For a feature space of arbitrary dimensionality K, w_0 can be included in $\mathbf{w}$ by extension of $\mathbf{m}$ into homogeneous coordinates (cf. (7.208)). Simultaneous normalization by the length of $\mathbf{w}$ gives[1]

$$d(\mathbf{m}) = \left| \widehat{\mathbf{w}}^{\mathrm{T}} \cdot \widehat{\mathbf{m}} \right| \quad \text{with} \quad \widehat{\mathbf{w}} = \begin{bmatrix} \dfrac{w_0}{\|\mathbf{w}\|} \\ \dfrac{\mathbf{w}}{\|\mathbf{w}\|} \end{bmatrix} = \begin{bmatrix} \widehat{w}_0 \\ \widehat{w}_1 \\ \vdots \\ \widehat{w}_K \end{bmatrix} \quad \text{and} \quad \widehat{\mathbf{m}} = \begin{bmatrix} 1 \\ \mathbf{m} \end{bmatrix} = \begin{bmatrix} 1 \\ m_1 \\ \vdots \\ m_K \end{bmatrix} . \tag{9.57}$$

Setting $d(\mathbf{m}){=}0$, an equation describing a decision hyperplane results analogously with (9.54). The problem to be solved is the optimization of position and orientation of this hyperplane. A first approach for optimization can be determined from the analysis of those elements from the training set, which were wrongly classified for the case of any given decision hyperplane. Let $\mathbf{m}_p^{(2|1)}$ be a feature vector which was mistakenly assigned into class S_2, though it is known to belong to S_1, and let $\mathbf{m}_p^{(1|2)}$ express the opposite case. The number of vectors assigned wrongly into classes S_1 and S_2 will be R_1 and R_2, respectively. The overall error of classification can then be expressed by the cost function

---

[1] The argument function $\widehat{\mathbf{w}}^T \widehat{\mathbf{m}}$ can be interpreted as the 'signed Euclidean distance' from the decision line, where the sign indicates on which side of the decision line a feature vector $\mathbf{m}$ is positioned in the feature space. In the example of Fig. 9.5, the sign will be positive in the partition of class S_2. This can be used as a very simple classification criterion in the 2-class decision.

$$\Delta(\widehat{\mathbf{w}}) = \sum_{p=1}^{R_1}\left|\widehat{\mathbf{w}}^{\mathrm{T}}\widehat{\mathbf{m}}_p^{(1|2)}\right| + \sum_{p=1}^{R_2}\left|\widehat{\mathbf{w}}^{\mathrm{T}}\widehat{\mathbf{m}}_p^{(2|1)}\right| = \sum_{p=1}^{R_2}\widehat{\mathbf{w}}^{\mathrm{T}}\widehat{\mathbf{m}}_p^{(2|1)} - \sum_{p=1}^{R_1}\widehat{\mathbf{w}}^{\mathrm{T}}\widehat{\mathbf{m}}_p^{(1|2)}. \qquad (9.58)$$

Similar to the gradient optimization algorithms described in chapter 8, an optimization is made by the direction of the negative gradient of the cost function, derived over $\widehat{\mathbf{w}}$. For the right side of (9.58), the argument $\widehat{\mathbf{w}}^T\widehat{\mathbf{m}}$ is always negative in the partition of class S_1. The derivative is

$$\frac{d}{d\widehat{\mathbf{w}}}\big(\Delta(\widehat{\mathbf{w}})\big) = \sum_{p=1}^{R_2}\widehat{\mathbf{m}}_p^{(2|1)} - \sum_{p=1}^{R_1}\widehat{\mathbf{m}}_p^{(1|2)}. \qquad (9.59)$$

Iterative linear optimization of the orientation vector $\widehat{\mathbf{w}}$ can then be performed as

$$\widehat{\mathbf{w}}^{(r+1)} = \widehat{\mathbf{w}}^{(r)} + \Delta\widehat{\mathbf{w}}^{(r)} = \widehat{\mathbf{w}}^{(r)} - \alpha^{(r)}\cdot\frac{d}{d\widehat{\mathbf{w}}}\big(\Delta(\widehat{\mathbf{w}})\big) \ \text{ with } \ 0 < \alpha^{(r)} \le 1. \qquad (9.60)$$

Typically, $\alpha^{(r)}$ must be decreased when convergence is approached, which can e.g. be achieved by proportionality with the percentage of mis-classified elements, $\Sigma R_l^{(r)}/Q$. The optimization can however show poor convergence behavior if the number of feature vectors which are close to the decision boundary is high, as these may arbitrarily change their class allocation with each iteration step. This can be avoided by introduction of a *rejection zone* (Fig. 9.6), which can be interpreted as a 'no man's land' stripe being parallel-symmetric on both sides of the decision boundary[1]. In principle, the 'accept/reject' decision lines are now individually defined for each class, but parallel and distant by the width of the rejection zone. Using a squared Euclidean distance norm and a decision hyperplane shifted by a rejection zone width $\|\mathbf{w}\|$ into the partition of the respective class, the following cost function results:

$$\Delta(\widehat{\mathbf{w}}) = \sum_{q=1}^{Q_1}\Big[\mathbf{w}^{\mathrm{T}}\mathbf{m}_q^{(1)} + w_0 + \|\mathbf{w}\|\Big]^2 + \sum_{q=1}^{Q_2}\Big[\mathbf{w}^{\mathrm{T}}\mathbf{m}_q^{(2)} + w_0 - \|\mathbf{w}\|\Big]^2. \qquad (9.61)$$

Optimization is done by computing the derivative of the cost function separately over w_0 and $\mathbf{w}$. The following conditions result:

$$\frac{d\Delta(\widehat{\mathbf{w}})}{dw_0} = 2\sum_{q=1}^{Q_1}\Big[\ \mathbf{w}^{\mathrm{T}}\mathbf{m}_q^{(1)} + w_0 + \|\mathbf{w}\|\ \Big] + 2\sum_{q=1}^{Q_2}\Big[\ \mathbf{w}^{\mathrm{T}}\mathbf{m}_q^{(2)} + w_0 - \|\mathbf{w}\|\ \Big] = 0 \qquad (9.62)$$

$$\frac{d}{d\mathbf{w}}\big(\Delta(\widehat{\mathbf{w}})\big) = 2\sum_{q=1}^{Q_1}\mathbf{m}_q^{(1)}\Big[\ \mathbf{w}^{\mathrm{T}}\mathbf{m}_q^{(1)} + w_0 + \|\mathbf{w}\|\ \Big] + 2\sum_{q=1}^{Q_2}\mathbf{m}_q^{(2)}\Big[\ \mathbf{w}^{\mathrm{T}}\mathbf{m}_q^{(2)} + w_0 - \|\mathbf{w}\|\ \Big] = 0$$

$$(9.63)$$

---

[1] This is a nonlinear optimization approach similar to the Levenberg-Marquardt method (cf. sec. 8.7); a one-step optimization approach is however assumed in the sequel. If performed iteratively, the width of the rejection zone could also be varied step by step.

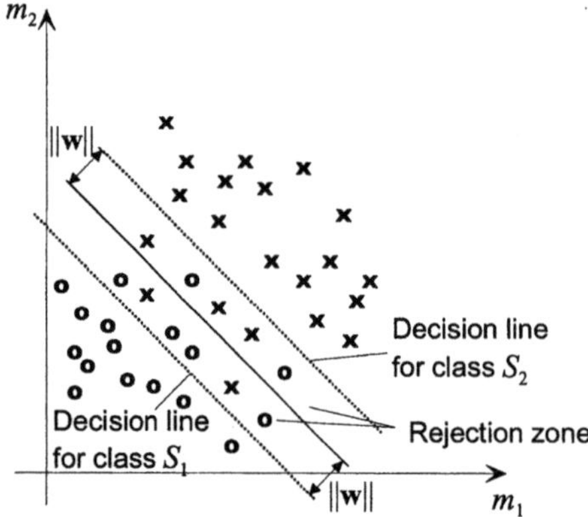

Fig. 9.6. Definition of a 'rejection zone' for optimization of linear classification

With the conditions for the relationships of class centroids and class occurrence distributions,

$$\frac{1}{Q_l}\sum_{q=1}^{Q_l}\mathbf{m}_q^{(l)} = \mathbf{z}^{(l)}; \ Q_1 + Q_2 = Q; \ P_l = \frac{Q_l}{Q} \tag{9.64}$$

the following formulation results by (9.62):

$$Q_1 \cdot \mathbf{w}^{\mathrm{T}} \cdot \mathbf{z}^{(1)} + Q_2 \cdot \mathbf{w}^{\mathrm{T}} \cdot \mathbf{z}^{(2)} + (Q_1 + Q_2) \cdot w_0 + (Q_1 - Q_2) \cdot \|\mathbf{w}\| = 0$$
$$\Rightarrow P_1 \cdot \mathbf{w}^{\mathrm{T}} \cdot \mathbf{z}^{(1)} + P_2 \cdot \mathbf{w}^{\mathrm{T}} \cdot \mathbf{z}^{(2)} + w_0 + (P_1 - P_2) \cdot \|\mathbf{w}\| = 0. \tag{9.65}$$

Defining the global centroid of the training set as $\mathbf{z} = P_1 \cdot \mathbf{z}^{(1)} + P_2 \cdot \mathbf{z}^{(2)}$, the final result is

$$w_0 = (P_2 - P_1) \cdot \|\mathbf{w}\| - \mathbf{w}^{\mathrm{T}} \cdot \mathbf{z}. \tag{9.66}$$

From (9.63),

$$\left[\sum_{q=1}^{Q_1}\mathbf{m}_q^{(1)} \cdot \left[\mathbf{m}_q^{(1)}\right]^{\mathrm{T}}\right] \cdot \mathbf{w} + \left[\sum_{q=1}^{Q_2}\mathbf{m}_q^{(2)} \cdot \left[\mathbf{m}_q^{(2)}\right]^{\mathrm{T}}\right] \cdot \mathbf{w}$$
$$+ w_0 \cdot \left[Q_1 \cdot \mathbf{z}^{(1)} + Q_2 \cdot \mathbf{z}^{(2)}\right] + \|\mathbf{w}\| \cdot \left[Q_1 \cdot \mathbf{z}^{(1)} - Q_2 \cdot \mathbf{z}^{(2)}\right] = 0. \tag{9.67}$$

Resolving the third term in (9.67) by using (9.64) and (9.66) gives

$$w_0 \cdot \left[Q_1 \cdot \mathbf{z}^{(1)} + Q_2 \cdot \mathbf{z}^{(2)}\right] = w_0 \cdot Q \cdot \mathbf{z} = (Q_2 - Q_1) \cdot \|\mathbf{w}\| \cdot \mathbf{z} - Q \cdot \left[\mathbf{z} \cdot \mathbf{z}^{\mathrm{T}}\right] \cdot \mathbf{w}. \tag{9.68}$$

If further the definition of the 'within-class' covariance matrix (9.42) is used, (9.67) can be re-formulated as

$$Q \cdot \mathbf{C}_{mm} \cdot \mathbf{w} = \left[Q_2 \cdot \left(\mathbf{z}^{(2)} - \mathbf{z} \right) - Q_1 \cdot \left(\mathbf{z}^{(1)} - \mathbf{z} \right) \right] \cdot \|\mathbf{w}\|, \qquad (9.69)$$

which finally results in the definition of the optimum vector $\mathbf{w}$ characterizing the direction of the decision line or plane,

$$\frac{\mathbf{w}}{\|\mathbf{w}\|} = \mathbf{C}_{mm}^{-1} \cdot \left[P_2 \left(\mathbf{z}^{(2)} - \mathbf{z} \right) - P_1 \left(\mathbf{z}^{(1)} - \mathbf{z} \right) \right]. \qquad (9.70)$$

For the special case of equal class probabilities ($P_1 = P_2$),

$$\frac{\mathbf{w}}{\|\mathbf{w}\|} = \frac{1}{2} \cdot \mathbf{C}_{mm}^{-1} \cdot \left[\mathbf{z}^{(2)} - \mathbf{z}^{(1)} \right]; \quad w_0 = -\mathbf{w}^{\mathrm{T}} \cdot \mathbf{z} = -\mathbf{w}^{\mathrm{T}} \cdot \frac{\mathbf{z}^{(1)} + \mathbf{z}^{(2)}}{2}, \qquad (9.71)$$

and for another special case of uncorrelated feature vectors being normalized by variances ($\mathbf{C}_{mm} = \mathbf{I}$), using condition $\mathbf{z} = P_1 \cdot \mathbf{z}^{(1)} + P_2 \cdot \mathbf{z}^{(2)}$,

$$\frac{\mathbf{w}}{\|\mathbf{w}\|} = \left[P_2 \left(\mathbf{z}^{(2)} - \mathbf{z} \right) - P_1 \left(\mathbf{z}^{(1)} - \mathbf{z} \right) \right] = 2 P_2 \cdot \left(\mathbf{z}^{(2)} - \mathbf{z} \right) = 2 P_1 \cdot \left(\mathbf{z} - \mathbf{z}^{(1)} \right). \qquad (9.72)$$

The decision hyperplane in the latter case will be perpendicular with the interconnecting line between the two class centroids. The position for $P_1 = P_2$ will be exactly at the center position of this line[1]. Otherwise, the decision line is shifted towards the centroid of the class with lower probability (see Fig. 9.7a)[2], where the amount of shift depends on the width of the rejection zone.

In case of correlated feature vectors, the vector Gaussian distribution is characterized by ellipsoids in the K-dimensional feature space which indicate points of equal probability density of feature vectors $\mathbf{m}$ that belong to a given class S_l. The principal axes of the ellipsoid are gained by eigenvector analysis of the class covariance matrix (see e.g. Figs. 3.5 and 9.2). The decision boundary will then be positioned such that those parts under the PDF volumes are minimized, which would be within the partition of the other class. In cases of correlated feature val-

---

[1] The interconnection line between two directly neighbored centroids is denoted as *Delaunay line*. It is divided into two pieces of equal length by the perpendicular *Voronoi boundary* (also a line in case of $K=2$, a plane or hyperplane for $K>2$). Points on the Voronoi boundary are at equal Euclidean distance from both of the neighbored centroids. In case $P_1 = P_2$ and $\mathbf{C}_{mm} = \mathrm{Diag}\{\sigma_k^2\}$, the class partitions are *Voronoi cells*, which are sub-spaces of the feature space, separated by Voronoi boundaries. Two centroids are regarded as direct neighbors, if the interconnection line (which then is a Delaunay line) does not traverse the Voronoi cell of a third centroid. For more background on Voronoi regions, Voronoi and Delaunay lines, which also play an important role in *Vector Quantization* (VQ), refer to sec. 11.5.1, Fig. 11.23.

[2] In principle, the latter case is very similar to *entropy-constrained VQ* (sec. 11.5.5), where however the optimized linear classification solution (9.70) is more general, taking into account correlations between the feature values within $\mathbf{m}$.

ues, the decision line (or plane) boundaries will hence no longer be perpendicular with the Delaunay line connecting the two class centroids (see Fig. 9.7b).

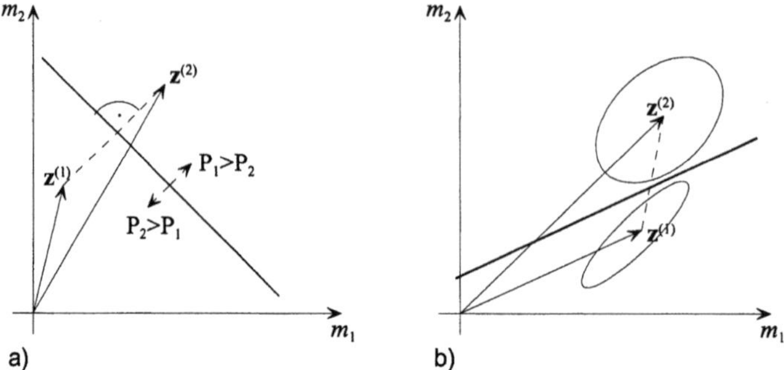

Fig. 9.7. Positions of class centroids and optimum decision lines in linear classification, $K=2$ **a** uncorrelated feature values   **b** different covariance characteristics in the two classes

9.4.2   Generalization of Linear Classification

The linear classifier is defined by infinitely-extended decision lines or hyper-planes. For $L>2$ classes, a total number of $(L-1)\cdot L/2$ pair-wise decisions between classes would have to be performed. However then, overlapping areas between the class partitions occur, where decisions can not be unique. This is shown in Fig. 9.8 for an example of $L=3$ classes.

A possible solution to this problem is *piecewise-linear classification*, where the extension of the decision lines or planes is bounded. This can also be applied for a two-class problem, where the class partitions are then described by a polygon of line segments as shown in Fig. 9.9a; this will obviously be advantageous when nonlinear dependencies are observed between feature values. For the example shown here, the correct class separation is clearly improved by defining three deci-sion lines of bounded validity. The complexity of the decision process is increased, as multiple conditions (depending on the number of lines) have to be tested to decide whether a feature vector falls into the region of a class[1]. In the case of mul-tiple classes, the decision process can be seen as a solution of the problem illus-trated in Fig. 9.8, where for each of the non-unique areas, a clear allocation to a specific class is set by rules. This is shown in Fig. 9.9b for a case of five different classes, where the dotted lines mark the extensions beyond the valid pieces of the decision boundaries.

The *optimization criteria* regarded so far are also based on the assumption that the decision boundary extends to infinity; only then, it is defined as a steady func-tion which can be differentiated for linear optimization. Hence, usage of the opti-mization methods described earlier would be sub-optimum. A possible solution is

---

[1] The number of decisions in the piecewise-linear two-class approach is equal to the num-ber of decision lines or planes which form the boundary between the class partitions.

to select only those class members from the training set for optimization, which are positioned in the vicinity of the valid piece of a decision boundary. This can be achieved by setting selection conditions related to the other decision boundaries involved.

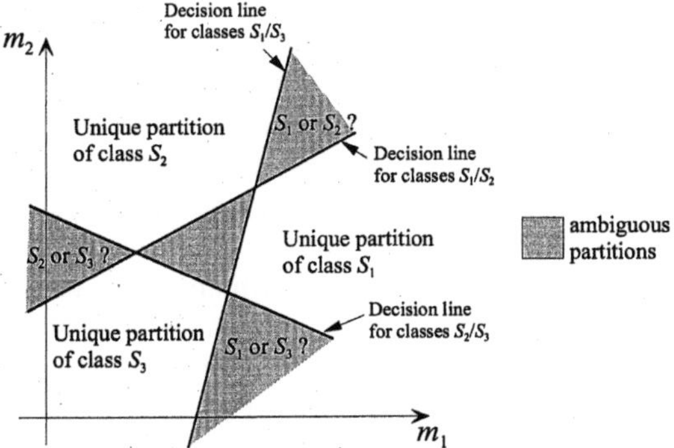

Fig. 9.8. Non-unique areas in a multiple-class decision problem

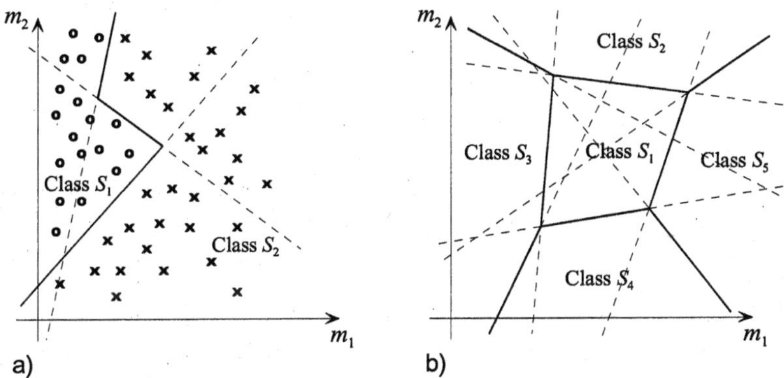

Fig. 9.9. Piecewise-linear classification **a** in case of two classes **b** in case of five classes

Support Vector Machines (SVM). The idea behind SVM is to map a nonlinear class decision problem into a higher-dimensional feature space, where it can be resolved by a linear classifier [BURGES 1998]. The basic approach still performs a two-class decision. Following (9.56) and assuming unity norm of **w**, the linear classification decision into class S_l, $l=1,2$ for a given vector **m** is made by the function

$$l(\mathbf{m}) = \varepsilon\left(w_0 + \mathbf{w}^{\mathrm{T}} \cdot \mathbf{m}\right) + 1\,, \tag{9.73}$$

where $\varepsilon(x)$ is the unit step function, $\varepsilon(x)=0$ for $x<0$, $\varepsilon(x)=1$ for $x>0$. In the SVM approach, a set of *I support vectors* $\mathbf{m}_i$ is defined, which are used in combination with a *kernel function* $k(\mathbf{x},\mathbf{y})$ to map the problem into a nonlinear decision function of type

$$l(\mathbf{m}) = \varepsilon\left(\sum_{i=1}^{I} v_i k\left(\mathbf{m},\mathbf{m}_i\right) + w_0\right) + 1 \ . \tag{9.74}$$

Low-complexity kernel functions suitable for analytic optimization are e.g. *polynomial kernels, radial basis function kernels*, and *sigmoid kernels*. For a given kernel function, the remaining problem is the determination of appropriate support vectors and associated weights v_i. Support vectors are typically positioned within a margin on both sides of the (nonlinear) decision boundary in the input feature space. They are then mapped by the chosen kernel function into the higher-dimensional space where they shall be located on hyperplanes which are co-planar at equal distances from the decision hyperplane. Under these constraints, it is possible to determine the support vectors $\mathbf{m}_i$ and the weights v_i, usually by nonlinear regression. SVM classifiers are perfectly suitable for binary ('yes/no') decision problems such as identification of face areas in images. Applicability to multiple-class problems with a larger number of classes is not yet fully resolved, or at least does not provide significant advantages as compared to other nonlinear classification methods.

9.4.3   Nearest-neighbor and Cluster-based Methods

In *nearest neighbor classification*, feature vectors from the training set are used for direct classification. For a given feature vector $\mathbf{m}$, the most similar vector (nearest neighbor) from the training set shall be found, e.g. using the criterion of Euclidean distance. As the a priori allocation of any training set vector $\mathbf{m}_q^{(l)}$ to a specific class is known, the vector $\mathbf{m}$ can simply be allocated to the same class as its nearest neighbor. This allows approximating almost arbitrary separation boundaries between classes, which can also be interpreted as a highly accurate polygon approximation of almost any class boundary shape. Fig. 9.10a shows an example where the Euclidean distance is used as nearest neighbor criterion[1]. Instead of this *one-neighbor method*, the k nearest training-set vectors can be sought for each feature vector $\mathbf{m}$; this is denoted as *k-nearest-neighbor* classification. The decision is then made in favor of the class for which the majority of neighbors is pre-assigned. The percentage of neighbors belonging to the same class can then further be used as a reliability criterion.

The complexity which results from these exhaustive search methods is hardly manageable in the case of a large training set. Fortunately, the training set is typically highly redundant in terms of the classification decision in particular for cases of training set vectors which are at higher distance from the decision boundary.

---

[1] The training set vectors can then be interpreted as centroids of Voronoi cells. Whenever a different class is assigned for two neighbored cells, the related Voronoi boundary establishes a section of the class separation polygon. See also footnote on p. 397.

The nearest neighbor comparison can consistently be limited to a small subset of training set vectors. The *Delaunay net* (cf. footnote on p. 397) defines an interconnection topology of all directly neighbored training set vectors. Then, all vectors can be discarded from the set, which exclusively have nearest neighbors belonging to the same class. In Fig. 9.10a, all Delaunay lines retained by this condition are drawn by dotted lines. The associated Voronoi boundaries establish the pieces of the actual class boundary. Further vectors may be removed from the nearest-neighbor set, when the associated segments of Voronoi boundaries are almost on identical lines (or hyperplanes for higher number of dimensions). Fig. 9.10b illustrates the example of a largely reduced set of vectors; the effect is indeed very similar to piecewise linear classification.

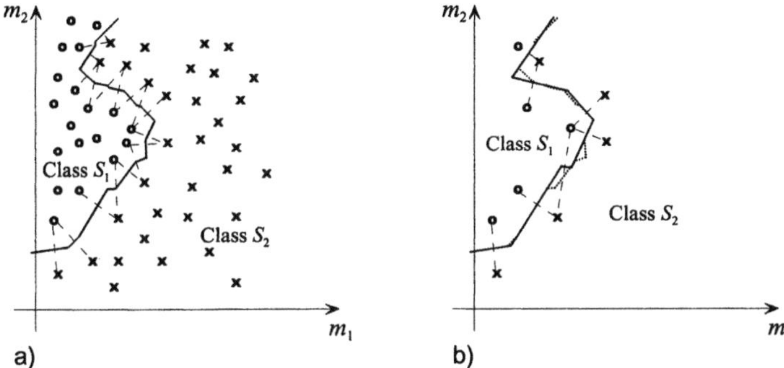

Fig. 9.10. Nearest neighbor classification (one-neighbor method) **a** class separation as resulting from the training set **b** approximation by reduced set of nearest-neighbor vectors

In the simplest method of cluster-based classification, a feature vector $\mathbf{m}$ is compared against L cluster centroids, and assigned to the class S_l where the Euclidean distance is minimum[1]:

$$l(\mathbf{m}) = \underset{\mathbf{z}^{(1)}...\mathbf{z}^{(L)}}{\arg\min} \sum_{k=1}^{K} \left(m_k - z_k^{(l)} \right)^2 = \underset{\mathbf{z}^{(1)}...\mathbf{z}^{(L)}}{\arg\min} \left(\left[\mathbf{m} - \mathbf{z}^{(l)} \right]^{\mathrm{T}} \left[\mathbf{m} - \mathbf{z}^{(l)} \right] \right). \quad (9.75)$$

To perform this comparison, cluster centroids must be given. When the allocations of feature vectors from the training set into the respective classes are known a priori, it is possible to compute the centroids $\mathbf{z}^{(l)}$ according to (9.11). The classification can then also be interpreted as a nearest-neighbor search related to these centroids.

---

[1] One cluster per class is typically sufficient if the classification problem is locally linear, such that the class membership can be expressed by a maximum-distance function around a class centroid; this is typically a spherical or elliptic function – the typical cases which can be interpreted by Gaussian PDF models. In other cases, it may also be appropriate to define several cluster centroids which map into the same class. In this case, the a priori PDF of the class can rather be interpreted by a *mixture of Gaussians* (9.30).

Fig. 9.11 shows an example of a two-dimensional feature space. Each dot marks a feature vector $\mathbf{m}=(m_1,m_2)$ from the training set. In this case, three different concentrations (clusters) are obvious, while some other feature vectors cannot uniquely be allocated to one of the clusters, as they are relatively far away from any of the cluster centroids. Such vectors would e.g. be positioned outside a *trust region*, which can best be determined based on the standard deviations among the feature vectors securely assigned to the respective cluster. If (9.75) is used for the decision on class allocation, the decision boundaries between clusters are in principle still Voronoi lines, where however the decision about values outside the trust regions can be flagged as potentially unreliable.

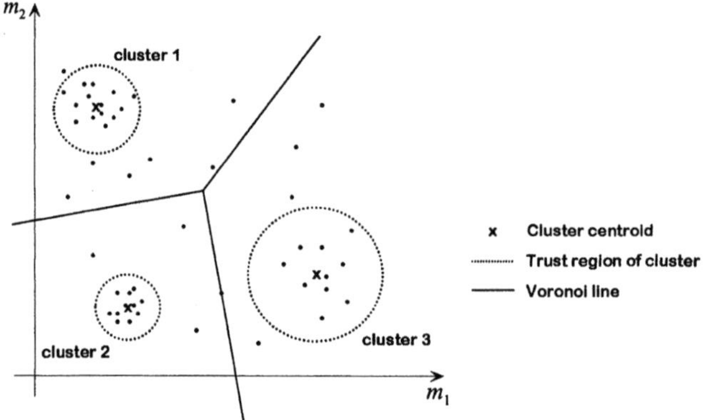

Fig. 9.11. Cluster-based classification in a K=2 dimensional feature space

One important advantage of clustering is the capability to be applied in *blind classification* processes, where no class allocation of training vector data must be given a priori. L' centroids must then be pre-defined, where the parameter L' relates to a hypothetical number of classes. Different strategies are applicable to determine the centroids initially. In *k-means clustering*, those vectors are selected from the training set which are as far apart as possible from each other. Random selection of vectors may also be reasonable in some cases. If initial centroids are defined, it is possible to form class-related subset partitions from the training set using (9.75). As a result, clusters $S_{l'}$ will have $Q_{l'}$ feature vectors $\mathbf{m}_q^{(l')}$ as members. When the quadratic Euclidean distance

$$d^{(l')} = \frac{1}{Q_{l'}} \sum_{q=1}^{Q_{l'}} \sum_{k=1}^{K} \left(m_{k,q}^{(l')} - z_k^{(l')} \right)^2 \tag{9.76}$$

is employed as a cost function, an optimized cluster centroid can be determined by derivation of (9.76) over $\mathbf{z}^{(l')}$. Assuming independency of the cost function in the different feature value dimensions, this can be performed separately for each dimension:

$$\frac{d}{dz_k^{(l')}}\left(d^{(l')}\right) = \frac{1}{Q_{l'}} \cdot \frac{d}{dz_k^{(l')}}\left[\sum_{q=1}^{Q_{l'}}\left(\left(m_{k,q}^{(l')}\right)^2 - 2 \cdot z_k^{(l')} \cdot m_{k,q}^{(l')} + \left(z_k^{(l')}\right)^2\right)\right]$$

$$= \frac{1}{Q_{l'}} \cdot \left[\sum_{q=1}^{Q_{l'}}\left(-2 \cdot m_{k,q}^{(l')} + 2z_k^{(l')}\right)\right] = 2z_k^{(l')} - \frac{2}{Q_{l'}} \cdot \sum_{q=1}^{Q_{l'}} m_{k,q}^{(l')} \overset{!}{=} 0. \tag{9.77}$$

Then, the optimized cluster centroids are given by[1]

$$\mathbf{z}^{(l')}{}_{\text{opt}} = \frac{\sum_{q=1}^{Q_{l'}} \mathbf{m}_q^{(l')}}{Q_{l'}} \quad ; \quad l' = 1, 2, \ldots, L' . \tag{9.78}$$

By any of the optimization steps, the mean distance between the training set vectors and their respective cluster centroids

$$d_{ges} = \frac{1}{Q}\sum_{l'=1}^{L'} Q_{l'} \cdot d^{(l')} \tag{9.79}$$

becomes lower. The optimization step (9.78) can however lead to a change of the cluster allocations, as another optimized centroid vector may now be a better choice for a given training set vector. Due to this fact, the cluster design must be performed iteratively until convergence is achieved. Regarding the mapping of clusters $S_{l'}$ into classes S_l, different strategies are possible. As the number L' is arbitrarily selected in the beginning, it is not useful to retain clusters separate which are hardly distinguishable[2]. These can be *merged* before a next iteration step is done, setting $L' \leftarrow L'-1$. Further, for clusters with high variation, it can be advantageous to perform *splitting*[3]. Finally, it is useful to eliminate clusters which do not have a pre-defined minimum number of members from the training set. These different methods must be iteratively applied until certain requirements (reasonable number of classes, distinguishable classes etc.) are fulfilled. By the end, the number of clusters will be equal to the number of distinguishable classes[4].

---

[1] In principle, except for selection strategies for initial centroids and strategies to reject outliers which are not in the trust region, the cluster optimization algorithm is equivalent to the Generalized Lloyd Algorithm (cf. sec. 11.5.3-11.5.5) for vector quantizer design.

[2] Usage of criteria similar to (9.43)-(9.46) is possible in this context.

[3] Cluster splitting is similar to splitting procedures in vector quantizer design (sec. 11.5.4). The cluster centroid $\mathbf{z}^{(l)}$ can be artificially modified into two different values $\mathbf{z}^{(l)}+\varepsilon$ and $\mathbf{z}^{(l)}-\varepsilon$. For these new centroids, allocation of training-set vectors must be determined again, followed by optimization (9.78).

[4] Complicated class constellations as mentioned in the footnote on p. 401 may be reasonably described by a mixture of several clusters. The decision whether this is reasonable or not can only be made by comparing the semantic meaning behind the different clusters, which typically must be done by human interaction.

At the boundaries between two class partitions, it is probable that certain feature vectors **m** are assigned to wrong classes both in nearest neighbor and in cluster-based methods. Classification is anyway unreliable outside the trust regions (or inside rejection zones). These decisions can either be flagged as 'undefined', or vectors are allocated to one of the clusters based on probability considerations. The *a priori PDF* $p(\mathbf{m}|S_l)$ can e.g. be derived by modeling the cluster PDF by a vector Gaussian distribution. If further the first-order probabilities for selection of the different classes are evaluated, the optimum MAP decision for such outliers can be made based on the Bayes rule (cf. sec. 8.5)

$$l(\mathbf{m}) = \arg\max_{S_1\ldots S_L} \left[P(S_l|\mathbf{m}) \right] = \arg\max_{S_1\ldots S_L} \left[p(\mathbf{m}|S_l) \cdot P(S_l) \right]. \tag{9.80}$$

In this case, the decision will eventually not be made for the cluster with the centroid nearest to **m**, but it will rather be considered

- which cluster has largest probability $P(S_l)$
- for which cluster the a priori probability $p(\mathbf{m}|S_l)$ is more wide-spread, such that also feature vectors which are relatively far from the centroid may be reasonably assigned as members.

9.4.4   Maximum a Posteriori (Bayes) Classification

Maximum a Posteriori (MAP) optimization methods are based on statistical criteria of *conditional probabilities*. In the context of estimation methods, the basic principle was introduced (see sec. 8.5), other applications are in classification and segmentation of multimedia signals. $P(S_l|\mathbf{m})$ indicates the probability by which a given feature vector **m** shall be assigned to class S_l. This *a posteriori* probability is not explicitly known in the beginning. It is a plausible assumption however, that the optimum classification decision will be made for the class where $P(S_l|\mathbf{m})$ becomes maximum. The *a priori* PDF $p(\mathbf{m}|S_l)$ defines the opposite relationship, expressing the probability that a certain feature vector **m** appears when a class S_l is given. This probability density can e.g. be approximated by analyzing the members of the training set which are pre-assigned to S_l. As an analytic model for this case, the vector Gaussian PDF (3.40) shall be used again. The parameters of the PDF are estimated from the vectors $\mathbf{m}_q^{(l)}$ by computing the class centroid vector and covariance matrix (9.33). The following relationship exists between the two conditional probabilities and the first-order probability functions $p(\mathbf{m})$ and $P(S_l)$:

$$P(S_l|\mathbf{m}) \cdot p(\mathbf{m}) = p(\mathbf{m}|S_l) \cdot P(S_l). \tag{9.81}$$

Re-formulation into the BAYES rule gives

$$P(S_l|\mathbf{m}) = \frac{p(\mathbf{m}|S_l) \cdot P(S_l)}{p(\mathbf{m})} = \frac{p(\mathbf{m}|S_l) \cdot P(S_l)}{\sum_{k=1}^{L} p(\mathbf{m}|S_k) \cdot P(S_k)}. \tag{9.82}$$

If (9.82) shall be maximized for a given $\mathbf{m}$, the denominator can be regarded as a constant. Using the vector Gaussian PDF as a model for the a priori probability of class S_l, the optimum MAP classification decision is

$$S_{l,opt} = \underset{S_1, S_2, \ldots, S_L}{\arg\max} \left[P(S_l) \cdot \frac{1}{\sqrt{(2\pi)^K \cdot \left| \mathbf{C}_{mm}^{(l)} \right|}} \cdot e^{-\frac{1}{2}\left[\mathbf{m}-\mathbf{z}^{(l)} \right]^T \left[\mathbf{C}_{mm}^{(l)} \right]^{-1} \left[\mathbf{m}-\mathbf{z}^{(l)} \right]} \right]. \tag{9.83}$$

By taking the logarithm of (9.83), the following function results, for which still the maximum must be found:

$$\ln\left[P(S_l) \cdot p(\mathbf{m}\,|\,S_l) \right]$$
$$= \ln P(S_l) - \frac{K}{2}\ln 2\pi - \frac{1}{2}\ln\left| \mathbf{C}_{mm}^{(l)} \right| - \frac{1}{2}\left[\mathbf{m}-\mathbf{z}^{(l)} \right]^T \left[\mathbf{C}_{mm}^{(l)} \right]^{-1} \left[\mathbf{m}-\mathbf{z}^{(l)} \right]. \tag{9.84}$$

To analyze this result, the optimum Bayes decision is now applied to a decision between two classes, even though it is clearly not restricted to this case. Assume that the MAP decision determines a sub-division of the feature space into two different partitions $\mathcal{R}_1$ and $\mathcal{R}_2$. The probability of a mis-classification within one class results by analysis of the volume under the a priori PDF, as far as it falls into the partition of the respective other class. The total probability of mis-classifications then results by summing the mis-classification probabilities of the single classes, weighted by the respective class probabilities:

$$P_{err} = P(S_2) \int_{\mathcal{R}_1} p(\mathbf{m}\,|\,S_2)d\mathbf{m} + P(S_1) \int_{\mathcal{R}_2} p(\mathbf{m}\,|\,S_1)d\mathbf{m}. \tag{9.85}$$

From (9.82), class S_1 will be selected if

$$p(\mathbf{m}\,|\,S_1)P(S_1) > p(\mathbf{m}\,|\,S_2)P(S_2) \quad \Rightarrow \frac{p(\mathbf{m}\,|\,S_1)}{p(\mathbf{m}\,|\,S_2)} > \frac{P(S_2)}{P(S_1)}. \tag{9.86}$$

Now, regard the case where both classes are modeled by a vector Gaussian PDF of identical covariance matrices, which then will be both also be identical to the mean 'within-class' covariance matrix $\mathbf{C}_{mm}$ (9.42). Taking the logarithm of (9.86) and substituting (9.84) results in the condition

$$\ln p(\mathbf{m}\,|\,S_1) - \ln p(\mathbf{m}\,|\,S_2)$$
$$= -\frac{1}{2}\left[\mathbf{m}-\mathbf{z}^{(1)} \right]^T \cdot \left[\mathbf{C}_{mm} \right]^{-1} \cdot \left[\mathbf{m}-\mathbf{z}^{(1)} \right] + \frac{1}{2}\left[\mathbf{m}-\mathbf{z}^{(2)} \right]^T \cdot \left[\mathbf{C}_{mm} \right]^{-1} \cdot \left[\mathbf{m}-\mathbf{z}^{(2)} \right]$$
$$= -\frac{1}{2}\mathbf{m}^T \left[\mathbf{C}_{mm} \right]^{-1}\mathbf{m} + \left[\mathbf{z}^{(1)} \right]^T \cdot \left[\mathbf{C}_{mm} \right]^{-1}\mathbf{m} - \frac{1}{2}\left[\mathbf{z}^{(1)} \right]^T \cdot \left[\mathbf{C}_{mm} \right]^{-1} \cdot \mathbf{z}^{(1)}$$
$$+ \frac{1}{2}\mathbf{m}^T \left[\mathbf{C}_{mm} \right]^{-1}\mathbf{m} - \left[\mathbf{z}^{(2)} \right]^T \cdot \left[\mathbf{C}_{mm} \right]^{-1}\mathbf{m} + \frac{1}{2}\left[\mathbf{z}^{(2)} \right]^T \cdot \left[\mathbf{C}_{mm} \right]^{-1} \cdot \mathbf{z}^{(2)}$$
$$= \left[\mathbf{z}^{(1)} - \mathbf{z}^{(2)} \right]^T \cdot \left[\mathbf{C}_{mm} \right]^{-1} \cdot \mathbf{m} - \frac{1}{2}\left[\mathbf{z}^{(1)} - \mathbf{z}^{(2)} \right]^T \cdot \left[\mathbf{C}_{mm} \right]^{-1} \cdot \left[\mathbf{z}^{(1)} + \mathbf{z}^{(2)} \right]$$

$$= \left[\mathbf{z}^{(1)} - \mathbf{z}^{(2)} \right]^{\mathrm{T}} \cdot \left[\mathbf{C}_{mm} \right]^{-1} \cdot \left[\mathbf{m} - \frac{\mathbf{z}^{(1)} + \mathbf{z}^{(2)}}{2} \right] > \ln \frac{P(S_2)}{P(S_1)}. \tag{9.87}$$

By this rule, an optimum decision boundary is defined where $p(\mathbf{m}|S_1) \cdot P(S_1)$ and $p(\mathbf{m}|S_2) \cdot P(S_2)$ are equal. This is shown for the example of one single (scalar) feature value (which could also be interpreted as a plane section of a PDF along the Delaunay line of a multi-dimensional feature space) in Fig. 9.12. A graphical interpretation of (9.87) for a case of two features is given in Fig. 9.13. Any vector $\mathbf{m}_0$ positioned on the separation line is described by the equation

$$\left[\mathbf{z}^{(1)} - \mathbf{z}^{(2)} \right]^{\mathrm{T}} \cdot \left[\mathbf{C}_{mm} \right]^{-1} \cdot \left[\mathbf{m}_0 - \frac{\mathbf{z}^{(1)} + \mathbf{z}^{(2)}}{2} \right] = \ln \frac{P(S_2)}{P(S_1)}. \tag{9.88}$$

First, regard the case $\mathbf{C}_{mm} = \mathbf{C}_{mm}^{-1} = \mathbf{I}$. From the multiplication of the two vector parenthesis expressions on the left side of (9.88), a constant value $\ln P(S_2) - \ln P(S_1)$ will result for all points $\mathbf{m}_0$, which are positioned on a line perpendicular with the Delaunay line between the two class centroids, $\mathbf{z}^{(1)}$-$\mathbf{z}^{(2)}$. For $P(S_1) = P(S_2)$, the separation line intersects the Delaunay line at the point $(\mathbf{z}^{(1)} + \mathbf{z}^{(2)})/2$, which means it is a Voronoi boundary. For $P(S_1) \neq P(S_2)$, the line is shifted from the center towards the class centroid of the class having less probability. For $\mathbf{C}_{mm} \neq \mathbf{I}$, the inverse covariance matrix plays the role of a rotation matrix, the separation line will no longer be perpendicular on the Delaunay line. This is reasonable, because then the vector Gaussian PDF also takes the shape of an ellipse around the class centroid, which is tilted by orientation of the principal axis.

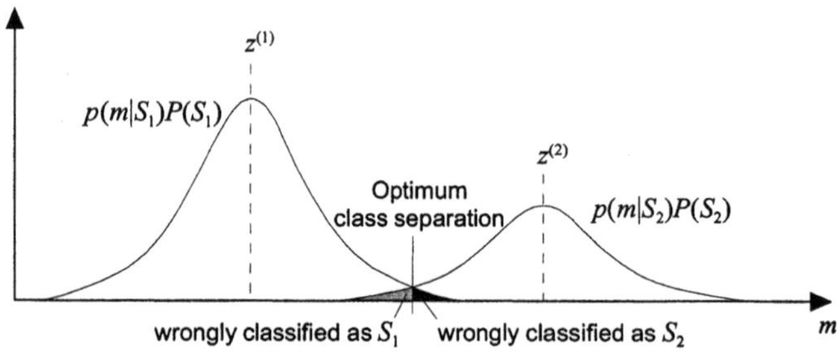

Fig. 9.12. Interpretation of the optimum MAP decision for the case of one single feature value and decision between two classes

For more than 2 classes, the problem of MAP decision is not fundamentally different. Still it will e.g. be possible to model a priori probabilities by vector Gaussian functions, such that finding the maximum value $P(S_i|\mathbf{m})$ can fully be solved analytically.

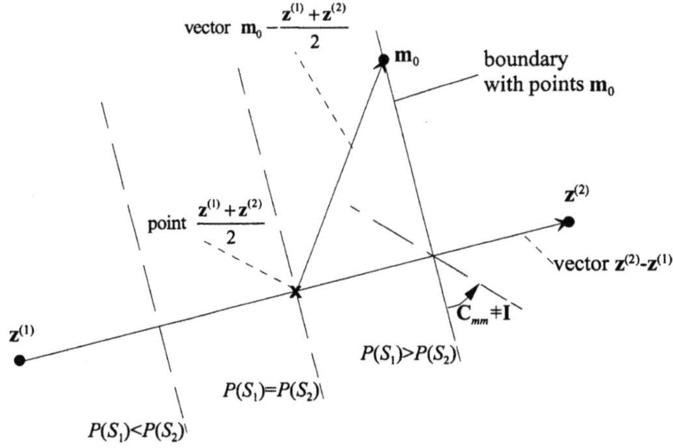

Fig. 9.13. Interpretation of the optimum MAP decision for the case of a 2D feature space and decision between two classes

9.4.5 Artificial Neural Networks

Artificial neural networks (ANN) are capable to capture nonlinear behavior of signals or features extracted from signals in a much wider sense, and can also be adapted to specific signal behavior by training procedures. The notion 'neural' reflects the similarity of the scheme with the functionality of neurons in the nervous system of humans and animals. These systems are inherently nonlinear, as excitations at the input of the neurons are only forwarded if a certain threshold level is superseded.

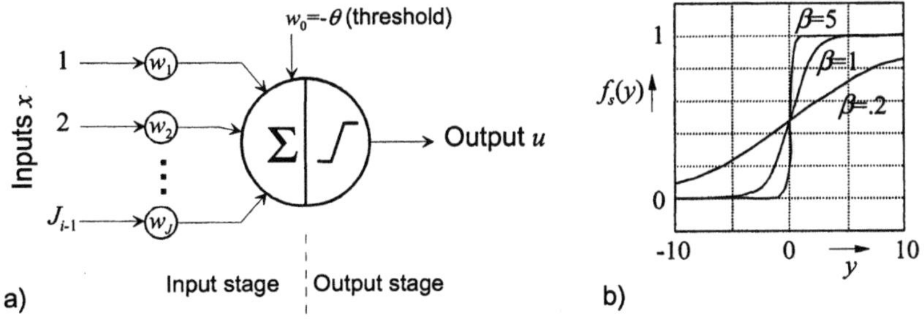

Fig. 9.14. a Typical basic node (perceptron) of an ANN **b** sigmoid function

A typical basic node of an ANN, the *perceptron*, is shown in Fig. 9.14. This was originally developed from a model of neurons in the retina (cf. sec. 6.1.1, Fig. 6.1). At the input stage, a linear combination of J input signals x_m is produced, and a bias value w_0 is added:

$$y = w_0 + \sum_{m=1}^{J} w_m x_m \, . \tag{9.89}$$

The bias shifts the effect of the subsequent nonlinear function by a value $\theta = -w_0$. The nonlinear function which is most commonly used in ANN is the *sigmoid function*

$$u = f_S(y) = \frac{1}{1 + e^{-\beta y}} \, . \tag{9.90}$$

The signal u can have values between 0 for $y \ll \theta$ and 1 for $y \gg \theta$; the factor β regulates the 'steepness' of the transition in the range around $y = \theta$: For the extreme case $\beta = 0$, the output is a constant value $u = 0{,}5$; for $\beta \to \infty$, the function is equivalent to a hard-threshold limiter (unit step function). The advantage of the sigmoid function is the steadyness, which makes it continuously differentiable. This is an important condition for optimization of neural networks. The derivative of the sigmoid is computed by

$$f_S(y) = \left(1 + e^{-\beta y}\right)^{-1} \Rightarrow \frac{df_S(y)}{dy} = -\left(1 + e^{-\beta y}\right)^{-2} \cdot (-\beta) \cdot e^{-\beta y} = \beta \cdot e^{-\beta y} \cdot f_S(y)^2 \, . \tag{9.91}$$

From (9.90), $f_S(y) \cdot [1 + e^{-\beta y}] = 1 \Rightarrow f_S(y) \cdot e^{-\beta y} = 1 - f_S(y)$, such that

$$\frac{df_S(y)}{dy} = \beta \cdot f_S(y) \cdot \left[1 - f_S(y)\right] . \tag{9.92}$$

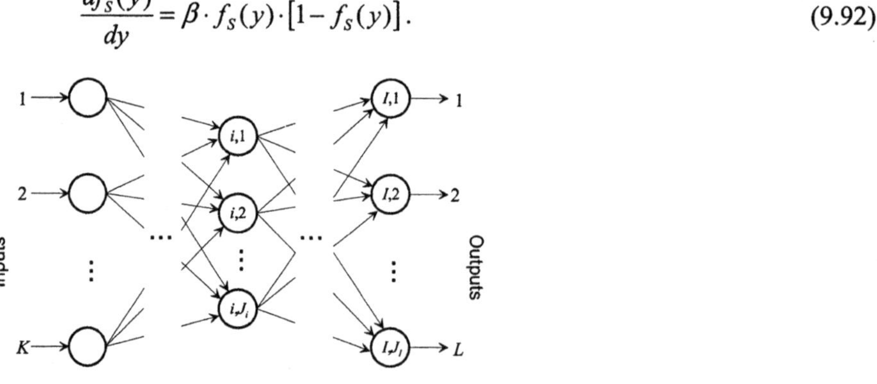

Fig. 9.15. Structure of a multi-layer perceptron

The real effect of the ANN is achieved by defining a topology of *interconnections* from many basic nodes. One of the commonly-used topologies is the *Multi-Layer Perceptron (MLP)* shown in Fig. 9.15. By the interconnection of nodes, it is structured similarly to the neuron structure in the human retina (cf. Fig. 6.1). The input layer of the MLP has K inputs, which will be the values from a feature vector **m** if the network is used for classification[1]. The nodes of the *input layer* are indeed no functional nodes of the network, but have rather a distribution functionality, forwarding the inputs towards all the nodes of the subsequent layer. The MLP consists of one or several *hidden layers* and one *output layer* of L outputs, from which the result of classification can be concluded, e.g. by selecting the class which corresponds to the maximum output[2]. The node structures of the hidden layers and the output layer are identical to the structure shown in Fig. 9.14a. In the MLP, each output of nodes at any layer of the network is connected to each input of nodes in the subsequent layer, where however input weights may be set to zero.

For usage as a classifier, the ANN must be trained by using a typical training set. For the MLP, the *Back Propagation Algorithm* (BPA) is the usual training procedure, which tracks possible errors in classification backwards from the output through all layers, thereby optimizing all weighting factors such that the classification error is minimized. Assume that an MLP consists of I active layers, where in the layer i a number of J_i perceptrons is used. For node j in layer i, $J_{i-1}+1$ weighting factors $w_{i,j,m}$, $m=0,...,J_{i-1}$, are needed for the connections from the outputs of layer $i-1$. The output signal of node (i,j) is

$$u_{i,j} = f_S\left[\sum_{m=0}^{J_{i-1}} w_{i,j,m} \cdot u_{i-1,m}\right]; \; u_{i-1,0} = 1. \tag{9.93}$$

The optimization of the network shall be performed using a training set with a total of Q vectors $\mathbf{m}_q$. The vector[3] $\mathbf{w}$ shall contain all weighting factors $w_{i,j,m}$ of all nodes of all layers of the network. Hence, the global criterion for the quality of the classification decision can be defined as

$$\Delta(\mathbf{w}) = \sum_{q=1}^{Q} \Delta_q(\mathbf{w}). \tag{9.94}$$

---

[1] The input values of the network could as well be *signal values*. In such a case, the output of the network can be a feature vector $\mathbf{m}$ or also a classification decision computed directly from the signal. The network is implicitly used as feature extractor in such a case. Additionally, ANN can also be applied for nonlinear estimation and prediction.

[2] Observe that the outputs of the ANN are continuous in amplitude. Hence, the amplitudes of the different outputs, which are in the range between 0 and 1, can be used as criteria for the reliability of a decision. A unique classification will be made if the network responds by only one strong output; the a posteriori probability of a class assignment could be derived from the output signals as $P(S_l \,|\, \mathbf{m}) = u_l \Big/ \sum_{k=1}^{L} u_k$.

[3] $\mathbf{w}$ could also be interpreted as a 3D tensor

As a quality criterion for classification of any training set member $\mathbf{m}_q$, the squared error in the classification resulting at the output of the network can be used, where $c_l(\mathbf{m}_q)$ shall be the ideal expected result for $\mathbf{m}_q$ at output l[1]:

$$\Delta_q(w) = \sum_{l=1}^{J_I} \left[u_{I,l}(\mathbf{m}_q) - c_l(\mathbf{m}_q) \right]^2 . \tag{9.95}$$

Optimization is performed by an iterative gradient algorithm, where the weighting factors in iteration $r+1$ are optimized by the direction of the negative gradient of the cost function as expressed by partial derivatives,

$$w_{i,j,m}^{(r+1)} = w_{i,j,m}^{(r)} - \varepsilon \cdot \left. \frac{\partial \Delta(\mathbf{w})}{\partial w_{i,j,m}} \right|_{\mathbf{w}^{(r)}} = w_{i,j,m}^{(r)} - \varepsilon \cdot \sum_{q=1}^{Q} \left. \frac{\partial \Delta_q(\mathbf{w})}{\partial w_{i,j,m}} \right|_{\mathbf{w}^{(r)}} . \tag{9.96}$$

The factor ε influences the speed of convergence in the iterative gradient optimization. The gradient can be decomposed into a linear and a nonlinear component

$$\frac{\partial \Delta_q(\mathbf{w})}{\partial w_{i,j,m}} = \frac{\partial \Delta_q(\mathbf{w})}{\partial u_{i,j}} \cdot \frac{\partial u_{i,j}}{\partial w_{i,j,m}} , \tag{9.97}$$

where the linear term is with (9.92)

$$\frac{\partial u_{i,j}}{\partial w_{i,j,m}} = \frac{\partial}{\partial w_{i,j,m}} \left[f_s \left(\sum_{n=0}^{J_{i-1}} w_{i,j,n} \cdot u_{i-1,n} \right) \right]$$

$$= \beta \cdot f_s \left(\sum_{n=0}^{J_{i-1}} w_{i,j,n} \cdot u_{i-1,n} \right) \cdot \left[1 - f_s \left(\sum_{n=0}^{J_{i-1}} w_{i,j,n} \cdot u_{i-1,n} \right) \right] \cdot \frac{\partial}{\partial w_{i,j,m}} \left(\sum_{n=0}^{J_{i-1}} w_{i,j,n} \cdot u_{i-1,n} \right)$$

$$= \beta \cdot u_{i,j} \cdot (1 - u_{i,j}) \cdot u_{i-1,m}. \tag{9.98}$$

The nonlinear term is re-written as

$$\frac{\partial \Delta_q(\mathbf{w})}{\partial u_{i,j}} = \sum_{n=1}^{J_{i+1}} \left[\frac{\partial \Delta_q(\mathbf{w})}{\partial u_{i+1,n}} \cdot \frac{\partial u_{i+1,n}}{\partial u_{i,j}} \right] = \sum_{n=1}^{J_{i+1}} \left[\frac{\partial \Delta_q(\mathbf{w})}{\partial u_{i+1,n}} \cdot \frac{\partial}{\partial u_{i,j}} \left(f_s \left(\sum_{p=0}^{J_i} w_{i+1,n,p} \cdot u_{i,p} \right) \right) \right]$$

$$= \sum_{n=1}^{J_{i+1}} \left[\frac{\partial \Delta_q(\mathbf{w})}{\partial u_{i+1,n}} \cdot \beta \cdot f_s \left(\sum_{p=0}^{J_i} w_{i+1,n,p} \cdot u_{i,p} \right) \cdot \left(1 - f_s \left(\sum_{p=0}^{J_i} w_{i+1,n,p} \cdot u_{i,p} \right) \right) \cdot \frac{\partial}{\partial u_{i,j}} \sum_{p=0}^{J_i} w_{i+1,n,p} \cdot u_{i,p} \right]$$

$$= \beta \cdot \sum_{n=1}^{J_{i+1}} \left[\frac{\partial \Delta_q(\mathbf{w})}{\partial u_{i+1,n}} \cdot u_{i+1,n} \cdot (1 - u_{i+1,n}) \cdot w_{i+1,n,j} \right]. \tag{9.99}$$

The following initial conditions are set at the output layer for the nonlinear term:

---

[1] Ideally, as said in the footnote on p. 409, $c_l(\mathbf{m}_q)=1$ for the true class $l=S(q)$, zero otherwise. If however the classification of a training set member is vague, other than ideally expected weights can be assigned.

$$\frac{\partial \Delta_q(\mathbf{w})}{\partial u_{I,j}} \overset{!}{=} u_{I,j}(\mathbf{m}_q) - c_j(\mathbf{m}_q) \quad ; \quad j = 1, 2, \ldots, L. \tag{9.100}$$

Now, the optimization can be started at the output, because according to (9.99), each previous layer nonlinear derivative term can recursively be computed from the subsequent layer result. Though algorithmically simple, the BPA is computationally costly when the number of perceptrons is high. After each iteration step, the cost function (9.94) must again be updated. Alternatively, it is possible to update the cost function in each iteration only for a subset of vectors, which should then represent sufficient variety over different classes.

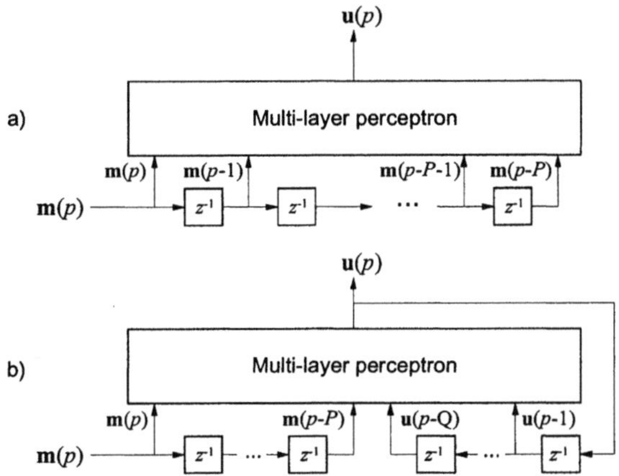

Fig. 9.16. Networks establishing relationships between subsequent classification results **a** Time-delay network **b** Recurrent network

The topology of neural networks can also be augmented such that a relationship is established between subsequent classification decisions. A *time delay network* can be interpreted as an ANN embedded into an FIR filter, replacing the linear superposition by the perceptron structure. The feature vectors are fed into a tapped delay line at the input, such that they take influence on several subsequent classification decisions (Fig. 9.19a). In a *recurrent network*, a similar approach combines FIR and IIR structures, such that also feedback of previous outputs (classification decisions) is used (Fig. 9.19b).

Radial Basis Functions (RBF) network. The RBF network in its most simple form (Fig. 9.17) consists of only 2 layers, where the first layer is not a perceptron, but a maximum-likelihood classifier[1] as defined by Gaussian a priori distribution

---

[1] An ML classifier follows the same principle as the maximum-likelihood estimator (8.35),(8.36). The function (9.101) is proportional with the a priori probability that a fea-

functions for J different classes. The output signal of node j in this layer, with feature vector **m** as input, will be

$$u_{1,j} = e^{-\frac{1}{2}\left[\mathbf{m}-\mathbf{z}^{(j)}\right]^{\mathrm{T}}\left[\mathbf{C}_{mm}^{(j)}\right]^{-1}\left[\mathbf{m}-\mathbf{z}^{(j)}\right]}.$$

(9.101)

Here, $\mathbf{z}^{(j)}$ and $\mathbf{C}_{mm}^{(j)}$ are defined as centroid vectors and covariance matrices in J different a priori defined classes; the respective output values $u_{1,j}$ tend against 1 for a good match with the expected distribution probability of the class, and will tend towards 0 otherwise. The class parameters can e.g. be determined by cluster analysis, or by direct computation of (9.11) and (9.33). The output layer consists of perceptrons, for which the weighting factors can be optimized by a procedure similar to the back propagation algorithm.

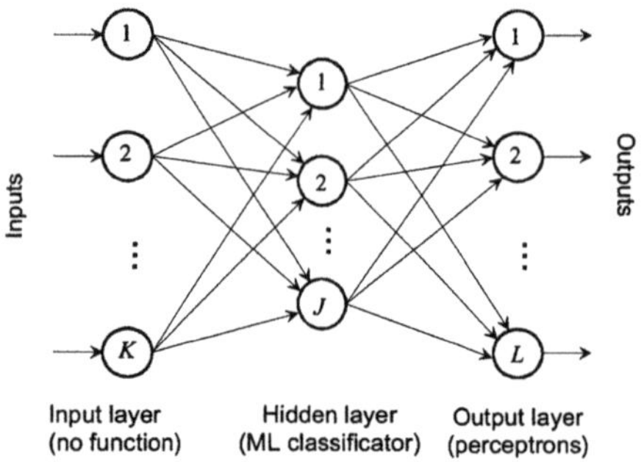

Fig. 9.17. Radial basis function (RBF) network

In general, the performance of a neural network highly depends on the topology, which was assumed to be given a priori so far. For an optimization of the topology itself, different strategies are applicable. It is e.g. possible to start with a network of high number of nodes, and prune out in an iterative process those nodes and/or complete layers which apparently only take low influence on the quality of the classification result. Due to the inherent nonlinear behavior of the network, there will however not be a linear dependency between the number of nodes/layers and the classification performance. Further, as in all classification algorithms, the final performance highly depends on the correct selection of the training set data used for optimization.

Self organizing feature maps (SOFM). SOFM, also known as *Kohonen maps* [KOHONEN 1982] or *topographic maps* are competitive neural networks, which allow mapping feature data of arbitrary dimension into a low-dimensional feature space, where items of similar characteristics will be clustered, and maximum dissimilar data will be located at a far topological distance. This allows for example *blind*

---

ture vector **m** is assigned to class S_j described by a Gaussian PDF with mean value $\mathbf{z}^{(j)}$ and covariance $\mathbf{C}_{mm}^{(j)}$.

sorting of any feature data sets $\mathcal{M}$ which have no prior class assignments; it can then also help to find class-related systematic properties.

A general structure of an SOFM is shown in Fig. 9.18. K-dimensional input feature vectors shall be clustered into L cells of a P-dimensional feature space. The matrix $\mathbf{W}=[\mathbf{w}_1\ \mathbf{w}_2\ \dots\ \mathbf{w}_L]^{\mathrm{T}}$ is a KxL matrix consisting of *synaptic weights* related to L neurons. The neurons are associated with cells of the feature space, competing for allocation of incoming vectors $\mathbf{m}_q$, $q=1,\dots Q$, where typically $Q \gg L$. Initially, the weights $\mathbf{w}_{k,l}$ are populated by a random selection of vectors $\mathbf{m}_q$ from $\mathcal{M}$. For any incoming vector $\mathbf{m}_q$, the winning neuron is the one with minimum squared or Euclidean distance regarding the synaptic weights,

$$l^* = \arg\min_{l=1,\dots,L} \sum_{k=1}^{K} \left(m_q(k) - w_{k,l} \right)^2 \tag{9.102}$$

The mapping matrix $\mathbf{P}=[\mathbf{p}_1\ \dots\ \mathbf{p}_L]^{\mathrm{T}}$ of size PxL receives the index l^*, and maps $\mathbf{m}_q$ to the position of the winning neuron $\mathbf{p}_{l*}$ in the feature space; in addition, the topological distance relative to all other neurons is evaluated, which is then used to compute an update for the synaptic weights for $l=1,\dots,L$ by

$$\Delta\mathbf{w}_l = \varepsilon_r \cdot \Gamma_r(l,l^*) \cdot \left[\mathbf{m} - \mathbf{w}_l \right] \ ; \ \ \mathbf{w}_l = \left[w_{1,l}\ \ \ w_{2,l}\ \ \ \cdots\ \ \ w_{K,l} \right]^{\mathrm{T}}, \tag{9.103}$$

where ε_r is a step size factor depending on the convergence status of the SOFM. $\Gamma_r(l,l^*)$ is a neighborhood weighting function, which typically applies a Gaussian-shaped weight depending on the topological distance between l and l^* in the feature space. The synaptic weights are then updated as $\mathbf{w}_l+\Delta\mathbf{w}_l$, and the process of competition continues until convergence is reached.

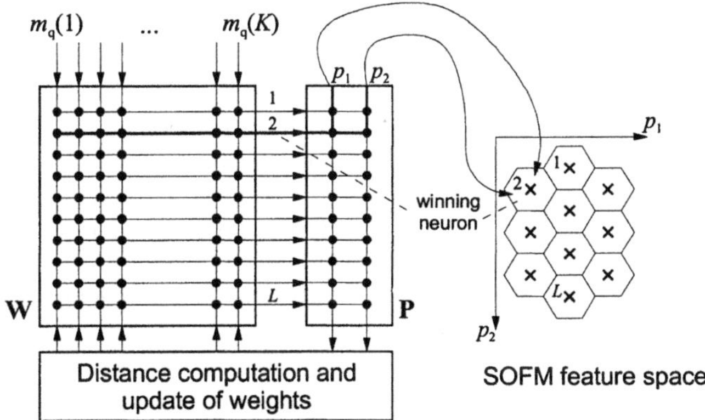

Fig. 9.18. Self-organizing feature map (SOFM) network, example $L=10$, $P=2$

9.4.6   Hidden Markov Models

Markov models (cf. sec. 3.4) are fully described by transition probabilities between states, where the most general definition (3.79) describes all transition probabilities $P(S_j|S_i)$ as elements of the matrix $\mathbf{P}$. Often, a semantic event can be described by the coincidence of different single observations which should occur by certain expected sequence(s). The observations are categorized beforehand into a finite set of observation symbols (e.g. phonemes in speech recognition). The number of possible events (e.g. spoken words) shall be finite as well. Only the observations are known. Furthermore, there shall be a degree of *uncertainty* about the association of an observation with a state of the Markov model, which could be due to the fact that the method of observation is inaccurate, that the source of events is biased etc. Typically, the sequence of observations $\mathcal{O}$ will also be finite. A *Hidden Markov Model*, if used for classification, will evaluate such sequences and determine the amount of certainty that the underlying sequence of observations corresponds to an expected event. This is made on the basis of the following parameter sets:

- The probabilities of initial states by which sequences start;
- The probabilities of transitions between the states, which are actually the parameters of the Markov model itself;
- The probabilities by which an observation will occur in association with a given state.

Assume that the entirety λ of these parameter sets shall characterize the HMM. Typical optimization problems are [RABINER 1989]:

1. Given the model parameters λ, determine $P(\mathcal{O}|\lambda)$, the probability of occurrence for an observation sequence. A common solution to this problem is the *forward-backward procedure*, which from a given state at any position in the sequence determines the probabilities once towards the beginning, and once towards the end of the sequence.
2. Given the model parameters λ, choose a state sequence S, such that the joint probability with the observation sequence $P(\mathcal{O},S|\lambda)$ is maximized. This is a typical classification problem, where one of different pre-defined state sequences S is expected (e.g. the sequence of phonemes constituting a spoken word, a sequence of gestures in sign language for deaf people, a sequence of turns that ends by a goal in a sports game). The usual solution to this is the application of the *Viterbi algorithm* (cf. sec. 11.6.1), which analyzes a likelihood metric for paths of state sequences. The maximum number of paths to be compared is then upper bounded by the number of states in the model.
3. Derive the model parameters λ, such that $P(\mathcal{O}|\lambda)$ or $P(\mathcal{O},S|\lambda)$ is maximized. While the former two problems were *analysis-related*, this one concerns the *synthesis* of the model, or training for a given classification problem. Typical solutions to this problem are the *segmental k-means algorithm*, which measures the variations of observations $\mathcal{O}$ from a training set against idealized

state sequences S, and the *Baum-Welch re-estimation method*, which adjusts the model parameters iteratively such that the probability is increased up to the maximum.

As HMMs are fully described by statistical parameters, it is possible to compute the distance between two models by statistical distance metrics. Of particular interest is the *Kullback-Leibler distance* (9.20)-(9.22), which could be re-formulated as follows, assuming e.g. that λ_1 are the parameters of a given reference model, and λ_2 the parameters related to an observation. Then,

$$\Delta(\lambda_1, \lambda_2) = \sum_{\mathcal{O}} P(\mathcal{O} \mid \lambda_1) \ln \frac{P(\mathcal{O} \mid \lambda_1)}{P(\mathcal{O} \mid \lambda_2)} \,. \tag{9.104}$$

Unfortunately, the direct solution by (9.104) would become overly complex, as minimization would require computing all possible state sequences. By re-formulating the distance function based on the likelihood metric provided by the Viterbi algorithm, more practical solutions can be implemented.

9.5   Problems

Problem 9.1

The following set of feature vectors, and related a priori assignments into two classes S_l, $l=1,2$ shall be given:

$$S_1: \quad \mathbf{m}_1^{(1)} = c \cdot \begin{bmatrix} 1 \\ 1 \end{bmatrix}, \quad \mathbf{m}_2^{(1)} = c \cdot \begin{bmatrix} 2 \\ 1 \end{bmatrix}, \quad \mathbf{m}_3^{(1)} = c \cdot \begin{bmatrix} 1 \\ 2 \end{bmatrix}, \quad \mathbf{m}_4^{(1)} = c \cdot \begin{bmatrix} 2 \\ 2 \end{bmatrix}$$

$$S_2: \quad \mathbf{m}_1^{(2)} = \begin{bmatrix} 4 \\ 3 \end{bmatrix}, \quad \mathbf{m}_2^{(2)} = \begin{bmatrix} 5 \\ 3 \end{bmatrix}, \quad \mathbf{m}_3^{(2)} = \begin{bmatrix} 6 \\ 4 \end{bmatrix}, \quad \mathbf{m}_4^{(2)} = \begin{bmatrix} 3 \\ 2 \end{bmatrix}$$

Two cases are regarded: *i)* c=0,5   *ii)* c=2.

a) Compute the covariance matrices and determine the achievable quality of feature selection for the class separation by (9.43) for both cases.

b) Determine the position of the separation line for both cases according to the MAP criterion (9.87); $\mathbf{m}_0$ shall be feature vectors which are positioned on the separation line, i.e.

$$\left[\mathbf{z}^{(1)} - \mathbf{z}^{(2)} \right]^{\mathrm{T}} \cdot \left[\mathbf{C}_{mm} \right]^{-1} \cdot \left[\mathbf{m}_0 - \frac{\mathbf{z}^{(1)} + \mathbf{z}^{(2)}}{2} \right] = \ln \frac{P(S_2)}{P(S_1)} \,.$$

c) Sketch for both cases the positions of the feature vectors $\mathbf{m}_q^{(l)}$, the class centroids $\mathbf{z}^{(l)}$ and the separation lines in the feature plane. Determine the number of wrongly classified cases. Interpret the results.

Problem 9.2

A continuous-value signal feature shall be given, where the probability density according to Fig. 9.19 can be interpreted as the superposition of probability densities from two classes S_1 and S_2, both of which are uniformly distributed in the respective ranges.

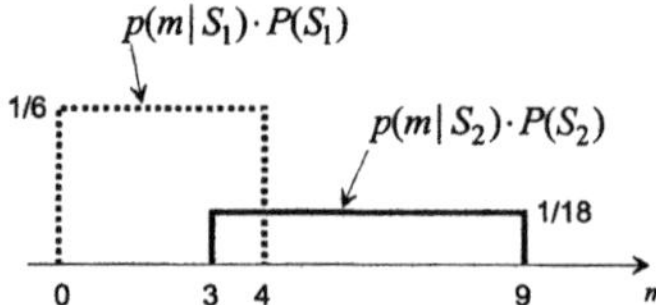

Fig. 9.19. Probability density by superposition of two classes

a)   Sketch the resulting PDF $p(m)$, and compute the class centroids $z^{(1)}$ and $z^{(2)}$.
b)   Compute the probabilities $P(S_1)$ and $P(S_2)$, and express the 'a priori' density functions $p(m|S_1)$ and $p(m|S_2)$.
c)   Determine and sketch the 'a posteriori' probabilities $P(S_1|m)$ and $P(S_2|m)$.
d)   The classification shall be performed according to a threshold method based on an MAP criterion. Determine the optimum threshold θ_{MAP} and the probabilities of a wrong classification, separately for each class and in total.
e)   For which threshold value $\theta_=$ would the numbers of wrong classifications in both of the classes become equal? Which would then be the overall probability of wrongly-classified elements?

Problem 9.3
The two features of three distinct classes S_1, S_2 and S_3 shall be uniformly distributed within the areas of the feature space as shown in Fig. 9.20. The probabilities shall be $P(S_1)=0.25$, $P(S_2)=0.6$ and $P(S_3)=0.15$.

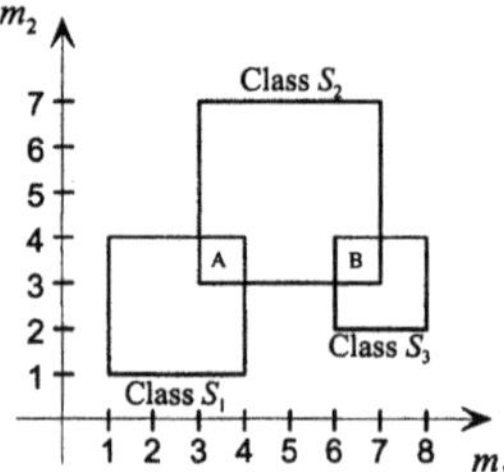

Fig. 9.20. Probability density by superposition of three classes

a)   Determine the class centroids $\mathbf{z}^{(l)}$ and the covariance matrices $\mathbf{C}_{mm}^{(l)}$, $l=1..3$ according to (9.33).
b)   Determine the contributions to the overall probability density, $p(\mathbf{m}|S_l)P(S_l)$ within the ranges of the three classes.
c)   Determine the a posteriori probabilities $P(S_l|m)$ within those ranges of the feature space where two classes intersect (marked by A/B).
d)   An MAP classification shall be performed. Determine the probabilities of classification errors $P(S_1|S_2)$, $P(S_2|S_1)$, $P(S_2|S_3)$, $P(S_3|S_2)$ and the overall probability of classification errors.

10 Signal Decomposition

The goal of signal decomposition is extraction and separation of signal components from composite signals, which should preferably be related to semantic units. Examples for this are distinct objects in images or video, video shots, melody sequences in music, spoken words or sentences in speech signals. Signal decomposition methods are closely related to classification of underlying features, which characterize the component to be separated. An important category of signal decomposition methods is segmentation of image, video, audio and speech signals, which is often a prerequisite for useful feature extraction and classification, but can also be used to improve the performance of compression algorithms. Segments are localized either over a time interval or over a spatial region; they will typically be defined by criteria that indicate homogeneity of certain features or semantic coherence. In segmentation, usually no overlap of the signal components to be isolated occurs. A more challenging approach of decomposition is separation of single components from mixed signals, where the composite signal consists of a sample-wise superposition from multiple components.

In image and video signal analysis, the following segmentation approaches can be distinguished:

- *Spatial segmentation:* Decomposition of the signal into distinct regions, for which in the ideal case a semantic meaning can be assigned (e.g. localization of faces, persons);
- *Temporal segmentation:* Decomposition of a video sequence into subsequences (e.g. shots);
- *Spatio-temporal segmentation:* Separation of regions by their temporal behavior (e.g. movement of a particular object through a video sequence within a certain period of time).

A superposition of components is rare in natural images (it can happen e.g. in case of reflections or in case of transparent objects); it is however often applied as artificial production effect (e.g. transitions, fades in a video). In audio signal analysis, also segmentation along the temporal axis is typically needed, e.g. for separation of spoken words, breaking a music piece into lyrics and refrain, breaking it further into times, breaks and notes. As polyphonic sounds are superpositions from different audio sources, separation of these composite signals is also of high importance for audio signal analysis, e.g. for detection and isolation of single instruments from an orchestra.

Separation is often done manually, in particular in media production and post-production (editing, cutting, manual separation of objects from movies). As this is time-consuming, development of fully automatic decomposition methods, or at least automatic aid in manual decomposition is an important task. The basic principle of signal decomposition is very similar to classification, i.e. identification of signal sections which are coherent regarding underlying features. Considering different types of signals, the optimum strategy may also differ largely. Due to this fact, a huge variety of signal-specific segmentation algorithms have been developed over the past decades. In the sequel, it is tried to characterize and categorize common approaches regarding the underlying strategies, even though a consistent theoretical framework for segmentation does not exist yet. It is further important to select segmentation algorithms for a given application, based on the computational complexity of these methods.

10.1 Segmentation of Image Signals

The goal of image segmentation is to identify regions of homogeneous or coherent features within an image or a frame from a video sequence. It is then assumed that these regions relate to objects shown in the image, which can further be categorized by their semantic meaning afterwards. Semantic recognition can e.g. be performed by analysis of shape or other features. In this context, the selection of correct combinations of features is important, which may indeed be context-dependent such as a priori assumptions about expected object sizes, color or contrast variations within the area of an object, or statistical assumptions about allowable parameter variations within closed segments. Important criteria in image segmentation are

- local amplitudes and colors of pixels, as well as their differences;
- statistical parameters characterizing the image globally or locally;
- Texture parameters characterizing object surfaces;
- Edges, corners etc. which can be found by edge detection.

10.1.1 Pixel-based Segmentation

In pixel-oriented segmentation, pixel-wise attributable features like luminance, color or local difference are subject to statistical analysis in a one- or multidimensional feature space. Based on this analysis, each pixel is assigned to a class, where it is then assumed that neighbored pixels belonging to the same class will be members of one segment. The histogram threshold segmentation described first can be regarded as a special case from a larger category of methods, which extend to multi-dimensional feature space analysis and more than two segmentation clusters.

Histogram-based thresholding. Thresholding separates segments by classification related to amplitude ranges. Often, the luminance (gray-value amplitude) of an image signal is used as an analysis criterion in this context. Assume that the objects to be separated (e.g. foreground/background) can be distinguished by their luminance levels, such that separation into just two luminance classes is sufficient. A good indicator for the suitability of such a simplified model is given by the histogram or probability distribution of such images, which will typically show two characteristic value ranges with concentration around significant peaks. If a threshold value is defined, this will effect a binarization of the image according to

$$b(m,n) = \begin{cases} F & \text{if} \quad x(m,n) \le \Theta \\ B & \text{if} \quad x(m,n) > \Theta. \end{cases} \tag{10.1}$$

Here, a semantic categorization is assumed where the background (B) is lighter, and foreground (F) is darker in the original image[1]. By assuming a gray-value signal $x(m,n)$ which is quantized into J amplitude levels, for which further the histogram or occurrence distribution $P(j)$ shall be available, the probability of a pixel to be classified as background or foreground under the constraint of the threshold level Θ in (10.1) will be

$$P(F) = \sum_{j=1}^{\Theta} P(j) \; ; \quad P(B) = \sum_{j=\Theta+1}^{J} P(j) \,. \tag{10.2}$$

Then, the mean values and variances within the foreground and background segments can be computed as

$$\mu_F(\Theta) = \frac{1}{P(F)} \sum_{j=1}^{\Theta} x_j \cdot P(j) \quad ; \quad \mu_B(\Theta) = \frac{1}{P(B)} \sum_{j=\Theta+1}^{J} x_j \cdot P(j) \tag{10.3}$$

$$\sigma_F^{\ 2}(\Theta) = \frac{1}{P(F)} \sum_{j=1}^{\Theta} (x_j - \mu_F(\Theta))^2 \cdot P(j) = \frac{1}{P(F)} \sum_{j=1}^{\Theta} x_j^{\ 2} \cdot P(j) - \mu_F^{\ 2}(\Theta)$$

$$\sigma_B^{\ 2}(\Theta) = \frac{1}{P(B)} \sum_{j=\Theta+1}^{J} (x_j - \mu_B(\Theta))^2 \cdot P(j) = \frac{1}{P(B)} \sum_{j=\Theta+1}^{J} x_j^{\ 2} \cdot P(j) - \mu_B^{\ 2}(\Theta) \tag{10.4}$$

The global mean and variance of the discrete-amplitude image are given by

$$\mu = \sum_{j=1}^{J} x_j \cdot P(j) = P(B) \cdot \mu_B(\theta) + P(F) \cdot \mu_F(\theta) \tag{10.5}$$

---

[1] This definition could as well be made vice versa; a typical example where the background is light and the foreground is dark are printed letters or drawings on a white sheet of paper.

$$\sigma^2 = \sum_{j=1}^{J}(x_j - \mu)^2 \cdot P(j) = \sum_{j=1}^{J} x_j^2 \cdot P(j) - \mu^2$$

$$= P(B) \cdot \left[\sigma_B^2(\Theta) + \mu_B^2(\Theta) \right] + P(F) \cdot \left[\sigma_F^2(\Theta) + \mu_F^2(\Theta) \right] - \mu^2 \tag{10.6}$$

If now the mean of the variances *within* the two segments is defined as

$$\overline{\sigma^2}(\Theta) = P(B) \cdot \sigma_B^2(\Theta) + P(F) \cdot \sigma_F^2(\Theta) \tag{10.7}$$

and the variance *between* segment mean values (deviation of class centroids from the global centroid) as

$$\sigma_\mu^2(\Theta) = P(B) \cdot (\mu_B(\Theta) - \mu)^2 + P(F) \cdot (\mu_F(\Theta) - \mu)^2 , \tag{10.8}$$

it can be shown that (10.6) is the sum of (10.7) and (10.8):

$$\sigma^2 = \overline{\sigma^2}(\Theta) + \sigma_\mu^2(\Theta) . \tag{10.9}$$

The probability of segmentation errors P_{err} will result from the percentage of pixels which either belong originally to the background and are larger than Θ, or belong to the foreground and are less or equal to Θ. This can be expressed in dependency of the conditional (a priori) probabilities that a foreground or background pixel have an amplitude x_j,

$$P_{\mathrm{err}} = P(B) \cdot \sum_{j=1}^{\Theta} P(j \mid B) + P(F) \cdot \sum_{j=\Theta+1}^{J} P(j \mid F) . \tag{10.10}$$

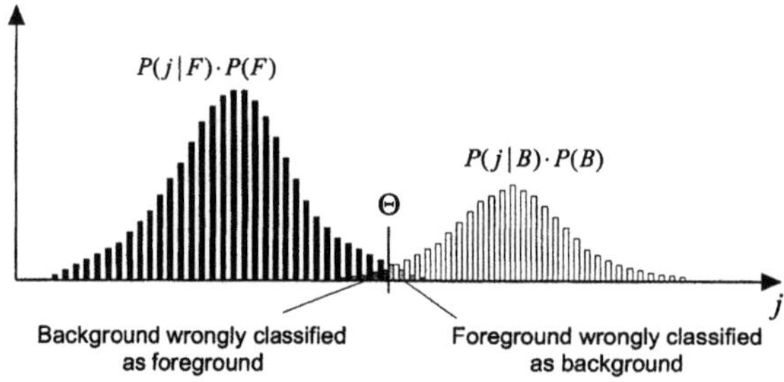

Fig. 10.1. Segmentation error contributions in a two-class threshold segmentation

The error can be interpreted as the percentage of occurrences expressed by those histogram lines which are shown gray in Fig. 10.1. If the distribution functions of the foreground and background classes are assumed to be Gaussian, the overall probability distribution of the image can be interpreted by a mixture of Gaussians (9.30). Minimization of the segmentation error can then analytically be achieved by derivation of (10.10) over Θ. If a mixture of Gaussians is assumed, the error will become minimum in a case when $\overline{\sigma^2}(\Theta)$ becomes as small as possible, while $\sigma_\mu^2(\Theta)$ becomes as large as possible. A useful optimization criterion can then be defined as[1]

$$\Theta_{opt} = \arg\max_\Theta \left[\frac{\sigma_\mu^2(\Theta)}{\sigma^2(\Theta)} \right], \tag{10.11}$$

where the constraint will be that the given probability function of the image $P(j)=P(j|V)+P(j|H)$ can be expressed as close as possible by a mixture of Gaussians. Mean values, variances and weights (probabilities) of the foreground and background classes must be fitted by joint minimization of a statistical test criterion (sec. 3.4) and maximization of (10.11). An example of a simple threshold segmentation using different threshold values is shown in Fig. 10.2.

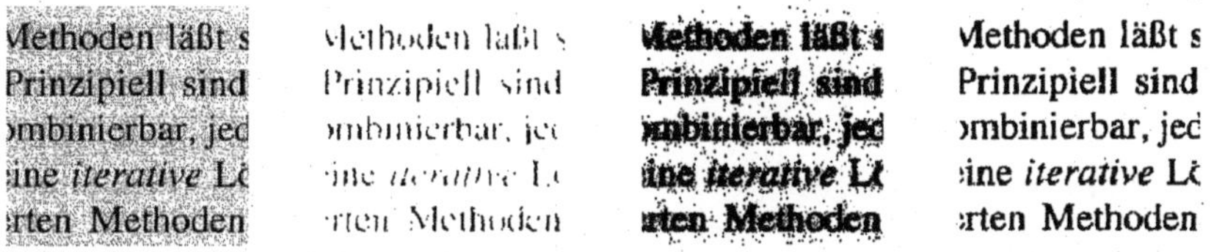

Methoden läßt s
Prinzipiell sind
 combinierbar, jec
ine iterative Le
rten Methoden
a
Methoden läßt s
Prinzipiell sind
combinierbar, jec
ine iterative Le
rten Methoden
b
Methoden läßt s
Prinzipiell sind
combinierbar, jec
ine iterative Le
rten Methoden
c
Methoden läßt s
Prinzipiell sind
combinierbar, jec
ine iterative Le
rten Methoden
d

Fig. 10.2. Binary image segmentation example: **a** Original 8-bit sub-image area and **b-d** Results of binary image segmentation using different thresholds, **b** 110 (too low) **c** 230 (too high) **d** 184 (optimum) [source : HÖYNCK]

If images of more complex gray-value statistics have to be segmented, the method using one single global threshold value is not suitable. Alternatively, the threshold-value optimization can be performed *locally*, e.g. by determining optimum threshold values within smaller blocks of an image.

Relationship with cluster-based methods. The basic assumption in the histogram-threshold approach is the expectation of significant agglomerations of values (clusters) which can be detected by statistical analysis in the feature space. So far, only one feature was considered, and only assignment into one of two possible classes was made[2]. This may only be suitable for images of very simple structure. In more challenging segmentation applications, multiple features must be regarded, e.g. several color components, texture, edges. In many cases where partitioning into two classes is not sufficient, color including chrominances is a much better classification criterion than luminance. Significant statistical agglomerations in a multidimensional feature space can be identified by *cluster analysis*. A cluster is a sub-space of the feature space, which is typically described by a centroid and a certain surrounding sub-space variation, e.g. described by a hypersphere or elliptic

---

[1] This criterion is a scalar expression closely related to vector criteria (9.43)-(9.46).

[2] An example of threshold-based segmentation into three different classes is treated in Problem 10.2.

volume. Cluster positions can be pre-defined, which is viable when image regions of certain features are expected[1]; automatic analysis and adaptation of clusters to a given image is also possible (see sec. 9.4.3). In the general case, K features extracted at one pixel position (m,n) are combined into a feature vector

$$\mathbf{m}(m,n) = \left[m_1(m,n), m_2(m,n), \ldots, m_K(m,n) \right]^{\mathrm{T}} . \tag{10.12}$$

To decide about cluster allocation for the pixels, a comparison against L different cluster centroids

$$\mathbf{z}^{(l)} = \left[z_1^{(l)}, z_2^{(l)}, \ldots, z_K^{(l)} \right]^{\mathrm{T}} \tag{10.13}$$

must be performed. A typical comparison criterion is the L_P norm (9.16) by which the pixel is allocated to the feature class

$$S(m,n) = \arg\min_{l=1,\ldots,L} \sum_{k=1}^{K} \left| m_k(m,n) - z_k^{(l)} \right|^P . \tag{10.14}$$

For the case of Euclidean distance ($P=2$) this will be the centroid which is a closest linear distance within the feature space. The clustering can also be an iterative process of classification and cluster optimization, e.g. following (9.75) and (9.78). Specifically in video (image sequence) segmentation, this iterative optimization of clusters can be performed recursively from one frame to the next [OHM, MA 1997].

It should be observed that the definition of the number of clusters is not independent of the goal of segmentation. When the number of clusters is too high, an *over-segmentation* occurs, resulting in a multitude of small segment patches, which have barely anything in common with semantic objects shown in the scene. This can be avoided, if clusters are not defined at the global image level, but rather over local areas, where it can be expected that a smaller variety of feature classes will be present, or by pre-filtering which eliminates detail (see sec. 5.1 and 10.1.3). The result of clustering will be an image consisting of cluster (segment) labels $S(m,n)$. In principle, this image can be further processed like any image, e.g. it is possible to eliminate small patches of labels which diverge from their local neighborhood. This can be done by application of median or morphological filters, or by statistical correction, e.g. deterministic relaxation (see sec. 10.1.4). The cluster labels however do not yet directly identify connected regions of homogeneous features, as indeed two or more disconnected areas may have been assigned to the same cluster. It is hence still necessary to perform *segment labeling* (see sec. 10.1.5).

---

[1] An example of cluster segmentation using only one class known a priori is chroma key segmentation (see sec. 15.1). In movie studios, foreground objects are frequently captured in front of a saturated blue background, a color rarely occurring in nature. Only for this one color a cluster definition must be made; all pixels *not* contained in this cluster are then assigned to the foreground object.

10.1.2   Region-based Methods

Region-based segmentation methods are implicitly designed to find connected regions of pixels which show similar features. They do not explicitly expect pre-defined clusters or feature classes, but can also be combined with statistical analysis approaches. In principle, they are suitable to blindly analyze where homogeneity or significant changes of feature characteristics can be found in images. By definition of the homogeneity criteria, the sensitivity of region-based methods in case of gradual changes can be flexibly tuned. Typical parameters to decide for presence of a region boundary are e.g. the maximum allowed feature value differences between a pixel and its neighbors or between a pixel and the feature value centroid of the region. Optimization of these parameters is often made on a heuristic basis, and optimum settings can be quite different depending on images properties.

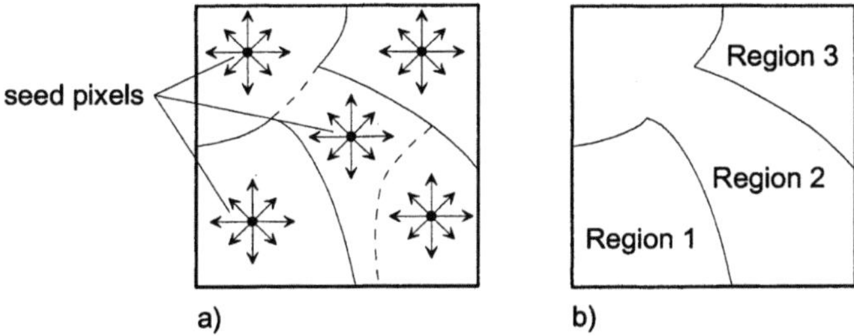

Fig. 10.3. Region growing: **a** Evolution of regions starting from *seed pixels*
b Merging of regions

Region growing. These methods start from single *seed pixels* which can either be selected randomly within an image, may be pixels with highest possible distance in feature space (like in k-means clustering), or pixels which are good representatives of expected segment statistics. Originating from these seed pixels, the region grows by checking feature homogeneity towards the nearest neighbors (see Fig. 10.3a). The homogeneity criterion can be based on luminance or color of pixels, local mean, variance, covariance, mean difference of pixels in a neighborhood, but can also include features like spectral characteristics signifying textures, input from edge-detection methods etc. If the homogeneity is violated, i.e. if the feature of a neighbor pixel deviates by more than a pre-defined maximum (threshold criterion), this neighbor will by itself become a seed pixel of another region. If in the growing-process two regions converge, they can be *merged* provided that the homogeneity criterion is not violated across the boundary. The principle of merging is indicated in Fig. 10.3a by the dotted lines; the final result of the segmentation is shown in Fig. 10.3b. Region growing can directly be coupled with segment labeling, as initially one distinct label is assigned to each seed pixel; in the sequel, if two regions are merged, one of the labels is eliminated, or if a new seed pixel is set, a new label is created[1].

Many variants of region growing algorithms have been proposed, which achieve improved segmentation results e.g. by adaptation of the homogeneity criteria depending on the region size. This is in particular advantageous to avoid small regions (over-segmentation) by supporting merge or elimination of small regions even in cases of higher deviation from homogeneity.

Split-and-merge methods. These methods work by a principle which is somehow antipodal with region growing. While the latter ones start from the level of pixels and then performs segmentation until the entire image is covered ('bottom up'), split and merge starts at the level of the entire image and then builds a hierarchy of sub-areas, which can proceed down to the level of single pixels ('top down'). The basic approach of split-and-merge schemes is shown in Fig. 10.4. During each step, rectangular blocks from the image[2] are *split* into sub-blocks, typically of same size. Splitting into four sub-block is often applied as in the example of Fig. 10.4b, but other sub-partitions can be made likewise. Then for each sub-block, feature criteria are computed such as mean, variance of colors etc. If the features of neighbored sub-blocks are found to be homogeneous, a *merge* is applied. If the merge leads to the same result which was available before the split, the sub-block never needs to be split again. A merge can however be performed across the boundary of a block which was previously split. In the example shown, no merge is possible yet after the first splitting step. Then, a second split is applied (Fig. 10.4c), after which a merge can be performed across the boundaries of adjacent larger-scale block units. This leads to the final segmentation result (Fig. 10.4d). In principle, blocks that could not yet be merged must provisionally be further subdivided in order to analyze whether they are sufficiently homogeneous internally.

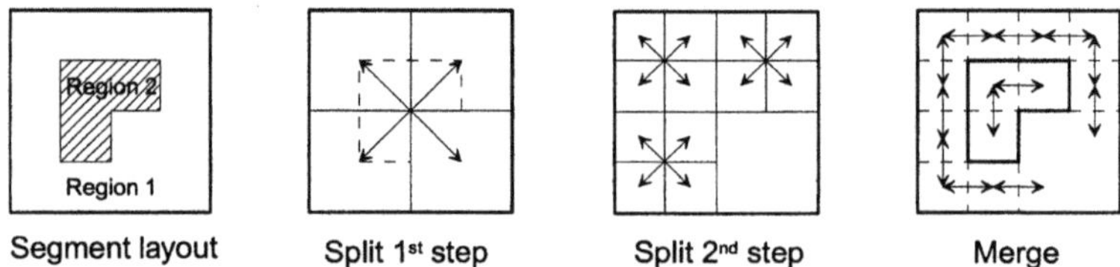

Fig. 10.4. Split-and-merge segmentation

A combination of region-based methods with other principles introduced before is possible, e.g. with the pixel-based feature classification methods from sec. 10.1.1. This is in particular advantageous in finding a compromise between statistical separation criteria which can be derived systematically from image characteristics,

---

[1] Methods of segment labeling are introduced in sec. 10.1.5.

[2] in the first step only splitting is possible, starting by one single block covering the entire image

and other goals of segmentation, e.g. the targeted resolution precision of the regions[1].

10.1.3   Texture Elimination

If texture features are used for the segmentation decision, it is possible to detect segments of homogeneous texture. Texture-related segmentation is however significantly more complex than segmentation based on simpler features like color, gradients etc.; it usually involves frequency analysis or statistical neighborhood analysis, which in addition means that texture features cannot be gained as true single-pixel features. Texture-related segmentation can be simplified, if the pixel-wise texture properties are already available by a compact representation, e.g. by using the wavelet coefficient signature patterns as extracted in the scheme shown in Fig. 7.11.

If segmentation is performed purely on the basis of luminance or color features, presence of structured textures can play a role which counteracts the goal of a good segmentation, as the amplitude variations within local texture structures can be mis-interpreted as being region boundaries. By systematic elimination of textures, this effect can be reduced, where however edges should be preserved. Texture elimination is either done before (as pre-processing) or during the segmentation process. In the ideal case, only homogeneous areas with clearly distinct boundaries shall remain; this can hardly be achieved by usage of linear filters, but nonlinear filters or combinations of adaptive linear and nonlinear methods can achieve this goal. An important constraint regarding the operation of the filters is the preservation of true region boundaries; this requires high accuracy if the goal shall be a pixel-wise accurate segmentation. Median filtering and more general rank order filters or morphological filters (cf. sec. 5.1) are preprocessing methods of good effect, while being of small or moderate complexity. If texture elimination methods are in the extreme case capable to produce *ideally-flat areas*, the remaining segmentation problem can entirely be solved by application of segment labeling (sec. 10.1.5) based on the amplitude plateau values of a perfectly detail- and noise-free signal.

Anisotropic diffusion. This is a *scale space* approach, which falls under a similar category as wavelet decomposition (sec. 4.4.4) and Gaussian pyramid decomposition (sec. 4.4.6), but is nonlinear in contrast to the previous methods. By reducing the resolution of the signal, texture details are inherently eliminated, but linear filtering would eliminate edge information simultaneously. The iterated application

---

[1] The region-based segmentation must not necessarily be performed down to the pixel level, but could as well stop at a level of small blocks, which can be realized very elegantly by the split and merge approach. Examples for application of 'block-based' segmentation are found in image and video coding schemes using variable block sizes (cf. sec. 12.6.3 and 13.2.6). Split and merge can also closely be combined with feature extraction methods, where feature homogeneity analysis is performed as part of the feature extraction process.

of Gaussian filter kernels in the Gaussian pyramid is also denoted as *isotropic diffusion*, which however loses the localization accuracy about prospective region boundaries. *Anisotropic diffusion* can also be based on Gaussian filter kernels, the strength and directional orientation of these filters are however additionally steered by properties of the signal [PERONA, MALIK 1990]. The optimized methods of filter adaptation can be quite complex, and it is beyond the scope here to give a detailed insight. Instead, a more practical method is presented, which adapts the smoothing filters by parameters gained from edge analysis. A Gaussian filter kernel is used which has an extension of –P/2...+P/2 in horizontal direction and –Q/2...+Q/2 in vertical direction around the respective position (m,n). In the following example, the absolute values of the differences in the second derivative[1], and deviations by intensity from the given central position are used to define criteria $d(p,q)$ and $e(p,q)$, where

$$d(p,q)=\left|\nabla x(m,n)-\nabla x(m+p,n+q)\right| \ ; e(p,q)=\left|x(m,n)-x(m+p,n+q)\right|$$
$$\text{for } -P/2 \le p \le P/2 \ ; \ -Q/2 \le q \le Q/2 . \tag{10.15}$$

The weighting of the 2D Gaussian-shaped filter impulse response (7.57) is then determined at position (m,n) by the following method:

$$w(p,q;m,n) = \left(f\left[d(p,q)\right] \cdot g\left[e(p,q)\right] \right) \cdot [1-s(m,n)]+s(m,n) . \tag{10.16}$$

Setting the switching function to a value $s(m,n)=1$ at the given position will invoke an isotropic diffusion process (which means that edges will be smoothed out as well). This can be based on additional rules or criteria, e.g. desirable minimum size of a segment, or other features gained independently like homogeneity of motion, when the method is used for moving-object segmentation. The functions $f(\cdot)$ and $g(\cdot)$ are nonlinear, ranging between zero and one for large and small argument values, respectively. An example is the following definition of a ramp hysteresis between two threshold values Θ_1 and Θ_2,

$$f\left[d(\cdot)\right] = \begin{cases} 1 & \text{for } d(\cdot) \le \Theta_1 \\ 1-\dfrac{d(\cdot)-\Theta_1}{\Theta_2-\Theta_1} & \text{for } \Theta_1 < d(\cdot) \le \Theta_2 \\ 0 & \text{for } d(\cdot) > \Theta_2. \end{cases} \tag{10.17}$$

From the corresponding unweighted (isotropic) Gaussian kernel $h(p,q)$ in (7.57), the anisotropic kernel at position (m,n) is computed by the formula

$$\tilde{h}(p,q) = \frac{w(p,q;m,n)\cdot h(p,q)}{\displaystyle\sum_p\sum_q w(p,q;m,n)\cdot h(p,q)} \ ; \ -P/2 \le p \le P/2 \ ; \ -Q/2 \le q \le Q/2 . \tag{10.18}$$

For the case $w(p,q;m,n)=1$, the anisotropic filter process is identical to the Gaussian filter. The anisotropic diffusion method can also be implemented iteratively,

---

[1] this is a similar criterion as the zero-crossings of the second derivative, cf. sec. 7.3.2

like a pyramid approach, using only short filter kernels by each iteration step. Fig. 10.5 shows 2D impulse response shapes of an unweighted and a weighted Gaussian kernel for the example of a significant image edge that shall be preserved.

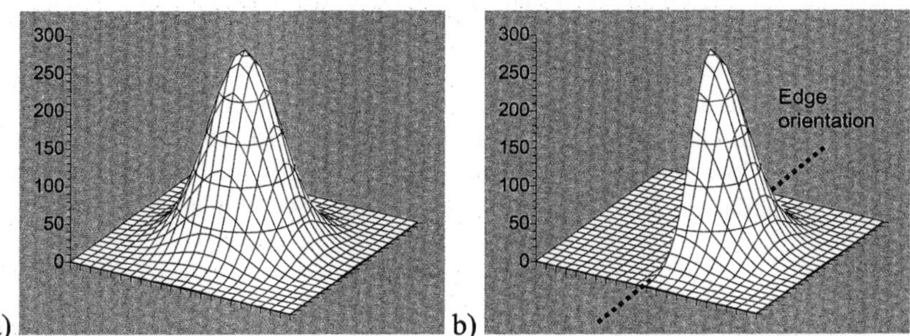

Fig. 10.5. Isotropic (**a**) and anisotropic-weighted (**b**) Gaussian filter kernels [source: IZQUIERDO]

Morphological smoothing and segmentation methods. Morphological filters (sec. 5.1.2) are often used for pre-processing and other tasks in the context of segmentation. Direct elimination of textures can be achieved by iterated application of *open* and *close* operations to gray-level images, which brings amplitudes to unique plateaus, while preserving edges. Likewise, median filters can be used to eliminate isolated labels from a segmentation result, or straighten inconsistent segment boundaries.

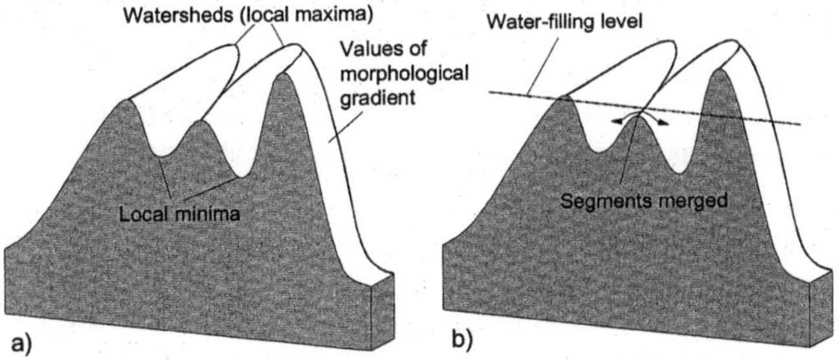

Fig. 10.6. Watershed algorithm **a** Interpretation of morphological gradient as surface shape topology **b** Merging of segments in case of 'water overflow'

Preferably after texture elimination e.g. by open/close operations, the morphological gradient can further be applied to localize segments within an image. The *watershed algorithm* is a useful method to evaluate morphological gradients for the purpose of segmentation. The amplitude shape of the gradient is interpreted like the 3D surface shape of a mountain (Fig. 10.6a). Local maxima of the gradient are interpreted as 'watersheds', which are assumed to be identical with the segment boundaries of objects. The search for watersheds starts however at local minima of the gradient, the lowest points in height where the 'filling of basins by water' starts[1]. If water would flow into a neighbored basin, a lowest elevation of a watershed is found, from where the watershed can further be tracked by the direction of the maximum gradient. If the watershed is yet found at a relatively low elevation, this indicates a weak gradient, such that it can be decided to discard this boundary; this means in principle that two basins (segments) are merged (Fig. 10.6b). In fact, by setting threshold levels as rules for segment merging or retaining the watersheds, the number of segments finally obtained can be influenced; from this, the watershed algorithm can be tuned by just one parameter to rather achieve 'over-segmentation' or a low number of segments. All watersheds finally retained establish segment boundaries. By guarantee, these boundaries will always be *closed contours*.

10.1.4   Relaxation Methods

As introduced in sec. 8.5, statistical methods are suitable to gain additional evidence about the *reliability* of a segmentation decision. In general, statistical reliability criteria can be invoked in any of the methods introduced so far. In *relaxation* methods, they are run in the context of pixel-based segmentation decision, where the reliability criterion is in principle a consistency check of segmentation information within a neighborhood. *Iterative methods* must be applied to achieve this. In a statistical approach, decisions are weighted as unreliable if a high deviation among a set of neighboring pixels is observed. This is a plausible assumption in the context of segmentation, as pixels near segment boundaries may eventually arbitrarily have been assigned to one or the other segment. As this is an open competition, the final assignment will be made to the segment which turns out to be the winner by statistical evidence. As in any statistical method, an underlying statistical model must be assumed, where it is the task of the optimization to align the parameters of the model as good as possible with the observation about the signal.

Markov Random Fields and stochastic relaxation. The concept of *Markov Random Fields* (MRF) was introduced earlier in the context of texture modeling (sec. 7.2.1). It is generally based on multi-dimensional conditional probabilities of pixel amplitudes, regarding a neighborhood context. For segmentation of image signals, the model can also be applied based on conditional probabilities for a neighborhood of pixels to belong to the same or different segments[2]. A useful model for this is the *Gibbs distribution* function, a type of exponential distribution,

---

[1] The water-filling process can be implemented by a gray-level dilation filter, which fills the troughs of a signal.

[2] In the segment relationship modeling, the property of Markov random fields as a state-transition model becomes more illustrative than for the case of texture modeling, where the number of states can indeed be rather high. For example, the case of binary segmentation as performed by the histogram thresholding method results in just one foreground and one background region class. This could well be modeled by a simple 2-state Markov chain (cf. sec. 3.4).

by which the probability of a pixel to belong to a given segment decays exponentially[1], when less pixels from the neighborhood hypothetically belong to the same segment. As a basis, a definition of boundaries within the sampling raster of a 2D signal must formally be made. As every pixel shall finally belong to only one segment, the segment boundaries must be positioned between pixels. Two different approaches for this are shown in Fig. 10.7. In the pixel raster (o) in Fig. 10.7a, segment boundaries can be found at the positions marked by '|' and '—'. In the alternative definition of Fig. 10.7b, boundary positions are marked as '+'. The number of boundary values to be described is identical to the number of pixels in the representation of Fig. 10.7b; however, both representations can uniquely be mapped into each other. Fig. 10.7c shows an example, where the boundary positions are drawn as bold elements, marking the transition from one segment (light pixels) to another (dark pixels). A label field of boundary positions $t(i,j)$ is thereby defined which shall have the value '1' if a boundary is present towards the respective neighbor, '0' otherwise.

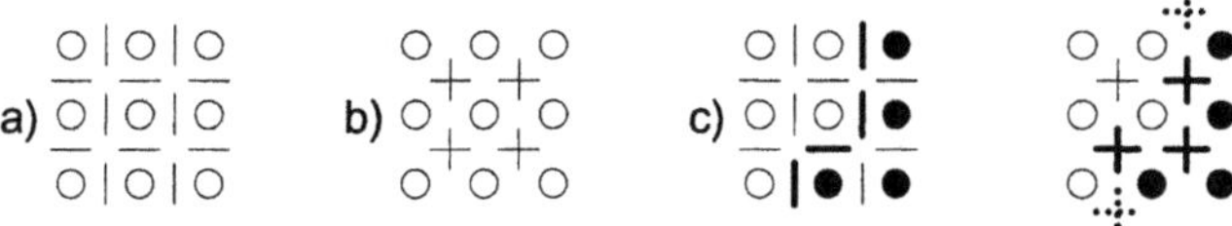

Fig. 10.7. Positions of pixels and boundary points for modeling of segments by Markov random fields.

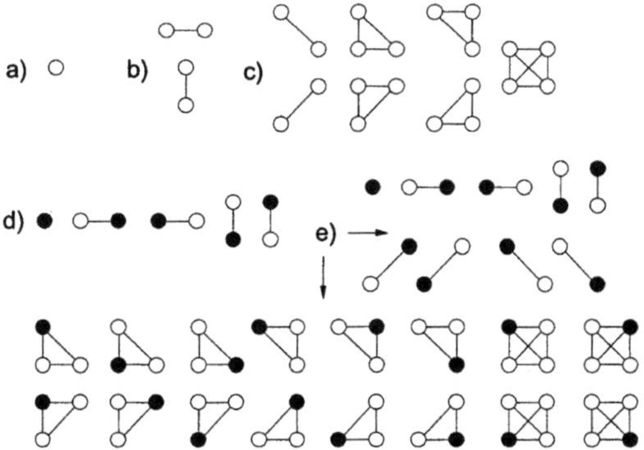

Fig. 10.8. Clique configurations of homogeneous neighborhood systems for $\mathcal{N}_c^{(2)}(m,n)$ **a** $c=0$ **b** $c=1$ **c** $c=2$, and clique ensembles for **d** $c=1$ **e** $c=2$.

---

[1] Exponentially decaying probabilities are typical for state durations in memoryless state models, where the time of rest in a state does not depend on the previous state.

To define a relationship between neighbored pixels, homogeneous neighborhood systems $\mathcal{N}_c^{(P)}(m,n)$ of (4.1) are used[1], which have the property of being *symmetric*, such that pixels are mutually members of their respective neighborhoods.

The concept of symmetry can even further be extended; a group of pixel positions, where *each* is a member of *each other's* neighborhood system $\mathcal{N}_c(m,n)$, is denoted as a *clique*. Fig. 10.8a-c shows the respective clique configurations for cases of homogenous neighborhood systems with $c=0$, $c=1$ and $c=2$. Connections of pixels within a clique are indicated here by graphs. The number of possible cliques increases drastically by an increased size of the neighborhood system: For a system of $c=8$, cliques of up to 9 pixels would be possible, and each possible subset of the largest clique can be regarded as a disjoint clique as well. Further, in a clique configuration of Q pixels, the actual pixel can be found at Q different *orientations*. A *clique ensemble* $\mathcal{C}_c(m,n)$, related to a homogeneous neighborhood system $\mathcal{N}_c(m,n)$, is the set of all possible cliques by all possible orientations. Fig. 10.8d-e shows $\mathcal{C}_c(m,n)$ including all possible orientations for cases $c=1$ and $c=2$.

An important property to be analyzed for the ensemble $\mathcal{C}_c(m,n)$ are the luminance differences between the pixels contained in the cliques, by which a *potential* $V(m,n)$ can be computed at each pixel position (m,n). This can be regarded as a criterion about the probability for presence of a segment boundary. $V(m,n)$ is influenced by information about amplitude differences Δ between the pixel $x(m,n)$ and its eight neighbors $x(i,j)$, further a hypothesis about presence of a segment membership of the different clique members, $\hat{S}(m,n)$ for the current pixel and $\hat{S}(i,j)$ for the eight neighbors is considered. The value of the boundary position field $t(i,j)$ is defined as follows:

$$t(i,j) = \begin{cases} 1 & \text{if } \hat{S}(m,n) \neq \hat{S}(i,j) \\ 0 & \text{if } \hat{S}(m,n) = \hat{S}(i,j) \end{cases} \quad ; \quad (i,j) \in \mathcal{C}_c(m,n). \tag{10.19}$$

The definition of the potential at position (m,n) is

$$V(m,n) = \sum_{(i,j) \in \mathcal{C}_c(m,n)} \lambda_t \cdot t(i,j) + \lambda_\Delta \cdot \left(1 - t(i,j)\right) \cdot \Delta\left(x(m,n), x(i,j)\right). \tag{10.20}$$

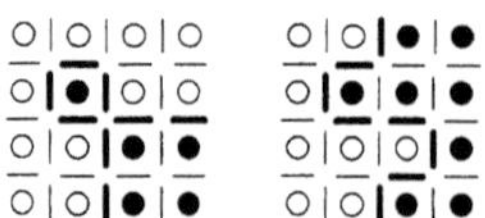

Fig. 10.9. Examples of irregular segment boundary configurations

The function $\Delta(\cdot)$ is a pixel-amplitude difference which can e.g. express the absolute or squared difference values between the pixels; alternatively, nonlinear func-

---

[1]In the sequel, $\mathcal{N}_c(m,n)$ is used with default order $P=2$, i.e. should read as $\mathcal{N}_c^{(2)}(m,n)$.

tions based on threshold criteria or adaptive functions can be used. The constants λ_t and λ_Δ effect a weighting regarding the influence of boundaries and local amplitude differences[1]. In addition, irregular or undesirable configurations of boundaries are penalized, as then a higher potential is assigned due to the higher number of boundary positions within the cliques (see examples in Fig. 10.9).

The principle of optimization is based on modeling the segmentation field $t(i,j)$ by an MRF. The Gibbs distribution signifies how large the expected segment size (rest of the MRF model in the same state) will be; the probability of smaller segments obviously becomes higher, when the potential increases, as then the exponential function decays more rapidly:

$$p(t) = \frac{1}{z} \cdot e^{-V(t)} \quad \text{with} \quad z = \sum_t e^{-V(t)} . \tag{10.21}$$

If this distribution function is assumed to be the a priori PDF, an optimization based on an MAP criterion can be applied (cf. sec. 8.5 and 9.4.4). The hypothesis about presence of segment boundaries is then judged in the posterior estimation, based on criteria such as the fit it gives with the prior model. The goal is minimization of the potential over the *entire image*, where interrelationships exist in principle between all pixels. If the λ parameters and the difference function $\Delta(\cdot)$ are appropriately selected, the optimum result will be achieved with a minimum number of segments under the constraint of the given amplitude differences. The solution can only be obtained iteratively, where the segmentation result from the previous iteration is modified such that the potential is further minimized. The statistical optimization algorithm used for this purpose is denoted as *stochastic relaxation* [GEMAN, GEMAN ET AL. 1985]. It is based on a model from thermodynamics, which assumes that elementary particles at high temperatures possess an almost-free capability to move. Movement capability of gas particles is usually modeled by a *Boltzmann distribution*[2]. For optimization of segmentation, the initial phase uses a 'high temperature', which means that segment allocation of pixels will have a high degree of freedom in the beginning, they may easily become member of one or another segment. This can be achieved if the parameter λ_Δ is set to a higher value initially. By gradual reduction during the subsequent iterations, the optimum of segmentation is finally reached[3]. The method has been proven to reach the global optimum, but typically requires a very high number of iterations.

---

[1] by presence of a boundary hypothesis, the contribution to the potential V becomes λ_t; if no boundary hypothesis is made, the contribution in V is $\lambda_\Delta \cdot \Delta(\cdot)$.

[2] which is actually similar to the Gibbs distribution given above, however the potential is temperature dependent.

[3] A similar principle is applied in annealing of metals – by exactly observing a 'temperature schedule' the desired result is achieved, e.g. to produce steel of high durability. In stochastic optimization, such methods have been denoted as *simulated annealing*, which are often applied to avoid convergence towards local minima in iterative optimization processes.

Deterministic Relaxation. While stochastic relaxation reaches an optimum from any given initial segmentation at high computational cost, *deterministic relaxation* is more simple but requires a good initial hypothesis about the segments contained in the image. An optimized result is then achieved by a relatively low number of iterations. An example of a deterministic relaxation algorithm, which is purely based on probability criteria determined from segment assignments within pixel neighborhoods, is described here. Deterministic relaxation can also be applied as statistically-based post-processing of a segmentation result originating from any other segmentation procedure.

Initially, for each pixel position (m,n), a probability value $P^{r=0}(m,n;l)$ must be available, which indicates by which certainty this position could be assigned to a segment or cluster S_l (out of L different). The sum over all values about assignment of a pixel at position (m,n) to any of the L segments must be unity[1]:

$$\sum_{l=1}^{L} P^r(m,n;l) = 1 . \tag{10.22}$$

The basic principle of deterministic relaxation is the iterative refinement of probability values by update values q^r, where the index r expresses the iteration step. The update process at position (m,n) is as follows:

$$P^{r+1}(m,n;l) = \frac{P^r(m,n;l) \cdot \left[1 + q^r(m,n;l) \right]}{\sum_{k=1}^{L} P^r(m,n;k) \cdot \left[1 + q^r(m,n;k) \right]} . \tag{10.23}$$

To determine the update value q^r for the $r+1^{\text{th}}$ iteration, the probabilities of neighborhood pixels are evaluated. As an example, the eight pixels belonging to neighborhood $\mathcal{N}_2(m,n)$ can be used. A procedure to determine the update is

$$q^r(m,n;l) = \frac{1}{8} \cdot \left[\sum_{(i,j)\in\mathcal{N}_2(m,n)} \left(\sum_{k=1}^{L} c(k,l) \cdot P^r(i,j;k) \right) \right] \tag{10.24}$$

with

$$c(k,l) = \begin{cases} 1 & \text{if } k = l \\ -1 & \text{if } k \neq l. \end{cases} \tag{10.25}$$

Other weighting functions $c(k,l)$ are possible. For larger neighborhood systems, it is useful to establish a weight which is adaptive by the distance, or take into account geometric criteria such that straight segment boundaries are given a preference. Relaxation increases the probability of a pixel to be member of a segment, if its neighbors are by high probability members of the same segment as well, and

---

[1] If the starting point is some arbitrary segmentation result where the segmentation label $S(m,n)=l'$ was assigned without any further knowledge about statistical reliability, set initially $P^{r=0}(m,n;l=l')=1$ and $P^{r=0}(m,n;l\neq l')=0$.

vice versa. If convergence is reached by iteration R, the pixel at position (m,n) is finally allocated to the segment which is found by highest probability

$$S(m,n) = \arg\max_l P^R(m,n;l) \ . \tag{10.26}$$

In principle, this is still an a posteriori decision, however the neighborhood relationships are taken into account here, such that the reliability of the segmentation results can be expected to be much higher than in pixel-separate decisions.

10.1.5 Image Region Labeling

The segment classification label $S(m,n)$ as e.g. provided by a histogram-threshold, cluster-based or other segmentation algorithms is not yet a unique criterion for pixels belonging to a closed region. In fact, pixels with same class labels assigned could be members of two disparate regions. As an example, an image may show different patches of grass which will be indexed with same class labels by a texture classification, but shall be treated as different regions[1]. Hence, an identification of those pixels is necessary, which have been annotated by identical class labels $S(m,n)$ and are connected in one region. Typically, the number L_R of distinct region labels $R(m,n)$ must be larger than the number L of different classification labels $S(m,n)$. Due to the fact that region shapes may be topologically difficult (contain holes, noses etc.), simple implementation covering all possible cases is an issue of high importance. This can still be done by a sequential scan of pixels, however the region assignment must eventually later be revised, as two patches that are disparate in an earlier phase of the scan may later be identified as connected. Assume that a sequential scan is performed row-wise, starting in the top-left corner of an image. A new region label is assigned, whenever a pixel does not have the same classification label as any of its previously processed neighbors. However, in the example shown in Fig. 10.10a, only during the scan of the last row it can be identified that the pixels which are marked by horizontal and vertical stripes actually belong to the same region. An example algorithm is presented here, which distinguishes the following cases and implements suitable decisions[2]:

1. $S(m,n)=S(m-1,n)$ and $S(m,n)\neq S(m,n-1)$ $\Rightarrow$ region label $R(m,n)=R(m-1,n)$
2. $S(m,n)\neq S(m-1,n)$ and $S(m,n)=S(m,n-1)$ $\Rightarrow$ region label $R(m,n)=R(m,n-1)$

---

[1] It however depends on the application and the semantic context, if eventually two disparate regions shall better be treated as one object.

[2] The algorithm given here excludes 'weak' connections of regions, which means that two pixels are not assigned to the same region when they are diagonally adjacent, while neither horizontal nor vertical neighbors belong to the same region. This would for example make any square on a checkerboard a separate region.

3. $S(m,n)=S(m-1,n)$ and $S(m,n)=S(m,n-1)$ and $S(m-1,n-1)\neq S(m,n-1)$ $\Rightarrow$ merge the regions previously assigned to the left and top neighbors, then assign the same region label $R(m,n)=R(m-1,n)=R(m,n-1)$ to the present pixel
4. $S(m,n)\neq S(m-1,n)$ and $S(m,n)\neq S(m,n-1)$ $\Rightarrow$ create a new region label R_{new} and assign $R(m,n)=R_{new}$.

Only three comparisons of classification indices are necessary at any pixel position by this procedure. Possible configurations corresponding to cases 1-4 are shown in Fig. 10.10b, where the dotted lines indicate that it is irrelevant for the decision whether a region boundary is found at the position or not.

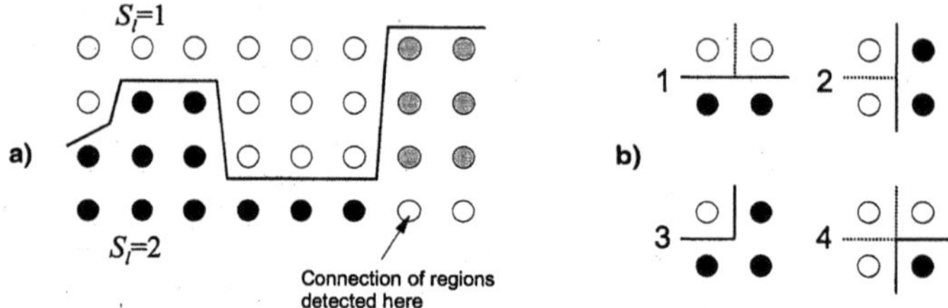

Fig. 10.10. a Illustration of a case 3, where two regions which are handled separate initially are identified as connected   **b** illustration of the 4 cases in the algorithm described above

The management of the segment labels can best be performed by table processing. Usually it will be sufficient to mark which regions have been merged, and run a second pass which assigns the final labels to pixel positions.

10.2 Segmentation of Video Signals

10.2.1   Temporal Segmentation for Scene Changes

Goal of a temporal segmentation is partitioning of longer video sequences into sub-sequences of unique characteristics. This is not only useful for recognition of movie structures, but is also very important in the context of encoding, where correct usage of temporal correspondences must be observed in case of scene changes. Typically, the similarity between adjacent frames also highly depends on the speed of motion, which could be either object motion or global (camera) motion. Scenes can change either abruptly (at shot boundaries) or gradually (e.g. by fade effects, camera focusing on different content, new objects entering the scene etc.). Basically, many different features can be used to analyze the changes within a video sequence. The simplest approach is analysis of frame differences, which may however lead to false scene cut hypotheses when the gradual change by motion is high. Instead of performing motion-compensated difference analysis, it is then simpler to analyze features which are invariant against motion translation or other forms of motion. Among the most simple (and most effective) features in this category is color analysis, where e.g. color histograms are largely invariant against motion effects, and can only change significantly by presence of larger covered/uncovered areas. To analyze the impact of gradual changes more thoroughly, motion features are nevertheless most important, as it is then e.g. possible to differentiate between fast and slow motion, analyze percentage of covered/uncovered areas etc.

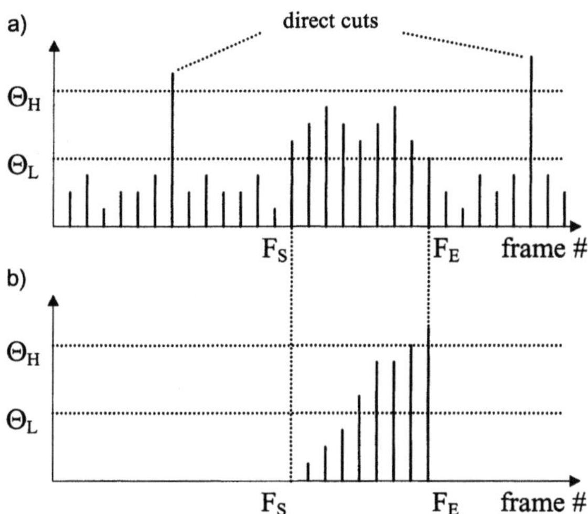

Fig. 10.11. Color histogram difference thresholding between subsequent frames: **a** Primary difference between adjacent frames to detect direct cuts **b** Secondary difference with reference to frame F_S for detection of gradual scene transitions (source : HÖYNCK)

A typical model assumption for shot-boundary detection is that global features like color distribution will only change gradually unless an abrupt scene change occurs. Assume that a color histogram analysis shall be used to detect scene changes. A shot boundary can be identified if two subsequent frames are clearly different, which can be concluded if histogram differences (9.23) supersede a pre-defined threshold value Θ_H. Direct cuts will be clearly detected by such an approach. The case of indirect cuts, e.g. by fades or wipes as frequently used in production, is more critical, as these could be mis-interpreted as gradual changes. Definition of a lower threshold value might solve this problem, but would increase the possibility of false scene cut hypotheses[1]. The problem can better be solved by a strategy introducing a *second threshold value*, $\Theta_L < \Theta_H$. When Θ_L is superseded, a hy-

---

[1] For general concepts of trade-offs between false hypotheses and missed hypotheses, see annex A.2.

pothesis about the potential presence of a gradual scene transition is made, but not finally decided yet. In the example shown in Fig. 10.11a, this will be the case at frame F_S. A secondary histogram difference is then computed for all subsequent frames, taking reference to frame F_S. If after a pre-defined maximum number of frames the difference threshold Θ_H is indeed superseded in the secondary difference, a gradual change is classified. Color histogram analysis is however quite sensitive in cases of sudden illumination changes, e.g. flashlights which may be falsely interpreted as scene cuts; such problems can however also be resolved by setting rules about typical relationships of differences between several adjacent frames. In any case, with more features used, the detection of true video segment boundaries will be more reliable.

In principle, segmentation of video based on frame-wise features can also be performed by *cluster analysis*. A contiguous video shot can in many cases be characterized by similarity of the frame features, which then establish a clear cluster in the feature space. As an extension of such a concept, it is also possible to find a characteristic *key frame*, which as a still image gives most relevant information about the content of the underlying video segment. This should be the frame which by its feature values is nearest to the centroid of the cluster, as computed from feature values of all frames belonging to the segment.

10.2.2   Combination of Spatial and Temporal Segmentation

In image sequence segmentation, the temporal continuity of segmentation results is of high importance, in particular if it shall be possible to track an object by its temporal changes. Further, in particular the motion information can be utilized to identify objects, where in many cases it will not be possible to classify an object as one entity by pure spatial segmentation, as only the homogeneity of motion will give sufficient evidence. Finally, it is also possible to improve the motion estimation itself by combination with segmentation information.

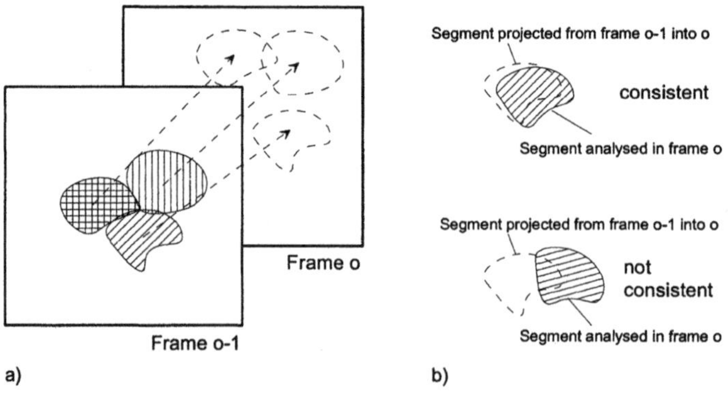

Fig. 10.12. Segment tracking **a** projection into subsequent frame **b** identification of corresponding segments

Object Tracking. The problem of object tracking is in principle a prediction problem. In a very simple method, forward-oriented motion vectors can be used to project segments from frame *o*-1 into frame *o* (Fig. 10.12a). Then, a comparison with the segmentation results obtained *independently* for that frame at the same position is made. If area, contour and other feature parameters are coherent, a continuity of segmentation can be concluded (Fig. 10.12b). There is however no consistent rule how to handle segments when no continuity is found, which could either be caused by erroneous segmentation or by erroneous motion estimation.

As it can be expected that for moving image content at least small deviations in the segmentation occur, a further stabilization is necessary, where in particular the segmentation can no longer be performed independently within the single frames. The following methods are often used in this context:

- *Kalman Tracking.* When the Kalman estimation method is used for tracking of motion trajectories as described in sec. 8.6, stabilization of the temporal variation of possible segmentation errors can be included. In this case, the estimate of the state vector $\hat{\mathbf{x}}_r$ of the Kalman filter can be defined such that it contains the coordinate positions of pixels exclusively from a segment to be tracked, the respective motion shifts and acceleration parameters. The matrix **A** from Fig. 8.3 and (8.67) then performs the prediction of segment positions from the previous state vector $\hat{\mathbf{x}}_{r-1}$.

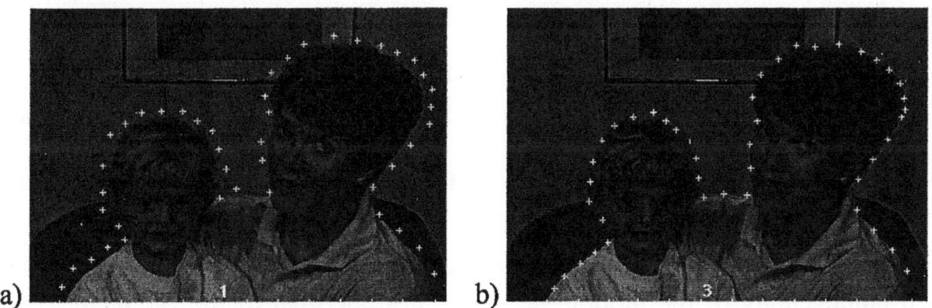

Fig. 10.13. Manually-marked outer shape of an object in a video frame (*a*) and update by an automatic ACM method (*b*) (source: KRUSE)

- *Active contour models (ACM).* In ACMs, also denoted as *snakes*, the control points describe the segment or object boundary by a contour, which can e.g. be gained from a contour approximation method, see sec. 7.4.1. Approximated contours are projected from one frame to the next using motion parameters. Then, an update is performed using an 'elasticity model', which means the control points are shifted by shortest possible distance under the constraint of approximating a contour shape in the new frame (as e.g. found by gradient detection) as good as possible. The snake method can also be used to improve the result of a raw segmentation, e.g. to find automatically a pixel-accurate closed contour in the vicinity of a polygon that was manually marked as the outer shape of an object (see Fig. 10.13).

10.2.3   Segmentation of Objects based on Motion Information

If a scene has been captured by a static camera (e.g. on a tripod), it is possible to identify static scene parts by very simple methods of *frame-difference computation*; when the difference is above a pre-set threshold level, the corresponding position can be identified as being probably part of a moving foreground object. High differences may however also appear in case of shadow projections from foreground objects; these can better be distinguished by analyzing chrominance differences instead of gray-value differences[1]. Granular local motion appearing in a scene may be another problem, e.g. caused by movement of a curtain or of leaves of a tree. This can be avoided by increasing the difference threshold within regions where random small-scale fluctuations of motion are found.

It can not be expected that a foreground object which has been detected and separated within a video sequence moves permanently. Hence, it is necessary to freeze an object in the segment memory when it stops moving and track it again when it restarts a motion action.

Combined Segmentation and Motion Estimation. In pixel-based motion vector estimation, the following criteria apply[2], which can be used to judge the reliability of the estimation, and can also be used as a basis for hypotheses about segment or object boundaries:

– Energy of displaced frame differences (DFD), except for occluded areas, where higher values may be allowed when a hypothesis about presence of an object boundary is made;
– Homogeneity of the motion vector field, except for positions of object boundaries;

Similar criteria are applied in matching-based motion estimation, in particular when homogeneity of the motion vector field is one goal of the estimation, see (7.185). Methods of motion estimation introduced so far did not systematically define exception conditions at segment boundaries. This can indeed only be achieved by combining motion estimation and segmentation. In the stochastic relaxation method for segmentation (sec. 10.1.4), it is possible to include the properties of the local motion vector field in the computation of the potential for the segmentation model. As a side effect, by introducing a condition of motion vector homogeneity, the number of actual hypotheses about segment boundaries can be-

---

[1] By a shadow that changes lighting conditions, the color hue and saturation are mainly unchanged, while the luminance changes significantly.

[2] cf. (7.168) for the case of optical-flow based methods

come much lower than by a criterion of gray-value homogeneity, in particular for cases of uniformly moved (object) or uniformly static (background) textured areas.

The segment boundary field $t(i,j)$ in (10.17) carries the actual hypothesis about the presence of a boundary by image-based criteria (amplitude, gradient), another field $v(i,j)$ is now introduced analogously for definition of possible boundaries in the motion vector field. Herein, a threshold parameter Θ_v indicates which maximum difference between neighbored-pixel motion vectors will be allowed without formulating a hypothesis about a motion boundary:

$$v(i,j) = \begin{cases} 1 & \text{if } \|\mathbf{k}(i,j)-\mathbf{k}(m,n)\| > \Theta_v \\ 0 & \text{if } \|\mathbf{k}(i,j)-\mathbf{k}(m,n)\| \leq \Theta_v \end{cases} \quad ; \quad (i,j) \in \mathcal{N}_c(m,n). \quad (10.27)$$

In addition, a field $u(m,n)$ is defined, which indicates the presence of an occlusion area. A useful criterion for this is a typical cost function relating to the deviation between the pixel and its correspondence in the reference frame, while considering the motion shift of the actual estimation. Here, the absolute difference criterion which should not supersede a threshold parameter Θ_u, is taken as an example:

$$u(m,n) = \begin{cases} 1 & \text{if } d(m,n) > \Theta_u \\ 0 & \text{if } d(m,n) \leq \Theta_u \end{cases}$$

$$\text{with } d(m,n) = |x(m,n,o) - x(m+k,n+l,o-1)|. \quad (10.28)$$

$d(m,n)$ establishes a simple criterion for the reliability of the motion estimation based on the DFD. The potential at position (m,n) then is

$$V(m,n) = V_1(m,n) + V_2(m,n) + V_3(m,n) \quad (10.29)$$

where the three components are

$$V_1(m,n) = \lambda_u \cdot u(m,n) + \lambda_d \cdot (1-u(m,n)) \cdot d(m,n) \quad (10.30)$$

$$V_2(m,n) = \sum_{(i,j)\in\mathcal{N}_c(m,n)} \lambda_v \cdot v(i,j) + \lambda_\Delta \cdot (1-v(i,j)) \cdot \Delta(\mathbf{k}(m,n),\mathbf{k}(i,j)) \quad (10.31)$$

$$V_3(m,n) = \sum_{(i,j)\in\mathcal{N}_c(m,n)} \lambda_t \cdot (1-t(i,j)) \cdot v(i,j). \quad (10.32)$$

The goal of the estimation is global minimization of the potential, both based on decision about motion vectors and motion boundaries. Herein, the first component is based on the DFD, where occluded areas are explicitly excepted; on the other hand, an occlusion hypothesis $[u(m,n)=1]$ is weighted by the factor λ_u. The second component is the typical potential used here for the motion-based segmentation hypothesis, analogously with (10.20); $\Delta(\cdot)$ again expresses an arbitrary motion vector difference function, where e.g. the function in (10.28) based on the L_P norm can be used. The third component penalizes a hypothesis about the presence of a segment boundary in the motion vector field, for the case where not *simultaneously* a hypothesis of a segment based on gray-value criteria is made. The field $t(i,j)$ is equivalent to the definition in (10.19), but could simply be determined from an edge-detection gradient filter with subsequent threshold decision (sec. 7.3.1), when no gray-value based segmentation information is available a priori. Minimization of (10.29) must then be performed iteratively to find an optimum joint constellation for

- the motion parameters $\mathbf{k}(m,n)$
- the occlusions positions $u(m,n)$ where motion parameters are undefined,
- the motion vector field discontinuities $v(i,j)$.

After convergence, motion parameters *and* a moving-object segmentation are available.

10.3 Segmentation and Decomposition of Audio Signals

In audio signal segmentation, the definition of features, by which the uniformity of segments shall be judged, is of prior importance as well. This highly depends on the goals of segmentation: In particular, different temporal granularity levels, e.g. segmentation into units of single tones/notes, time-based units, or larger units like melody sequences and detection of song structures have to be considered:

- Regarding separation of single, separable tones or notes, the analysis of the temporal envelope (hull) is of highest importance. This can be achieved by threshold analysis of the loudness, similar to the method illustrated in Fig. 7.80, if the tones are clearly separated.
- For single (monophonic) tones with bindings, as they will frequently appear in harmonic musical instruments, pitch analysis of sufficient granularity will enable detection of tone transitions. This can also be done by spectral analysis, where it must however be observed that increased resolution of frequency analysis leads to decreased time-domain resolution. Shorter transform block lengths are advantageous to detect the point of tone transition exactly.
- The analysis or separation of polyphonic tones, or analysis of time structure in case of polyphonic sound events, is more challenging. Possible criteria can be the analysis of signal power over a short window of samples, which will often show some periodic characteristics. Criteria testing the homogeneity of spectral properties also allow a separation, where however the precision of the time position is low when the window length used for the spectrogram analysis is too long.
- For segmentation of the global structure of music pieces, the temporal granularity is less important, while changes of global features like tempo, loudness or timbre are more important. Best results will be achieved when different features and rules are combined in the decision-making process.

Separation of mixed sound sources is another challenging problem in audio signal decomposition, which is still far from development of fully satisfactory solutions. Prospective applications are sound identification, speech recognition in noisy or multi-speaker environments, post processing of single sounds from mixed signals in music production, and separate analysis of mixed sound sources for analysis/synthesis audio coding methods (sec. 14.3).

In *blind separation*, most successful approaches are based on SVD (sec. 8.4), ICA (sec. 9.1.2) or similar methods like *independent subspace analysis* [CASEY, WESTNER 2000]. A basic prerequisite of these methods, which will not necessarily be fulfilled in reality (e.g. when an orchestra is playing), is the assumption that the single sound sources are *statistically independent*. Another group of methods uses acquisition by multiple microphones, which allows localization by combination with beamforming, and *single input, multiple output* modeling in the ICA context [SARUWATURI, KAWAMURA, SHIKANO 2001]. Also this approach seems to be problematic when the number of sound sources exceeds the number of microphones available.

For cases where blind separation is an ill-posed problem, separation of mixed audio components originating from different sound sources could additionally be based on a priori assumptions about the properties of the individual sources. Human listeners utilize knowledge about additional features to distinguish instruments within an orchestral sound. This will in particular be characteristic temporal changes of spectra, identification of melodies, localization of different instruments by binaural hearing, a priori knowledge about instrument characteristics or even about the music piece itself, where it may be known which instrument takes which part.

10.4 Problems

Problem 10.1
The following gray-value image shall be binarized by a thresholding segmentation (cf. sec. 10.1.1).

$$\mathbf{X} = \begin{bmatrix} 1 & 2 & 4 & 5 & 5 \\ 1 & 1 & 2 & 4 & 5 \\ 1 & 2 & 4 & 5 & 5 \\ 1 & 1 & 2 & 4 & 5 \\ 1 & 1 & 1 & 2 & 4 \end{bmatrix}$$

a) Determine the allocations of pixels with coordinates (m,n) into two regions S_l, $l=1,2$ for three cases $\Theta=\{1,5\;;\;3\;;\;4,5\}$. The region index shall be computed according to the formula

$$l(m,n) = \begin{cases} 1 & \text{if } x(m,n) < \Theta \\ 2 & \text{if } x(m,n) \geq \Theta. \end{cases}$$

b) For all three cases of Θ, compute the mean values μ_{S_l} and variances $\sigma^2_{S_l}$ of the pixels $x(m,n)$ allocated to the different regions. Which threshold value would be optimum according to the criterion (10.11)?

c) Determine for the best case found in b) the segmentation error under the assumption, that the mean value and variance are representative parameters of a Laplacian probability density functions, which represent the amplitude values in the two regions,

$$p(x\,|\,S_l) = \frac{1}{\sqrt{2}\sigma_{S_l}} \cdot e^{-\frac{\sqrt{2}|x-\mu_{S_l}|}{\sigma_{S_l}}} \quad .$$

Give a graphical interpretation.

Problem 10.2

The following gray-value image $\mathbf{X}$ shall be processed by threshold segmentation. Using a threshold set $\Theta = [\ \Theta_1\ \Theta_2\ ]^T$, an allocation into three different region classes S_l, $l=1,2,3$ is performed. The region allocation shall be performed according to the following rule:

$$l(m,n) = \begin{cases} 1 & \text{if } x(m,n) < \Theta_1 \\ 2 & \text{if } \Theta_1 \le x(m,n) \le \Theta_2 \\ 3 & \text{if } \Theta_2 < x(m,n) \end{cases} \quad \text{with} \quad \mathbf{X} = \begin{bmatrix} 1 & 2 & 3 & 4 & 5 \\ 1 & 2 & 3 & 4 & 5 \\ 1 & 2 & 3 & 4 & 5 \end{bmatrix}.$$

The following sets of threshold levels can alternatively be used: $\Theta_A = [\ 2\ 4\ ]^T$, $\Theta_B = [\ 3\ 3\ ]^T$.

a) Determine for Θ_A and Θ_B the matrices of index images.

b) Determine for Θ_A and Θ_B the mean values μ_{S_l} and variances $\sigma^2_{S_l}$ as well as the occurrences of pixels within the three region classes.

c) Determine the 'within class' variance $\overline{\sigma^2}(\theta)$ and the 'between classes' variance $\sigma_\mu^2(\theta)$. Which of the two threshold sets is the better selection, according to a criterion to maximize $\sigma_\mu^2(\theta)/\overline{\sigma^2}(\theta)$?

d) Give an example for a set of threshold values, where all pixels will fall into the same segment. Give reasons, why this is the worst possible case of segmentation, according to the criterion tested under c).

MIX
Papier aus verantwortungsvollen Quellen
Paper from responsible sources
FSC® C105338

If you have any concerns about our products,
you can contact us on
ProductSafety@springernature.com

In case Publisher is established outside the EU,
the EU authorized representative is:
Springer Nature Customer Service Center GmbH
Europaplatz 3, 69115 Heidelberg, Germany

Printed by Libri Plureos GmbH
in Hamburg, Germany

Jens-Rainer Ohm

Multimedia Communication Technology

Springer-Verlag Berlin Heidelberg GmbH

Jens-Rainer Ohm

Multimedia Communication Technology

Representation, Transmission and Identification of Multimedia Signals

With 441 Figures

Springer

Professor Jens-Rainer Ohm
RWTH Aachen University
Chair and Institute of Communications Engineering
Melatener Str. 23
52074 Aachen
Germany

Cataloging-in-Publication Data applied for

ISBN 978-3-642-62277-9   ISBN 978-3-642-18750-6 (eBook)
DOI 10.1007/978-3-642-18750-6

Typesetting: Digital data supplied by author
Cover-Design: Design & Production, Heidelberg
Printed on acid-free paper   62/3020 Rw 5 4 3 2 1 0

Preface

Information technology provides a plenty of new ways to process, store, distribute and access audiovisual information. Beyond traditional broadcast and telephone channels and analog storage media like film or tapes, the emerging Internet, mobile networks and digital storage are going to revolutionize the terms of distribution and access. This development is ruled by the convergence of audiovisual media technology, information technology and telecommunications technology. By capabilities of digital processing, established media like photography, movie, television and radio are changing their roles and are becoming subsumed by new integrated services which are *mobile, interactive, pervasive,* usable from anywhere, giving freedom to play with, and penetrating everyday life. Multimedia communication establishes new forms of communication between people, between people and machines, allows also communication between machines using audiovisual information or related feature parameters. Intelligent media interfaces are becoming increasingly important, and machine assistance in accessing media, in acquiring, organizing, distributing, manipulating and consuming audiovisual information becomes inevitable in the future.

This book intends to provide a deep insight into important enabling technologies of multimedia communication systems, which are methods of multimedia signal processing, analysis, identification and recognition, and schemes for multimedia signal representation, compression and expression by features or other properties. All these are lively and highly innovative areas at present, where this book reviews state-of-the-art technology and its scientific foundations, but shall primarily support systematic understanding of underlying methods, algorithms and their theoretical foundations. It is strongly believed that this is the best approach to contribute to future improvements in the field.

In part, the book is a substantially upgraded translation of my German language textbook on digital image and video coding, which was published by the mid '90s. Since then, the progress that was made in compression of audiovisual data has been breath-taking, and consequently newest developments are reflected, including the Advanced Video Coding standard and motion-compensated Wavelet coding. The second basis for this book are my lectures on topics of multimedia communications held regularly at RWTH Aachen University. These treat all aspects of image, video and audio compression, including networking interfaces, and also include multimedia signal identification and recognition. These latter aspects, topically related to the MPEG-7 multimedia content description standard, establish a profound basis for intelligent multimedia systems.

Most chapters are supplemented by homework problems, for which solutions are available from **http://www.ient.rwth-aachen.de**.

The book would not have been possible without contributions of numerous students and many other people who have worked with me on topics of image, video and audio processing, encoding and recognition over more than 15 years. These are (in alphabetical order) Sven Bauer, Michael Becker, Markus Beermann, Sven Brandau, Nicole Brandenburg, Michael Brünig, Ferry Bunjamin, Kai Clüver, Emmanuelle Corne, Holger Crysandt, Sila Ekmekci, Christoph Fehn, Ingo Feldmann, Oliver Fromm, Karsten Grüneberg, Karsten Grünheit, Jens Güther, Hafez Hadinejad, Konstantin Hanke, Guido Heising, Hans Dieter Höhne, Michael Höynck, Laetitia Hue, Ebroul Izquierdo, Peter Kauff, Jörg Kramer, Silko Kruse, Patrick Laurent, Thomas Ledworuski, Wolfram Liebsch, Oliver Lietz, Phuong Ma, Bela Makai, Claudia Mayer, Bernd Menser, Domingo Mery, Karsten Müller, Patrick Ndjiki-Nya, Bernhard Pasewaldt, Andreas Praatz, Lars Prokop, Oliver Rockinger, Katrin Rümmler, Thomas Rusert, Mihaela van der Schaar, Ansgar Schiffler, Oliver Schreer, Holger Schulz, Aljoscha Smolic, Frank Sperling, Peter Stammnitz, Jens Wellhausen, Mathias Wien and Detlef Zier. Please forgive me if I forgot anybody.

Very special thanks are also directed to my scientific mentors Peter Noll, Hans Dieter Lüke and Irmfried Hartmann, all people of IENT and to my family.

Aachen, August 15, 2003
Jens-Rainer Ohm

Table of Contents

Part C: Coding of Multimedia Signals

11 Quantization and Coding

After sampling, multimedia signal are discrete in space and/or time. Transformed equivalents of these signals, e.g. frequency coefficients or prediction error signals, and feature values or parameters related to these signals are often represented by continuous (or large sets of fine-granular discrete) amplitude values. To obtain a more compact digital representation, quantization is necessary. Quantization typically effects a distortion, which depends on the chosen quantizer step size and the number of quantization levels. On the other hand, the number of levels has direct influence on the rate that is necessary for representation. The relationship between rate and distortion is an important topic of information theory, by which the rate-distortion function and the source coding theorem were developed. The rate-distortion function defines the minimum rate to be provided for representation when a maximum distortion shall not be superseded, and depends on the statistic properties of a source. An analytic derivation is however only possible for stationary signal models. Statistical dependencies within source processes can implicitly be integrated into the rate-distortion dependency formulations. Following the statements of the source coding theorem, the minimum rate expressed by the rate-distortion function can be approached, if a large number of source symbols are encoded jointly, which can be implemented by block codes or sliding-block codes. Concrete methods to approach the rate-distortion bound for minimum-rate lossless representation of continuous and discrete signals are entropy coding, vector quantization, tree and trellis coding.

11.1 Scalar Quantization

To process and transmit multimedia signals digitally, they must be sampled and transformed into a discrete set of values, which is done by *quantization*. If the signal has continuous values, the process of quantization can be regarded as a unique mapping procedure, where exactly *one reconstruction value* from a discrete source alphabet can be selected if the continuous value is within a certain range of amplitudes[1]. For representation, it is then sufficient to provide an index i, which uniquely identifies the value from the discrete alphabet. The continuous signal shall be denoted as $x(\mathbf{n})$, the quantized (or reconstructed) signal as $y(\mathbf{n})$. The procedure of quantization can then be regarded as a mapping $x(\mathbf{n}){\rightarrow}i(\mathbf{n})$, the reconstruction is the mapping $i(\mathbf{n}){\rightarrow}y(\mathbf{n}){=}y_{i(\mathbf{n})}$, which is not exactly the inverse, as in general the quantization mapping is *non-invertible*[2] if $x(\mathbf{n})$ is truly continuous by value. If it is a discrete signal of high-precision integer or float representation, a comparably high number of bits must be spent for the representation; this is the case of lossless coding of discrete sources. If high-precision discrete signals are re-quantized into a lossy representation, the basic effects are not significantly different from the quantization of truly continuous signals; consequently, this is not treated as a separate case here.

Uniform quantization. In uniform quantization, the distances between reconstruction values which are neighbors on the amplitude scale are *quantization intervals* of constant width Δ. Each reconstruction value y_j is positioned at the center of a quantization interval, and all continuous-amplitude signal values within this interval are mapped into y_j. If however a continuous value x is beyond the minimum or maximum reconstruction values, an effect of *overload* occurs; the quantization intervals at the boundaries of the value range supported by the quantizer are in principle open and reach to infinity. The effect of overload shall be neglected for the moment; it is further assumed that the quantization intervals are sufficiently narrow, such that the PDF of the input signal can be regarded as constant within the interval[3]. The *quantization error* $q(\mathbf{n}){=}x(\mathbf{n}){-}y(\mathbf{n})$ can then be modeled as being uniformly distributed within the range $[-\Delta/2,\Delta/2]$ and have zero mean. Then, the variance of this quantization error is

$$\sigma_q^{\ 2} = \frac{1}{\Delta} \int\limits_{-\Delta/2}^{\Delta/2} q^2 dq = \frac{\Delta^2}{12} . \tag{11.1}$$

The properties of quantizers are fully described by the *quantizer characteristic*, which defines the nonlinear mapping $x(\mathbf{n}){\rightarrow}y(\mathbf{n})$. It is important to properly adapt the minimum-maximum range, considering the expected statistics of the input signal. Fig. 11.1 shows different examples of uniform quantizers, and also illustrates the relationship between an input signal amplitude $x(\mathbf{n})$ and the quantization error amplitude $q(\mathbf{n})$, which results as the difference between the quantizer characteristic and a function $y{=}x$, the zero-distortion case.

---

[1] This statement is true when quantization is regarded as an independent process. In the context of *rate-distortion optimization* (cf. sec. 11.3), re-mapping into different reconstruction values is intentionally performed to achieve a lower rate.

[2] For interpretation of quantization from a more general view of amplitude-mapping functions, see also sec. 5.3.1.

[3] These assumptions are strictly true only for signals which have a uniform PDF and perfectly fit by their amplitude range with the range of the quantizer.

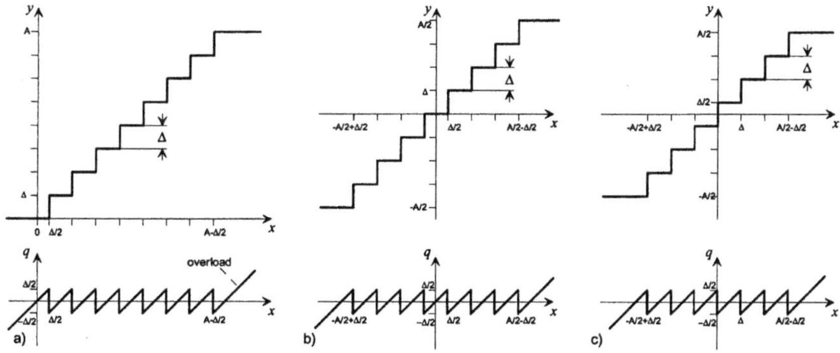

Fig. 11.1. Uniform scalar quantizer characteristics, and characteristics of quantization errors in dependency of the signal amplitude variable: **a** Quantizer for positive-only signal and **b/c** Quantizers which are symmetric about value zero (**b** with, **c** without zero reconstruction value)

If a uniform quantizer is used, the processes of quantization and reconstruction ('de-quantization') are extremely simple due to the linear relationship between the reconstruction values and the indices. For computation of the index i from the signal value x, it is sufficient to divide the continuous value by Δ and perform nearest-integer rounding. Optionally, an *offset* shift can be compensated in the quantization step. To compute the reconstruction value y, scaling of the index by Δ and reverse offset shift must be performed. For indices within the range $j=1,...,J$, a uniform quantization process determines the optimum index i and reconstruction y_i as follows[1]:

$$i = \mathrm{nint}\left(\frac{x - \mathit{offset}}{\Delta}\right) + 1 \quad ; \quad i = \max(i,1) \quad ; \quad i = \min(i,J)$$

$$y = (i-1) \cdot \Delta + \mathit{offset}.$$

(11.2)

For the examples of Fig. 11.1, $\Delta=A/J$ in all cases, *offset*=0 for Fig. 11.1a, and *offset*=A/2 for Figs. 11.1b/c. In case of Fig. 11.1a, J can be an even or odd number, while due to symmetry reasons J must be odd for Fig. 11.1b and even for Fig. 11.1c.

Non-uniform quantization. The application of a quantizer with non-uniform step size (Fig. 11.2a) can be useful if the signal $x(\mathbf{n})$ has a non-uniform PDF. Further, quantizers of non-uniform step sizes can be used to implement nonlinear amplitude-mapping functions as a by-product of quantization. This can e.g. be useful if quantization errors are perceived as more severe at low amplitude ranges, where

---

[1] Other methods of indexing can be used which address the sign and the magnitude separately. For the case of the quantizer in Fig. 11.1b, it would not be necessary to convey the sign in the case of the zero reconstruction value.

more accurate quantization should be performed. The classical method for the design of non-uniform quantizers is the *Lloyd Algorithm* [LLOYD 1957], which is based on a criterion for minimization of the squared error in the case of a given signal PDF. If a signal value x is mapped by an arbitrary quantizer into the nearest reconstruction value y_j from a codebook $\mathcal{C}=\{y_j; j=1,..,J\}$, the index is

$$i = \arg\min_{y_j \in \mathcal{C}}(x - y_j)^2. \tag{11.3}$$

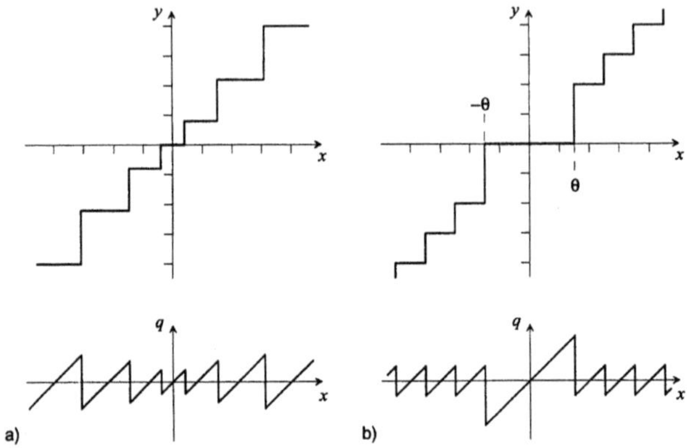

Abb. 11.2. Quantizer characteristics **a** of a non-uniform **b** of a dead zone quantizer

All amplitude variables x which would be mapped into an index j by this procedure shall be denoted as $x^{(j)}$. For these values, the expected squared distortion is

$$D_j = E\{(x^{(j)} - y_j)^2\}. \tag{11.4}$$

The mean squared distortion over all j will be

$$D = \sum_{j=1}^{J} P(j) \cdot D_j . \tag{11.5}$$

To minimize D_j, which is dependent on the choice of values y_j, the following derivation is performed:

$$D_j = E\{(x^{(j)})^2\} - 2E\{x^{(j)}\} \cdot y_j + y_j^2$$

$$\Rightarrow \frac{\partial D_j}{\partial y_j} = -2E\{x^{(j)}\} + 2 \cdot y_j \quad \text{and} \quad \frac{\partial D_j}{\partial y_j} = 0 \text{ for } y_j = y_{j,opt} \tag{11.6}$$

$$\Rightarrow y_{j,\text{opt}} = E\{x^{(j)}\}.$$

The updated value $y_{j,\text{opt}}$ can be determined if either a sufficiently large training set of typical values x is available, or if an appropriate model of the PDF $p(x)$ or a discrete approximation thereof is known. In analogy with (11.6), for the case of a continuous PDF,

$$D_j = \frac{1}{P(j)} \int_{x \in x^j} p(x) \cdot (x - y_j)^2 dx ,$$ (11.7)

for which the optimized value is

$$y_{j,opt} = E\left\{x^j\right\} = \frac{1}{P(j)} \int_{x \in x^j} p(x) \cdot x dx .$$ (11.8)

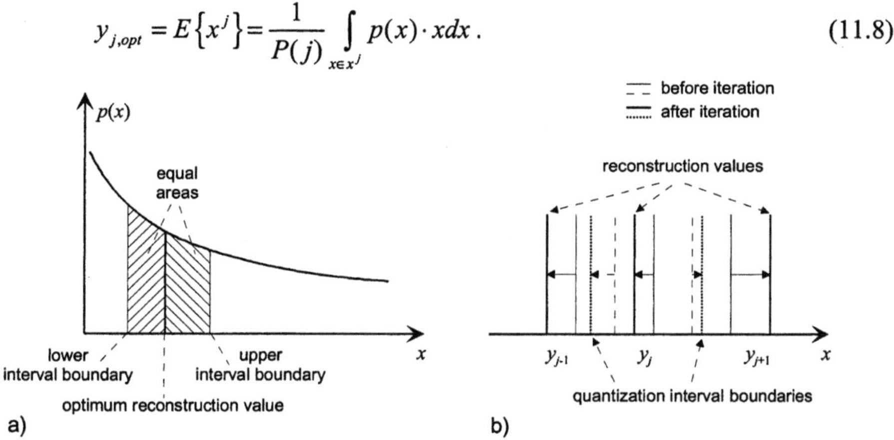

Fig. 11.3. **a** Optimum position of a reconstruction value, depending on the shape of the PDF within a quantization interval   **b** update of reconstruction values and interval boundaries by one iteration

Unlike uniform quantization, the optimized reconstruction value is no longer at the center position of the interval, but it is the mean value of all signal values $x^{(j)}$ which are allocated to this interval, or the mass centroid of the respective PDF partition (Fig. 11.3a). For a scalar quantizer, the distance between each of two neighbored reconstruction values and the interval boundary in between must be equal[1]. When a set of new values $y_{j,opt}$ is computed, the optimum boundaries between the quantization intervals will implicitly move (Fig. 11.3b). As a consequence, the optimum solution can only be approached iteratively, where it is guaranteed by each iteration step that the result can never become worse: Optimization of the reconstruction values $y_{j,opt}$ guarantees a lower distortion; then, by the re-definition of the quantization interval boundaries other reconstruction values may provide quantization by even less distortion than before for some given x. However then, the set of values $y_{j,opt}$ that was computed in the previous iteration is no longer optimum, as the centroids of the quantization intervals again are shifted. Another optimization step is necessary. This process of iterative optimizations by guarantee converges; by guarantee, the overall distortion can never increase by any iteration step, but the

---

[1] According to (11.3), the interval boundaries are at positions where the distortion of a signal value is equal compared to both adjacent reconstruction values.

improvement will also continuously decrease. A quantizer with a finally optimized set of reconstruction values is obtained, if the overall distortion D is not changing significantly any more as compared to the previous iteration step.

A specific category in between uniform and non-uniform quantizers is defined by the *dead zone quantizer characteristic* shown in Fig. 11.2b. Here, the innermost quantization zone (for values near zero) is enlarged. Beyond this dead zone, the quantizer characteristic is of uniform step size. The purpose of the dead-zone quantizer is an increase of the probability for the zero-quantized values. This can be advantageous when the zero values can be encoded by a very low rate as compared to non-zero values.

11.2 Coding Theory

11.2.1   Source Coding Theorem and Rate Distortion Function

The source coding theorem [SHANNON 1959] can be summarized as follows:

—   For the encoding of a discrete, memoryless source, where a distortion smaller or equal to D shall be allowed, a block code of minimum rate $R=R(D)+\varepsilon$ with $\varepsilon > 0$ exists;
—   To approach $R(D)$, the block length K of the code must be sufficiently large.

The determination of $R(D)$, the *Rate-Distortion Function* (RDF) for signals of arbitrary PDF is in general quite complex and only possible by numeric approximations[1]. An analytic solution can be given for the case of an uncorrelated, stationary Gaussian signal $z(n)$, which is white noise of variance σ_z^2. For a given squared distortion metric D,

$$R(D) = \frac{1}{2}\log_2 \frac{\sigma_z^{\,2}}{D} .\tag{11.9}$$

D can e.g. be expressed by the quantization error variance according to (11.1). The uncorrelated Gaussian model is important, as in terms of the rate-distortion behavior, it is at the verge of the *worst case* among all memoryless sampled signals of same variance σ^2 [BERGER 1971]. However, (11.9) only applies to cases $D \leq \sigma_z^2$, as otherwise the logarithmic function would bring the rate into the negative range. A more general expression can hence be formulated as[2]

---

[1] An algorithm to compute the approximation of the RDF from a set of data samples was proposed in [BLAHUT 1987]

[2] For $R=0$, no information about the signal is conveyed; the squared distortion metric is upper bounded by σ_z^2, the maximum possible uncertainty about the signal.

$$R(D) = \max\left(0, \frac{1}{2}\log_2 \frac{\sigma_z^2}{D}\right) \quad \Rightarrow \quad D(R) = \sigma_z^2 \cdot 2^{-2R}, \, R \geq 0. \tag{11.10}$$

By using a block code, the properties of the multi-dimensional PDF of a signal can be exploited. The K-dimensional vector PDF of an uncorrelated Gaussian signal is the product from the PDFs of the contained random variables, according to (3.28) $p_K(\mathbf{x}) = p(x_1) \cdot p(x_2) \cdot \ldots \cdot p(x_K)$. The vector $\mathbf{x}$ usually consists of K subsequent samples $x(n)$, even though for uncorrelated sources any samples of the signal could be combined for encoding. Practical realizations of block codes are the methods of *entropy coding* (sec. 11.4), *vector quantization* (sec. 11.5) and *sliding block coding* (sec. 11.6). The advantage of block coding results from the fact that even for uncorrelated sources, certain 'untypical' vector constellations exist which are of extreme low probability. For a PDF like the zero-mean Gaussian, where amplitudes concentrate around the value zero, combinations where *all* values have a *high amplitude* are of extremely low probability. Typical combinations are vectors with many low and some higher amplitude values, which can be considered accordingly in the design of codes.

As $R(D)$ is always a convex function, a unique inverse function $D(R)$ – the *Distortion-Rate Function* (DRF) – is defined, which also 'inverts' the source coding theorem as follows:

– If a rate $R \geq 0$ is available for encoding of a discrete, memoryless source, the distortion can never become lower than $D(R)$.

11.2.2   Rate-Distortion Function for Correlated Signals

For a correlated (AR) Gaussian process, the RDF is

$$R(D_\Theta) = \frac{1}{4\pi} \int_{-\pi}^{\pi} \max\left(0, \log_2 \frac{S_{xx}(\Omega)}{\Theta}\right) d\Omega. \tag{11.11}$$

Here, the correlated process is interpreted as being decomposed into an infinite number of orthogonal frequency components, represented by statistical properties of the power spectrum. As an argument function of the spectral integration, (11.10) is used[1]. If the spectral power at a given frequency is above the threshold Θ, the rate of (11.9) is assigned, otherwise the rate will become zero. In the latter case, the spectral distortion for a given frequency Ω is equal to the power spectrum itself; in the former case, the distortion equals Θ. When not all power spectrum values within the range $0 \leq |\Omega| < \pi$ are above the threshold, the integration in (11.11) can only be solved piece-wise. If all spectral values are above the threshold, the

---

[1] Similar to Parseval's theorem (3.71), where the variance is computed by integration over the values of the power spectrum, (11.11) interprets the rate by the rate contributions of the orthogonal frequency components.

resulting overall distortion D equals Θ. For the more general case, the distortion $D_\Theta \leq \Theta$ must also be determined by piece-wise integration:

$$D_\Theta = \frac{1}{2\pi} \int_{-\pi}^{\pi} \min\left[\Theta, S_{xx}(\Omega)\right] d\Omega .\qquad (11.12)$$

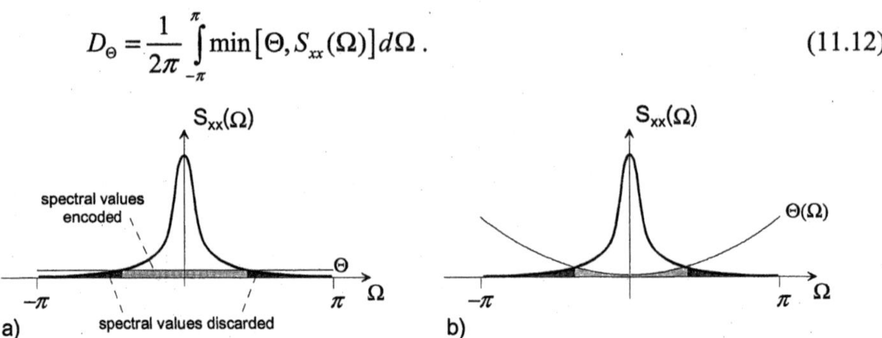

Fig. 11.4. Interpretation of the rate distortion function for an AR(1) process: **a** Equal distortion over all spectral components **b** Spectrally weighted distortion threshold function

Fig. 11.4a illustrates the totally resulting distortion D_Θ by the shaded areas below the argument function of (11.12). The distortion shall never become larger than the signal power at a specific frequency. The spectral components of the dark-shaded areas are hence not encoded at all: The signal loses high-frequency components in the given example. Fig. 11.4b in addition shows the example of a *frequency-weighted threshold function*[1].

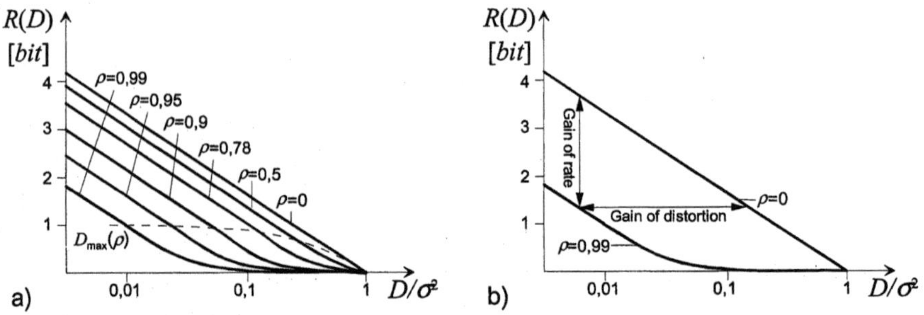

Fig. 11.5. a $R(D)$ of an AR(1) process parametrized over different parameters ρ **b** Gain of rate and gain of distortion by exploitation of correlation properties (Graphical presentation after [CLARKE 1985])

Example : R(D) of an AR(1) process. The AR(1) process is described by the correlation coefficient ρ and variance of the innovation signal σ_z^2. The power spectrum

---

[1] Such functions are often used in encoding of multimedia signals. Due to spectral masking inherent in human senses of vision and hearing, higher distortion can be allowed for certain components. In principle, this also allows to define distortion metrics and RDF definitions which are adapted to human perception.

$S_{xx}(\Omega)$ is given by (4.58). If $S_{xx}(\Omega)>\Theta$ throughout the range $0\leq|\Omega|<\pi$, the rate distortion function is

$$
\begin{aligned}
R(D) &= \frac{1}{4\pi} \int_{-\pi}^{\pi} \log_2 \frac{\sigma_x^2 (1-\rho)^2}{D\cdot(1-2\rho\cos\Omega+\rho^2)} d\Omega \\
&= \frac{1}{4\pi} \int_{-\pi}^{\pi} \log_2 \frac{\sigma_x^2 (1-\rho)^2}{D\cdot(1+\rho^2)} d\Omega - \frac{1}{4\pi} \int_{-\pi}^{\pi} \log_2 \left(1-\frac{2\rho\cos\Omega}{1+\rho^2}\right) d\Omega \\
&= \frac{1}{2}\log_2 \frac{\sigma_x^2 \left(1-\rho^2\right)}{D} = \frac{1}{2}\log_2 \frac{\sigma_z^2}{D}.
\end{aligned}
\tag{11.13}
$$

It turns out that $R(D)$ for the AR(1) process[1] of variance σ_x^2 can directly be expressed as the RDF of its white-noise innovation signal which has variance σ_z^2. This is however only valid for the case where all spectral components are above Θ. As the power spectrum (4.58) of the AR(1) process steadily decreases up to the minimum value at $|\Omega|=\pi$, the maximum allowable distortion for validity of (11.13) is

$$
D\leq D_{max} = \frac{1-\rho}{1+\rho}\cdot\sigma_x^2.
\tag{11.14}
$$

If this condition holds, also $D=\Theta$. For larger D, the parametrization over Θ in (11.11) and (11.12) can be resolved as follows for the AR(1) process. By definition the frequency Ω_Θ where $S_{xx}(\Omega)<\Theta$ for $|\Omega|>\Omega_\Theta$,

$$
R(D_\Theta) = \frac{1}{2\pi} \int_{0}^{\Omega_\Theta} \log_2 \frac{S_{xx}(\Omega)}{\Theta} d\Omega \quad ; \quad D_\Theta = \frac{1}{\pi}\left[\int_{0}^{\Omega_\Theta} \Theta d\Omega + \int_{\Omega_\Theta}^{\pi} S_{xx}(\Omega)d\Omega\right]
\tag{11.15}
$$

$$
\text{with} \quad \Theta \overset{!}{=} \frac{\sigma_z^2}{1-2\rho\cos\Omega_\Theta+\rho^2} \quad \Rightarrow \quad \Omega_\Theta = \arccos\frac{1}{2\rho}\left(1+\rho^2 - \frac{\sigma_z^2}{\Theta}\right).
$$

Fig. 11.5a illustrates $R(D)$ graphs for AR(1) processes of different parameters ρ. The D axis is shown on a logarithmic scale, such that the $R(D)$ relationship (11.9) is linear for the uncorrelated process. The dotted line interconnects the values D_{max} according to (11.14) for different values of ρ; above D_{max}, the $R(D)$ curves of the correlated processes are lines in parallel to $R(D)$ of the uncorrelated process ($\rho=0$). The corresponding value in rate is

$$
R(D_{max}) = \log_2 \left(1+\rho\right).
\tag{11.16}
$$

The distance between the $R(D)$ plots of $\rho=0$ and $\rho>0$ can be interpreted as either a gain in terms of rate (vertically) or in terms of distortion (horizontally), see Fig. 11.5b. This gain however decreases when the distortion grows larger than D_{max}.

---

[1] This applies for any AR(P) process.

The drastic breakdown of rate-distortion performance can indeed be observed in encoding of multimedia signals at low rates. By using a coding scheme which utilizes the correlation between samples of an AR(1) process while observing the constraint (11.14), a maximum *coding gain*

$$G = \frac{1}{\gamma_x^{\,2}} = \frac{\sigma_x^{\,2}}{\sigma_z^{\,2}} \left(= \frac{1}{1-\rho^2} \quad \text{for AR(1) model} \right) \tag{11.17}$$

can be achieved. The coding gain expresses the factor by which the distortion decreases when a coding scheme utilizes the correlation, as compared to independent (PCM) encoding of samples at the same rate. Alternatively, keeping the same distortion as for the case of independent encoding of samples, the rate can in the case of AR(1) be reduced by the factor

$$R_G = -\frac{1}{2} \cdot \log_2 \left(1-\rho^2\right). \tag{11.18}$$

By substituting (11.14) into (11.13), it is found that this gain can be realized when $R \geq R(D_{\max})$. The coding gain can be interpreted here as the factor, by which a decorrelating coding scheme[1] is able to reduce the distortion as compared to a PCM scheme. The coding gain is also the reciprocal value of the *Spectral Flatness Measure* (SFM) [JAYANT, NOLL 1984]

$$\gamma_x^{\,2} = \frac{2^{\left[\frac{1}{2\pi}\int\limits_{-\pi}^{\pi}\log_2 S_{xx}(\Omega)d\Omega\right]}}{\sigma_x^{\,2}} = \frac{2^{\left[\frac{1}{2\pi}\int\limits_{-\pi}^{\pi}\log_2 S_{xx}(\Omega)d\Omega\right]}}{\frac{1}{2\pi}\int\limits_{-\pi}^{\pi} S_{xx}(\Omega)d\Omega}, \tag{11.19}$$

which is the ratio between the *geometric mean* over the power spectrum compared to the variance or *arithmetic mean*. Both values are equal and the coding gain becomes unity, if all spectral components contribute equally to the overall power, which is the case of white Gaussian noise. In *any other case* (non-uniform spectral distribution), the geometric mean is lower than the arithmetic mean, such that a coding gain can be realized. Observe that the SFM can only be determined for processes where in the range $[-\pi,\pi]$ all spectral components are larger than zero, and is strictly valid only for Gaussian processes.

11.2.3   Rate Distortion Function for Multi-dimensional Signals

The RDF for a 2D Gaussian (e.g. AR) model is an extension of (11.11)

---

[1] In case of Gaussian (AR) processes, all statistical dependencies are linear and can be expressed as correlations.

$$R_{2D}(D_\Theta) = \frac{1}{8\pi^2} \int\limits_{-\pi}^{\pi}\int\limits_{-\pi}^{\pi} \max\left(0, \log_2 \frac{S_{xx}(\Omega_1,\Omega_2)}{\Theta}\right) d\Omega_1 d\Omega_2 , \tag{11.20}$$

and (11.12)

$$D_\Theta = \frac{1}{4\pi^2} \int\limits_{-\pi}^{\pi}\int\limits_{-\pi}^{\pi} \min\left[\Theta, S_{xx}(\Omega_1,\Omega_2)\right] d\Omega_1 d\Omega_2 . \tag{11.21}$$

If the AR model and its spectrum are separable, the separability of $R(D)$ does not follow implicitly except within the range of low distortion, where no spectral components are discarded by encoding. A 2D separable AR(1) process is specified by horizontal and vertical correlation coefficients ρ_h and ρ_v, where according to the relationship between the variances of the process and its innovation signal

$$R_{2D}(D) = \frac{1}{2}\log_2 \frac{\sigma_x^2\left(1-\rho_h^2\right)\left(1-\rho_v^2\right)}{D} = \frac{1}{2}\log_2 \frac{\sigma_z^2}{D} . \tag{11.22}$$

The validity of (11.22) is now constrained to cases of low distortions[1]

$$D \le D_{\max,2D} = \frac{\left(1-\rho_h\right)\cdot\left(1-\rho_v\right)}{\left(1+\rho_h\right)\cdot\left(1+\rho_v\right)}\cdot\sigma_x^2 , \tag{11.23}$$

which by identical correlation parameters results in a lower value than for the 1D case (11.14). The maximum coding gain by exploiting correlations along both horizontal and vertical directions is

$$G_{2D} = \frac{1}{\left(1-\rho_h^2\right)\cdot\left(1-\rho_v^2\right)} , \tag{11.24}$$

but can only be realized for rates $R \ge R(D_{\max,2D}) = \log_2(1+\rho_h)+\log_2(1+\rho_v)$. The reduction in rate then is $R_{G,2D} = -\frac{1}{2}\log_2(1-\rho_h^2)-\frac{1}{2}\log_2(1-\rho_v^2)$. Within the low distortion range, the correlations in both dimensions contribute independently to the overall gain in the case of the separable model.

Example. For a 2D separable AR(1) source with $\rho_h=\rho_v=.95$, a further reduction of distortion by 10.1 dB or reduction of rate by 1.68 bit/sample can be achieved in comparison to the 1D case. As compared to PCM (not exploiting correlation at all), the gain is 20.2 dB in distortion, or 3.36 bit/sample in terms of rate. This model analysis matches fairly well the gain which can e.g. be achieved in *lossless* coding of still images; typically, an 8 bit/sample PCM image representation can be lossless encoded (without any modification of samples) into 3-5 bit/sample by methods of linear predictive coding.

---

[1] cf. Problem 11.1

11.3 Rate-Distortion Optimization of Quantizers

A continuous-amplitude signal is mapped by quantization into a signal of discrete amplitude levels. Another case is a signal having discrete amplitudes originally, which is *re-quantized*, i.e. mapped into a signal of less amplitude levels. To represent the indices i which express the reconstruction values in case of digital transmission, the rate of *entropy* (3.94) is necessary at minimum, where the set of indices establishes an alphabet with finite number of symbols. As coding method, any of the entropy coding schemes described in sec. 11.4 could be applied. The approach to regard quantization and entropy coding as separate entities is however only useful in the range of high rates, where the number of symbols in the alphabet is large. In particular at low rates, the optimization of *quantization and encoding* must rather be regarded as a *combined problem*. For example, it may be useful to select a reconstruction causing slightly higher distortion, if this leads to significantly reduced rate as compared to another value giving slightly lower distortion. In the process which optimizes the quantizer (e.g. (11.4)-(11.8)) or during the quantization itself it is then necessary to observe an additional rate constraint. The goal should be to allocate the available bit budget in an optimum way, such that the overall distortion is kept as low as possible under the rate constraint. Assume that $R(j)$ is the number of bits to be spent when quantization index j is selected. Then, the overall rate is

$$R = \sum_{j=1}^{J} P(j) \cdot R(j) \, . \tag{11.25}$$

Modifying (11.5) by formulation of a constrained problem that can be solved by *Lagrangian optimization* includes the rate into the 'distortion' criterion

$$D^* = D + \lambda \cdot R = \sum_{j=1}^{J} P(j) \cdot \left[D_j + \lambda \cdot R(j) \right] . \tag{11.26}$$

Depending on the factor λ, the influence of D and R can be adapted, and obviously $\lambda=0$ gives exactly (11.5). The optimum factor λ can be found, if D^* is minimized depending on the rate by derivation of (11.26),

$$\frac{\partial D^*}{\partial R} = \frac{\partial D}{\partial R} + \lambda_{\mathrm{opt}} \overset{!}{=} 0 \;\Rightarrow\; \lambda_{\mathrm{opt}} = -\frac{\partial D}{\partial R} \, . \tag{11.27}$$

The optimum λ for a given rate is determined by the negative slope of the DRF $D(R)$, which is the inverse function of $R(D)$[1]. Indeed, for multimedia signals, $R(D)$ is usually unknown, such that this slope must be estimated. This can be achieved by introducing an *Operational Rate-Distortion Function* (ORDF), which is tracked

---

[1] As $R(D)=D^{-1}(R)$, the optimum λ an also be interpreted as the reciprocal negative slope of the RDF for a given distortion point D.

during the iterative optimization of a quantizer or during the quantization decisions. If a number of values is quantized and encoded by the output entropy of the quantizer, a certain pair of D (11.5) and R (11.25) results which can be localized in the R-D plane. An assumption that generally applies is the convexity of the RDF. By testing different configurations of quantizers, all those with R/D pairs which contradict the assumption of convexity can be rejected as sub-optimum. This can be done by linear interpolation between known R/D pairs, eliminating all pairs which are above the lowest resulting interconnection curve, which is the ORDF (see Fig. 11.6). Then, a new optimization trial can be performed, adjusting λ according to the local slope as estimated from the recent ORDF approximation.

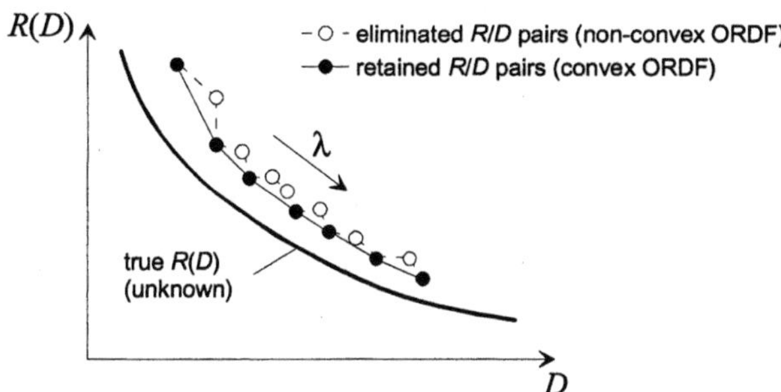

Fig. 11.6. Development of the ORDF based on convexity constraint

This method is denoted as *Entropy-constrained Quantization* (ECQ). During the process of quantizer optimization, the self information $i(j)=-\log_2 P(j)$ can be used as an approximation for the expected rates $R(j)$. It can be estimated from the occurrence counts of the quantizer cell selections or the area below the PDF within the range of the quantization cell. When rate constraints are employed during encoding by a given quantizer/encoder configuration, the actual resulting rate allocation or an entropy-based estimate thereof can be used.

For high rates (low distortion), quantizer designs optimized by ECQ are approaching uniform characteristic. If a quantizer has to be optimized for lower rates, it is advisable to start by a uniform quantizer with a high number of reconstruction values J, and increase λ during the iterative optimization process, until an optimum around the target rate (point on the ORDF) is approached. By increasing the rate cost, an effect occurs that certain reconstruction values are not used anymore; these can be eliminated from the set of reconstruction values. Asymptotically, for $\lambda \to \infty$ only one reconstruction value remains ($J=1$), which is the mean value determined from the training set or the source PDF used for optimization; the encoding rate is zero. This is the point where $R(D)$ approaches the D axis, such that $D=\sigma_x^2$ (see Fig. 3.12).

For higher-rate encoding of stationary sources such as Gaussian processes, it would be possible to approach any point on the rate-distortion curve by performing a uniform quantization and fine-tune the step-size Δ such that a discrete output is produced for which the rate of the entropy is $R(D)$. For real-world multimedia sources, the stationarity assumption does not apply. These could rather be interpreted by a *switched* or *composite source models*, where statistics vary locally. For example, clearly disparate regions of high detail and regions of low detail co-exist in images. The steepness of the 'localized' rate-distortion functions will be very different for a given overall rate/distortion point; refer to Fig. 11.5, where some areas of an image may follow the graph for ρ=0.9, other less-detailed areas may better be characterized by the graph for ρ=0.99. If a certain rate budget can be spent, it is best invested for components which give the highest value in terms of distortion minimization. This is guaranteed by choosing a fixed value of λ in (11.26) for rate-constrained quantization of all components, regardless of their local statistics. This leads to following consequences:

- Assume an amplitude value which is near the boundary between two quantization cells. If a rate constraint is applied, this value would better be represented by the reconstruction value which requires the lower rate, even though the distortion (e.g. according to squared error criterion) may be slightly higher. This can also be interpreted such that the *quantizer characteristic* is modified depending on λ, where the decision boundary is no longer at the centered position between two neighbored reconstruction values y_j. A more thorough analysis for the more general case of vector quantization is given in sec. 11.5.5.
- Assume that different quantizer configurations are available for selection, which relate to different rate-distortion points. A straightforward example would be a uniform quantizer of variable step size Δ. Again, the decision can be obtained by adding the resulting λ-weighted rate as a penalty over the distortion metric of the quantizer. The decision for the optimum quantizer configuration should be performed for groups of samples from local neighborhoods or from signal components of similar rate-distortion behavior. This would lead to a selection of different quantizers for components of different statistical properties in the composite source.

Embedded quantization. The distortion which occurs due to quantization is directly related to the quantizer step size Δ, see e.g. (11.1). Assume now that the residual error shall be *re-quantized* by a quantizer which has exactly half of the step size, $\Delta/2$. Then, the number of reconstruction values J will be roughly doubled. The layout of the second quantizer shall divide each of the previous quantization cells into two halves of equal width. If the reconstruction values are assumed to be positioned at the centers of quantization intervals, each reconstruction value y_j for the quantizer of step size Δ will branch for the $\Delta/2$-quantizer into two values at positions $y_j \pm \Delta/4$. Two different quantizer characteristics of this type are shown in Fig. 11.7; the dotted lines represent the quantizers of half step size. Fig. 11.7a shows a configuration for quantization of only positive values, while Fig. 11.7b gives the example of a dead zone quantizer (cf. Fig. 11.2b); the dead zone width is 2Δ, and the zero-quantization threshold is Δ. The dead zone is divided into three intervals, one of which is a new dead-zone of exactly half width Δ, the other two are 'normal' quantization intervals of width $\Delta/2$.

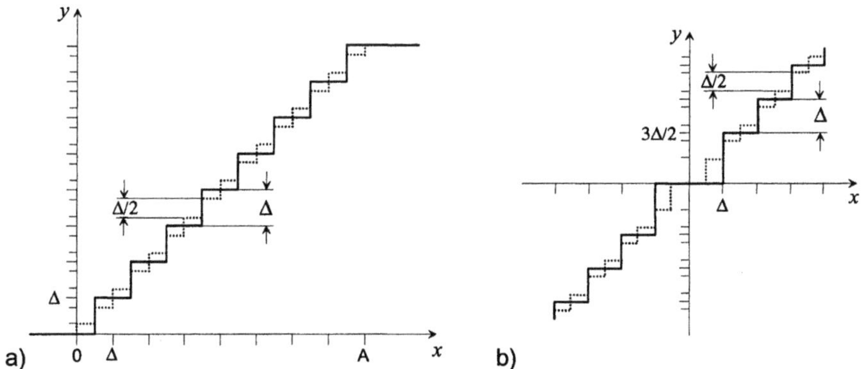

Fig. 11.7. Quantizer characteristics for the case of re-quantization by step size $\Delta/2$
a Asymmetric positive-only range of quantizer **b** Symmetric dead-zone quantizer

Fully embedded quantizers perform the method of re-quantization over multiple levels, starting with only two quantization intervals for the case of positive-value signals, and three intervals for the case of the dead-zone quantizer. This is shown in Fig. 11.8; the reconstruction values are always positioned at the centers of the quantization intervals. The final result of the quantization is an integer-value index which has a word length equal to the number of levels; the first level relates to the *Most Significant Bit* (MSB), the last level to the *Least Significant Bit* (LSB) of the integer word. This index can then be truncated starting from the LSB, which directly relates to a coarser quantization with lower number of levels. The entire quantization could indeed be performed by the binary tree decision as explained here. A much simpler way is to employ an ordinary uniform quantization using the interval width (quantizer step size) of the last level (LSB), where the total number of quantization intervals must be a power of two, 2^B. Implicitly then, each level of the embedded quantizer corresponds exactly to one bit at the related significance position of a positive integer of word length B. This corresponds exactly to the scheme in Fig. 11.8a which shows the quantization of positive-amplitude signals. The quantized representation of the amplitude range between MIN and MAX is becoming more accurate with each additional bit becoming available. This is denoted as a *bit-plane representation*, where a bit-plane is constructed by the respective bits of a certain significance level from all available samples of a signal. One single bit-planes can then be interpreted as binary signal of same number of samples, by combination of all bit-planes a multi-level representation is established.

The dead-zone scheme in Fig. 11.8b must allow zero, positive and negative reconstruction values at any step of quantizer refinement. The zero value does not require a sign bit. In a magnitude/sign representation, the sign must only be conveyed for the non-zero amplitude levels.

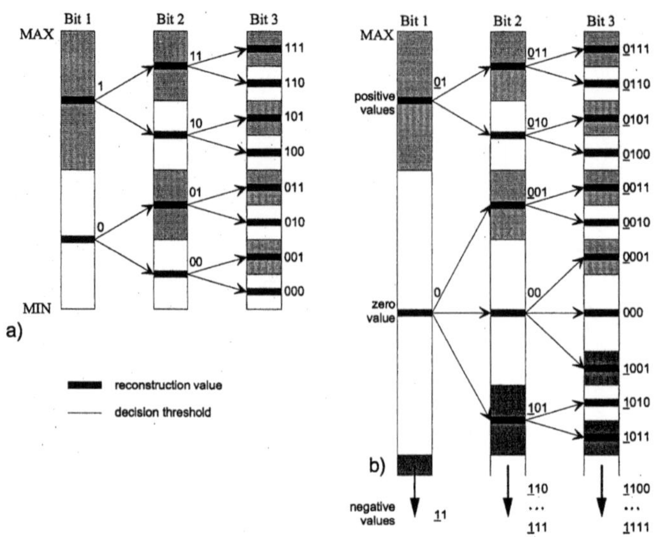

Fig. 11.8. Embedded quantizer structures and associated bit-plane code **a** for positive-valued signals **b** with dead zone for sign/magnitude signals; sign bit underlined

Embedded quantization and bit-plane coding are useful for *scalable representation* of quantized signals, where the more significant bits are a representation which allows reconstruction by a higher distortion at a lower rate. It is however not possible to move freely on the rate-distortion curve by adding or omitting bits from the bit-plane representation. More specifically, even when efficient entropy coding is applied to the bit-plane signals, the rate-distortion curve can *exactly* be mapped only at those rate points where full bit-planes are available, allowing to decode all samples of the signal by exactly the same bit-plane resolution. This phenomenon is known as the problem of *fractional bit-planes*, and can be interpreted as follows:

- Assume that it is possible to approach the rate-distortion bound by adjusting some quantizer step size Δ, which maps into a distortion D_Δ, and the rate $R(D_\Delta)$ is the entropy of the quantizer output.
- Towards higher rates, the embedded quantizer will not allow to select arbitrary smaller step-sizes Δ, but it will again exactly perform equal to a freely adjustable quantizer at rate-distortion points that relate to step sizes $\Delta/2$, $\Delta/4$, $\Delta/8$,...., etc. (see Fig. 11.9). At these points with rates $R(D_{\Delta/2^b})$, *all* samples of the signal can be quantized by the same accuracy.
- For points between the rates $R(D_{\Delta/2^b})$, the available bit budget should be used to encoding at least some samples by the higher accuracy of the next finer bit-plane. As the squared distortion D results as a mean value over all samples, it will statistically decrease *linearly* with each additional sample (bit) encoded by the next bit-plane. Hence, the phenomenon is observed that for bit-plane coding the rate-distortion function is a *polygon approximation* of the original RDF, which connects the exactly-matching rate-distortion points $R(D_{\Delta/2^b})$ of

full bit-plane populations by straight lines (see Fig. 11.9). The optimum performance could only be achieved if all samples are treated equally, which is simply impossible for all rates in between.

By rate-distortion optimized bit-plane truncation, re-ordering of information in the bit planes can be performed, such that samples or components which contribute by highest reduction of distortion at minimum cost of bits are conveyed first. Such strategies partially compensate the phenomenon of fractional bit-planes. This is however only achievable if the method of bit-plane encoding allows clustering the information into components of more or less significance. Such strategies are e.g. applied by the EBCOT algorithm used in the JPEG 2000 standard (cf. sec. 12.4.2).

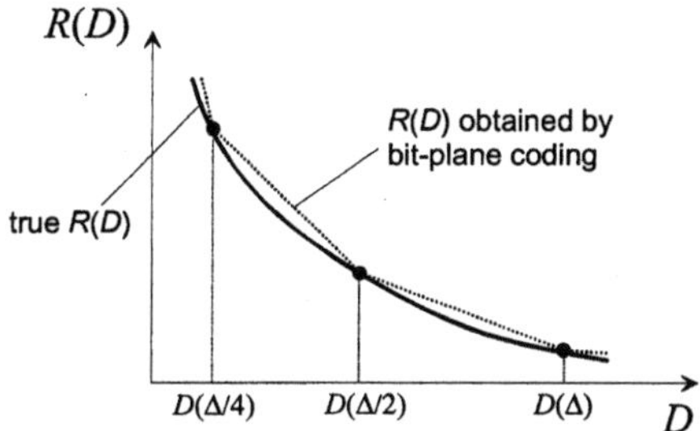

Fig. 11.9. Problem of *fractional bit-planes* in bit-plane truncation

11.4 Entropy Coding

11.4.1 Properties of Variable-length Codes

The *entropy* $H(\mathcal{A})$ of (3.93) and (3.94) is the minimum rate, by which a discrete source with source alphabet $\mathcal{A}=\{a_1,a_2,..,a_A\}$ can be lossless encoded. The goal of *entropy coding* is to define a code $\mathcal{C}$, which allows this encoding by approximately the rate of entropy. In principle, this is possible by using *Variable-Length Codes* (VLC). One important constraint of binary codes is the prerequisite of *integer bit allocation*. Each element of the alphabet shall be assigned a codeword (binary string) consisting of z_j bits, which as good as possible should match the self information of a_j, i.e. $z_j \approx i(j) = -\log_2 P(j)$.

Codes of variable length must be decodable not only codeword by codeword, but also if a *stream of codewords* is received by the decoder. This incurs the following constraints:

— *Uniqueness of codewords* (same condition applies to fixed-length codes);

–   *Uniqueness of codeword lengths*, which means that during decoding, it must be possible to identify the start bit and the terminating bit belonging to each codeword within the stream.

The latter condition could be realized by using a specific *terminating symbol* or *separation symbol* after each codeword inserted into the stream. Such an approach is e.g. implemented in the *Morse code*, where the break (no beep) between the codewords indicates their separation. Unfortunately, this introduces an overhead, which can make it impossible to approach the lower bound of the entropy rate. The most important class of codes fulfilling the condition of unique decodability without overhead are the *prefix codes*. These are constructed such that never a valid entire string of one codeword shall be the prefix of another codeword's bit string. Regard two codes of 4 codewords each, $\mathcal{C}$=[0,10,110,111] which is uniquely decodable, the other $\mathcal{C}$=[0,10,100,111] which is not unique, as the string '10' can either be the entire second codeword, or the prefix of the third. Prefix codes are the only class of uniquely decodable codes which allow *direct* or *instantaneous decoding*, which means that decoding can immediately be performed if the entire bit string characterizing the codeword is available at the decoder. Prefix codes can be mapped onto the structure of a *code tree* (Fig. 11.10). In case of binary coding, the tree consists of binary (two-way) *branches*, each of which represents one bit. The branches connect *nodes*; specific nodes are the *root* and the *leaves*, which represent the start and termination of valid codewords[1]. The entire *path* of branches between root and leaf signifies the bit string associated with one codeword.

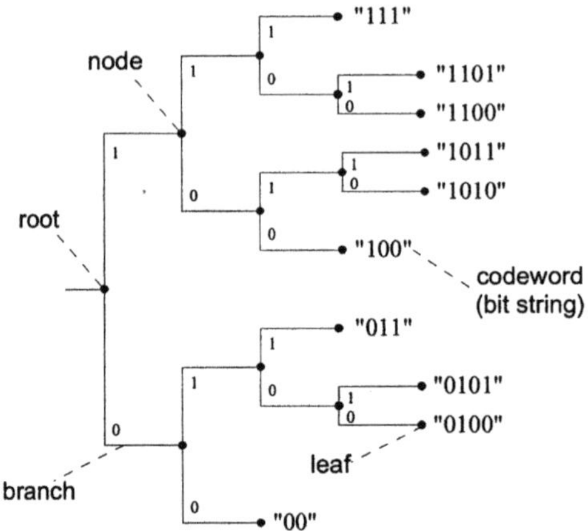

Fig. 11.10. Code tree representing a variable-length code of 10 symbols

---

[1] If parsing of an entire *stream of codewords* is regarded, the leaf terminating one codeword establishes the root of the subsequent codeword.

For a given set of binary codeword (bit string) lengths z_j, it is always possible to construct a valid prefix code if *Kraft's inequality* applies [COVER, THOMAS 1991][1]

$$\sum_{j=1}^{J} 2^{-z_j} = c \leq 1. \tag{11.28}$$

For other uniquely-decodable codes, which do not belong to the class of prefix codes, decoding must impose additional constraints (e.g. forbid certain codewords if they are not part of a clearly-defined sequence of codewords; disallow certain codewords at the beginning or at the end of a stream etc.).

The expected mean rate by using a variable length code to encode a source alphabet $\mathcal{A}=\{a_j; j=1, \dots, J\}$ is

$$R = E\{z_j\} = \sum_{j=1}^{J} P(j) \cdot z_j \geq H(\mathcal{A}) = -\sum_{j=1}^{J} P(j) \cdot \log_2 P(j). \tag{11.29}$$

The self information $i(j) = -\log_2 P(j)$ is in general *not an integer number*, while bits cannot exist by fractional units, i.e. z_j *must* be an integer number. This means that only for the specific case where all $P(j)=2^{-m}$ (integer powers of two), the rate of entropy (3.85) can exactly be approached. A simple design for a code of variable length is the *Shannon Code*, where

$$z_j = \begin{cases} \text{int}[i(j)] & \text{if } P(j)=2^{-m} \\ \text{int}[i(j)]+1 & \text{else.} \end{cases} \tag{11.30}$$

A Shannon code can by guarantee approach the entropy by a margin[2]:

$$H(\mathcal{A}) \leq R < H(\mathcal{A})+1 \tag{11.31}$$

A very effective method to improve the performance of a variable-length code is combination of several source symbols (letters) into vectors. From (11.31), in the most inconvenient case of z_j value configurations, the rate could be almost one bit above the entropy. In case of vector combinations, this penalty is spread over all K source symbols which are combined for encoding. Then, a Shannon-type prefix code exists by guarantee which has a rate per symbol (letter)

$$\frac{1}{K} \cdot H_K(\mathcal{A}) \leq R < \frac{1}{K} \cdot H_K(\mathcal{A}) + \frac{1}{K} \leq H(\mathcal{A}) + \frac{1}{K}, \tag{11.32}$$

---

[1] In (11.28), condition $c=1$ holds true, if the code tree is fully occupied, i.e. no 'open leaves' remain in the code tree. If $c<1$, strings exist that are not used to represent letters from the alphabet; codes are indeed sometimes constructed such that additional strings may be reserved for re-synchronization or other purposes.

[2] (11.31) is indeed a guaranteed upper bound for any method of entropy coding which tries to approximate the self information of source symbols by the binary length of code symbols as close as possible.

where $H_K(\mathcal{A})$ denotes the entropies of vectors $\mathbf{a}=[a(1)\ a(2)\ ...\ a(K)]^{\mathrm{T}}$, assembled from K source symbols of values from the alphabet $\mathcal{A}$. The equality in the right-most equation of (11.32) applies for the case where the source symbols $a(k)$ within $\mathbf{a}$ are statistically independent. Then, following (3.24),

$$P_K(\mathbf{a}) = P\left[a(1)=a_{j_1}\right] \cdot P\left[a(2)=a_{j_2}\right] \cdot ... \cdot P\left[a(K)=a_{j_K}\right] \quad \text{with} \quad a_{j_k} \in \mathcal{A} \quad (11.33)$$

and $H_K(\mathcal{A})=K{\cdot}H(\mathcal{A})$. In any other case, the entropy of the vector will be *lower than* the K-fold entropy of single source symbols, as joint statistical properties can be implicitly exploited by the design of the code. Hence, in any case, the entropy of the source can be approached by a margin $\varepsilon=1/K$, in case $K{\to}\infty$ it is possible to get arbitrary close, which is exactly the message of Shannon's source coding theorem. The real problem is the complexity of such codes (in terms of code tables to be managed, decisions to be made etc.), which grows *exponentially* by the length of the vectors. Hence, it can be concluded that Shannon codes are an impractical solution. The real challenge in entropy coding will be to find coding mechanisms, which are able to approach the rate of the entropy closely, while still keeping the complexity low. Such schemes should also be capable to utilize statistical dependencies between source samples to be encoded, and should be adaptable to arbitrary source statistics.

(11.29) can be re-written as

$$R = -\sum_{j=1}^{J} P(j)\cdot\log_2 2^{-z_j} = -\sum_{j=1}^{J} P(j)\cdot\log_2\left[c\cdot R(j)\right] \quad ; \quad R(j)=\frac{2^{-z_j}}{c}, \quad (11.34)$$

where the constant c shall be as defined in (11.28). Now, the difference between the rate that a variable-length code produces and the lower bound of entropy can be expressed as

$$R-H(\mathcal{A}) = -\sum_{j=1}^{J} P(j)\cdot\log_2\left[c\cdot R(j)\right] + \sum_{j=1}^{J} P(j)\cdot\log_2 P(j)$$

$$= \sum_{j=1}^{J} P(j)\cdot\log_2\frac{P(j)}{R(j)} + \log_2\frac{1}{c}. \quad (11.35)$$

For codes which fulfill the upper bound of Kraft's inequality (11.28), the right term will be zero, and the left term is the *relative entropy* (9.21). In fact, this code would be optimum for a source with probability distribution $R(j)$, and the differential entropy exactly expresses the rate penalty which has to be paid if this is not the case.

11.4.2 Huffman Codes

Huffman codes [HUFFMAN 1952] are one of the most widely used design methods for variable-length coding. The Huffman code design is more efficient than the Shannon code, as it allows to choose some of the symbol lengths as $z_j<i(j)$, while retain-

ing the property of a decodable prefix code[1]. The principle of Huffman codes can best be understood by the following code design procedure, for which it is only necessary to know the probabilities of the symbols a_j in the source alphabet $\mathcal{A}$:

1. Construct a list $\mathcal{L}$ of probabilities $P(a_1),P(a_2),...,P(a_J)$. Associated with each item in the list are indices j of the source symbols, where initially each list field relates to exactly one source symbol. These indices further point to the bit strings $\{c_j; j=1,...J\}$ of the code $\mathcal{C}$. Initially, the bit strings consist of zero bits each.

2. In $\mathcal{L}$, search for the two list fields of smallest probabilities. Extend all bit strings associated with these two fields; all strings related to the first field are e.g. extended by a '0'-bit, all associated with the second field by a '1'-bit. The code grows 'from the tail', i.e. the first bit allocated will become the terminating bit of the actual code string; expressed in terms of a code tree, this means that the code design starts at the leaves and propagates towards the root.

3. Within the list $\mathcal{L}$, the two fields processed before are removed. The sum of their probabilities is associated with a new list field. This new field is further associated with all source symbol indices and bit strings which were previously associated with the two deleted fields.

4. If $\mathcal{L}$ contains only one field, the code is readily designed. Otherwise, the procedure iterates over steps 2-4 again.

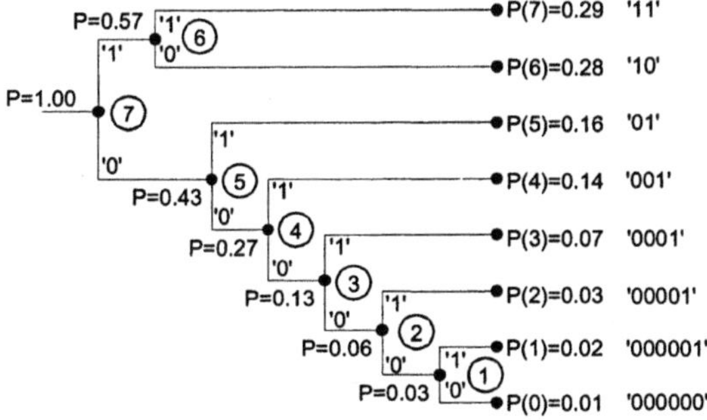

Fig. 11.11. Construction of a Huffman code tree by a total of 7 steps

Fig. 11.11 shows the design of a Huffman code with an example of 8 source symbols of different probabilities, which can best be interpreted by steps of a code tree construction. Each node in the tree is marked by the probability value as stored in the list, which determines the decision about the next two paths to be extended. In this example, 7 iterations are necessary. While the design starts from the *leaves of the tree*, decoding must start from the root. Any step produces a fully populated code tree, such that by guarantee $c=1$ in (11.28). In principle, the upper bounds

---

[1] The Shannon design often leads to bit allocations where (11.28) gives a value $c<1$, which indicates that certain branches of the tree would not be occupied (see Problem 11.2). An additional penalty occurs which can e.g. be characterized by (11.35)

(11.31) and (11.32) still apply, however it can be shown that the upper bound is never met. The following approximation gives an estimate of the penalty over the entropy rate [COVER, THOMAS 1991]:

$$R < \frac{1}{K}\left(H_K(\mathcal{A}) + P_{max}\right) \qquad \text{if } P_{max} < 0.5$$
$$R < \frac{1}{K}\left(H_K(\mathcal{A}) + P_{max} + 0.086\right) \text{ if } P_{max} \geq 0.5 \qquad ; \; P_{max} = \max\left[P_K(\mathbf{a})\right] \quad (11.36)$$

The number of code symbols in $\mathcal{C}$ and hence the complexity grows exponentially with K. The code table must typically have the size JK. The entropy is better approached even for lower K than it is the case by the Shannon code. The usage of vectors however affects the complexity of

- the design process, where the list $\mathcal{L}$ becomes rather very large;
- the coding process, where it is necessary to use a large memory containing all different code strings;
- the decoding process, where not only a large memory is required, but also the parsing for valid code strings becomes increasingly complex due to the irregularity of the code tree.

11.4.3   Systematic Variable-length Codes

Of particular concern in the case of Huffman codes are long code strings which must be provided for untypical source symbols or source symbol vectors of extremely low probability. From the complexity point of view, it may be preferable to not consider these cases at all in the design of VLCs. Alternatively, an upper bound can be imposed to the length of the codeword strings. Some methods following this paradigm are described in this and the subsequent sections, such that both good performance of the code and moderate complexity can be compromised. Besides a reduction of the code table size, the decoding complexity can also be reduced by introducing more *regularity* in the code design. For example, subsets of symbols can be identified which have common prefixes to allow simplified parsing of the code string. Another aspect is the *universality* of the code design, which means that the same code can be adapted to sources of different probability distributions. This will in particular be important when adaptive entropy coding shall be used; it is however also favorable in terms of complexity, when the same decoding unit (hardware or software) is able to decode different source components, and is not 'hard-wired' to parse only one specific VLC.

Escape codes. A common solution to reduce the complexity in terms of code table size is made by *not assigning* variable-length codes to source symbols which are *untypical* (expected to occur with a very low probability). In *escape codes*, a certain VLC symbol is reserved which indicates that one of the untypical source symbols is expressed by a *fixed-length code suffix*. If J' untypical symbols are identified, the number of bits to be attached to the escape symbol is $R'=\log_2 J'$ if J' is a power-of-two number, $R'=\log_2 J'+1$ else. The escape code can still be expressed by a code tree, where a regular binary tree of depth R' unfolds beyond the path leading to the escape symbol. The rate penalty to be paid is usually rather low, depending on the total percentage of untypical source symbols. The VLC length for the escape codeword is determined by adding all probabilities of the untypical symbols.

Table 11.1. VLC table for encoding of 'AC' coefficients and separate encoding of exceptional RUN/LEVEL combinations not contained in the VLC table (excerpt from the MPEG-1 standard)

1. VLCs for run length and quantizer index, 's'=sign

*) differently defined for first and last coefficient
) largest LEVEL value in the table   *) largest RUN value in the table

RUN	LEVEL	code	RUN	LEVEL	code
EOB	--	10	9	1	0000101s
0	1	1s / 11s *)	0	5	00100110s
1	1	011s	...	...	...
0	2	0100s	13	1	00100000s
2	1	0101s	0	7	0000001010s
0	3	00101s	...	...	...
3	1	01111s	21	1	000000010110s
4	1	00110s	0	12	0000000011010s
1	2	000110s	...	...	...
5	1	000111s	0	40**)	000000000010000s
...	...	...	...	...	...
ESCAPE	--	000001	31***)	1	0000000000011011s

2. Fixed-length code, separate coding of run-length and quantizer level index. Invoked by ESCAPE codeword: 6 bit express run length, 8 bit for |index|<128, 16 bit for |index|<256.

RUN	code	LEVEL	code
0	000000	-255	1000000000000001
1	000001	...	...
2	000010	-128	1000000010000000
...	...	-127	10000001
...	...	...	...
...	...	127	01111111
...	...	128	0000000010000000
...	...	...	...
63	111111	255	0000000011111111

As an example for such a strategy, an excerpt of a VLC table from the MPEG-1 standard is shown in Table 11.1. This is a so-called '2D VLC', which combines source symbols RUN and LEVEL into a vector (cf. sec. 12.4.1). Combinations of short RUN and small LEVEL are expected to occur most frequently due to the statistical properties of transform coefficients and are represented by the shortest codewords. In case of ESCAPE (6 bit code), the RUN and LEVEL symbols are encoded independently, one by a fixed-length code, the other by a primitive VLC supporting only two different codeword lengths. The length of codewords in the case of ESCAPE is 20 or 28.

Universal variable-length codes. Source models of Gaussian distribution and its generalized extension (3.20) shows an exponential decay of probabilities towards higher amplitude levels. Markov chains are typical models for correlated binary sources, and also have an exponential decay of probabilities for increasing length of same-value runs (3.78). Optimum VLCs for such sources will typically show an increasing VLC length towards source symbols of higher values. According to (3.10), the probability of a quantized source symbol is equivalent to the area under the continuous PDF within the quantization interval. The optimum bit allocation in a variable length code usually approximates the self information as good as possible, which means that one bit more shall be assigned to a source symbol of roughly half probability compared to another one. In exponential distributions, the tangential slope of the PDF decreases towards higher amplitudes. Hence, it can be expected that the number of adjacent quantized source symbols which have to be encoded by using the same number of bits in the VLC codewords increases exponentially towards higher amplitude levels when uniform quantization is used (see Fig. 11.12).

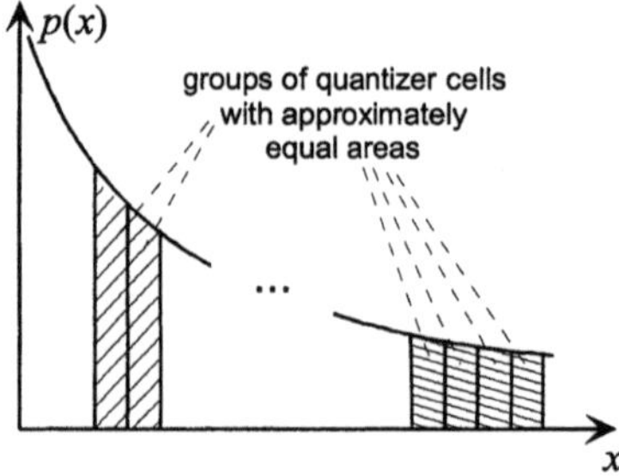

Fig. 11.12. Groups of quantizer cells which will be assigned an equal numbers of bits in a VLC design for the case of exponentially decaying source distribution and uniform quantization

Codes which are appropriate to encode sources of increasing code-length property by systematic codeword assignment are the *Golomb-Rice codes* (GR) and the *Exponential Golomb codes* (EG). While the numbers of codewords having the same length is constant for GR codes, it is exponentially increasing with the codeword length for EG codes. To comply with the phenomenon illustrated in Fig. 11.12, the EG codes are the better choice. An example of a *Universal VLC* (UVLC) constructed on the basis of an EG code is shown in Fig. 11.13. This code is highly regular. For any possible codeword length, a prefix 1, 01, 001 etc. expresses the number of bits b_i to follow in the codeword. The prefixes are however interleaved with the b_i, which has the advantage that most probably two subsequent

'1' bits (when appearing in the code stream) indicate the end of a codeword. This way the code provides an additional capability of detection and eventual recovery from transmission errors.

		Source symbol	Codeword
		0	1
	1	1	001
		2	011
a) $\quad$ 0 $\ b_0$ 1 $\qquad$ b)		3	00001
		4	00011
0 $\ b_1$ 0 $\ b_0$ 1		5	01001
		6	01011
0 $\ b_2$ 0 $\ b_1$ 0 $\ b_0$ 1		7	0000001
		8	0000011
0 $\ b_3$ 0 $\ b_2$ 0 $\ b_1$ 0 $\ b_0$ 1		9	0001001
		10	0001011
...	...	...	...

Fig. 11.13. Universal VLC construction based on an EG code design **a** Construction schema **b** Code table

Table 11.2. Construction of RVLC from GR and EG codes

Source symbol	Reversible GR code		Reversible EG code	
	Prefix	Suffix	Prefix	Suffix
0	0	00	0	0
1	0	01	0	1
2	0	10	101	0
3	0	11	101	1
4	11	00	111	0
5	11	01	111	1
6	11	10	10001	0
7	11	11	10001	1
8	101	00	10011	0
9	101	01	10011	1
10	101	10	11001	0
11	101	11	11001	1
12	1001	00	11011	0
13	...	...	11011	1
...	...	...	...	...

Reversible variable-length codes. Variable-length codes as discussed so far are typically prefix codes. In decoding, the string has to be parsed in a unique direction, corresponding to the path in the code tree which originates from the root. If a bit error occurs, a wrong path may be selected which has a different length, such that the starting bit of the next codeword will be misinterpreted, and also subsequent decisions can be affected. To overcome this problem, *Reversible Variable-Length Codes* (RVLC) were developed, which can also be parsed in reverse direction [TAKASHIMA, WADA, MURAKAMI 1995][WEN, VILLASENOR 1997]. In addition to the prefix condition, a *suffix condition* is formulated, which says that no codeword's suffix shall coincide with any longer codeword's suffix, such that the code can instantaneously be decoded by reverse order. RVLC design methodologies are known for Huffman, GR and EG codes, where the latter two have the most efficient solutions.

A systematic construction can be made for codewords of length z_j which are partitioned into a prefix of length z_j-m and an m-bit suffix. For GR codes the suffix condition is fulfilled, if the prefix starts and ends by a '1', all other bits of the prefix being '0'. For EG codes, this condition can be formulated such that the prefix shall also start and end by a '1', imposing the additional zero constraint only to the odd-indexed prefix bits. A prefix of length one shall always be '0'[1]. Bit strings can be found which use GR or EG codewords as prefixes fulfilling these conditions. Examples of a GR code with $m=2$ and an EG code with $m=1$ are shown in Table 11.2[2].

Application of RVLC (as other systematic code designs) imposes a penalty on efficiency due to the constraints imposed on the codeword length allocations, which can numerically be expressed by (11.35). More recently, an RVLC design was proposed which can largely close this gap [GIROD 1999].

11.4.4   Arithmetic Coding

Arithmetic coding is a method of entropy coding which has found wide acceptance in the newest generations of standards for multimedia signal compression. In contrast to Huffman or systematic VLCs, it can no longer uniquely be identified how bits from a bit stream relate to specific source symbols; arithmetic decoding can rather be interpreted as a *sliding window* method, where encoder and decoder are in a well-defined state at each step, depending on past source symbols. The decoder reconstructs a source symbol by combining the information carried by bits newly arriving with the decision rules valid by the present state. Decoded source symbols are output whenever this is uniquely possible by the bits received. The notion *arithmetic coding* is related to the fact that the underlying computation must be performed by a high arithmetic precision, as otherwise the performance will suffer.

The simplest type of an arithmetic code is the *Elias code* [ELIAS 1975]. It shall be presented here to explain the case of arithmetic compression for an unequally distributed binary source producing symbols 'A' and 'B' without correlation between subsequent source outputs.

P_A and $P_B=1-P_A$ shall be the probabilities of the two source symbols. For sequences 'AA', 'BB', 'BA' and 'AB', the respective probabilities then are P_A^2, $(1-P_A)^2$, $(1-P_A)\cdot P_A$ and $P_A\cdot(1-P_A)$. To perform the code mapping, *probability intervals* of widths equal to the probabilities of the associated source sequences are constructed. Initially (at the 'root' of the code), only one interval $I_{N=0}=[0,1]$ of width 1 exists, which means that nothing is known about the state of the source yet. This interval is then sub-divided into decreasingly smaller intervals, each expressing by its width the probability of a possible source symbol sequence. For exam-

---

[1] The suffix condition is of course also fulfilled, if the role of '0' and '1' bits is interchanged.

[2] The prefixes of the RVLC are complete EG codewords. To fulfill the suffix condition, the code pyramid of Fig. 11.13 must be modified as '0', '$1b_0 1$', '$1b_1 0b_0 1$', '$1\,b_2 0b_1 0b_0 1$' etc.

ple, if the first source symbol is q_1='A', the associated interval is $I_{N=1}=[0,P_A]$, while for the source symbol q_1='B', the interval is $I_{N=1}=[1-P_B,1]=[P_A,1]$. If e.g. the first source symbol is q_1='B' and the second is q_2='A', the next step of sub-division will be an interval $I_{N=2}=[P_A,P_A+P_B\cdot P_A]=[P_A,P_A+(1-P_A)\cdot P_A]$. In general, the interval which is associated with a sequence of N source symbols $Q=[q_1,q_2,q_3, ..., q_N]$ has lower and upper limits

$$I_N = \left[W_1\cdot r_1 + W_2\cdot r_2 + ... + W_N\cdot r_N, \right.$$

$$\left. W_1\cdot r_1 + W_2\cdot r_2 + ... + W_N\cdot r_N + \frac{W_N}{P_A}\cdot\left(P_A\cdot(1-r_N)+r_N\cdot(1-P_A)\right)\right] \qquad (11.37)$$

$$= \left[\sum_{n=1}^{N}W_n\cdot r_n, \sum_{n=1}^{N}W_n\cdot r_n + \underbrace{\frac{W_N}{P_A}\cdot\left[P_A\cdot(1-r_N)+r_N\cdot(1-P_A)\right]}_{\text{width of the interval}}\right]$$

with

$$W_n = \begin{cases} P_A & \text{for } n=1 \\ P_A\cdot\prod_{v=1}^{n-1}P_A\cdot(1-r_v)+r_v\cdot(1-P_A) & \text{else} \end{cases}$$

$$r_v = \begin{cases} 0 & \text{for } q_v = 'A' \\ 1 & \text{for } q_v = 'B' \end{cases} \qquad (11.38)$$

The factor W_n continuously decreases the interval widths by increasing n, according to the probabilities of previous source symbols. The last term in the interval's upper limit expression is the actual width of the interval, related to the probability of the entire sequence. The central mechanism in *arithmetic encoding* is comparison of the *non-uniform grid* of probability intervals against a *uniform grid* of *code intervals* I_M. The grid of code intervals associated with a possible code string sequence $B=[b_1,b_2,b_3, ..., b_M]$ of length M is defined as

$$I_M = \left[0.5\cdot b_1 + 0.25\cdot b_2 + ... + 2^{-M}b_M , \; 0.5\cdot b_1 + 0.25\cdot b_2 + ... + 2^{-M}b_M + 2^{-M}\right]$$

$$= \left[\sum_{m=1}^{M}2^{-m}\cdot b_m, \sum_{m=1}^{M}2^{-m}\cdot b_m + 2^{-M}\right]. \qquad (11.39)$$

The rule of encoding is to release all bits of this sequence for transmission, whenever an interval I_N is completely contained in I_M, i.e. if the following conditions hold true; $I=[I_{lo},I_{up}]$ representing intervals by their lower and upper boundaries:

$$I_{N,\text{lo}} \geq I_{M,\text{lo}} \quad \text{and} \quad I_{N,\text{up}} \leq I_{M,\text{up}} . \qquad (11.40)$$

As the width of interval I_N is identical with the probability of the source symbol sequence of length N, the associated self information can be expressed as

$$i_N = -\log_2\left(I_{N,\text{up}} - I_{N,\text{lo}}\right) \geq -\log_2\left(I_{M,\text{up}} - I_{M,\text{lo}}\right) = -\log_2 2^{-M} = M \ . \qquad (11.41)$$

From (11.40), the code interval I_M is always larger or equal as compared to the probability interval I_N. From (11.41), the number of bits sent up to this step is always less than or equal to the self information of the source sequence. Hence,

- The number of bits sent is smaller or equal to the accumulated self information of the source symbols;
- The sequence of source symbols can not yet fully be decoded from the bits sent so far, unless the probability intervals perfectly match with the code intervals, in which case the number of bits sent equals the self information.

A unique decoding of the source sequence associated with interval I_M is only possible, if the number of bits received is at least equal to the self information of the source sequence to be decoded,

$$-\log_2\left(I_{M,\text{up}} - I_{M,\text{lo}}\right) + \log_2\left(I_{N,\text{up}} - I_{N,\text{lo}}\right) \geq 0 \ . \qquad (11.42)$$

In principle, decoding is performed by the reverse principle of encoding. It is tested whether the interval I_M, associated with the received sequence of code bits, fully fits into an interval I_N associated with a source sequence:

$$I_{N,\text{lo}} \leq I_{M,\text{lo}} \quad \text{and} \quad I_{N,\text{up}} \geq I_{M,\text{up}} \ . \qquad (11.43)$$

If the sequence of source symbols grows sufficiently long, the intervals will get arbitrarily small, such that the entropy can also be approached arbitrarily close. This however is the constraint of arithmetic coding: As computation of the interval boundaries can only be performed by finite arithmetic precision, it is not possible to approach the entropy perfectly. However, as will be discussed below, even for case of finite arithmetic precision the margin over the entropy rate will be very small.

Fig. 11.14 graphically shows the grids of probability intervals and code intervals for the case $P_A=1/4$. As an example, regard two sequences of source symbols, which are actually the cases of most rare and most probable occurrence among all sequences of length $N=3$:

- Sequence 'AAA' : For q_1='A', the probability interval $I_{N=1}=[0,1/4]$ falls exactly into the code interval $I_{M=2}=[0,1/4]$. Bits $b_1=0$ and $b_2=0$ are sent immediately. Likewise, $I_{N=2}=[0,1/16]$ falls exactly into $I_{M=4}=[0,1/16]$ and $I_{N=3}=[0,1/64]$ falls exactly into $I_{M=6}=[0,1/64]$, which means that in total the code bit sequence '000000' is sent up to this point. This is the special case where both encoding and decoding can be performed instantaneously, which is caused by the fact that probability and code intervals are exactly equal. Here, the probability $P_A=1/4$ is an integer power of 2, for which also Huffman and Shannon codes give optimum performance.
- Sequence 'BBB' : Neither $I_{N=1}=[1/4,1]$ nor $I_{N=2}=[7/16,1]$ fall entirely into any code interval, both expressing self information of less than one bit. The next

sub-division $I_{N=3}=[37/64,1]$ is contained in $I_{M=1}=[1/2,1]$. With q_3='B', b_1='1' has to be sent. This intuitively shows that a sequence of higher probability causes significantly less bits to be sent; however receiving b_1='1', it will only be possible to decode the first two source symbols q_1='B' and q_2='B': $I_{M=1}=[1/2,1]$ falls entirely into probability intervals $I_{N=1}=[1/4,1]$ and $I_{N=2}=[7/16,1]$, but the third source symbol can not be decoded yet. The partial information about q_3 which was carried by b_1 is however retained in the state of the decoder, such that it can uniquely be decoded when future bits are received.

This example illustrates that in arithmetic coding a *variable number of source symbols* is mapped into a *variable number of code bits*. It can be observed that untypical sequences of source symbols, such as long chains of the less probable letter 'A', are encoded almost instantaneously without imposing a critical burden on the complexity, as it is not necessary to parse code trees of large depth. It is further interesting to note that due to the uniform code interval raster, the probabilities of '0' and '1' code bits will be equal. The code bits will not have any statistical dependency, such that further entropy coding of the code bit sequence would be useless. Any efficient entropy coder would show this property.

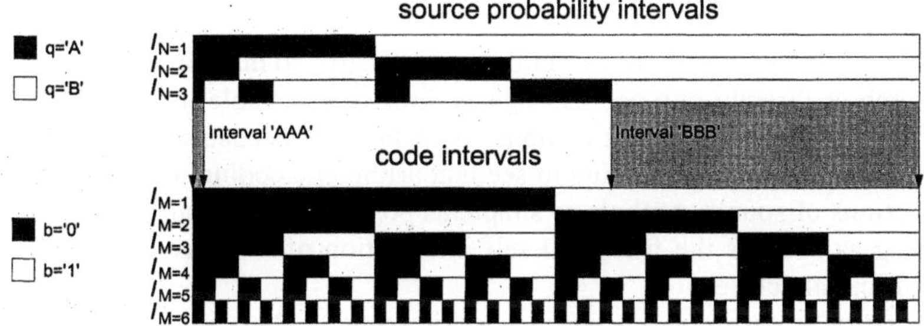

Fig. 11.14. Code intervals and probability intervals in the example discussed above

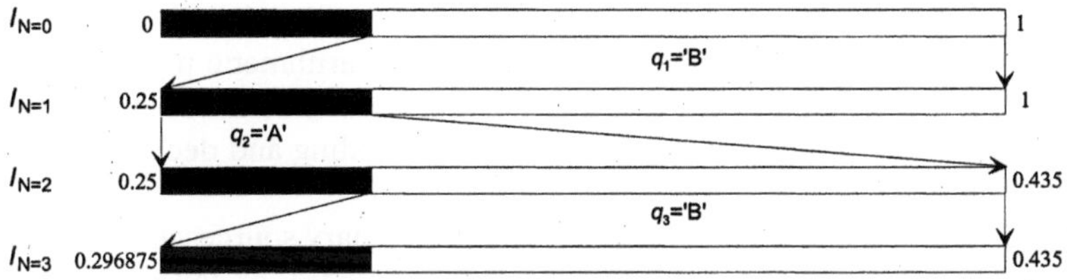

Fig. 11.15. Expansion of probability intervals into the full range between 0 and 1 after each encoding step; the source sequence is 'BAB'

In practical realizations, neither the encoder nor the decoder need to construct the complete set of intervals; it is only necessary to track the path corresponding to the source symbols arriving at the encoder and of the associated bit sequence received by the decoder. An interpretation is that after each successful encoding or decoding step, the recently-active probability interval is expanded to the full scale [0,1] (see Fig. 11.15). This also implies that the factor W_n in (11.37) must not explicitly be computed.

The interval computations must actually be performed by integer arithmetic; if an integer word length of B bits is used, the probability interval boundaries can be approximated by an accuracy of 2^{-B}. The step of interval expansion described above can in fact implicitly be combined with the integer rounding of probability intervals. It is further not even necessary to compute the code intervals at all, because the integer-number representation inherently carries the information about the code interval mapping[1]. After a probability interval corresponding to a source symbol has been computed by an integer representation of its lower and upper boundaries, all leading bits that are equal in representation of both boundaries can be sent as code bits. As an example, with a 6 bit positive integer representation, let the lower boundary of the current probability interval be represented by an integer number '010000', the upper boundary by '011100'. Then, the leading bit string '01' can be sent, the interval is expanded into full six bits by left-shifting the remaining bits, such that the lower boundary becomes '000000', the upper boundary '110000'. The decoder receives the bits by the sequence they are sent, shifts them in from the left, and compares the leading bits continuously with the boundaries of probability intervals, where the same interval expansion is performed each time a source symbol can be decoded.

Due to the rounding, it even turns out that it is possible to map finite sequences of source symbols directly into code strings as directly decodable units (cf. Problem 11.4). Even though this direct mapping is in fact not needed to perform the encoding or decoding, it is interesting to see that arithmetic coding handles untypical combinations of source symbols as simple as possible by releasing the associated bit string as soon as the fractional self information of a source event is approached sufficiently close.

If computation is performed using integer arithmetic of B bit precision, the entropy can by guarantee be approached within a margin of $\log_2 2^{-B+2}$ bit [PENNEBAKER ET AL. 1988]. The 'loss' of two bits as compared to the smallest integer step size 2^{-B} can be explained due to the facts that rounding errors occur at both lower and upper interval boundaries[2]. Nevertheless, using sufficient arithmetic precision, the entropy is approached by a much smaller margin than usually achievable by Huffman or systematic code designs, while the arithmetic encoding and decoding processes are still quite regular.

The method of arithmetic coding is not restricted to binary sources as discussed so far. When a source alphabet of an arbitrary finite number J of source symbols shall be encoded, the binary code intervals remain identical, whereas for a source

---

[1] For more detail on efficient implementation of arithmetic encoders and decoders, refer to [WITTEN, MCNEAL, CLEARY 1987] and [PENNEBAKER ET AL. 1988].

[2] This seems indeed to be a strong worst case assumption. Regarding the rounding inaccuracy of the algorithm described above, and by statistical expectation of equally distributed rounding errors, a bound of $R \geq H + \log_2 2^{-B+1}$ bit appears more realistic.

alphabet of J symbols each step of sub-division will generate J new probability intervals. Fig. 11.16 shows an example of a three-letter alphabet with probabilities $P("A")=0.4$, $P("B")=P("C")=0.3$.

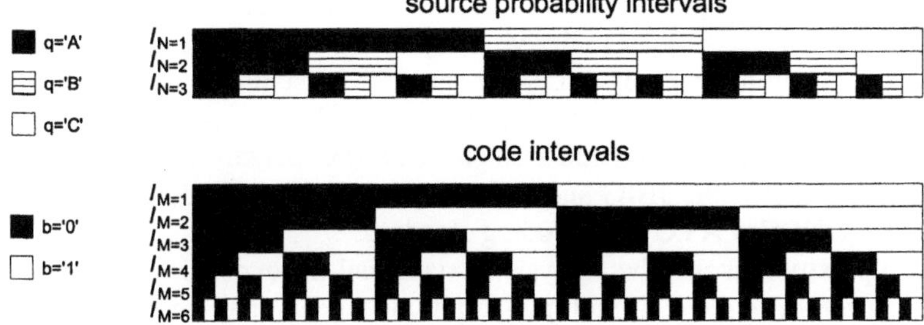

Fig. 11.16. Example of probability and code interval grids for a source with 3 different symbols

Unlike Huffman coding or other types of prefix-based variable length codes, it is not necessary to use a code table. Moreover, the probability values are *directly used* to run the procedures of interval subdivision in encoding and decoding. This largely simplifies usage of *dynamically changing* probability values, which means that arithmetic encoding and decoding devices are *universally applicable* to any source statistics and alphabets, and *easily adaptable* to changing source statistics.

11.4.5   Context-dependent Entropy Coding

In all algorithms for entropy coding introduced so far, first-order probabilities, joint or conditional probabilities can be used to design the codes. If conditional probabilities are used, it is possible to achieve a more efficient coding of sources in which subsequent symbols or subsequent vectors of symbols are *statistically dependent*. This even applies if they are not encoded as joint units. Assume that a *context* $\mathcal{K}$ is established from previously decoded source symbols. In the simplest case, this context could be the state of the directly preceding source symbol, which is then used to pre-determine the probabilities of values in the present state. The goal is then to approach the *conditional entropy* of source alphabet $\mathcal{A}$ relating to the known context $\mathcal{K}$, which gives according to (3.91)

$$H(\mathcal{A}\,|\,\mathcal{K}) = H(\mathcal{A}) - I(\mathcal{A};\mathcal{K}) \leq H(\mathcal{A}).\tag{11.44}$$

Realization of context-dependent entropy coding can nevertheless be of considerable complexity. The number of conditional probability values that have to be considered is increasing exponentially by the size of the context $\mathcal{K}$. In particular for Huffman coding, this may be difficult to realize, as then a huge number of code tables would have to be managed and stored. In case of arithmetic coding, only tables of conditional probabilities have to be provided which are then used for sub-division of the probability intervals. The concepts and implementations of coding and decoding procedures are not changed in principle.

The conditional probabilities to be used for context-dependent entropy coding could be based on linear dependencies[1], or can also relate to nonlinear dependencies. The latter category is difficult to derive analytically or on the basis of models, but can be captured by analysis of a training set or from statistics of previously-decoded samples of the current signal. Context-dependent entropy coding has become more widely used recently, and is in particular extensively exploited in the arithmetic coding parts of the JBIG and JPEG 2000 still image coding standards, and in the H.264/MPEG-4 AVC video coding standard.

Due to complexity reasons, the analysis of statistical dependencies is often reduced to *binary context models*. If multiple-amplitude signals shall be encoded, the first step to be performed is a *binarization*, which is often done by one of the following approaches:

- Integer representation of a signal $x(\mathbf{n})$ is separated into sign and magnitude as $|x(\mathbf{n})|=b_0(\mathbf{n}){\cdot}2^0+b_1(\mathbf{n}){\cdot}2^1+...+ b_{B-1}(\mathbf{n}){\cdot}2^{B-1}$, $x(\mathbf{n})=|x(\mathbf{n})|{\cdot}\mathrm{sgn}(\mathbf{n})$. The B bit-plane values $b_i(\mathbf{n})$ establish a binary representation (cf. sec. 11.3). The binary context can then be defined by values from previously decoded values of the next-higher bit plane at or around the current location. This method is e.g. applied in the EBCOT algorithm of the JPEG 2000 standard (cf. sec. 17.3.1).
- Representation by a simple VLC which systematically leaves statistical dependencies after the mapping of source symbols into the binary codewords. An example for this is the universal VLC method described in sec. 11.4.3. By defining contexts either within or across binary codewords, a context-dependent arithmetic coder is able to further compress such representations significantly. *Context-adaptive Binary Arithmetic Coding* (CABAC) as implemented in the MPEG-4 AVC standard [MARPE ET AL. 2003] additionally uses a selectable initialization and adaptation of contexts, such that the scheme can universally be used for different source components.

The binary representation can be modeled as a Markov source of vector context as described in (3.81)-(3.82). For a binary context definition consisting of K previously decoded bits, the specification of 2^K transition probabilities is necessary. As the arithmetic decoding process can be interpreted as a finite-state machine anyway, the inclusion of state-dependent context probabilities is straightforward.

11.4.6 Adaptive Entropy Coding

In the previously described methods of entropy coding, the probability distribution of the source was assumed to be known a priori and should not be subject to

---

[1] For example, autocovariance values estimated from a signal can be used with a vector Gaussian distribution to determine conditional probabilities (3.29) of adjacent samples or in a vector context.

change during encoding. For the case of stationary sources, this allows the design of an efficient code approaching the entropy rate. As multimedia signals include instationary components, this can lead to sub-optimum solutions. In the worst case, it may happen that the rate is considerably increased by a badly designed variable-length code which does not fit the statistics of the source. This problem can be solved by *adaptive entropy coding*. The core approach for compression is not different from the methods described so far, but the code tables or probability parameters have to be adapted using the actual source characteristics. This can be done by one of the following principles:

- *Forward adaptation* (Fig. 11.17a) : For a given segment of the source, which can be a window over a certain number of values or an entire signal like an image, occurrences are measured and used for code adaptation. Either the probability maps resulting from the measurements, or the related code parameters must separately be conveyed to the decoder as side information, which means they are part of the encoded representation. The disadvantages of this method are two-fold: The source signal must be stored temporarily to determine the occurrence values, which causes a delay; the rate is increased, as the parameters for code adaptation have to be transmitted separately.
- *Backward adaptation* (Fig. 11.17b) : As entropy coding is lossless, both the encoder and the decoder can measure the occurrences of source symbols over a defined window *from the past*. If the signal does not show too much variation of statistical behavior, this can well reflect the current statistics and be used for adaptation of the entropy coder and decoder. A disadvantage occurs in the case of transmission errors, where encoder and decoder might run into states where they use different parameters for adaptation, which will typically lead to uncontrollable errors during decoding. On the other hand, in contrast to the forward-adaptive approach, no transmission of side information is necessary.

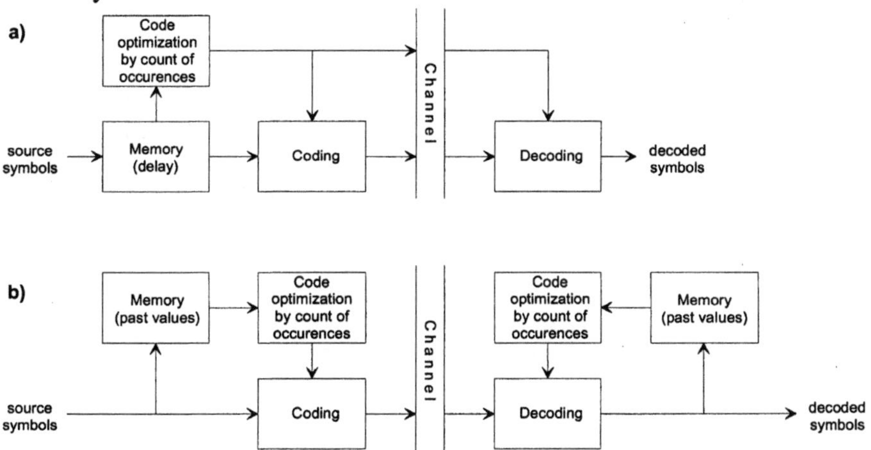

Fig. 11.17. Adaptive Entropy coding. **a** Forward adaptation **b** Backward adaptation

A method for probability estimation. Probabilities must typically be estimated by counting the occurrences of the different events. In the case of binary context-related coding decisions, the probability $P(b(\mathbf{n})=0|\mathbf{b})=1-P(b(\mathbf{n})=1|\mathbf{b})$ must be estimated for each of 2^K different contexts. Assume that $C_{0|\mathbf{b}}$ and $C_{1|\mathbf{b}}$ are the zero- and one-counts found for the conditioning context $\mathbf{b}$, either analyzed in the window of forward estimation, or over a past window in case of backward estimation. The following *scaled-count estimate* can then be made:

$$\hat{P}(b=0|\mathbf{b}) = 1 - \hat{P}(b=1|\mathbf{b}) = \frac{C_{0|\mathbf{b}} + \Delta \cdot P'_{0|\mathbf{b}}}{C_{0|\mathbf{b}} + C_{1|\mathbf{b}} + \Delta \cdot \left(P'_{0|\mathbf{b}} + P'_{1|\mathbf{b}}\right)} \quad ; \quad \Delta > 0. \qquad (11.45)$$

The values $P'_{0|\mathbf{b}}$ and $P'_{1|\mathbf{b}}$ effect an initialization of the probability estimate, which will give a reasonable result even when no single instance of a certain value $\mathbf{b}$ was found previously. These are initial guesses about the probabilities of zero- and one-values. The factor Δ plays a role regarding the speed of the adaptation. If Δ is set to a higher value, the adaptation will only take significant effect when a certain number of counts is found; otherwise the probability estimate will be roughly retained by the initial guess.

For context-independent encoding of an alphabet having J different source symbols, J probabilities $P(i)$ must be estimated. The same method is applicable and extends as follows:

$$\hat{P}(i) \equiv \hat{P}(a=a_i) = \frac{C_i + \Delta \cdot P_i'}{\sum_{j=1}^{J} C_j + \Delta \cdot P_j'} \quad ; \quad \Delta > 0. \qquad (11.46)$$

11.4.7   Entropy Coding and Transmission Errors

Variable length codes are rather sensitive in case of transmission errors. By deletion or flipping of one single bit, a mis-interpretation can occur such that a different prefix is recognized, which may belong to a codeword of different length. The problem likewise affects Huffman-type VLCs and arithmetic codes; systematic VLCs (cf. sec. 11.4.3) can be designed in a way that effects of error propagation are minimized, or that even partial recovery can be performed at the decoder. In principle, the mis-interpretation can propagate from one code symbol to the next, until finally an invalid code symbol is identified, or until it is recognized that the number of decoded source symbols is larger than expected. In both cases, it is almost impossible to identify the position of the original erasure, such that often the whole decoded sequence must be considered as invalid. The following example explains different possible cases in the context of a simple prefix code.

Example : Propagation of errors in case of a variable-length code of prefix type. The code string representation of four source symbols shall be 'A'⇔'0', 'B'⇔'10', 'C'⇔'110' und 'D'⇔'111'. The original encoded sequence shall be 'ABCD', such that the code sequence '0|10|110|111' is sent. In the following cases, the respective distorted bit is underlined.

Case A: Receive stream '011̲110111', interpretation of the decoder: '0|111|10|111' ⇔ 'ADBD'. 2 letters are wrong, however the error is occasionally compensated, i.e. the last letter 'D' is decoded correctly. No error is detected by the decoder, as the code stream can be separated into valid codewords. The number of decoded source symbols is correct as well.

Case B: Receive stream '1̲10110111', interpretation of the decoder: '110|110|111' ⇔ 'CCD'. All decoded symbols are wrong, the error is not detected as the code stream can be separated into valid codewords, unless the decoder would know to expect 4 symbols. In fact, the last letter 'D' is decoded correctly, but it will be mis-interpreted as it is found at the wrong position.

Case C: Receive stream '0100̲10111', interpretation of the decoder: '0|10|0|10|111' ⇔ 'ABABD'. In principle the same case as *B*, however the number of letters is by one too high.

Case D: Receive stream '010110101̲', interpretation of the decoder: '0|10|110|10|1' ⇔ "ABCB?". The last letter is wrong, however the error is recognized, as the code stream can not entirely be separated into valid codewords, the last bit remains isolated.

By a certain degree of probability, the decoding process may catch up again after a number of code symbols decoded. Then, it synchronizes again with a correct starting bit of a valid codeword string, as in cases *A-C*. In the meantime however, it is also quite probable that a wrong number of source symbols are decoded (cases *B* and *C*). To detect and at least partially recover from errors, one of the most effective counteractions is the introduction of *resynchronization* mechanisms. A resynchronization code consists of a bit string which must be unique, i.e. will not be allowed to be part of any other string of the variable length code, nor be reproducible by concatenation of different code strings. For systematic construction, it is often defined as a long string of only '0' or '1' bits, which can easily be forbidden within other codewords by systematic construction of the code. By using resynchronization, it is possible to detect the occurrence of transmission errors with a high reliability, and re-start the decoding process from that point. If RVLCs as described in sec 11.4.3 are used, it is also possible to recover source symbols *backwards* from the resynchronization point up to the position where the error actually occurred, or where a divergence is found between the forward and backward decoding processes.

11.4.8   Run-length Coding

Run length coding is a specific type of entropy coding applicable on *two-level* (binary) signals, by which positions of *changes* between the two levels are encoded. It is likewise applicable for encoding of any on-off decisions. In principle, run-length coding is a lossless, reversible transformation of a two-level signal into a multi-level signal with lower number of samples. If typical signal statistics let expect that not all values of the multi-level signal are equally probable, application of variable-length codes to the run-length information attains an even more efficient encoding of the two-level signal. An advantage of the run-length transformation is the implicit joint encoding of binary samples, where however the complexity of the code only grows linearly by the maximum run-length[1]. The run length indicates the number of subsequent samples having an equal level in the binary signal. Two different methods are commonly used, which are illustrated in Fig. 11.18; it depends on the properties of the binary signal which of these methods would be the better choice:

–   If both levels are approximately equally probable, and adjacent samples are highly statistically dependent: The run-length code signifies the number of subsequent samples of same value, i.e. in principle the positions of transitions ('0'→'1', '1'→'0') are marked (Fig. 11.18a). It is in addition necessary to convey the starting level (here: '(1)') to the decoder. The smallest possible run length is one.

–   If one of the two levels is much more probable than the other : The run-length code describes the numbers of samples instantiated by the level of higher probability between two samples of the lower-probability type (Fig. 11.18b). A run-length value of zero signifies that another sample of low-probability level follows immediately. It is not necessary to convey the starting level: This is by default assumed to be the level of higher probability; otherwise, the first run-length will be a zero value, as in the example shown.

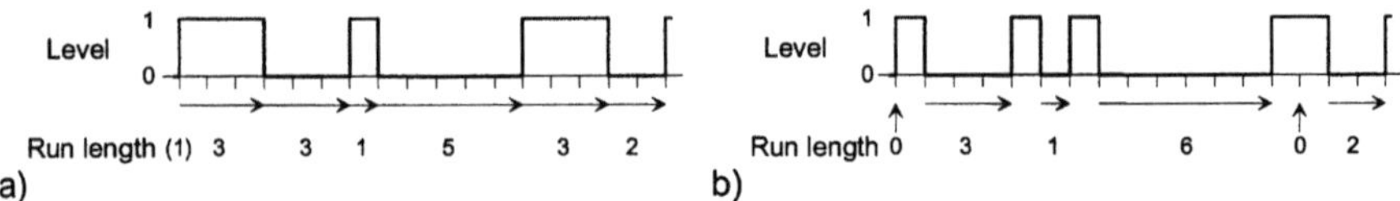

Fig. 11.18. Run-length coding **a** Method assuming equal probability levels **b** Method assuming unequal probability levels $[P(0)>P(1)]$

Run-length coding of 1D binary sequences can be interpreted as an approach to optimally encode binary processes generated by the 2-state Markov chain model described in sec. 3.4. The entropy is determined by (3.96). Accordingly, the entropy of the two-level signal, determined separately for the two states is

$$H(b)\big|_{b=1} = -P(0\,|\,1)\cdot\log_2 P(0\,|\,1) - P(1\,|\,1)\cdot\log_2 P(1\,|\,1) \tag{11.47}$$

$$H(b)\big|_{b=0} = -P(1\,|\,0)\cdot\log_2 P(1\,|\,0) - P(0\,|\,0)\cdot\log_2 P(0\,|\,0)\,, \tag{11.48}$$

---

[1] In contrast, if binary samples are combined into vectors for the purpose of entropy coding, the complexity would grow exponentially with the vector length. The design of a run-length code, even in tandem combination with VLC of the run length, has a much lower complexity than direct VLC encoding of the binary sequence.

where the total entropy results from weighting by the probabilities of the two states (3.77)

$$H(b) = P(0) \cdot H(b)\big|_{b=0} + P(1) \cdot H(b)\big|_{b=1} \ . \tag{11.49}$$

The transition probabilities of the Markov chain model can be adapted from statistics of a given signal by counting the occurrences of run-lengths of '0' and '1' sequences from a training set, and then using (3.78) to determine best-fitting parameter sets. This can in principle also be interpreted as an approach to adapt a hidden Markov model (cf. sec. 9.4.6). Extension to more efficient coding of run-length values for 2D or multi-dimensional signals can be made by defining a binary context **b** as in (3.81), which includes more values to determine the transition probabilities, e.g. pixels from the line above in a binary image.

A coding scheme approaching the entropy rate according to (11.49) can be designed using the expected run length probabilities as determined by (3.78). Any common entropy code design method can be applied. Observe that (3.78) defines an exponential distribution of probabilities, such that longer run lengths will have higher self information and will be encoded by longer code symbols. Usually, this will lead to code constructions which can excellently be matched by systematic codes such as Golomb-Rice or Exp-Golomb codes (cf. sec. 11.4.3).

Due to the relative addressing inherent in run length coding, a high impact of transmission losses results. This can even get more severe, if the run-length values are variable-length coded. Hence, the introduction of resynchronization mechanisms, as discussed for the case of error-prone transmission of variable length codes, is very important in the case of run-length coding as well. In fax transmission standards, which extensively use run-length methods, a resynchronization is typically set at the beginning of each line.

11.4.9   Lempel-Ziv Coding

Lempel-Ziv (LZ) codes are widely used for universal lossless compression. Due to flexible adaptation mechanisms, these are in principle applicable for compression of any kind of digital data. For generic compression of multimedia signals, LZ codes are of limited value however, as mechanisms that are using models of signal statistics can typically achieve still better results; further, for *lossy* compression, a tight integration of quantization and entropy coding is required.

LZ codes are frequently applied to compress text files, the variant LZ78 is also used in the modem standard V.42 and in the graphics/image representation format GIF. Technology-wise, LZ codes nicely complete the review of entropy coding methods. *Fixed-length codewords* are used to encode source symbol vectors of *variable length*, the contrary principle of Huffman coding. This is achieved by installing a *dictionary* which contains source symbol vectors of high expected probability. In particular the adaptive modifications of LZ are interesting: LZ77 (having fixed length of the dictionary, but automatic adaptation of its content) [ZIV, LEMPEL 1977], LZ78 [ZIV, LEMPEL 1978] and LZW [WELCH 1984], supporting dynamic adaptation and growth of dictionary size. Here, a sliding window over past values is used to update the dictionary from occurrence counts of recently decoded source symbols (see Fig. 11.19). Source symbols which cannot be found in the dictionary are encoded uncompressed.

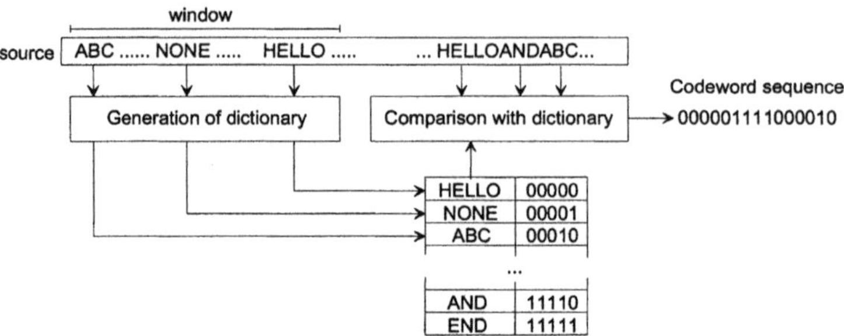

Fig. 11.19. Principle of dictionary adaptation in Lempel-Ziv-Welch coding

Due to the fixed length of codewords, error propagation in case of transmission errors can not occur by wrong identification of codeword starting points, but the number of decoded source symbols may be wrong, and a wrong adaptation of the dictionary can be caused at the decoder. Further, as source symbols are of variable length, the reconstructed source sequence may indeed have less or more elements than the original one.

Table 11.3. Properties of different entropy-coding methods

Codewords	Source vectors	
	fixed-length	variable-length
fixed-length	Fixed-length coding (FLC), systematic VLC with 'escape' condition	Run-length coding with FLC, Lempel-Ziv coding
variable-length	Shannon & Huffman coding, systematic VLC with 'typical' condition	Run-length coding with VLC, arithmetic coding

It is interesting at this point to summarize the different entropy coding methods by the way *how* they implement codes which as close as possible match the self-information of the source symbols by the codeword lengths. Generally, it is possible to use *fixed or variable length of source-symbol vectors* and *fixed or variable length of codewords*. The overview given in Table 11.3 shows that the methods of run-length coding and arithmetic coding exploit all possible flexibility, as both the input and output of the code mapping are based on the variable-length method.

This gives one interpretation for the good performance of these methods, while relatively low complexity is retained.

11.5 Vector Quantization

Vector quantization (VQ) is another method to combine samples in block-wise encoding. Unlike the method of 'scalar quantization plus entropy coding', the combination of samples into vectors is applied during the quantization step. Fig. 11.20 illustrates how vectors can be formed from 1D or 2D signals. If groups of K samples are combined into vectors, a sequence of vectors $\mathbf{x(n')}$ is available for quantization and encoding. In case of 1D signals, the vectors are often composed by natural sample order, $\mathbf{x(n')}=[x(n'K),x(n'K+1),...,x(n'K+K-1)]^T$, where $n'=\mathrm{int}(n/K)$ as shown in Fig. 11.20a. Other methods of vector formation are possible, e.g. alternating or interleaved packing of samples into vectors[1]. In vector quantization of 2D image signals, vectors are mostly defined as rectangular or quadratic blocks of dimension $K=M'\mathrm{x}N'$ (see Fig. 11.20b). Vector quantization can not only be applied directly to the signal, but can also be used for representation of transform coefficients, prediction error signals, filter coefficients, motion vectors or other parameters which have to be encoded.

11.5.1 Basic Principles of Vector Quantization

A typical block diagram of a vector quantizer is shown in Fig. 11.21. A *codebook* $\mathcal{C}=\{\ \mathbf{y}_j\ ;\ j=1,..,J\ \}$ contains a set of J different K-dimensional *codewords* (reconstruction vectors) $\mathbf{y}_j=[y_j(0),y_j(1),...,y_j(K\text{-}1)]^T$. The task of the encoder is to find the codeword $\mathbf{y}_i$ which is most similar when compared to an incoming vector $\mathbf{x}\equiv\mathbf{x(n')}$. The output of the encoder is the index i associated to the codeword; if necessary, the discrete index values can be subject to additional entropy coding[2] according to their probability distribution. If a squared-error criterion is used as a distortion metric, the optimum vector from the codebook is found as

---

[1] This can be advantageous for purposes of scalability, robust transmission etc., but will reduce the compression efficiency in case of signals where statistical dependency between direct neighbors is high.

[2] Additional constraints can be imposed during codebook design, such as equal probability of codeword usage over the whole set, or optimization based on entropy constraints (cf. sec. 11.5.5). If fixed-length coding of indices is used, high robustness is achieved in particular when transmission is made over error-prone channels. Transmission errors will then not take influence on subsequent source symbol reconstructions, as it would usually be the case when variable-length codes are used.

$$i = \arg\min_{y_j \in \mathcal{C}} \left[d(\mathbf{x}, \mathbf{y}_j) \right]; \quad d(\mathbf{x}, \mathbf{y}_j) = \sum_{k=0}^{K-1} (x(k) - y_j(k))^2 = \left[\mathbf{x} - \mathbf{y}_j \right]^{\mathrm{T}} \cdot \left[\mathbf{x} - \mathbf{y}_j \right]. \quad (11.50)$$

The decoder simply uses the received index i to pick the vector $\mathbf{y}_i$ from an identical codebook; the output is then placed in the reconstructed signal at the respective position $\mathbf{n}'$. The vectors $\mathbf{x}$ and $\mathbf{y}$ can be interpreted as points within a K-dimensional signal space $\mathcal{R}^K$ (cf. sec. 3.2) with K orthogonal coordinate axes. The Euclidean norms of the vectors, $\|\mathbf{x}\|_2$ and $\|\mathbf{y}\|_2$ express the distance from the origin of $\mathcal{R}^K$. The square root of (11.50) is the *Euclidean distance* or L_2 *norm distance* between points in $\mathcal{R}^K$, which is identical to the *linear geometric distance*. For the case of two-dimensional vectors (see Fig. 11.22), this is obvious by Pythagoras' theorem.

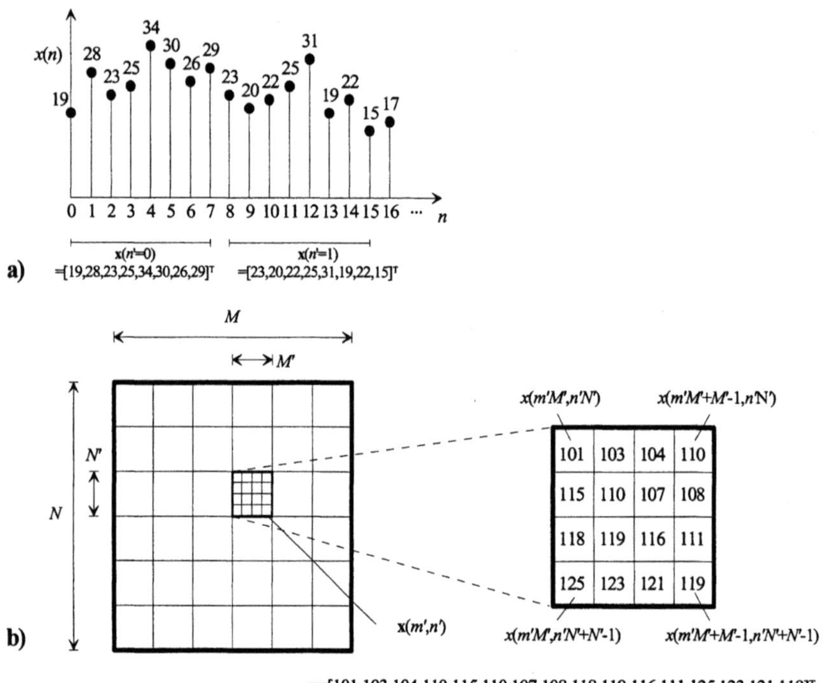

x=$[101,103,104,110,115,110,107,108,118,119,116,111,125,123,121,119]^{\mathrm{T}}$

Fig. 11.20. Examples for combinations of samples into vectors **a** in 1D signals **b** in 2D signals

The most elaborate task of the VQ encoder is the search for the optimum reconstruction vector $\mathbf{y}$. The complexity of the decoder is comparably low. Hence, VQ is quite attractive for applications where low decoding complexity is mandatory, while the encoder complexity is not as important e.g. in cases where the content is encoded only once, and decoded multiple times.

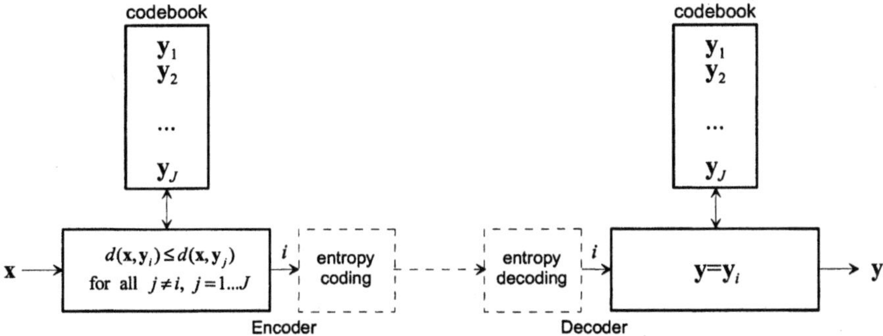

Fig. 11.21. Structure of a vector quantizer

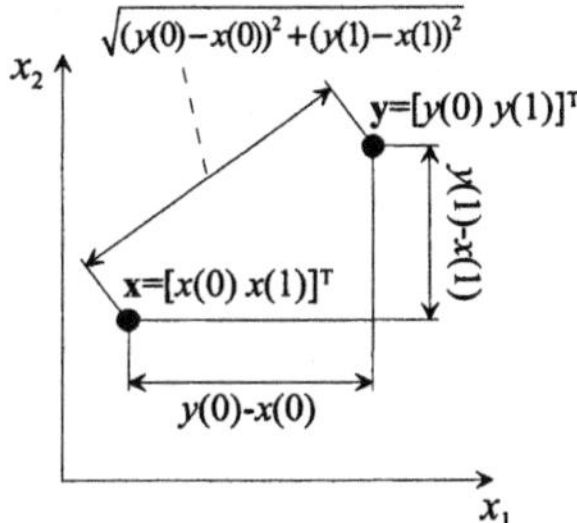

Fig. 11.22. Interpretation of the Euclidean distance between vectors **x** and **y** in $\mathcal{R}^2$.

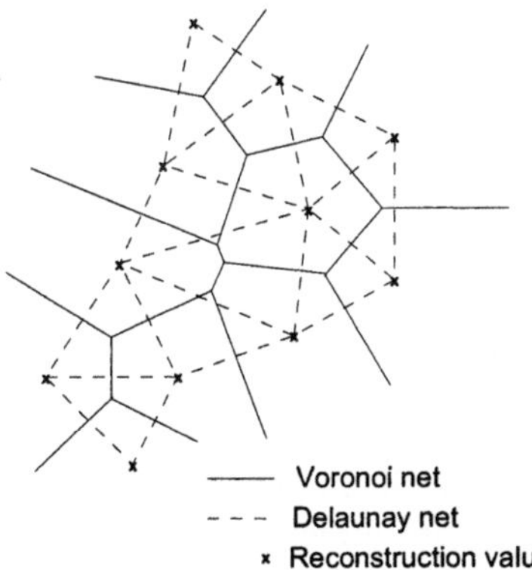

Fig. 11.23. Voronoi net and Delaunay net in $\mathcal{R}^2$, example codebook of J=9 codewords $\mathbf{y}_j$

Fig. 11.23 shows the layout of a codebook by the positions of the reconstruction values $\mathbf{y}_j$ in $\mathcal{R}^2$. From (11.50) it is possible to determine the boundaries of J distinct regions in $\mathcal{R}^2$, which are the quantization cells of the codewords $\mathbf{y}_j$. For the case of an Euclidean distance criterion, these partitions are denoted as the *Voronoi regions* in $\mathcal{R}^K$ (also *Dirichlet* or *nearest-neighbor regions*). The net of boundaries is the *Voronoi net*. It is a complementary representation of the *Delaunay net*, which is constructed from the connecting lines between direct neighbors y_j (cf. footnote on p. 397). The lines of the Voronoi net are perpendicular with the lines of the Delaunay net, and divide them into equal halves. It is obvious that the Euclidean distance from any x positioned on the Voronoi net towards the directly neighbored y_j is exactly equal.

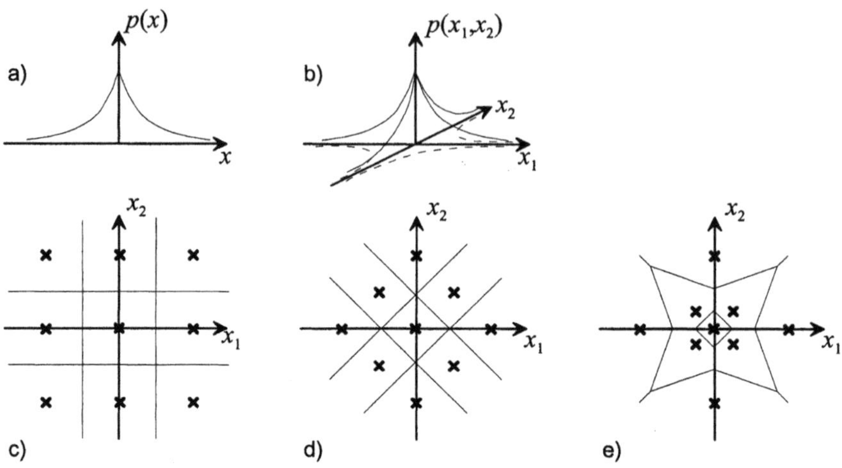

Fig. 11.24. Interpretation about the efficiency of VQ: **a** PDF of a signal **b** related joint PDF of two samples **c** Voronoi regions as realizable by a scalar quantizer **d-e** Voronoi regions of codebooks with same number $J=9$ of uniformly distributed (**d**) and non-uniformly distributed (**e**) reconstruction values.

The advantage of VQ, as compared to a scalar quantization using the same number of reconstruction values, is illustrated in Fig. 11.24. Fig. 11.24a shows the non-uniform PDF $p(x)$ of a signal. Assuming that samples are statistically independent, and each two samples over variables x_1 and x_2 are combined into vectors x, the resulting joint (vector) PDF is $p_2(x)=p(x_1,x_2)=p(x_1){\cdot}p(x_2)$, as shown in Fig. 11.24b. Now, a scalar quantizer shall be used allowing $J=3$ different reconstruction values per sample; if applied to the vectors, the degree of freedom in terms of reconstruction value selection is limited to the configuration of Voronoi regions as shown in Fig. 11.24c. A vector quantizer with a codebook constellation as shown in Fig. 11.24d places 9 different reconstruction values by more preference at positions where higher occurrence can be expected; concentration of higher values also along the axes of $\mathcal{R}^2$ appears reasonable, as signal constellations where either both values are of low amplitude, or one value is lower and the other is higher are typical. Even though the layout of Fig. 11.24d is still regular, having equal distances between any sub-sets of neighbored reconstruction values, it can no longer be described as a scalar quantizer[1]. Such types of codebook constellations with equally-shaped Voronoi cells will be denoted as *uniform VQ codebooks*, in analogy with uniform scalar quantization. The configuration in Fig. 11.24e goes one step beyond, as reconstruction values are irregularly distributed (*non-uniform VQ codebook*). More reconstruction values are concentrated around the origin of $\mathcal{R}^2$, the area of highest probability according to the joint PDF.

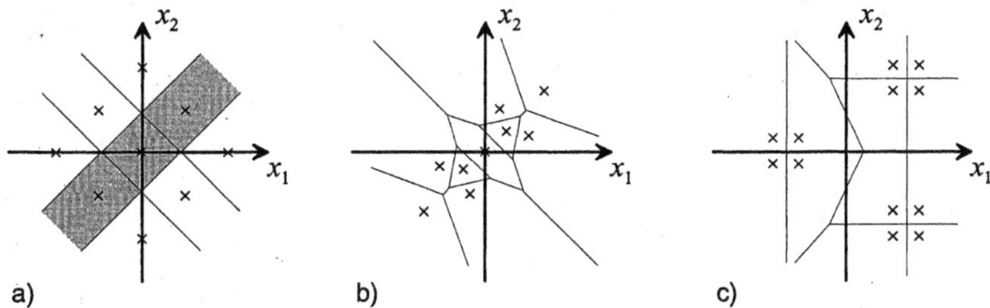

Fig. 11.25. Capability of VQ to exploit statistical dependencies: **a** Layout of a uniform codebook (Voronoi regions of highest probability in case of correlated signals are shaded) **b** Layout of a non-uniform codebook for correlated signals   **c** Layout of a non-uniform codebook for an example of nonlinear statistical dependencies

If the values within the vector are correlated (linearly statistically dependent), concentrations around the principal axis with an angular orientation of 45^0 within $\mathcal{R}^K$ can be expected (see Fig. 3.7). This can be considered in the design of entropy coding for the uniform codebook VQ in Fig. 11.24d, where shorter VLC codewords can be used to encode the indices associated with the shaded Voronoi regions in Fig. 11.25a. A non-uniform codebook layout can directly be optimized such that the correlation within the signal is taken into account (Fig. 11.25b). Another advantage of non-uniform vector quantization methods is the capability to exploit *nonlinear statistical dependencies* as well. These are related to concentrations of values which are disparate from the principal axis, but contradict the condition for statistical independency, $p(x_1,x_2)=p(x_1){\cdot}p(x_2)$; an example of a non-uniform codebook optimized for such a case is shown in Fig. 11.25c.

If a codebook of J reconstruction values is used, it is possible to represent the information by a rate of $\log_2 J$ bit/vector or $\log_2 J/K$ bit/sample, not considering entropy coding yet. In the subsequent sections, the principles to design vector quantizer codebooks of uniform and non-uniform characteristics are described.

---

[1] In fact, the same set of reconstruction values would be possible by application of a Haar transform, which rotates the coordinate axis of $\mathcal{R}^2$ by 45^0, performing subsequent scalar quantization of transform coefficients; see Problem 11.7

11.5.2   Vector Quantization with Uniform Codebooks

If the codebook describes a regular grid of reconstruction values, the VQ scheme is denoted as *uniform-codebook* or *lattice VQ*[1]. For the following illustrations the two-dimensional space $\mathcal{R}^2$ is chosen again. Fig. 11.26a shows a rectangular grid of reconstruction values, the lattice $\mathbf{Z}_2$. Reconstruction values represent a 2-dimensional grid of integers. This expresses the vector counterpart of a uniform scalar quantizer, where the axes have to be scaled by the quantizer step size Δ. Fig. 11.26b shows the lattice $\mathbf{A}_2$, which for $K=2$ describes the optimum uniform VQ, showing Voronoi regions of hexagonal structure[2].

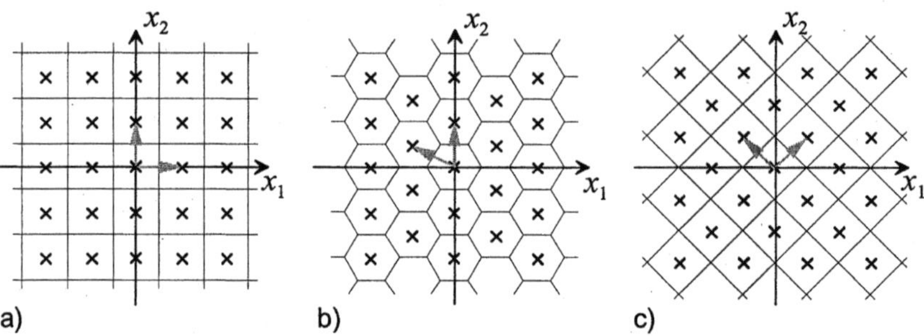

Fig. 11.26. Lattice structures for $K=2$ **a** $\mathbf{Z}_2$ lattice **b** hexagonal $\mathbf{A}_2$ lattice **c** $\mathbf{D}_2$ lattice

A clear analogy exists between lattice VQ and regular multi-dimensional sampling (see sec. 2.2.3). In fact, lattice VQ can be interpreted as *regular sampling*[3] of the vector signal space $\mathcal{R}^K$. $\mathbf{Z}_2$ is equivalent with rectangular sampling, $\mathbf{A}_2$ relates to hexagonal sampling. Another group are the $\mathbf{D}_K$ lattices, which consist of the set of all integer numbers in $\mathcal{R}^K$ where the elements of the vector have an even sum. $\mathbf{D}_2$ (Fig. 11.26c) relates to quincunx sampling in $\mathcal{R}^2$.

For any vector size K, the optimum lattice is given by the densest packing of identically-shaped Voronoi cells. Likewise, this is also the best possible approximation of a K-dimensional hypersphere (where the hypersphere-shaped cells would *not* allow truly dense packing, as empty spaces remain between the non-planar surfaces). For $K=3$, the $\mathbf{D}_3{}^*$ lattice is optimum. Here, the packing of Voronoi cells can be interpreted as a 'pile of balls' (Fig. 11.27a), which find their densest packing system by gravity, provided that the bottom layer is densely packed. While the centroids of the Voronoi cells are identified by the centers of

---

[1] An excellent reference on lattice VQ is [CONWAY, SLOANE 1988].

[2] This lattice, for a given number of reconstruction values (or given area of the elementary Voronoi regions) describes the densest packing within a 2D plane, such that the mean distance between the centroid of the Voronoi region and the points falling into this region is smaller than in the case of $\mathbf{Z}_2$.

[3] Also scalar quantizers perform sampling of amplitude variables x. The mass distribution function $p(x_i)$ can straightforwardly be interpreted as a sampled expression of the PDF $p(x)$.

the balls, the actual shape of the cells is the dodecaeder shown in Fig. 11.27b. Assuming that the Voronoi cell has a unity volume, the approximation error as compared to a hypersphere is decreased by increasing vector dimension K. Optimum lattices are often determined from optimum error-protection block codes, which should have the property of equal Hamming distances between any pairs of nearest valid code words. For example, the lattice Λ_{16} which is optimum for $K=16$ is constructed from a Reed-Muller code, and the lattice Λ_{24} is derived from an extended Golay code.

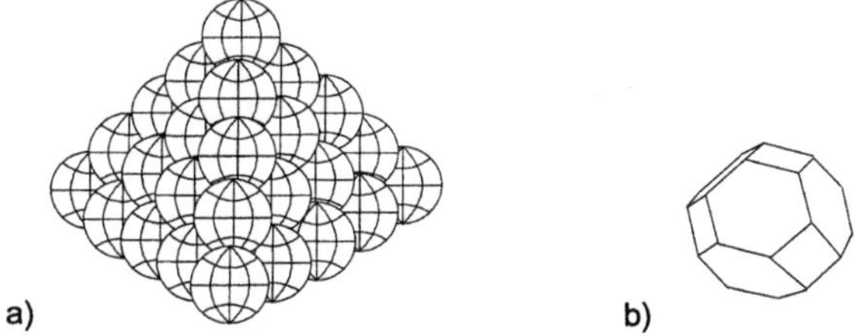

Fig. 11.27. a Illustration of $\mathbf{D}_3{}^*$ lattice packing as 'pile of balls' **b** geometry of a single Voronoi cell

The regular lattice structure of reconstruction values can be described by a *generator matrix* $\mathbf{G}=[\mathbf{g}_0\ \mathbf{g}_1]$, which for the examples of lattices $\mathbf{Z}_2$, $\mathbf{A}_2$ and $\mathbf{D}_2$ can be defined as[1]

$$\mathbf{G}_{\mathbf{Z}_2} = \begin{bmatrix} 1 & 0 \\ 0 & 1 \end{bmatrix} = \mathbf{I} \ ; \quad \mathbf{G}_{\mathbf{A}_2} = \frac{2}{\sqrt{3}}\begin{bmatrix} 0 & -\dfrac{\sqrt{3}}{2} \\ 1 & \dfrac{1}{2} \end{bmatrix} \ ; \quad \mathbf{G}_{\mathbf{D}_2} = \frac{\sqrt{2}}{2}\begin{bmatrix} 1 & 1 \\ -1 & 1 \end{bmatrix} . \quad (11.51)$$

$\mathbf{G}$ is equivalent to the sampling matrix $\mathbf{D}$ of (2.51). The columns of $\mathbf{G}$ are the *basis vectors* $\mathbf{g}_k$ of the lattice, for which the set of reconstruction values can be obtained by a K-dimensional vector of integer index values i_k, $k=0...K-1$ by

$$\Lambda = \{\mathbf{y} : \mathbf{y} = i_0\mathbf{g}_0 + i_1\mathbf{g}_1 + ... + i_{K-1}\mathbf{g}_{K-1}\} \Rightarrow \mathbf{y} = \mathbf{G}\cdot\mathbf{i}$$

$$\text{with}\quad \mathbf{i} = \begin{bmatrix} i_0 & i_1 & \cdots & i_{K-1} \end{bmatrix}^{\mathrm{T}}. \quad\quad (11.52)$$

As the number of discrete points defined by Λ could in principle be infinite, it is necessary to limit the range of values i_k for encoding. Then, a codebook with finite

---

[1] Several definitions of *dual lattices* exist, which are addressing the same lattice grids in principle, but would lead to another range of quantization when addressed by a finite set of index values $\mathbf{i}$ as in (11.52). Observe that $\det|\mathbf{G}|$ determines the areas or volumes of the Voronoi cells, which have been normalized to unity in all cases of (11.51). Fast algorithms exist for quantization, which do not actually need to perform the scaling by the irrational numbers.

number of reconstruction values $\mathbf{y}_i$ is obtained. The encoding step of lattice VQ is rather simple, as in principle the generator matrix can be interpreted to define a transformation of the coordinates in $\mathcal{R}^K$. The inverse transformation $\mathbf{G}^{-1}$ is applied to the signal values, which can then be quantized in a $\mathbf{Z}_K$ lattice (i.e. scalar), which directly gives the index values. For some types of lattices (in particular for the optimum lattices in K=2,3,4,8,12,16 und 24), more simple algorithms exist for direct quantization. Decoding is performed directly by (11.52).

A lattice VQ with highest packing density of Voronoi cells for a given vector dimension K is an optimum quantizer for encoding of signals with a uniform PDF. As it is known from scalar quantization, the uniform quantizer in combination with entropy coding is optimum for any PDF at least in the range of higher rates[1]. To still better adapt a lattice quantizer for encoding of signals with non-uniform PDF it is possible

–   to adapt the 'outer shape' of the codebook; the 'spherical' shape in Fig. 11.28a is asymptotically optimum for uncorrelated signals of Gaussian PDF, the 'pyramid' shape in Fig. 11.28b is better suitable for signals which by their statistical behavior follow the Laplacian PDF [FISHER 1986, 1989].
–   *Companding* of signal values can be performed, where nonlinear amplitude-mapping functions (cf. sec. 5.3.1) are applied to the signal values prior to quantization, and the inverse mapping is then applied after reconstruction from the lattice points [SIMON 1998]. In fact, this is equivalent to a vector quantizer with non-uniform distribution of the reconstruction values as shown in Fig. 11.28c.

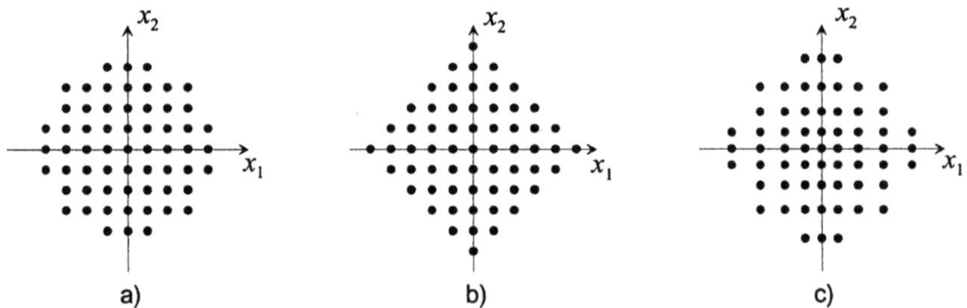

Fig. 11.28. Structures of lattice codebooks.
a spherical **b** pyramid **c** spherical with non-uniform expansion

Lattice vector quantizers also provide nearly-optimum solutions for *embedded vector quantization* at least in the range of higher rates. Assume that a representation of the signal shall be provided in multiple amplitude resolution levels, such that re-quantization of the Voronoi cells is necessary. Due to the equal shape of the cells, the same finer-level cell partitioning can be used everywhere, which should

---

[1] In general, VQ with uniform codebooks is mainly useful in higher-rate encoding, while non-uniform codebooks are better suitable for low rates.

as good as possible approximate the shape of the larger cells of the previous level. Fig. 11.29a shows an example for the A_2 lattice, where each cell is re-quantized by a package of three smaller cells; the leftmost part shows a root cell, which marks the entire range of quantization. This approach is conceptually very similar to embedded scalar quantization and bit-plane coding, however ternary symbols are required to represent each next-finer resolution level. As three additional symbols are required to encode the refined information about two samples, the 'non-fractional' points on the rate-distortion curve (cf. sec. 11.3) will be apart of each other by distances of ½ $\log_2 3 \approx 0.79$, which is closer than in the case of scalar quantization[1]. The distances between the 'non-fractional' rate-distortion points can be further decreased by increasing the vector length. Fig. 11.29b shows the implementation of a dead zone lattice VQ; again, an embedded sub-partitioning can be realized which reduces the size of the dead zone into one third for the next finer resolution level.

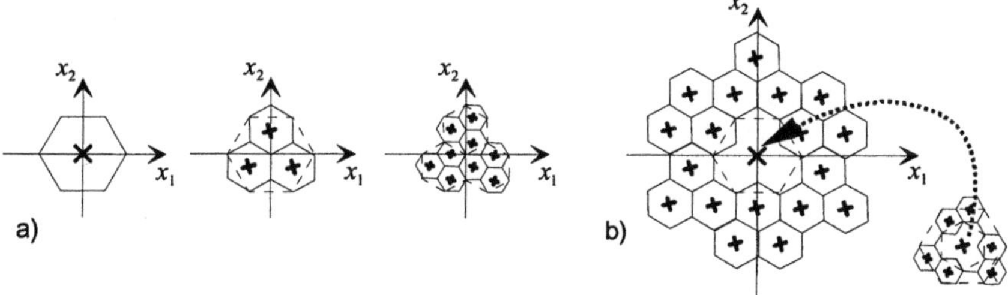

Fig. 11.29. **a** Embedded A_2 lattice VQ, root-cell partition and first two levels of 3 and 9 reconstruction values in the codebook **b** Embedded A_2 lattice VQ with a dead zone

11.5.3   Vector Quantization with Non-uniform Codebooks

To design codebooks for non-uniform VQ, the *Generalized Lloyd Algorithm* (GLA), also known as *Linde-Buzo-Gray algorithm* [LINDE, BUZO, GRAY 1980] or one of its numerous variants are usually applied. The principle is a vector extension of the Lloyd algorithm for the design of non-uniform scalar quantizers as described in (11.4)-(11.8). For VQ design, a *training set* is commonly used instead of a model distribution, which gives the advantage that eventually linear or nonlinear statistical dependencies between the samples of a signal can implicitly be exploited, even if they cannot be described analytically[2]. The training set must be selected carefully, such that the signals contained are statistically typical representatives of the signals which later shall be encoded by the quantizer. As in the (scalar) Lloyd algorithm, an iterative procedure is applied. Within one iteration, let $\mathbf{x}^j$ denote

---

[1] A performance penalty is caused however by the mis-match between the shapes of the re-quantizing Voronoi cells and the larger cell. As compared to scalar embedded quantization, this is not severe, as lattice VQ achieves a lower-distortion encoding anyway.

[2] When a training set is used, the similarity between the GLA and cluster-based classification algorithms (cf. sec. 9.4.3) becomes obvious.

those N_j vectors, which fall into the Voronoi cell of $\mathbf{y}_j$. Instead of using statistical expectation as in (11.4) and (11.6) the values from the sets $\{\mathbf{x}^j\}$ are used, the occurrence counts approximate the probabilities, such that

$$D_j = \frac{1}{N_j}\sum_{N_j}\sum_{k=1}^{K}(x^j(k)-y_j(k))^2 \quad ; \quad \hat{P}(j)=\frac{N_j}{\sum\limits_{j}N_j}. \tag{11.53}$$

By deriving D_j over $\mathbf{y}_j$, the elements $y_{j,\mathrm{opt}}(k)$ of the reconstruction vector which minimizes the distortion are determined as

$$y_{j,\mathrm{opt}}(k)=\frac{\sum\limits_{N_j}x^j(k)}{N_j} \quad\Rightarrow\quad \mathbf{y}_{j,\mathrm{opt}}=\frac{\sum\limits_{N_j}\mathbf{x}^j}{N_j}. \tag{11.54}$$

$\mathbf{y}_{j,\mathrm{opt}}$ is the arithmetic mean over $\{\mathbf{x}^j\}$ or asymptotically, $\mathbf{y}_{j,\mathrm{opt}}=E\{\mathbf{x}^j\}$. An example how the algorithm works with a small training set is illustrated in Fig. 11.30 for a vector quantizer of $K=2$ and $J=4$. 25 samples ('•') are used as training data. Reconstruction values $\mathbf{y}_j$ are marked as '$\mathbf{x}$'. Initially, a symmetric uniform codebook is used (Fig. 11.30a); in this case, the boundaries of the Voronoi regions are the coordinate axes of $\mathcal{R}^2$. In the next step, the $\mathbf{y}_j$ are shifted towards the mean values (centroids) as derived by the sets $\{\mathbf{x}^j\}$. This modifies the Voronoi net (Fig. 11.30b). As a consequence, some of the training vectors can become members of a different set $\{\mathbf{x}^{j'}\}$, relating to a more similar reconstruction vector $\mathbf{y}_{j'}$ which has been generated by the past iteration. Hence, in the next iteration (Fig. 11.30c) it is again possible to reduce the overall distortion D by a further optimization of the reconstruction vectors. In this whole process, it is guaranteed that D can never increase from one iteration to the next[1].

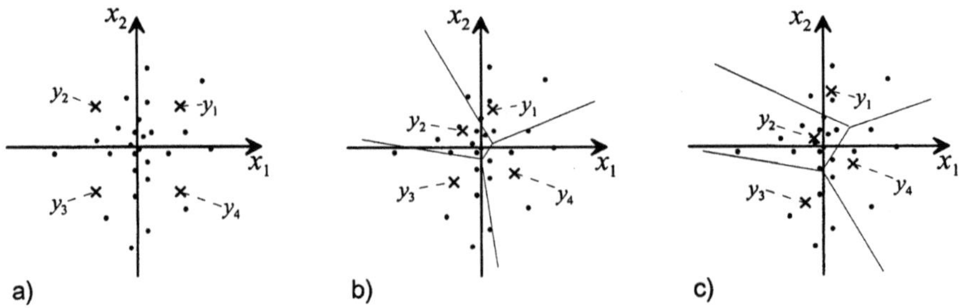

Fig. 11.30. Graphical illustration of two iterations of the GLA

---

[1] Also refer to sec. 9.4.3, optimization of cluster centroids

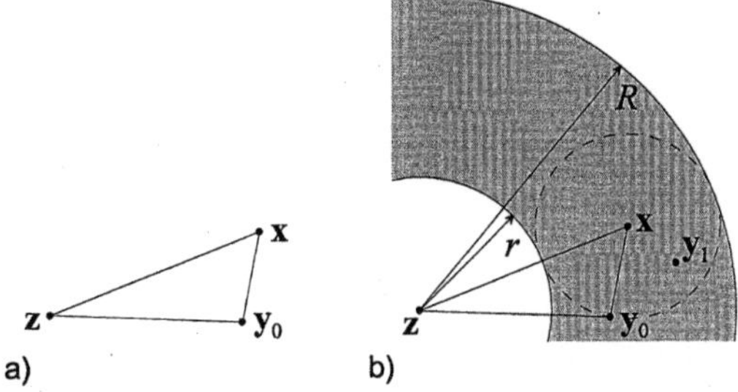

Fig. 11.31. Graphical interpretation **a** of the triangular inequality **b** of a method to limit the search to a specific range of the codebook, according to (11.58) (after BRÜNIG)

No guarantee exists for any regularity in the structure of non-uniform codebooks. Hence, it is necessary to compare a vector **x** to be coded against *each* $\mathbf{y}_j$ e.g. by the criterion (11.50). The resulting complexity in terms of multiplications and additions will be J operations per sample, or $K \cdot J$ operations per vector. As the rate grows logarithmically by J (if expressed as rate per sample, $R=(\log_2 J)/K \Rightarrow J=2^{RK}$), the complexity of encoding grows exponentially by the rate. Hence, it is impractical to apply VQ algorithms with unstructured non-uniform codebooks, when a large rate is required. A possible solution would be the reduction of K, which however may have a negative effect on the performance, according to the postulates of the source coding theorem[1]. Algorithms exist which can at least partially resolve this problem by excluding those $\mathbf{y}_j$ from the search which most probably are no good candidates to encode a given **x**. For this, a *fix point* **z** is defined, and distances $\|\mathbf{z}\text{-}\mathbf{y}_j\|$ between the fix point and all $\mathbf{y}_j$ are computed in advance (which is possible if the codebook is frozen after it has been designed). Further, an arbitrary *initial vector* shall be selected from the codebook, e.g. the vector $\mathbf{y}_0$. Then, according to the *triangular inequality* (see graphical interpretation in Fig. 11.31a) the following conditions apply regarding the vector **x**:

$$\|\mathbf{z}-\mathbf{y}_0\| \le \|\mathbf{z}-\mathbf{x}\| + \|\mathbf{x}-\mathbf{y}_0\| =: R$$
$$\|\mathbf{z}-\mathbf{y}_0\| \ge \|\mathbf{z}-\mathbf{x}\| - \|\mathbf{x}-\mathbf{y}_0\| =: r \ .$$

$$(11.55)$$

Further, for the best match $\mathbf{y}_i$ available in the codebook

$$\|\mathbf{x}-\mathbf{y}_i\| \le \|\mathbf{x}-\mathbf{y}_0\| ,$$

$$(11.56)$$

must give by combination with (11.55)

---

[1] A small vector length K is more disadvantageous if the signal samples contained in the vector are statistically dependent. Further, for instationary sources, it may be advantageous to use different codebooks or also different rates and vector lengths for different characteristics of data to be encoded. Methods which can be applied based on such ideas are more specifically discussed in sec. 12.2.

$$\left\| \mathbf{z} - \mathbf{y}_i \right\| \le \left\| \mathbf{z} - \mathbf{x} \right\| + \left\| \mathbf{x} - \mathbf{y}_i \right\| \le R$$
$$\left\| \mathbf{z} - \mathbf{y}_i \right\| \ge \left\| \mathbf{z} - \mathbf{x} \right\| - \left\| \mathbf{x} - \mathbf{y}_i \right\| \ge r. \qquad (11.57)$$

Consequently, it is sufficient to constrain the search to all those codebook vectors $\mathbf{y}_j$ for which the following condition holds:

$$r \le \left\| \mathbf{z} - \mathbf{y}_j \right\| \le R . \qquad (11.58)$$

This can easily be checked by once computing r and R, and compare the range they express against the distances relative to the fix point, which are pre-computed. However, the efficiency of such an approach, i.e. the percentage of $\mathbf{y}_j$ which can actually be excluded, depends on proper selection of $\mathbf{z}$ and the initial vector $\mathbf{y}_0$. A graphical interpretation is given in Fig. 11.31b. Methods which use several fix points are advantageous when large codebooks are used, but the principle remains similar.

11.5.4   Structured Codebooks

The VQ methods described in this section also use non-uniform codebooks, but lead to implementations of lower encoding complexity. In contrast to the method described before, the performance will however be inferior if compared against full-search results gained on monolithic non-uniform codebooks. The best way for complexity reduction is to break down the decision process into a sequence of partial decisions on a decision tree, which are related to a pre-defined structure in the codebook. A by-product of some of these methods is *scalability*, where it is possible to structure the resulting stream of code symbols such that parts relating to the first decisions can be used as a raw approximation, the remaining parts as refinement or enhancement information to achieve lower distortion.

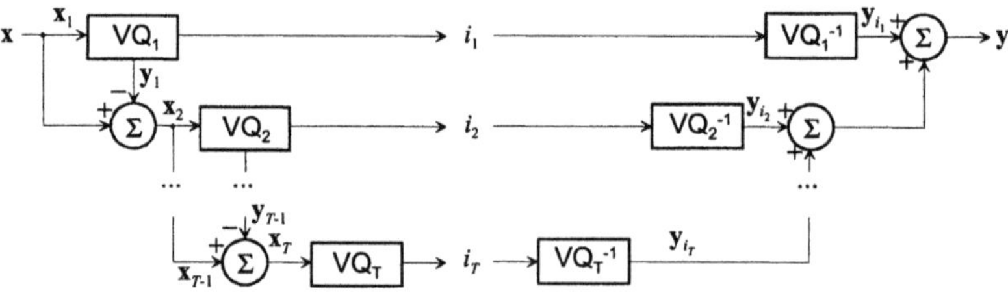

Fig. 11.32. Multi-level vector quantization

The *multi-level VQ* method shown in Fig. 11.32 consists of T levels or stages of vector quantizers which work sequentially. The subsequent levels encode residuals to reduce the coding errors of previous levels. By initial settings $\mathbf{y}_0{=}\mathbf{0}$ and $\mathbf{x}_0{=}\mathbf{x}$, the input signal of level t is described as $\mathbf{x}_t = \mathbf{x}_{t-1}{-}\mathbf{y}_{t-1}$. For each level, a specific codebook $\mathcal{C}_t$ containing J_t reconstruction vectors is provided. For $t{\ge}2$, these are residual signal vectors. The entire information is the concatenation of indices i_t, relating to the partial codebooks. The overall data rate is (without entropy coding of indices)

$$R = \frac{1}{K} \sum_{t=1}^{T} \log_2 J_t \quad \text{[bit/sample]}. \tag{11.59}$$

The reconstruction vector **y** results as the sum of all vectors,

$$\mathbf{y} = \sum_{t=1}^{T} \mathbf{y}_{it} . \tag{11.60}$$

Decoding could be terminated at any level, as all previous levels represent more raw approximations of the signal (scalability property). For optimality of multi-level VQ it would have to be assumed that statistical properties of the residual signals in the respective levels are independent, such that no statistical dependency should exist between the $\mathbf{y}_{it}$. This is apparently not the case, as selection of a vector with a relatively small Voronoi cell in a previous stage will typically lead to selection of small residual vectors in subsequent stages. The particular values J_t have to be chosen *before* codebook generation. In this process, the codebooks for each of the single levels, starting from level one, must be optimized successively. Hence, multi-level VQ will be inferior in terms of coding performance as compared to a any single-level VQ.

A *tree-structured codebook* layout is shown in Fig. 11.33a. Here, the search for the best reconstruction vector is also performed by several steps related to levels of a *code tree*. The tree-like nature of decisions is more obvious here than in multi-level VQ. In each step, the new choices depend on the result of the previous step. If the tree consists of T levels, and each node within the tree at level t can branch into I_t different nodes at the subsequent level, the number of vector comparison operations to be performed is

$$M = \sum_{t=1}^{T} I_t , \tag{11.61}$$

while the total number of different reconstruction vectors which could be reached by any tree path up to level t is

$$J_t = \prod_{t'=1}^{t} I_{t'} . \tag{11.62}$$

The entire number of possible choices at the last level is $J=J_T$, such that $R=(\log_2 J)/K$ is the number of bits/sample necessary to represent the information (without considering entropy coding). In case of a *binary tree*, $I_t=2$ over all levels, and only $M=2\cdot\log_2 J$ comparisons (instead of J in case of full search with the same total number of reconstruction vectors) are necessary. In a binary-tree decision, the complexity of VQ only grows *linearly* with the rate, in contrast to the case of exhaustive (full) search, where an exponential growth is observed.

The multi-level VQ described above can also be interpreted by a code tree, and is hence a specific type of tree-structured VQ. By constraint of the multi-level approach, reconstruction vectors are defined as superpositions of all residuals along a tree path; the extensions of the paths at the same level of the tree are not defined independently, as the same residual codebooks are used at any node to extend into the next level. The advantage of multi-level VQ is the reduced memory for codebook storage; in fact, for binary tree-structured VQ, a memory of $J+J/2+J/4+\ldots\approx 2J$ vectors must be provided at the encoder, while the decoder must only store J vectors of the highest level T. This means that the codebook size and thereby the memory complexity of tree-structured VQ still grows *exponentially* with the rate. The refinement of quantization performed at the subsequent tree levels perfectly matches the residual errors of the previous levels, whereby it can be concluded that tree-structured VQ compensates the major drawback of multi-level VQ as described above[1]. On the other hand, the number of vectors to be stored in multi-level VQ is only ΣJ_T, which is drastically lower. Hence, due to memory constraints, multi-level VQ is a better candidate for higher-rate applications, whereas scalability is supported in a less optimum sense.

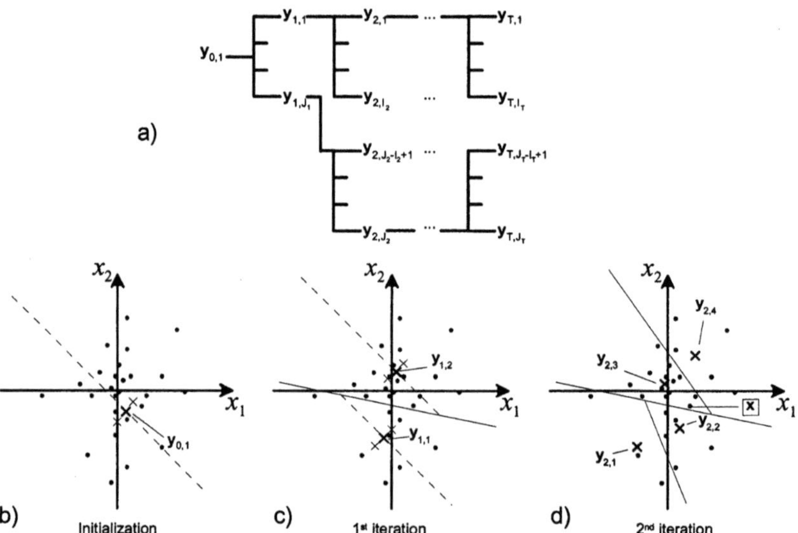

Fig. 11.33. a Tree-structured codebook **b-d** First iteration steps illustrating the 'splitting' of quantizer partitions for the case of a binary tree structure

Codebooks with a symmetric tree structure can be designed by a *splitting algorithm* (see Fig. 11.33b). The numbers of branches I_t at the respective levels have to

---

[1] Indeed, tree-structured VQ provides a fully-embedded quantizer solution, where each quantizer cell is individually split into finer partitions by the next step. This will become

be defined in advance. In Fig. 11.33b, the example for a binary tree structure with $I_f{=}2$ over all levels is shown. From the training set $\mathscr{X}{=}\{\mathbf{x}(n');\ n'{=}0,1,...,N'{-}1\}$, the initial reconstruction vector $\mathbf{y}_{0,1}$ at the root of the tree, is computed as the global centroid, which is the mean over all vectors from the training set. The associated Voronoi region is the entire space $\mathscr{R}^K$, which is then arbitrarily divided into two sub-spaces (see Fig. 11.33b)[1]. Within each hemisphere, the two optimized vectors $\mathbf{y}_{1,1}$ and $\mathbf{y}_{1,2}$ are then computed according to (11.54), see Fig. 11.33c. Each of the two new Voronoi regions is again split, such that during the second iteration which generates the second level of the codebook tree, 4 reconstruction vectors are generated (Fig. 11.33d). As a splitting procedure, *additive splitting*

$$\mathbf{y}_{\text{split},1} = \mathbf{y} - \varepsilon \cdot \mathbf{1} \quad ; \quad \mathbf{y}_{\text{split},2} = \mathbf{y} + \varepsilon \cdot \mathbf{1} \quad ; \quad 0 < \varepsilon \ll 1 \tag{11.63}$$

was used here; another procedure is *multiplicative splitting*

$$\mathbf{y}_{\text{split},1} = \mathbf{y} \cdot (1 - \varepsilon) \quad ; \quad \mathbf{y}_{\text{split},2} = \mathbf{y} \cdot (1 + \varepsilon) \quad ; \quad 0 < \varepsilon \ll 1 . \tag{11.64}$$

If it is necessary to split into more than two vectors, it is useful to employ a set of orthogonal basis function, e.g. those of the Hadamard transform (4.125) (also weighted by ε) in (11.63)/(11.64) instead of the unity vector. The optimum split directions could further be determined by an eigenvector analysis of the second-moment mass distribution in the given quantizer partition, which is equivalent to a covariance analysis between the elements within the training-set vectors quantized into this partition[2]. The separations should then be done by adding and subtracting the ε-weighted first eigenvectors of the covariance matrix. This will also have an effect of newly-formed Voronoi cells to be orthogonally orientated, such that most likely each covers a similar-size subspace in $\mathscr{R}^K$. The split procedure described here is also frequently used to generate *initial codebooks* for a VQ design of an exhaustive-search VQ encoder. Further, it is possible to run several iterations for codebook optimization after each of the splitting steps to gain better optimized structured codebooks.

In a tree-structured codebook search, it can nevertheless occur that at the last level $t{=}T$, a vector $\mathbf{y}_{T,j}$ exists, which is actually more similar (by the Euclidean distance criterion) to the signal vector $\mathbf{x}$ than the vector selected by the tree-based decision. This is caused by the fact that configurations of quantizer partitions at previous decision levels take influence on the subsequent results of the best-vector search. In Fig. 11.33d, the signal vector '⊠' is originally closest to the reconstruction vector $\mathbf{y}_{2,2}$; indeed $\mathbf{y}_{2,3}$ is selected, because at the first level of the tree, $\mathbf{y}_{1,2}$ was chosen. A more thorough analysis of this behavior shows that the decision boundary between the quantizer partitions of $\mathbf{y}_{2,2}$ and $\mathbf{y}_{2,3}$ is *not a Voronoi line* (not per-

---

[1] In case of correlated samples within the vector, it is best to perform the sub-division orthogonal to the principal axis of the mass distribution of the training set (cf. Fig. 3.7a). This is in principle done both by the additive and multiplicative splitting procedures described below.

[2] cf. e.g. (9.33), which is an equivalent case for feature vectors allocated to a class

pendicular and centered on the Delaunay line). It can be concluded that tree-structured VQ cannot be optimum in the sense of rate-distortion minimization. A possible method to circumvent this problem is tracking of multiple tree paths, which could actually be done similar as in the *M-algorithm* described in sec. 11.6.2.

Both the multi-level VQ and the tree-structured VQ can be interpreted in a more general sense as *product codes*. If the reconstruction of a VQ method is described by a combination from indexing of T codebooks, the set of possible combinations establishes a *super codebook*, which can be expressed as the outer product of all sub codebooks:

$$\mathcal{C} = \mathop{\mathsf{X}}_{t=1}^{T} \mathcal{C}_t .$$

(11.65)

This super codebook could only be equally efficient as one monolithic codebook addressed by same number of freely-selectable indices, if no statistical dependencies are effective between the partial information encoded by the sub codebooks, and if the rate allocations to sub codebooks are optimum in a rate-distortion sense. From this point of view, proper design of sub codebooks turns out to be a very difficult optimization problem, and definitely none of the methods presented so far is capable to fulfill these conditions.

Another simple form of a product code is *Gain/Shape VQ* [SABIN, GRAY 1984]. This uses reconstruction vectors $\mathbf{y}_j^*$ which are normalized by unity norm $\|\mathbf{y}_j^*\|_2 = 1$. To generate the final reconstruction $\mathbf{y}$, these normalized vectors are multiplied by a gain factor σ_i. The task is to find the pair $(\sigma_i, \mathbf{y}_j^*)$ which minimizes the Euclidean distance compared to an incoming vector $\mathbf{x}$. This metric can be expressed by

$$d_2(\mathbf{x}; \mathbf{y}_j^*, \sigma_i) = \left\| \mathbf{x} - \sigma_i \cdot \mathbf{y}_j^* \right\|_2 = \mathbf{x}^\mathsf{T}\mathbf{x} - 2 \cdot \sigma_i \cdot \mathbf{x}^\mathsf{T}\mathbf{y}_j^* + \sigma_i^2 \cdot \left[\mathbf{y}_j^*\right]^\mathsf{T} \mathbf{y}_j^* . \quad (11.66)$$

This expression can first be minimized independently of the choice for the gain factor, if the term $\mathbf{x}^\mathsf{T}\mathbf{y}_j^*$ becomes maximum; this is the *cross correlation* between the input vector and the respective codebook vector. Finally, by deriving (11.66) over σ_i and due to $[\mathbf{y}_j^*]^\mathsf{T}\mathbf{y}_j^*=1$, the optimum gain factor results as $\sigma_i = \mathbf{x}^\mathsf{T}\mathbf{y}_j^*$.

11.5.5   Rate-constrained Vector Quantization

Methods which are similar to schemes for optimization of scalar quantizers considering rate constraints (sec. 11.3) can also be applied in the context of VQ. In the GLA, *minimization of distortion* during each iteration step is guaranteed; it might happen however that the *rate increases* (as e.g. measured by entropy) during the optimization of the codebook. This lead to cases where the result, regarding the gap towards the rate-distortion bound, is worse than before the iteration. This can of course only be relevant if variable-rate encoding is performed, e.g. by entropy coding of index values. Again, a modified distance metric is used in analogy with

(11.26), taking into account the rate $R(j)$ to be spent for encoding of the code symbol by a Lagrangian multiplier weight[1]:

$$d^{*}_{2}(\mathbf{x},\mathbf{y}_{j}) = \left[\mathbf{x}-\mathbf{y}_{j}\right]^{T}\left[\mathbf{x}-\mathbf{y}_{j}\right] + \lambda \cdot R(j). \tag{11.67}$$

The parameter λ expresses the weight by which the rate shall influence the decision, the optimum is given by (11.27). At the decision boundary between two quantizer partitions, this distance metric must be equally balanced regarding the two reconstruction vectors. Hence, depending on λ, the boundary between the quantizer cells related to the reconstruction vectors of indices j_1 and j_2 is now placed according to the following condition, where $\mathbf{x}_b$ denotes any vectors residing on the boundary[2]

$$\left[\mathbf{x}_{b}-\mathbf{y}_{j_1}\right]^{T}\left[\mathbf{x}_{b}-\mathbf{y}_{j_1}\right] + \lambda \cdot R(j_1) = \left[\mathbf{x}_{b}-\mathbf{y}_{j_2}\right]^{T}\left[\mathbf{x}_{b}-\mathbf{y}_{j_2}\right] + \lambda \cdot R(j_2), \tag{11.68}$$

which can be re-written as

$$\mathbf{y}_{j_1}^{T}\mathbf{y}_{j_1} - \mathbf{y}_{j_2}^{T}\mathbf{y}_{j_2} + 2\mathbf{x}_{b}^{T}\left(\mathbf{y}_{j_2}-\mathbf{y}_{j_1}\right) = 2\left(\mathbf{x}_{b} - \frac{\mathbf{y}_{j_1}+\mathbf{y}_{j_2}}{2}\right)^{T} \cdot \left(\mathbf{y}_{j_2}-\mathbf{y}_{j_1}\right)$$
$$= \lambda \cdot \left(R(j_2)-R(j_1)\right). \tag{11.69}$$

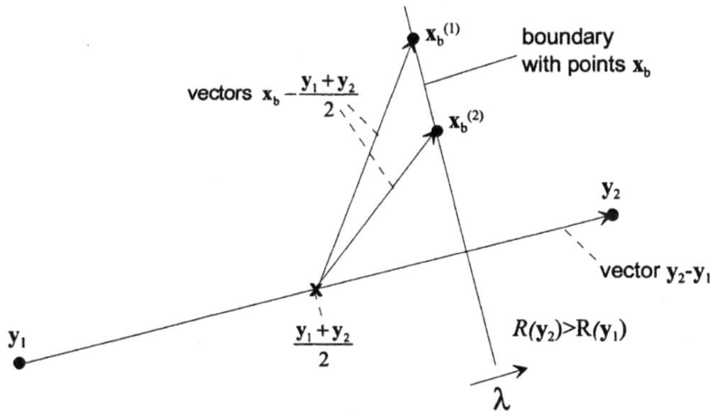

Fig. 11.34. Positions of two neighbored reconstruction vectors, and boundary between the quantizer regions in rate-constrained VQ (after BRÜNIG)

An illustration for the case of $K=2$ is given in Fig. 11.34. For $\lambda=0$ ('ordinary' VQ without rate-constraint criterion), this results in a line which crosses the point

---

[1] During codebook design, the rate can best be estimated by the self information (3.84), as computed from the normalized occurrence counts; if rate constraints are applied during encoding, the actual rate of a VLC used to encode the codebook indices can be used.

[2] This expresses a line in $\mathcal{R}^2$, a plane or hyperplane in higher-dimensional $\mathcal{R}^K$

$(\mathbf{y}_{j1}+\mathbf{y}_{j2})/2$ (the center of the Delaunay line between the vectors), and is perpendicular to the vector $\mathbf{y}_{j2}-\mathbf{y}_{j1}$ (which actually is a vector of length and orientation of the Delaunay line). In this case, the decision boundary is the Voronoi line. For $\lambda>0$, the boundary is shifted towards the reconstruction vector requiring larger rate, but remains perpendicular with the Delaunay line. For the case $R(j_1)=R(j_2)$, the boundary will also be a Voronoi line, regardless of λ. Else, if the vector $\mathbf{x}$ is at approximately equal distance from both reconstruction vectors, the vector requiring lower rate is selected, as this choice helps to approach the optimum in the rate-distortion sense.

An optimization of the VQ for a specific entropy rate can also be done by tuning the parameter λ. For growing λ, the rate decreases, and for $\lambda\to\infty$, $R\to0$. Then, all vectors $\mathbf{x}$ from the training set are exclusively allocated to only one reconstruction vector, which is the global centroid of the set.

Rate-distortion optimization in tree-structured codebooks. Tree-structured VQ as described in sec. 11.5.4 always leads to symmetric-tree codebooks with equal path lengths of the code tree. As variable-length coding is typically related to asymmetric code trees of different path length (see sec. 11.4.1), it is obvious that asymmetric-tree codebooks would certainly provide better results in the rate-distortion sense. In the splitting procedure described earlier, in fact no guarantee exists that simultaneous splitting of all quantizer partitions will lead to the appropriate reduction in distortion, as could be expected by spending one more bit (in the case of binary trees). If e.g. the quantizer cell of $\mathbf{y}_{t,j}$ is relatively small and populated by a small number of training-set vectors, the effect by further splitting may be minimum. One possible solution to this problem is made by application of *tree pruning algorithms* [CHOU, LOOKABAUGH, GRAY 1989][RISKIN, GRAY 1991], where branches are removed from a full regular tree depending on their effectiveness, which means that nodes that have least effect on additional distortion as compared to the saved bit budget are removed first; this again relates to the slope of an operational rate-distortion function (cf. sec. 11.3). In principle, *growing* a tree step by step is a similar but antipodal design approach. In growing, the quantizer partitions are not split simultaneously for all $\mathbf{y}_{t,j}$, but rather step by step, where the next splitting is made such that the *largest decrease of distortion* is achieved. Let D_j be the distortions (11.53), and $P(j)$ the probabilities of codebook members accepted so far by the growing algorithm. Both values can be determined by analysis of the training set. Further, $t(j)$ expresses the depth of the tree at the given position. If partition j would be split into 2 sub-partitions j' with indices as given below, the decrease of distortion is

$$\Delta D(j) = D_j - \sum_{i=1}^{2} \frac{P(j')}{P(j)} \cdot D_{j'} \tag{11.70}$$

with

$$j' = 2(j-1)+i ; \quad t(j') = t(j)+1 . \tag{11.71}$$

The two new index values j' are extended by one bit as compared to their ancestor index j. Splitting is performed for a cell with maximum of $\Delta D(j)$ at any position of the tree. Then, the two new vectors are integrated into the codebook by supplementing indices j' and retaining the ancestor j as an intermediate step for tree-based search. The procedure is continued until some pre-set termination criterion is reached. Observe that this results both in an asymmetric tree structure and in a variable-length code representation of the vectors residing at the leaves of the tree. Indices j relating to the leaves of the tree can readily be used as variable-length code strings and represent a unique prefix code. The rate related to the indices at any phase of the splitting procedure or by the time of termination will be

$$R = \sum_{j} P(j) \cdot t(j) \,. \tag{11.72}$$

As the probabilities of the resulting new cells are not necessarily equally distributed by each splitting step, it can still be useful to apply entropy coding of the index values; optimum growing of the tree can then be made by including a rate constraint in the decision for splitting (11.70).

11.6 Sliding Block Coding

In VQ, each vector or block of samples is encoded separately, i.e. independently of its neighbors[1]; a unique code symbol (index value) is generated, which can be used for direct decoding, where a one-to-one mapping is performed. This is the typical case of a block code. In *sliding block coding*, several code symbols are needed for decoding of each output, where each of these code symbols takes also influence on the decoding of several other outputs. The principle of decoding in a sliding-block method is shown in Fig. 11.35.

Each code symbol $i(n')$ is combined with its $L-1$ predecessors $i(n'-1)...i(n'-L+1)$ to generate a reconstruction (output) $y(n')$, which can be a scalar or a vector value. Linear or nonlinear mapping functions can be used. The current code symbol $i(n')$ is also used in the following $L-1$ decoding steps, such that in total it takes influence on L subsequent outputs[2]. The parameter L is the *constraint length* of the sliding block code. Important sub-classes of sliding block codes used in source coding are

---

[1] This statement is not fully true for the methods of Predictive VQ and Finite-state VQ which will be introduced in sec. 12.2. A major difference as compared to sliding-block coding is still that the quantization decision is made instantaneously in VQ, such that a code symbol (index) can always be directly related to a source symbol (vector). In sliding-block coding, the decision about the optimum code symbol can only be made by evaluating sequences of source symbols.

[2] If the mapping function is an IIR filter, L could theoretically approach infinity, but usually the effect of a code symbol decays gradually.

trellis and *tree coding*. If linear mapping functions (filters) are used, sliding block coding is also denoted as *convolutional coding*.

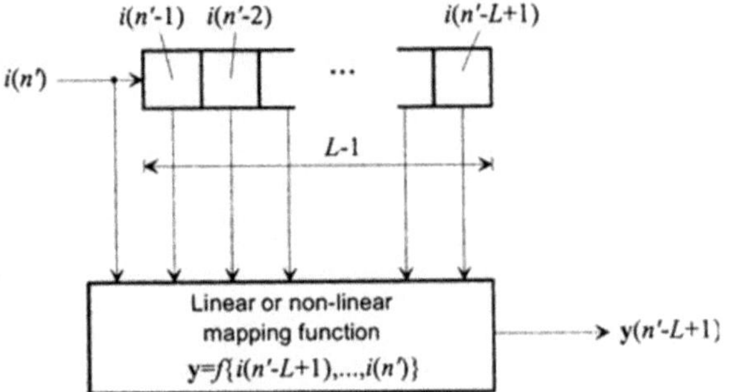

Fig. 11.35. Decoder in sliding block coding

11.6.1   Trellis Coding

If the mapping function is a codebook, the subsequent code symbols are interpreted as a concatenated codebook index. Then, the number of possible reconstruction values is *finite*. If each code symbol can be instantiated by J different values, the codebook $\mathcal{C}=\{y_j; j=1,...,J^L\}$ will output one of J^L different reconstruction vectors $y(n')=f[i(n'),i(n'-1),...,i(n'-L+1)]$. Each $y(n')$ can be a vector relating to K source samples in the most general case. The configuration of possible decoder states is shown in Fig. 11.36 for an example of $J=2$, $L=2$. The regular structure of state transition graphs as shown in Fig. 11.36a is a *trellis diagram*. In each decoding step n', one of 4 different reconstruction vectors $y(n')$ is generated, which are related to the *nodes* representing possible states at each level of the state diagram. The mapping table is shown in Fig. 11.36b. It follows e.g. if $y(n'-1)=y_1$ or $=y_2$, $y(n')$ can only be $=y_1$ or $=y_3$ etc. The code symbols $i(n')$ are related to the *branches* that lead towards the nodes[1]. Due to the finite-state property of $L=2$, the code symbol $i(n'-2)$ will no longer have any direct influence on $y(n')$; however it is obvious that the selection of $i(n'-1)$ influences which two nodes at level n' are addressed by selection of $i(n')$. The task of the encoder is the search for the *optimum sequence* of codewords, such that the overall reconstruction error over the path in the trellis is minimized.

The task of the trellis encoder can best be interpreted, if the sequences of code symbols $i(n')$ are mapped to a *path* in the trellis diagram (Fig. 11.36a). It would in principle be necessary to compare all possible paths in the trellis to find the optimum code symbol sequence. The decision is based on the *accumulated distortion* along the different paths aimed to find one that minimizes the overall path distor-

---

[1] In this case, binary code symbols are used, hence there are two branches emanating from or leading to each node.

tion. In principle this means to run as many decoders in parallel as paths exist[1]. Before the optimum code symbol $i(n')$ can be determined, at least L-1 further decoding steps have to be evaluated. After these L-1 steps, it may however still not yet be possible to come to a final decision. Even though a certain code symbol *directly* influences only L reconstruction values, there is an indirect influence as the subsequent steps are not necessarily independent from the initial state. A final decision about the optimum path in the trellis can hence only be made if an $i(n')$ to be selected is the code symbol $i(n^*)$ which is the common *root* of all paths (Fig. 11.37). By the *Viterbi algorithm* described in the next sub-section, the number of paths to be compared in trellis coding is always *upper bounded* by J^{L+1}, even though a code symbol may take indirect influence on a considerably larger number $L^* \gg L$ of reconstruction vectors.

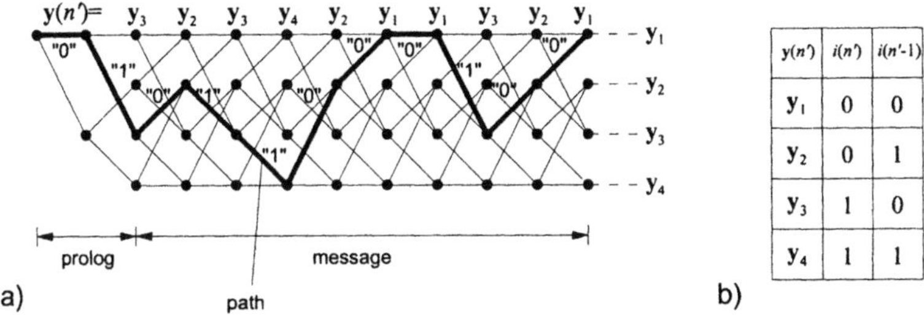

$y(n')$	$i(n')$	$i(n'-1)$
y_1	0	0
y_2	0	1
y_3	1	0
y_4	1	1

Fig. 11.36. a Trellis diagram for J=2, L=2 and **b** related mapping table

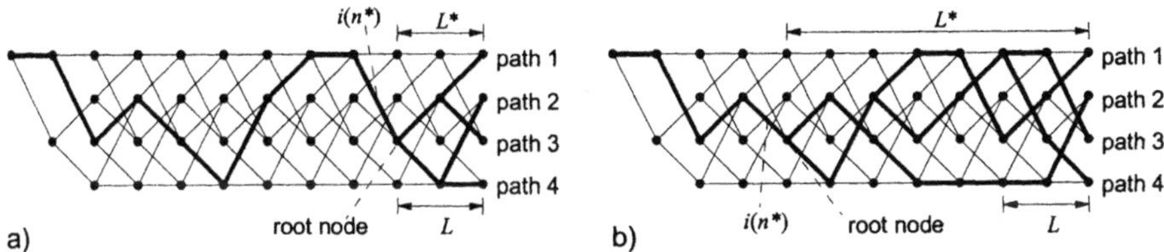

Fig. 11.37. Viterbi algorithm. Retained paths from the trellis diagram and position of the root node **a** in case of minimum possible $L^*=L=2$ **b** in case $L^*>L$

Viterbi algorithm. In the trellis diagram (Figs. 11.36 and 11.37), each of the J^L nodes can be reached by any of J different paths. Emanating from the node, each of the paths also has J possible extensions. As however each of these extensions will add the same amount of distortion to any of the incoming paths, the path with minimum distortion occurring up to the position of the node will by guarantee be the optimum among all incoming paths of this node. All other paths can be excluded from further decisions. The *globally optimum path* must be one of the optimum paths leading towards the J^L different nodes. According to the algorithm

---

[1] The task of a VQ encoder can be interpreted similarly: In an exhaustive search of the codebook, J different decoders are run in parallel to find the optimum result. The VQ can also be interpreted as Trellis coding of constraint length L=1.

proposed in [VITERBI 1967], it is only necessary at each node to compare J incoming paths and select the best during one encoding step. If then all $M=J^L$ optimum paths of the single nodes at this encoding step originate from *one identical root node*, the code symbol $i(n^*)$ which leads to this root node is by guarantee a member of the globally optimum path and can finally be released. The root node resides L^* steps in the past, wherein the number L^* is variable (see the examples in Fig. 11.37a/b). By consequence, the *search depth* over the trellis and the *encoding delay* are variable as well. The computational complexity of the encoding process is constant, as in total per encoding step (and hence per code symbol released) never more than $J{\cdot}M=J^{L+1}$ comparison operations will be necessary.

Codebook optimization. Optimization of codebooks for trellis-coded quantization (TCQ) can be performed by a variant of the GLA described in sec. 11.5.3. It is simply necessary to use the Viterbi algorithm during the encoding step of each iteration, while the optimization of code words is not modified [STEWART, GRAY, LINDE 1982], [AYANOGLU, GRAY 1986]. Without significantly modifying the methods described in sec. 11.5.5, it is also possible to run codebook optimization for TCQ using *entropy* or *rate constraints* [FISCHER, WANG 1992].

11.6.2   Tree Coding

In multimedia signal coding/decoding algorithms, *recursive processing* is frequently used. For example, IIR filters are employed in linear-prediction synthesis, or backward-related conditional decisions are made by entropy coders. If a linear filter with infinite impulse response supplements the convolutional decoder described above, the number of states and therefore also the constraint length will in principle be asymptotically infinite. The same is the case for any nonlinear context-related decisions of entropy coding. It is very unlikely then that paths frequently merge; at least this will not happen systematically as it was the case in a trellis. The related graph structure is a genuine tree with the number of nodes at each level exponentially growing by the depth. The number of branches emanating from each node will still be the number J of code symbols available, assuming that a constant J is used by each encoding step. An interesting example is *tree coding of correlated sources*, where an AR synthesis filter is used as code generator (see Fig. 4.11), which is fed by an innovation signal from a codebook; this could be represented by scalar or vector values. If J different code symbols are used at each encoding/decoding step, the possible states of the decoder can be interpreted by a Jary code tree (Fig. 11.38).

No algorithm of finite complexity would be able to determine the globally optimum path through this tree, but sub-optimum algorithms typically are sufficiently powerful for this purpose. An example is the *M-algorithm* described below, some other tree-search algorithms are e.g. compared in [ANDERSON, MOHAN 1984]. All these algorithms could also be applied to trellis structures, if the number of paths to be compared by the Viterbi algorithm would grow too large due to complexity limitations. Further, these algorithms are applicable in any encoding methods with re-lated decisions, e.g. tree-structured VQ, but likewise in motion estimation, contour or motion trajectory tracking etc. If a coding method is based on a decoder with recursive decisions, a multiple-path tree decision will usually improve the performance at the cost of increased complexity only of the encoder, without modifying the decoder at all. Examples for this are delayed-decision DPCM [MODESTINO, BHASHKARAN 1981] or predictive VQ [OHM, NOLL 1990].

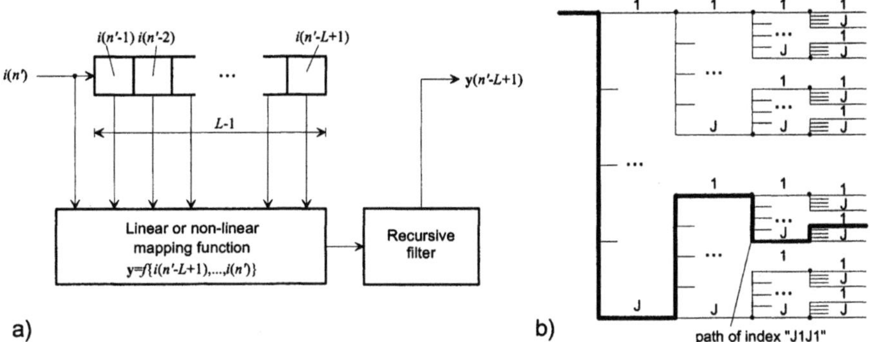

Fig. 11.38. a Decoder and **b** code tree in tree coding

***M*-algorithm.** The steps performed by this algorithm are described here by the example of the state where the signal vector $\mathbf{x}(n')$ arrives. The encoding delay introduced shall be constant and is defined a priori by a parameter L^*. It does *not* influence the search complexity of the algorithm, except for the size of the parameter tables to be stored. These contain at position n', for any of M paths existing in the table with temporary-assigned table indices m,

- path indices $\mathbf{i}_m(n')=[i_m(n'-L^*+1), i_m(n'-L^*+2),..., i_m(n'-1)]$;
- accumulated coding distortions of the paths d_m such that $d_1 \leq d_2 \leq ... \leq d_M$;
- previous reconstruction values $\mathbf{Y}_m(n')=[y_m(n'-L^*+1), y_m(n'-L^*+2),..., y_m(n'-1)]$.

The decoder mapping function is given for path m and a code symbol j emanating from this path as

$$\mathbf{y}_{j,m}(n') = f\left[\mathbf{i}_m(n'), \mathbf{Y}_m(n'), j\right]. \tag{11.73}$$

Herein, the interdependency of preceding reconstruction values is effected by the recursive filter. The algorithm consists of the following steps:

1. Output (final decision) of index $i(n'-L^*)=i_1(n'-L^*)$ related to the *root node* of the code tree, which relates to the minimum distortion path as retained prior to this encoding step at table position $m=1$. Hereby, the reconstruction vector $y(n'-L^*)=y_1(n'-L^*)$ is finally determined.
2. Checking of all remaining M-1 paths from the table ($m=2,3,...,M$) for identity of their own root node index with the root node index decided under 1. Elimi-

nation of all paths with $i_m(n'-L^*)\neq i_1(n'-L^*)$ is made. $M^*\leq M$ paths are retained after this step.

3. Coding of $\mathbf{x}(n')$ for all retained M^* paths, extending them by each J possible reconstruction values, and accumulation of the resulting distortions to the path distortions:

For $m=1,2,...,M$; $j=1,2,...,J$:

$$\mathbf{y}_{j,m}(n') = f\left[\mathbf{i}_m(n'),\mathbf{Y}_m(n'),j\right]$$

$$d_{j,m} = d_m + d\left(\mathbf{y}_{j,m}(n'),\mathbf{x}(n')\right)$$

4. Among the M^*J paths available now, search M paths with smallest accumulated distortion; update the *Mary* table for the next encoding step, which again results in a list ordering by decreasing distortion:

For $k=1,2,...,M$:

$$d_k = \min d_{j,m} \quad ; \quad j=1,...,J \quad ; \quad m=1,...,M$$

$$(i,m) = \arg\min d_k \quad ; \quad d_{i,m} = d_{\max}{}^1$$

$$\mathbf{i}_k(n'+1) = \left[i_m(n'-L^*+1),i_m(n'-L^*+2),...,i\right]$$

$$\mathbf{Y}_k(n'+1) = \left[\mathbf{y}_m(n'-L^*+1),\mathbf{y}_m(n'-L^*+2),...,\mathbf{y}_{i,m}(n')\right]$$

The search complexity of the M-algorithm is only dependent on M and J. For larger values of M, better results can be expected, as the probability is higher to find a path which is near to the global optimum. The algorithm can flexibly be adapted to the requirements set by a specific tree- or trellis-based encoding approach, and is also well suitable for hardware realizations [MOHAN, SOOD 1986].

11.7 Problems

Problem 11.1

a) Show for the AR(1) process the relationship of the spectral flatness measure $\chi_x^2=\sigma_z^2/\sigma_x^2$ under assumption of 'low distortion' according to (11.14),.

b) Show the validity of (11.18). Which gain in rate can be achieved in coding a 2D separable AR(1) process, if the 2D correlation properties are exploited ? Compute the values of the gains in the 1D and 2D cases with $\rho_h=\rho_v=0.95$.

c) Which is the rate at the boundary of the range of 'low distortion' according to (11.23) for the separable 2D AR(1) process? Compute the value of this rate for $\rho_h=\rho_v=0.95$.

d) The distortion parameter Θ in (11.11) shall be chosen such that in coding of a 1D AR(1) process, a band limitation to *i)* $\Omega_{max}=\pi/4$, *ii)* $\Omega_{max}=\pi/2$ and *iii)* $\Omega_{max}=\pi$ is achieved. Determine the Θ values for these frequencies Ω_{max}, where the AR(1) process is defined by *I)* $\rho=0.5$ and *II)* $\rho=0.95$.

---

[1] d_{max} is an arbitrary high value, which prevents that a path which was already accepted for the table is investigated again in the loop.

Problem 11.2
The probabilities of the 4 symbols of a discrete source alphabet are $P(1)=0.4$; $P(2)=P(3)=P(4)=0.2$. Determine the code symbol lengths of a Shannon code design, and sketch the code tree. Further, design a Huffman code. Compute the entropy and the rates that are necessary for both codes. Why is the Huffman code more efficient ?

Problem 11.3
A memoryless binary process [0,1] is characterized by the first order probabilities $P(0)=0.25$ and $P(1)=0.75$.
a) Compute the entropy.
b) Determine the joint probabilities for pairs of samples $P(0,0)$, P(0,1), $P(1,0)$ and $P(1,1)$ and for triplets of samples $P(0,0,0)$, $P(0,0,1)$, ... , $P(1,1,1)$.
c) Determine the resulting rates in entropy coding by the Huffman method, using first order probabilities, then for vectors of samples the second order and third order joint probabilities as determined in b). Compute the differences compared to the entropy.

The samples shall now be statistically dependent (same first order probabilities), which is characterized by a binary Markov chain model (Fig. 3.10). One of the two transition probabilities is $P(1|0)=0.5$.

d) Determine the probabilities $P(0|1)$, $P(1|1)$ and $P(0|0)$.
e) Find the probability values of sample pairs $P(0,0)$, P(0,1), $P(1,0)$ and $P(1,1)$.Then, determine the code symbol lengths for a Huffman code, compute the rate and give an interpretation in comparison to the result of c).
f) Compute the entropy according to (11.49), and give an interpretation in comparison with the results of c) and e).

Problem 11.4
A discrete source is arithmetic encoded. The source can output two different letters 'A' and 'B' of probabilities $P(A)=0.8$, $P(B)=0.2$.
a) Compute the entropy.
b) Sketch the layout of probability intervals within the range [0,1] for all combinations of 3 subsequent source symbols, assuming statistical independency. For each interval, compute the lower and upper boundaries.

Encoding shall be performed by an arithmetic unit which has a precision of 3 bit.

c) Sketch the code interval raster for three subsequent code bits. Perform rounding of the probability interval boundaries from b) according to the 3 bit precision. In this case, do not further subdivide probability intervals which exactly match a code interval, as it can be assumed that the encoding process again starts independently afterwards. Sketch a code tree relating to all intervals that are retained. For each path within the tree, identify the string of code bits that would be transmitted along the path, and the sequence of source symbols that is expressed.
d) Which bit rate (per source symbol) results ? Compare against the entropy, and check whether the value is within the guaranteed bounds of arithmetic encoding, regarding the given precision of arithmetic operations.

Problem 11.5
A binary source with memory shall be encoded by a run-length code. An analysis of the source has given the parameters of a 2-state Markov chain model: $P(0|1)=0.8$; $P(1|0)=0.4$.

a)   The source outputs the sequence '0110000101011000'. Determine the run-length codes for both methods illustrated in Fig. 11.18a/b.
b)   Compute the probabilities for '0' and '1' sequences of lengths 1,2 and 3.
c)   Determine the probabilities for run lengths 0,1,2 and 3 for the method according to Fig. 11.18b. The run lengths shall be encoded by a Huffman code, where an ESCAPE symbol shall signal run lengths larger than 3. Determine the code string lengths for the run lengths 0-3 and for the ESCAPE symbol.

Problem 11.6
In a VQ constellation of vector length K=2, and J=2 reconstruction values, the codebook shall be optimized using the Generalized Lloyd algorithm (see sec. 11.5.3) The training set is $\mathcal{T}$={$[-1\ 0]^T$;$[-3\ -2]^T$;$[\ 1\ 1]^T$;$[-5\ -4]^T$;$[0\ 1]^T$;$[1\ 0]^T$;$[2\ 2]^T$}. Initially, a codebook with vectors $\mathbf{y}_0$=$[-1\ -1]^T$ and $\mathbf{y}_1$=$[1\ 1]^T$ shall be used.
a)   Determine the optimized $\mathbf{y}_0$ and $\mathbf{y}_1$ in two iteration steps.
b)   For the second iteration, determine the optimized vectors for an entropy constrained VQ method (see sec. 11.5.5) with value λ=65. The self information $i(j)$ shall be determined from the occurrence counts of the $\mathbf{y}_j$ selections in the first iteration. From the first iteration, determine the boundary line between the two quantizer cells for cases λ=0 and λ=65, and sketch the boundary lines as well as the reconstruction values in $\mathcal{R}^2$.

Problem 11.7
The vector PDF $p(x_1,x_2)$ of a signal shall be uniform within the area sketched in Fig. 11.39.

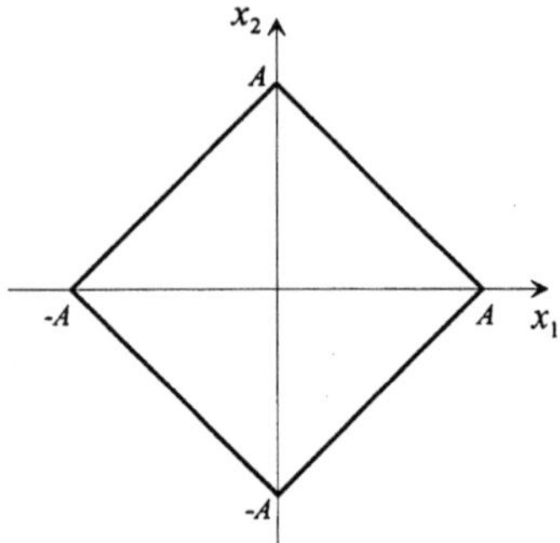

Fig. 11.39. Layout of a 2D uniform PDF

a)   Express the PDF in dependency of x_1 and x_2.
b)   Determine the probability of signal values with $|x_1| \geq A/2$.

A vector quantizer with the following set of reconstruction values is used for quantization: $\mathbf{y}_1$=$[-a,0]^T$; $\mathbf{y}_2$=$[a,0]^T$; $\mathbf{y}_3$=$[0, -a]^T$; $\mathbf{y}_4$=$[0,a]^T$.

c)   Sketch the Voronoi regions and determine the optimum value of a.
d)   Before quantization, a linear block transform shall be performed, such that the vector quantizer can be replaced by a scalar quantizer applied to the transform coefficients. Which linear orthonormal block transform is suitable for this purpose? Sketch the Voronoi regions in the transformed signal space.
e)   Determine the variance of the quantization error for the optimum value of a.

12 Still Image Coding

Still image coding includes compression of binary images and multiple-amplitude level (gray scale or color) images. Substantially different methods are applied for these two cases. This chapter gives a broad overview on different methods, which are in principle combinations of methods for signal decorrelation and analysis, quantization and coding, optimized for the specific characteristics of image signals. For binary images, run-length methods and methods related to conditional entropy coding are most relevant. For multiple-amplitude image signals, vector quantization, predictive coding, transform coding and fractal coding are presented in more detail. Transform coding methods can further be clustered into block transform, filterbank and wavelet transform related methods. These methods are presented mainly for examples of luminance (gray-level) compression, as typically the chrominance components are compressed by the same techniques, but are less challenging in terms of structure and hence will allow higher compression ratios. Building blocks which are necessary to understand the principles of still image coding standards like JPEG and JPEG 2000 are discussed in detail. Further important aspects relate to the robustness of still-image compression methods in the case of transmission losses, and to content-related encoding, which allows to further improve the quality by adaptation to the content properties and signal structure. The basic methods for still image coding are also important as elements within video compression methods, which will be further discussed in chapter 13. When applied to video, still image coding is also denoted as intraframe coding, expressing that compression of a sequence of video frames is performed without exploiting the interframe redundancies.

12.1 Compression of Binary Images

A binary image is an image of only two amplitude levels, e.g. a bi-level black and white image; typical binary images are scanned text pages or two-tone prints. Also, shape masks expressing binary shapes (7.90) are binary images. Transmission of binary images has a long history in telecommunications. Long-distance transmission of bi-level image material was first performed in telegraphy during the 19th century. Fax services were the first wide-spread application for transmission of personalized picture information. As many image sources, in particular photographs, have more than 2 gray-amplitude levels, it is first necessary to transform them into binary images if required for transmission, storage or processing. The simplest method is a thresholding operation, where the binary image is processed from the gray-level image by the operation

$$b(m,n) = \begin{cases} 0 & \text{if } x(m,n) < \Theta \\ 1 & \text{if } x(m,n) \geq \Theta. \end{cases} \tag{12.1}$$

By determination of optimum threshold values Θ, it can be avoided that the relevant picture structures disappear by thresholding. Methods to compute best choices of threshold values are in principle similar to histogram-based thresholding described for the purpose of segmentation in sec. 10.1.1. Concerning the statistical behavior, it can be expected that the binarized image inherits properties of statistical dependency between adjacent samples (e.g. correlation) from the original gray-value image (cf. Problem 7.5). The coherence between adjacent samples in the binary signal is best modeled by Markov chains (cf. sec. 3.4).

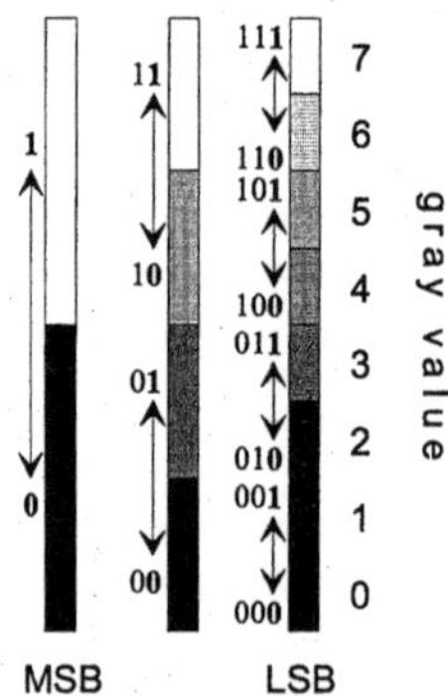

Fig. 12.1. Contributions of single bits to overall gray levels of a 3 bit discrete amplitude representation.

Encoding of a gray-level image can also be performed by binary-image coding techniques, if the signal is interpreted by a series of *bit planes* as obtained by embedded quantization (cf. Fig. 11.8). In an unsigned-number code (e.g. positive integer numbers for PCM representation of an image), each bit interprets a certain difference of gray level (see. Fig. 12.1). In the example of a $B=3$ bit representation, $2^3=8$ different gray levels can be represented, where the most significant bit (MSB) signifies a change of the amplitude level by 2^{B-1} (=4), the second bit effects a value difference by 2, and the least significant bit (LSB) a change by $2^0=1$. The respective (0,1) bit values from each pixel position are combined into B separate binary images, which does not increase the total number of bits as compared to a B bit PCM representation. Fig. 12.2 shows an example of the first three bit planes of a gray-level image. It is interesting to observe that the structural detail increases by decreased significance of the bit plane; most structure is usually observed in the LSB, where also the effect of local amplitude fluctuations caused e.g. by texture structures or additive noise is highest. As a consequence, binary coding methods which exploit the spatial coherence of the signal are becoming less efficient when applied to bit planes of lower significance; direct bit plane coding of images is typically not useful for more than 4 bit-plane levels, which in total would correspond to a representation of 16 different gray-level amplitudes[1].

Fig. 12.2. Left to right: Bit planes 1 (LSB), 2 and 3 (MSB) of a PCM signal.

Halftoning is another method which is often used (in particular in printing systems) to map a multiple-amplitude signal into a binary signal, while retaining a gray-level illusion. This is achieved by increasing the binary sample density. For the case of printing on white paper, the percentage of black dots is made higher in dark areas and lower in light areas. If the packaging of dots is sufficiently dense, the observer's visual system, which has a lowpass amplitude transfer function, performs an interpolation such that a mixture of black dots and white areas appears to be a certain gray-level[2]. By halftoning, high-frequency components are introduced systematically into the bi-level signal, which effects frequent changes between levels '0' and '1'. Consequently, half-toned signals do not expose the typical nearest-neighbor coherence behavior usually expected from simple binary image models. As halftoning patterns however usually show periodic structures or structures that can be broken down into elementary groups of equal pixel constellations, application of larger-area coherence or context models can better be used for systematic data compression. Halftoning is a typical example, how reduced amplitude resolution can be traded against increased sample resolution, thereby preserving the relevant information in the signal[3].

---

[1] This is the usual approach taken by a fax transmission with 'photo quality' setting.

[2] Halftoning methods are known, which are optimum for subjective impression, see e.g. [MEŞE, VAIDYANATHAN 2000].

[3] Interestingly, run-length coding is in a certain sense the antipodal approach, where a binary signal is transformed into a multi-level signal of less samples.

Run-length coding. The method of run-length coding as introduced in sec. 11.4.8 is widely used in binary signal coding. For 2D signals, row-wise processing is usually performed. However, additional methods can be employed in case of 2D signals to utilize the redundancy (correlation) between adjacent lines. The following methods can be applied:

- *Prediction*: Previously-encoded samples either from the current row, or from the row above can be used to predict the current sample **X**. It is necessary to use only samples already known to the decoder, which must perform the same prediction. In Fig. 12.3a, it is assumed that four samples **A-D** are used, which can have a total of 16 different configurations. For each configuration, a prediction rule must be defined; examples of three different rules are also illustrated in Fig. 12.3a. The prediction error will then be set to '0', when the actual value fits with the prediction and '1' otherwise. A high occurrence of zero-values can be expected if the prediction is successful, such that it is advantageous to apply the run-length method of Fig. 11.18b afterwards[1].

- *Relative Address Coding* (RAC) and *Relative Element Address Designate* (READ): The addresses of transitions (white→black, black→white) are encoded relative to the address of a transition in the previous row (Fig. 12.3b); only for the first row, direct run-length coding is necessary. Different variants of this principle are possible, which are e.g. implemented in the fax standards of classes G3/G4.

- *Line skipping*: If a row is completely identical with the previous, this is indicated by a specific code.

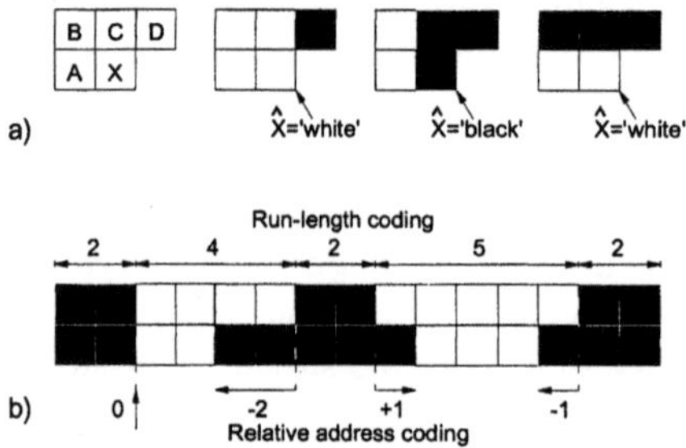

Fig. 12.3. a Prediction of binary samples, prediction topology (left) and examples of three different prediction rules **b** Relative address coding

Template prediction. Run-length methods have a disadvantage if applied to fine-structured drawings and raster graphics, as these systematically show frequent changes between white and black pixels. In such cases, the method of *template prediction* can achieve significantly better results (see Fig. 12.4a): A template is a

---

[1] Application of such prediction schemes is not limited to combination with run-length coding; any other entropy-coding method can be used as well to encode the prediction error signal.

pattern from the direct neighborhood of the current pixel position. Within a certain surrounding range of pixels previously transmitted, the best match for this template (excluding the current position itself) is searched. The identical search is then possible at the encoder and the decoder. The prediction for the current pixel will be the value which is found at the related position of the best match. If the actual value of the current pixel deviates from the prediction, a '1' is encoded, otherwise a '0'. In the case of complex and periodic structures as they appear in rastered images and shaded graphics, the prediction result will be much better than the simple nearest-neighbor prediction shown in Fig. 12.3a. Any entropy coding can be used to encode the prediction errors; adaptive arithmetic coding is one of the best suitable methods here, a combination which is e.g. used in the JBIG standard (see sec. 17.3.1).

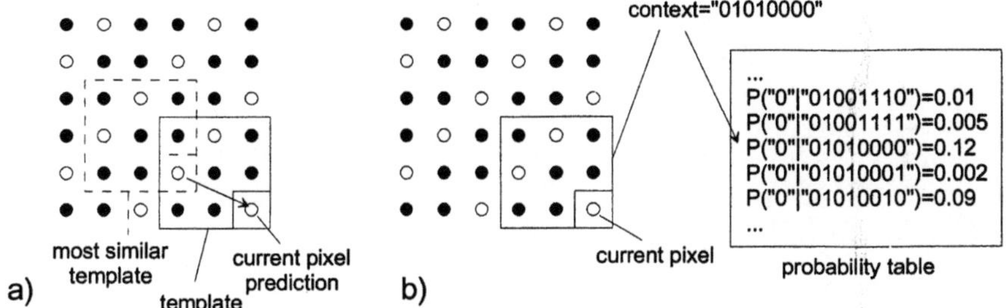

Abb. 12.4. a Template prediction **b** context-based arithmetic encoding

Context-based arithmetic encoding (CAE). CAE is typically based on exploitation of conditional probabilities over a local area of samples (cf. sec. 11.4.5). In principle, the effect is similar with prediction methods, however a direct coupling is made between the (nonlinear, i.e. conditional probability based) prediction step and the step of entropy coding. An example is shown in Fig. 12.4b. The context $\mathcal{K}$ is a pre-defined mask of pixels. If the neighborhood is defined over P pixels, 2^P different configurations can occur; for Fig. 12.4b, $P=8$. For each of the possible configurations, the conditional probability must be given, by which the current pixel will have a value '0' or '1'. A fixed probability table can be used, where only one of the two probabilities must be stored in the table due to the relationship $P(1|\mathcal{K})=1-P(0|\mathcal{K})$. Arithmetic coding is directly employed using this probability table; additionally, adaptation of probability values can be used, as discussed in sec. 11.4.6. CAE is likewise applicable to binary signals or to bit-planes. CAE is used in the MPEG-4 standard to encode the binary shape of segmentation information, based on fixed probability tables, but with context configurations that can be switched depending on shape orientation, temporal prediction etc. In JPEG 2000, it is applied for bit plane coding of wavelet coefficients, where also different contexts are used and the probability tables are continuously adapted.

12.2 Vector Quantization of Images

VQ can be applied directly to image signals, as shown in Fig. 11.20b. If the values within the vector blocks are statistically dependent, this is implicitly exploited when codebooks are designed by the Generalized Lloyd Algorithm (GLA, see sec. 11.5.3).

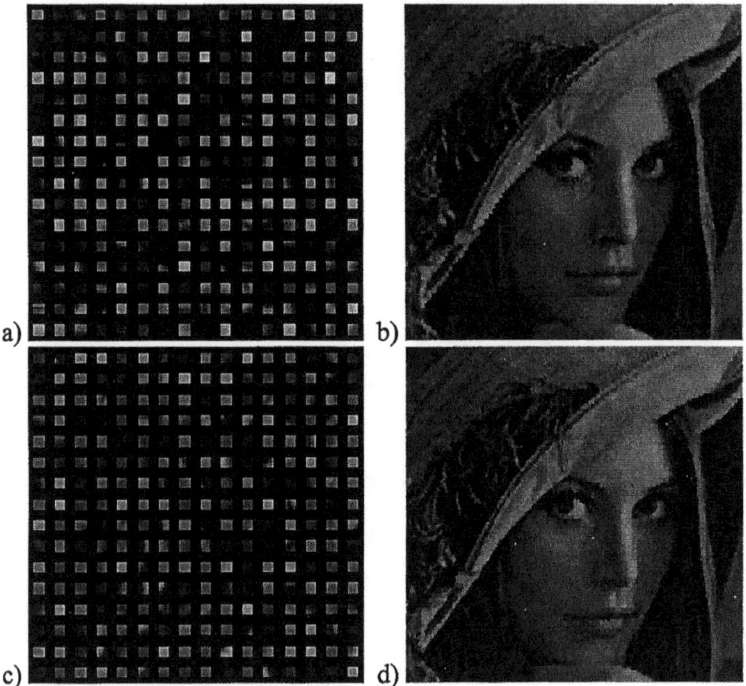

Fig. 12.5. VQ in image domain, codebook size J=256 **a** Codebook generated by GLA procedure from a training set of 15 images **b** VQ reconstruction of an image not contained in training set **c** codebook generated from single image **d** VQ reconstruction of this image

An important aspect is proper selection of the training set. If the set is not statistically representative, the characteristics of the codebook may be biased to the properties of the set, and quality may suffer when different images outside of the training set are encoded[1]. Alternatively, it is possible to use *adaptive VQ*, which has a similar effect as adaptive entropy coding. If the overhead needed to transmit the codebook is not too high, a gain in coding performance can be achieved by using a codebook specifically designed for one image. This also applies for image sequences, where images within one shot can be expected to be quite similar, such

---

[1] This statement is true for any optimization of coding algorithms where fixed code tables are used, including methods of entropy coding.

that the rate needed for transmission of the entire codebook or some adapted vectors can be held reasonably small as compared to the overall rate[1].

The effect of the codebook design on the quality of the reconstruction is illustrated in Fig. 12.5. Codebooks designed for an image domain VQ with J=256 vectors and K=4x4 pixels vector block size are shown in Figs. 12.5a/c; each of the small squares is one reconstruction value. The codebook in Fig. 12.5a was generated using a training set of 15 images. The reconstruction result of an image not contained in the training set is shown in Fig. 12.5b. In contrast, Fig. 12.5c shows a codebook which was specifically designed for this one image. Observe that some vectors exactly show structures like the brim of the hat. Obviously, the reconstruction image in Fig. 12.5d shows much less blocking artifacts.

The block-separate VQ method described here has the following disadvantages:

- *Blocking artifacts* become visible in the reconstructed signal, most typically in flat areas and around edges;
- Due to relatively small block sizes which have to be used to keep codebook size and algorithmic complexity low, a high redundancy can exist between neighboring blocks.

In the remaining part of this section, some methods are described to solve these problems[2].

Mean separating VQ. Amplitude discontinuities in images (e.g. at edges) result in a *locally variable* mean value. In the codebook examples from Fig. 12.5, many vectors are apparently optimized purely for quantization of approximately flat areas. Separating the mean value of a block prior to VQ encoding will hence enforce a higher percentage of codebook vectors to represent details within the block. In *mean separating VQ* (Fig. 12.6a), the block mean value of the vector $\mathbf{x}(m',n')$

$$\mu(m',n') = \frac{1}{M' \cdot N'} \sum_{m=m'M'}^{(m'+1)M'-1} \sum_{n=n'N}^{(n'+1)N'-1} x(m,n) \tag{12.2}$$

is computed, subtracted from the elements of the vector and encoded separately. After decoding, the reconstructed mean value is added to the pixels of the zero-mean reconstruction vector. If a rate R_M is used to encode the block mean values, the total rate per vector (without entropy coding) becomes $R_M + \log_2 J$. As high correlation can be expected between amplitude levels of neighbored blocks, pre-

---

[1] This means that the codebook either should be small, or can be represented by a reasonably low number of bits; alternatively, it is also possible to adapt only a sub-set of vectors from the codebook.

[2] Most of these methods are likewise applicable to other types of multimedia signals. They are however more effective and hence mainly used in image coding due to the specific properties like non-zero mean, high correlation of nearest neighbors etc.

dictive (DPCM) coding of mean values (see sec. 12.3) allows to allocate even less rate, such that a better compression performance is achieved.

Classifying VQ. In classifying VQ [RAMAMURTHY, GERSHO 1986], a *feature classification* of the vectors is performed prior to encoding (Fig. 12.6b). This is in principle based on a composite-source model, assuming that the image consists of different types of sub-sources like edges, textured areas etc.; optimization of encoding is then done separately for each sub-source type. Useful classification criteria are *block variances* and dominant *edge directions* within the blocks. Based on the classification result, each vector is assigned a class label S_l, indicating one out of L classes. For each class, a specific *sub-codebook* $\mathcal{C}_l=\{\mathbf{y}_{j,l}; j=1,2,...,J_l\}$ is provided; these sub-codebooks can be generated when an identical classification is applied before running the codebook-generation algorithm, such that the training sequence is partitioned a priori into the given classes. The *vector index j* and the *class index l* must be conveyed to the decoder. The reconstruction vector is then addressed from a codebook which is the unification of all sub-codebooks

$$\mathcal{C} = \bigcup_{l=1}^{L} \mathcal{C}_l = \left\{\mathbf{y}_{j,l}; j=1,2,...,J_l; l=1,2,...,L\right\}. \tag{12.3}$$

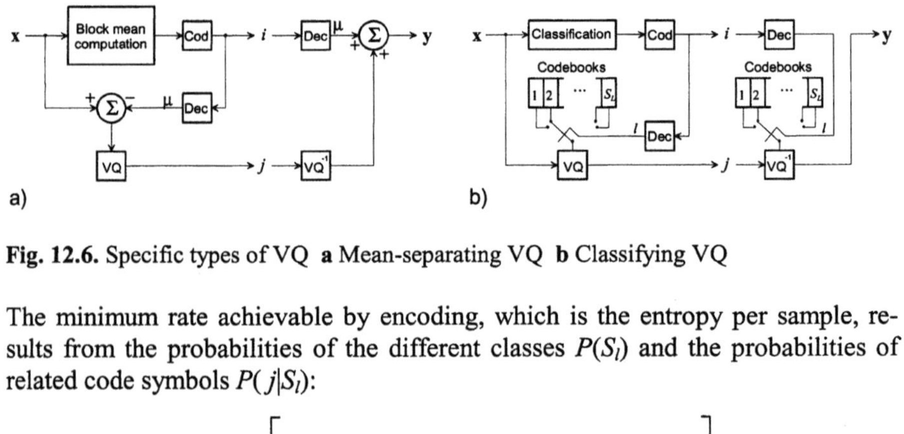

Fig. 12.6. Specific types of VQ **a** Mean-separating VQ **b** Classifying VQ

The minimum rate achievable by encoding, which is the entropy per sample, results from the probabilities of the different classes $P(S_l)$ and the probabilities of related code symbols $P(j|S_l)$:

$$R \le -\sum_{l=1}^{L} \frac{P(S_l)}{K_l} \cdot \left[\underbrace{\log_2 P(S_l)}_{\text{class index entropy}} + \underbrace{\sum_{j=1}^{J_l} P(j|S_l) \cdot \log_2 P(j|S_l)}_{\text{vector index entropy}} \right]$$

$$\le \sum_{l=1}^{L} \frac{P(S_l)}{K_l} \cdot \log_2 \frac{J_l}{P(S_l)}. \tag{12.4}$$

Here, also the case is considered where different block sizes (vector lengths K_l) are used in the classes. If the distortion D_l occurs in class S_l, the total distortion over all classes will be

$$D = \sum_{l=1}^{L} P(S_l) \cdot D_{S_l} .$$

(12.5)

Classifying VQ allows to use a higher rate for classes which are more complex, e.g. more detailed or edge blocks. Due to the classification, better adaptation of the codebooks to locally changing (instationary) source statistics is achieved.

Block-overlapping VQ. Block overlap techniques use information from two or more decoding block units at a given pixel position, which typically achieves reduction of blocking artifacts[1]. The principle is illustrated in Fig. 12.7a. The top-left position of a block with index (m',n') is at the coordinate $(m' \cdot M', n' \cdot N')$, the block size (of the vectors) is $K=M''\text{x}N''$. The block overlaps with eight direct neighbors by a width $M''-M'$ horizontally and $N''-N'$ vertically. A separable window function $w(m'',n'')$ can be used to realize the block overlap and the composition of the respective neighbored blocks, such that

$$w(m'',n'') = \begin{cases} w(m'') \cdot w(n'') & \text{for } 0 \le m'' < M'' \text{ and } 0 \le n'' < N'' \\ 0 & \text{for } m'' < 0 \text{ or } m'' \ge M'' \text{ or } n'' < 0 \text{ or } n'' \ge N'' \end{cases}$$

(12.6)

with

$$m'' = m - m'M' \quad ; \quad n'' = n - n'N' ,$$

(12.7)

and the 1D window functions defined such that the following conditions hold:

$$w(m'') = w(M'' - m'' - 1) \quad ; \quad w(m'') + w(m'' + M') = 1$$
$$w(n'') = w(N'' - n'' - 1) \quad ; \quad w(n'') + w(n'' + N') = 1.$$

(12.8)

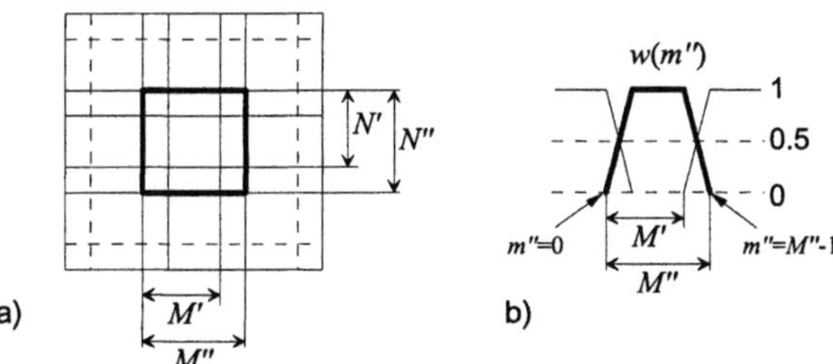

Fig. 12.7. VQ with block overlap **a** Overlap geometry in the 2D spatial domain
b Example of a trapezoid window function, with overlap positions for neighbored blocks

An example of a trapezoid window function which increases/decreases linearly within the range of overlap is shown in Fig. 12.7b. It is also possible to realize non-separable window functions or complete block overlaps (e.g. $M''=2M'$ will

---

[1] Similar block-overlapping methods are also applied in context of other block-based coding schemes, like block transforms and block-based motion compensation.

give an overlap up to the block center position). The distortion after decoding results by the superposition of coding errors from all blocks contributing to one sample position, weighted by the respective window function. If the weighting by the window function shall be applied prior to encoding of $\mathbf{x}(m',n')$, also the vectors within the codebook must be appropriately windowed:

$$\mathbf{x}(m',n') = \left[x(m'M', n'N') \cdot w(0,0), ..., x(m'M' + M'' - 1, n'N' + N'' - 1) \cdot w(M'' - 1, N'' - 1) \right]^{\mathrm{T}} . \quad (12.9)$$

This allows to stick to the VQ procedures as introduced previously; also design of the codebook must then be performed from a training sequence of windowed vectors. The search for optimum reconstruction vectors can be performed as in conventional VQ, separately for the single blocks[1]. According to (12.8)

$$\sum_{m''=0}^{M''-1} \sum_{n''=0}^{N''-1} w(m'',n'') = M' \cdot N' , \quad (12.10)$$

however it is further straightforward to show that for any window function that fulfills (12.8),

$$\sum_{m''=0}^{M''-1} \sum_{n''=0}^{N''-1} w^2(m'',n'') \le M' \cdot N' , \quad (12.11)$$

where equality is exactly and only obtained for the case of non-overlapping rectangular blocks, $M''=M'$ and $N''=N'$. From this, it is interesting to observe that, even though the area of the vector is extended, the squared vector norm of the windowed vectors is apparently decreased. Hence, while the vector size K increases by the percentage overlapping block areas, the codebook size J needs not to be increased or can even be made lower for comparable performance. Introduction of block overlap in VQ *will not result in an increased data rate*, but will help to reduce the unnatural blocking artifacts. Observe however that block-overlapping VQ as introduced here does not take into account the statistical dependencies between adjacent blocks; all vectors are encoded independently, and the overlap processing is done prior to encoding, or in the reconstruction after decoding.

Finite State VQ (FSVQ). When the vector $\mathbf{x}(\mathbf{n}')$ is processed, coder and decoder are in a certain state $s(\mathbf{n}')$, which can e.g. be determined from the constellation of reconstruction vectors selected previously in the neighborhood. If the number of possible values s instantiated in $s(\mathbf{n}')$ is limited to S, encoder and decoder can be interpreted as *finite-state machines*. In any state s, only a subset of possible reconstruction vectors, which are organized within a *state codebook* $\mathcal{C}_s$, are allowed to be addressed. The unification of all state codebooks forms the *super codebook*

---

[1] The 'normal' block-wise VQ can be interpreted as a special case, where the window functions are non-overlapping rectangles, such that $M''=M'$, $N''=N'$.

$$\mathcal{C} = \bigcup_{s=1}^{S} \mathcal{C}_s \quad ; \quad \mathcal{C}_s = \left\{ \mathbf{y}_{j,s}; j = 1, 2, ..., J_s \right\}. \tag{12.12}$$

By the formulation in (12.12), it is not yet determined whether reconstruction vectors can be members of different state codebooks. If the $\mathcal{C}_s$ are overlapping sets, the total number of reconstruction vectors will be lower than the sum over all J_s from the single states. The computation of the code symbols $i(\mathbf{n}')$ and the follow-up state $s(\mathbf{n}'+1)$ are performed according to the relations[1]

$$i(\mathbf{n}') = \arg\min_{\mathbf{y}_j \in \mathcal{C}_{s(\mathbf{n}')}} d(\mathbf{x}(\mathbf{n}'), \mathbf{y}_j) \quad ; \quad s(\mathbf{n}'+1) = f(i(\mathbf{n}'), s(\mathbf{n}')) . \tag{12.13}$$

In state s, J_s different code symbols can be selected. This requires a significantly lower rate as compared to the address space necessary for the entire super code-book $\mathcal{C}$. The performance highly depends on the definition of the *next-state function* $f(j,s)$. By this function – dependent on the previous state and the previous transmitted code symbol – the state codebooks $\mathcal{C}_s$ are defined. These should consist of those subsets of reconstruction vectors, which are most likely to be selected within a certain state configuration. General methods for optimization of FSVQ codebooks are described in [DUNHAM, GRAY 1985], [FOSTER, GRAY, DUNHAM 1985].

The block diagram of an FSVQ encoder and decoder is shown in Fig. 12.8a. Observe the structural similarity with classifying VQ (Fig. 12.6b) – FSVQ could also be interpreted as a backward-controlled version of classifying VQ, not requiring encoding and transmission of the information about the sub-codebook which is actually selected.

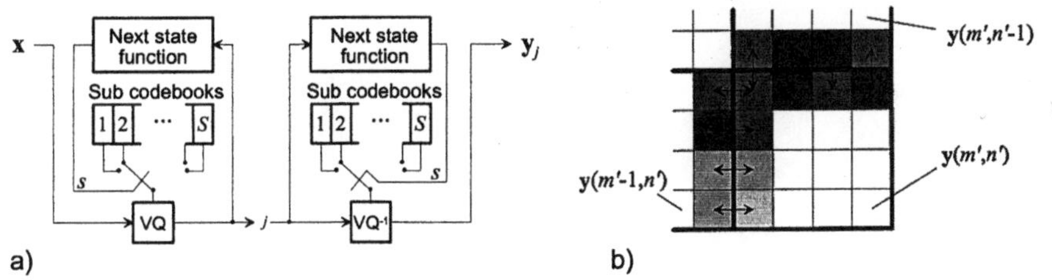

Fig. 12.8. **a** FSVQ encoder and decoder
b Example for definition of the next-state function in side-match VQ

Side-match FSVQ. In this method which was introduced in [KIM 1992], a super codebook of J reconstruction vectors is designed as in conventional VQ. The number $J_s = J^*$ of code symbols shall be constant in all states. Within one state, those J^* reconstruction vectors are selected, for which the *lowest differences* in amplitude occur at the boundaries between the actual vector and its left and top already decoded neighbors (see arrows and shaded areas in Fig. 12.8b). The maximum possible number of states is then $J \times J$. The state selection function is

---

[1] For $\mathbf{n}'=(m',n')$, $\mathbf{n}'+1$ means e.g. $(m'+1,n')$ in the case of row-wise sequential processing.

$s(m',n')=f[\mathbf{y}(m'-1,n'),\mathbf{y}(m',n'-1)]$, whereby the reconstruction vectors again result as a function of previous states and code symbol indices.

Combination of FSVQ with entropy coding. The side-match method or other next-state functions can also be used for conditional entropy coding of the code symbols (indices) as output by a common VQ encoder. Assume a codebook has size J, where only $(J^*+1)\ll J$ code symbols are represented by a variable-length code. These are J^* vectors best fitting according to the side-match method, and one 'escape' code indicating that the best match from the codebook is not one of the J^* side-match vectors. As it can further be expected that the vectors are selected by highest probability when the amplitude difference across the block boundary is lowest, the entropy rate can in fact be significantly below $\log_2 J^*$. For typical images, the escape condition is used by less than 10%. Only by these exceptional cases, the index allowing to select one of J vectors of the super codebook directly must be encoded by $\log_2 J$ bits. With $J=256$, the encoded data rate of a typical 2D block VQ can be reduced by a factor of almost two without any loss. This again shows the high amount of redundancy still inherent in methods of separate-block encoding. Fig. 12.9 shows reconstructed results generated by entropy-encoded FSVQ.

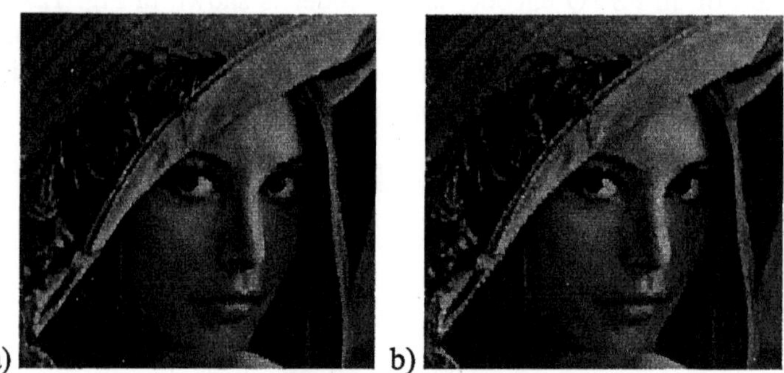

Fig. 12.9. Side-Match-FSVQ used for conditional entropy coding
a $R=0{,}31$ *bpp* $(J=1024, J_s=64)$ **b** $R=0{,}21$ *bpp* $(J=256, J_s=16)$

FSVQ is based on a context-dependent *nonlinear prediction*, or in the latter method can also be interpreted as a conventional VQ with subsequent context-adaptive entropy coding. Vector quantization in combination with *linear prediction* will be treated with more detail in sec. 12.3.3.

12.3 Predictive Coding

Predictive coding has been applied in speech, image and video compression over decades. Even though the main stream in still image coding performs transform coding today, even the most recent developments of multimedia compression standards make extensive use of predictive components. Besides the image information (color, texture) itself, parameters such as motion vectors, and also transform coefficients are often predictive encoded. Predictive coding is also frequently applied in the area of lossless compression. In particular by combination with more advanced quantization and entropy coding techniques, still more improvements seem to be achievable.

12.3.1 DPCM Systems

In Fig. 4.14, the general principle of linear prediction was shown. If the signal $e(\mathbf{n})$ can be represented by a discrete set of amplitude values, entropy coding can directly be applied to the prediction error signal. This will lead to a reduction in data rate, compared to independent sample-wise encoding of the (likewise discrete) $x(\mathbf{n})$, which is by the order of magnitude of the gains derived in (11.17), (11.18) and (11.24) for the case of AR processes. A different situation occurs however, when $e(\mathbf{n})$ is quantized into reconstruction values $v(\mathbf{n})$, as shown in Fig. 12.12a[1]. Here, a quantization error $q(\mathbf{n})=e(\mathbf{n})-v(\mathbf{n})$ occurs. The resulting effect can be expressed in the spectral domain for any κ-dimensional prediction system as follows. Frequency spectra over $\Omega=[\Omega_1,\Omega_2,..,\Omega_\kappa]$ shall be defined for the original signal $X(\Omega)$, the reconstructed signal $Y(\Omega)$, the prediction error signal $E(\Omega)$, the quantized prediction error signal $V(\Omega)$, the predictor filter $H(\Omega)$ and the quantization error $Q(\Omega)$. Then, the spectrum difference between the signals $x(\mathbf{n})$ and $y(\mathbf{n})$ can be determined as

$$Y(\Omega) = \frac{V(\Omega)}{1-H(\Omega)} = \frac{E(\Omega)}{1-H(\Omega)} - \frac{Q(\Omega)}{1-H(\Omega)} = X(\Omega) - \frac{Q(\Omega)}{1-H(\Omega)}, \qquad (12.14)$$

by which

$$X(\Omega) - Y(\Omega) = \frac{Q(\Omega)}{1-H(\Omega)} \quad \Rightarrow \quad x(\mathbf{n}) - y(\mathbf{n}) = q(\mathbf{n})*b(\mathbf{n}). \qquad (12.15)$$

In (12.15), $b(\mathbf{n})$ is the impulse response of the recursive synthesis filter with transfer function $B(\mathbf{z})=1/[1-H(\mathbf{z})]$. The effect of convolving the quantization error by the impulse response of the synthesis filter is denoted as *drift*. Due to the high correlation properties of image signals, synthesis filters will typically be recursive

---

[1] The system shown in Fig. 12.12a is also denoted as D*PCM [JAYANT, NOLL 1984].

filters with poles near to or even on the unit circle/hypersphere in the **z** domain; in such cases, the drift becomes extremely critical and uncontrollable.

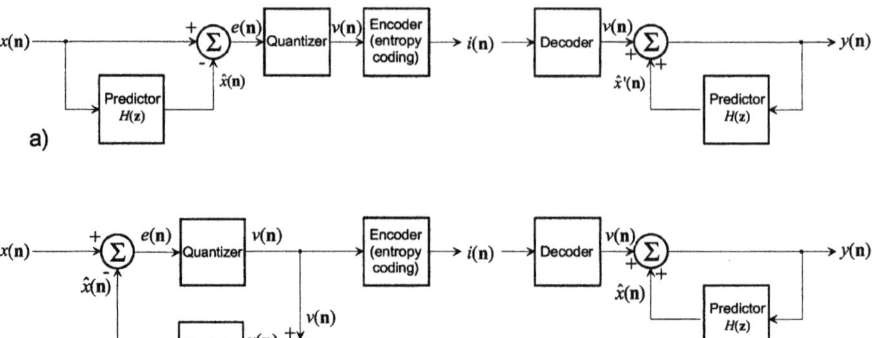

Fig. 12.10. Encoders and decoders of predictive systems with quantization of the prediction error signal **a** Open loop (D*PCM) **b** Closed loop (DPCM)

The system from Fig. 12.12a has an *open loop*, where the recursion at the decoder is not under control from the encoder side. As a consequence, the estimates $\hat{x}(\mathbf{n})$ and $\hat{x}'(\mathbf{n})$ diverge. The block diagram of a *Differential Pulse Code Modulation* (DPCM) system, which is a closed-loop approach and compensates this crucial defect, is shown in Fig. 12.10b. The predictor computes an estimate $\hat{x}(\mathbf{n})$ for the pixel $x(\mathbf{n})$. The *prediction error*

$$e(\mathbf{n}) = x(\mathbf{n}) - \hat{x}(\mathbf{n}) \tag{12.16}$$

is quantized into the value $v(\mathbf{n})$, which is then encoded and transmitted. At the receiver side, the decoder performs the following operation to generate the reconstructed value

$$y(\mathbf{n}) = v(\mathbf{n}) + \hat{x}(\mathbf{n}) \, . \tag{12.17}$$

As only the *quantized value* $v(\mathbf{n})$ is available at the receiver, both the encoder and the decoder must use identical estimates $\hat{x}(m,n)$. This can be achieved if the decoding step (12.17) is identically performed at the encoder side, which is the basic difference compared to the D*PCM scheme of Fig. 12.10a. As a result, the quantization error between $e(\mathbf{n})$ and $v(\mathbf{n})$ becomes identical with the reconstruction error (coding error) between $x(\mathbf{n})$ and $y(\mathbf{n})$:

$$q(\mathbf{n}) = e(\mathbf{n}) - v(\mathbf{n}) = \left[x(\mathbf{n}) - \hat{x}(\mathbf{n})\right] - \left[y(\mathbf{n}) - \hat{x}(\mathbf{n})\right] = x(\mathbf{n}) - y(\mathbf{n}) \, . \tag{12.18}$$

Only for the case $e(\mathbf{n})=v(\mathbf{n})$, which means when lossless encoding of $e(\mathbf{n})$ is performed, DPCM and D*PCM give identical results. This also requires that the signal $x(\mathbf{n})$ and the prediction $\hat{x}(\mathbf{n})$ are both quantized into a set of discrete amplitude values, such that $e(\mathbf{n})$ will also be instantiated from a discrete set. In any other case, DPCM encoding is lossy, such that $y(\mathbf{n}) \neq x(\mathbf{n})$, and $y(\mathbf{n})$ must be used to compute the prediction.

The transfer function of the predictor filter is unchanged as compared to (4.92), however the input-output relationship of a DPCM encoder loop can no longer be described as a linear system as in (4.88), as the mapping $x(m,n) \rightarrow v(m,n)$ includes the nonlinear operation of quantization. The coding gain of a DPCM system is defined as the ratio of variances between original signal and prediction error signal in analogy with (11.17),

$$G = \frac{\sigma_x^{\,2}}{\sigma_e^{\,2}} .$$

(12.19)

The solution to the drift problem made by DPCM does however not come without a penalty. The related effect is known as *quantization error feedback*, which is derived here again generically for any κ-dimensional DPCM system. By formulating (12.16)-(12.18) in the frequency domain, we get

$$Q(\Omega) = E(\Omega) - V(\Omega) = X(\Omega) - Y(\Omega)$$

$$Y(\Omega) = \frac{V(\Omega)}{1 - H(\Omega)}; \quad E(\Omega) = X(\Omega)[1 - H(\Omega)],$$

(12.20)

which gives

$$V(\Omega) = \left[X(\Omega) - Q(\Omega) \right] \cdot \left[1 - H(\Omega) \right]$$

$$E(\Omega) = \underbrace{X(\Omega) \cdot \left[1 - H(\Omega) \right]}_{E'(\Omega)} + Q(\Omega) \cdot H(\Omega).$$

(12.21)

When $e'(\mathbf{n})$ is defined as the prediction error of the lossless or D*PCM cases, the prediction error in DPCM can be expressed by

$$e(\mathbf{n}) = e'(\mathbf{n}) + q(\mathbf{n}) * h(\mathbf{n}) ,$$

(12.22)

which means that the DPCM prediction error signal contains an additional component, which is the convolution of the quantization error by the impulse response of the predictor filter. Hence, it is unavoidable that the prediction gain of DPCM (12.19) is lower than the maximum possible gain (11.17) whenever the encoding gets lossy. This means that lossy DPCM can never realize the maximum possible coding gain, unless specific encoding mechanisms would be used that take into account the characteristics of quantization error feedback and the resulting correlation in the prediction error. The spectrum of the prediction error signal, even in the case of ideal prediction of a stationary Gaussian process (AR model) would not be flat (white noise) anymore. This means that further compression of the prediction error could be made by utilizing the feedback phenomenon, which is however not easy to achieve due to the recursive relationships in the DPCM loop. Further, it can be concluded that this effect increases towards lower rates, where $E\{q(\mathbf{n})\}$ grows larger. Regard the extreme case of quantizing all samples into zero, $V(\Omega)=0$, $Q(\Omega)=X(\Omega)$ and hence $E(\Omega)=X(\Omega)$. This will be the point $D=\sigma_x^2$, $R=0$ where all rate-distortion graphs in Fig. 11.5 converge; prediction is useless, as the decoder does not hold any information about previous samples, either.

12.3.2   Predictor filters in 2D DPCM

The computation of estimates must only use values already known to the decoder due to causality considerations. This is guaranteed in the DPCM scheme shown in Fig. 12.10, when a causal filter is used in the prediction loop. For simplicity, it shall be assumed that the sequence of processing starts in the top left corner and proceeds row-wise over an image. The prediction equation in DPCM using a quarter-plane filter of order PQ-1 is by modification of (4.91)

$$\hat{x}(m,n) = \sum_{\substack{p=0 \\ (p,q)\neq(0,0)}}^{P-1} \sum_{q=0}^{Q-1} a(p,q)\cdot y(m-p,n-q). \tag{12.23}$$

The choice of the predictor filter highly influences the performance of the DPCM system. The predictor filters may be chosen according to separable or non-separable AR models, but sometimes are designed under more heuristic considerations. Examples are the following non-separable predictor filters, where the first performs the prediction from the nearest horizontal and vertical neighbors

$$\hat{x}(m,n) = a_h \cdot x(m-1,n) + a_v \cdot x(m,n-1) \quad \Rightarrow H(z_1,z_2) = a_h \cdot z_1^{-1} + a_v \cdot z_2^{-1}, \tag{12.24}$$

where e.g. $a_h=a_v=0{,}5$ (averaging) is a typical choice of values. The other example is the asymmetric half-plane filter of four nearest causal neighbors

$$\hat{x}(m,n) = a_h \cdot x(m-1,n) + a_v \cdot x(m,n-1) + a_{d^+} \cdot x(m-1,n-1) + a_{d^-} \cdot x(m-1,n+1)$$
$$\Rightarrow H(z_1,z_2) = a_h \cdot z_1^{-1} + a_v \cdot z_2^{-1} + a_{d^+} \cdot z_1^{-1} \cdot z_2^{-1} + a_{d^-} \cdot z_1^{-1} \cdot z_2^{1}. \tag{12.25}$$

For natural images, predictors with fixed settings of predictor coefficients will not generally be optimum, as image signals are non-stationary. Hence, schemes using an *adaptation of predictors* can be expected to result in a performance gain. For example, correlation will be lower in edge regions than it is in flat areas, further a dependency on edge direction or texture properties can be observed. Components of predictor adaptation are included in the block diagram of Fig. 12.11. The following methods are typically applied for adaptive prediction:

–   Forward adaptation: On the basis of local analysis and estimation of the auto-covariance statistics of the image, predictor coefficients can be determined that would be optimum for an autoregressive model having same statistical properties (see sec. 4.2.1). In case of forward adaptation, the estimation of the parameters is performed from the area to be encoded, and it is necessary to convey the prediction filter parameters as side information to the decoder.

- Backward adaptation: Adjustment of predictor coefficients is performed from signal values and prediction error values which at this step are already known to the decoder. In contrast to (4.95) and (4.96), the LMS algorithm must now use quantized prediction error values $v(m,n)$. It is not necessary to encode and transmit predictor parameters, since the operation can identically be performed at the encoder and decoder sides; no overhead information increases the bit rate.

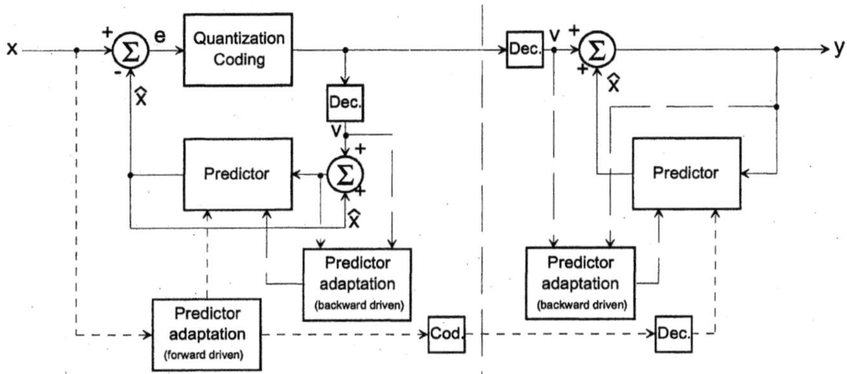

Fig. 12.11. DPCM encoder and decoder, integrating components for predictor adaptation

Switched prediction. Here, a small set of different predictor filters (e.g. specific predictor geometries for horizontal, vertical, diagonal prediction) is pre-defined, from which the most appropriate is selected according to criteria like minimization of the prediction error. Switched prediction is the most simple type of adaptive prediction, and can also be realized either by backward or forward adaptation. In backward-adaptive switching, a typical approach is made by analysis of the causal neighborhood for edge orientations to adapt the predictor geometry by pre-set rules [RICHARD, BENVENISTE, KRETZ 1984]. An example is given in Fig. 12.12. The pixel **X** shall be predicted by a combination of the neighbored decoded pixels **A-D**, which corresponds to (12.25). Depending on the local edge orientation, different rules for the computation of the estimate $\hat{x}$ are set.

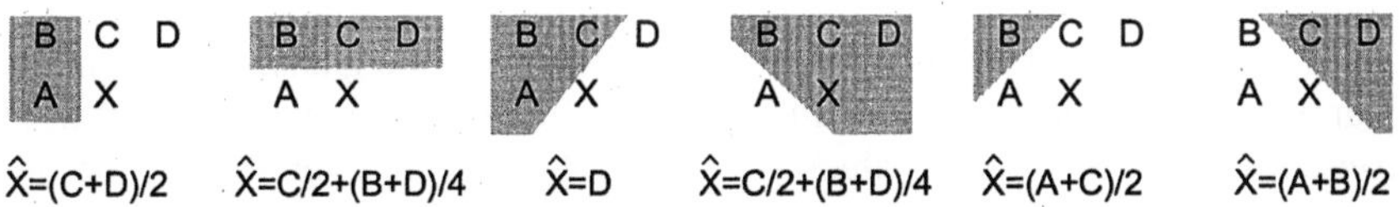

Fig. 12.12. Examples of edge directions and appropriate prediction rules

In forward-adaptive switching, the settings have to be encoded as side information. This can either be done for groups of pixels, or in the extreme case for each pixel individually. For example, in the lossless mode of the JPEG standard, a predictor setting is valid for a group of subsequent pixels, until a new setting is made. Typically, forward-adaptive switching should also include a case where *no prediction* is performed, which then degenerates the scheme into PCM coding. A backward adaptive switched prediction scheme is implemented in the lossless and near-lossless coding standard JPEG-LS.

If DPCM is employed for lossless encoding of discrete-value (PCM) signals, the prediction error values shall be integer numbers, which can only be the case if the estimates $\hat{x}$ are integer as well. This can either be achieved by appropriate setting of the predictor filters, or by a direct integer rounding of the estimates.

Nonlinear predictors. In edge regions, prediction often fails. This can be interpreted such that assumptions of stationarity and linear statistical dependencies between samples do not apply. Nonlinear prediction methods can eventually tackle this problem. Variants of nonlinear predictors are

- *Median predictors*: The median value of a set of neighbored samples results in a simple but nevertheless efficient nonlinear estimate $\hat{x}$. Typically, relatively small filters of three or five input samples are used; weighted median filters or hybrid median/FIR filters can be applied as well.
- *Volterra predictors*: These polynomial filters introduced in sec. 5.1.3 can straightforwardly be applied for prediction, if filter inputs are defined from a causal neighborhood. These filters are however extremely sensitive in case of changing signal statistics, and must be implemented as adaptive filters in any case.
- *Neural network predictors*: Pixels from a causal neighborhood are used as inputs to an ANN, e.g. a multi-layer perceptron (cf. Fig. 9.15). The number of pixels to be used for prediction corresponds to the K inputs of the network, and one output node ($J=1$) gives the prediction result[1]. The adaptation is performed by training the network accordingly. In addition to the better prediction properties at positions of signal discontinuities, ANNs are less sensitive against transmission errors and noise in the signal [DIANAT ET AL. 1991]. The overall procedure is however significantly more complex than for the case of simple linear or median predictors.

12.3.3 Quantization and Encoding of Prediction Errors

Fig. 12.13 shows PDF estimates of one-step prediction error signals, as determined by occurrence counts from a set of natural images. The significant concentration of occurrences for low amplitude values is obvious. In general, even though the image signal is entirely positive, prediction error signals can have positive and negative values and tend to be of zero mean. In the simplest case, a scalar quantization

---

[1] unless a *vector prediction* of several samples shall be performed.

and subsequent entropy coding can be applied. Due to the non-uniform PDF with high concentration around e=0, an entropy $H(\mathcal{V}) \ll \log_2 J$ can be expected when a quantizer with a set of J reconstruction values $\mathcal{V}=\{v_j\,;j=1,\ldots J\}$ is used. For example, let the original signal be represented by 8 bit PCM. The range of possible integer prediction error values is $-255 \leq e \leq 255$, which are 511 different values requiring roughly 9 bit/sample for direct representation. Nevertheless, the entropy of the prediction error signal will be much lower than 9 bit/sample, depending on detail and noise in the original signal. Natural images can typically be lossless encoded by DPCM at rates in the range between 3 and 5 bit/pixel. DPCM using scalar quantization and entropy coding is still a dominating technique in lossless encoding of images, which is e.g. applied in the standards JPEG (lossless mode) and JPEG-LS (see sec. 17.3.2). For the case of B-bit integer precision of the PCM original and usage of integer estimates, the value range of prediction errors must apparently be represented by B+1 bit precision integer values. If B bit positive-integer arithmetic units are used for the computation of the prediction error difference, an underflow of the value range can occur at the encoder during the subtraction in (12.16); the carry flag can however simply be ignored, as the complementary addition in the decoder (12.17) will exactly compensate this by an overflow. This has been denoted as *folded quantization* [BOSTELMANN 1974]. As a consequence, the binary representation of DPCM prediction errors $e(\mathbf{n})$ requires only the same bit-depth precision as given for the original $x(\mathbf{n})$. In fact, for each *concrete* value $\hat{x}(\mathbf{n})$, only 2^B different values of $e(\mathbf{n})$ are possible. For the B=8 bit example and $\hat{x} = 100: -155 \leq e \leq 100$, if $0 \leq x \leq 255$.

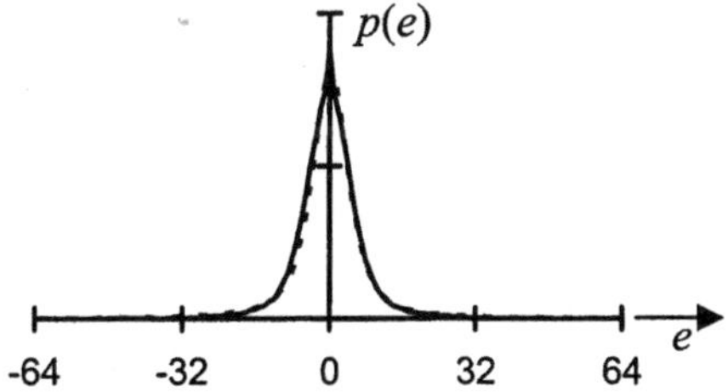

Fig. 12.13. PDF estimate of a prediction error signal in separable 2D prediction (——) , and in 1D horizontal prediction (···)

By combination of scalar quantization and entropy coding of the prediction error signal, encoded rates of less than 1 bit/pixel can be realized. For example, with a set of J=3 reconstruction values $\mathcal{V}=\{-V,0,+V\}$, the probability of selecting v=0 is increased when V is set to a higher level. As all values $|e|<V/2$ would be mapped into the zero reconstruction, the entropy becomes lower as well. For $V \rightarrow \infty$, $P(v$=$0) \rightarrow 1$, such that $H(\mathcal{V}) \rightarrow 0$. The whole image will then be decoded by the value that is used as an initialization of the predictor memory before the decoding starts.

This is indeed the case of maximum possible distortion according to the properties of the rate-distortion function for a rate of zero (sec. 3.5 and 11.2).
Fig. 12.14 shows reconstructed images obtained from DPCM coding by different numbers of quantizer reconstruction values J and resulting entropy rates $H(\mathcal{V})$. Fig. 12.14a represents the case of lossless coding.

Abb. 12.14. DPCM using scalar quantization and entropy coding.
a J=511, $H(\mathcal{V})$=4.79 *bpp* **b** J=15, $H(\mathcal{V})$=1.98 *bpp* **c** J=3, $H(\mathcal{V})$=0.88 *bpp*

Predictive vector quantization. An alternative way to obtain rates of less than one bit/sample by DPCM-based coding schemes is *Predictive VQ* (Fig. 12.15). A *vector predictor* computes an estimate, which is then subtracted from a signal vector, resulting in a *prediction error vector*. A vector quantizer, however now using a codebook of prediction error vectors, is then used within the prediction loop.

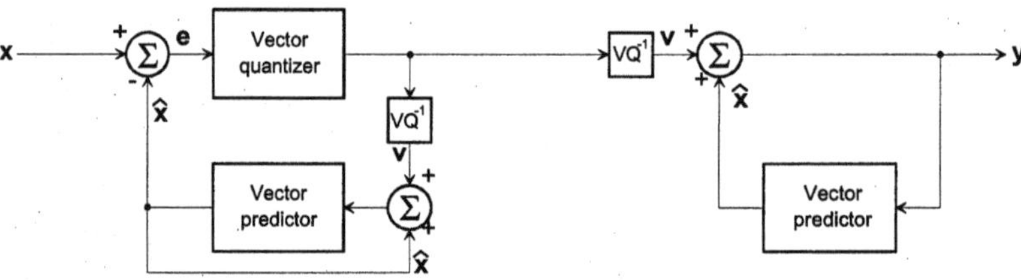

Fig. 12.15. Predictive vector quantization (encoder and decoder)

In predictive VQ, the entire vector must be predicted at once, because otherwise the simultaneous vector quantization would not be possible. As a consequence, samples must not be predicted from other samples within the same vector, which implies as a consequence that the prediction error samples are no longer obtained by *one-step* predictions, where a filter operates recursively sample by sample[1].

---

[1] An alternative way to combine vector codebooks with one-step linear prediction synthesis filters is *code-excited linear prediction* (CELP), which is widely used in speech coding (cf. sec. 14.1). In CELP, the encoder cannot perform the operations of prediction and VQ sequentially in a prediction loop, but must compute the reconstruction from each vector

This shall be shown here for the example of a 1D vector prediction of samples from an AR(1) process. Assume that a K-dimensional vector $\mathbf{x}=[x(n+1)...x(n+K)]^T$ shall be predicted and mapped into a prediction error vector $\mathbf{e}=[e(1)...e(K)]^T$. Obviously, all samples will best be predicted by the nearest neighbor $x(n)$ which is available from the previous vector[1]. The optimum prediction will then be

$$\hat{x}(n+k) = \rho^k \cdot x(n) \Rightarrow e(k) = x(n+k) - \rho^k \cdot x(n) \; ; \; k > 0 . \tag{12.26}$$

This is an optimum one-step prediction from a direct neighbor $x(n)$ only for the first sample in the vector, $x(n+1)$. Now, (12.26) can be used to determine the auto-covariance values of $\mathbf{e}$, which could e.g. be used to set up an autocovariance matrix $E\{\mathbf{ee}^T\}$:

$$\begin{aligned}
E\{ e(l) \cdot e(k)\} &= E\{ \left(x(n+l) - \rho^l x(n) \right) \cdot \left(x(n+k) - \rho^k x(n) \right) \} \\
&= \underbrace{E\{ x(n+l)x(n+k)\}}_{\sigma_x^2 \cdot \rho^{|k-l|}} - \rho^l \underbrace{E\{x(n)x(n+k) \}}_{\sigma_x^2 \cdot \rho^k} \\
&\quad - \rho^k \underbrace{E\{x(n)x(n+l) \}}_{\sigma_x^2 \cdot \rho^l} + \rho^l \rho^k \underbrace{E\{x^2(n) \}}_{\sigma_x^2} \\
&= \sigma_x^2 \cdot \left(\rho^{|k-l|} - \rho^{|k+l|} \right).
\end{aligned} \tag{12.27}$$

Obviously, the samples within the prediction error vector are correlated, even though the correlation is weaker than for samples within the original signal vector $\mathbf{x}$. This effect similarly applies to the case of 2D vector prediction.

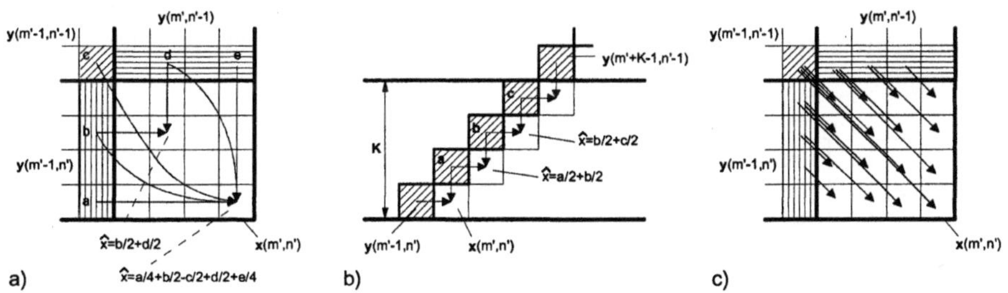

Fig. 12.16. Vector structures in predictive vector quantization: **a** block-wise **b** diagonal **c** block-wise directional

Fig. 12.16 shows three different methods of 2D vector prediction. In the scheme of Fig. 12.16a, a block vector arrangement (block size K=4x4 pixels) is used [HANG,

available in the codebook. This makes both the encoding procedure and the codebook design strategy substantially different compared to predictive VQ.

[1] Actually, the prediction should be performed from the reconstructed value $y(n)$. We assume lossless compression here, which allows to study the effect of vector prediction without the additional effect of quantization error feedback.

WOODS 1985][BHASHKARAN 1987]. In Fig. 12.16b, a group of diagonally-neighbored samples is combined into a vector (in this example, K=4). This technique allows the usage of a set of quarter plane predictor filters as vector predictor combination, such that in fact one-step (nearest-neighbor) prediction is possible for any sample, which minimizes the prediction error [OHM, NOLL 1990]. Fig. 12.16c shows a configuration of *directional prediction*, where the direction itself can be adapted by a switched prediction[1].

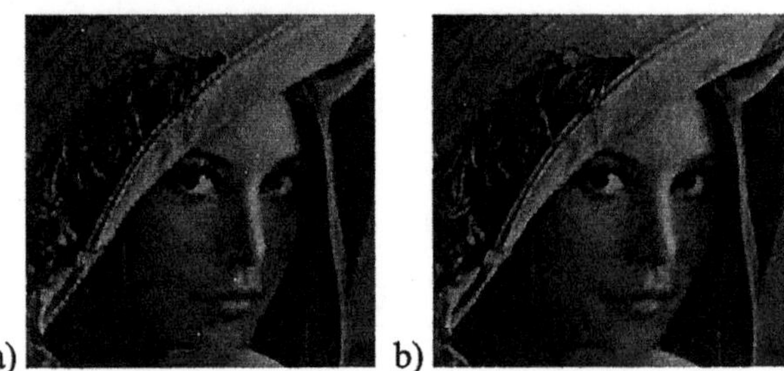

Fig. 12.17. Predictive VQ **a** with block vector structure, R=0,5 b/p
b with diagonal vector structure and delayed decision coding, R=0,3 b/p

In fact, the correlation within the prediction error vectors is not critical for compression performance, as the VQ codebook design can inherently take into account any statistical dependencies. From this point of view, predictive VQ also partially resolves the problem of quantization error feedback, as the codebook statistics will exactly follow the statistics of signals that are used in training. Fig. 12.17 shows reconstruction examples obtained from predictive vector quantization schemes, clearly showing the improved quality over simple DPCM at rates below one bit/sample.

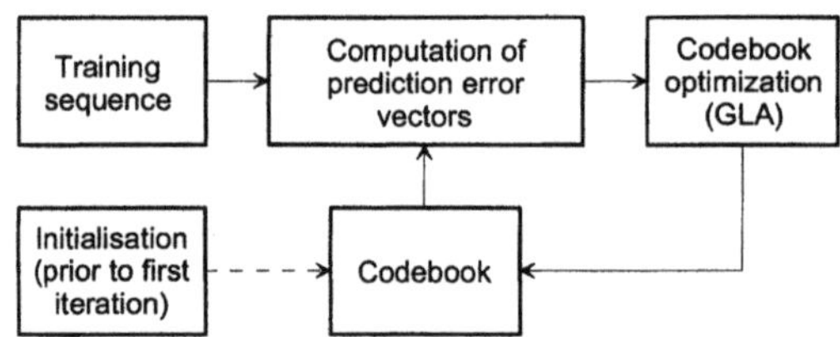

Fig. 12.18. Codebook design in vector DPCM

Codebooks for predictive VQ can also be generated using a training sequence of image signals and the Generalized Lloyd algorithm (cf. sec. 11.5.3). For best adaptation of the codebook to the prediction error signal statistics (where in particular

---

[1] This type of vector prediction is used for intraframe coding in the Advanced Video Coding standard, even though not VQ is used for encoding of the residual, but a block transform is applied to remove the remaining redundancy.

the properties of within-vector correlations and the rate-dependent quantization error feedback are important), the *closed-loop design* shown in Fig. 12.18 should be used. The mapping from original image signal values into prediction error vectors is again performed by each iteration, using as a reference the codebook which was optimized in the previous iteration step.

Tree encoding and DPCM with delayed decision. Scalar and vector DPCM schemes can also be combined with tree encoding, exploiting recursive dependencies over subsequent encoding steps (cf. sec. 11.6.2). In principle, the final decision about the quantization is withheld until it is known which effect it would have on the prediction of subsequent samples. Additionally, the variability of the quantizer can largely be extended by a *sliding block code* (cf. sec. 11.6). A scheme in combination with vector prediction was introduced in [HANG, WOODS 1985]. If in addition an *entropy constrained* codebook design is employed (cf. sec. 11.5.5), encoding by variable bit rate better approaches the entropy rate [COHEN, WOODS 1989].

A specific case, where identical quantizers are used by each step, is *DPCM with delayed decision*. The quantization of the prediction error sample into the reconstructed sample $v(\mathbf{n})$ has a direct effect on the reconstruction value $y(\mathbf{n})$, and can straightforwardly be selected to be optimum for this one value. As however $y(\mathbf{n})$ is again used for prediction of subsequent values, the question arises whether not another $v(\mathbf{n})$ would be a better choice with regard to *future* estimates and reconstructions. If the effect of the recursive DPCM decoder is tracked by a multiple-path comparison within a tree, the task is to determine the optimum path. This means an *optimum sequence* of coded prediction error values $v(\mathbf{n})$ shall be selected, such that the error over the sequence of all related reconstruction values $y(\mathbf{n})$ is minimized. In this context, the M-algorithm has been applied successfully both for scalar DPCM [MODESTINO, BHASKARAN 1981] and for vector DPCM [OHM, NOLL 1990]. The gain of SNR as compared to methods without decision delay is reported to be around 4 dB. The decision delay implicitly counteracts the effect of quantization error feedback in the prediction loop, as the prediction error energy as well as the quantization error energy are minimized over sliding windows of samples. While the encoder complexity grows significantly by the introduction of the decision delay, the DPCM or vector DPCM decoder structures must not be changed at all.

12.3.4 Error propagation in DPCM

The effect of drift generally occurs in cases where different predictions are used at the encoder and decoder sides of DPCM, as then the encoder side makes wrong assumptions about the information which is available for prediction at the decoder. Another severe source of drift are *transmission losses*, as occurring in error-prone channels. To avoid interrelationships between quantization error feedback and transmission error propagation in the subsequent analysis, lossless encoding is assumed, such that $E(\Omega)=V(\Omega)$ and $X(\Omega)=Y(\Omega)$ if no errors occur in the channel. Assume that by a data loss a component $K(\Omega)$ is superimposed to the prediction error before being fed into the synthesis filter. The reconstruction signal is then influenced as follows:

$$V(\Omega) = E(\Omega) + K(\Omega)$$

$$Y(\Omega) = \frac{E(\Omega)}{1-H(\Omega)} + \frac{K(\Omega)}{1-H(\Omega)} = X(\Omega) + \frac{K(\Omega)}{1-H(\Omega)}. \qquad (12.28)$$

The additional interference component in the reconstructed signal is the convolution of the channel error component $K(\Omega)$ by the impulse response of the (recursive) synthesis filter. If $K(\Omega)$ has the characteristics of impulse noise (single bit errors), the signal is augmented by impulse responses of the synthesis filter, originating from the positions where the errors occur. Fig. 12.19 shows examples of error propagations in cases of 1D and 2D predictors. While in the first case, propagation can only be observed along one of the horizontal/vertical directions, in the latter case the interferences propagate over both dimensions.

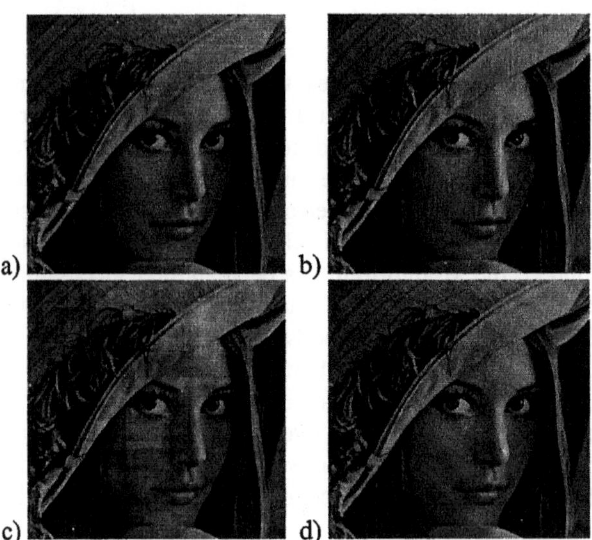

Fig. 12.19. Transmission error propagation in DPCM, random bit error probability $P_{err}=10^{-3}$, no VLC coding: **a** 1D prediction, horizontally **b** 1D Prediction, vertically **c** separable 2D prediction (3.15) **d** non-separable 2D prediction (3.17)

As compared to the separable predictor, the non-separable predictor in Fig. 12.19d effects less visible interferences, because due to the choice of coefficients $a=a_h=a_v=0{,}5$, the influence decays exponentially by a^l over both dimensions

$(0.5; 0.25; 0.125; ...)$[1]. For the separable filter using $\rho_h, \rho_v \approx 1$, a much slower decay can be observed. Indeed, the non-separable filter has worse performance in terms of data compression[2], while its properties regarding resilience against channel errors are significantly better. This is a typical example where the design of data compression methods should take into account the entire transmission chain. In the case given here, it would be advisable to select the best filters by knowing the specific error properties of the channel, and eventually go for a compromise regarding compression performance in the error-free case. Drift and error propagation are crucial problems in DPCM systems. Similar effects occur in motion-compensated DPCM encoding of video signals, which will further be discussed in sec. 13.2.8 and 13.2.9.

12.4 Transform Coding

Transform coding (TC) comprises encoding methods based either on orthogonal block transforms, or based on filterbank / wavelet transforms. The basic principles of these transforms were introduced in sec. 4.3 and 4.4; this section now concentrates on the aspects of *quantization* and *encoding* of transform (frequency) coefficients. These methods are to a certain extent different for cases of block transforms and filterbank transforms; it is tried to show the commonalities whenever possible.

12.4.1   Block Transform Coding

The decorrelating effect of discrete linear transforms was analyzed in detail in sec. 4.3.3. The general method is to perform the transform over a finite block of samples, by which property the scheme is denoted as *block transform coding*. The following derivations are made in particular for orthonormal transforms. If the transform is orthogonal in general, it is necessary to consider the normalization factor A, as e.g. used in the analysis of the correspondence between distortion in the signal and frequency domains (4.161)[3]. To encode the coefficient $c_{u,v}$ by an expected distortion D_{TC}, the rate

---

[1] See Problem 12.2c

[2] See Problem 4.4

[3] For non-orthogonal transforms (e.g. bi-orthogonal wavelet bases), (4.161) does not apply, which means that it is not straightforward to map the distortion introduced by encoding of the coefficients directly to the distortion appearing in the reconstructed signal. In such cases, it is mostly assumed that the transform is 'almost' orthogonal, which may in fact introduce additional errors. More exact solutions for optimum definition of quantization in case of non-orthogonal transforms are discussed in sec. 12.4.2.

$$R_{u,v} = \max\left[0, \frac{1}{2}\log_2 \frac{E\left\{c_{u,v}^{\;2}\right\}}{D_{\mathrm{TC}}}\right] \qquad (12.29)$$

has to be provided according to (11.10). If the max($\cdot$) function which cannot be described analytically is neglected[1], and constant distortion is assumed over all $U \cdot V$ coefficients, the mean rate is

$$R_{\mathrm{TC}} = \frac{1}{2 \cdot U \cdot V}\sum_{u=0}^{U-1}\sum_{v=0}^{V-1}\log_2 \frac{E\left\{c_{u,v}^{\;2}\right\}}{D_{\mathrm{TC}}} = \frac{1}{2 \cdot U \cdot V}\log_2\left[\prod_{u=0}^{U-1}\prod_{v=0}^{V-1} \frac{E\left\{c_{u,v}^{\;2}\right\}}{D_{\mathrm{TC}}}\right]. \qquad (12.30)$$

If the image is encoded without exploiting the correlation properties (i.e. by PCM, and for strict definition, also assuming a Gaussian model), the following relationship exists between rate and distortion, which can be determined by (11.9):

$$R_{\mathrm{PCM}} = \frac{1}{2}\log_2 \frac{\sigma_x^{\;2}}{D_{\mathrm{PCM}}} \Rightarrow D_{\mathrm{PCM}} = \frac{\sigma_x^{\;2}}{2^{2 \cdot R_{\mathrm{PCM}}}}. \qquad (12.31)$$

An interesting question is by which factor the distortion D_{TC} will be decreased in transform coding as compared to D_{PCM}, if the same rate $R_{\mathrm{TC}}=R_{\mathrm{PCM}}$ is used. Substituting (12.30) into (12.31) gives

$$D_{\mathrm{PCM}} = \frac{\sigma_x^{\;2}}{2^{2 \cdot \frac{1}{2 \cdot U \cdot V}\log_2 \prod_{u=0}^{U-1}\prod_{v=0}^{V-1}\frac{E\left\{c_u^{\;2}\right\}}{D_{\mathrm{TC}}}}}. \qquad (12.32)$$

The ratio of D_{PCM} over D_{TC} can be interpreted as *coding gain*, which for a 2D transform of block size $U \mathrm{x} V$ turns out to be the ratio of *arithmetic* and *geometric* mean values of the expected energies over the discrete set of coefficients[2],

$$G_{\mathrm{TC}} = \frac{D_{\mathrm{PCM}}}{D_{\mathrm{TC}}} = \frac{\dfrac{1}{U \cdot V}\sum_{u=0}^{U-1}\sum_{v=0}^{V-1}E\left\{c_{u,v}^{\;2}\right\}}{\left[\prod_{u=0}^{U-1}\prod_{v=0}^{V-1}E\left\{c_{u,v}^{\;2}\right\}\right]^{1/U\cdot V}}. \qquad (12.33)$$

The coding gain of block transform coding (12.33) can also be interpreted as a discrete counterpart of (11.17) and the related spectral flatness measure (11.19), in which the geometric mean of a continuous spectrum is derived by integrating over the logarithmic spectrum. The coding gain becomes larger, when the spectral power is better concentrated in a low number of coefficients. From the definition

---

[1] This is valid in the case of high rates where $E\{c_{u,v}^{\;2}\} \geq D_{\mathrm{TC}}$.

[2] In an orthonormal transform, the arithmetic mean over the expected coefficient energies within a block is identical with the power (variance) of the (zero-mean) signal. The following condition holds for any orthogonal transform, as then the factor A from (4.161) must be considered in the computation of the coding gain.

of the spectral flatness measure, it is straightforward to show that for a 2D AR process

$$\log_2 \sigma_z^2 = \frac{1}{4\pi^2} \int\limits_{-\pi}^{\pi}\int\limits_{-\pi}^{\pi} \log_2 S_{xx}(\Omega_1,\Omega_2)\, d\Omega_1 d\Omega_2 \,. \tag{12.34}$$

In (12.33), the numerator term is equal to σ_x^2 for the case of an orthonormal transform, as can be concluded from (4.161). The denominator can be rewritten as

$$\begin{aligned}
\log_2 \left[\prod_{u=0}^{U-1}\prod_{v=0}^{V-1} E\{ c_{u,v}^2 \} \right]^{1/U\cdot V} &= \frac{1}{UV}\sum_{u=0}^{U-1}\sum_{v=0}^{V-1} \log_2 E\{ c_{u,v}^2 \} \\
&\geq \frac{1}{4\pi^2} \int\limits_{-\pi}^{\pi}\int\limits_{-\pi}^{\pi} \log_2 S_{xx}(\Omega_1,\Omega_2)\, d\Omega_1 d\Omega_2.
\end{aligned} \tag{12.35}$$

The inequality is concluded by the fact that even for the optimum transform (4.149), the values $E\{c_{u,v}^2\}$ are transformed from a *finite* autocovariance sequence, which must be periodic or symmetrically extended. $S_{xx}(\Omega_1,\Omega_2)$ relates to an autocorrelation sequence of *infinite* extension, which decays towards zero and will clearly lead to a lower 'spectral entropy'. This penalty becomes larger for smaller block length and higher correlation of the signal[1]. The reason for this penalty can also be interpreted by the fact that block transform coding by itself can not take into account the statistical dependencies between adjacent blocks. It can be concluded that for the transform of an AR process

$$\begin{aligned}
&\lim_{U,V\to\infty} \left[\prod_{u=0}^{U-1}\prod_{v=0}^{V-1} E\{ c_{u,v}^2 \} \right]^{1/U\cdot V} = \sigma_z^2 \\
&\Rightarrow \left[\prod_{u=0}^{U-1}\prod_{v=0}^{V-1} E\{ c_{u,v}^2 \} \right]^{1/U\cdot V} \geq \sigma_z^2 \quad \Rightarrow \quad G_{\mathrm{TC}} \leq G_{\mathrm{AR}},
\end{aligned} \tag{12.36}$$

where G_{AR} is the maximum possible coding gain of an AR model as given by (11.17). This conclusion can however not be generalized unconditionally to the case of image signals, as it assumes stationarity of the signal; in fact, images show highly instationary properties with frequent changes of the local amount of detail, change of mean, edges etc. From this point of view, usage of smaller transform block sizes is justified, however the reason for the penalty (12.36), the lack of exploitation of statistical dependencies between adjacent blocks, should be tackled as well.

In 2D transform coding, the image signal is typically partitioned into small blocks of size $M'\mathrm{x}N'$ (Fig. 12.20a). The block with index (m',n') has the starting coordinate (top-left pixel) at position $(m,n)=(m'\cdot M',n'\cdot N')$ of the image. For the case of a linear block transform, $U\cdot V$ 2D frequency coefficients are computed per

---

[1] see also Problem 12.3

block, where $U=M'$, $V=N'$. Meaningful block sizes in natural-image transform coding are between 4x4 and 32x32 pixels. The choice of the optimum block size depends mainly on the spatial correlation of the signal; for highly correlated signals large blocks are advantageous, while in case of lower correlation smaller blocks are preferable:

– If the signal is highly correlated and the block size is too small, a high redundancy may still exist between coefficients of adjacent blocks;
– if the signal has low correlation, or if the transform block consists of image sub-areas with different statistical behavior (e.g. at edges), it is better to select smaller block sizes which may improve the concentration of information in a smaller number of coefficients.

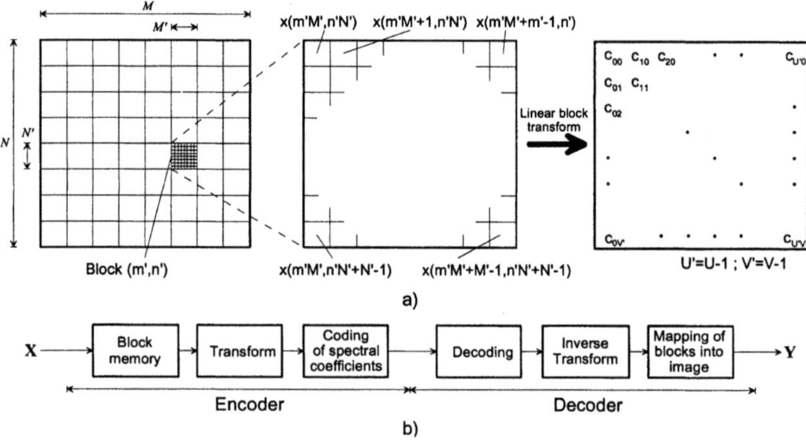

Fig. 12.20. a 2D transform coding of sub-image block partitions
b Overall scheme of block-based transform coding and decoding

To minimize the problem of inter-block correlation, DPCM coding is often applied in combination with transform coding, where prediction is made by information from previously decoded blocks. If fixed block sizes are used, a choice of 8x8 pixels can be regarded as a good compromise, as it is e.g. defined in the DCT-based block transform coding as applied in JPEG and other standards. Smaller block sizes such as 4x4 pixels are only appropriate if sophisticated inter-block prediction techniques are applied additionally. Fig. 12.20b shows a typical block diagram of a transform coding scheme. After the transform has been performed, coefficients are quantized and encoded. At the decoder, the reconstruction of the quantized coefficients is followed by the inverse transform. Application of DPCM or other prediction techniques are interpreted by this high-level view as part of the encoding and decoding processes. Specific prediction techniques include:

– *Mean ('DC') prediction*: The coefficient c_{00} bears most information about the local mean value of the signal, which typically varies only gradually within a local neighborhood. It can be predicted using the respective DC coefficients of any previously decoded blocks, e.g. as shown in the configuration of Fig. 12.21a.

- *Detail ('AC') prediction*: In specific image configurations, e.g. horizontal or vertical edge structures which extend beyond the block boundaries, also structures can be predicted. This will in particular apply to the first row (in case of vertically-oriented structures) *or* the first column (in case of horizontally-oriented structures) of 'AC' coefficients. The first row is predicted from the block above; the first column is predicted from the left-handed block. As it can usually not be expected that both horizontally and vertically oriented structures play a dominant role simultaneously, switching can be performed (see Fig. 12.21b).

- *Directional prediction*: For separable 2D transforms, it is difficult to directly analyze the effect of structures different from horizontal and vertical orientations by the values of transform coefficients. Moreover, as the transforms are typically not shift invariant[1], it is almost impossible to predict coefficients by the values from a neighbored block in the case of diagonally-oriented structures. Alternatively, a *directional prediction* can be performed on the block directly in the image domain, where then the transform is applied to the residual signal. In principle, this is a method of *vector prediction*, as it was analyzed in sec. 12.3.3; the block transform will then exploit the remaining correlation in the prediction error signal.

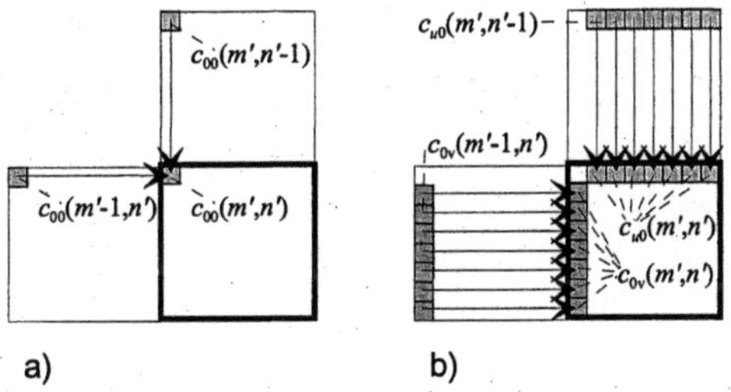

Fig. 12.21. Prediction of block transform coefficients **a** DC coefficient **b** AC coefficients

Fig. 12.22 shows probability distributions of coefficients $c_{0,0}$, $c_{1,0}$ and $c_{2,0}$ as analyzed from a 2D DCT of size 8x8. For a sufficiently large set of images, the block-mean coefficients $c_{0,0}$ show a wide range of amplitude distributions, while for individual images, distributions with characteristic peaks of amplitude concentrations occur (Fig. 12.22a). In contrary, the higher-frequency coefficient statistics show a concentration of values around zero, and show this statistical behavior quite consistently, when analyzed either over large sets or single images.

---

[1] Very different amplitudes of coefficients can result, if the block position is shifted.

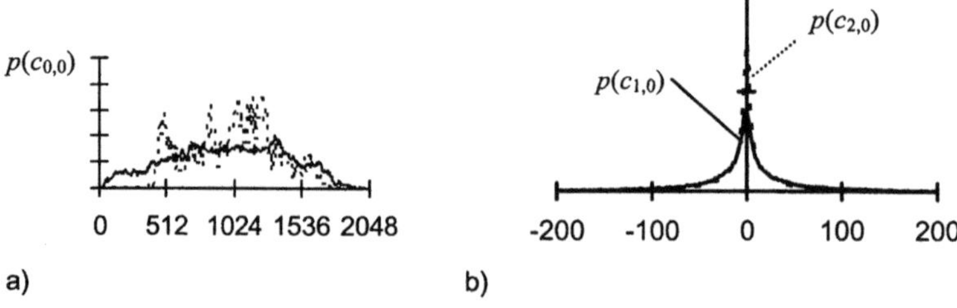

Fig. 12.22. PDF of DCT coefficients **a** Coefficient $c_{0,0}$ of one single image ($\cdots$) and averaged over 15 images (—) **b** Coefficients $c_{1,0}$ (—) und $c_{2,0}$ ($\cdots$)

Quantization of transform coefficients. The transform coefficients as resulting by block transforms like the DCT have an extended value range as compared to the original image signal. For sinusoidal basis function transforms, the coefficients can hardly be mapped onto a discrete set of values, even when the input is a discrete image with a finite number of amplitude levels. As a consequence, sinusoidal-basis block transforms are not usable for *lossless coding*. This case is different for transforms over integer-defined basis function sets, as e.g. the rectangular-basis transforms (4.119)-(4.126), or the integer-basis transforms (4.136)-(4.138). However even in these cases, higher bit precision is needed for the transform coefficients as compared to the original image samples. Only by sophisticated exploitation of inter-relationships between values of different coefficients, efficient lossless coding is achievable, which then is considerably more complex than DPCM or similar solutions. Hence, block transform coding is mainly useful in the range of lossy representation of image signals. The coefficients have to be quantized, which introduces the distortion D_{TC}. In principle, any quantization method can be used, such as uniform, dead-zone or non-uniform scalar quantizers. Due to the concentration of transform coefficient values around zero, quantizers allowing a zero-reconstruction value should be selected. Vector quantization can also be applied, which is discussed separately in sec. 2.4.3.

Of particular importance are schemes employing *frequency-dependent quantization*, which is justified under psycho-visual aspects (cf. sec. 6.1.2). An individual quantizer attenuation factor $w_{u,v}$ must then be defined for each coefficient, while a step size reference Δ shall apply globally for all coefficients. The quantizer step size to be used for the quantization of coefficient $c_{u,v}$ then is $\Delta_{u,v}=w_{u,v}\cdot\Delta$. Normally, all coefficients of values less than $\Delta_{u,v}/2$ will be quantized into the zero reconstruction value. For the case of dead-zone quantizers, all coefficients which are below the threshold Θ are quantized into zero (cf. Fig. 11.2). Neither in cases of frequency-weighted quantization nor for dead-zone quantizers, improvement can be expected with regard to squared-error distortion metrics; in contrary, cases can be found where frequency-weighted quantization increases the squared reconstruction error significantly without having a beneficial effect in terms of rate reduction[1].

After quantization, entropy coding has a significant effect to minimize the encoded rate, in particular when many coefficients are quantized into zero. An important difference between natural images and a statistical model like AR(1) is the lack of stationarity, where in natural images a local variation between blocks of lower and higher detailed content is typical. This could best be modeled as a *composite source*, which makes it however difficult to design unique codes of variable length which are applicable to approach the entropy rate for any category of blocks within an image. In addition, particularly at low bit rates, a high percentage of coefficients must be quantized into the zero reconstruction value (see Fig. 12.22b). Hence, alternative methods should be used in entropy coding of zero-quantized coefficients, or to signal groups of coefficients which shall entirely be set to zero.

Zonal coding and classification based coding. Early methods in transform coding of images were based on *zonal coding*, where non-zero coefficients were assumed to occur only in sub-partitions or zones of transform blocks. Block layouts in the following examples are arranged as in Fig. 12.20a right, with coefficient $c_{0,0}$ in the top-left corner. Fig. 12.23a shows an example of a zone separation into 'quantized' and 'not quantized' coefficients, as it can e.g. be determined from a separable AR(1) model for a given rate-distortion point. Here, all those coefficients which statistically fall below the spectral threshold parameter Θ according to Fig. 11.4 are assigned into the 'not quantized' zone. Such an approach would however only be reasonable for stationary signals. When the amount of detail varies,

- many coefficients would be present in the 'quantized zone' which actually can be set to the zero value in blocks of low detail;
- relevant coefficients would be found in the 'not quantized' zone, such that detail is lost in high-detailed blocks.

Both cases are sub-optimum for best rate-distortion performance, as either rate is wasted, or unnecessarily high distortion occurs.

As extensions of zonal coding methods, being better adaptable to composite sources of different block activity, classification-based methods of adaptive block transform coding [CHEN, SMITH 1977] were introduced. The best fitting zonal separation scheme is then selected for each block from a set of different schemes available. Examples of zone configurations for blocks with horizontal and vertical structures, low and high detail are shown in Fig. 12.23b-e. Such layouts must however be adapted to the characteristics of individual images to achieve the best performance.

---

[1] See Problem 12.4; this comparison is however not balanced, as the SNR is a squared distortion metric which assumes uniform frequency weighting of errors over all transform coefficients due to the validity of Parseval's theorem (3.70). Unequal quantization of coefficients is based on reasoning about perceptually weighted distortion.

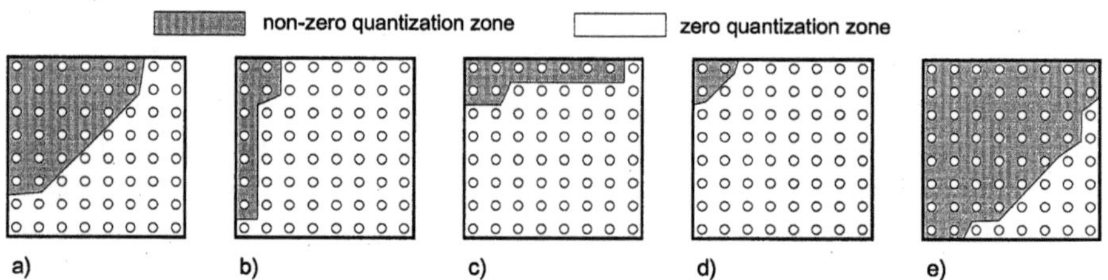

Fig. 12.23. a Zonal transform coding b-e Different configurations of classified transform coding with adapted zero-quantization zones: b Highly detailed horizontal structure c Highly detailed vertical structure d Low detail e High detail without specific direction

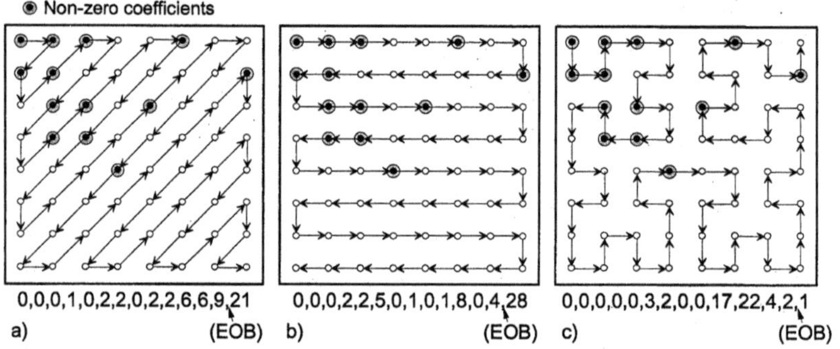

Fig. 12.24. Scanning schemes in run-length coding within a transform block:
a zigzag scan b row-wise scan c Peano-Hilbert scan

Run-length coding. The most-widely used scheme to describe positions of relevant (non-zero quantized) coefficients individually for each block is based on *run-length coding*. As however run-length values relate to a unique 1D technique, it is necessary to 'unwrap' the 2D block of coefficients in a first step by defining of a *scan sequence*. The number of runs is equal to the number of coefficients quantized into non-zero values. The most commonly used sequence is the *zigzag scan* shown in Fig. 12.24a[1]. It can be expected that coefficients of higher frequencies will by higher probability be quantized into the zero value. In the zigzag scheme, these are visited last, regardless of directional orientation (horizontal, vertical, diagonal). Thus, a distribution of many short, and only few long run-lengths can be expected which is more favorable for the subsequent entropy coding. Even though the zigzag scan is the best scheme on average, *switching of the scan mode* for individual blocks can be useful in some cases. The row-wise scan sequence of Fig. 12.24b (if turned by 90^0: column-wise) is advantageous if the block shows vertical (horizontal) edge structures, as then high-energy coefficients can mainly be expected in the horizontal orientation. The Peano-Hilbert scan (Fig. 12.24c) is the

---

[1] The zigzag-scan method was first proposed in [CHEN, PRATT 1984]. Many variants exist, which e.g. give preference to visit coefficients in one of the two directions earlier.

best approach when statistical dependencies (concentrations of high-energy coefficients) can be expected in local areas of the coefficient matrix. This can be the case in specific transforms like the Discrete Hartley transform (4.135). In the case of transform coefficients which encode the residual from a directional spatial block prediction (Fig. 12.16c), also a *double scan* method is used in the Advanced Video Coding standard (cf. sec. 17.4.4). The DC coefficient c_{00} is mostly excluded from the run. As was shown previously, its statistical properties differ significantly, such that it can better be encoded separately.

Over the scan sequence, it must be checked whether coefficient amplitudes fall below the threshold of the zero quantizer cell (see Fig. 12.25). The run-length values in Fig. 12.25 indicate the numbers of zero-quantized coefficients between two coefficients quantized into non-zero values. This is equivalent to the run-length coding scheme of Fig. 11.18b. Following this scheme, also the run-lengths for the different scan modes are given in the examples of Fig. 12.24. If all non-zero coefficients have been visited, this is often signaled by an 'end of block' (EOB) symbol.

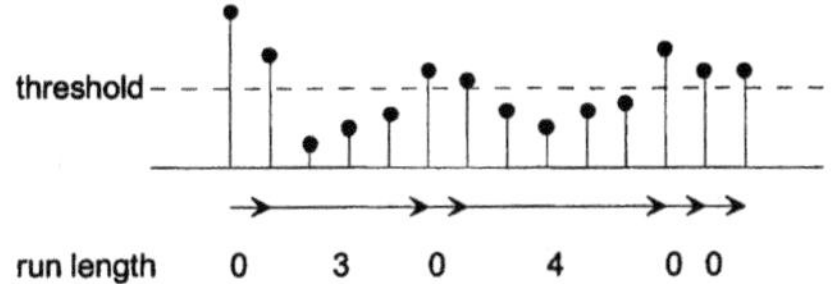

Fig. 12.25. Generation of run-length code sequence based on a zero-threshold decision

Even with entropy coding, the bit rate to encode the run-length information can be unacceptably high. This is in particular the case when many coefficients have to be transmitted, i.e. in high-quality encoding (small quantizer step size). In schemes for run-length coding as widely used in image and video coding standards[1], entropy codes are used which combine the *run-length information* (which is basically a position information) and the *quantizer level information* for the non-zero coefficients in a {RUN;LEVEL} code[2]. Similar methods with a combined code for {RUN;LEVEL;EOB} are also used. The latter is usually denoted as *3D VLC*, the former as *2D VLC*, indicating the number of sources symbols (vector length K) which are combined in entropy coding. For AC coefficients, amplitudes of positive and negative signs can be expected to occur by approximately equal probability. The information about the sign is therefore simply appended to the VLC as an additional bit.

Bit-plane coding of transform coefficients. Embedded quantization and bit-plane coding as described in sec. 11.3 can also be applied to transform coefficients. Again, the DC coefficients c_{00} should be treated separately, as they have only posi-

---

[1] E.g. JPEG, MPEG, H.26x, see sec. 17.3 and 17.4.
[2] See Table 11.1.

tive values and are different by their statistical behavior compared to the AC coefficients. As AC coefficients can have positive or negative values, it is useful to encode amplitude and sign separately. A simple method is the combination of run-length coding with bit-plane coding; it is then necessary to perform one run per bit plane, starting from the MSB, proceeding down to the LSB. In principle, no genuine quantization step is necessary, when the whole procedure is applied to an integer representation of the transform coefficient magnitudes, where the full representation of all bit planes relates to the lowest distortion level that can be achieved. Positive and negative signs can be expected to be of same probability if the PDF is symmetric; hence, entropy coding of the sign would not have significant effect. However the significance of the sign depends on the significance of the highest bit-plane where a '1' value is found in the amplitude representation. As the PDF of the coefficients indicates concentration around the value zero, it can further be expected that less '1' bits are found in the more significant bit planes.

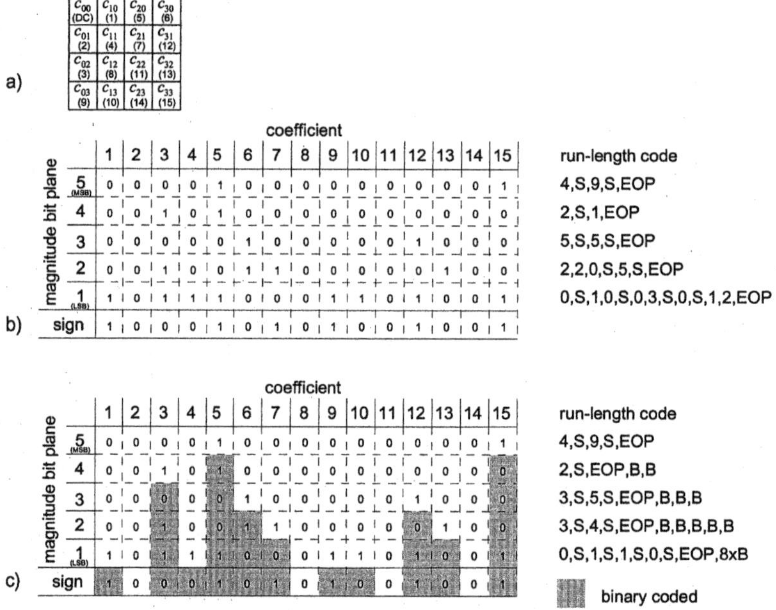

Fig. 12.26. Bit-plane coding of transform coefficients using **a** sequence of coefficients in a 4x4 transform block **b** run-length coding **c** combined run-length/binary coding

Two different methods of run-length/bit-plane coding realizations are shown in Fig. 12.26. Both examples relate to a zigzag scan of a 4x4 block, where only the bit-plane coding of AC coefficients 1-15 is considered here (Fig. 12.26a). The end of each bit-plane scan is indicated by an *End of Plane* (EOP) symbol, which fulfills a similar functionality as the 'EOB' described above. In the method of Fig. 12.26b, numbers indicate run lengths of zero values, and 'S' indicates that a sign bit is encoded whenever the highest significant bit related to a coefficient is found. It can be expected that runs are getting shorter towards the lower bit-planes. In the method of Fig. 12.26c, the run-length code only marks the highest-significance non-zero bit of each coefficient. All less significant bits of the same coefficient are encoded in a binary representation 'B' within their respective bit-plane position[1].

The advantage of this scheme is that the run-length statistics are less different for the individual bit-planes, such that the design of entropy coding of run-length values must not be made separately for each plane, as it would be advisable for the first method. On the other hand, the values '0' and '1' of the binary coded part may not be uniformly distributed over all bit planes. This could deteriorate the coding performance, unless additional entropy coding is used for this part of the information.

The method of bit-plane coding is in particular of high importance for *scalable* or *embedded* coding of transform coefficients. When a specific rate shall be output, the process of encoding can stop anywhere as soon as the bit budget is consumed. As the significance of the bit planes directly relates to the amount of distortion which will be introduced, this allows to approach a reasonable rate-distortion behavior at different operational points of quantization[2].

For block transform coding, aspects which can further improve the compression performance include the utilization of all statistical dependencies between the different transform coefficients, both within the blocks, and between the blocks. To some extent, it can be concluded that the choice of the transform, even the choice of the transform block size, becomes less important if sophisticated methods are applied which can make use of any remaining linear (correlation) or nonlinear dependencies[3]. Even though this is a problem of extreme dimensionality, methods like context-based and context-adaptive entropy coding, in particular implemented by flexible arithmetic coding, can be expected to provide further progress in compression. Second, due to the choice of the transform basis functions, certain limitations exist which introduce systematic errors to the recon-

---

[1] It is also a common practice to perform two scans, one searching for coefficient positions which are encoded for the first time ('significance scan'), the other refining the magnitude representation of previously known coefficients by on more bit (the values 'B'). Similar strategies are applied in bit-plane coding of wavelet coefficients

[2] Note that the problem of *fractional bit-planes* (cf. sec. 11.3) applies likewise in bit-plane coding of transform coefficients.

[3] From statistical models of power spectra like the AR(1) model, it can e.g. be expected that *when* within a specific location (transform block) spectral energy is present in certain frequency bands, it is highly probable that energy also resides in neighbored coefficients expressing similar directional orientation etc. This dependency will not be expressible as a correlation, as the total energy will result from coefficients of either positive or negative sign; the dependency can however be utilized in entropy coding, when signs are encoded separately from magnitudes. Statistical dependencies between magnitudes, which are indeed generated by a (nonlinear) absolute-value computation, are a typical example of nonlinear coherence.

structed signal. In the case of block transforms, typically blocking artifacts are observed when encoding is performed at low data rates. By a more systematic removal of these errors, both on the basis of statistical signal models and by distortion models including characteristics of basis functions and quantization, a much better improvement would be possible than by the simple methods of postprocessing as introduced in sec. 5.2.

12.4.2   Subband and Wavelet Transform Coding

In general, quantization of subband coefficients can follow the same principles as developed for block transform methods. If a block-overlapping or subband transform is used which results in frequency bands of equal width, the resulting coefficients can be interpreted exactly like block transform coefficients, just produced by a different set of basis functions. This is indeed the typical point of view when block-overlapping transforms are used for 2D image compression [MALVAR, STAELIN 1989][MALVAR 1991]. In contrast, the first methods for 2D subband coding of images were designed such that the subbands were regarded as independent, where only intra-band dependencies were further taken into consideration [WOODS, O'NEILL 1986][WESTERINK ET AL. 1988][GHARAVI, TABATABAI 1988]. In particular, as for the 'DC' coefficient of a block transform, predictive encoding (DPCM) of the scaling coefficients can significantly improve the compression performance; this signal, in spite of sub-sampling, still exhibits high spatial correlation along both horizontal and vertical directions. Significant *directional correlations* can also be observed within any frequency bands where lowpass filtering is applied along one spatial direction. This can be exploited by horizontal/vertical directional prediction, which is conceptually comparable with the 'AC prediction' for the first row or first column of coefficients in a block transform. Due to the frequency-inversion principle in sub-sampling of highpass bands (cf. sec. 4.4.1), slight *negative correlations* typically appear between samples of these bands, which are however relatively low such that exploitation for encoding is not worthwhile.

Strictly spoken, both of the aforementioned approaches are simplistic, and best compression performance can only be obtained if both *inter-band* and *intra-band* statistical dependencies are exploited.

A fundamental barrier, which eventually inhibits utilization of statistical dependencies between adjacent samples of one subband, is the interference of alias components, which are unavoidable during sub-sampling. This alias is also a general problem in reconstruction, as it cannot be compensated (see (4.175)), when a complementary component required for alias cancellation is suppressed during encoding. While this loss of information results in blocking artifacts for the case of block transforms, a smooth transition over block boundaries is guaranteed by block-overlapping transforms due to the longer basis functions. For the case of more general filterbanks, blocking artifacts are also inhibited if the filter length is larger than the sub-sampling factor. For the case of longer filters, the problem is different; the suppression of highpass coefficients typically causes *ringing artifacts* at amplitude discontinuities of the signal, e.g. at edge positions[1].

During the last decade, *Wavelet transform coding* has evolved as one main direction in subband coding of 2D images. The principal reasons for this are as follows:

- Very efficient methods exist to utilize inter-band and intra-band dependencies between scaling and wavelet coefficients simultaneously;
- In the vicinity of edges, the Wavelet transform which uses shorter basis functions with better localization property for the high-frequency components, is advantageous both in terms of compression performance and minimization of the ringing problem;
- The Wavelet transform allows usage of *bi-orthogonal filters* (cf. sec. 4.4.2-4.4.4) with perfect reconstruction and other attractive properties, still leading to a quasi-orthogonal frequency representation when a sufficient depth of the wavelet tree is used. Subband decompositions of equal bandwidth should definitely be implemented using *orthogonal filters* (e.g. QMF), as otherwise the iterative decomposition of highpass bands would cause problems of irregularity. This typically imposes constraints in terms of filter length, perfect reconstruction capability etc.
- The Wavelet transform provides a naturally scalable representation of image signals in a resolution pyramid, while still being a critically-sampled representation (cf. sec. 4.4.6).

Zero-tree coding. To illustrate what statistical dependency *between* frequency bands in wavelet transform means, Fig. 12.27a shows the result of a 2D wavelet transform, where at the top-left position the lowest frequency band (scaling band) appears. In 2D wavelet transform, a decomposition of the previous scaling band into one scaling and three wavelet bands of quarter size is performed again by each subsequent level of the wavelet tree (cf. sec. 4.4.5). This procedure is iterated on the respective sequence of scaling-coefficient images, which are in fact images being down-scaled by factors of 4, 16, 64, ..., etc. as compared to the original. From Fig. 12.27a, the *localization property* of the wavelet transform becomes obvious. Each wavelet coefficient represents its respective frequency component at a specific position within the image. Herein, the coefficients from different frequency bands which relate to the same positions are obviously not statistically independent: If high energy is observed in a lower-frequency band at a specific location, it is very likely that the next higher band also contains relatively high

---

[1] This can also be interpreted by the *Gibbs phenomenon*, where the reconstruction of a signal with amplitude discontinuities from the coefficients of a Fourier expansion is shown to produce a ringing of constant amplitude at this discontinuity, whenever Fourier coefficients are discarded; for other transform bases than Fourier, ringing is less severe, fortunately.

energy around the same location, however with quadruple number of samples[1]. This effect can be observed in Fig. 12.27a in particular at vertical, diagonal and horizontal edges. The statistical dependencies between the wavelet bands follow in general by the horizontal, vertical and diagonal orientation characteristics, as it is shown in Fig. 12.27b[2]. Hence, in entropy coding it will be useful to determine the probabilities under a condition on whether the directionally adjacent wavelet band of next-lower frequency contains significant energy at the respective position or not. This implies that a higher rate must be spent for encoding of less probable configurations such as a low value of a lower-frequency coefficient accompanied by a higher amplitude of a high-frequency coefficient at any co-located position.

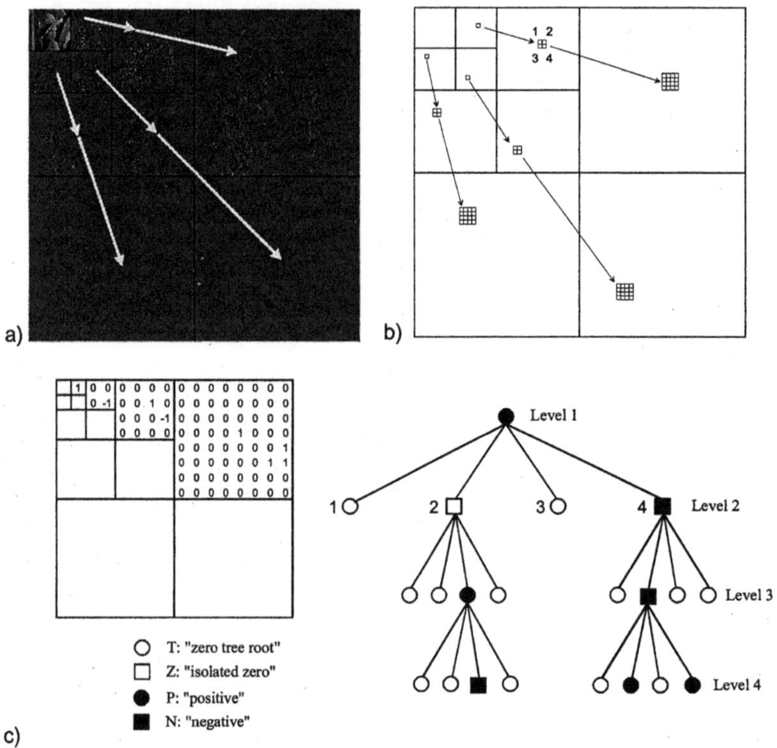

Fig. 12.27. a 2D wavelet decomposition **b** geometric referencing in zero tree coding **c** structure of the tree relationships within the MSB bit plane of the horizontal direction

[1] Amplitudes were amplified by a factor of 4 in Fig. 12.27 to make structures in the wavelet components visible at all.

[2] The same property can be observed in block transform coding; if a block covers an area of high detail, many different coefficients of high amplitude will be present in specific orientations, which is e.g. utilized by best scanning sequences of run-length coding, or by the adaptation mechanisms in classified coding.

One of the earliest approaches to use a tree-like data structure over the entire wavelet tree was proposed in [LEWIS, KNOWLES 1992]. A key step was then made by, the development of *Embedded Zero-tree Wavelet* (EZW) coding [SHAPIRO 1993]. As in Fig. 12.27b, trees are defined separately over coefficient groups of horizontal, diagonal and vertical orientations. The method integrates the concept of embedded quantization (cf. sec. 11.3), and also has a high similarity with bit-plane coding as introduced for the case of block transforms in sec. 12.4.1, where however instead of the zigzag scan, a scan of coefficients over the zero tree is performed. Each parent branches into four children at the next resolution level, where the ordering is made as shown by numbers 1-4 in Figs. 12.27b/c. An example concerning the scan of the most significant bit plane over the horizontal orientation is given in Fig. 12.27c. Possible decisions of the quantizer as shown in Fig. 11.8b are positive, negative or zero values, which can occur for the MSB at any coefficient position. As however it is highly expected that a zero-value parent node will have zero-value children as well, two cases are treated differently in the symbols to be quantized into zero. The following symbols are defined:

- *Zero-tree root* ('T'): The value is zero at the given parent position, and at *all* subsequent children down the entire tree. Encoding of all levels below is terminated.
- *Isolated zero* ('Z'): The value is zero at the given parent position, but may be positive or negative significant at a child anywhere in the entire tree.
- *Positive* ('P'): The value is of positive amplitude, and above the given threshold of the embedded quantizer level.
- *Negative* ('N'): The value is of negative amplitude, and below the given (negative) threshold of the embedded quantizer level.

All coefficients found as positive or negative are regarded as implicitly relevant at the subsequent passes to be performed for the next, less significant bit plane. Observe that the encoding of the sign, which was regarded as separate in the block transform bit-plane coding methods presented in the previous section, is integrated in the information of these four symbols. It must never again be encoded when a coefficient was once found as relevant. Hence, for the subsequent bit planes, the method described so far is called the *dominant pass*, which is only performed for those positions which were not yet found as relevant from upper bit-planes. For all relevant positions, a *sub-ordinate* pass is performed, which only encodes the branching into the upper or lower half of the next-finer embedded quantizer cell by '0' and '1' symbols; this exactly corresponds to the method of binary-coded information in Fig. 12.26c.

The information symbols 'T', 'Z', 'P', 'N', '0' and '1', as occurring at the respective coefficient positions within one bit plane, are re-ordered to obtain sequences which proceed from coarser to finer spatial resolution levels. A typical scan sequence is shown in Fig. 12.28. The resulting sequence of symbols is then arithmetically encoded, taking into account the expected probabilities at the respective bit-plane levels.

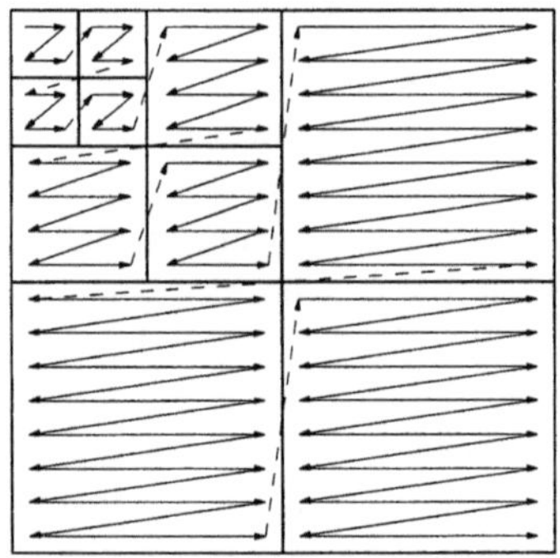

Fig. 12.28. Typical scanning sequence of zero-tree information in EZW

EZW has been the ancestor for a number of subsequent algorithm designs, which further improved its excellent performance. One of the most popular zero-tree approaches is *Set Partitioning into Hierarchical Trees* (SPIHT), presented in [SAID, PEARLMAN 1996]. The key concepts are partial ordering of the transformed coefficients by magnitude using a set partitioning sorting algorithm, and ordered bit-plane transmission of refinement bits. The re-ordering of the information, which aims to send the most relevant bits first, introduces a certain amount of overhead information. Nevertheless, it has been shown that SPIHT in most cases clearly outperforms EZW. In particular, by the 'relevant first' paradigm, the so-called problem of *fractional bit planes* (cf. sec. 11.3) can be reduced to a minimum effect, when the algorithm is used to implement a scalable representation of image signals. However, SPIHT does not provide an implicit mechanism to truncate bit streams for a reconstruction by reduced spatial resolution, discarding higher-frequency wavelet coefficients.

Context-adaptive Binary Arithmetic Coding (CABAC), which is now used as an arithmetic encoder in the Advanced Video Coding standard, was originally developed for adaptive entropy coding of wavelet coefficient over zero trees. This scheme allows flexible exploitation of intra- and inter-band dependencies in the bit planes of embedded-quantized wavelet coefficients, and also allows bit-stream truncation by reduced spatial resolution [MARPE, CYCON 1999].

Nevertheless, context-based codecs *not* using zero-tree correspondences over wavelet trees have also been shown to provide excellent compression performance, in particular when proper context modeling and sophisticated rate-distortion optimization is performed [LI, LEI 1999]. An algorithm of this class (EBCOT, see below) has become the basis of the JPEG 2000 compression standard. Complete omission of inter-band contexts is attractive due to the low complexity and simpler realization of spatially scalable streams. Nevertheless, for certain classes of images, better performance can be expected if the coherence between subbands can be exploited, similar as in the zero-tree approach. A scheme which combines both of these aspects is *Embedded Zero Block Coding* (EZBC) proposed in [HSIANG, WOODS 2000]. EZBC performs a quadtree-based set partitioning within subbands of a wavelet tree, signifying zero blocks in a hierarchical structure. The underlying context modeling allows flexible exploitation of statistical dependencies between coefficients *within one subband* or *across subbands*. A method of bit-plane interleaving additionally improves the compression performance. Due to the block structure, EZBC provides a fully embedded solution supporting both spatial and bit-plane scalability, similar to EBCOT.

Embedded Block Coding with Optimum Truncation (EBCOT). The EBCOT algorithm [TAUBMAN 2000] is the method of embedded entropy coding of wavelet coefficients used in the JPEG 2000 standard[1]. It is based in principle on the same methods of embedded quantization of the coefficients represented by a wavelet tree as described earlier; however, EBCOT does not belong to the class of zero-tree methods, as it encodes the coefficients over the different scales of the wavelet representation basically independent of each other. Instead, EBCOT makes extensive usage of intra-band redundancies by employing highly sophisticated context-based and adaptive arithmetic coding. The motivation to give up the zero-tree paradigm is a more flexible access to coefficients from different wavelet bands. In principle, though not implemented in JPEG 2000, this gives a sufficient degree of flexibility to support *wavelet packet* representations (cf. sec. 12.4.4). The method enforces an artificial block structure to groups of wavelet coefficients, where all information relating to certain local areas can be uniquely accessed and decoded independently. These structures are entitled as *code blocks*, which span over the same areas for all subbands of the wavelet pyramid, which means that the number of coefficients contained in a code block increases by factors of four with each level towards the higher-frequency bands. The entropy coding related to bit planes of the embedded quantizer performs different passes. The first is *significance coding*, which has again the target to find coefficient positions which were still insignificant (all zero) in the already encoded higher bit planes. The conditional probabilities of a coefficient to become significant in a bit plane are determined from spatial neighborhood contexts of the next-higher bit plane. The contexts themselves are defined by horizontal and vertical primitives, which are combinable into larger contexts. This is performed differently for subbands of horizontal, vertical or diagonal directionality. Depending on the number of significant coefficients in the neighborhood, an expectation is determined for the current coefficient to become significant by the next step. This is called the *normal mode* of the significance scan. In addition, a *run mode* is defined, which allows to efficiently encode cases of multiple adjacent coefficients to stay insignificant, which is indeed very similar to the run-length coding of bit planes in block transform coding in Fig. 12.26c. The remaining two steps are again *sign coding* for newly found significant coefficients, which is invoked immediately after the significance was found, and

---

[1] It is possible here to give only a rough idea how EBCOT works, and how it is related to other methods discussed so far. An excellent treatise on all detailed aspects is [TAUBMAN, MARCELLIN 2001].

magnitude refinement coding of coefficients, which were already found to be significant in higher bit planes.

The important core of the EBCOT algorithm is a re-ordering of the information into streams using rate-distortion optimization criteria, which is denoted as *post compression RD optimization*. Given a certain bit budget, it has to be checked for each code block whether it is reasonable to consume additional bits from this budget, which must typically be spent for units of completely decodable information from one of the coding passes. For example, at low rates, the magnitude refinement is typically found to provide a significant contribution to minimize the distortion with lowest amount of bits to be spent, while for higher rates, the significance propagation seems to become more important. With a relatively low amount of overhead information, the bit streams can be re-ordered such that the optimum performance is achieved at a given number of operational rate-distortion points. The reasoning behind is similar as in the re-ordering made by SPIHT, however EBCOT may performs optimum due to the consistent application of rate-distortion criteria and optimization. Clearly, the problem of 'fractional bit planes' is solved by this method likewise.

A clear advantage of wavelet coding in general and of its combination with bit-plane based entropy coding in particular is the free scalability of the coded information, both in terms of spatial resolution and quality. Bit rate savings achievable by wavelet-based methods as implemented in the JPEG 2000 standard, as compared to the block-DCT method of the previous JPEG standard, are in the order of magnitude of 25-50% at low rates, depending on the characteristics of images. It is however difficult to say which percentage of this gain is made due to the transform, and how much is contributed by more sophisticated entropy coding methods, in particular context-related and adaptive arithmetic coding.

Lossless coding using the Wavelet transform. The compression performance of wavelet coders highly depends on the specific transform basis which in chosen. In principle, by using the bit-plane representation, it will be possible to achieve lossless compression, if the output of the filters provides a finite set of integer values. This can e.g. be realized if short-kernel PRFs (4.193) or integer-precision biorthogonal filters (4.194) are used. On the other hand, such filters are less efficient at lower rates due to worse frequency separation between the bands, which cause higher aliasing errors when subbands are discarded. Nevertheless, it is remarkable that Wavelet transform methods are able to keep their excellent performance up to the range of lossless or near lossless coding, a range which traditionally was dominated by DPCM based methods, and that this can be integrated as part of a unique scalable representation from lossless quality down to lossy levels.

Quantization adjustment and coding gain in wavelet coding. The considerations about the coding gain, which were introduced in the context of block transform coding in sec. 12.4.1, are not directly applicable to the case of Wavelet coding due to the following reasons:

- The direct mapping of the coding error in the frequency coefficients towards the reconstruction error in the image domain as introduced in (4.161), is strictly valid only for the case of orthogonal transforms.
- The formulation of the coding gain as made in (12.33) assumes frequency representations of equal bandwidth and equal number of samples.

To study the effect of a quantization or coding error which is added to the signal in the domain of wavelet or scaling coefficients, it is useful to formulate the mapping of the quantization errors $Q_0(\Omega)$ and $Q_1(\Omega)$ into the reconstruction error $E(\Omega)$, as occurring in the lowpass and highpass bands of a 2-band system as

$$E(\Omega) = G_0(\Omega)Q_0(\Omega) + G_1(\Omega)Q_1(\Omega) \ . \tag{12.37}$$

Taking the squared absolute values and assuming statistical independency of the two transform quantization errors gives

$$\left|E(\Omega)\right|^2 = \left|G_0(\Omega)\right|^2 \left|Q_0(\Omega)\right|^2 + \left|G_1(\Omega)\right|^2 \left|Q_1(\Omega)\right|^2 \ . \tag{12.38}$$

From this, and due to the validity of Parseval's theorem we can conclude that

$$\sigma_e^2 = \sigma_{q_0}^2 \underbrace{\sum_k g_0^2(k)}_{w_0} + \sigma_{q_1}^2 \underbrace{\sum_k g_1^2(k)}_{w_1} = \sigma_{q_0}^2 \, w_0 + \sigma_{q_1}^2 \, w_1 \ , \tag{12.39}$$

which means that the quantization errors are weighted by the squared norms of the reconstruction filter impulse responses. Now, it has to be considered that in a wavelet tree, the number of samples is different within the particular wavelet bands, and that the coefficients residing at the deeper levels of the tree will be weighted multiple times as they pass through several reconstruction filters. For a 1D wavelet transform with T levels of the wavelet tree, scaling coefficient c_0 and wavelet coefficients $c_1 \ldots c_T$, the resulting reconstruction error will be

$$\sigma_e^2 = \sigma_{q_0}^2 \, w_0^T + \sum_{t=1}^{T} \sigma_{q_t}^2 \, w_0^{T-t} w_1 \ . \tag{12.40}$$

Similarly, for a 2D wavelet tree of T four-band decomposition levels, assuming equal filter pairs for the horizontal and vertical steps and a frequency layout with indexing scheme as shown in Fig. 4.49d, the result is

$$\sigma_e^2 = \sigma_{q_{0,0}}^2 \left(w_0^2\right)^T + \sum_{t=1}^{T} \left[\sigma_{q_{t,0}}^2 \, w_1 w_0 + \sigma_{q_{0,t}}^2 \, w_1 w_0 + \sigma_{q_{t,t}}^2 \, w_1^2 \right] \left(w_0^2\right)^{T-t} \ . \tag{12.41}$$

A signal-flow scheme which allows to interpret the respective contributions of the quantization errors into the reconstruction error is shown in Fig. 12.29.

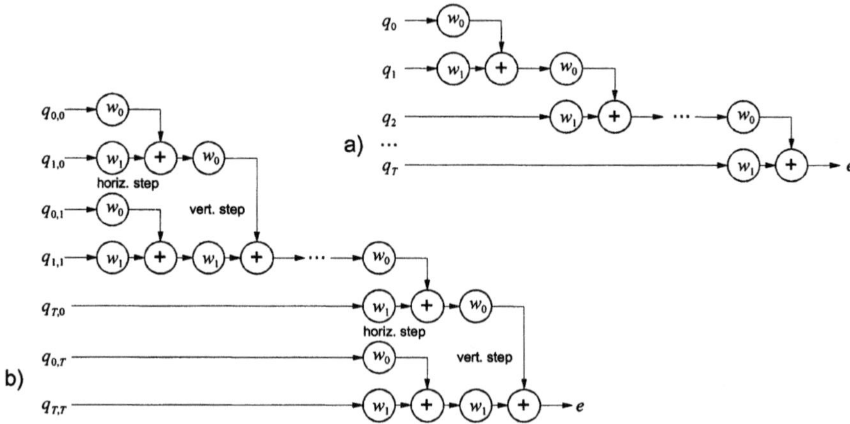

Fig. 12.29. Signal flow and weighting of quantization errors from wavelet bands into the reconstruction **a** for a 1D wavelet tree **b** for a 2D wavelet tree

It is now straightforward to determine how the quantization step sizes have to be chosen for the particular wavelet bands, such that the overall reconstruction error is minimized. As the squared quantization step size is proportional with the quantization error variance, it is necessary to adapt the quantization step size by the reciprocal square root of the respective subband's weighted influence in (12.40), which gives the quantizer step size for the single subbands, normalized by a global step size factor Δ

$$\Delta_0 = \frac{\Delta}{\sqrt{w_0^{\,T}}} \quad ; \quad \Delta_t = \frac{\Delta}{\sqrt{w_0^{\,T-t} w_1}} = \Delta_0 \cdot \sqrt{\frac{w_0^{\,t}}{w_1}} \text{ for } t > 0 . \tag{12.42}$$

By the last equality, it is possible to establish a relationship between the quantization step sizes of all wavelet bands with the step size used for quantization of the scaling band. For the case of a 2D wavelet tree, the related dependency follows from (12.41) as

$$\Delta_{0,0} = \frac{\Delta}{\sqrt{\left(w_0^{\,2}\right)^T}} \quad ; \quad \Delta_{t,0} = \Delta_{0,t} = \frac{\Delta}{\sqrt{\left(w_0^{\,2}\right)^{T-t} w_1 w_0}} = \Delta_{0,0} \cdot \frac{w_0^{\,t-\frac{1}{2}}}{\sqrt{w_1}}$$

$$\Delta_{t,t} = \frac{\Delta}{\sqrt{\left(w_0^{\,2}\right)^{T-t} w_1^{\,2}}} = \Delta_{0,0} \cdot \frac{w_0^{\,t}}{w_1} \text{ for } t > 0 . \tag{12.43}$$

Observe that for the case $w_0 = w_1 = 1$, all quantization step sizes shall be chosen equal. This would exactly be the case for an orthonormal transform. Some biorthogonal filters exist, e.g. the 9/7 filter (4.195), which indeed almost approach the orthonormality case.

It is now possible to determine the coding gain of wavelet coding in a more general formulation. A factor $M^*(t)$ is introduced which describes the number of coefficients in a specific wavelet band by proportionality with the scaling band, where $M^*(t)=1$ for $t=0$, and $M^*(t)=2^{t-1}$ for the wavelet bands, $0<t\leq T$. Assume further that the distortion will be dependent on the squared quantizer step size, where D_t may e.g. be $\Delta_t^2/12$ according to (11.1). The mean of the rate is

$$R_{\mathrm{WC}} = \frac{1}{2^T}\cdot\frac{1}{2}\sum_{t=0}^{T} M^*(t)\cdot\log_2\frac{E\{c_t^2\}}{D_t} = \frac{1}{2\cdot 2^T}\log_2\left[\prod_{t=0}^{T}\left(\frac{E\{c_t^2\}}{D_t}\right)^{M^*(t)}\right]. \quad (12.44)$$

For PCM, the rate is assumed to be (e.g. with $D_{\mathrm{PCM}}=\Delta^2/12$)

$$R_{\mathrm{PCM}} = \frac{1}{2}\log_2\frac{\sigma_x^2}{D_{\mathrm{PCM}}} \Rightarrow D_{\mathrm{PCM}} = \frac{\sigma_x^2}{2^{2\cdot R_{\mathrm{PCM}}}}. \quad (12.45)$$

Substituting (12.44) into (12.45) with $R_{\mathrm{WC}}=R_{\mathrm{PCM}}$ gives

$$D_{\mathrm{PCM}} = \frac{\sigma_x^2}{\left[\prod_{t=0}^{T}\left(\frac{E\{c_t^2\}}{D_t}\right)^{M^*(t)}\right]^{\frac{1}{2^T}}}. \quad (12.46)$$

If now the quantizer step sizes are separated from D_t, assuming that all wavelet coefficients will contribute to the total reconstruction error in (12.39), the ratio of D_{PCM} over D_{WC} can again be expressed as the *coding gain*,

$$G_{\mathrm{WC}} = \frac{D_{\mathrm{PCM}}}{D_{\mathrm{WC}}} = \frac{\sigma_x^2}{\left[\prod_{t=0}^{T}\left(\frac{E\{c_t^2\}}{\Delta_t^2}\right)^{M^*(t)}\right]^{\frac{1}{2^T}}}. \quad (12.47)$$

The related formulation for a 2D wavelet tree is

$$G_{\mathrm{WC}} = \frac{D_{\mathrm{PCM}}}{D_{\mathrm{WC}}} = \frac{\sigma_x^2}{\left[\frac{E\{c_{0,0}^2\}}{\Delta_{0,0}^2}\prod_{t=1}^{T}\left(\frac{E\{c_{t,0}^2\}}{\Delta_{t,0}^2}\cdot\frac{E\{c_{0,t}^2\}}{\Delta_{0,t}^2}\cdot\frac{E\{c_{t,t}^2\}}{\Delta_{t,t}^2}\right)^{M^*(t)}\right]^{\frac{1}{4^T}}}, \quad (12.48)$$

where now $M^*(t)=1$ for $t=0$, and $M^*(t)=4^{t-1}$ for the wavelet bands, $0<t\leq T$.

12.4.3   Vector Quantization of Transform Coefficients

From the aspects discussed in the previous sections, a crucial aspect in frequency coding is the best exploitation of any existing statistical dependencies, both *within* and *between* frequency bands. This is the motivation behind the usage of prediction methods and context-based entropy coding. Vector quantization of frequency coefficients (resulting from block transforms, block-overlapping, subband or wavelet transforms) is another method to approach this goal. By combining a larger number of spectral coefficients during quantization yet, a better performance in the rate-distortion sense can be expected than by separate scalar quantization and entropy coding. This will in particular be true at low data rates, where the disadvantage of pre-determined scalable quantization thresholds may not fully be compensated by sophisticated entropy coding. Different methods of transform vector quantization are e.g. compared in [BLAIN, FISCHER 1991], [SENOO, GIROD 1992], [COSMAN, GRAY, VETTERLI 1996]. *Sliding block codes*, like methods of tree and trellis coded quantization, have also been shown to provide compression advantages [OHM 1990], [NANDA, PEARLMAN 1992], [PEARLMAN, JAKATDAR, LEUNG 1992], [XIONG, WU 1999]. In principle, it is possible to combine vectors from spectral coefficients of only *one frequency band*

$$\mathbf{c}_{uv} = \left[c_{uv}(m',n'), c_{uv}(m'+1,n'), ... \right]^{\mathrm{T}} , \tag{12.49}$$

or from *different frequency bands* at the same spatial position (m',n')

$$\mathbf{c}(m',n') = \left[c_{u,v}(m',n'), c_{u+1,v}(m',n'), ... \right]^{\mathrm{T}} . \tag{12.50}$$

The method (12.49) shall be denoted as *intra-band VQ*, and the method (12.50) as *inter-band VQ*.

Intra-band VQ. For subband transforms of constant bandwidths, a decomposition of a size MxN image is made into UxV subband images of sizes M/UxN/V each (see Fig. 12.30a). For the case of the wavelet transform, the subband images are of dyadically increasing size. Small spatial blocks of sizes $M'_{u,v}$ x $N'_{u,v}$ are extracted from the subband images as vectors to be encoded. The case of block transform coding is shown in Fig. 12.30b. Here, the coefficients of identical frequency are extracted from different transform blocks and then arranged as vectors. This allows to exploit any spatial dependencies between the samples from the same frequency band by a proper design of the codebooks. The vectors should be orientated along directions where most statistical dependencies are expected, e.g. for subbands of horizontal low frequency it is appropriate to use vectors of large horizontal extension. As it can be expected by (12.29) that higher rates have to be spent on the lower frequency coefficients, the vector (block) sizes should be variable over the different frequency bands, such that approximately codebooks of constant size are obtained. The codebook size and vector dimension for a particular subband with an expected rate $R_{u,v}$ can be determined as

$$J_{u,v} = 2^{K_{u,v} \cdot R_{u,v}} \quad \Rightarrow K_{u,v} = M'_{u,v} \cdot N'_{u,v} = \frac{\log_2 J_{u,v}}{R_{u,v}} . \qquad (12.51)$$

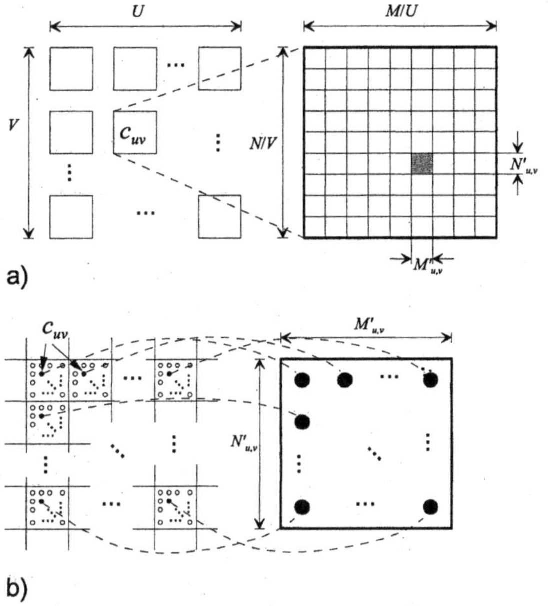

Fig. 12.30. Arrangement of intraband vectors
a for Subband/VQ coding **b** for block Transform/VQ coding

Due to the instationary statistical behavior of image signals, it is however not reasonable to allocate constant rates anywhere in the image. Methods which can be used to solve this problem are

- Classified VQ, switching of codebooks depending on the amount and characteristics of local detail;
- switching of codebooks or selection of sub-codebooks depending on contexts gained from other frequency bands;
- combination with run-length or zero-tree coding to employ VQ only for those coefficients which are above a given threshold.

Both types of dependencies between samples, *inter-band* and *intra-band*, can be exploited by these methods or a combination thereof.

Inter-band VQ. Inter-band VQ methods combine coefficients from different spatial bands of the same spatial location into vectors, and will hence inherently exploit inter-band dependencies when an appropriate codebook design is made. For block transform coding, simply coefficients from the same transform block are combined as vectors. For wavelet coding, the combination can be performed similar to the zero-tree correspondences. When applied to a larger transform block which is split into smaller vector block partitions, transform VQ coding can indeed also be interpreted as a *product code* (cf. sec. 11.5.4) with the entire transform block as a larger vector entity to be encoded from sub-codebooks [CUPERMAN 1989].

If the entire transform block of size $K=U \cdot V-1$ (excluding the DC coefficient) is regarded as one block vector $\mathbf{x}$, this is first transformed into a coefficient vector $\mathbf{c}$. Subsequently, $\mathbf{c}$ is subdivided into T sub-vectors $\mathbf{c}_1, \mathbf{c}_2, ..., \mathbf{c}_T$ of lengths $K_1, K_2, ..., K_T$. Similarly, sub-vectors can be constructed from partial elements of a wavelet tree. At one sampling position of the scaling-band image in a wavelet tree of T levels, a total of 2^T-1 wavelet coefficients exist for the case of a 1D transform, or 4^T-1 for a 2D transform. These can be arranged into sub-vectors, e.g. for the three directional orientations as in the zerotree method, each of a vector length $K_i=(4^T-1)/3$ or smaller sub-vector partitions. In general, it should be observed that coefficients are combined into vectors for which highest statistical dependencies are expected. If codebooks $\mathcal{C}_t$ of sizes J_t are used for the sub-vectors, the following conditions hold:

$$K = \sum_{t=1}^{T} K_t \quad ; \quad \mathcal{C} = \mathbf{X}_{t=1}^{T} \mathcal{C}_t \quad ; \quad R = \frac{1}{K} \sum_{t=1}^{T} ld\, J_t . \tag{12.52}$$

Here, R expresses the rate per sample. The unified codebook $\mathcal{C}$ (representing one transform block or one full scan of the wavelet tree) is established by the *outer product* of the single codebooks, where all possible combinations of sub-vectors can be used. To achieve rate allocations depending on the amount of detail in the signal, entropy coding or classifying mechanisms with codebook switching can be used. As both intra-band and inter-band statistical dependencies will exist, combination with spatial prediction mechanisms or entropy coding based on spatial contexts can be used. In general however, it must be concluded that VQ of frequency coefficients has insufficient flexibility with regard to rate allocation, or will become overly complex when a sufficient degree of flexibility is needed, which could e.g. be achieved by discarding of samples, dynamic change of vector length etc. Further, implementation of scalability by methods like tree-structured codebooks will hardly allow to cover comparably broad ranges of rates as with scalar embedded quantizers. Possible applications of transform VQ coding can mainly be identified for ranges of extremely low data rates, where scalar quantizers suffer from an insufficient degree of variety by their constrained sets of reconstruction values.

Trellis-coded quantization. A lack of variety in quantizer reconstruction values occurs at low data rates when scalar quantizers are used. This problem can also be tackled by *trellis coded quantization* (TCQ) of transform coefficients, which follows the basic scheme as introduced in sec. 11.6.1. The set of allowable reconstruction values is expanded, however restrictions are made which values may be used in combination by subsequent encoding steps, as defined by the paths in the related trellis diagram. This can be interpreted similar to a VQ codebook. TCQ can also be combined with embedded quantizers, where improvements over conventional quantization techniques have been shown (see e.g. [BILGIN, SEMENTILLI, MARCELLIN 1999]).

12.4.4 Adaptation of transform bases to signal properties

Local statistical properties of image signals vary largely. For example, the amount of local detail influencing correlation and frequency characteristics can abruptly change at boundaries of objects. For large flat areas or large areas with regular structure (texture), frequency transforms can best exploit the correlation properties of the signal by using large contexts (e.g. large blocks in transform coding). For areas with instationary signal behavior like edges, it is better to use narrow contexts. For Wavelet transform methods, this problem is partially solved by the *localization property*. Nevertheless, it may happen that a dyadic Wavelet representation loses coding gain as compared to a transform which performs a more narrowband frequency analysis also in the high-frequency range; for example, in the case of signals containing high-frequency periodic components, a representation by longer basis functions similar to the STFT (sec. 4.4.4) could be advantageous. Better compression performance can eventually be expected if *variable block sizes* are allowed in a block transform (see Fig. 12.31a). For the case of a wavelet transform, the strategy would be adaptation of the *transform decomposition depth*, which leads to frequency bands of freely variable bandwidth and variable spectral resolution; this is denoted as *Wavelet Packet decomposition* (see Fig. 12.31b).

Variable block size transforms (VBST). A VBST method based on the DCT was proposed in [VAISEY, GERSHO 1992]. Quadratic block sizes in the range between 4x4 and 32x32 pixels are considered, where the block layout structure is encoded by a quad-tree (see Fig. 12.42). The optimization problem is the decision about the best local block sizes. Starting from the largest block size, this can be done by joint analysis of mean value and variance differences between the next smaller sub-blocks, where typically large variations indicate that a significant change of detail structure occurs, such that a division into smaller sub-blocks will be useful. It is however necessary to track occurrences of mean/variance deviations between sub-blocks down to the smallest block size before the decision over all levels can be made; cases occur where no deviation is found between larger block partitions e.g. of size 16x16, while significant deviations are observed between blocks of smaller sizes.

In [WIEN 2003], a VBST method is developed which is based on an integer-value transform (4.136)-(4.138), in combination with sophisticated spatial prediction methods and rate-distortion criteria. The method has also been applied to motion-compensated frame differences in video coding. Block partitions are either quadratic with sizes ranging between 4x4 and 16x16, or rectangular of sizes 4x8 and 8x16 with horizontal or vertical orientation.

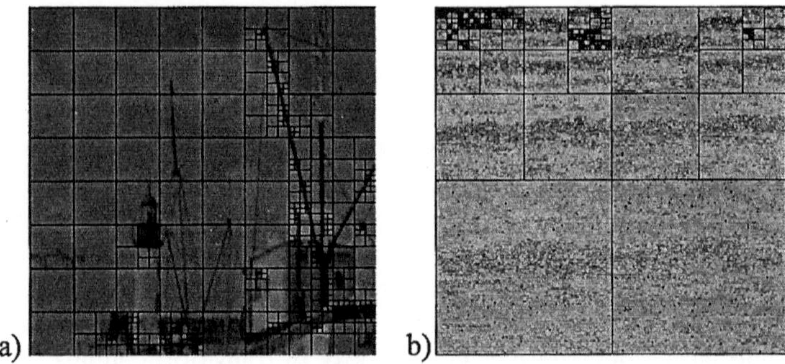

Fig. 12.31. a Spatial block layout in variable block size transform coding
b Wavelet Packet transform result using variable bandwidth of frequency decomposition
[Source: SPERLING (a), VAN DER SCHAAR (b)]

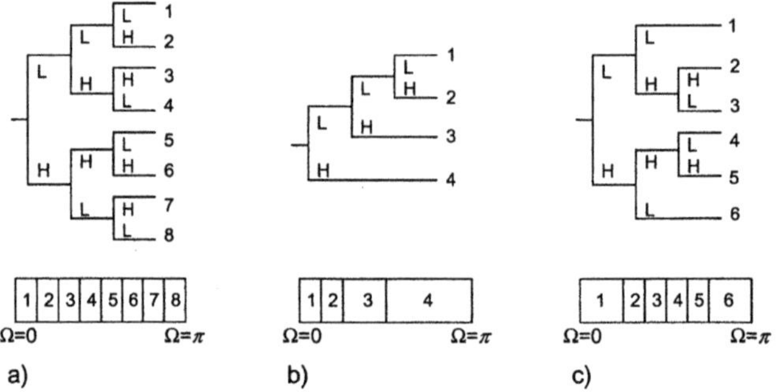

Fig. 12.32. Wavelet splitting trees and related frequency bandwidth[1] layouts: **a** Binary
(uniform-bandwidth) **b** Dyadic (Wavelet) **c** Irregular (Wavelet Packet)

Wavelet Packet transform. Subband decomposition of equal-bandwidth frequency resolution over all bands and wavelet transform of octave-bandwidth frequency resolution can be interpreted as binary tree and dyadic tree structures; both
are highly regular. In between, other irregular decomposition structures are possible, which correspond to cases where high-frequency bands from the elementary 2-
band split are decomposed whenever useful (Fig. 12.32).

In the case of the 2D Wavelet transform, the elementary split steps can be
based on quartile decompositions as shown in Fig. 12.31b, where again a quad tree
code can be used to represent the frequency layout; otherwise it would also be
possible to regard the horizontal/vertical decomposition steps as two independent
two-band splits, such that non-quadratic frequency layout cells would be realizable
as well.

---

[1] Frequency bands are shown in Fig. 12.32a/c by correct (increasing) frequency ordering,
where the frequency-axis reversal occurring in H band sub-sampling is compensated by
interchange of L and H decomposition filters (cf. sec. 4.4.1).

As a criterion for the split/non-split decision, the formulations of coding gain (12.47)-(12.48) can be used by slight modification[1] to examine whether at any level of the tree an additional split will increase the coding gain. It is however possible that no increased gain is observed by growing the tree one level, but a significant gain could be realized if a split was performed at a higher level. From this point of view, strategies are preferable which start from the full decomposition as in Fig. 12.32a, and prune the tree by removing any splits which do not contribute to the coding gain. In [RAMCHANDRAN, VETTERLI 1993], tree-pruning methods are studied in combination with rate-distortion criteria. In fact, coding gain criteria will only be valid at low distortion or high data rates (cf. sec. 11.2.2), and it turns out that significantly different best packet bases are found depending on the operational point on the rate-distortion curve. The approach in general is very similar to rate-distortion optimization of quantizers (cf. sec. 11.3), where now band-split decisions are preferred which contribute the steepest decrease of the rate-distortion curve.

Adaptation of basis functions. Even though certain standard types of wavelet filter kernels are widely used, their energy compaction properties will hardly be optimum for any characteristics of signals. In particular, bi-orthogonal filter designs offer a considerable degree of freedom when performed under the constraint of a given signal's frequency characteristics. Another goal of adaptive filter designs could be stronger removal of details from the scaled versions of the signal, which indeed would shift additional energy into the higher-frequency wavelet bands [YANG, RAMCHANDRAN 2000][MAYER, HÖYNCK 2001].

Completely different types of filter bases exist which adapt very specifically to detail characteristics of 2D signals like edges, corners etc., and will eventually leave the paradigm of dyadic decomposition. Some examples of this category have been proposed under names like *edgelets*, *curvelets*, *ridgelets*, *wedgelets*, or *chirplets*. Usability for image coding is not fully clear yet; nevertheless it can be expected that adaptive and more flexible designs of filter bases will remain as a hot research topic in the future.

12.4.5   Transform coding and transmission losses

Transform coding schemes are generally more robust in the case of transmission losses than predictive coding methods. This is in particular caused by the fact that the basic principle is non-recursive, in contrast to the fully recursive structure of DPCM decoders, such that error propagation is less severe when independent frequency components are decoded. Where in DPCM the effect is propagation of

---

[1] In (12.47)-(12.48), the index t relates to the position in a wavelet tree which can uniquely be mapped to the respective wavelet bands. For wavelet packets, it is now necessary to use an index which indicates each node of the tree individually. The principle of computation and the meaning of parameters Δ and M^* is not changed.

the error by the impulse response of the recursive synthesis filter, in transform coding the effect are erroneous components of synthesis basis functions which are superimposed to the decoded signal. This effect occurs for any types of basis functions, irrespective of block transform, overlapping-block transform or wavelet/subband types. Consequently, bits which contribute most to the reduction of distortion in the rate-distortion sense will also have the highest impact on distortion when channel errors occur. E.g. in a bit plane representation, the most significant bit planes will have the highest effect in deteriorated visual quality when affected by transmission losses. Likewise, it can be expected from statistical models that low-frequency coefficients have highest energy and would contribute most to the visibility of errors if corrupted by channel errors. Hence, most emphasis for error-resilient transmission must be put on DC and lowest-frequency AC coefficients in transform coding, on scaling and lowest-frequency wavelet coefficients in wavelet coding.

A very efficient protection to guarantee highest attainable quality in presence of data losses can be made by appropriate error protection of these most relevant components. Embedded quantization and bit-plane coding offer a natural solution to this problem. If higher-significance bit-planes are protected accordingly, a minimum reconstruction quality is guaranteed; errors in less significant bit-planes, which usually take the highest percentage of rate, can eventually be tolerated[1].

However, even for cases of quite conventional compression methods like block transforms with one-step quantization and entropy coding, certain parts of information can be identified which are more important than others. It can statistically be expected that errors in the lower-frequency coefficients have more impact on the reconstruction result, such that the associated run/level entropy coded values have to be protected. Splitting of the bit stream into sub-streams which represent signal components of different relevance is denoted as *data partitioning*. An example which is applicable in a zigzag run-length scanning scheme is shown in Fig. 12.33.

Similarly, in wavelet compression, the information residing within the wavelet tree may be cut at a certain resolution level, such that a signal of lower spatial resolution is more safely transmitted. Even though data partitioning is attractive due to its simplicity, there is no guarantee that the information which is contained in the gray-shaded area of Fig. 12.33 is upper bounded to a maximum percentage of the entire information. Typically, low-frequency coefficients consume a considerable amount of the entire data rate, and in general, significant error protection should be applied only to a relatively small part of the information, as otherwise the effort to be made for channel coding would significantly increase the rate to be transmitted. If this principle is not observed, the quality decreases substantially if no errors are present, as less bits are available for source coding. From this point of view, coding schemes are preferable which allow to clearly identify which parts of the information shall be protected by which priority. This is an eminent advantage of transform coding schemes in general over predictive schemes. From this

---

[1] Methods of *unequal error protection*, which can fulfill this requirement, are discussed in sec. 15.2.2.

point of view, transform coding schemes with flexible priority identification properties such as wavelet and bit-plane coding are preferable.

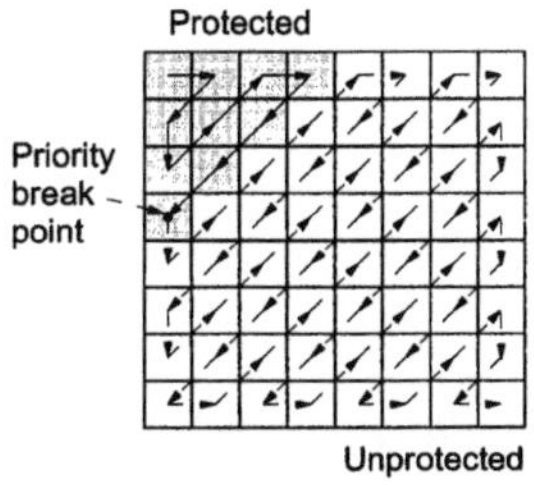

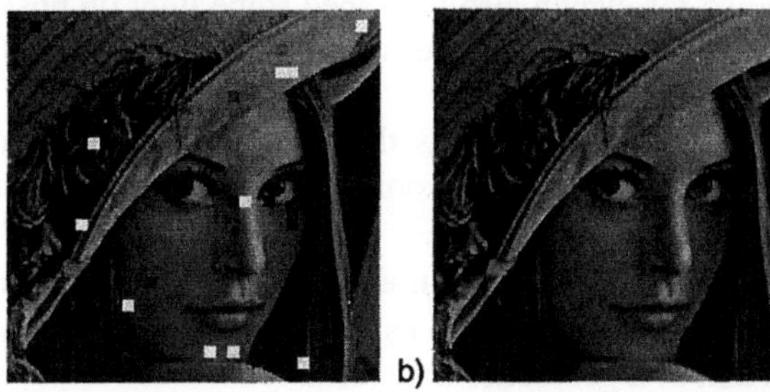

Fig. 12.33. Principle of 'data partitioning' in a zigzag scan of a transform block

As the analysis in this section has shown, substantial increase in compression efficiency can be expected when *inter-band* (or *intra-block*) and *intra-band* (or *inter-block*) statistical dependencies are largely exploited. On the other hand, regarding error resilience of the encoded representation, this is not very useful, as eventually infinite error propagation may occur as it was the case in prediction-based coding. Clearly at this point, any approach to separately optimize source and channel coding is obsolete, as eventually a small gain in source compression would require a considerable amount of error-protection overhead to recover from problems in case of transmission losses. In such cases, joint optimization of compression and error protection methods is necessary.

a)   b)

Abb. 12.34. DCT with bit plane coding at a rate R = 1 *bpp*, channel error rate of statistically independent errors $P_{err} = 10^{-2}$: **a** Transmission without error protection **b** Transmission with error protection in significant bit planes

Fig. 12.34a/b shows reconstructed images for the case of an error-prone transmission. The encoding scheme in these examples is a DCT with embedded quantization and a bit-plane coding scheme. In Fig. 12.34a, no error protection is done[1], in 12.34b, 10% of the bits belonging to the more significant bit planes are error pro-

---

[1] except for resynchronization after an error, which can be made at the starting coordinate of any block; this comes at the cost of an additional overhead in the encoded information.

tected with no losses remaining[1]. Observe that errors affecting the DC coefficients are most visible, which is consistent with the assumption that they contain the largest percentage of overall energy.

12.5 Fractal Coding

The theory of fractals is based on the assumption of *self-similarity* (similarity between different scales and resolutions), which is inherent to many natural processes and most probably appears in image signals as well. A typical example is the similarity of structures appearing in the directional wavelet bands at different resolution levels (see e.g. Fig. 12.27a). To exploit this for image compression, it is necessary to define a *fractal transform* for entire images or at least smaller areas from images, which would be able to describe the self-similarity. The *collage theorem* establishes the foundations, by which an image signal can be reconstructed with sufficient accuracy by the parameters of a fractal transform. The efficiency of a fractal transform highly depends on the strategy by which the parameters of the transform are determined and encoded. In principle, matching is applied for computation of the fractal code, which can also be interpreted as a specific form of vector quantization, with a codebook that is extracted from the image itself. At the end of this section it will be discussed how fractal coding can be combined with other coding methods introduced so far, which also may shed some light on possible ways to further improve encoding e.g. of wavelet or block transform coefficients.

The theory of *iterated function systems*, which is the most common basis of image coding methods denoted as *fractal*, is however only weakly connected to the theory of complex fractals like the Mandelbrot sets[2]. Moreover, it is tried in a brute force approach to describe the assumed self-similarity of an image by parameters of a linear or nonlinear transform $f(\cdot)$, which usually includes components of an amplitude-value transform and of a geometric transform.

---

[1] Random bit errors (no burst characteristic) are assumed here, such that the latter assumption is realistic when channel coding is applied for error protection.
2 Mandelbrot sets can be visualized by a 2D graphical presentation of infinitely renewable and refinable image structures. Even though being chaotic processes, convergence is guaranteed up to infinitely fine resolution. For applications in image coding, it is interesting that in principle images with extremely fine resolution and infinitely increasable amount of detail can be generated by a finite set of parameters.

12.5.1 Principles of Fractal Transforms

Regard a certain area (e.g. a block) from an image to be encoded. It shall be tried to describe this area (target) as good as possible from *another* area (origin) of the same image. For this purpose, the origin area may be modified in *geometry, brightness and contrast*.

First, it is important that *any* change is performed at all, it would not be reasonable to select origin and target to be identical.

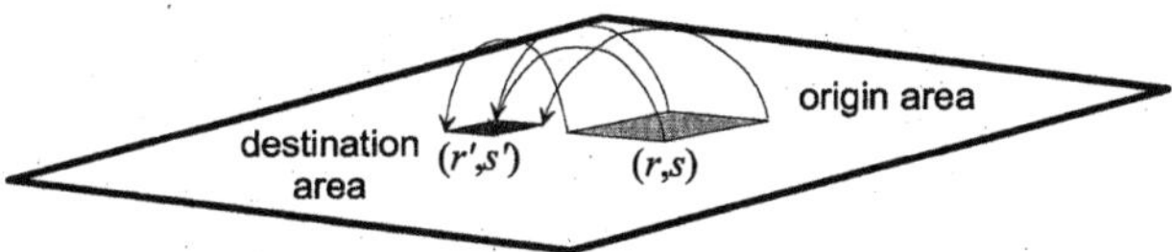

Fig. 12.35. Principle of a fractal block transform

The principle of a fractal block transform is shown in Fig. 12.35. In this case, an origin area described by coordinates (r,s) is mapped by a *geometric transform* $\gamma(\cdot)$ into a block-shaped target area described by coordinates (r',s'). In the example shown here, the geometric transform consists of a translation and a scaling, such that the origin area will also be block-shaped, and 3 parameters are needed to describe the transform for the given block. In general, any geometric transform (typically a parametric coordinate transformation, as e.g. the affine transform) can be used. In addition, an *amplitude transform* $\lambda(\cdot)$ is applied, by which mean value and variation (contrast) of the origin block amplitudes are adapted to the actual values of the target block in the image to be decoded. The entire fractal transform for the given block area is then the functional combination of geometric and amplitude transform, $f(\cdot)=\gamma(\cdot)\circ\lambda(\cdot)$.

12.5.2 Collage Theorem

Contractivity of the geometric transform $\gamma(\cdot)$. The geometric transform performs a mapping of coordinates (r,s) from an origin area into coordinates (r',s') of the target area. This mapping is *contractive*, if no geometric distance between any two points in the target system (r',s') is larger than the distance between the corresponding points in the origin system (r,s). Herein, k is the contractivity factor, such that

$$(r,s)\xrightarrow{\;\gamma(\cdot)\;}(r',s')$$

$$d\big((r_1',s_1'),(r_2',s_2')\big)\le k\cdot d\big((r_1,s_1),(r_2,s_2)\big) \quad;\quad k<1. \tag{12.53}$$

Regarding the degrees of freedom as shown for the affine transform in Fig. 7.37, translation and rotation will not effect any change of the distance at all, these are non-contractive. Scaling will be contractive when the scaling factor is less than 1.

If other transform modifications like shear are used, a much lower scaling factor must eventually be used to still guarantee a contractive transform.

Contractivity of the amplitude transform $\lambda(\cdot)$. The amplitude transform modifies the variation between the pixels by a gain factor a, and shifts the mean by an offset value b; in case of color images, separate factors can be applied to the different components. The transform is contractive, if the variation of amplitudes is not increased during the mapping, which means the gain factor must be less than unity:

$$x(r',s') \xrightarrow{\;\lambda(\cdot)\;} y(r',s') = a \cdot x(r',s') + b \quad ; \quad k = |a| \overset{!}{<} 1 \tag{12.54}$$

The offset value should be constrained such that the output image needs not to be clipped into the allowable amplitude range $0 \ldots A_{\max}$.

Collage Theorem. The *Collage Theorem* [BARNSLEY 1988] states that an iterative generation of an image from an arbitrary source image is possible by a fractal transform $f(\cdot)$ if

1.   The description of the image from itself can be achieved by the transform with sufficiently small distortion;
2.   The complete transform $f(\cdot) = \gamma(\cdot) \circ \lambda(\cdot)$ is contractive.

Proof of the Collage Theorem. Let $\mathbf{X}$ be an original image, and $f(\mathbf{X})$ be the image which results when the fractal transform is applied *once* to this image. This can be seen as the optimum case of decoding from the fractal transform parameters $f(\mathbf{X})$, which have to be designed such that the image can be described as good as possible from itself. Let $\mathbf{Y}^r = f^{(r)}(\mathbf{Z})$ be the reconstructed image after the rth iteration of the fractal transform applied to an arbitrary initial image $\mathbf{Z}$. Further, $f^{(r)}(\mathbf{X})$ shall be the result after the rth application of the transform to the original image $\mathbf{X}$; observe that $f^{(r)}(\mathbf{X})$ probably will have a larger deviation than $f(\mathbf{X})$ from the original image $\mathbf{X}$, as in the iteration $2 \ldots r$ the original is not used. By the triangular inequality, the following relationship applies (where $d(\cdot,\cdot)$ expresses the Euclidean distance):

$$d(\mathbf{X},\mathbf{Y}) \le d\left(\mathbf{X}, f^{(r)}(\mathbf{X})\right) + d\left(f^{(r)}(\mathbf{X}), f^{(r)}(\mathbf{Z})\right). \tag{12.55}$$

Further,

$$d\left(\mathbf{X}, f^{(r)}(\mathbf{X})\right) \le d\left(\mathbf{X}, f^{(1)}(\mathbf{X})\right) + d\left(f^{(1)}(\mathbf{X}), f^{(2)}(\mathbf{X})\right) + \ldots + d\left(f^{(r-1)}(\mathbf{X}), f^{(r)}(\mathbf{X})\right)$$
$$\le (1 + k + \ldots + k^{r-1}) \cdot d\left(\mathbf{X}, f^{(1)}(\mathbf{X})\right) \le (1-k)^{-1} \cdot d\left(\mathbf{X}, f(\mathbf{X})\right) \tag{12.56}$$

and

$$d\left(f^{(r)}(\mathbf{X}), f^{(r)}(\mathbf{Z})\right) \le k \cdot d\left(f^{(r-1)}(\mathbf{X}), f^{(r-1)}(\mathbf{Z})\right) \le \ldots \le k^r \cdot d(\mathbf{X},\mathbf{Z}). \tag{12.57}$$

Using (12.55), (12.56) and (12.57), the following upper bound condition holds for the resulting distortion:

$$d\left(\mathbf{X},\mathbf{Y}^r\right) \leq \underbrace{(1-k)^{-1}\cdot d\left(\mathbf{X},f(\mathbf{X})\right)}_{\varepsilon} + k^r\cdot d(\mathbf{X},\mathbf{Z}) . \tag{12.58}$$

As according to (12.53) and (12.54) k must be smaller than 1, the distance between the original image $\mathbf{X}$ and the output $\mathbf{Y}^r$ of this iteration will converge towards ε after a sufficient number of iterations r. This distance is lower bounded (for $k{\rightarrow}0$) by the distance between $\mathbf{X}$ and $f(\mathbf{X})$. The convergence will be faster for smaller k. Due to the many inequalities contained in this proof, the distortion in the reconstructed signal will usually be *much smaller* than the upper bound given by (12.58).

Fig. 12.36. Convergence of fractal decoding, first three iterations:
a Generated from uniform gray image **b** Generated from different image

12.5.3   Fractal Decoding

Fig. 12.35 shows the procedure how the collage theorem is applied in fractal decoding of one area (block) of an image, using parameters of the fractal transform. It is now necessary to provide parameters for any area of the image. If the procedure is applied iteratively, the initial source image must not be in any way related to the image to be decoded, and an arbitrary image can be used as starting point. Whenever the fractal transform is contractive, convergence is guaranteed, and even after one iteration the result will be more similar to the image to be decoded according to (12.58). Further iterations are performed, until convergence is achieved, i.e. until no or no more significant changes occur from one iteration to the next. This is the decoding process. The parameters of the fractal transform can hence be interpreted as a code describing the image. The encoder has to determine these parameters by analysis of the original image, which means to find the optimum fractal mapping of the image into itself. The result of decoding, as shown above, can however not become better than the result from applying the fractal transform *once* to the original image.

The convergence of the fractal decoding process is illustrated in Fig. 12.36. Same parameters are used in both cases, however the initial source image is once a uniform gray image, and once a different image than the image to be decoded. It is nevertheless obvious that the difference decreases significantly after a small number of iteration steps in both cases. Convergence is apparently faster, when the initial image has uniform gray-level values.

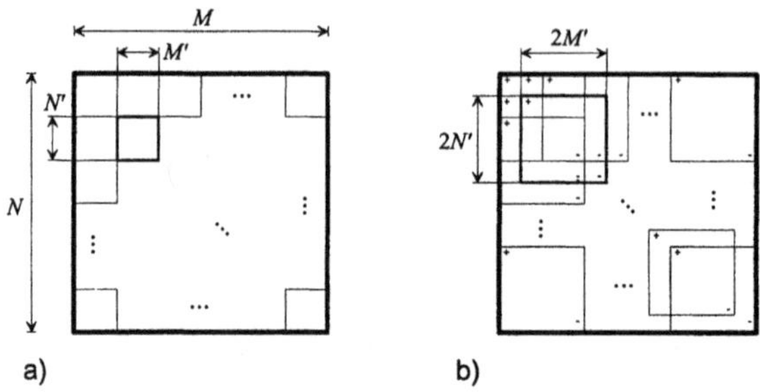

Fig. 12.37. Fractal block coding: **a** Positions of target blocks where parameters of the fractal transform must be defined (non overlapping) **b** Example of possible selection of origin blocks (overlapping)

Fractal block coding as described in [JACQUIN 1992] is a simple realization to determine and encode parameters of a fractal transform. The principle is shown in Fig. 12.37. Certain limitations are imposed to the geometric transform, which helps to speed up the search for the fractal parameters and make their representation more compact:

- The target block areas are of equal size M'xN' (denoted as *range blocks* in [JACQUIN 1992]);
- Contractivity of the geometric transform is fixed to a scaling factor $\Theta=0,5$;
- The free parameters of the geometric transform are translations, rotations of blocks by 0°, 90°, 180° and 270° and horizontal or vertical (mirrored) reflections about the central axes.

The origin blocks (denoted as *domain blocks* in [JACQUIN 1992]) must possess a size of $2M'$x$2N'$. To strictly fulfill the contractivity criterion and avoid alias in the reconstructed signal, they must be lowpass filtered and sub-sampled by a factor of two when mapped into the target blocks. To limit the number of code symbols, the number J of allowable source blocks is limited; J is the product of translational and rotational variants to be checked.

Computation of transform parameters. Let **x** be the block of the original image to be coded, $\mathbf{y}_j$* a member from a set of source blocks to be used for the geometric transform, sub-sampled to a size $M'\text{x}N'$, $j=1,2,...,J$. The parameter a of the amplitude transform shall be applied to zero-mean source blocks, where the actual mean value shall be $\mu_{\mathbf{y}_j\!*}$. If the source block of index j is selected, the reconstruction result after geometric and amplitude transforms is

$$\mathbf{y}_j = a\cdot(\mathbf{y}_j\!*-\!\mu_{\mathbf{y}_j\!*})+\mathbf{b} \quad ; \quad \mu_{\mathbf{y}_j\!*}=\mu_{\mathbf{y}_j\!*}\cdot\mathbf{1} \quad ; \quad \mathbf{b}=b\cdot\mathbf{1}\,. \tag{12.59}$$

In addition, a constant offset value a_0 can be used, which effects that in the case of similar mean of $\mathbf{x}$ and $\mathbf{y}_j\!*$ (a case which is very likely if the source and destination of the mapping are close), the parameter b of the amplitude transform can be kept small:

$$\mathbf{y}_j = a\cdot(\mathbf{y}_j\!*-\!\mu_{\mathbf{y}_j\!*})+a_0\cdot\mu_{\mathbf{y}_j\!*}+\mathbf{b}\,. \tag{12.60}$$

The optimum source block to be selected shall have maximum cross covariance between source and destination

$$r_{xy_j\!*}{}' = \mathbf{x}^{\mathrm{T}}\mathbf{y}_j\!*-M'\cdot N'\cdot\mu_x\cdot\mu_{\mathbf{y}_j\!*}\,. \tag{12.61}$$

This criterion is similar to Gain/Shape VQ (11.66), however the covariance is used in (12.61) as zero mean is assumed in (12.59).

Optimum parameters a and b of the amplitude transform are determined as in the procedure for (11.66), again observing the mean subtraction:

$$a_{\mathrm{opt}} = \frac{r'_{xy_j\!*}}{\sigma_{y_j\!*}{}^2} = \frac{\mathbf{x}^{\mathrm{T}}\mathbf{y}_j\!*-M'\cdot N'\cdot\mu_x\cdot\mu_{y_j\!*}}{\left[\mathbf{y}_j\!*\right]^{\mathrm{T}}\mathbf{y}_j\!*-M'\cdot N'\cdot\mu_{y_j\!*}{}^2} \quad ; \quad b_{\mathrm{opt}}=\mu_x-a_0\cdot\mu_{y_j\!*}\,. \tag{12.62}$$

If the parameter a_{opt} resulting from (12.62) violates the condition of contractivity, it must be modified for $|a_{\mathrm{opt}}|<1$. The search for the optimum transform to be performed at the encoder side turns out to be a problem of *matching* (cf. sec. 7.5).

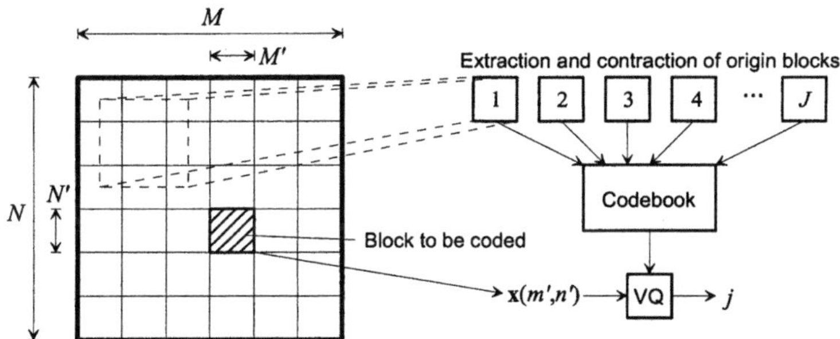

Fig. 12.38. Interpretation of fractal encoding as 'VQ with virtual codebook'

Interpretation as 'VQ with virtual codebook'. The set of allowable sub-sampled source blocks can be interpreted as a codebook, where the optimum reconstruction

y_j has to be found according to the criterion of cross covariance. The vectors contained in this codebook however depend on the image content (in the decoding they also change by each iteration). The interpretation of fractal encoding as a VQ using a *virtual codebook* is shown in Fig. 12.38.

Choice of parameter a_0. A typical case in fractal block transform is the choice of an origin block at the *same position* or in the near neighborhood of the target block to be encoded. In this case, $\mathbf{x}$ and $\mathbf{y}_j{}^*$ will have almost identical mean values. The introduction of the parameter a_0 then even allows to exploit the fractal self-similarity in encoding of the offset parameter b. The image shall be decoded, starting iterations from an image with a constant gray value c. For $a_0=0$, the mean value of the blocks would be readily reconstructed after only one iteration of decoding, however $b=\mu_x$ has to be encoded separately. Setting a constant $a_0>0$, only $b=\mu_x-a_0\cdot\mu_{y_j}{}^*$ has to be encoded according to the relationships given in (12.60). The reconstructed mean μ_y after one iteration will be $b+a_0\cdot c$, after two iterations $b(1+a_0)+a_0{}^2\cdot c$, and after the rth iteration (cf. (12.56))

$$\mu_y = b\cdot(1+a_0+...+a_0{}^{r-1})+a_0{}^r\cdot c . \tag{12.63}$$

Using $a_0=1$, no reconstruction would be possible at all; useful choices are in the range $a_0=0.5 ... 0.6$ [BARTHEL, VOYÉ 1994]. In the following examples, some typical configurations of images are investigated, by which the benefit of the fractal transform becomes obvious.

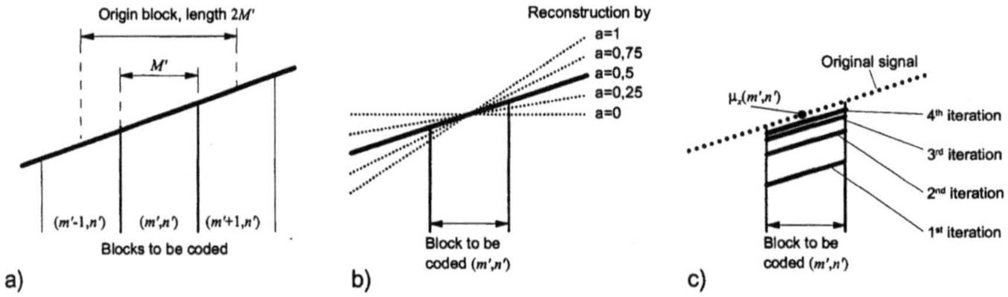

Fig. 12.39. **a** Image signal with ramp increase of amplitude **b** effect of parameter a **c** iterations of fractal decoding using $a=0.5$, $a_0=0.5$, $b=\mu_x/2$

Example: Linear ramp increase of amplitude. Fig. 12.39b shows the effect of a fractal transform with geometric contractivity $k=0.5$, where the original signal has a ramp-type increase of amplitude (Fig. 12.39a). The value $a=0,5$ will result as an optimum amplitude gain,; the contractivity condition is hence fulfilled for both amplitude and geometric transforms. Observe that the increasing ramp of amplitude can only be reconstructed when neighbored blocks have different mean values; the contractive mapping of the geometric transform implicitly effects a utilization of neighbored-block redundancies, which could also be interpreted as a mutual prediction process. In the evolution of iterations, Fig. 12.39c shows the results of the first four iterations of fractal decoding; fast convergence towards the actual amplitude values of the signal is observed both for the ramp-like characteristics and for the block mean values.

Example: Step and ramp edges. In the case of step and ramp edges (cf. Fig. 7.14), the origin block is again usually found at the position or in the direct neighborhood of the target block to be encoded. If an edge is located at the center of the block, a geometric transform of zero translation will be optimum (cf. Fig. 12.40). The parameter a of the amplitude transform will however now be near to 1; this also implies that more iteration steps are necessary to achieve an adequate reconstruction of edges. Accurate decoding of edges by a minimum number of parameters is a strong aspect of fractal coding as compared to transform coding[1].

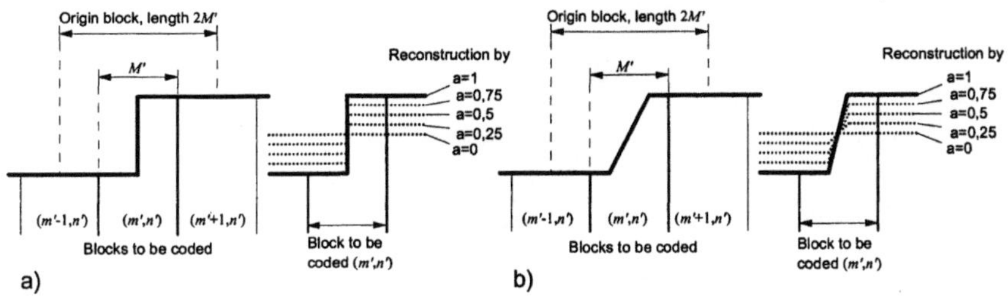

Fig. 12.40. Effect of fractal block transform **a** at a step edge **b** at a ramp edge

Encoding of fractal transform parameters. The capability for compression in fractal coding highly depends on smart strategies to encode the parameters of the transform. Due to the highly nonlinear interdependencies, the optimization is not straightforward. When a higher number J of geometric transform variations is supported, the resulting data rate increases, but the quality will increase as well. As not all positions will be selected by equal probability, application of entropy coding is useful; from the examples above, origin blocks in the vicinity of the target blocks are likely to be selected by higher probability. Evidence for a suitable statistical model in entropy coding of fractal parameters is however not shown so far. Classification of blocks into flat, texture, ramp and edge types can help to overcome this problem. As typical in still image coding, more variety of transform parameters will be required to obtain sufficient quality in highly-detailed areas, while areas of low structure are less critical and can be represented by a lower number of bits (fractal parameters with less variation or valid for larger block sizes). The offset parameters b of the luminance transform play a similar role as the DC coefficient in block transform coding and are likely to be redundant between neighboring blocks. DPCM coding can be applied to these components.

---

[1] Recall that both in block and wavelet transforms, typically a large number of coefficients will be necessary for accurate representation of edges, while still running into the danger of blocking or ringing artifacts.

In general, the performance of fractal coders saturates for higher data rates. Fractal coding is not universally applicable over a wide range of rates, nor is it straightforwardly possible to scale the information of the fractal transform. A solution to this problem is *hybrid fractal/transform coding*, where the fractal coder performs an encoding at very low data rates, and a transform coder encodes the residual error of fractal decoding towards higher rates with arbitrary precision. Conceptually, this is very similar to hybrid MC prediction/transform video coding (cf. sec. 13.2.2), where the residual of motion-compensated prediction is encoded by a transform coder [MURPHY 1993]. In this sense, fractal coding could also be interpreted as 'motion compensated prediction of the image by itself'.

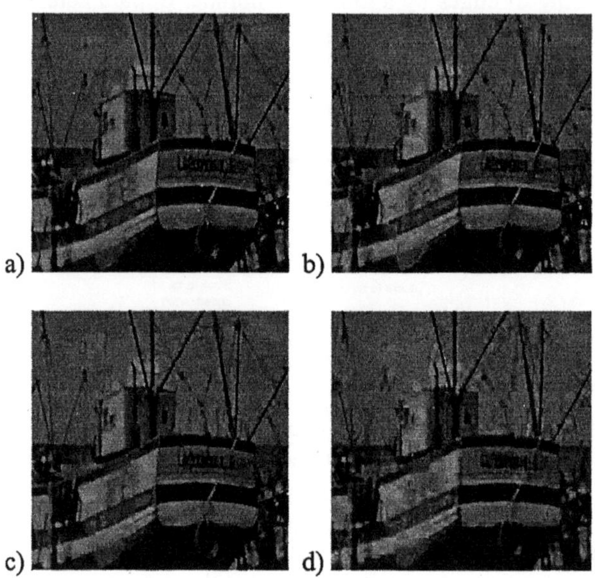

Fig. 12.41. Decoded images using fractal block coding (*a/b*) and hybrid fractal/DCT transform coding (*c/d*) at rates **a** R=0,4 *bpp* **b** R=0,22 *bpp* **c** R=0,22 *bpp* **d** R=0,12 *bpp*

In a method presented in [BARTHEL ET AL. 1994], the complete amplitude transform (after geometric transform) is performed in the DCT frequency domain. The gain parameters a and offset parameters b are individually adapted for each frequency component by defining individual gain factors a_{uv} and offset values b_{uv}, whenever necessary to obtain a sufficiently low reconstruction error. This allows to adapt much better to local details in the image in cases of low self-similarity[1]. In most cases, a *constant* a_{uv} can be used, which due to the linearity of the DCT is exactly equivalent to the fractal block coding methods described so far. Another extreme case would be to set *all* $a_{uv}=0$. The b_{uv} will then reflect the entire information about the block to be encoded, such that the method is equivalent to pure block trans-

---

[1] Observe that a decomposition of the gain factor into a DC component a_0 and AC component a was already applied in the modification from (12.59) into (12.60).

form coding. More interesting are the cases in between, because there is potential to adaptively exploit advantages of both block transform and fractal coding. The resulting reconstruction quality is reported to be significantly better over a wider range of rates than the quality of standalone fractal or standalone DCT transform coders.

Fig. 12.41 shows decoded results of fractal block coding at 0.4 and 0.22 *bpp*, and of hybrid fractal/DCT transform coding at 0.22 and 0.12 *bpp*.

12.6 Region-based coding

The subjective relevance of different regions in an image will not be equal. For example, a human typically first looks at faces if present in an image, to recognize the person that is visible. In many cases, the most relevant content can be found in the center area of an image. Consequently, it is reasonable to use compression schemes which adapt their properties according to the content of specific areas or regions. This appears in particular useful if regions are homogeneous by color and texture, for which good models are available, and also the perception by humans is well known. It is not unrealistic to replace such areas by synthetically generated pixel values, even though early methods of this kind have never really met the high expectation that were initially formulated (see e.g. [KUNT ET AL. 1985]).

If the adaptation to the properties of the region shall be exact in terms of a pixel-accurate shape, it is necessary to convey shape parameters for decoding as well. Such an accurate representation of the shape is inevitable if specific manipulations shall be applied to the image signal, e.g. replacement of areas in production, or if interactive applications shall be enabled.

In general, properties of regions can be expressed by their *shape* (typically a binary mask expressing *which* pixels belong to the region) or by their *contour* (positions of boundary pixels of the region). In the subsequent sections, typical shape coding methods are explained first. Then, modifications to coding methods introduced earlier are described, such that encoding of color and texture information within regions of *arbitrary shape* can be performed consistently and efficiently. The aspects of contour/shape coding and representation are also treated in sec. 7.4.1-7.4.3 from a recognition-related perspective. Here, we concentrate on methods which will allow an exact (lossless) representation of contours and shapes.

12.6.1 Binary Shape Coding

The rough position of a region is typically described by a bounding rectangle given by corner position, width and height. If a more exact location and rough description of the global shape features are required, any shape approximation method (cf. sec. 7.4.1) can be employed. The exact information about the shape of an arbitrary region is defined as a binary image within the area of the bounding rectangle or other circumscribing approximation,

$$b(m,n) = \begin{cases} 1 : \text{pixel is contained in the region} \\ 0 : \text{pixel is not contained in the region.} \end{cases} \qquad (12.64)$$

Methods for binary image coding as presented in sec. 12.1 can directly be used for shape coding. For example, binary *context arithmetic coding* (CAE) is defined in the MPEG-4 standard to describe the shape of arbitrary regions, either in single images or in an image sequence, where in the latter case the context is extended to the third (temporal) dimension.

Another efficient method of shape coding is the *quad tree representation* [SAMET 1984], which is in principle a block-based approximation of a region, but can also reach the granularity of a pixel-accurate representation. This method or variants thereof are also perfectly suitable to express the properties of block-based coding with variable block size (e.g. variable block-size transform coding, variable block-size motion compensation), or frequency layouts in wavelet-packet decomposition[1]. In the original quad-tree approach, the code expresses whether a block of size $K \cdot L$ is sub-divided into 4 smaller blocks of size $K/2 \cdot L/2$, or is left by its original size. Any of the smaller blocks can again be split into four children etc., such that a hierarchy of quad-block sub-partitions results, which is mapped onto a hierarchical tree structure. An example of a region shape to be expressed is shown in Fig. 12.42a. The associated quad tree is outlined in Fig. 12.42b. Each node of the tree has 4 branches, if the associated block is further sub-divided. If no sub-division is made, the tree terminates, i.e. a leaf is set instead of a node.

The information about the sub-division is represented by the *quad tree code*, where in the example of Fig. 12.42 nodes (quadrangles) are represented by '0' bits, while leaves (circles) are represented by '1' bits. This code is uniquely decodable like a prefix code, as each '0' must be followed by at least four additional bits. In the example shown, the decoder would interpret the tree as follows, where the parentheses express the levels of parsing: 0(10(1111)10(1110(1111))). At the last level of sub-division, additional bits can be saved, as it is known they can not be further split. In the example shown, the bit string '01011111101110' would be sufficient, if it is clear that the blocks $J...M$ belong to the smallest unit which can be generated. If e.g. an image shall be sub-divided into blocks of maximum size 32x32 pixel, minimum size 4x4 pixel, the smallest possible code consists of 1 bit (a single '1' would indicate that the largest-size block is not sub-divided any more), while the maximum number of bits would be 21 (1x '0' for size level 32x32, 4x '0' level 16x16, 16x '0' for level 8x8; then, it is uniquely indicated that *all* blocks shall

---

[1] Also the zero-tree concept used for wavelet coding is conceptually very similar to the quad tree.

be sub-divided into smallest size of 4x4). Such a range of sub-divisions may be a typical example for variable block size processing, e.g. in block transforms. In principle, the quad-tree splitting can continue until the level of pixels is reached.

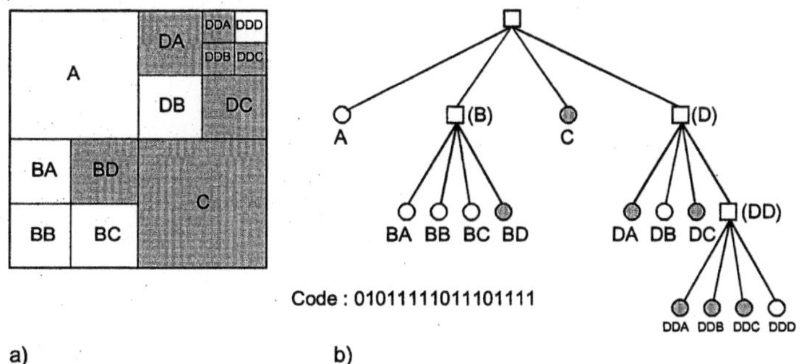

Fig. 12.42. a Quad-tree sub-division of an image b associated quad tree with code

If the quad tree shall be used for binary shape representation, it is necessary to assign the information about the region membership to the leaves. One single bit attached to the leaf then expresses whether pixels belonging to the related sub-block are or are not member of the region. Alternatively, multiple labels can be used which allow to express segmentation information related to multiple regions. This is illustrated by the gray-shaded areas and nodes in Fig. 12.42.

The method of sub-dividing regular rectangular blocks is not limited to usage in image shape or structure representation; An extension into the third (temporal) dimension are *octree* codes for representation of volumetric data. Other hierarchical tree codes, e.g. allowing non-quadratic layouts like horizontal/vertical 2-block splits, are possible as well.

12.6.2 Contour shape coding

Chain codes. A lossless description of discrete contours (Fig. 12.43a) can be encoded by *chain codes* [FREEMAN 1970] from a start point to an end point. A chain code points from one contour pixel to the next, which in the simplest case will be restricted to be one of the four or eight nearest neighbors (Fig. 12.43b)[1]. If start and end points are direct neighbors by themselves, the contour is *closed*. The co-ordinate reference of at least one point (usually the starting point) has to be encoded in addition, if the exact position must be given.

Contour descriptions based on interconnections in 4-neighborhood systems ($\mathcal{N}_1^{(2)}$ of (4.1)) can also be used. It can however be expected that the number of contour pixels increases due to the fact that no direct interconnections by diagonals are supported (cf. Problem 7.10). Assuming that all contour directions are equally

---

[1] This corresponds to the system $\mathcal{N}_2^{(2)}$ of (4.1).

probable, the $\mathcal{N}_1$ system will require $\log_2(4) = 2$ bit per contour pixel for encoding, while in the $\mathcal{N}_2$ system $\log_2(8) = 3$ bit per contour pixel are required. The expected mean rate will be the same, as for case of diagonal interconnections (which could be half of all cases when equal probability of directions is assumed) two contour points must be encoded in the $\mathcal{N}_1$ system. For direct chain coding, application of *entropy coding* will only be useful if certain contour directions have higher probability, which would definitely be the case for regular geometric forms such as rectangles.

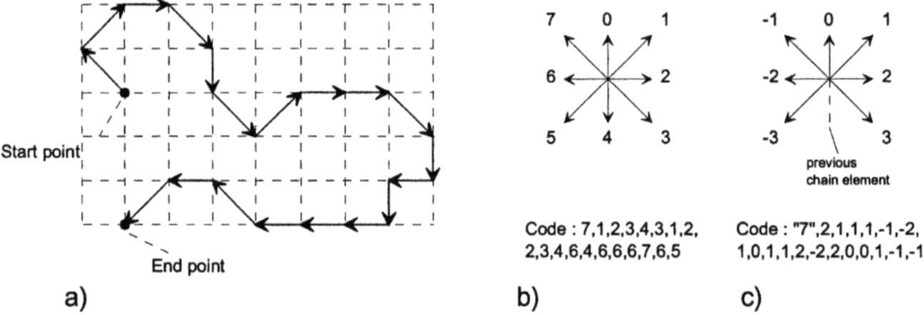

Fig. 12.43. **a** Discrete contour and its chain description; interconnnection directions for an 8-neighborhood **b** in direct chain coding **c** in differential chain coding

Differential chain codes. *Differential chain codes* describe changes of the contour direction from one element of the chain to the next [KANEKO, OKUDAIRA 1985]. If no $180°$ (back) turns are allowed, the differential code must support 3 different continuation directions in the case of the $\mathcal{N}_1$ system, and 7 different directions in the case of the $\mathcal{N}_2$ system (Fig. 12.43c). In particular for smooth contours, the combination of differential chain coding and entropy coding is highly efficient, where the $0°$ direction (straight continuation) is expected by highest probability, while directions '-3' and '+3' would only be selected at extreme corner locations which can be expected to occur rarely. Typically, in the case of the $\mathcal{N}_2$ system, discrete contours can be represented by differential chain/entropy coding using around 1.5 bit per contour pixel.

To further reduce this rate, chain codes can be extended to encode combinations of subsequent elements. Also in this case, more smooth contours will require less bits for representation, as e.g. chains of double or triple continuations in the same direction would occur by higher probabilities. Systematic smoothing of contours could be employed as part of the encoding process to achieve lower rates, however the contour will then no longer be lossless encoded.

Other methods which can be applied for lossy contour encoding are based on *interpolation* of contours from a number of control points, which must no longer be direct neighbors (cf. sec. 7.4.1). By these methods, also hierarchical representations with variable accuracy of the contour representation can be realized[1].

12.6.3 Coding within arbitrary-shaped Regions

In region-based coding, it is still necessary to encode the image information (color and texture) within the regions, unless the values are generated by color or texture synthesis. The following modifications of coding schemes described earlier can be used for this purpose:

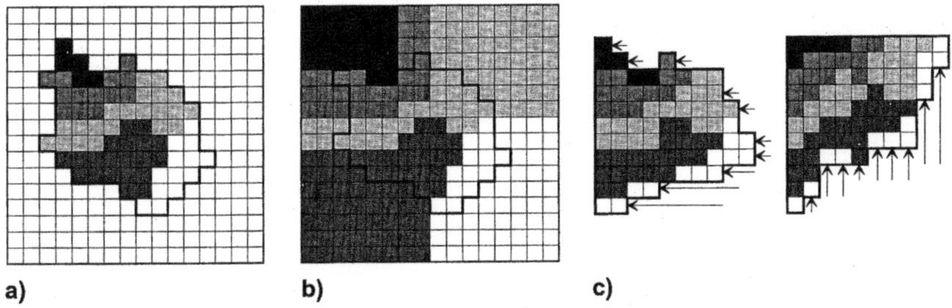

Fig. 12.44. a Arbitrary-shaped region, embedded within a larger rectangular block **b** Padding for extension of the signal to the size of a 16x16 block transform **c** Shift of rows and columns in the computation of a shape-adaptive transform

Block or wavelet transform coding with signal extension. Due to the inherently sub-sampled representation of transform coefficients in critically sampled transforms, a conflict occurs when an arbitrary number of pixels shall define regions of arbitrary shape. This can be resolved by a properly defined extension of the signal beyond the boundaries of the region. The signal outside the region is filled up by copying pixels within the region, such that a size of the signal is obtained that allows the processing of the transform. For example, in the case of a block transform, the area outside the region must be filled up to next-higher block size. This method is generally entitled as *padding*. As the number of transform coefficients will become larger than the number of samples within the region, the transform is in principle over-complete; by appropriate definition of the padding process, as many coefficients as possible are becoming irrelevant. This can usually be achieved if amplitude discontinuities in the extended signal are minimized; mostly, replication of pixels from the region boundaries is performed[2]. Fig. 12.44b shows

---

[1] A typical criterion to judge the similarity of two contours or shapes is the *area between*, which can be determined from an exclusive-or combination of the underlying binary shape images.

[2] This is equivalent to a constant-value extension of rectangular images, as it was shown in Fig. 4.2c. Symmetric extension (Fig. 4.2b) is also useful, while periodic extension is impractical for the case of arbitrary shaped regions. For symmetric wavelet filters, boundary extension methods like constant-value or symmetric extension can be implemented very

an example of a region padded up to a size of rectangular blocks, such that subsequent processing by a block transform of size 16x16 pixels can be employed. It is also possible to decompose larger regions into a number of smaller blocks, where the blocks in the inner part of the region are rectangular by nature, and padding will only be necessary for boundary blocks. In such a case, a rectangular area must be defined around the shape, which has at least the size of the bounding rectangle (7.92), but shall be a multiple of the smaller block side lengths (see Fig. 12.45).

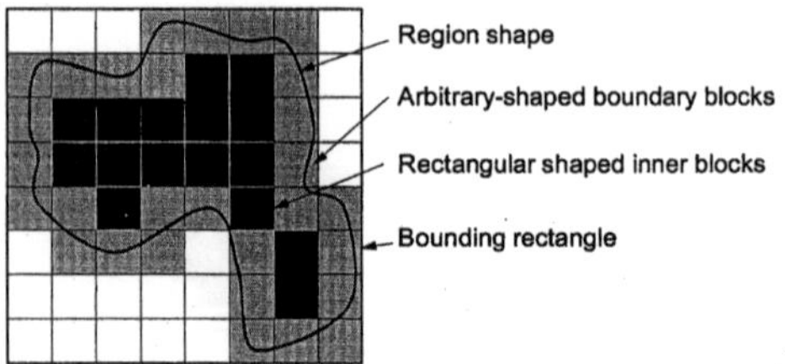

Fig. 12.45. Partitioning of a larger region into rectangular-shaped inner blocks and arbitrary-shaped boundary blocks

Transform coding of arbitrary-shaped regions. Orthogonal 2D transforms have been proposed for which the basis images are not rectangular, but of arbitrary shape [GILGE ET AL. 1989]. The computation of optimum basis images is similar to KLT eigenvector computation, but imposes additional constraints due to the non-regular region structure. The advantage of these transforms is that they are not inherently over-complete. From a complexity point of view, the usage of rectangular blocks in the inner part of a region as shown in Fig. 12.45 is advantageous as well, because transforms of arbitrary shape must then only be performed for boundary blocks which are of limited size. A simple variant which is nearly optimum for small blocks is the *shape-adaptive DCT* as defined in the MPEG-4 standard [SIKORA, MAKAI 1995]. The principle is shown in Fig. 12.44c. First, all rows of the region area are left-aligned, and vertical transforms are computed over the individual length of each column. Subsequently, the columns are top-aligned followed by a horizontal transform. This transform is indeed separable, as it can be broken down into a series of 1D transforms, individually adapted by the lengths of the columns and rows. In the example shown, first a series of vertical 1D transforms of lengths 11,10,8,8,8,7,7,7,4,2 is performed. Subsequently, horizontal 1D transforms of lengths 10,10,9,9,8,8,8,5,2,2,1 are applied to the intermediate result[1]. The design is slightly inconsistent by the fact that the DCT coefficients gained by transforms of different block lengths do not express identical frequen-

---

efficiently without padding; for a detailed discussion, refer to [KARLSSON, VETTERLI 1989][BAMBERGER ET AL. 1994].

[1] Remark that the shape-adaptive DCT as defined in the MPEG-4 standard allows a maximum transform length of 8, such that the region shown here would have to be separated into 4 blocks of size 8x8 each.

cies in the signal. Further, the orthonormality of the transform is not guaranteed. It is possible to uniquely reconstruct the region by application of the inverse transform, followed by the inverse vertical/horizontal shift operations, if the shape of the region is known to the decoder.

Predictive coding, vector quantization and fractal block coding. As scalar DPCM methods use pixel-based processing, they can directly be used for encoding of arbitrary-shaped regions. Adaptation of predictor filters can then be made based on the autocovariance function computed over the region. In general, the performance of predictive coding at lower rates is however inferior to transform coding schemes. Vector quantization (VQ) is less suitable for arbitrary-shape region encoding, as it would be necessary to provide a large set of different codebooks to support different region shapes; the same restriction applies to Predictive VQ. As fractal coding can however be interpreted as VQ with self-generating codebooks, this limitation is released, the shape of the origin area can flexibly be adapted to the shape of the target area.

Instead of representing color and texture information by waveform coding methods, synthesis can be performed, where however a highly natural impression must be achieved. Regions of exactly constant color will hardly be found in natural images. Typically, slight amplitude fluctuations occur between neighbored pixels, which could be modeled by additive noise; using noise overlay in synthesis is called *dithering*[1]. Further, large-scale color amplitude variations are typically present within larger region patches, as e.g. caused by gradual shade transitions on curved object surfaces, which could be modeled by a low number of parameters using polynomial functions fitted to the color amplitude surface. If more structure is present within a region, *texture synthesis* must be performed. Different methods exist to generate naturally-looking textures by random processes. These include autoregressive and ARMA synthesis [CHELAPPA, KASHYAP 1985], random feeding of wavelet synthesis filters [PORTILLA, SIMONCELLI 2000] and generation from Markov random field models [PAGET/LONGSTAFF 1998].

When different region patches are generated separately, an unnatural impression can still occur at the region boundaries. This problem can be avoided by application of transition or blending functions (cf. sec. 16.1).

---

[1] The synthesis of some special types of (non-white) noise, such as signal-dependent film grain, may however require more sophisticated models.

12.7 Problems

Problem 12.1

A common distortion measure in image and video coding is the *Peak Signal-to-Noise Ratio* (PSNR) (A.2). In case of an 8 bit digital signal as original source, value ranges are $0...255$, such that the maximum amplitude is $A=255$. Let an original signal be given by an image matrix

$$\mathbf{X} = \begin{bmatrix} 20 & 17 & 18 \\ 15 & 14 & 15 \\ 19 & 13 & 14 \end{bmatrix}.$$

Reconstruction images are output by two different coding schemes as

$$\mathbf{Y}_1 = \begin{bmatrix} 19 & 18 & 17 \\ 16 & 15 & 14 \\ 18 & 14 & 13 \end{bmatrix} \; ; \quad \mathbf{Y}_2 = \begin{bmatrix} 20 & 17 & 18 \\ 15 & 23 & 15 \\ 19 & 13 & 14 \end{bmatrix}.$$

For both reconstruction images, compute the mean absolute difference of pixels $|x(m,n)-y(m,n)|$ and the PSNR. Give an interpretation of the result. In particular, which of the distortions would be more clearly visible ?

Problem 12.2

Let an image matrix be given, $\mathbf{X} = \begin{bmatrix} 0 & 0 & 0 & 0 & 0 \\ 0 & 1 & 1 & 1 & 0 \\ 0 & 1 & 1 & 1 & 0 \\ 0 & 1 & 1 & 1 & 0 \\ 0 & 0 & 0 & 0 & 0 \end{bmatrix}.$

a) Compute the associated matrix $\hat{\mathbf{X}}$ of the prediction estimate within the marked area for the cases

  i) $\hat{x} = x(m-1, n)$ ii) $\hat{x} = 0{,}5 \cdot x(m-1, n) + 0{,}5 \cdot x(m, n-1)$

  iii) $\hat{x} = x(m-1, n) + x(m, n-1) - x(m-1, n-1)$.

b) Determine the values e in the following prediction error matrix $\mathbf{E} = \mathbf{X} - \hat{\mathbf{X}}$ for the 3 cases from part a) of this problem:

$$\mathbf{E} = \begin{bmatrix} 0 & 0 & 0 & 0 & 0 \\ 0 & e(1,1) & e(2,1) & e(3,1) & e(4,1) \\ 0 & e(1,2) & e(2,2) & e(3,2) & e(4,2) \\ 0 & e(1,3) & e(2,3) & e(3,3) & e(4,3) \\ 0 & e(1,4) & e(2,4) & e(3,4) & e(4,4) \end{bmatrix}$$

c) Compute the reconstruction matrices $\mathbf{Y} = \mathbf{E'} + \hat{\mathbf{X}}'$ provided that $\mathbf{E'} = \mathbf{E}$ at all positions except for the value $e'(1,1)=0$. The prediction $\hat{\mathbf{X}}'$ shall recursively be computed from $\mathbf{Y}$. Then, compute the difference between $\mathbf{X}$ and $\mathbf{Y}$ and give an interpretation of the result.

d)   To quantize the prediction error signal, a quantizer with 3 reconstruction values $\mathcal{V}=\{-1/3,0,1/3\}$ is available. Compute the matrices of the reconstruction $\mathbf{Y}$, the related prediction $\hat{\mathbf{X}}'$, the prediction error $\mathbf{E}=\mathbf{X}-\hat{\mathbf{X}}'$ and the quantized prediction error $\mathbf{V}$. Then, compute the differences $\mathbf{X}$-$\mathbf{Y}$ and $\mathbf{E}$-$\mathbf{V}$, and interpret the results by comparing against the results from b).
[Hint: Recursive processing is necessary in computing the solutions of c) and d). Assume that the predictor memory is initialized by zero values.]

Problem 12.3
Compute the coding gain for DCT transform coding of an autoregressive model with correlation parameters ρ=0.5 and ρ=0.95. Four cases of 1D and 2D-separable transforms shall be considered: U=2, U=3, UxV=2x2, UxV=3x3. Compare the results against the maximum achievable coding gain for an AR(1) model for the 1D and 2D cases.
[Hint : Use the result from Problem 4.7 to determine the variances of coefficients for the DCT block lengths $\{U,V\}$=3.]

Problem 12.4
By computation of a 2D DCT over a block size of 4x4, the following matrix of transform coefficients results, coefficients ordered as in Fig. 12.20a:

$$C = \begin{bmatrix} 235 & 35 & 15 & 3 \\ -67 & 3 & 5 & -9 \\ -17 & 13 & -7 & 9 \\ 5 & 37 & 2 & 1 \end{bmatrix}$$

a)   Perform a quantization by the following matrices $\mathbf{Q}_1$ und $\mathbf{Q}_2$, which characterize the quantization step sizes related to the particular coefficients. To compute the quantizer indices i from coefficients $c_{u,v}$ and for reconstruction of the quantized coefficients $c^{(q)}_{u,v}$, use the following method

$$i = \mathrm{nint}\left(\frac{c_{u,v}}{q_{u,v}}\right) ; \quad c^{(q)}_{u,v} = i \cdot q_{u,v}$$

[nint ( ) expresses rounding to the nearest integer value.]

$$\mathbf{Q}_1 = \begin{bmatrix} 8 & 8 & 8 & 8 \\ 8 & 8 & 8 & 8 \\ 8 & 8 & 8 & 8 \\ 8 & 8 & 8 & 8 \end{bmatrix} ; \quad \mathbf{Q}_2 = \begin{bmatrix} 4 & 6 & 8 & 12 \\ 6 & 8 & 12 & 16 \\ 8 & 12 & 16 & 20 \\ 12 & 16 & 20 & 24 \end{bmatrix}$$

b)   Run-length coding shall be applied to the zigzag-scanned coefficients (cf. Fig. 12.23a) for both cases of quantization. Determine the run-lengths if coding according to the method illustrated in Fig. 11.18b is used.
c)   Entropy coding of quantizer indices (without using run-length coding) shall be performed by using a systematic VLC according to the following bit allocation table. Compute the resulting bit rates for both cases of quantization.

	Index						
	0	±1	±(2..3)	±(4..7)	±(8..15)	±(16..31)	±(17..63)
Number of bits	1	3	5	7	9	11	13

d)   Construct the table of a valid code that results in the given bit allocation.
e)   Compute the distortion expressed by the Peak SNR (A.2), related to a peak amplitude $A=255$ for both cases of quantization. Which difference in rate could theoretically be expected for encoding of an AR(1) process, assuming that both quantizations are still in the linear range of the RDF (11.23)?

Problem 12.5

An AR(1) process described by correlation coefficient $\rho_h=\sqrt{3}/2$ and variance $\sigma_x^2=4$ is coded by DPCM. Fig. 12.46 shows the case of one-dimensional linear prediction; in the two-dimensional case, a separable AR(1) process with $\rho_h=\rho_v$ and appropriately optimized prediction shall be used accordingly.

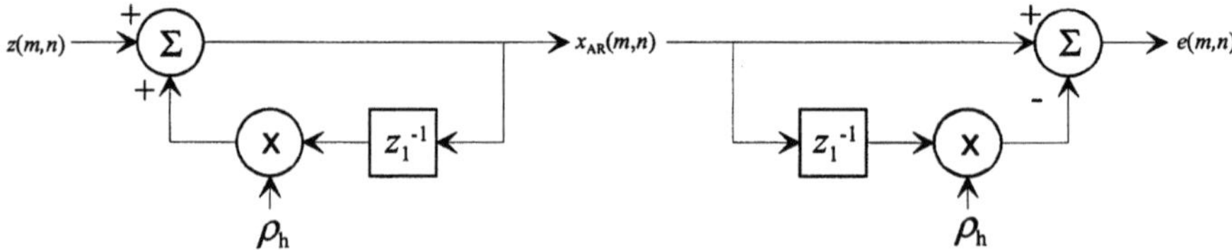

Fig. 12.46. Prediction and synthesis of an AR(1) process

a)   Compute the variance of the prediction error $e(m,n)$ and the coding gain for the 1D and the 2D cases.
b)   By how many bit/pixel can the rate be reduced in the low distortion range fulfilling *i*) (11.14) and *ii*) (11.23), when *i*) one-dimensional *ii*) two-dimensional DPCM is applied?
c)   Instead of the conventional DPCM encoder structure (Fig. 12.10b), a modified structure in Fig. 12.47 shall be used. Supplement the "+" and "-" signs at the summations, such that both structures produce exactly the same signals $i(\mathbf{n})$.

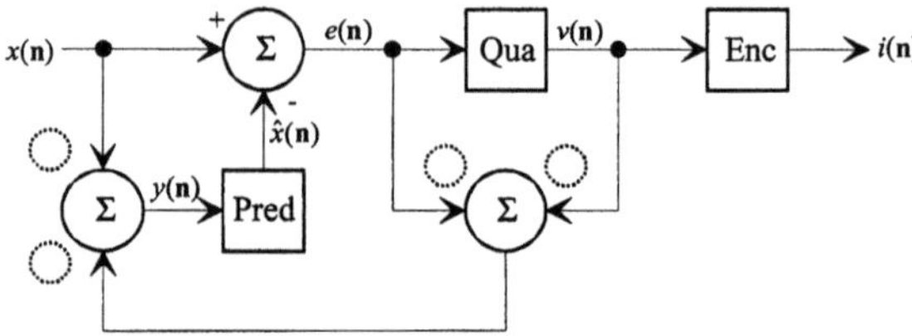

Fig. 12.47. Alternative DPCM encoder structure

Problem 12.6

Black and white (binary) image patches of size 4x4 pixels shall be compressed.

a)   Determine the maximum entropy rate per sample. Which number of bits should be used at maximum to encode the entire patch?

Now, the image patch shown in Figure 12.47 shall be compressed.

b)   Determine the occurrences of black and white pixels. Then, determine the entropy under the assumption that the occurrence values map exactly into probabilities.

c)   Statistical independency of neighbored pixels shall be assumed. Design a Huffman code, which combines each two samples into a vector before encoding. Which rate per sample results?

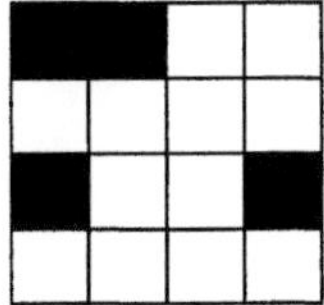

Fig. 12.48. Binary image patch to be compressed

Vectors can be configured either from horizontally-adjacent or from vertically-adjacent pixels, as shown in Figure 12.49. In an adaptive entropy coding method, one additional bit is needed to signal which configuration is used. The code table from Figure 12.49 is used for encoding.

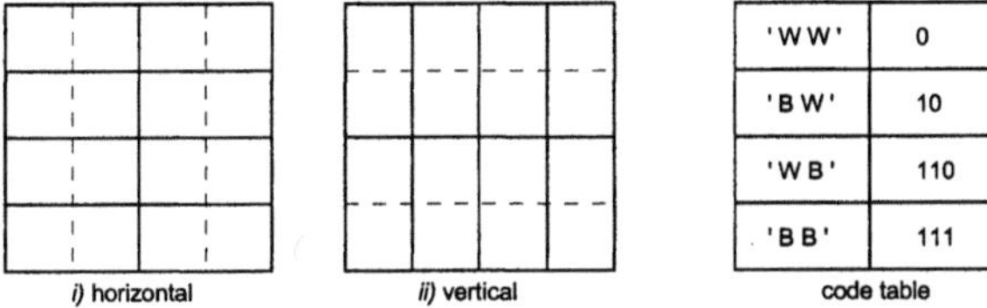

'WW'	0
'BW'	10
'WB'	110
'BB'	111

Fig. 12.49. Combinations of two pixels into vectors, and code table

d)   Determine the occurrences of the different black/white combinations for both configurations *i)* and *ii)*, when the image from Figure 12.47 is encoded.

e)   Determine for both configurations the number of bits needed to compress the image, using the code table. Discuss the advantage of adaptive coding.

Problem 12.7
An AR(1) process is characterized by parameters $\rho=0.75$ and $\sigma_x^2=16$. For encoding, a rate $R=2$ bit/pixel is available.

a)   Determine theoretically the minimum possible distortion and the maximum coding gain, if all correlation properties are exploited properly.

b)   For encoding, a Haar transform kernel with $U=2$ is employed. Compute the expected variances of coefficients c_0 and c_1. Determine the coding gain. Which distortion results by a rate $R=2$?

c)   The distortion D shall have a variance not higher than 1. Which is the necessary rate, *i)* according to the rate distortion function of an AR(1) process *ii)* if the Haar transform is used?

d)   A 2D separable AR(1) process of same variance with $\rho_h=\rho_v=\rho$ shall be coded. Determine the variances of the coefficients, using the results from b), and the coding gain of the separable transform for $U=V=2$.

13   Video Coding

In encoding of video sequences, similarity between subsequent frames can be exploited for compression. By many aspects, video coding schemes can be regarded as extensions of still image coding schemes as presented in the previous chapter, however techniques of motion compensation play a key role to achieve high compression performance. Highest redundancy between adjacent frames will be found along the motion trajectory; also coding errors will be less visible if they are consistent with the motion of the scene. This chapter starts by introducing more simple schemes which are aiming to exploit the temporal redundancy without motion compensation. Then, the combination of motion-compensated prediction with spatial transform coding is investigated, which is denoted as hybrid coding. This is the type of video coding scheme which is implemented in most of today's video coding standards, for which the foundations are investigated in detail. Methods of advanced motion compensation are discussed, scalable and multiple-description coding concepts are introduced. Hybrid coders are based on a temporally recursive DPCM loop, which imposes a number of constraints in encoding and enforces provisions to prevent possible error propagations. As a possible alternative, motion-compensated wavelet methods are introduced, which extend methods of wavelet coding into the dimension of the temporal axis. Recent research has shown that no limitation exists in principle to combine wavelet filtering along motion trajectories with any type of motion compensation. This in particular provides flexible scalability of video streams and gives significantly higher robustness in case of data losses as compared to the hybrid solution. The chapter concludes by investigating methods for motion vector encoding, which is of particular importance as the data rate to be spent for motion information can reach a significant amount in the more sophisticated classes of video coding schemes.

13.1 Methods without Motion Compensation

In principle, a movie can be encoded by processing each frame separately using one of the still image coding methods described in chapter 12. The processing should then be performed as fast that real-time constraints (e.g. for video presentation by 25 or 30 frames per second) are met. Such methods are denoted as *intraframe coding*, where only the spatial redundancies within single frames of video are exploited for compression. An example for this is *Motion JPEG*[1], which is built by the same compression algorithm as the DCT based still image coding standard, but in addition conveys synchronization information (e.g. for syncing video and audio, or syncing video with the display). A disadvantage of intraframe coding methods is the potential of high visibility of coding errors in video due to inconsistency. *Temporal variation of coding errors* occurs when motion or other differences (e.g. noise) between subsequent frames are present in the scene. This is most critical if the underlying principle of intraframe coding is shift variant; e.g. the results from DCT or wavelet transforms can be significantly different if the block grid or the sub-sampling phases are shifted. Application of still image coding to video compression hence appears only useful in the high-rate and high-quality ranges, where such effects can be ignored.

Single frames from a decoded sequence processed by intraframe compression are often perceived to provide a good quality, however when displayed as moving video, coding errors are becoming clearly visible due to temporal fluctuation. If in contrary *interframe motion-compensated coding* is employed, coding errors are typically shifted along the motion path, such that they are not visually annoying. Due to this phenomenon, motion-compensated methods typically perform perceptually more pleasing than intraframe methods, even if quality metrics like SNR seem to indicate equivalence in quality. On the other hand, motion-compensated methods have also disadvantages, in particular a higher complexity of encoder and decoder.

More simple methods of interframe coding are presented in sections 13.1.1-13.1.3, which exploit the temporal redundancy in a video sequence without using motion compensation. They are still more complex than pure intraframe coding methods, specifically due to the need of frame memories. Other general disadvantages of interframe coding methods with or without motion compensation are lack of capability for direct decoding (access) of arbitrary frames, and necessity of transcoding if editing or cuts are made. Due to such considerations, usage of intraframe coding methods can be justified for certain video coding applications. If only low motion appears in the scene, *interframe schemes without motion compensation* will be significantly more efficient than intraframe schemes, e.g. if the background is static due to a fixed camera. Such properties can typically be found in applications of videoconferencing, surveillance etc.

---

[1] Motion JPEG is not an international standard format, but originally a proprietary solution which has found wide acceptance in products.

13.1.1 Frame Replenishment

Frame replenishment is a video coding method which can be combined with any still image (intraframe) coding technique, exploiting cases when no changes occur from one frame to the next. The first frame of a sequence is fully intraframe encoded, while from subsequent frames only those areas are encoded which are significantly changing compared to the predecessor. Small changes should be ignored, which are potentially caused by noise, jitter in sampling positions etc. It must be signaled to the decoder which areas are subject to change. This can effectively be achieved by performing a decision on a block-by-block basis, such that the amount of side information is kept low. Useful decision criteria are based on cost functions as e.g. the variance or mean absolute value of the differences between the original image $X(o)$ and a decoded copy of the previous image $Y(o-1)$, which is stored in a frame memory. The principle is shown by the block diagram of Fig. 13.1.

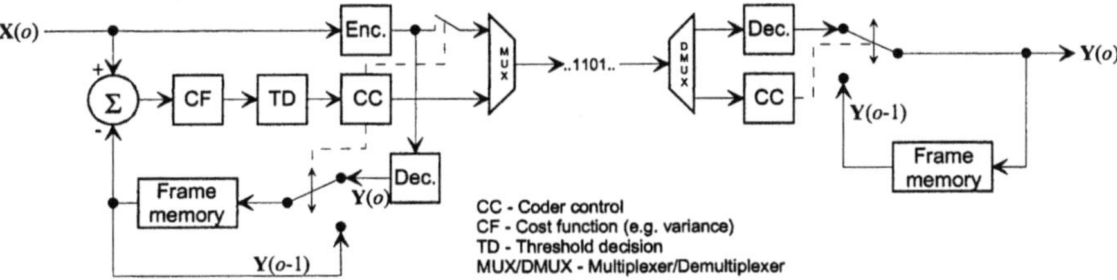

Fig. 13.1. Frame replenishment, coder (left) and decoder (right); switch shown in a position where new intraframe encoding is performed

For interlaced video sequences, the non-replenished parts should refer to the field of same parity from the previous frame, as this is located at the identical spatial position. This means that a full frame memory (not only a field memory) must be provided for interlaced video. In case of motion, replenishment can be expected to apply in most areas, which means that no substantial difference in performance could be expected then, as compared to pure intraframe coding of same type.

Also in the case of motion-compensated hybrid coding (see sec. 13.2), the goal is to obtain as much information as possible from the previous frame, and mechanisms are provided to apply intraframe coding on single blocks. From this point of view, frame replenishment can be interpreted as a special case and early predecessor of the interframe prediction methods, where however no motion compensation is performed and no prediction error signal is ever transmitted. Most hybrid coding schemes can be operated in modes which are very similar to replenishment; this may in particular be useful, when the encoder complexity shall be kept low.

A method of replenishment is defined in ITU-T recommendation H.120, designed for 2 Mbit/s video transmission rate in conferencing applications. Here, 2D DPCM is used as intraframe coding method. Further, some proprietary formats defined for storage and replay of video on computer systems are based on replenishment.

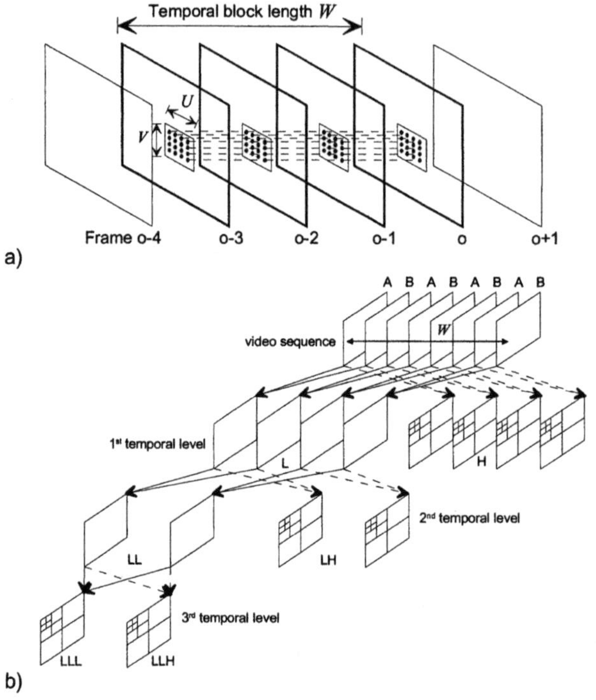

Fig. 13.2. a Extension of a block transform into the temporal dimension by definition of a spatio-temporal cube   **b** Spatio-temporal wavelet decomposition using T=3 levels of a temporal wavelet tree

13.1.2   3D Transform and Subband coding

It is also possible to extend (block-based, subband or wavelet) transform coding methods to the third (temporal) dimension [NATARJAN, AHMED 1977] [KARLSSON, VETTERLI 1987]. As an example, regard a 3D frequency decomposition defined over W frames from a video sequence. The transform shall be a 3D DCT, which is then applied to cubes of side lengths U,V and W in the two spatial dimensions and the temporal dimension, respectively:

$$c_{u,v,w} = C_0 \cdot \sqrt{\frac{8}{U \cdot V \cdot W}}$$

$$\cdot \sum_{m=0}^{U-1} \sum_{n=0}^{V-1} \sum_{o=0}^{W-1} x(m,n,o) \cdot \cos\left[u(m+0.5)\frac{\pi}{U}\right] \cdot \cos\left[v(n+0.5)\frac{\pi}{V}\right] \cdot \cos\left[w(o+0.5)\frac{\pi}{W}\right] \quad (13.1)$$

$$\text{with} \quad C_0 = \begin{cases} \frac{\sqrt{2}}{4} & \text{for } (u,v,w)=(0,0,0) \\ \frac{1}{2} & \text{for } (u,v,w)=\left[(0,0,c) \text{ or } (0,b,0) \text{ or } (a,0,0)\right] \\ \frac{\sqrt{2}}{2} & \text{for } (u,v,w)=\left[(0,b,c) \text{ or } (a,0,c) \text{ or } (a,b,0)\right] \\ 1 & \text{for } (u,v,w)=(a,b,c) \end{cases} \quad ; \quad a \neq 0, b \neq 0, c \neq 0$$

Fig. 13.2a shows the placement of a transform cube, having spatio-temporal extensions (U,V,W) within a video sequence. Fig. 13.2b illustrates a 3D Wavelet transform, where in the simplest case a Haar basis can be used for wavelet decomposition along the temporal axis[1]. This can be interpreted as decomposition of frame pairs into average (lowpass) and difference (highpass) frames

$$L(m,n) = \frac{1}{2}\left[A(m,n) + B(m,n)\right] \quad ; \quad H(m,n) = A(m,n) - B(m,n), \quad (13.2)$$

where pairs of lowpass frames are then again combined, iterating the procedure over the different levels of the wavelet tree. At the end nodes of the temporal decomposition, a 2D spatial wavelet transform is applied. A specific advantage as compared to a block transform comes due to the localization property, which better allows to identify positions of temporal changes, which otherwise would spread as high energy over all coefficients of the temporal block extension. With a number of T wavelet tree levels temporally, the resulting temporal block length[2] in 3D wavelet transform is $W=2^T$.

The 'temporal' frequency characterizes temporal changes between the frames. The following interpretations can be made (cf. sec. 2.1.2) :

–   In the case where no motion occurs, all information will be concentrated in the lowest-frequency coefficients, the temporal-axis 'DC' or scaling coefficient. If only the temporal-axis transform is performed, an image constituted from the low-frequency coefficients at their respective spatial position is the average over all W frames of the block, typically scaled by an amplitude factor, depending on the normalization of the transform basis.

---

[1] In fact, a non-orthonormal version of the Haar basis is used in the following explanations, which allows to interpret the temporal lowpass and highpass filter outputs as average and difference frames, respectively.

[2] Indeed, for the case of a Haar basis, this can still be interpreted as a block transform of block length W over the temporal axis.

- As the 3D spectrum is sheared depending on the amount of translational motion shift, the positions where relevant 'temporal' frequency coefficients are found depend both linearly on the presence of spatial high frequency components (detail) and on the strength of motion.
- If both spatial detail and motion are high, this will effect a spreading of energy over a broad range of spatio-temporal bands.

Best strategies for encoding of 3D frequency coefficients require careful consideration of these effects. In particular 3D wavelet schemes allow utilization of contexts which implicitly relate to the shear effects of the 3D spectrum. To illustrate this, the 3D wavelet decomposition of Fig. 13.2b is re-interpreted as a *wavelet transform cube* in Fig. 13.3. 3D zero-tree methods have been proposed, where typically correspondences between spatial bands are unchanged as compared to conventional 2D zero-tree structures (Fig. 12.27). Additional correspondences exist however over the bands of the temporal-axis wavelet decomposition, which implicitly reflect the fact that the 'temporal' frequency linearly increases with 'spatial' frequency if translational motion occurs. This means that once a directional correspondence relating to the sheared spectrum is found, it is most probable that the same shear direction can further be traced towards higher frequencies, with high probability to find zero-tree correspondences. A 3D version of SPIHT has been introduced in [KIM, XIONG, PEARLMAN 2000]. A 3D extension of EBCOT (which is not a zero-tree method, indeed) was proposed in [XU, LI, ZHANG 2000].

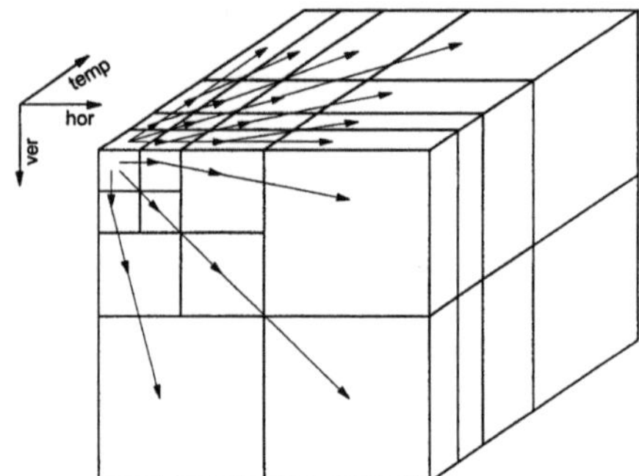

Fig. 13.3. 3D Wavelet transform cube with possible zero-tree correspondences

In any interframe video coding scheme, it is reasonable to perform a *scene change detection* and switch off temporal-axis processing where correspondences would not exist beyond scene boundaries. In 3D frequency coding, high energy in temporal-axis high-frequency coefficients would otherwise result. This can be avoided by proper adaptation of temporal analysis block lengths in cases where scene changes are detected.

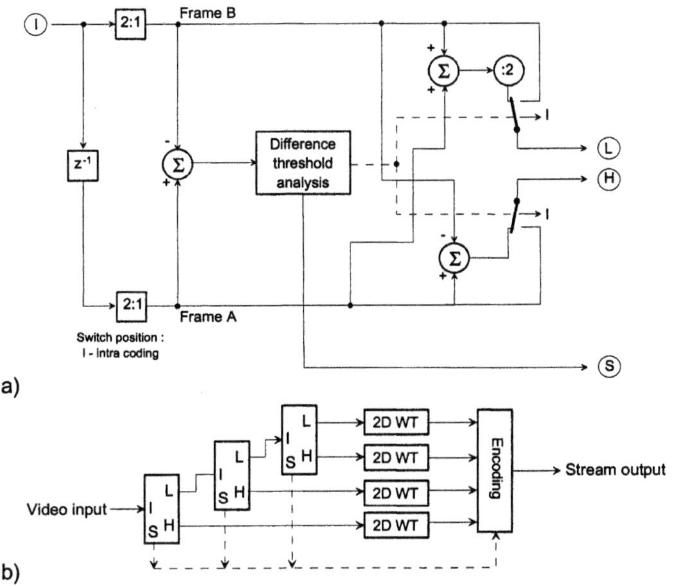

Fig. 13.4. 3D Wavelet decomposition with temporal-change adaptation
a Temporal 2-band split element **b** Cascaded structure

A next step would be implementation of a *local change* analysis. It is also useful to switch off the temporal-axis frequency decomposition locally, if large changes are detected [PODILCHUK, JAYANT, NOLL 1990], [QUELUZ 1992]. This can be interpreted as an *adaptation to the presence of motion*, which is simple to integrate into the temporal-axis wavelet decomposition. In particular when Haar filter kernels are used, the analysis can be performed in any local area for any frame pair, such that the decision on whether to use the temporal redundancy can be made very flexible. A block diagram of a motion-adaptive method is shown in Fig. 13.4[1]. After difference threshold analysis between a pair of frames, which can be performed locally or globally, the temporal-axis filtering is discarded in the 'intraframe' position. A non-orthonormal set of Haar filters using different norms for the lowpass and highpass basis functions as in (13.2) is used here in the 2-band split element, which is implemented by a polyphase structure (cf. sec. 4.4.3).

Further discussion of these concepts can be found in the context of their extension to fully motion-compensated Wavelet coding in sec. 13.4.

---

[1] This is indeed a very similar concept as in frame replenishment, where switching to intraframe coding is performed when the temporal similarity compared to the corresponding area from the previous frame is low.

13.2 Hybrid Video Coding

13.2.1 Motion-compensated Hybrid Coders

The notion *hybrid coding* was introduced in multi-dimensional image and video coding to characterize the combination of different encoding principles which are applied along the different dimensions of the signal [ROESE, PRATT, ROBINSON 1977]. For video coding, this could e.g. be combinations of DPCM (in temporal dimension) with DCT or wavelet transform coding or vector quantization in the spatial dimensions. Today, the term *Hybrid Video Coding* is mainly used to express the combination of motion-compensated DPCM with block transform coding, which has become the basis for most video coding standards existing today.

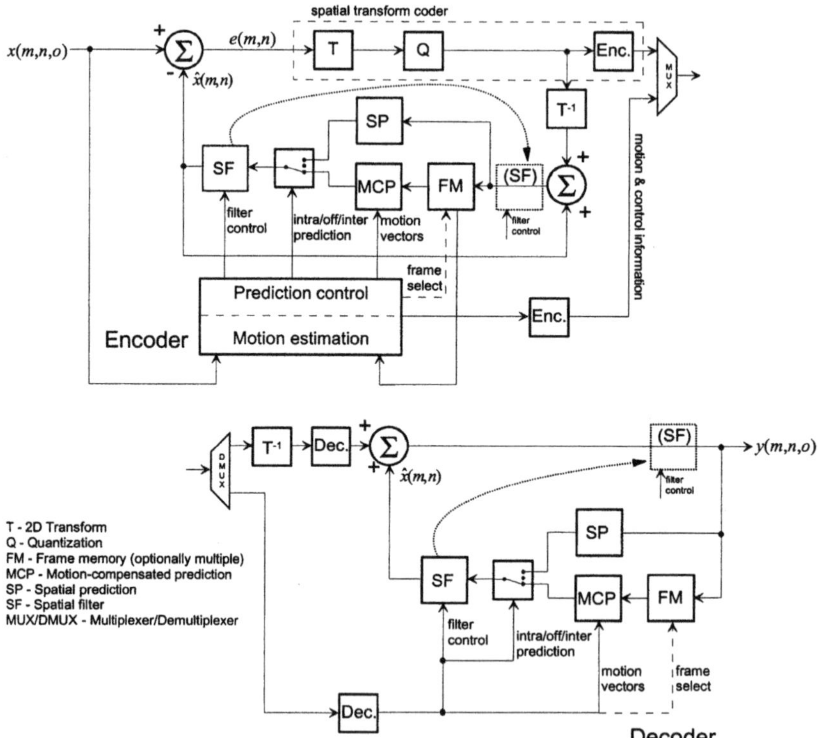

Fig. 13.5. Hybrid (temporal predictive + 2D transform) interframe coding with motion compensation

The structure of a hybrid video coder, combining motion-compensated prediction (cf. sec. 4.2.2) and a 2D spatial transform coding of the motion-compensated prediction error signal, is shown in Fig. 13.5. Regarding the temporal dimension, this could be interpreted as a conventional DPCM with forward predictor adaptation performed by the motion vector parameters[1]. The estimation of the motion vectors can either take into account the comparison between two original frames, or the comparison between the present original frame and the previous decoded frame, which is available in the frame memory for computation of the prediction anyway.

The functional block 'prediction control' allows optional switching of different options in the prediction loop, including to switch off the prediction entirely. The following signals can optionally be fed into the 2D (spatial) block transform coder[2]:

- Current block minus motion-compensated prediction from previous frame;
- Current block minus spatial prediction from previously decoded blocks; this is a directional block prediction as introduced in sec. 12.4.1;
- Both previous options, spatially filtered;
- Original frame only (switch mid position 'prediction off'); in this case, pure intraframe coding is used.

Different variants are possible, e.g. spatial prediction is often performed in the transform domain by the 'DC' and 'AC' prediction methods introduced in sec. 12.4.1. The spatial loop filter operation can alternatively be applied before writing into the frame memory (see dotted arrows), such that it can simultaneously be used to improve the quality of the reconstructed signal $Y(o)$[3].

For the case of a scene change, global switching to intraframe coding should be performed. Global intraframe coding can also be enforced for other purposes such as capability of direct access to a frame. Frames which are not predicted from other frames are denoted as *I-type* frames. Settings of prediction control and adaptation of motion parameters are typically performed locally (e.g. for units of blocks), which still includes the possibility to encode only selected areas by intraframe coding.

The hybrid coding scheme as shown in Fig. 13.5 has a low delay; only one previously decoded frame is used for temporal prediction. The encoding and decoding delays will in principle not be longer than in intraframe coding. Frame memories are necessary both at encoder and decoder sides, which must be randomly accessible as the motion shift is locally varying. At the encoder side, the motion estimation often is the element of highest computational cost, which also imposes a high number of memory accesses. Random decodability of frames is difficult due to the fact that each frame relies on its predecessors in the recursive prediction loop; in

---

[1] Backward-adaptive motion-compensated prediction has also been studied (see e.g. [NETRAVALI, ROBBINS 1979], [LIM, CHONG, DAS 2001], but is often regarded to be less robust and to cause an additional burden in decoder complexity.

[2] Not all of these options are supported in any realizations of hybrid coders as e.g. defined in video coding standards.

[3] This configuration is used in the Advanced Video Coding standard (cf. sec. 17.4.4).

principle, decoding must always start at the most recent *I*-type frame. From this point of view, and also for the purpose of increased stability in case of potential transmission failures or decoding inaccuracies[1], intraframe encoded information should be inserted within certain time intervals.

13.2.2 Characteristics of Interframe Prediction Error Signals

In the following analysis, the case of lossless coding is regarded first, which means that original frame $\mathbf{X}(o)$ and reconstructed frame $\mathbf{Y}(o)$ are assumed for simplicity to be identical. If a motion shift by k pixels horizontally and l pixels vertically occurs from one frame to the next, the optimum motion-compensated prediction generates a prediction error signal

$$e(m,n,o) = x(m,n,o) - x(m-k,n-l,o-1) = r(m,n,o). \qquad (13.3)$$

The signal $r(m,n,o)$ characterizes all components which can not be predicted even by perfect motion compensation, e.g. newly uncovered parts of a scene[2]. If no motion compensation is applied in the prediction, which means that prediction is always performed using the identical coordinate position from the previous frame as estimate, the prediction error signal is

$$\begin{aligned} e(m,n,o) &= x(m,n,o) - x(m,n,o-1) \\ &= r(m,n,o) + x(m,n,o) - x(m+k,n+l,o). \end{aligned} \qquad (13.4)$$

To analyze the coding gain achievable by motion-compensated hybrid coding, it is useful to compare the 2D power spectra of the original frame, and of the prediction error frames with and without motion compensation. As these 2D signals are subject to subsequent spatial transform coding, the spatial frequency characteristics take high influence on the overall effectiveness. For the case of 'ideal' motion compensation, the spatial prediction error spectrum of frame o follows by (13.3)[3]

---

[1] Decoding inaccuracies can e.g. be caused by slight deviation in the arithmetic precision, by which the inverse transform is computed at the encoder and decoder. Some video coding standards define a mandatory refresh by intra coding after 132 frame cycles to circumvent this problem.

[2] Observe that this residual will not be a white noise signal, such that it is not a type of an innovation process. It can rather be modeled like an image signal, however in the case of ideal motion compensation only consisting of small non-zero patches which are the newly-uncovered areas. These components are spatially correlated. In reality, additional components will appear which are caused due to noise which is temporally uncorrelated, as well as effects of imperfect motion compensation. In principle, both correlated and uncorrelated components will be present in $e(m,n,o)$, but the spatial correlation is generally lower than in original images. For an investigation on these properties, refer to [SHISHIKUI 1992].

[3] This statistical modeling analysis asymptotically assumes infinitely extended frames, which are moving by constant translational shift.

$$E(j\Omega_1, j\Omega_2, o) = R(j\Omega_1, j\Omega_2, o) \,, \tag{13.5}$$

such that the power spectrum is

$$S_{ee,\mathrm{mc}}(\Omega_1, \Omega_2, o) = S_{rr}(\Omega_1, \Omega_2, o) \,. \tag{13.6}$$

'Ideal' motion compensation may be interpreted such that just switching to intra coding must be performed whenever no motion compensation is possible, and otherwise the prediction would be perfect. If *no* motion compensation is applied at all, the prediction error spectrum can be determined by the difference of actually shifted pictures,

$$E(j\Omega_1, j\Omega_2, o) = X(j\Omega_1, j\Omega_2, o) - X(j\Omega_1, j\Omega_2, o) \cdot e^{jk\Omega_1} \cdot e^{jl\Omega_2} + R(j\Omega_1, j\Omega_2, o), \tag{13.7}$$

which is re-formulated into

$$\begin{aligned}
E(j\Omega_1, j\Omega_2) &= X(j\Omega_1, j\Omega_2, o) \cdot (1 - e^{j(k\Omega_1 + l\Omega_2)}) + R(j\Omega_1, j\Omega_2, o) \\
&= X(j\Omega_1, j\Omega_2, o) \cdot (-2j) \cdot e^{j\frac{k\Omega_1 + l\Omega_2}{2}} \cdot \sin\left(\frac{k\Omega_1 + l\Omega_2}{2}\right) + R(j\Omega_1, j\Omega_2, o).
\end{aligned} \tag{13.8}$$

The result is a power spectrum[1]

$$\begin{aligned}
S_{ee,\mathrm{no_mc}}(\Omega_1, \Omega_2) &= S_{xx}(\Omega_1, \Omega_2) \cdot 4 \cdot \left[\sin\left(\frac{k\Omega_1 + l\Omega_2}{2}\right)\right]^2 + S_{rr}(\Omega_1, \Omega_2) \\
&= 2 \cdot S_{xx}(\Omega_1, \Omega_2) \cdot \left[1 - \cos\left(k\Omega_1 + l\Omega_2\right)\right] + S_{rr}(\Omega_1, \Omega_2).
\end{aligned} \tag{13.9}$$

This prediction error spectrum is highly coherent with the spectrum of the frame $X(o)$, which means that the prediction error picture will be correlated with the original picture in the case where no motion compensation is performed. The characteristics of this spectral component depend on the actual motion shift that parametrizes the 'sin^2' function which is modulating the signal spectrum. In addition, the component S_{rr}, which can not be predicted from the previous frame, is still present. It is interesting to observe that specific power spectrum components of the uncompensated prediction error can be larger by a factor of four as compared to the original signal spectrum. This means that temporal prediction without motion compensation can even perform *worse than intraframe coding*, depending on the characteristics of the signal and the strength of motion. This will be the case in particular for images with a high amount of spatial detail. In the case of 'ideal' motion compensation, the performance can never be worse than for intraframe coding.

---

[1] Assuming for simplicity that $r(m,n,o)$ would be uncorrelated with $x(m,n,o)$, which in fact would not be the case. The additional components of cross power spectra can however be ignored for the subsequent analysis.

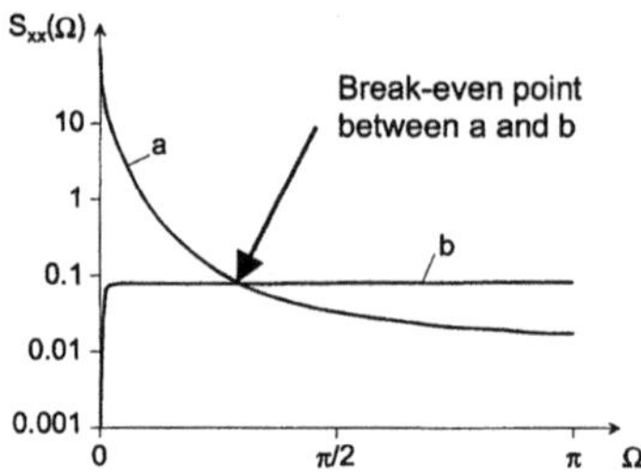

Fig. 13.6. 1D spectrum of an AR(1) model, ρ=0,96 (a) and related interframe prediction error spectra (without motion compensation, without considering component S_{rr}) for the case of horizontal motion shift k=1 (b)

As an example to illustrate this phenomenon, Fig. 13.6 shows a 1D power spectrum of an AR(1) model, and the prediction error spectrum according to (13.9) for k=1. If the deviation from the true motion is significant, the application of interframe prediction is advantageous only for very low spatial frequencies. Even for the case k=1, the break-even point is roughly at one third of the maximum spatial frequency. This means that interframe prediction of all higher-frequency components would be useless and would only deteriorate the coding performance. The same effect applies if motion compensation is used, but the motion is estimated erroneously. By more accurate motion compensation, the 'break even frequency' will be increased.

In this context, it must be observed that in motion estimation and compensation a certain deviation from the optimum often occurs. In the sequel, the deviation error is denoted as $[k_e, l_e]$. In principle, the characteristics of the motion-compensated prediction error signal are then similar to the uncompensated case, but $[k_e, l_e]$ will usually be relatively small. In analogy with (13.9),

$$S_{ee,\mathrm{mc}}(\Omega_1,\Omega_2) = 2 \cdot S_{xx}(\Omega_1,\Omega_2) \cdot \left[1 - \cos\left(k_e\Omega_1 + l_e\Omega_2\right)\right] + S_{rr}(\Omega_1,\Omega_2). \quad (13.10)$$

In practice, the deviation from true motion will not be constant as assumed in (13.10). Moreover, a superposition of different cosine components occurs in the spectrum. The contributions can then be determined by the PDF of the motion deviation $p(k_e, l_e)$ [GIROD 1987], respectively the real part of the Fourier transform $\mathscr{F}$ thereof[1],

---

[1] The Fourier transform of the PDF is also called its *characteristic function* (cf. sec. 3.2). In the case regarded here the interpretation is as follows: If only *one* constant motion deviation of value $[k', l']$ exists, $p(k_e, l_e) = \delta(k_e - k', l_e - l')$, for which the Fourier transform is $e^{j(k'\Omega_1 + l'\Omega_2)}$. The real part thereof is the cosine function in (13.10). If several discrete values of motion errors are present by certain probabilities, the spectral error results as a superposition of their corresponding cosine functions (weighted by probabilities), which constitute the real part of the Fourier transform. The limit transition for any PDF of continuous-valued motion deviations then gives (13.11).

$$S_{ee,\mathrm{mc}}(\Omega_1,\Omega_2) = 2 \cdot S_{xx}(\Omega_1,\Omega_2) \cdot \left[1 - \mathrm{Re}\left\{ \mathscr{F}\big(\mathrm{p}(k_e,l_e)\big) \right\} \right] + S_{rr}(\Omega_1,\Omega_2) \,. \qquad (13.11)$$

The PDF of deviations from true motion can e.g. be modeled by a two-dimensional Gaussian distribution, the Fourier transform thereof is a real-valued Gaussian function as well.

13.2.3   Quantization error feedback and error propagation

The effects of quantization error feedback and propagation of transmission errors, as analyzed in the context of 2D DPCM (see sec. 12.3.1 and 12.3.4), are likewise present in the case of motion-compensated interframe prediction. Quantization error feedback is expressed by the component $Q(\Omega) \cdot H(\Omega)$ from (12.21), where $H(\Omega)$ is described from the z-transfer function in (4.100) for motion-compensated prediction. As the 2D power spectrum is invariant against motion shifts of the signal at least in the case of translational motion[1], the 2D quantization error spectrum resulting by encoding of the previous frame can be modeled by the argument of the function in (11.21),

$$S_{qq}(\Omega_1,\Omega_2) = \min\left[\Theta, S_{xx}(\Omega_1,\Omega_2) \right] . \qquad (13.12)$$

This spectrum is superimposed to the spectrum of the motion-compensated prediction error (13.11). For the case of lossy coding with higher distortion, the quantization error spectrum is not white, but identical to the spectrum of the signal in particular for higher frequencies (refer to the example of the AR(1) model, sec. 11.2.2). If e.g. the first frame is intraframe coded and the high spatial frequencies are eliminated due to lossy coding, these can never be used in prediction, such that they will again fully appear in the motion-compensated prediction error of the next frame[2]. Finally, the application of a spatial loop filter $A(z_1,z_2)$ from (4.100), which could either be a noise reduction filter or an interpolation filter, must be considered as part of the predictor filter. Then, the prediction error signal for the case of constant deviation from the true motion is

$$E(j\Omega_1, j\Omega_2) = X(j\Omega_1, j\Omega_2) \cdot (1 - A(j\Omega_1, j\Omega_2) \cdot e^{j(k_e\Omega_1 + l_e\Omega_2)}) , \qquad (13.13)$$

such that

---

[1] Translational motion of otherwise unchanged components effects a pure phase shift of the spectrum.

[2] In practice, this effect may even be worse, as the spectrum $S_{qq}(\Omega_1,\Omega_2)$ can include components of non-linear and spatially variant coding artifacts (e.g. blocking), which would be superimposed to the signal $e(m,n,o)$.

$$S_{ee}(\Omega_1,\Omega_2) = E^*(j\Omega_1,j\Omega_2)\cdot E(j\Omega_1,j\Omega_2)$$

$$= S_{xx}(\Omega_1,\Omega_2)\cdot\Big[\ \big[1-\mathrm{Re}\{A(j\Omega_1,j\Omega_2)\}\cdot\cos(k_e\Omega_1+l_e\Omega_2)$$

$$+\mathrm{Im}\{A(j\Omega_1,j\Omega_2)\}\cdot\sin(k_e\Omega_1+l_e\Omega_2)\big]^2$$

$$+\big[\mathrm{Re}\{A(j\Omega_1,j\Omega_2)\}\cdot\sin(k_e\Omega_1+l_e\Omega_2) \tag{13.14}$$

$$+\mathrm{Im}\{A(j\Omega_1,j\Omega_2)\}\cdot\cos(k_e\Omega_1+l_e\Omega_2)\big]^2\ \Big]$$

$$= S_{xx}(\Omega_1,\Omega_2)\cdot\big[1+|A(j\Omega_1,j\Omega_2)|^2 -2\,\mathrm{Re}\{A(j\Omega_1,j\Omega_2)\}\cdot\cos(k_e\Omega_1+l_e\Omega_2)$$

$$+2\,\mathrm{Im}\{A(j\Omega_1,j\Omega_2)\}\cdot\sin(k_e\Omega_1+l_e\Omega_2)\big]\ \big].$$

By taking the same approach as in (13.11), a generalization for the case of arbitrary PDF in the deviation of motion gives

$$S_{ee}(\Omega_1,\Omega_2) = S_{xx}(\Omega_1,\Omega_2)\cdot\Big[1+|A(j\Omega_1,j\Omega_2)|^2 -2\cdot\mathrm{Re}\{A^*(j\Omega_1,j\Omega_2)\cdot\mathcal{F}(p(k_e,l_e))\}\Big]. \tag{13.15}$$

The overall power spectrum of the motion-compensated prediction error is then

$$S_{ee}(\Omega_1,\Omega_2) = S_{xx}(\Omega_1,\Omega_2)\cdot\Big[1+|A(j\Omega_1,j\Omega_2)|^2 -2\cdot\mathrm{Re}\{A^*(j\Omega_1,j\Omega_2)\cdot\mathcal{F}(p(k_e,l_e))\}\Big]$$
$$+\min\big[\Theta,S_{xx}(\Omega_1,\Omega_2)\big]\cdot|A(j\Omega_1,j\Omega_2)|^2 +S_{rr}(\Omega_1,\Omega_2). \tag{13.16}$$

In case of high deviation from the true motion and in case of high distortion (low rates), this spectrum can have a higher density of power than the spectrum of the original signal; this will mainly be the case for high spatial frequency components. For an optimum solution, prediction should only be applied for those spectral components, where the prediction error signal would indeed have a lower power spectrum than the original signal. Referring to $S_{ee}(\Omega_1,\Omega_2)$ of (13.16), an ideal filter would have an amplitude transfer function

$$|A(j\Omega_1,j\Omega_2)| = \begin{cases} 0 & \text{if} \quad S_{xx}(\Omega_1,\Omega_2) < S_{ee}(\Omega_1,\Omega_2) \\ 1 & \text{if} \quad S_{xx}(\Omega_1,\Omega_2) \geq S_{ee}(\Omega_1,\Omega_2), \end{cases} \tag{13.17}$$

which would however be an ideal filter that cannot practically be realized. A typical linear optimization could use a Wiener filter that minimizes the area under S_{ee} in (13.16) under the given constraints of the loop model [GIROD 1993]. In practice, in particular if the motion estimation is relatively inaccurate anyway, much simpler lowpass filters can be used. ITU-T Rec. H.261 defines the separable binomial loop filter, which can selectively be turned on or off:

$$\mathbf{H} = \begin{bmatrix} 1/16 & 1/8 & 1/16 \\ 1/8 & 1/4 & 1/8 \\ 1/16 & 1/8 & 1/16 \end{bmatrix}. \tag{13.18}$$

If on the other hand motion compensation can be expected to be highly precise, but feedback of coding errors turns out to be the major problem (in particular at low data rates), other strategies can be applied. For this purpose, the AVC/H.264 standard defines a nonlinear de-blocking filter as loop filter, which adapts by the quantizer step size (reflecting the expected coding error) and local signal properties.

If motion compensation is performed by using sub-pixel accuracy, an interpolation of intermediate values is necessary. This is also performed by a filter $A(z_1,z_2)$ residing in the loop, which is typically a spatial lowpass filter. Even though lowpass filtering may be advantageous as indicated above, this may not always be the case, in particular if the encoding quality is high and motion estimation is sufficiently accurate. Assume an almost perfect motion compensation which means $p(k_e,l_e) \approx \delta(k_e,l_e)$, such that the Fourier transform of the PDF is almost constant. Following (13.15), this leads to a power spectrum of the prediction error signal

$$S_{ee}(\Omega_1,\Omega_2) = S_{xx}(\Omega_1,\Omega_2) \cdot \left[1 - \left|A(j\Omega_1,j\Omega_2)\right|^2\right], \qquad (13.19)$$

which means that for a lowpass interpolation filter with strong roll-off characteristics, a considerable part of the high frequency components would be unpredictable and will appear in the prediction error signal. This effect has led to the definition of high-quality sub-pixel value interpolation filters in the most recent generations of video coding standards.

If *channel losses* occur, the power spectrum of the reconstructed signal is distorted by a component which results as $K(\Omega)/(1-H(\Omega))$ according to (12.28). This effect shall now be analyzed for the case of a motion-compensated prediction using a spatial filter $A(z_1,z_2)$ in the prediction loop. Assume that a loss occurs at frame o_K, where an error spectrum component $S_{kk}(\Omega_1,\Omega_2)$ appears in the prediction error frame, which will likewise appear in the reconstructed frame $\mathbf{Y}(o)$. Subsequent reconstructed frames will then recursively be distorted by a component which is fed back by the prediction loop

$$S_{kk}(\Omega_1,\Omega_2,o) = S_{kk}(\Omega_1,\Omega_2) \cdot \prod_{p=o_K}^{o-1} \left[\left|A(j\Omega_1,j\Omega_2,p)\right| \cdot \left|H(j\Omega_1,j\Omega_2,p)\right|\right]^2 \qquad (13.20)$$

$$\text{for } o > o_K.$$

Herein, the transfer function $H(j\Omega,p)$ expresses the motion shift occurring by the respective past frames, and the transfer function $A(j\Omega,p)$ expresses all elements of spatial filtering, including interpolation of sub-pixel positions in motion compensation. The model in (13.20) cannot fully reflect reality, as again constant motion shift is assumed over frames which are infinitely extended. In this case, occlusion effects would never occur in the motion-compensated prediction process, and $|H(j\Omega,p)|^2=1$. Provided that no spatial filtering is applied ($|A(j\Omega,p)|^2=1$ likewise), an *infinite propagation* of errors will occur.

If spatial filtering is performed in the loop, the effect of the loss at frame o_K is gradually smoothed out over time, however mainly the spatial high-frequency components of the error will be suppressed. In the more realistic case where the motion compensation is spatially variant, in addition the effect can be observed that some of the lost pixels will no longer be used for future prediction. This will in particular be the case at positions where the motion is discontinuous, which happens at boundaries between differently moving objects or at the boundaries of the frame. Fig. 13.7 shows this effect of error propagation as it evolves within a time span of 10 frames. It was assumed here that motion vectors are decoded without a loss. As the motion vectors are a dominating factor influencing the predictor filter's impulse response, the effect on the reconstruction error will be much more severe if wrong motion vectors are used for reconstruction.

It can be concluded that most attention must be paid to the error protection of motion vectors and other prediction control parameters, and to the lower-frequency components of the prediction error signal. The effect of visibility after a loss is clearly dependent on the importance of the lost component with regard to subsequent predictions. The inherent problem is the deviation of the prediction between the encoder and the decoder after an error occurs. This effect was previously introduced as the *drift problem* (sec. 12.3.1. and 12.3.4). By systematic provisions, e.g. error protection for bits which would cause a large reconstruction error in case of losses, it can be tried to keep the amount of drift to a minimum. In any case, if it can be expected that errors occur occasionally, secure mechanisms to guarantee recovery from errors are systematic insertion of *I*-type frames, systematic refresh of smaller areas or other mechanisms which reset the prediction loop into a controllable state (see sec. 13.2.8, 13.2.9 and 15.2.2);

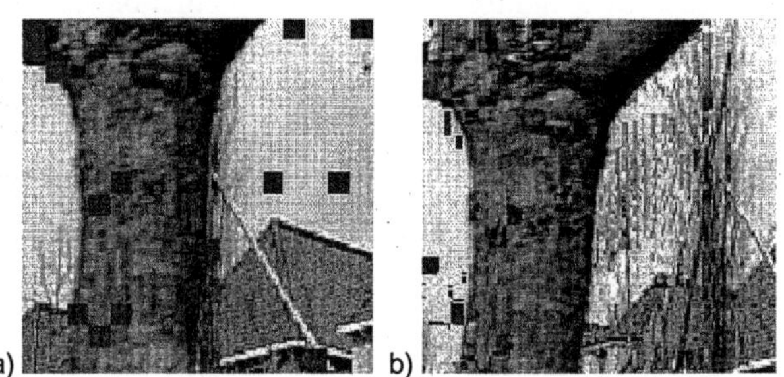

Fig. 13.7. Error propagation in case of motion-compensated prediction
a Reconstructed frame o_K where loss occurs   **b** Reconstructed frame o_{K+10}

13.2.4   Forward, Backward and Multiframe Prediction

In motion estimation, the definitions of *current frame* and *reference frame* were introduced. Typically, the coordinate system of the motion vector field to be estimated relates to the current frame, and the best match is searched in the reference frame. In the special case of block matching, blocks are fixed in the current frame, while the highest-similarity blocks are floating in the reference frame (cf. Fig.

7.55). Hence, not all pixels of the reference frame are used for prediction of the current frame; this is even reasonable, as areas may exist which are actually not visible in the current frame due to occlusion effects. Less reasonable are double references which typically also occur and indicate an inconsistency in the motion definition. If the reference frame precedes the current frame (as it was assumed in the principles of motion compensated prediction discussed so far), motion estimation and compensation are denoted as *forward oriented* (Fig. 13.8a). If the reference frame follows the current frame, motion definitions are *backward oriented* (Fig. 13.8b)[1].

In motion-compensated prediction, the difference computation to obtain the prediction error signal is usually performed between the pixels of the current frame, which is the frame to be encoded, and the estimate thereof which is generated from the reference frame by observing the estimated motion shift. Both the prediction frame $\hat{\mathbf{X}}$ and the prediction error frame $\mathbf{E}$ are aligned with the current frame coordinate system in this case. To obtain a causal sequence of frame-recursive processing, forward-oriented motion compensation must exclusively be used. This shall be denoted as the *conventional motion compensation* method.

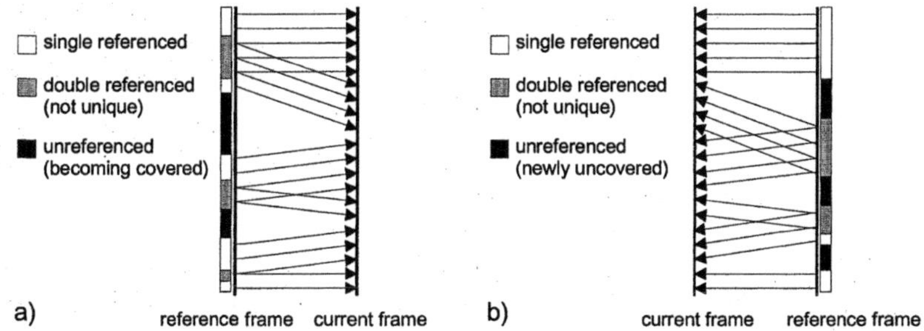

Fig. 13.8. Forward oriented (**a**) and backward oriented (**b**) motion compensation

Usage of backward-oriented motion estimates is nevertheless possible in strictly frame-recursive prediction processing, however the 'current' frame of motion estimation can then no longer be the frame which is encoded. Assume that the previously decoded frame takes the 'current' role in motion estimation, while the frame to be encoded is the reference frame. The prediction frame $\hat{\mathbf{X}}$ must then be generated by *back-projecting* pixels from the previous frame into the present frame coordinate system. In this case, *empty holes* will be generated where no projection is made, which are in fact the unreferenced positions as illustrated in Fig. 13.8b. Existence of these holes is reasonable, as they may typically relate to newly-uncovered positions in the frame to be encoded. To generate a prediction for these positions, methods of spatial interpolation/prediction can be applied (cf. sec.

---

[1] A consistency relationship between backward and forward oriented motion vectors was introduced in sec. 7.6.6 for methods to detect discontinuities in the motion vector field.

5.4.5), or they are otherwise intraframe encoded. Nevertheless, this method of *projection-based motion compensation* is rather applicable in context of highly-accurate motion compensation, warping grid motion compensation or other types of motion compensation using dense motion vector fields.

Forward-oriented vectors can also be used for prediction of the current frame in the conventional motion compensation method, if a *future reference frame* is already available at the decoder. For this, it is however necessary to break the natural order of the frame-by-frame processing sequence, and transmit certain future frames earlier. This is used in the method of *bi-directional* prediction (see sec. 13.2.5).

Neither in forward-oriented nor in backward-oriented motion compensation the reference frame must necessarily be a direct neighbor of the current frame. Often cases occur where a better prediction is found from a frame which is at a farther distance. This freedom to choose one out of an available set of reference frames for prediction is denoted as *multiframe prediction* [WIEGAND, GIROD 2001]. As a consequence, multiple frames have to be stored in the encoder and decoder, and also motion estimation becomes more complex, as it is necessary to determine the best match among the multiple candidate frames available. Considerable gain in prediction quality is achieved by this technique in certain cases. As an example, assume that a person shown in a video closes the eye or opens the mouth for a short moment in the directly preceding frame. Then, a better prediction can clearly be obtained from an earlier frame, where the eye was likewise open or mouth was shut as in the current frame.

13.2.5   Bi-directional Prediction

So far, a strictly-recursive approach of motion-compensated prediction was regarded, enabling frame-wise sequential processing. A better compression efficiency – even though at the expense of increased computational and memory complexity and increased delay – could be expected if usage of future frames was additionally allowed for prediction. If two reference frames are used for prediction, where typically one is a predecessor and one a successor of the current frame, this method is denoted as motion-compensated *bi-directional prediction*[1].

The bi-directional principle as implemented in MPEG standards is illustrated in Fig. 13.9. To guarantee a causal processing sequence at the decoder, it is necessary to first process (encode/decode) those frames which are intended to be used for bi-

---

[1] Prediction could also be made using *two past* or *two future* frames as references, which would however not be *bi-directional* in the strict sense of the word. This generalized mode, as implemented in the AVC standard, is denoted as *bi-predictive*. In principle, there is no limitation to further extend these concepts into a true multi-frame prediction, where an arbitrary number of frames with arbitrary weights is used as in (4.99). It is however questionable how efficient this would be in video compression, in particular regarding the fact that motion parameters and weight adaptation parameters would have to be encoded for each frame used for prediction.

directional prediction. The prediction of those frames (drawn shaded in Fig. 13.9) is hence restricted to the uni-directional sequential principle. This can be interpreted as a sub-sampled video sequence of lower frame rate which is processed by the conventional method. The members of this sequence are denoted as P-type frames. If preceding and subsequent decoded P-type frames are available, the remaining frames can be encoded by bi-directional prediction (these are denoted as B-type frames). Two sets of motion vectors (forward-oriented and backward-oriented) must be estimated for the B-type frames. The sub-sampled sequence needs only to be processed in advance as far as needed for causal recursive processing. In the example shown, the necessary causal sequence of frame encoding steps would be ...1-...-4-2-3-7-5-6-... . This imposes a delay, where the frames of B-type have to be stored intermediately until the next P-type frame needed for prediction is available and readily processed. Additional frame memory is necessary at the decoder side as well, as both the previous and the subsequent P-type frames have to be available for the prediction of the B-type frames, and have to be stored until they can be output by correct display order.

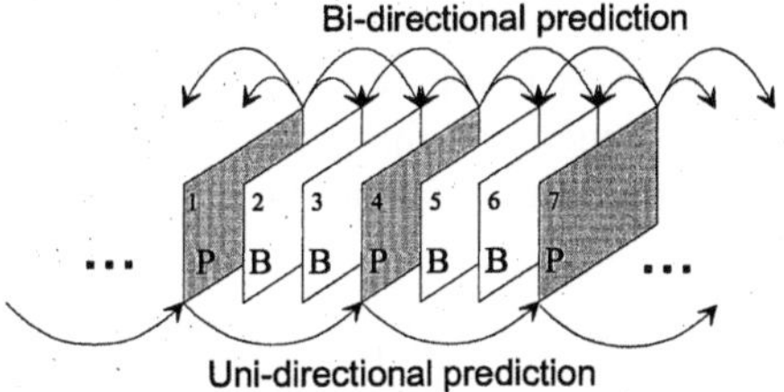

Fig. 13.9. Causal processing sequence by combination of uni-directional and bi-directional prediction

The block diagram of a hybrid video encoder modified for support of bi-directional prediction is shown in Fig. 13.10. It is useful to have a choice of different prediction modes for the B-type frames, which are forward- or backward-only and bi-directional prediction, as well as intraframe coding. Typically, these prediction modes can be switched on a block-by-block basis.

To analyze the coding gain which can be achieved by bi-directional prediction, we assume a simple case where exactly one B-type is between two P-type frames, and the predictions obtained from both neighbors are averaged:

$$\hat{x}(m,n,o) = 0.5 \cdot x(m+k_f,n+l_f,o-1) + 0.5 \cdot x(m+k_b,n+l_b,o+1) \ . \qquad (13.21)$$

Herein, subscripts b and f indicate backward-oriented and forward-oriented motion compensation, respectively. Both partial compensations will most probably not be perfect, where the bi-directional prediction error signal results as

$$e_{\mathrm{bi}}(m,n,o) = x(m,n,o) - \hat{x}(m,n,o) \ . \qquad (13.22)$$

In analogy with (13.9) and (13.10), taking into account motion compensation inaccuracies (k_e,l_e), the following 2D spectrum results:

$$E_{\text{bi}}(j\Omega_1, j\Omega_2) = X(j\Omega_1, j\Omega_2) \cdot \left[1 - 0,5 \cdot e^{-j(k_{e,f}\Omega_1 + l_{e,f}\Omega_2)} - 0,5 \cdot e^{j(k_{e,b}\Omega_1 + l_{e,b}\Omega_2)} \right]$$

$$= X(j\Omega_1, j\Omega_2) \cdot \left[j \cdot e^{-j\frac{k_{e,f}\Omega_1 + l_{e,f}\Omega_2}{2}} \cdot \sin\left(\frac{k_{e,f}\Omega_1 + l_{e,f}\Omega_2}{2}\right) - j \cdot e^{j\frac{k_{e,b}\Omega_1 + l_{e,b}\Omega_2}{2}} \cdot \sin\left(\frac{k_{e,b}\Omega_1 + l_{e,b}\Omega_2}{2}\right) \right]$$

$$+ R(j\Omega_1, j\Omega_2).$$

$$(13.23)$$

The resulting power spectrum

$$S_{ee,\text{bi}}(\Omega_1, \Omega_2) = S_{xx}(\Omega_1, \Omega_2) \cdot \left[\sin\left(\frac{k_{e,f}\Omega_1 + l_{e,f}\Omega_2}{2}\right) - \sin\left(\frac{k_{e,b}\Omega_1 + l_{e,b}\Omega_2}{2}\right) \right]^2 + S_{rr}(\Omega_1, \Omega_2) \quad (13.24)$$

will only be identical with the corresponding components caused by inaccurate motion compensation in uni-directional prediction (13.8), if $k_{e,f} = -k_{e,b}$ and $l_{e,f} = -l_{e,b}$; it will occasionally become zero, if $k_{e,f} = k_{e,b}$ and $l_{e,f} = l_{e,b}$. If the forward- and backward oriented motion vectors are estimated independently, the motion compensation inaccuracies can also be regarded as statistically independent. As then the extreme cases mentioned above may occur by same probability as well as all cases in between, it can be expected from a statistical point of view that the energy of the motion-compensated bi-directional prediction error, as caused by inaccurate motion compensation, will be reduced by guarantee as compared to uni-directional prediction (P type frames). The actual gain will depend on the spatial spectrum of the current frame and the PDF of the MC compensation inaccuracies.

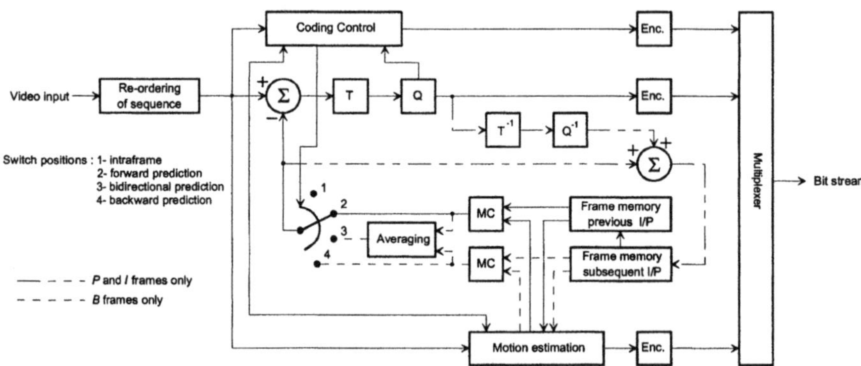

Fig. 13.10. Hybrid encoder supporting bi-directional prediction

Due to the fact that two statistically independent motion-compensated prediction errors are averaged, partial extinction can occur. This can also be interpreted as a temporal lowpass filtering (smoothing) of those components in the prediction error spectrum which are caused by inaccurate motion compensation.

Another component relates to the percentage of unpredictable areas[1], which appear as component S_{rr} in (13.24). Here, bi-directional prediction has another eminent advantage: If it is possible to switch to uni-directional prediction of backward or forward directions, it can be expected that areas being uncovered are better predictable from the subsequent frame. If on the other hand an area is being covered from the current frame to the next, it can better be predicted using only the previous frame as reference. This will further reduce the prediction error energy.

It should however be observed that the gain to be expected by bi-directional prediction is accompanied by probably *worse* prediction in the sequence of unidirectionally predicted P-type frames. As the temporal distance between these frames is higher than in a frame-wise prediction, it can be expected that current frame and reference frame are more dissimilar. This effects both a higher probability of unpredictable areas, and a higher possibility of erroneous motion compensation. The best tradeoff between the gain made by bi-directional prediction and the loss due to the underlying unidirectional prediction is highly sequence-dependent and can hardly be described analytically. As a rule of thumb, it is usually advantageous to use up to three B-type frames between P-type frames, if motion is slow and continuous, and high spatial detail appears in the sequences, which might be a source of high-energy spectral components in the prediction error signal. In case of high motion, it can sometimes be observed that the usage of bi-directional prediction does not give any advantage, as the penalty of worse prediction in the longer-distance P-type sequence is too high.

A disadvantage of bi-directional prediction which should also be considered is the double number of motion parameters. This increases the complexity of motion estimation and compensation, but also potentially increases the data rate. The forward and backward motion vectors are typically not much different with exception of the sign. Hence, joint encoding can be useful; in addition, if motion trajectories are consistent over the sequence, it will be possible to determine the motion vectors in a specific area of the B-type frame from the motion vector of the (motion-wise) 'co-located' area of the adjacent P-type frame. This is exploited by the *direct mode* as implemented in MPEG-4 and other video coding standards (see sec. 17.4.3).

The effect of *quantization error feedback* according to (13.12) is present in the prediction of B-type frames as well. Quantization error components which are fed back from the previous and the subsequent frame may however be correlated, when the latter is a P-type frame which is also predicted from the first. On the other hand, the temporal averaging of two components will usually have a smoothing effect, in particular for high-frequency artifacts like blocking. Further, B-type frames are typically not used to predict other frames[2]. This means that neither

---

[1] This may be the more important component in the case where high-quality motion estimation and compensation are performed, such that motion vector inaccuracies are less severe.

[2] Exceptions could be made in the case of B-frame pyramids, see sec. 13.2.8. However even then, no long recursions occur.

critical quantization error feedback nor critical channel error propagation will originate from of a decoded B-type frame. Actually, it can be argued that the recursive decoder structure of frame-by-frame MC prediction is broken up in the case of B-type frames. As side effect, these can usually be quantized coarser than P-type frames, and must not necessarily be strongly protected against transmission losses.

The case of switching to intraframe coding in single block areas will occur fairly rare in B-type frames, because a good reference will typically be found either in the previous or in the subsequent reference frame even in the case of an occlusion. Exceptions would be made for areas which appear just for one frame and are not visible before and after, which could be e.g. be a short closing of an eye; also sudden changes of lighting conditions by flashlights or reflections could enforce an instantaneous intraframe mode.

Fig. 13.11 shows prediction error frames for cases of uni-directional and bi-directional prediction (compare also with Fig. 4.16). Generally, decreased structural detail of the prediction error can be observed; at the right edge of the tree the switching to backward prediction reduces the prediction error for an uncovered area.

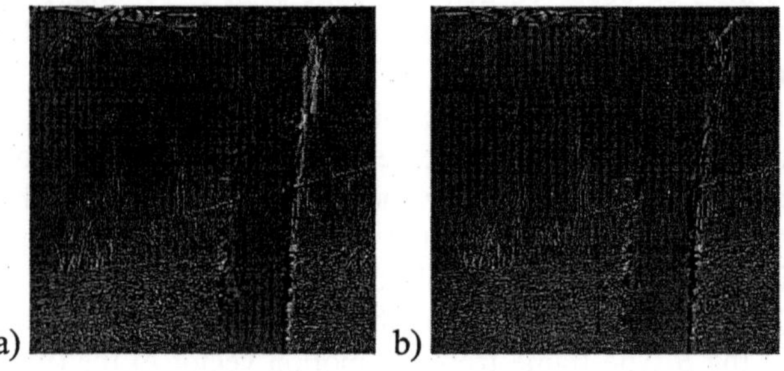

Fig. 13.11. Prediction error frame for motion compensation with 1/2 pixel accuracy **a** unidirectional prediction (P-type frame) **b** bi-directional prediction (B-type frame)

13.2.6 Improved Methods of motion compensation

Block-based motion compensation is a poor model, as it is hardly possible to support smooth variations and discontinuities in the motion vector field properly. This is in particular true when motion vectors defined over relatively large block sizes like 16x16 pixel are used. On the other hand, it should be avoided that too much rate is spent for the motion information unless it is necessary. This must be one design paradigm of the methods described subsequently. Particular attention must however also be paid to the motion estimation itself. Straightforward usage of minimum frame difference criteria as in (7.184) may not be the appropriate solution with regard to rate-distortion criteria.

The motion-compensated prediction error signal does *not* have a white spectrum, as was shown in (13.16). Many algorithms for motion estimation do not take into account the spectral properties of the MC prediction error regarding the effectiveness of the subsequent transform. It is not guaranteed that the motion vector chosen by criteria of prediction error energy minimization is also optimum with respect to the rate that is needed for the encoding of the resulting transform coefficients. Eventually, a prediction error constellation having a better capability for spectral concentration may be a better choice than another signal which has less energy, but is spectrally more flat (see Problem 13.2). A possible solution would be to select the motion vector which gives the minimum logarithmic energy of coefficients as resulting from the transform of a prediction error block. This would takes into account the spectral flatness of the prediction error residual (cf. (11.19) and (12.30)). A less complex approximation of spectral-flatness based criteria is the *Sum of Absolute Transformed Differences* (SATD)[1]. Herein, a 2D transform is applied to prediction error blocks resulting by usage of a subset of best-candidate motion vectors. Then, the vector is selected which provides the minimum L_1 norm (sum of absolute values) computed over the transform coefficients. This approach favors prediction error constellations which concentrate energy in a low number of coefficients, that will normally also be the best choice for encoding by lowest rate[2]. Low-complexity transforms like the Hadamard or Walsh transforms (4.123)-(4.126) are sufficient for this purpose. The decision about the best motion compensation constellation becomes more complex when different modes can be selected, e.g. variable MC block sizes, which also highly influence the rate to be spent for the vectors. Typical methods of rate-distortion optimization are applied in this context, where variants of the metric (7.186) are used in motion estimation based on a rate/quality dependent Lagrangian multiplier [WIEGAND, GIROD 2001]. In principle, motion estimation and coding mode selection algorithms targeting for minimization of the rate could be extended into extremely complex methods. Testwise encoding of prediction error candidate blocks of different vectors and different coding modes could be performed to search exhaustively for the configuration that gives best rate-distortion performance.

Variable block size MC. In cases of smooth variation and of discontinuities in the motion vector field, a general flaw of block based motion compensation results by the fact that constant motion vectors are enforced over the entire area. A reduction of the block size could largely improve the performance; It should however only be applied when indeed changes of the motion vector field are present, as otherwise the rate for more accurate motion parameters would be spent in vain. A solution is variable block size motion compensation as shown in Fig. 13.12. Useful methods are based on splitting a larger block into two or four equal sub-partitions whenever justified. The rate to be spent for the finer resolution of the motion vec-

---

[1] This method is used in reference encoders for the AVC standard.

[2] For an example, see Problem 13.2

tor field must be more than compensated by the expected savings in bit rate for encoding of the motion-compensated prediction error, which however is rate dependent. Rate-distortion optimization criteria as described above must be applied.

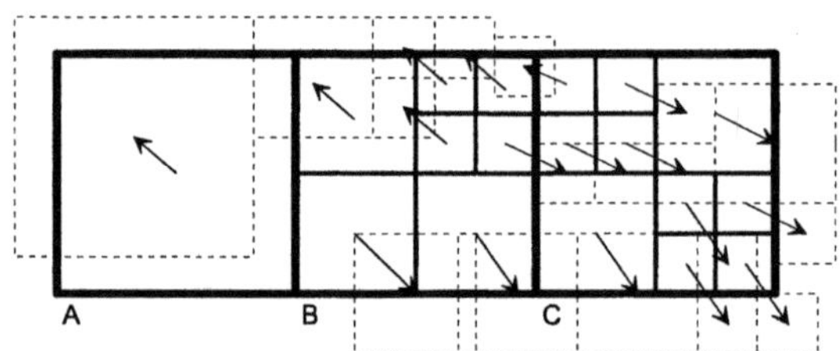

Fig. 13.12. Example of motion compensation using variable block sizes

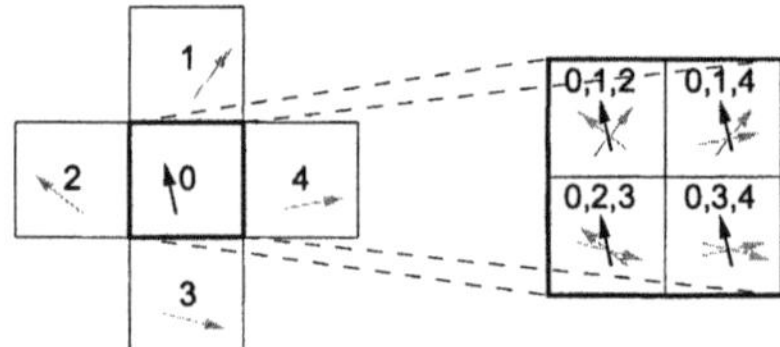

Fig. 13.13. OBMC: usage of neighbored-block motion vectors 1-4 for weighted superposition of different prediction errors in quarter tiles of the current block '0', as described by the weighting functions (13.26)

Overlapping-block MC. At object boundaries, discontinuities between adjacent block motion vectors occur. As it is very unlikely that the actual boundaries of differently-moving objects coincide with block boundaries, unpredictable components of high energy and structure will typically appear in the prediction error images at such positions. A possible solution to this problem is *overlapping block motion compensation* (OBMC) [ORCHARD, SULLIVAN 1994]. Herein, a weighted superposition of different motion-compensated predictions is performed. These are generated by application of additional motion vectors within the current block that were actually allocated to neighbored blocks. This effects a smoothing of the prospectively erroneous components in the motion compensated predictions. An example is given in Fig. 13.13. The prediction in the current block (0) is generated using its own motion vector $\mathbf{k}^{(0)}$ and the four motion vectors $\mathbf{k}^{(p)}$ of its horizontal and vertical neighbors ($p=1\ldots4$), using an appropriate weighting function such that

$$\hat{x}(m,n,o) = \sum_{p=0}^{4} w_p(m,n) \cdot x(m+k^{(p)}, n+l^{(p)}, o-1) \quad ; \quad \sum_{p=0}^{4} w_p(m,n) = 1. \qquad (13.25)$$

An example of weighting functions is given by the following weighting matrices for a block size of 8x8 pixels[1]:

$$\mathbf{W}_0 = \frac{1}{8}\begin{bmatrix} 4 & 5 & 5 & 5 & 5 & 5 & 5 & 4 \\ 5 & 5 & 5 & 5 & 5 & 5 & 5 & 5 \\ 5 & 5 & 6 & 6 & 6 & 6 & 5 & 5 \\ 5 & 5 & 6 & 6 & 6 & 6 & 5 & 5 \\ 5 & 5 & 6 & 6 & 6 & 6 & 5 & 5 \\ 5 & 5 & 6 & 6 & 6 & 6 & 5 & 5 \\ 5 & 5 & 5 & 5 & 5 & 5 & 5 & 5 \\ 4 & 5 & 5 & 5 & 5 & 5 & 5 & 4 \end{bmatrix}$$

$$\mathbf{W}_1 = \frac{1}{8}\begin{bmatrix} 2 & 2 & 2 & 2 & 2 & 2 & 2 & 2 \\ 1 & 1 & 2 & 2 & 2 & 2 & 1 & 1 \\ 1 & 1 & 1 & 1 & 1 & 1 & 1 & 1 \\ 1 & 1 & 1 & 1 & 1 & 1 & 1 & 1 \\ 0 & 0 & 0 & 0 & 0 & 0 & 0 & 0 \\ 0 & 0 & 0 & 0 & 0 & 0 & 0 & 0 \\ 0 & 0 & 0 & 0 & 0 & 0 & 0 & 0 \\ 0 & 0 & 0 & 0 & 0 & 0 & 0 & 0 \end{bmatrix} \quad \mathbf{W}_2 = \frac{1}{8}\begin{bmatrix} 2 & 1 & 1 & 1 & 0 & 0 & 0 & 0 \\ 2 & 1 & 1 & 1 & 0 & 0 & 0 & 0 \\ 2 & 2 & 1 & 1 & 0 & 0 & 0 & 0 \\ 2 & 2 & 1 & 1 & 0 & 0 & 0 & 0 \\ 2 & 2 & 1 & 1 & 0 & 0 & 0 & 0 \\ 2 & 2 & 1 & 1 & 0 & 0 & 0 & 0 \\ 2 & 1 & 1 & 1 & 0 & 0 & 0 & 0 \\ 2 & 1 & 1 & 1 & 0 & 0 & 0 & 0 \end{bmatrix}$$

$$\mathbf{W}_3 = \frac{1}{8}\begin{bmatrix} 0 & 0 & 0 & 0 & 0 & 0 & 0 & 0 \\ 0 & 0 & 0 & 0 & 0 & 0 & 0 & 0 \\ 0 & 0 & 0 & 0 & 0 & 0 & 0 & 0 \\ 0 & 0 & 0 & 0 & 0 & 0 & 0 & 0 \\ 1 & 1 & 1 & 1 & 1 & 1 & 1 & 1 \\ 1 & 1 & 1 & 1 & 1 & 1 & 1 & 1 \\ 1 & 1 & 2 & 2 & 2 & 2 & 1 & 1 \\ 2 & 2 & 2 & 2 & 2 & 2 & 2 & 2 \end{bmatrix} \quad \mathbf{W}_4 = \frac{1}{8}\begin{bmatrix} 0 & 0 & 0 & 0 & 1 & 1 & 1 & 2 \\ 0 & 0 & 0 & 0 & 1 & 1 & 1 & 2 \\ 0 & 0 & 0 & 0 & 1 & 1 & 2 & 2 \\ 0 & 0 & 0 & 0 & 1 & 1 & 2 & 2 \\ 0 & 0 & 0 & 0 & 1 & 1 & 2 & 2 \\ 0 & 0 & 0 & 0 & 1 & 1 & 2 & 2 \\ 0 & 0 & 0 & 0 & 1 & 1 & 1 & 2 \\ 0 & 0 & 0 & 0 & 1 & 1 & 1 & 2 \end{bmatrix}. \tag{13.26}$$

The prediction error $e(m,n,o) = x(m,n,o) - \hat{x}(m,n,o)$ is then computed by (13.25). When adjacent-block motion vectors are different, the variant of OBMC shown here is by a factor of approximately 4-6 more complex than conventional block-based MC; at any position of the block, weighted superposition of three different prediction signals must be performed. It should be observed that OBMC must be applied as well during the motion estimation (cf. (7.187)), as otherwise the result of the prediction would not be optimized for minimization of the prediction error resulting from the superposition.

While being advantageous in case of clear motion discontinuities, the OBMC method is not fully consistent for the case of *smooth variations* in the motion vector field. It will then usually effect a spatial blurring or smearing of the estimate, which can cause higher prediction errors than the warping MC described in the subsequent subsection.

Warping MC and Global MC. When a control grid (7.188) or a parametric model (cf. sec. 7.4.4) are used to describe the motion vector field, individual motion vectors can be determined for each pixel position, but the description of the motion is still very compact[2]. The estimate is then computed as[3]

$$\hat{x}(m,n,o) = \tilde{x}\big[m + \tilde{k}(m,n), n + \tilde{l}(m,n), o - 1\big]. \tag{13.27}$$

It is necessary to consider the warping function during the optimization of the

---

[1]Other weighting functions are possible, e.g. including all eight neighbored blocks. In general, at each position of the block, the weights must sum to unity.

[2] E.g. no higher number of motion vectors is required as compared to block matching when a warping grid is used (cf. sec. 7.6.4), and even less in case of global motion parameters.

[3] In (13.27), the tilde symbol '~' is used to express interpolated signals or motion vectors

motion vectors, see (7.190). The warping model suffers in the case of motion discontinuities, where however solutions exist as described in sec. 7.6.4.

A special case of warping-based MC is *Global Motion Compensation* (GMC), where the warping function is defined by a global motion model, e.g. an affine or perspective transformation. Both in cases of small-scale and global warping MC, it should be observed that usage of sub-pixel shifts is the typical case, which must be provided with sufficient accuracy. This also implies that high-quality spatial interpolation must be performed for the generation of the estimates.

Sprites. The technique of mosaicing has been introduced in sec. 7.8. If a mosaic is used for the purpose of encoding, it is commonly denoted as a *sprite*. This is a large image assembled by patches from different frames of a video (typically a moving background), from which single frames can be reconstructed by inverse projection. *Static sprites* as defined in the MPEG-4 standard are based on a fixed picture model assumption where no changes except for motion shall occur during the entire usage. This is not applicable to real-time applications, as all frames have to be available before the sprite can be generated. In contrast to that, *dynamic sprites* allow changes, which can also be encoded as differential information [SMOLIC, SIKORA, OHM 1999]. Typically, sprites are used as one element of *object-based coding* as described subsequently; only in special cases like global zoom or pan over an otherwise unchanged scene, sprite mapping could be used as a stand-alone methodology for encoding of a video sequence.

Object-based coding. The general flaws of block-based motion compensation were described above; also none of the techniques described so far provides a consistent solution to the problem of *occlusions* which necessarily occur when objects are moving differently within a video scene. The ultimate goal of *object-based coding* is to describe the objects contained in a sequence *separately* by as low number of parameters as possible. The meaning of these objects will not be relevant for the purpose of encoding, but could be useful in the context of their automatic extraction. The background or other large areas, which are either static or subject to consistent (e.g. global) motion, are treated like any other object in the scene. In particular, if a part of the background is temporarily occluded by other objects, it is very efficient to recall it from a background memory when it appears again [THOMA, BIERLING 1989]. A more general type of background memory would be a sprite which can be moving as well, e.g. due to camera motion.

For any object within a scene, it is necessary to provide the *shape/position* and the *color/texture* information for any time instance (frame). The *motion* information will be the key to keep the representation as compact as possible, which means that changes in the other parameters need only be described if they cannot be determined by the motion. A block diagram of the original approach in object-based video coding as proposed in [MUSMANN, HÖTTER, OSTERMANN 1989] is shown in Fig. 13.14. This method is more related to replenishment methods than to hybrid coding, as it was postulated that the object motion can either be modeled perfectly or fails, such that no prediction difference must be encoded at all. The color/texture information of an object once found is frozen, until it changes in reality such that it needs to be newly transmitted. This latter case is entitled as *model failure*. In the other case of *model compliance*, the new frame can be described accurately enough mainly by motion information, which should combine with mechanisms to change the shape. The underlying motion model is assumed to play a pre-dominant role. Both 2D [HÖTTER 1990] and 3D motion models [OSTERMANN 1994],[KAMPMANN 2002] have been investigated.

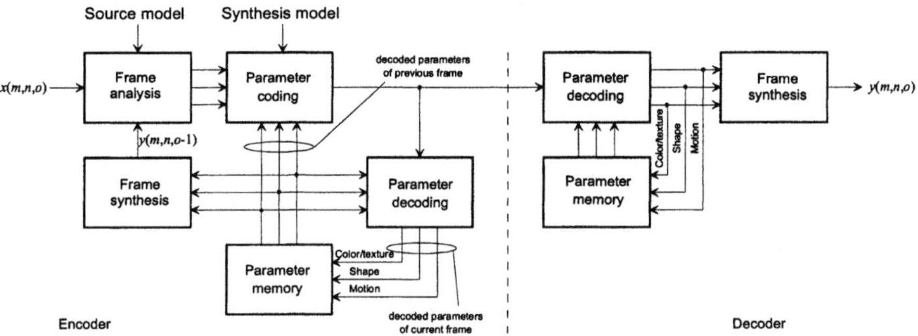

Fig. 13.14. Block diagram of an object-based video encoder and decoder as proposed in [MUSMANN, HÖTTER, OSTERMANN 1989]

The method in Fig. 13.14 is of recursive nature as hybrid coding, both analysis and synthesis relate to the previous frame. Motion and shape parameters are encoded as differential information as well; in [HÖTTER 1994], the method is interpreted as a generalization of conventional hybrid coding, the latter one using block-based shape and motion models, and a block transform for texture coding.

The restriction imposed previously, postulating that texture changes shall exclusively be encoded by the intraframe 'model failure' information, performs often worse than encoding of prediction errors. This fact was considered in the definition of object-based video coding as it is implemented in the MPEG-4 standard. The decoding process for a scene consisting of *video objects* is shown in Fig. 13.15. The scene description information (cf. sec. 17.2) defines the positions of the objects, and also the information which of two objects shall be visible as foreground in the case of a spatial overlap. For each single object, information components of *motion, shape* and *texture* must be decoded. The motion information is used to exploit the temporal redundancy in both shape and texture information:

– *Context Arithmetic Encoding* (CAE) as introduced in sec. 12.1 is used for representation of the binary shape, where for exploitation of the temporal redundancy the context uses both spatially and temporally preceding neighbors. The motion shift is taken into account to determine the temporally co-located positions from the previous frame.

– Texture encoding is based on motion-compensated prediction differences encoded by a block-wise 8x8 DCT transform coding for the inner part of objects, and DCT based on *block padding* or *shape adaptive* methods for the boundary blocks, as explained in sec. 12.6.3.
– In addition to the binary shape, a tool is available to encode a *gray-level shape* which allows to define *transparency* of objects, expressed by a so-called *alpha map* which describes a blending factor for each pixel. The alpha map is encoded by the same method that is used for the texture, a hybrid motion-compensated prediction and DCT transform coding.

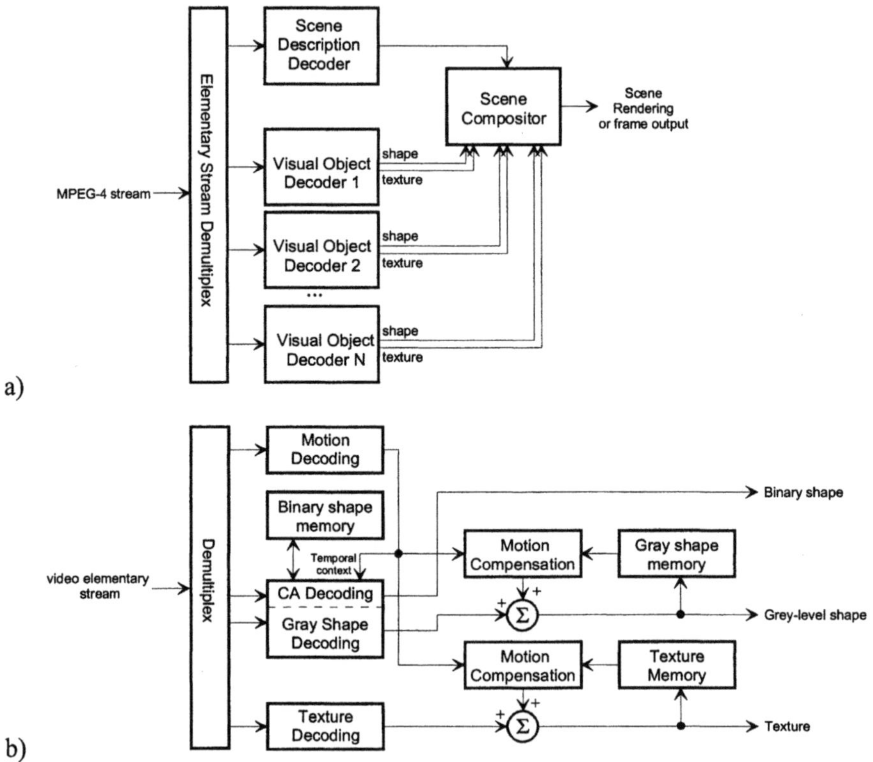

Fig. 13.15. Object-based decoding in MPEG-4 **a** Composition of N video objects into a scene or a frame **b** Structure of one single video object decoder

Object-based coding has not yet reached a maturity status which might allow application for any video scene. This mainly relates to insufficient methods of scene analysis and exact segmentation which is necessary before the coder can be run. For certain scenes, in particular with single objects in front of static background, good results can be obtained, in particular for efficient encoding of temporarily occluded but otherwise unchanged areas. From the complexity point of view, the combination of a conventional (block-based) hybrid video coding with an object-based hybrid coding, both consisting of common building blocks, is advantageous. Simple switching between the two cases can be made to whatever is optimum. Regardless of potential increase in coding performance, object based video coding is needed for interactive applications, where e.g. the scene content shall be influenced by the user, or shall be tailored to specific user needs by the content provider.

13.2.7 Hybrid Coding of Interlaced Video Signals

Most content acquired by video cameras is originally captured in interlaced formats (see sec. 1.2 and 2.2.4). Hence, hybrid encoding and motion compensation in general must also be adaptable to the interlaced sampling structure. This concerns both the motion compensation and the spatial 2D transform. Methods for interlaced coding were in particular developed for the MPEG-2 standard and establish one fundamental extension of MPEG-2 video coding as compared to MPEG-1. Similar methods have also been adopted by subsequently defined video coding standards like MPEG-4. It can typically be decided at a global level if the fields shall be encoded independently ('field mode') or jointly combined as one frame ('frame mode'). In particular in the frame mode, it is then possible to select between different approaches for motion compensated prediction on a block by block basis as described below.

Depending on the properties of motion (i.e. whether no motion occurs, whether the motion is by an even or odd number of pixels vertically etc.), it may be advantageous to predict the even and odd fields of an interlaced video frame from the respective fields either of same or of opposite parity in the previous frame. In the case of fast and inhomogeneous motion, it can also be useful to define different motion vectors for the two fields. For homogeneous motion and in the interest of a low rate, it is better to use only one common vector for the related block area of both fields. Three prediction modes for interlaced video are shown in Fig. 13.16[1]:

- In *frame prediction*, only one motion vector is used for both fields. The fields are in principle combined into a frame, regardless of the time shift that is actually effective between their sampling time points. This means that prediction of a current field could be performed either using the field of same or opposite parity of the reference frame, depending on whether the (frame-related) vertical motion shift is by an even or odd number of pixels. As however is evident from Fig. 13.16a, this mode is inconsistent whenever an odd vertical motion shift occurs, as then the vertical orientations of the motion vectors are different for both of the fields (same motion vector is used regardless of the fact that either one or three field periods are between the actual field and the field it is predicted from). Further, this method is usually disadvantageous in the case of fast motion.

---

[1] These are exactly the modes defined by the MPEG-2 standard.

- In *field prediction*, independent motion vectors are used for the two fields. Further, it is selectable if any field is predicted from a field of either same or opposite parity in the reference frame. Eventually, depending on the size of the motion vectors, vertical interpolation of lines missing from the selected reference field will be necessary. In Fig. 13.16, such interpolated prediction pixels are indicated by opaque circles or squares. In addition to the configuration illustrated in Fig. 13.16b, it is possible that the second field of a frame uses the first field of the frame for prediction reference when fields are encoded independently (in global 'field mode').

- In *dual prime prediction*, any field of the current frame is predicted using an interpolation from both fields of the reference frame. Common motion vectors are used, but different from the frame prediction described above, these are scaled to make the motion shift consistent with the temporal distance between the respective current field and the fields it is predicted from. In principle, this can also be interpreted as if by a first step a full-frame picture is interpolated from both reference fields, taking into account the motion shift. This 'prediction picture' is then used to predict the current fields, where the respective temporal distance influences the scaling of the motion vector.

The blocks for the computation of the DCT (or other transform) can then also be applied either by frames or by fields, depending on whether the frame mode or the field mode is selected (see Fig. 13.17). The latter case is advantageous for presence of high motion, where the similarity between the even and odd lines can be low, such that a DCT of frame-wise arranged blocks would output vertical frequency coefficients of high energy. In case of no motion, it is better to combine both fields into a frame, as then the vertical correlation is highest, and less vertical high-frequency coefficients need to be encoded. This is in particular relevant in cases where motion compensation fails or local intraframe coding is performed.

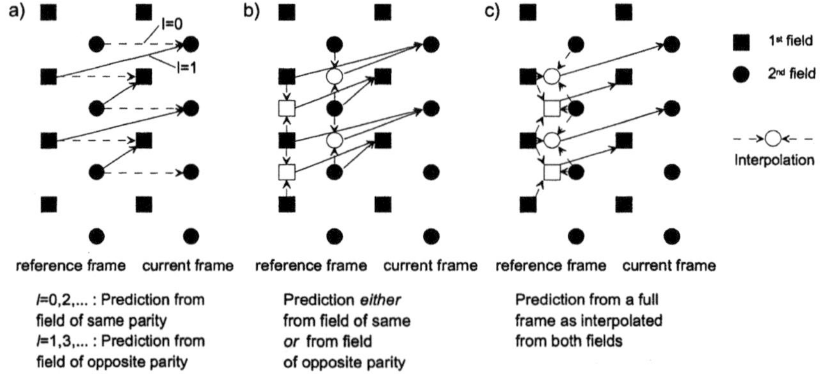

Fig. 13.16. Motion-compensated prediction schemes for fields in interlaced video: **a** Frame prediction **b** Field prediction **c** 'Dual prime' prediction

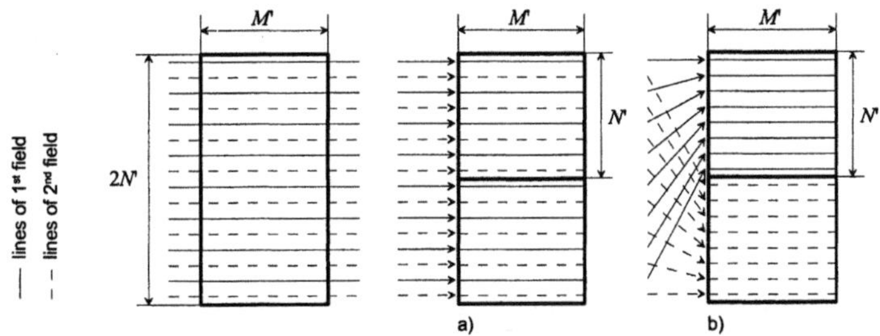

Fig. 13.17. a DCT in frame mode **b** separate DCT in the fields

13.2.8 Scalable Hybrid Coding

Scalable representation of still image frames, allowing flexible adaptation of spatial resolution and quality (quantizer) resolution independent of the encoding process, is efficiently solved by a combination of wavelet transform and embedded quantization with bit-plane representation (see sec. 12.4.2). Scalable coding is also highly desirable for video signal representation, in particular regarding error-resilient transmission and transmission over networks with variable bandwidth allocation (cf. sec. 15.2). In video representation, one more degree of freedom is *temporal scalability*, i.e. variable frame rate by different scalable resolution levels.

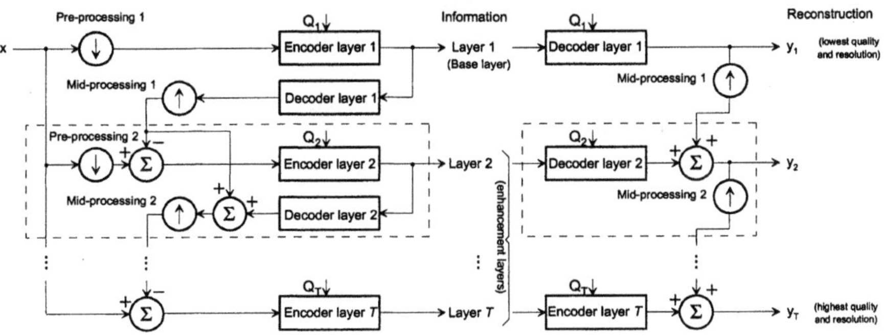

Fig. 13.18. Principle of scalable coding with T layers

A very general principle of scalable coding and decoding is shown in Fig. 13.18, where by supplementing further building blocks of the intermediate-level type (highlighted by a dotted rectangle), an arbitrary number of scalable layers can be realized. The spatio-temporal signal resolution to be represented by the base layer is first generated by decimation (pre-processing). In the subsequent encoding stage, an appropriate setting of the quantizer will then lead to a certain overall quality level of the base information. The base layer reconstruction is an approximation of all the higher-layer resolution levels, and can be utilized in the decoding of the subsequent layers. The *mid-processing* unit performs up-sampling of the subsequent layers. The *mid-processing* unit performs up-sampling of the next-lower layer signal to the subsequent layer resolution. Typically, pre- and mid-processing are performed by decimation and interpolation throughout all stages, whereas the particular action to be taken can be quite different depending on the dimension of scalability, e.g. motion-compensated processing can be implemented for frame-rate up-sampling. In principle, the information is propagated from the lower into the higher resolution layers both during encoding and decoding.

In all types of scalability (temporal, spatial or quantization/quality), the constraints imposed by the frame-recursive processing of hybrid video coding have to be carefully considered. Any lower layer should in the ideal case be self-contained, which means that the prediction should not use any decoded information from higher layers. Otherwise, different estimates would be used at the encoder and decoder sides, and a similar *drift effect* would occur as it was observed in the presence of channel losses (cf. sec. 13.2.3). The prediction of the base layer information will be worse than it could be if all enhancement layer information was allowed in the prediction, a phenomenon that was already expressed in (12.22). This does not penalize the operational point of the base layer, which will implicitly perform like a single layer coder at the same rate; as however the base layer information is used for prediction of the enhancement layer, the rate-distortion performance towards higher rates will be worse than it could be in a single-layer coder. Alternatively, the full enhancement information could blindly be used for prediction[1]; in this case, the reconstruction quality of the highest enhancement layer could approach the performance of a single-layer coder, while the reconstruction quality of the base layer and all intermediate layers would eventually suffer dramatically due to drift (12.15). This basic dilemma to *penalize either enhancement or base layer* performance is inherently caused by the recursive nature of the DPCM structure integrated in the hybrid coding concept.

In the following sub-sections, different principles of scalable hybrid coding will be discussed in detail, which also will include solutions trying to find a trade-off between the two extrema described above. In most cases, two-layer configurations are explained, however extension to more layers as in Fig. 13.18 is straightforward by using multiple enhancement layer coder/decoder building blocks.

Scalability of quantizers. Quality scalability is implemented in the block diagram of Fig. 13.18 by introducing different quantization parameters Q_t. In a scalable quantizer configuration, it has to be observed that the quantizer of the enhancement layer re-quantizes the quantization error from the base layer, such that the quantization range has to be adapted properly. If base-layer stability shall be observed, it is inevitable to implement a separate prediction loop for the base layer both at the encoder and decoder sides[2]. In the simplest case, the quantization of the

---

[1] This is a mode of *SNR scalability* as defined in the MPEG-2 standard.

[2] Most following diagrams omit the blocks of transform/inverse transform and the blocks of entropy coding/decoding from the loops to make the structures more simple to read. Straightforward augmentation by the respective blocks of transform and inverse transform

residual error from the base layer can be performed directly as intraframe coding (Fig. 13.19a)[1]. However then, the method is strongly sub-optimum as significant temporal correlation may exist between the residual signals of subsequent frames when the base layer is operating at low quality level[2]. As an alternative solution, interframe prediction with a separate loop can be applied in the enhancement layer coding (Fig. 13.19b). By this principle, the residual error to be encoded in the enhancement layer will typically become lower than for the case of base-only prediction, as detail information available in the previous enhancement frame is utilized. Identical motion parameters can be used in both prediction loops. Both configurations of Fig. 13.19b are interchangeable, i.e. each coder/decoder structure can be combined with the other[3]. In the decoder diagram at the top, only one prediction loop is needed, if only the best available enhancement information shall be reconstructed, which is indeed the normal case in a single-purpose receiver. The decoder structure at the bottom reconstructs both base and enhancement layers in parallel.

Sub-optimality of the double-loop method. By utilizing the full-resolution signal in the enhancement-layer prediction, the performance will still be inferior as compared to a non-scalable coder with only one loop, which can be proven as follows. Assume the usage of a uniform quantizer of step size Δ_B. Then, the residual error from the base layer will be within the range $\pm\Delta_B/2$, if the effect of overload is neglected. In the system of Fig. 13.19a, an embedded quantizer of two steps of size $\Delta_B/2$ would actually be usable in the next layer. However, in the system shown in Fig. 13.19b, the effect of quantization error feedback occurs in the enhancement prediction loop additionally. As a result, the range of the residual signal to be fed into the enhancement quantizer is in fact not limited by guarantee to $\pm\Delta_B/2$. As a consequence, established methods for embedded quantizer scalability (see sec. 11.3) can not straightforwardly be applied here. As on the other hand only embedded quantizers could approach the rate-distortion performance of non-scalable coding, the system of Fig. 13.19b will necessarily be inferior as compared to a single-layer method at the full rate of the enhancement layer. The base layer encoder is equivalent to a self-contained single-layer coder, such that no quality drop occurs at the base-layer rate.

---

is shown e.g. in Fig. 13.26, and entropy coding does not effect the loop of signal processing at all.

[1] This is the method of *Fine Granularity Scalability* as implemented in the MPEG-4 Visual standard [Li 2001].

[2] The correlation between residuals will be higher for cases of no or only small change of the video scene, where all finer detail structures must again and again be encoded for each enhancement layer frame.

[3] See problem 13.3

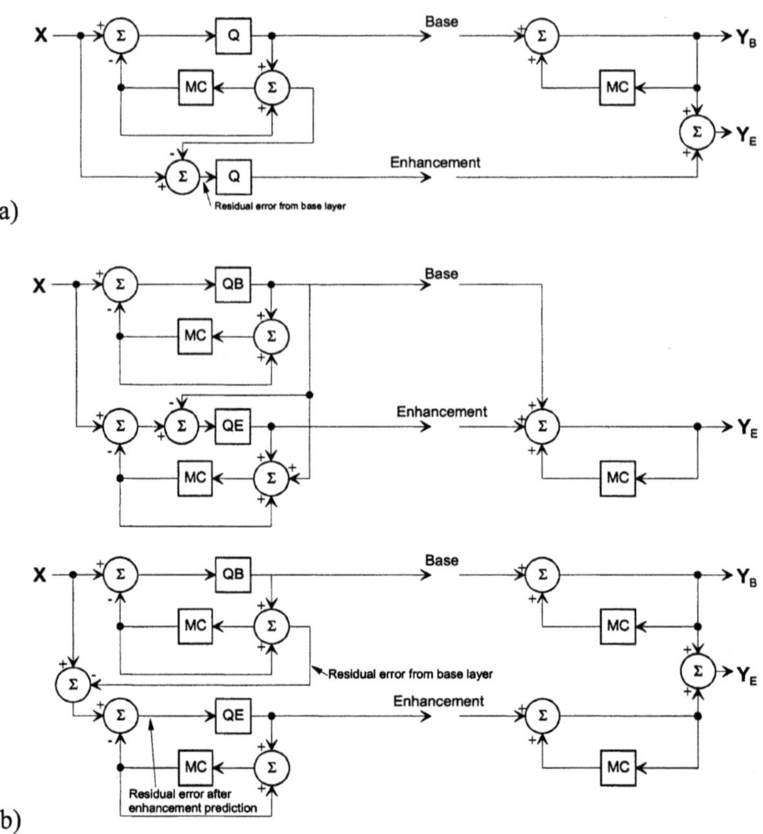

Fig. 13.19. Quantizer scalability in an MC prediction coder
a without prediction in the enhancement layer **b** with prediction in the enhancement layer
(QB : quantizer of base layer; QE : quantizer of enhancement layer)

To enforce better compression performance at the full enhancement rate, prediction at the encoder side could *always* be performed from the highest enhancement-layer information, irrespective of what information the decoder has available (Fig. 13.20a). This is a method of 'SNR scalability' as defined in the MPEG-2 standard. While this approach does not lead to any penalty at the rate of the full enhancement layer, it can penalize the lower layer quality severely due to the *drift*, which is caused by the fact that different predictions are used at encoder and decoder when only partial information is received. In Fig. 13.20b, a slightly different coder structure with identical output is shown. The idea is here to track the drift which would occur in a decoder only receiving the base-layer prediction error EB within the local loop of the encoder. The quantizers in both parts of Fig. 13.20 are arranged in a tandem configuration, where the base layer quantizer (having larger step size) performs re-quantization of the quantized enhancement information. This can e.g. be realized by embedded quantization. The quantized signals EB (base layer) and EE (enhancement layer) then correspond to representations of the prediction error signals using different bit-depths. These are now formally split to trace the signal flows. The prediction PE corresponds to the full enhancement layer, and can be produced as a sum of the base-layer prediction PB and a drift component D, which is generated recursively from the quantization differences between base and enhancement layers.

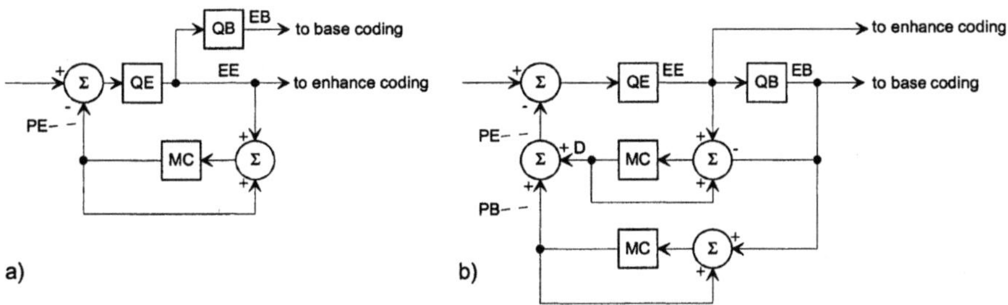

Fig. 13.20. Hybrid SNR scalability structures incurring drift at the base layer
a Direct realization **b** Method for tracking the decoder drift

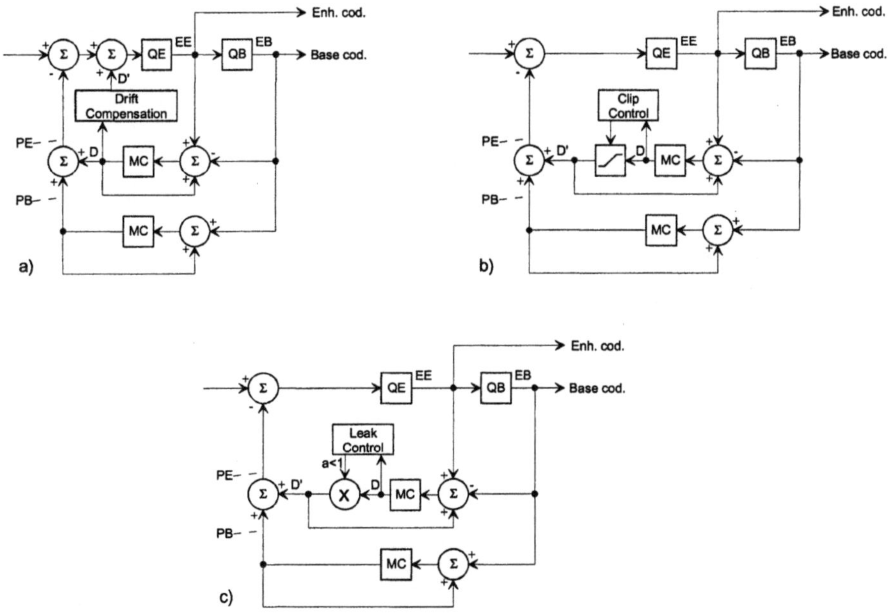

Fig. 13.21. SNR scalability structures for limitation of drift
a Drift compensation **b** Drift clipping **c** Drift leaking

Control of drift. It is obvious that a drift-free base-layer prediction would be possible when only PB was used. Turning off the path of D would be identical to the system of Fig. 13.19a, which is clearly sub-optimum. Alternatively, it can be tried to *control* D, to avoid that drift runs out of control. Due to the recursion, D could eventually become much larger than the difference between base and enhancement layer step sizes, when a number of subsequent differences EE-EB of same sign are accumulating. This will however not be the typical case, but can rather occur in extreme situations of high temporal changes in the video signal. Different derived concepts to keep the drift under control are shown in Fig. 13.21:

- In *drift compensation* (Fig. 13.21a), a value D' is added to the prediction error prior to quantization [MAYER, CRYSANDT, OHM 2002]. For the case D'=−D, no drift would occur, but also no usage of enhancement information would be made. For D'=0, the drift would be fully present. It is assumed here that the decoder side is not aware of the drift compensation made, which means that the usual MC decoder loop could be used without any modifications. The tricky part is to optimize the component D' at the encoder side, such that a good balance between the penalties for the base and enhancement decoding is achieved; this depends on the operational target (whether more advantage should be achieved when operating decoders near base or near enhancement rates). Even though it is a clear advantage to keep the decoder unchanged and leave the drift control as an issue of encoder optimization, this method performs worse than the following solutions, where the drift control is symmetrically run in the decoder loop as well.
- In *drift clipping* (Fig. 13.21b), the drift is dynamically limited if a maximum value D_{max} is reached. A good choice for D_{max} is approximately the base-layer quantizer step size. Strategies are then either to set D'=0 or D'=−D; the latter method would immediately resynchronize the standalone base layer loop to the drift-free case, while the first method imposes less penalty on the quality of the enhancement layer. It has been shown that the first method gives a better SNR performance over a broad range of rates, when compared to the double-loop configuration of Figure 13.19b [MAYER, CRYSANDT, OHM 2002]. Identical drift clipping must be performed at the decoder side. The clipping rules could in principle also be adapted to the signal characteristics, which would then require to transmit clipping mode parameters as side information. $D_{max}=0$ would correspond to the drift-free case, and $D_{max}=\infty$ is the case of unlimited drift.
- In *drift leaking* (Fig. 13.21c), the accumulation of drift is limited by multiplying D'=a·D with a leak coefficient a<1. Here, for a=0 the drift-free case and for a=1 the unlimited-drift case are given. The best selection of the drift coefficient is also dependent on the operational target and sequence characteristics, such that an adaptive setting is appropriate. Usage of a similar method has been studied in [AMON, ILLGNER, PANDEL 2002].

Another method of drift control, which is a hybrid with the double-loop method of Fig. 13.19b and guarantees unconditional base layer stability, is denoted as *Progressive FGS* [WU, LI, ZHANG 2000]. Here, enhancement layer information of a bit-plane representation is partially used for prediction by a sophisticated prediction mechanism which terminates error propagation after a fixed number of frames in cases where not the full enhancement information is available at the decoder.

Extension to multiple layers. The structures shown in Fig. 13.21b actually do not require multiple-loop MC implementations; in fact, implementation can be much simpler than for the method of Fig. 13.19b. This is illustrated in a multiple-bit-plane example of Fig. 13.22, where the base layer is enhanced by two motion-compensated bit planes, each controlled by the clipping method. Additional bit planes can be supplemented as intraframe enhancements without motion compensation, similar to the method of Fig. 13.19a. When clipping is made exactly into the corresponding quantizer step size of the respective bit plane, one-bit frame memories are sufficient for each additional bit plane, assuming separate handling of the sign, which is valid for all bit planes (cf. Fig. 11.8b). Further, when the motion compensation is identical in all bit planes, the entire dotted rectangle in Fig. 13.22 can be implemented by a monolithic frame memory from which the motion compensation is performed by one operation, simultaneously for all bit planes. In the receiver loop, clipping must be made into the range of the highest significant bit plane received, but also here, one MC loop is sufficient.

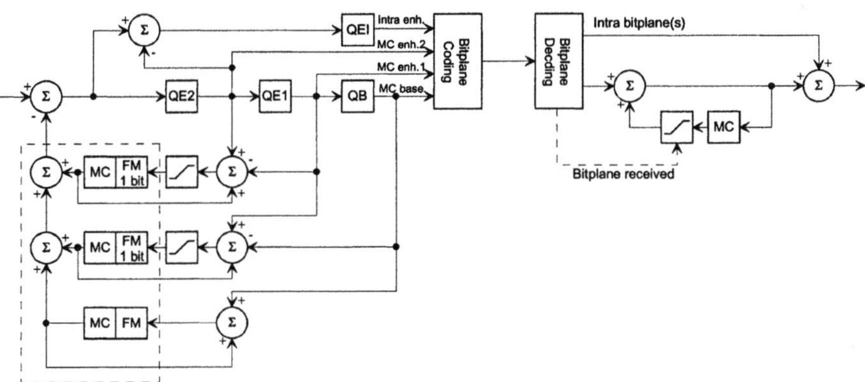

Fig. 13.22. Drift-limiting structure based on the clipping approach with multiple bit-planes and additional intraframe enhancement layer(s)

The intraframe enhancement difference is used for further refinement by quantizing the difference between the MC prediction error and the finest motion-compensated bit plane. The sign bit is always the sign of the unquantized prediction error, and is only encoded when non-zero information is conveyed by one of the bit planes, like in the original embedded quantization shown in Fig. 11.8. Unlike the methods presented in the previous paragraphs, the principle for combined motion compensation of bit planes can no longer be realized if a 2D block transform is embedded into the prediction loop. As embedded quantization is then performed in the transform domain, a 1:1 mapping of the differences between the different layers after the inverse transform would not apply. If however MC prediction is performed in the transform domain (cf. sec. 13.3.2), the method can fully be utilized.

In summary, implementation of SNR (quality) scalability within hybrid video coders will always cost a penalty as compared to single-layer coders, which is mainly due to the recursive structure and the drift problem. The amount of penalty is sequence dependent, but by the different adaptation mechanisms described above it would be possible to optimize the performance for different sequence characteristics under the constraints of expected rate targets. Qualitatively, the rate-distortion behavior of the different methods discussed is sketched in Fig. 13.23, which is in coincidence with measurements that were e.g. reported in [MAYER, CRYSANDT, OHM 2002] and [V.D. SCHAAR, RADHA 2001]. It is assumed here that rates are flexibly scalable with fine granularity between lowest and highest rate points. Any of the schemes can in principle be tuned to a best rate point where the performance of a single-layer coder is either exactly obtained or at least nearly approached. It can be concluded that the methods of drift control establish a reasonable compromise, showing moderate penalty towards both lower and higher rates. When these methods shall cover a sufficient wide range of rates with sufficient compression performance and stability, implementation of multiple loops is necessary, which results in systems of considerable complexity both at the encoder and decoder sides.

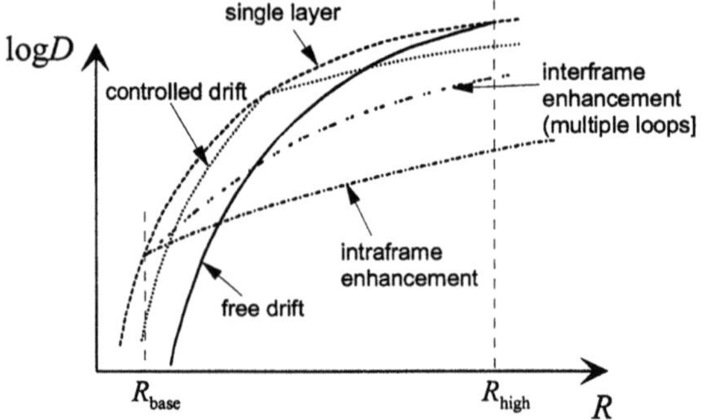

Fig. 13.23. Qualitative rate-distortion behavior of different scalable hybrid coders

Temporal Scalability. Assume that the base layer shall relate to a reconstructed sequence of lower frame rate. If the base information shall only be predicted from itself, these frames must establish a self-contained sub-sequence. This can e.g. be achieved, if the base layer is encoded as a sequence from which frames are skipped, while the enhancement layer supplements these frames for the higher frame rate, which are then predicted from the base-layer frames (Fig. 13.24). In principle, this will not lead to an increased frame rate as compared to a single-layer coder, if e.g. a frame sequence of ..PBBPBBP.. is used, where the B-type frames establish the enhancement layer. In this case, the up-sampling filter can be interpreted as a motion-compensated lowpass prediction or interpolation filter.

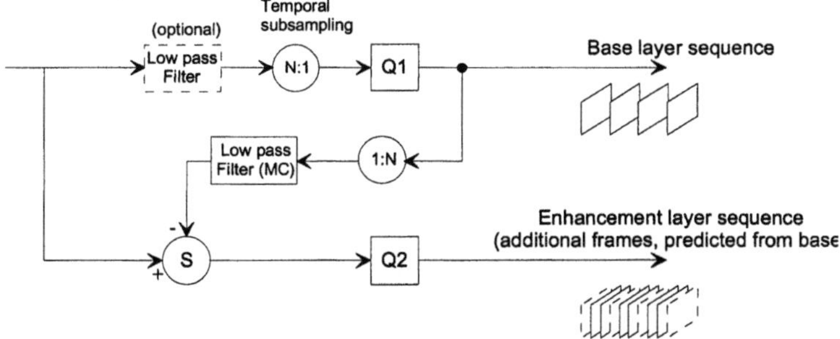

Fig. 13.24. General concept of temporal scalability based on MC difference (prediction) encoding of enhancement layer

Temporal scalability based on B-type frames is the only case where scalable hybrid coding may not be inferior as compared to a single-layer hybrid coder, as scalability does not change in principle the single-layer compression method. This is the case, as no prediction recursion originates from the B-type frames which establish the enhancement layer, such that the drift problem does not apply. If P-type frames shall be used for the enhancement layer, the normal frame-recursive processing sequence must be broken and replaced by a hierarchy of self contained layers. This necessarily leads to constellations where the distance between the frames and their prediction references becomes larger, which will cause a loss in performance (see Fig. 13.25a for an example of temporal scalability over three different frame rates). A similar scalability configuration can also be realized using B-type frames (Fig. 13.25b). Here, unlike in the bi-directional prediction definition introduced earlier, the B-type frames 'B1' of the first enhancement layer are indeed used to predict the frames 'B2' of the second layer; as the number of frames increases by two with each additional level. This shall be denoted as a *B-frame pyramid*[1].

This method improves the compression performance as compared to the conventional constellation, where B-type frames are exclusively predicted from surrounding P-type frames (cf. sec. 13.2.5). Observe that in any case, only frames from lower layers are used to predict a frame in a higher layer. As the B_1 frames are partially used to predict the B_2, drift may occur whenever these are not available to the full resolution or are affected by transmission losses. This drift is however less severe, as it is by guarantee restricted to one or two steps, and is further diminished by the fact that for the subsequent prediction of another B-type frame the drift propagates only by a factor of 0.5[2].

---

[1] Such constellations of B-type frames are indeed supported in the AVC/H.264 standard (cf. sec. 17.4.4)

[2] A similar effect was observed in spatial prediction for the case of averaging prediction filters, see e.g. Fig. 12.19d.

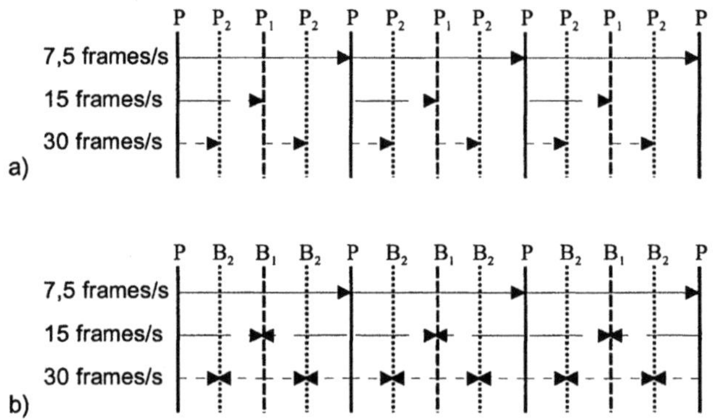

Fig. 13.25. Temporal scalability with two enhancement layers (1/2) supporting three different frame rates **a** based on *P*-type enhancement frames **b** based on *B*-type enhancement frames

Scalability of Spatial Resolution. Wavelet transform methods have evolved as an optimum solution for highly efficient scalable coding of still images. In video coding, incompatibilities of wavelet basis functions with block-based motion compensation exist (cf. sec. 13.3.1), such that block transforms are mostly preferred as spatial transforms in hybrid coding. This also influences the implementation of spatial scalability, where the operations of decimation and interpolation must be applied outside of the prediction loop. Spatial scalability is then realized similar to a differential pyramid, where motion compensated prediction can be applied within each pyramid level in addition to the coarse-to-fine prediction. These methods can also be interpreted as extensions of the principles for quantizer scalability from Fig. 13.19. Systems either without (Fig. 13.26a) or with (Fig. 13.26b) motion compensation in the enhancement layer can be realized. In the latter case, it has to be observed that the values of motion parameters have to be scaled by the same factor by which the frames are scaled spatially. Unlike the cases of quantizer scalability, it would also be useful to estimate different motion vectors for the base and enhancement layers due to the different spatial resolution accuracy.

In the scheme of Fig. 13.26b, the enhancement layer prediction loop is applied to a residual signal, similar to the lower part of Fig. 13.19b. This implies that *always* first the residual from the base layer is computed, and then prediction of the enhancement layer is performed.

There may be cases however, where using *only* the previous frame enhancement layer reconstruction allows better prediction of the actual enhancement frame, without referencing to the current base layer frame. Such a more flexible structure is depicted in Fig. 13.26c. Here, the enhancement layer frame can either be predicted entirely from the up-sampled base layer (which would be the same as in Fig. 13.26a), from the previous enhancement layer reconstruction, or from the mean value of both. In MPEG-4 video coding, this latter case is also denoted as 'bi-directional' prediction, even though it is somewhat different from the original concept of *B*-type frames.

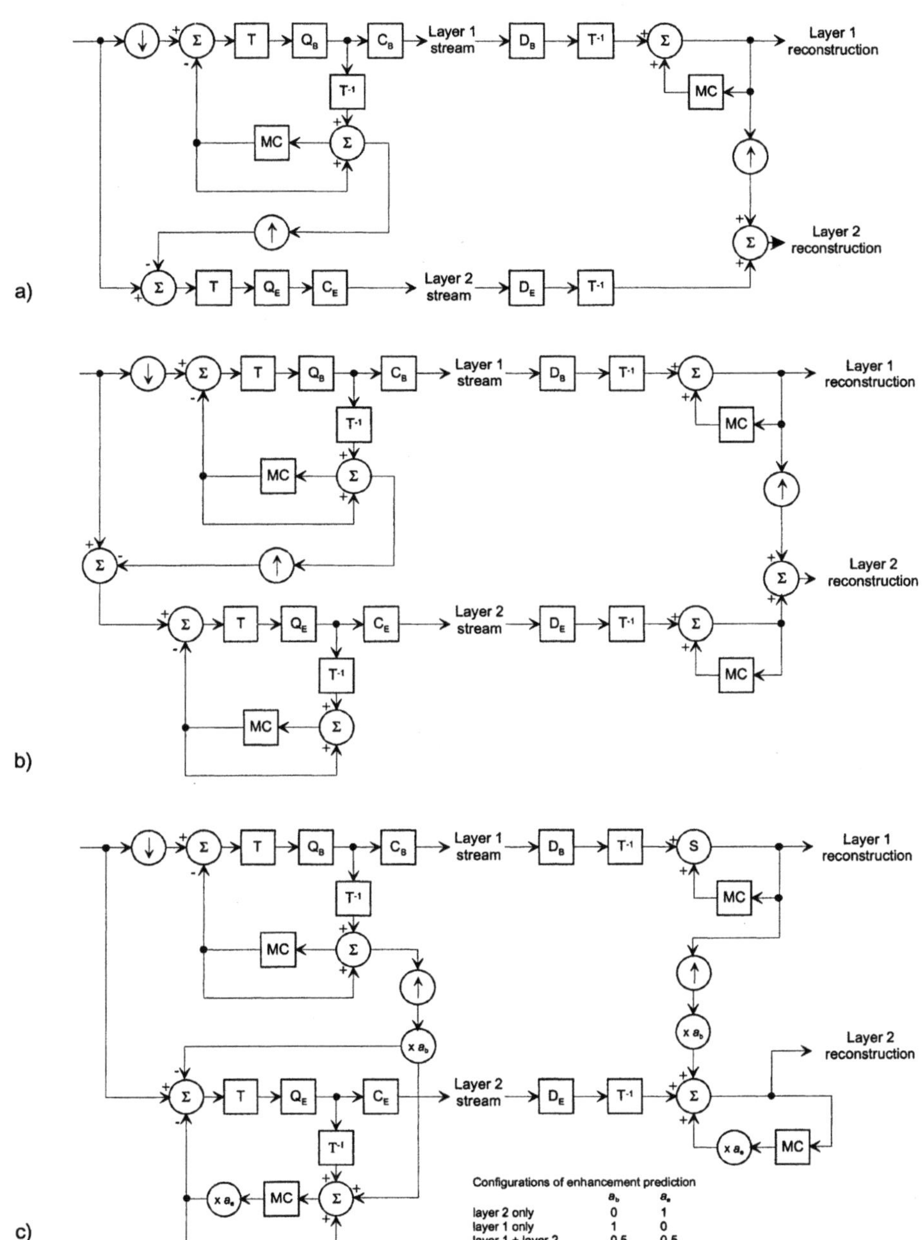

Fig. 13.26. Spatial scalability combined with quantizer scalability in a hybrid coder: **a** Without MC prediction in enhancement layers **b** With double loop, supporting MC prediction in enhancement layer residual **c** With double loop, supporting switchable MC prediction in the enhancement layer [T: Transform; $\downarrow/\uparrow$: decimation/interpolation; for all other acronyms see caption of Fig. 13.19]

13.2.9   Multiple-description Video Coding

Multiple description coding (MDC) relates to source-channel models where the information can reach the receiver by different paths (*diversity*). In wireless communications, *Multiple Input / Multiple Output* (MIMO) sender/channel/receiver constellations could highly improve the system reliability. In wired networks (e.g. for Internet connections), different transmission routes can also establish a *path diversity*. The basic idea of MDC is to transmit a signal by multiple self-contained streams, all of which can be linked such that the quality improves when more streams are actually received. A certain level of quality is nevertheless guaranteed by any of the streams received standalone. Hybrid video coding gives good reasons for application of MDC, as severe data losses can occur by the recursion relationships. The three basic methods which can be utilized are *MDC quantizers*, *MDC transforms* and *MDC temporal prediction* methods. Only examples of 2-stream MDC methods are presented here, but extensions to higher numbers of partial descriptions are possible.

MDC quantizers. The idea behind MDC quantization is to quantize samples by two quantizers with coarser quantization step sizes but different offsets [VAISHAMPAYAN 1993]. If the information from both is combined, more exact conclusions can be drawn about *where* the original position of the value has been in the quantizer cells (e.g. upper half or lower half). This is equivalent to the information conveyed by a single quantizer of finer step sizes. Systematic design of such quantizers can be made, for which an example is given by the matrix in Fig. 13.27. The equivalent finer quantizer has 22 steps, which can be uniquely addressed when both code symbols i_1 and i_2 are received. Each standalone MDC stream conveys only information about 8 coarser steps. For example, if only i_1=2 is received, the actual finer-quantizer value could have been 5,7, or 8. It then depends on the source statistics which will be the best reconstruction; for example, in case of uniformly distributed sources, the average value 6.66 should be the optimum output. The efficiency of the MDC quantizer in terms of source compression (as compared to a single quantizer) relates to the number of cells that are populated in the matrix. In the given example, 2x3=6 bit/sample have to be transmitted in MDC, while only $\log_2 22 \approx 4.45$ bit/sample would be necessary if a monolithic representation was used, which shows an overhead of this MDC scheme of roughly 33%. The extreme cases would be population of the main diagonal only (which means duplicate transmission of information over the two paths) and full population (which does not cost anything in compression efficiency, but neither provides much evidence about the value if only one path is received). Optimum design algorithms are described in [VAISHAMPAYAN 1993]. In general, the reconstruction quality in case of channel losses suffers, if more values off the main diagonal are allowed. The source statistics and the expected channel error statistics must both be observed in the optimization of the design.

index i_1 received by first stream

index i_2 received by second stream	0	1	2	3	4	5	6	7
0	1	3	X	X	X	X	X	X
1	2	4	5	X	X	X	X	X
2	X	6	7	9	X	X	X	X
3	X	X	8	10	11	X	X	X
4	X	X	X	12	13	15	X	X
5	X	X	X	X	14	16	17	X
6	X	X	X	X	X	18	19	21
7	X	X	X	X	X	X	20	22

Fig. 13.27. Schema of an MDC quantizer designed for two separate streams; numbers in bold are the quantization indices of the corresponding single-description quantizer

MDC transforms. The basic idea is to generate transforms which output an uncorrelated set of transform coefficients within each description, where however pair-wise correlation of coefficients will be observed across different descriptions. An extreme case fulfilling this paradigm is made by sending the same description over both channels, which obviously doubles the total rate, but guarantees full reconstruction if only one stream is received. Another simple example would be to generate polyphase (even/odd indexed) sequences of samples from a signal, and apply a transform coder to both of them separately for transmission over the two MDC channels. This means that each second sample will be missing if only one stream is received. Further, the two transforms will be less efficient than one, as the correlation is lower within the separate polyphase sequences[1]. In [WANG ET AL. 2001], optimum *pair-wise decorrelating transforms* are developed for this purpose from the background of a KLT design. Nevertheless, due to the remaining interstream correlation that cannot be exploited, MDC transforms cannot reach the compression efficiency of a transform optimized for a single stream transmission.

MDC prediction strategies. If a transmission error occurs at any position in a video sequence, the remaining frames up to the next *I*-type frame can be affected (Fig. 13.28a). A typical approach to cope with this problem is frame freezing, if the loss is severe. Alternatively, if separate prediction sequences are established for the even- and odd-indexed frames and conveyed by two separate MDC streams [APOSTOLOPOULOS 2000], at least half of the frames can be decoded (which means by half frame rate), if one of the streams is received properly (Fig. 13.28b). This method has also been denoted as *multiple state encoding*. Alternatively, it can be tried in such exceptional situations to use the properly received stream for partial recovery of the other stream, which could e.g. be done by frame interpolation. Like the other two MDC methods described, the prediction strategy of Fig. 13.28b is inferior as compared to Fig. 13.28a.

---

[1] For an analysis of correlation statistics in even/odd polyphase sequences, refer to Problem 4.12.

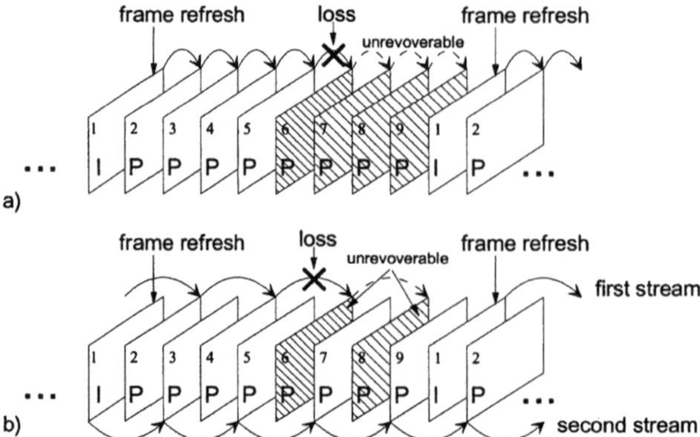

Fig. 13.28. a Error propagation in case of a transmission loss **b** Benefit of MDC prediction, where one of the streams is not affected by the error

Simulcast. Simulcast (for *Simul*taneous broad*cast*) is a specific method of decoding a video or other multimedia signal from separate self-contained streams. Simulcast streams are typically not *symmetric* as the different MDC streams are, and are not intended to *complement each other*. A typical example is a video signal which is encoded in various spatial (frame size) or temporal (frame rate) resolutions or by variable data rates. The typical goal in simulcast is to serve devices of different resolution capability or networks of different transmission capacities, where the exact streaming rate and the terminal properties are not known beforehand when encoding is done. The inefficiency of simulcast as compared to a single stream of highest quality and resolution comes by the additional rate that is needed to encode the lower-quality streams (see also sec. 17.1).

Switching frames. *Switching frames* (*S*-type frames) were recently introduced to allow transition between the prediction processes in multiple streams which contain information about the same video signal [GIROD, FÄRBER, HORN 1997]. The idea is that the *S*-type frame contains the differential information that is necessary to change from the decoder state (which means the state of the frame memory in the recursive loop) of one stream to that of the other stream. This can be useful in the context of multiple-description or simulcast coding, but also in the case of *stream switching* (cf. sec. 15.2), where instantaneous switching to lower or higher data-rate streams shall be allowed by request without insertion of an *I*-type frame. The principle is shown in Fig. 13.29. If the *S*-type frame is inserted at the position of a *P*-type frame from the other stream, it is denoted as *SP* frame; if it contains the upgrade information needed to achieve the state of an *I*-type frame in the frame memory, it is denoted as *SI* frame. The information related to *S*-type frames adds overhead (additional data rate) as compared to a single-stream coder.

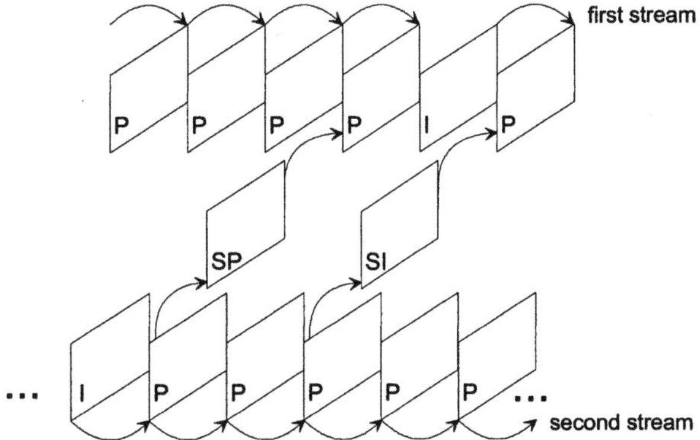

Fig. 13.29. Placement of switching frames in a multiple-stream representation of a video sequence

It can be observed that any of the techniques regarded in this section, as well as most of the scalable coding methods from the previous section[1] perform their purpose by a rate penalty. One general difference between MDC and scalable coding should be notified: MDC assumes to generate a number of *equally important sub-streams*, assuming that equal impact of losses applies to all. Scalable coding is based on a *hierarchical prioritization* of sub-streams, which follows from the rate-distortion relevance of different parts of the full stream. It can be concluded that in case of lossy channels, scalable coding without additional protection mechanisms for the most important sub-streams is useless, while scalable coding augmented by redundancy (unequal error protection, multiple transmission or even MDC coding of important components) can well be interpreted as an efficient MDC scheme. It will depend mainly on the penalty that any of said methods imposes over single-stream transmission, and also on the flexibility it offers for adaptation to any transmission conditions, which is the optimum solution in the context of a given application scenario.

13.2.10 Optimization of Hybrid Encoders

Standards for video compression typically define only the bitstream syntax and semantics and the operation of decoders (cf. sec. 17.1). Hence, optimization of encoders offers freedom for improvement in quality, being fully compatible with existing standards-compliant decoding devices. One of the key issues is the optimization of motion compensation, which must mainly be regarded by the effect on other coding elements, like the rate resulting by entropy coding of motion vectors and transform coefficients. The optimum selection of motion vectors is typically *rate dependent*, and may not necessarily relate to the true motion.

---

[1] except for certain configurations of temporal scalability

The first group of methods to improve the compression performance is *rate distortion optimization*. In principle, a selected operational point is targeted, such that for a given rate budget the distortion gets as low as possible. Numerous configurations could in principle be selected, as different components contribute to the overall rate. For example, the rate to be spent for encoding of transform coefficients in a hybrid coder relies on the selection of motion parameters and on mode switching (e.g. selection of intraframe coding modes, of uni-directional or bi-directional prediction). The typical solution for rate/distortion problems is made by Lagrangian optimization (cf. sec. 11.3), which however assumes that the rate-distortion behavior changes smoothly around the given operational target. In particular mode selection is however a highly nonlinear optimization problem, for which no consistent solution exists except for running multiple coding configurations for selection of the best result, which is highly complex.

The second group of optimization strategies relates to the *recursive dependencies* in the MC prediction loop. Not all picture types to be encoded (*I*, *P*, *B*) are similarly relevant by their contribution to the overall distortion. It is usually a good strategy to give more relevance to an *I* picture in the beginning of a GOP, as this is recursively used to predict many other pictures; if strongly distorted, the prediction of subsequent pictures would be worse, which would increase the overall rate. On the other hand, this depends on the amount of change in the sequence; with fast changes, the influence of encoding quality to subsequent frames typically decays faster. If *B* pictures are not used to predict any other, they can typically be quantized more coarsely. Reconstructed P frames are used for prediction of other frames, and should be encoded with higher quality as well, but this may be traded versus the length of the dependency chain. Methods to use such dependency chains in the context of rate-distortion optimization bear high potential in hybrid encoder improvement, but are not well developed yet due to their enormous complexity. Manageable solutions with restriction to selected cases are described in [ORTEGA, RAMCHANDRAN, VETTERLI 1994][RAMCHANDRAN, ORTEGA, VETTERLI 1994] and [BEERMANN, WIEN, OHM 2002]. In principle, the error feedback of the prediction loop (12.21) must be fully analyzed to decide which is the optimum way to jointly minimize rate and distortion in combination with frame look-ahead decisions. However even *within a frame*, sophisticated video encoders use multiple feedbacks such as spatial prediction from block to block, prediction of motion vectors from block to block, adaptation mechanisms in entropy coding. A brute-force strategy to approximate best rate-distortion behavior under constraint of frame dependencies is *multi-pass encoding*, where one frame is encoded at different rates e.g. by different quantizer settings, and only after the rate/distortion implications on subsequent frames have been tested, the final decision is made which of the pre-encoded streams shall be used.

The third group of optimizations is related to *rate control*. In applications of constant-rate transmission, or for best transmission policies in packet-switching networks, it is necessary to employ rate regulation mechanisms, which are usually based on variations of quantizer step size (see sec. 15.1.2). The interdependency between rate control and rate-distortion optimization mechanisms, in particular best strategies to hold variations in the quantizer step size as low as possible, are of high importance in this context.

The fourth group of optimization strategies relates to *pre-processing*. For a given rate and encoding of a given sequence, it must be decided which is the best spatio-temporal resolution by which the sequence should be encoded. This highly influences the perceptual reconstruction quality. In principle, methods of frame skipping, spatial sub-sampling to smaller image format, or resolution-reduction preprocessing by lowpass filters can be applied. Further, semantic criteria can be taken into account by identification of *regions of interest* which should be encoded by higher quality[1]. This can likewise be done by blurring (resolution reduction) of less relevant regions by pre-processing. It is also possible to take even more advantage of perceptual weighting functions than possible in still image coding; for example, human observers are hardly able to detect detailed structures in the case of fast motion, which means that effects such as motion blur could be mimicked by pre-processing as well, which would on the other hand directly improve the performance of motion-compensated prediction.

All methods described so far are purely methods for maximization of *compression performance*. To constrain the optimization by aspects of source coding without taking into account the properties of transmission channels is a blind-eyed attitude, as the overhead necessary for protection may become significant, even more when the source is compressed to the limits. This problem will further be discussed in sec. 15.2.

13.3 MC Prediction Coding using the Wavelet Transform

The advantages of using wavelets for still image compression have been well recognized, especially the high coding efficiency in conjunction with the inherent scalability they provide is attractive. Even though different trials have been made to extend wavelet technology to video coding, the more complicated nature of spatio-temporal signals, in particular the need to apply motion compensation for high coding efficiency, have prevented usage of wavelet-based methods so far. Only recently, wavelet-based video coding has become an interesting trend. This and the following section will give an analysis on the problems to be solved for efficient implementation of wavelet video coding, and give insight to some of the newer techniques.

---

[1] A typical example is detection of faces. It can also simply be assumed that the most relevant area is found in the center of the scene, which at least is true when the camera tracks this content.

13.3.1   Wavelet Transform in the Prediction Loop

The direct way to combine wavelet coding with motion compensation would be to replace the block transform by a wavelet transform in the encoding and prediction loop (Fig. 13.30). Unfortunately, this is penalized by the following problems:

1. By using block-based motion compensation, high spatial frequencies occur at the block boundaries when adjacent block motion vectors diverge. As the wavelet transform has a block-overlapping effect (lengths of filter impulse responses are longer than block sizes in transform coding), a worse effect can be expected here as compared to the case of a block transform which coincides with positions of MC blocks[1]; in particular, this can cause severe ringing artifacts when single wavelet coefficients are discarded.
2. Block-wise intraframe mode switching introduces amplitude discontinuities in the 2D signal which is fed into the transform. This will again cause problems in combination with basis functions which are applied beyond block boundaries, with similar effects as described under 1.
3. Replacement of the transform in the hybrid prediction loop does not solve the drift problem, which is the main reason preventing implementation of efficient scalable coding, as was discussed above.
4. It appears to be straightforward to integrate the spatial resolution pyramid of the wavelet tree with the down-sampling steps that are necessary in the spatial scalable hybrid coding methods of Fig. 13.26. However, the scaling band of wavelet-transformed frames is not by guarantee alias-free, as certain constraints apply to the filters by which a critically sampled perfect reconstruction pyramid representation is guaranteed. Motion compensated prediction of those frames may be less efficient than for the case of specifically optimized low-pass down-sampling filters.

Solutions have been developed to the first two problems. In particular, wavelet transforms perform well in the hybrid prediction loop when block-based motion compensation is replaced by grid-based or overlapping block MC (sec. 13.2.6); blending functions reduce the problem of amplitude discontinuities in case of intra-mode switching (sec. 16.1) [MARTUCCI ET AL. 1997][HEISING ET AL. 2001]. In Fig. 13.30, this is expressed by the capability to adapt a factor α which allows smooth transition between intraframe coding (α=0) and full MC prediction (α=1)[2]. Performance gain over block-transform based hybrid coding is reported, however the complexity is also significantly increased. No flexible scalability solution comparable to the properties of still image wavelet coders is provided. Hence, the additional value of such methods over traditional hybrid coding appears rather limited.

---

[1] It could even be argued that one of the reasons to use a transform of block size 4x4 in the AVC standard is the better harmony with the 4x4 MC blocks that must be supported at minimum.

[2] This block diagram is highly simplified. Other elements of state of the art hybrid codecs (cf. Fig. 13.5) can be integrated.

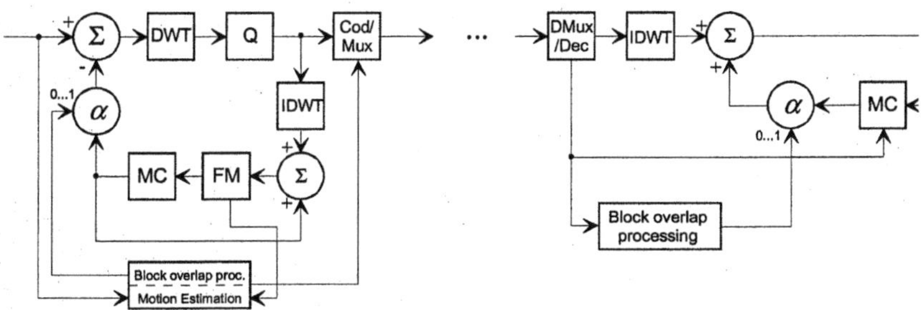

Fig. 13.30. MC prediction codec using DWT in place of block transform

13.3.2 Frequency Coding with In-band Motion Compensation

Motion-compensated frame prediction as presented so far is mostly applied to video frames in the original resolution by which the signal shall be decoded. In the case of spatial scalability encoding of Fig. 13.26, motion compensation is applied to different resolution levels in parallel, but the entire information to be encoded is *overcomplete*, as it is based on a difference pyramid representation. Frequency coding, in particular by using the wavelet transform, provides inherent mechanisms for spatially scalable representation without being overcomplete.

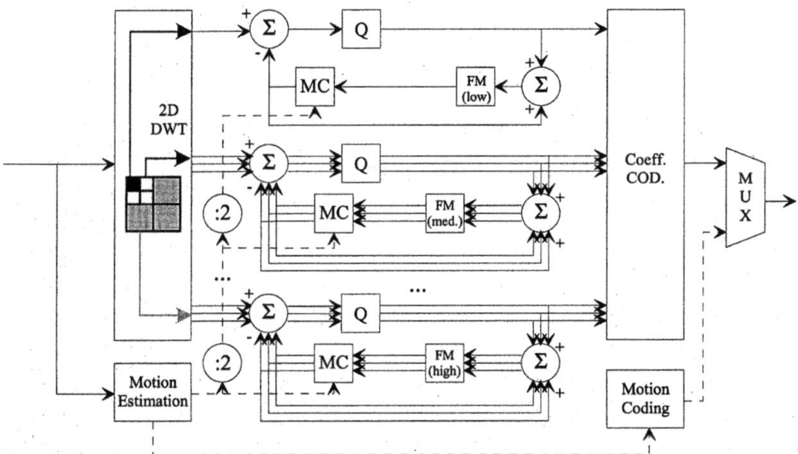

Fig. 13.31. MC prediction scheme with in-band motion compensation, inverting the sequence of MC prediction and spatial wavelet transform

An extension to video coding would be desirable, for which a possible configuration is shown in Fig. 13.31. Here, the frequency transform over the spatial coordinates is performed *first* (not in the prediction loop), and then motion-compensated temporal DPCM encoding is applied to each spatial sub-image of wavelet coefficients. Different methods have been proposed both for motion compensation in the wavelet transform domain and in the DCT domain in [BOSVELD ET AL. 1992][YANG, RAMCHANDRAN 1997][SHEN, DELP 1999][BENZLER 2000].

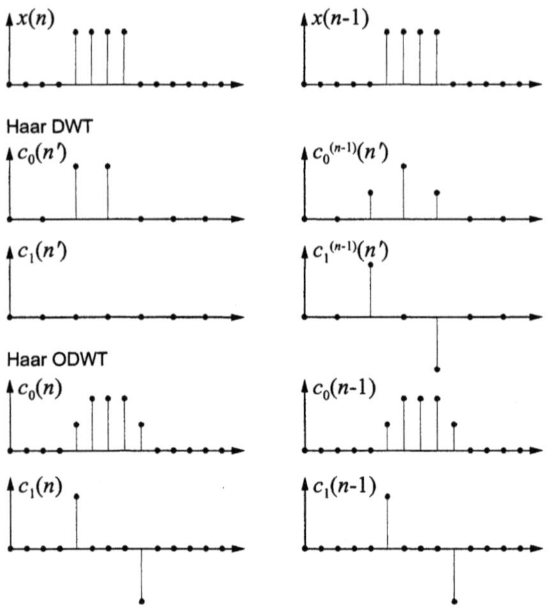

Fig. 13.32. Shift variance of the Haar Wavelet transform. Right signal shifted by one sample to the right, lowpass and highpass coefficients in Haar DWT and Haar ODWT

None of these methods has shown to be highly efficient in compression performance. The reason behind is the *shift variance* of the transforms used for frequency decomposition, which prevents a good prediction. In particular, the alias components inherent in the sub-sampled frequency coefficients are highly variant and unpredictable, depending on the relative position between signal and transform sub-sampling phase. This can best be explained for the example of the Haar basis wavelet transform. The top part of Fig. 13.32 shows two identical signals, where the right one is shifted by one sample, a typical case of motion shift. The center part of the Figure shows the down-sampled representations of low and high frequency band signals after a Haar DWT applied to the original signals. It is obvious that the low-frequency component $c_0^{(n-1)}(n')$ of the shifted signal may be perfectly predictable by linear interpolation from the unshifted $c_0(n')$, but not so vice versa. For the high-frequency signals, the prediction will fail completely in any case[1].

---

[1] If longer filters are used, the signals $c_0^{(n-1)}(n')$ and $c_0(n')$ are becoming more similar and better predictable from each other. On contrary, no highpass signals will be consistently predictable from the other sub-sampling position. This can be interpreted in different ways: Shift of the sub-sampling phase by one sample is critical, as it relates to a phase shift of

A possible solution to this problem is the application of an *overcomplete* wavelet transform (ODWT), which is in principle a representation of the lowpass and highpass filtered signals without sub-sampling. As obvious from the lower part of Fig. 13.32, full prediction capability is retained in this case also for the highpass component. In principle, it is sufficient to retain the overcomplete representation for the highpass component, which is similar to pyramid methods (cf. sec. 4.4.6).

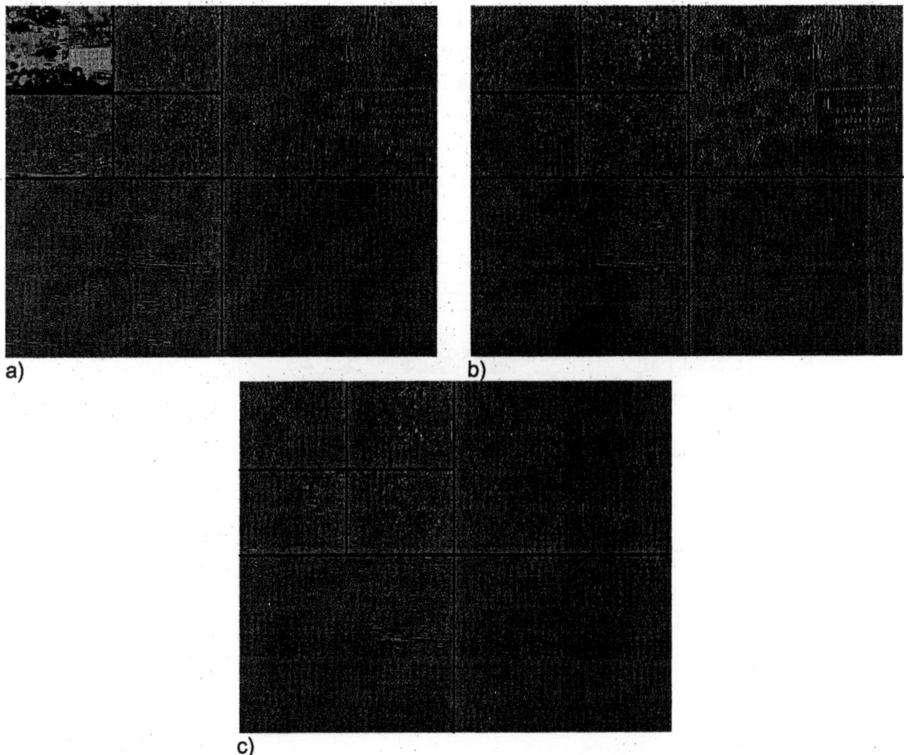

Fig. 13.33. **a** Wavelet decomposition of a video frame **b** Motion-compensated in-band prediction **c** Motion-compensated in-band prediction using ODWT

Unlike pyramid schemes however, in the overcomplete wavelet representation only the *critically sampled* signal must be transmitted [ZACIU ET AL. 1996]. When it is needed in the prediction loop of the decoder, the overcomplete representation can be generated locally from the critically-sampled representation. An implementation of this overcomplete scheme in an in-band motion-compensated wavelet video

---

almost half a period (π) for the case of a signal which bears components near half sampling frequency; the sub-sampling of highpass signals involves frequency inversion, which does not allow straightforward alias-free interpolation of the signal towards the other sampling position; finally, the percentage of alias energy is typically higher for the highpass component, at least if a signal model with spectral decay towards higher frequencies is assumed.

codec is shown in Fig. 13.34. For the prediction of each spatial resolution scale, only information from this scale and from lower scales is needed, such that spatial scalability is fully guaranteed without drift. Fig. 13.33 shows the 2D wavelet decomposition of a video frame, and the related in-band prediction errors without and with overcomplete transform computation.

The ODWT representation as needed for MC prediction is generated locally by application of the inverse DWT. As the motion shift is known, it can even be decided a priori which phases of the ODWT are needed, such that the complexity can be reduced [AUWERA ET AL. 2002]. Likewise, it is possible to combine the steps of IDWT and ODWT into just one interpolation filtering operation [LI, KEROFSKY 2001].

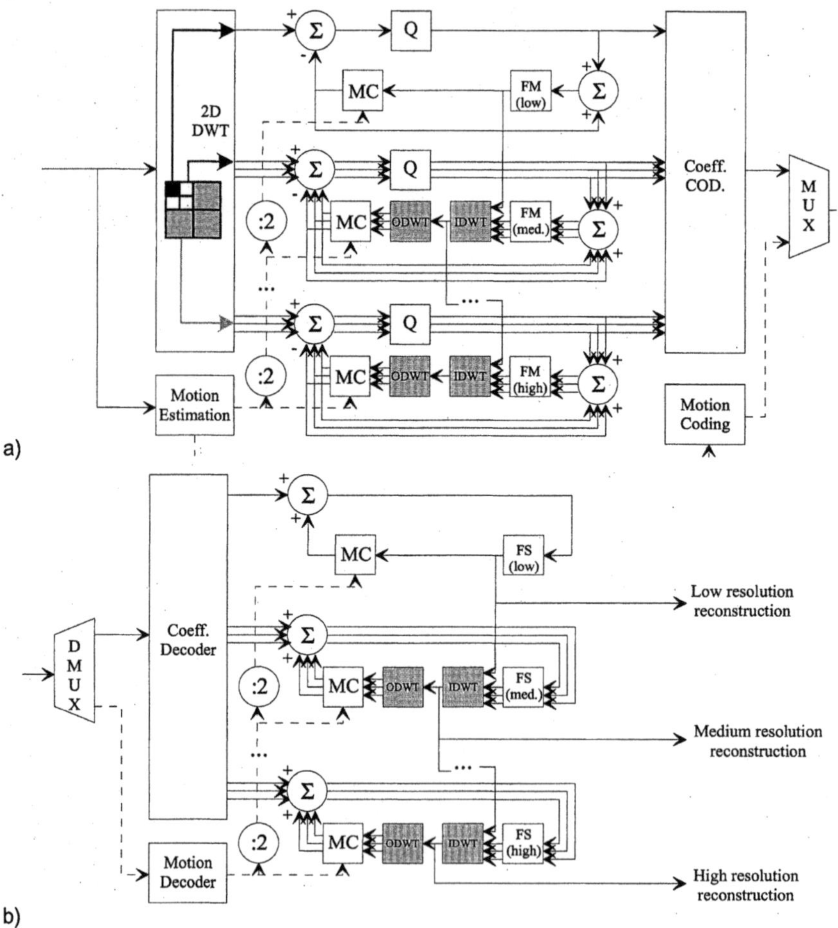

Fig. 13.34. Video codec using an overcomplete wavelet transform in the prediction loop: **a** Encoder **b** Decoder   [DWT: Discrete Wavelet Transform; IDWT: Inverse DWT; ODWT: Overcomplete DWT; MC: Motion Compensation; Q: Quantization; FM: Frame memory]

In the structure of Fig. 13.32, the motion estimation is performed using the frames of original (full) resolution, once for the entire encoding process. For each level of the wavelet transform, it is then necessary to scale the motion vectors by a factor of two, as the pixel shift effected by the same physical motion has to be scaled down simultaneously with the frame resolution. Any motion compensation scheme, including block matching[1], can be applied. If the motion is estimated for the original signal resolution however, *no scaling of the motion information* is possible, which causes an unnecessary overhead of motion information for the lower spatial resolutions. As an example, assume that a block matching is used and the motion estimation is performed at full resolution, using a block size of 16x16 pixels and full-pixel accuracy. The same vectors after one spatial scaling level will have an equivalent meaning of half-pixel accuracy and a block size of 8x8, after another spatial scaling it will be quarter-pixel accuracy and block size 4x4 etc. The natural increase of accuracy of motion vectors – both in spatial resolution of the motion vector field and by the sub-pixel accuracy – effects that the percentage of bits to be spent for motion information would largely increase when the image size is reduced. This is in fact not desirable and most likely the high resolution of the motion vector field will not be needed at all; in fact, there is no guarantee that the motion vectors estimated for the full resolution are suitable at all to predict the spatially down-sampled wavelet bands.

Alternatively, a separate motion estimation refinement can be performed within each of the spatial levels, which can then be implemented similar to *hierarchical motion estimation* (cf. sec. 7.6.3). This means that the motion estimation result from the next-lower resolution is used as an initial estimate for the finer resolution. A modified encoder structure, implementing hierarchical motion estimation, is shown in Fig. 13.35. Now, it is possible to keep both the spatial resolution (block size) and the accuracy (sub-pixel resolution) constant over all levels, such that the amount of motion information will scale down consistently with the frame size. Scalable encoding of motion information can be implemented in this configuration, typically by differential encoding of the next finer resolution with reference to the coarser resolution. Consequently, it is useful to start the estimation process at the lowest spatial resolution level, and use the result in a scaled-up version also as an initial estimate for the next finer level. This can be done straightforwardly, as down-scaling of the signal is performed beforehand in the wavelet pyramid anyway.

Results have been reported which show that the ODWT based in-band MC prediction scheme performs similar to the full resolution MC prediction approach [ANDREOPOULOS ET AL. 2003] under comparable conditions of motion compensation accuracy. This means that at least spatial scalability can be realized now without significant loss in compression efficiency. On the other hand, the basic structure is

---

[1] In fact, block matching will not cause any problems here with the transform bases, as the *original* signal is transformed, and not the prediction error residual. Also, switching to intraframe mode is performed in the wavelet bands when necessary, which can also be done on a block-by-block basis without an interference to wavelet basis functions.

still based on a recursive decoder, such that the problem of drift inherent in MC prediction schemes is not resolved. This means that also the ODWT scheme cannot straightforwardly be combined with efficient quantizer or SNR scalability. On one hand, the problem is simplified, because no transform resides within the loop, which allows e.g. simple combination with the drift compensation methods discussed in sec. 13.2.8. On the other hand, much more flexible and wider-range quantizer scalability will be needed.

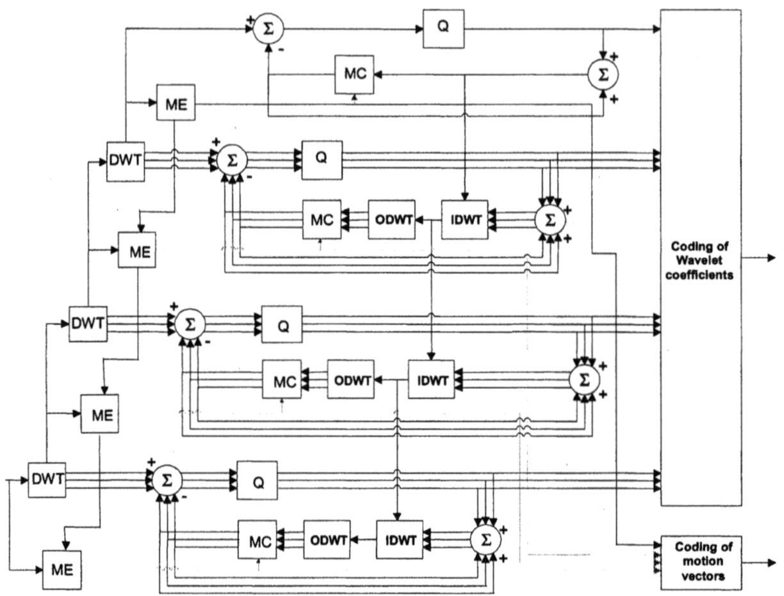

Fig. 13.35. ODWT-based video encoder with hierarchical motion estimation and scalable motion vector encoding

When decoding of higher resolution in spatial scalability shall be made, also the lower-resolution (scaling and wavelet) bands must be represented more accurately than for the purpose of low-resolution reconstruction – coding errors in the lower resolution bands are becoming more visible as they spread over a larger number of pixels in full resolution. To circumvent this problem, the multiple-loop approaches described in sec. 13.2.8 can be employed. As however already multiple loops exist for the different frequency bands, and each of these must be extended by multiple loops for SNR scalability, the entire structure could become rather complex. The necessity to implement one separate loop (at least on the encoder side) for each rate point that shall be free of drift is a fundamental restriction caused by the recursive nature of MC prediction decoders. It can be concluded in general that the drift problem prohibits the realization of efficient solutions for scalability. A different approach is taken in the spatio-temporal frequency coding methods described in the subsequent section, where the recursive processing along the temporal axis is replaced by non-recursive, motion-compensated subband filters.

13.4 Spatio-temporal Frequency Coding with MC

In Section 13.1.2, spatio-temporal (denoted as *3D* or *2D+t*) frequency coding methods without motion compensation were introduced. Application of motion compensation (MC) so far was only regarded for prediction schemes. There is indeed no constraint restricting usage of MC to the case of prediction, as it can rather be regarded as a method to align a filtering operation which is then performed along the temporal axis on an assumed motion trajectory [KRONANDER 1989]. In the case of MC prediction, the filters are in principle LPC analysis and synthesis filters, while in cases of transform or wavelet coding, transform basis functions are subject to MC alignment[1].

The predictor filter (4.98) describes the shift of coordinates in the previous frame (z_3^{-1}) by k pixels horizontally (z_1^k) and l pixels vertically (z_2^l). The decorrelating operation is prediction error filtering $1-H(z_1,z_2,z_3)$, which is by principle a temporal filter operation aligned with the motion trajectory. *Motion-compensated temporal filtering* (MCTF) can be interpreted very similar. As transform and subband/wavelet methods are also described by linear filter operations, they can probably likewise be applied along the motion trajectory. As an example, regard the basis functions of the Haar transform, which can be expressed by the filter pair (4.180). Motion-compensated versions of these filters, applying motion shift $\mathbf{k}=[\ k\ \ l\ ]^{\mathrm{T}}$ in the previous frame, have a z-transfer function

$$H_0(z) = \frac{\sqrt{2}}{2} + \frac{\sqrt{2}}{2} \cdot z_1^{\ k} \cdot z_2^{\ l} \cdot z_3^{-1}$$

$$H_1(z) = -\frac{\sqrt{2}}{2} + \frac{\sqrt{2}}{2} \cdot z_1^{\ k} \cdot z_2^{\ l} \cdot z_3^{-1}. \tag{13.28}$$

As this Haar basis can be interpreted as a block transform of temporal block length $W=2$, the effect of the transform can be regarded as a decomposition of a pair of frames $\mathbf{X}(o)$ and $\mathbf{X}(o+1)$ (o even) into a 'lowpass' frame $\mathbf{C}_0(o')$ and a 'highpass' frame $\mathbf{C}_1(o')$, where $o'=o/2$:

$$c_0(m,n,o') = \frac{\sqrt{2}}{2} \cdot x(m,n,2\cdot o'+1) + \frac{\sqrt{2}}{2} \cdot x(m+k,n+l,2\cdot o')$$

$$c_1(m,n,o') = -\frac{\sqrt{2}}{2} \cdot x(m,n,2\cdot o'+1) + \frac{\sqrt{2}}{2} \cdot x(m+k,n+l,2\cdot o'). \tag{13.29}$$

If k and l are integer numbers which are constant over the entire frame[2], a perfectly invertible analysis/synthesis relationship exists. Periodic extension of frames could

---

[1] Motion-compensated frame interpolation (sec. 16.4) has indeed lots of commonalities with the latter case, where usually filters of finite response are applied along the motion trajectory

[2] e.g. in case of global translational motion

resolve the missing correspondences where vectors point beyond the frame boundaries; alternatively, intraframe switching as shown in Fig. 13.4a can be applied in such areas.

If however motion vectors are *spatially varying*, isolated areas may be present, which are not member of any uniquely connected motion trajectory. Upon unique trajectories (Fig. 13.36a), all pixels can ideally be reconstructed by the respective synthesis filtering, which must include inverse MC mapping. In case of inhomogeneous motion vector fields (Fig. 13.36b), as they e.g. occur when objects move differently, motion trajectories can diverge, such that certain pixels or entire areas may not be members of any motion trajectory; these are related to newly uncovered areas, and these positions as *unconnected*. Another case happens when motion trajectories converge or merge, which e.g. happens when areas are being covered. Here, certain coordinate references are *multiple connected*[1]. In the latter case, information would be contained duplicate in the transform coefficients of (13.29), while in the former case, information would be missing and reconstruction would be impossible.

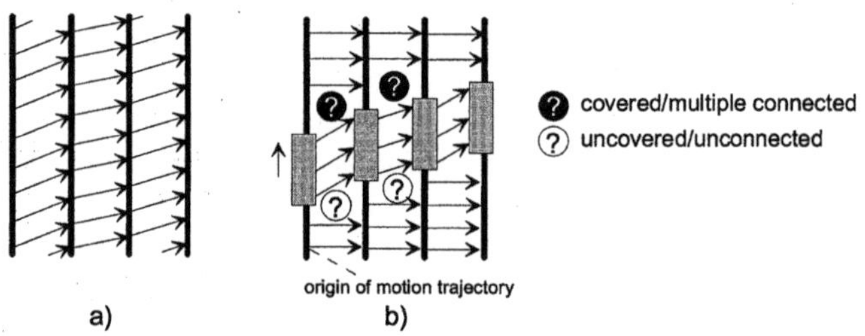

Fig. 13.36. Forward motion trajectories in case of **a** homogeneous **b** inhomogeneous motion vector fields

13.4.1 Temporal-axis Haar Filters with MC

A solution to the problem of unreferenced pixels in case of Haar filters can be made as follows by re-defining the coordinate references with regard to the motion shifts [OHM 1991]. Regard a motion-compensated non-orthonormal Haar filter pair with z transform

$$H_0(z) = \frac{1}{2}\left(1 + z_1^{\tilde{k}} \cdot z_2^{\tilde{l}} \cdot z_3^{-1}\right)$$
$$H_1(z) = -z_1^{k} \cdot z_2^{l} + z_3^{-1}.$$

(13.30)

---

[1] Note that these relationships are closely tied to the definitions of *forward motion* and *backward motion* as discussed in sec. 7.6.6. Fig. 13.36 shows the case of forward-directed motion. For the case of backward motion definitions, the semantic association would be 'becoming covered' for the unconnected case and 'newly uncovered' for the multiple-connected case.

The effect of this modification shall again be interpreted by transforming pairs of even/odd indexed frames A and B into each one 'lowpass' frame L and one 'highpass' frame H, such that

$$\underbrace{c_0(m,n,o')}_{L(m,n)} = \frac{1}{2}\cdot\underbrace{x(m,n,2\cdot o'+1)}_{B(m,n)}+\frac{1}{2}\cdot\underbrace{x(m+\tilde{k},n+\tilde{l},2\cdot o')}_{A(m+\tilde{k},n+\tilde{l})}$$

$$\underbrace{c_1(m,n,o')}_{H(m,n)} = \underbrace{x(m,n,2\cdot o')}_{A(m,n)}-\underbrace{x(m+k,n+l,2\cdot o'+1)}_{B(m+k,n+l)}. \tag{13.31}$$

Obviously here, for the case of temporal-axis Haar filters, the L frame is the motion-compensated *average*, and the H frame is the motion-compensated *difference* between the two frames. The motion vector $[k,l]^{\mathrm{T}}$ shall characterize the forward motion originating from frame A towards frame B, while $[\tilde{k},\tilde{l}]^{\mathrm{T}}$ describes the backward motion from B towards A[1]. If a unique motion trajectory exists, both motion vectors cannot be independent of each other, as they shall connect corresponding pixels. If e.g. estimation of $[k,l]^{\mathrm{T}}$ is performed at all positions (m_A,n_A) in frame A, parameters $[\tilde{k},\tilde{l}]^{\mathrm{T}}$ can uniquely be defined, whenever corresponding positions (m_B,n_B) are neither unconnected nor multiple connected:

$$\begin{aligned}\tilde{k}(m_B,n_B) &= -k(m_A,n_A)\\ \tilde{l}(m_B,n_B) &= -l(m_A,n_A)\end{aligned} \quad \text{with} \quad \begin{cases}m_B = m_A + k(m_A,n_A)\\ n_B = n_A + l(m_A,n_A).\end{cases} \tag{13.32}$$

In case of multiple-connected mappings, it is still possible to determine a value for $[\tilde{k},\tilde{l}]^{\mathrm{T}}$ by setting *selection rules*, e.g. to use the smaller of two or more vectors targeting one pixel. All remaining positions (m_B,n_B) then belong to unconnected areas in B, where parameters $[\tilde{k},\tilde{l}]^{\mathrm{T}}$ cannot be determined from (13.32) or by selection rules. For these latter positions, original values from B are filled into L. For the not-selected multiple connected positions, i.e. those which violate (13.32) and were also rejected by the selection rules, it is possible to fill a motion-compensated prediction error in H, as will be described in detail in the following paragraphs. In total, this procedure does not produce any overhead or spatial discontinuity in the motion-compensated L and H subband frames, except for possible effects which are caused by discontinuous or erroneous motion vector fields.

In contrast to (13.29), a non-orthonormal pair of Haar filters is used. The motivation behind is avoidance of spatial discontinuities in amplitude between the temporally-filtered (connected) and substituted (unconnected or multiple-connected) pixels in L and H. The analysis equations for the normal case of unique connections are, recalling (13.31):

---

[1] In the sequel, we will generally assume that the coordinate system of H is related to the positions of A, while the coordinate system of L relates to positions of B. These relationship definitions are arbitrary and can be made vice versa without any restriction.

$$L(m,n) = 0.5 \cdot B(m,n) + 0.5 \cdot A\big(m + \tilde{k}(m,n), n + \tilde{l}(m,n)\big)$$
$$H(m,n) = A(m,n) - B\big(m + k(m,n), n + l(m,n)\big). \tag{13.33}$$

The information about remaining 'multiple connected' pixels from frame A is integrated as prediction difference into the highpass frame, while the unconnected pixels from frame B are embedded into the lowpass frame (see Fig. 13.37):

$$L(m,n) = B(m,n) \qquad \text{if 'unconnected'}$$
$$H(m,n) = A(m,n) - \hat{A}(m,n) \quad \text{if 'multiple connected'.} \tag{13.34}$$

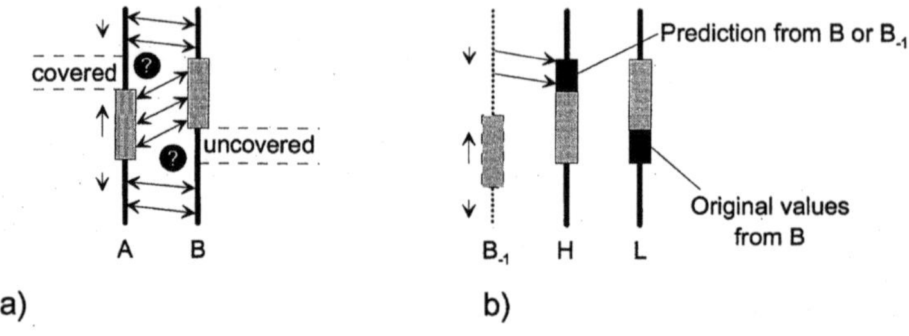

Fig. 13.37. a Covered and uncovered areas in case of frame pairs **b** Substitution of predictive coded areas into the 'highpass' frame, original frame areas into the 'lowpass' frame

The prediction reference $\hat{A}(m,n)$ can in principle refer to the (subsequent) frame B or to the preceding frame B_{-1}. As vectors $[k,l]^{\mathrm{T}}$ are defined for any position[1], irrespective of multiple connections occurring, it is straightforward to select between the following two modes, where the mode switching information must be conveyed to the decoder,

$$\hat{A}(m,n) = B\big(m + k(m,n), n + l(m,n)\big) \qquad \text{'backward mode'}$$
$$\hat{A}(m,n) = B_{-1}\big(m - k(m,n), n - l(m,n)\big) \qquad \text{'forward mode'.} \tag{13.35}$$

All operations defined in (13.33)-(13.35) are then fully invertible. For normally-connected pixels the synthesis equations are:

$$\tilde{A}(m,n) = L\big(m + k(m,n), n + l(m,n)\big) + 0.5H(m,n)$$
$$\tilde{B}(m,n) = L(m,n) - 0.5H\big(m + \tilde{k}(m,n), n + \tilde{l}(m,n)\big), \tag{13.36}$$

while for the exceptional cases of unconnected or multiple-connected pixels

$$\tilde{B}(m,n) = L(m,n) \qquad \text{if 'unconnected'}$$
$$\tilde{A}(m,n) = \hat{A}(m,n) + H(m,n) \quad \text{if 'multiple connected'.} \tag{13.37}$$

---

[1] This must not necessarily mean that individual vectors must be defined differently for any position; in fact, block-based definition of motion vector fields is often used in MCTF systems.

The second of these operations can only be performed after the first is executed over the allowable motion compensation range of the respective pixel position. Perfect reconstruction is strictly possible, when full-pixel accuracy of motion compensation is implemented. Motion compensation using sub-pixel motion shift will lead to lossy reconstruction, as then sub-pixel position interpolations would be necessary in analysis *and* synthesis steps, which could never be perfect unless an ideal interpolator was used. Nevertheless, it was shown that *arbitrary methods of motion compensation* can be used and that the reconstruction error can be made reasonably small when interpolators of high quality are used to compute the sub-pixel positions [OHM 1994A].

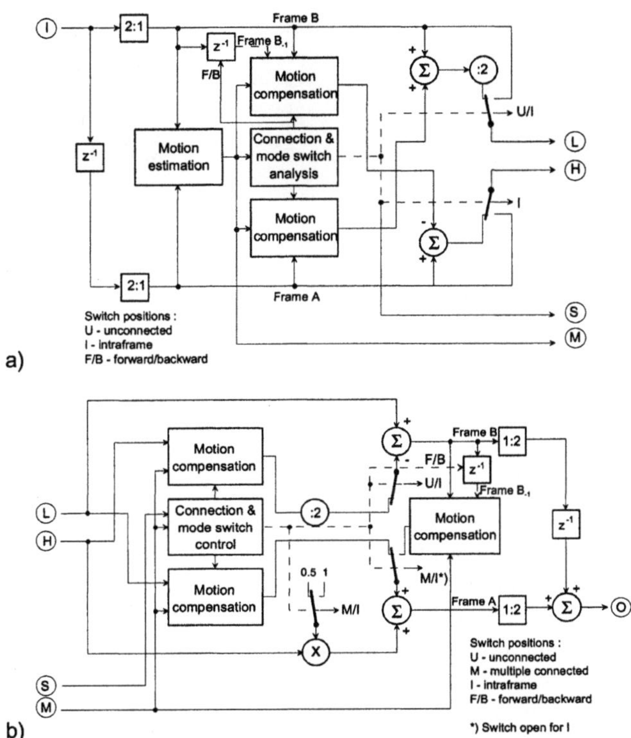

Fig. 13.38. Motion-compensated Haar filter kernel, **a** analysis **b** synthesis

Fig. 13.38a/b shows block schematics of the analysis and synthesis stages described by (13.33)-(13.37). These are similar polyphase-filter implementations as used in Fig. 13.4a; in addition to the aspects discussed so far, also intra mode switching is again integrated.

Several of these building blocks can now be combined in a cascaded structure, which realizes a temporal-axis wavelet filter tree. This is in principle identical to the method shown in Fig. 13.4b, but the motion information must be conveyed in addition. If the same rules are applied at the encoder and decoder sides for the definition of unconnected and multiple connected pixels, this information can completely be gained from the motion parameters. Intra and forward/backward mode switching information must additionally be encoded as side information. Motion parameters could be estimated independently within the single levels of the temporal wavelet tree, must be encoded and conveyed to the decoder for synthesis. It is more consistent however to use an estimation and encoding which sets a relationship between the different levels of the wavelet tree; this aspect will further be discussed in sec. 13.4.3. The frames at the end nodes of the temporal wavelet tree are finally decomposed by a spatial wavelet transform (as in Fig. 13.2b), such that the final result is a 3D motion-compensated spatio-temporal wavelet transform.

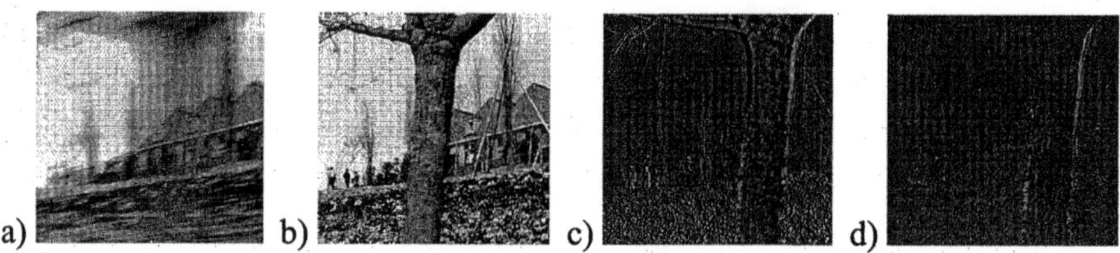

Fig. 13.39. Frames resulting by temporal-axis wavelet tree over T=4 levels: **a** Lowpass frame (LLLL) without motion compensation and **b** with motion compensation **c** Highpass frame (H) without motion compensation and **d** with motion compensation

Fig. 13.39 shows frames processed by the motion-compensated temporal axis wavelet filtering, employing four levels of temporal-axis transform, which are compared against the result of processing without motion compensation as introduced in sec. 13.1.2. It is obvious that without motion compensation, the low-frequency frame LLLL is becoming heavily blurred, while the high-frequency frame H carries a lot of detail information yet. In principle, the highpass frame shows the same behavior as a prediction error frame without motion compensation, where it was already shown to be disadvantageous in terms of compression efficiency when no MC is applied[1]. In the motion-compensated case, the lowpass frame LLLL contains all relevant image information; it looks like an original frame, but indeed is an average over 16 frames here; such a frame can well be used as a member of a temporally sub-sampled sequence which can be displayed at lower frame rate. It is obvious that only spatio-temporal wavelet coding *with MC* is useful for the purpose of temporal scalability.

---

[1] Observe that the information from the highpass frame is weighted by a factor of 0.5 during synthesis, see (13.36). Hence, the penalty by using no MC is in fact less than in prediction without MC; theoretically, temporal-axis transforms without MC could never become worse than intraframe coding, which in fact is not the case for temporal prediction without MC, as was shown earlier.

13.4.2   Temporal-axis Lifting Filters for arbitrary MC

Motion-compensated Lifting filters. Any pair of biorthogonal wavelet filters can be implemented in a lifting structure similar as shown in Fig. 4.40. This lifting scheme can also be used to give a different interpretation of the motion-compensated transform between a pair of frames A and B of a video sequence, which shall be transformed into one lowpass frame L and one highpass frame H. Herein, the frames A and B are again interpreted as the even and odd polyphase components of the temporal-axis transform. Assume that A^* and B^* establish a pair of pixels which is unambiguously 'connected'. This means that unique, invertible correspondences exist by $B^*=B(m,n) \Leftrightarrow A^*=A(m+\tilde{k},n+\tilde{l})$, respectively $B^*=B(m+k,n+l) \Leftrightarrow A^*=A(m,n)$; A^* and B^* are still related by integer motion shift $\mathbf{k}=[k,l]^{\mathrm{T}}$, where typically $\tilde{\mathbf{k}}=-\mathbf{k}$. The lifting structure inherently enforces the spatial coordinate relationships as defined in the previous section, where positions in B shall be mapped into identical positions of the lowpass frame L, while positions in A shall map into the coordinate reference positions of highpass frame H. With pixels connected by unique integer shift, this can be interpreted as the pair of non-orthonormal Haar filters in lifting implementation (4.212), where the prediction and update filters are in fact now 3D filters integrating the motion shift, such that $P(\mathbf{z}) = -z_1^{\,k} z_2^{\,l}$ and $U(\mathbf{z}) = \tfrac{1}{2} z_1^{\,\tilde{k}} z_2^{\,\tilde{l}}$,

$$H(m,n) = A(m,n) - B(m+k,n+l)$$

$$L(m,n) = B(m,n) + \frac{1}{2} H(m+\tilde{k},n+\tilde{l}) = \frac{1}{2}\Big[B(m,n) + A(m+\tilde{k},n+\tilde{l}) \Big]. \qquad (13.38)$$

The equivalence with (13.33) is obvious. The consequence of re-defining the motion-compensated Haar filters by a lifting structure are however more fundamental, as the lifting structure is able to guarantee perfect reconstruction in any case, when the same prediction and update filters are used during the reverse operation of synthesis. This means that it will now be possible to release the restriction of full-pixel shifts and gain *perfect reconstruction for arbitrary motion vector fields*. The interpretation by lifting filters was first made in [PESQUET, BOTTREAU 2001],[LUO ET AL. 2001] and [SECKER, TAUBMAN 2001]. A special case was previously developed in [OHM, RÜMMLER 1997], where it was shown that the polyphase kernels of 1D or 2D biorthogonal filter pairs can be used as perfect-reconstructing interpolation filters in the case of a half-pixel accurate motion compensation with temporal-axis Haar filters; the gain achievable by this method in an operational MCTF coding system was first reported in [CHOI, WOODS 1999].

Assume that in addition to the pixel-wise shift (k,l), a sub-pixel displacement shall be compensated, such that the actual shift will be $k+\alpha$ horizontally, and $l+\beta$ in vertical direction, $0 \leq (\alpha,\beta) < 1$. For simplified explanation, the sub-pixel shift is for the first example applied in vertical direction only, where a Haar lifting filter with linear interpolation of sub-pixel positions in the prediction and update steps is used. Further, the forward an backward motion shall be assumed to match the typical case of correct motion flow, $\tilde{\mathbf{k}} = -\mathbf{k}$. A flow diagram is shown in Fig. 13.40, where it is assumed that the full-pixel shift component has already been considered by aligning the positions of corresponding pixel pairs $A*$ and $B*$, which makes the diagram easier to read. For the case $\beta{=}0$, the resulting lowpass and highpass samples are identical to (13.38). For $\beta{>}0$, H samples are generated in the prediction step using two-tap linear interpolation (5.34), which means that two vertically-adjacent pixels from frame B are weighted by factors β and $(1\text{-}\beta)$ to gain the prediction of the A pixel. In the update step, the L pixels are generated from two adjacent H pixels, weighted by $\beta/2$ and $(1\text{-}\beta)/2$. The prediction and update filters can then be described as

$$P(\mathbf{z}) = -z_1^{k} z_2^{l} \cdot A(z_2) \quad ; \quad U(\mathbf{z}) = \tfrac{1}{2} z_1^{-k} z_2^{-l} \cdot A(z_2^{-1})$$

$$\text{with} \quad A(z_2) = (1 - \beta) + \beta z_2 \tag{13.39}$$

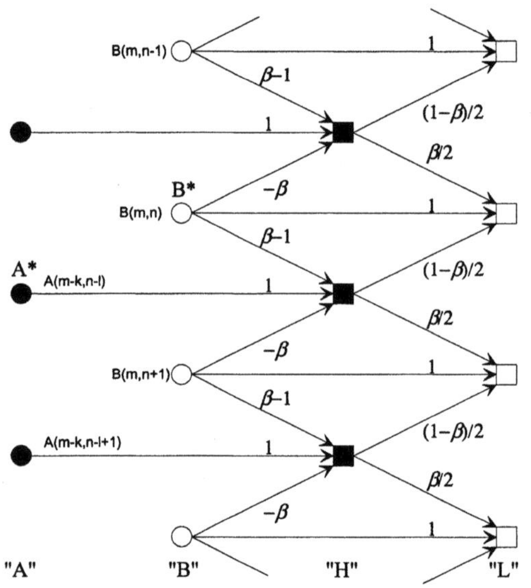

Fig. 13.40. Signal flow diagram of a motion-compensated lifting filter with sub-pixel shift in one (vertical) dimension; correspondence $A*/B* : B*$ is shifted by k pixels horizontally and $l{+}\beta$ pixels vertically relative to $A*$

The operations to generate the H and L frames are

$$H(m,n) = A(m,n) - (1 - \beta)B(m+k,n+l) - \beta B(m+k,n+l+1)$$

$$L(m,n) = B(m,n) + \frac{\beta}{2} H(m-k,n-l-1) + \frac{(1-\beta)}{2} H(m-k,n-l). \tag{13.40}$$

By inversion of this principle in synthesis, perfect reconstruction is guaranteed, where first the entire frame B must be reconstructed:

$$\tilde{B}(m,n) = L(m,n) - \frac{\beta}{2} H(m-k,n-l-1) - \frac{(1-\beta)}{2} H(m-k,n-l)$$
$$\tilde{A}(m,n) = H(m,n) + (1-\beta)\tilde{B}(m+k,n+l) + \beta\tilde{B}(m+k,n+l+1). \tag{13.41}$$

The lowpass analysis in (13.40) can further be re-written as

$$L(m,n) = \frac{1}{2}\big[\beta A(m-k,n-l-1) + (1-\beta)A(m-k,n-l)$$
$$-\beta(1-\beta)B(m,n-1) + (1+2\beta-2\beta^2)B(m,n) - \beta(1-\beta)B(m,n+1)\big]. \tag{13.42}$$

This can be interpreted such that L is the mean from a linearly-interpolated value of A and a vertically contrast-enhancement filtered value of B^1, which gives a picture of original sharpness if the motion match fits. The result of the prediction step H is identical to frame difference computation in MC prediction, with sub-pixel values generated by linear interpolation. The effect in synthesis will however be different from the prediction synthesis, as the values from H spread equally over both frames A and B. For $\beta=0$, (13.40) equals (13.38). For **k=0** and $\beta=1/2$, frames A and B can be interpreted as if they were fields of an interlaced frame of double height. Then, (13.40) is equivalent to the 5/3 filter (4.194) applied verti-cally to this 'big frame'^{2}. The concept can now be extended to 2D, where

$$P(\mathbf{z}) = -z_1^{k}z_2^{l} \cdot A(z_1,z_2) \quad ; \quad U(\mathbf{z}) = \tfrac{1}{2} z_1^{-k}z_2^{-l} \cdot A(z_1^{-1},z_2^{-1})$$
$$\text{with} \quad A(z_1,z_2) = \big[(1-\alpha)+\alpha z_1\big] \cdot \big[(1-\beta)+\beta z_2\big] \tag{13.43}$$

The corresponding operations are

$$H(m,n) = A(m,n) - (1-\alpha)(1-\beta)B(m+k,n+l) - (1-\alpha)\beta B(m+k,n+l+1)$$
$$-\alpha(1-\beta)B(m+k+1,n+l) - \alpha\beta B(m+k+1,n+l+1)$$
$$L(m,n) = B(m,n) + \frac{\alpha\beta}{2} H(m-k-1,n-l-1) + \frac{\alpha(1-\beta)}{2} H(m-k-1,n-l) \tag{13.44}$$
$$+ \frac{(1-\alpha)\beta}{2} H(m-k,n-l-1) + \frac{(1-\alpha)(1-\beta)}{2} H(m-k,n-l)$$

Perfect-synthesis inversion is straightforward. Re-writing the lowpass decomposi-tion similar to (13.42) gives

---

^{1} For contrast enhancement filters, see (5.18).

^{2} This is exactly the solution proposed in [OHM, RÜMMLER 1997] for perfect reconstruction MCTF with half-pixel accuracy.

$$
\begin{aligned}
L(m,n) = \frac{1}{2}\big[& \alpha\beta A(m-k-1,n-l-1) + \alpha(1-\beta)A(m-k-1,n-l) \\
& +(1-\alpha)\beta A(m-k,n-l-1) + (1-\alpha)(1-\beta)A(m-k,n-l) \\
& -(1-\alpha)(1-\beta)\alpha\beta B(m-1,n-1) - (\beta-\beta^2)(1-2\alpha+2\alpha^2)B(m,n-1) \\
& -(1-\alpha)(1-\beta)\alpha\beta B(m+1,n-1) - (\alpha-\alpha^2)(1-2\beta+2\beta^2)B(m-1,n) \\
& +(2-(1-2\alpha+2\alpha^2)(1-2\beta+2\beta^2))B(m,n) - (\alpha-\alpha^2)(1-2\beta+2\beta^2)B(m+1,n) \\
& -(1-\alpha)(1-\beta)\alpha\beta B(m-1,n+1) - (\beta-\beta^2)(1-2\alpha+2\alpha^2)B(m,n+1) \\
& -(1-\alpha)(1-\beta)\alpha\beta B(m+1,n+1)\big].
\end{aligned}
\tag{13.45}
$$

Now, bilinear interpolations (5.35) considering horizontal/vertical sub-pixel shift factors $\alpha\equiv h$ and $\beta\equiv v$ are performed in the prediction and update steps. Again, for $\alpha=\beta=0$, (13.44) equals (13.38); further, it is interesting to note that for the case $\alpha=\beta=0.5$ and $\mathbf{k=0}$, A and B complement in a 'big frame' which has a quincunx sampling grid, and is decomposed into the rectangular sampled L and H frames. Here, (13.44) is a $45°$-rotated version of (4.243) which was indeed designed for the purpose of quincunx-to-rectangular sub-sampling systems.

The bilinear interpolation as applied in (13.44) is attractive due to its low complexity. The operations can be implemented by only 4 multiplications/sample. Nevertheless, as in MC prediction, usage of higher-quality interpolation filters is attractive. This could be realized by integration of higher-order interpolators into the filters $P(\mathbf{z})$ and $U(\mathbf{z})$ of the Haar lifting structure, as proposed in [Secker, Taubman 2001]. Alternatively, higher-quality interpolation can also be realized by additional lifting and update steps in the structure discussed above, where in each of the steps a simple bilinear interpolation with sub-pixel shift factors α and β must be applied. Often, symmetric bi-orthogonal filter pairs are factored into lifting steps, which then have symmetric prediction and update filters as $P_i(z)=p_i z^{-1}+1+p_i z^1$ and $U_i(z)=u_i z^{-1}+1+u_i z^1$. One coefficient value p_i or u_i will completely define each lifting step (cf. Fig. 4.41c). Fig. 13.41 shows an example, where the equivalent β weighting mechanism as in Fig. 13.40 is used in a lifting structure with each 2 prediction and update steps. For $\beta=0.5$, exactly the lifting factorization of the 9/7 filter (4.195) results. Remark that for $\beta=0.5$, Fig. 13.40 relates to the bi-orthogonal 5/3 filter which is known to have a sub-optimum effect in compaction of frequency bands due to its relatively large deviation from orthogonality. On the other hand, the 9/7 filter is known to approximate orthogonality fairly well (cf. sec. 12.4.2), such that it can directly be concluded that higher-quality interpolation will have a beneficial effect to improve data compression. The optimum choice for a combination between MCTF and spatial interpolation filters, in particular providing consistent filter norms which would be relevant for encoding, is not yet well investigated.

The interpretation of sampling position shifts by introduction of branch-weight factors in the lifting flow diagrams in principle allows re-definition of the sub-pixel shifts at each sampling position, though still guaranteeing perfect reconstruction. Further development of such systems would enable a generic wavelet decomposition of irregularly-sampled signals. The scope of applications is much wider, and could include the implementation of arbitrary geometric mapping as part of the wavelet synthesis filter. This would for example also allow arbitrary (non-dyadic) sampling resolutions of the synthesis output.

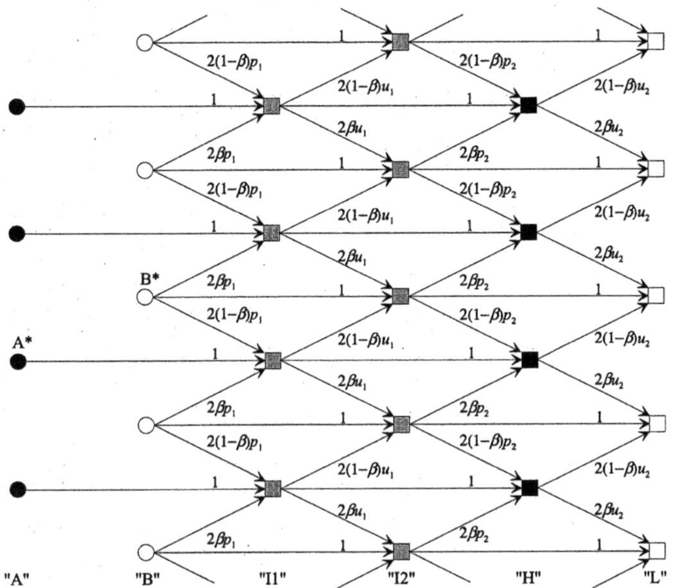

Fig. 13.41. Flow diagram of a motion-compensated lifting filter with sub-pixel shifts integrated into a 9/7 lifting filter

Non-homogeneous motion vector fields. So far, it was assumed that the integer motion shift $[k,l]$ shall be constant at least over the area which is analyzed. A basic concept to cope with the problem of discontinuities in the motion vector field, by defining reasonable substitutions at the positions of unconnected and multiple-connected pixels, was given in (13.34). It will now be shown how this integrates seamlessly with the lifting filters. Fig. 13.42a shows the case of an 'unconnected' pixel. Notice that the references A^*/B^* and also the sub-pixel shifts β change at the motion boundary (discontinuity). In this case, the backward motion vector field diverges from the view of a position in A, such that one pixel from B (highlighted by '#') stays isolated.

The multiple-connected case is shown in Fig. 13.42b. Here, the motion vector field converges from the view of a position in A. Again, the references A^*/B^* change at the motion boundary, and one pixel from A (indicated by '#') stays isolated, which means that it is not used in the update filtering step. Pixels from B around the motion boundary will become multiple-connected[1].

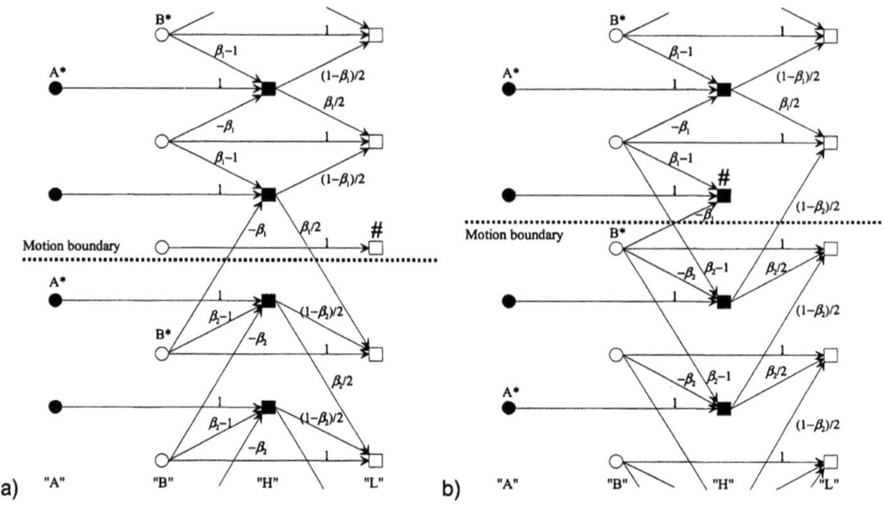

Fig. 13.42. Lifting structure with cases of **a** unconnected **b** multiple-connected pixels

Mode switching. As mentioned in sec. 13.1.2 and 13.4.1, it is not necessary to enforce all possible references between pixels A^* and B^* from frames A and B. It can even be advantageous to embed more 'unconnected' pixels in the L frame when no unique match is found between the two frames. Typical reliability criteria for motion estimation, in the simplest case unexpectedly high frame differences, can be used as decision criteria. This would establish an equivalent case as switching to intraframe coding in the case of MC prediction. A possible method for separate intraframe coding of pixels from frames A and B is shown in Fig. 13.43a. It would not be reasonable here to encode the frame A by prediction from frame B, as indeed the normal connection has been discarded due to the missing correspondence[2]. Hence, intraframe coding has to be applied in both frames A and B. This will however mean that a signal component which is not of zero mean will have to be inserted in the 'highpass' frame. To avoid an amplitude discontinuity, which might be problematic for the subsequent spatial wavelet decomposition, a smooth

---

[1] Remark that in principle also the pixel above the hash-marked element could be defined as 'isolated' logically; by definition here, the pixel with the smallest motion vector was set to the 'connected' status for the case of multiple connections. Also, usage of sub-pixel weight factors is not fully consistent across motion boundaries, such that other variants would be possible.

[2] Indeed, a *plane prediction*, where pixels in A are predicted from the average of a larger number of pixels in B, might be useful in some cases, and can be implemented as a specific mode.

transition can be realized by gradually adapting the weights of the prediction step between the areas of different type [HANKE, RUSERT, OHM 2003]. 'Transition' pixels which are indeed a mixture between intraframe coding and the normal prediction step of the lifting filter, are marked by 'T'. It is reasonable to adapt weights symmetrically in the lifting flow, such that complementary transition pixels will also be inserted in the lowpass frame between the normally filtered and the intraframe coded pixels. Another approach of mode switching is the usage of forward motion vectors instead of backward vectors for the multiple-connected pixels in frame A, cf. (13.35).

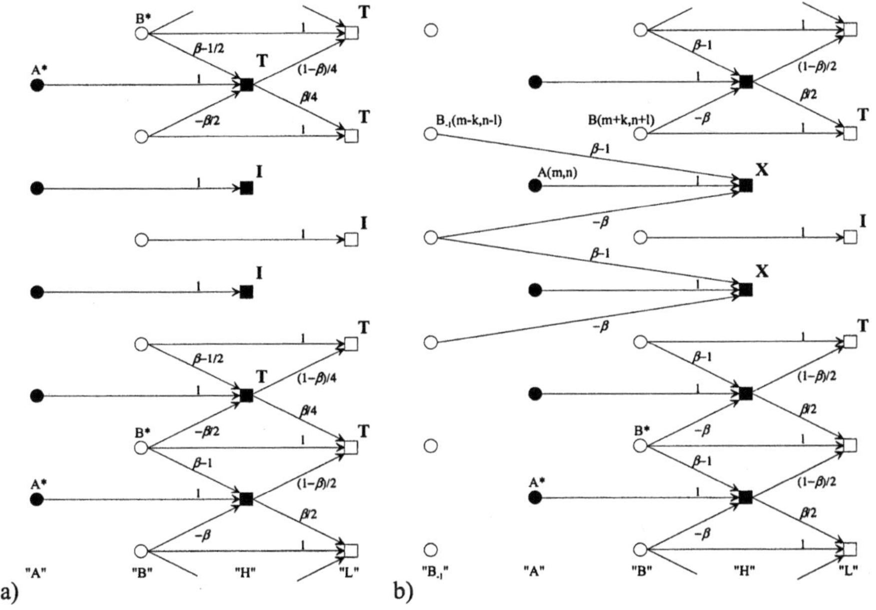

Fig. 13.43. Embedding of intra-coded (**a**) and backward predicted pixels (**b**) in the lifting structure [**I** intra **T** transition **X** backward predicted]

As multiple connected pixels can semantically be categorized as 'becoming covered' (see introduction to sec. 13.4), the backward switching is rather the preferable mode. Such a method can also be applied alternatively for those positions in A where intraframe coding is applied at the related positions in B (see Fig. 13.43b). Both intraframe and prediction direction mode switching must be signaled by side information, e.g. on a block-by-block basis.

Lifting Filters extended over the temporal axis. One single analysis level of the wavelet tree, again by view of a pair-wise frame decomposition, is illustrated in Fig. 13.44a, giving yet another interpretation of the motion-compensated Haar filters. As it was shown above, the motion-compensated prediction step in the lifting filter structure (resulting in the H frame) is almost identical to conventional motion-compensated prediction. However, at any transform level, no further recursion is performed evolving from positions of predicted frames A/H, such that the motion-compensated wavelet scheme is naturally non-recursive, and it is not necessary to reconstruct frames at the encoder side. The interpolation mechanisms included in the lifting filter structures from previous sections are now illustrated as simple 'MC' blocks. In fact, for the purpose of sub-pixel accurate motion compensation, arbitrary spatial interpolation filters can be used here.

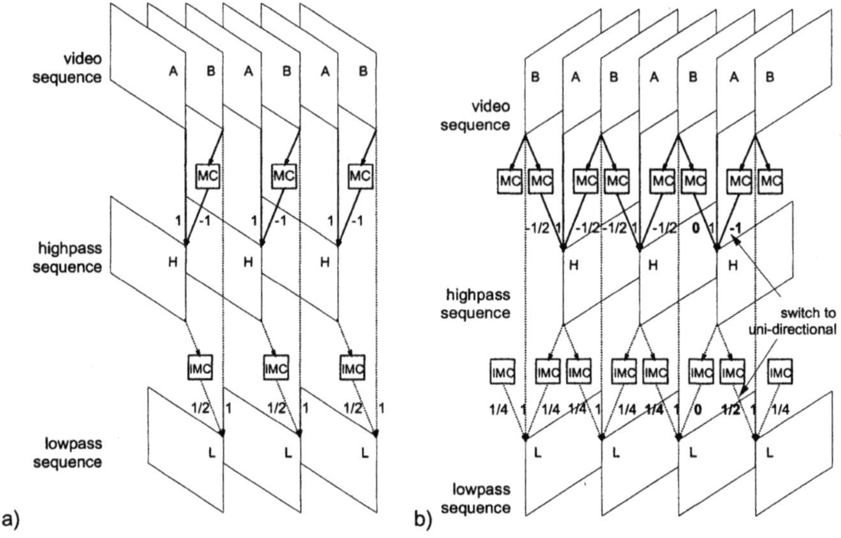

Fig. 13.44. MC wavelet transformation step $A/B{\rightarrow}H/L$ in lifting structure **a** Haar filter with uni-directional prediction and update **b** 5/3 filter with bi-directional prediction and update

An example how MC operates is given in Fig. 13.45 for a block-based motion compensation scheme. Here, the block positions are fixed with regard to the coordinates of frame A, which are identical to the coordinates of frame H. Hence, it is possible to predict any pixel, regardless of overlapping motion vector fields. The second step of the lifting filter is the update step, which generates the L frame. If this shall be performed in a consistent way in combination with MC, any pixel being mapped from frame B into frame H during the prediction step must be projected back to its original position in the L frame during the update step. This appears reasonable, as the L positions are defined by the same coordinate reference as for pixels in B. Hence, the MC applied to H, which is used to generate L during the update step, should as close as possible be the inverse of the MC (IMC) that was used during the prediction step. If this is not observed, ghosting or other artifacts could appear in the lowpass frame, and it would not be usable for temporal scalability. As typical in block-based MC, the blocks are fixed in A and H but floating in B and L, which has two consequences (see Fig. 13.45b) :

- Pixels which remain blank after IMC are the 'unconnected' pixels. As then the information mapped from H into L is zero[1], original values from B are automatically filled in.
- For duplicate mappings by IMC, a rule must be defined which one is valid; this is the case of "multiple connections".

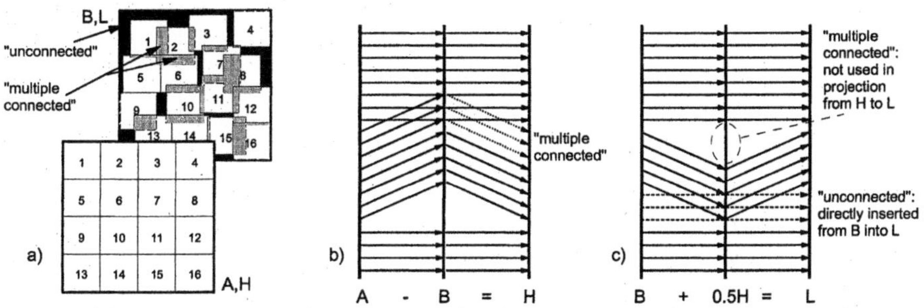

Fig. 13.45. a Unconnected and multiple-connected areas in block matching **b** backward MC in prediction step **c** projection-based IMC in update step

It is now also straightforward to extend this scheme to bi-directional frame prediction concepts, which have been proven to achieve higher coding efficiency than uni-directionally predicted frames for MC prediction coders (cf. sec. 13.2.5). The principle is shown in Fig. 13.44b. Here, also the update step is performed bi-directionally, wherein still the reverse correspondence between MC and IMC must be observed due to the reasons given above. Similar to the case of MC prediction coders, it is also possible to switch dynamically between forward, backward and bi-directional prediction, or implement an intraframe mode. If for example an H frame shall only be computed by the prediction of A from the subsequent B, the left-branch weight of the prediction step generating that frame must be set to 0, and the right-branch weight will be set to -1. To observe symmetry of the update step, the branch weight corresponding to the zero weight within the prediction step must also be set to 0, which is reasonable as no motion vector exists. An example is shown for the processing of the rightmost H frame in Fig. 13.44b. It should be emphasized that in principle the MC in the prediction and the IMC in the update step could be independent. This still would guarantee perfect reconstruction by the inherent properties of the lifting structure, as was shown in [SECKER, TAUBMAN 2001]). Nevertheless, the match between MC and IMC seems to be important to guarantee undistorted L frames[2].

---

[1] This is similar to the methods of *projection-based MC* (based on backward vectors) as introduced in sec. 13.2.4.

[2] The symmetry of prediction and update steps, both by the relationship of MC/IMC and with regard to the lifting-filter weights is important for optimum encoding performance, cf. sec. 13.4.4.

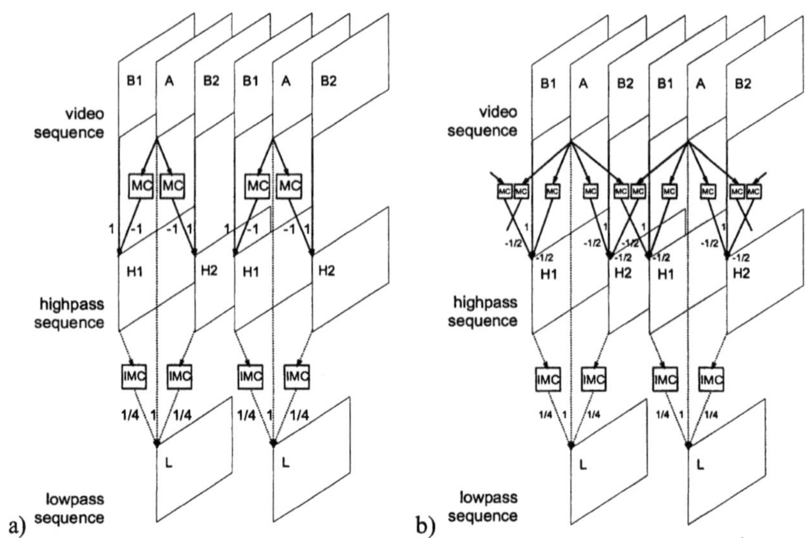

Fig. 13.46. MC wavelet with non-dyadic transformations A/B→L/H, ratio of L:H is 1:2
a Uni-directional prediction and update **b** Bi-directional prediction and update

The filter which is realized in Fig. 13.44b is exactly a 5/3 biorthogonal filter operating along the temporal axis. An advantage as compared to Haar filters is the *symmetry*, which means that neither the forward nor the backward direction is favored in any way unless explicitly activated by mode switching. By usage of Haar filters, a well-defined frame grouping structure is implicit, where a Group of Pictures (GoP) of length 2^T establishes a self-contained access unit[1]. For the case of 5/3 temporal filters, the concept of a temporal-axis block transform can finally be given up; useful decoding could start at any position and would guarantee availability of full information after a limited number of frames, depending on the depth of the temporal wavelet tree. Finally, it must be observed that the number of motion parameters is doubled by introduction of the bi-directional scheme. This can however be avoided by usage of proper motion vector field encoding such as the *direct mode* used in MC prediction (cf. sec. 13.2.5).

If the schemes as shown in Fig. 13.44 are arranged in a wavelet tree as in Fig. 13.2b, the flexibility of temporal scalability would be constrained to dyadic levels, when only the lowpass output shall be used. Cases like down-scaling from 30 *Hz* to 10 *Hz* sequences, which is often used in temporal scalability by MC prediction coding, would not be possible. As Fig. 13.46 shows, an even more generalized view of the MCTF lifting concept can overcome this limitation, allowing non-dyadic temporal-axis decompositions. In the example of Fig. 13.46a, two frames *B*1 and *B*2 are uni-directionally predicted from one *A* frame, such that two frames

---

[1] This is mainly related to the fact that the Haar transform is indeed a block transform.

$H1$ and $H2$ are generated in a group with one L. Following the MC/IMC principles explained above, the L processing is a binomial filter, $\frac{1}{4} B1 + \frac{1}{2} A + \frac{1}{4} B2$.

The principle of non-dyadic decomposition can also be extended to the case of bi-directional prediction, as it is shown in Fig. 13.46b. Regarding the H frames, this is actually very similar to the widely-used structure of *IBBPBBP..* in conventional MC prediction coders, as shown in Fig. 13.9. Dyadic and non-dyadic structures can be combined within a wavelet tree, such that a significant degree of freedom exists to define the temporal scalability levels of the decomposition. Nevertheless, in a strict sense, such a decomposition could no longer be denoted as being a wavelet transform, and will eventually not lead to a fully-decorrelated representation of video frames[1].

13.4.3 Improvements on Motion Compensation

Block-based motion compensation, even though being attractive by low computational complexity, is unnatural in the context of spatial wavelet decomposition, and possible source of artifacts. Nevertheless, good compression results have been reported with block-based MC in motion-compensated wavelet coders, which are competitive with state of the art hybrid coder designs. Most of the concepts for improved motion compensation, as developed in sec. 13.2.6 for the case of hybrid coders, can be applied to 3D wavelet coders likewise. They are even more beneficial here as being a more natural choice for the combination with wavelet basis functions, while block transforms seem to be better in harmony with block-based motion compensation. Usage of variable block size motion compensation was proposed in [OHM 1994B][GOLWELKAR, WOODS 2003]. Results on 3D subband and wavelet coders with warping MC were reported in [OHM 1994B] and [SECKER, TAUBMAN 2001]. Alternatively, overlapping-block methods can be used [HANKE, RUSERT, OHM 2003], which in principle means that weighted superpositions are performed at motion boundaries. Fig. 13.47 illustrates example cases of very short weighted overlaps, where two different predictions and updates are averaged at the positions directly beneath the motion boundaries, each weighted by a factor of 0.5[2]. This is done by extending the respective motion interconnection relationships by positions beyond the boundary. The modified weights in the prediction step are complemented with the update step[3]. A comparison against the cases of Fig. 13.42 shows

---

[1] The same is true in the case of MC prediction using several adjacent B-type frames. Due to the fact that they are predicted from the same P-type frames, they will most probably be correlated, a similar phenomenon as in vector prediction (12.26)/(12.27).

[2] Larger overlaps are advantageous in practice.

[3] Again here, the weighting factors of the lifting branches at the discontinuous positions of the motion vector field can not be fully consistent, as the filters are non-orthonormal in general, and the degree by which actually energy preservation within the L and H signals is approached also varies with the sub-pixel shift. Optimization of the lifting weights under

that now previously unconnected and multiple-connected samples are more consistently integrated in the subband decomposition process. The block overlapping method blurs prediction differences in the *H* frame in the vicinity of motion boundaries, but will also produce more blurred areas in the *L* frame where the motion is inconsistent. This is beneficial for higher compression efficiency and for the usage of *L* frame sequences in temporal scalability, achieving better subjective quality (see Fig. 13.48).

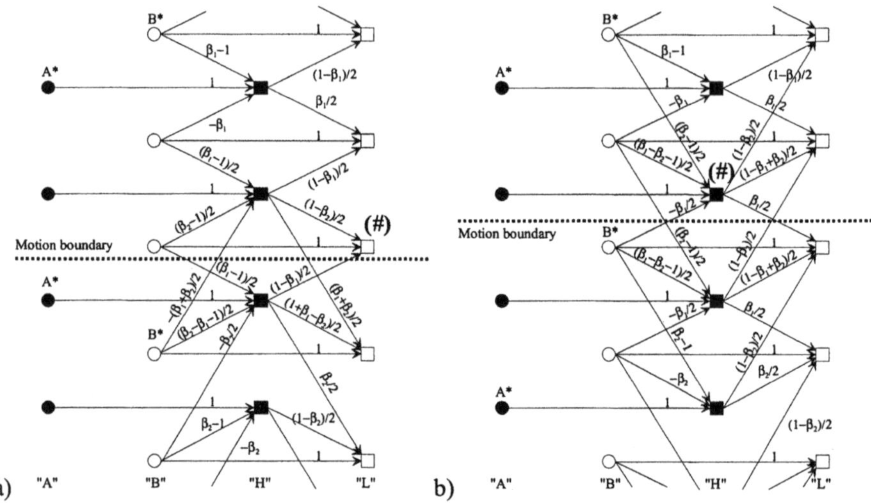

Fig. 13.47. Effect of overlapping-block motion compensation in the case of previously "unconnected" pixels (*a*) and "multiple connected" pixels (*b*) in the lifting structure

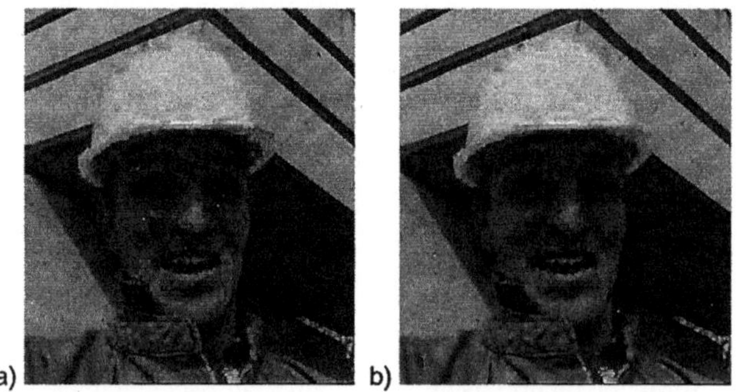

Fig. 13.48. Reconstruction of video frames from an MCTF wavelet coded representation with block-based MC (*a*) and with OBMC/de-blocking (*b*)
[source: HANKE/RUSERT/THULL]

---

the aspect of minimization of distortion will be an important topic for future developments. This aspect is further discussed in sec. 13.4.4.

In general, 3D wavelet schemes will take more advantage by *true motion* estimation than hybrid coders do. This can be justified by the fact that for high compression ratios it is very likely that most information contained in the H frames will be suppressed, such that the reconstruction of the original frames is more or less a motion projection from the information contained in the L frames. As no prediction loop exists, it would also consistently be possible to improve the reconstruction quality by integrating methods of format conversion (see sec. 16.4) into the synthesis process at the different levels of the wavelet tree. Methods for motion estimation as applied in existing 3D wavelet coders have mainly been developed from related hybrid coders, which are typically optimized for the prediction step, but not necessarily *jointly for prediction and update steps*. A first approach to solve this problem was a combined forward/backward motion estimation [OHM 1994A]. Further, criteria can be applied which prefer motion vector fields that are spatially and temporally consistent over the levels of the wavelet pyramid [OHM 1994B]. Rate constraints for variable block size motion vector fields have been introduced, but optimum motion estimation in a rate-distortion sense, where the vector should be applicable over a broad range of rates in a scalable representation, is a problem which is yet unresolved. Solutions on the following aspects must be found:

– Unlike still-image wavelet coders where the respective orthonormality weighting determined from the synthesis basis functions (12.40)-(12.43) is constant over the entire image, it is variable in motion-compensated 3D wavelet coders. This is due to the fact that the synthesis basis varies with the motion shift, e.g. influenced by sub-pixel interpolation filters, such that the effect of quantization in the subbands on the resulting distortion is not necessarily constant. This should in fact be considered in motion estimation. Motion-compensated wavelet bases which are at least invariant by their Euclidean norm would solve the problem, however it is not clear yet if a solution exists for this. If the spatial transform follows the temporal transform (which shall be denoted as a $t+2D$ scheme in the sequel), the motion estimation should in principle take into account the spectral flatness or concentration in spatial frequency components as well[1].

– The influence of the rate to be spent on the motion vectors must be considered, which is not straightforward to solve for a scalable representation. The percentage of the motion information rate typically largely increases towards lower total rates. The best trade-off between more accurate motion description and optimization of the energy compaction effected in the 3D transform must be found not for a single rate point, but over an entire envisaged range of rates. The effect of the motion compensation on the rate-distortion performance can be assumed to diminish towards higher rates. Practically, the best so-

---

[1] This would be similar to motion estimation based on criteria in the transform domain, such as the SATD (cf. sec. 13.2.6).

lution would be to clearly associate different resolution levels of the motion vector field with related spatio-temporal resolution levels of the signal.

- Novel methods for encoding of the motion vector field must be developed, which rather regard the motion information itself as a 3D entity of motion trajectories (see also sec. 13.5).

13.4.4   Quantization and Encoding of 3D Wavelet Coefficients

The transforms introduced so far in motion-compensated wavelet filtering are not orthonormal. For quantization, it is important to investigate the effect of expected transform domain quantization error to the expected variance of the decoding error. It shall be assumed that a spatial transform is used where orthonormality applies (or at least approximately, e.g. for the 9/7 bi-orthogonal filters (4.195)). Then, the aspect of optimum quantization can be analyzed separately for the temporal transform, which could then be apply by linear weights to any spatial coefficient within a given temporal band of the 3D representation. The case of Haar filters is regarded in detail.

Quantization of L and H frames from motion-compensated filters. To obtain an orthonormal representation from (13.38), normalization must be performed such that H frames are multiplied by $K_H = 1/\sqrt{2}$, while L frames must be multiplied by $K_L = \sqrt{2}$ according to (4.212). In practice, these up- and down-scaling operations of amplitude ranges must not actually be performed, but can be implemented by proper definitions of quantization step sizes during encoding, which is in particular beneficial for integer implementations. Orthonormality by $K_H = 1/\sqrt{2}$ means that the unnormalized H frames in (13.38) can be encoded with *less accuracy* than a prediction error frame in conventional hybrid MC prediction, while by $K_L = \sqrt{2}$ the unnormalized L frames must be encoded with higher accuracy than I frames. This phenomenon can also be interpreted by the signal flow in the lifting filter structure, as illustrated for the Haar filter case in Fig. 13.49a. A quantization error q_H of the H component is first weighted by a factor ½ and then subtracted from the quantization error q_L of the L component. The error injected into B is then $\Delta_B = q_L - \tfrac{1}{2}q_H$. Now, in the synthesis of A, the prediction is generated from the reconstructed B and added to H. The reconstruction error in A is $\Delta_A = q_L + \tfrac{1}{2}q_H$, which shows that the *back path* of the lifting filter compensates half of the error in H. The actual quantization step sizes must be chosen as if the quantization errors were normalized by the factors K_L and K_H, i.e. q_L/K_L and q_H/K_H. Under the assumption of uncorrelated quantization errors in L and H, the following contributions are then made to the reconstruction error variance:

$$\sigma^2_{\Delta_B} = E\left\{\left(\underbrace{\frac{\sqrt{2}}{2}}_{1/K_L}\cdot q_L\right)^2\right\} + E\left\{\left(\underbrace{-\sqrt{2}}_{1/K_H}\cdot\frac{1}{2}q_H\right)^2\right\} = \frac{1}{2}\left[\sigma^2_{q_L} + \sigma^2_{q_H}\right]$$

$$\sigma^2_{\Delta_A} = E\left\{\left(\underbrace{\frac{\sqrt{2}}{2}}_{1/K_L}\cdot q_L\right)^2\right\} + E\left\{\left(\underbrace{\sqrt{2}}_{1/K_H}\cdot\frac{1}{2}q_H\right)^2\right\} = \frac{1}{2}\left[\sigma^2_{q_L} + \sigma^2_{q_H}\right]$$

$$(13.46)$$

Obviously, the lowpass component must be quantized using double accuracy or half quantization step size as compared to the highpass component. In (13.46), $\sigma^2_{q_L}$ and $\sigma^2_{q_H}$ denote the orthonormally adjusted quantization error variances, which *are equally spread* into the reconstruction error variances of frames A and B, respectively. This is however only true if the assumption made above comes true, which means that the back path indeed compensates half of the quantization error. The inherent assumption must be that the operations of MC and IMC are *exactly inverses* of each other. If this would not be the case, the back path would produce another quantization error $\frac{1}{2}q_H'$ which originates from another sample, and the reconstruction error in A would become $\Delta_A=q_L+q_H-\frac{1}{2}q_H'$. However, except for the case of full-pixel shifts, MC and IMC can not perfectly match, as sub-pixel interpolations are involved. It follows that the quality of the interpolation filters has a direct influence on the minimization of the reconstruction error[1].

Similarly, optimum quantization step sizes for other types of filters can be derived by (12.42). For example, in the case of 5/3 filters applied along the temporal axis, the lowpass component should be quantized by a factor of

$$\sqrt{\frac{\left(\frac{1}{2}\right)^2 + 1^2 + \left(\frac{1}{2}\right)^2}{\left(\frac{1}{8}\right)^2 + \left(\frac{1}{4}\right)^2 + \left(-\frac{3}{4}\right)^2 + \left(\frac{1}{4}\right)^2 + \left(\frac{1}{8}\right)^2}} = \sqrt{\frac{48}{23}} \tag{13.47}$$

more accurate as compared to the highpass component, which means that the quantizer step size for the lowpass signal should be smaller by this factor.

The cases of unconnected and multiple-connected pixels. The synthesis flow in the cases of unconnected and multiple-connected pixels is illustrated in Fig. 13.49b. Clearly, for the unconnected case, $\Delta_B=q_L$. For the other exception case of multiple connections, pixels are embedded into the H frame as prediction errors (13.34). The worst case will then be that the pixel B which is used for prediction is connected and contains a reconstruction error $\Delta_B'=q_L-\frac{1}{2}q_H'$, wherein q_L', q_H' and

---

[1] Practically, it can be found that a sub-pixel accuracy of 1/8 pixel in motion compensation still effects a remarkable gain in motion-compensated (MCTF) wavelet coding, while for hybrid coding usually a saturation is reached with about 1/4 pixel accuracy.

the 'own' quantization error q_H are uncorrelated[1]. Then, $\Delta_A = \Delta_B' + q_H = q_L' - \tfrac{1}{2}q_H' + q_H$. For these cases, it is necessary to use different normalization factors $K_L = K_H = 1$, which means that these components should be quantized by the same quality that would be used for intraframe coding of the signal:

$$\sigma^2_{\Delta_B} = E\left\{(q_L)^2\right\} = \sigma^2_{q_L}$$

$$\sigma^2_{\Delta_A} = E\left\{\left(\frac{\sqrt{2}}{2}\cdot q_L'\right)^2\right\} + E\left\{\left(-\sqrt{2}\cdot\frac{1}{2}q_H'\right)^2\right\} + E\left\{(q_H)^2\right\} = \frac{1}{2}\sigma^2_{q_L} + \frac{3}{2}\sigma^2_{q_H}. \tag{13.48}$$

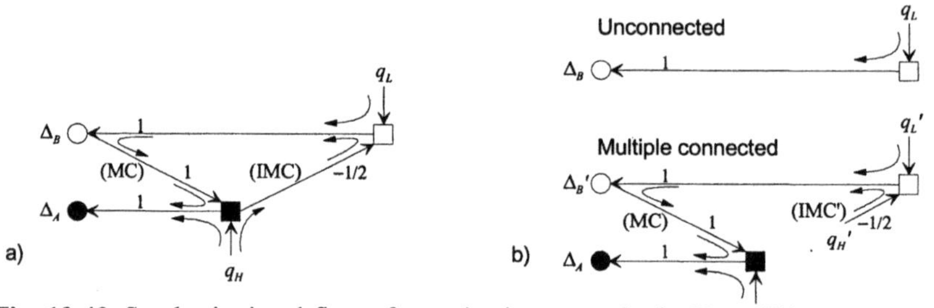

Fig. 13.49. Synthesis signal flow of quantization errors in the Haar lifting structure (*a*) and in the case of unconnected / multiple-connected pixels (*b*)

For the multiple-connected positions, this problem is critical, as they suffer both from the error that is intruding via the back path[2] (q_L',q_H') and from the 'direct' quantization error q_H. This asymmetric behavior is the reason for quality fluctuations that are often observed at specific frame positions in the decoded result of 3D wavelet coders based on Haar filters [RUSERT, HANKE, OHM 2003]. Indeed, if the prediction of the multiple-connected pixels would be performed from *reconstructed* values of B, the quantization error feedback from B is eliminated such that $\Delta_A = q_H$. Even then, it is still necessary to adjust the quantization weighting (or the normalization factors) depending for the positions of unconnected and multiple-connected pixels[3]. In principle, to determine the effect of quantization errors accurately, it is necessary to track the evolution of the errors through the entire wavelet tree (temporally and spatially); the additional cost for this optimization is one additional

---

[1] This case is not unrealistic, in particular when forward prediction from B_{-1} is used.

[2] In principle this a similar problem as the quantization error feedback in open-loop prediction systems, but without infinite propagation. This effect is however becoming worse when superpositions from different levels of the temporal wavelet tree occur.

[3] It could in principle similarly be useful to consider the effect of the sub-pixel shift in motion compensation in this context; as it was shown before, interpolation in the lifting branches may modify Haar filters into bi-orthogonal filters, for which different normalization factors K_L and K_H have to be applied.

spatio-temporal transform to be performed at the encoder side [OHM 1994A][RUSERT, HANKE, OHM 2003][1].

The unequal weighting of L and H samples in (13.46) is one of the main reasons effecting that 3D transform schemes can be superior in performance compared to hybrid (prediction based) coders. Even though at the first sight the high-pass frame H seems to be very similar to an MC prediction frame, the partial compensation of coding errors via the back path, and the systematic spreading over different adjacent frames give an advantage regarding the total balance of the squared error. A theoretical analysis of this gain is given in [FLIERL, GIROD 2003], which comes to the result that up to 40% reduction in rate, as compared to a hybrid coder with same methods of motion compensation, can be achieved.

As the L frames (scaling band of the temporal wavelet transform) have to be quantized by increasingly finer quantizers when stepping up the levels of the wavelet tree, this effect increases exponentially by the number of levels[2]; for example, the LLLL frame from a four-level temporal Haar wavelet tree contains all relevant information of $2^4{=}16$ frames, and should be quantized by a factor of $\sqrt{2}^4 = 4$ more accurately than a single intra-coded frame. As a by-product, noise, sampling inconsistencies etc. are discarded at lower rates by the temporal filtering process. From this point of view, motion-compensated wavelet coding can realize advantages of joint multiple-frame compression straightforwardly, which in a hybrid coder could only be achieved by extremely complex look-ahead decisions over a large number of frames.

Methodologies to encode the motion-compensated 3D wavelet coefficients as developed until now are not much different yet from 2D wavelet coding (sec. 12.4.2) or 3D wavelet coding without MC (sec. 13.1.2). Embedded quantizers are used, which can straightforwardly be applied without penalty, as the synthesis filter structure is still non-recursive by principle[3]. Conventional 2D wavelet coders can directly be applied to the subband frames resulting by the temporal wavelet tree processing; this is particularly suitable in a configuration where the entire temporal transform is performed first. This case shall be denoted as a t+2D transform, corresponding to the scheme shown in Fig. 13.2b.

---

[1] The quantization of a given frequency coefficient must be made more or less accurate depending on the squared norm of the overall 3D synthesis basis (including all steps of spatial, temporal and MC interpolation synthesis filters).

[2] This however depends on the sequence characteristics; when no correspondences can be found between two frames, no gain can be realized by 3D MC Wavelet coding. For sequences with fast motion, saturation is often reached after 2 or 3 levels of the temporal wavelet tree, while for slow-moving sequences or sequences with static background, even an increase to 7 or 8 levels gives an advantage and further increases the coding performance.

[3] Certain restrictions apply however for the prediction method made for the multiple-connected positions, as discussed above.

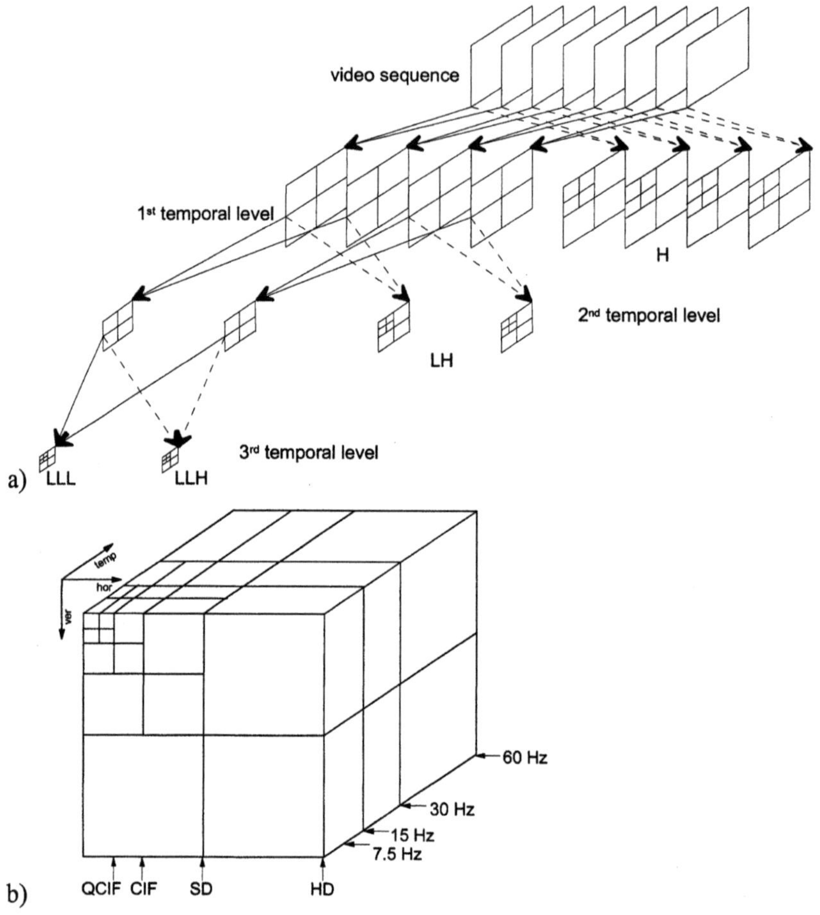

Fig. 13.50. a Wavelet tree with reduction of spatial size throughout the temporal levels
b corresponding wavelet cube

The optimum strategy of spatio-temporal decomposition is a significant topic of further exploration. The scalability property of the spatial/temporal wavelet transform may e.g. be utilized to reduce the size of the frame memory necessary to perform encoding and decoding. An example is shown in Fig. 13.50a, where the spatial size of the frames is reduced by a factor of 4 (implemented by one level of spatial 2D wavelet decomposition) with each temporal level. Inherently, the depth of the spatial tree is now much lower for the higher temporal frequency bands, which is also reasonable as these signals have less spatial correlation anyway. The related wavelet cube is shown in Fig. 13.50b. The best spatio-temporal decomposition structure could be found by wavelet-packet design criteria (cf. sec. 12.4.4), where the next split in the 3D wavelet tree can be made spatially or temporally, depending on best effect in coding gain. This would implicitly include criteria of temporal similarity between frames and scene cut detection, as the gain by further splitting in temporal direction at the deeper levels of the tree is clearly highest for sequences of low motion[1]. Additional constraints must be made by scalability requirements, such that at least splits which support the required operational ranges of spatial or temporal scalability must be provided by default. As an example, the wavelet cube shown in Fig. 13.50b would allow spatial scalability between sub-QCIF and HD resolutions spatially, and temporal scalability for frame rates between 7.5 and 60 *Hz* temporally; for HD indeed, no lower frame rates than 15 *Hz* would be supported, which appears reasonable.

Alternatively, a spatial/temporal decomposition can be realized where the spatial wavelet filtering is done as the first step, and the temporal filtering is applied to the coefficients of the spatial transform ('2D+t' transform). Within the lifting filters which are used for the temporal transform, it is then possible to use an *overcomplete DWT* (ODWT) which can generate any sub-sampling phase which matches with the motion trajectory locally, similar to the methods described in sec. 13.3.2 [ANDREOPOULOS ET AL. 2003]. In terms of coding efficiency, this method is reported to be not inferior compared to the 't+2d' transform, but has a number of advantages:

1. Spatial scalability of motion vector fields, combined with consistent hierarchical motion estimation can be applied (see sec. 13.3.2). This is particularly important when a fully-scalable representation over several spatial transform levels shall be realized. In the case of the 't+2D' transform, even though spatial scaling can be applied, no consistent way exists to decode the lower resolution frames by using a lower-resolution motion vector field as well.
2. Even though spatial scalability can be applied to the 't+2D' transform, methods to explicitly optimize for shift invariance of the 2D transform are not straightforward. In fact, if the sub-sampled resolutions are generated by conventional down-sampling in the spatial wavelet pyramids of *L* and *H* frames, the inverse motion compensation may implicitly mix different sub-sampling phases; this can be a cause of artifacts in the vicinity of motion boundaries, but also may lead to temporally-varying alias effects in the reconstructed signal.
3. As discussed before, spatially varying properties of filters, including partial violations of the constant norm principles at unconnected and multiple-connected positions, are inherent to MCTF. In order to keep these effects under control by appropriate quantization, it is advantageous to introduce them only by the last step of the 3D wavelet decomposition. It is then possible to adapt the quantization step sizes sample by sample, where the adaptation pa-

---

[1] Ideally, the steps of spatial and temporal decomposition would be arbitrarily interchangeable with the same result. This is however not achievable with the solutions of motion-compensated 3D wavelet coding known so far due to the shift variance of the MCTF process.

rameters can directly be determined directly from the motion vectors with minimum effort.

3D motion-compensated wavelet coding is a field of high potential for improvements and further new developments, overcoming many disadvantages of traditional (hybrid) video coding. The non-recursive structure of the decoder avoids problems like error feedback, drift and sensitivity to data losses. The wavelet-based (3D) spatio-temporal coding methods have the property of being freely scalable in spatial and temporal resolution as well as quality (SNR or quantizer scalability). Best exploitation of temporal redundancy by motion compensation is fully supported, such that the overall compression performance is high. Due to these properties, this new class of video compression methods is an excellent candidate for future video coding systems. Ideal support is given for transmission of video signals in environments of heterogeneous, error-prone networks, where any fluctuation of bandwidth can be served instantaneously. The capability to implement flexible spatial and temporal scalability is suitable to serve terminals of different resolution and different computational capabilities from only one encoded representation (bit stream). A clear disadvantage – if not allowable by an application – is the higher encoding delay which is inherent in motion-compensated temporal-axis frequency coders[1]. In the next section, a rough analysis of complexity and delay of such coders is given.

13.4.5   Delay and Complexity of 3D Wavelet Coders

Delay. If T levels of temporal-axis wavelet transform are used (refer to the example of Fig. 13.2b with $L=3$), and the Haar filter structure of Fig. 13.44a is invoked, a delay of 2^T-1 frames occurs at minimum until the information necessary to decode the first frame of a Group of Pictures (GOP) of length 2^T can be conveyed to the decoder. This delay is mainly caused by the temporal lowpass filtering, the update step in the lifting structure.

For the particular structures introduced previously, the delays d_t occurring at *one* level of the wavelet decomposition tree can be analyzed as follows:

---

[1] Low delay – however at the expense of deteriorated coding efficiency – can be achieved if no update step is performed in MCTF. The construction of the temporal wavelet tree would then be identical to the temporal scalability pyramids in Fig. 13.25. In fact, the configuration in Fig. 13.25b can be interpreted as a wavelet tree based on a (1,3) biorthogonal filter (only using a 'lazy wavelet' in 'lowpass' filtering) to realize a low-delay mode allowing instantaneous encoding and decoding. The temporal scalability of hybrid coders based on P-type frames as shown in Fig. 13.25a corresponds to a Haar filter with omitted update step. Due to the quantization properties explained in the previous section, such a mode is inferior in terms of coding efficiency. The general trade-off which exists between compression performance and latency that was already pointed out in section 1.1.1 is again in effect here.

- Fig. 13.44a: d_t=1 frame;
- Fig. 13.44b: d_t=4 frames;
- Fig. 13.46a: d_t=2 frames;
- Fig. 13.46b: d_t=6 frames.

(13.49)

If any of these structures is used at a specific level t (t=1,…,T) within a wavelet tree, the delay (in numbers of original frames) contributed by that level to the overall delay is $d_t \cdot 2^{t-1}$. The overall minimum delay (assuming that all processing can be done sufficiently fast) can then be determined as the accumulation of delay contributions over all levels:

$$\mathrm{min\_delay} = \sum_{t=1}^{T} d_t \cdot 2^{t-1} \, .$$

(13.50)

In contrast to encoding, the delay at the decoder side, provided that all necessary information is available e.g. on a hard disc, is only given by the processing time of decoding T frames through the levels of the wavelet pyramid. Within this time, random access to any frame is guaranteed, which is an important advantage as compared to hybrid coders, where decoding always must start at an I-type frame and proceed through the entire sequence of P-type frames. In a streaming or broadcast application a raw resolution of the most relevant information (e.g. *LLLL* frames of low spatial resolution) can be sent iteratively at short time distances without adding much rate overhead. Then, entry points for decoding can be provided much more frequently than with hybrid prediction coders. For storage applications, the forward-backward symmetry of the decoding process allows efficient implementation of reverse play, and the different levels of temporal scalability also allow fast forward/reverse play at different speeds. From this point of view, no delay penalty exists at all for the case of pre-recorded streams in broadcast and storage, which establishes one important class of applications for fully-scalable video codecs.

Memory Complexity. Investigating the memory complexity of the motion compensated temporal wavelet transform, it can be found that it is not necessary to store the entire GOP during encoding or decoding. In fact, all transform steps $B\leftrightarrow H$ and $A\leftrightarrow L$ during encoding (and vice versa during decoding) can be performed in-place in the lifting structure, as common coordinate systems are used. For example, in the case of combining Figs. 13.2b and 13.44a, the first processing step is the prediction part of the lifting transform, $A,B\rightarrow H$, which requires one frame memory for B and one in common for A/H. The next step is $B,H\rightarrow L$, using one common memory for B and L. Afterwards, H can be encoded, such that it must not be stored anymore. However, L must be retained for further processing within the wavelet tree until it is processed in combination with a neighbor at the next level. This leads to the conclusion that as a minimum (if all processing can be performed in sufficiently short time), only $T+1$ frame memories are needed in a Haar-based temporal wavelet tree consisting of T levels[1].

In general, the minimum numbers of frame memories necessary per level are directly related to the frame delays d_t given in (13.49) for the different temporal filter structures. Assuming that d_t is the delay occurring at level t out of T levels, the overall minimum number of frame memories necessary to perform motion-compensated wavelet analysis or synthesis will be

$$\min\_\text{frame\_memory} = 1 + \sum_{t=1}^{T} d_t \, , \tag{13.51}$$

or $1+T{\cdot}d$ for the case of a constant $d=d_t$ (same filter length) over all temporal wavelet tree levels. The one additional frame memory is needed to store the newest input picture in addition to the lowpass subband frames which must be held in memory at the different levels for further processing. The scalability property of spatial/temporal wavelet transform can also be utilized to reduce the size of the frame memory necessary to perform encoding and decoding. In the structure shown in Fig. 13.50, the spatial size of the frames is reduced by a factor of 4 (implemented by one level of the spatial 2D wavelet decomposition) with each temporal level. In this case, the number of necessary full-frame memories reduces to only

$$\min\_\text{frame\_memory} = 1 + \sum_{t=1}^{T} \left(\tfrac{1}{4}\right)^{t-1} d_t < 1 + d_t \left(\frac{1}{1-1/4}\right), \tag{13.52}$$

which is upper bounded to $1+4d/3$ in the case of constant $d=d_t$ over all levels, even for very large T. The *amount of frame memory* is not the only indicator for memory complexity. Another factor, often regarded as more important, is the *number of memory accesses*. Memory accesses are in general a severe bottleneck for both hardware (specific processor) and software (general-purpose processor) video encoder and decoder implementations. The criticality of memory accesses is first related to the number of pixel operations, but the regularity, i.e. the predictability about the locations of pixels which shall be used in a processor pipeline is also important. As will become clear from the following sub-section, MCTF wavelet coding is still in the same order of magnitude of complexity when compared to hybrid MC prediction coding when viewed under this aspect. Due to the regularity of the temporal wavelet tree, it may be even more regular for predictability of accesses as compared to techniques like multiframe prediction, which are used in standard MC prediction coders of today.

Computational Complexity. The computational complexity of the MC wavelet lifting filter implementations is mainly influenced by the motion compensation steps. The aspect of motion estimation shall be excluded here; due to the usage of same motion parameters in both prediction and update stages, a similar degree of

---

[1] An additional frame memory will be required if prediction of the multiple-connected positions shall use frames B_{-1} as in (13.35).

complexity as in MC prediction coders will be in effect, which means that estimation of one or two motion vector fields per frame, depending on usage of uni-directional or bi-directional processing, is necessary.

First, the case of Haar filters, which means uni-directional prediction and update in the lifting filters is regarded, using linear interpolation for sub-pixel accuracy in MC. From (13.44), 4 multiplications and additions (M&A) per pixel and frame are necessary to perform the 2D linear interpolations at one level of the MC wavelet transform. If iterated over the temporal wavelet tree of arbitrary depth, where the number of frames is reduced by a factor of 2 at each level, the total number of M&A per pixel of the original sequence is

$$\frac{\text{mul+add}}{\text{pixel}} = 4 \cdot \sum_{t=1}^{T} \left(\tfrac{1}{2}\right)^{t-1} , \tag{13.53}$$

which is upper-bounded to a value of 8, both at the encoder and decoder, irrespective of the number of levels. This is slightly less than double the number of computations as compared to a unidirectional (P-type frame only) MC prediction using the same bilinear interpolation for sub-pixel values. For the memory-friendly structure of Fig. 13.50, the number is

$$\frac{\text{mul+ add}}{\text{pixel}} = 4 \cdot \sum_{t=1}^{T} \left(\tfrac{1}{2}\tfrac{1}{4}\right)^{t-1} = 4 \cdot \sum_{t=1}^{T} \left(\tfrac{1}{8}\right)^{t-1} , \tag{13.54}$$

which is upper bounded to 32/7 or approximately 4.5 M&A per pixel. This is only a marginal increase as compared to the case of uni-directional MC prediction, and is only slightly more than half the number of operations that are necessary for the case of MC prediction using B-type frames.

For the cases of Figs. 13.44b and 13.46b, where bi-directional predictions and updates are used in the lifting filters, the number of M&A per pixel as well as the memory accesses are exactly doubled as compared to the uni-directional cases given in (13.53) and (13.54), if still the linear sub-pixel interpolation filter is used. For the non-dyadic decomposition of Fig. 13.46b, a factor of 4/3 applies as compared to (13.53) and (13.54), as bi-directional MC is applied in the update step. If longer sub-pixel value interpolation filters (see e.g. Fig. 13.41) are used, the number of M&A per pixel increases linearly with the order of the filter, when the 2D interpolations are computed by a separable system. Qualitatively, the effect of higher complexity due to longer filter kernels is not different from the increased complexity by longer interpolation filter kernels as used in state of the art MC prediction coders.

All complexity contributions are scaled down by memory-friendly structures as shown in Fig. 13.50. This directly show the favorable applicability of the fully-scalable motion-compensated 3D wavelet methods for the purpose of *complexity scalability*.

13.5 Encoding of Motion Parameters

Motion parameters are a component of side information which is of prior importance in encoding of video sequences. Motion vector fields as extracted from natural scenes are typically *spatially* and *temporally coherent*: An object moves through a series of subsequent frames either by constant velocity or accelerated. As a consequence, the spatio-temporal correlation in the motion vector field is high, which can not only be utilized in motion estimation, but also for encoding of motion parameters by lower data rate. If motion parameters are estimated over discrete value sets and for discrete positions, they are typically lossless encoded. Even though lossy encoding is applicable likewise, it depends on the stability of the motion-compensated compression algorithm whether this is desirable.

The motion is perfectly characterized by the motion vector field, which describes the translational motion of each single pixel. This motion vector field can by itself be interpreted as a correlated *signal*. Hence, PCM encoding of motion parameters would lead to an unnecessary waste of bits.

Most methods for encoding of locally varying motion vector fields are based on a sub-sampled representation. This e.g. naturally follows if motion is estimated from block matching or warping grid procedures (see sec. 7.6.3). In the case of global motion, it is typically sufficient to describe the spatial variation of motion by only one set of parameters relating to a parametric motion model (cf. sec. 7.4.4).

The method of motion parameter encoding must be known to the estimation procedure, when rate-distortion optimization shall be applied, considering the rate to be spent for the motion information. A method which preferably estimates a spatio-temporally continuous motion vector field usually also requires a lower rate for encoding due to the smooth predictability of the local motion vectors.

13.5.1 Spatial Contexts in Motion Coding

Spatial predictive coding. The most simple method to utilize the coherence within the motion vector field is spatial predictive coding. Only motion vectors belonging to previously encoded blocks may be used for prediction. Different strategies for prediction are applicable, e.g. 1D or 2D linear prediction, or median prediction are applied in video compression standards of today. The difference (prediction error) between the prediction and the current value of the motion vector is entropy-coded. This is much more efficient than entropy coding of the original motion vector values. Fig. 13.51 illustrates how values of motion vector prediction differences are highly concentrated around the value zero; this is clearly not the case for the original motion vector values. The statistical distribution of motion vector differences is often modeled by a negative exponential (e.g. Laplacian) PDF.

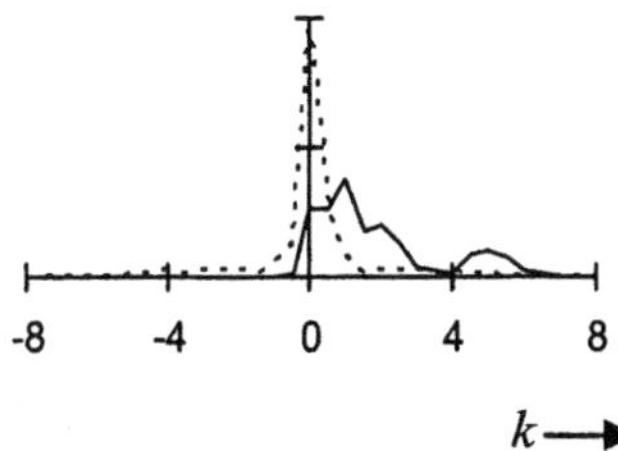

Fig. 13.51. Occurrences of the horizontal motion vector component in one video sequence (—) and associated difference component in case of prediction (···)

Predictive coding is mostly applied separately to the horizontal and vertical translation components. For more complex motion models, e.g. including rotational motion, it can no longer be expected that motion components in horizontal and vertical directions are statistically independent, such that a joint prediction based on a model is more appropriate. This is in principle achieved by methods of *Global Motion Compensation* (GMC), where local parameters are only used in addition if the local motion is not represented sufficiently accurate by the global model.

Spatial hierarchical coding. Multi-resolution motion estimation algorithms (cf. sec. 7.6.3) support smooth motion vector fields. During the hierarchical process of estimation, an increasingly higher-accuracy representation of the motion vector field is achieved.

This can directly be utilized for hierarchical encoding of the motion vector field. In fact, the progression from a coarser to a finer resolution of the motion block grid (Fig. 13.52a) exactly relates to the approach of *pyramid encoding* (sec. 4.4.6). Parameters within the hierarchy can best be encoded differentially (Fig. 13.52b). Advantages are as follows:

- The differential pyramid has a decorrelating effect;
- Motion parameters within the different levels of the pyramid naturally have different levels of relevance, which can be important for the definition of protection mechanisms, and inhibits error propagation, which can occur in motion vector encoding using the recursive spatial prediction methods described above;
- Motion parameters relating to the different levels of the pyramid can be applied to video frames of different resolution levels (e.g. SD and HD).

Efficient encoding combines the pyramid representation with a tree description of the pyramid, by which it is signaled if it is at all necessary to proceed to a finer resolution, or if motion vectors within a local area are homogeneous[1]. If entropy coding is applied to the difference values, in general the rate will be lower when

---

[1] A similar method is used to encode variable-size motion blocks in the Advanced Video Coding (AVC) standard (see sec. 17.4.4).

the motion vector field is varying smoothly, because the prediction from one pyramid level to the next will then typically give only small differences, to be represented by codewords of short lengths.

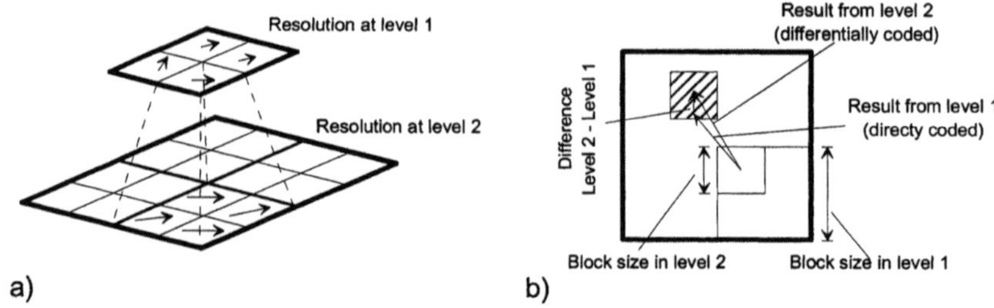

Fig. 13.52. Hierarchical coding of a block-based motion vector field
a resolution accuracy within the hierarchy levels **b** differential encoding

13.5.2 Temporal Contexts in Motion Coding

Temporal predictive coding. For most types of motion, the motion vector field also shows a high coherence along the temporal direction, which can be expressed by a continuous *motion trajectory*. If motion information from a previous frame position is used for prediction, it must be observed that the motion parameters themselves describe the direction along which the motion trajectory evolves, where largest temporal coherence can be expected. Temporal prediction is in general more problematic than spatial prediction due to the following reasons:

- The motion vector estimated for the previous frame can be erroneous, such that not only the prediction fails, but is also made for the wrong position in the current frame;
- the prediction can be non-unique, if several motion vectors from the previous frame point into one position of the current frame;
- The motion vector field can be subject to accelerations or de-accelerations, in particular if several types of motion are mixed. Such a behavior of the motion trajectory could better be predicted if the motion evolution over more than one frame is considered.

An example for the latter case is a rolling wheel, where the center of rotation moves coupled with the rotational movement. If a point is just at the wheel base (touching the ground), its motion will be zero. In the next moment, this point starts accelerating again until it reaches the top of the wheel. Application of temporal prediction of motion is also necessary when objects shall be tracked within a scene. In this context, by analyzing the evolution of motion parameters or motion trajectories over more than one frame, it is possible to decide at which positions occlusions (covered and uncovered areas) are present.

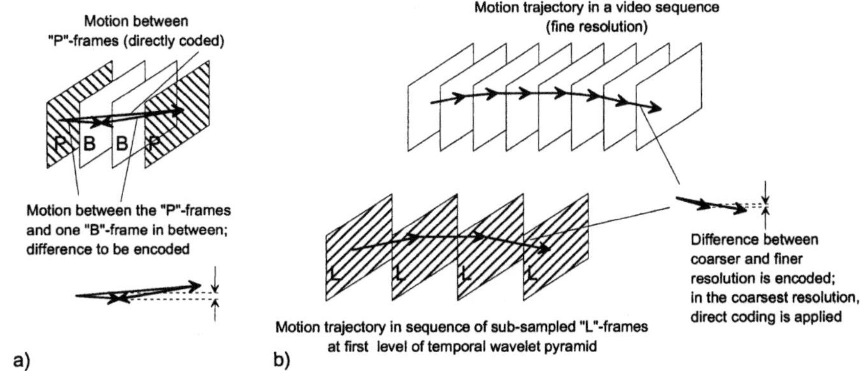

Abb. 13.53. Temporal-hierarchical difference coding of motion parameters
a for case of *B*-type frames **b** for case of motion-compensated wavelet coding

Temporal hierarchical coding. If continuous motion is present over several subsequent frames, hierarchical coding along the temporal dimension is possible as well. Unlike temporal prediction which is an extrapolation, this method is rather related to representation of motion at different temporal resolutions and interpolative description of the temporal changes in the motion vector field. If e.g. an object moves within the scene by 2 pixels from frame o to $o+1$, its motion will be expected to be 4 pixels from o to $o+2$, 6 pixels from o to $o+3$ etc. A temporally-hierarchical coding of motion parameters is in particular useful in combination with methods in which also frames are encoded in a temporal hierarchy. It must be observed that the values of the motion vectors scale linearly with the frame distance. For example, in the context of bi-directional prediction (sec. 13.2.5) the motion information for B-type frames can be predicted and if necessary be encoded differentially using information from corresponding P-type frame areas after appropriate scaling (Fig. 13.53a). An efficient application of temporal motion vector prediction is the temporal *direct mode* which is implemented in some video coding standards (H.263, MPEG-4, H.264/AVC). Here, the motion vectors for macroblock units in B-type frames can be determined from the motion vectors in the *co-located* (by direction of the motion trajectory) macroblock of an adjacent P-type frame.

In motion-compensated wavelet coding (sec. 13.4), several temporal hierarchy levels exist for the motion information. An example where the temporal resolution is increased by a factor of two is shown in Fig. 13.53b. In general, investigations made so far have shown that temporal prediction and temporal hierarchy trees are slightly less powerful in compression of motion vector fields than spatial prediction or spatial hierarchical coding[1].

---

[1] This may however well be caused by the fact that the motion estimation methods as used today barely optimize for temporal continuity of motion trajectories.

13.5.3   Fractal Video Coding

Fractal block coding as introduced in sec. 12.5 was constrained to still images. It shall shortly be revisited here, as certain commonalities with motion compensation were found by the mapping models applied in the fractal transform. If fractal coding shall be applied to video, 3D cubes must be used instead of the coding blocks (see Fig. 13.54a). The contractivity of the transform can then either be performed over the spatial or over the temporal axis; the latter case is eventually advantageous when scene changes occur, as these result in amplitude discontinuities over the temporal axis, which is somehow equivalent to edges within a 2D image, where the fractal transform is known to have particular strengths. If the content is moving, combination with motion compensation could be made; in fact, if the content of subsequent frames can fully be mapped by motion parameters, motion compensation can be implemented as part of the fractal transform [BARTHEL, VOYÉ 1995]. Two cases may be regarded, where either moving content establishes the origin volume and unmoved content the destination volume of the fractal transform, or vice versa (Fig. 13.54b): In the first case, the geometric transform to be applied to the origin volume can be determined *directly* from the motion mapping, in the second case the *inverse* motion mapping has to be applied. Of particular interest is the case where a continuously moved area is mapped by temporal and spatial contraction into its own position, which means that the center of the origin and destination volumes is identical. Here, at least for the case of arbitrary translational motion, including the case of constant acceleration, *no motion parameters* are required at all. The motion is scaled along with the video frames, such that the self-similarity fully applies (Fig. 13.54c). The same would be the case for rotational motion when the geometric transform allows rotational mapping over time.

These are idealized views, which in practice would be constrained by many exceptions. For example, if motion vector fields are inhomogeneous, occlusions occur, and the typical spatio-temporal contractive mapping will fail. Nevertheless, the properties of 3D fractal transforms can shed some interesting light on joint properties of image and motion information, which eventually may be useful for more efficient compression of either of these components in the future.

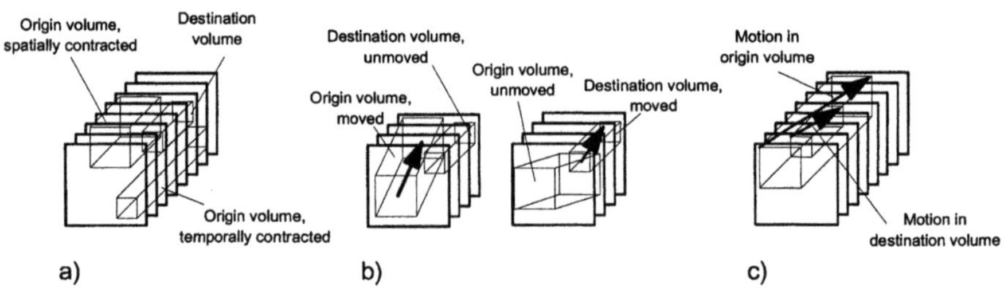

Fig. 13.54. The principle of 3D fractal coding

13.6 Problems

Problem 13.1
A motion-compensated prediction of a video signal is performed. The misalignment of motion compensation shall follow a uniform distribution into values $[k_e, l_e] \in \{-1, 0, 1\}$, and shall be statistically independent between horizontal and vertical dimensions.

a)   Determine the power spectrum $S_{ee}(\Omega_1, \Omega_2)$ of the prediction error frame, if the corresponding original frame from the sequence has a power spectrum $S_{xx}(\Omega_1, \Omega_2)$.
b)   Which would be the optimum cutoff frequency of a lowpass filter in the prediction loop, such that the variance of the prediction error signal is minimized?
c)   The power spectrum of the image frame shall be described in horizontal direction as $S_{xx}(\Omega_1) = A \cdot |\pi - \Omega_1|$. Compute the coding gain that is achievable by motion-compensated prediction, both without and with the lowpass filter from b).

Problem 13.2
In motion estimation, it shall be decided which of two motion vectors $\mathbf{k}_1$ and $\mathbf{k}_2$ shall be chosen. The following prediction error matrices have been computed:

$$\mathbf{E}(\mathbf{k}_1) = \begin{bmatrix} 10 & 10 \\ 0 & -10 \end{bmatrix} \quad ; \quad \mathbf{E}(\mathbf{k}_2) = \begin{bmatrix} 20 & 20 \\ 20 & 20 \end{bmatrix}$$

a)   Determine the best vector following the criterion of minimum error variance [(7.184) with $p=2$].
b)   Apply a 2D Walsh transform of block size 2x2. Determine the resulting bit rate, using the table from problem 12.4c. Based on a criterion to minimize the rate, which of the vectors would now be the preferable choice ? Give an interpretation.

Problem 13.3
In the block schematics for scalable hybrid coders as given in Fig. 13.19b, identical reconstructed signals shall result if the entire information (base and enhancement layer) is used. Further, it is assumed that the motion compensation (MC) uses identical shift in all MC blocks of base and enhancement layers. Show that also the prediction errors at the respective quantizer inputs (QE und QB) are identical in both schemes then.

[Hint : Use the relationships between the respective signals in the frequency domain, similar to (12.14) and (12.15) to simplify the proof. Use $H(\Omega)$ for the entire block of motion compensation without explicitly expressing the sizes of the motion vectors.]

Problem 13.4
The spatially characteristic properties of the first frame from a video sequence of 10 frames are modeled by a separable 2D AR(1) process of parameters $\rho_h = \rho_v = \sqrt{3}/2$ and $\sigma_x^2 = 16$. By a camera pan, a horizontal translational movement of exactly 10 pixels/frame leftwards occurs. The frame size shall be $M \times N = 90 \times 50$ pixels.

a)   The signal is encoded using $R=6$ bit/pixel, without utilizing any correlation between samples. Determine the minimum possible distortion D.

In the following points, the signal coded according to a) is regarded as the original reference for subsequent compression.

b)   By which amount [bit/pixel] can the rate be reduced without introducing additional distortion, if *i)* 1D and *ii)* 2D spatial correlation in the signal is utilized?

c)   The exact motion shift shall be known to the decoder, such that perfect motion-compensated prediction can be performed. Sketch those areas in the second frame which have to be newly coded and those which can be perfectly predicted. By this, determine the possible reduction in bit rate due to MC prediction.

d)   Compute the necessary total bit rate for encoding of the sequence with distortion D, if all known spatial and temporal correlation properties are used.

Due to a finite length of a transmission buffer used for rate regulation, the maximum number of bits to be used for encoding of any single frame is limited to 9000.

e)   Compute the minimum additional distortion D_1 (compared to the encoding of a)-d)), which must now be introduced for the first frame of the sequence.

f)   How will the distorted representation of the first frame affect the prediction of the second frame? Which power spectrum of the motion-compensated prediction error will be found within those areas of the second frame which could perfectly be predicted under the previous assumptions?

14 Audio Coding

Speech and audio signals can be characterized as waveforms, which often include components of quasi-periodic signals, but also aperiodic and noise-like components may appear. Such characteristics can be mapped to signal generation models, and also imply that analysis of longer-term behavior will improve the performance of compression algorithms. The importance of periodic components is related to the harmonic properties of many audio sources; harmonic spectra appear in cases of voiced speech and many types of musical instruments, which can directly be related to the physical process of sound generation. Most speech coding schemes are based on linear prediction methods. Synthesis models for speech generation have also reached an advanced status and can be used for parametric coding. In the sector of speech, understandability of the reconstructed signal is an important design criterion. In general-purpose waveform coding of audio signals, utilization of psycho-acoustic properties of hearing is a key factor to achieve what transparent quality, compression without noticeable difference as compared to a digital (PCM) original. The typical core of these methods is transform coding based on block-overlapping or subband transforms. Sound synthesis, enabling synthetic generation of natural or newly-created sounds, is a very mature technology as well. It is however mostly based on nonlinear models, universal usage in analysis/synthesis audio coding is still an area requiring further development.

14.1 Coding of Speech Signals

Most speech codecs are based on principles of *Linear Predictive Coding* (LPC) (sec. 4.2.2), having tight relationship with autoregressive synthesis (sec. 4.2.1). The speech generation in the human vocal tract can excellently be modeled by excitation of a tube with varying perimeters, which then by its acoustic properties of reflection and resonance is mapped into parameters of an autoregressive synthesis filter [RABINER, SCHAFER 1978]; the z-poles of such a filter will correspond to the resonance frequencies of the tube, which again are the physical reason for harmonics in particular in voiced speech. To determine optimum filter parameters, the Wiener-Hopf method (4.74) can be used, which will then typically use autocorrelation values as measured from segments of a speech signal. If the filter is fed by an excitation signal of flat spectral properties (white noise or impulse-like signals), the spectral properties of the output are similar to the properties of the original signal as modeled by the filter. For audible perception, properties of the amplitude spectrum are of high importance. For speech synthesis however, in particular within voiced segments of the signal, the phase information is highly relevant as well, where arbitrary phase of the excitation signal may lead to an unnatural impression. Hence, linear-prediction based speech coding schemes can be classified mainly according to the chosen excitation strategy into the categories described in the following sub-sections.

DPCM and Adaptive DPCM methods. These use direct coding of the prediction error signal, by which the true waveform shall be approximated as close as possible. Due to their simplicity, these methods are still the best choice for higher data rates, and achieve a reduction of rate as compared to PCM-coded (64 kbit/s) speech signals by a factor of 2-3 while retaining good quality. The basic principle is very similar to the DPCM concepts as developed in sec. 12.3.1. In *Adaptive DPCM* (ADPCM), methods with forward and backward adaptation of predictor filters are employed (see sec. 4.2.2), where forward adaptation typically uses filter lengths of 12-16 taps, while backward adaptation uses shorter filters of 4-8 taps. An extremely simplified DPCM method is *Delta Modulation* (DM), which is also applicable to speech compression at higher rates. This is based on a one-tap prediction filter ($a=1$) and a one-bit quantizer which only can reconstruct the prediction error by values $\pm\Delta/2$. In principle, the quantizer is mostly operated in the overload (clipping) zone. Nevertheless, due to the effect of quantization error feedback, it is possible to consistently track the shape of the signal, in particular when DM operates with *over-sampling*. Alternatively, an adaptive quantizer can be used, which increases the steps size Δ in cases where several subsequent quantization decisions have the same sign, and decreases Δ when equal distribution of positive/negative signs is observed. Similar methods of *backward quantizer adaptation* are also used in ADPCM of speech.

Analysis-by-synthesis methods. In this class of speech coders (Fig. 14.1a), the LPC synthesis filter is fed by vectors of innovation signal samples (excitation sequences) from a codebook. The LPC filter is usually determined by a conventional forward-adaptation method, e.g. the Wiener-Hopf equation (4.72) and efficient solutions thereof. Unlike the vector prediction methods presented in sec. 12.3.3, the synthesis filter is a one-step recursive filter, which means that prediction relies on a number of previous samples including the directly preceding sample. No drift-free closed-loop analysis as in DPCM can then be implemented, because reconstruction values related to excitation signal samples in one vector have mutual influences in the reconstruction process. The methodology to find the best excitation signal is an analysis/synthesis approach, where the available excitation vectors are fed into the synthesis filter and the reconstruction sequences are compared against the original. In the process of comparison, signal-dependent spectral weighting is often applied. This introduces a distortion criterion based on *spectral noise shaping*, which is assumed to give best understandability of the synthesized speech signal.

Analysis-by-synthesis (A/S) methods are most widely used for speech coding at low bit rates, where still a natural impression of the speech (retaining the individual characteristics of a speaker as well) shall be achieved. For unvoiced speech, generation of a good synthesis result is uncritical from white-noise (random) type innovation vectors, where the synthesis filter performs the required spectral shaping; the phase information seems to be less important in this case. Different A/S methods can better be characterized by their strategies on voiced speech segments, where a unique pitch property in the signal can only be obtained when a train of excitation pulses is fed into the synthesis filter. The following methods are typically employed:

- *Multi-Pulse Excitation* (MPE), which is based on optimum placement of non-equidistant pulses, where the number of pulses shall be as low as possible;
- *Regular Pulse Excitation* (RPE), similar to MPE, but with equidistant pulses; in MPE and RPE, the set of different excitation configurations (of which only a finite number exists) establishes the excitation codebook;
- *Code Excited Linear Prediction* (CELP), in which excitation signals are selected from a codebook without specifically differentiating between voiced and unvoiced speech segments (in fact, the codebook is usually designed such that both cases are appropriately supported); a variant of CELP is *Vector Sum Excited Linear Prediction* (VSELP).

In particular for encoding at very low bit rates (below 3 kbit/s) also mixtures of these methods are used, e.g. *Mixed Excitation Linear Prediction* (MELP), which is a hybrid between CELP and MPE.

Another method for analysis-synthesis coding with specific handling of voiced speech segments is *Harmonic Vector Excitation Coding* (HVXC), in which the excitation signal is characterized only by its pitch and the envelope of its spectrum (the latter by vector quantization encoding of LPC parameters). This method is implemented as a speech codec in the framework of the MPEG-4 standard, and is claimed to deliver at a rate of 2 kbit/s a comparable quality as the best CELP defined so far (FS 1016) at 4,8 kbit/s. For an overview on different speech coding standards, refer to sec. 17.5.1.

If forward adaptation of LPC parameters is used, the percentage of bit rate to be spent for the side information can become quite significant in low-rate speech coding. Different representations of LPC parameters can be used for the purpose of encoding, e.g. *parcor coefficients* [ITAKURA, SAITO 1972], *log area coefficients* [MAKHOUL ET AL. 1985] or *line spectrum pairs* [ITAKURA, SUGAMURA 1979], which are in principle one-to-one mappings of the direct filter coefficients, but are less sensitive in deviation from original spectral properties, when quantization for low-rate encoding is performed.

Vocoder methods. Vocoders (Fig. 14.1b) are speech synthesis devices, which can e.g. be used to output a written text as a synthetic speech signal, when a mapping is available from letter sequences into phonemes into associated parameters of an LPC synthesis model. In addition, tone height, spectral shape etc. can be varied, such that e.g. either male or female voices can be produced as desired. Vocoders in general lack naturalness, but still provide a good understandability of speech. Only a rough classification into voiced/unvoiced speech segments is made. The excitation signal is switched between a pulse train with a pulse distance equal to the pitch period within voiced speech segments, and a white noise signal within unvoiced speech segments. If applied for encoding, the parameters to be transmitted are the LPC parameters, the information about voiced/unvoiced classification, the pitch period for the voiced segments, and the amplitude (variance) of the excitation signal.

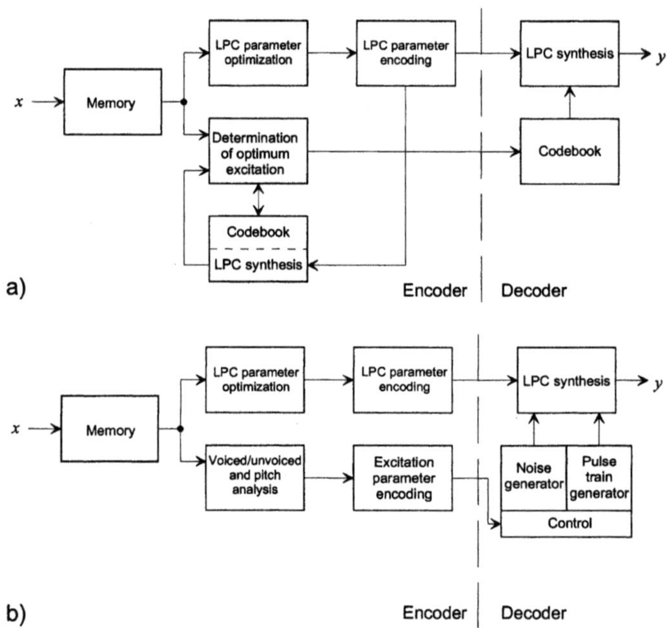

Fig. 14.1. Speech coding: **a** Analysis by synthesis method **b** LPC vocoder method

14.2 Waveform Coding of Audio signals

Audio signals are mainly composed from different sound sources of different nature, such that it is difficult to characterize the generation by a unique physical model. Hence, methods of *waveform coding* establish the most appropriate audio compression technology available today, as they allow reconstruction by good quality for arbitrary audio signals[1]. For waveform coding methods, it is straightforward to employ mechanisms which take into account the psycho-acoustic properties of hearing (sec. 6.2), mainly related to masking. It is in particular simple to integrate such mechanisms into transform coding methods, where a frequency-weighted quantization leads to the desired effect of perceptual weighting. A general method for audio transform coding is illustrated in Fig. 14.2. The psycho-acoustic model will mainly be based on the frequency-dependent masking thresholds, which are the most effective masking mechanisms of hearing; in addition, temporal masking can be used. For signal coding, block-overlapping transforms and subband filter banks are the most suitable methods. For the determination of frequency-dependent masking thresholds, an additional windowed DFT is usually employed in parallel, as the optimum analysis for perception-related frequency groups may be different from the optimum frequency decomposition performed for the purpose of encoding. The thresholds are signal dependent, because instantaneously dominant spectral components can mask out coding noise within other frequency groups. The frequency coefficients are quantized and encoded according to the psycho-acoustic weighting function as determined by the psycho-acoustic model, such that coding noise is preferably shifted into spectral bands where it can hardly be perceived.

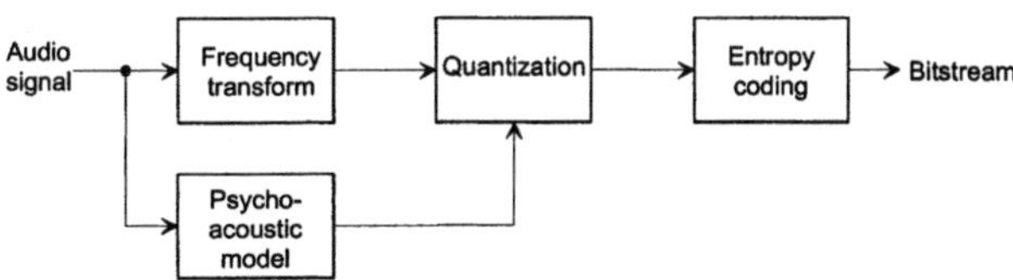

Fig. 14.2. General structure of an audio encoder based on a psycho-acoustic model

32 frequency bands of equal width have been used in the first digital audio codecs which still were mainly optimized for minimization of distortion, e.g. using SNR criteria[2]. To achieve a sufficiently good approximation observing the frequency sensitivity properties of psycho-acoustic weighting functions (see sec. 6.2), this frequency resolution is by far not sufficient. While transforms typically use equal-bandwidth frequency channels, the perceptually adjusted Bark scale is nonlinear, such that variable numbers of transform bands have to be combined into frequency groups. In audio coders which make more extensive use of such weighting functions, between 128 and 1024 frequency bands are typically needed in the transform

---

[1] In contrast to that, e.g. music signals are perceived strangely distorted when encoded by an LPC method which is originally designed for speech signals.

[2] Reference is made here e.g. to the MPEG-1 audio compression standard, layer 1 (cf. sec. 15.5.2)

analysis. The shorter of these numbers is only useful, if a better resolution on the time axis is required which is the case if temporal masking functions shall be employed additionally. Polyphase filterbanks or block-overlapping transforms (sec. 4.3.5) are usually employed for frequency decomposition in these cases.

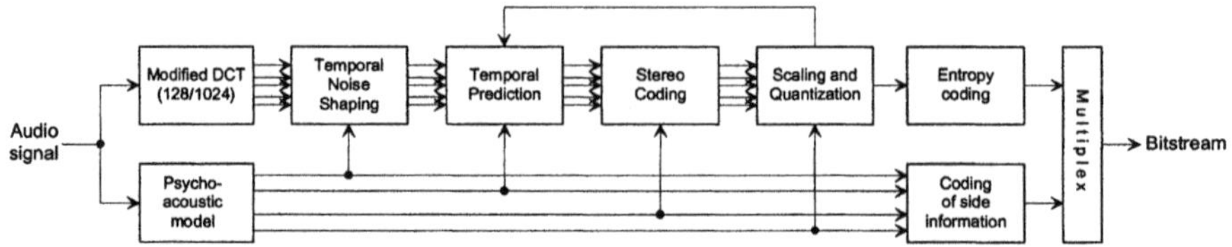

Fig. 14.3. MPEG-2/-4 Advanced Audio Coding (AAC)

Advanced Audio Coding (AAC) is one of the best-performing audio coding methods available today. As compared to the widely used 'MP3' codec (MPEG-1 layer 3, see sec. 17.5.2), it provides further reduction of the data rate by approximately a factor of two without loss of perceptual quality. Originally starting as an amended part of the MPEG-2 standard, AAC has meanwhile been supplemented by a number of additional tools to further increase the coding efficiency; the structure gives a good basis to understand what methodologies are used in state-of-the-art audio waveform coding. A block diagram of the main building blocks is shown in Fig. 14.3.

Transform with switchable resolution. As a monolithic transform with block overlap, a *Modified DCT* (MDCT) is used for frequency decomposition. The block length and hence the frequency resolution is switchable between 128 and 1024 spectral lines. Switching of the filter bank for the shorter analysis block length is in particular useful when the short-duration temporal pre-masking effect shall be utilized (see sec. 6.2), or when instantaneous transitions between sounds or notes are present. When the signal is quasi-stationary over the transform segment, it is advantageous to use the longer block length with better frequency resolution, which allows a more precise adaptation to the frequency masking functions.

Temporal prediction of transform coefficients. In the case of 'uniform tone' signals, a significant statistical dependency exists between the transform coefficients of subsequent blocks. This can be exploited by invoking a DPCM loop, by which a prediction is made from the coefficients of previous transform blocks. Two types of prediction are defined, where the *long-term prediction* is attractive due to the significantly lower complexity.

Advanced quantization and entropy coding. Improved entropy coding of spectral coefficients is supported by a flexible adaptation capability for the VLC tables of the entropy coder. For extremely low rates (<40 kbit/s) a method of *Transform*

Domain Weighted Interleaved Vector Quantization (TWIN-VQ) is additionally defined, which then replaces the separate quantization and encoding of transform coefficients. The other building blocks of AAC are retained without changes. The selection of the vectors is also controlled by the perceptual model. The notion *interleaving* reflects the method of ordering the spectral coefficients into vectors, which are preferably assembled from subbands of similar properties. Further, a scalable mode is defined, which optionally replaces the one-layer encoding step by a multi-layer method, which is based on embedded quantization and *Bit Slice Arithmetic Coding* (BSAC), an entropy-coding method performing bit-plane coding very similar to the methods described in sec. 11.3, however without sophisticated adaptation or context utilization mechanisms. As it is difficult to utilize prediction in scalable coding due to the drift problem (cf. sec. 13.2.8), the BSAC method is inferior as compared to a single-layer coder. Scalable coding can also be employed in combination with TWIN-VQ. Here, the information encoded by TWIN-VQ establishes the base layer, and a residual error is encoded as enhancement layer. Therefore, compression performance at higher rates is worse than with the fully-embedded BSAC method. Additional mechanisms are defined for error-resilient transmission of AAC streams.

Stereo redundancy. For exploitation of *stereo redundancy*, it must be observed that directional identification in stereophonic signals or in binaural perception is not only based on amplitude differences between the left an right channels ('stereo by intensity'). The delay between the signals arriving at the left and right ears and the acoustic transfer functions (characterized by the room properties and the shape of the head) play an important role as well. The latter aspects are in particular important in 'artificial head' stereophony. Also in cases where the stereophonic signal is recorded using different microphones at distant positions, delays between the sounds arriving by left and right channels are not negligible. This will cause phase shifts between the left and right channels, which prevents direct similarity matching of the left and right signals. In addition, these shifts can be frequency dependent due to the room and ear transfer functions; when different sound sources located at different positions are recorded, the resulting phase shifts will be variable, and the left and right channel signals which result from mixtures of all sources could become relatively dissimilar. A universal method in stereo signal encoding, which allows to exploit redundancy as far as possible under different conditions, is shown in Fig. 14.4. The system is switchable between the following modes:

- Separate encoding of left and right channel signals;
- 'Mid/side' coding, i.e. encoding of a (monophonic) sum signal and a difference signal from the two stereo channels; this method is very similar to the method used in analog (FM) stereo audio broadcast;
- Intensity coding, i.e. coding of one channel, and use of an intensity factor to synthesize the second channel as an amplitude-modified copy of the first. Intensity coding is mainly useful for synthetically mixed sources where both ste-

reo channels are originating from a monophonic source and have been placed on the stereo basis by amplitude balance adjustment.

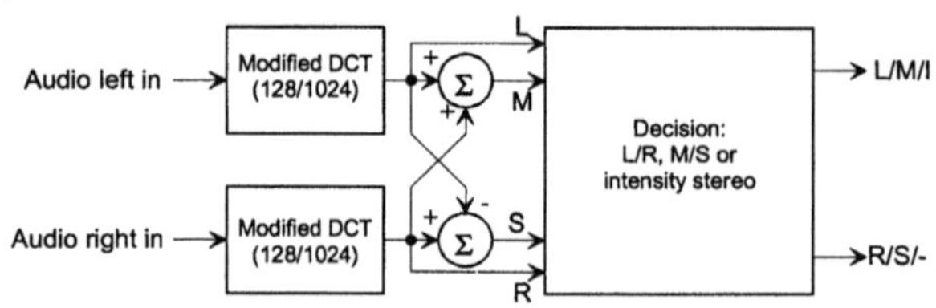

Fig. 14.4. Stereo signal coding modes in Advanced Audio Coding (AAC)

Perceptual noise substitution and spectral band replication. Analyzing psycho-acoustic properties of human hearing, certain noise-like (non-harmonic) signal components exist, for which exact waveform coding, including preservation of the phase relationships, is irrelevant. These components can be substituted within their respective frequency bands by noise signals of same variance. This principle of synthetic generation of signal components has been further developed towards more general mechanisms of *spectral band replication* (SBR), which also allow to substitute higher-frequency harmonics by synthetic signal components. Typically then, only a narrowband signal of 8 *kHz* bandwidth is encoded, and all frequency components above are synthetically reproduced, which is done differently for harmonic and non-harmonic components.

Another improvement of quality is achieved by *temporal noise shaping*, by which the frequency weighting functions are not freely adapted within each transform block, as sudden fluctuations of the spectral characteristics of coding noise can have an annoying effect. By enforcing consistency of coding noise between neighbored analysis blocks, a significant perceptual quality improvement is achieved.

It is interesting to note that many of these methods have their direct counterparts in techniques that are applied in still image and video coding. Switchable transform resolution is analogous with variable block-size transforms in image coding (sec. 12.4.4), where shorter sizes are also applied in cases of transitional signal behavior. The temporal prediction in the context of transforms over finite bases had been identified as a key to improve compression performance in image coding as well. The goal of noise shaping is very similar to a central goal in video coding, to keep coding distortions as consistent as possible over frame sequences.

The AAC with bandwidth extension allows practically transparent (without audible impairment) encoding of stereo audio signals at a rate as low as 48 kbit/s. Using TWIN-VQ, a quality which is better than AM radio is achieved at a data rate of 20 kbit/s. This allows applications like Internet radio streaming for extremely narrow-bandwidth channels, e.g. via telephone modems or mobile terminals. Perceptually adapted audio coding schemes can eventually impose high *measurable* distortions (e.g. by SNR criteria) to the signal, while the effect may not be *audible* at all. This is a typical approach to exploit the principle of irrelevance. If however the encoded bit rate is too low, artifacts may become audible quite drastically, in particular as characterized by

- Non-harmonic distortions (observe that clipping distortions as known from analog audio systems impose mainly harmonic distortions, which may in fact be more convenient to the ear);
- Frequency-selective and non-stationary noise;
- 'Roughness' of the signal, which becomes in particular audible due to temporal fluctuations between the analysis blocks;
- Loss of high frequencies.

An important aspect in *encoder optimization* for audio signals is the best adjustment of the perception models, but rate-distortion criteria can be used as well (see sec. 6.2 and 11.3). For example, it can be observed that different MP3 encoders which are available in the market are significantly different in terms of quality of the decoded signal, for streams encoded by identical bit rates.

Lossless coding of audio signals. Recently, some interest has developed in *lossless audio coding*, which means that each bit in the PCM representation, which can range up to 24 bit/sample/channel and 192 *kHz* sampling rates, must be preserved. The methods used for this purpose are quite conventional. Predictive (open-loop) coding is used in combination with systematic entropy coding methods like Rice codes [LIEBCHEN 2003].

14.3 Parametric Coding of Audio and Sound Signals

Audio signals as they typically have to be encoded not only consist of natural sound recordings, synthetically generated sounds have become more important in music production during the last decades. Synthesis methods exist which are capable to imitate natural sounds such that they are almost indistinguishable. These synthesis methods can often be characterized by a small set of parameters (e.g. filters, oscillator frequencies, noise source properties, coupling and modulation of different sources, temporal envelope characteristics). Audio synthesis description formats such as MIDI are supported by many devices like PC soundcards, mobile phones etc. This implies the possibility to encode both natural and synthetic signals by the characterization of a suitable synthesis model and a small set of synthesis parameters. However many unresolved problems exist to directly map a sound available as a waveform into a parametric synthesis model. Indeed, for most musical instruments useful synthesis models exist, which are based on acoustical principles of sound generation, e.g. related to properties like attack and decay characteristics, waveform modulation etc. For the latter aspect, frequency modulation

(FM) has established as a method which generates very natural instrument imitations or new synthetic sounds; FM is however nonlinear, such that it is not straightforward to determine the parameters which are necessary to obtain a desired synthesis result.

The viable principle to find best settings in parametric coding is analysis-by-synthesis optimization. This could be highly complex, in particular if the parameters are sensitive, if the parameter space is multi-dimensional, and if optimization includes the selection of the synthesis method as well. This problem already applies to the case of single sound sources, even though it is solvable in this case using exhaustive analysis-by-synthesis optimization. For composite sources, such as polyphonic (e.g. orchestra) sounds, additional problems are as follows:

- Signal decomposition and identification is necessary, i.e. before the analysis of a pure sound can be made, it must be determined which components exist in the composite signal. Even though an orchestra sound could also be imitated by a single synthesis model, a more natural impression will be achieved if several synthesis models fine-tuned to the properties of the single instruments are mixed. Such a method must also include an identification of the sound sources within the orchestra sound to decide for the appropriate synthesis model.
- When recording is not made in a reverberation-free environment, room modeling or canceling out of room influences will become an important part of the analysis problem. This could include reverberation and echo analysis and compensation.

These problems turn out to be at least similarly complex as for the case of content-based encoding of image and video signals. As a perfect separation of sound sources from any given recording seems to be impossible, again the properties of human hearing should be exploited, which means that an analysis-by-synthesis approach should be able to differentiate mixed sound sources by their properties as good as a human would be able to. This must involve a priori knowledge about feature characteristics of individual sound sources and signal estimation methods.

If the requirement is not a *perceptually lossless* coding, application of parametric audio coding is already realistic today. Possible applications can be identified in the range of extremely low bit rates, where waveform coding will fail or produce unacceptable coding distortions. A method for parametric coding of audio signals is defined in the audio part of the MPEG-4 standard, which gives good synthesis quality for cases of monophonic sources (single musical instruments) or polyphonic sounds of low complexity. A block schematic of this *Harmonic and Individual Lines plus Noise* (HILN) scheme is shown in Fig. 14.5. The first step is a decomposition of the signal into harmonic components, sinusoidal components and noise components, all of which are parametrically encoded based on a perception model using relevance detection and weighting. At the decoder end, the reverse principle of composition of the synthesized signal from these three components is performed. The method has been developed for encoded stream rates between 4 and 16 kbit/s. For monophonic signals, reconstruction of a signal from a

6 kbit/s stream delivers a fairly acceptable quality, but is by far not transparent when compared to the original. The most important properties of the signal are retained, which would at such low rates never be possible with one of the waveform coding methods described in sec. 14.2. Recently, new extensions to parametric coding were proposed which include *transients* as an additional synthesis element. This is better capable to model temporal-envelope characteristics such as attack properties, and significantly improves the synthesis quality also in case of more complex polyphonic sounds.

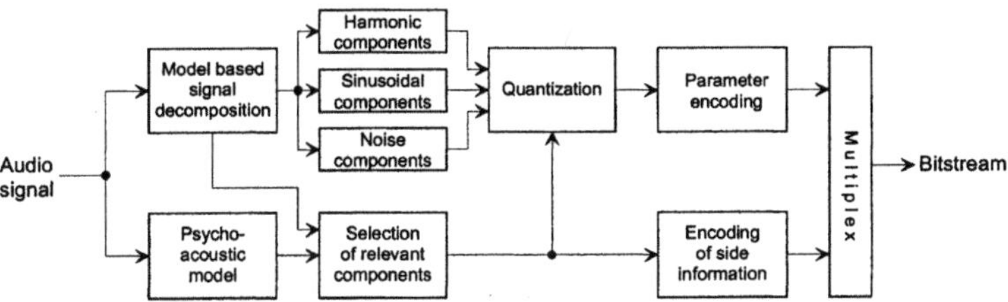

Fig. 14.5. Block diagram of the parametric audio coding scheme HILN as defined by the MPEG-4 standard

Another important aspect of natural audio synthesis is related to the room environment in which signals are typically generated. Sound field modeling, e.g. by room transfer functions (see sec. 7.10.4) will have a high impact on the degree of naturalness by which a synthetic sound is perceived. Even though this can mostly be seen as a problem of signal composition (cf. sec. 16.6), it could also be used for parametric audio compression, if the properties of rooms can be extracted from a given signal and be modeled by a low number of parameters, e.g. early echo positions and reverberation time.

Part D: Applications and Standards

15 Transmission and Storage

Transmission and storage are core parts of multimedia communication systems. Before multimedia content can be consumed, it has to be made available either via a transmission network or locally on a storage medium. Digital techniques firstly are advantageous compared to traditional (analog) solutions, as they can provide better quality consuming less bandwidth. In contrast to analog technology, where transmission channels and storage media were designed for monolithic media types as video or speech signals, digital multimedia representation allows to convey content by any transmission and storage medium, provided that sufficient bandwidth is available. This chapter introduces solutions for digital media transmission and storage in scenarios where traditional services are replaced, and also where new services are evolving, like internet streaming and mobile access to media sources. Traditional monolithic-media systems are replaced by completely heterogeneous network environments and diversified presentation devices (terminals) for different situations supporting mobile access, home access etc., such that requirements for flexible media adaptation are becoming dominant. On the other hand, a unification is achieved, as any media types can be accessed anywhere with a given terminal in the ideal case. In addition, multimedia storage and transmission services can be augmented by additional functionalities like content-driven or customized access to media.

15.1 Convergence of Digital Multimedia Services

Digital techniques allow a mass dissemination of multimedia signals. A driving force of this development is the convergence of traditionally separate sectors of *telecommunications*, *computers*, audiovisual *entertainment*, and personal *photography/video*. New types of terminals are being developed which allow to access all types of multimedia information everytime and everywhere, and also to personally acquire audiovisual signals for storage or for up-stream sending over networks.

Digital techniques using compressed formats for storage and transmission have a number of advantages over traditional analog media[1], which are going to be replaced either entirely, or kept as originals, while digital media allow simplified mass dissemination. The most eminent advantages of digitization are as follows:

- Copying without quality loss, prevention from physical/chemical ageing of carriers by lossless backup copies;
- Less bandwidth consumption, which allows personalized transmission of multimedia signals even over narrow-bandwidth channels;
- Simple integration of different media for usage in multi-functional devices;
- Simple transformation from one format into another;
- Lower susceptibility against losses in storage and transmission, when optimized accordingly;
- Simple augmentation by additional information (metadata);
- Usage of random-access recording media.

On contrary, disadvantages are:

- If losses occur, the degradation is often ungraceful;
- Fast 'ageing' of formats and devices can be observed due to continuous innovations. In this context, also *transcoding* of signals may become necessary to provide them in the new format or make them replayable by a new type of device, which is often a lossy process.
- Without proper protection, unauthorized access is simplified.

In short, two reasons can be identified for digital multimedia representations to take over the previous analog-dominated domains: They are cheaper and they provide more functionality. Audiovisual information is becoming omnipresent in professional and private sectors, and the time we live in has already been entitled as the *Visual Information Age*[2]. Audiovisual media are changing the ways how we communicate and cooperate, acquire information, learn and work, how we interact with environments and form our imagination about the outside world. Last but not least, they are intended to be sold as articles of commerce. The latter aspect is critical, as the simplified universal access to multimedia sources makes control over dissemination more complicated, and any security mechanisms typically are open for cracking.

Mobility, *personalization*, *interactivity* and *spatialization* are central new functionality aspects, which elevate digital media over analog predecessors. Mobility covers the aspect of ubiquitous multimedia information access and dissemination,

---

[1] The term 'traditional analog' covers recording and transmission of audiovisual signals on canvas, paper, film, magnetic tapes, records etc. in a very wide sense.

[2] Without doubt, audiovisual information has largely influenced the human society in the past century and is proceeding to do so maybe even more in the present century. For more background on these aspects, interested readers are referred to the works of MARSHALL MCLUHAN, VILEM FLUSSER and PAUL VIRILIO.

and enables usage for personalized purposes from anywhere. In the context of mobility, further integration of traffic systems (e.g. vehicles, intelligent traffic analysis and routing systems) with multimedia systems can be expected. On the other hand, digital multimedia systems enable *virtual mobility*, where by interactivity and spatialization mechanisms (e.g. 3D displays or 3D navigation), the illusion of being in a different place is provided. This eventually will decrease the amount of physical mobility in the world, even though presently such tendencies can not be verified; it may well be the case that virtual mobility will just supplement the different ways how we experience the world and how we communicate. Catchwords in this context are virtual meetings and teleconferences, virtual schools and universities, virtual visits of experts to solve a specific problem, virtual shopping, journeys etc.

Different requirements result by the nature of applications. At the consumer end, a typical postulate is the satisfactory perception of multimedia content. *Perceptual quality* and *quality of content* are equally important requirements. *Latency* is critical in applications with time-critical interaction or in bi-directional communications, where typically round-trip times of more than 200 ms are perceived as annoying. In time-critical surveillance applications, low latency is of high importance as well. Switching of programs in broadcast, switching between different sources in a surveillance application or access to stored material without incurring long delays is also more convenient, even though this will be less critical than the previous case. The grade of compliance with such requirements is often weighted against the *cost*. This relates to the cost of transmission, cost of devices and cost of the content itself a user is willing to pay.

For recording and transmission of digitally represented audio, image and video signals, different storage media and transmission channels have very different characteristics regarding *capacity* (storage, access and transmission bandwidth), *latency* (access and transmission delay), and *transmission quality* (error characteristics, fluctuations etc.). This will directly or indirectly map into the level of user satisfaction. From the viewpoint of the user, it is irrelevant where the content is residing, provided that the service is sufficiently fast and of sufficient quality. We can stick to this abstracted point of view, and investigate the impact that the characteristics of any channel will have on the quality perceived by the user.

It is not necessary here to make any difference between transmission and storage any more; both cases are viewed as abstract channels with specific characteristics. In fact, communication networks today are a mixture of transmission paths and storage devices, e.g. providing temporary storage at proxy servers or anywhere else on the route, optimized for the goals of lowest possible latency and best quality to be delivered to the user. Supplementing transmission chains by temporary local storage, providing *variable latency* at the receiver end, is even more advantageous as it grants additional functionality like pause, reverse and forward replay modes at least over a limited range of the timeline.

Aspects related to the adaptation of digital multimedia representations for transmission over channels of different characteristics are discussed in the subsequent section. In sec. 15.3 and 15.4, more specific methods of transmission for the cases of broadcast and internet/mobile streaming are introduced.

15.2 Adaptation to Channel Characteristics

With the advent of digital multimedia signal transmission, traditional ways of transmitting *one type of data over one network* specifically designed for that purpose are either gradually or radically broken up. It depends on best strategies for transition toward fully digital solutions and other reasoning, if temporary co-existence of analog and digital transmission shall be considered. Traditional analog services like radio and TV broadcast use monolithic systems, which specify the entire chain from the acquisition device (camera, microphone) output, all parts of the transmission system (including exact signal specifications) and the receiver. In principle, also the first generations of digital media broadcast systems (*Digital Audio Broadcast*, DAB; *Digital Video Broadcast*, DVB – see sec. 15.3) follow a similar paradigm. For a principal category of applications the encoded data formats, transport mechanisms etc. are specified, even though flexible transport of other types of data is possible. The transport mechanisms of these networks are designed to convey digital information by a pre-defined rate in real time, which means that the network QoS at least as relating to transmission bandwidth and latency can be guaranteed. Further guarantees are usually made by broadcast service providers to install a sufficient infrastructure, such that under normal conditions a reasonably error-free transmission is achieved.

As another example, the Internet is not homogeneous, it can rather be regarded as a *network of networks*. Even more, the original design approach of the Internet protocol stack was not made for real-time streaming of media data. Both transmission bandwidth and latency are variable. The transmission mechanism is based on *packet switching*, which means that the available bandwidth is in principle dynamically allocated to different users, depending on the frequency by which they release data packets into the network. This can cause network overload, which is the main reason for data losses and variable latency within the network; in principle, always entire data packets are then lost during transmission.

Digital personalized mobile transmission networks as installed today (GSM or similar) are based on *circuit switching*, which means they reserve a fixed transmission capacity with guaranteed maximum latency for each user, however the bandwidth is rather low. These networks were primarily designed for speech communication. Next generations of mobile networks such as UMTS/IMT-2000 or beyond use packet switching as well, where however better control mechanisms with regard to packet losses exist than they are available for the Internet of today.

Due to the high flexibility of digital transmission, it appears quite realistic to achieve interoperability and adaptation of transmission mechanisms with any kind of channels. A high-level view of this problem is shown in Fig. 15.1. In the ideal case, it should be possible to define adaptation mechanisms *completely independent* of the source coding/decoding algorithm, such that a flexible adaptation can be made *after encoding is finalized*, depending on the type of network and instantaneous network conditions. This could further take into account the *replay format* of the terminal, such that no unnecessary information is transmitted which is not used at the receiver end. This ideal case can not be achieved for any kind of source coding algorithms. If networks are operated in a chain, this chain will always be as weak as its weakest link, e.g. if a narrow-bandwidth and error-prone channel path is present anywhere in the entire transmission chain, this will constrain the overall quality.

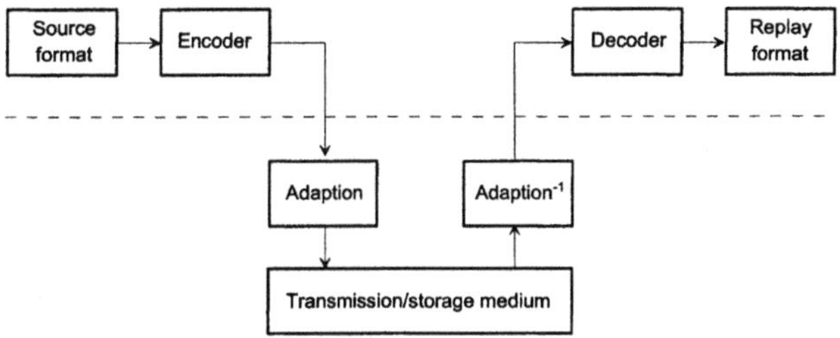

Fig. 15.1. Adaptation of a generic multimedia signal encoder/decoder to a transmission or storage medium

Regarding transmission characteristics, channels are classified into *Variable Bit Rate* (VBR; e.g. Internet, advanced mobile channels, hard-discs, DVDs) or *Constant Bit Rate* (CBR; e.g. broadcast or GSM mobile channels, tape drives) types. In principle, VBR transmission is ideally synergetic with multimedia sources, as most coding algorithms by their nature produce variable-rate streams. In real-time transmission, full advantage of VBR can only be achieved by a multiplex system, where it can statistically be expected that some sources instantaneously produce high rate while other sources produce medium or low rate[1]. Most variable-rate channels use packet switching, where data access units as generated by sources are bundled in packets and transmitted only when necessary[2]; In packet switching, delays and packet losses occur if the total amount of data comes up to the entire capacity of the underlying physical channels. This is in particular critical in real-time streaming. As a consequence, even in VBR channels it will often be necessary to adapt the data rate which the source produces, in order to prevent unacceptable delays and packet losses. For CBR channels, the rate must be limited to the rate

---

[1] This effect is denoted as *statistical multiplexing*.

[2] A data packet generally consists of a *header*, which contains all necessary information to identify the packet, and a *payload* which contains the usable information. For typical cases, the overhead data rate by the header information is below 10 percent, which highly depends on the packet length and the compactness of the header.

that can be conveyed over the channel anyway. From the viewpoint of source coding, *rate control* mechanisms must be designed.

A second criterion to be observed in adaptation are the *error characteristics* of the network. Typical errors in packet-switching networks are *packet losses*[1], and it is often assumed that other (e.g. isolated bit) errors are irrelevant in these types of networks anyway. For synchronous transmission over VBR networks, bit losses are the more typical errors occurring, which can either be *random errors* (e.g. in case of cable networks or CDs) or have the nature of *burst errors* (e.g. in circuit switching mobile networks or on magnetic tapes).

A distinction criterion for best strategies on error control relates to the property of whether a *back channel* exists in the network, which can be used to inform the sender about problems in transmission. Typically, back channels are not practical for a *point-to-multipoint* operation, as occurring in broadcast applications or multicast[2] over the Internet, because it is then impossible that the transmitter side reacts for each receiver individually. For *point-to-point* connections, usage of back channels is highly practical and allows to implement powerful error control mechanisms.

Finally, in optimization of transmission strategies it is important to distinguish between live streaming, streaming of stored media and media file transfer applications. The case of *file transfer* is uncritical regarding transmission errors, as retransmissions are requested in typical file transfer protocols until the file is available at the receiver end without errors. On the other hand, this has the highest latency in a sense that it is necessary to wait until the entire file has been transferred, before the signal(s) represented in the file can be presented to the user. In *streaming of stored media*, presentation of content typically starts immediately after a sufficient amount of data has been received. Two aspects have to be considered here:

– Encoding is done before storing. If either rate or error control have any impact on the operation of the encoding algorithm, a guess (estimate) of the network conditions at the time of transmission must be made. Alternatively, different streams supporting different conditions can be stored, or real-time *transcoding* can be made by the time of transmission, which however increases the server complexity and most likely decreases the quality. Ideally, the nature of an encoding algorithm would allow independent adaptation of a once-encoded stream, which could be a quite realistic case for scalable or embedded-stream solutions.

– Latency is typically less critical in streaming of stored content, as the user will only notice an initial delay between the time of request until the signal starts replaying. Further, encoding delay is at all uncritical here, the entire latency

---

[1] Also in hard disc storage, which is often assumed to be practically error free, entire data blocks are lost or discarded in case of defects, giving a typical packet-loss error characteristic.

[2] depending on the number of connections simultaneously served in multicast.

perceived results by transmission, buffering at the receiver end and decoding. This is important, as increased latency can typically help to improve the quality of transmission and encoding.

For live streaming, real-time encoding is indeed necessary. This means that any useful adaptation of the encoding process as related to network conditions can be made instantaneously. In the case of multicast live streaming, individual adaptation for different transmission chains can further become necessary. In such a case, scalability of streams can save a considerable amount of bandwidth, which is in particular important in case of narrow-bandwidth upstream channels.

Due to complexity considerations, real-time encoders often provide a lower reconstructed quality than highly optimized off-line encoders which are specifically designed for a scenario 'encode once, decode multiple times with highest quality achievable'. The aspect of latency is application dependent. If real-time interaction of users is mandatory, latency should be low. If this is not the case, it is well possible to employ the same policy as stated above under the second bullet.

For unicast transmission over variable bandwidth networks without QoS guarantee, *stream switching* is a widely used method. In particular for streaming server applications, a number of pre-encoded single layer streams must be stored in parallel that are encoded by data rates fitting with the expected network bandwidth variations. At the time of transmission, the respective stream which best matches the instantaneously available network bandwidth is selected for transmission.

15.2.1   Rate and Transmission Control

The bit rate resulting by the data compression methods described in chapters 11-14 highly depends on the complexity[1] of the sources; for example, if the same quantizer setting (quantization step size) is used, signals with less variability in time or space will cause less bit rate output, typically due to the use of VLCs. This means if the quantizer setting (distortion) is fixed, variable bit rate (VBR) will be the normal case for multimedia sources. If a transmission or storage medium shall be used where the *bandwidth is fixed*, the distortion must be varied for constant bit rate (CBR). Typical examples where this is necessary are modem channels (ISDN, xDSL), broadcast channels (DVB, DAB) and storage media with a constant replay rate (tape recorders, CD players). As the bit rate resulting for a specific signal can not be foreseen a priori, methods to produce a fixed bit rate output of an encoder use rate control mechanisms, where the bits are not output instantaneously but written into a temporary first-in-first-out (FIFO) buffer, which is filled by variable rate from the encoder output, but streamed out by fixed rate towards the channel input (see Fig. 15.2). A model for this is the *leaky bucket* [REIBMAN, HASKELL 1992], assuming a bucket of given volumetric capacity being filled with water which is not flowing in continuously, but dropping out by a constant number of drops per time unit. The task is then to regulate the input flow such that the bucket never

---

[1] resolution, detail properties etc.

overflows, but that also at least a minimum amount of information always is retained within the bucket.

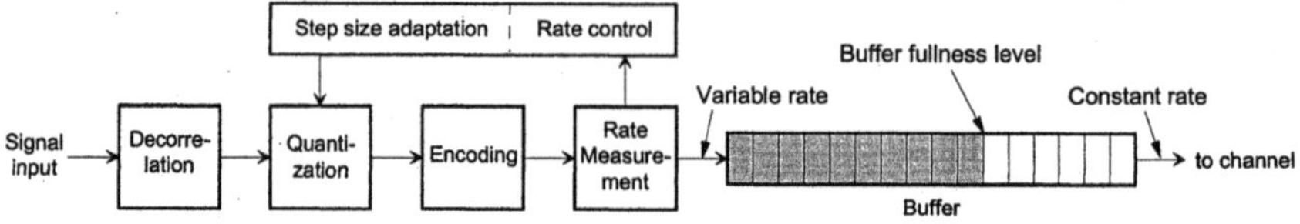

Fig. 15.2. Rate control by quantizer step size adaptation

The problem is more tricky than it looks at first sight. In particular, it is not sufficient to only serve the channel by a fixed rate without taking into regard the schedule of the decoder. If for example a video frame is scheduled to be readily decoded for display at a certain point in time, the information being related to this frame must have been readily available for decoding by a certain time span in advance, which is at minimum the time required for decoding. This can only be achieved in the given scenario, when a complementary buffer is implemented at the decoder. This decoder buffer works symmetrically to the leaky bucket: It is filled by a constant amount of bits per time unit, but is flushed by regular distances, where the flushing stops when all bits necessary to decode one unit of access (e.g. a frame of video) have been released from the buffer[1]. The buffer model can be described by three parameters R, B and F, where R is the (constant output) rate transmitted over the channel, B is the buffer size and F is an initial buffer fullness which shall be reached before decoding starts [RIBAS-CORBERA ET AL. 2002]. An example timing diagram related to the decoder buffer fullness is shown in Fig. 15.3. An initial delay F/R occurs before decoding starts.

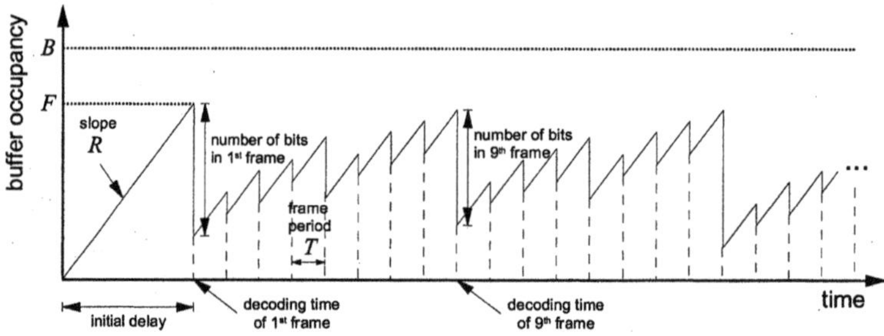

Fig. 15.3. Decoder buffer timing diagram, case of synchronous frame decoding and display

---

[1] The example of video is regarded here, as it establishes one of the most challenging cases, due to the high fluctuations in data rate that are inherent by statistical properties of video signals.

Both cases of buffer overflow and buffer underflow have to be prevented[1]. In principle, the rate control algorithm has to track the (hypothetical) state of the decoder buffer for this purpose. Typically, a regulation of quantizer step size must be executed when the buffer occupancy either approaches a certain margin towards 0 or B. Algorithms for buffer control can become quite complex, e.g. taking into account expected bit budgets for I-, P- and B-type frames[2]. Simple regulation of the quantization step size Δ for the purpose of rate control may also be sub-optimum in the sense of rate-distortion optimization (RDO). Combinations of rate control with RDO are however straightforward [CHOU, MIAO 2001], as increasing or decreasing the Lagrangian multiplier λ in RDO algorithms is directly related to a decrease or increase of rate, respectively. In both cases, adjustment of Δ or adjustment of λ, strong fluctuations should be avoided as far as possible, which again would be sub-optimum in the sense of rate-constrained quality at the global level of the entire sequence of frames.

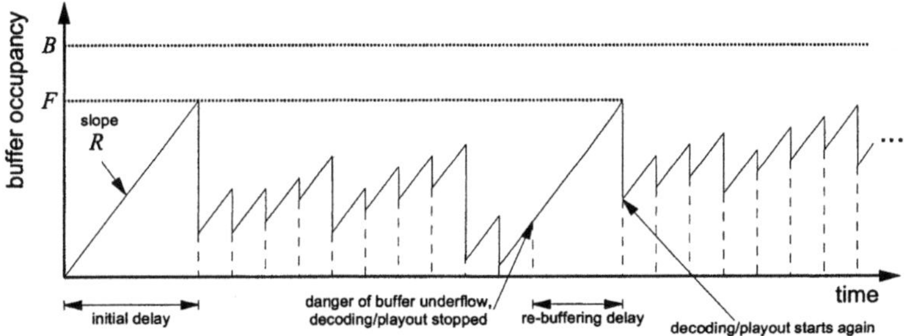

Fig. 15.4. Decoder buffer timing diagram with re-buffering

Typically, if the decoder buffer runs into underflow, losses occur because the information required for decoding is not available. This is in particular true for real-time applications with scheduled timing of outputs, e.g. synchronous frame rates of video or audio signals. In cases where a low delay is required, but synchronous frame play-out is not important (e.g. low-cost videoconferencing), the decoder could perform decoding whenever all bits related to a new frame are available, but the appearance may become jerky. In other cases, *re-buffering* can be used, for which decoding is stopped for a short period until the buffer is sufficiently full again. This incurs an additional delay and has the consequence that the replay time

---

[1] The case of decoder buffer *overflow* could simply be avoided if irrelevant *stuffing bits* are sent over the channel, however then a mechanism must be implemented to prevent these stuffing bits from being written into the decoder buffer.

[2] In Fig. 15.3, an example is shown where each 8th frame could be an I-type frame consuming more data rate. In the time between two I-type frames, the buffer fills up gradually, such that sufficient headroom is available when the bits for the next I-type frame are flushed. In principle, this means that the transmission of bits related to the I-type frames takes significantly longer time than a normal frame period.

e.g. of a movie is extended by the time of the re-buffering periods[1]. An example is shown in Fig. 15.4.

In practice, the method of re-buffering is not useful for constant-rate/constant-delay networks, where a sufficiently stable rate control can be implemented at the encoder which guarantees that decoder buffer underflows will never occur. Re-buffering is a decoder-side emergency mechanism which can prevent from completely losing control in cases of transmission over variable-rate/variable-delay networks, where the time by which the information arrives at the decoder cannot be controlled by the encoder side. It can nevertheless be tried at the encoder to implement transmission strategies by which the need for re-buffering is kept to a minimum. For this case, it is necessary that the transmitter has knowledge about *network parameters*, which can be gained in different ways:

- If a VBR transmission over a *network with QoS guarantees* is made, typical negotiated parameters are mean rate, peak rate, maximum peak duration, maximum delay, delay jitter and loss-related parameters. As these are guarantees, transmission strategies can be tuned to the given bounds. These networks would eventually reject acceptance of packets from transmitters who do not stick to the negotiated parameters, which however is automatically prevented by the described strategies such as leaky bucket control.
- If a *best-effort network* is used, no guarantees about transmission parameters are made. It is however possible to gain feedback information about the instantaneous transmission quality either directly from the network or from the client (receiver) side.

In the case of VBR transmission, the decoder buffer timing diagrams are varied such that the rate-related slope is not constant anymore. In this case, the fullness level B can also be approached arbitrarily close, and the transmitter simply would stop sending information when this point is reached. This corresponds to the fact that in VBR transmission it is allowed to send with a *peak rate R_{max}* for a short time period. The advantage is that the initial delay becomes shorter, and that the buffer resources can be managed more flexibly. Unfortunately, networks which support VBR transmission usually also cannot guarantee a constant delay. This is simply due to the fact that the network itself contains buffers to shape the traffic which is multiplexed from different sources. If many sources instantaneously transmit by higher rates, more buffering is needed such that the delay increases. This requires the implementation of more intelligent *scheduling mechanisms* in the decision about optimum timing for transmitting and decoding the information. In the case of packet-based networks, the schedules for packetization at the transmitter and de-packetization at the receiver, taking into account expected delays and delay jitter, data losses and re-transmission (see subsequent section) are a crucial part of such strategies, and receiver/decoder buffers have to be extended such that the overall transmission chain can cope with more extreme situations.

---

[1] In case of audio, slowing down the play-out speed is more complicated (cf. sec. 16.6)

With regard to rate control, embedded and scalable coding are ideal solutions, as the streams are readily organized as multiple-rate streams with clear priority which information has to go first when only a certain bit-rate budget can be transmitted. If scalability is combined with scalable stream arrangement in a rate-distortion optimized approach, a near optimum result is achievable, depending on the granularity of scalable layers. Implementation of constant-rate or variable-rate streaming becomes extremely simple, as no encoder has to be invoked any more. This opens the path for much more flexible solutions, where decisions about rate control strategies can be done at any place anywhere in the transmission path, when it turns out that the assumptions made at the transmitter side are in conflict with the actual delay or with the instantaneous bandwidth that can be supported. This will however only take full effect when combined with appropriate error control strategies.

15.2.2   Error Control

Some source coding algorithms provide inherent tools for resilience against transmission losses (cf. sec. 11.4.3, 11.4.7, 12.3.4, 12.4.5, 13.2.8 and 13.2.9)[1]. Implementation of *resynchronization* is of high importance, which means that entry points must be inserted in the media streams where the decoding process can restart in a case when an unrecoverable error has occurred previously. Often, error-resilient and re-synchronization methods are less efficient in terms of compression, but on the other hand, regarding the event of network errors, can indeed provide better quality for a given transmission situation. A typical example for this is the application of intraframe coding or interframe prediction in the case of video sequence coding. Obviously, when all frames are intraframe coded, no propagation of errors from one frame to the next will ever occur, but less compression will be achieved than in interframe coding. On the other extreme, if only the first frame is intraframe coded, and all subsequent frames are coded by interframe prediction, compression will be good, but in principle infinite propagation of errors can occur, when only some small part of the information is lost. It is hence necessary to find an optimum between both extremes, which is often made by *cyclic intraframe coding*, not affecting the compression performance largely, but also preventing infinite propagation of errors[2]. In principle, these strategies replace the IIR synthesis filter of the MC prediction decoder loop by an FIR filter. Two different methods are shown in Fig. 15.5. The partial refresh over selected areas which are changed from frame to frame (Fig. 15.5b) is advantageous if data rate fluctuations shall be kept low e.g. for a simpler buffer control and lower overall delay. The

---

[1] Some of these methods, e.g. data partitioning and scalable coding, assume that a part of the information can be conveyed by 'first class' priority, where less transmission errors shall occur; this aspect will be further discussed below.

[2] Alternatives are methodologies which perform a non-intra update of the decoder frame memory, e.g. by usage of switching frames (cf. sec. 13.2.9).

full-frame refresh as a by-product gives well-defined access or entry points for decoding, and is more appropriate for broadcast or storage applications. The optimum refresh rate depends on the actual probability and other characteristics (e.g. burst behavior) of transmission errors.

On the other hand, the capacity of any channel is limited, and the probability of errors increases when more information shall be transmitted over the channel. Finding optimum solutions in such cases are typical decision problems of *joint source and channel coding*.

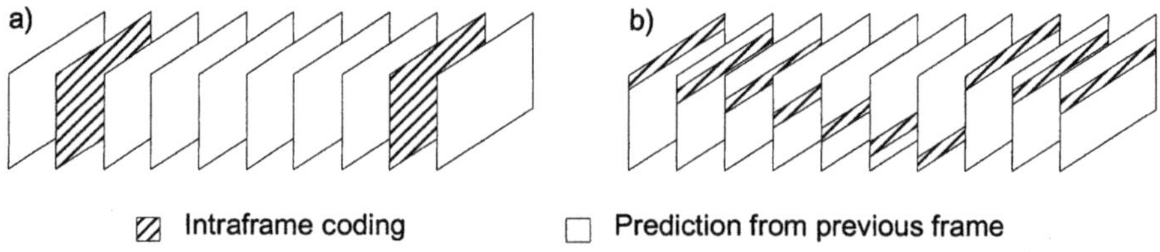

Fig. 15.5. Cyclic intraframe coding to avoid infinite propagation of transmission error effects **a** performed over entire frames **b** performed over parts of frames

Not necessarily the conclusion is correct that better error resilience of a coding method implies a decrease in compression performance. As an example, for still images or intra-coded video frames, 2D transform coding schemes show both better compression *and* better error resilience if compared to prediction schemes. Likewise, by using B-type frames in MC prediction coders, a better error resilience is achieved without sacrificing (or often even increasing) compression performance. This can be interpreted by an analysis of *dependencies* in the decoding process. Both in 2D DPCM and in frame-to-frame prediction long dependency chains can exist; any new information is useless if it relies on previous information that was received erroneously. Such dependency chains are typically much shorter in 2D transform coding and in frame-type sequences like IBBPBBP.. . In wavelet coders (both 2D and 3D variants), dependency chains are of variable length relating to the depth of the wavelet pyramid, where in addition the number of elements with longer dependencies *decreases exponentially* with the dependency length (in DPCM, it only decreases linearly). This gives an additional advantage with regard to error resilience, when a combination with methods of unequal error protection or priority-weighted transmission is made. In most schemes with shorter dependency chains (except for the methods of intraframe refresh), the penalty will however be an increased encoding or decoding delay, which again shows that a tradeoff between efficiency and latency can be made.

When no error resilient encoding tools are available, or when certain parts of the streams must undergo a specifically strong protection, additional mechanisms for error control have to be used in transmission over error-prone networks. In principle, error-free transmission does not exist, but for some network types errors are as rare that it is not necessary to take care. For example, if R [bit/s] is the transmission rate, and P_{err} is the probability of bit errors, typically $P_{err}R$ bit errors will occur per second, or one error after each $1/(P_{err}R)$ s. It is interesting to note here, that by a requirement to have a low number of errors, e.g. one per hour, the bit error probability is reciprocally dependent on the data rate, such that higher-rate services need higher quality channels from this point of view. This is however not the entire truth, because it does not take into account the relevance or effect of errors: A single bit error will be hard to detect when it affects a high-frequency transform coefficient in an HDTV video sequence, while it may have drastic implications when a motion vector in a QCIF resolution sequence is affected.

Best strategies for error control can again best be clustered around different network characteristics. Networks providing *Quality of Service* (QoS) typically give guarantees about mean error rate and error variations like burst duration in addition to the other QoS parameters already discussed in the previous section. Another important distinction criterion is to be made for networks *with* or *without* *back channels* providing the transmitter with information about the state of the receiver.

QoS networks typically allow to define different QoS classes (*differentiated services*). This would be ideal in cases where different priority levels can be assigned to different sub-stream parts, e.g. in scalable coding of sources or in data partitioning. If however no network QoS is guaranteed, as in case of best-effort networks, it is still possible to achieve a similar effect by implementing *application layer QoS*, which is often done by providing *Unequal Error Protection* (UEP) for different sub-streams. For correction of single bit errors (as far as these are not simply ignored), and also for correction of any errors in application cases where no back channel exists or where low end-to-end delay is required, *Forward Error Correction* (FEC) is the most appropriate method for error control. In case of networks providing a back channel, *Automatic Repeat Request* (ARQ) is often applied. It has also been shown that hybrid solutions between FEC and ARQ provide additional advantages [CHANDE ET AL. 1999][ZHANG, KASSAM 1999].

Forward error correction. FEC is a common approach of channel coding[1], where redundancy is supplemented to the information bits, such that detection or correction of errors becomes possible. In block-based channel coding, often denoted as *Cyclic Redundancy Check* (CRC) coding, N information bits are protected by K redundancy bits, which typically allows to detect up to K errors occurring in the block of length $N+K$, or correct up to $K/2$ errors (in case of even K). Typical CRC codes which are often applied in multimedia signal transmission are *Bose-Chaudhury-Hocquenghem* (BCH) codes and *Reed-Solomon* (RS) codes. The other large class are the *convolutional codes*, which can usually be described by a trellis graph (cf. sec. 11.6.1), here defining valid sequences of bits that would arrive at the decoder when no bit error has occurred. The Viterbi algorithm is then applied to find the valid sequence which was most probably sent, having closest distance with the sequence received. Convolutional codes are typically described

---

[1] For a general reference on channel coding, see e.g. [BOSSERT 1999].

by the *constraint length* and the *code rate r*, which specifies the percentage of the information bits carried within the total rate sent over the channel.

Convolutional codes can almost optimally be combined with multi-symbol modulation, e.g. in *trellis-coded modulation* (TCM) [UNGERBOECK 1974]. One of the most efficient classes of convolutional codes existing today are the turbo codes [BERROU, GLAVIEUX 1993] and *Rate-Compatible Punctured Convolutional Codes* (RCPC) [HAGENAUER 1988].

FEC is a perfect method to avoid single bit errors. For burst error characteristics, the effect of error correction can only be obtained if the influence length of the error-correction code becomes longer than the error burst phase. This can be achieved by *interleaving*, which is an arrangement of a longer sequence of bits in a row-wise loaded matrix, where FEC is then applied column-wise by supplementing further rows. As the bits in the matrix are transmitted row by row, burst errors will mainly affect rows, but can most probably be recovered by the column-wise FEC. In this context, a CRC code and a convolutional code are often combined as *outer* and *inner codes*. Fig. 15.6 shows an example of interleaving in the context of packet-based transmission. As typically entire data packets are lost, this can be solved by application of an FEC column-wise *across packets*. A row-wise FEC could be applied additionally to correct single bit errors. In the specific case of packet drops, the positions of losses are known, but neither right nor wrong information is available. It is possible to reconstruct the information when a maximum of K losses occurs out of $N+K$ information packets sent, as any valid codeword blocks are different for at least K bit positions, which is the minimum *Hamming distance* in a well-constructed code. Consequently, only one valid codeword can be found in each case which perfectly must match at the positions of bits from packets that were received.

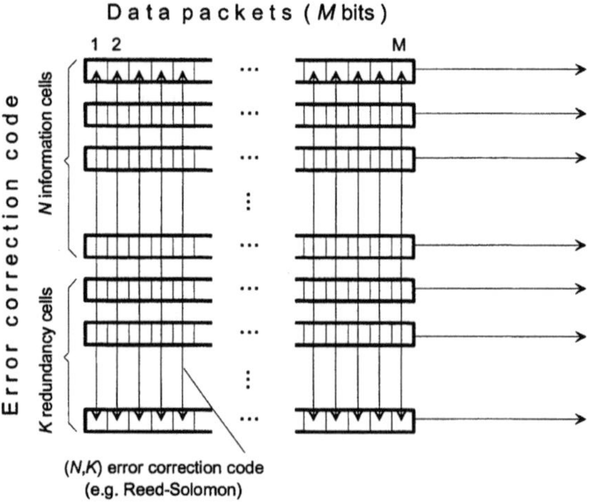

Fig. 15.6. Interleaving combined with error correction coding for error protection in case of burst errors

In the case of single or burst-like bit errors, all bits are received but some may be wrong. If the received sequence of bits does not match a valid codeword, an error *syndrome* is computed by the channel decoder, which shows the not matching bit positions compared to the most similar valid codeword.

Exploitation of latency and ARQ. If a back channel exists from the transmitter to the receiver, and sufficient latency is allowed by the application layer, error control can in addition invoke re-transmission of lost information. In [CHANDE ET AL. 1999] it is shown that this significantly increases the error resilience as compared to interleaving FEC techniques, when only as much additional redundant information is sent to allow recovery from losses[1]. In [CHOU, MIAO 2001], rate-distortion optimal streaming over a channel with packet losses and a complementary back channel is studied. For a given set of packets and a time delivery deadline, it must be decided which packet to send by the next opportunity. The goal would be to achieve minimum distortion under the given rate constraint, which in particular for the case of video coding must also include analysis of signal dependencies (see above), as it is e.g. useless to transmit information about a B-type frame when a P-type frame on which it relies has been discarded. In any case, it is necessary to identify those packets which must be re-sent with high priority, as re-transmission of all packets which are lost might incur even higher losses due to additional network overload. For the case of ARQ applications, the scheduling mechanisms regarded in the previous section under pure transmission delay aspects must be revised, including in addition the ARQ round-trip delay and the probability of remaining errors after ARQ was performed. ARQ approaches are one of the most important error control strategies used for Internet streaming, for 3G mobile networks and wireless LAN. Again it is observed that whenever the application allows (which is the case typically when no real-time reaction is necessary) it is advisable to allow as much latency as acceptable, to support recovery from errors by appropriate control mechanisms. This is true for both cases – FEC, where interleaving introduces additional delay, and ARQ which relies on a minimum amount of round-trip latency anyway.

Error concealment. Error concealment reduces the effect of errors through specific processing performed at the receiver side. In principle, concealment is a post-processing which is not part of the normative decoder behavior and needs not to be standardized. Some methods of concealment will however not be completely independent of error resilient transmission methods, but will rather try to utilize them as smart as possible. One example for this is the best strategy to proceed with the information gained by reversible VLC decoding, e.g. how to establish best hypotheses which part of the information is likely to have been decoded correctly etc.

---

[1] This is e.g. possible by using RCPC, due to the rate-compatibility property, where a weaker code is a sub-set of the stronger code.

Another example is utilization of *signal interleaving*[1]. If data packets are assembled from coded information of locally neighbored signal areas, larger concatenated areas may be undecodable in case of packet losses, as it is shown for the example of an image signal in Fig. 15.7a, top. By interleaving, the errors can be spread over different areas of the signal (Fig. 15.7a, bottom). A typical concealment approach in this case is interpolation of information from correctly received neighbored areas, which is most likely to be successful (Fig. 15.7b).

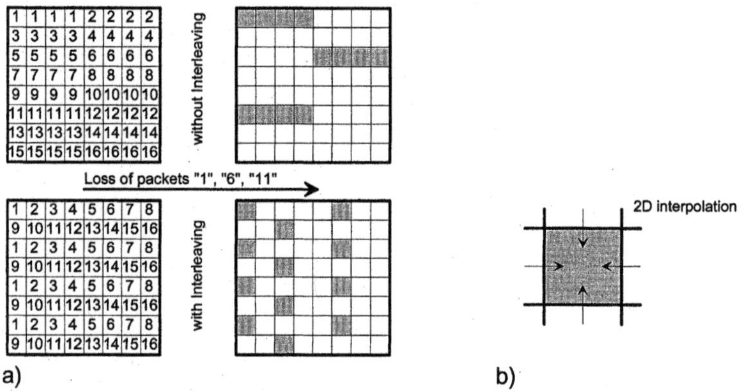

Fig. 15.7. Signal interleaving where signal blocks are combined into packets 1-16 (**a**) combined with interpolation for error concealment (**b**)

For video transmission, methods of temporally consistent concealment are of prior importance. The most simple concealment method to avoid visibility of strange errors is *frame freezing*, where the decoding process is stopped until the next re-synchronization point is reached. Even though the subjective perception in case of frame freezing may be less annoying than error propagation which would occur otherwise (cf. Fig. 13.7), it is not very convenient for subjective impression if applied frequently. An alternative are techniques of motion-compensated concealment [KAISER, FAZEL 1999], where it is tried to get as much as possible advantage from motion vectors correctly received to map previously correctly decoded areas into subsequent frames. In principle, these methods have a certain similarity with frame interpolation (cf. sec. 16.4). In cases where motion vectors are lost, techniques of motion vector field interpolation can be applied for concealment; temporally adjacent and spatially adjacent correctly received motion vectors can be used for this purpose. Both linear (e.g. bilinear) or nonlinear (e.g. median filtered) motion vector concealment is possible. Nevertheless, there is no guarantee that motion-compensated concealment will indeed lead to improved quality; in particular in cases where the motion vector field is inhomogeneous, the danger of introducing unnatural artifacts is considerably high. From this point of view, the optimum solution would likely be between the simple freeze-frame concealment and more

---

[1] This is e.g. supported by *slice data partitioning* in the AVC standard (cf. sec. 17.4.4).

advanced methods, where the latter ones should only be applied where the motion information is classified as being consistent and reliable.

More advanced approaches for concealment could be based on *synthesis* of lost information. This has successfully been applied in case of transmission losses in speech transmission, where vocoder methods can be used to substitute missing samples. For image and video signals, in particular color and texture synthesis are promising methods in this context; for homogeneous textured areas, it is almost impossible to distinguish synthesized and natural texture; likewise, areas homogeneous by color can usually be filled by interpolation or extrapolation. With only minimum overhead, *signal-based metadata* (e.g. low-level signal feature information as encoded by a standardized MPEG-7 representation) would also be usable to perform synthesis with a relatively high similarity to the original signal. Such methods could open up completely new ways of error protection in future multimedia systems.

15.3 Digital Broadcast

As a typical example for transmission methodologies in broadcast services defined for digital media, Digital Video Broadcast (DVB) is explained in more detail here, which is built around the signal coding and systems parts of the MPEG-2 standard (see sec. 17.2, 17.4.2 ad 17.5.2). DVB was designed to use previous analog TV channels for digital services. With the same bandwidth capacity that was originally reserved for one analog program, the introduction of digital video compression allows to increase the number of programs, to increase the quality or resolution (e.g. expand from SD to HD), or provide a combination of these improvements. Originally defined as a European standard by the European Telecommunication Standards Institute (ETSI), DVB has meanwhile been adopted by a number of countries outside Europe for implementation of digital TV either built upon an existing analog infrastructure or from scratch. Besides some extensions to the MPEG-2 transport stream (TS) by 'Program Specific Information' (PSI) which allows e.g. to embed electronic program guides (EPG), and by definition of conditional access mechanisms as necessary for Pay TV services, DVB specifications are mainly concerned with transmission principles over satellite TV channels (DVB-S), cable TV channels (DVB-C) and terrestrial TV channels (DVB-T).

For each of these configurations, definitions of modulation and error protection mechanisms are specifically made. The first building blocks are identically used in DVB-S, DVB-C and DVB-T as shown in Fig. 15.8. Following the base-band interface, which defines the access to the packetized bytes residing in an MPEG-2 transport stream, an 'energy dispersal' is performed which has the goal to suppress the carriers of the subsequent modulation as far as possible for avoidance of interchannel interference; in principle, this is a scrambling process which is in particular effective if longer runs of '0' or '1' bits are present. Channel coding is a combination of an outer encoder based on a Reed-Solomon [204,188] CRC code[1], an interleaver and an 'inner' encoder, which is a punctured convolutional encoder of rates selectable as r=1/2, 2/3, 3/4, 5/6 or 7/8. The inner encoder adds at maximum one bit of redundancy to each bit sent, but is very flexible in terms of the amount of redundancy that can be added.

The modulation principles are different for the types of channels. DVB-S uses *Quadrature Phase Shift Keying* (QPSK), which allows transmission of 2 bits by each modulation symbol sent. If the bandwidth of a satellite transponder is 36 *MHz*, the effective source data rate which can be conveyed by these methods is variable between 26.1 Mbit/s (for convolutional code rate 1/2) and 45.6 Mbit/s (for convolutional code rate 7/8). Cable channels provide much better transmission quality (in terms of channel bit errors) than satellite channels. Hence, DVB-C uses *Quadrature Amplitude Modulation* (QAM) as modulation method, which provides more effective utilization of bandwidth, carrying a higher number of bits by each modulation symbol. This can be selected as either 16-QAM (4 bits/symbol), 32-QAM (5 bits/symbol) or 64-QAM (6 bits/symbol). Differential encoding is used to enable receiver synchronization of the phase. DVB-T, which is used for reception of digital TV via terrestrial TV channels, is the most complex method. The error characteristics of terrestrial TV channels exhibit a frequency-selective behavior, which is even worse for the case of mobile reception. Due to this reason, a spreading of the coded binary information over multiple carriers is performed. This is achieved by Orthogonal Frequency Division Multiplex (OFDM), which provides a very efficient implementation of multi-carrier modulation, where the kernel of the algorithm is an inverse DFT at the transmitter and a DFT at the receiver [VAN NEE, PRASAD 2000]. OFDM is also combined with quadrature modulation, where QPSK (2 bits/symbol), 16-QAM (4 bits/symbol) and 64-QAM (6 bits/symbol) can be used as modulation methods, the selection being adaptable to the actual channel characteristics.

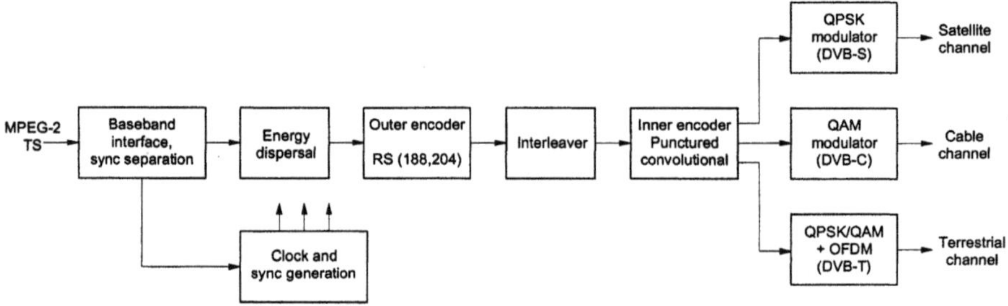

Fig. 15.8. Block diagram of DVB encoders

---

[1] For each PES packet of 188 bytes, 16 bytes are added as redundancy; PES = Packetized Elementary Stream, see sec. 17.2.

In a hierarchical variant of DVB-T modulation, different quadrature modulation methods can be mixed, such that different parts of the bit stream can be transmitted providing variable protection (as e.g. QPSK is much more resilient against transmission errors than 64-QAM). After OFDM signal generation, a guard interval must be inserted, which allows to suppress interference from delayed components of the modulated signal, which are expected to arrive from neighbored transmission stations (TV towers). Without the guard interval, the carriers would lose the orthogonality property. The necessary length of the guard interval depends on the expected distance of neighbored transmitters, and can be set to a maximum of 224 μs (which allows a maximum of 67 km distance between transmitters, a value which is reasonable due to the earth surface curvature). By using OFDM, neighbored transmitters can use the same TV channels (transmission frequencies) in DVB-T, which allows a much more economic bandwidth utilization than in analog TV where such cases must be forbidden due to interference problems. The measurable spectrum of the modulated signal in DVB-T appears nearly as white noise, such that interference with neighbored channels, but also interference with analog channels which are still to be run in parallel with DVB for a certain period of time, are minimized. Examples of usable data rates which can be transmitted over one former analog TV channel are given in Table 15.1 for different combinations of modulation methods and convolutional code rates. If hierarchical modulation is used, combinations in between are possible, such that e.g. using QPSK (code rate 1/2) embedded into 16-QAM (code rate 5/6), a rate can be achieved which results from adding (R_{QPSK} + $R_{16\text{-}QAM}$). This way, it is possible to jointly transmit data of high priority at 4.7 Mbit/s and data of low priority at 15.67 Mbit/s, both over the same channel.

Table 15.1. Usable data rates (Mbit/s) per terrestrial TV channel for different combinations of code rates and modulation types in DVB-T (source : REIMERS)

Code rate $r=$	QPSK	16-QAM	64-QAM
1/2	4.7	9.4	14.10
2/3	6.27	12.53	18.8
3/4	7.05	14.10	21.15
5/6	7.83	15.67	24.68
7/8	8.23	16.45	24.68

DVB is typically used to transmit 5 programs within the equivalent bandwidth of one analog TV channel. The MPEG-2 standard is used for video and audio encoding, where the quality by using an optimized source encoder implementation is reported to be still slightly better than for analog TV. It has further been shown that DVB-T in cases of sufficiently strong signal reception (e.g. with near by TV towers), or by using the more reliable coding/modulation modes such as $r=1/2$ and QPSK is usable for mobile reception from vehicles traveling by speeds of up to

180 km/h (approx. 110 mph). This is an additional functionality which was not achievable with analog TV[1].

DVB was originally designed for stationary reception, and the possibility for mobile reception in DVB-T is a surplus value that was detected to be feasible when the first experiments within an operational network were made. For *Digital Audio Broadcast* (DAB), the capability for mobile reception was an inherent design goal from the beginning. DAB is also usable for reception from vehicles moving with higher velocity, e.g. high-speed trains. The basic transmission methods are very similar to DVB-T, and DAB shall not be treated in more detail here.

15.4 Media Streaming

For finite-size or short-duration media (e.g. still images, short video or sound clips), the first practical way for transmission is definition of a file format, where the file contains the entire bit stream. The file is then transmitted as an entire unit, before the replay starts at the receiver end. For videos, movies or music pieces of longer duration this method is not practical, because the user would not like to wait until the transmission is finished, the receiving device may not have sufficient storage capacity etc. In *streaming* transmission, decoding and replay should start almost immediately after transmission has begun. For packetized transmission of streams, network protocols play a central role. These can for the case of the Internet roughly be clustered into *network-layer protocols*, *transport protocols* and *session control protocols*. Of course, additional specific targeted protocols can be defined in the application layer.

Internet Streaming. The *Internet Protocol* (IP) is based on a concept of a 'network of networks' which is abstracted from the underlying physical layer. The Internet was originally designed for file transfer, but today real-time media streaming has reached a significant amount of the entire Internet traffic. On top of IP, the *Transport Control Protocol* (TCP) is widely used, which guarantees reliable transmission by acknowledgement of IP packet reception at each link in the transmission chain (between switches, routers). Unfortunately, TCP is too slow for real time streaming applications due to its link-to-link hand shaking, re-transmission mechanisms and other protocol overhead such as congestion-dependent flow control[2]. As an alternative lower-layer transport protocol, the *Unified Datagram Pro-*

---

[1] In Germany, terrestrial analog TV transmission is planned to be terminated within the next years, where in the region of Berlin / Brandenburg transition to fully digital terrestrial services has been finished by autumn 2003. Coexistence of analog and digital transmission can in general be seen as an episode of limited duration.

[2] In local networks without many hops and with limited congestion, TCP is well usable for streaming.

tocol (UDP) was defined, which does not use any acknowledgement mechanism at all, but encounters the danger that packets become entirely lost in cases of network congestion. Losses typically occur, when overflow of input or output buffers happens on the network path, e.g. in switches. Both TCP and UDP also employ simple CRC based error detection and will discard packets which are affected by bit errors. As a higher-layer transport protocol, the *Real-Time Transport Protocol* (RTP) is primarily used on top of UDP for Internet media streaming. RTP provides timing and synchronization information associated with the data, and is responsible to re-order packets into their original sequence. Timing information conveyed in RTP can e.g. be used for synchronization or to decide whether a packet is outdated and may be discarded. Payload type identification is conveyed for certain common types such as video and audio streams according to ISO/IEC standards or ITU-T recommendations. RTP can be used in conjunction with the *Real-Time Control Protocol*, which is primarily intended to provide feedback about the instantaneous network QoS (delay, packet loss rate etc.), e.g. to be used to optimize transmission for best error resilience.

Newer versions of the Internet protocol such as IP v6 also foresee implementation of QoS and classified (differentiated) services. On the other hand, as the Internet is a 'network of networks' in which many older-generation routers and switches exist, global implementation of QoS mechanisms is unrealistic by the short term. Implementation of QoS and differentiated services is however quite realistic in certain network domains, private and local area networks which are consistently equipped by devices supporting QoS.

A big advantage can be taken by assigning different network QoS levels or priority classes for sub-streams of different relevance, which can ideally be combined with scalable (layered) media representations or data partitioning methods. To guarantee QoS, it is necessary by the network side to perform bandwidth reservation, and perform *Call Admission Control* (CAC) to prevent network overload. In cases where sub-streams of higher relevance exist, it is typically only necessary to reserve bandwidth for these parts (e.g. the base-layer stream in scalable coding), and convey the remaining parts by best effort subject to availability of unused bandwidth capacity, but without any guarantees. This approach is sometimes denoted as *Available Bit Rate* (ABR) transmission, which indicates that it is conceptually positioned somewhere in between CBR and VBR[1].

For bandwidth reservation over Internet connections, the *Reservation Protocol* (RSVP) was defined, by which it can be checked whether the foreseen capacity is available on the transmission path to convey a UDP/RTP connection, such that the requested bandwidth can be reserved. The transmitting device is then informed about the bandwidth granted for the connection. This principle can only work

---

[1] The ideas behind are very similar to the methods originally projected for Broadband ISDN (B-ISDN) with its underlying Asynchronous Transfer Mode (ATM) packet switching technology. Presently, rudiments of ATM technology actually exist, which are used below the IP layer in high-bandwidth backbone connections serving the Internet; it will not be possible however to really make use of their priority switching in end-to-end connections.

satisfactorily, if the source observes the negotiated parameters, for which the simplest way is the implementation of a leaky bucket (see sec. 15.2.1) as virtual buffer, which counts the number of packets that is produced by the source within a certain time interval. The length of this interval corresponds to the negotiated *peak duration*, during which the source is allowed to send by a rate higher than the negotiated mean. If the leaky bucket runs full, the rate must be reduced by rate control mechanisms as described in sec. 15.2.1.

Even in cases where no actual reservation is provided by the network, a transmitter can estimate the presently available network bandwidth by collecting feedback from the receiver side about the network transmission quality. Then, a useful approach is preventive reduction of the transmission rate on a voluntary basis.

The principle of optimum adaptation of rate regarding the available bandwidth requires the application of rate control methods. However, as most traffic in media streaming relates to transport of *pre-encoded data* from a server to a client, independent rate control is impossible unless transcoding is applied or a scalable (embedded) bitstream is stored. A third approach is to encode and store streams of different rates in advance, which are then selected based on the instantaneously available bandwidth. *Stream switching* is a technique widely used in this context. The number of pre-encoded streams must however not be too high, as otherwise the encoding effort and the storage capacity to be provided at the server side would become unacceptably high. On the other hand, with a limited number of streams[1], the exact transmission rate of the network will hardly be matched. To avoid network overload, the rate must then usually be held lower than the rate which could in principle be transmitted, which is penalized by a quality loss.

An alternative methodology to achieve this goal is usage of scalable coding with fine-granular scalable embedded bit streams. Here, the problem of rate control is totally decoupled from the encoding process, but at least for the case of predictive video coding, fine-granular scalability is penalized by a significant loss in coding efficiency (cf. sec. 13.2.8). This situation probably looks different for the cases of temporal scalability based on B-type frames and non-recursive methods such as intraframe wavelet coding and motion-compensated interframe wavelet coding (sec. 13.4).

In spite of provisions taken as described, packet losses frequently occur in Internet streaming. Under severe network load conditions, packet loss ratios of around 10% are realistic in streaming applications.

Wireless Networks. Mobile multimedia communication is getting more important now and expectedly even more in the future. This development is accompanied by the growing usage of wearable computers (laptops), Personal Digital Assistants (PDA) and mobile phones with multimedia capabilities, e.g. built-in cameras, color LCD screens etc. In the future, when wireless networks of higher bandwidth capacity will be available, general multimedia data (beyond unimodal speech signals or

---

[1] A number of 3 streams is typically used in stream switching applications today.

text message interfaces) could establish most traffic over wireless links. For digital wireless networks, it can be distinguished between

–   Local networks, e.g. wireless LAN or DECT (Digital Enhanced Cordless Telephone), the latter with data transmission services based on Bluetooth;
–   Mobile networks, e.g. GSM and associated enhanced data services like GPRS (General Packet Radio Structure), third-generation networks like UMTS (IMT-2000) and beyond;
–   Broadcast distribution networks, e.g. DAB, DVB-T, which were already discussed in sec. 15.3.

While the first two categories provide individual communication services, broadcast distribution networks are not originally designed for that purpose. Distribution networks are unidirectional by nature, serving only the *down-stream*. For feedback from the user to the network, an *up-stream* must be available as well, which is only provided by the first two categories of networks. For terminals which can connect to different networks, it is however irrelevant whether the same connection is used for down- and up-streams. From this point of view, it can be expected that available bandwidth resources in distribution networks may be used for additional personalized *multicast* services. Individual content could be sent when requested by a sufficiently large group of people, such that it is economically justified. Presently, new scenarios are also discussed for usage of DVB-T channels in narrow-bandwidth broadcast transmission directly to mobile devices such as cellular phones, where a much higher number of TV programs could be served due to the lower spatial resolution which is required.

A main purpose of wireless local networks is simplification of connection for *migrating* terminal devices, which are frequently moved from one place to another e.g. within a building. Unlike 'true' mobile communication, it is often expected that during the connection the devices are not moved or at least only moved very slowly (e.g. a PDA user may walk around during a fair). This is important, as the bandwidth which can be made available highly depends on the stationarity of network conditions to allow reliable adaptation of modulation and error protection methods. Typical wireless LAN (WLAN) technology as available today, e.g. designed in compliance to the standard IEEE 802.11, provides transmission bandwidth of up to 11 Mbit/s; follow-up networks providing up to 100 Mbit/s and more are under development. Typically, different users have to share this bandwidth, where the number of simultaneous accesses also depends on the density of base stations. This leads to similar error characteristics based on congestion as in wired packet-switching networks, where however the WLAN is more prone to additional single-bit or burst errors caused by the fading characteristics of radio transmission. These are detected and if possible corrected at the physical and network link layer, otherwise re-transmission is initiated.

In mobile communication, where the receiver is moving (e.g. built in a vehicle), transmission and reception characteristics are usually changing fast, which leads to much more severe errors at the physical layer. The most relevant mobile network existing today is based on GSM (Global Standard for Mobile Communication). GSM defines a connection-oriented circuit switching, where physical transmission capacity in space and time is assigned to individual users. For one GSM channel, the data transmission rate will usually not exceed 9.6 kbit/s per direction (duplex). As originally designed mainly for the purpose of speech communication, transmission of other data like images, video etc. is relatively slow, and real-time streaming is almost impossible unless extremely low quality is acceptable.

The first step towards mobile services providing higher rates is made by GPRS, which is based on dynamic multiplexing of GSM channels, where also the number of channels can be variable depending on availability and resources requested. The multiplexed transmission capacity is shared by several users in a packet-based methodology. GPRS is well suitable for mobile services based on the Internet Protocol, however additional provisions must be taken such as usage of short packets and additional error control mechanisms. In principle, if media streaming is performed over these channels, the error characteristics are not too much different from other packet switching technology, where random or short-burst errors occurring in the underlying GSM channel have to be corrected by the network link layer.

The Universal Mobile Telecommunication System (UMTS) is the third generation ('3G') of mobile networks, a framework which was developed under the roof of the ITU-T. An alternative denotation is IMT-2000 (International Mobile Telecommunications 2000). In fact, UMTS is not homogeneous, but comprises an entire group of standards, some of which are newly defined radio transmission methods and protocol mechanisms, but also integrates existing networks like GSM and GPRS, which are used when no other connection is available. The projected data rate is up to 2 Mbit/s, which however will barely be available to single users; a more realistic range of rates would be 64-384 kbits/s or even below. Packet switching is used, where packets are much smaller in size than typical IP packets; if an IP connection is established via UMTS, the IP packets are decomposed into sub-packets of smaller size. Packet losses frequently occur, which can in many cases also be caused by severe burst errors occurring at the physical level. The usual method to compensate for packet losses is re-transmission. Quality of Service mechanisms are defined. The available bandwidth and the probability of transmission losses depend on the number of users which share the same channel.

Even though the higher layers of transmission in mobile networks are typically IP-oriented, in fact it is necessary to set up a specific protocol stack for the network link layer. Both FEC and ARQ are employed for this purpose. Unlike previous ARQ approaches which were made over the entire transmission chain, ARQ performed between the mobile and the base station has a very short round-trip time. On the other hand, it is still necessary to carefully decide in a mobile streaming application which packets are worthwhile to re-transmit, because re-transmission increases the network load, which again decreases the transmission quality etc. This can however be managed efficiently due to the availability of QoS negotiation mechanisms.

In the future, mobile devices will support multiple transmission protocols, and select the best alternative path available in the given environment. For example, 'hot spots' like airports, trains, fairs will be equipped by WLAN or successors thereof, which provide cheap transmission and high bandwidth, but are based on a picocellular topology. Each base station covers a radius of some 10 meters only, such that access is extremely limited to one location. Satellite-based services are most universally available from any point of the world, but are less suitable for mobile transmission and reception, as the antenna (or satellite dish) needs exact adjustment. New technology will also evolve in the medium-range radio networks. On this background, discussion about a future generation of mobile networks ('4G') has already started, which will most likely integrate all these approaches and eventually support new error-prone modulation, coding and multiplexing methods to achieve an even higher bandwidth. Of high importance for further improvements are *multiple-input multiple-output* technologies (MIMO) which take advantage of diversity mechanisms, such that a receiver can combine information which is available from multiple transmitters, and *adhoc networks*, where any mobile device can take over tasks such as routing towards other mobile devices or towards a base station.

15.5 Content-based Media Access

The ever improving methods for signal analysis, including content analysis, allow to augment signal data by *metadata*, which can be used for the purpose of *indexing*. In a certain sense, the information flood and in particular the flood of audiovisual signals which have become available due to digital services, directly enforce the usage of more intelligent search and retrieval mechanisms. The same is true for any information residing on world-wide networks, which would hardly be valuable without intelligent retrieval aids. While however indexing information needed by text-based search engines can well be automatically extracted by parsing texts for key words, satisfactory indexing of audiovisual data must often be done by manual annotation. Some information such as date and place of capturing/production, illumination conditions etc. could also be generated automatically; further, if editing scripts exist, these can also be used to automatically generate index data. This section rather concentrates on automatic procedures which can gain indexing information directly from signal features, for which the extraction processes were described in chapter 7.

Features of multimedia signals can in many cases be represented much more compactly than the signal itself, but still give relevant information about the content. For applications of similarity-based search for signals, feature based descriptions are more expressive than signal samples. Table 15.2 contains selected examples of applications for which signal-based features are useful.

Table 15.2. Applications where indexing by features is required

Application	Items searched for	Feature types
Retrieval from movie and video archives	Genre, Author, actors, specific scene content and events, mood of scenes, keyframes, change characteristics	All visual and audio basic features, feature constellations characterizing persons, objects, events
Retrieval from image and photo archives/databases	Images of certain color, texture, shape, localization; specific objects or persons	All visual basic features except motion, feature constellations characterizing persons and objects,
Retrieval from audio archives and databases	Genre of music pieces, sound characteristics, search for melodies or melody fragments	All audio basic features, features in time-line behavior, feature constellations characterizing genres
Multimedia production	Segmentation, cut detection, key-frame extraction, all other retrieval purposes	All visual and audio basic features
Digital broadcast, Electronic Program Guides, Intelligent set-top boxes	Genre, Programs of specific content, authors, actors etc., relationships (linking) with other programs or web links, events in programs	Mainly textual features, linking and timing information
Surveillance, security control	Expected or unexpected events in pre-defined scenes; identification of persons, objects or events	Audiovisual basic features in particular motion, face features, localization, silence; feature constellations characterizing persons, objects or events
Web cams	persons, objects, events	feature constellations characterizing persons, objects or events
Audiovisual communication	Appearance and actions of persons (talk/no talk); tracking of persons	motion, silence, face features, localization features
Sports event augmenting	Automatic analysis of distances, comparison of time behavior of different runners, scene augmentation	motion, time-line and spatial localization features, feature constellations characterizing persons, objects or events
Automation, inspection, service	Identification of objects or events; unexpected events	feature constellations characterizing states, objects or events
Signal identification for copyright protection	Similarity with reference items, which must be stable under modifications	Signal footprints, fingerprints, watermarks

Table 15.2. (continued)

| Smart cameras and microphones | Optimum scene properties for capture, focus on preferred objects; trigger acquisition in case of pre-defined events | Adjustment of illumination, color, tracking of camera motion or ego-motion of objects; localization of objects or persons; characterization of objects or events |

Content description for multimedia signals can be used to filter the information flood which comes as a by-product of increasingly cheap storage and transmission of digital media. However, the content description by itself can significantly contribute to reduce the necessary bandwidth in transmission and storage, as signals which can clearly be identified as undesirable by a compact feature description must not be transmitted or stored at all. In some cases, *real-time analysis* of feature data is necessary (e.g. in surveillance, smart cameras, communication); in other cases, more complex analysis methods may be performed off-line. In particular when retrieval from archives or databases is made, the index data relating to the items in the database are mostly pre-computed and also stored in this or another database.

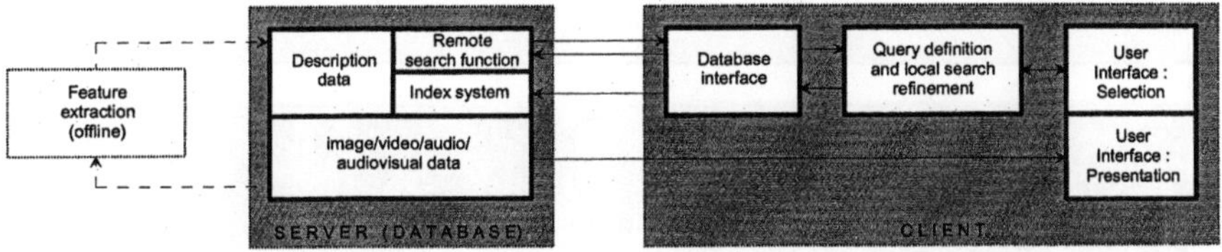

Fig. 15.9. Block diagram of a distributed retrieval application

Fig. 15.9 shows the block diagram of a retrieval application, where media data shall be found in a remote database. As databases usually provide powerful and efficient search functions, it is necessary to map the feature-based distance function as resulting from a specific query definition to the respective remote search function of the database, which could be defined as an application program interface (API). Typical search criteria and strategies are as described in sec. 9.3. If not all desired comparison methodologies are supported by the given database system, it is also possible to perform a search refinement locally at the client, after a remote pre-selection made at the server (database) side. It is indeed not useful to perform exhaustive search at the client side, as even transmission of compact feature-related data could cause a high amount of data to be transmitted, if the number of items in the database is large. After the indices of a limited number of most similar items have been determined, the signals themselves can be retrieved from the database system and be presented to the user for final selection or consumption.

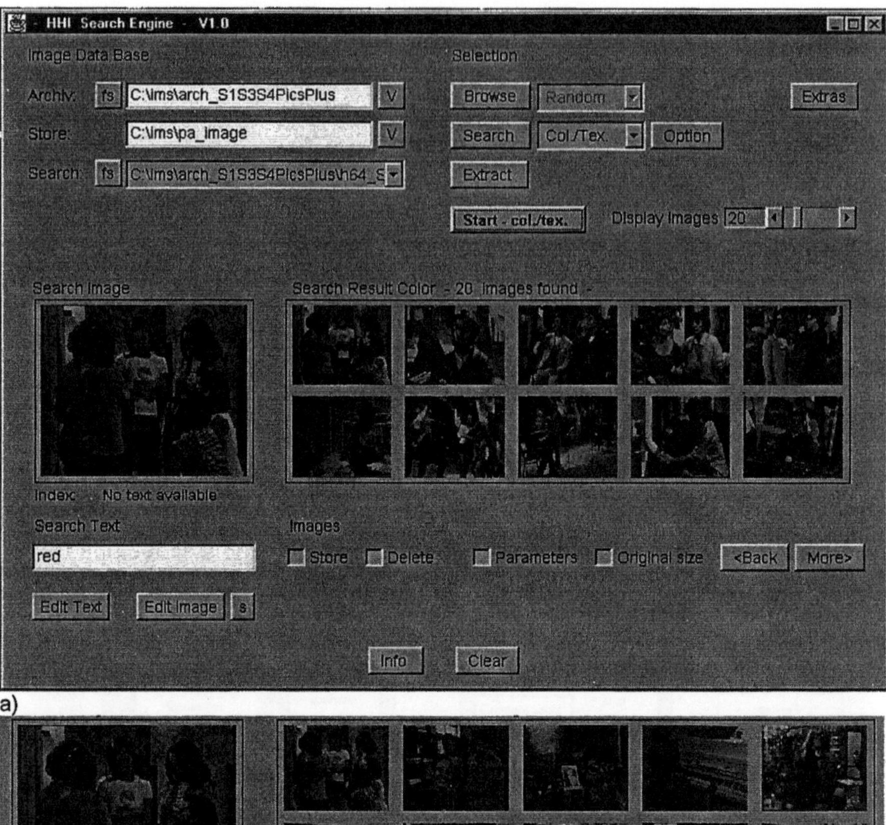

Fig. 15.10. a User interface of an image search engine for 'query by example' using combined color histogram and wavelet-based texture feature, and **b** using only the color histogram feature (source : HHI)

Fig. 15.10a shows the user interface of an image search engine, based on a *query by example* methodology. Feature data related to one given signal item are extracted at the client side, and transmitted to the server (database) system with the mission to find more similar items. Here, the example image is shown on the left-hand side[1]. The ten images at the right-hand side are the most similar entries found

---

[1] Query by example can quite universally be used for different media types by selecting an example signal, extract the respective feature data and use them for database search. Other signal-specific query types are *query by sketch* for image signals, where the user draws a raw figure of the image layout or shape which is searched, and *query by humming* for the case of audio signals, where a rough example of the melody is conveyed to the machine.

in the database, where the example image itself appears at first position, as it was indeed contained in the database (which must not necessarily be the case). A global color histogram and a wavelet-based texture feature (see sec. 7.2, Fig. 7.11) were used for comparison. Fig. 15.10b shows another query by example result, where color only was used for comparison. It is obvious that the general structure of the images retrieved is clearly different from the query, as no structure-related feature (as in Fig. 15.10a) is used.

15.6 Content Protection

Digital Rights Management is of crucial importance in business models for the delivery of digital content. As distribution of digital media can in principle be performed by anybody, perfectly without losses, it is necessary to integrate mechanisms which prohibit unauthorized or illegal usage and manipulation of content. The basic goals to be achieved are shortly described in the subsequent subsections.

Protection of content from being used or altered without authorization. A classical approach to achieve this is *encryption*. Bit streams describing the content are disturbed or scrambled in a systematic way that is pre-determined by a key which should only be known to the authorized user[1]. Unfortunately, development of de-crypting methods for 'cracking' any encrypted information is simply a matter of time. The most reliable mechanisms are related to usage of several keys, some of which may be public, some private. A useful means is combination of the key with individual hardware properties, e.g. the serial number of a machine or device on which the content is allowed to be presented. Further, methods can be used where authorization keys must be newly acquired (e.g. via the Internet) each time the content shall be replayed. The disadvantage of these latter methods is that usage of digital media can definitely become rather difficult even for the authorized user. Even if this problem is resolved, one fundamental problem is that any authorized user would be capable to record the content – when necessary by transcoding into a different format – as soon as the play-out is possible. From this point of view, the methods described in the next sub-section seem to be at least equally important.

Detection of illegal usage of content. Illegal usage can be detected if simple methods exist to uniquely *identify* digital content. This can either be achieved

---

[1] It is beyond the scope here to go into detail with encryption methods. The interested reader is referred to [SALOMON, EWENS 2003].

- by augmentation of a piece of content with imperceptible *tags* which cannot be removed without destroying the content;
- by defining *features* which can be extracted from any piece of content, and can uniquely be matched against features held in a database identifying content that shall be protected.

The first method is mainly implemented by methods of *watermarking*, the second is mostly denoted as *fingerprinting*.

Watermarking methods[1] exist both for the spatial/time domain signals and for frequency domain signals (transform coefficients). Watermarks should be imperceptible and robust against different types of manipulations, such as filtering, transcoding and geometric modifications (e.g. scaling, rotation) of the signal, and should also resist against systematic attacks. By watermarking, it would be possible to uniquely identify content which is used illegally, and eventually it may even be possible to determine the identity of (authorized) users who gave away the content for illegal distribution.

Fingerprints (also denoted as *signatures*) are features extracted from a piece of content which should allow unique identification by minimum computational effort. Fingerprint identification must also be possible when only a part (e.g. one area from a larger image signal, segment of some seconds from a longer audio track) are available. By matching the fingerprints extracted from an (unknown) piece of content against fingerprints collected in a database, it is possible to identify the usage authorization status of that content. For an example of fingerprints (signatures) related to audio signals, refer to sec. 17.6.4.

Combination of media protection with media representation. Scalable coding or any other method of coding where a small part of the information (e.g. the motion vectors in video coding) is of high importance for the reconstruction quality, can flexibly be combined with protection methods. In principle, this can even be implemented in a way that the most important part of the content is used as an authorization key. Scalable coding methods are more flexible by this regard, as they can provide graceful degradation over different protection levels. Two scenarios could be:

- *Protection of usage for the entire piece of content:* This is achieved by protecting only an innermost base layer, or by using this base layer itself as a key to use the content by full resolution. It will not be possible to replay the signal even at the lowest quality level without this base layer
- *Provide the base layer for free usage, protect the enhancement layer(s):* It may be desirable to provide the content of low-quality approximation (base layer) for free preview or pre-listening, while the usage of the full-resolution signal (base plus enhancement layers) shall be charged. This can be achieved by protection of any part of the bit stream which is above the layer provided for free usage.

---

[1] For a reference on watermarking, see [COX, BLOOM, MILLER 2001].

16   Signal Composition, Rendering and Presentation

Signal composition and rendering methods are applied in the context of generation (production) and presentation (consumption) of multimedia signals. This involves mixing of signals which originate from different sources, adaptation and projection as required by the format of the composite signal or the output device.

Viewing the entire chain from the generation to the presentation of multimedia signals, the following steps are typically invoked[1]:

1. Signal acquisition: This can either be the capture of a natural signal by a camera or microphone, or the generation of a 'synthetic' signal (computer graphics, sound synthesis);
2. Signal adaptation and enhancement: noise reduction, dynamics companding, manipulation of the spectrum (e.g. blurring, contrast enhancement), resolution adaptation (interpolation/decimation), size adaptation (cropping/cutting);
3. Signal decomposition: Separation of single elements from the complete signal;
4. Signal mixing: Composition of different single signal elements into a composite signal (e.g. insertion of a foreground object separated out of a video sequence into another background, composition of an image from photographic elements and graphical elements, mixing of different natural or of natural and synthetic sounds);
5. Rendering: Preparation of a signal for projection into an output format which can be replayed on a given device;
6. Output of the rendered signal to the given output device (e.g. display, loudspeaker).

The boundaries between these steps are fuzzy, and the practical implementation and combination also depends on specific application requirements. *Multimedia signal production* typically includes the steps 1-5, where *editing* is an important part of the production process (working with previously generated sources), which involves steps 2-4. When applications are interactive, the multimedia production

---

[1] We do not consider the processes of transmission and storage here, as these were already extensively discussed in previous chapters. Techniques related to steps 1-3 were also treated in chapters 2, 5 and 10.

will have to provide mechanisms that allow execution of steps 4 and 5 at the receiver (consumption) end, which then must also be described as part of the media representation. In these cases, mixing and rendering may actually not be done finally during the production process. When the content shall be provided to the consumer in a finalized form (e.g. a movie), everything but step 6 will be executed during production, but may eventually be done by off-line processing. *Real-time mixing and rendering* of signals are important in interactive applications, where the user is allowed to take direct influence on the presentation, which must then invoke steps 4-6.

This chapter in particular concentrates on the tasks of multimedia signal processing in signal mixing, and introduces some more aspects of rendering and output. Often, the capabilities of the output device, e.g. regarding display size, computational power, graphics capabilities etc. will constrain the methods that have to be applied within the entire chain. This is however less rigid in the case of digital media than e.g. in analog TV, where any step from acquisition to consumption is made under the constraint of a given resolution format. In general, rendering of 3D scenes is much more challenging, and if simplification is made by using geometric manipulation and projection exclusively from 2D images, also the signal processing steps which are made in advance to rendering can be largely simplified. In particular, image warping and image-based rendering for viewpoint adaptation (sec. 16.2 and 16.3) play an important role in this context. These are treated with more detail here, while 3D computer graphics is not a focus of this book. In some cases, even extremely simple methods for image/video scene mixing and blending (sec. 16.1) can be employed to generate surprising three-dimensional effects. For the generation of arbitrary-resolution output formats, conventional methods of decimation and interpolation (sec. 4.4.1, 5.4) can be employed. An interesting aspect not yet treated before is *video frame rate conversion* (sec. 16.4). This is necessary if the output device displays by a lower or higher frame rate than used in the original source representation, if different video sources have to be mixed which were acquired by different sampling rates, or if the signal shall be displayed by a different speed (e.g. slow motion) than in original acquisition. For *rendering*, only selected topics are presented which are in the closer focus of multimedia signal processing, whereas many other aspects of rendering are more related to computer graphics techniques.

16.1 Composition and Mixing of Visual Signals

A general scheme for composition of a 2D output signal (which could also be one frame of a video) from N different sources is shown in Fig. 16.1. The sources are first subject to individual coordinate-mapping transforms $\Gamma(\mathbf{n})$, which could also imply projection of a 3D volumetric signal into the 2D image plane, or adjustment of size, orientation, position in space and time. A 2D linear filter $H(z_1,z_2)$ is invoked, and then each signal is weighted by a position-dependent function $\alpha(m,n)$, which determines by which amount the amplitude of the signal shall appear at the respective position of the output. All these operations are determined by the composition information, which must be provided beforehand by user settings, by automatic or semi-automatic extraction from signals.

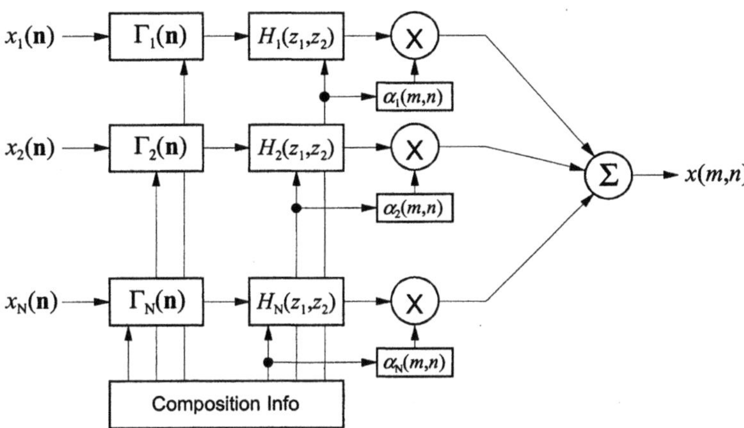

Fig. 16.1. Composition of a 2D output signal from N sources

When different patches from image, video or computer graphics signals are mixed, these will typically be opaque, such that only one of the signals normally appears at one position. In this context, *boundary transitions* between the different signal patches which are 'glued together' is of high importance for a natural impression. Natural images do not typically possess extremely abrupt changes in amplitude, which however depends on lighting conditions; in particular for diffuse lighting, transitions are typically rather smooth. Gradually-changing amplitude transitions can be achieved by *blending techniques*, where one signal decays over the width W of a transition period, while another increases. The signal $x(m,n)$ as composed from two or more signals $x_i(m,n)$ will then result as

$$x(m,n) = \alpha_1(m,n) \cdot x_1(m,n) + \alpha_2(m,n) \cdot x_2(m,n) + \dots \tag{16.1}$$

Typically, the weighting functions should be defined such that they sum up to unity, $\Sigma \alpha_i(m,n)=1$. The increasing and decaying behavior should be implemented *perpendicular* by the boundary orientation. This way, simple weighting can be defined on the basis of 1D functions, which are expanded along a particular direction. An example of blending two 1D signals using 1D ramp functions of width $W=3$ is shown in Fig. 16.2. A general definition of 1D ramp function pairs for an actual boundary exactly at position n' in case of an *even number W* is

$$\alpha_1(n) = \begin{cases} 1 & \text{for } n \le n' - \dfrac{W}{2} \\[2ex] \dfrac{n'-n}{W} + \dfrac{1}{2} & \text{for } n' - \dfrac{W}{2} < n < n' + \dfrac{W}{2} \\[2ex] 0 & \text{for } n \ge n' + \dfrac{W}{2}. \end{cases} \quad ; \quad \alpha_2(n) = 1 - \alpha_1(n) \qquad (16.2)$$

For the case of an *odd number W* (in this case, the actual boundary is between pixel positions n' and $n'+1$), the linear-ramp transition function is

$$\alpha_1(n) = \begin{cases} 1 & \text{for } n \le n' - \dfrac{W-1}{2} \\[2ex] \dfrac{n'-n}{W} + \dfrac{1}{2} & \text{for } n' - \dfrac{W-1}{2} < n < n' + \dfrac{W+1}{2} \\[2ex] 0 & \text{for } n \ge n' + \dfrac{W+1}{2}. \end{cases} \qquad (16.3)$$

Weighting functions of this type can be applied in different cases of one- and multi-dimensional signal blending. Application over the temporal axis include fade-in and fade-out effects for video sequences or audio signals.

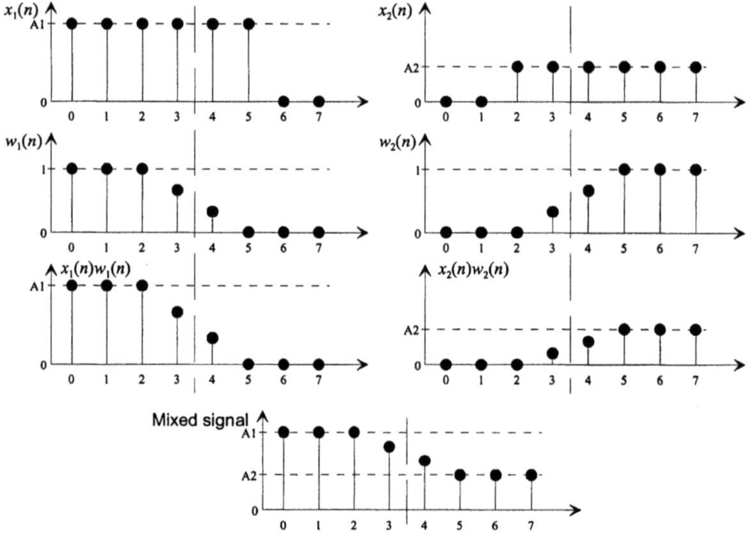

Fig. 16.2. 1D blending of two signals using ramp weighting functions, $W=3$

Other blending functions fulfilling the unity condition, being more natural than the ramp (cf. edge models in Fig. 7.14) are the *tangent hyperbolic function* and the *sigmoid function* (9.90). Examples of horizontally- and diagonally-orientated 2D blending functions derived from the linear ramp (here with a ramp of width $W=4$) are given as

$$\mathbf{W}^h_1 = \begin{bmatrix} \ddots & \vdots & \vdots & \vdots & \vdots & \vdots & \iddots \\ \cdots & 1 & \frac{3}{4} & \frac{1}{2} & \frac{1}{4} & 0 & \cdots \\ \cdots & 1 & \frac{3}{4} & \frac{1}{2} & \frac{1}{4} & 0 & \cdots \\ \cdots & 1 & \frac{3}{4} & \frac{1}{2} & \frac{1}{4} & 0 & \cdots \\ \iddots & \vdots & \vdots & \vdots & \vdots & \vdots & \ddots \end{bmatrix} \quad ; \quad \mathbf{W}^h_2 = \begin{bmatrix} \ddots & \vdots & \vdots & \vdots & \vdots & \vdots & \iddots \\ \cdots & 0 & \frac{1}{4} & \frac{1}{2} & \frac{3}{4} & 1 & \cdots \\ \cdots & 0 & \frac{1}{4} & \frac{1}{2} & \frac{3}{4} & 1 & \cdots \\ \cdots & 0 & \frac{1}{4} & \frac{1}{2} & \frac{3}{4} & 1 & \cdots \\ \iddots & \vdots & \vdots & \vdots & \vdots & \vdots & \ddots \end{bmatrix} \quad (16.4)$$

$$\mathbf{W}^d_1 = \begin{bmatrix} \ddots & \vdots & \vdots & \iddots & \iddots & \iddots & \iddots \\ \cdots & 1 & 1 & \frac{3}{4} & \frac{1}{2} & \frac{1}{4} & \iddots \\ \cdots & 1 & \frac{3}{4} & \frac{1}{2} & \frac{1}{4} & 0 & \cdots \\ \iddots & \frac{3}{4} & \frac{1}{2} & \frac{1}{4} & 0 & 0 & \cdots \\ \iddots & \iddots & \iddots & \iddots & \vdots & \vdots & \ddots \end{bmatrix} \quad ; \quad \mathbf{W}^d_2 = \begin{bmatrix} \ddots & \vdots & \vdots & \iddots & \iddots & \iddots & \iddots \\ \cdots & 0 & 0 & \frac{1}{4} & \frac{1}{2} & \frac{3}{4} & \iddots \\ \cdots & 0 & \frac{1}{4} & \frac{1}{2} & \frac{3}{4} & 1 & \cdots \\ \iddots & \frac{1}{4} & \frac{1}{2} & \frac{3}{4} & 1 & 1 & \cdots \\ \iddots & \iddots & \iddots & \iddots & \vdots & \vdots & \ddots \end{bmatrix} .$$

These examples assume straight edge lines. A 2D version applicable to non-straight boundary shapes is the *feathering filter*. Here, the weight is determined as linearly increasing/decreasing ramp depending on the shortest distance from the object boundary, which can be determined by a distance transform (cf. sec. 7.4, Fig. 7.34).

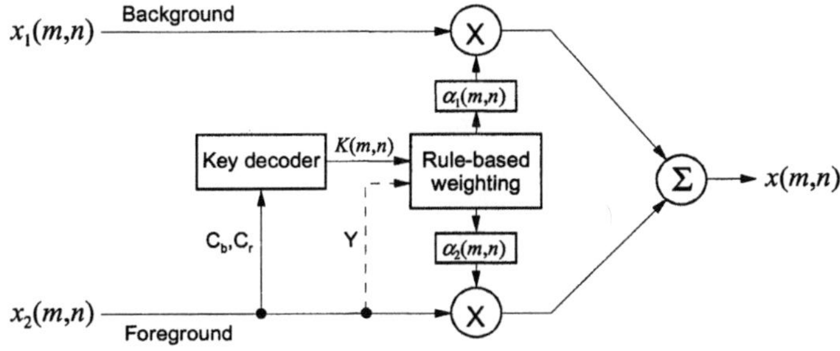

Fig. 16.3. Principle of a digital chroma-key unit

Chroma keying. The chroma key or 'blue box' method is widely used for trick mixing of movie or video signals. A foreground object is positioned in front of a sufficiently saturated, uniform-colored background. Then, a key mixer which in principle is a color classifier tuned to the blue color, can exchange the blue area by another arbitrary background from a video, still image or synthetic graphics signal. This can also be interpreted as *controllable segmentation* where criteria are defined a priori at the time of production. Analog chroma-key mixers have mostly been replaced by more advanced digital methods, which allow a more accurate adjustment and also pre- or post-processing of the key signal. In particular, the mixing at boundaries between the foreground and background can use soft transition filtering which allows a more natural impression and reduces artifacts. Using digital methods, it is possible to individually adjust the weights by which foreground and background are mixed individually for each pixel by certain rules, which e.g. allow to enforce transition weighting functions as introduced above. A block diagram of a typical digital chroma key unit is shown in Fig. 16.3.

The key signal K is mostly generated from chrominance components C_b and C_r only, which has the effect that the entire range of luminance can be utilized in the area of the foreground signal. A blue key color is advantageous, as it occurs less frequently in nature than other colors and has clearly distinguishable characteristics. By rough segmentation (clustering) of the video into blue background and colored foreground, it is possible to determine the exact key color automatically by statistical analysis. If the key color is uniquely detected by the key decoder, a background pixel is unconditionally switched to the output without any modification. If the color is identified *not* to belong to the range of the key color, *a* pixel of the foreground is mapped to the output. When blue color values are within a transition range (unreliable classification), K can also have a value between 0 and 1, which can further be adjusted by the rule-based weighting unit. This can also include analysis of neighbored key signal positions or spatial filtering of the key signal. By setting both α-values to intermediate amplitudes between 0 and 1, transparency between foreground and background is realized, which helps to improve the visual quality of the mixing result at object edges; fine structures like hair can in fact be partially transparent. If in addition analysis of luminance variations within the blue background area is made, transparency effects like shadows and smoke can be passed from the 'blue box' into the synthesized scene, which provides very natural impressions.

Z keying. If a *depth-based* segmentation of parts from a background scene is available, so-called *Z-key* methods are also applicable. Here, the depth (Z distance) information of foreground and background objects is used to decide which parts shall become visible after scene composition (Fig. 16.4). Hence, Z-key mixing allows a simultaneous composition of scenes from several sources (several independent objects), where in case of overlap the visibility is simply decided in favor of the object of smallest depth distance. In contrast, chroma keying is only capable to include exactly one object at a time in front of a full-frame background, such that several objects may only be included one after another. This effects that always the object which is most recently included covers any other. In Z keying, a newly included object can also be covered by parts of the background scene or other objects. Fig. 16.4 shows a simple example, where depending on the depth information the rider will either disappear behind the house or in front. This requires the availability of *pixel-accurate* depth information, which can be obtained automatically by disparity estimation combined with segmentation, or must be set manually. In Z-key based composition, blending filters introduced above should also be used to achieve more natural effects at object or segmentation boundaries.

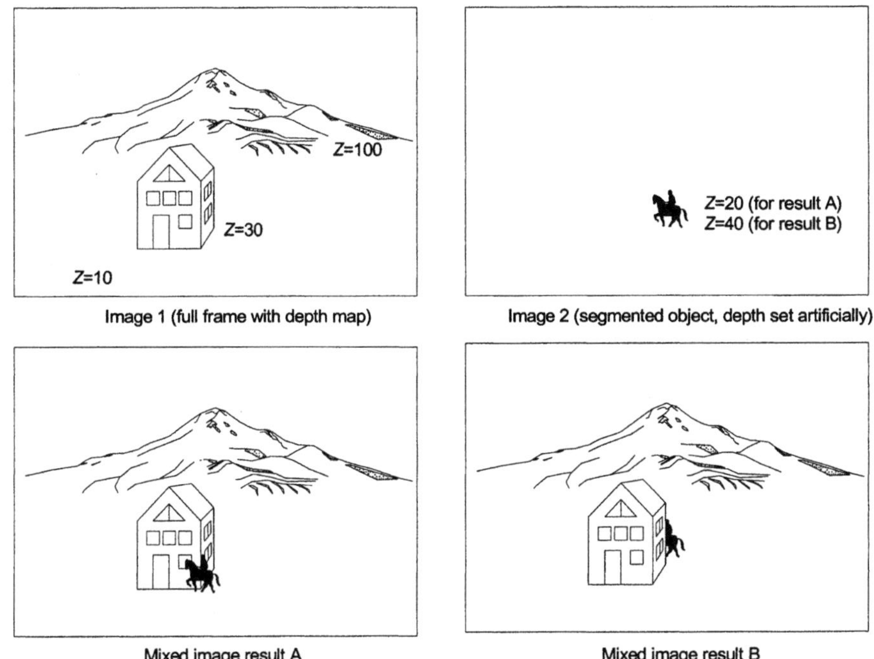

Fig. 16.4. Principle of Z-Keying with different depth assignments for the horse rider.

Representation of key signals. Key signals must typically be provided in pixel-accurate resolution, even though they usually do not show much spatial variation. Nevertheless, compression of key maps is important in interactive applications and also in media production. In fact, the α-values can be represented jointly with the video content as so-called gray-scale shapes (see Fig. 13.15) having a non-binary amplitude scale as needed for purposes of scene composition. In the MPEG-4 video codec, the gray-scale shape representation can support up to three of these 'auxiliary components', which could also hold depth data, warping data or other data types usable for scene manipulation on a pixel-by-pixel basis. As an extension of the conventional color sampling (see sec. 1.3.1), formats which support an additional alpha channel are denoted as '4:4:4:4' (which typically is an alpha weighting map in addition to RGB color components) or '4:2:2:4' (alpha in addition to luminance/chrominance components).

Scene composition as described here exclusively supports 2D image plane outputs. A depth (3D) impression can be achieved by exploiting the effect of *motion parallax* shift. If a camera position is moved, the effective motion of an object within the image plane will be larger for objects which are positioned nearer to the camera (the same effect as parallax shift in a stereoscopic camera system, cf. sec. 7.7). It is important to observe this effect when a foreground object is mixed into a background scene which was acquired by a moving camera.

16.2 Warping and Morphing

Warping is a method of image manipulation in which an image, a video frame or a region from one of these is geometrically aligned ('warped') when it is mixed into the output. *Morphing* subsumes methods, where in addition the amplitude (gray-level or color) value is adjusted during the mapping[1]. Typical applications of these techniques are e.g. replacement of faces, transformations of figures or banners which are artificially imprinted onto surfaces, being geometrically modified as necessary due to motion or perspective projection. To apply the geometric manipulations it is first necessary to describe the geometric modification by defining a number of *reference point pairs*. These points may establish either a regular or an irregular warping grid. At least one of the structures (source or destination of the geometric mapping) will usually be irregular[2]. It is also possible that both are irregular, in particular when characteristic feature points shall be mapped between two images. For example, when warping a face into another person's face in another image, eye must map to eye, nose to nose, mouth to mouth etc. Only occasionally, all of these feature points would be expected to be positioned on a regular grid. Fig. 16.5 shows an example of a triangular regular grid which is mapped into an irregular grid, and another example of irregular-to-irregular grid mapping.

For manipulation of images, the definition of reference points allows practical implementation of locally-varying warping grids. This introduces sufficient degrees of freedom in deformation which would not be achievable by global parametric shift models (sec. 7.4.4). Individual shifts are applied to the pixels, where it is assumed that for small local patches parametric shift models are sufficiently consistent. The basic relationships to map the shifts of the warping-grid feature points into pixel shift vectors at any position in between are given in (7.191) and (7.192) for the case of triangular patches. The definition of the triangular patch composition is however not unique. Typically, a densest interconnection topology exists, which is defied by the *Delaunay lines* (cf. sec. 9.4.3). Observe however that with arbitrary allowable mappings of grid points, the warped grid may not necessarily have a Delaunay topology any more. This may cause unreasonable figure mappings like mirrored patches, where also the interpolation of individual shifts and pixel amplitudes would not be reasonable. This often corresponds to areas which are becoming closed (occluded) within the warped image. A similar problem has been discussed in the context of warping-based motion compensation in sec. 7.6.4. The problem can be avoided by imposing constraints to the neighbored grid shifts, such that the difference between the shift values shall never override the distance of the grid positions.

---

[1] For more background on warping and morphing techniques, refer to [WOLBERG 1990]

[2] Regularity of one of the grids is for example given by the case of motion warping models as described in sec. 7.6.4.

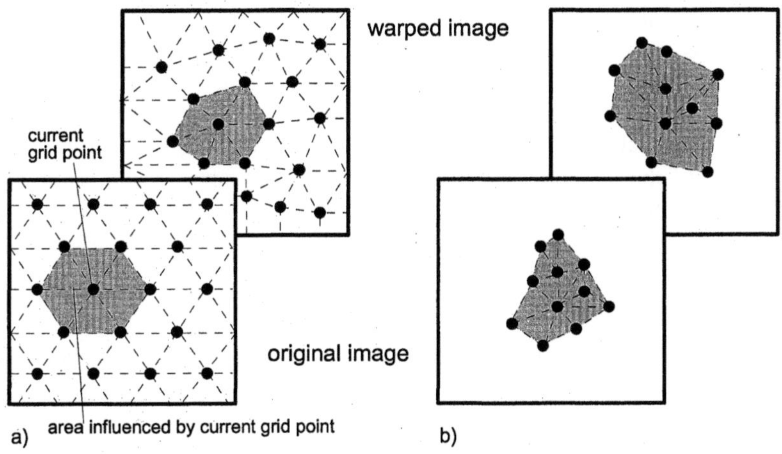

Fig. 16.5. Warping of different types of grids **a** Regular hexagonal grid defining triangular patches **b** Irregular-to-irregular grid mapping over a bounded area

Morphing is an artificial transformation of one initial image patch $x_i(m,n)$ into a destination patch $x_d(m',n')$, including both gradual change of geometry and of pixel amplitudes (texture). If applied for trick modes in movies, this will typically be performed as a blending over a time period of F frames. If the change shall proceed linearly, the geometric transformation occurs by F equal step widths. Simultaneously, a linearly progressing modification (smooth blending) of the associated pixel amplitude values is performed, such that in an intermediate frame $x_f(m'',n'')$, $f=1,...F-1$, the result is

$$x_f(m'',n'') = \frac{F-f}{F} \cdot x_i(m,n) + \frac{f}{F} \cdot x_d(m',n')$$

$$\text{with} \quad m'' = m + \frac{f}{F}(m'-m) \; ; \; n'' = n + \frac{f}{F}(n'-n). \tag{16.5}$$

The position (m'',n'') is typically not an integer-valued sampling position. Therefore, methods for interpolation on irregular grids (cf. sec. 5.4.5) must be applied additionally to produce an output picture which is regularly sampled.

16.3 Viewpoint Adaptation

A special case of warping and morphing applies if a scene is captured by several cameras. Assume that two camera views $x_1(m,n)$ and $x_2(m,n)$ are available, e.g. the left and right cameras of a stereoscopic system. The point-wise references between these pictures are given by the disparity vectors $\mathbf{d}(m,n)=[d_h(m,n),d_v(m,n)]$, which can be determined automatically by disparity estimation (sec. 7.7). A parameter f shall be defined as in (16.5), which shall have a value range between 0 and 1, such that normalization by F is not necessary. A value $f=0$ expresses the view as captured in image x_1, $f=1$ shall relate to x_2. An arbitrary intermediate view, which corresponds to a virtual camera position on the interconnecting line between the two focal points of the available camera views, results by appropriate scaling of the disparities and interpolation of texture values from the two original frames:

$$x_f\left[m+f\cdot d_h(m,n),n+f\cdot d_v(m,n)\right]$$
$$=(1-f)\cdot x_1(m,n)+f\cdot x_2\left[m+d_h(m,n),n+d_v(m,n)\right]. \tag{16.6}$$

Fig. 16.6 shows an example where a 'mid view' was generated from the left and right views of stereoscopic image pairs. The simple linear weighting function as described in (16.6) is not necessarily the optimum solution. When occlusions occur between the left and right positions, higher-quality interpolation results are obtained, if the respective pixel is *projected* uni-directionally from the image where it is visible. This gives best quality if nonlinear weighting functions are used to superimpose the left and right pictures, in particular when occlusion areas are present [IZQUIERDO, OHM 2000].

Fig. 16.6. Generation of an intermediate view (mid) from the original left and right views of a stereoscopic frame pair by disparity-compensated interpolation

Another method which is exclusively based on projection is the *Incomplete 3D* method [OHM, MÜLLER 1999]. In a first step, preprocessing is performed to gather all available pictorial information in the so-called *incomplete 3D plane*, which is associated with a dense disparity map (Fig. 16.7 top). For synthesis, it is only necessary to project this plane, using disparity vectors which are scaled according to the desired view point. Fig. 16.7 bottom shows synthesis results, including views which are beyond the original left and right camera positions (outside the baseline of a stereoscopic camera system).

All methods described here require availability of a *dense disparity field*. This is of particular importance, if the scene includes objects which are close to the cameras, because in such cases the disparities of neighbored pixels can have high differences (cf. sec. 7.7). In cases where all parts of the scene are relatively far from the cameras, while the cameras are near to each other (in the extreme case: at the same position, but with different orientations), the perspective model (7.109) is sufficient to describe the mapping. Such simplifications of the problem are applied for systems which allow panoramic look-around views. Methods supporting camera view adaptation directly by interpolation or projection from one or several images are subsumed under the term of *image based rendering* [SEITZ, DYER 1995]. These methods produce more natural-looking images than it is possible by genuine computer graphics methods. The complexity is relatively low, which allows implementation of real-time view adaptation even for video signals [OHM ET AL. 1998].

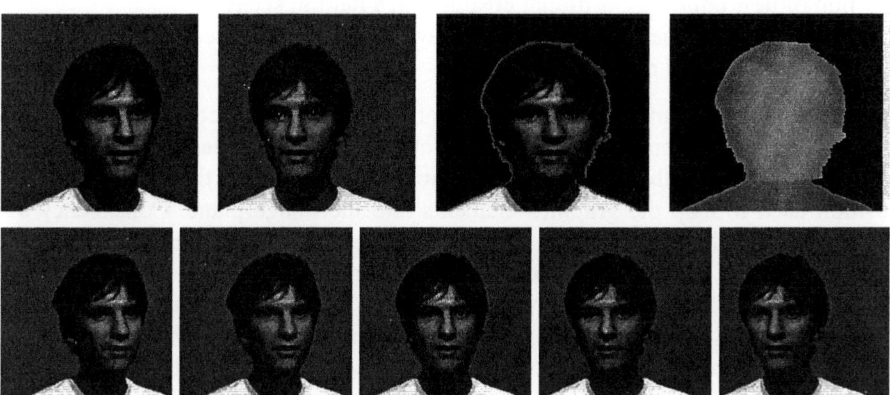

Fig. 16.7. 'Incomplete 3D' technique. Top row– Original left and right camera views, incomplete 3D plane combined from both views and associated disparity map. Bottom row– Synthesized view positions (from left to right): Beyond left camera, left camera, central view, right camera, beyond right camera.

To obtain good quality for larger ranges of view adaptation, e.g. look-around views in a room, in a stadium or over a landscape, it is necessary to provide a larger number of views, where for a 360^0 look-around view each camera should approximately cover an angle of not more than 15^0–30^0, to allow reliable correspondence matching between pictures from neighbored cameras. The tremendous amount of data rate necessary to store or transmit the information of such a high number of video streams would require specific compression methods which are able to flexibly exploit the redundancies between the different camera views. These methods are often denoted as *light field compression*, where however most approaches to date only deal with compression of static (still picture) multiple camera views. Different methods have been proposed in this context which are extensions of known video coding methods such as disparity-compensated hybrid coding [OHM 1999] or disparity-compensated wavelet coding [GIROD ET AL. 2003]. These methods perform compression based on the original image or video frames. Examples of other representation spaces are

– the *ray space representation*, which provides a dense packing of different (original and synthesized) camera views [FUJII, TANIMOTO 2002];
– representations which combine as much information as possible into one or a limited number of image planes. Examples for this are the I3D technique de-

scribed above and methods similar to mosaics (cf. sec. 7.8), but applied to multiple camera views.

In many applications, natural content (captured by cameras) is mixed with synthetic (computer graphics) content. This will be the case when

- synthetic objects shall be included in a natural scene (trick effects, e.g. simulation of new buildings, artificial creatures in a natural scene);
- natural objects shall be included in a synthetically generated scene (e.g. people in an artificial room).

To achieve a natural impression of mixed natural and synthetic content, it is important that the projection relationships are consistent, which means that the actual camera position for the natural scene shot must match with the virtual camera position used to map the graphics elements into the image plane. Consistency of illumination conditions (direction of light source etc.) should be observed as well. Blind estimation of camera parameters from a projected image is an ill-posed problem; camera calibration information (cf. sec. 7.7) is needed in this context. If such information is not available, the mixing of natural and synthetic content or mixing of content from different natural visual sources must follow a trial-and-error approach, by tuning parameters manually until the subjective impression is appropriate, such that size, view angle, illumination etc. fit plausibly.

16.4 Frame Rate Conversion

In analog video, the formats (frame/field rates, number of lines etc.) are assumed to be identical over the entire chain of acquisition, storage, transmission and replay, except for cases where different standard definitions like NTSC and PAL formats are used. When multimedia systems shall be used to jointly present signals of different sources (e.g. scanned film, interlaced and progressive video, video of different frame rates, still images, still or animated graphics), a high variability of acquisition and representation formats applies. Additionally, the content shall be replayable on different devices. For presentation on computer monitors and projectors, the frame frequency (refresh rate) could be almost arbitrary, while these devices only support a progressive (full frame) scan (cf. sec. 1.2). This requires one more signal processing step to convert the interlaced video signal into another display format before the output is performed. While the modification of the spatial resolution can be performed by spatial interpolation or decimation as presented in sec. 4.4.1 and 5.4, the change of frame rate and the conversion between interlaced and progressive formats are more challenging, in particular when motion occurs.

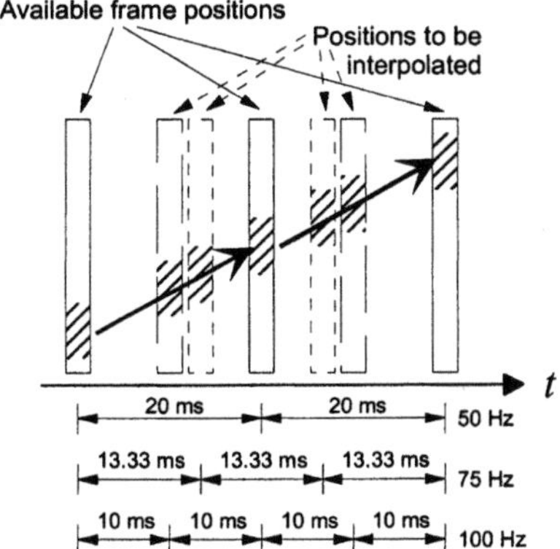

Fig. 16.8. Temporal positions of intermediate frames and instantaneous positions of an object moving along a trajectory for cases of different frame rates

To illustrate the role of motion in the case of intermediate frame interpolation, Fig. 16.8 shows the temporal sampling positions of frames for cases of three different sampling rates (50 *Hz*, 75 and 100 *Hz*). If parts of the scene (or the entire scene in case of global motion) are moving, it is necessary to determine the correct positions at which local patches must appear in an intermediate frame. Simple temporal-axis interpolation will fail, as motion incurs aliasing, such that filtering must be done along the motion trajectory. Only then, objects will appear at correct positions, and correct flow of motion is guaranteed. Regard the example of an up-conversion from a 50 *Hz* sequence into a 75 *Hz* sequence. As shown in Fig. 16.8, two out of three frames must be displayed at newly generated temporal positions, only one third of the frames can remain unchanged. As a consequence, in cases when the interpolation is performed by erroneous motion information, perceptible geometric distortions can occur.

For the generation of new frames, it is hence useful to implement a *fallback mode*, which is switched on whenever the motion vectors might be unreliable. In the most simple case, this could be a projection of the temporally nearest original frame (without using motion information), in principle a 'freeze frame' method, which can also be applied locally only for those parts of the frame with unreliable motion vectors. A smooth transition between motion-compensated (MC) and non-MC areas in the interpolated frame is then necessary, which can be achieved by spatial blending similar to the scheme shown in Fig. 16.2[1]. The block diagram of such a method is illustrated in Fig. 16.9. Reliability criteria can be used as described in sec. 7.5 and 7.6, in particular a high value of motion-compensated displaced frame difference (DFD) or high divergence in the motion vector field are good indicators for unreliable motion vectors, e.g. in case of occlusions. As the inputs are fields from an interlaced video sequence, the first step is de-interlacing to generate a full-frame sequence which has a frame rate equal to the former field

---

[1] This is also similar to the prediction approach of overlapping-block MC, see sec. 13.2.6.

rate. In a second step, progressive frames of arbitrary rate are generated by temporal interpolation; the nearest de-interlaced frame is used in the fallback mode. The composition step mixes the frame-interpolated and de-interlaced (fallback) areas, and also provides the smooth transition filtering at the boundaries between normal and fallback areas.

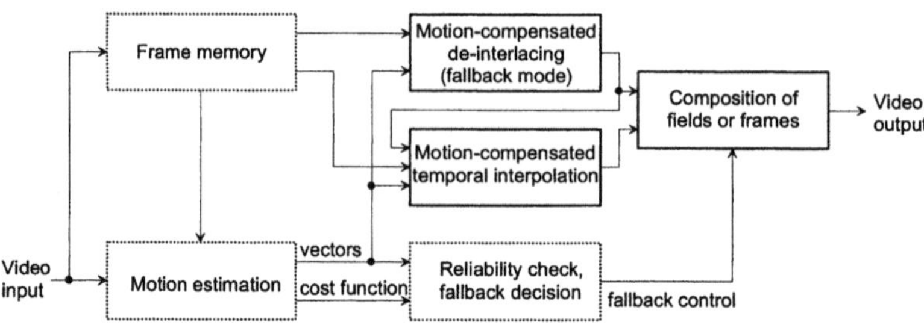

Fig. 16.9. Block diagram of a system for error-tolerant motion-compensated frame-rate up conversion (inputs are fields of an interlaced video sequence)

For the purpose of up-conversion from interlaced fields to progressively-sampled video frames (de-interlacing), only *intermediate lines* have to be generated. Combination of the two fields into one frame will not be a viable solution unless the scene is unmoved. Otherwise, the motion shift occurring between the different temporal sampling positions of the fields would cause a displacement between even and odd lines. Nearest spatial positions of the current field can be used in combination with information from at least one other field, wherein motion shift should be analyzed to achieve highest quality. If the motion estimation fails, only spatially-neighbored information from the current field should be used, which can however cause a loss of spatial detail or alias effects.

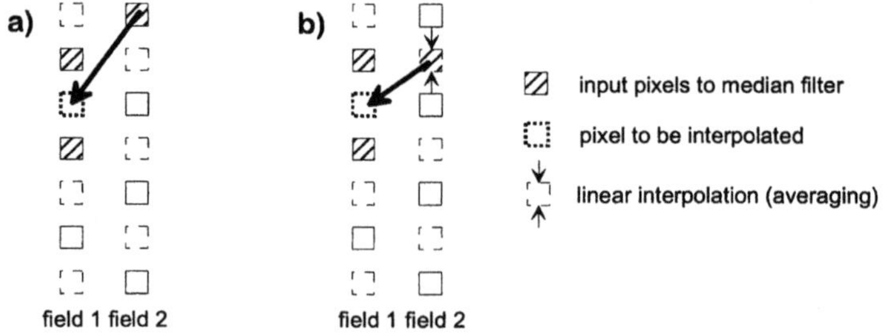

Fig. 16.10. De-interlacing using a 3-tap median filter **a** Motion vector points to an existing line of the opposite-parity field **b** Motion vector points to a non-existing line

For de-interlacing and intermediate frame interpolation, *median filters* (sec. 5.1.1 and 5.4.5) show excellent performance [BLUME 1996]. These guarantee preservation of image sharpness, and suppress outliers from the interpolated result. Fig. 16.10 shows a method of motion-compensated de-interlacing based on a 3-tap median filter, for which one input value is selected from the opposite parity field, depending on the direction of the motion vector. When the motion vector points to an existing line in the other field, an original values is used as filter input, otherwise the average value from two adjacent lines is computed first.

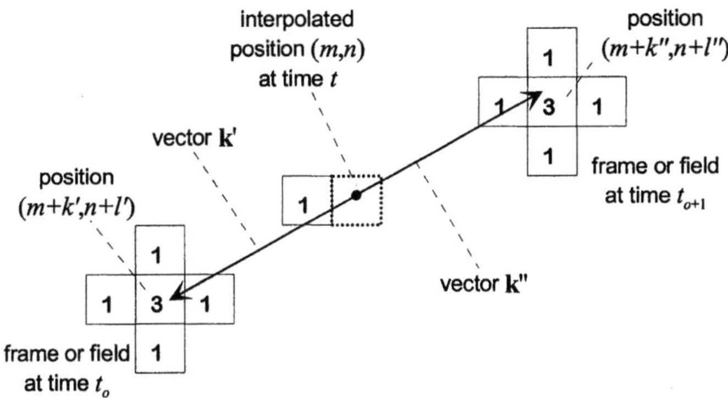

Fig. 16.11. Example of a weighted median filter used for intermediate frame interpolation

In the case of intermediate frame interpolation, pixels of the new frame must be generated for a time point t, when $t_o = oT$ and $t_{o+1} = (o+1)T$ are the sampling times associated with the two available frames, and $t_o < t < t_{o+1}$. Fig. 16.11 shows the application of a weighted median filter for this purpose [BRAUN ET AL. 1997]. The input values are gathered from two horizontal/vertical neighborhoods $\mathcal{N}_1^{(1)}$ as defined in (4.1), which are centered at positions $(m+k',n+l')$ and $(m+k'',n+l'')$ and interconnected by the motion vector $\mathbf{k}$ which has been estimated between these positions. The center pixels are weighted by a factor 3. The interpolated pixel is generated at position (m,n) which is traversed by the vector; one additional input to the median filter is gathered from a previously-computed position of the interpolated frame. The motion vectors $\mathbf{k'}=[k'\ l']^T$ and $\mathbf{k''}=[k''\ l'']^T$ are scaled versions

$$\mathbf{k'} = \mathbf{k} \cdot \frac{t_o - t}{T} \quad ; \quad \mathbf{k''} = \mathbf{k} \cdot \frac{t_{o+1} - t}{T} \tag{16.7}$$

of the (backward) motion vector $\mathbf{k}$, which must be checked for validity at the local position in the intermediate frame[1].

---

[1] Two or multiple vectors may cross the position (m,n) in the intermediate frame, or in other cases, *no* motion vector may be identified at all. In addition to the reliability check during estimation, the validity of a motion vector for a given position can be based on its size or the consistency with neighbored vectors. In a simplified approach, motion vectors can be used that are available at the same coordinate position (m,n) of the nearest existing frame, which then should be the 'current' frame of motion estimation.

16.5 Rendering of Image and Video Signals

Rendering of image or video signals is the preparation for a projection to the output medium (screen). Depending on the resolution of the system, image or frame sizes must eventually be scaled proportionally. Geometric distortion can occur due to properties of the projection system (e.g. if a video projector's optical axis is not perpendicular with the screen); if these distortions are known, they can be pre-compensated by methods of signal processing such as geometric transformations. Rendering also can include processes of view adaptation by user request and simple methods of post-mixing such as picture-in-picture overlays. In the latter case, adaptation of the sizes of the single elements can also be regarded as part of the rendering process. The result of rendering then is a newly composed image plane as it is physically fed into the output device.

Viewpoint adaptation is of particular importance in interactive applications. Usage of disparity-compensated processing for *image based rendering* was introduced in sec. 16.3. Other simple methods of view adaptation can be based on the *parallax shift*, where the impression of a view point change is stimulated by shifting the position of a (segmented) foreground object relative to the background. No further modification must then be applied to the 2D image patches of foreground and background, except for a scaling when the viewpoint changes by depth. The latter class of methods can also be subsumed under the category of *2D rendering*, while disparity or depth dependent image projection is often denoted as *2½D rendering*.

In genuine *3D rendering*, the full 3D (volumetric) information about the scene to be projected onto the screen must be available. For general image or video data, 3D rendering is unrealistic, as the information that can be gained about the 3D space by a limited number of camera views will always be incomplete. 3D rendering is well applicable for 'image plus volumetric' information originating from 3D acquisition devices for single objects (e.g. sensors in computer tomography, laser range samplers, single objects photographed by multiple camera positions). In computer graphics applications, complete volumetric 3D scenes can be stored with associated surface textures. The projection which then has to be applied for rendering onto 2D screens is in principle not much different from the projection of the 3D exterior world into a camera image plane (cf. sec. 7.4.4). The camera is however not an existing optical system, but rather a *virtual camera* where the projection equations are applied numerically for each pixel to be rendered, which can be a computationally expensive task. To simplify the process, methods of 3D rendering are often based on *orthographic projection* (7.101). The disadvantage is the underlying assumption that all objects to be projected are relatively far from the camera, or are quite flat in depth. From this point of view, orthographic projection is comparable with simple parallax-based 2D rendering methods, and indeed more correct depth impressions could be generated by manipulation of the 3D scenes. The precise and correct projection of 3D volumes onto 2D screens is the *perspec-*

tive projection (7.100). Here, the projection ray must explicitly be computed for each pixel, which includes a division operation.

For image, video and graphics signals, stereoscopic rendering and projection is important. It is used to generate a depth impression by the phenomenon of the stereoscopic parallax (cf. sec. 7.7). In natural vision, the parallax is effective mainly in the near range (up to 6 *m*). As however stereoscopic screens are often operated within an even lower range of viewing distances, the parallax effect can indeed be important; the depth impression is generated here by stimulating left and right eyes with slightly different images. The inherent problem of stereoscopic display systems is the necessity for simultaneous projection of left and right images, where it must be guaranteed that the two signals are sufficiently separated such that the image to be viewed by the left eye does not interfere with the right eye and vice versa. In principle, this is achieved by multiplexing, where the penalty is a *loss of resolution* in another dimension, e.g. temporal, spatial or color resolution. Common methods for stereoscopic display are:

- *Head mounted displays.* These consist of two small screens close to the eyes, where each is physically only visible to one eye. Perfect separation is guaranteed, as no common optical path exists. No loss of resolution occurs, but a disadvantage is the limited usage, isolating the viewing persons from the real exterior world.
- *Red/green glasses.* The left and right signals are displayed by different primary colors and rendered into only one image. Glasses using red and green optical filters are used for separation. This is maybe the most simple method for stereoscopic display, which is usable for print media, TV screens, projection systems etc. The restriction is the loss of color information, the perceived signal appears to be almost monochrome.
- *Shutter glasses.* The left and right frames of the stereoscopic signal are displayed in alternating order (temporal multiplex) on a screen, where specific glasses must be worn which close a shutter, synchronized with the display to make any frame invisible which is not intended for viewing by the respective eye. These systems have a very good signal separation property, but display systems with high temporal refresh rate are needed, and also the shutter glasses themselves are rather expensive.
- *Polarization glasses.* The left and right images are projected using optical polarization filters (which retain the color), such that one signal is projected by horizontally- and the other by vertically-polarized light. The observer also wears glasses with same polarization filter properties, which separate the signal channels. The method is applicable only with double-lens projection systems, not for TV or computer monitors, channel separation is less efficient than for shutter glasses. A loss occurs in brightness, color and contrast due to the polarization filters.
- *Autostereoscopic displays.* These are systems without glasses, which are usually based on *spatial multiplex* of left and right signals, where typically alter-

nating columns of the display are assigned to the left/right channels. For separation of the signals, such that the light rays of the respective column only reach the correct eye destination, lens rasters [BÖRNER 1993] or stripe masks [EZRA ET AL. 1995] are used.

The stereoscopic illusion is achieved, when certain areas of the image are not perceived as positioned *on the screen*, but rather appear *before* or *behind* the screen. For stereoscopic vision, one important criterion is the perceived parallax (cf. sec. 7.7). Due to focusing of the lens and the high density of receptors in the area of the fovea centralis (cf. sec. 6.1), an eye can see only a small area around the fixated point by full resolution; the 'viewing rays' of the two eyes will usually converge at that point in case of stereoscopic vision. For evaluation of distances, this experience is of similar importance as the parallax. From the principles shown in Fig. 16.12, it is obvious that a certain point in stereoscopic presentation is observed *behind the screen*, if the right eye perceives the point further right on the screen than the left eye sees the corresponding point; obviously then, the viewing rays seem to converge behind the screen (Fig. 16.12a). Likewise, in the opposite case where the right eye looks farther to the left than the left eye, the convergence point will be interpreted as being positioned *in front of the screen* (Fig. 16.12b).

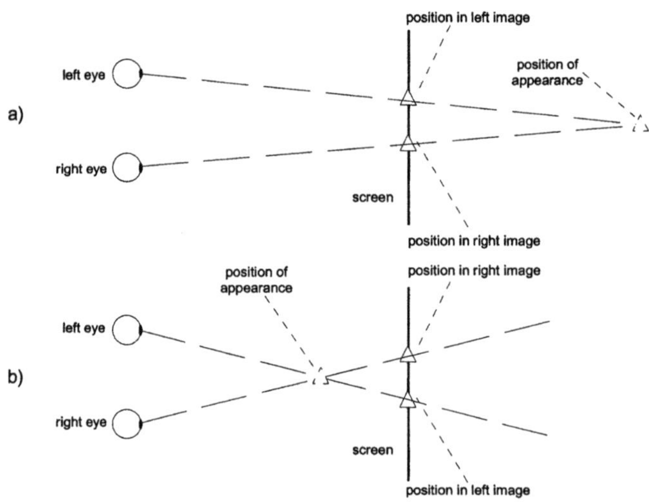

Fig. 16.12. Stereoscopic projection for appearance of an element of a scene
a behind the screen   **b** in front of the screen

From these interpretations, it can further be concluded that a perfect spatial illusion is hardly achievable using a stereoscopic system[1]: It would be required that

---

[1] In fact, usage of stereoscopic systems requires some training to overcome the unnatural aspects of stereoscopic display. Indeed, a trained person can well extract the additional 3D spatial information; in the brain, it is processed in a more abstracted sense anyway.

the capturing cameras possess identical properties (in particular regarding the baseline distance) with the eyes, and that the viewing distance is also very similar to the original distance between the captured object and the camera (or that at least a natural scaling of sizes is observed). The stereoscopic parallax however establishes only one component in the depth perception part of the human visual system. The motion parallax as experienced by egobody motion seems to be at least similarly important. In stereoscopic systems, the perfect spatial illusion can in fact be violated if the viewer changes the viewing position or distance. Based on such reasoning, systems have been developed where disparity- or depth-based view projection is used to change the presented image slightly[1]. This gives a more natural depth illusion than 'static' stereoscopic systems.

Output systems giving spatial impression beyond stereoscopy are still in early stages of development. An example are holographic systems, which are quite mature for still images, where however for video no monolithic acquisition systems exists, and video holographies must be artificially generated by mapping of multiple simultaneously captured camera feeds. These systems still suffer from low resolution which does not in general outweigh the spatial illusion[2].

16.6 Composition and Rendering of Audio Signals

Composition of audio signals is mainly performed as *mixing by linear superposition principles*; optionally an equalization (filtering) for frequency spectrum tuning is applied to the individual signals. Further, spatial localization is applied, depending on the capabilities of the output system. If in a multimedia application audio signal elements can be associated with visible elements (e.g. an object which is visible *and* audible), a natural illusion also requires adaptation of the loudness depending on the apparent distance, such that the volume of the audio signal should become lower if the object is visually disappearing. For sound composition where an illusion of room properties shall be produced, the inclusion of room transfer functions or an approximation thereof by echo and reverberation effects is important (cf. sec. 7.10.4). Other typical effects applied in audio composition are nonlinear amplitude manipulations by compression and expansion (cf. sec. 5.3), nonlinear spectral manipulations like *harmonization* (artificial generation of har-

---

[1] This simulates the motion parallax by the expected change of a scene that would occur when the observer is moving.

[2] To some extent, this remark applies to any output technique which tries to mimic a spatial (volumetric) illusion. Understanding of 3D spatial contexts belongs to categories of higher abstraction of the brain, where humans are permanently coping with incomplete information gathered by the eyes. From this point of view, methodologies are questionable where the provision of stereoscopic or other spatial illusion comes by a penalty in other information components such as spatial resolution.

monics), artificial generation of modulations as vibrato and tremolo (cf. sec. 7.10.3) or phase variations. Some of these effects add non-linear components into the mixing process.

According to the definition given above, *audio rendering* is the preparation of the signal(s) for an output to a loudspeaker system. This must take into account the interrelationship of the loudspeaker positions, the room characteristics and the sound field to be generated. Systems with multiple loudspeakers, such as stereophonic and surround sound systems, are most commonly used. Depending on the spatial localization assigned in scene composition, the audio rendering must then determine the signal components to be fed into the separate loudspeakers, such that the desired effect is achieved. This is quite straightforward for multi-channel systems, which serve the different loudspeakers separately with a clear directional assignment (front left, back right etc.).

Sound field generation. More sophisticated systems are able to generate almost arbitrary 3D *sound fields* by a limited number of signal feed channels. One example for this is the *ambisonics system* [GERZON 1977], which decomposes the 3D soundfield into one sum channel and three orthogonal difference channels related to directionally-bipolar orientated acquisition over orthogonal axes of the 3D room coordinate system. This is in principle an extension of the sum/difference encoding in stereophony (see sec. 14.2, which only analyzes the left-right axis), by the remaining two axes front-back and top-down. By appropriate mixing of the different signals, which relates to weighting by trigonometric functions of the azimuth and elevation angles, it is either possible to feed a loudspeaker at the respective position, or to add more directional sound sources to the composite ambisonics signal. Loudspeakers are typically arranged at equidistant positions on a sphere, where the optimum listening position is at the center of the sphere. In principle, to retain the full spatial sound illusion, it will be necessary to track the position of the listener and adjust the signal accordingly.

More ambitious approaches are related to *virtual sound field generation*. A sound field can be generated for one particular listening position in a room, where it is necessary to include sound source positions, room transfer function and ear transfer function of the listener in the model. The output can be listened by earphones, where the effect is then similar to *artificial head stereophony*. Alternatively, the *cross talk cancellation* method uses two loudspeakers with appropriately pre-processed inputs to provide an exact copy of a true sound field for one specific listener's position [WARD, ELKO 2000]. This method is perfectly applicable only within reflection-free rooms, and is also quite sensitive in case of listener movements; in case of single users, tracking of movements and appropriate real-time adjustment of the generated sound fields is necessary.

Methods of *wavefield synthesis* are not restricted to single locations. By using a large number of loudspeakers in a surround arena, a soundfield can in principle be generated as it would actually occur based on natural sound events [BOONE, VER-

[BOONE,VERHEIJEN 1993]. If pre-computed, a large number of audio channels, relating to the feed of a number of loudspeakers, must be stored.

Audio speed adaptation. Methods of video frame rate conversion can be used to generate slow-motion or also faster motion video, when replay is performed by a different frame rate than originally intended. Also for audio signals, it may be desirable to adjust the speed during the rendering process, e.g. if the audio signal has to be synchronized with a video signal that is altered in replay speed. Simple sampling rate conversion will not perform this purpose, as it also will change the frequency of the replayed signal. The challenge is to change the speed but keep the tone height constant. An application scenario from the context of audiovisual streaming is introduced in [STEINBACH ET AL. 2001].

A viable method for audio speed adaptation with unchanged tone height is based on application of a short-time Fourier Transform (STFT). The spectral coefficients are decimated or interpolated over the time axis, resulting in either more or less STFT blocks than for the original signal. By inverse transform, the tone height is in fact not changed, but the replay speed will become longer if additional transform blocks have been generated, and will become shorter in the opposite case. In addition, it is necessary to perform phase shift alignment of adjacent blocks, as this otherwise will lead to perceivable phase switching or other inconsistencies at the block boundaries.

17 Multimedia Representation Standards

The primary purpose of standardization in the area of multimedia communication systems is to guarantee interoperability between devices and representation formats. Wide acceptance of a standard will help to achieve product life cycles of reasonable duration, and hence will also contribute to the availability of products at reasonable cost. For this, it is important that standards are open, which means that anybody has full access to all background information about technology implemented in the standard, and is allowed to use them by fair conditions. Compliance with standards must be strongly postulated, but it is often possible to further improve the efficiency within the framework of an existing standard, e.g. by optimizing encoders. Standards have to be extended when new application areas emerge, or when progress of technology allows significant improvement. In such cases, it should be targeted to retain compatibility with previous versions. In the design of a complex multimedia communication system, it is necessary to observe definitions made by many different standards e.g. from the domains of networking and transmission, source representation, coding and display technology. It is beyond the scope of this book to give a complete overview about this field. Moreover, it is shown how the methods introduced in previous chapters are implemented in the multimedia coding and content description standards of the International Standardization Organization (ISO/IEC) and the International Telecommunication Union (ITU).

17.1 Interoperability and Compatibility

Interoperability can be achieved if multimedia communication systems follow the same rules and can understand (communicate with) each other. Simple examples are interoperability between transmitter and receiver, encoder and decoder. The encoder must know what the operational capabilities of the decoder are, and the decoder must be able to interpret the bitstream sent by the encoder. These concepts extend to much more complex configurations including multiple transmitters and receivers, interfaces to networks and storage, user interfaces etc.

Interoperability in image, video and audio coding standardization is mostly achieved by defining the *bitstream syntax*, the *bitstream semantics* and the *decoding process*. The syntax relates to the structure of a bitstream, such that it can be parsed and interpreted as a sequence of codewords, which is typically the first step of the decoder operation. The semantics specifies the way by which these codewords are to be interpreted during the decoding process, e.g. the mapping of VLC codewords into reconstructed source symbols, interpretation of side information parameters etc. The decoding process itself generates the output signal by a well-defined concatenation of processing steps.

It is important to note that it is not necessary here to exactly define the design of encoders, as it is uniquely possible by the given specification to build an encoding device which produces a bitstream of valid syntax that can be transformed into a usable output signal by a standard compliant decoding device. This enables plenty of freedom degrees to further *optimize an encoder*, retaining full compatibility with an existing standard, allowing competition between different encoder manufacturers for the best advanced encoder technology in the market. On the other hand, a low-cost encoder could be built e.g. for real-time applications with low power consumption constraints, which might not show the optimum performance that theoretically could be achieved within the framework of the standard. Encoders are of course constrained by the bitstream and decoder definitions. For example, most video coding standards define block-based motion compensation, where certain restrictions may be made to the maximum range of motion compensation. It would be useless if an encoder finds a best motion vector for a shift of 17 pixels, if the specification of the standards-compliant bitstream semantics only allows a range of ±15 pixels.

In speech coding, a different approach is often taken. As it is understood to be important that intelligibility of speech is guaranteed, also encoders are fully specified by their algorithmic operations, thus guaranteeing a specified perceptual quality. On the other hand, this also limits the capability for post standardization improvements[1].

Decoder conformance. Standards conformance is typically postulated only for decoders for the case of standards with open encoder solutions. It is simply assumed that encoding devices, which produce bit streams of poor reconstruction quality or bit streams that crash when fed into standards-compliant decoders, will not survive in the market.

However, only decoding devices of limited complexity can be implemented in certain application environments, and it is e.g. reasonable that a decoder in a mo-

---

[1] As an example for such improvements, the development that occurred in the context of the MPEG-2 standard is impressive. The last experimental version of the MPEG committee was the *Test Model 5* (TM5). Commercial products using sophisticated high-end encoders today typically achieve comparable quality as TM5 at only half data rate. This means that the compression factor is roughly doubled, while the streams are indeed still replayable by decoders that were built years ago.

bile phone would not be able to decode an HD resolution video signal or complex 3D graphics elements. Such reasonable limitations of decoder complexity within the framework of a more generic standard are imposed by definition of *profiles* and *levels*, where a certain profile/level combination describes a *decoder conformance point*[1]:

- A *profile* specifies a collection of algorithmic elements (tools)[2] which must be supported in the decoder to produce a reasonable output;
- A *level* is related to a maximum resolution of the signal which must be supported by the decoder at a given conformance point, e.g. expressed in numbers of samples that must be processed per second, or in terms of maximum bit rates.

The profile/level indication is usually conveyed in a header part of the bitstream, such that a decoding device can immediately decide if it will be able to process this stream. Most audiovisual media coding standard provide normative mechanisms for *conformance testing*[3], where typical standard-compliant bitstreams are provided which must be decodable by a conformant decoder, trying as much as possible to drive the decoding device to its limits.

Compatibility principles. When new standards are defined or when extensions (amendments) to existing standards are made, one important question will be whether these are *compatible* with previous solutions. The following compatibility principles apply[4]:

- *Upward and downward compatibility*: A receiver supporting higher resolution (e.g. HD video) should be upward compatible, able to replay a signal from a bitstream which was originally generated for a receiver of lower resolution (e.g. SD video). By the reverse principle of downward compatibility, also a receiver designed for lower resolution would be able to replay a signal – at least partially or by reduced quality – from a bit stream which was originally generated for a receiver of higher resolution. From the principles given above, upward compatibility would always be guaranteed for a conformant decoder within the same profile at higher level, since the algorithmic elements (syntax,

---

[1] This approach has been taken systematically in all MPEG standards since MPEG-2. MPEG-1 was originally designed with restriction to one specific application, and hence did not need the flexibility of defining multiple conformance points. Some other standards (e.g. JPEG) are more vague in definition of conformance points, where it can however be noticed that standards elements which go beyond the baseline structure are hardly accepted in the market.

[2] In the MPEG-4 standard, another entity of *object type* is introduced, which allows a more systematic grouping of tools. Profiles are then defined as combinations of object types.

[3] See the respective Part 4 of the standards MPEG-1, MPEG-2, MPEG-4 and JPEG 2000, Part 2 of JPEG.

[4] Most of these principles were introduced in [OKUBO 1992].

semantics etc.) which must be supported are identical over all levels. Downward compatibility is more difficult to achieve, as it may break the complexity limitations of lower-resolution decoders. It could straightforwardly be realized for the case of embedded bit streams and scalable decoders. Otherwise, *complexity scalable decoders* can be implemented, where a lower-resolution decoder performs a best effort approach for signal reconstruction from a higher-resolution stream. It is however clear that this could not be a standards-compliant decoding process.

– *Forward and backward compatibility*: In forward compatibility, the bitstream generated according to an older standard shall also be interpreted by decoders operating according to either an extension (newer generation) of this standard or a completely new standard (Fig. 17.1). Backward compatibility is the reverse principle, where a new standard is defined such that a bitstream can at least for partial or low-quality reconstruction be interpreted by a decoding device which operates according to an earlier standard. The former case can easily be achieved if the new standard or the extension of a standard by a new profile defines a *superset* in terms of algorithmic elements. Backward compatibility is more difficult to realize, and either will sacrifice the compression efficiency of a newer generation of standards, or will enforce higher complexity of decoding devices. On the other hand, backward compatibility can be rather important for acceptance in the market.

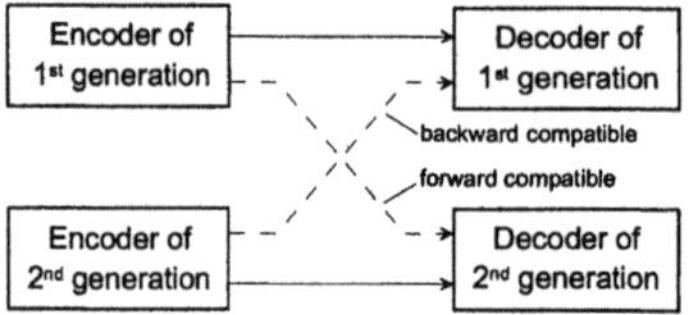

Fig. 17.1. Forward and backward compatibility

Interestingly, realizations for the compatibility principles are very similar to solutions where content shall be transmitted over networks by various rates or resolutions. The different methods can be summarized as follows:

– Simulcast: Bit streams related to different resolution levels, quality levels or standard generations are transmitted simultaneously and can be regarded as completely independent of each other except for the fact that they are multiplexed within one stream. The decoder selects the appropriate stream according to its capabilities. In this case, decoders need not to be compatible at all, however a penalty comes by increased transmission/storage bandwidth resulting from the sum of the different streams. A stand-alone decoder of higher resolution or second generation may be capable to decode the first generation streams, if it is upward/forward compatible. Simulcast is only useful if multiple decoders of different capabilities must be served simultaneously.

–   Embedded (scalable or layered) coding: The bit stream is structured such that
    it is fully used at the highest resolution or the more advanced version of a
    standard, but the lower resolution uses a sub-set of the stream. In principle, the
    encoder consists of two (or multiple) stages, each of which produces one level
    of the bit stream. However the information already available from the lower
    layers is not discarded as in simulcast. Instead, the higher layer performs some
    kind of differential encoding starting from the information provided by the
    lower layer decoder. In more elegant solutions for embedded coding like
    wavelet coding (sec. 12.4.2 and 13.4.4) or embedded quantization (sec. 11.3),
    the generation of the different resolution levels is inherently included in a sin-
    gle-step encoding process.

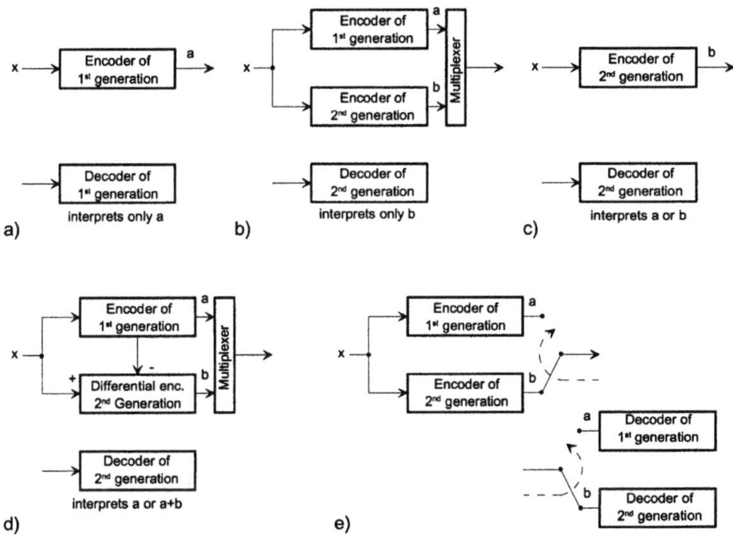

Fig. 17.2. Compatibility principles **a** Encoder and decoder of first generation without
backward compatibility **b** Simulcast **c** Syntax extension (superset) **d** Embedded coding
e Switchable coding   (adopted from [Okubo 1992])

–   Syntax extension: A new generation is defined in a way that it implicitly con-
    tains all options from one or several previous generation(s), i.e. it establishes a
    superset of all previous syntax and semantics. This implies that any bit
    streams generated according to the old standard generation(s) can implicitly
    be decoded by devices built according to the new generation as well. Back-
    ward compatibility is not necessarily supported. As an example, MPEG-2
    video coding defines a superset of MPEG-1, i.e. MPEG-2 video decoders are
    usually capable to decode MPEG-1 streams, but not vice versa. On the other
    hand, if syntax extension is made by using syntax elements that were defined
    as *reserved for future usage* in the older generation, an older decoder may

simply ignore the functionality expressed by the new syntax elements, but still may work correctly on the remaining syntax.

– Switchable coding: Switching must at least be implemented at the encoder *or* at the decoder. If the encoder is switchable, it negotiates the standard configuration to be used in the communication with the decoder at the receiver end. If the decoder is switchable, it selects its decoding mechanism according to the stream which is sent by the encoder. Only in the latter case, the method is still applicable when one encoder shall serve multiple decoders, e.g. in broadcast or multicast applications. From this point of view, switching is no good solution in point-to-multipoint applications, as any older decoding devices that were existing before the extensions were defined would be outdated.

Tab. 17.1. Compatibility principles mapped to the methods from Fig. 7.3

	Upward/forward compatibility	Downward/backward compatibility	Multiple receivers
Simulcast	optional	yes	yes
Syntax extension	yes	limited	yes
Embedded coding	yes	yes	yes
Switchable coding	yes, if encoder *or* decoder is switchable		yes, if *all* decoders are switchable

Schematic diagrams for all four methods are shown in Figs. 17.2b-e. Table 17.1 summarizes which compatibility principles are achieved by the methods. It is evident that all compatibility mechanisms can best be fulfilled by the embedded coding method. Whether this is indeed the optimum solution for a given application depends on the compression performance which often is penalized by scalable source coding methods (cf. sec. 13.2.8 and 14.2), and on the implementation complexity. I exceptional cases, when a new extension or a new standard is designed for an entirely new application domain which was not yet served by any older generation standard, or if the new standard gives drastic performance improvement over prior existing solutions, compatibility principles may also be ignored. Due to economical considerations, it would however even then be reasonable to stick to older solutions for which full or partial implementations are available. Acceptance of an entirely new technology should be justified by the benefit that can be expected globally, e.g. by saving storage or transmission cost, by achieving significantly improved service quality, new functionality etc.

General principles of standardization. Standardization in the area of multimedia systems in first place requires evidence that it is useful and supported by market needs to develop a normative specification of technology. *Requirements* must then be set up that can be matched by foreseen application and use cases. It is advantageous to develop standard frameworks which are as generic as possible, able to support different application domains, provided that this generic approach does not lead to unnecessarily complex systems[1].

If the requirements to be fulfilled are clear, and if it is evident that prospective technology exists that would be able to fulfill these requirements, it is necessary to evaluate such technology and select a method or a combination of different methods proposed as starting point of standardization[2]. Further improvements are then made by detailed investigations, typically accompanied by rigid experimentation, testing and verification procedures, which enable objective comparison in the context of a (software) model. The whole process is ruled by an increasing degree of convergence, which is reflected by different draft versions[3], accurately documenting the technology which is foreseen to appear in the prospective standard.

17.2 Definitions at Systems Level

To realize a complete interoperable system for multimedia data compression and decoding, aspects have to be observed which go beyond the genuine properties of encoding algorithms. For example, unique identification of the data streams to be decoded, multiplexing, synchronization of streams and decoded content (e.g. synchronous presentation of video and audio elements) have to be guaranteed. Further, interoperability in adaptation to specific properties of a network, error protection etc. play an important role in multimedia signal transmission. Hence, specifications are needed for access-level systems syntax and related mechanisms, which includes identification, access structure, multiplexing and synchronization of data streams, adaptation to network characteristics, aspects of user interaction and presentation of the decoded content by output devices.

---

[1] Of course, within the framework of a generic standard it is always possible to circumvent the complexity problem by defining sub-sets of elements for specific application solutions via the profiling approach described above. It should however be guaranteed that this does not affect the readability of the standard at large.

[2] If standardization of a larger framework shall be started, a *Call for Proposals* is usually issued by standardization bodies. This clearly states the desirable requirements to be fulfilled by the technology that is called for, and defines testing conditions or evaluation criteria by which different competing technologies shall be compared. In the case of coding algorithms, this testing employs quality evaluation methods as described in Appendix A.1. The evaluation will typically investigate the performance, universal applicability and complexity of different methods proposed.

[3] In the case of ISO/IEC standards, these phases are *Working Draft* (WD), *Committee Draft* (CD), *Final Committee Draft* (FCD) and *Final Draft International Standard* (FDIS). In the case of amending an existing standard, the CD is replaced by *Proposed Draft Amendment* (PDAM), the FCD by *Final Proposed Draft Amendment* (FPDAM), and the FDIS by *Final Draft Amendment* (FDAM).

An important aspect in this context is the structure of streams, where decomposition into smaller access units is required, which can be identified and treated separately. Without actually starting to interpret (decode) these pieces, sufficient information must be given by additional syntactic elements e.g. to identify whether all data necessary to decode a certain frame of video were already received. In addition, mechanisms must be provided which allow to combine different substreams (video, audio, supplemental information) for transport over a common stream.

A typical example for such an approach is the *transport stream hierarchy* of MPEG-2. As shown in Fig. 17.3, each encoding device (e.g. video and audio encoders) generates *Elementary Streams* (ES), which can directly be decoded by the complementary decoding device. To make the ESs piecewise accessible, decomposition into *Packetized Elementary Streams* (PES) is performed. The *Transport Stream* (TS) is a concatenation of PES, as it will actually be transmitted over the network[1]. Additionally, higher-level multiplexes of several TS can be constructed, which are e.g. related to different programs. Supplemental information at this level are the *Program Association Tables* (PAT), containing a list of all TS contained in the multiplex, including indexes for the *Program Map Tables* (PMT) which are associated with the particular program units and used to manage the TS packets. In addition, *Service Information* (SI) tables can be provided, which e.g. can be used to convey electronic program guides.

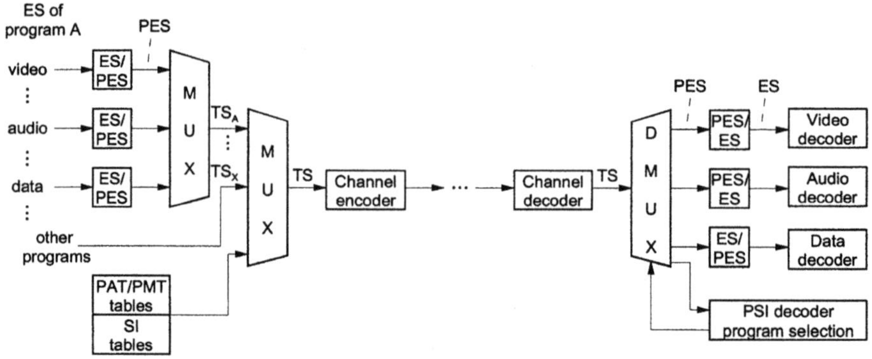

Fig. 17.3. Transport stream multiplex of MPEG-2

In the PES, packet sizes are variable with the length indicated in the header. Each packet also contains the ES identification which is needed to re-assemble the ES. Other optional parts of the PES header contain information about priority, copy-

---

[1] For some cases, such as Internet transport, the MPEG-2 TS would again be transported via RTP packets; in such cases however, it is tried to keep the overhead to a minimum. For other networks, such as DVB, the MPEG-2 TS is the primary transport mechanism which directly interfaces with the network layer (cf. sec. 6.3).

right, scrambling for access protection, additional error detection for the previous packet and synchronization information.

TS packets have a fixed length of 188 bytes[1], the payload data are filled from PES packets. A transport stream contains (in principle in arbitrary sequence) PES data from all associated elementary streams, which belong to a program of a given *Program Identifier* (PID). In addition, the TS-packet header also contains information about priority, scrambling etc. *Program Clock Recovery* (PCR) information is used for synchronization between different transport streams in the TS multiplex. In addition, private header data can be conveyed, which can e.g. be used for the purpose of encryption and to identify access rights in Pay TV.

Synchronization. For time-accurate presentation of the decoded data, like synchronous replay of video and audio, it is necessary to provide synchronization mechanisms at the ES level. To support synchronization, the ES are supplemented by *time stamps*. These can be defined based on a relative reference (e.g. the start of a movie) or based on absolute time. Different types of time stamps exist in MPEG standards:

- *Decoding Time Stamps* (DTS) indicate the time by which the decoding of the related frame shall be finished;
- *Composition Time Stamps* (CTS, only available in MPEG-4, where several objects can be composed into a scene before rendering) indicate the time by which the composition result shall be available;
- *Presentation Time Stamps* (PTS) indicate the time by which the presentation shall be made.

Additional requirements related to synchronization, in particular in MPEG-4, can refer to user interaction, where e.g. certain parts of the ESs are only activated (decoded) if initiated by the user.

Further important aspects that must be defined at the systems level are *file formats* for storage, *network interfaces, user control functions* and *rights management mechanisms*. The standards related to multimedia coding are typically sub-divided into several parts; the systems related specifications can be found as follows:

- ISO/IEC 11172 (MPEG-1), is entitled '*Coding of Moving Pictures and associated Audio for Digital Storage Media at up to about 1.5 Mbit/s*'. All systems-related definitions are made in '*Part 1: Systems*'.
- ISO/IEC 13818 (MPEG-2) is entitled '*Generic Coding of Moving Pictures and Associated Audio Information*'. The systems-related definitions are pro-

---

[1] The length of 188 bytes perfectly fits with the payload size that was once specified for ATM cells in broadband ISDN (see footnote on p. 707). At the time when MPEG-2 was developed, it was the original plan to transport most payloads of MPEG-2 data via ATM networks.

vided in *'Part 1: Systems'*, *'Part 6: DSM-CC'*[1], *'Part 9: Extension for real time interface for systems decoders'* and *'Part 11: IPMP on MPEG-2 systems'*[2].

- ISO/IEC 14496 (MPEG-4) is entitled *'Coding of Audiovisual Objects'*. The systems-related specifications are provided in *'Part 1: Systems'*, *'Part 6: Delivery Multimedia Integration Framework (DMIF)'*, *'Part 8: Carriage of ISO/IEC 14496 contents over IP networks'*, *'Part 11: Scene description and application engine'*, *'Part 12: ISO base media file format'*, *'Part 13: Intellectual Property Management and Protection (IPMP) extensions'*, *'Part 14: MP4 file format'*, *'Part 15: Advanced Video Coding (AVC) file format'* and *'Part 16: Animation Framework eXtension (AFX)'*.

Similar specifications for the systems level are also available for conversational multimedia services and applications in recommendations of the ITU-T, where the different aspects are however found in separate recommendations. For example,

- *Adaptation to various types of networks* is defined in the series of ITU-T Recommendations H.220ff; H.222 is identical with the respective part of MPEG-2 Systems;
- *Multiplexing of streams* for different types of networks is addressed in the series of ITU-T Recommendations H.240ff;
- *Audiovisual sessions*, including all synchronization aspects between different media types, are specified in ITU-T Recommendations H.320ff.

The MPEG-4 standard encompasses a complete framework for description of natural and synthetic audiovisual content, e.g. video and graphics, natural and synthetic audio. It enables to combine different types of *audiovisual objects* into a *scene context*. This allows realization of content-based and interactive applications. For natural media coding, MPEG-4 mostly builds on similar technology as its predecessors MPEG-1 and MPEG-2, but supplements new functionality (cf. sec. 17.4.3 and 17.5.2). While e.g. MPEG-1 and MPEG-2 were only capable to encode rectangular video frames and composite audio channels, MPEG-4 supports multiple *audiovisual objects* which are encoded and decoded independently. The signal flow chain is supplemented by a new functional stage of *scene composition* combining the different objects. The separate representation allows systematic manipulation and interaction with objects and scenes at the receiver end. Possible application areas of MPEG-4 are beyond pure audiovisual data compression. For example, tools are provided for three-dimensional object and scene description and scene navigation, as typically used in computer games. It is possible to fully

---

[1] DSM-CC = Digital Storage Media – Command and Control. This provides networking interfaces and control mechanisms for play-out of temporarily stored media streams, mainly usable for cyclic transmission of broadcast streams in so-called *data carousels*, which allow implementation of *near video on demand* services with simple pause, forward and rewind functions.

[2] IPMP=Intellectual Property Management and Protection.

integrate video and graphics signals, natural audio and synthetic sounds in a unique standardized representation.

The systems layer of MPEG-4 not only defines the *access* to units of the streams (e.g. video and audio, auxiliary data), their *identification* and *synchronization*, but also the methods for scene description (see below). Interfaces between multiplex, decoder(s) and scene composition are much more flexible than in MPEG-2. The interoperation between the different components of an MPEG-4 based system can be interpreted in a *layered model* as follows:

– The *compression layer* contains the building blocks related to data compression and representation for the different types of objects and scene related elements (e.g. video, audio, graphics, object and scene interrelationships). Specific decoders must be invoked which are optimized for best compression performance of the different media types. The related binary representations are provided in elementary streams separately for each audiovisual object or scene description element.

– In the *synchronization layer*, elementary streams are decomposed into access units and supplemented with time stamps, which allows a synchronization of the system processes at the level of streams.

– In the *transport layer*, the packetized units are assembled for transport over a network or for storage in a file. This also includes multiplex mechanisms for combination of multiple streams. Depending on the type of network that is used for transport, multiplex or timing mechanisms which are already defined by the network can be re-used to reduce possible overhead.

The actual method to *output* the composed scene is not defined by the standard, as the mechanisms which could be supported highly rely on the capabilities of the available hardware; e.g. the realizable visual effects depend on the real-time rendering power of the graphics chip. The scene description format, which defines the positions of different objects within a scene, is the *BInary Format for Scene description* (BIFS), a superset and extension of the *Virtual Reality Modeling Language* (VRML) (standard ISO/IEC 14772). BIFS can also be interpreted as a binary compressed format of VRML (which is only defined as a file format), with most important extensions related to dynamic (time-dependent) behavior of scenes and objects. 2D and 3D graphics primitive element representations (e.g. for circles, spheres, rectangles, quadrics) are included. Textures (still textures or videos, i.e. animated textures) can be mapped onto surfaces of any 2D or 3D graphics elements. Coordinate references can be defined for the 2D image plane or the 3D-world space, projection methods from a 3D scene onto 2D planes support virtual cameras at freely-definable viewpoints. The scene description is dynamic, e.g. the scene can change by a single object changing its position or shape over time, or by new objects appearing. To simplify this, audiovisual objects can be grouped hierarchically, where the entire group is defined as one unit concerning position and orientation in space and behavior over time; this includes changes of signal-related properties such as color, brightness and loudness. Another important property of

BIFS going beyond the capabilities of VRML is the *streaming support*, which allows immediate presentation of scenes from a media stream without transmitting a complete file.

An example of a BIFS scene graph is shown in Fig. 17.4. The 3D scene contains one 2D audiovisual object of rectangular shape, which is projected onto a screen, one arbitrary-shaped audiovisual object (person, arranged to be sitting behind the table), and synthetic (graphics) 3D elements.

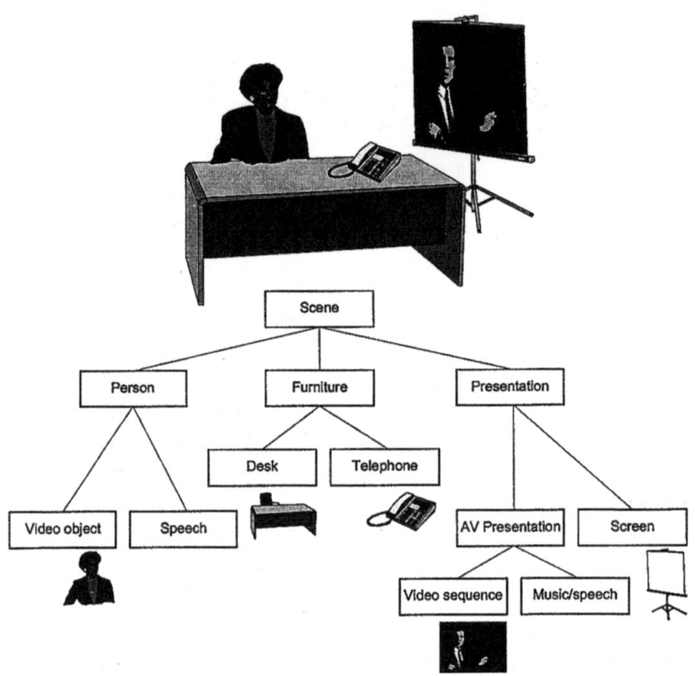

Fig. 17.4. Structure of a scene graph in BIFS scene description

Another important aspect to be supported at the systems level are *Application Program Interfaces* (APIs), or other downloadable pieces of software. Several multimedia application frameworks have established mechanisms for plug-ins of software applets. Examples for this are an extension of MPEG-4 systems entitled MPEG-J and the *Multimedia Home Platform* (MHP), a new part in the DVB standards, providing download capability to run applets in set-top boxes.

Downloadable software in general bears interesting potential, as it allows permanent improvements of algorithms and support for new applications. Downloadable media decoding software is already used today in a number of proprietary products, and can indeed resolve the problem of interoperability between devices almost perfectly. On the other hand, negative aspects of such approaches are quick outdating of formats, and considerably higher hardware cost than in custom-built decoding devices. It must be carefully decided which parts of a multimedia communication system should rather be implemented by software with update and download capability, and which parts have still a favorable and more cost-efficient solution by direct hardware implementation. Definitely, the percentage of operations that can more favorably be performed by software plugins can be expected to significantly increase in the future.

17.3 Still Image Coding

17.3.1 The JBIG Standards

The Joint Bilevel Images Group (JBIG) has defined the standard ISO/IEC 11544 (identical with ITU-T Rec. T.82), entitled *'Coded representation of picture and audio information – Progressive bi-level image compression'*. This standard allows lossy-to-lossless progressive transmission of bi-level images, and typically compresses for the lossless case by an approximately 40% rate reduction as compared to the widely used G.3/G.4 fax standards [HAMPEL ET AL. 1992]. The most important algorithmic elements of JBIG are

–   usage of 'line skipping', signaling that subsequent lines that are identical;
–   arithmetic coding for all elements;
–   usage of different prediction methods, in particular template prediction (sec. 12.2);
–   progressive transmission capability, which means that a binary image can first be decoded and reconstructed by a lower resolution; by further bits arriving, the resolution can then gradually be increased.

For progressive transmission, a method is applied which is very similar to pyramid coding (see sec. 4.4.6), providing a lowpass filtered (hence alias reduced) representation at the lower resolution. The structure of a JBIG encoder[1] is shown in Fig. 17.5a. The number of *differential encoders* (relating to the difference levels of a Laplacian pyramid) is selectable. If T resolution levels are used, T-1 differential encoders must be run. Fig. 17.5b shows the structure of a single differential encoder. The *typical prediction* is an up-sampling by a (zero-order) hold interpolator. Here, it is examined whether 4 adjacent samples at higher resolution are identical to the sample at the same spatial position in the next lower resolution layer, already available at the decoder. In *deterministic prediction*, an improved spatial interpolation is used, which is able to predict more complex binary structures as e.g. raster elements in graphics images. If both predictions fail, the template prediction as described in sec. 12.1 is employed. The prediction errors and the infor-

---

[1] Most multimedia compression standards only define normative *decoders*. This nevertheless prescribes typical encoder structures which are capable to output conformant streams.

mation about usage of the prediction modes are encoded by an arithmetic encoding scheme. Fig. 17.5c shows the structure of the coder for the lowest resolution level (base layer). In this layer, the typical prediction consists only of the line skip step. Encoding starts at the highest differential layer, while decoding starts at the base layer, and propagates through the differential layers by increased order $2...T$.

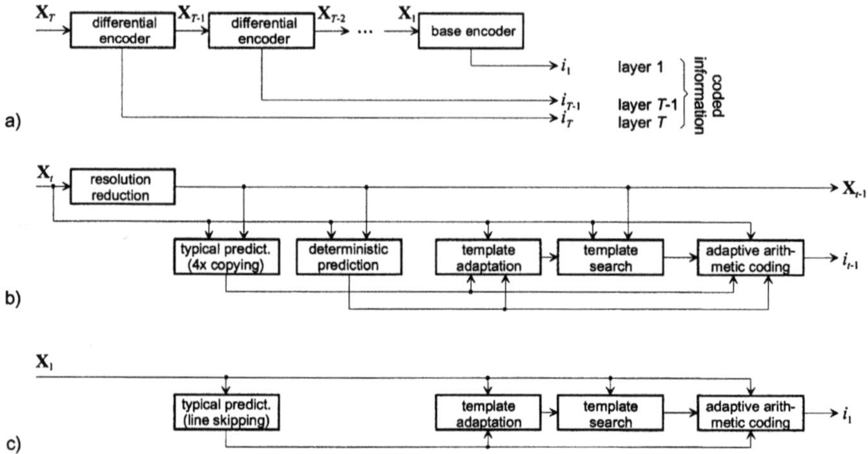

Fig. 17.5. JBIG encoder: **a** hierarchical structure **b** differential encoder stage **c** base encoder stage

Further improvement was made in the definition of the successive standard JBIG–2 (ISO/IEC 14492, ITU-T Rec. T.88) entitled *'Lossy/lossless coding of bi-level images'*, which in particular was designed for cooperative processing of documents in internet applications and document transmission over narrow-bandwidth wireless channels. The improvements are based on classification of areas from the image to be treated by specific encoding tools, such as text areas, halftone areas, line graphics areas etc.

17.3.2   The JPEG Standards

The Joint Photographic Experts Group (JPEG) is a working group which was established jointly by the International Standardization Organization (ISO) and the International Telecommunication Union (ITU). In 1990, JPEG finalized the first version of the still image coding standard ISO/IEC 10918 (same text as ITU-T Recommendation T.81ff.). This is usually entitled as *the JPEG standard* and has probably become the most widely used still image compression method, based on a block transform (DCT) scheme. Subsequently, the standard JPEG-LS was defined for lossless and near-lossless coding (ISO/IEC 14495-1, ITU-T Rec. T.87/T.870), which is mainly based on spatial prediction methods. The newest in the series of JPEG standards is JPEG 2000 (ISO/IEC 15444, ITU-T Rec.

T.800ff.), which defines a fully embedded still image codec based on wavelet transform technology. All standards produced by JPEG are applicable to digital images of different horizontal and vertical resolutions, different amplitude resolutions (bit/pixel in the PCM original), different color representations, e.g. gray-level images, *RGB* or YC_bC_r.

The JPEG standard is officially entitled ISO/IEC 10918 *'Digital compression and coding of continuous-tone still images'*, and is sub-divided into the following parts:

- ISO/IEC 10918-1 (identical with ITU-T Rec. T.81) : *Requirements and guidelines*;
- ISO/IEC 10918-2 (identical with ITU-T Rec. T.83): *Compliance testing*;
- ISO/IEC 10918-3 (identical with ITU-T Rec. T.84): *Extensions*.

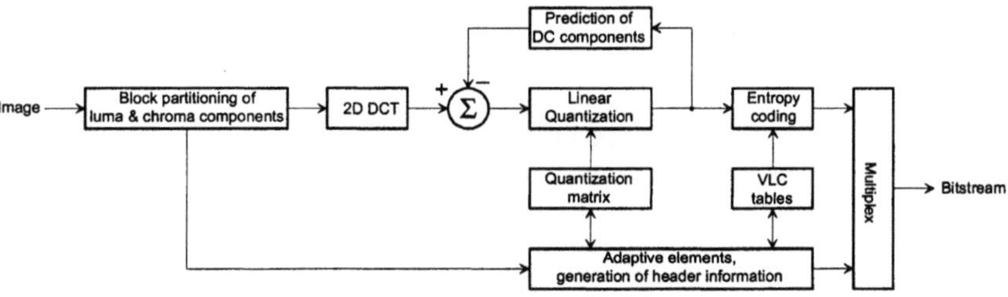

Fig. 17.6. Encoder of the DCT-based JPEG standard (sequential DCT mode)

The structure of a typical JPEG encoder is shown in Fig. 17.6. The core approach is adaptive block transform coding, using a 2D DCT of block size 8x8 pixels. The DC coefficients can either be predictive encoded by DPCM (using the DC coefficient from the left-hand and top-above blocks to compute the prediction), or can be directly encoded after uniform quantization. The AC coefficients are scanned in zigzag order as shown in Fig. 12.24a, the non-zero values are quantized uniformly.

A *frequency weighting* can optionally be used, by which an individual quantization step size can be defined for each transform coefficient. Based on visual perception properties (cf. sec. 6.1.2), low-frequency coefficients should be quantized by using smaller step sizes than for the higher-frequency coefficients. As a consequence, the threshold value by which it is decided whether a coefficient shall be set to zero, is also variable depending on the frequency. Table 17.2 shows an example of a frequency-dependent quantization table, also denoted as *quantizer weighting matrix*[1]. The positions of non-zero coefficients are expressed by a run-

---

[1] This example is the quantizer weighting matrix as defined as default configuration in the standards MPEG-1 and MPEG-2 for the case of intraframe coding of a transformed block, which follows the same basic principle as JPEG. The default in JPEG is to use no frequency weighting, whereas an example quantization table is included in an informative annex, which however seems to be less consistent with perceptual models than the MPEG table. The values as shown in the table have to be divided by 16 and are then multiplied by

length code, where both the run-lengths and the quantization level indices are encoded jointly by a variable length code.

Tab.17.2. Example of a quantizer weighting matrix as default-wise defined in the MPEG-1 standard for intraframe-coded DCT blocks

weighting G_{uv} for c_{uv} $v=$	$u=$ 0	1	2	3	4	5	6	7
0	8	16	19	22	26	27	29	34
1	16	16	22	24	27	29	34	37
2	19	22	26	27	29	34	34	38
3	22	22	26	27	29	34	37	40
4	22	26	27	29	32	35	40	48
5	26	27	29	32	35	40	48	58
6	26	27	29	34	38	46	56	69
7	27	29	35	38	46	56	69	83

Two methods of entropy coding are employed in the *baseline version* of JPEG, which defines fixed VLC tables separately for the luminance and the chrominance part of the information:

- *Coding of DC coefficients* (either predictive or non-predictive): A systematic VLC is used, where the prefix designates the number of bits necessary to represent the quantizer index, and the suffix is the index itself, encoded as an integer number of this given bit depth.
- *Coding of AC coefficients*: A '2D VLC' (cf. sec. 12.4.1) similar to the construction shown in Table 11.1 is employed. This is an escape code, where only typical (frequently occurring) combinations of RUN and LEVEL are mapped into VLC symbols. Any combination which is not supported by the VLC table must be encoded by a larger number of bits, consisting of the ESCAPE symbol and separate representations of the RUN and LEVEL values. If the zigzag scan detects no further non-zero coefficient up to the end of the block, an *End of Block* (EOB) symbol is encoded.

In extended coding process definitions of the JPEG standard[1], the encoder can determine optimized VLC tables for individual images; the parameters necessary for this adaptive entropy coding are then conveyed in the preamble of the bitstream. Further, adaptive binary arithmetic coding is defined which can further improve the compression performance.

---

the quantizer step size factor ('*Q factor*'), which then gives the individual quantizer step size to be used for a transform coefficient. In the JPEG standard, the quantizer step size factor is fixed over the entire image, while it can be varied on the basis of 16x16 macroblocks in the MPEG standards. The output rate actually produced mainly depends on the pre-set Q factor, but can also be influenced by the selection of the weighting matrix.

[1] These extended methods are in fact hardly used in practice.

The different modes by which JPEG encoders and decoders can be operated are as follows:

- *Sequential DCT mode*: This is the most frequently used mode of JPEG, where the transform blocks are scanned sequentially in a left-to-right, top-down order. All information about coefficients contained in one block is written to the bit-stream at once as defined by the zigzag scan. If a stream is received, the decoding process will work by the same sequence, and the reconstructed image can only appear row by row if the transmission is not finished yet. Different coding processes are specified, including baseline sequential, extended sequential with adaptive Huffman and adaptive arithmetic coding; 8 or 12 bit sample precision are supported.
- *Progressive DCT mode*: In principle, the compression algorithm is identical with the sequential mode, but the information bits relating to DCT coefficients are re-ordered before they are written to a file or bit stream. In *spectral-selection progression*, this is done by increasing frequency order, such that first the bits relating to DC coefficients of *all* blocks, then bits for all AC coefficients of first, second, third etc. diagonal of the zigzag scan are written. If the progressive stream is successively transmitted, a first decoding can already be performed when only a low number of information bits have been received, giving a raw approximation of the *entire image*. This method of *progressive transmission* and decoding is in particular useful when narrow-bandwidth channels are used. The data rate as compared to the sequential DCT mode is not increased in principle (except for the overhead to signal the ordering of progression). It is however necessary to execute several decoder runs, depending on the number of progression steps. The progressive DCT mode is also suitable for error protection by data partitioning, where the need for protection of bits naturally results by the order they appear in the bit stream. Beside the spectral-selection progression, a *full progression* operation is defined, which in addition allows refinement of the quantization of previously transmitted coefficients during subsequent progression steps. Different coding processes specified include Huffman or arithmetic coding, spectral selection or full progression, and 8 bit or 12 bit sample precision.
- *Lossless mode* : The DCT modes are not suitable for lossless encoding. This is mainly caused by the fact that the DCT requires high (in principle: float) precision for the computation of the sinusoidal basis functions. In lossless mode, the JPEG standard defines a 2D DPCM with forward-adaptive switchable predictors and entropy coding of the residual signal. This allows a reduction of the rate (depending on image characteristics) by approximately a factor of two as compared to an 8-bit PCM representation. Coding processes are specified for adaptive Huffman or arithmetic coding of residuals.
- *Hierarchical mode*. In the progressive DCT mode, no guarantee can be given that the quality reconstructed by the first steps of progressive transmission will be rather high; the low-frequency DCT coefficients in fact do not represent an appropriately lowpass filtered version of the signal. Typically, severe blocking

artifacts will occur in the first decoding steps of the progressive DCT mode. To circumvent this problem, a differential (Laplacian) pyramid method (cf. sec. 4.4.6) is provided in the hierarchical progressive mode. A block diagram of this method is shown in Fig. 17.7, where the differential signals at the particular levels are encoded either by sequential or progressive DCT mode encoders. The process typically starts at a coarsest level in which the image is sub-sampled by a factor of four in both horizontal and vertical directions (factor 16 in total). Quantization step sizes can be selected independently and influence the bit rate that results in each of the levels. Typically, this mode increases the total bit rate for the full-resolution signal by about 30% as compared to the sequential DCT mode, which is the main reason (along with the increased decoding complexity) why it has not widely been used so far.

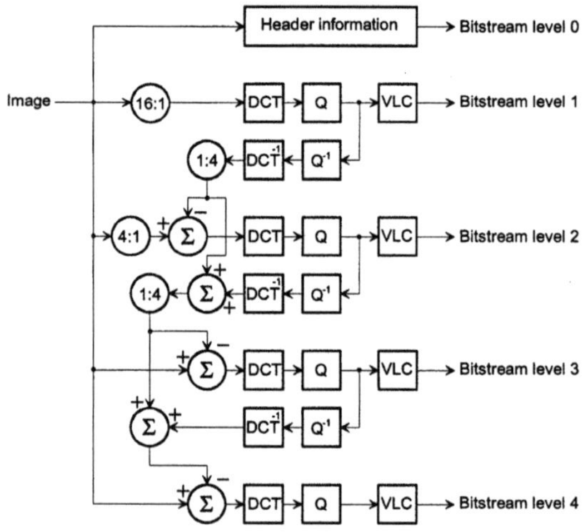

Fig. 17.7. Hierarchical-progressive mode of JPEG

JPEG-LS. Even though the JPEG standard provides capability for lossless coding, this has in fact rarely been used. Even more, there is a huge gap between the lossy range where the DCT modes reach highest efficiency, and the strictly lossless case. Both for *lossless* or *near lossless* coding of PCM signal representations, predictive methods or methods of entropy coding which utilize spatial contexts within the signal achieve better compression performance than transform coders[1]. This was the motivation to define a new standard ISO/IEC 14495, entitled '*Lossless and near-lossless compression of continuous-tone still images*', shortly denoted as JPEG-LS. It is sub-divided into the following parts:

– ISO/IEC 14495-1 (identical with ITU-T Rec. T.87): *Baseline*;

---

[1] This statement is strictly true for transforms which require high arithmetic precision to obtain orthogonal bases. Transforms with integer precision can achieve almost the same compression performance as DPCM methods at high data rates.

– ISO/IEC 14495-2 (identical with ITU-T Rec. T.870: *Extensions*.

For lossless coding, a reduction of the rate by approximately a factor 2-4 as compared to a PCM representation can be achieved, where the actual reduction highly depends on image characteristics such as amount of detail or noise, and also typically decreases with larger bit-depth of the original signal. JPEG-LS defines two coding modes, for which the selection is determined from a neighborhood context:

– *Run mode* : This assumes a number of subsequent pixels of identical amplitude values (or nearly identical in the case of near-lossless mode); run-length coding is performed, where the amplitude value has only to be encoded once per run.
– *Prediction mode* : This mode is rather preferred if significant fluctuations in amplitude occur in the neighborhood. Backward-adaptive switchable predictor configurations are used. The residual (prediction error signal) is encoded by a VLC.

For entropy coding, a systematic VLC based on an Exp-Golomb design (cf. sec. 11.4.3) is defined in the baseline version. The extended version further also specifies context-based arithmetic coding.

JPEG 2000. This newest product in the series of JPEG standards is based on a fully-embedded wavelet codec design, which in principle inherently comprises progressive, low-rate up to lossless coding within one single framework, and adds additional functionality that was not supported by other still image coding standards so far. The official name is ISO/IEC 15444 '*JPEG 2000 image coding system*'; the standard is sub-divided into the following parts[1]:

– ISO/IEC 15444-1 (same text as ITU-T Rec. T.800): *Core coding system*;
– ISO/IEC 15444-2 (same text as ITU-T Rec. T.801): *Extensions*;
– ISO/IEC 15444-3: *Motion JPEG 2000*;
– ISO/IEC 15444-4 (same text as ITU-T Rec. T.803): *Conformance testing*;
– ISO/IEC 15444-5 (same text as ITU-T Rec. T.804): *Reference software*;
– ISO/IEC 15444-6: *Compound image file format*;
– ISO/IEC 15444-8: *Secure JPEG 2000*;
– ISO/IEC 15444-9: *Interactivity tools, APIs and protocols*;
– ISO/IEC 15444-10: *3-D and floating point data*;
– ISO/IEC 15444-11: *Wireless*;
– ISO/IEC 15444-12: *ISO base media file format* (common text with MPEG-4 Part 12).

The baseline mode allows selection of two different bi-orthogonal wavelet filter pairs, both of which are specified by their equivalent lifting structures (see sec. 4.4.3):

---

[1] The numbers beyond 5 were still in the progress of development by the time this book was published.

–   The 9/7 filter (4.195) provides good data compression efficiency at low bit rates, however does not allow lossless reconstruction due to the floating point arithmetic precision that must be used, even in case of very accurate quantization;

–   The 5/3 filter (4.194) has integer precision (only shifts and additions necessary in the implementation of the lifting transform, see Fig. 4.41). This allows synthesis achieving perfect lossless reconstruction from a well-defined finite set of quantization levels. On the other hand, due to the poor frequency separation properties of this filter, the compression performance at lower rates is worse than with the 9/7 filter.

The wavelet representation supports *spatial scalability,* which means that partial information from the file or stream can be used to perform decoding of lower spatial resolution. The discrete representation of wavelet coefficients is generated by embedded quantizers with dead zone (see Fig. 11.8b). This additionally enables *quality scalability,* the bit stream can also be scaled to reconstruct a signal of same spatial resolution at lower rate but with higher distortion. This leads to the definition of a generic coding scheme which is capable to cover the entire range from low rates up to lossless encoding in a unique consistent way, implicitly including features for highly-efficient hierarchical and progressive transmission. A certain inconsistency in this approach results by the fact that different filter types have to be used to achieve best efficiency in the low- and high-rate ranges.

The coding of wavelet coefficients is performed by the EBCOT algorithm as described in sec. 12.4.2. Unlike zero-tree methods, EBCOT does not utilize statistical dependencies between different frequency bands of the wavelet representation, and re-orders coefficients in a block-like fashion. This allows to define mechanisms for extremely flexible access to partial information referenced by pointers to sub-stream entries. These sub-streams are structured into segments and packets, which can uniquely be related e.g. to selective decoding of certain areas of the image, specific spatial resolution or quality levels. Within *code blocks,* all binary information is concatenated which is necessary to decode groups of wavelet coefficients. Several code blocks are composed into a higher-level structure called *precinct.* By this it is e.g. possible to separately decode particular areas of images. Typically, these areas are rectangular blocks. Each of the units is self contained including all header information, which actually increases the overall data rate of the representation, but is necessary to achieve the desired flexibility. In spite of this overhead, JPEG 2000 still achieves a reduction at low rates of around 30–50 % as compared to the DCT-based JPEG standard. This random access functionality is e.g. important in applications where a user enlarges (or zooms in) a part of an image, which can then flexibly be combined with an increased spatial resolution decoding of the selected area.

To encode images by JPEG 2000 into an exact number of bits, it is in principle not necessary to adapt the quantizer step size. If the bits from the particular bit-planes are packed in a bitstream according to their sequence of relevance, it is possible to approximately approach a desired rate by simple truncation of the stream. The best sequential arrangement of these streams, even though supporting the other features like localized or variable-resolution access, is the core element of the EBCOT algorithm (cf. sec. 12.4.2). As the selection process is in principle non-normative, it bears a high potential for further improvements in JPEG 2000 encoders. The JPEG 2000 syntax fully supports such a flexible approach, as it can be signaled which bit plane from which code block provides the next relevant information in the pack stream.

The extensions beyond the baseline mode support the following features:

– Usage of *arbitrary wavelet filters* by definition in the header;
– *More flexible quantization* of wavelet coefficients, including *psycho-visual weighting functions* and local adaptation of the embedded quantizer step size;
– Improved compression by *trellis-coded quantization* (cf. sec. 11.6.1);
– Nonlinear pre-processing of images by *gamma correction* and *lookup tables*;
– Support of *additional color spaces*, including *multi-spectral* components (more than three);
– *Region of interest* (ROI) definition, which allows decoding of non-rectangular regions of arbitrary shape, also in combination with the scalability features; it is possible to increase the spatial resolution in decoding only within a specific ROI, or decode the ROI by higher quality of the embedded quantizer.

Psycho-visually weighted quantization. The two-dimensional wavelet decomposition in JPEG 2000 is performed according to Fig. 12.27, typically using 5 levels of the wavelet tree. Sub-sampling by a factor of 4 within each level produces a scaled image at the last level with a number of pixels lower by a factor of $4^5=1024$ as compared to the full-resolution image. An advantage of the embedded representation is also the higher flexibility with regard to frequency weighting. In the DCT-based JPEG standard, weighting must always finally be determined by the time of encoding. Using JPEG 2000, the weighting matrix can be defined by user preferences when a stream is retrieved. This is advantageous, as the optimum psycho-visual weighting highly depends on the viewing distance (cf. sec. 6.1.2). Examples of distance-dependent weighting are given in Table 17.3. Obviously, if the viewing distance is doubled, the perceived spatial frequency is also increased by a factor of two, such that the same weighting factors should then be used at a lower level of the wavelet tree.

Part 3 of the standard defines *Motion JPEG 2000*, which supports applications where a video sequence is encoded frame by frame by a JPEG 2000 coder, which is a pure intraframe approach (see sec. 13.1). This also includes mechanisms for accurate timing in video presentation and synchronization with other media types such as audio[1]. At the systems level, the mechanisms are very similar to the methods of MPEG-4 systems. In the meantime, the MPEG and JPEG committees have

---

[1] Many implementations of so-called *Motion JPEG* exist in the market. These use the DCT-based JPEG standard for synchronized video encoding, but are proprietary solutions never officially defined by the JPEG committee.

defined a common file format (part 12 of the respective standards), which achieves a compatibility for most systems-related aspects between both standards.

Table 17.3. Frequency weighting for wavelet bands (viewing distance expressed as multiples of pixel sampling distance on the screen) (source: JPEG)

Level	Viewing distance 1000			Viewing distance 2000			Viewing distance 4000		
	HL	LH	HH	HL	LH	HH	HL	LH	HH
1	1	1	1	1	1	1	1	1	1
2	1	1	1	1	1	1	1	1	0.731
3	1	1	1	1	1	0.727	0.564	0.564	0.285
4	1	1	0.727	0.560	0.560	0.284	0.179	0.179	0.043
5	0.560	0.560	0.284	0.178	0.178	0.043	0.014	0.014	0.0005

The highly-flexible bitstream structure of JPEG 2000 also allows to provide efficient mechanisms for transmission error protection. Firstly, it is straightforward to protect the more relevant parts of the embedded stream selectively, e.g. by unequal error protection (cf. sec. 15.2.2). Further, as the information within code blocks is self-contained (separately decodable), resynchronization is guaranteed even in the case of partial data losses.

17.3.3   MPEG-4 Still Texture Coding

The Visual part of the MPEG-4 standard (ISO/IEC 14496-2) also defines a method of wavelet coding for still textures. The primary purpose is flexible encoding of textures which shall be mapped onto surfaces of 2D or 3D graphics elements in MPEG-4 scenes. Unlike JPEG 2000, this is a zero-tree method (cf. sec. 12.4.2), but a fully embedded stream representation is provided as well. Localized access, which is useful in streaming of MPEG-4 scenes where a texture may eventually only partially be visible, is provided by a coding tool entitled as *wavelet tiling*. This separates a larger texture image into smaller *tiles* which can be decoded independently.

17.4 Video Coding

The most important series of recommendations and standards for video compression were defined by groups of the ITU (ITU-T Rec. H.261/262/263/264) for application domains of telecommunications, and by the *Moving Pictures Experts Group* (MPEG) of the ISO/IEC (ISO standards MPEG-1/-2/-4) for applications in computers and consumer electronics, entertainment etc. There are commonalities between the standards defined by the both groups, and partially work has been done jointly. For example, ITU-T rec. H.262 is identical to the ISO standard

MPEG-2, and ITU-T rec. H.264 is identical to part 10 of the ISO standard MPEG-4, 'Advanced Video Coding'.

Proprietary solutions beyond open standards are not described here, as it is in most cases difficult to obtain sufficient information about the underlying techniques. It may be worthwhile to mention that some older standard solutions exist which are meanwhile outdated, e.g.

- ITU-T Rec. H.120 '*Codecs for Videoconferencing using Primary Digital Group Transmission*', which is a frame-replenishment method based on spatial 2D DPCM intended for a transmission rate of 2 Mbit/s;
- ITU-R Rec. BT.721 '*Transmission of Component Coded Digital Television Signals for Contribution-Quality Applications at Bitrates near 140 Mbit/s* ', which is designed for near-lossless coding of SD resolution video, and uses a switchable PCM/DPCM solution;
- ITU-R Rec. BT.723 '*Transmission of Component Coded Digital Television Signals for Contribution-Quality Applications at the Third Hierarchical Level of ITU-T Rec. G.702*' which is intended for encoded rates around 34 Mbit/s, using motion-compensated hybrid coding. Meanwhile, MPEG-2 is available in the same application domain, providing a higher degree of flexibility.

17.4.1   ITU-T Recommendations H.261 and H.263

The first normative specification of video coding following the principle of a hybrid motion compensated prediction with block transform coding of the residual was defined in ITU-T Rec. H.261 '*Video Codec for Audiovisual Services at p x 64 kbit/s*'. It was primarily designed for usage in conversational (videophone and video conferencing) applications, where a short encoding and decoding delay is of primary importance. A suitable encoder structure is shown in Fig. 17.8. The decoder defined by H.261 supports the following features:

- Application to CIF or QCIF resolution pictures for different frame rates between 8 1/3 *Hz* and 30 *Hz*;
- Block-based translational motion compensation for macroblocks of size 16x16 pixel, with allowable ranges of motion vectors of ±15 pixel horizontally/vertically, accuracy of motion shift by full-pixel steps;
- Motion-compensated prediction loop with access to one decoded previous frame, optionally a lowpass loop filter (13.18) can be switched at the macroblock level;
- Switching between intraframe and interframe coding at the macroblock level;
- Block DCT of size 8x8 pixels applied separately to luminance and chrominance components; the coding of DCT coefficients is similar to the method of the JPEG standard, including the zigzag scan method within the DCT blocks;
- Within a macroblock, a *coded block pattern* (CBP) code signals which 8x8 DCT transform blocks are encoded (not entirely set to zero value);

– Uniform, non-weighted quantization of coefficients, adaptation of quantization step sizes is possible at each macroblock;
– DPCM encoding of motion parameters and DC transform coefficients.

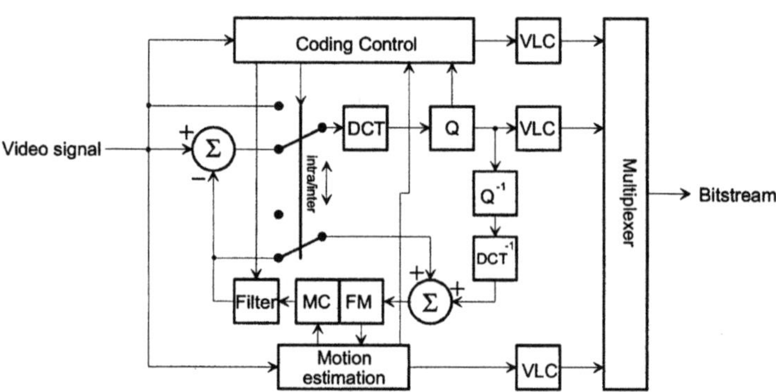

Fig. 17.8. Functional block diagram of a video encoder which is compatible to ITU-T Rec. H.261 [MC: Motion compensation; FM: Frame memory; Q: Quantizer]

All parameters (such as motion vectors, quantizer step sizes, DCT block RUN/LEVEL information etc.) are encoded by specially-defined codes of variable length. The bit stream structure of ITU-T Rec. H.261, even though it is clearly less flexible than in newer standards, shall be presented by some more detail here, as it is a typical example for video stream structures and simple to understand as a *strictly hierarchical* approach. Fig. 17.9 shows the different layers of the hierarchy, starting by the *sequence layer*. Each layer provides re-synchronization or entry points, such that even after a data loss a decoder is capable to catch up, if it starts parsing for a re-synchronization codeword, and starts to decode the bitstream again at the next entry point. Below the *picture layer* which relates to single frames, a partitioning of the respective frame into a *group of blocks* (GOB) structure is made; the number of GOBs depends on the frame size. Each GOB consists of 33 *macroblocks* (MB) of size 16x16 pixels each; finally, one single MB combines 4 luminance blocks and 2 chrominance blocks of size 8x8 pixels, which are the units that are processed by the DCT. A special run-length code denoted as *next macroblock address* allows to skip (setting all motion parameters and transform coefficients to zero) entire MBs, which is a very efficient mechanism if the prediction from the previous frame is sufficiently good.

H.261 (like other hybrid DCT-based video coding standards) forbids infinite frame prediction loops. The motivation behind this is to break the propagation of errors in the case of lossy transmission, but also to prevent possible drift effects which may be caused by using different implementations of the DCT at the encoder and decoder sides (different rounding in fast algorithms, different precision of integer or float arithmetic units etc.). Hence, it is prescribed that after at most 132 frame cycles an intraframe-coded area must be interspersed[1].

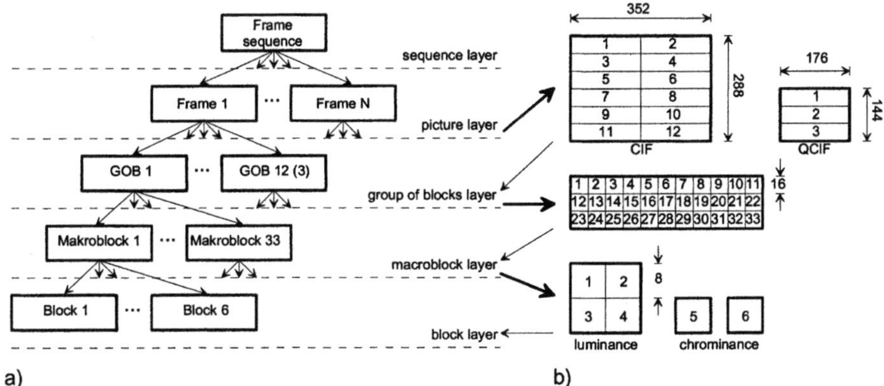

Fig. 17.9. a Hierarchy definition in the ITU-T Rec. H.261 syntax **b** Partitioning of a frame into GOBs in the cases of CIF and QCIF frame sizes, MBs within one GOB, DCT coding blocks within one MB

Transmission at the lowest rate of 64 *kbit/s*[2] is usually performed using the QCIF format with 10 frames/s or less. When stronger action is shown in the scene to be encoded, the reconstruction quality substantially degrades at this rate. Good quality at full CIF resolution and 25 or 30 frames per second can typically be achieved using a compressed data rate of 384 *kbit/s*, however still assuming scenes of teleconference type, where the amount of motion usually is lower than in general-purpose video. As better quality was desirable in particular for the lower-rate range, e.g. for ISDN, modem-based or wireless transmission, a more advanced framework for usage in visual conversational applications was defined in ITU-T Rec. H.263, *'Video Coding for Low Bit Rate Communication'*. One of the key issues for better compression was the improvement of motion compensation. Using H.263, the quality achievable for QCIF frame size is significantly better than with H.261 for an encoded rate of 64 *kbit/s*, and even at lower transmission rates (e.g. 28 *kbit/s* for low-speed modem) a fair quality can still be achieved. In some cases of easily encodable sequences, H.263 can still be used at extremely low rates such as 16 *kbit/s*. This however requires further reduction of the frame rate to 5 frames/s or below.

The most important new encoding tools of H.263 which have allowed a significant increase in quality as compared to H.261 are:

---

[1] This approach has in fact been adopted by all subsequent video coding standards, and only recently was given up in the definition of the AVC/H.264 standard (sec. 17.4.4), where an integer-precision transform has replaced the DCT.

[2] A rate control as described in sec. 6.2.1 must be employed to transmit by a fixed rate.

- Motion compensation can be performed by half-pixel accuracy, the motion block size is variable, and can be selected as either 8x8 or 16x16 pixels. The strict limitation of motion vector ranges is abandoned;
- To predict motion vectors for the purpose of encoding, information from the blocks above the current block can be used (in H.261, only the motion vector from the left-positioned block is used as predictor). Median prediction provides both better coding performance and better stability in cases of transmission losses;
- DC *and* AC coefficient prediction (see Fig. 12.21) are defined;
- H.263 supports a structure of 'PB picture. This is a unit consisting of a *B*-type picture and a subsequent *P*-type picture (see sec. 13.2.5). As this allows only one *B*-type picture between two *P*-type pictures, the delay is still relatively short, as it is important for conversational applications.
- A '3D VLC' (RUN,LEVEL,LAST) is used for encoding of the block transform coefficients;
- The frame rate is variable, which allows to guarantee a certain minimum level of reconstruction quality at the expense of temporal resolution, and also reduces some buffering overhead which would be necessary to guarantee synchronous presentation.
- More different picture formats are supported than in H.261, and the 'group of blocks' structure is more flexible, allowing a definable number of macroblocks to be included.

A number of additional annexes have been amended since the first release of H.263, most of them intending to further improve compression performance or provide other functionality, e.g. by improved prediction mechanisms such as overlapping block MC (sec. 13.2.6), multi-frame prediction, and a quantization-adaptive loop filter. Additional methods for error-resilient transmission and modes for temporal, spatial and SNR scalability are defined. These enhancements are often subsumed under the names of 'H.263+' and 'H.263++', but are barely used in full combination[1].

17.4.2   MPEG-1 and MPEG-2

In contrast to conversational applications in video communication, a number of requirements applies in the context of storage and replay of stored data, which mainly are related to random access:

- Editing (e.g. extracting or replacement of frames) shall be possible;
- The video sequence shall be replayable forward and backward;
- Fast forward/reverse modes are desirable.

---

[1] As the relevant additional tools of H.263+/++ are likewise found in the more recent standards MPEG-4 and H.264 (see sec. 17.4.3 and 17.4.4), no more detailed description shall be given here. The interested reader may e.g. refer to [CÔTÉ ET AL. 1998].

If video is transmitted over distribution channels (e.g. digital broadcast channels, Internet multicast or broadcast), the requirement for random access is also very important, as subscribers may switch into a channel at any time, where they should not be forced to wait for an unnecessarily long time before the signal replay can be started. On the other hand, the requirement for extremely low encoding delay is released. In replay of stored data, the encoding is completely de-coupled, which is also true for many cases in broadcast transmission scenarios; only for live broadcast, on-line encoding must be done, which however will not have as rigid delay requirements as for conversational applications.

The random-access requirements are fully supported if all frames of a video sequence would be coded in intraframe mode. This however would lead to a clear disadvantage in compression performance, as the similarity between successive frames could not be exploited. As an alternative, periodic intraframe coding can be applied (e.g. by insertion of one intra coded frame within periods of 0.5 or 1 s), which still allows *near* random access with an acceptable delay. Decoding for random access must then start at the position of such a frame; to realize fast forward or reverse, the sub-sampled sequence of intra-coded frames could be re-played.

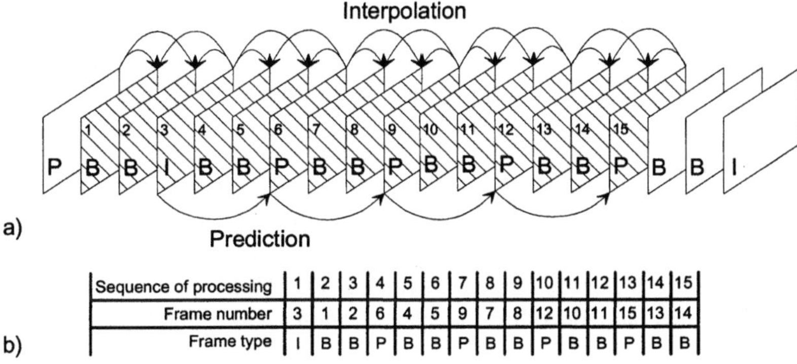

Sequence of processing	1	2	3	4	5	6	7	8	9	10	11	12	13	14	15
Frame number	3	1	2	6	4	5	9	7	8	12	10	11	15	13	14
Frame type	I	B	B	P	B	B	P	B	B	P	B	B	P	B	B

b)

Fig. 17.10. a GOP structure typical for MPEG-1 and MPEG-2 (case of 'open GOP')
b Sequence of processing

Additionally, it must be observed that the requirement for good image quality is often higher in video broadcast or storage applications than it is for conversational services. Also, the statistical behavior of movie-like video sequences is considerably different from videotelephony-like sequences, which makes them more difficult to compress:

- More and faster motion occurs;
- Higher spatial detail is often present;
- Non-static cameras (in many cases not on a tripod) are used, zoom, pan and other camera motion occurs;
- Frequent scene changes are typical.

Part 2 of the MPEG-1 standard (ISO/IEC 11172-2 *'Coding of Moving Pictures and associated Audio for Digital Storage Media at up to about 1.5 Mbit/s, Part 2: Video'*) defines a hybrid video compression method which was specifically designed for storage applications[1]. Historically created after H.261 but for a different application domain, a number of new tools and mechanisms was added aiming in particular at higher reconstruction quality for video sequences of general nature and more complex content, and to fulfill the requirements of random access:

– Another access entity is defined as *Group of Pictures* (GOP). Three picture (frame) types are defined, which are intraframe-encoded (type *I*) as well as the *P* and *B* types (see sec. 13.2.5). Within a GOP, at least one picture must be an *I* picture which must be positioned such that the remaining pictures of the GOP can uniquely be decoded. An exception is allowed in the *open GOP mode* for *B* pictures at the beginning of the GOP, which may eventually refer to at most one reconstructed picture from the previous GOP (Fig. 17.10a). Flexible prediction mode switching (forward, backward etc.) is further enabled on a macroblock by macroblock basis, with selectable modes depending on the picture type.

– Accuracy of motion compensation can be half pixel, the maximum motion shift range is in principle unlimited, e.g. for half pixel accuracy, shifts of up to ±512 pixels could theoretically be used in horizontal and vertical directions. This will however reasonably be limited by the frame size; further, the data rate to be spent for motion vectors will be considerably lower when a lower range is used.

– The positioning of *I*-, *P*- and *B*-type pictures within a GOP is definable, however a causal sequence of processing must be observed. In particular those *I* and *P* pictures which are used to predict a *B* picture must be encoded and decoded before the *B* picture can be processed (Fig. 17.10b). This results in an increased coding and decoding delay and additional memory requirements when *B* pictures are used.

– Psycho-visual weighting of DCT coefficients can be applied in quantization. The default quantization matrix as provided for *I* frames was already shown in Table 17.2. It is possible to define alternative quantization tables (e.g. adapted to properties of a sequence), which must then be transmitted as side information in the header.

– Picture sizes, color representations, pixel aspect ratio, picture rates and similar parameters are conveyed in the header, which makes MPEG-1 quite flexible for usage with different types of sources.

– A decoder buffer and timing model is defined (*Video Buffer Verifier*, VBV), which allows to design encoders supporting normative decoder timing behavior. Necessary parameters such as expected buffer size are included in the video bitstream (refer to sec. 15.2.1 for these concepts).

---

[1] The original purpose of MPEG-1 was storage of compressed video plus audio on CDs.

No loop filter is defined. At the macroblock level, encoding methods are similar to H.261, including a coded block pattern (CBP) code to indicate the transform blocks that are encoded at all, and capability for setting of quantizer step sizes. The bit stream defined in MPEG video coding standards is also structured into a number of syntactic hierarchy layers. In MPEG-1 video, these are *video sequence layer, group of pictures layer, picture layer, slice layer, macroblock layer* and *block layer*. The syntax of each layer contains the related characteristic information (e.g. length and structure of a GOP, specific prediction or intra modes invoked etc.). For the slice layer and above, start codes are defined which allow re-synchronization. The slice layer plays a similar role as the 'group of blocks' layer in H.261, but is more flexible by combining variable numbers of macroblocks in row-wise sequential scan order. The quantization and encoding of DCT coefficients uses the typical zigzag scan and RUN/LEVEL encoding scheme as described in sec. 12.4.1.

The original goal of MPEG-1 compression was to allow recording of movies on Video CDs, in which a data rate of 1.1 Mbit/s would be available for the video signal. For sequences of CIF resolution, with a fair amount of spatial detail and content changing moderately over time, the reconstruction quality is comparable with that of analog consumer-quality video tape (e.g. VHS). At least with critical material (high temporal change, high spatial detail) distortions are yet becoming visible, in particular blocking artifacts caused by the DCT and other artifacts caused by the block-wise motion compensation, e.g. noise around moving structures[1].

MPEG-1 does not support specific coding tools which take into account the characteristics of *interlaced video sources*, which are however of particular importance for higher quality video services e.g. in TV broadcast applications. Extension of the basic methods of MPEG-1 for more generic classes of video sources and applications is the purpose of the subsequent MPEG-2 standard, where again part 2 defines the video compression (ISO/IEC 13818-2 *'Generic Coding of Moving Pictures and Associated Audio Information – Part 2: Video'*, which has an identical text with ITU-T Recommendation H.262). This standard covers a wide range of applications and data rates of typically up to about 40 Mbit/s for storage and transmission, or even higher for applications in video production. Larger frame sizes of up to HD resolution (cf. sec. 1.2) are supported. MPEG-2 is also forward compatible with MPEG-1 (which means that MPEG-1 streams observing typical constraints e.g. in frame sizes and data rates can be decoded by MPEG-2 decoders). In terms of video encoding tools, specific provisions are made for interlaced video. Further, MPEG-2 defines tools for scalable video coding, e.g. to embed streams which can be used to either decode with CIF+SD, SD+HD resolution. The main technical extensions as compared to MPEG-1 can hence be summarized as follows:

---

[1] The actual quality highly depends on the degree of encoder optimization (cf. sec. 13.2.10), which is beyond the normative issues of these standards.

- To support the properties of interlaced video, different methods for field/frame adaptive motion compensation (*frame based*, *field based* and *dual prime* prediction modes), as well as switching between field-/frame-wise DCT are specified (see sec. 13.2.6). Further, it is possible to switch into a 16x8 prediction mode, where separate motion vectors can be defined for the top and bottom halves of a macroblock.
- Methods of *scalable* coding are defined, which provide SNR scalability and spatial scalability over a limited number of layers each (cf. sec. 13.2.8). In this context, the bit stream is sub-divided into two or three parts, which means that by retaining or receiving only core parts of the stream, it is e.g. possible to re-construct frames at lower resolution. To encode DCT coefficients related to the different resolution levels, differently optimized variable length codes are provided in the spatial scalable mode.
- Methods to encode sequences with 4:2:2 chrominance sampling are defined by allowing additional 8x8 transform blocks to be subsumed within a macro-block.
- A method of temporal scalability is defined, which allows prediction of addi-tional inserted frames either from a base-layer sequence or from another frame of the enhancement layer sequence. This method can also be used for encod-ing of stereoscopic sequences with an LRLRLR... interleaving of left and right pictures, as e.g. required for shutter-glass display (cf. sec. 16.5)
- All scalability modes can also be used to achieve a forward and backward compatible combination of MPEG-1 and MPEG-2, when MPEG-1 syntax is exclusively used in the base layer.
- A method of data partitioning for DCT coefficients is defined, similar to the scheme shown in Fig. 12.33.

While MPEG-1 was originally designed to be used in just one application domain (video for CD storage), the number of application domains and necessary combi-nations of elementary tools is manifold for the case of MPEG-2. As it appears not to be useful that any MPEG-2 device may support all elements of the standard, MPEG-2 defines different *profiles*. Further, within each *profile*, *levels* are speci-fied, which describe maximum sizes or image formats which must be decodable. Each bit stream carries the information indicating the profile implementation ex-pected at the decoder, as well as the format (level) which the decoder must be able to process in real time. From this information, any MPEG-2 compliant decoder can decide immediately whether it will be able to process the stream; within certain application domains, specific 'profile@level' configurations have established as mandatory, e.g. 'main profile@main level' is typically required for digital TV broadcast or DVD storage applications. Four levels are defined as 'low', 'main' (SD), 'high-1440' and 'high' (HD), where however not each level is combinable with each profile. The profiles defined by MPEG-2 video are as follows:

- *Simple profile*: This is for *low cost* and *low delay* applications, allowing frame sizes up to SD resolution (ITU-R Rec. 601) and frame rates up to 30 *Hz*. Us-age of *B* frames is not allowed.

- *Main profile*: This is the most widely used MPEG-2 profile, defined for SD and HD resolution applications in storage and transmission, without providing compatible decoding of different resolutions. All interlaced-specific prediction modes as well as *B* frames are supported.
- *SNR scalable profile*: Similar to Main profile, but allows in addition SNR scalability invoking drift at the base layer (cf. sec. 13.2.8); resolutions up to SD are supported.
- *Spatial scalable profile*: Allows usage of (drift free) spatial scalability, also in combination with SNR scalability.
- *High profile*: Similar to Spatial Scalable profile, but supporting a wider range of levels, and allowing 4:2:2 chrominance sampling additionally. This profile was primarily defined for upward/downward compatible coding between SD and HD resolutions, allowing embedded bit streams for up to 3 layers with a maximum of two different spatial resolutions.
- *4:2:2 profile*: This profile extends the Main profile by allowing encoding of 4:2:2 chrominance sampling, allows larger picture sizes and higher data rates.
- *Multiview profile*: This allows encoding of stereoscopic sequences provided in a left-right interleaved multiplex, using the tool of temporal scalability for exploitation of left-right redundancies. The displacement between left and right pictures is estimated and used for *disparity compensated prediction*, which follows the same principles as motion-compensated prediction. In addition, camera parameters can be conveyed in the stream.

MPEG-2 is mainly used for consumer-level video broadcast (e.g. DVB) and storage (e.g. DVD), and for professional applications like video storage in studios.

17.4.3   MPEG-4 Visual

The standard ISO/IEC 14496-2: *'Coding of Audiovisual Objects – Part 2: Visual'* includes the specification of MPEG–4 video coding. Methods are partially based on proven technology from previous standards (e.g. block-based motion compensation, DCT), but tools for new functionalities are specified as well. While MPEG–1 and MPEG–2 are only capable to encode rectangular video frames, MPEG–4 defines encoding of *video objects* which can have either rectangular or arbitrary shape. Video scenes can be composed from several objects which may change in position, appearance, size etc. independent of each other, but it is also possible to embed video elements into 3D scenes (cf. sec. 17.2). Frame-based MPEG–4 video coding is also used for storage of video content on CDs and for video streaming over the Internet and mobile channels. Network interfaces, including interfaces with Internet protocols, are better adaptable than in previous MPEG standards, more efficient and flexible degrees of scalability are provided, and better mechanisms for error protection and error resilience in case of transmission losses are defined in the video syntax.

As the video objects to be processed by the video codec are not necessarily of arbitrary shape in any case, the approach taken in MPEG-4 is the definition of more simple decoder configurations by combinations of tool subsets in object types and profiles. Such low-complexity configurations are only able to process rectangular video frames, while a generic extension of the syntax and semantics allows processing of arbitrary-shaped video objects. The rudimentary codec is in principle very similar to the hybrid MC prediction / DCT scheme of previous standards. With a specific *short header* format, an MPEG-4 video decoder is also capable to decode streams generated by an H.263 baseline encoder. Coding of arbitrary-shaped video objects is enabled by supplementing a shape coding mechanism, where subsequent steps of the decoding procedure are restricted to the area which is defined by the object shape mask (Fig. 17.11). For unique definition of both rectangular and arbitrary-shaped video objects in the bitstream syntax, the entity of *Video Object Plane* (VOP) is introduced, which can likewise be used to represent a rectangular-plane frame or arbitrary-shaped object plane. Motion-compensated prediction processing is defined for *B-VOP* (bi-directional prediction) and *P-VOP* (unidirectional prediction). Within an *I-VOP* (intraframe encoded), only spatial prediction of DC and first row/column of AC transform coefficients can be performed.

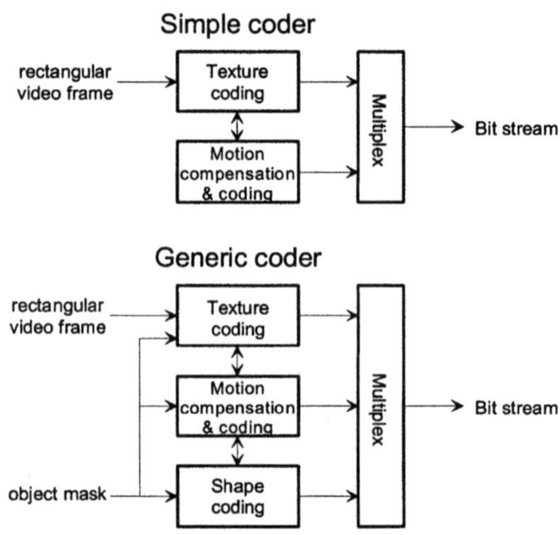

Fig. 17.11. Evolutionary MPEG-4 video coding concept : conventional hybrid codec as basis for frame-based coding, supplemented by shape coding for the generic VOP case

The shape is represented by a binary mask which is compressed by *Context Arithmetic Encoding* (CAE, see sec. 11.4.5 and 12.1). Binary shape parameters are also encoded by utilization of motion shift, where motion-compensated samples from the reference frame can be used within the context of CAE. As the basic concept is block-based (DCT, motion compensation), the shape is aligned with a block grid, where then blocks of rectangular shape and boundary blocks of non-rectangular shape co-exist (cf. Fig. 12.45). It is also possible to define transparency of an object by either assigning a constant transparency (alpha) value, or by providing spatially variable transparency by a *gray-scale shape* (cf. sec. 13.2.6); the encoding is performed by the same motion-compensated DCT algorithm that is used for the texture information.

The MPEG-4 video codec invokes a set of new tools, some of which are intended for better compression performance, and some support object-based or other functionality:

- Quarter-pixel accuracy and variable block size of 8x8 or 16x16 can be used in motion compensation;
- Global motion compensation is defined, which allows to express e.g. the effect of camera motion by using only a low number of parameters;
- Padding or shape-adaptive DCT (sec. 12.6.2) is used to encode boundary blocks of non-rectangular shape, where the padding must also be performed in a normative way when a motion vector points to a position which is not part of the reference VOP (e.g. beyond the VOP boundary of the previous frame);
- Different VLC tables can be selected, where the codes are designed for more efficient encoding at ranges of lower or higher rates; the optimum choice depends on the target rate;
- In the *short header* mode, bitstream-level compatibility with the H.263 baseline syntax is realized;
- The direct mode can determine the motion vectors within B-VOPs from the co-located P-VOP motion vectors without rate overhead (see sec. 13.5);
- Interlace video encoding tools are specified which are similar to those defined in MPEG-2, but have been modified for the special case of arbitrary-shape object encoding;
- Different options of scalability, including temporal scalability modes, spatial scalability modes and fine-granular SNR scalability are defined (cf. sec. 13.2.8);
- Static sprites are defined, which are mosaic-like images that can geometrically be aligned by 2D global warping, such that reverse mapping into the frames of a video sequence can be performed (cf. sec. 13.2.6).
- Coding of signals with up to 12 bit amplitude resolution and different color sampling formats (including 4:4:4) is supported.
- For high-quality studio storage and inter-studio transmission applications, a different method of DCT coefficient encoding is introduced[1]. This is based on grouping of DCT coefficients by similar amplitude values instead of the conventional zigzag run-length and entropy scheme. Recursive selection of VLC tables is applied to groups of coefficients, where the selection function relies on previously coded groups. Coded data are the group indicator and a fixed-length code determining the actual coded value.
- For high-quality studio storage, a lossless coding method based on switching between DPCM and PCM is defined.

---

[1] This and the subsequent bullet apply only for the two *Studio Object Types*.

The concept of profiles and levels already used by MPEG-2 is also implemented in MPEG-4. Below the entity of profiles, MPEG-4 defines *object types*. These are combinations of *tools* (basic coding methods like B VOPs, interlaced coding etc.). For the visual part of the standard, usually a profile exists with the same name as an object type, but the profile may support more object types than the one with the same name. This is in particular useful, because object types follow a hierarchy, e.g. the advanced simple object type is an extension (superset) of the simple object type. Until now, 13 video-related object types have been defined, which are (without claim for complete description of properties):

- *Simple* and *Simple Scalable*: Support for only rectangular VOPs, no B-VOPs, half-pixel accuracy of motion compensation, tools for error resilience. Simple scalable further allows spatial and temporal scalability (including B-VOPs).
- *Advanced Simple*: Superset of Simple, allowing B-VOPs, quarter-pixel accuracy of motion compensation, global motion compensation, interlaced video coding tools.
- *Core* and *Core Scalable*: Superset of Simple and Simple Scalable, respectively. These allow arbitrary binary-shape video objects, B-VOPs, different quantization methods. Another type very similar to Core is the *N-Bit* object type, in addition allowing variable amplitude resolution of 4-12 bit.
- *Advanced Coding Efficiency*: Superset of Advanced Simple, allowing arbitrary shaped video objects with binary or gray-scale shape.
- *Fine Granularity Scalable*: Superset of Advanced Simple, providing fine-granular SNR and temporal scalability.
- *Main*: Superset of Core, invoking most of available MPEG-4 video tools such as sprites, gray-scale shape, interlaced coding.
- *Simple Studio* and *Core Studio*: These are defined specifically for high resolution and high quality in applications of studio production and materials exchange. Studio-typical color sampling formats as 4:4:4 and up to 12 bit amplitude resolution are supported. Additional tools useful for production are included such as lossless coding, sprite coding and multiple alpha channels for auxiliary data.
- *Advanced Real-Time Simple*: Invokes additional error-resilience functionality, such as encoder/decoder re-synchronization in case of transmission errors, and resolution reduction.
- *Error Resilient Simple Scalable*: Superset of Simple Scalable with additional error resilience tools, in particular resynchronization mechanisms for the enhancement layer.

Fig. 17.12 shows the relationships of MPEG-4 video-related object types. It is evident that the *Simple* object type and profile establishes the *kernel* of the standard, and most other types are forward-compatible extensions. Consequently, e.g. a decoder being conformant to the *Advanced Simple Profile* will implicitly be able to decode streams originally generated for the *Simple Profile*. The only exception are the *Simple Studio* and *Core Studio* types, which are designed for a completely different application domain, and establish a forward-compatible hierarchy by themselves.

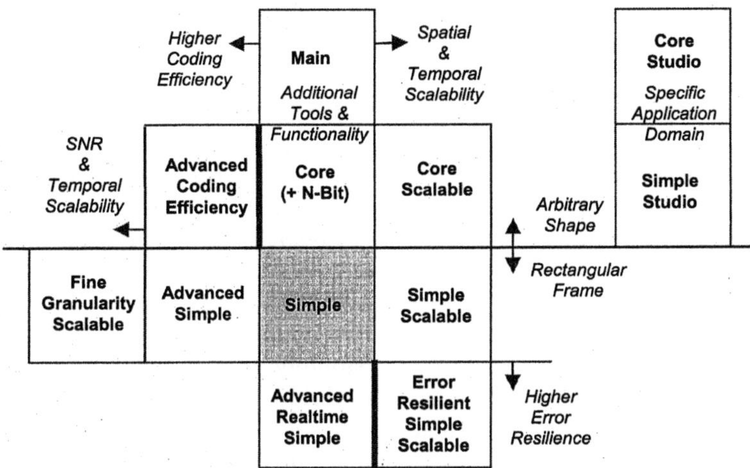

Fig. 17.12. Structure of MPEG-4 video object types

Several tools for error resilience mechanisms are provided. These allow to vary the strength of error protection mechanisms (which usually require additional data rate), such that flexible adaptation to the error characteristics of specific networks can be achieved:

– *Resync markers* can be embedded at different points of the bit stream, down to the level of 'group of blocks' (GoB), which are units containing a variable (defined by encoder) number of macro-blocks. The GoB header then contains position information which is necessary to recover after data losses and re-start decoding at the correct position.
– *Data partitioning* allows to separate the motion data (for which a loss would be quite critical, as the reconstructed image may appear severely geometrically distorted otherwise) from the less relevant texture data (DCT coefficients).
– *Reversible variable-length codes* allow to reconstruct the information *backwards* from a resynchronization point in the case of data losses (cf. sec. 11.4.3 and 11.4.7).

Other tools and objects defined in the Visual part of MPEG-4 are:

– A fully scalable still texture codec based on a 2D wavelet transform with zero-tree encoding (cf. sec. 17.3.3);
– A codec for human face and body animation based on face/body models and additional semantic parameters;

- Tools for animation of 2D warping grids with uniform and irregular (Delaunay) topology (cf. sec. 16.2).

These tools are likewise combined into object types and profiles. which are *Animated 2D Mesh, Basic Animated Texture, Advanced Scalable Texture, Simple Face Animation* and *Simple Face/Body Animation*.

17.4.4   H.264/MPEG-4 Part 10 Advanced Video Coding (AVC)

The demand for ever increasing compression performance has recently led to the definition of a new part of the MPEG-4 standard, ISO/IEC 14496-10: *'Coding of Audiovisual Objects – Part 10: Advanced Video Coding'*, which is identical by text with ITU-T Rec. H.264. The development of AVC was performed by the *Joint Video Team* (JVT), which consists of members of MPEG and of the ITU-T Video Coding Experts Group. To achieve highest possible compression performance, any attempt to achieve backward or forward compatibility with earlier standards was given up. Indeed, the basic approach is again a hybrid video codec (MC prediction + 2D transform). A high-level block diagram of AVC includes all options illustrated in Fig. 13.5. The most relevant new tools and elements are as follows:

- Motion compensation using variable block sizes (denoted as sub-macroblock partitions) of 4x4, 4x8, 8x4, 8x8, 8x16, 16x8 or 16x16, motion vectors encoded by hierarchical prediction starting at the 16x16 macroblock level;
- Motion compensation is performed by quarter-sample accuracy, using high-quality interpolation filters;
- Usage of the integer transform (4.136) of block size 4x4. This is not a DCT, but could be interpreted as an integer approximation thereof. The transform is orthogonal when the appropriate normalization is observed during quantization. For the entire building block of transform and quantization, implementation by 16-bit integer arithmetic precision is possible both for encoding and decoding. In contrast to previous standards based on the DCT, there is no dependency on a floating point implementation, such that in principle no drift between encoder and decoder can occur in normal operation. To compensate for the drawback of relatively small block sizes, extensive usage of spatial inter-block prediction must be made, and a sophisticated deblocking filter must be applied in the loop.
- Intraframe coding is performed by first predicting the entire block from boundary pixels of adjacent blocks. Prediction is possible for 4x4 and 16x16 blocks, where for 16x16 blocks only horizontal, vertical and DC prediction (each pixel predicted by same mean value of boundary pixels) is allowed. In 4x4 block prediction, also directional prediction is specified (see Fig. 12.16c), where eight different prediction directions are selectable. As typically the result of full-block (vector) prediction does not decorrelate the pixels (12.27), the 4x4 integer transform is applied to the residual of intra prediction for best compression performance. A special double scan method is

defined for this case, replacing the ordinary zigzag scan. The set of luma DC coefficients from a macroblock of size 16x16 (which contains 16 blocks of size 4x4 each) can additionally be processed by a 2D Walsh transform of block size 4x4[1] to achieve better inter-block decorrelation.

– An adaptive deblocking filter (cf. Fig. 13.5) is applied in the prediction loop. As shown in (13.16), the optimum selection of this filter highly depends on the amount of quantization error fed back from the previous frame. The adaptation process of the filter is nonlinear, with lowpass strength of the filter steered firstly by the quantization parameter (step size). Further parameters considered in the filter selection are the difference between motion vectors at the respective block edges, the coding mode used (e.g. stronger filtering is made for intra mode), the presence of coded coefficients and the differences between reconstruction values across the block boundaries. The filter itself is a linear 1D filter with directional orientation perpendicular to the block boundary, impulse response lengths are between 3 and 5 taps depending on the filter strength.

– Multiple reference picture prediction allows to define references for prediction of any macroblock from one of up to F previously decoded pictures; the number F itself depends on the profile/level definition, which specifies the maximum amount of frame memory available in a decoder. Typically, values around $F=5$ are used, but for some cases $F=15$ could be realized.

– Instead of B-type, P-type, I-type *pictures*, type definitions are made slicewise, where a slice may at maximum cover an entire picture.

– New types of *switching slices* (S-type slices, with *SP* and *SI* sub-types, cf. sec. 13.2.9) allow controlled transition of the decoder memory state when stream switching is made; this information however causes a rate overhead as compared to a single stream.

– The B-type slices are generalized compared to previous standards, denoted as *bi-predictive* instead of *bi-directional*. This in particular allows to define structures of prediction from *two previous* or *two subsequent* pictures, provided that a causal processing order is observed. Furthermore, prediction of B-type slices from other B-type slices is possible, which e.g. allows implementation of a B-frame pyramid (cf. Fig. 13.25). Different *weighting factors* can be used for the reference frames in the prediction (formerly, this was restricted to averaging by factors 0.5 each). In combination with multiframe prediction, this allows a high degree of flexibility, but also incurs less regular memory accesses and increased complexity of the encoder and decoder.

– Two different entropy coding mechanisms are defined, one of which is *Context-adaptive VLC* (CAVLC), the other *Context-adaptive Binary Arithmetic Coding* (CABAC). Both are universally applicable to all elements of the code syntax, which is based on a systematic construction of variable-length code

---

[1] The chroma DC coefficients, which are only quarter the number of luma samples for case of 4:2:0 sampling, are subject to a 2x2 2D block transform equivalent to the Haar transform.

tables. By proper definition of the contexts, it is possible to exploit non-linear dependencies between the different elements to be encoded. As CABAC is a coding method for binary signals, a binarization of multi-level values such as transform coefficients or motion vectors must be performed before it can be applied; methods which can be used are *unary codes* or *truncated unary codes* (VLCs consisting of '1' bits with a terminating zero), Exp-Golomb codes or fixed-length codes. Four different basic context models are defined, where the usage depends on the specific values to be encoded [MARPE ET AL. 2003].

– Additional error resilience mechanisms are defined, which are *Flexible Macroblock Ordering* (FMO – allowing macroblock interleaving), *Arbitrary slice ordering* (ASO), data partitioning of motion vectors and other prediction information and encoding of *redundant pictures*, which e.g. allows duplicate sending or re-transmission of important information.

– Other methods known from previous standards, such as frame/field adaptive coding of interlaced material, direct mode for *B*-slice motion vector prediction, predictive coding of motion vectors at macroblock level etc. are implemented as usual.

– A *Network Abstraction Layer* (NAL) is defined for simple interfacing of the video stream with different network transport mechanisms, e.g. for access unit definition, error control etc.

The key improvements are again made in the area of motion compensation, which approaches almost the degree of flexibility as originally defined in (4.99), with the additional option to use subsequent frames as reference when they are decoded earlier. As the decoding process is more or less defined independently of the display timeline, it would even be possible to define and encode additional pictures which are just used as reference for motion compensation, but by themselves are never displayed; these could e.g. be used as memory for temporarily occluded background areas. The sophisticated loop filter allows a significant increase of subjective quality at low and very low data rates, and fully compensates the potential blocking artifacts produced by the 4x4 block size transform. State-of-the-art context-based entropy coding drives compression to the limits. On the other hand, the high degrees of freedom in mode selection, reference-frame selection, motion block-size selection, context initialization etc. will only provide significant improvement of compression performance when appropriate optimization decisions, in particular based on rate-distortion criteria, are made (cf. sec. 13.2.6 and 13.2.10). Some elements of this type have been included in the reference encoder software used by the JVT during the development of the standard.

The combination of all different methods listed has led to a significant increase of the compression performance compared to previous standard solutions. Reduction of the bit rate at same quality level by up to 50% as compared to H.263 or MPEG–4 Simple Profile, and up to 30% as compared to MPEG–4 Advanced Simple Profile have been reported [WIEGAND, BJONTEGAARD, SULLIVAN, LUTHRA 2003]. On the other hand, the complexity of encoders must significantly be increased as well to achieve such performance. Regarding decoder complexity, the integer transform and usage of systematic VLC designs clearly reduces the complexity as compared to previous (DCT based) solutions, while memory accesses are also becoming more irregular due to usage of smaller motion block sizes and possibility of multiframe access. The loop filter and the arithmetic decoder also add complexity.

The concept of profile and level definitions for decoder conformance points is also implemented in the AVC standard. Presently, the following profiles are defined:

- *Baseline profile*: Constraint to usage of *I*- and *P*-type slices, no weighted prediction, no interlace coding tools, no CABAC, no slice data partitioning, some more specific constraints on the number of slice groups and levels to be used with this profile.
- *Main profile*: Only *I*-, *P*- and *B*-type slices, all major error resilience tools such as slice data partitioning, arbitrary slice order, multiple slice group per picture are disabled; some more specific constraints are made on levels to be used with this profile.
- *Extended profile*. No CABAC, all error resilience tools used (including *SP* and *SI* slices), some more specific constraints imposed to the direct mode, number of slice groups and levels to be used with this profile.

The profile/level construction is made in a way such that – unlike the approach of MPEG-4 video – no forward or backward compatibility between the different profiles is achieved. A total of 5 major levels and 15 sub-levels is defined. Level restrictions relate e.g. to the maximum number of macroblocks per second, maximum frame size, maximum decoded picture buffer size (imposing constraints on multiframe prediction), maximum bit rate, maximum coded picture buffer size and vertical motion vector ranges. These parameters can be mapped to a model of a *Hypothetical Reference Decoder* (HRD), which mainly relates to buffer and timing behavior similar to the models discussed in sec. 16.2.1.

Presently, an extension is prepared to support usage of AVC in professional applications (e.g. studio storage, digital cinema), which will define encoding based on color sampling formats like 4:4:4 and 4:2:2[1], and up to 12 bit amplitude resolution precision.

---

[1] The other profiles only support encoding at 4:2:0 color sampling resolution.

17.5 Audio Coding

17.5.1   Speech Coding

Table 17.4 gives an overview about most common speech coding standards available today, most of which are based on the principle of linear prediction, which perfectly fits with physical models of the speech tract (cf. sec. 14.1). These standards are used for speech encoding and transmission either standalone or combined with other media types in multimedia communication systems.

Table 17.4. Overview of standards for encoding of speech signals[1]

Standard	Codec type	Bit rate [kbit/s]
ITU-T G.711	PCM	64
ITU-T G.726	ADPCM	16, 24, 32, 40
ITU-T G.727	Embedded ADPCM	16, 24, 32, 40
ITU-T G.728	Low-Delay CELP	16
ITU-T G.729	Algebraic CELP (ACELP)	8
ITU-T G.723.1	ACELP / MP-LPC	5.3 / 6.3
GSM 06.10	RPE-LTP	13
GSM Enhanced Full Rate	ACELP	12.2
GSM Half Rate 06.20	VSELP	5.6
MPEG-4 Audio Part	CELP scalable narrow band	3.85-12.2 (steps ~ 2 kbit/s)
MPEG-4 Audio Part	CELP scalable wide band	10.9-23.8
MPEG-4 Audio Part	HVXC scalable	2 / 4
FS 1015 (US Fed. Secure)	LPC-10E	2.4
FS 1016	CELP	4.8
FS 1017	MELP	2.4

---

[1] PCM:*Pulse Code Modulation*, ADPCM:*Adaptive Differential PCM*, CELP:*Code Excited Linear Prediction*, LPC:*Linear Predictive Coding*, MP-LPC:*Multi-Pulse LPC*, RPE-LTP:*Regular Pulse Excitation / Long Term Prediction*, VSELP:*Vector Sum Excited Linear Prediction*, HVXC:*Harmonic Vector Excitation Coding*, MELP:*Mixed Excitation Linear Prediction*

For speech synthesis, the Audio part of the MPEG-4 standard further defines a *Text-to-Speech Interface* (TTSI), which is designed for transmission rates between 200 and 1200 bit/s.

17.5.2 Music and Sound Coding

For the series of MPEG audio waveform coding standards, an overview of configurations and achievable stereo data rates is given in Table 17.5. Both cases of CD quality (perceptually lossless) and FM radio quality (lossy) are shown. These numbers clearly show the technical progress that was made in audio compression over the last decade.

Table 17.5. Overview of MPEG standards for waveform coding of audio signals

Standard	Transform	Frequency bands	Stereo data rate (CD quality) [*kbit/s*]	Stereo data rate (FM quality) [*kbit/s*]
MPEG-1 Layer 1	Polyphase filterbank	32	384	256
MPEG-1 Layer 2	Polyphase filterbank	32	256	192
MPEG-1 Layer 3 ('MP3')	Polyphase filterbank (32) + MDCT (6/18)	192 / 576 switchable	180	128
MPEG-2 AAC	MDCT	128 / 1024 switchable	128	64
MPEG-4 AAC	MDCT	128 / 1024 switchable	<128	40-48

The MPEG-1 standard (ISO/IEC 11172-3 *'Coding of Moving Pictures and associated Audio for Digital Storage Media at up to about 1.5 Mbit/s, Part 3: Audio'*) defines the syntax and semantics of three different audio codecs, which follow a forward-compatible principle over three layers. Layer 1 and 2 codecs both use the same polyphase (subband) filterbank processing a signal into 32 (uniform-bandwidth) frequency bands, adaptive bit allocation for fixed-length codes is applied in encoding of the frequency coefficients instead of entropy coding. MPEG-1 Audio Layer 2 is the first codec making more extensive use of perceptual weighting functions in encoding, which is however limited due to the usage of 32 uniform frequency bands, which cannot be mapped directly with the nonlinear Bark scale of 24 frequency groups (cf. sec. 6.2). In the improved scheme of MPEG-1 Audio Layer 3 (better known as *MP3*), a second filter bank is supplemented which is realized by a block-overlapping transform (*modified DCT*, MDCT) switchable for sub-division of each of the former 32 bands into 6 or 18 bands; this allows a spectral decomposition into a total of either 6x32=192 or 18x32=576 frequency bands. With a sampling frequency of 48 *kHz*, this results in a lowest (bandwidth) resolution of 24/576 *kHz* =41.67 *Hz*. It depends on the principle of adaptation to the signal properties, whether it is better to select the lower or higher number of frequency bands[1]. As the 32-band polyphase filterbank is still used, a certain level of forward compatibility with Layers 1 and 2 is retained, even though the genuine quantization and encoding steps are also different in Layer 3; for decoding of Layer 1 or Layer 2 streams, a Layer 3 decoder must switch to a different coefficient decoding algorithm, but can use the same (partial) filterbank. A typical block diagram of an MP3 encoder is shown in Fig. 17.13.

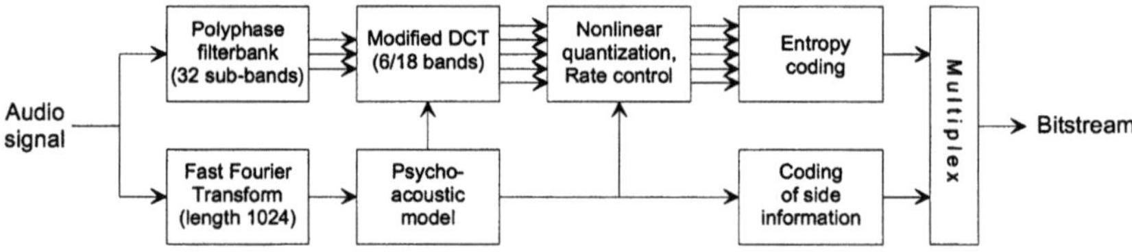

Fig. 17.13. Typical MPEG-1 Layer 3 ('MP3') Audio encoder

The codecs of MPEG-1 Layers 1-3 were identically defined for the MPEG-2 standard in ISO/IEC 13818-3 *'Generic Coding of Moving Pictures and Associated Audio Information - Part 3: Audio'*, such that full compatibility between the two standards is retained at the level of audio signal coding. After finalization of the first edition of the MPEG-2 standard, it became evident that significantly better compression performance could be achieved when the tandem configuration of transforms in Layer 3 would be replaced by a monolithic transform, and different other tools would be supplemented (cf. sec. 14.2). This has led to the definition of a new part of MPEG-2 (ISO/IEC 13818-7 *'Generic Coding of Moving Pictures and Associated Audio Information - Part 7: Advanced Audio Coding'*)[2]. Advanced Audio Coding (AAC) is also the primary technology for audio waveform coding in the MPEG-4 standard (ISO/IEC 14496-2: *'Coding of Audiovisual Objects – Part 3: Audio'*). As an open standard, AAC is presently the best scheme available for waveform coding of audio signals. Compared to MP3, a reduction of the data rate by a factor of 2 can roughly be achieved, depending on the data rate; in general, this gain is more remarkable for lower data rates (range of lossy compression). The most important novel elements to achieve better compression performance are

---

[1] In general, longer transform blocks are better suitable for signal segments of more stationary behavior, e.g. for sustained harmonic sounds, while usage of shorter blocks is advantageous in cases of tone transitions.

[2] An encoder block diagram is shown in Fig. 14.3.

- Spectral resolution switchable for 128 or 1024 subbands, using a monolithic MDCT transform;
- Optional temporal prediction of spectral coefficients (determination of an estimate from previously decoded spectral components), which significantly increases coding efficiency for the case of 'uniform tone' signals;
- Improved encoding of spectral coefficients by more flexible adaptation capability for the VLC;
- Utilization of stereo redundancy;
- Scalable sampling rate (SSR);
- Switching of the filterbank into a shorter analysis block length, which in particular allows better utilization of the pre-masking effect (sec. 6.2.2);
- Introduction of 'temporal noise shaping', which allows to achieve better consistency of the coding noise between neighbored analysis blocks and prevents perceptible temporal blocking artifacts.

In the MPEG-4 extensions of AAC, the following mechanisms are further defined:

- A method for *perceptual noise substitution* (PNS), which identifies noise-like (e.g. non-harmonic) signal components and substitutes the signal within appropriate frequency bands by a noise signal of same variance;
- *Long term prediction* (LTP), which drastically reduces the complexity as compared to the previous (MPEG-2) coefficient prediction method – even though it does not result in quality improvement in general;
- For extremely low rates (40 *kbit/s* and below) *Transform Domain Weighted Interleave Vector Quantization* (TWIN-VQ) is specified for quantization and encoding of transform coefficients. The term *interleaving* reflects the specific approach of ordering the spectral coefficients into vectors. The selection of vectors (quantization) is again controlled by the perceptual model, while the other elements of AAC are retained without changes.
- Scalability: Optional replacement of the one-layer encoding by a multi-layer encoding. MPEG-4-AAC allows data rate scalability with minimum steps of 8 kbit/s. To achieve fine granularity of the bit rate scalability, *Bit Slice Arithmetic Coding* (BSAC), similar to the embedded quantization methods described in sec. 11.3, is employed instead of conventional VLC coding. Scalable coding is also possible in combination with TWIN-VQ, but this penalizes the compression performance at higher rates.
- Additional mechanisms for *error resilience* (ER) and error protection are specified.
- A low-delay mode is introduced by using shorter analysis block length, providing a lowest delay of only 20 *ms*.
- A method for *Spectral Band Replication* (SBR) is defined, where only a narrow-band signal of typically 8 *kHz* bandwidth is encoded, and all frequency components above are synthetically reproduced at the decoder, which is done by distinction between harmonic and non-harmonic components.

The MPEG-4 AAC with SBR allows transparent (without perceptible impairment) encoding of stereo audio signals at rates of 96 *kbit/s* and below. Using TWIN-VQ, a quality which is better than AM radio is achieved at a data rate of 20 *kbit/s* for the stereo signal. This allows applications like Internet radio streaming for extremely narrow-band channels like telephone modems.

As conformance points, MPEG-2 AAC defines three different *profiles*, which are *AAC LC* (low complexity), *AAC Main* (extension of LC by the temporal prediction tool) and *AAC SSR* (extension of LC by the SSR tool). In MPEG-4 audio, the entity of *object types* is defined additionally (similar as in MPEG-4 video coding). This allows to define useful combinations of tools by a more granular level. Forward-compatible relationships are retained as far as possible. A topology of these relationships is shown in Fig. 17.14, where the MPEG-2 AAC LC (low complexity) profile is the common core method, which would still be decodable by all extended-version decoders. Level-related conformance in audio defines sampling rates and number of objects (e.g. audio channels) that a decoder should be able to process.

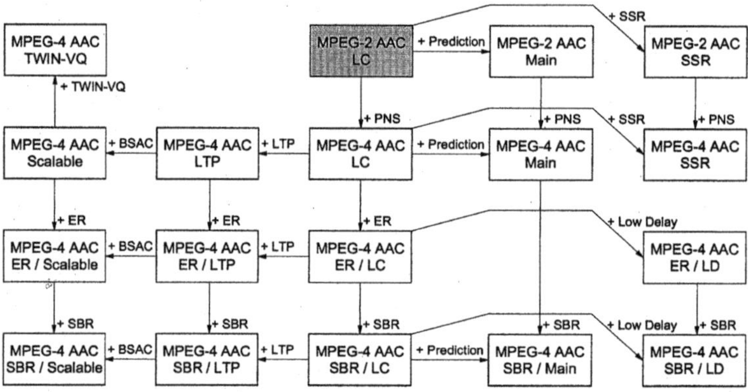

Fig. 17.14. AAC profile/object type definitions in MPEG-2 and MPEG-4 (after GRILL)

Waveform coding of audio signals as defined in the MPEG standards allows a considerable degree of freedom to optimize the encoder[1]. The most important aspect is the best exploitation of perception models (sec. 6.2), but this may further be combined with rate-distortion criteria (sec. 11.3), then however taking into account the perceptual distortion level. The effects of frequency masking, also of temporal post-masking, can directly be integrated in the quantization control. For utilization of temporal pre-masking, it is necessary to switch to shorter analysis blocks. For certain tools, e.g. for temporal noise shaping and prediction methods

---

[1] Different MP3 encoders available in the market differ significantly in terms of decoded signal quality, even when streams of the same bit rate are produced.

the best parameter selection can only be made by combined encoding of subsequent transform blocks.

In addition to the waveform coding methods, MPEG-4 audio defines tools, object types and profiles for *synthetic audio* representation. These consist of

– A flexible music synthesis language denoted as *Structured Audio Orchestra Language* (SAOL);
– An efficient wavetable synthesis based on the *Structured Audio Sample Bank Format* (SASBF);
– An adoption of General MIDI (Musical Instrument Digital Interface) format which is widely supported by sound synthesis devices;
– Capability for flexible combination with audio-related BIFS nodes in the MPEG-4 systems part, which allows flexible mixing of audio objects, insertion of audio effects and modeling of sound propagation in 3D rooms;
– The *Harmonic Individual Lines with Noise* (HILN) parametric audio coder (see sec. 14.3), which can also be combined with the HVXC parametric speech coder (see sec. 14.1).

Presently, a new amendment is prepared for advanced parametric coding which includes transient (temporal envelope) behavior. Such methods can be expected to provide significantly better results than waveform codecs in the range of low bit rates, where only perceptually lossy coding can be realized.

17.6 Multimedia Content Description Standard MPEG-7

Metadata information is becoming increasingly important for multimedia signal representations. These *metadata* can be clustered into:

– Content-abstracted (high level) metadata: Information about copyright, authoring, acquisition conditions (time, place etc.), storage/transmission formats and locations, and abstracted content categories (e.g. genre of movies or music pieces);
– Content-related conceptual (mid level) metadata: Concrete descriptions of image, video or audio content, e.g. scene properties, objects contained, events, actions and interactions etc.; summaries of the content, information about available variations (versions) of the same content or related content e.g. in content collections;
– Content-related structural (low level) metadata: Features which can be extracted by analysis directly from the audiovisual signals, e.g. length or size of segments, motion properties or trajectories in video, image related properties like color, edges or texture description.

By using standardized metadata description formats it is possible to build *interoperable* and *distributed systems* for content-related applications. For example, multimedia signal retrieval applications (sec. 15.5) could simultaneously look up multiple databases, media or web servers, each of which would preferably accommodate the same schema of feature description. If this is not the case, transcoding of the metadata format must be applied, which typically is costly and time-consuming, such that fast responses of the retrieval system are impossible. Examples of multimedia related metadata description standards are the *Resource Description Framework* (RDF) of the Internet Engineering Task Force (IETF), the *Metadata Dictionary* of the Society of Motion Picture and Television Engineers (SMPTE), and MPEG-7 (ISO/IEC 15938: *Multimedia Content Description Interface*).

Regarding the focus of this book in multimedia signal processing, the MPEG-7 standard is most interesting, as it directly includes methods to describe low-level features of audiovisual signals as introduced in chapter 7. In general, standardized definitions of metadata are required if distributed systems shall be capable to exchange data. Regarding low-level features, MPEG-7 could also be interpreted as an 'audiovisual language', which fills a gap that exists due to the fact that signal features can hardly be described by text. This can straightforwardly be used e.g. in content-related storage and streaming applications as described in sec. 15.5. The scope for future intelligent systems utilizing audiovisual information resources is however much wider. For example, a standardized description framework is necessary to define *ontologies*, to support a common understanding between automatic systems about the meaning of certain signal features and the way how they can provide semantic classification. This could e.g. be used to install lexical databases of audiovisual feature constellation examples to be mapped into a certain semantic meaning, or for identification of a certain piece of content. The translation between different modalities (text/sound/image related information) expressing the evidence about an event or the nature of an object appearing in audiovisual scenes is also a topic of high importance in this context.

The MPEG-7 standard normatively specifies only the representation of the content description, while generation of descriptions and consumption of descriptions are regarded as application specific topics (see Fig. 17.15). The normative representation must relate both to the syntactic structure of a description and to the semantic meaning of description elements. Further, mechanisms must be provided to identify and access MPEG-7 descriptions at a systems interoperability level.

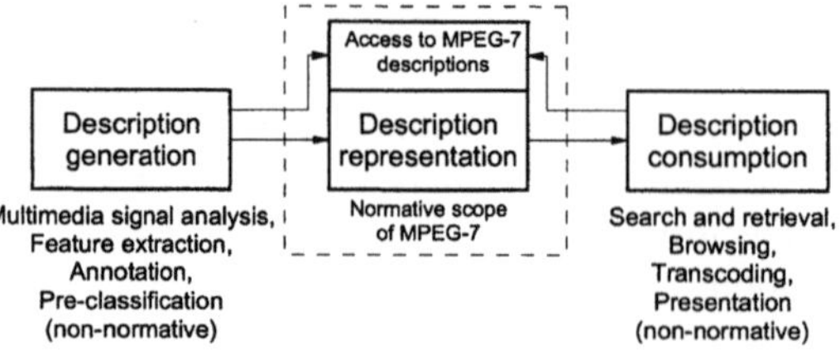

Fig. 17.15. Normative and non-normative elements in an MPEG-7 application

17.6.1 Elements of MPEG-7 Descriptions

The relationship of the different elements specified in the MPEG-7 standard is illustrated in Fig. 17.16. The structural elements of a description according to the MPEG-7 standard are *Description Schemes* (DS) and *Descriptors* (D). A DS is typically a combination of several Descriptors, but can also contain further subordinate DSs. In the example of Fig. 17.16, the DS "A" at the top level has two Ds "A" and "B" directly attached, as well as another DS "B", which again consists of Ds "C" and "D". Descriptors are the lowest-level elements of the description, which are described in the standard by the syntax of *descriptor values* and the associated semantic meaning (e.g. which value instantiation would express red or green color in a color descriptor). The syntactical structure both of descriptors and description schemes is expressed by the *Description Definition Language* (DDL).

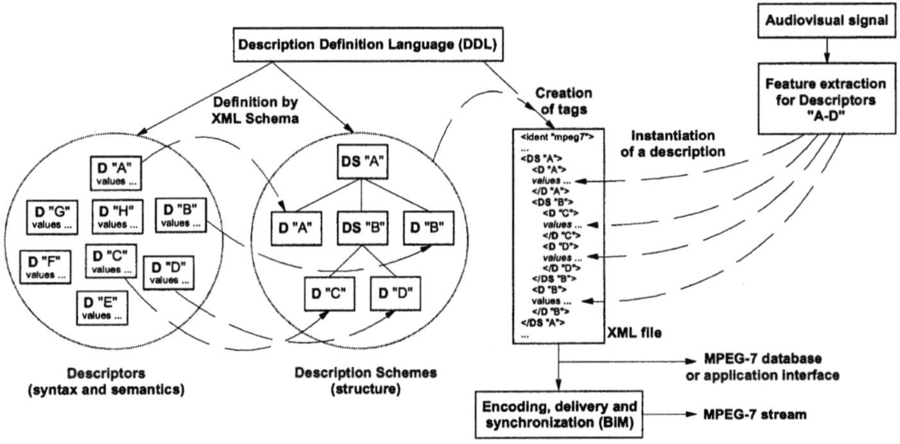

Fig. 17.16. Elements of the MPEG-7 standard and their relationship

The DDL is derived from the *Extensible Markup Language* (XML)[1], and uses the XML Schema language which allows to specify MPEG-7 specific extensions which are necessary to define Ds and DSs. These extensions mainly relate to simple and complex type definitions, attribute declarations and element declarations as e.g. necessary to identify and classify the normative elements; they also allow specific definitions of data types which are needed by some descriptors such as vectors, matrices or very compact binary mappings of descriptor values. The description itself is then instantiated e.g. by performing feature extraction, and stored in an XML file. XML however is a pure file format not directly suitable to support

---

[1] XML and XML Schema are language standards defined by the World Wide Web Council (W^3C)

real-time applications such as streaming of descriptions, dynamic change of single descriptor values etc. Further, the representation is not very compact, as XML is text-based. To overcome these limitations, MPEG-7 also specifies a *Binary Format for MPEG-7 Descriptions* (BiM), which can uniquely be mapped with any XML based descriptions, and can indeed be regarded as a generic binary encoding for any XML files guided by schema information. Typical compression factors achievable over the source text format are around 98%. Further, BiM simplifies access to partial descriptions by specific sub-tree pointer mechanisms, which allows to process XML files piece-wise. This can e.g. be used for efficient signaling, when only a part of the description changes dynamically over time. In principle, BiM allows more efficient parsing of descriptions than XML, in particular when only a part of a larger description shall be used by an application. As however XML is widely used and many application program interfaces exist which are based on XML (e.g. for database storage and query), it depends on the application whether it is more useful to employ the text-based (XML) or binary (BiM) description representations of MPEG-7.

The MPEG-7 standard (ISO/IEC 15938: *Multimedia Content Description Interface*) consists of the following parts:

- 15938-1: Part 1 – Systems. This defines mechanisms for efficient transport, storage and access of MPEG-7 descriptions.
- 15938-2: Part 2 – Description Definition Language. This specifies the grammar of the XML-based language, including MPEG-7 specific extensions, as used to define the syntax of MPEG-7 description tools.
- 15938-3: Part 3 – Visual. This defines description elements and tools which exclusively relate to visual information features.
- 15938-4: Part 4 – Audio. This defines description elements and tools which exclusively relate to audio information features.
- 15938-5: Part 5 – Multimedia Description Schemes. This specifies description elements and tools for generic feature descriptions and DS structures in general.
- 15938-6: Part 6 – Reference Software. This is a collection of software tools, integrating implementations of the normative parts of the standard, and also non-normative examples for feature extraction and description usage.
- 15938-7: Part 7 – Conformance. This describes guidelines and procedures for syntax-related conformance testing of MPEG-7 implementations.
- 15938-8: Part 8 – Extraction and Usage of Descriptions. This is a Technical Report which contains informative material about non-normative procedures related to extraction and usage of MPEG-7 descriptions.

17.6.2   Generic Multimedia Description Concepts

In principle, by using the DDL it is possible to define specific description structures beyond the normative elements of the MPEG-7 standard. Even though the *syntax* of such application-specific extensions can fully be described by DDL, it would not be possible to assign a *semantic meaning*, such that in fact no interoperability could be achieved except if the meaning of values is negotiated privately between systems. For visual and audio information types, the parts 3 and 4 of the standard define specific description elements which can generically be used for different applications, covering all relevant audiovisual features (see sec. 17.6.2 and 17.6.3). In addition, part 5 specifies a sufficiently rich set of general-purpose description elements (D and DS) which can be clustered into the following categories:

- *Basic elements*: This category contains a number of schema tools which allow formation, packaging and annotation of MPEG-7 descriptions, basic data types such as vectors and matrices, methods for linking and localization of media and primitive textual annotations e.g. relating to time, place, persons.
- *Content management*: Description of *creation information* such as title, author, date, classification by genre, purpose, recommended audience etc.; description of *media information* such as formats, compression methods and original sources; description of *usage information* such as usage rights and linkage to rights holders.
- *Content description*: Specific DSs for *structural description* (e.g. segment or region properties, association of low level features with certain locations in space and time), and *conceptual description* (e.g. semantic relationships of objects and events). These are key concepts directly related to signal analysis, for which a more detailed explanation will be given below.
- *Navigation and access*: This category shall facilitate browsing, quick preview and retrieval of multimedia content. It comprises description of *summaries* by different levels of detail, *partitions and decompositions* e.g. for multi-resolution (scalable) access and *variations* such as down-scaled or extended versions and availability in different modalities.
- *Content organization*: This category allows organization and modeling of content collections or description collections. In particular, collections can be described based on their common properties, or be mapped to analytic models. Furthermore, probability based classification models, state transition models (Hidden Markov) and statistical description models are provided. This allows to establish classified sets of audiovisual data and description of classification mechanisms.
- *User interaction*: Description of user preferences, which is usable for personalization of content access, presentation and consumption.

Content description structures. Typically, spatial sub-regions of an image, segments of a music piece or a video can be described by *hierarchical structures*. For this type of structural description of audiovisual signals, MPEG-7 specifies a generic class of *Segment DS* with specific realizations as

- *Video Segment DS*: This is intended to describe segments (to be specified by start time and duration) from a video sequence. A video segment may not be contiguous in time, it can also consist of overlapping or unconnected sub-segments. In addition to the position on the time axis, low-level visual features of categories color, texture and motion can be described. A video segment is always related to full-frame (rectangular) video.

- *Still Region DS*: This relates to one single image or a region from an image. The low-level visual features which can be attached to this DS are color, texture and shape. A still region can again consist of sub-regions, which may be overlapping, contiguous or unconnected.
- *Moving Region DS*: Unlike the Video Segment DS, this relates to smaller regions within video frames, such that it could e.g. describe the behavior of an object and its features over time. The additional visual feature which is needed (in addition to the features described for the Video Segment DS) is shape. A moving region segment can consist of sub-segments/-regions, which may be overlapping, contiguous or unconnected either temporally or spatially. Hence, the moving region is the most generic visual-related DSs, because it can be interpreted as an extension either of the Video DS or the Still Region DS.
- *Mosaic DS*: This is an extension of the Still Region DS, where the region is a mosaic generated from different images (either frames of a video sequence, or still images which were combined, cf. sec. 7.8). In addition to the low-level features combinable in the Still Region DS, this supports global motion information, such that the relationship between the mosaic and its source pictures can be described.
- *Audio Segment DS*: This allows specification of a segment from an audio signal, is conceptually very similar to the Video Segment DS, however related to audio features like spectral properties, timbre, loudness, tempo etc.
- *Audio-Visual Segment DS*: This is a combination which e.g. allows to describe video jointly with synchronized audio.

Description of relationships. Segments can be allocated in a hierarchical structure, where any segment can have one or more sub-segments. Such a construction can be used to convey increasingly detailed information. Any sub-segment will usually describe a part of the signal which is shorter in time or smaller in area than the parent segment. A parent segment can be decomposed into concatenated sub-segments, but sub-segments of the same parent segment can also be overlapping or unconnected. Sub-segments must not be of the same type as the parent segment. For example, a *Moving Region* sub-segment may describe the detailed behavior of a single object within a *Video* segment, or a *Still Region* sub-segment may describe a key-frame from a *Video* segment.

Even though the hierarchical structure, which can be described by a tree, is very efficient regarding access and parsing of descriptions or scalability of descriptions, it may not be fully appropriate for any case. For a more flexible approach, a *Segment Relation DS* is defined, which can establish links (relational graphs) between segments at arbitrary positions in a description representation.

For *conceptual* (semantic) descriptions, the focus is rather on events, objects and concepts, state, place and time in real or narrative worlds. This is supported by the *Semantic Base DS*, which can e.g. be instantiated as *Object DS*, *Event DS*, *Concept DS* etc. These purely semantic descriptions will typically be text based. *Description of abstractions* allows to combine similar instances, and replace specific objects, events or other semantic entities by corresponding *classes*. By a *Semantic Relation DS*, linking of abstract classes with related instantiations of *Segment DSs*, and with *Analytic Model DSs* describing statistical properties is also possible. This way, concrete mappings of primitive features (automatically extractable from multimedia signals) into semantic classification criteria can indeed be established.

17.6.3 Visual Descriptors

The group of *Visual* description tools specified in the part 3 of the MPEG-7 standard[1] covers all low-level features of visual information, and includes in addition mechanisms for visual-specific structural description, as well as higher-level information for specific visual signals such as human faces. The different categories of Visual Ds are described in the following sub-sections. In most cases, descriptor values are quantized into very compact binary representations of a low number of bits, often using non-uniform quantizers. This leads to very compact descriptions of visual features, whereas the quantizations have been carefully designed such that the descriptors would not lose expressiveness.

Basic structures. These are not descriptors in the original sense, but rather visual-specific low-level structural description tools, which allow to attach or combine other descriptors efficiently within a specific spatio-temporal context:

- *Grid Layout*: This defines a grid of $P \cdot Q$ equally-sized rectangular block partitions (P blocks horizontally, Q blocks vertically). It is either possible to use the same configuration of Ds for all blocks (such that just the descriptor values have to be conveyed for each block individually), or to attach individual Ds to each block. A 1 bit flag is used to express whether a descriptor value is conveyed for a specific block or not.
- *Time Series*: This provides a definition of equidistant or non-equidistant temporal positions, where for each time instance, the descriptor values for one or several Descriptors are attached. This can be interpreted as a regular or non-regular temporal sampling of dynamically changing description information.
- *2D/3D Multiple View*: This is a combination of 2D (image plane related) descriptors which provide information about the same scene or an object that was shot from different view angles.
- *Spatial 2D Coordinates*: A generic definition to establish two-dimensional reference coordinate systems which can be used in combination with Visual Ds. Coordinate definitions are made either by pixel units, by pixel units nor-

---

[1] The subsequent list already includes Visual description elements which are defined in the first amendment to 15938-3, which is going to be finalized in 2003.

malized to image width/height or by length units. In case of *local* coordinate systems, the reference is fixed, while in case of *integrated* coordinate systems, reference can be made by mapping into a coordinate system of a different picture (e.g. another frame in a video).

- *Temporal Interpolation*: Definition of a linear or quadratic interpolation function, by which descriptor values at intermediate temporal positions can be approximated; this allows a compact and accurate representation of the temporal change in a description from a low number of time instances; this is e.g. suitable to describe motion trajectories in combination with a set of parametric descriptions.
- *GoF/GoP Feature*: This relates to application of a descriptor to a group of frames from a video sequence or a group of pictures, where the descriptor values are combined into one instantiation describing the entire group. Methods of combination are average, median or intersection (minimum value) of the respective descriptor values from the single pictures.

Color. Color features are supported by a variety of descriptors, which are suitable to describe localized or global color features. The first sub-category are not descriptors in the strict sense, but rather supporting tools for color description:

- *Color Space*: Specification of the color space, supporting *RGB*, YC_bC_r, *HSV*, *HMMD* (a mapping derived from HSV), linear matrix transform, optionally also a white reference (cf. sec. 7.1.1).
- *Color Quantization*: Definition of quantization methods in the color space, resulting in a discrete set of indices for color values. Methods of uniform and non-uniform quantization are supported.
- *Illumination Invariant Color*: This is a structure which wraps the *Dominant Color*, *Scalable Color*, *Color Structure* and *Color Layout* Descriptors as given below, with the property that the combined information is less sensitive against illumination changes.

The different types of color descriptors are:

- *Dominant Color*: Definition of up to eight dominant colors by mean and variance (cf. (7.15)). Extraction of this descriptor is typically performed by cluster analysis.
- *Scalable Color*: This D is based on a histogram computed from an HSV color space representation quantized into 256 color values. The histogram values are nonlinearly mapped, and the histogram is then transformed by a Haar basis applied to pairs of neighbored colors in a hierarchical (wavelet-like) tree, which allows to represent the histogram by a minimum of 16 and a maximum of 256 coefficients dyadically. The coefficients are encoded by an embedded quantizer, such that scalability can be applied both by number of coefficients and accuracy of the coefficient representation (see also Fig. 7.3). If the full resolution is used, the descriptor is similar expressive as an original histogram; by using the coarser resolutions, much faster feature comparison can be

realized. The number of bits necessary for the representation of the histogram can range between 22 and 832.

- *Color Structure*: This descriptor is based on a color histogram computed from a representation in the HMMD color space, using a 7x7 structure element for computation (see (7.17) and Fig. 7.5). Instead of pixel-based statistics, this rather considers occurrence of color values within pixel neighborhoods, which relates semantically to the 'local purity' of colors. The accuracy of the descriptor can be set by quantization of the HMMD space into 32-256 bins in dyadic steps. Eight bits are used for the representation of (non-uniformly quantized) histogram values.

- *Color Layout*: This descriptor expresses the local distribution of dominant colors within a regular grid of 64 rectangular blocks. The matrix of dominant color values extracted from the single blocks is compressed by a 2D DCT (cf. Fig. 7.4). By retaining only a small number of coarsely quantized coefficients, a minimum of 64 bits are necessary to express the color layout of an image. The precision can be increased by more accurate representation of the coefficients, which may require up to 971 bits.

- *Color Temperature*: This is a very compact descriptor requiring only 8 bits for the representation of the color feature, expressed in categories such as 'hot', 'warm', 'moderate' and 'cool'. This is e.g. useful for quick browsing and pre-selection from huge databases.

- *GoF/GoP Color*: This is a combination of the GoF/GoP feature as described above with the Scalable Color D.

Texture. This group contains descriptors which can express properties of homogeneous textures or edge-like structures:

- *Homogeneous Texture*: Texture analysis is performed by directional Gabor wavelet basis functions. The feature description is based on absolute moments of first and second order (7.39). The Homogeneous Texture D supports a maximum of 30 frequency channels , which are arranged by 6 angular orientations and 5 scaling resolutions (cf. Fig. 7.13).

- *Texture Browsing*: This D is defined at a rather semantic level, adapted to texture perception by humans. The Descriptor contains 3 parameters of regularity, angular orientation and detail content. Conceptually, to determine these features from real-world material, it is recommended to gather the information by using the same Gabor basis functions as in the homogeneous texture D. The descriptor is very compact, only 12 bits are needed for representation.

- *Edge Histogram*: This descriptor expresses occurrences of the four main orientations of edges (horizontal, vertical and the two 45 degree orientations) and 'non-directional' edges (cf. Fig. 7.21). The descriptor optionally includes a simple localization mechanism similar to the grid layout.

Shape. This group contains descriptors for two-dimensional (contour, binary) and three-dimensional (surface) shapes. All these descriptors have the property to be rotation and scale invariant, and are also quite insensitive against small deformations of shape:

- *Contour Shape*: This descriptor is applicable to shapes which are described by a unique contour line. It is a combination of a few global shape parameters such as circularity (7.98), eccentricity (7.135) both for the original and convex-smoothed contour. The remaining part is based on the *Curvature Scale Space* (CSS) representation, where the number of relevant CSS peaks and the CSS peak amplitudes/positions are expressed (cf. Fig. 7.31).
- *Region Shape*: This descriptor is applicable to shapes where the contour description is not efficient, e.g. binary shapes containing holes. For extraction of the descriptor values, the binary shape is centered within a square window of normalized size. Then, a transform using the ART basis functions (cf. Fig. 7.41) is applied, where the absolute values of transform coefficients establish the descriptor values. A total of 35 coefficients is used, where however the descriptor also keeps sufficient expressiveness by a lower number of coefficients.
- *Shape Variation*: This is an extension related to the Region Shape D, which allows to describe shape variations within a collection of binary shapes (e.g. temporal variations within a video sequence) in terms of two variation maps and standard deviations of ART coefficients. The variation maps are constructed by accumulating the binary shapes $b(m,n)$ of the collection (static map) and accumulating the inverses $1-b(m,n)$ (dynamic map), then computing ART coefficients from both maps. In addition, the standard deviations of the ART coefficients (as extracted from each binary shape of the collection individually) are computed.
- *Shape 3D*: This is a histogram related to curvatures of a 3D surface. The strength of the curvature fluctuations, denoted as *shape index*, can be analyzed by comparing the angles between orientations of neighbored triangular patches in the wireframe representation of a shape. The descriptor is scalable by number of histogram bins and by precision of histogram values.

Motion. This group of Descriptors supports different types of motion which can occur within video sequences:

- *Camera Motion*: Qualitative and quantitative description of the extrinsic motion of a camera, including translations and rotations about the 3 axes of the coordinate system. These are the translational movements *track* (left-right), *boom* (up-down) and *dolly* (forward/backward) and the rotational movements *pan* (left-right), *tilt* (up-down) and *roll* (around the Z axis). In addition, zoom is described as the only type related to intrinsic camera parameters. The quantitative description is relatively coarse, it is not possible to use this D for accurate camera calibration.
- *Motion Trajectory*: Definition of a motion trajectory from reference points (each point expressed by the time and spatial coordinates), the respective reference coordinate system (including the spatial 2D coordinate basic function)

and an interpolation function (including the temporal interpolation basic function). By universal use of the coordinate reference, trajectories can be defined between video frames or within the 3D world. It can further be signaled whether the camera follows an object. The motion trajectory is independent of spatial resolution of frames and frame rates.

- *Parametric Motion*: Description of values related to different parametric motion models is defined, ranging from 2 parameters (translation), 4 parameters (translation, rotation and scaling), 6 (affine), 8 (perspective) up to 12 (parabolic) parameters (cf. sec. 7.4.4). Optionally, a reference coordinate system can be described by including the basic function for spatial 2D coordinates.
- *Motion Activity*: By analyzing the motion vector fields (where e.g. block-based motion vectors can be used as an approximation of the motion vector field properties), the general level of motion activity in a video sequence can be determined. Descriptor values express the intensity of motion (by 6 different levels), and the dominant direction of motion (by 8 directions). The spatial distribution of different motion properties is expressed by analysis of the run lengths over similar motion vectors (where a long run means that a large area is subject to a similar motion). Optionally, a better localization can be achieved by a combination with a simplified grid layout, which allows to describe global and local motion properties separately. The temporal variation within a sequence of frames can also optionally be described by a histogram of motion intensity values.

Localization. This group of features is intended to describe spatio-temporal localization of content:

- *Region Locator*: This describes the position of a region within a *single image* either by a bounding box or by a polygon approximation.
- *Spatio-Temporal Locator*: This describes the position of a region spatio-temporally, which is done either by combining a motion trajectory with a parametric motion model, or by defining the trajectory of vertex coordinates relating to a bounding box, a polygon or an ellipsoid. The Spatio-Temporal Locator can describe both spatially connected and non-contiguous regions.

Face description. This comprises descriptor structures which are intended for the purpose of face recognition, e.g. to retrieve face images which match a query face image:

- *Face Recognition*: This is based on projection of normalized and geometrically adjusted face images into a coefficient representation, using a set of basis images (similar to eigenfaces from a mean face, see sec. 7.9; the basis images are fixed and fully specified in the standard). The description then consists of 48 coefficients, each quantized into 5 bits.
- *Advanced Face Recognition*: This descriptor is more robust against variations in pose and illumination conditions. It is based on Linear Discriminant Analysis (LDA), which seeks to find a linear transformation by maximizing the be-

tween-class variance and minimizing the within-class variance (9.48). LDA is also applicable to *principal components* (e.g. eigenvalues or other transform expansions), then denoted as PCLDA. To extract the Advanced Face Recognition D, 2D Fourier coefficients are computed globally and in a multi-block mode[1] from normalized and geometrically adjusted face images. PCLDA is then applied separately to the Fourier absolute values of the global transform and a merged scan of the multi-block representation. Finally, both results are normalized, and the joint vector of values is again processed by LDA. The bases for both PCLDA transforms and for the LDA transform have fixed values which are fully specified in the standard. The quantized result of the LDA projection is denoted as *Fourier feature*, for which the number of coefficients can be variable.

17.6.4   Audio Descriptors

The group of *Audio* description tools specified in the part 4 of the MPEG-7 standard[2] covers all low-level features of audio information. It defines in addition mechanisms for audio-specific structural description and higher-level information related to audio semantics. The different categories of audio description elements are listed in the following sub-sections; most of the basic concepts were already introduced in sec. 7.10.

Structures. The temporal evolution of low-level audio features can either be described by sampling (extracting descriptions) at regular intervals, or by using segments which should demark periods of similar sound properties. Both methods are applicable to scalar and vector types of descriptor values. The result is a series of descriptor instantiations. In addition, a *scalable series* is defined, which allows to progressively down-sample the series of values and use a variable degree of temporal granularity, which can directly be weighted versus the compactness of the description. The resulting *scale tree* can be augmented by summary values such as minima, maxima, mean, variances, covariances of descriptor values hidden in the next-finer levels of the series.

Basic. This group comprises two descriptors, which are the *Waveform Envelope D* (expressing minimum and maximum value of amplitudes) and the *Audio Power D* which expresses the power, both descriptors relating to time intervals with a default duration of 10 ms.

---

[1] The multi-block extraction is performed using a so-called *holistic image*, which is a cropped center part of the face. In addition to the holistic image itself, partitions into 4 and 16 sub-blocks are made, where Fourier coefficients are extracted from all three representations.

[2] The subsequent list already includes Audio description elements which are defined in the first amendment to 15938-4, which is going to be finalized in 2003.

Basic spectral. This group of descriptors allows to express short-term spectral properties of audio signals, which relate to a finite spectral band. The default bandwidth uses a lower band edge of 62.5 *Hz* and a higher band edge of 16 *kHz*, logarithmically spaced. The following descriptors are defined:

- *Audio Spectrum Envelope*: This describes the power spectrum of an audio signal. The number of bands between lower and upper edges is definable by a *resolution* parameter, whereas all energy residing beyond the edges is expressed by one coefficient each.
- *Audio Spectrum Centroid*: This describes the center of gravity of the log-frequency power spectrum as defined in (7.221).
- *Audio Spectrum Spread*: The spectrum spread is defined as the root mean square deviation of the logarithmic-frequency power spectrum from the centroid, as given by (7.222).
- *Audio Spectrum Flatness*: This describes the flatness of the spectrum within a given number of frequency bands, as specified in (7.224).

Spectral Basis. The spectral basis descriptors relate to low-dimensional projections of spectrograms, which are computed by Independent Component Analysis (ICA, see sec. 9.1.2). The *Audio Spectrum Basis D* contains the basis functions as generated by SVD analysis, while the *Audio Spectrum Projection D* represents the low-dimensional features (or coefficients) which are generated by applying (9.7). In fact, the procedure described below (9.7) is used for generation of the most compact description. The procedure is applied to a time series of *Audio Spectrum Envelope D*, where the amplitude values are first logarithmically scaled and normalized to unity.

Signal parameters. These descriptors are applicable mainly to periodic and quasi-periodic signals. The *Audio Harmonicity D* expresses values of the harmonic ratio (7.226) and the upper limit of harmonicity (7.228). The *Audio Fundamental Frequency D* describes the fundamental frequency or pitch of a harmonic or near-harmonic signal.

Temporal timbre descriptors. This group contains the *Log Attack Time D* with extraction method (7.230) and the *Temporal Centroid D* with extraction according to (7.231).

Spectral timbre descriptors. Five spectral timbre descriptors are defined, where the first four operate on regularly-spaced spectral components, and are primarily applicable to harmonic sounds:

- *Harmonic Spectral Centroid D*, relating to the extraction method of (7.233);
- *Harmonic Spectral Deviation D*, relating to the extraction method in (7.234);
- *Harmonic Spectral Spread D*, relating to the extraction method in (7.236);
- *Harmonic Spectral Variation D*, relating to the extraction method in (7.237);

- *Spectral Centroid D*, relating to the extraction method of (7.238).

Silence. The *Silence D* is a simple binary (on/off) descriptor, with a possible extraction method as shown in Fig. 7.80.

High-level description tools. These are pre-defined combinations of low-level descriptors, typically optimized for specific application domains:

- *Audio signature*: The Audio Signature D allows identification of audio content, and is robust against possible modifications such as lossy coding or transcoding, filtering, transmission noise or acoustic transfer by loudspeakers and microphones. Matching must be performed against a database of descriptor values extracted from known music pieces, which e.g. allows to locate remote metadata related to a music piece. Highly robust signatures of this type are based on the *Audio Spectral Flatness D*, which is provided in a scalable time series, expressing mean and variance of descriptor values. Identification is typically possible with high reliability when only some seconds of a song are played.

- *Musical instrument timbre description*: Two different combinations are made from low-level timbre descriptors for specific classes of instruments. The *Harmonic Instrument Timbre D*, which is best suitable for sustained harmonic sounds, combines the *Log Attack Time D* with the four harmonic spectral timbre descriptors. The *Percussive Instrument Timbre Descriptor* combines both temporal timbre descriptors with the *Spectral Centroid D*.

- *Sound recognition and indexing*: This is the main application of the spectral basis descriptor group (ICA based methods), but other descriptors may be used as well. The *Sound Model DS* is based on a continuous Hidden Markov Model (cf. sec. 9.4.6), which signifies transition probabilities related to the temporal changes of descriptor values for a given sound. The basis functions related to the *Audio Spectrum Basis D* are stored along with the model, and are used to compute the ICA projection of an audio segment which shall be matched against the model. The *Sound Classification Model DS* is a collection of different sound models which can be used for automatic classification of audio segments, employing appropriately trained sound models; applications include genre classification, instrument and voice recognition. Similarity comparison can be based on the most likely state path using the Viterbi algorithm (*Sound Model State Path D*), or on the relative frequency by which the respective states are selected with the given audio segment (*Sound Model State Histogram D*).

- *Spoken content description*: These description tools allow description of spoken words, similar to the representation at the output of automatic speech recognition systems (before the final classification step). This way, it is possible to systematically track inaccuracies of recognition systems, and retain intermediate levels of the recognition process for later verification. The first functional unit is the *Spoken Content Lattice DS*, representing a series of nodes (supplemented with timing information) and links which are related to phonemes or words. The *Spoken Content Header DS* provides additional information such as speaker info, confusion count, word and phoneme lexicon. As speech is an important element of multimedia applications, reliable recognition and combination of recognition with other modalities (e.g. search for spoken keywords in a video or movie) are highly relevant in MPEG-7 applications.

- *Melody description*: The two main elements for melody description are the *Melody Contour DS* and the *Melody Sequence DS*, where the first is a compact representation of a melody extracted from a segment, expressing the change of melody by 5 grades of ascent/descent; rhythmic information such as beats are also supported. The *Melody* Sequence is a more verbose representation based on very precise interval (change in tone height) and timing information related to single notes.

- *Audio signal quality description*: This is a set of descriptors related to different aspects of audio quality, which can e.g. be used to decide which of different versions of the same piece of music would be preferable. Parameters which can be expressed are e.g. background noise level, bandwidth, correlation between different audio channels, information about transmission channels (including storage medium).

- *Audio tempo description*: Besides the accurate descriptions of tone lengths and variations which are available in the melody description category, a more global description of the tempo of a music piece is expressed in BPM (beats per minute).

17.7 Multimedia Framework MPEG-21

The vision for MPEG-21 is to define an open *multimedia framework* that will enable transparent and augmented use of multimedia resources across a wide range of networks and devices. This shall cover the entire chain that is traversed during the lifecycle of multimedia content, encompassing creation, production, manipulation, management, delivery, trading and consumption. However, the notion of content is too simplistic to express the richness of all benefits that are provided by multimedia applications. One central focus of MPEG-21 is the definition of *Digital Items* (DI). A DI is a hierarchical container of resources, metadata and other sub-ordinate DIs. It can best be defined as the digital representation of a 'work', which is the resource being managed, described, traded, consumed etc. in a multimedia application. For seamless interoperability, it will be necessary to identify all necessary resources which will make the DI operational. In principle, it would be consistent to subsume all these resources under the DI itself, which will however not consistently be possible in all cases. From this point of view, it is necessary to define mechanisms for *Digital Item Adaptation* (DIA) which would make a DI run under varying resource environments.

The second important paradigm in MPEG-21 is user-centricity. In general, users create, modify, protect, adapt, trade and consume digital items. All participants of the content value chain are regarded as users (regardless if they are operators, service providers, content creators or end-users), while they can simultaneously have different roles, interacting with each other or with the DIs that are the purpose of their interaction. This evolving framework addresses all possible actions on DIs including Digital Rights Management (DRM), Digital Item Adaptation (DIA) and Digital Item Processing (DIP).

The current status of technology provided in MPEG-21 allows static adaptation of Digital Items with respect to resource constraints and user profiles. The standard is entitled *ISO/IEC 21000 "Multimedia Framework (MPEG-21)"*, and is presently consisting of the following sub-parts[1]:

- *TR 21000-1: Vision, Technologies and Strategy.* This (non-normative) Technical Report develops a vision about the framework, which shall enable transparent usage of augmented multimedia resources across different networks and devices, serving the needs of all potential users.
- *21000-2: Digital Item Declaration* (DID). This part specifies a set of abstract terms and concepts to establish a useful model for definition of DIs. The model shall be as general and flexible as possible, while providing mechanisms to enable higher-level functionality. Declaration elements are containers, items, components, anchors, descriptors, conditions, choices, selections, annotations, assertions, resources, fragments, statements and predicates. These are defined by normative syntax and semantics, which is represented in XML. Further, a normative XML schema definition is provided which comprises the entire grammar of the DID representation.
- *21000-3: Digital Item Identification* (DII). The scope of this part is unique identification of DIs and parts thereof, identification of intellectual property rights related to DIs, identification of Description Schemes (DS) and different types of DIs, usage of identifiers to link DIs with related information such as metadata. For this purpose, DIs and parts thereof are identified by encapsulation of *Uniform Resource Identifiers* (URI) into the Identification DS.
- *21000-4: Intellectual Property Management and Protection* (IPMP). Even though IPMP hooks for content protection are provided in other standards such as MPEG-2, MPEG-4 and MPEG-7, the non-proprietary interworking of such methods is still a major challenge, where a consistent solution shall be provided by MPEG-21, including e.g. standardized ways to retrieve IPMP tools from remote locations, exchange messages between IPMP tools and terminals etc.
- *21000-5: Rights Expression Language* (REL). The REL is seen as a machine-readable language by which it is possible to declare rights and permissions us-

---

[1] Parts 4, 7-12 still have draft status by the time this book is published.

ing the terms as defined in a dictionary (RDD). A central concept is *granting* of rights, which identifies the *principal* to whom the grant is issued, the *right* that the grant specifies, the *resource* to which the right applies and the *condition* that must be fulfilled before the right can be exercised. The REL is also defined using the XML Schema language.

- *21000-6: Rights Data Dictionary* (RDD). The RDD comprises a set of uniquely identified terms to support the REL. The specification relates to the structure of the dictionary and the methodology to create the dictionary. It also specifies provisions to track additions, amendments and deletions to terms and their attributes.

- *21000-7: Digital Item Adaptation* (DIA). The goal of DIA is to provide the user with interoperable transparent access to multimedia content by different types of terminals and networks without taking care of any installation, management and implementation issues. The specific scope targeted for DIA includes *user characteristics*, including preferences for specific media resources and their presentation, and mobility characteristics of the user; *terminal capabilities*, e.g. encoding and decoding capability, hardware / software limitations and communication (network interconnect) capabilities; *network characteristics*, which are mainly related to network QoS like bandwidth utilization, delay and error characteristics; *environment characteristics* such as location of a user and ambient conditions like noise and illumination levels; *resource adaptability tools*, which assist in the adaptation of resources including binary resources (e.g. bit streams) and metadata, including best trade-offs regarding complexity and QoS, and *session mobility tools*. DIA also defines a framework for description of bitstream structures.

- *21000-8: Reference Software*. This is the reference software implementation of the technical elements of MPEG-21.

- *21000-9: File Format*. The specification of a file format for storage of MPEG-21 information, including possible 'multi-purpose' combinations with media file formats.

- *21000-10: Digital Item Processing* (DIP). DIP turns a DI from a data structure to an object that can be processed for manipulation in an intended manner. A typical usage of this would be at the terminal side, where e.g. it would be necessary to crop an image when the display is too small; by DIP, the content provider could prescribe recommended methods of such manipulations.

- *21000-11: Evaluation Methods for Persistent Association*. The term *persistent association* categorizes techniques to uniquely tie and manage identification and description methods with the content. This includes carriage of identifiers in file or transport formats, embedding identifiers in the content by watermarks, and protection of identifiers against unauthorized removal and modification.

- *21000-12: Test Bed for MPEG-21 Resource Delivery*. This shall be defined as a non-normative part (Technical Report) of the standard, including a software package for a fully operational test bed which provides capabilities such as network and terminal adaptation of multimedia resource streams.

For media and metadata representation, MPEG-21 intends to use existing standards like MPEG-4 and MPEG-7. It is however one goal to identify missing links within such standards, and provide better solutions regarding the *entire chain* of multimedia communication systems, giving maximum benefit to all users at reasonable and fair cost. It is obvious that the attitude of MPEG-21 is highly related to the (nevertheless still simplistic) layout of a multimedia communication system as initially presented in Fig. 1.1b. Multimedia communication systems of the 21st century are diversified, based on multiple users interacting with multiple media resources, where the actual amount of information provided to the user highly relies on the performance of the systems which serve for this purpose. For example, an intelligent system which perfectly adapts to the user's needs with regard to the content that is offered and the desired form of presentation, can clearly save immense network bandwidth resources, as no unnecessary information must be transmitted. From this point of view, indeed a close interrelationship exists between the different variants of QoS initially introduced (Network, Perceptual and Semantic QoS).

Communication media have taken influence on societal aspects during the entire history of mankind. The underlying technology has always been a driving force in this development. The last century was dominated by centralized mass media, where most individuals are more or less degraded to be only recipients of normalized information (with certain freedom of selection, whatsoever). This paradigm starts to change with the advent of multimedia communications technology, and by the capabilities that are provided in modern telecommunication networks. A part of this process are media that are becoming more individualized, where more individuals can actively participate in the process of information generation and distribution. It is not fully clear yet what the impact on societal development will be: May this lead to a more democratic role of media, or will it rather lead to uncontrollable manipulations? *Seeing is believing*, but on the other hand digital technology enables simplified manipulations such that it becomes possible to see whatever one wishes. From this point of view, a critical attitude of human beings may be more important in the age of multimedia than it was ever before, which does not mean to reject, but rather to accept information suspiciously. In addition, technology is highly needed which helps to detect and track manipulations. If this is guaranteed, multimedia communications will turn out to be highly beneficial, opening up completely new and more effective ways of cooperating, learning and living. The destination of this development is hard to predict, but it is rather clear that multimedia systems are one important factor to change this world, and will make it appear substantially different within 20 years from now.

Appendices

A   Quality Measurement

A.1   Signal Quality

A.1.1   Objective Signal Quality Measurements

Objective measurement of signal quality is of high importance to compare algorithms designed for multimedia signal coding or signal enhancement. In many optimization approaches, the goal is the *minimization of the squared error*, which relates to the shortest Euclidean distance between signal amplitudes. If the difference $q(\mathbf{n})=x(\mathbf{n})-y(\mathbf{n})$ between the signal and the reconstruction is interpreted as a zero-mean noise process[1], the related quality measurement can be expressed as follows by the *Signal-to-Noise Ratio* (SNR), which is usually mapped into a logarithmic scale:

$$SNR\left[dB\right]=10\cdot\log_{10}\frac{E\left\{\left(x(\mathbf{n})-\mu_x\right)^2\right\}}{E\left\{q^2(\mathbf{n})\right\}}=20\cdot\log_{10}\frac{\sigma_x}{\sigma_q}. \tag{A.1}$$

(A.1) refers to the variance of the signal. SNR measurements relating to the signal power or energy (for finite signals) are less usual, because the result would be highly biased by the signal mean value. For audio and speech signals, the zero-mean property applies anyway. Image and video signals are strictly positive-valued[2] and bounded by a maximum amplitude. The mean of a given image or video frame can by itself be regarded to be a random variable. Deviations in mean could indeed play an important role, and would also lead to a non zero-mean sig-

---

[1] The zero-mean property of $q(\mathbf{n})$ is a reasonable assumption, unless $y(\mathbf{n})$ is systematically biased by an amplitude modification.

[2] In fact, this is not true for chrominance components such as C_b and C_r. In practice however, digital chrominance components are often represented as positive integer values by adding a bias compensating the maximum negative amplitude; e.g. in case of an 8-bit representation, a value of 128 is added. Chrominance components of a given image are typically not zero-mean signals, as they are biased by the individual color characteristics: In case of a 'reddish' image, the component C_r of (1.6) would clearly have a positive mean value.

nal $q(\mathbf{n})$. A metric which is not biased by the mean of individual images is the *Peak SNR* (PSNR)

$$PSNR[dB] = 10 \cdot \log_{10} \frac{A_{\max}^2 \sum_{m=0}^{M-1} \sum_{n=0}^{N-1} b(m,n)}{\sum_{m=0}^{M-1} \sum_{n=0}^{N-1} b(m,n) \cdot \left(x(m,n) - y(m,n) \right)^2} . \qquad (A.2)$$

This has established as the most widely used metric for image distortion comparison. The binary mask $b(m,n)$ relates to the definition in (7.90) and allows comparison of two images over an arbitrary-shaped area, smaller than the rectangular frame of size MxN. $A_{\max}$ is the maximum amplitude, or could also be interpreted as the maximum absolute difference that can occur between two distinct amplitude values from signals x and y[1]. Consequently, the PSNR can never be less than zero. The PSNR fits surprisingly well with subjective quality assessment, if results of algorithms are compared which introduce distortion of similar nature. Usually then, a PSNR difference of around 1 dB also indicates a perceptually noticeable quality difference. This must however be taken cautiously, as the following effects can occur:

– SNR measurements in general are highly sensitive against phase shifts. It must be noted that in case of images, linear phase shift are uncritical for high visual quality. For example, an observer would hardly notice a difference if an image is shifted by half a pixel using a high-quality interpolator, while the SNR criterion might indicate a poor quality.
– The PSNR does not take into account masking of noise by highly-detailed structures. Typically, images of low detail are judged to have a good quality only with much higher PSNR values than images of high detail. From this point of view, the measure as given in (A.1) would be favorable, but would be too simplistic to reflect the complex spatio-temporal masking relationships in image and video signals. For example, coding noise that varies over time is observed as very annoying, even though it may hardly be noticeable within a single frame.

Segmental SNR measurements. SNR criteria are average measures, which do not take into account *fluctuations* of quality. Often, a signal with only one highly distorted segment or region will be judged to be of bad quality, while the SNR computed over the entire signal may be extremely high. As possible workaround solutions, additional criteria such as SNR measurements over smaller segments, computation of minima, maxima or other criteria which express fluctuations have been proposed. The *segmental SNR*, which is the mean or expected value of SNR results from equal-size segments [JAYANT, NOLL 1984], penalizes fluctuations in SNR. This is however only true when the definition (A.1) is used, and when the fluctuations are mainly caused by variations of the instantaneous signal power. Fluctuations in

---

[1] E.g. $A_{\max}$=255 for signals originally represented by 8-bit PCM.

the instantaneous power of the error (while the power of the signal remains constant), will have the inverse effect, as will be shown below for the case of the Peak SNR averaged over a sequence of video frames.

Mean of PSNR measurements. When the PSNR over a video sequence of O frames is computed, this is often done by averaging the PSNR contributions of the individual frames, which seems to be similar to the computation of the segmental SNR. The result will be a criterion

$$
\begin{aligned}
PSNR_{mean} &= \frac{1}{O}\sum_{o=0}^{O-1}10\log_{10}\frac{A_{max}^2}{\frac{1}{MN}\sum_{m=0}^{M-1}\sum_{n=0}^{N-1}(x(m,n,o)-y(m,n,o))^2} \\[2ex]
&= 10\log_{10}\left[\prod_{o=0}^{O-1}\frac{A_{max}^2}{\frac{1}{MN}\sum_{m=0}^{M-1}\sum_{n=0}^{N-1}(x(m,n,o)-y(m,n,o))^2}\right]^{\frac{1}{O}} = -10\log_{10}\left[\prod_{o=0}^{O-1}\frac{\frac{1}{MN}\sum_{m=0}^{M-1}\sum_{n=0}^{N-1}(x(m,n,o)-y(m,n,o))^2}{A_{max}^2}\right]^{\frac{1}{O}} \\[2ex]
&\geq -10\log_{10}\left[\frac{1}{O}\sum_{o=0}^{O-1}\frac{\frac{1}{MN}\sum_{m=0}^{M-1}\sum_{n=0}^{N-1}(x(m,n,o)-y(m,n,o))^2}{A_{max}^2}\right] = 10\log_{10}\left[\frac{A_{max}^2}{\frac{1}{MNO}\sum_{o=0}^{O-1}\sum_{m=0}^{M-1}\sum_{n=0}^{N-1}(x(m,n,o)-y(m,n,o))^2}\right]
\end{aligned}
$$

$$\tag{A.3}$$

The inequality follows from the relationship between arithmetic and geometric mean; both will be equal only for the case of identical values being averaged. This means that by averaging the PSNR over a sequence of video frames clearly an algorithm which produces SNR fluctuations is favored. Similarly, the average PSNR computed over a set of still images may not be suitable to compare different algorithms, when one algorithm performs better only for a subset of the images under test.

Perceptually weighted criteria. As mentioned above, SNR measurements do not fully coincide with subjective impression, since they do not take into account perceptual masking functions. Development of better perceptually-weighted quality measurement criteria has been a topic of research over decades; for image quality evaluation criteria see e.g. [MIYAHARA 1988], [XU, HAUSKE 1994]. Typically, such criteria would include frequency weighting functions (cf. chapter 6), but due to the different types of possible signal degradations, it will further be necessary to classify errors into categories.

A.1.2 Subjective Assessment

When a reliable evaluation about the perceived signal quality shall be achieved, it is inevitable to perform formal subjective assessment tests involving a representative ensemble of test subjects. For some categories of signal compression algorithms, such as audio compression attempting to approach transparent (audible lossless) quality, this is the only viable assessment method, objective measures like SNR would not be useful at all.

In subjective assessment, an ensemble of test subjects are asked to give their opinions about the perceived quality. The answers are statistically evaluated, where criteria as mean value and variance of opinions are computed to determine the average results and their significance. Rejection of clear outliers is usually part of the process. The qualification of the test subjects (expert or non-expert) is also of high importance, but it also depends on the goal of the test how the subjects shall be selected.

Further, any methodology for quality assessment must define a quality scale. For the case where the goal of transparent (perceptually lossless) quality is investigated, a binary decision is sufficient. When the material under test is expected to exhibit quality losses, the five-grade quality/impairment attribute scale as defined in Table A.1 is widely used. By evaluating the answers gained from the ensemble of test subjects, it is possible to achieve results of *Mean Opinion Score* (MOS) assessments.

Table A.1. Score and related subjective assessment criteria used in MOS tests

Score	Quality	Impairment
5	*Excellent*	Imperceptible
4	*Good*	Perceptible, but not annoying
3	*Fair*	Slightly annoying
2	*Poor*	Annoying
1	*Bad*	Very annoying

Subjective tests must be reproducible, otherwise their value would be limited. Definition of assessment conditions for different types of multimedia signals has a long history. In this context, some methodologies defined by ITU-R recommendation BT.500-11 shall be shortly described; this is the latest from the BT.500 series of recommendations which actually started in 1974, entitled *'Methods for the Assessment of the Quality of Television Pictures'*[1]:

---

[1] ITU-R BT.500 covers the cases of SD resolution quality assessment. For HD resolution, the corresponding recommendation is ITU-R BT.710, which however defines exactly the same testing methodologies except for the setup of viewing and ambient conditions.

- Viewing conditions for assessment are defined differently for home viewing and lab environments; this includes ambient lighting conditions, preferred viewing distances and angles, properties of displays such as luminosity, resolution and aspect ratio;
- Quality and impairment scales are defined as in Table A.1;
- An ensemble of 15 non-expert subjects is specified as the minimum number to perform the assessment;
- Methodologies to evaluate the results of the assessment are described.

Depending on the material to be evaluated, different test methods are defined; three which are more frequently used in practice are:

- *Double Stimulus Continuous Quality Scale* (DSCQS): 10 second test sequences A and B are presented in a schema A-*break*-B-*break*-A-*break*-B-*break*. During the breaks of 3 seconds, a mid gray level is shown, the last break is longer when necessary for voting. Voting can start when the second round begins. One of A and B is the original source, the other is processed material, but the observer is unaware which is which. Attributes for scores are given by the *quality scale* of Table A.1.
- *Double Stimulus Impairment Scale* (DSIS): Viewers are shown multiple sequence pairs consisting of a *reference* and a *test* sequence. Subjects are asked to score the amount of impairment in the test sequence as compared to the reference by using the *impairment scale* of Table A.1. This kind of test is often used e.g. to evaluate coding algorithms in the lower range of data rates, where it would be clear that the impairment as compared to the original source would be severe. The reference can then be a standard coding algorithm with a reasonable level of quality, used as a common 'anchor' when different other algorithms shall be compared.
- *Single Stimulus Continuous Quality Evaluation* (SSCQE): This test methodology is usable on live material or can also be used to judge the quality of video streaming e.g. over the Internet, where rapid fluctuations of quality may occur. A common reference sequence is not explicitly available. A series of video sequences is presented once to the viewer, who evaluates the picture quality instantaneously (in real time) by moving a slider with position corresponding to the scores on the quality scale. Samples of the slider position are taken twice a second. This test can either be used to evaluate the quality fluctuations of one algorithm, or also to compare different algorithms when the slider positions of corresponding time points, where the same material was shown, are matched.

In all cases, before the actual test starts, the subjects are introduced to the test environment by a short training period; different material than shown later in this test must be used for this purpose. Further, the test sessions must be time limited to avoid fatigue behavior of the test subjects.

A.2 Classification Quality

To judge the quality of a classification algorithm or a retrieval method, it is necessary to rely on a pre-defined *ground truth*. This is determined by the classification result that a human subject normally would expect, such that it can be decided whether the algorithm works well. For cases where a classifier is optimized by a training set, the ground truth could usually be defined as being identical to the a priori class assignments made for the items in the training set (cf. sec. 9.4). However, for objective evaluation of classification performance, it is not reasonable to let the ground truth (which is used as a means to judge the algorithm) overlap with the training set that is used to optimize the algorithm. As however the pre-classified training set would be an ideal ground truth, a solution for objective evaluation is separation of a certain part of data from the training set, which are exclusively used as ground truth and not for classifier optimization. If e.g. 10% of the data belonging to each class from the training set are separated as ground truth, the algorithm can still be optimized by using the remaining 90%, but the quality of classification would be judged by the 10% which were not used for training. This procedure can then be iterated by separating another sub-set of 10% etc., and finally the quality is judged by averaging the single results.

Assume that a given class S_l has Q_l members in the training set. Then, misclassifications consist of two different cases. Firstly, not all of the original members of the class may be identified by the classification, which means they are missed. This leads to the *precision rate* which is numerically expressed as

$$\text{PREC} = \frac{N_{\text{correct}}}{N_{\text{correct}} + N_{\text{missed}}} . \tag{A.4}$$

Second, a certain number of elements may have been falsely assigned to the class, which originally belong to other classes or are just members of the population of items which are randomly present and do not uniquely belong to any class. This leads to the *recall rate*,

$$\text{REC} = \frac{N_{\text{correct}}}{N_{\text{correct}} + N_{\text{false}}} . \tag{A.5}$$

Observe that for any class, it is possible to tune an algorithm e.g. by shifting the decision boundaries in the feature space, such that either the precision or the recall rate is maximized. For example, in a security-relevant application, it may be desirable to detect all critical cases, while it is acceptable that some false alarms are made which must then be sorted out manually. Here, the algorithm should rather be tuned such that PREC is maximized. On the other hand, in an application where automatic admission control is made by a face detection algorithm, it may rather be preferable to reject (miss) an authorized person instead of leaving an unauthorized (false) person pass. In this case, the algorithm should be tuned for maximization of REC. In general, it can be argued that in a multiple-class decision problem one will penalize REC of other classes by artificially increasing PREC of one class and vice versa. In such cases, the best decision would be to keep both numbers of missed and false elements in balance, which will be realized when the harmonic average of precision and recall is maximized:

$$F = \frac{PREC \cdot REC}{\frac{1}{2}(PREC + REC)} \overset{!}{=} \max .$$
(A.6)

To judge the quality of retrieval applications, similar concepts can be applied. Here, *queries* are formulated, and the retrieval algorithm is expected to respond on each query by finding the related ground truth items which are e.g. contained in a database. The definition of precision and recall is however a bit more vague, as e.g. a user would be fully satisfied if the correct retrievals are interspersed with some other items that can easily be sorted out. Assume that a number $NG(q)$ of ground truth items are available for a query q. In the perfect case, the algorithm should find these items by the first $NG(q)$ ranks of the retrieval result. A user might however still be satisfied looking at $K(q){>}NG(q)$ ranks[1]. Let the number $N_{found}{\le}NG(q){\le}K(q)$ be the number of items belonging to the ground truth set of the query, which are actually found within the first $K(q)$ ranks of the retrieval. A measure which is indeed very similar to (A.4) (but less rigid), is the *retrieval rate* $RR(q)$ related to a query q. From this, the *average retrieval rate ARR* is determined over all NQ queries:

$$RR(q) = \frac{N_{found}}{NG(q)} \quad \Rightarrow ARR = \frac{1}{NQ} \sum_{q=1}^{NQ} RR(q) .$$
(A.7)

During the development of MPEG-7 Visual descriptors, a new quality measure for descriptor expressiveness has been defined, which is based on a *query by example* retrieval scenario. It is necessary to specify a database, a query set and the ground-truth data from the database corresponding to each of the queries. The ground-truth data is a set of images which are visually similar to a given query image, considering the properties expressed by a descriptor under test. Typically, at least 5000 images should be contained in the database, and the number of queries should be about 1% of the number of images in the database, such that sufficient statistical evidence of results is guaranteed. The database must contain a sufficient variety of items, such as images from stock photo galleries, screen shots of television programs and animations. The query and corresponding ground truth image definitions must be manually established through a process of visual inspection and cross verification by different independent people.

It is then necessary to weight the query results based on a unique numeric measure. Assume that $NG(q)$ is the size of the ground truth set for a query q, which can indeed be very different for individual queries, depending on how many

---

[1] Typically, $K(q){=}2NG(q)$ is a reasonable value.

ground truth images are found in the dataset. Further, the ground truth sets may or may not be constituted as clearly separated classes. Hence, the definition of a retrieval measure has to take into account the following points:

– Normalization of the measure irrespective of sizes of ground truth sets is necessary;
– Actual ranks shall be obtained from the retrieval, to favor algorithms that retrieve ground truth items in highest ranks;
– Consideration of the number of missed ground truth items by assignment of a 'penalty', where beyond a certain limit of ranks $K(q)$ any item is counted as missed.

In principle, it was a goal to find an integrated metric which reflects both the precision and the recall, but is more 'soft' by not unconditionally postulating that the ground truth items are found in the first $NG(q)$ ranks. The following solution was adopted in the MPEG-7 core experiment process. Consider a query q ; assume that as a result of the retrieval, the kth ground truth image for this query q is found at $RANK(k)$. Further, $K(q) \geq NG(q)$ is defined to specify the 'relevant ranks', which are the ranks that would still count as feasible in subjective evaluation of the retrieval. For relatively large $NG(q)$ (20-25 items), subjects would judge the retrieval results as still useful if items are found with ranks around $2NG(q)$, while for smaller ground truth sets, even more tolerance would be allowed. The penalty rank assigned to each missed item should be $\geq K(q)$, but a penalty just equaling $K(q)$ would put retrievals with too many misses in advantage. A good compromise derived from this reasoning was found by defining a $RANK^*(k)$ as:

$$RANK*(k) = \begin{cases} RANK(k) & \text{if } RANK(k) \leq K(q) \\ 1.25 \cdot K(q) & \text{if } RANK(k) > K(q) \end{cases}$$

$$K(q) = \min\{4 \cdot NG(q), 2 \cdot \max[NG(q) \;\forall\; q]\}$$

(A.8)

From (A.8), the *Average Rank* for query q is

$$AVRANK(q) = \frac{1}{NG(q)} \sum_{k=1}^{NG(q)} RANK*(k)$$

(A.9)

However, with ground truth sets of different size (actually, NG varied between 3 and 32 in some of the MPEG-7 experiments), the $AVRANK$ counted from ground truth sets with small and large $NG(q)$ values would largely differ. To eliminate influences of different $NG(q)$, the *Modified Retrieval Rank*

$$MRR(q) = AVRANK(q) - 0.5 \cdot [1 + NG(q)]$$

(A.10)

is defined, which is always larger than 0, but with upper margin still dependent on NG. This finally leads to the *Normalized Modified Retrieval Rank*

$$NMRR(q) = \frac{MRR(q)}{1.25 \cdot K - 0.5 \cdot [1 + NG(q)]} \qquad (A.11)$$

which can only take values between 0 (indicating that the whole ground truth found) and 1 (indicating that nothing was found), irrespective of the size of NG in the query. From (A.11), it is now straightforward to compute the *Average Normalized Modified Retrieval Rank* (*ANMRR*), giving just one number to indicate the retrieval quality over all queries:

$$ANMRR = \frac{1}{NQ} \sum_{q=1}^{NQ} NMRR(q), \qquad (A.12)$$

The *ANMRR* metric approximately coincides with the results of subjective evaluation about retrieval accuracy of search engines. Interestingly enough, it was found in MPEG-7 Core Experiments that a clear interrelationship exists between the compactness of descriptors which can be defined for different precision levels and the retrieval accuracy; the compactness is counted in numbers of bits needed for the representation. This allows to use *rate-accuracy* plots expressing the quality of content descriptions, which can play a similar role as the rate-distortion plots widely used in image and video coding.

B Vector and Matrix Algebra

Vector and matrix operations are frequently used in multimedia signal processing, as this allows to express groups of signal samples in a very efficient way (e.g. K samples from an audio signal form a vector of length K, an image consisting of M columns and N rows is a matrix of size MxN). Linear mathematical operations can easily be expressed by vector and matrix algebra, as used in many parts of this book for the purpose of a simplified notation. The underlying conventions are summarized in this section.

A vector is a one-dimensional structure of K scalar values. We exclusively use column vectors, i.e. vertical structures. A KxL matrix is a two-dimensional structure with L rows and K columns. Higher-dimensional structures are called *tensors*. Matrices and vectors are indicated by bold letters, where in principle the denomination of value types is retained (see Appendix C). For example, $\mathbf{x}$ and $\mathbf{X}$ are a vector and a matrix consisting of samples x. The most important rules used in vector and matrix operations are summarized in the remaining part of this section.

— The transpose of a vector is

$$\mathbf{a}^{\mathrm{T}} = \begin{bmatrix} a_1 \\ \vdots \\ a_K \end{bmatrix}^{\mathrm{T}} = \begin{bmatrix} a_1 & \cdots & a_K \end{bmatrix}. \tag{B.1}$$

— The transpose of a matrix is performed by exchange of rows and columns,

$$\mathbf{A}^{\mathrm{T}} = \begin{bmatrix} a_{11} & \cdots & a_{1K} \\ \vdots & \ddots & \vdots \\ a_{L1} & \cdots & a_{LK} \end{bmatrix}^{\mathrm{T}} = \begin{bmatrix} a_{11} & \cdots & a_{L1} \\ \vdots & \ddots & \vdots \\ a_{1K} & \cdots & a_{LK} \end{bmatrix}. \tag{B.2}$$

— The inner product of two vectors of same length K is a scalar value

$$\mathbf{a}^{\mathrm{T}} \cdot \mathbf{b} = \begin{bmatrix} a_1 & \cdots & a_K \end{bmatrix} \cdot \begin{bmatrix} b_1 \\ \vdots \\ b_K \end{bmatrix} = a_1 \cdot b_1 + a_2 \cdot b_2 + ... + a_K \cdot b_K. \tag{B.3}$$

— The outer product of two vectors of length K is a KxK matrix

$$\mathbf{a} \cdot \mathbf{b}^T = \begin{bmatrix} a_1 \\ \vdots \\ a_K \end{bmatrix} \cdot \begin{bmatrix} b_1 & \cdots & b_K \end{bmatrix} = \begin{bmatrix} a_1 b_1 & a_1 b_2 & \cdots & a_1 b_K \\ a_2 b_1 & a_2 b_2 & & \vdots \\ \vdots & & \ddots & \\ a_K b_1 & \cdots & & a_K b_K \end{bmatrix}. \tag{B.4}$$

– The product of a vector of length K and a KxL matrix is a vector of length L

$$\mathbf{A} \cdot \mathbf{x} = \begin{bmatrix} a_{11} & \cdots & a_{1K} \\ \vdots & \ddots & \vdots \\ a_{L1} & \cdots & a_{LK} \end{bmatrix} \cdot \begin{bmatrix} x_1 \\ \vdots \\ x_K \end{bmatrix} = \begin{bmatrix} a_{11}x_1 + \ldots + a_{1K}x_K \\ \vdots \\ a_{L1}x_1 + \ldots + a_{LK}x_K \end{bmatrix}. \tag{B.5}$$

– The inner matrix product (first matrix is KxL, second matrix MxK) is an MxL matrix

$$\begin{aligned} \mathbf{A} \cdot \mathbf{B} &= \begin{bmatrix} a_{11} & \cdots & a_{1K} \\ \vdots & \ddots & \vdots \\ a_{L1} & \cdots & a_{LK} \end{bmatrix} \cdot \begin{bmatrix} b_{11} & \cdots & b_{1M} \\ \vdots & \ddots & \vdots \\ b_{K1} & \cdots & b_{KM} \end{bmatrix} \\ &= \begin{bmatrix} a_{11}b_{11} + \ldots + a_{1K}b_{K1} & \cdots & a_{11}b_{1M} + \ldots + a_{1K}b_{KM} \\ \vdots & \ddots & \vdots \\ a_{L1}b_{11} + \ldots + a_{LK}b_{K1} & \cdots & a_{L1}b_{1M} + \ldots + a_{LK}b_{KM} \end{bmatrix}. \end{aligned} \tag{B.6}$$

If two (non-quadratic) matrices of equal size are multiplied, one of them must be transposed. Then, the following relationship holds:

$$\mathbf{A}^T \mathbf{B} = \begin{bmatrix} \mathbf{B}^T \mathbf{A} \end{bmatrix}^T. \tag{B.7}$$

– The outer (Kronecker) product of two matrices (sizes KxL and MxN) is performed such that each element of the first matrix is multiplied by each element of the second. The result is a matrix of size KMxLN, which can be partitioned into KL sub-matrices, each of size MxN:

$$\begin{aligned} \mathbf{A} \times \mathbf{B} &= \begin{bmatrix} a_{11} & \cdots & a_{1K} \\ \vdots & \ddots & \vdots \\ a_{L1} & \cdots & a_{LK} \end{bmatrix} \times \begin{bmatrix} b_{11} & \cdots & b_{1M} \\ \vdots & \ddots & \vdots \\ b_{N1} & \cdots & b_{NM} \end{bmatrix} \\ &= \begin{bmatrix} a_{11}b_{11} & \cdots & a_{11}b_{1M} & \cdots & a_{1K}b_{1M} \\ \vdots & \ddots & \vdots & & \vdots \\ a_{11}b_{N1} & \cdots & a_{11}b_{NM} & \cdots & a_{1K}b_{NM} \\ \vdots & & \vdots & & \vdots \\ a_{L1}b_{N1} & \cdots & a_{L1}b_{NM} & \cdots & a_{LK}b_{NM} \end{bmatrix}. \end{aligned} \tag{B.8}$$

– The determinant of a quadratic matrix of size $K \mathrm{x} K$ is the sum over $K!$ possible permutations $(\alpha, \beta, .., \omega)$ of numbers $(1, 2, .., K)$, where k is the number of inversions within a permutation (sequence $a_{1,\alpha} a_{1,\beta}$, $\alpha > \beta$):

$$D = |a_{ij}| = \begin{vmatrix} a_{11} & a_{12} & \cdots & a_{1K} \\ a_{21} & a_{22} & & a_{2K} \\ \vdots & \vdots & \ddots & \vdots \\ a_{K1} & a_{K2} & \cdots & a_{KK} \end{vmatrix} = \sum_{(\alpha,\beta,..,\omega)} (-1)^k a_{1\alpha} a_{2\beta}...a_{K\omega} . \tag{B.9}$$

This can better be interpreted as computing the sum of 'products over diagonals' within the periodically extended matrix. All products of diagonals oriented in parallel with the primary (trace) axis contribute by positive sign, all secondary-oriented diagonals (top right to bottom left) by negative sign; e.g. for cases $K=2$ and $K=3$:

$$|\mathbf{A}| = \begin{vmatrix} a_{11} & a_{12} \\ a_{21} & a_{22} \end{vmatrix} = a_{11}a_{22} - a_{12}a_{21}$$

$$|\mathbf{A}| = \begin{vmatrix} a_{11} & a_{12} & a_{13} \\ a_{21} & a_{22} & a_{23} \\ a_{31} & a_{32} & a_{33} \end{vmatrix} = a_{11}a_{22}a_{33} + a_{12}a_{23}a_{31} + a_{13}a_{21}a_{32}$$

$$-a_{11}a_{23}a_{32} - a_{12}a_{21}a_{33} - a_{13}a_{22}a_{31}. \tag{B.10}$$

– Inverting a matrix, $\mathbf{A}^{-1}$ is useful in many places, e.g. to solve linear equation systems. Additional conditions are $[\mathbf{A}^{-1}]^{-1}=\mathbf{A}$ and $\mathbf{A}^{-1}\mathbf{A}=\mathbf{A}\mathbf{A}^{-1}=\mathbf{I}$, i.e. the matrix multiplied by its inverse results in the *identity matrix*

$$\mathbf{I} = \begin{bmatrix} 1 & 0 & \cdots & & 0 \\ 0 & 1 & 0 & & \vdots \\ \vdots & 0 & 1 & \ddots & \\ & & \ddots & \ddots & 0 \\ 0 & \cdots & & 0 & 1 \end{bmatrix}. \tag{B.11}$$

If the determinant of the matrix and all determinants of sub-matrices are unequal zero, the matrix is invertible, otherwise it is called to be singular. The inversion of matrices of sizes 2x2 and 3x3 is performed as follows:

$$\mathbf{A}^{-1} = \begin{bmatrix} a_{11} & a_{12} \\ a_{21} & a_{22} \end{bmatrix}^{-1} = \frac{1}{|\mathbf{A}|} \cdot \begin{bmatrix} a_{22} & -a_{12} \\ -a_{21} & a_{11} \end{bmatrix} \tag{B.12}$$

$$
\mathbf{A}^{-1} = \begin{bmatrix} a_{11} & a_{12} & a_{13} \\ a_{21} & a_{22} & a_{23} \\ a_{31} & a_{32} & a_{33} \end{bmatrix}^{-1}
$$

$$
= \frac{1}{|\mathbf{A}|} \cdot \begin{bmatrix} a_{22}a_{33} - a_{23}a_{32} & a_{13}a_{32} - a_{12}a_{33} & a_{12}a_{23} - a_{13}a_{22} \\ a_{31}a_{23} - a_{21}a_{33} & a_{11}a_{33} - a_{13}a_{31} & a_{13}a_{21} - a_{11}a_{23} \\ a_{21}a_{32} - a_{31}a_{22} & a_{12}a_{31} - a_{11}a_{32} & a_{11}a_{22} - a_{12}a_{21} \end{bmatrix}. \tag{B.13}
$$

Inversion of larger size matrices can be reduced to recursive inversion sub-matrices by the following formula, where any $\mathbf{A}_{11}$ and $\mathbf{A}_{22}$ must be quadratic:

$$
\mathbf{A} = \begin{bmatrix} \mathbf{A}_{11} & \mathbf{A}_{12} \\ \mathbf{A}_{21} & \mathbf{A}_{22} \end{bmatrix} \Rightarrow \mathbf{A}^{-1}
$$

$$
= \begin{bmatrix} \left[\mathbf{A}_{11} - \mathbf{A}_{12}\mathbf{A}_{22}^{-1}\mathbf{A}_{21}\right]^{-1} & -\mathbf{A}_{11}^{-1}\mathbf{A}_{12}\left[\mathbf{A}_{22} - \mathbf{A}_{21}\mathbf{A}_{11}^{-1}\mathbf{A}_{12}\right]^{-1} \\ -\mathbf{A}_{22}^{-1}\mathbf{A}_{21}\left[\mathbf{A}_{11} - \mathbf{A}_{12}\mathbf{A}_{22}^{-1}\mathbf{A}_{21}\right]^{-1} & \left[\mathbf{A}_{22} - \mathbf{A}_{21}\mathbf{A}_{11}^{-1}\mathbf{A}_{12}\right]^{-1} \end{bmatrix}. \tag{B.14}
$$

Further,

$$
[\mathbf{AB}]^{-1} = \mathbf{A}^{-1}\mathbf{B}^{-1} \quad ; \quad [c\mathbf{A}]^{-1} = \frac{1}{c}\mathbf{A}^{-1}. \tag{B.15}
$$

- The conjugate $\mathbf{A}^*$ of a matrix containing complex values is computed by multiplication of all imaginary parts by -1.

- The eigenvector of a quadratic matrix is a vector which multiplied by the matrix results in a scaled version of itself. The scaling factor is the associated eigenvalue λ. A non-singular quadratic matrix of K rows has K different eigenvectors $\mathbf{\Phi}_k$ and K eigenvalues λ_k:

$$
\mathbf{A} \cdot \mathbf{\Phi}_k = \lambda_k \cdot \mathbf{\Phi}_k, \ 1 \leq k \leq K. \tag{B.16}
$$

- The trace of a quadratic matrix of size $K \mathrm{x} K$ is the sum of the elements along its primary diagonal axis:

$$
\mathrm{tr}[\mathbf{A}] = \sum_{k=1}^{K} a_{k,k}. \tag{B.17}
$$

Additionally,

$$
\mathrm{tr}[\mathbf{A} \cdot \mathbf{B}] = \mathrm{tr}[\mathbf{B} \cdot \mathbf{A}] \ ; \quad \mathrm{tr}[\mathbf{A} \times \mathbf{B}] = \mathrm{tr}[\mathbf{A}] \cdot \mathrm{tr}[\mathbf{B}]. \tag{B.18}
$$

- The *Euclidean norm* of a vector results from (B.3) by multiplication with itself and square-rooting

$$\sqrt{\mathbf{a}^{\mathrm{T}} \cdot \mathbf{a}} = \sqrt{[a_1 \quad \cdots \quad a_K] \cdot \begin{bmatrix} a_1 \\ \vdots \\ a_K \end{bmatrix}} = \sqrt{\sum_{k=1}^{K} a_k^{\ 2}} \tag{B.19}$$

- *Orthogonality* means that the product of two vectors is zero. An *orthogonal set* of several vectors is structured such that the product of any two different vectors is zero, while the Euclidean norm of all vectors is equal. A stronger criterion is *orthonormality*; here, the Euclidean norm of all vectors is one (unity). For example, an orthonormal set of two vectors **a** and **b** fulfills the conditions

$$\mathbf{a}^{\mathrm{T}}\mathbf{b} = \mathbf{b}^{\mathrm{T}}\mathbf{a} = 0 \; ; \; \mathbf{a}^{\mathrm{T}}\mathbf{a} = \mathbf{b}^{\mathrm{T}}\mathbf{b} = 1. \tag{B.20}$$

- Orthogonality as a relationship between two vectors is a special case of *biorthogonality*. If two vectors establish a *basis system*, they can be arranged as columns of a matrix **A**. Then, a *dual basis system* $\tilde{\mathbf{A}}$ must exist, which fulfills the following conditions:

$$\mathbf{A} = \begin{bmatrix} \mathbf{a} & \mathbf{b} \end{bmatrix} \; ; \; \tilde{\mathbf{A}} = \begin{bmatrix} \tilde{\mathbf{a}} & \tilde{\mathbf{b}} \end{bmatrix}^{\mathrm{T}} \; ; \; \tilde{\mathbf{A}}\mathbf{A} = \mathbf{I}$$
$$\Leftrightarrow \tilde{\mathbf{a}}^{\mathrm{T}}\mathbf{b} = \tilde{\mathbf{b}}^{\mathrm{T}}\mathbf{a} = 0 \; ; \; \tilde{\mathbf{a}}^{\mathrm{T}}\mathbf{a} = \tilde{\mathbf{b}}^{\mathrm{T}}\mathbf{b} = 1. \tag{B.21}$$

The orthonormal basis system is only a special case thereof,

$$\tilde{\mathbf{a}} = \mathbf{a} \; ; \; \tilde{\mathbf{b}} = \mathbf{b} \; \Leftrightarrow \; \tilde{\mathbf{A}}^{\mathrm{T}} = \mathbf{A} \; \Leftrightarrow \; \mathbf{A}^{-1} = \mathbf{A}^{\mathrm{T}}. \tag{B.22}$$

It follows that inversion of orthogonal matrices is simply done by transposing and dividing by their quadratic (squared Euclidean) norm.

C Symbols and Variables

Unique symbols are used in equations whenever possible. If duplicate symbols are used, usually no ambiguity will occur due to the context (for example, the continuous speed variables u,v are never used jointly with the discrete frequency indices u,v).

Table C.1. Running variables and signal sizes over different dimensions in signal and spectral domains

	horizontal	vertical	depth	temporal
World coordinates $\mathbf{X}$	X	Y	Z	t *)
Speed $\mathbf{V}$ in 3D world space	V_X	V_Y	V_Z	-
continuous coordinate $\mathbf{r}$	r	s	-	t *)
Speed $\mathbf{v}$ in image plane	u	v	-	-
discrete coordinate $\mathbf{n}$	m *)	n	-	o
Sampling interval	R	S	-	T *)
Sub-sampled signal	m' *)	n'	-	o'
Size of a discrete signal	M *)	N	-	O
Size of a sub-sampled signal or block	M', M'' *)	N', N''	-	O'
Frequency of a continuous signal ω	ω_1	ω_2	-	ω_3, ω *)
Frequency of a discrete signal Ω	Ω_1	Ω_2	-	Ω_3, Ω *)
z transform	z_1	z_2	-	z_3, z *)
Discrete-spectrum frequency	u *)	v	-	w
Number of discrete spectral coefficients	U *)	V	-	W
index of filter coefficient, filter order	p, P *)	q, Q	-	r, R
Number of blocks	P *)	Q	-	-
Index of impulse response, correlation function or shift, length of impulse response	k, K *)	l, L	-	-

*) also used for 1D signals, typically with temporal dependency

Multimedia signals are often *multi-dimensional* (e.g. row, column and temporal directions in video). Table C.1 lists the *running variables*, *sizes* or *limits* and spectral denotations for each of the up to four dimensions. Table C.2 lists the *signal types* and their representation in the signal domain and in the spectral domain, where for the vector running variables ($\mathbf{n}$, $\mathbf{z}$, $\mathbf{\omega}$ etc.); the respective index variables from the Table C.1 have to be supplemented, depending on the number of dimensions.

Table C.2. Signal types in signal and spectral domains

	Signal domain	Spectral domain
Original signal, continuous	$x(\mathbf{r})$	$X(j\omega)$
Original signal, sampled	$x(\mathbf{n})$	$X(j\Omega), X(\mathbf{z})$
Signal estimate	$\hat{x}(\mathbf{n})$	$\hat{X}(\Omega)$
Binary signal	$b(\mathbf{n})$	$B(j\Omega)$
Noise signal	$z(\mathbf{n})$	$Z(j\Omega)$
Reconstructed signal	$y(\mathbf{n})$	$Y(j\Omega), Y(\mathbf{z})$
Prediction error signal, residual signal	$e(\mathbf{n}), r(\mathbf{n})$	$E(j\Omega), R(j\Omega)$
Reconstructed prediction error signal	$v(\mathbf{n})$	$V(j\Omega)$
Window function	$w(\mathbf{n})$	$W(j\Omega)$
Quantization error signal	$q(\mathbf{n})$	$Q(j\Omega)$
Interference signal	$k(\mathbf{n})$	$K(j\Omega)$
Filter coefficient (non-recursive, recursive)	$a(\mathbf{p})$ $b(\mathbf{p})$	$A(j\Omega), A(\mathbf{z})$ $B(j\Omega), B(\mathbf{z})$
Filter impulse response	$h(\mathbf{n}), g(\mathbf{n})$	$H(j\Omega), H(\mathbf{z}), G(j\Omega), G(\mathbf{z})$
Transform coefficient	$c_{\mathbf{u}}(\mathbf{n}')$	$C_{\mathbf{u}}(\mathbf{z})$

$S_{xx}, S_{yy}, \ldots$	Power spectra of $x, y, \ldots$
$r_{xx}, r_{yy}, \ldots$	Autocorrelation functions of $x, y, \ldots$
S_{xy}, r_{xy}	Cross power spectra, cross correlation
r_{xx}', r_{xy}'	Auto covariance, cross covariance
ρ_{xx}', ρ_{xy}'	Correlation coefficient (normalized)
$a_{\mathrm{h}}, r_{xx,\mathrm{h}}, \rho_{\mathrm{h}}$	Filter coefficient, autocorrelation function, correlation coefficient in horizontal direction
$a_{\mathrm{v}}, r_{xx,\mathrm{v}}, \rho_{\mathrm{v}}$	Filter coefficient, autocorrelation function, correlation coefficient in vertical direction
$a_{\mathrm{t}}, r_{xx,\mathrm{t}}, \rho_{\mathrm{t}}$	Filter coefficient, autocorrelation function, correlation coefficient in temporal direction
$\alpha(\cdot)$	Alpha map, blending function
$\delta(\cdot)$	Dirac impulse, unit impulse (Delta function)
$\varepsilon(\cdot)$	Unit step function

$\gamma(\cdot)$	Geometric transformation mapping
λ_u	Eigenvalue
ϕ_u	Eigenvector
$\mathbf{t}_u$	Basis vector
$\mathbf{T}$	Transform matrix
$\mathbf{H}, \mathbf{G}$	Filter matrix
$\mathbf{I}$	Identity matrix
Δ	Quantizer step size, distance function
$i(\cdot)$	Self information
$p(\cdot)$	Probability density function
$P(\cdot)$	Probability, cumulative probability
$H(\cdot), C(\cdot)$	Histogram, cumulative histogram
$f(\cdot)$	Amplitude mapping function
$p(\cdot,\cdot), P(\cdot,\cdot)$	Joint probability density function, joint probability
$p(\cdot\|\cdot), P(\cdot\|\cdot)$	Conditional probability density function, conditional probability
$p_K(\cdot)$	Vector probability density function
σ_x^2, σ_y^2	Variance of $x, y, \ldots$
μ_x, μ_y	Mean value of $x, y, \ldots$
L_x, L_y	Quadratic mean value (energy, power) of $x, y, \ldots$
$m_x^{(P)}, m_x'^{(P)}$	P^{th} order moment, central moment
$E\{\cdot\}$	Expected value
$L\{\cdot\}$	System transfer function
$\text{Im}\{\cdot\}$	Imaginary part of function
$\text{Re}\{\cdot\}$	Real part of function
$H(\mathcal{S})$	Entropy over set $\mathcal{S}$
$H(\mathcal{S}_1,\mathcal{S}_2)$	Joint entropy of sets $\mathcal{S}_1,\mathcal{S}_2$
$H(\mathcal{S}_1\|\mathcal{S}_2)$	Conditional entropy of sets $\mathcal{S}_1,\mathcal{S}_2$
$I(\mathcal{S}_1;\mathcal{S}_2)$	Mutual information of sets $\mathcal{S}_1,\mathcal{S}_2$
$d(\cdot,\cdot)$	Difference function, distance function, distortion
i,j	Code symbol index
k	Index of value in vector
l	Class or segment index
m_k	Feature value
r	Iteration number
t	Index of level
u_k	Output of nonlinear system
w_k	Weight factor, element of normal vector
w_k	Weight factor, element of normal vector
z_k	Centroid related to feature value
$A, A_{\text{max/min}}$	Amplitude value, maximum and minimum amplitude
B	Number of bits
F	Focal length
I, J	Number of codewords, code symbols or letters in discrete alphabet
K	Vector length

L	Constraint length
R	Number of iterations
S_l	Class of index l
T	Number of levels
$\mathbf{C}_{mm}$	Covariance matrix
$\mathbf{x}, \mathbf{y}, ...$	Vector from elements $x, y, ...$
$\mathbf{X}, \mathbf{Y}, ...$	Matrix from elements $x, y, ...$
$\mathbf{A}, \mathbf{D}, \mathbf{F}$	Coordinate mapping or sampling matrices in signal and frequency domains
$\mathbf{R}, \mathbf{t}$	Rotation matrix, translation vector
$\mathbf{R}_{xx}, \mathbf{r}_{xx}$	Autocorrelation (or autocovariance) matrix, vector
H, S, V	Hue, saturation, value color components
R, G, B	Red, green, blue (primary colors)
Y, C_b, C_r	Luminance and color difference components
$\mathcal{A}, \mathcal{B}$	Discrete alphabet
$\mathcal{C}$	Code, codebook
$\mathcal{F}\{\cdot\}$	Fourier transform
$\mathcal{K}$	context set
$\mathcal{S}$	Set, state
$\mathcal{V}$	Scale space
$\mathcal{W}$	Wavelet space
$\mathcal{R}^K$	Vector space of K dimensions
ε	Step size factor, small incremental value
η_x	Efficiency parameter
κ	Number of dimensions in signal
λ	Wavelength, Lagrangian multiplier
ϕ, θ	Angles (e.g. of rotation)
τ	Period
ξ_x	Geometric feature parameter
$\varphi(\tau)$	Scaling function, interpolation function
$\psi(\tau)$	Wavelet function
$\tilde{\omega}, \tilde{\Omega}$	Oscillation frequency
Θ	Threshold, scaling factor
Λ	Eigenvalue matrix, shape of an area
Π	Search range, shape of filter mask, projection profile
Φ, Ψ	Transform matrices of optimized linear transforms

D Acronyms

AAC	Advanced audio coding
ABR	Available bit rate
AC	Alternating current (transform coefficients of frequency higher than zero)
ACF	Auto correlation function
ADPCM	Adaptive DPCM
AM	Amplitude modulation
ANMRR	Average normalized modified retrieval rank
ANN	Artificial neural network
AR	Autoregressive process
ARQ	Automatic repeat request
ARR	Average retrieval rate
ART	Angular radial transform
ATM	Asynchronous transfer mode
AVC	Advanced video coding
BCH	Bose-Chaudhury-Hocquenghem (channel code)
BIBO	Bounded input, bounded output (stability)
BIFS	Binary format for scenes
BIM	Binary format for MPEG-7 descriptions
BPA	Back propagation algorithm
BSAC	Bit slice arithmetic coding
CABAC	Context-adaptive binary arithmetic coding
CAC	Call admission control
CAE	Context arithmetic encoding
CAVLC	Context-adaptive VLC
CBP	Coded block pattern
CBR	Constant bit rate
CCD	Charge coupled device
CCIR	Comitee Consultatif International de Radiodiffusion - now ITU-R
CCITT	Comitee Consultatif International de Telegraph et Telephone - now ITU-T
CD	Compact disc
CELP	Code excited linear prediction
CGI	Control grid interpolation
CIE	Commission Internationale d'Éclairage
CIF	Common Intermediate Format
COD	Coding device, encoder
CRC	Cyclic redundancy check

CSS	Curvature scale space
CTS	Composition time stamp
D	Descriptor
DAB	Digital audio broadcast
DC	Direct current (transform coefficient of frequency zero)
DC	Digital cinema
DCT	Discrete cosine transform
DDL	Description definition language
DEC	Decoding device, decoder
DECT	Digital enhanced cordless telephone
DFD	Displaced frame difference
DFT	Discrete Fourier transform
DHT	Discrete Hartley transform
DI	Digital item
DIA	Digital item adaptation
DID	Digital item declaration
DII	Digital item identification
DIP	Digital item processing
DM	Delta modulation
DMIF	Delivery multimedia integration framework
DMUX	Demultiplexer
DPCM	Differential PCM
DPD	Displaced pixel difference
DRM	Digital rights management
DS	Description scheme
DSCQS	Double stimulus continuous quality scale
DSIS	Double stimulus impairment scale
DSL	Digital subscriber loop
DSM-CC	Digital Storage Media – Command and Control
DST	Discrete Sine transform
DTS	Decoding time stamp
DVB	Digital video broadcast
DVD	Digital versatile disc
DWT	Discrete Wavelet transform
EBCOT	Embedded block coding with optimum truncation
ECQ	Entropy constrained quantization
EG	Exponential Golomb (code)
EOB	End of block
EOL	End of line
EPG	Electronic program guide
ER	Error resilience
ES	Elementary stream
ETSI	European Telecommunications Standards Institute
FAU	Facial action unit
FBA	Face and body animation
FEC	Forward error correction
FFT	Fast Fourier transform
FGS	Fine granularity scalability
FIFO	First in, first out (buffer)
FIR	Finite impulse response (filter)

FLC	Fixed-length coding
FM	Frequency modulation
FoE	Focus of expansion
GMC	Global motion compensation
GoB	Group of blocks
GoP	Group of pictures (within a video sequence)
GPRS	Generalized packet radio structure
GR	Golomb-Rice (code)
GSM	Global standard for mobile communication
HHR	Half horizontal resolution
HMM	Hidden Markov model
HRD	Hypothetical reference decoder
HSC	Harmonic spectral centroid
HSD	Harmonic spectral deviation
HSS	Harmonic spectral spread
HSV	Harmonic spectral variation
HT	Haar transform
ICA	Independent component analysis
IETF	Internet Engineering Task Force
IIR	Infinite impulse response (filter)
IP	Internet protocol
IPMP	Intellectual property management and protection
ISDN	Integrated services digital network
ISO	International Standardization Organization
ITU	International Telecommunication Union
JBIG	Joint Bilevel Images Group (working group of ISO and ITU-T)
JPEG	Joint Photographic Experts Group (working group of ISO and ITU-T)
JVT	Joint Video Team (working group of ISO and ITU)
HDTV	High definition TV
HSV	Hue – saturation – value
KLT	Karhunen-Loève transform
LAN	Local area network
LC	Low complexity
LCD	Liquid crystal display
LD	Low delay
LOT	Lapped Orthogonal Transform
LPC	Linear predictive coding
LSB	Least significant bit
LTP	Long term prediction
MA	Moving average (process)
MAD	Minimum absolute difference
MAT	Medial axis transform
MC	Motion Compensation
MCP	Motion-compensated prediction
MCTF	Motion-compensated temporal filtering
MDCT	Modified DCT (block-overlapping filterbank)
ME	Motion Estimation
MIDI	Musical instrument digital interface
MIMO	Multiple input, multiple output
ML	Maximum likelihood

MLP	Multi layer perceptron
MOS	Mean opinion score
MPE	Multi pulse excitation
MPEG	Moving Pictures Experts Group (working group of ISO)
MRF	Markov random field (statistical model)
MSB	Most significant bit
MSD	Minimum squared difference
MSE	Mean squared error
MUX	Multiplex, multiplexer
MVF	Motion vector field
NAL	Network abstraction layer
NN	Nearest neighbor
NTSC	National Television Standards Committee
OBMC	Overlapping block motion compensation
OFDM	Orthogonal frequency division multiplex
PAL	Phase alternating lines
PAT	Program association table
PCM	Pulse code modulation
PCR	Program clock recovery
PDA	Personal digital assistant
PDF	Probability density function
PES	Packetized elementary stream
PID	Program identifier
PRF	Perfect reconstruction filter
PSI	Program specific information
PSNR	Peak signal-to-noise ratio
PVQ	Predictive vector quantization
Q	Quantization
QAM	Quadrature amplitude modulation
QCIF	Quarter CIF
QMF	Quadrature mirror filter
QoS	Quality of service
QPSK	Quadrature phase shift keying
RAC	Relative address coding
RBF	Radial basis function (neural network)
RCPC	Rate-Compatible Punctured Convolutional Codes
RDD	Rights data dictionary
RDF	Resource description framework
RDO	Rate-distortion optimization
READ	Relative element address designate (coding)
REL	Rights expression language
RGB	Red – green – blue
RLC	Run-length coding
RPE	Regular pulse excitation
RR	Retrieval rate
RS	Reed-Solomon (channel code)
RSVP	Reservation protocol
RTP	Real-time transport protocol
SADCT	Shape-adaptive DCT
SAOL	Structured audio orchestra language

SAT	Symmetry axis transform
SBC	Subband coding
SBR	Spectral band replication
SDH	Synchronous digital hierarchy
SDTV	Standard definition TV
SFM	Spectral flatness measure
SIF	Standard intermediate format
SNR	Signal-to-noise ratio
SOFM	Self organizing feature map
SQ	Scalar quantization, scalar quantizer
SSCQE	Single stimulus continuous quality evaluation
STFT	Short time Fourier transform
SVD	Singular value decomposition
SVM	Support vector machine
T	Transform
TC	Transform coding
TCM	Trellis coded modulation
TCP	Transport control protocol
TDAC	Time domain aliasing cancellation
TS	Transport stream
TTSI	Text to speech interface
UDP	Unified datagram protocol
UMTS	Universal mobile telecommunication system
URI	Uniform resource identifier
URL	Uniform resource locator
VBR	Variable bit rate
VBV	Video buffering verifier
VCEG	Video Coding Experts Group (Working group of ITU-T)
VCV	Video complexity verifier
VLC	Variable-length coding
VMV	Video memory verifier
VOP	Video object plane
VQ	Vector quantization
VRML	Virtual Reality Modeling Language
WAN	Wide area network
WLAN	Wireless LAN
WT	Wavelet transform
W^3C	World Wide Web Council
XML	Extensible markup language
1D	One-dimensional
2D	Two-dimensional
3D	Three-dimensional
3G	Third generation (mobile network)
3GPP	Third Generation Partnership Project
4G	Fourth generation (mobile network)

References

Aach, T. , Kaup, A. , Mester, R. : Statistical model-based change detection in moving video. *Signal Process.* 31 (1993), pp. 165-180

Adolph, D. , Buschmann, R. : 1.15 Mbit/s coding of video signals including global motion compensation. *Signal Process. : Image Commun.* 3 (1991), pp. 259-274

Aggarwal, J. K. , Nandhakumar, N. : On the computation of motion from sequences of images - a review. *Proc. IEEE* 76 (1988), pp. 917-934

Ahmed, N. , Natarjan, T. , Rao, K. R. : Discrete cosine transform. *IEEE Trans. Comp.* 23 (1974), pp. 90-93

Aign, S. : A temporal error concealment technique for I-Pictures in an MPEG-2 video decoder. *Proc. SPIE VCIP* (1998), vol. 3309, pp. 405-416

Aizawa, K. , Choi, C. S. , Harashima, H. , Huang, T. S. : Human facial motion analysis and synthesis with applications to model-based coding. *Motion Analysis and Image Sequence Processing*, M. I. Sezan and R. L. Lagendijk (eds.), Dordrecht : Kluwer 1993

Aizawa, K. , Harashima, H. , Saito, T. : Model-based analysis/synthesis image coding system for a person's face. *Signal Process. : Image Commun.* 1 (1989), pp. 139-152

Alexander, S. T. , Rajala, S. A. : Image compression results using the LMS adaptive algorithm. *IEEE Trans. Acoust., Speech, Signal Process.* 33 (1985), pp. 712-714

Amon, P. , Illgner, K. , Pandel, J. : SNR scalable layered video coding. *Proc. Packet Video Workshop* (2002)

Anandan, P. , Bergen, J. R. , Hanna, K. J. , Hingorani, R. : Hierarchical model-based motion estimation. *Motion Analysis and Image Sequence Processing*, M. I. Sezan and R. L. Lagendijk (eds.), Dordrecht : Kluwer 1993

Anderson, J. B. , Mohan, S. : Sequential coding algorithms : A survey and cost analysis. *IEEE Trans. Commun.* 32 (1984), pp. 169-176

Anderson, J. B. , Mohan, S. : Source and Channel Coding. Dordrecht : Kluwer 1991

Andreopoulos, Y. , van der Schaar, M. , Munteanu, A. , Barbarien, J. , Schelkens, P. : Complete-to-overcomplete discrete wavelet transforms for fully-scalable video coding with MCTF. *Proc. SPIE VCIP* (2003), vol. 5150, pp. 719-731

Andrew, J. : A simple and efficient hierarchical image coder. *Proc. IEEE ICIP* (1997), vol. 3, pp 658-661

Antonini, M. , Barlaud, M. , Mathieu, P. , Daubechies, I. : Image coding using wavelet transform. *IEEE Trans. Image Process.* 1 (1992), pp. 205-220

Apostolopoulos, J. : Error-resilient video compression through the use of multiple states. *Proc. IEEE ICIP* (2000), pp. 2585-2588

Aravind, R. , Gersho, A. : Low rate image coding with finite-state vector quantization. *Proc. IEEE ICASSP* (1986), pp. 4.3.1-4.3.4

Aravind, R. , Civanlar, M. R. , Reibman, A. R. : Packet loss resilience of MPEG-2 scalable video coding algorithms. *IEEE Trans. Circ. Syst. Video Tech.* 6 (1996), pp. 426-435

Van der Auwera, G. , Munteanu, A. , Schelkens, P. , Cornelis, J. : Bottom-up motion compensated prediction in the wavelet domain for spatially scalable video coding. *IEE Electron. Lett.* 38 (2002), pp. 1251-1253

Ayanoglu, E. , Gray, R. M. : The design of predictive trellis waveform coders using the generalized Lloyd algorithm. *IEEE Trans. Commun.* 34 (1986), pp. 1073-1080

Ballard, D. H. , Brown, C. M. : Computer Vision. Englewood Cliffs: Prentice Hall, 1985

Bamberger, R. H. , Eddins, S. L. , Nuri, V. : Generalized symmetric extension for size-limited multirate filter banks. *IEEE Trans. Image Process.* 3 (1994), pp. 82-87

Barnsley, M. : Fractals everywhere. San Diego : Academic Press 1988

Barnsley, M. , Anson, L. F. : The Fractal Transform. Peters 1993

Barnsley, M. , Hurd, L. P. : Fractal Image Compression. Peters 1993

Barr, A. : Superquadrics and angle-preserving transformations. *IEEE Trans. Comput. Graphics Appl.* 1 (1981), pp. 1-20

Barthel, K.-U. , Voyé, T. , Noll, P. : Improved fractal image coding. *Proc. PCS* (1993), no. 1.5, Lausanne : EPFL 1993

Barthel, K.-U. , Schüttemeyer, J. , Voyé, T. , Noll, P. : A new image coding technique unifying fractal and transform coding. *Proc. IEEE ICIP* (1994), vol. III, pp. 112-116

Barthel, K.-U. , Voyé, T. : Three-dimensional fractal video coding. *Proc. IEEE ICIP* (1995), vol. III, pp. 260-263

Beermann, M. , Wien, M. , Ohm, J.-R. : Look-ahead coding considering rate/distortion-optimization. *Proc. IEEE ICIP* (2002), vol. I, pp. 93-95

Belfor, R. A. F. , Lagendijk, R. L. , Biemond, J. : Subsampling of digital image sequences using motion information. *Motion Analysis and Image Sequence Processing*, M. I. Sezan and R. L. Lagendijk (eds.), Dordrecht : Kluwer 1993

Bellifemine, F. , Capellino, A. , Chimienti, A. , Picco, R. , Ponti, R. : Statistical analysis of the 2D-DCT coefficients of the differential signal for images. *Signal Process. : Image Commun.* 4 (1992), pp. 477-488

Benzler, U. : Spatial scalable video coding using a combined subband-DCT approach. *IEEE Trans. Circ. Syst. Video Tech.* 10 (2000), pp. 1080-1087

Berger, T. : Rate Distortion Theory. Englewood Cliffs : Prentice-Hall 1971

Berrou, C. , Glavieux, A. , Thitimajshima, P. : Near Shannon limit error-correcting coding and decoding: Turbo-codes. *Proc. IEEE ICC* (1993), vol. II, pp. 1064-1070

Besag, J. : On the statistical analysis of dirty pictures. *J. Roy. Stat. Soc. B* 48 (1986), pp. 259-302

Bhaskaran, V. : Predictive VQ schemes for grayscale image compression . *Proc. IEEE GLOBECOM* (1987), pp. 436-441

Bierling, M. : Displacement estimation by hierarchical block matching. *Proc. Visual Commun. Image Process.* (1988), SPIE vol. 1001, pp. 942-951

Bierling, M. , Thoma, R. : Motion compensated interpolation using a hierarchically structured displacement estimator. *Signal Process.* 11 (1986), pp. 387-404

Bilgin, A. , Sementilli, P. J. , Marcellin, M. W. : Progressive image coding using trellis coded quantization. *IEEE Trans. Image Proc.* 8 (1999), pp. 1638-1643

Blahut, R. E. : Computation of channel capacity and rate-distortion functions. *IEEE Trans. Inf. Theor.* 18 (1972), pp, 460-473

Blahut, R. E. : Principles and Practice of Information Theory. Reading : Addison-Wesley 1987

Blain, M. E. , Fischer, T. R. : A comparison of vector quantization techniques in transform and subband coding of imagery. *Signal Process. : Image Commun. 3* (1991), pp. 91-105

Blume, H. : Vector based nonlinear upconversion applying center weighted medians. *Proc. SPIE Nonlin. Image Proc.* VII, (1996), vol. 2662, pp. 142-153

Bober, M. : MPEG-7 visual shape descriptors. *IEEE Trans. Circ. Syst. Video Tech.* 11 (2001), pp. 716-719

Börner, R. : Autostereoscopic 3-D imaging by front and rear projection on flat panel displays. *Displays* 14 (1993), pp. 39-46

Boone, M.M. , Verheijen, E.N.G. : Multi-channel sound reproduction based on wave field synthesis, *Proc. 95th AES Convention* (1993)

Bose, N. K. : Applied Multidimensional Systems Theory. New York : Van Nostrand Reinhold, 1983

Bossert, M. : Channel Coding for Telecommunications New York: Wiley, 1999

Bostelmann, G. : A simple high quality DPCM-codec for video telephony using 8 Mbit/s. *Nachrichtentechnische Zeitschrift* 28 (1974), Heft 3

Bosveld, F. , Lagendijk, R. L. , Biemond, J. : Hierarchical coding of HDTV. *Signal Process. : Image Commun.* 4 (1992), pp. 195-225

Bosveld, F. , Lagendijk, R. L. , Biemond, J. : Compatible spatio-temporal subband encoding of HDTV. *Signal Process.* 28 (1992), pp. 271-289

Bottreau, V. , Benetiere, M. , Felts, B. , Pesquet-Popescu, B. : A fully scalable 3D subband video codec. *Proc. IEEE ICIP* (2001), vol. 2, pp. 1017-1020

Bouman, C. A. , Shapiro, M. : A multiscale random field model for Bayesian image segmentation. *IEEE Trans. Image Process.* 3 (1994), pp. 162-177

Bozdagi, G. , Tekalp, A. M. , Onural, L. : 3-D motion estimation and wireframe adaptation including photometric effects for model-based coding of facial image sequences. *IEEE Trans. Circ. Syst. Video Tech.* 4 (1994), pp. 246-256

Braun, M. , Hahn, M. , Ohm, J.-R. , Talmi, M. : Motion-compensating real-time format converter for video on multimedia displays. *Proc. IEEE ICIP* (1997), vol.I, pp. 125-128

Brünig, M. , Niehsen, W. : Fast full-search block matching. *IEEE Trans. Circ. Syst. Video Tech.* 11 (2001), p. 241-247

Buhmann, J. , Lange, J. , von der Malsburg, C. , Vorbrüggen, J. C. , Würtz, R. P. : Object recognition with Gabor functions in the dynamic link architecture - Parallel implementation on a Transputer network. In B. Kosko (ed.), *Neural Networks for Signal Processing*, pp. 121-159, Englewood Cliffs: Prentice Hall, 1992.

Burges, C. J. C. : A tutorial on support vector machines for pattern recognition. *Data Mining and Knowledge Discovery* 2 (1998), no.2, pp. 121-167

Burt, P. J. , Adelson, E. H. : The Laplacian pyramid as a compact image code. *IEEE Trans. Commun.* 31 (1983), pp. 532-540

Cadzow, J. A. : Least squares, modeling, and signal processing. *Dig. Signal Process.* 4 , pp. 2-20. San Diego : Academic Press 1994

Casey, M. A. , Westner, A. : Separation of mixed audio sources by independent subspace analysis. *Proc. Int. Computer Music Conf.*, Berlin 2000

Chahine, M. , Konrad, J. : Estimation of trajectories for accelerated motion from time-varying imagery. *Proc. IEEE ICIP* (1994), vol. II, pp. 800-804

Chai, B. B. , Vass, J. , Zhuang, X. : Statistically adaptive wavelet image coding. In *Visual Information Representation, Communication, and Image Processing*, pp. 73-95, Marcel Dekker, 1999

Chan, M.-H. , Yu, Y.-B. : Variable size block matching motion compensation and application to video coding. *Proc. IEE* 137 (1990), pp. 202-212

Chan, W.-Y. , Gersho, A. : Generalized product code vector quantization : A family of efficient techniques for signal compression. *Dig. Signal Process.* 4 (1994), pp. 95-126

Chande, V. , Farvardin, N. , Jafarkhani, H. : Image communication over noisy channels with feedback. *Proc. IEEE ICIP* (1999), vol. II, pp. 540-544

Chang, T. , Kuo, C.-C. J. : Texture analysis and classification with tree-structured wavelet transform. *IEEE Trans. Image Process.* 2 (1993), pp. 429-441

Chelappa, R. , Kashyap, R.L.: Texture synthesis using 2-D noncausal autoregressive models. *IEEE Trans. Acoust., Speech, Signal Process.* 33 (1985), pp. 194-203

Chelappa, R. , Chatterjee, S. : Classification of textures using Gaussian Markov random fields. *IEEE Trans. Acoust., Speech, Signal Process.* 33 (1985), pp. 959-963

Chellappa, R. , Jain, A. (eds.) : *Markov Random Fields: Theory and Applications.* Academic Press, 1996

Chen, C.-T. , Wong, A. : A self-governing buffer control strategy for pseudoconstant bit rate video coding. *IEEE Trans. Image Process.* 2 (1993), pp. 50-59

Chen, H.-H. , Chen, Y.-S. , Hsu, W.-H. : Low-rate sequence image coding via vector quantization. *Signal Process.* 26 (1992), pp. 265-283

Chen, W.-H. , Pratt, W. K. : Scene adaptive coder. *IEEE Trans. Commun.* 32 (1984), pp. 225-232

Chen, W.-H. , Smith, C. H. : Adaptive coding of monochrome and color images. *IEEE Trans. Commun.* 25 (1977), pp. 1285-1292

Chen, W.-H. , Smith, C. H. , Fralick, S. : A fast computational algorithm for the discrete cosine transform. *IEEE Trans. Commun.* 25 (1977), pp. 1004-1009

Chen, Y. , Pearlman, W. : Three-dimensional Subband Coding of Video using the zero-tree method. Proc. SPIE VCIP (1996), vol. 2727, pp. 1302-1309

Cheng, P.-Y. , Li, J. , Kuo, C.-C. J. : Multiscale video compression using wavelet transform and motion compensation. *Proc. IEEE ICIP* (1995), vol. I, pp. 606-609

Chin, T. M. , Luettgen, M. R. , Karl, W. C. , Willsky, A. S. : An estimation theoretic perspective on image processing and the calculation of optical flow. *Motion Analysis and Image Sequence Processing*, M. I. Sezan and R. L. Lagendijk (eds.), Dordrecht : Kluwer 1993

Choi, C. S. , Aizawa, K. , Harashima, H. , Takebe, T. : Analysis and synthesis of facial image sequences in model-based image coding. *IEEE Trans. Circ. Syst. Video Tech.* 4 (1994), pp. 257-275

Choi, S.-J. , Woods, J. W. : Motion-compensated 3-D subband coding of video. ", *IEEE Trans. Image Process.* 8 (1999), pp. 155-167

Chou, P. A. , Lookabaugh, T. , Gray, R. M. : Entropy-constrained vector quantization. *IEEE Trans. Acoust., Speech, Signal Process.* 37 (1989), pp. 31-42

Chou, P. A. , Lookabaugh, T. , Gray, R. M. : Optimal pruning with applications to tree-structured source coding and modeling. *IEEE Trans. Inf. Theor.* 35 (1989), pp. 299-315

Chou, P. A. , Miao, Z. : Rate-distortion optimized streaming of packetized media. *Microsoft Research Tech. Rep.* MSR-TR-2001-35, February 2001

Chowdhury, M. F. , Clark, A. F. , Downton, A. C. , Morimatsu, E. , Pearson, D. E. : A switched model-based coder for video signals. *IEEE Trans. Circ. Syst. Video Tech.* 4 (1994), pp. 216-227

Chuang, G. C.-H. , Kuo, C.-C. J. : Wavelet descriptor of planar curves : Theory and applications. *IEEE Trans. Image Proc.* 5 (1996), pp. 56-70

Chung, Y.-S. , Kanefsky, M. : On 2-D recursive LMS algorithms using ARMA prediction for ADPCM encoding of images. *IEEE Trans. Image Process.* 1 (1992), pp. 416-422

Commission Internationale d'Éclairage (CIE) : Industrial Colour-Difference Evaluation. CIE Publication No. 116, Vienna, 1995

Clarke, R. J. : Transform Coding of Images. London : Academic Press 1985

Cohen, R. A. , Woods, J. W. : Sliding block entropy coding of images. *Proc. IEEE ICASSP* (1989), pp. 1731-1734

Cohn, D. , Riskin, E. A. , Ladner, R. : Theory and practice of vector quantizers trained on small training sets. *IEEE Trans. Patt. Anal. Mach. Intell.* 16 (1994), pp. 54-65

Connor, J. T. , Martin, R. D. , Atlas, L. E. : Recurrent neural networks and robust time series prediction. *IEEE Trans. Neur. Netw.* 5 (1994), pp. 240-254

Conway, J. H. , Sloane, N. J. A. : Voronoi regions of lattices, second moments of polytopes, and quantization. *IEEE Trans. Inf. Theor.* 28 (1982), pp. 211-226

Conway, J. H. , Sloane, N. J. A. : Fast quantizing and decoding algorithms for lattice quantizers and codes. *IEEE Trans. Inf. Theor.* 28 (1982), pp. 227-232

Conway, J. H. , Sloane, N. J. A. : A fast encoding method for lattice codes and quantizers. *IEEE Trans. Inf. Theor.* 29 (1983), pp. 820-824

Conway, J. H. , Sloane, N. J. A. : Sphere Packings, Lattices and Groups. New York : Springer 1988

Cosman, P. , Gray, R. M. , Vetterli, M. : Vector quantization of image subbands – A survey. *IEEE Trans. Image Proc.* 5 (1996), pp. 202-225

Côté, G. , Erol, B. , Gallant, M. , Kossentini, F. : H.263+: Video Coding at Low Bit Rates. IEEE Trans. Circ. Syst. Video Tech. 8 (1998) pp. 849-866

Cover, T. M. , Thomas, J. A. : Elements of Information Theory. New York: Wiley 1991

Cox, I. , Bloom, J. , Miller, M. : Digital Watermarking: Principles & Practice. Morgan Kaufmann, 2001

Crochiere, R. E. , Rabiner, L. R. : Multirate Digital Signal Processing. Englewood Cliffs : Prentice-Hall 1983

Cross, G. R. , Jain, A. K. : Markov random field texture models. *IEEE Trans. Patt. Anal. Mach. Intell.* 5 (1983), pp. 25-39

Cuperman, V. : On adaptive vector transform quantization for speech coding. *IEEE Trans. Commun.* 37 (1989), pp. 261-267

Cuperman, V. , Gersho, A. : Vector predictive coding of speech at 16 kbits/s. *IEEE Trans. Commun.* 33 (1985), pp. 685-696

Daubechies, I. : Orthonormal basis of compactly supported wavelets. *Comm. Pure Applied Math.* 41 (1988), pp. 909-996

Daubechies, I. : The wavelet transform, time-frequency localization and signal analysis. *IEEE Trans. Inf. Theor.* 36 (1990), pp. 961-1005

Delogne, P. , Macq, B. : Universal variable length coding for an integrated approach to image coding. *Annales Télécommun.* 46 (1991), pp. 452-459

Dempster, A. P. : A generalization of Bayesian inference, *J. Roy. Stat. Soc.* 30 (1968), Series B

Dempster, A. P. : Upper and lower probabilities induced by a multivalued mapping, *Annals Mathem. Statist.* 38 (1976)

Dianat, S. A. , Nasrabadi, N. M. , Venkataraman, S. : A nonlinear predictor for differential pulse-encoder (DPCM) using artificial neural networks. *Proc. IEEE ICASSP* (1991), pp. 2793-2796

Diehl, N. : Object-oriented motion estimation and segmentation in image sequences. *Signal Process. : Image Commun.* 3 (1991), pp. 23-56

Dubois, E. : Motion-compensated filtering of time-varying images. *Multidim. Syst. Signal Process.* 3, pp. 211-240. Dordrecht : Kluwer 1992

Dubois, E. , Konrad, J. : Estimation of 2-D motion fields from image sequences with application to motion-compensated processing. *Motion Analysis and Image Sequence Processing*, M. I. Sezan and R. L. Lagendijk (eds.), Dordrecht : Kluwer 1993

Dudgeon, D. E. , Mersereau, R. M. : Multidimensional Digital Signal Processing. Englewood Cliffs : Prentice-Hall 1984

Dugelay, J.-L. : Toward an estimation of three-dimensional motion in a 3D TV image sequence. *Proc. Visual Commun. Image Process.* (1991), SPIE vol. 1605, pp. 688-693

Dugelay, J.-L. : Toward an estimation of three-dimensional motion in a 3D TV image sequence II : Extension to curved surfaces. *Proc. Visual Commun. Image Process.* (1992), SPIE vol. 1818, pp. 688-693

Dunham, M. O. , Gray, R. M. : An algorithm for the design of labeled-transition finite-state vector quantizers. *IEEE Trans. Commun.* 33 (1985), pp. 83-89

Efstratiadis, S. N. , Katsaggelos, A. K. : An adaptive regularized recursive displacement estimation algorithm. *IEEE Trans. Image Process.* 2 (1993), pp. 341-352

Elias, P. : Universal codeword sets and representations of the integers, *IEEE Trans. Inf. Theor.* 21 (1975), pp. 194-203

Enkelmann, W. : Investigations of multigrid algorithms for the estimation of optical flow fields in image sequences. *Comp. Graph. Image Process.* 43 (1988), pp. 150-177

Ezra, D. , Woodgate, G. , Omar, B. , Holliman, N. , Harrold, J. , Shapiro, L. : New auto-stereoscopic display system. *Proc. SPIE Stereosc. Displ. Apps.* (1995), vol. 2409, pp. 31-40

Farkash, S. , Pearlman, W. A. , Malah, D. : Transform trellis coding of images at low rates with blocking effect removal. *Proc. EUSIPCO* (1988), *Signal Process. IV : Theor. and Appl.*, pp. 1657-1660, Amsterdam : Elsevier 1988

Farrelle, P. M. , Jain, A. K. : Recursive block coding - a new approach to transform coding. *IEEE Trans. Commun.* 34 (1986), pp. 161-179

Fischer, T. R. : A pyramid vector quantizer. *IEEE Trans. Inf. Theor.* 32 (1986), pp. 568-583

Fischer, T. R. : Geometric source coding and vector quantization. *IEEE Trans. Inf. Theor.* 35 (1989), pp. 137-145

Fischer, T. R. : On the rate-distortion efficiency of subband coding. *IEEE Trans. Inf. Theor.* 38 (1992), pp. 426-429

Fischer, T. R. , Marcellin, M. W. , Wang, M. : Trellis-coded vector quantization. *IEEE Trans. Inf. Theor.* 37 (1991), pp. 1551-1566

Fischer, T. R. , Wang, M. : Entropy-constrained trellis-coded quantization. *IEEE Trans. Inf. Theor.* 38 (1992), pp. 415-426

Flierl, M. , Girod, B. : Investigation of motion-compensated lifting wavelet transform. *Proc. Picture Coding Symposium* (2003), pp. 59-62, Saint-Malo, April 2003

Flierl, M. , Girod, B. : Generalized B pictures and the draft H.264/AVC video-compression standard. *IEEE Trans. Circ. Syst. Video Tech.* 13 (2003), pp. 587- 597

Forchheimer, R. , Kronander, T. : Image coding - from waveforms to animation. *IEEE Trans. Acoust., Speech, Signal Process.* 37 (1989), pp. 2008-2023

Forney, G. D., jr. : The Viterbi algorithm. *Proc. IEEE* 61 (1973), pp. 268-278

Foster, J. , Gray, R. M. ; Dunham, M. O. : Finite-state vector quantization for waveform coding. *IEEE Trans. Inf. Theor.* 31 (1985), pp. 348-359

Freeman, H. : Boundary encoding and processing. *Picture Processing and Psychopictorics*, B. S. Lipkin and A. Rosenfeld (eds.), New York : Academic Press 1970

Fujii, T. , Tanimoto, M. : Free viewpoint TV system based on ray-space representation. *Proc. SPIE Three-Dimens. TV, Video, and Display* (2002), vol. 4864, pp. 175-189

Gabor, D. : Theory of communication. *J. Inst. Elect. Eng.* 93 (1946), pp. 429-457

Gardos, T. et al. : Efficient Receiver-Driven Layered Video Multicast Using H.263+ SNR Scalability. *Proc. IEEE ICIP* (1998), vol. 3, pp. 32-35.

Geman, D. , Geman, S. , Graffigne, C. , Dong, P. : Boundary detection by constrained optimization. *IEEE Trans. Patt. Anal. Mach. Intell:* 12 (1990), pp. 609-628

Geman, S. , Geman, D. : Stochastic relaxation, Gibbs distribution, and the Bayesian restauration of images. *IEEE Trans. Patt. Anal. Mach. Intell.* 6 (1984), pp. 721-741

Gersho, A. : Optimal nonlinear interpolative vector quantization. *IEEE Trans. Commun.* 38 (1989), pp. 1285-1287

Gersho, A. , Gray, R. M. : Vector Quantization and Signal Compression. Dordrecht : Kluwer 1992

Gerzon, M.A. : Design of ambisonic decoders for multi speaker surround sound. *Proc. 58th AES Convention*, New York, 1977

Ghanbari, M. : Two-layer coding of video signals for VBR networks. *IEEE J. Sel. Areas Commun.* 8 (1989), pp. 771-781

Ghanbari, M. : The cross search algorithm for Motion Estimation. *IEEE Trans. Commun.* 38 (1990), pp. 950-953

Ghanbari, M. , Azari, J. : Effect of bit rate variation of the base layer on the performance of two-layer video codecs. *IEEE Trans. Circ. Syst. Video Tech.* 4 (1994), pp. 8-17

Gharavi, H. , Tabatabai, A. : Subband coding of monochrome and color images. *IEEE Trans. Circ. Syst.* 35 (1988), pp. 207-214

Gibson, J. D. , Sayood, K. : Lattice Quantization. *Advan. Electron. and Electron. Phys.* 72 (1988), pp. 259-330

Gilge, M. , Engelhardt, T. , Mehlan, R. : Coding of arbitrarily shaped image segments based on a generalized orthogonal transform. *Signal Process. : Image Commun.* 1 (1989), pp. 153-180

Gimlett, J. I. : Use of "activity" classes in adaptive transform image coding. *IEEE Trans. Commun.* 23 (1975), pp. 785-786

Girod, B. : The efficiency of motion-compensating prediction for hybrid coding of video sequences. *IEEE J. Sel. Areas Commun.* 5 (1987), pp. 1140-1154

Girod, B. : Psychovisual aspects of image communication. *Signal Process.* 28 (1992), pp. 239-251

Girod, B. : Motion compensation : Visual aspects, accuracy and fundamental limits. *Motion Analysis and Image Sequence Processing*, M. I. Sezan and R. L. Lagendijk (eds.), Dordrecht : Kluwer 1993

Girod, B. : Rate-constrained motion estimation. *Proc. Visual Commun. Image Process.* (1994), SPIE vol. 2308

Girod, B. : Image sequence coding using 3D scene models. *Proc. Visual Commun. Image Process.* (1994), SPIE vol. 2308

Girod, B. : Bidirectionally decodable streams of prefix code-words. *IEEE Comm. Lett.* 3 (1999), pp. 245-247

Girod, B. , Färber, N. , Horn, U. : Scalable codec architectures for Internet video-on-demand. *Proc. 1997 Asilomar Conf. Signals and Syst.*, Pacific Grove, 1997

Girod, B. , Färber, N. , Stuhlmüller, K. : Internet video streaming. In *Multimedia Communications*, pp. 547-556, Springer, 1999

Girod, B. , Chang, C.-L. , Ramanathan, P. , Zhu, X. : Light field compression using disparity-compensated lifting. *Proc. IEEE ICASSP* (2003)

Golwelkar, A. , Woods, J. W. : Scalable video compression using longer motion compensated temporal filters. *Proc. SPIE VCIP* (2003), vol. 5150, pp. 1406-1417

Goyal, V.K. , Kovacevic, J. , Arean, R. , Vetterli, M. : Multiple description transform coding of images. Proc. IEEE ICIP (1998), vol. 1, pp 674 – 678

Gray, R. M. : Sliding block source coding. *IEEE Trans. Inf. Theor.* 21 (1975), pp. 357-368

Gray, R. M. : Vector quantization. *IEEE ASSP Mag.* 1 (1984), no. 2, pp. 4-29

Gray, R. M. : Source Coding Theory. Dordrecht : Kluwer 1990

Gray, R. M. , Karnin, E. D. : Multiple local optima in vector quantizers. *IEEE Trans. Inf. Theor.* 28 (1982), pp. 256-261

Gray, R. M. , Linde, Y. : Vector quantizers and predictive quantizers for Gauss-Markov sources. *IEEE Trans. Commun.* 30 (1982), pp. 381-389

Greiner, T. , Eberle, E. , Pandit, M. : Design of signal-matched wavelets - theory and results. *Proc. EUSIPCO* (1992), *Signal Process. VI : Theor. and Appl.*, pp. 929-932, Amsterdam : Elsevier 1992

Gröchenig, K. , Madych, W. R. : Multiresolution analysis, Haar bases, and self-similar tilings of R^n. *IEEE Trans. Inf. Theor.* 38 (1992), pp. 556-568

Gross, M. H. , Staadt, O. G. , Gatti, R. : Efficient triangular surface approximations using wavelets and quadtree data structures. *IEEE Trans. Visual. Comp. Graph.* 2 (1996), pp. 130-143

de Haan, G. , Biezen, P. W. A. C. , Huijgen, H. , Ojo, O. A. : True-motion estimation with 3-D recursive search block matching. *IEEE Trans. Circ. Syst. Video Tech.* 3 (1993), pp. 368-379

Hagenauer, J. : Rate-compatible punctured convolutional codes (RCPC codes) and their applications. *IEEE Trans. Commun.* 36 (1988), pp. 389-400

Hajek, B. : Cooling schedules for optimal annealing. *Math. Oper. Res.* 13 (1988), pp. 311-329

Haley, G. M. , Manjunath, B. S. : Rotation-invariant texture classification using a complete space-frequency model. *IEEE Trans. Image Process.* 8 (1999), pp. 255-269

Hampel, H. et al. : Technical features of the JBIG standard for progressive bi-level image compression. *Signal Process. : Image Commun.* 4 (1992), pp. 103-111

Han, S.C. , Woods, J.W. : Object-based subband/wavelet video compression. In *Wavelet Image Compression,* P. Topiwala (ed.), Kluwer Academic Press, 1998

Hang, H.-M. , Haskell, B. G. : Interpolative vector quantization of color images. *IEEE Trans. Commun.* 36 (1988), pp. 465-470

Hang, H.-M. , Woods, J. W. : Predictive vector quantization of images. *IEEE Trans. Commun.* 33 (1985), pp. 1208-1219

Hanke, K. , Rusert, T. , Ohm, J.-R. : Motion-compensated 3D video coding using smooth transitions. Proc. SPIE VCIP (2003), vol. 5022, pp. 933-940

Heising, G. , Marpe, D. , Cycon, H. L. : A wavelet-based video coding scheme using image warping prediction. *Proc. IEEE ICIP* (1998), vol. I, pp. 84-86

Heising, G. , Marpe, D. , Cycon, H. L. , Petukhov, A. P. : Wavelet-based very low bit-rate video coding using image warping and overlapped block motion compensation, *IEE Proceedings - Vision, Image and Signal Processing* 148 (2001), no. 2, pp. 93-101

Heising, G. : Efficient and robust motion estimation in grid-based hybrid video coding schemes. *Proc. IEEE ICIP* (2002),Vol. I, pp. 697-700

Heitz, F. , Bouthemy, P. : Multimodal estimation of discontinuous optical flow using Markov random fields. *IEEE Trans. Patt. Anal. Mach. Intell.* 15 (1993), pp. 1217-1232

Hepper, D. : Efficiency analysis and application of uncovered background prediction in a low bit rate image coder. *IEEE Trans. Commun.* 38 (1990), pp. 1578-1584

Herley, C. , **Vetterli, M.** : Wavelets and recursive filter banks. *IEEE Trans. Signal Process.* 41 (1993), pp. 2536-2556

Hlawatsch, F. , **Boudreaux-Bartels, G. F.** : Linear and quadratic time-frequency signal representations. *IEEE Signal Process. Mag.* 9 (1992), no. 2, pp. 21-67

Ho, Y.-S. , **Gersho, A.** : Variable-rate multi-stage vector quantization for image coding. *Proc. IEEE ICASSP* (1988), pp. 1156-1159

Ho, Y.-S. , **Gersho, A.** : Classified transform coding of images using vector quantization. *Proc. IEEE ICASSP* (1989), pp. 1890-1893

Hötter, M. : Differential estimation of the global motion parameters zoom and pan. *Signal Process.* 16 (1989), pp. 249-265

Hötter, M. : Object-oriented analysis-synthesis coding based on moving two-dimensional objects. *Signal Process. : Image Commun.* 2 (1990), pp. 409-428

Hötter, M. : Optimization and efficiency of an object-oriented analysis-synthesis coder. *IEEE Trans. Circ. Syst. Video Tech.* 4 (1994), pp. 181-194

Horn, B. P. , **Schunck, B. G.** : Determining optical flow. *Artif. Intell.* 17 (1981), pp. 185-204

Horn, U. , **Stuhlmüller, K. W.** , **Link, M.** , **Girod, B.** : Robust Internet video transmission based on scalable coding and unequal error protection. Signal Proc.: Image Commun. 15 (1999), pp. 77-94

Hotelling, H. : Analysis of a complex of statistical variables into principal components. *Journal of Educational Psychology* 24 (1933), pp. 417-441,498-520

Hsiang, S.-T. , **Woods, J. W.** : Invertible three-dimensional analysis/synthesis system for video coding with half-pixel accurate motion compensation. *Proc. SPIE VCIP* (1999), vol. 3653, pp. 537-546

Hsiang, S.-T. , **Woods, J. W.** : Embedded image coding using zeroblocks of subband/ wavelet coefficients and context modeling. *Proc. IEEE ISCAS* (2000), vol. 3, pp. 662-665

Hsiang, S.-T. , **Woods, J. W.** : Embedded video coding using invertible motion-compensated 3D subband/wavelet filter bank. *Signal Proc.: Image Commun.* 16 (2001), pp. 705-724

Hsieh, C.-H. , **Chuang, K.-C.** , **Shue, J.-S.** : Image compression using finite-state vector quantization with derailment compensation. *IEEE Trans. Circ. Syst. Video Tech.* 3 (1993), pp. 341-349

Hu, M. K. : Visual pattern recognition by moment invariants. *IRE Trans. Inf. Theor.* 8 (1962), pp. 179-187

Huang, C.-L. , **Hsu, C.-Y.** : A new motion compensation method for image sequence coding using hierarchical grid interpolation. *IEEE Trans. Circ. Syst. Video Tech.* 4 (1994), pp. 42-51

Hubel, D. H. , **Wiesel, T. N.** : Receptive fields, binocular interaction and functional architecture in the cat's visual cortex. *J. Physiol. (London)* 160 (1962), pp. 106-154

Huber, P. J. : Robust Statistics. New York: Wiley 1981

Hürtgen, B. , **Büttgen, P.** : Fractal approach to low rate video coding. *Proc. Visual Commun. Image Process.* (1993), SPIE vol. 2094

Huffman, D. A. : A method for the construction of minimum redundancy codes. *Proc. IRE* 40 (1952), pp. 1098-1101

Hush, D. R. , **Horne, B. G.** : Progress in supervised neural networks. *IEEE Signal Process. Mag.* 10 (1993), no. 1, pp. 8-39

Husøy, J. H. , Ramstad, T. A. : Application of an efficient parallel IIR filter bank to image subband coding. *Signal Process.* 20 (1990), pp. 279-292

Itakura, F. , Saito, S. : On the optimum quantization of feature parameters in the PARCOR speech synthesizer. *Proc. 1992 Conf. Speech Commun.*, pp. 434-437

Itakura, F. , Sugamura, M. : LSP speech synthesizer, its principle and implementation. *Trans. Committee on Speech Res.*, ASJ, vol. S79-46, Nov. 1979

Izquierdo M., E. , Ohm, J.-R. : Image-based rendering and 3D modeling : A complete framework. *Signal Proc.: Image Commun.* 15 (2000), pp. 817-858

Jacobs, E. W. , Fisher, Y. , Boss, R. D. : Image compression : A study of the iterated transform method. *Signal Process.* 29 (1992), pp. 251-263

Jacquin, A. E. : Image coding based on a fractal theory of iterated contractive image transformations. *IEEE Trans. Image Process.* 1 (1992), pp. 18-30

Jähne, B. : Digital Image Processing. New York/Berlin: Springer 2002 (5th edition)

Jain, A. K. : A fast Karhunen-Loève transform for a class of random processes. *IEEE Trans. Commun.* 34 (1976), pp. 1023-1029

Jain, A. K. : Fundamentals of Digital Image Processing. Englewood Cliffs : Prentice-Hall 1989

Jain, J. R. , Jain, A. K. : Displacement measurement and its application in interframe image coding. *IEEE Trans. Commun.* 29 (1981), pp. 1799-1808

Jayant, N. S. , Noll, P. : Digital Coding of Waveforms. Englewood Cliffs : Prentice-Hall 1984

Jelinek, F. , Anderson, J. B. : Instrumentable tree encoding of information sources. *IEEE Trans. Inf. Theor.* 17 (1971), pp. 118-119

Jeong, D. G. , Gibson, J. D. : Uniform and piecewise uniform lattice vector quantization for memoryless Gaussian and Laplacian sources. *IEEE Trans. Inf. Theor.* 39 (1993), pp. 786-804

Johnston, J. D. : A filter family designed for use in quadrature mirror filter banks. *Proc. IEEE ICASSP* (1980), pp. 291-294

Kaiser, S. , Fazel, K. : Comparison of error concealment techniques for an MPEG-2 video decoder in terrestrial TV-broadcasting. *Signal Proc.: Image Commun.* 14 (1999), pp. 655-676

Kampmann, M. : Automatic 3-D face model adaptation for model-based coding of videophone sequences. *IEEE Trans. Circ. Syst. Video Tech.* 12 (2002), pp. 172-182

Kaneko, T. , Okudaira, M. : Encoding of arbitrary curves based on the chain code representation. *IEEE Trans. Commun.* 33 (1985), pp. 697-707

Karczewicz, M. , Kurceren, R. : The SP- and SI-frames design for H.264/AVC. *IEEE Trans. Circ. Syst. Video Tech.* 13 (2003), pp. 637-644

Karlsson, G. , Vetterli, M. : Subband coding of video signals for packet switched networks. *Proc. Visual Commun. Image Process.* (1987), SPIE vol. 845, pp. 446-456

Karlsson, G. , Vetterli, M. : Extension of finite length signals for subband coding. *Signal Process.* 17 (1989), pp. 161-168

Katto, J. , Ohki, J. , Nogaki, S. , Ohta, M. : A wavelet codec with overlapped motion compensation for very low bit-rate environment. *IEEE Trans. Circ. Syst. Video Tech.* 4 (1994), pp. 328-338

Kiang, S.-Z. , Baker, R. L. , Sullivan, G. J. , Chiu, C.-Y. : Recursive optimal pruning with applications to tree structured vector quantizers. *IEEE Trans. Image Process.* 1 (1992), pp. 162-169

Kim, B.-J. , **Pearlman, W.** : An embedded wavelet video coder using three-dimensional set partitioning in hierarchical trees. *Proc. Data Compression Conference* (1997), pp. 251-260

Kim, B.-J. , **Xiong, Z.** , **Pearlman, W. A.** : Low bit-rate scalable video coding with 3D SPIHT. *IEEE Trans. Circ. Syst. Video Tech.* 10 (2000), pp. 1374-1387

Kim, J.-S. , **Park, R.-H.** : Local motion-adaptive interpolation technique based on block matching algorithms. *Signal Process. : Image Commun.* 4 (1992), pp. 519-528

Kim, J. T. , **Lee, H. J.** , **Choi, J. S.** : Subband coding using human visual characteristics for image signals. *IEEE J. Sel. Areas Comm.* 11 (1993), pp. 59-64

Kim, T. : Side match and overlap match vector quantizers for images. *IEEE Trans. Image Process.* 1 (1992), pp. 170-185

Kim, Y. H. , **Modestino, J. W.** : Adaptive entropy coded subband coding of images. *IEEE Trans. Image Process.* 1 (1992), pp. 31-48

Kohonen, T. : Self-organized formation of topologically correct feature maps. *Biological Cybernetics* 43 (1982), pp. 59-69

Kohonen, T. : Self-Organization and Associative Memory. Berlin: Springer 1984 (3rd edition 1989)

Koilpillai, R. D. , **Vaidyanathan, P. P.** : Cosine-modulated FIR filter banks satisfying perfect reconstruction. *IEEE Trans. Signal Process.* 40 (1992), pp. 770-783

Konrad, J. , **Dubois, E.** : Bayesian estimation of motion vector fields. *IEEE Trans. Patt. Anal. Mach. Intell.* 14 (1992), pp. 910-927

Kovacevic, J. , **Vetterli, M.** : Nonseparable perfect reconstruction filter banks and wavelet bases for $\mathcal{R}^n$. *IEEE Trans. Inf. Theor.* 38 (1992), pp. 533-555

Kronander, T. : Some aspects of perception based image coding. *Dissertation*, Universität Linköping 1989

Kruse, S. : Scene segmentation from dense displacement vector fields using randomized Hough transform. *Signal Proc.: Image Commun.* 8 (1996), pp. 29-41

Kunt, M. , **Ikonomopoulos, A.** , **Kocher, M.** : Second-generation image coding techniques. *Proc. IEEE* 73 (1985), pp. 549-574

van Laarhoven, P. , **Aarts, E.** : Simulated Annealing : Theory and Applications. Dordrecht : Reidel 1987

Lazar, M. S. , **Bruton, L. T.** : Fractal block coding of digital video. *IEEE Trans. Circ. Syst. Video Tech.* 4 (1994), pp. 297-308

Leduc, J.-P. , **Odobez, J.-M.** , **Labit, C.** : Adaptive motion-compensated wavelet filtering for image sequence coding. *IEEE Trans. Image Proc.* 6 (1997), pp. 862-878

Le Gall, D. : MPEG : A video compression standard for multimedia applications. *Comm. ACM* 34 (1991), pp. 47-58

Le Gall, D. : The MPEG video compression algorithm. *Signal Process. : Image Commun.* 4 (1992), pp. 129-140

Le Gall, D. , **Tabatabai, A.** : Subband coding of digital images using symmetric short kernel filters and arithmetic coding techniques. *Proc. IEEE ICASSP* (1988), pp. 761-764

Lempel, A. , **Ziv, J.** : Compression of two-dimensional data. *IEEE Trans. Inf. Theor.* 32 (1986), pp. 2-8

Lewis, A. S. , **Knowles, G.** : Image compression using the 2D wavelet transform. *IEEE Trans. Image Proc.* 1 (1992), pp. 244-250

Li, J. , **Lei, S.** : An embedded still image coder with rate distortion optimization. *IEEE Trans. Image Proc.* 8 (1999), pp xx-xx (7, July 1999)

Li, W. : Overview of Fine Granular Scalability in MPEG-4 Video Standard. *IEEE Trans. Circ. Syst. Video Tech.* 11 (2001), pp. 385-398

Li, X. , Kerofsky, L. , Lei, S. : All-phase motion compensated prediction in the wavelet domain for high performance video coding. *Proc. IEEE ICIP* (2001), vol. 3, pp. 538-541

Li, X. , Kerofsky, L. : High performance resolution scalable video coding via all-phase motion compensated prediction of wavelet coefficients. *Proc. SPIE VCIP* (2002), vol. 4671, pp. 1080-1089

Liebchen, T. : MPEG-4 Lossless Coding for High-Definition Audio. *Proc. 115th AES Convention*, New York, October 2003

Lim, J. S. : Two-Dimensional Signal and Image Processing. Englewood Cliffs : Prentice-Hall 1990

Lim, K. P. , Chong, M. N. , Das, A. : Low-bit-rate video coding using dense motion field and uncovered background prediction. *IEEE Trans. Image Process.* 10 (2001), pp. 164–166.

Limb, J. O. , Murphy, J. A. : Measuring the speed of moving objects from television signals. *IEEE Trans. Commun.* 23 (1975), pp. 474-478

Lin, J.-N. , Unbehauen, R. : 2-D adaptive nonlinear equalizers. *Proc. EUSIPCO* (1992), *Signal Process. VI : Theor. and Appl.*, pp. 1381-1384. Amsterdam : Elsevier 1992

Lin, Y. , Astola, J. , Neuvo, Y. : A new class of nonlinear filters - Neural filters. *IEEE Trans. Signal Process.* 41 (1993), pp. 1201-1222

Linde, Y. , Buzo, A. , Gray, R. M. : An algorithm for vector quantizer design. *IEEE Trans. Commun.* 28 (1980), pp. 84-95

Ling, F. , Li, W. , Sun, H. : Bitplane coding of DCT coefficients for image and video compression. *Proc. SPIE VCIP* (1999), vol. 3653, pp. 500-508

Lippman, A. : Video coding for multiple target audiences. *Proc. SPIE VCIP* (1999), pp. 780-782

List, P. , Joch, A. , Lainema, J. , Bjontegaard, G. , Karczewicz, M. : Adaptive deblocking filter. *IEEE Trans Circ. Syst. Video Tech.* 13 (2003), pp. 614-619

Liu, B. , Zaccarin, A. : New fast algorithms for the estimation of block motion vectors. *IEEE Trans. Circ. Syst. Video Tech.* 3 (1993), pp. 148-157

Lloyd, S. P. : Least squares quantization in PCM. (1957) Reprint : *IEEE Trans. Commun.* 30 (1982), pp. 129-137

Lu, J. : Signal processing for Internet video streaming: A review. *Proc. SPIE VCIP* (2000), vol. 2974, pp. 246-259

Luettgen, M. R. , Karl, W. C. , Willsky, A. S. : Efficient multiscale regularization with applications to the computation of optical flow. *IEEE Trans. Image Process.* 3 (1994), pp. 41-64

Luo, L. , Li, J. , Li, S. , Zhuang, Z. , Zhang, Y.-Q. : Motion compensated lifting wavelet and its application in video coding. *Proc. ICME 2001*, July 2001

Maglaris, B. , Anastassiou, D. , Sen, P. , Karlsson, G. , Robbins, J. D. : Performance models of statistical multiplexing in packet video communications. *IEEE Trans. Commun.* 36 (1988), pp. 834-844

Mallat, S. : A theory for multiresolution signal decomposition : The wavelet representation. *IEEE Trans. Patt. Anal. Mach. Intell.* 11 (1989), pp. 674-693

Mallat, S. : Multifrequency channel decompositions of images and wavelet models. *IEEE Trans. Acoust., Speech, Signal Process.* 37 (1989), pp. 2091-2110

Malvar, H. S. : Signal Processing with Lapped Transforms. Norwood : Artech House 1991

Malvar, H. S. , Staelin, D. H. : The LOT : Transform coding without blocking effects. *IEEE Trans. Acoust., Speech, Signal Process.* 37 (1989), pp. 553-559

Malvar, H.S. , Hallapuro, A. , Karczewicz, M. , Kerofsky, L. : Low-complexity transform and quantization in H.264/AVC. *IEEE Trans. Circ. Syst. Video Tech.* 13 (2003), pp. 598- 603

Mandelbrot, B. B. : The Fractal Geometry of Nature. San Francisco : Freeman 1982

Maragos, P. A. , Schafer, R. W. : Morphological skeleton representation and coding of binary images. *IEEE Trans. Acoust., Speech, Signal Process.* 34 (1986), pp. 1228-1244

Maragos, P. A. , Schafer, R. W. , Mersereau, R. M. : Two-dimensional linear prediction and its application to adaptive predictive coding of images. *IEEE Trans. Acoust., Speech, Signal Process.* 32 (1984), pp. 1213-1229

Marcellin, M. W. , Fischer, T. R. : Trellis coded quantization of memoryless and Gauss-Markov sources. *IEEE Trans. Commun.* 38 (1990), pp. 82-93

Marcellin, M. W. , Sriram, P. , Tong, K.-L. : Transform coding of monochrome and color images using trellis coded quantization. *IEEE Trans. Circ. Syst. Video Tech.* 3 (1993), pp. 270-276

Marpe, D. , Cycon, H. L. : Very low bit-rate video coding using wavelet-based techniques. *IEEE Trans. Circ. Syst. Video Tech.* 9 (1999), pp. 85–94

Marpe, D. , Blättermann, G. , Ricke, J. , Maaß, P. : A two-layered wavelet-based algorithm for efficient lossless and lossy image compression. *IEEE Trans. Circ. Syst. Video Tech.* 10 (2000), pp. 1094-1102

Marpe, D. , Schwarz, H. , Blättermann, G. , Heising, G. , Wiegand, T. : Context-based adaptive binary arithmetic coding in JVT/H.26L. *Proc. IEEE ICIP* (2002), Vol. II, pp. 513-516

Marpe, D. , Schwarz, H. , Wiegand, T. : Context-based adaptive binary arithmetic coding in the H.264/AVC video compression standard. *IEEE Trans. Circ. Syst. Video Tech.* 13 (2003), pp. 620-636

Marr, D. , Nishihara, K. : Representation and recognition of the spatial organization of 3D shapes. *Proc. Royal Soc. London B* 200 (1978), pp. 269-294

Martucci, S. A. , Sodagar, L. , Chiang, T. H. , Zhang, Y.-Q. : A zerotree wavelet coder. *IEEE Trans. Circ. Syst. Video Tech.* 7 (1997), pp. 109-118

Mathews, M. : Introduction to timbre. In *Music, Cognition and Computerized Sound*, P.R. Cook (ed.), Cambridge: MIT Press, 1999

Max, T. : Quantizing for minimum distortion. *IRE Trans. Inf. Theor.* 6 (1960), pp. 7-12

Mayer, C. , Höynck, M. : Linear phase perfect reconstruction filter banks with alias-reduced lowpass filters for spatially scalable video coding. *Proc. 10th Aachen Symp. Signal Theory* (2001), pp. 397-402, Aachen, September 2001

Mayer, C. , Crysandt, H. , Ohm, J.-R. : Bit plane quantization for scalable video coding. *Proc. SPIE VCIP* (2002), vol. 4671, pp. 1142-1152

McCanne, S. , Vetterli, M. , Jacobson, V. ; Low-complexity video coding for receiver-driven layered multicast. *IEEE J. Sel. Areas Comm.* 16 (1997), pp. 983-1001

McMillan, L. , Bishop, G. ; Plenoptic modeling: An image-based rendering system. *Proc. SIGGRAPH* (1995), Los Angeles, Aug. 1995

Meiri, A. Z. , Yudilevich, E. : A pinned sine transform image coder. *IEEE Trans. Commun.* 29 (1981), pp. 1728-1735

Meer, P. , Sher, C. A. , Rosenfeld, A. : The chain pyramid : Hierarchical contour processing. *IEEE Trans Patt. Anal. Mach. Intell.* 12 (1990), pp. 363-376

Meşe, M. , Vaidyanathan, P. P. : Optimized Halftoning using dot diffusion and methods for inverse halftoning. *IEEE Trans. Image Process.* 9 (2000), pp. 691-709

Miyahara, M. : Quality assessments for visual service. *IEEE Commun. Mag.*, Oct. 1988, pp. 51-60

Modestino, J. W. , Bhaskaran, V. : Robust two-dimensional tree encoding of images. *IEEE Trans. Commun.* 29 (1981), pp. 1786-1791

Mokhtarian, F. , Bober, M. : Curvature Scale Space Representation: Theory, Applications and MPEG-7 Standardization. Dordrecht: Kluwer Academic Publishers, 2003

Mohan, S. , Sood, A. K. : A multiprocessor architecture for the (M,L)-algorithm suitable for VLSI implementation. *IEEE Trans. Commun.* 34 (1986), pp. 1218-1224

Moorhead, R. J., II , Rajala, S. A. , Cook, L. W. : Image sequence compression using a pel-recursive motion-compensated technique. *IEEE J. Sel. Areas Commun.* 5 (1987), pp. 1100-1114

Morikawa, H. , Harashima, H. : 3-D structure extraction coding of image sequences. *Proc. IEEE ICASSP* (1990), pp. 1969-1972

Müller, K. , Ohm, J.-R. : Contour description using wavelets. *Proc. WIAMIS'99*, pp. 77-80, Berlin, May 1999

Muirhead, R. J. : Aspects of Multivariate Statistical Theory. New York: Wiley, 1982

Muirhead, R. J. , Chen, Z. : A comparison of robust linear discriminant procedures using projection pursuit methods. In *Multivariate Analysis and its Applications,* T. W. Anderson, K. T. Fang and I. Olkin (eds.), pp. 163-176, Hayward: IMS Monograph Series, 1994

Munteanu, A. , Cornelis, J. , Van der Auwera, G. , Cristea, P. : Wavelet-based lossless compression scheme with progressive transmission capability. *Int. J. Imaging Syst. Technol.* 10 (1999), pp. 76-85

Murakami, H. , Hashimoto, H. , Hatori, Y. : Quality of band-compressed TV services. *IEEE Commun. Mag.*, Oct. 1988, pp. 61-69

Murphy, N. : Fractal block coding in image and video compression. *COST 211ter Workshop : New techniques for coding of video signals at very low bitrates*, Hannover 1993

Musmann, H. G. , Pirsch, P. , Grallert, H. J. : Advances in picture coding. *Proc. IEEE* 73 (1985), pp. 523-548

Musmann, H. G. , Hötter, M. , Ostermann, J. : Object-oriented analysis-synthesis coding of moving images. *Signal Process. : Image Commun.* 1 (1989), pp. 117-138

Nam, K. M. , Kim, J.-S. , Park, R.-H. : A fast hierarchical motion vector estimation algorithm using mean pyramid, *IEEE Trans. Circ. Syst. Video Tech.* 5 (1995), pp. 341-351

Nanda, S. , Pearlman, W. A. : Tree coding of image subbands. *IEEE Trans. Image Process.* 1 (1992), pp. 132-147

Nasrabadi, M. N. , King, R. A. : Image coding using vector quantization : A review. *IEEE Trans. Commun.* 36 (1988), pp. 957-971

Natarajan, T. , Ahmed, N. : On Interframe Transform Coding. *IEEE Trans. Comm.* 25 (1977), pp. 1323-1329

Naveen, T. , Woods, J. W. : Motion compensated multiresolution transmission of high definition video. *IEEE Trans. Circ. Syst. Video Tech.* 4 (1994), pp. 29-41

van Nee, R. , Prasad, R. : OFDM for Wireless Multimedia Communications. Boston: Artech House, 2000

Netravali, A. N. , Haskell, B. G. : Digital pictures - Representation and compression. New York and London : Plenum Press 1988

Netravali, A.N. , Robbins, J. D. : Motion compensated television coding, Part I. *Bell Syst. Tech. J.* 58 (1979), pp. 631-670

Netravali, A. N. , Robbins, J. D. : Motion-compensated coding : Some new results. *Bell Syst. Tech. J.* 59 (1980), pp. 1735-1745

Neuhoff, D. L. , Lee, D. H. : On the performance of tree-structured vector quantization. *Proc. IEEE ICASSP* (1991), pp. 2277-2280

Ngan, K. N. , Chooi, W. L. : Very low bit rate video coding using 3D subband approach. *IEEE Trans. Circ. Syst. Video Tech.* 4 (1994), pp. 309-316

Ngan, K. N. , Koh, H. C. : Predictive classified vector quantization. *IEEE Trans. Image Process.* 1 (1992), pp. 269-280

Ngan, K. N. , Leong, K. S. , Singh, H. : Adaptive cosine transform coding of images in perceptual domain. *IEEE Trans. Acoust., Speech, Signal Process.* 37 (1989), pp. 1743-1750

Nicoulin, M. , Mattavelli, M. , Li, W. , Basso, A. , Popat, A. C. , Kunt, M. : Image sequence coding using motion-compensated subband decomposition. *Motion Analysis and Image Sequence Processing*, M. I. Sezan and R. L. Lagendijk (eds.), Dordrecht : Kluwer 1993

Nikias, C. L. , Mendel, J. M. : Signal Processing with higher-order spectra. *IEEE Signal Process. Mag.* 10 (1993), no. 3, pp. 10-37

Ohm, J.-R. , Noll, P. : Predictive tree encoding of still images with vector quantization. *Annales Télécommun.* 45 (1990), pp. 465-470

Ohm, J.-R. : A Hybrid Image Coding Scheme for ATM Networks Based on SBC-VQ and Tree Encoding. *Proc. 4th Intern Workshop Packet Video*, pp. B.2-1 - B.2-6, Kyoto, August 1991

Ohm, J.-R. : Temporal domain subband video coding with motion compensation. *Proc. IEEE ICASSP* (1992), vol. III, pp. 229-232

Ohm, J.-R. : Advanced packet-video coding based on layered VQ and SBC techniques. *IEEE Trans. Circ. Syst. Video Tech.* 3 (1993), pp. 208-221

Ohm, J.-R. : Three-dimensional subband coding with motion compensation. *IEEE Trans. Image Process.* 3 (1994), pp. 559-571 [OHM 1994A]

Ohm, J.-R. : Motion-compensated 3-D subband coding with multiresolution representation of motion parameters. *Proc. IEEE ICIP* (1994), vol. III, pp. 250-254 [OHM 1994B]

Ohm, J.-R. : Motion grid interpolation with simple contour adaptation. *Proc. Picture Coding Symposium* (1996), Melbourne, March 1996

Ohm, J.-R. , Rümmler, K. : Variable-raster multiresolution video processing with motion compensation techniques. *Proc. IEEE ICIP* (1997), vol.I, pp. 759-762

Ohm, J.-R. , Ma, P. : Feature-based cluster segmentation of image sequences. *Proc. IEEE ICIP* (1997), vol.III, pp. 178-181

Ohm, J.-R. , Grüneberg, K. , Hendriks, E. , Izquierdo M., E. , Kalivas, D. , Karl, M. , Papadimatos, D. , Redert, A. : A realtime hardware system for stereoscopic videoconferencing with viewpoint adaptation. *Signal Process.: Image Commun.* 14 (1998), pp.147-171

Ohm, J.-R. , Müller, K. : Incomplete 3D - Multiview representation of video objects. *IEEE Trans. Circ. Syst. Video Tech.* 9 (1999), pp. 389-400

Ohm, J.-R. : Stereo/Multiview Encoding Using the MPEG Family of Standards. *Proc. SPIE Stereosc. Displ. Virt. Real. Syst.* (1999), vol. 3639, pp. 242-253

Ohm, J.-R. : Encoding and reconstruction of multiview video objects. *IEEE Signal Process. Mag.* (1999), pp. 47-54

Ohm, J.-R. , Bunjamin, F. , Liebsch, W. , Makai, B. , Müller, K. , Smolic, A. , Zier, D. : A set of visual feature Descriptors and their combination in a low-level Description Scheme. *Signal Process.: Image Comm.* 16 (2000), pp. 157-179

Ohm, J.-R. , Cieplinski, L. , Kim, H. J. , Krishnamachari, S. , Manjunath, B. S. , Messing, D. S. , Yamada, A. : Color Descriptors. In *Introduction to MPEG-7*, B.S. Manjunath, P. Salembier, T. Sikora (eds.), pp. 187-212, New York: Wiley 2002

Okubo, S. : Requirements for high quality video coding standards. *Signal Process. : Image Commun.* 4 (1992), pp. 141-151

Oppenheim, A. V. , Willsky, A. S. , Young, I. T. : Signals and Systems. 2nd edition Englewood Cliffs : Prentice-Hall 1997

Orchard, M. T. : Predictive motion field segmentation for image sequence coding. *IEEE Trans. Circ. Syst. Video Tech.* 3 (1993), pp. 54-70

Orchard, M. T. , Sullivan, G. J. : Overlapped block motion compensation : An estimation-theoretic approach. *IEEE Trans. Image Proc.* 3 (1994), pp. 693-699

Ortega, A. , Ramchandran, K. , Vetterli, M. : Optimal trellis-based buffered compression and fast approximations. *IEEE Trans. Image Process.* 3 (1994), pp. 26-40

Ostermann, J. : Object-based analysis-synthesis coding based on the source model of moving rigid 3D objects. *Signal Process. : Image Commun.* 6 (1994), pp. 143-161

Ostermann, J. : Object-based analysis-synthesis coding (OBASC) based on the source model of moving flexible 3-D objects. *IEEE Trans. Image Proc.* 3 (1994), pp. 705-710

Pang, K. K. , Tan, T. K. : Optimum loop filters in hybrid coders. *IEEE Trans. Circ. Syst. Video Tech.* 4 (1994), pp. 158-167

Paget, R. , Longstaff, I. D. : Texture synthesis via a noncausal nonparametric multiscale Markov random field. *IEEE Trans. Image Process.* 7 (1998), pp. 925-931

Papoulis, A. : Probability, Random Variables and Stochastic Processes. New York: McGraw Hill 1984

Park, H. W. , Lee, Y. L. : A postprocessing method for reducing quantization effects in low bit-rate moving picture coding. *IEEE Trans. Circ. Syst. Video Tech.* 9 (1999), pp. 161-171

Pearlman, W. A. : Performance bounds for subband coding. *Subband Image Coding*, J. W. Woods (ed.), Dordrecht : Kluwer 1991

Pearlman, W. A. , Jakatdar, P. : A transform tree code for stationary Gaussian sources. *IEEE Trans. Inf. Theor.* 31 (1985), pp. 761-768

Pearlman, W. A. , Jakatdar, P. , Leung, M. M. : Adaptive transform tree coding of images. *IEEE J. Sel. Areas Commun.* 10 (1992), pp. 902-912

Pearson, D. E. : Texture mapping in model-based image coding. *Signal Process. : Image Commun.* 2 (1990), pp. 377-395

Peleg, S. , Rousso, B. , Rav-Acha, A. , Zomet, A. : Mosaicing on adaptive manifolds. IEEE Trans. Patt. Anal. Mach. Intell. 22 (2000), pp. 1144-1154

Pennebaker, W. B. , Mitchell, J. L. : JPEG Still Image Compression Standard. New York: Van Nostrand Reinhold 1993

Pennebaker, W. B. , Mitchell, J. L. , Langdon, G. G. , Arps, R. B. : An overview of the basic principles of the Q-coder adaptive binary arithmetic coder. *IBM J. Res. Develop.* 32 (1988), pp. 717-752

Perkins, M. G. , Lookabaugh, T. : A psychophysically justified bit allocation algorithm for subband image coding systems. *Proc. IEEE ICASSP* (1989), pp. 1815-1818

Perona, P. , Malik, J. : Scale-space and edge detection using anisotropic diffusion. *IEEE Trans. Patt. Anal. Mach. Intell.* 12 (1990), pp. 629-639

Pesquet-Popescu, B. , Bottreau, V. : Three-dimensional lifting schemes for motion-compensated video compression. *Proc. IEEE ICASSP* (2001), pp. 1793-1796

Plataniotis, K. N. , Venetsanopoulos, A. N. : Color Image Processing and Applications. New York/Berlin: Springer, 2000

Platt, S. M. , Badler, N. I. : Animating facial expressions. *IEEE Comput. Graph.* 13 (1981), pp. 242-245

Po, L.-M. , Ma, W.-C. : A Novel Four Step Search Algorithm For Fast Block Motion Estimation. IEEE Trans. Circ. Syst. Video Tech. 6 (1996), pp 313-317

Podilchuk, C. I. , Jayant, N. S. , Noll, P. : Sparse codebooks for the quantization of non-dominant subbands in image coding. *Proc. IEEE ICASSP* (1990), pp. 2837-2840

Portilla, J. , Simoncelli, E. P. : A parametric texture model based on joint statistics of complex wavelet coefficients. *Intern. J. Comp. Vision* 40 (2000), no. 1

Princen, J. P. , Bradley, A. B. : Analysis/synthesis filter bank design based on time domain aliasing cancellation. *IEEE Trans. Acoust., Speech, Signal Process.* 34 (1986), pp. 1153-1161

Puri, A. , Aravind, R. : Motion-compensated video coding with adaptive perceptual quantization. *IEEE Trans. Circ. Syst. Video Tech.* 1 (1990), pp. 351-361

Puri, A. , Aravind, R. , Haskell, B. G. : Adaptive frame/field motion compensated video coding. *Signal Process. : Image Commun.* 5 (1993), pp. 39-58

Puri, A. , Aravind, R. , Haskell, B. G. , Leonardi, R. : Video coding with motion-compensated interpolation for CD-ROM applications. *Signal Process. : Image Commun.* 2 (1990), pp. 127-144

Queluz, M. P. : A 3-dimensional subband coding scheme with motion-adaptive subband selection. *Proc. EUSIPCO* (1992), *Signal Process. VI : Theor. and Appl.*, pp. 1263-1266, Amsterdam : Elsevier 1992

Rabiner, L. R. , Schafer, R. W. : Digital Processing of Speech Signals. Englewood Cliffs: Prentice-Hall 1978

Rabiner, L. : A tutorial on Hidden Markov Models and selected applications in speech recognition. *Proc. IEEE* 77 (1989), pp. 257-286

Ramamurthy, B. , Gersho, A. : Classified vector quantization of images. *IEEE Trans. Commun.* 34 (1986), pp. 1105-1115

Ramasubramanian, V. , Paliwal, K. K. : Fast K-dimensional tree algorithms for nearest neighbor search with application to vector quantization encoding. *IEEE Trans. Signal Process.* 40 (1992), pp. 518-531

Ramchandran, K. , Vetterli, M. : Best wavelet packet bases in a rate-distortion sense. *IEEE Trans. Image Process.* 2 (1993), pp. 160-175

Ramchandran, K. , Ortega, A. , Vetterli, M. : Bit Allocation for dependent quantization with applications to multiresolution and MPEG video coders. *IEEE Trans. Image Process.* 3 (1994), pp. 533-545

Ranganath, S. , Jain, A. K. : Two-dimensional linear prediction models - Part I : Spectral factorization and realization. *IEEE Trans. Acoust., Speech and Signal Process.* 33 (1985), pp. 280-299

Reed, I. S. , Solomon, G. : Polynomial codes over certain finite fields. *SIAM J. Appl. Math.* 8 (1960), pp. 300-304

Reibman, A. R. : DCT-based embedded coding for packet video. *Signal Process. : Image Commun.* 3 (1991), pp. 231-237

Reibman, A. R. , Haskell, B. G. : Constraints on variable bit-rate video for ATM networks. *IEEE Trans. Circ. Syst. Video Tech.* 2 (1992), pp. 361-372

Reininger, R. C. , Gibson, J. D. : Distribution of two-dimensional DCT coefficients for images. *IEEE Trans. Commun.* 31 (1983), pp. 835-839

Reusens, E. : Sequence coding based on the fractal theory of iterated transformation systems. *Proc. Visual Commun. Image Process.* (1993), SPIE vol. 2094

Ribas-Corbera, J. , Chou, P. A. , Regunathan, S. L. : A flexible decoder buffer model for JVT video coding. *Proc. IEEE ICIP* (2002), vol. II, pp. 493-496

Ribas-Corbera, J. , Chou, P. A. , Regunathan, S. L. : A generalized hypothetical reference decoder for H.264/AVC. *IEEE Trans. Circ. Syst. Video Tech.* 13 (2003), pp. 674-687

Richard, C. , Benveniste, A. , Kretz, F. : Recursive estimation of local characteristics of edges as applied to ADPCM coding. *IEEE Trans. Commun.* 32 (1984), pp. 718-728

Rioul, O. , Vetterli, M. : Wavelets and signal processing. *IEEE Signal Process. Mag.* 8 (1991), no. 4, pp. 14-38

Riskin, E. A. , Gray, R. M. : A greedy tree growing algorithm for the design of variable rate vector quantizers," *IEEE Trans. Signal Process.* 39 (1991), pp. 2500-2507

Robson, J. G. : Spatial and temporal contrast sensitivity functions of the visual system. *J. Opt. Soc. Amer. A* 56 (1966), pp, 1141-1142

Roese, J. , Pratt, W. , Robinson, G. : Interframe Cosine Transform Image Coding. *IEEE Trans. Commun.* 25 (1977), pp. 1329- 1339

Rosenfeld, A. , Kak, A. C. : Digital Picture Processing. Orlando : Academic Press 1982

Rotem, E. , Malah, D. : Subband coding of images using a trellis coder. *Proc. EUSIPCO* (1988), *Signal Procress. IV : Theories and Applications*, pp. 1153-1156, Amsterdam : Elsevier 1988

Rusert, T. , Hanke, K. , Ohm, J.-R. : Transition filtering and optimized quantization in interframe wavelet video coding, Proc. SPIE VCIP 2003, vol. 5150, pp. 682-694

Rydfalk, M. : CANDIDE, a parameterized face. *Report* No. LiTH-ISY-I-866, *Dept. of Electrical Engineering*, Linköping University, Sweden, 1987.

Sabin, M. , Gray, R. M. : Product code vector quantizers for waveform and voice coding. *IEEE Trans. Acoust., Speech, Signal Process.* 32 (1984), pp. 474-488

Sabin, M. , Gray, R. M. : Global convergence and empirical consistency of the generalized Lloyd algorithm. *IEEE Trans. Inf. Theor.* 32 (1986), pp. 148-155

Safranek, R. J. , Johnston, J. D. : A perceptually tuned subband image coder with image dependent quantization and post-quantization data compression. *Proc. IEEE ICASSP* (1989), pp. 1945-1948

Said, A. , Pearlman, W. : A new fast and efficient image codec based on set partitioning in hierarchical trees. *IEEE Trans. Circ. Syst. Video Tech.* 6 (1996), pp. 243-250

Saito, H. , Kawarasaki, M. , Yamada, H. : An analysis of statistical multiplexing in an ATM transport network. *IEEE J. Sel Areas Commun.* 9 (1991), pp. 359-367

Sakrison, D. J. : Image coding applications of vision models. In *Image Transmission Techniques*, W. K. Pratt (ed.), London : Academic Press 1979

Salembier, P. , Pardàs, M. : Hierarchical morphological segmentation for image sequence coding. *IEEE Trans. Image Proc.* 3 (1994), pp. 639-651

Salomon, D. , Ewens, W. J. : Data Privacy: Encryption and Information Hiding. Berlin: Springer 2003

Samet, H. : The quadtree and related hierarchical data structures. *Comput. Surv.* 16 (1984), pp. 187-260

Saruwaturi, H. , Kawamura, T. , Shikano, K. : Blind source separation for speech based on fast convergence algorithm with ICA and beamforming. Proc. Eurospeech (2001), pp.2603-2606

Sayood, K. , Gibson, J. D. , Rost, M. C. : An algorithm for uniform vector quantizer design. *IEEE Trans. Inf. Theor.* 30 (1984), pp. 805-814

van der Schaar, M. , Radha, H. : A hybrid temporal-SNR fine-granular scalability for internet video. *IEEE Trans. Circ. Syst. Video Tech.* 11 (2001), pp. 318-331

van der Schaar, M. , Chen, Y. , Radha, H. : Embedded DCT and Wavelet Methods for Fine Granular Scalable Video: Analysis and Comparison. *Proc. SPIE VCIP* (2000), vol. 2974, pp. 643-653

Shafer, G. : A mathematical theory of evidence. Princeton: Princeton University Press 1976

Secker, A. , Taubman, D. : Motion-compensated highly-scalable video compression using an adaptive 3D wavelet transform based on lifting. *Proc. IEEE ICIP* (2001), pp. 1029-1032

Seitz, S. M. , Dyer, Ch. R. : Physically-valid view synthesis by image interpolation. *Proc. IEEE Worksh. on Represent. of Visual Scenes* pp. 18-25, Cambridge 1995

Senoo, T. , Girod, B. : Vector quantization for entropy coding of image subbands. *IEEE Trans. Image Process.* 1 (1992), pp. 526-533

Serra, J. : Image Analysis and Mathematical Morphology, vols. 1 and 2. San Diego : Academic Press 1982/1988

Servetto, S. , Ramchandran, K. , Orchard, M. T. : Wavelet based image coding via morphological prediction of significance. *Proc. IEEE ICIP* (1995), pp. 530-533

Shannon, C. E. : A mathematical theory of communication. *Bell Syst. Tech. J.* (1948), reprint : University of Illinois press 1949

Shannon, C. E. : Coding theorems for a discrete source with a fidelity criterion. *IRE Nat. Conv. Rec.* (1959), reprint : *Information and Decision Processes*, R. E. Machol (ed.), New York : McGraw Hill 1960

Shannon, C. E. : The Collected Papers. Piscataway : IEEE Press 1992

Shapiro, J. M. : Embedded image coding using zerotrees of wavelets coefficients. *IEEE Trans. Signal Process.* 41 (1993), pp. 3445-3462

Shariat, H. , Price, K. E. : Motion estimation with more than two frames. *IEEE Trans. Patt. Anal. Mach. Intell.* 12 (1990), pp. 417-434

Shen, K. , Delp, E. : Wavelet based rate scalable video compression. IEEE Trans. Circ. Syst. Video Tech. 9 (1999), pp. 109-122

Shepard, R. : Pitch perception and measurement. In *Music, Cognition and Computerized Sound*, P.R. Cook (ed.), Cambridge: MIT Press, 1999

Shishikui, Y. : A study on modeling of the motion compensation prediction error signal. *IEICE Trans. Commun.* 75-B (1992), pp. 368-376

Sikora, T. , Makai, B. : Shape-adaptive DCT for generic coding of video. *IEEE Trans. Circ. Syst. Video Tech.* 5, pp. 59-62, 1995

Silsby, P. L. , Bovik, A. C. , Chen, D. : Visual pattern image sequence coding. *IEEE Trans. Circ. Syst. Video Tech.* 3 (1993), pp. 291-301

Simon, S. F. : On suboptimal multidimensional companding. *Proc. IEEE Data Compress. Conf.* (1998), pp. 438-447

Simoncelli, E. P. , Adelson, E. H. : Non-separable extensions of quadrature mirror filters to multiple dimensions. *Proc. IEEE* 78 (1990), pp. 652-663

Simoncelli, E. P. , Adelson, E. H. : Subband Transforms. *Subband Image Coding*, J. W. Woods (ed.), Dordrecht : Kluwer 1991

Smith, M. J. T. : IIR analysis/synthesis systems. *Subband Image Coding*, J. W. Woods (ed.), Dordrecht : Kluwer 1991

Smith, M. J. T. , Barnwell, T. P. , III : Exact reconstruction techniques for tree-structured subband coders. *IEEE Trans. Acoust., Speech, Signal Process.* 34 (1986), pp. 434-441

Smith, M. J. T. , Eddins, S. L. : Analysis/synthesis techniques for subband image coding. *IEEE Trans. Acoust., Speech, Signal Process.* 38 (1990), pp. 1446-1456

Smolic, A. , Sikora, T. , Ohm, J.-R. : Long-term global motion estimation and application for sprite coding, content description and segmentation. *IEEE Trans. Circ. Syst. Video Tech.* (1999), pp. 1227-1242

Smolic, A. , Wiegand, T. : High-Resolution Video Mosaicing. *Proc. IEEE ICIP* (2001), vol. III, pp. 872-875

Snyder, M. A. : On the mathematical foundations of smoothness constraints for the determination of optical flow and for surface reconstruction. *IEEE Trans. Patt. Anal. Mach. Intell.* 13 (1991), pp. 1105-1114

Soman, A. K. , Vaidyanathan, P. P. : On orthonormal wavelets and paraunitary filter banks. *IEEE Trans. Signal Process.* 41 (1993), pp. 1170-1183

Steinbach, E. G. , Färber, N. , Girod, B. : Adaptive playout for low latency video streaming. Proc. IEEE ICIP (2001), vol. I, pp. 962-965

Stewart, L. C. , Gray, R. M. , Linde, Y. : The design of trellis waveform coders. *IEEE Trans. Commun.* 30 (1982), pp. 702-710

Stiller, C. , Hürtgen, B. : Combined displacement estimation and segmentation in image sequences. *Proc. EUROPTO Image Commun. Video Compr.* (1993), SPIE vol. 1977, pp. 276-287

Stollnitz, E. , DeRose, T. , Salesin, D. : Wavelets for Computer Graphics. San Francisco: Morgan Kaufmann, 1996

Storer, J. A. : Image and Text Compression. Dordrecht : Kluwer 1993

Strintzis, M. G. : Test of stability of multidimensional filters. *IEEE Trans. Circ. Syst.* 24 (1977), pp. 432-437

Sullivan, G. J. , Baker, R. L. : Motion compensation for video compression using control grid interpolation. *Proc. IEEE ICASSP* (1991), pp. 2713-1716

Sullivan, G. J. , Wiegand, T. : Rate-Distortion Optimization for Video Compression. *IEEE Signal Proc. Mag.* 15, pp. 74-90

Sun, M.-T. , Reibman, A. R. (editors) : Compressed Video over Networks. Marcel Dekker, 2000

Sweldens, W. : The lifting scheme : A new philosophy in biorthogonal wavelet constructions. *Proc. SPIE* vol. 2569 (1995), pp. 68-79

Takashima, Y. , Wada, M. , Murakami, H. : Reversible variable length codes. *IEEE Trans. Commun.* 43 (1995), pp. 158-162

Tan, T. K. , Pang, K. K. , Ngan, K. N. : A frequency scalable coding scheme employing pyramid and subband techniques. *IEEE Trans. Circ. Syst. Video Tech.* 4 (1994), pp. 203-207

Tan, W. , Zakhor, A. : Real-time Internet video using error resilient scalable compression and TCP-friendly transport protocol. *IEEE Trans. Multimedia* (1999), pp. 172-186

Tanabe, N. , Farvardin, N. : Subband image coding using entropy-coded quantization over noisy channels. *IEEE J. Sel. Areas Commun.* 10 (1992), pp. 926-943

Taubman, D. , Zakhor, A. : Multirate 3-D subband coding of video. *IEEE Trans. Image Proc.* 3 (1994), pp. 572-588

Taubman, D. : High performance scalable image compression with EBCOT. *IEEE Trans. Image Proc.* 9 (2000), pp. 1158-1170

Taubman, D. S. , Marcellin, M. W. : JPEG 2000: Image Compression Fundamentals, Standards and Practice. Dordrecht: Kluwer Academic Publishers, 2001

Terzopoulos, D. , Waters, K. : Analysis and synthesis of facial image sequences using physical and anatomical models. *IEEE Trans. Patt. Anal. Mach. Intell.* 15 (1993), pp. 569-579

Tewfik, A. H. , Sinha, D. , Jorgensen, P. : On the optimal choice of a wavelet for signal representation. *IEEE Trans. Inf. Theor.* 38 (1992), pp. 747-765

Thoma, R. , Bierling, M. : Motion compensating interpolation considering covered and uncovered background. *Signal Process. : Image Commun.* 1 (1989), pp. 191-212

Tubaro, S. , Rocca, F. : Motion field estimators and their application to image interpolation. *Motion Analysis and Image Sequence Processing*, M. I. Sezan and R. L. Lagendijk (eds.), Dordrecht : Kluwer 1993

Turk, M. , Pentland, A. : Eigenfaces for Recognition. *J. Cognitive Neuroscience* 3 (1991), pp. 71-96

Ungerboeck, G. : Adaptive Maximum Likelihood Receiver for Carrier Modulated data Transmission Systems. *IEEE Trans. Commun.* 22, 624-636

Unser, M. , Aldroubi, A. , Eden, M. : B-spline signal processing. *IEEE Trans. Signal Process.* 41 (1993), pp. 821-833 (Part I-Theory), pp. 834-848 (Part II-Efficient design and applications)

Uz, K. M. , Ramchandran, K. , Vetterli, M. : Combined multiresolution source coding and modulation for digital broadcast of HDTV. *Signal Process. : Image Commun.* 4 (1992), pp. 283-292

Uz, K. M. , Vetterli, M. , Le Gall, D. J. : Interpolative multiresolution coding of advanced television with compatible subchannels. *IEEE Trans. Circ. Syst. Video Tech.* 1 (1991), pp. 86-99

Vaidyanathan, P. P. : Quadrature mirror filter banks, M-band extensions and perfect-reconstruction techniques. *IEEE ASSP Mag.* 4 (1987), no. 3, pp. 4-20

Vaidyanathan, P. P. : Multirate digital filters, filter banks, polyphase networks, and applications : A tutorial. *Proc. IEEE* 78 (1990), pp. 56-93

Vaidyanathan, P. P. : Multirate Systems and Filter Banks. Englewood Cliffs: Prentice Hall 1993

Vaidyanathan, P. P. , Regalia, P. A. , Mitra, S. K. : Design of doubly-complementary IIR digital filters using a single complex allpass filter, with multirate applications. *IEEE Trans. Circ. Syst.* 34 (1987), pp. 378-389

Vaisey, J. , Gersho, A. : Image compression with variable block size segmentation. *IEEE Trans. Signal Process.* 40 (1992), pp. 2040-2060

Vaishampayan, V. A. : Design of multiple description scalar quantizers. *IEEE Trans. Inf. Theor.* 39 (1993), pp. 821-834

Vetterli, M. : Multi-dimensional subband coding : some theory and algorithms. *Signal Process.* 6 (1984), pp. 97-112

Vetterli, M. : Multirate filter banks for subband coding. *Subband Image Coding*, J. W. Woods (ed.), Dordrecht : Kluwer 1991

Vetterli, M. , Herley, C. : Wavelets and filter banks: Theory and design. *IEEE Trans. Signal Proc.* 40 (1992), pp. 2207-2232

Vetterli, M. , Kovacevic, J. : Wavelets and Subband Coding. Englewood Cliffs: Prentice Hall, 1995

Vetterli, M. , Uz, K. M. : Multiresolution coding techniques for digital television : A review. *Multidim. Syst. Signal Process.* 3, pp. 161-188. Dordrecht : Kluwer 1992

Viswanathan, R. , Makhoul, J. : Quantization properties of transmission parameters in linear predictive systems. *IEEE Trans. Acoust., Speech, Signal Proc.* 23 (1975), pp. 309-321

Viterbi, A. J. : Error bounds for convolutional codes and an asymptotically optimum decoding algorithm. *IEEE Trans. Inf. Theor.* 13 (1967), pp. 260-269

Viterbi, A. J. , Omura, J. K. : Principles of Digital Communication and Coding. New York : McGraw Hill 1985

Wallace, G. K. : The JPEG still picture compression standard. *Comm. ACM* 34 (1991), pp. 31-44

Wang, J. Y. A. , Adelson, E. H. : Representing moving images with layers. *IEEE Trans. Image Proc.* 3 (1994), pp. 625-638

Wang, Y. , Lee, O. : Active mesh - a feature seeking and tracking image sequence representation scheme. *IEEE Trans. Image Proc.* 3 (1994), pp. 610-624

Wang, Y. , Orchard, M. , Vaishampayan, V. , Reibman, A. R. : Multiple description coding using pairwise decorrelating transform. *IEEE Trans. Image Proc.* 10 (2001), pp. 351-366

Wang, Y. , Ostermann, J. , Zhang, Y.-Q. : Video Processing and Communications. Englewood Cliffs: Prentice Hall 2002

Wang, Y. , Wenger, S. , Wen, J. , Katsaggelos, A. G. : Error resilient video coding techniques. *IEEE Signal Process. Mag.* 17(4) (2000), pp. 61-82

Wang, Y. , Zhu, Q.-F. : Error control and concealment for video communication: A review. *Proc. IEEE* 86 (1998), pp. 974-997

Ward, D. B. , Elko, G. W. : Virtual Sound Using Loudspeakers: Robust Acoustic Cross-Talk Cancellation. *Acoustic Signal Processing for Telecommuncation*, Chapter 14, pp. 303-317, Kluwer Academic Publishers, 2000

Wedi, T. , Musmann, H.G. : Motion- and aliasing-compensated prediction for hybrid video coding. *IEEE Trans. Circ. Syst. Video Tech.* 13 (2003), pp. 577-586

Welch, T. A. : A Technique for High Performance Data Compression. IEEE Comput. 17 (1984), pp. 8-19

Wen, J. , Villasenor, J. : A class of reversible variable length codes for robust image and video coding. *Proc. IEEE ICIP* (1997), vol. II, pp. 65-68

Westerink, P. H. , Biemond, J. , Boekee, D. E. : Subband coding of color images. *Subband Image Coding*, J. W. Woods (ed.), Dordrecht : Kluwer 1991

Westerink, P. H. , Biemond, J. , Boekee, D. E. , Woods, J. W. : Subband coding of images using vector quantization. *IEEE Trans. Commun.* 36 (1988), pp. 713-719

Westerink, P. H. , Biemond, J. , Muller, F. : Subband coding of image sequences at low bit rates. *Signal Process. : Image Commun.* 2 (1990), pp. 441-448

Wiegand, T. , Girod, B. : Multi-Frame Motion-Compensated Prediction for Video Transmission. Dordrecht: Kluwer, 2001

Wiegand, T. , Sullivan, G. J. , Bjontegaard, G. , Luthra, A. : Overview of the H.264/AVC video coding standard. *IEEE Trans. Circ. Syst. Video Tech.* 13 (2003), pp. 560-576

Wiegand, T. , Schwarz, H. , Joch, A. , Kossentini, F. , Sullivan, G.J. : Rate-constrained coder control and comparison of video coding standards. *IEEE Trans. Circ. Syst. Video Tech.* 13 (2003), pp. 688-703

Wien, M. : Variable block-size transforms for H.264/AVC. *IEEE Trans. Circ. Syst. Video Tech.* 13 (2003), pp. 604-613

Willemin, P. , Reed, T. , Kunt, M. : Image sequence coding by split and merge. *IEEE Trans. Commun.* 39 (1991), pp. 1845-1855

Witten, I. H. , Neal, R. M. , Cleary, J. G. : Arithmetic coding for data compression. *Comm. ACM* 30 (1987), pp. 520-540

Wolberg, G. : Digital Image Warping. Washington : IEEE Computer Society Press 1990

Woods, J. W. , O'Neill, S. D. : Subband coding of images. *IEEE Trans. Acoust., Speech, Signal Process.* 34 (1986), pp. 1278-1288

Wu, F. , Li, S. , Zhang, Y.-Q. : DCT-prediction based progressive fine granularity scalable coding. *Proc. IEEE ICIP* (2000), pp. 1903-1906

Wu, S. F. , Kittler, J. : A differential method for simultaneous estimation of rotation, change of scale and translation. *Signal Process. : Image Commun.* 2 (1990), pp. 69-80

Xie, K. , van Eycken, L. , Oosterlinck, A. : Estimating motion vectorfields with smoothness constraints. *Proc. EUROPTO Image Commun. Video Compr.* (1993), SPIE vol. 1977, pp. 238-247

Xiong, Z. , Ramachandran, K. , Ortega, M. T. : Space-frequency quantization for wavelet image coding. *IEEE Trans. Image Proc.* 6 (1997), pp. 677-693

Xiong, Z. , Wu, X. : Wavelet image coding using trellis coded space-frequency quantization. Proc. SPIE VCIP (1999), vol. 3653, pp. 306-314

Xu, J. , Li, S. , Zhang, Y.-Q. : A wavelet codec using 3-D ESCOT. *Proc. IEEE-PCM2000*, Dec. 2000

Xu, W. , Hauske, G. : Picture quality evaluation based on error segmentation. *Proc. Visual Commun. Image Process.* (1994), SPIE vol. 2308

Yang, X. , Ramchandran, K. : Hierarchical backward motion compensation for wavelet video coding using optimized interpolation filters. *Proc. IEEE ICIP* (1997), vol. 1, pp. 85-88

Yang, X. , Ramchandran, K. : Scalable wavelet video coding using alias-reduced hierarchical motion compensation. *IEEE Trans. Image* Process. 9 (2000), pp. 778-791

Young, R. W. , Kingsbury, N. G. : Frequency domain motion estimation using a complex lapped transform. *IEEE Trans. Image Process.* 2 (1993), pp. 2-17

Yu, P. , Venetsanopoulos, A. N. : Hierarchical multirate vector quantization for image coding. *Signal Process. : Image Commun.* 4 (1992), pp. 497-505

Yu, Y.-B. , Chan, M.-H. : Low bit rate video coding using variable block size model. *Proc. IEEE ICASSP* (1990), pp. 1096-1099

Zaciu, R. et al. : Motion estimation and motion compensation using the overcomplete discrete wavelet transform. *Proc. IEEE ICIP* (1996), vol. I, pp. 973-976

Zhang, Q. , Kassam, S. A. : Hybrid ARQ with selective combining for fading channels. *IEEE J. Select. Areas Commun.* 17 (1999), pp. 867-880

Zhang, R. , Tsai, P.-S. , Cryer, J. E. , Shah, M. : Shape from shading: a survey. *IEEE Trans. Patt. Anal. Mach. Intell.* 21 (1999), pp. 690-706

Zhang, Y.-Q. , Zafar, S. : Motion-compensated wavelet transform coding for color video compression. *IEEE Trans. Circ. Syst. Video Tech.* 2 (1992), pp. 285-296

Zhang Z. , Xu G. : Epipolar Geometry in Stereo, Motion and Object Recognition. Dordrecht: Kluwer, 1996

Ziv, J. , Lempel, A. : A universal algorithm for sequential data compression, *IEEE Trans. Inf. Theor.* 23 (1977), pp. 337-343

Ziv, J. , Lempel, A. : Compression of individual sequences via variable-rate coding. *IEEE Trans. Inf. Theor.* 24 (1978), pp. 530-536

Zurmühl, R. : Matrizen. Berlin : Springer 1964

Zwicker, E. : Psychoakustik. Berlin : Springer 1982

Index

MIX
Papier aus verantwortungsvollen Quellen
Paper from responsible sources
FSC® C105338

If you have any concerns about our products,
you can contact us on
ProductSafety@springernature.com

In case Publisher is established outside the EU,
the EU authorized representative is:
Springer Nature Customer Service Center GmbH
Europaplatz 3, 69115 Heidelberg, Germany

Printed by Libri Plureos GmbH
in Hamburg, Germany